Agricultural, Biosystems, anc Engineering Educati

AF760648

Agricultural engineering, developed as an engineering discipline underpinned by physics, applies scientific principles, knowledge, and technological innovations in the agricultural and food industries. During the last century, there was exponential growth in engineering developments, which has improved human wellbeing and radically changed how humans interact with each other and our planet. Among these, "Agricultural Mechanization" is ranked among the top 10 in a list of 20 Top Engineering Achievements of the last century that have had the greatest impact on the quality of life.

While many success stories abound, the problems of low appeal among students, identity crises, and limited job opportunities in many climes continue to trouble the discipline's future in many parts of the world. Yet agriculture and agricultural engineering remain fundamental to assuring food and nutrition security for a growing global population.

Agricultural, Biosystems, and Biological Engineering Education provides the first comprehensive global review and synthesis of different agricultural, biosystems, and biological engineering education approaches, including a detailed exposition of current practices from different regions.

Key Features:

- Describes novel approaches to curriculum design and reform
- Outlines current and emerging epistemology and pedagogies in ABBE education
- Provides a framework to grow agricultural engineering in Africa and other developing regions
- Highlights the role of ABBE education in the context of the SDGs

Presented in 3 parts and containing 42 chapters, this book covers the historical evolution of agricultural engineering education and discusses the emergence of biological and biosystems engineering education. It will appeal to engineers and other professionals, education planners and administrators, and policy makers in agriculture and other biological industries.

Contemporary Food Engineering

Series Editor

Professor Da-Wen Sun, Director

Food Refrigeration & Computerized Food Technology,
National University of Ireland, Dublin
(University College Dublin)
Dublin, Ireland
http://www.ucd.ie/sun/

The **Contemporary Food Engineering Series** consists of edited books addressing engineering processes related to food production. Classical unit operations in engineering applied to food manufacturing are covered as well as such topics including the transport and storage of liquid and solid foods; heating, chilling and freezing of foods; mass transfer in foods; dehydration, thermal processing, non-thermal processing, extrusion, liquid food concentration, membrane processes and applications of membranes in food processing; packaging, cleaning, and sanitation. The books will be aimed at professional food scientists, academicians researching food engineering problems, and graduate level students.

Enhancing Extraction Processes in the Food Industry, *edited by Nikolai Lebovka, Eugene Vorobiev, and Farid Chemat*

Thermal Food Processing: New Technologies and Quality Issues, Second Edition, *edited by Da-Wen Sun*

Advances in Fruit Processing Technologies, *edited by Sueli Rodrigues and Fabiano Andre Narciso Fernandes*

Biosensors in Food Processing, Safety, and Quality Control, *edited by Mehmet Mutlu*

Edible Oils: Extraction, Processing, and Applications, *edited by Smain Chemat*

Engineering Aspects of Membrane Separation and Application in Food Processing, *edited by Robert W. Field, Erika Bekassy-Molnar, Frank Lipnizki, and Gyula Vatai*

Engineering Aspects of Food Emulsification and Homogenization, *edited by Marilyn Rayner and Petr Dejmek*

Advances in Meat Processing Technology, *by Alaa El-Din A. Bekhit*

Engineering Aspects of Food Biotechnology, *edited by Jose A. Teixeira and Antonio A. Vicente*

High Pressure Processing of Fruit and Vegetable Juices, *edited by Milan Houška and Filipa Vinagre Marques da Silva*

Trends in Fish Processing Technologies, *edited by Daniela Borda, Anca I. Nicolau, and Peter Raspor*

Food Biofortification Technologies, *edited by Agnieszka Saeid*

Advances in Postharvest Fruit and Vegetable Technology, *edited by Ron B.H. Wills and John Golding*

Computational Fluid Dynamics in Food Processing, Second Edition, *edited by Da-Wen Sun*

Food Nanotechnology: Principles and Applications, *edited by C. Anandharamakrishnan and S. Parthasarathi*

Ultraviolet Light in Food Technology: Principles and Applications, by *Tatiana Koutchma*

Advances in Processing Technologies for Bio-based Nanosystems in Food, *edited by Oscar Leandro da Silva Ramos, Ricardo Nuna Correia Pereira, Miguel Angelo Parente Ribeiro Cerqueria, Jose Antonio Couto Teixeira, and Antonio Augusto Vicente*

Innovative Technologies in Seafood Processing, *edited by Yesim Ozogul*

Advances in Vinegar Production, *edited by Argyro Bekatorou*

Advances in Food Bioproducts and Bioprocessing Technologies, *edited by Monica Lizeth Chavez Gonzalez, Nagamani Balagurusamy, Christobal N. Aguilar*

Advances in Noninvasive Food Analysis, edited by *Muhammad Kashif Iqbal Khan*

Agricultural, Biosystems, and Biological Engineering Education: Global Perspectives and Current Practice, edited by *Umezuruike Linus Opara*

For more information about this series, please visit: https://www.crcpress.com/Contemporary-Food-Engineering/book-series/CRCCON-FOOENG

Agricultural, Biosystems, and Biological Engineering Education

Global Perspectives and Current Practice

Edited by
Umezuruike Linus Opara

CRC Press
Taylor & Francis Group
Boca Raton London New York

CRC Press is an imprint of the
Taylor & Francis Group, an **informa** business

Cover by Umezuruike Linus Opara

First edition published 2024
by CRC Press
2385 NW Executive Center Drive, Suite 320, Boca Raton FL 33431

and by CRC Press
4 Park Square, Milton Park, Abingdon, Oxon, OX14 4RN
CRC Press is an imprint of Taylor & Francis Group, LLC

© 2024 Taylor & Francis Group, LLC with the exception of Chapter 5 Enhancing the Role of Agricultural and Biological Engineering through Education: Perspective from FAO. Chapter 5 copyright is retained by the Food and Agriculture Organization of the United Nations. The views expressed in this publication are those of the author(s) and do not necessarily reflect the views or policies of the Food and Agriculture Organization of the United Nations.

Reasonable efforts have been made to publish reliable data and information, but the author and publisher cannot assume responsibility for the validity of all materials or the consequences of their use. The authors and publishers have attempted to trace the copyright holders of all material reproduced in this publication and apologize to copyright holders if permission to publish in this form has not been obtained. If any copyright material has not been acknowledged please write and let us know so we may rectify in any future reprint.

With the exception of Chapters 4, 11, 19, and 41, no part of this book may be reprinted, reproduced, transmitted, or utilized in any form by any electronic, mechanical, or other means, now known or hereafter invented, including photocopying, microfilming, and recording, or in any information storage or retrieval system, without written permission from the publishers.

Chapters 4, 11, 19, and 41 of this book are freely available as downloadable Open Access PDFs at http://www.taylorfrancis.com. They are available under a Creative Commons Attribution-Non Commercial-No Derivatives (CC-BY-NC-ND) 4.0 license (Chapters 4, 11, and 41) and a Creative Commons Attribution-NonCommercial-ShareAlike (CC-BY-NC-SA) license (Chapter 19).

For permission to photocopy or use material electronically from this work, access www.copyright.com or contact the Copyright Clearance Center, Inc. (CCC), 222 Rosewood Drive, Danvers, MA 01923, 978-750-8400. For works that are not available on CCC please contact mpkbookspermissions@tandf.co.uk

Trademark notice: Product or corporate names may be trademarks or registered trademarks and are used only for identification and explanation without intent to infringe.

Library of Congress Cataloging-in-Publication Data

Names: Opara, U. L. (Umezuruike Linus), editor.
Title: Agricultural, biosystems, and biological
engineering education: global perspectives and current practice /
edited by Umezuruike Linus Opara.
Description: First edition. | Boca Raton, FL: CRC Press, 2024. |
Series: Contemporary food engineering series |
Includes bibliographical references and index.
Identifiers: LCCN 2023053970 (print) | LCCN 2023053971 (ebook) |
ISBN 9781032488776 (hardback) | ISBN 9781032496078 (paperback) |
ISBN 9780429150111 (ebook)
Subjects: LCSH: Agricultural engineering--Study and teaching. |
Bioengineering--Study and teaching. Classification: LCC S678.5 .A47 2024 (print) |
LCC S678.5 (ebook) | DDC 630.71--dc23/eng/20240314
LC record available at https://lccn.loc.gov/2023053970
LC ebook record available at https://lccn.loc.gov/2023053971

ISBN: 978-1-032-48877-6 (hbk)
ISBN: 978-1-032-49607-8 (pbk)
ISBN: 978-0-429-15011-1 (ebk)

DOI: 10.1201/9780429150111

Typeset in Times
by Deanta Global Publishing Services, Chennai, India

The Open Access version of Chapter 4 and 41 was funded by National Research Foundation, South Africa.
The Open Access version of Chapter 11 was funded by University of Western Australia.
The Open Access version of Chapter 19 was funded by Center for Research on Engineering Applications in Tropical Agriculture (CREATA) and Institut Pertanian Bogor University.

Dedicated to

My parents
Pápá Julius Uzôma and Mámá Glady Ôkaraonyemma
for the opportunity to learn agricultural engineering by doing.
My wife Gina (nee Malifa)
and our children
Ijeôma Chidi Uwaezuoke Ônumaraekwu
Ôkaraonyemma Lanuola Ulumma
for supporting and encouraging me to complete this book.

Contents

Foreword

Cultivating Minds, Nurturing Futures: A Journey in Agricultural, Biosystems, and Biological Engineering Education

In the vast expanse of human endeavour, few pursuits hold the promise of sustenance and progress like agriculture. From ancient grains sown in fertile soil to the marvels of modern bioengineering, the story of agricultural engineering is a tale of innovation, adaptation, and a steadfast commitment to feeding the world.

This book, *Agricultural, Biosystems, and Biological Engineering Education: Global Perspectives and Current Practice*, stands as a testament to the enduring legacy of agriculture, tracing its historical roots and celebrating its evolution into a vibrant discipline with biological sciences at the core, and leveraging engineering advances for the human wellbeing. Professor Umezuruike Linus Opara's edited volume provides an account of how the nurturing of minds has paved the way for a sustainable and prosperous future and offers a comprehensive resource for educating the next generation of agricultural, biosystems, and biological engineers curated from experts around the world.

A Journey Through Time and Innovation

To understand the present and envision the future, one must first appreciate the past. This book embarks on a journey through the annals of agricultural engineering, unearthing the seeds of wisdom that have sprouted into the diverse and dynamic field of Agricultural, Biosystems, and Biological Engineering as we know it today. From rudimentary plows to precision agriculture, every chapter in this story is marked by the indomitable spirit of progress.

Biosystems/Biological Engineering: A New Dawn

As the agricultural landscape has evolved, so have the challenges it presented. The emergence of biological and biosystems engineering heralded a paradigm shift, where the principles of biology and the rigour of engineering converged to meet the demands of a changing world. Agriculture today is practiced in a world where society and technology are changing faster than ever – where population growth, limited natural resources, and global warming create enormous challenges, and technological breakthroughs present an abundance of opportunities. This book delves deep into this transformative phase, exploring the emergence of a discipline uniquely equipped to tackle the complexities of modern agriculture.

Innovations in Curriculum Design and Reform

Education is the crucible where ideas are forged into action. The chapters within these pages illuminate novel approaches to curriculum design and reform from leading institutions around the world, reflecting the dynamic nature of agricultural, biosystems, and biological engineering (ABBE) education. From integrating cutting-edge technologies to fostering a holistic understanding of sustainability, these cutting-edge approaches to ABBE curriculum support transformative experiences where future leaders and innovators are shaped.

Epistemology and Pedagogies: Forging New Frontiers

In a world where knowledge is constantly evolving, it is imperative that our educational approaches keep pace. This book ventures into the realm of epistemology and pedagogies, unveiling the contemporary evidence-based methods that empower students to grapple with the intricate challenges of agricultural and biological systems. It is a testament to the adaptability and resilience of educators and learners alike.

A Framework for Growth: Empowering Developing Regions

The promise of agricultural progress extends far beyond the confines of well-established regions. This book provides a thoughtful framework for nurturing agricultural engineering in Africa and other developing and emerging areas. It is a call to action, a blueprint for empowering communities with the knowledge and skills they need to build sustainable agricultural futures.

ABBE Education and the Sustainable Development Goals

In the quest for a more equitable and sustainable world, agricultural and biological systems education takes center stage. This book underscores the pivotal role ABBE education plays in the context of the Sustainable Development Goals put forth by the United Nations. From ensuring food security to mitigating climate change, the contributions of this discipline reverberate through every facet of global progress.

As you embark on this journey through the tapestry of agricultural and biological systems education, may you be inspired by the resilience of the human spirit and the boundless potential of innovation. For within these pages lies not just rich historical perspective, but a roadmap for a future where agriculture thrives and communities flourish.

It is my hope that this book will serve as a beacon, illuminating the path forward for educators, students, and practitioners alike. Together, let us continue to cultivate minds and nurture futures, for in doing so, we shape a world that sustains and enriches us all.

Stephanie Farrell
Professor and Founding Department Head, Experiential Engineering Education, Rowan University, USA
President, International Federation of Engineering Education Societies (2022-2024)

Foreword

The discipline of agricultural engineering developed as an academic frontier servicing the rural sector. With initial focus on rural electrification, drainage, farm structures, mechanization, etc., the discipline has transformed during the past century to include modern concepts and practice in automation and robotics, digitalization of agriculture, postharvest technology and engineering, processing and value addition, environmental control, bioresources conservation,, and sustainable use, among others. With dwindling enrolment in many regions and countries of the world, the discipline has evolved and transformed through curriculum changes and refocused to align with technological trends and socio-economic challenges and opportunities.

This required a new vision for agriculture and agricultural engineering education, including the transition into sustainable agri-food systems in line with the United Nations Agenda 2030 for sustainable development (Transforming our World) as symbolized by the Sustainable Development Goals (SDGs) and the African Union's Agenda 2063 – the Africa We Want. Following these agendas, various countries are taking on the challenge of solving extremely complex problems in our food system by announcing solid agricultural policies that balance sustainability and productivity. Against this backdrop, it is becoming increasingly important to develop human resources with a deep understanding of current food issues and who will boldly take on the challenge of realizing a sustainable food system accompanied by technological innovation.

This book, edited by Distinguished Professor Umezuruike Linus Opara, a globally renowned agricultural and biosystems engineer, provides a critical look at the important matter of how we educate and train the global agricultural engineer for sustainable development. The editor has successfully brought together, in one book and the first of its kind, leading educators and thought leaders to document the historical evolution of agricultural engineering from its rural origin in North America, to the transformation of the discipline from physics and mathematics as the foundation to the emergence of biology as the new frontier. Through case studies from countries in different continents, several book chapters narrate the development of unique degree programmes that have been adapted to address local human capacity development needs. Chapters in the book also address the challenges of growing agricultural, biological, bioresources, and biosystems engineering education in Africa and other developing regions. To successfully train the truly global citizen for sustainable development, the book challenges educators in agricultural engineering and other disciplines to embrace the tenets of the United Nations Agenda 2030 and the aspirations of the African Union's Agenda 2063.

The depth and width of the topics covered position the book as a valuable resource for teachers, students, policymakers, development professionals, and industry practitioners interested in the education, training, research, and application of engineering principles and technologies for sustainable production and sustainable consumption of food, feed, fibre, and other biomaterials, as well as environmental protection and bioresources conservation. As the world grapples with the challenges of sustainably feeding 9 billion people by 2050, the role of agricultural, biosystems, and biological engineering has come to the fore more than ever before. The publication of this new book on *Agricultural, Biosystems, and Biological Engineering Education: Global Perspectives and Current Practice* is, therefore, timely.

This book is expected to be a valuable guide for developing human resources to achieve sustainable agriculture and food systems. It is a welcome and unique addition to digital and physical libraries of knowledge on agriculture, engineering and technology, education, agri-food systems transformation, and integrated rural development.

Seishi Ninomiya

Professor Emeritus, The University of Tokyo, Japan
President, International Commission of Agricultural and Biosystems Engineering, CIGR (2023-2024)

Preface

Agricultural engineering profession and agricultural engineering education play an essential role in assuring a secure and sustainable food system and biomaterials for industrialization. There is also increasing recognition of the contribution of agricultural engineering to environmental protection and natural resources conservation and use. During its formative years as an academic discipline and engineering profession, the primary mission of agricultural engineering was to improve lives and the environment in rural areas through increased farm productivity and provision of rural infrastructure such as electricity, roads, and drainage. The emerging discipline and academic programme were commonly referred to as "Rural Engineering", which later became 'Agricultural Engineering'. Physics and mathematics were the fundamental sciences of agricultural engineering.

Reasons for the focus of agricultural engineering on the rural environment and on-farm activities during its formative years are encapsulated in the definition of agricultural engineering by the American Society of Agricultural Engineers in the 1950s: "agricultural engineering is the application of any and all branches of engineering to the extent that they may be used in farming, in rural living, rural processing of farm products, and such allied activities as malaria control and wildlife conservation." While promoting agricultural engineering as a profession, the forebears emphasized that it calls for a knowledge of many other kinds of engineering, notably, civil, hydraulic, mechanical, electrical, and structural engineering.

As the human population surged and new territories were occupied in search of a better life, the demand for improved engineering technologies to clear forests, till the land, and process agricultural products for trade continued to rise. With mounting labour shortages and harsh farming conditions, improved and new power sources and machinery were needed.

Developments in animal-drawn implements using animals such as horses and oxen as power sources led to the first revolution in mechanized agriculture. The transformation and modernization of agriculture accelerated even faster in profound ways with the development and introduction of the 4-wheeled tractor running on an internal combustion engine. This marked a golden era of agricultural engineering – both in agricultural engineering education and practice. Drudgery was drastically reduced, the timeliness of agricultural operations increased significantly, and consequently, with other technological inputs such as irrigation and high-yielding plant and animal materials, productivity (including yield) and total factor production skyrocketed, especially in the industrialized world. The apocalyptic Malthusian prediction of mass global starvation and death arising from the human population outgrowing food production was averted – thanks in large part to the success of agricultural mechanization, which has been recognized among the top 10 technological advancements in the 19th Century.

These developments had push-pull effects on the education of agricultural engineers, including curriculum design and content, instruction materials and methods, etc. The demand for engineering in agriculture was loud. Governments worldwide responded by setting up agricultural engineering departments in tertiary institutions as well as dedicated agricultural engineering research and extension Institutes, Centres or Services under relevant Ministries or Departments but mainly the Ministry of Agriculture. Leading international development agencies in the agriculture and food domain, such as the Food and Agriculture Organization (FAO) of the United Nations (UN), set up an Agricultural Engineering Service (AES). With bright prospects ahead, student enrolments continued to rise or remained stable for decades. Funding for agricultural engineering education, research, and extension services boomed.

Towards the end of the 20th Century, cracks began to show on the walls of agricultural engineering education. The combination of declining student enrolments and funding of agricultural engineering programmes posed an existential challenge to the discipline and profession. Governments responded first by cutting budgets and later closing many agricultural engineering institutes and centres. Academic managers and administrators responded in more nuanced and varied ways, including renaming departments and programmes, reforming and restructuring academic programmes and/or departments, merging departments, merging programmes, etc. In severe cases, academic departments and/or programmes were gradually closed. These changes occurred to different degrees in each of the universities that I have worked at during the past three decades. In New Zealand, the Agricultural Engineering Institute was closed and some of its staff and programmes merged into a new R&D company called Lincoln Technology, when I eventually got my first employment as a Research Engineer after completing my PhD. At Massey University, New Zealand, the Department of Agricultural Engineering was closed as part of a restructuring exercise and the major was retained as an academic program for some time. At Sultan Qaboos University, Sultanate of Oman, the department was closed, and its staff merged with the Department of Soils and Water in the Department of Soils, Water and Agricultural Engineering, while retaining the academic programme. The agricultural engineering program at Stellenbosch University, South Africa, was closed long before I joined.

Overall, while the responses to the challenges facing agricultural engineering education varied considerably across continents, countries, and institutions in the same country, one thing that has emerged is the widespread acceptance of biology as the fundamental science which underpins the philosophy and approach. Unlike other engineering disciplines and professions, agricultural engineers are constantly required to work with biological systems, and agriculture is one of those. Agricultural machines interact with biological systems such as plants, animals, micro-organisms, and soils. Major reforms that have taken place include the incorporation of biology into the agricultural engineering curricula, and as a result, the names of many academic departments and programmes changed to include 'biosystems', 'biological', 'bioresources', 'bio-environmental', etc. For instance, while the term 'biological engineering' is dominant in North America, especially in the USA, most institutions in Europe prefer 'biosystems engineering', while 'agricultural engineering' and 'bioresources engineering' are common in Africa.

While recognizing and celebrating the success of agricultural intensification, including mechanization and other related technologies, as important in assuring the security of supply of food, feed, and fibre, it is also now widely recognized that current production and consumption practices and patterns must be changed to assure a better and equitable world – present and future. The sustainability of life and planet Earth are under severe threat due to various grand challenges, including climate change, greenhouse gas emissions, rising temperatures, deforestation, biodiversity loss, and environmental degradation. Agriculture is well known as a major contributing factor to these challenges and must, therefore, be part of the solution. The United Nations has echoed the call for sustainable development through the 17 Sustainable Development Goals (SDGs) as well as the African Union's Agenda 2063 – The Africa We Want. Both the SDGs of Agenda 2030 and the Aspirations of Agenda 2063 embody the call for sustainable agricultural production and consumption. In response to these challenges, the FAO has spearheaded the calls for the adoption of sustainable intensification and sustainable mechanization strategies. These present new opportunities for agricultural engineering education.

This edited book provides a historical perspective on the development of agricultural engineering education and discusses the different approaches to the education and training of current and future agricultural, biosystems, biological and bioresources engineers. With 82 authors contributing a total of 42 chapters from 44 different institutions in 42 cities, 21 countries, and 5 continents, the book is certainly a comprehensive compendium on how to educate the global agricultural engineer for sustainable transformation and modernization of the agri-food system. The book is divided into three parts.

Part 1 discusses the evolution of the agricultural engineering discipline and agricultural engineering education, including perspectives on its role in achieving the SDGs and Agenda 2063. As the leading agency of the United Nations on agriculture and food systems, authors from the Food and Agriculture Organization (FAO) present an instructive perspective on how to enhance the role of agricultural and biological engineering through education.

The 23 chapters in Part 2 focus on curriculum design and reform, including case studies from several countries. This is a rich source of information for those contemplating the development of new agricultural engineering programmes or restructuring existing ones.

The chapters in Part 3 present and discuss different approaches and technological tools to support teaching and learning. One chapter focuses on strategies to grow agricultural engineering in Africa based on empirical insights from undergraduate students in East, West, and Southern Africa. Other chapters highlight prospects for agricultural, biosystems, biological, and bioresources engineering education, especially in light of new and emerging global trends and technological advancements for achieving sustainable agricultural development.

With the variety and depth of materials covered, this book will benefit a wide array of audiences, including academics, students, researchers, educationists, governments, and other policymakers, as well as development practitioners interested in rural development and agriculture and food systems transformation. It will also serve as a readily accessible reference material for developing the emerging sub-discipline of agricultural, biosystems, biological, and bioresources engineering education.

Umezuruike Linus Opara
Stellenbosch 7600
South Africa

Acknowledgements

The idea to edit a book on agricultural engineering education followed the successful publication of a three-part Special Issue on the subject matter in the *International Journal of Engineering Education*, which I initiated and co-edited nearly two decades ago at Sultan Qaboos University in the Sultanate of Oman. Following the release of the Special Issue, my friend and colleague, Professor Da-Wen Sun, at the University College, Dublin, Ireland, supported my idea to advance the project into a book for a wider audience, especially given its critical contribution to the education and training of the next generation of agricultural, biosystems, biological, and bioresources engineers.

The invitation to contribute chapters to this book received tremendous global response and support from academics, researchers, and international agricultural and rural development practitioners. I am grateful to the numerous authors who contributed to this collection on global trends and current practices in agricultural, biosystems, biological, and bioresources engineering education. By sharing your expertise and experiences on the various aspects of agricultural engineering education – from the design and development of new academic programmes to curriculum reform, epistemology and pedagogy, and innovative teaching and learning tools, you have contributed to the production of important reference materials for the continuous improvement of agricultural engineering educational programmes which is necessary for the proper training of truly global agricultural engineers.

To the CRC Press/Taylor and Francis staff, thank you for standing by me, trusting, and urging me to complete this book. A special mention is due to Mr Stephen Zollo – Senior Editor, who has been with me from the inception to the completion of this project. To Ms Ashley Weinstein and Ms Amanda Parida, who are no longer with the publisher, thank you for your support and assistance during the early stages. To Ms Laura Piedrahita – Editorial Assistant, you stepped in at a critical time when I had a renewed commitment and enthusiasm to finish this book, and I thank you for your guidance and timely support. I appreciate you all for accommodating my busy schedule which delayed this publication.

Various colleagues and students assisted me with reviewing and providing editorial assistance on some of the chapters. Thank you Dr Robert Lufu, Kasiemobi Ezeora, Siphumle Jama, Buhle Maphosa, Esther Obioma, ThankGod (Okwute) Opara, Israel Ogra, Kenneth Opara, Tochi Ahumibe, Adejo Prince, and Kelechi Ekechi.

Finally, the completion of this book owes a lot to the dedication and hard work of Mrs Lizanne De Kock (MSc in Food Science), who has been my erstwhile Research and Project Assistant over the years. A special thanks to you, Lizanne, for your diligence, admirable attention to detail, superb editing skills, and perseverance towards the completion of this book. I could not have wished for a better assistant and administrator to enable me to achieve the publication of this book.

About the Editor

Umezuruike Linus Opara is a Distinguished Professor at Stellenbosch University, South Africa, where he also holds the Chair of the South African Research Chairs Initiative (SARChI) in Postharvest Technology, and the founding Director of the Africa Institute for Postharvest Technology (AIPhT). He graduated with degrees in Agricultural Engineering (BEng, First Class Honors; MEng, *cum laude*) from the University of Nigeria, Nsukka, a PhD in Agricultural Engineering from Massey University, Palmerston North, New Zealand and a Certificate in Innovation for Economic Development, from the Executive Education programme at Harvard Kennedy School, Harvard University, USA. He is a Chartered Engineer, Certified Food Scientist, member of the Academy of Science of South Africa, and has leading memberships of several professional agricultural engineering societies, including being a Fellow of the International Academy of Agricultural and Biosystems Engineering, the Institution of Agricultural Engineers (UK), the South African Institution of Agricultural Engineers, the Nigerian Institution of Agricultural Engineers, and having Life Membership of the American Society of Agricultural and Biological Engineers, the Asian Association for Agricultural Engineering, and the Indian Society of Agricultural Engineers.

He has published over 480 peer-reviewed journal articles, book chapters and peer-reviewed conference papers, and graduated over 70 MSc/MEng and PhD students in universities in South Africa, the Sultanate of Oman, and New Zealand. He was recognized in 2019 and 2021 as a "Highly Cited Researcher" by the Web of Science Group, distinguishing the "world's most influential researchers of the past decade". With a current H-Index of 61 (Scopus) and 69 (Google Scholar), Prof. Opara is a leading global expert on agricultural engineering and postharvest technology and also the leading global individual researcher on pomegranates – covering the value chain from fruit development to postharvest technology, agro-processing, and impacts on human health. He was lead co-author of the Mechanization Component of the 1989–2004 Agricultural Development Plan of Nigeria (1989), the first Medium-term Agricultural Development Plan of Timor-Leste (2009), and the recent report on "Agricultural Engineering in Africa: A Key Driver for Transforming Agriculture to Deliver Food Security and to Support Economic Prosperity" (2021).

Among many international, continental, national, and institutional awards for his sustained research productivity and impact, he was honoured as Laureate of the African Union's Kwame Nkrumah Continental Award for Life and Earth Sciences – Africa's highest research award for senior researchers, Winner of the Impact Research and Science in Africa (IMPRESSA) Award for "excellence in research and building Africa's human resources capacity", and the Distinguished Researcher Award at Sultan Qaboos University, Sultanate of Oman. In 2019, Stellenbosch University bestowed its highest honour on Prof. Opara – the Chancellor's Award, for his "sustained excellence in research". In 2022, he was honoured with the Distinguished Alumnus Award for academic and research excellence by the Department of Agricultural and Bioresources Engineering, University of Nigeria, Nsukka.

He was the founding President of the Pan African Society for Agricultural Engineering, past Executive Secretary of the International Academy of Agricultural and Biosystems Engineering, and past President of the International Commission of Agricultural and Biosystems Engineering. He was the founding Editor-in-Chief (now Honorary Editor) of the *International Journal of Postharvest Technology and Innovation* (Inderscience), and currently Co-Editor-in-Chief of *Food Safety and Health* (Wiley), Associate Editor of *Food and Energy Security* (Wiley), Associate Editor of *Frontiers in Plant Science*, Associate Editor of *Frontiers in Bioengineering and Biotechnology*, Review Editor for *Frontiers in Food Science and Technology*, Regional Editor of *Agricultural Mechanization in Asia, Latin America and Africa*, and serves on the Editorial Board of several other international journals. He is also co-editor of a recent book on *Mechanical Damage in Fresh Produce – Measurement, Analysis and Control* (Springer). He has served and continues to serve on the governing board and advisory board of many international agencies, including universities and research entities and has consulted for UN agencies such as the FAO, WHO, and UNIDO on agricultural engineering and postharvest technology for development. His ongoing activities focus on building Africa's research capacity in engineering and biosciences for agri-food systems transformation, wealth creation, and sustainable development.

Prof. Opara is currently developing the UNESCO International Centre for Biotechnology in Nsukka, Nigeria, a Category II Centre under the auspices of UNESCO, for advancing biotechnology research and high-level capacity building in the Africa region in the areas of food security, tropical diseases, and bioresources conservation.

List of Contributors

Joseph C. Adama
Michael Okpara University of Agriculture Umudike
Umudike, Nigeria

Chioma V. Agu
National Root Crops Research Institute
Umudike, Nigeria

Yaseen Ahmed Al-Mulla
Sultan Qaboos University
Al Seeb, Sultanate of Oman

Akindele Folarin Alonge
University of Uyo
Uyo, Nigeria

Chinenye Chukwuemeka Anyadike
University of Nigeria,
Nsukka, Nigeria

Hermann Auernhammer
Technical University of Munich
Freising, Germany

Stephen O. Babalola
Michael Okpara University of Agriculture Umudike
Umudike, Nigeria

László Baranyai
Hungarian University of Agriculture and Life Sciences
Budapest, Hungary

Pilar Barreiro Elorza
Polytechnic University of Madrid
Madrid, Spain

Manuel Berenguel
University of Almería
Almería, Spain

Remigio Berruto
University of Turin
Turin, Italy

Emmanuel Yaovi Hunnuor Bobobee
Kwame Nkrumah University of Science and Technology (KNUST)
Kumasi, Ghana

Thomas J. Brumm
Iowa State University
Ames, Iowa USA

Patrizia Busato
University of Turin
Turin, Italy

Adrián-Enrique Chavarría-Vidal
Technological Institute of Costa Rica
Cartago, Costa Rica

Ogbonnaya Chukwu
Federal University of Technology
Minna, Nigeria

Grant Clark
McGill University
Quebec, Canada

Ahmet Colak
Ankara University
Ankara, Turkey

Antonio Comparetti
University of Palermo
Palermo, Italy

Eva Cristina Correa Hernando
Polytechnic University of Madrid
Madrid, Spain

Nugrahaning Sani Dewi
National Research and Innovation Agency
Jakarta, Indonesia

Belén Diezma Iglesias
Polytechnic University of Madrid
Madrid, Spain

Can Ertekin
Akdeniz University
Antalya, Turkey

Frank P. Ferrie
McGill University
Montreal, Canada

Kehdinga George Fomunyam
Durban University of Technology
Durban, South Africa

F. García-Mañas
University of Almería
Almería, Spain

J.D. Gil
University of Almería
Almería, Spain

Ayub N. Gitau
University of Nairobi,
Nairobi, Kenya

Natalia Gómez-Calderón
Technological Institute of Costa Rica
Cartago, Costa Rica

Isabel Guzmán-Arias
Technological Institute of Costa Rica
Cartago, Costa Rica

José Luis Guzmán
University of Almería
Almería, Spain

Andrew L. Guzzomi
The University of Western Australia
Perth, Australia

Yasumaru Hirai
Kyushu University
Fukuoka, Japan

Kenji Hiyoshi
University of Miyazaki
Miyazaki, Japan

Muhammad Zakaria Hossain
Bangladesh Agricultural Research Institute
Gazipur, Bangladesh

Karim Houmy
Food and Agriculture Organization of the United Nations
Rabat, Morocco

Ángeles Hoyo
University of Almería
Almería, Spain

Anthony C. Iheonye
McGill University
Quebec, Canada

Shigeki Inaba
Saga University
Saga, Japan

Arthur T. Johnson
University of Maryland
College Park, Maryland USA

Deepak Keshwani
University of Nebraska
Lincoln, Nebraska USA

Jennifer Keshwani
University of Nebraska
Lincoln, Nebraska USA

Josef Kienzle
Food and Agriculture Organization of the United Nations
Rome, Italy

Masakazu Komatsuzaki
Ibaraki University
Ibaraki, Japan

Marybeth Lima
Louisiana State University
Baton Rouge, Louisiana USA

Tinashe Madzivanzira
University of Zimbabwe
Harare, Zimbabwe

Ronaldo Maghirang
University of Illinois
Urbana, Illinois USA

Emmanuel Manzungu
University of Zimbabwe
Harare, Zimbabwe

Duncan O. Mbuge
University of Nairobi
Nairobi, Kenya

Karolina Villagra Mendoza
Technological Institute of Costa Rica
Cartago, Costa Rica

Steven K. Mickelson
Iowa State University
Ames, Iowa USA

Hirofumi Mino
Chiba University
Chiba, Japan

Joseph Mpagalile
Food and Agriculture Organization of the United Nations
Rome, Italy

Manuel Muñoz
University of Almería
Almería, Spain

Brighton Mvumi
University of Zimbabwe
Harare, Zimbabwe

Patrick Obi Ngoddy
University of Uyo,
Uyo, Nigeria

Lien Le Phuong Nguyen
Hungarian University of Agriculture and Life Sciences
Budapest, Hungary

Joel Nwaeze Nwakaire
African Technology Policy Studies Network
Nairobi, Kenya

Yukiharu Ogawa
Chiba University
Chiba, Japan

Takashi Okayasu
Kyushu University
Fukuoka, Japan

John Oluranti Olajide
Ladoke Akintola University of Technology
Ogbomoso, Nigeria

Umezuruike Linus Opara
Stellenbosch University
Stellenbosch, South Africa
UNESCO International Centre for Biotechnology,
Nsukka, Nigeria

Valérie Orsat
McGill University
Quebec, Canada

Hadi K Purwadaria
IPB University
Bogor, Indonesia

Vijaya Raghavan
McGill University
Quebec, Canada

Francisco Rodríguez
University of Almería
Almería, Spain

Peter Rwakatiwana
University of Zimbabwe
Harare, Zimbabwe

Jorge Antonio Sánchez-Molina
University of Almería
Almería, Spain

John Schueller
University of Florida
Gainesville, USA

Takeshi Shikanai
University of the Ryukyus
Okinawa, Japan

Yoshiyuki Shinogi
Kyushu University
Fukuoka, Japan

Milton Solórzano-Quintana
Technological Institute of Costa Rica
Cartago, Costa Rica

Andrea Soto-Grant
Technological Institute of Costa Rica
Cartago, Costa Rica

Terumi Sugawara
Kyushu Okinawa Agricultural Research Center
Fukuoka, Japan

Fumihiko Tanaka
Kyushu University
Fukuoka, Japan

Fumina Tanaka
Kyushu University
Fukuoka, Japan

K.C. Ting
University of Illinois
Urbana, Illinois USA

James P. Trevelyan
The University of Western Australia
Perth, Australia

Rejoice Tsheko
Botswana University of Agriculture and Natural Resources (BUAN)
Gaborone, Botswana

Y. Yuwana
University of Bengkulu
Bengkulu, Indonesia

Fedro S. Zazueta
University of Florida
Gainesville, Florida USA

Part 1

Historical Evolution and Importance of Agricultural Engineering

1

Introduction – Revisiting and Rethinking Agricultural, Biosystems, and Biological Engineering Education for Sustainable Development

Umezuruike Linus Opara

Introduction

The application of engineering principles and related technologies to agriculture can be traced to prehistoric times and the use of simple hand tools powered by humans, which was followed by advancements in animal-drawn implements and the raising of livestock as a source of power for agricultural operations including the processing of food, feed, and fibre. The advent of mechanical power in the early part of the 19th century heralded the emergence of agricultural engineering as a practice and agricultural engineering (AE) education as an academic discipline. Key functions of AE during this era were to enhance agricultural extensification (clearing forests and virgin landraces), to bring more land under cultivation, and agricultural intensification, by producing more per unit of land through multiple croppings per year, as well as increasing stocking/planting density. Physics and mathematics were the fundamental sciences underpinning agricultural engineering education during those formative years.

There is now overwhelming evidence that while these practices contributed to massive increase in crop yield and total production, they have had long-term damaging impacts on the climate, the environment, and natural resources. The boom in AE education began to dwindle in the mid-1970s to the late-1980s in developed countries in the global north and in some parts of the southern hemisphere (notably, Australia, New Zealand and South Africa), leading to the closure of many AE academic and research programmes due to low and economically unviable student enrolments and an associated loss of funding. While student enrolments continued to rise or remained steady in the developing world, student interest in and preference for the major remained low, with the result that most AE students reported that the major was not their first choice.

In response to this growing dilemma around the future of AE education and the discipline, the names of academic programmes and departments have been changed to include words like 'Food', 'Environment', and variants of the word 'Biology' such as 'Bioresources', 'Bio-Environmental', 'Biosystems', and 'Biological' (Brumm & Mickelson, 2023, in this book; Clark & Orsat, 2023, in this book; Johnson, 2023, in this book; Furry, 2007; Opara, 2003a, b; Opara, 2006; Opara, 2004a,b; Opara et al., 2004a,b; Opara et al., 2006). As the profession grapples with these changes, AE education must also address the current and emerging grand challenges facing humanity, including sustainable development, food and nutrition insecurity, rising urbanisation and populations especially in developing regions, pollution, rapid environmental and ecosystems degradation, and widespread bioresource depletion, etc. Agriculture is not only highly vulnerable to these environmental threats, but is also a significant contributor. Facing these scenarios, and as humanity enters a new epoch – the Anthropocene – we are now confronted with new challenges in the transition towards sustainable futures, including agriculture and security for the Anthropocene (Newman et al., 2023).

The future of agri-food systems for the Anthropocene and the future of agricultural engineering are intertwined. This future requires the rethinking of AE education for sustainable development. The challenge for AE education in the Anthropocene is to determine how to train and equip present and future AE engineers to oversee the transitioning of agri-food systems towards more ecologically, socially, and economically sustainable futures, including dealing with the question of how the entire food value chain and ecosystem services can sustainably feed and nourish a growing global population under changing and dynamic environmental conditions. Practising agricultural engineers also need continuing professional development under this new regime of agricultural engineering for sustainable development. Professional AE societies and academic programmes in tertiary institutions should play important roles in designing and providing the relevant training required (Opara, 2012; Taylor et al., 2021).

Agricultural engineering (AE) developed as an engineering discipline, underpinned by physics, which applies scientific principles, knowledge, and technological innovations to the agricultural and food industries. During the last century, there has been an exponential growth in engineering developments, which has improved human wellbeing and radically changed the way humans interact with each other, as well as with the planet Earth. Among these, "Agricultural Mechanization"

DOI: 10.1201/9780429150111-2

ranked among the top 10 in a list of the 20 Top Engineering Achievements which have had the greatest impact on quality of life (www.greatachievements.org/). Agricultural mechanization, as one of the key missions of agricultural engineering, contributed to reducing drudgery and improving productivity in agriculture, thereby transforming agriculture from subsistence to commerce and modernizing the rural economy.

The importance of mechanization in agricultural transformation can be viewed in terms of three periods of industrialized agricultural development: the period leading up to 1850, characterised as the human-power era in agriculture, 1850–1910, characterised as the animal-power era, and the period from 1910 to the present, characterised as the mechanical-power era. In this regard, it should be noted that the transition from hand labour to machine operations involved more than the substitution of the tractor for human muscle. On the contrary, "it was the adaptation of the internal-combustion engine for agricultural operations (automobiles, trucks, auxiliary engines, and stationary power) and the use of the electric motor to agricultural operations (on-farm and postharvest) that contributed to progress in mechanization which has made the modern farmer a director of power rather than its source" (McColly and Martins, 1955).

Today, with several grand existential challenges facing humanity – especially climate change, rising populations particularly in the developing regions of the world, rising temperatures and greenhouse gas emissions, and deforestation and the degradation of the environment and ecosystems – the future of agriculture as we know it looks bleak. Many scholars argue that we have now entered a new planetary era, the Anthropocene, which is characterised by the possibility of human-induced effects crossing critical thresholds that could undermine key functions of the Earth system. Agriculture for the Anthropocene requires a new vision for agri-food systems, including education and the development and application of technology in multi-functional agriculture. This calls for a transition towards sustainable agri-food systems in line with the United Nations Agenda 2030 for sustainable development (Transforming our World) and the African Union's Agenda 2063 – The Africa We Want, Africa's masterplan for transforming the continent into a global powerhouse of economic prosperity. Given that Sub-Saharan Africa remains largely rural, with the economies of most countries depending on agriculture for livelihood under smallholder subsistence agriculture, it is well established that the modernization and transformation of agri-food systems into commercial value chains is critical to achieving the aspirations of Agenda 2063. Agricultural engineering education and research and agricultural engineers have a crucial role to play (Opara, 2012; Taylor et al., 2021).

At the same time, significant changes have occurred regarding the way we educate and train engineers, including agricultural engineers. With pressure on academic departments and their programmes due to declining student enrolment, many have closed, reformed, or repositioned, so as to remain relevant. The changes made have included revised curricula and changes to the name of the discipline and/or department to include variants of 'Biology' as the new fundamental science underpinning agricultural, biosystems, and biological engineering (ABE) education. While success stories abound, the problems of low appeal among students, identity crises, and limited job opportunities continue to trouble the future of the discipline in some parts of the world. Yet, agriculture remains fundamental to assuring adequate food and nutrition for a growing global population. Despite these challenges, considerable progress has been made at the global level in developing novel technologies to improve productivity and to assist in mitigating the negative impacts of agriculture on the environment and the impact of climate change and other environmental challenges on agriculture. With the present challenges, and the increasing possibility of new threats, a rethinking of AE education is warranted.

Revisiting Agricultural Engineering Education – A Humble, Rural Beginning

The origin of agricultural engineering practice and agricultural engineering education can be seen as a response to the rural development challenges faced by farmers who needed improved cultivation methods to reduce drudgery, improve productivity, enhance profitability, and develop rural infrastructure. This is exemplified by the development of agricultural engineering in the USA. The last half of the 1800s witnessed the development of colleges of agriculture and engineering, and in the land-grant colleges, many departments of civil and rural engineering. A typical description of such majors is as follows (ABET, 1932): "The study embraces the application of the engineering art to agriculture as seen in improved machinery, well-constructed buildings, conveniently arranged fields, etc.". Given the rapid development of the country, demand for enrolments was high but teachers were in limited supply. Few engineers from the already well-established engineering disciplines – mining, mechanical, civil, electrical – were willing or interested to serve in agriculture. The fact was that "many people were rather more interested in getting away from agriculture" (Edwards, 1992).

According to Furry (2007): "What began as a farm-oriented discipline, directed by bright American men with rural backgrounds, has broadened into a theoretical and applied discipline made of men and women from around the world who work to discover how best to make use of and preserve the resources of the Earth".

As an example, Table 1.1 shows the evolution of the departmental name for this discipline at Cornell University, USA,

TABLE 1.1

Agricultural engineering department name changes at Cornell University, USA.

Name	Year
Division of Rural Engineering and Architecture	1907
Department of Farm Mechanics	1910
Department of Rural Engineering	1913
Department of Agricultural Engineering	1930
Department of Agricultural and Biological Engineering	1988
Department of Biological and Environment Engineering	2001

which began as Rural Engineering, and is now Biological and Environmental Engineering. The description of the new academic programmes at Cornell University aptly captures this transition from rural and agricultural engineering to biological and environmental engineering (Furry, 2007):

> "Biological Engineering integrates engineering practice and quantitative biology, with a focus on food systems, life sciences, human health, and the environment. Environmental Engineering is aimed at combining engineering and environmental sciences in a coordinated manner so as to include a balance of basic, developmental, and applied investigative efforts. Once concerned primarily with the rural environment, the programme now addresses a wide range of environmental issues in both the private and public sectors."

Similarly, Table 1.2 shows the names of the technical sections/divisions in the American Society of Agricultural and Biological Engineers. Starting with a strong focus on rural infrastructure in the early 1900s, the emphasis shifted to power, machinery, farm structures, and processing in the mid-1900s. The current themes now focus on systems thinking, underpinned by biology and applied to facilities, machinery, processing, energy, natural resources and the environment, and information and communication technologies. The inclusion of education, outreach, and professional development is worthy of note, and relevant to the goal of publishing this book.

According to Hall (1992a), it was in 1852 in England that the first book in English with the words "agricultural engineering" was published. In the 1950s, McGraw-Hill released the following publications as part of a series in Agricultural Engineering, with Quincy C. Ayres as Consulting Editor:

- *Soil Erosion and Its Control*
- *Land Drainage and Reclamation*
- *Rural Electrification Engineering*
- *Farm Service Buildings*
- *Farm Engines and Tractors*
- *Farm Gas Engines and Tractors*
- *Farm Shop Practice*
- *Introduction to Agricultural Engineering*
- *Moisture Requirements in Agriculture*
- *Engineering for Agricultural Drainage*
- *Farm Machinery and Equipment*
- *Planning Farm Buildings*

In the preface to their seminal book *Introduction to Agricultural Engineering*, McColly and Martin (1955) called for increased mechanisation of American agriculture and increased attention to the discipline of agricultural engineering:

> "The American farm is being mechanized, and the time is approaching when muscles will be almost entirely replaced by power from electric motors, internal -combustion engines, or hydraulic units. Man-hours are expensive, and if they can be reduced a real saving results. The U.S. Department of Agriculture states that agricultural production per man hour has increased 35 per cent since 1940. The production per man hour for industry gained only 11 per cent. The Oklahoma Agricultural Extension Service claims that, if present trends continue, each farmer will feed and clothe 55 people in 1975. To be in a position to keep pace with this trend, students and workers in the business of agriculture must give increased attention to the field of agricultural engineering."

Nearly a century and half after the publication of the first book on agricultural engineering in England, Hall and Olsen (1992)

TABLE 1.2

Technical Sections/Divisions/Communities of the American Society of Agricultural Engineers for selected years to show changes in technical focus.

1907 ASAE Founding Charter Areas of Activity*	1920s ASAE Technical Sections**	1950s ASAE Technical Divisions**	1980s ASAE Technical Divisions	2023 ASABE Technical Communities	
Drainage and irrigation Rural highway construction Design and construction of farm building Improving farm implements and mechanizing the farm Use of electrical power on the farm	Farm power and machinery Farm buildings Reclamation Rural electric	Power and machinery Soil and water Farm structures Electric power and processing	Electric and electronic systems Food and process engineering Power and machinery Soil and water Structures and environment Emerging technologies***	Plant, animal, and facility systems Machinery systems Natural resources and environmental systems Processing systems	Information technology, sensors, and control systems Applied science and engineering Education, outreach, and professional development Energy systems Ergonomics, safety, and health

*H.W. Riley, charter member ASABE, Contribution during the first ASABE conference in Madison, Wisconsin. ** Hall (1992b). ***According to C.W. Hall (1992b), this was included "to identify new technologies, mistakenly considered by some as home for new division, although that was not the objective".

published *The Literature of Agricultural Engineering*, which listed, analysed, and summarised "the books and journals that have made important contributions to the evolution of the subject".

Agricultural engineering as a discipline and profession took hold in Europe much later than in North America. Similar to the developments that occurred in North America, the 'rural' foundation and focus of the early days of agricultural engineering in Europe is reflected in the name of the first pan-European scientific meeting on the subject, the "Congrès International du Génie Rural", which was held at the University of Liège, Belgium, in August 1930. It was during this congress that the apex global network of agricultural engineers was formed – the "Commission Internationale du Génie Rural (CIGR)", which was translated into English as the "International Commission of Agricultural Engineering" while retaining the original abbreviation in French (CIGR). This network of adhering national and regional societies, associations, and institutions has recently been renamed as the "International Commission of Agricultural and Biosystems Engineering" (CIGR), in response to the recognition and global consensus among professionals in the discipline that agricultural engineers principally work with biological systems.

Pioneer Programmes and Graduates

Literature evidence suggests that the development of agricultural engineering as an academic course of learning started in the USA. According to McColly and Martin (1955), the first listing of a course in agricultural engineering in the USA appeared at the University of Nebraska:

> "In the 1896 catalogue of the University of Nebraska, the School of Agriculture listed two courses – agricultural engineering, under O.V.P. Stout, and practical mechanics, under C.R. Richards. Subjects under agricultural engineering were farm survey with chain, drainage levelling, topographical survey for irrigation or landscape gardening, measurement and division of water, application to crops, and minor irrigation structures. Agricultural mechanics included draft of vehicles and implements, wind wheels and pumps, strength of timber, ventilation of buildings, and bearing power of soils, while elementary study of heat, carpentry, and blacksmithing were listed under practical mechanics."

There is currently no recorded evidence showing whether students were enrolled in response to this catalogue, whether they graduated, and if so, under which department.

However, the first agricultural engineering department and undergraduate degree programme in the USA were established at Iowa State College (now Iowa State University), headed by J. Brownlee Davidson, who had joined the university in 1906. Subsequently, Iowa State University became the first institution to confer a degree in Agricultural Engineering (Bachelor of Science) in 1910. The goal of this programme was to educate and train graduate engineers who were well equipped with the skills needed to solve problems in the agricultural sector. Both the curriculum and graduate attributes were like those of the other engineering disciplines – notably, rural and civil engineering in the 1800s – except that they focused on meeting the needs of the agricultural sector.

The first PhD in Agricultural Engineering in the USA was awarded by Cornell University to Earl Archibald White in 1917 in the Department of Rural Engineering (Furry, 2007) based on his dissertation on the tillage mechanics of the plow bottom upon the furrow slice. The second PhD in Agricultural Engineering, Forest B. Wright, graduated in 1933, eleven years after his BS degree from Cornell, and 26 years after the first PhD in agricultural engineering was conferred.

Outside North America, the first agricultural engineering department was set up in India at Allahabad in 1921, followed by an undergraduate curriculum that was developed in 1942 and led by Mason Vaugh who had graduated earlier in the USA (Kumar and Tiwary, 1989; Kumar et al., 1989). Today, agricultural engineering and related majors (biological, biosystems, bioresources, food, environmental, natural resources, etc.) are offered in universities and polytechnics around the world, producing graduates at both the undergraduate and postgraduate levels, in Africa, Asia, the Middle East, North America, Latin America and the Caribbeans, Oceania, and Europe.

Epistemological Foundations of Agricultural, Biosystems, and Biological Engineering Education

In general terms, AE is basically the utilisation of science and technology in agriculture. Today, the fields of agriculture and engineering cover many broad areas. While agriculture identifies a rapidly growing number of problems in production, handling, and processing which need to be solved, engineering on the other hand offers a rapidly growing number of options for solving these problems. Given its original subject matters, such as farm mechanics and construction of rural infrastructure including drainage, roads, and buildings, physics was the foundational science that underpinned agricultural engineering during its formative years. Although the recognition of biology as a foundational science in agricultural engineering has long been recognised – since the 1960s – it is only during the past two decades that this understanding has been incorporated into AE curricula, as well as into the names of both the relevant majors/programmes and academic departments (e.g., bioresources, biological, biosystems, natural resources, etc.). Afterall, agricultural processes which produce food, feed, and fibre are biological processes. It is therefore not surprising that many of the earliest programmes in agricultural engineering, as well as its predecessors (rural engineering, farm mechanics, agricultural mechanisation, etc.), were started in faculties of agriculture and in some other departments such as agronomy.

The driving force for these predecessors (e.g., mechanisation) was mainly the desire to reduce the drudgery associated with field operations such as land clearing, swamp draining, land preparation for cultivation, and crop harvesting (Hall, 1992a,b). On the other hand, the driving force behind

AE development was led by processes and devices that were already in existence or had been imagined. While some disciplines start with a foundation in theory, AE started in the opposite manner, with movements from those already practising agriculture, such as tool makers, craftsmen and women, and manufacturers of machines and implements. According to Hall (1992, a,b), the knowledge base of science was added later to improve upon these tools and devices.

Today, AE is deeply rooted in both engineering and agriculture. On the other hand, agricultural, biosystems, and biological engineering (ABE) education is at the interface of engineering, agriculture, and education (Figure 1.1). The delivery of quality ABE education requires a good understanding of basic and advanced modules in engineering, an adequate understanding of biology and agriculture, and the understanding and application of novel teaching and learning techniques. This requirement was echoed more than 60 years ago by McColly and Martin (1955) when they opined that: "In order to understand the fundamental role of engineering in agriculture, it is necessary to know the basic meanings of the two sciences, agriculture and engineering. They involve considerably more than just growing crops or operating and repairing equipment". To educate the global agricultural, biosystems, and biological engineer with an eye to sustainable development, educators must embrace the latest technological advances in engineering, agriculture, and education, including pedagogy, ethics, and the principles and practice of sustainable development.

Rethinking Agricultural Engineering Education

Today, the world is facing multiple grand challenges which threaten the future of life on planet Earth: climate change, food and nutrition insecurity, energy insecurity, environmental degradation, pollution, rising populations and urbanisation, the threat of pandemics, the depletion of natural bioresources at an alarming rate, etc. There is a general agreement among scientists and policymakers that the agriculture that we have practised can no longer be sustained, and it has been implicated in contributing to and accelerating the rate of these climatic and environmental hazards and emergencies. Our agri-food systems must embrace sustainable practices across the value chain!

In response to the climate crisis and the imminent danger of agriculture and food systems for the Anthropocene (Newman et al., 2023), we must rethink agricultural engineering education and practice with an eye to sustainable development. Agricultural engineering technologies – including products and processes – must be designed to enhance productivity with minimal negative impacts, while assuring a reliable supply of food of sufficient quantity, quality, nutrition and safety under profitable and equitable agri-food systems. Part of the strategy requires revisiting the philosophy and ethics of the profession to embolden the commitment of agricultural, biosystems, and biological engineering education to the SGDs of Agenda 2030 of the United Nations and the Aspirations of Agenda 2063 – The Africa We Want. Revisiting curricula, devising novel and impactful methods of teaching and learning, and increasing

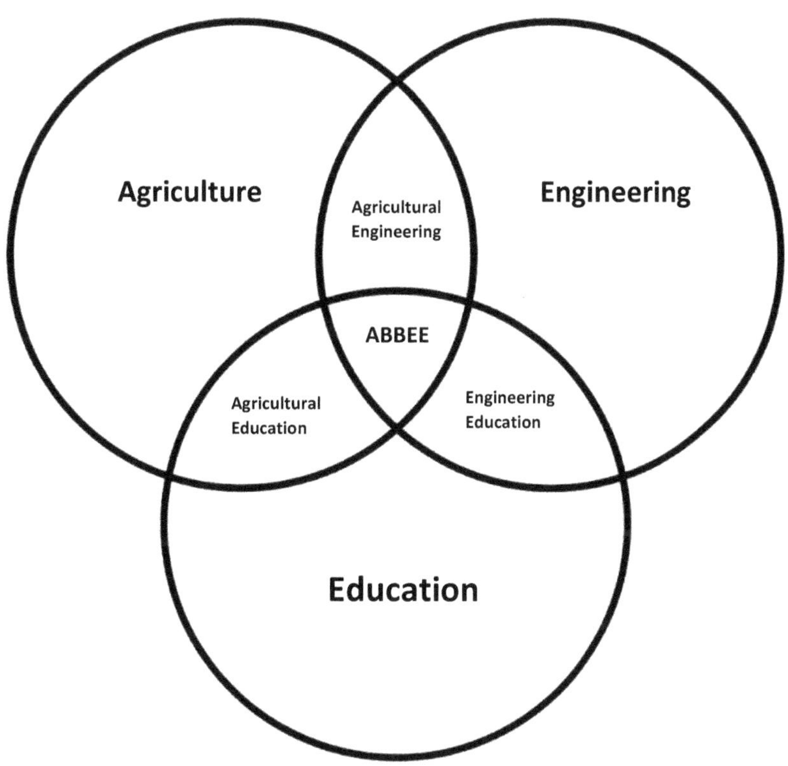

FIGURE 1.1 Epistemological foundation of Agricultural and Biosystems/Biological Engineering (ABE) Education.

the exposure of students to real-world challenges (and opportunities!) will be crucial. Implementing project-based learning, especially in the developing world where many students lack exposure to industry and policy, will become critical.

The grand challenges facing humanity are complex, vexing, and transcend geo-political boundaries, disciplinary boundaries, and time zones. Solutions to these problems require systems thinking, collaboration, and high levels of cooperation and partnership at all levels – communities and institutions local, national, regional, and global. Therefore, agricultural engineering education must equip graduates with relevant knowledge and technical skills, but also soft skills like ethical reasoning, empathy, creativity, teamwork, leadership, communication, critical thinking, and entrepreneurial thinking. Future agricultural engineers must become thought leaders who are able to understand global challenges but also be sufficiently equipped to solve local problems. These attributes are particularly important given that technological innovations often occur at the boundaries of knowledge, and the future work and careers of graduates are not necessarily linked to their majors/discipline. Studies in the USA showed that on average only 27% of graduates work in the field of their major (Abel and Deitz, 2013).

Many cross-cutting, advanced, and emerging scientific and technological innovations offer considerable promise with regard to facilitating and improving the quality of agricultural engineering education. Data science and computational thinking, artificial intelligence, and related technologies such as machine learning and deep learning, offer new prospects for harnessing the wealth of data and information inherent in agri-food systems and the environment. The technological innovation triad of biotechnology, nanotechnology, and communication technologies (Opara, 2004c), as well as digitisation, offer tremendous opportunities for AE education and research. Collaboration and cooperation will be essential for harnessing these opportunities in AE education.

In rethinking agricultural, biosystems, and biological engineering education, our goal is to educate students to become well-rounded agricultural engineers "who understand how to Conceive-Design-Implement-Operate complex, value-added engineering products, processes, and systems in a modern, team-based environment" (Crawley et al., 2014). This approach addresses three critical goals, which are to:

> Master a deeper working knowledge of technical fundamentals.
> Lead in the creation and operation of new products, processes, and systems.
> Understand the importance and strategic impact of research and technological development on society.

The field of agricultural, biosystems, and biological engineering education research has just begun, as evidenced in the chapters published in part 2 of this book. As the world grapples with the urgent challenge of transforming and modernising agriculture and food systems for a sustainable future under the Anthropocene, the future of agricultural, biosystems, and biological engineering education is bright. Just as agricultural engineering (including agricultural mechanisation) successfully led the transformation of rural areas and subsistence farming in the developed economies and many emerging economies into technology-enabled and knowledge-intensive modern living areas and industrial agribusiness hubs, this feat can be replicated to achieve sustainable agri-food systems transformation in the 21st century. In sub-Saharan Africa, which has missed the benefits of the much-acclaimed Green Revolution, agricultural, biosystems, and biological engineering is expected to play far more crucial and urgent roles in transforming the predominantly subsistent smallholder agri-food systems into commercial value chains for economy-wide development and industrialisation. Given that the majority of the population of Sub-Saharan Africa depends on agri-food systems for livelihood, being mainly subsistent smallholders and petty traders, it is clear that Africa's economic development will depend on what happens to agriculture and food systems in Africa. Therefore, to realise this aspiration of Africa as a global economic powerhouse, as encapsulated in Agenda 2063 – The Africa We Want (AUC, 2014), Africa needs an industrial agri-food system that is sustainable, profitable, and competitive for it to have a chance of addressing many of the socio-economic challenges facing the continent – from food and nutrition insecurity to rising populations, urbanisation, and youth unemployment.

Futurists and techno-economic forecasters have corroborated the prospect of a bright future for agricultural engineering education, as evidenced by the inclusion of Agricultural Engineering among "The Top (10) Majors For The Class Of 2022", only preceded by Math and Robotics, and ahead of Hospitality Management, Health and Biotechnology, Pre-Law, Quantum Engineering, 3-D Printing, The Liberal Arts, and Aerospace Engineering (Knapp, 2012). In this regard, the education of the Class of 2030 and Class of 2063 agricultural, biosystems, and biological engineers must begin today – first, through the conception of new and sustainable agri-food systems, and second, through the conception of a new ABE education for sustainable development.

This book is the first to focus on the critical subject matter of educating the agricultural, biosystems, and biological engineer, and serves as a reference material for present and future academics and other professionals interested in new programme design and evaluation, and the reform of existing ones. Together, the chapters in this book provide a rich source of information on current philosophies, curricula designs, and global practices in agricultural, biosystems, and biological engineering education. They also lay the foundation for future research in agricultural, biosystems, and biological engineering education, including curriculum development and reformation, effective teaching methods and models for better understanding how students learn, quality assurance, and impact assessment, etc.

Conclusions

The history of agricultural engineering practice predates that of agricultural engineering education. Over the past 100 years, the discipline has evolved from a rural base to a mainstream academic discipline, which was originally rooted in physics and mathematics and which has transformed into a bio-inspired academic discipline and profession. As industrialisation and globalisation take hold across the world, with fewer people engaged in primary agriculture, and new 'hitech' industries continuing to emerge, agricultural engineering education continues to face serious challenges of identity, poor image, low profile, and limited employment opportunities. In turn, these factors contribute to low student enrolment and challenges to the economic viability of programmes. Measures have been taken to rename and/or merge programmes, with varying degrees of success. In many parts of the developing world with sustained enrolment, most students did not select agricultural engineering as their preferred choice, and this raises the additional and highly important issue of the quality of the 'raw material' – the students that are enrolled. The major also has problems with student retention.

The challenge for AE education in the Anthropocene is to determine how to train and equip tomorrow's AE engineers with skills, competence, and compassion, engineers who can oversee the transition of agri-food systems towards more ecologically, socially, and economically sustainable futures in line with the Sustainable Development Goals (SDGs) of the United Nations Agenda 2030 and the Aspirations of the African Union Agenda 2063 – The Africa We Want.

Rethinking agricultural engineering education for sustainable development and for Anthropocene agriculture offers new opportunities for repositioning the discipline. Rapid advancements in cross-cutting and high-impact technological innovations in data science and computational thinking, artificial intelligence, digitisation, biotechnology, nanotechnology, un-manned aerial vehicles, etc, offer new opportunities to transform and modernise agri-food systems for sustainable development. The incorporation of novel teaching and learning methods and tools which enhance effective learning in the digital era offers great opportunities. The opportunities are tremendous for Sub-Saharan Africa and other rural economies to harness these innovations for teaching and learning, and, in research, to leapfrog into a new Green Revolution that embodies the SDGs of Agenda 2030 and the Aspirations of Agenda 2063. To succeed, connecting this new thinking with stakeholders is crucial: prospective and present students, academia, industry, policy, development agencies, and others.

REFERENCE LIST

Abel, J.R. & Richard Deitz, R. (2013). Do big cities help college graduates find better jobs? Federal Reserve Bank of New York. https://Libertystreeteconomics.Newyorkfed.Org/2013/05/Do-Big-Cities-Help-College-Graduates-Find-Better-Jobs/. Accessed 20 August 2023.

ABET. (1932). Annual report. Baltimore, MD: Accreditation Board for Engineering and Technology.

AUC. (2014). Agenda 2063. The Africa we want, 2nd edition, Popular Version. African Union Commission: Addis Ababa. 20 pages.

Brumm, T.J. & Mickelson, S.K. (2023). Chapter 10: Reshaping agricultural and biosystem engineering degree programs for student success and retention. In: Opara, U.L. (ed.). *Agricultural, Biosystems, and Biological Engineering Education – Global Perspectives and Current Practice.* CRC Press/Taylor & Francis Group.

Clark, O.G. & Orsat, V. (2023). Chapter 8: Bioresource engineering curriculum and reform at Mcgill University, Montreal, Canada. In: Opara, U.L. (ed.). *Agricultural, Biosystems, and Biological Engineering Education – Global Perspectives and Current Practice.* CRC Press/Taylor & Francis Group.

Crawley, E.F., Malmqvist, J., Ostlund, S., Brodeur, D.R. & Edstrom, K. (2014). *Rethinking Engineering Education: The CDIO Approach*, 2nd edition. New York: Springer, 311 pages.

Edwards, D.M. (1992). Chapter 3. Agricultural engineering education programs. In Hall, C.W. and Olsen, W.C. Editors. *The Literature of Agricultural Engineering*, pp. 25–42. Cornell University Press: Ithaca.

Furry, R.B. (2007). *A Pioneering Department: Evolution from Rural Engineering to Biological and Environmental Engineering at Cornell University, 1907–2007*, 186 pages. Ithaca: The Internet-First University Press.

Hall, C.W. (1992a). Preface. In Hall, C.W. and Olsen, W.C. Editors. *The Literature of Agricultural Engineering*, pp. vii–viii. Ithaca: Cornell University Press.

Hall, C.W. (1992b). Chapter 2. The influence and scope of the Agricultural Engineering profession. In Hall, C.W. and Olsen, W.C. Editors. *The Literature of Agricultural Engineering*, pp. 11–24. Ithaca: Cornell University Press.

Hall, C.W. & Olsen, W.C. (1992). Editors. *The Literature of Agricultural Engineering.* Ithaca: Cornell University Press.

Johnson, A.T. (2023). Chapter 29: Educating the compleat biological engineer. In: Opara, U.L. (ed.). *Agricultural, Biosystems, and Biological Engineering Education – Global Perspectives and Current Practice.* CRC Press/Taylor & Francis Group.

Knapp, A. (2012). The top majors for the class of 2022. Forbes. https://www.forbes.com/sites/alexknapp/2012/05/09/the-top-majors-for-the-class-of-2022/?sh=7f36961d322f. Accessed 20 August 2023.

Kumar, A. & Tiwary, S. (1989). Agricultural engineering teaching, research, and extension in India: Present state and future trends. *Proceedings of the Eleventh International Congress on Agricultural Engineering, Dublin, Ireland.* Rotterdam: A.A. Balkerman.

Kumar, A., Tiwary, S. & Rai, R.K. (1989). Response of undergraduate agricultural engineering students to education and evaluation. *Proceedings of the Eleventh International Congress on Agricultural Engineering, Dublin, Ireland.* Rotterdam: A.A. Balkerman.

McColly, H.F. & Martin, A.W. (1955). *Introduction to Agricultural Engineering.* New York: McGraw-Hill.

Newman, L., Newell, R., Dring, C., Glaros, A., Fraser, E., Mendly-Zambo, Z., Green, A.G. & Bahadur, K.K.C. (2023). Agriculture for the anthropocene: Novel applications of technology and the future of food. *Food Security*, 15, 613–627.

Opara, L.U. (2003a). Image is everything: Undergraduate students' contributions towards better understanding and improvement of public attitudes and perceptions of agricultural engineering. *World Transactions on Engineering and Technology Education*, 2(3), 457–460.

Opara, L.U. (2003b). From farmers to blue-collar professionals: Agricultural Engineering students' perspectives on their career prospects, critical skills, and discipline name change. *World Transactions on Engineering and Technology Education*, 2(3), 399–402.

Opara, L.U. (2004a). emerging technological innovation triad for smart agriculture in the 21st century. Part I. Prospects and impacts of nanotechnology in agriculture. *Agricultural Engineering International: The CIGR Journal of Scientific Research and Development.* Invited Overview Paper, VI, 27 pages.

Opara, L.U. (2004b). Outlook on agricultural engineering education and research and prospects for developing countries. *Outlook on Agriculture*, 33(2), 101–112.

Opara, L. (2004c). Emerging technological innovation triad for smart agriculture in the 21st century. Part I. Prospects and Impacts of Nanotechnology in Agriculture. Invited Overview Paper. *Agricultural Engineering International: The CIGR Journal of Scientific Research and Development*, VI(July), 1–27.

Opara, L.U. (2006). A comparison of students' insights and attitudes towards agricultural engineering education: Evidence from New Zealand and The Sultanate of Oman. *Proceedings of the 34th International Symposium on Actual Tasks in Agricultural Engineering*, Zaghreb, Croatia, pp. 13–19.

Opara, U.L. (2012). The need for a professional network of agricultural and biosystems engineers in Africa. *International Journal of Postharvest Technology and Innovation*, 2(4), 426–432.

Opara, L.U., Al-Adawi, S.S. & Al-Shukeili, T.Y. (2004a). Students' perceptions of agricultural engineering education at Sultan Qaboos University in Oman, Part 1: Academic program selection. *World Transactions on Engineering and Technology Education*, 3(1), 139–142.

Opara, L.U., Al-Adawi, S.S. & Al-Shukeili, T.Y. (2004b). Students' perceptions of agricultural engineering education at Sultan Qaboos University in Oman, Part 2: Curriculum content and career preferences. *World Transactions on Engineering and Technology Education*, 3(1), 143–146.

Opara, L.U., Perret, J.S. & Ampratwum, D.B. (2006). The evolution of agricultural engineering education at Sultan Qaboos University. *International Journal of Engineering Education*, 22(6), 1139–1148.

Taylor, A.J., Opara, U.L. & Smithers, J.C. (2021). *Agricultural Engineering in Africa: A Key Driver for Transforming Agriculture to Deliver Food Security and to Support Economic Prosperity.* Pretoria: Pan African Society for Agricultural Engineering (PASAE/AfroAgEng).

2

Emergence and Advancement of Agricultural Engineering Education, Research and Practice – A Historical Perspective

Umezuruike Linus Opara

Introduction

Engineering techniques have long been used to increase agricultural productivity (both on and off the farm). They can be traced back to when early human gatherers settled to cultivate the land and stored some of the produce for seasons of bad weather. Today's agriculture, on the other hand, is dramatically different in many ways, and the move from subsistence to market-oriented commercial agriculture is unquestionably one of the major characteristics of human development. With the rise of other industries supplying non-agricultural goods and services, farmers were able to produce not just enough for themselves and their immediate families but also others involved in large-scale food and raw material processing and non-agricultural sectors of the economy. This necessitated both substantial increases in cultivated areas as well as improvements in agricultural productivity, which required increased energy consumption and more efficient energy sources for land cultivation, harvesting, transportation, and processing. The pioneers of the agricultural engineering profession were motivated by these difficulties to pursue their goals.

In the past 20th century, advances in agricultural engineering science and technology have drastically altered how we farm and how agriculture contributes to society. New machinery and process techniques for the production, postharvest management, and use of food and agricultural raw materials have come from advances in agricultural engineering. These advancements were synonymous with the terms "tractorisation" and "mechanisation" in the early years of this century, reflecting the need for increased power input to agriculture. With the evolution of agricultural engineering as a profession and academic discipline, its activities have expanded dramatically to include information systems technology for production, handling, and processing, as well as improved methods for efficient and sustainable land, water, and other natural resource utilisation.

Nowadays, agricultural engineering is undoubtedly at a crossroads, and there is constant worldwide debate about the future of the discipline and profession. Some forecasts foresee the decline and/or extinction of agricultural engineering as an academic discipline, while many others have proposed remedial measures to enhance student enrolment and professional membership. Concerns about the future relevance, legitimacy, vitality, and growth of agricultural disciplines (Just & Rausser, 1993), as well as non-agricultural disciplines (Alha et al., 2000; Kilsby, 1999; Becker, 2000; Colander, 2000; Maroni, 1999; Younes, 2000; Vohra, 2000) are not limited to agricultural engineering. Paarlberg (1992) argued that "the agricultural disciplines – agricultural economics, agricultural engineering, agricultural chemistry, agricultural business, and agricultural statistics – will gradually be absorbed by their parent disciplines".

While we have witnessed the closure of many agricultural engineering departments and programmes (most notably in developed countries), often concurrent with the dismantling of Faculties/Colleges of Agriculture in tertiary institutions, we must not sit back and accept the eventual demise of both the discipline and our professional association. We must prepare future agricultural engineers to respond adequately to global problems such as environmental degradation and sustainable natural resource use, as well as strengthen our traditional roles in improving food security and poverty alleviation by developing and applying more efficient production and agro-processing technologies.

In response to dwindling student enrolments and the need to address these critical global challenges, several colleges, particularly in North America, have modified their names and/or programmes to emphasise Biology, Mathematics, and the Environment. Using the terms "Bioengineering", "Biological engineering", "Biosystems", "Bioresource", and "Natural Resources", alone or in combination with others (such as Food, Agricultural Engineering), is a prevalent and developing trend. As a result of these trends, Biology, Engineering, and Systems have been championed as the foundation of our profession's future.

While the developed (Western) world has been debating how to ensure the survival of the agricultural engineering profession, most developing countries are still struggling to feed themselves, partially due to insufficient food production, inefficient postharvest handling and distribution systems, and man-made disasters. Agriculture is still practised at the subsistence level, and adequate mechanisation technologies for improving

DOI: 10.1201/9780429150111-3

the productivity and efficiency of rural family farms are still needed. Traditional courses such as farm mechanisation and rural processing make up a significant portion of agricultural engineering programmes in many of these countries' tertiary institutions, and many developing countries continue to rely on Western universities for postgraduate agricultural engineering qualifications. Agriculture globalisation and agricultural engineering education pose new problems for both developed and developing countries in the twenty-first century.

The objectives of this chapter are to (a) discuss the evolution of the agricultural engineering discipline and examine its role in socio-economic development over the last century; (b) highlight external factors influencing the professional future; and (c) highlight some critical tasks in addressing major global challenges facing society.

Origin of Agricultural Engineering

The first agricultural engineers were engineering experts (mostly military, civil, and mechanical) and other scientists who utilised their skills and abilities to tackle practical problems in agriculture and the broad rural environment. During the American Civil War, Col. Samuel Walker Shattuk, credited as the first Agricultural Engineer, was thought to be an army engineer. When the University of Illinois at Urbana first admitted students on March 2, 1868, he was hired as an Assistant Professor of Civil Engineering and Instructor in Mathematics (Anon, 1965). The next year, Col. Shattuck was designated as a Professor of Civil Engineering, Instructor in Mathematics, and Professor of Agricultural Engineering at the College of Agriculture.

During the same period, a vacant Rural Engineering position was noted on the academic roster. Early records at the University of Illinois indicated that Col. Shattuck conducted drainage and sewerage research and outreach activities to meet the requirements of local farmers. In the April 1932 edition of *Agricultural Engineering*, George Washington, the first President of the USA, was cited as an agricultural engineer because he tackled practical challenges on his Mount Vernon estate as an engineer. Agricultural engineering's aims and objectives during these early years were "to boost the productive capacity of the farm worker and to aid him in utilising the income from the surplus produced for those things which contribute to the general wellbeing of rural life" (Anon, 1965). The American Society of Agricultural Engineers was founded in 1907 by a group of engineers with various professional backgrounds who were engaged in agricultural engineering.

Agricultural engineering was first listed as a course of study in any academic institution, including universities, back in 1896 in a catalogue of the University of Nebraska (Stewart, 1979). Thereafter, in 1906 at the University of Illinois, a conference was held by F.R. Crane (host), C.A. O'Cock (University of Wisconson), and J.B. Davidson (Iowa State College) to deliberate teaching techniques and the development of educational materials for a new profession of agricultural engineering. It was during the second meeting in the agricultural engineering building at the University of Wisconsin in Madison (27-28 December 1907) that the American Society of Agricultural Engineering (now the American Society of Agricultural and Biological Engineering) was founded. Subsequently, Iowa State College (now Iowa State University) became the first institution to confer a degree in Agricultural Engineering (Bachelor of Science) in 1910.

As the profession grew and evolved to meet the engineering challenges of contemporary agriculture, new academic and research departments of agricultural engineering were founded, initially and primarily in the United States and eventually in other areas of the world. In the early 1960s, diminishing agricultural price supports (and profits) prompted some authors to rethink agriculture's economic function (Whitten, 1965). At the same time, it was discovered that farmers' abilities to produce had outpaced their ability to efficiently harvest, store, and process the many food and fibre commodities at a cost equivalent to that required in other industries (Hobgood, 1964).

Notwithstanding these challenges and opportunities, the agricultural engineering profession soon encountered several issues and questions ranging from low enrolment to future course content and curricula needs, obsolescence, and public relations. Addressing the problems associated with a low profile and poor image was deemed especially important by J.W. Borden, president (1963–1964) of the American Society of Agricultural Engineers (ASAE),: "I believe that we are a meek, modest group dedicated to technical excellence and to achieving this objective – yet nobody knows we are in town" (cited by Hobgood, 1964). Many authors also questioned the viability of existing professional agricultural engineering programmes, as well as the utility of the many agricultural para-engineering qualifications available at the time, such as Agricultural Equipment Science, Agricultural Mechanics, Agricultural Engineering Technology, and Agricultural Mechanisation (Stewart, 1965; Hunt, 1965).

Many authors have explored these concerns over the last decade, and the debate over the exact future of the agricultural engineering profession, including the name (Cuello, 1992; Opara, 2000a) and curriculum content, remains unresolved (Wall, 1996). Nowadays, the term "Biological Engineering" appears to be the most generally used term in North America and across several institutions in Western Europe, and it represents the profession's new order (Cuello, 1992; Wells & Taylor, 1995, 1996). It should be noted that in regions of Europe and Francophone Africa, some agricultural engineering academic/research departments or programmes have retained the historical name of Rural Engineering ("Génie Rural"). agricultural engineering has thus successfully grown and evolved through the millennium from a rural-based engineering profession with agriculture at its core to a diverse discipline that includes engineers with skills and expertise in designing and managing mechanical, biological, and expert systems for agriculture and other process industries.

What Is Agricultural Engineering?

One of the significant dilemmas that would-be agricultural engineers have is determining whether the profession is agricultural or engineering. I had this issue as an undergraduate agricultural engineering student over 40 years ago! Even though my department (of Agricultural Engineering) was part of the Faculty of Engineering at the University of Nigeria, this unsettling feeling did not go away. Although the compulsory courses in Animal Science, Crop Science, and Soil Science (offered by the Faculty of Agriculture) made us better agricultural engineers, they also enhanced our sense of 'a bit of agriculture and a bit of engineering' as students. Most of the time, the profession's impression (and even emphasis) was on 'agriculture' rather than 'engineering'. Among other factors, this dilemma has contributed to the emergence of Biological as the new order of the profession, allowing the discipline to break through its restrictive and oppressive "agricultural ceiling" and liberate itself to ascend and explore new heights of opportunities and challenges (Cuello, 1992). Apart from the emotional implications of the agricultural engineering identity crisis, the principles and definitions of the new term "Biological Engineering" will be examined further in the following section. But first, basic definitions of engineering and agricultural engineering are considered.

The ultimate goal of the engineering profession is to improve human welfare. It is defined as "the profession in which knowledge of the mathematical and natural sciences obtained through study, experience, and practise is applied with judgement to find ways to economically utilise the materials and forces of nature for the benefit of mankind" (ABET, 1986). Engineering is clearly a diverse profession that has traditionally been separated into numerous divisions such as Civil, Mechanical, Electrical, Chemical, and so on. In response to its ever-expanding technological basis and scientific frontiers, new sub-divisions and specialised fields of engineering continue to arise. Several agricultural engineering educational and research programmes were part of traditional and well-established engineering disciplines, such as Civil or Mechanical Engineering, throughout their formative years as a profession.

Agricultural engineering arose in reaction to the ever-increasing need to deploy efficient engineering systems to boost agricultural productivity, first focusing on designing and applying mechanical systems to reduce the drudgery of human muscle. agricultural engineering, in a nutshell, is the application of engineering in agriculture, both on and off the farm. For diverse purposes, several distinct definitions of agricultural engineering and agricultural/farm mechanisation have been proposed. For example, in Australia, the Commonwealth Council for Rural Research and Extension (1981) defined agricultural engineering as "the application of engineering principles and practices to the solution of agricultural production problems in the major areas of Soil and Water, Power and Machinery, Farm Structures, and Environment". Farm mechanisation, on the other hand, was defined as "the effective application of machines, systems, structures, and instrumentation in agricultural production systems". While agricultural engineering is oriented towards design and emphasises mathematics and physical sciences, farm mechanisation places a much greater emphasis on agricultural science, technology, and economics due to its focus on the farming business and its operations (O'Shea & Manson, 1985). The study and understanding of complex interactions between machines and biological systems, as well as the design of devices and processes capable of working under these conditions, distinguishes agricultural engineering from other engineering disciplines.

Agricultural mechanisation was a key aim of agricultural engineering during the past century. The rationale for this focus was summarised by Professor E.U. Odigboh (1985) during his Inaugural Professorial Lecture at the University of Nigeria, Nsukka: "My hands will bleed if I dig into the earth's crust with my hands. If I place a spade between my hand and the earth, the crust breaks, and my hands remain entire as a result of my body's labour ". The author expanded on this inspiring logic, saying, "Let a machine be placed between the man and the spade; the spade cuts, the machine labours, the field is tilled, and the man is spared to devote his attention to other duties, to greater levels of human endeavour".

During the early development of the discipline, the various technical specialisations within the agricultural engineering profession could be broadly classified as (a) farm power and machinery, and (b) soil and water engineering. Numerous new specialised areas are becoming increasingly important, such as postharvest and processing engineering, sustainable energy systems, information and expert systems, and environmental engineering, resulting in an overlap between the profession and the traditional engineering disciplines from which it originated.

The Emergence of Biological and Biosystems Engineering

Faced with declining undergraduate intakes and funding, as well as a generally negative public perception of the profession, the debate that arose in the 1960s, primarily in the United States, on the future of the agricultural engineering profession resurfaced during the final decade of the twentieth century. The recent debate centred primarily on an appropriate name for the profession and the necessity of developing agricultural engineering (like other engineering professions) on the basis of fundamental science (such as physics, chemistry, or biology) rather than on agriculture only (Cuello, 1992). As potential alternatives to "Agricultural", the terms "Biological" and "Environmental" dominated the debate, although there were some strong opponents to the idea of a name change (Boyce, 1996). The term "Biological Engineering" was well received in North America and some parts of Western Europe, and as a result, many agricultural engineering departments in these parts of the world have included the terms biological or environmental engineering in their names, either additionally or as an alternative. Some educational institutions have likewise revised their curricula to reflect these new developments.

In North America, one of the reasons for the predominance of "Biological Engineering" as the new order of the agricultural engineering profession is that biology becomes its scientific base, just as Civil and Mechanical Engineering have physics and Chemical Engineering has chemistry. Agriculture was thought to be a limited area of applied biology. It has also been suggested that the new name lacks internal overlaps between its sub-areas, allowing the discipline to break beyond its limiting "agricultural ceiling" as biological science incorporates agriculture (Cuello, 1992). Biological engineering has been defined as a biologically based engineering discipline that integrates traditional engineering themes, engineering design, and applied biological sciences into a unified curriculum (Wells and Taylor, 1995). It applies basic, applied, and engineering sciences to solving problems involving biological systems and their environments. In practice, Biological Engineering consists of two major components: (a) the characterisation, measurement, and representation of systemic processes within an organism or patterns of relationships between organisms and their environment, and (b) engineering design to develop processes and systems that influence, control, or use biological materials and organisms for the benefit of society (Wells and Taylor, 1995). Individual plants and animals that compose a biological domain, as well as processes that use prokaryotic and/or eukaryotic organisms, are examples of such biological systems. These definitions serve as the foundation for creating a disciplinary curriculum in biological engineering.

Biosystems and Bioresources Engineering in Europe and Africa

It will take time to determine whether the new name and planned new curricula (Wells and Taylor, 1996) will have the desired impact on the profession's student numbers, R&D funding, and public image. Nonetheless, it does allow for the profession's constant evolution in response to the external environment (within academic governance structures and society at large). Yet, certain facets of this new order of the agricultural engineering profession necessitate an immediate plan to address both the educational and professional aspects. Worldwide, the bulk of undergraduate agricultural engineering programmes and students are found in developing and transitional countries where the name and curriculum content are generally aligned with the traditional agricultural engineering areas of specialisation. This raises the following questions: Is there a need for international harmonisation of names and curricula? Is this realistic while there are still major variations in agricultural development between the global North and South? What would the effect of current and planned reforms and redesigns have on the global mobility of young professional agricultural engineers?

To adequately address these challenges, a debate on the future role of agricultural engineering in sustainable development is required in developing countries and those in transition. Given the growing impact of globalisation on agriculture, engineering, and other sectors of the economy, a dialogue between agricultural engineers in both the North and South is also required to identify opportunities for possible curricula harmonisation and strengthening of collaborations in research and educational programmes. According to Wall (1996), the curriculum in agricultural engineering programmes has been under significant pressure; it is on the verge of becoming overcrowded if additional content is simply added to account for the growth in knowledge (particularly in the biological sciences) and to accommodate pressures from the profession and society at large. The International Commission of Agricultural and Biosystems Engineering (CIGR) is uniquely positioned to promote and facilitate these initiatives.

A Century of Agricultural Engineering Success

According to the growing scientific literature on agricultural engineering, what began as a mere utilitarian term has blossomed over the last century into a vibrant adolescent discipline with rich connections to fundamental (mathematics, physics, biology, chemistry) and applied (agriculture, engineering) sciences. Agricultural engineering approaches have transformed both conceptual and empirical interpretations of agricultural production systems and engineering processes for the transformation of food and other agricultural raw resources. Despite its unappealing reputation, it sits undeniably among the legitimate engineering disciplines that have greatly impacted human life in the twentieth century and beyond, primarily through agricultural mechanisation.

Faced with the threat of starvation because of the rapidly increasing human population and low agricultural productivity per labour force, agricultural mechanisation assisted in transforming vast and unproductive hectares of land into highly efficient and productive estates, removing the drudgery of manual labour. It was so fitting that agricultural engineering shared the limelight at the close of the twentieth century when Agricultural Mechanisation was recognised as one of the top 20 engineering achievements of the previous millennium (Anon, 2000; Shirley, 2000). Agricultural Mechanisation was ranked 7th on the list, alongside such engineering accomplishments as electricity, automobiles, computers, and safe and abundant water supplies. Numerous agricultural machines, such as the tractor, reaper, and combine harvester, greatly boosted farm efficiency and output, freeing up the majority of the developed world's people to industrialise the world. Hence, Agricultural Mechanisation's success was not restricted to agriculture.

In a separate survey, the American Society of Mechanical Engineers ranked Agricultural Mechanisation fourth among the top ten engineering feats of the twentieth century, alongside the automobile, the Apollo space missions, power generation, the aeroplane, integrated circuit mass production, air conditioning, and refrigeration (Falcioni, 2000). Nowadays, the abundance of high-quality, low-cost food in industrialised countries is attributed to better mechanical systems and processes in food production, handling, processing, and distribution. The challenge of the twenty-first century is to create sustainable systems that can be implemented indefinitely to eliminate food insecurity, poverty, and malnutrition in developing countries.

Advancing Agricultural Engineering in a Competitive Global Environment

Several significant trends influence the competitive conditions in which agricultural engineering professionals work. Some of these trends are more prevalent in developed countries than others. This section of the essay identifies these key trends briefly, and in the following section, I will discuss the issues confronting the agricultural engineering profession in the new millennium within the context of these competitive trends.

Declining Terms of Trade Affecting Agriculture. Several indicators, such as low farm earnings, a decreased share of farming in the national GDP, and a declining labour force involved in agriculture, illustrate how the declining terms of trade are harming agriculture, particularly in developed countries. The non-farm sector is also becoming more significant to the economy. Using data from UK agriculture, McInerny (1999) accurately described the nature of these developments. Karl Marx recognised during the nineteenth century that as economies expand, the value of agricultural products declines steadily as a share of both weekly household expenditure and the country's GDP. Farmers will see diminishing terms of trade because, while returns on farm produce tend to keep up with general economic growth, the price of agricultural inputs does not.

Supply Chain Management. The idea that farmers are the only ones who generate our food has been thrown into doubt by advancements in supply chain management and integrated food supply systems. From farm to plate, the food supply chain has become more networked and integrated. Just around 25% of the total value is contributed by agriculture, including the 10% that the input industries provide. The term "supply chain management" refers to the control of the complete system of manufacturing, distribution, and marketing procedures used to provide a consumer with a desired good. It offers resources for controlling quality, unpredictability, and particular customer values. It offers resources for managing client values, quality, and variability. Innovative solutions that preserve product integrity, quality, and safety at all chain levels are necessary for successful supply chain management.

Food Quality. Customers are increasingly selective about the quality of goods and services they purchase, and there is a general shift away from quantity and towards quality. The consumer is unquestionably the king! This trend has also resulted in a significant paradigm change away from supply-push agriculture and towards demand-pull agriculture. Consumers are becoming more willing to pay for desired product quality characteristics as well as other perceived quality indices, including the production process, effects on the environment, and place of origin. For the online evaluation of critical quality criteria, new technologies are needed to meet these needs.

Food Safety and Traceability. Public concern over how we produce and handle food has grown due to the prevalence of food-related illnesses like Bovine Spongiform Encephalopathy (BSE) in Europe and other food poisoning incidents. Consumers increasingly demand openness and traceability in the food supply chain, in addition to the challenge of finding a solution to the problem (Opara & Mazaud, 2001). The major challenges facing agriculture and the food business are preventing food safety hazards, identifying such hazards in the food chain, and the degree of responsiveness in adopting corrective actions.

Urbanisation and New Lifestyles. People are increasingly congregating in cities, and this trend is accelerating in developing countries. This has consequences for how and what we consume, as well as when and where we buy our food. Convenience is becoming increasingly important in our food consumption patterns, with the number of meals eaten outside the home constantly growing. Technology for ready-to-cook and ready-to-eat foods has been created, and future advancements in this area are expected. Because of the growing population density in major cities, urban agriculture and landscaping have expanded, providing additional sources of food and leisure activities.

Environmental Issues. It is now widely acknowledged that contemporary agriculture must be based on environmentally friendly practices that do not harm the environment or the ecological system. One of the critiques levelled at large-scale agricultural mechanisation is that it promotes the destruction of natural ecosystems and, if handled incorrectly, can lead to soil erosion. Agriculture in the next millennium must seek to generate enough food for humanity while minimising negative environmental and worker impacts.

Ethical Issues. Consumers, special interest organisations, and certain governments have worked hard to improve agricultural ethics. Many consumers and governments are now concerned about animal welfare, worker welfare and safety, and the exploitation of child labour. Some of these difficulties impact how farm animals are fed, housed, transported, and slaughtered. These critical components must be addressed in future engineering breakthroughs in these areas.

New and Emerging Technologies. The past half-century has seen astounding breakthroughs in information technology and computer science. Agricultural engineering has harnessed the Information Revolution through global positioning systems and geographic information systems (GPS/GIS) and utilised it to achieve precision farming. Technological advancements have transformed how we apply fertiliser and seed, as well as yield and maturity control. The World Wide Web (WWW) has aided in the dissemination and availability of scientific knowledge and data. Nanotechnology and biotechnology are two technological advancements with tremendous opportunities for future agricultural transformation for widespread economic development. Agricultural engineers cannot ignore these changes and their possible effects on the plants and environments with which we operate.

Globalisation. The world is now commonly referred to as a small village. Practically every aspect of our lives is becoming more global. So, what exactly is globalisation? It is the emergence of a set of sequences and processes that are more unhindered by territorial or jurisdictional constraints and, as a result, promote the expansion of transnational practices in economic, political, cultural, and social spheres (CSGR, 1998). As trade grows, so does competition for commodities and professional services. Transnational agriculture enterprises and worldwide supermarkets are becoming increasingly important in our food system's globalisation. Globalisation impacts everything from agricultural supplies and food products to farm labour. It also impacts agricultural engineering professionals' mobility, notably from developing to developed countries. This latter scenario relates to the topic of agricultural engineering education and curriculum quality. Because of significant curriculum similarities, many graduate students from developing nations have been able to complete their postgraduate studies in developed countries. Others have also gotten professional jobs in the global labour market. This flow of agricultural engineers and other professionals will undoubtedly continue. Nonetheless, it is expected that the ongoing changes in the names and curriculum of agricultural engineering programmes in most parts of North America and Western Europe will make a truly global profession challenging.

In addition to these trends, negative images and an uncertain outlook dominate agricultural commentary in the Western media: the controversy over the impact of genetically modified organisms and biotechnology on agriculture and human health, the emergence of food-related diseases such as mad cow disease in humans and other food safety issues, concerns about animal welfare, and the negative environmental impacts of agriculture. These developments and difficulties, taken combined, pose more challenging tasks for agricultural engineers working in the area.

Future Crucial Roles for Agricultural Engineering

The ongoing global competitive environment offers both problems and opportunities for agricultural engineering. The primary goal of agriculture in the previous millennium was to boost productivity, particularly on farms, through the mechanisation of production processes. Today and into the future, new existential challenges have compounded the historical tasks facing agricultural engineering, as expounded in the Sustainable Development Goals (SDGs) and Agenda 2063 – the Africa We Want. Considering the role of engineering in the sustainable agricultural transformation and modernisation necessary to meet the goals and aspirations enshrined in these global blueprints for socio-economic development, the following objectives will contribute to a meaningful impact:

a) increasing agricultural productivity in a way that is effective and helps to reverse the deteriorating terms of trade of agriculture;
b) design and development of novel mechanical systems and processes that enable environmental preservation and sustainable agriculture, especially on marginal lands;
c) development of measurement methods and criteria for evaluating new agricultural technologies' long-term effects on the environment and ecosystems;
d) development of reliable, non-invasive and cost-effective methods for the rapid evaluation of food safety and quality along the supply chain;
e) development of engineering processes and technologies that take advantage of the possibilities of new fields of study and technologies (such as nanoscience, information and computer technology, and biotechnology) in agriculture;
f) develop engineering systems and cost-effective technologies suitable for the difficulties facing agricultural and food supply in resource-poor developing nations. It is appropriate to draw attention to the pressing need for technologies to aid in the growth and commercialisation of these regions' agro-industrial and postharvest sectors. One of the issues preventing overall economic development, food insecurity, and poverty in the least developed nations is the high postharvest losses of food material and the absence of a thriving agro-industrial sector (Opara, 2000b).

Agricultural engineers will need to combine strong engineering abilities with an in-depth knowledge of the interactions between biological systems and physical and process technologies to carry out these tasks. Agricultural engineers must cooperate increasingly with other disciplines to accomplish the intended goals considering growing consumer (and society at large) concerns for food and environmental quality and safety. Many consumers are no longer willing to sacrifice the environment to increase agricultural production, unlike in the early days of agricultural mechanisation when there was limited understanding and awareness of the long-term negative impacts (such as erosion, pollution, and health hazards) of mechanisation on the environment and people. This astonishing progress obliges agricultural engineers to adopt new technologies, competencies, and attitudes.

What Does the Future Hold for Agricultural Engineering?

There is no question that agriculture (and the provision of food, feed and fibre) is necessary for human survival. So, it follows that engineering will keep performing challenging roles in agriculture in both developed and developing nations. Nonetheless, it should be noted that agricultural engineering's prospects are intimately linked with the sectors it supports and with other competitive trends in society at large. A consistent fall in undergraduate enrolment and a chronic reduction or lack of funding for programmes are two of the most obvious indicators of difficult times ahead for agricultural engineering education and research. When relevant academic departments, research institutions and/or their programmes are renamed or eliminated, the problem of the poor image, profile, and public awareness of the discipline and profession is worsened even more. These are all issues that agricultural engineering is currently dealing with in many parts of the world.

To remain relevant and survive into the next century, new approaches are needed to boost student enrolment, professional society membership, finance, and public perception of agricultural engineering. Many institutions (mostly in North America) have established a new focus on biological engineering. This brave step requires the backing of our professionals. Secondly, this step addresses the issues of identity, ambiguity, and image, allowing the discipline to embrace biology as a fundamental science for the agricultural engineering profession. The other basic sciences – physics, chemistry, and mathematics – will undoubtedly continue to play a significant role in agricultural and biosystems/biological engineering education. Second, the transition to biology allows us to play a prominent and visible part in the current biological revolution, which is exemplified by the enormous advancements in and prospects of biotechnology in agriculture and other areas.

While it is envisaged that changing names and redesigning curricula will help attract more talented students, it is also vital to establish solutions to address low-profile challenges in society at large. This necessitates that we seize the opportunities given by current communication technology while simultaneously taking an interest in and getting involved in non-technical issues that are important to the larger community. With many agricultural engineering universities renaming, restructuring, and/or closing, we must ensure that prospective students have clearly identified professional bodies with which they may identify and strive to belong. Renaming training and research programmes without carefully considering the possibility of new names for the adhering professional associations and networks, and the focuses of these professional societies, could further undermine the future of agricultural engineering.

Regarding the breadth of topics the discipline covers, we must capitalise on agriculture's multifunctionality and expand into the postproduction sector. Certainly, our profession's current link with farming, in general, has not served us well in the public sphere. The on-farm sector is widely seen as filthy, low-paying, and only suitable for the less academically gifted! Efforts concentrated in this area are less likely to lure bright young stars into the agricultural engineering profession, especially given the on-farm sector's diminishing terms of trade.

Given the evident variations in agricultural systems and overall economic development between developed and developing countries, a distinct strategy is required in determining the future of agricultural engineering in various regions (Opara, 2000a). Agriculture in developed countries (DCs) is highly modernised and mechanised, with large and medium farm sizes and a small number of farmers. As a result, many crops and goods are overproduced in these locations, with a higher emphasis on quality than on quantity. On the other hand, agriculture is still primarily practised for subsistence in most of the least developed countries (LDCs), with small and scattered farm sizes. Crop productivity and product quality remain low, and food insecurity, poverty, and malnutrition are serious obstacles to long-term development. These socioeconomic issues would shape agricultural engineering's future and its contribution to these societies.

Consider the keynote lectures at the International Conference, AgEng 2000, 2–7 July, held at the University of Warwick, UK, which highlighted the necessity of diverse tactics for agricultural engineering development in DCs and LDCs. Several of the papers discussed "advanced engineering" issues that were prevalent in European and North American research and education programmes, such as biotechnology, microelectromechanical systems, magnetoresistance technology, large-scale precision agriculture, advanced integration systems, and postharvest innovations (Witney, 2000). However, one speaker stated that agricultural engineering had produced few concrete contributions to the wellbeing of resource-poor farmers in developing countries because too many programmes and tools have been too expensive, too difficult, too huge, or too fragile (Bunch, 2000). To be effective, the author proposed that agricultural engineers must become more aware of the needs of poorer farmers and focus specific technologies on this group through participatory or farmer-led research. In the same year, on the other hand, the themes of an International Conference on Agricultural Engineering held in Ghana ("Engineering contributions to food security in developing countries") and Nigeria ("Engineering challenges in agriculture and poverty alleviation in the new millennium") pointed to the nature of the problems confronting agricultural engineers in many other LDCs. While domestic and regional challenges may differ, future agricultural engineers must become lifelong learners that are equipped with sufficient technical knowledge as well as relevant soft skills to address both local and global challenges. This presents a challenge and an opportunity for agricultural and biosystems/biological engineering education for sustainable development.

Conclusions

Agricultural engineering grew as a genuine and mature engineering discipline over the twentieth century. Historically, it arose from other schools of engineering (most notably civil and mechanical) and concentrated on the mechanisation of agriculture as its primary domain in the formative years. Today, the profession encompasses a wide range of on-farm and off-farm sub-disciplines, including postharvest engineering, forestry engineering, food engineering, renewable energy, waste and water engineering, bioprocess and biochemical engineering, computers and electronics in agriculture, etc. The profession has produced enormous gains in agriculture and the rural environment during the last 100 years, but the mechanisation of agriculture has been recognised as one of the top 20 greatest engineering achievements of the twentieth century.

While these wonderful accomplishments are acknowledged, it is also appropriate to evaluate how the profession may need to evolve to thrive in the face of competitive trends in the 21st century environment. Some of these trends include agriculture's declining importance in the global economy, the globalisation of the world economy, society's demand for sustainable agricultural practices and environmental protection, the emergence of new technologies and commerce such as information/computer technology and biotechnology, and a shift in the agrifood sector from supply-push to demand-pull. If these trends continue, they have the potential to significantly transform agriculture and food systems, including with regard to what and how we eat and how society utilises natural resources for food and nutrition security. As a result, these trends pose a significant challenge and opportunity for agricultural engineers in the new millennium. The ongoing discussion in DCs about the future of agricultural engineering has resulted in a significant shift towards biological engineering as a new order of agricultural engineering. A similar debate is needed in LDCs and around the world about the influence of these competitive tendencies on the future role of agricultural engineering in these nations, including desirable educational objectives and research tactics.

One of the most significant problems facing humanity in the current millennium is to further improve sustainable agriculture that satisfies the rising world population's food insecurity, and to meet these raw material needs without negatively impacting the environment and ecosystems. This is an important task for agricultural engineering in the new competitive global environment. Given the negative messages and general confusion about where agriculture is heading in the Anthropocene era, as well as a sense among farming communities that something new and unpleasant is happening, this is no easy task. To make this a reality, we must work closely and jointly with colleagues and other disciplines across the world. This includes the pursuit of a relevant agricultural and biosystems/biological engineering education which addresses the socio-economic, environmental, and ecological challenges facing humanity, now and into the future.

REFERENCE LIST

ABET. (1986). *Annual Report 1985.* Accreditation Board for Engineering and Technology, New York.

Alha, K., Holliger, C., Larsen, B.C., Purcell, P. & Rauch, W. (2000). Environmental engineering education - Summary report of the 1st European seminar. *Water Science and Technology*, 41(2), 1–7.

Anon. (1965). First agricultural engineer? *Agricultural Engineering*, 46(8), 414.

Anon. (2000). Agricultural mechanisation named among top 20 engineering achievements. *Resource*, 7(4), 17.

Becker, W.E. (2000). Teaching economics in the 21st century. *Journal of Economic Perspectives*, 14(1), 109–119.

Boyce, D.S. (1996). An important debate - The future of agricultural engineering. *Landwards* (Summer), 20–21.

Bunch, R. (2000). Keeping it simple: What resource-poor farmers will need from Agricultural Engineers during the next decade. *Journal of Agricultural Engineering Research*, 76(3), 305–308.

Colander, D. (2000). New millenium economics: How did it get this way, and what way is it? *Journal of Economic Perspectives*, 14(1), 121–132.

Centre for the Study of Globalisation and Regionalisation. (1998). *Research Agenda.* Centre for the Study of Globalisation and Regionalisation, University of Warwick: Warwick.

Commonwealth Council for Rural Research and Extension. (1981). *Agricultural Engineering in Australia.* Australian Government Publishing Service, Canberra, pp. 2–3.

Cuello, J.L. (1992). Toward a new engineering order: Biological engineering breaks through the 'agricultural ceiling'. *Agricultural Engineering* (Nov.), 17–19.

Falcioni, J.G. (2000). 20th century engineering feats. Editorial. *Mechanical Engineering*, 122(1), 4.

Hobgood, P. (1964). Profession of the future. *Agricultural Engineering*, 46(7), 379, 393.

Hunt, D.R. (1965). Is there is need for a curriculum in agricultural mechanisation? *Agricultural Engineering*, 46(11), 621, 646.

Witney, B.D. (2000). Editorial. Special issue. *Journal of Agricultural Engineering Research*, 76(3), 217. doi:10.1006/jaer.2000.0582

Just, R.E. & Rausser, G.C. (1993). The governance structure of agricultural science and agricultural economics: A call to arms. *American Journal of Agricultural·Economics*, 75, 69–83.

Kilsby, D.C. (1999). Food microbiology: The challenges for the future. *International Journal of Food Microbiology*, 50, 59–63.

Maroni, M. (1999). Toward an internationally harmonised, multiprofessional educational program in indoor air sciences: Needs and strategies. In: N. Bosch (ed.), *Education and Training in Indoor Air Sciences*, Vol. 60, pp. 21–26. Budapest: Kluwer Academic Puhl.

Mclnerney, J. (1999). Agriculture at the crossroads. *Journal of the Royal Agricultural Society of England*, 160, 8–27.

O'Shea, J.A. & Manson, J.L. (1985). The contribution of science to Australian agricultural engineering. 11. Education for agricultural engineering and farm mechanisation. *The Journal of the Australian Institute of Agricultural Science*, 51(4), 272–274.

Odigboh, E.U. (1985). Mechanisation of cassava production and processing: A decade of design and development. Inaugural Lecture Series No. 8, University of Nigeria, Nuskka, 52 pages.

Opara, L.U. (2000a). Current trends and future prospects for agricultural engineering and implications for developing countries. Paper No. RD001, presented at the International Conference, AgEng 2000, 2–7 July, University of Warwick.

Opara, L.U. (2000b). The role of post-harvest management in food security and poverty alleviation. AGSI Seminar, 28 November. FAO, Rome.

Opara, L.U. & Mazaud, F. (2001). Food traceability from field to plate. *Outlook on Agriculture*, 30(4), 239–258.

Paarlberg, D. (1992). The land grant college system in transition. *Choices*, (Third Quarter), 45.

Shirley, M. (2000). Agricultural mechanisation among century's top engineering achievements. *Institutions of Agricultural Engineers - Newsletter* (spring), 3–5.

Stewart, R.E. (1965). Is there a need for a curriculum in agricultural mechanisation? *Agricultural Engineering*, 46(9), 493.

Stewart, R.E. (1979). *7 Decades that changed America: A history of the American Society of Agricultural Engineers 1907–1977.* Michigan: MI: ASAE Publication. St. Joseph.

Vohra, F.C. (2000). Changing trends in biology education: An international perspective. *Biology International*, 39, 49–55.

Wall, G.L. (1996). *Curriculum Over-Crowding - A Common Problem or Not?* Department of Agricultural Engineering, Massey University, Palmerston North, 8 pages.

Wells, J.H. & Taylor, T.A. (1995). A contemporary vision of biological engineering education. *Resource* (July), 13–15.

Wells, J.H. & Taylor, T.A. (1996). First principles in biological engineering education. *Resource* (April), 12, 22.

Whitten, J.L. (1965). Agriculture's role in the economy. *Agricultural Engineering*, 46(10): 553.

Younes, T. (2000). Biological education: Challenges of the 21st century. *Biology International*, 39, 8–13.

3

Agricultural Engineering in the Context of SDGs: A Quest for Global Relevance

Kehdinga George Fomunyam and Umezuruike Linus Opara

Introduction

The last century was marked by rapid progress in all human endeavours. From discoveries in medicine which reduced and, in some cases, averted widespread death from diseases and pandemics to improvements in housing, water supplies, sanitation, transport, and space travel, developments in science, engineering, and technology have enabled us to improve our quality of life. However, it is now well established that these glorious achievements and general economic progress have come at huge cost, one that threatens human existence on planet earth. The over-exploitation of natural resources – below and above ground – and wasteful consumption have resulted in massive destruction of forests and other ecosystems which are essential in sustaining the lives of humans, animals, and plants. Widespread environmental degradation, pollution of air, water resource depletion and contamination, greenhouse gas emissions (GHGEs) and rising temperatures, and biodiversity loss and extinction are now major challenges facing the world. Above all, climate change, which is the manifestation of these negative impacts of humans on planet earth, has become an existential problem confronting humanity. In all these challenges, agriculture is both a contributor and victim. Therefore, agricultural, and biosystems/biological engineering education and research has an important function in mitigating these challenges.

These existential challenges have led to the global call for a new development paradigm as articulated by the Member States of the United Nations in the 17 Sustainable Development Goals (SDGs) of Agenda 2030. The 7 Aspirations of Africa's Agenda 2063 – The Africa We Want are also underpinned by the tenets of sustainable development. Both Agenda 2030 and Agenda 2063 recognise the pivotal role of agriculture in sustainable development, especially in assuring food and nutrition security and eradicating poverty in rural economies. To achieve this, agricultural and biosystem engineering has critical roles to play, through the education and training of skilled workforce and in carrying out relevant research to produce new knowledge and technologies for sustainable agri-food systems modernisation and transformation. Climate-smart agriculture has become an imperative.

Therefore, the question currently begging for answers among agricultural engineers is: how can our discipline and profession achieve global relevance within the context of 21st century SDGs and Agenda 2063 – The Africa We Want? This chapter attempts to answer this question from multiple perspectives. We begin by reviewing the global agenda for sustainable development as encapsulated in Agenda 2030 and Africa's Agenda 2063 – The Africa We Want, both in the context of the relevance of agricultural engineering.

Source: From Millennium Development Goals (MDGs) to Sustainable Development Goals (SDGs): What Role for Agricultural and Biosystems Engineering Education?

The MDGs: 2000–2015

The initial global framework for collective and joint responsibility for development was established in the Millennium Development Goals (MDGs). In September 2000, the Member States of the United Nations signed the historic Millennium Declaration, outlining a set of 8 goals aimed at various objectives, including reducing extreme poverty and hunger, achieving universal primary education, and ensuring environmental sustainability, with a target date of 2015 (Table 3.1). The MDGs were hailed as ground-breaking when they were first introduced because they gave people a language through which to communicate and agree on global goals. They also included a clear system for measuring and monitoring progress. (SDGF, 2023). The fundamental purpose of the MDGs was to inspire the world community and offer a system of responsibility for measures to help millions of disadvantaged people improve their standard of living. The focus of the MDGs was clearly on the developing world.

The relevance of agricultural engineering to the MDGs can be clearly linked to Goals that address agriculture (Goal 1) and education (Goal 2) while recognising their cross-cutting impacts on the other Goals. Agriculture plays a pivotal role in eliminating extreme poverty and hunger (MDG 1), especially in developing countries and other land-based economies (Denning, 2023). Throughout history, improved agricultural

DOI: 10.1201/9780429150111-4

TABLE 3.1
The Millennium Development Goals (Source: SDGF, 2023)

Goal 1	Eradicate extreme poverty and hunger
Goal 2	Achieve universal primary education
Goal 3	Promote gender equality and empower women
Goal 4	Reduce child mortality
Goal 5	Improve maternal health
Goal 6	Combating HIV/AIDs, malaria, and other diseases
Goal 7	Ensure environmental sustainability
Goal 8	Develop a global partnership for development

productivity has been a fundamental driver of social and economic transformation in human societies. Asia provides the most convincing evidence of agriculture-driven poverty alleviation. This region has gone through remarkable structural change and economic growth during the last three decades (IFPRI ADB, 2007; Otsuka, 2007; Evenson & Golin, 2003). Hunger decreased from 30% to 16% over the same time span, while poverty decreased from 50% in the 1970s to 18% in 2004. Hunger decreased from 30% to 16% over the same time span, while poverty decreased from 50% in the 1970s to 18% in 2004. These achievements are largely related to increased agricultural productivity brought about by technological advancement and market liberalisation. Research conducted in four countries in Asia over the last twenty years discovered that households were able to escape poverty by diversifying their sources of income, moving away from rice, the region's main staple food, and towards other crops, animals, and non-farm sources. By investing in education and the growth of non-farm rural firms, increasing income from increased agricultural production created significant new prospects for higher incomes and employment in rural areas.

Africa's agricultural production increase over the previous 40 years lagged behind population growth, contrary to what was obtainable in Asia. For instance, over this period cereal yields in the region remained relatively constant at 1 metric ton per hectare (MT/ha), although they more than quadrupled in East Asia (IFPRI ADB, 2007; Otsuka, 2007; Djurfeldt, 2005). While there have been some advancements since 2000, Africa's agricultural gross domestic product (GDP) experienced negative growth rates during the 1980s and 1990s. The primary food crops, such as maize and root crops, saw nearly all of their production growth arising from an expansion of the cultivated area, with minimal contributions from increases in yield per hectare. The low agricultural productivity in Africa has been attributed to various factors, including inadequate infrastructure, high transportation costs, insufficient institutional support, political instability, diverse agro-ecological complexities, limited availability of suitable high-yielding crop varieties, and low use of fertilisers (Evenson & Golin, 2003; Djurfeldt, 2005). To reverse the stagnant or perhaps very low performance of agriculture for progressive socio-economic development in Africa, especially in reducing poverty and improving food and nutrition security, there is a need to boost agricultural productivity while protecting the environment. Agricultural engineering technologies including mechanisation, irrigation, and techniques and procedures for reducing preharvest and postharvest losses and food waste are key to achieving this goal.

The SDGs: 2015-2030

By the target year of 2015, some progress was made in achieving the MDGs; however, the primary goal of achieving widespread economic development in poor countries remained elusive. Agricultural productivity and overall production have remained at low levels, particularly in Africa and certain other developing regions, hindering substantial grassroots economic development and poverty reduction. As the MDGs' timeline neared its end, discussions on a post-2015 agenda gained momentum, emphasising the creation of a sustainable world where environmental sustainability, social inclusion, and economic development would be equally prioritised. Additionally, the focus shifted from predominantly poor developing countries to encompassing the entire world. Acknowledging the pressing and essential call for action, the United Nations General Assembly approved Resolution A/RES/70/1 on 25 September 2015 during its 70th Session. This resolution represents a new framework for reshaping our world: the 2030 Agenda for Sustainable Development. As stated by the UN (2015), "Sustainable development acknowledges the interconnectedness and interdependence of eradicating poverty in all its forms and dimensions, addressing inequality within and between nations, conserving the environment, fostering inclusive and sustainable economic growth, and promoting social inclusion".

UN's Agenda 2030 embodies 17 Sustainable Development Goals (SGDs) (Box 3.1). Through its provision of goods such as food, feed, fuel, and fibre, agriculture has direct functions to perform in multiple SDGs including Goal 1 (ending poverty), Goal 2 (eliminating hunger, food, and nutrition insecurity), Goal 8 (sustained economic growth and productive employment), and Goal 12 (sustainable consumption and production). To meet the targets for SDG 2 (Box 3.2), for instance, agri-food systems must provide families and communities with diverse and more nutrient-dense food to meet their food and nutrition requirements for healthy living and overall wellbeing. Value chains for foods high in micronutrients must be strengthened to increase consumer access to and affordability of such foods. In this regard, agricultural and biosystems engineering education must equip the workforce with the relevant knowledge and technical expertise in nutrition-sensitive agriculture and food systems transformation, from field to fork.

Other roles for agriculture that impact the SGDs include the provision of ecosystem services. Agriculture (including agronomy, horticulture, fisheries, and aquaculture) is the source of food and feed for human nutrition. The link between agriculture and human health has long been recognised, hence the renewed call for nutrition-sensitive agriculture (NSA) and nutrition-sensitive value chains (NSVCs) – the study and practice of raising plants and animals, along with postharvest handling, marketing, and distribution, in a way that promotes optimal human health and nutrition (Ruel et al., 2018; Allen & de Brauw, 2018). The history of economic development has demonstrated that where properly designed and implemented, agriculture has been the foundation for the successful rural development and industrialisation of nations and communities, especially where agriculture was the primary economic activity and means of livelihood (Johnson, 1993; Johnston & Mellor, 1961; Meijerink & Roza, 2007; Kjeldsen-Kragh, 2007; Timmer, 2002). The emergence of the so-called Asian Tigers in the 1960s–1970s, such as China, India, Thailand, Malaysia, South Korea, and Singapore, is a case in point.

BOX 3.1: SUSTAINABLE DEVELOPMENT GOALS (SDGS). SOURCE: UN (2015).

Goal 1 End poverty in all its forms everywhere

Goal 2 End hunger, achieve food security and improved nutrition and promote sustainable agriculture

Goal 3 Ensure healthy lives and promote wellbeing for all at all ages

Goal 4 Ensure inclusive and equitable quality education and promote

Goal 5 Achieve gender equality and empower all women and girls

Goal 6 Ensure availability and sustainable management of water and sanitation for all

Goal 7 Ensure access to affordable, reliable, sustainable and modern energy for all

Goal 8 Promote sustained, inclusive and sustainable economic growth, full and productive employment and decent work for all

Goal 9 Build resilient infrastructure, promote inclusive and sustainable industrialization and foster innovation

Goal 10 Reduce inequality within and among countries

Goal 11 Make cities and human settlements inclusive, safe, resilient and sustainable

Goal 12 Ensure sustainable consumption and production patterns

Goal 13 Take urgent action to combat climate change and its impacts

Goal 14 Conserve and sustainably use the oceans, seas and marine resources for sustainable development

Goal 15 Protect, restore and promote sustainable use of terrestrial ecosystems, sustainably manage forests, combat desertification, and halt and reverse land degradation and halt biodiversity loss

Goal 16 Promote peaceful and inclusive societies for sustainable development, provide access to justice for all and build effective, accountable and inclusive institutions at all levels

Goal 17 Strengthen the means of implementation and revitalize the Global Partnership for Sustainable Development

BOX 3.2: SDG 2. END HUNGER, ACHIEVE FOOD SECURITY AND IMPROVED NUTRITION AND PROMOTE SUSTAINABLE AGRICULTURE.

2.1 By the year 2030, achieve the goal of eradicating hunger and ensuring that all individuals, particularly those in impoverished and vulnerable circumstances, including infants, have continuous access to safe, nutritious, and adequate food throughout the year.

2.2 By 2030, put an end to all forms of malnutrition, which includes meeting the internationally agreed-upon targets for reducing stunting and wasting in children under the age of 5 by 2025, and addressing the nutritional needs of adolescent girls, pregnant and lactating women, and older individuals.

2.3 By 2030, double the agricultural productivity and incomes of small-scale food producers, with a particular focus on empowering women, indigenous peoples, family farmers, pastoralists, and fishers, through equitable and secure access to resources like land, inputs, financial services, knowledge, markets, and opportunities for value addition and non-farm employment.

2.4 By 2030, establish sustainable food production systems and adopt resilient agricultural practices that enhance productivity and output while also promoting the preservation of ecosystems, which will strengthen the capacity for adaptation to climate change, extreme weather events, drought, flooding, and other disasters, and simultaneously lead to the progressive improvement of land and soil quality.

2.5 By the year 2020, conserve the genetic diversity of seeds, cultivated plants, and farmed and domesticated animals, along with their related wild species, through the sound management and diversification of seed and plant banks at national, regional, and international levels, and promote equitable access to and fair sharing of benefits arising from the use of genetic resources and associated traditional knowledge, as agreed upon internationally.

2.A Increase investment, including through strengthened international collaboration, in rural infrastructure, agricultural research, extension services, technology development, and the establishment of plant and livestock gene banks to enhance agricultural productivity in developing countries, with a particular focus on the least developed nations.

2.B Rectify and prevent trade restrictions and distortions in global agricultural markets, including the simultaneous elimination of all forms of agricultural export subsidies and any export measures with comparable effects, in alignment with the objectives of the Doha Development Round.

2.C Implement measures to ensure the proper functioning of food commodity markets and their derivatives, facilitating timely access to market information, including data on food reserves, to mitigate extreme food price volatility.

SDG 12 (sustainable consumption and production patterns) is directly related to agriculture. Abundant scientific evidence confirms that the three major global challenges of climate change, biodiversity loss, and pollution are all consequences of unsustainable production and consumption practices. These issues pose existential threats to human wellbeing and jeopardise the achievement of the Sustainable Development Goals. The associated environmental degradation exacerbates the severity of these crises. Agriculture is a major user of natural resources (Castle, 1982) including land, water, and plant and animal genetic materials, and therefore agricultural practices, including related engineering technologies, can exert far-reaching impacts on both agriculture itself and the natural resources used in agriculture and food systems. This explains, in part, why some academic departments (and degree programmes) have changed their name and revised curriculum from Agricultural Engineering to include bioresources. Examples include the Department of Agricultural and Bioresources Engineering at the University of Nigeria, Nsukka, and the former Department of Bioresources and Agricultural Engineering at Sultan Qaboos University, Sultanate of Oman.

Agriculture also has wide-ranging environmental impacts and consequences (Lichtenberg, 2002) through deforestation, waste generation, greenhouse gas emissions, pollution, rising temperature, and climate change. Recognising the multidimensional impacts of agriculture on the environment (Clay, 2013) and climate change (Anwar et al., 2013; Pielke et al., 2007) and the need for agricultural engineering to become relevant in addressing these critical challenges, many academic departments and educational programmes have changed their name and revised their curricula to incorporate the environment (such as the Department of Agricultural and Environmental Engineering at the University Ibadan, Nigeria). The former Kenya Society of Agricultural Engineers has rebranded and reformed to become Kenya Society of Environmental, Biological and Agricultural Engineers (KeSEBAE).

With regards to SDG 12 (Box 3.3), the link between climate change, biodiversity loss, and pollution, both in the air and underground (affecting soil and water), is widely acknowledged to originate from unsustainable production and consumption practices. Postharvest losses and food waste continue to pose significant challenges in both developed and developing economies, leading to substantial consequences for the environment, climate change, and bioresources (Opara, 2006a, b; Munhuweyi et al., 2016). Approximately one-third of the world's food production, equivalent to over 1.3 billion tonnes, is lost or wasted during postharvest handling (FAO, 2019), resulting in substantial economic consequences exceeding USD 1 trillion globally. SDG 12.3 explicitly tackles this issue of high postharvest losses and food waste, aiming to reduce per capita global food waste (at retail and consumer levels) by half before the year 2030. However, while this target is specific on food waste (the major contributor to total waste in the global North), there are no quantitative targets for postharvest losses (up-stream in the value chain), which is the major contributor to food wastage in poor developing countries. It is also in these parts of the world where preharvest and postharvest losses are most prevalent. Efforts are being made at the local level to address this imbalance by implementing policy directives that establish targets and by conducting research to quantify the extent and consequences of postharvest losses. For instance, within Goal 5 (Modern Agriculture for increased productivity and production) of the African Union's Agenda 2063 - The Africa We Want, which aligns with both Goal 2 (End hunger, achieve food security and improved nutrition and promote sustainable agriculture) and Goal 12 (Sustainable production and consumption) of UN Agenda 2063, one of the measurable indicators is Target 5.4: to reduce postharvest losses by 50%. To improve understandings of the magnitude of the postharvest loss problem as a critical step towards reducing the losses, research has been carried out using the track and trace approach to quantify losses in the horticultural industry in the farm (Opara et al., 2021a,b; Blanckenberg et al., 2021, 2022a,b), in the packhouse (Opara et al., 2021a,b), and in retail (Munhuweyi et al., 2016). These studies highlight the contributions of agricultural and biosystems engineering research and education towards evidence-based policy to attain the objectives and targets of the SDGs.

BOX 3.3: SDG 12: ENSURE SUSTAINABLE CONSUMPTION AND PRODUCTION PATTERNS.

Target 12.1: All countries, with developed nations leading the way, should adopt and carry out the 10-Year Framework of Programmes on Sustainable Consumption and Production Patterns, considering the developmental status and capabilities of developing countries.

Target 12.2: By the year 2030, achieve the sustainable management and effective utilization of natural resources.

Target 12.3: By the year 2030, decrease per capita global food waste at the retail and consumer levels by fifty percent and mitigate food losses throughout production and supply chains, including postharvest losses.

Target 12.4: By the year 2020, reach the objective of environmentally responsible handling of chemicals and wastes throughout their entire life cycle, in alignment with agreed-upon international frameworks. Additionally, substantially decrease their emission into the air, water, and soil to minimize their detrimental effects on human health and the environment.

Target 12.5: By the year 2030, significantly decrease waste production by employing methods such as prevention, reduction, recycling, and reuse.

Target 12.6: Promote the adoption of sustainable practices, particularly among large and transnational companies, and encourage them to incorporate sustainability information into their reporting cycles.

Target 12.7: Encourage sustainable public procurement practices in line with national policies and priorities.

Target 12.8: By the year 2030, guarantee that individuals worldwide possess the necessary information and awareness to embrace sustainable development and lead lifestyles in harmony with nature.

Target 12.a: Assist developing countries in enhancing their scientific and technological capabilities to transition towards more sustainable models of consumption and production.

Target 12.b: Create and apply monitoring tools to assess the impacts of sustainable tourism, focusing on job creation, promotion of local culture, and the utilization of local products.

Target 12.c: Streamline ineffective fossil-fuel subsidies that foster wasteful consumption by eliminating market distortions. This should be done based on each country's individual circumstances, including restructuring taxation and gradually phasing out harmful subsidies where they exist, considering their environmental consequences. This process should be conducted while fully acknowledging the distinct requirements and situations of developing nations, with the goal of minimizing potential adverse effects on their development and safeguarding the welfare of impoverished and affected communities.

During the past four decades, considerable advances in postharvest science, engineering, and technologies have been made, contributing to improved techniques for reducing losses and waste, and maintaining and safeguarding fresh and processed foods along their value chains. Led by agricultural engineers and scientists in other disciplines such as plant science and food science, these advances have resulted in innovations for the better management of harvesting, postharvest handling, processing, storage, transport, distribution, preparation, and consumption (Opara, 2013, 2009, 2006a, b). Advancements in postharvest technology are widely recognised as essential to the reduction of food losses and waste to feed a hungry world (Opara, 2006a, b, 2010, 2015).

Agricultural Engineering in Sustainable Development – A Quest for Global Relevance

Agricultural engineering as we have come to know it, focuses on the business of agriculture and how to improve it. As such the understanding of it is far and wide and differs from author to author. For example Bello (2012), postulating on agricultural engineering, defines it from two perspectives: scientific and professional. The scientific understanding, according to him, is the utilisation of knowledge from all fields of science to improve agricultural operations with multiple processes. Meanwhile, the professional understanding, according to Bello, deals with the application of technological processes in dealing with the transformation of natural resources into agricultural land, machinery, structures, processes, and systems, ultimately for the betterment of humankind. This means that agricultural engineering by and large focuses on the improvement of plant and animal production. Wall (2006), on the other hand, sees agricultural engineering as process of creating, utilising, and managing the engineering systems required to deliver agricultural products. As such agricultural engineering deals with or focuses on the improvement of agriculture. Fomunyam (2019c, p. 40) presents an alternative interpretation of agricultural engineering as encompassing all actions that facilitate the production and processing of food, feeds, fibre, and other outputs, while also focusing on reducing laborious tasks, enhancing effectiveness and efficiency, and maximising the use of resources. All the above definitions, regardless of their difference, have one thing in common: the improvement of agricultural processes. This understanding makes agricultural engineering of strategic importance, particularly in the 21st century, as the world is battling food deficits and millions are dying of hunger.

Fomunyam (2019b, p. 172) argues that "it is noteworthy to keep in mind that one of the major challenges to be faced in this epoch is the need to have at least 60% increase in the current food production level so as to keep up with the increasing population". To this end, agricultural processes must be altered or improved upon if we are to meet up with the ever-increasing demand for agricultural products. This is further necessitated by the fourth industrial revolution which is changing the world of work and agricultural processes in particular. This drive is further fuelled by the drive to accomplish the SDGs, one of which primarily focuses on zero hunger. Since hunger can directly affect the achievement of all the SDGs, the improvement of agricultural processes becomes paramount. Since agricultural engineering doesn't function in isolation but functions in tandem with the other branches of engineering and science, there is a quest for global relevance, a yearning for uniqueness and pedigree which would particularly showcase agricultural engineering and its endeavours in transforming the world rather than being lost in the general conflux of engineering. The question currently begging for answers among agricultural engineers is: how can agricultural engineering achieve global relevance within the context of 21st century SDGs? This chapter attempts to answer this question from multiple perspectives.

Re-Engineering the Agricultural Engineering Curriculum

The first step towards achieving global relevance for agricultural engineering is the re-engineering of the agricultural curriculum. Senzanje (2003) opines that agricultural engineering primarily deals with the use of knowledge and principles in alternative fields of engineering to solve problems around agricultural production and by extension the production and delivery of food substances in a sustainable way. With our knowledge of climate change and its resulting impact constantly and consistently on the increase, there is an incessant need for the re-engineering of the curriculum. This would ensure that the curriculum doesn't just address sustainable food production, but also the effects of the same on the environment. According to Kolmos, Hadgraft, and Holgaard (2016, p. 391), there has been a widespread demand over the past 25 years for updated engineering competencies, leading to a gradual transformation in engineering education curricula and teaching methods. This trend has been observed globally, starting in the US, Europe, Australia, and now spreading to other regions. Academics and engineering experts agree that there is a need for changes in or the complete re-engineering of curricula, and for agricultural engineers this need is particularly important as they strive for global relevance. The fact that there is food scarcity in the world at the moment is a testament to the fact that the current agricultural engineering culture, which dictates how agricultural engineers are trained, is flawed to a certain extent, and if we are going to address this challenge we must re-engineer the curriculum.

This re-engineering of the curriculum is not supposed to be superficial, as has been reported in some places around the world. A reconfiguration of the core beliefs, theories, approaches, and context is important if this change is to yield results. Kolmos et al. (2016) confirm this when they posit that three approaches to curriculum change have emerged in engineering over the past few years. The first is the add-on approach, where academics simply add some materials or content to existing curricula and move on from there. The second character is an integration strategy, where certain things are integrated into the curriculum so as to achieve particular goals. The third strategy is rebuilding, which demands a total overhaul of the existing curriculum aimed towards the rebuilding of another for a better experience. The first two approaches are the most widely used in agricultural engineering, while the latter, which is what is needed, is yet to be fully engaged within the discipline. Kolmos et al. (2016) add that most engineering institutions encounter significant difficulties in completely overhauling their curricula, leading them to opt for smaller changes instead. This is a huge challenge for the growth of the discipline because the world and the economies of today are completely different from those of ten years ago, particularly because of Covid 19 and the spread of the fourth industrial revolution.

There is no doubt that universities across the globe are struggling to maintain the value and quality of higher education. Kähkönen and Hölttä-Otto (2022) contend that engineering is also part of this effort for change. They argue that the present higher education system, developed over centuries, was designed to equip engineers to solve intricate problems. However, they emphasise the need for new skills and a more interdisciplinary education to keep up with the evolving demands. The quote "We are currently preparing students for jobs that don't yet exist, using technologies that haven't been invented, in order to solve problems we don't even know are problems yet" holds even greater truth in the current context (Kähkönen & Hölttä-Otto, 2022, p. 516). To adequately equip students for jobs that are yet to emerge and empower them to address unknown challenges, a curriculum re-engineering is imperative. It is this re-engineering that will empower agricultural engineering to achieve global relevance.

However, how this re-engineering is supposed to unfold is another question begging for an answer. Mitchell, Nyamapfene, Roach, and Tilley (2021, p. 48) shed light on this issue by stating that "traditional engineering schools face difficulties in reconciling the need to offer innovative engineering education that aligns with the expectations of graduates and employers, while also dealing with the limitations and established structure of their current curriculum". A shift from the existing curriculum demands that responsiveness be taken into consideration in the re-engineering process.

Curriculum responsiveness focuses on the ability of the curriculum to address issues from multiple perspectives and provide solutions to current and future challenges in the society (Fomunyam & Teferra, 2017). Responsiveness, as it pertains to curricula, borders on four key aspects: economic, disciplinary, cultural, and pedagogical or learning responsiveness. Economic responsiveness, according to Moll (2004), focuses on the character of the curriculum to aid in the training of graduates who are highly skilled professionals and who can fully function in different parts of society. It is not simply about offering a degree in agricultural engineering in this case; rather the focus is on how agricultural engineers produced by this programme are about to revolutionise the business of food production and the strive to end world hunger. If these engineers can move beyond circling around problems in the field to working out new and sustainable solutions, then it can be purported that the curriculum is economically responsive. The re-engineering of the curriculum towards achieving global relevance within the scope of this chapter, would then focus on innovative ways of dealing with world hunger and the increasing demand for food products. As such revolutionary engineers are needed who would not only fit into the current and future job markets but would respond to the economic confrontations of the time through innovation, job creation, and resource management.

Disciplinary responsiveness, as the second construct of responsiveness in curricula, according to Moja (2004), deals with the curriculum's responsiveness to the research and knowledge in the field. Ferdinand (2009), supported by Fomunyam and Teferra (2017), argue that most curriculum documents are years behind the research in the field, leading to graduates who are unable to actually be caught up with what is happening in the discipline. Since a field like agricultural engineering is a community of experts constantly working to produce knew knowledge in the field, it is ironic that the curriculum does not follow suit. This deviation from the 'current' knowledge, though not deliberate, has resulted in a form of scholarship that evades everyday life, and the challenges humanity is facing today. The curriculum must be aligned with the latest research in the field of study, and this can't be done in agricultural engineering by adding on or integrating, but rather through a reconstruction which would not only bring in new content but ensure that over time the curriculum is always in time with the discipline. According to Moll (2004, p. 8), "disciplinary responsiveness" refers to a curriculum that is attuned to the nature of the underlying knowledge discipline. This is achieved by closely connecting the methods of knowledge production with the way students are educated and trained in that specific disciplinary area. This means that global relevance can't be achieved in agricultural engineering, and particularly through agricultural engineering education, if the curriculum is not consuming or speaking to the current research in the discipline.

Cultural responsiveness as a character of the curriculum connotes the propensity of the curriculum to deal with cultural exigencies, not only within the classroom but also within society. Wedekind (2019) argues that responsiveness, and particularly cultural responsiveness in this case, must be at the centre of curriculum endeavours. To put this more adequately, the re-engineering of curricula in agricultural engineering can't be successful without cultural responsiveness. Culture doesn't only speak to lifestyle but to language and choices, and particularly choices orchestrated by the nature of society. Cultural dissonance not only within the classroom but also within society must be a focus if global relevance is to be achieved. Since the world of work or the society in which graduates are to function is characterised by cultural dissonance, setting it as the focal point of curriculum re-engineering is primary to achieving global relevance. Gamble (2003), supported by Mathur and Rodriguez (2022) argues that, given the continuous polarisation of the world and the cultural dissonance this creates, curricula must be made to respond to the cultural proclivities both within and outside the classroom, so that graduates can be relevant everywhere they go. Gay (2010) adds that over the course of time, the task of creating a culturally responsive curriculum has evaded the academic world. Only the re-engineering of the curriculum can ensure that this is achieved. And to Wright, Williams, Starling, Reynolds, and Garcia-Lavin (2023, p. 22), the "significance of critical consciousness in instigating curricular changes lies in recognizing whose perspectives and methodologies are present or missing in our teaching and training approaches" is vital for global relevance and effectiveness in every educational endeavour. Martinez-Taboada et al. (2017) conclude that academic performance is adversely affected by students' perception of cultural dissonance and if this must be addressed, curriculum re-engineering is vital.

Learning responsiveness as a character of curriculum responsiveness presupposes that curricula must respond to and deal with student needs. It is the curriculum's capacity to deal with these needs that make the products of such curricula relevant. If their needs are not met, their development would be incomplete. Effective learning cannot occur if the curriculum fails to address the individual challenges of all students. Wedekind (2019) suggests that the one-size-fits-all approach in education has proven unsuccessful and will continue to be inadequate in meeting the diverse needs of students without curriculum reform. To achieve learning responsiveness, the curriculum must be re-engineered to encompass various approaches to design, instructional strategies, assessment methods, and student support that consider the characteristics and contexts of specific student groups (Ogude et al., 2005, p. 13). This learning responsiveness is crucial for allowing aspiring agricultural engineers to make a meaningful impact.

Pedagogy in Agricultural Engineering

Since the curriculum is not the only thing vital in the drive to achieve global relevance in the era of the fourth industrial revolution as far as agricultural engineering education is concerned, a critical look at pedagogy is vital. D'Souza and Rodrigues (2015, p. 828) argue that conventional instructor-centred, lecture-based teaching approaches in engineering education have faced criticism for being excessively linear, rigid, systematic, and limiting. Such approaches have resulted in the ineffectiveness or inability of agricultural engineers to solve issues around agriculture globally. D'Souza and Rodrigues (2015, p. 833) reach the conclusion that a significant portion of teaching in engineering colleges relies on lecture-based and teacher-centred methods. Faculty members often prioritise completing the pre-designed syllabus without adequately considering students' needs, learning styles, paces of learning, and desired learning outcomes. Consequently, students tend to become passive listeners who memorise subject content to achieve good exam grades, but they may not fully develop the necessary skills. This assessment of engineering pedagogy speaks to the need for a change in pedagogy alongside that of the curriculum. The reconfiguration of the curriculum without a similar change in pedagogy would be a waste of time, as no meaningful change can or will be achieved. There is a need to look at pedagogy in agricultural engineering, particularly in the era of the fourth industrial revolution. And this we posit should be centred around the different dimensions or levels of context: place, space, and time.

Context is critical in pedagogy, and this is even more critical when the current industrial revolution is taken into consideration. Fomunyam (2019d) posits that there are three dimensions of context, and when these are taken into consideration within the confines of pedagogy, their impact can be far reaching. Context as place, treated in relation to pedagogy, will centre around the place where the said pedagogical constructs are being applied, as well as where the results of the learning would be felt. Polyakova (2020) views pedagogy in engineering education as the establishment of a scientific basis, through which to convert the societal need for professional engineers into successful outcomes through the engineering education system. As such, regardless of the drive for global relevance and the fight against global hunger, the individual context, the institutional context, the immediate societal context, and the global stratosphere all constitute different levels or magnitudes of context which must be engaged if meaningful engineers are to be trained. This is because learning is a function of capital and capital can't be useful outside of context because, be it social, cultural, or political capital, what gives it meaning is the context from which the capital was developed. So, the global and the local context are vital aspects which must be built into the pedagogy. As such, regardless of the pedagogical style chosen – teacher, student or learning centred – context as place must be taken into consideration.

Secondly, context as space is also of great importance in all pedagogical endeavours in agricultural engineering. McAuliffe, Hargreaves, Winter, and Chadwick (2009) posit that true pedagogy seeks to create in the student a desire to transcend the simple acquisition of knowledge and skills, instead creating learning experiences which can be reconstructed in any setting. So, the focus is on the learning and the how of the learning, and not necessarily on what is being learnt. This would ensure that, as the fourth industrial revolution is constantly transforming the society and the world of work, the engineering graduate can adapt and fit because he or she is not simply focused on what was taught but on how to learn. The creation of these learning experiences can only happen within the confines of context as space since there is a need for the creation of such a space by both the lecturer and the student. This space, though it does constitute the physical classroom, goes beyond that to deal with the mental and the disciplinary. In this sense, the space in question is a limitless sphere where the student can activate agency and function at optimal levels. The disciplinary space, on the other hand, is the situating of both the student and their learning within the broader space of the discipline to explore its idiosyncrasies beyond the content of the course being taught, its current challenges, its current and future direction, and the nature of knowledge within such a space. However, this is not the focus of the teaching but the *approach* of the teaching, or its underpinning pedagogy.

Lastly, context as time within the confines of pedagogy in agricultural engineering deals with the age and how education relates to the current age. Fomunyam (2019a) argues that although we are in the age of the fourth industrial revolution, most engineering programmes are yet to fully take it into consideration, along with how it is altering and will continue to alter education and the world of work. Time is critical in pedagogy, not only because it dictates the pace of and approach to learning, but because it helps situates learning in context, thereby determining its usefulness. Context as time also informs the sequencing of learning and how the learning can be used to respond to global issues. The time or age in which we live determines not only what should be studied and how it should be studied but, more importantly, how to situate the learning experience within its context. For example, context as time in agricultural engineering today speaks not only to the fourth industrial revolution but to the fight against hunger. A neglect of this or a move away from it would consistently put agricultural engineering in obscurity, as relevance is contextual. As such the engagement of context as time in the process of learning is at the heart of achieving global relevance in the era of the fourth industrial revolution.

Driving Areas of Global Relevance for Agricultural Engineering

Since relevance is not a construct that can stand for itself in this case but must be attributed to others to make sense, there are certain driving areas to which it must be attributed for global relevance to be possible. While different aspects of agricultural engineering are the focus here, it is of primary importance to look at certain forces or charges that inject life into these areas, programming them for relevance. These charges are at work in these different areas to produce certain levels of empowerment.

Education charges or curricular charges (for they are generally used in curriculum theorising and the business of education) determine not only how learning unfolds but also how it is experienced. Depending on the kind of charge at work, what actions should be taken, who should take them and when it should be taken are some of the concerns at play. Fomunyam and Teferra (2017) postulates that these charges are a determining factor or a driving force towards achieving unique goals. As such, charges like responsibilities, consciousness, commitments, and projects among others can work to foster global relevance for agricultural engineering. The roles, duties and responsibilities of lecturers and students alike, as well as managers and support staff within a university, for example, would determine how and when relevance can be achieved.

Fomunyam and Khoza (2020), adding to this, posit that these charges would dictate the level of responsiveness which would in turn dictate how context is seen and understood in the fight for global relevance. As such these charges take effect from the re-engineering of the curriculum, as well as the pedagogical proclivities, to situate them in context to ensure that, whatever aspect of agricultural engineering you are looking at, life can be injected into it. So then, responsibilities (who bears the responsibility for everything required for agricultural engineering to achieve global relevance?), consciousness (how aware are those responsible not only about the different levels of context but also of how to navigate it?), commitments (how committed are those with responsibility to ensuring that the purpose, in this case the SDGs, are achieved?), and projects (what are the responsible parties doing to bring the plans to reality). With these questions in mind, we look at research and set key areas for relevance in agricultural engineering.

A Case for Agricultural Engineering in the Achievement of Sustainable Development

The SDGs provide a global rallying point for agricultural and biosystems engineering to demonstrate relevance with regard to societal needs, particularly in achieving improved agricultural productivity and reducing losses and waste through sustainable agricultural intensification. This involves three key areas of intervention: a) education, b) research and innovation, and c) thought leadership.

Agricultural Engineering Education and Training

Education is the key to equipping the labour force with the knowledge requisite to gainfully participate in and contribute to a better society. It is also education that prepares humans for life-long learning in an ever-changing world. To be effective in meeting these expectations, agricultural engineering education and training must equip students with relevant knowledge in the basic sciences, engineering and related technologies, soft skills, ethics, and an understanding of agriculture (in the wide sense) and of entrepreneurship. The SDGs are quite complex and interact with each other, and to meet the targets, a systems thinking approach is necessary to deal with the inherent complexities of the nexus between human development, natural resources, and sustainability. "To make a significant contribution towards the attainment of the SDGs," agricultural and biosystems engineering curricula must produce graduates who become system integrators – with relevant skills and competencies which enable them to meet the competing demands of productivity and profitability gains, while ensuring sustainable production and enhancing sustainable consumption. The education and training of the new agricultural engineer must deliver graduates with the ability to harness the power of existing, new, and emerging technologies such as biotechnology, nanotechnology, information technology, artificial intelligence, robotics, etc., to transform and modernise agriculture to feed and nourish a hungry world sustainably.

To reposition agricultural and biosystems engineering education to achieve the SDGs, there is a need to critique current education and training curricula to align them with the Goals of the SDGs – Agenda 2030 and the Aspirations of Agenda 2063 – The Africa We Want. A crucial goal in this respect would be to transform and modernise agri-food systems to achieve sustainability and resilience, and to become climate-smart. Given agriculture's transformational role as a veritable instrument for rural development and socio-economic development, a modernised, profitable, and sustainable agriculture can also act as an instrument for achieving equity and social justice. The agricultural and biosystems engineer of today and tomorrow should be equipped with essential knowledge and expertise through education and training to be an agent of positive change through work and service to society.

Agricultural Engineering Research and Innovation

Advances in agricultural engineering technologies made through research and innovation have contributed significantly to the massive global increases in productivity and production which averted the apocalyptic Malthusian prediction of widespread death and misery from hunger due to geometric increases in human population. Thanks to the so-called Green Revolution technologies such as high-yielding plant and animal genetic resources, the use of agri-chemicals including fertiliser, and irrigation, all powered and propelled by agricultural mechanisation, farmers achieved unprecedented increases in both yield and production. This resulted in rapid and widespread reduction of poverty and food insecurity, especially in countries in Asia and Latin America, where the Green Revolution has had a remarkable success and impact. Today, the world is still confronted by these same problems – widespread poverty and a high incidence of food and nutrition insecurity. These challenges have led to calls to increase food production by up to 70% to feed a world of 9.5 billion people by 2050. Juxtaposing the existential problem of climate change, it is clear that the business-as-usual approach to agriculture and food systems is no longer acceptable. This calls for the transformation and modernisation of current agri-food systems into a new agriculture that is efficient, productive, profitable, and sustainable. It must also be equitable and deliver social justice. While recognising this as a tall order, the SDGs demand these attributes of a new agriculture. Consequently, new frontiers in agricultural and biosystems research and innovation are required to contribute novel solutions to these grand challenges.

Subsistent smallholder agriculture remains dominant in many parts of the world, often in those parts with the highest population growth rate and with the highest incidence of poverty. Transforming and modernising the smallholder agri-food systems to higher productivity remains a challenge. Crop yields are the lowest compared to the rest of the world. Research and innovation are needed to attain optimal and sustainable productivity gains in smallholder agriculture, and to develop new value chains linking production to local, regional, and international markets. Technological advances such as modern agricultural machinery, precision agriculture, drones, postharvest handling, and processing offer new prospects for research, teaching and learning towards sustainable mechanisation of intensive agri-food systems. Research in these areas can provide new teaching and learning tools for agricultural and biosystems engineering education.

Advancements in innovative and revolutionary technologies like biotechnology and nanotechnology present vast prospects for education and research in agricultural and biosystems engineering for sustainable food systems transformation, bioresources conservation and utilisation, and environmental protection. Artificial Intelligence (AI) techniques, including Machine Learning (ML) (Nturambirwe and Opara, 2020) offer tremendous opportunities in agricultural and biosystems engineering education. Given the huge datasets that are generated at all stages along the agricultural value chain, converting these data into information and knowledge which can be used to drive and support technological innovation is an opportunity for agricultural engineers. Indeed, Data Science for agriculture and food systems transformation and modernisation must be embraced by agricultural engineering. Applications of Data Science in agriculture and related land-based industries include crop monitoring, water management and optimisation irrigation systems, precision agriculture, soil analysis for improved farming practices, fertiliser application and irrigation scheduling, crop forecasting for harvest management and supply chain optimisation, maintaining postharvest quality, and food safety and traceability. As system integrators, agricultural and biosystems engineers must understand the complexity of SDGs and their targets, and build robust system models with the appropriate trade-offs that are necessary to achieve sustainable agri-food systems among the competing demands. The technical, soft, and managerial skills and competencies needed for these over-arching integrative responsibilities must be incorporated into the educational curricula of agricultural and biosystems engineers. Furthermore, cutting-edge research is necessary to drive the advancement of new knowledge in agricultural and biological/biosystems engineering systems. No other discipline is better equipped to fulfil this role.

Thought Leadership and SET Policy, Diplomacy, Advocacy, and Activism

The UN Agenda 2030 for changing our world and the AU Agenda 2063 for inclusive and sustainable development each provide a masterplan and strategic framework for achieving a better world, with a focus on 5 Ps: people, planet, prosperity, peace, and partnership. By its nature, agriculture performs an important function in realising the aspirations of Agenda 2063 and the SDGs of Agenda 2030. Agriculture in particular, and agri-food systems in general, are the key vehicles and instruments for ending all forms of hunger, protecting the environment and planet, and ensuring prosperity (socio-economic and technological) in harmony with nature, especially in economies where agriculture remains the main source of livelihood and most critical contributor to gross domestic product (GDP), such as those of sub-Saharan Africa, Southeast Asia, and parts of Latin America and the Small Island Nations. Even in more advanced countries where the contribution of agriculture to the national economy is lower, such as in North America, Europe, and some countries in Asia and Africa, agriculture (including horticulture, forestry, fisheries, and aquaculture, etc.) remains the source of food, feed, and fibre, regardless of whether the crop is produced locally or imported. Given that food is a fundamental human need, agriculture as the source of food also plays an important role in fostering peace in a just and inclusive society. To achieve the SDGs and Aspirations of Agenda 2063, building partnerships through collaboration and cooperation to mobilise the means to implements the necessary actions is essential.

Despite the importance of agriculture and its interconnectedness to the SDGs, the agriculture industry alone cannot realise its potential without transforming our world for a sustainable future and realising the aspirations of The Africa We Want. To realise this potential, thought leadership is required at all levels – including agricultural engineering education, research, policy, and practice. Agricultural and biosystems engineering education must produce graduates as thought leaders of today and tomorrow. As we drive towards transforming agri-food systems for a sustainable, equitable, and inclusive future, we need a new envisioning of a future agriculture. This requires thought leadership that transcends beyond one discipline – leadership that is bold and visionary in imagining what is possible in the light of Agenda 2030 and Agenda 2063.

To succeed, such thought leadership requires collaboration, networking, innovation, entrepreneurship, learning individuals, and learning organisations. These dimensions of thought leadership must be incorporated into our agricultural and biosystems engineering education and research programmes and professional practice. The transformational thought leadership espoused here for the agricultural engineering discipline is about the championing of new ideas, rather than being anything to do with managing people or helping to achieve a specific personal goal. It promotes competition based on innovation towards teaching, research, and service to society for sustainable development. Such thought leadership is based on the premise that novel ideas can transform the way we think about research, teaching, and service – including the way we run our professional societies and associations.

There is also a case for the contributions of agricultural and biosystems engineering to science, engineering and technology policy, science diplomacy, science advocacy, and science activism. The integrated nature, inter-linkages, and complexity of the SDGs and aspirations of Agenda 2063 are so important that these must be considered to ensure that the goals are realised. Therefore, partnerships, collaborations, and cooperation at individual, organisation, community, regional, national, and global levels are necessary. Developing the necessary policies and setting targets at these levels for each SDG and aspiration becomes paramount. Given the critical role of science, engineering, and technology in transforming our lives, as articulated in both Agenda 2030 and Agenda 2063, agricultural engineers have a role in science diplomacy, advocacy, and activism to promote greater investment in education, research, and innovation for sustainable development in general and for the advancement of the agricultural engineering discipline and profession (Opara, 2023). For instance, investments in agricultural development, including donor aid, have historically focused on production agriculture and far less on post-farm gate, with the former often accounting for over 90%. This anomaly has led to sustained neglect of postharvest operations and the development of value chains for trade and commerce. It is argued that this disproportionate focus in promoting on-farm agriculture has contributed to the high occurrence of postharvest losses and food waste, which has been estimated to be valued at over US$ 1 trillion per annum in direct and indirect costs. In Africa alone, postharvest losses amount to over US$ 48 billion per annum, excluding the socio-economic, resource, and environmental impacts. Strong, focused, and sustained diplomacy, advocacy, and activism are required by agricultural engineers and other role players to change this and similar narratives to assist in the realisation of the SDGs and aspirations of Agenda 2063.

Agenda on Research Priorities and Key Areas for Relevance in Agricultural Engineering

Research is a critical nexus connecting education and service to society (Figure 3.1). Research produces new knowledge that is necessary to develop innovative curricula and new teaching and learning tools. Research is also necessary to guide evidence-based policies and to develop practical solutions to the developmental challenges facing society. The discipline of agricultural engineering is at the forefront of leading research to advance technological innovations in agriculture and other land-based industries. The question that often arises then is: What are some of the research priorities and key areas for relevance in agricultural engineering in transforming our lives

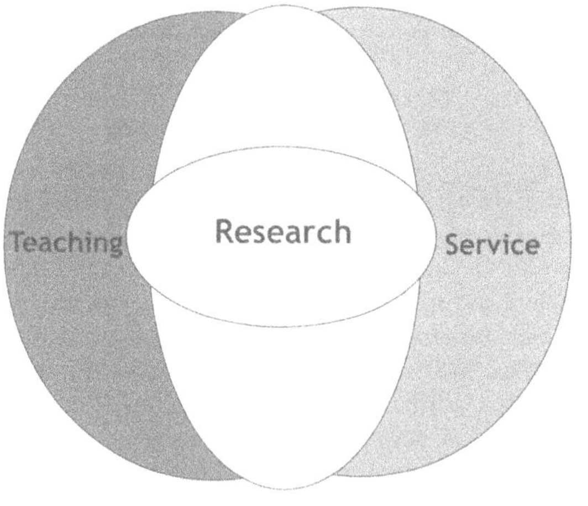

FIGURE 3.1 The nexus connecting research with teaching and learning, and service in agricultural engineering.

with the SDGs and Aspirations of Agenda 2063? These are summarised below:

Priority 1 – transform agricultural, biosystems, and biological engineering education for sustainable development. This requires rethinking the curricula, with more emphasis on complexity, inter-disciplinarity, and systems thinking. The future agricultural and biosystems engineer must be equipped, through education, training, and mentoring, to develop critical soft skills for the workplace and for functioning in an increasingly globalised society.

Priority 2 – transform and modernise agriculture for sustainable agri-food systems. This requires rethinking the role of subsistence smallholder agriculture in assuring food security and livelihoods of resource poor farmers, especially in sub-Saharan Africa. Despite considerable donor aid and other investments by national governments to improve the productivity of smallholder farmers during the past century, these farmers and their families still suffer the highest incidence of poverty and food insecurity compared to the rest of the population. Harnessing the power of modern agricultural technologies and innovation to push the frontiers of crop yields and the productivity of smallholder farms remains a challenge.

Priority 3 – promote and advance improved postharvest systems and agroprocessing for supply chain management and reduction of postharvest losses and food waste. This requires the mapping of existing postharvest systems of key agri-food commodities, including food, feed, and fibre, quantifying the magnitude of postharvest losses and food waste and their impacts, developing novel cold chains and dry chains to preserve the safety and quality of food materials, including maturity indexing, harvesting technologies, quality control and assurance, storage, drying, packaging, transport and logistics, and renewable energies for applications along the value chain.

Priority 4 – develop new agri-food value chains and promote market access. This requires developing novel technologies to measure, monitor, and control quality and safety along the value chain; developing new value-added products from agri-food materials with proven health benefits; developing standards for marketing and trade at local and international levels; and developing traceability systems to determine origin, composition, location, and the overall status of food products in the supply chain.

Priority 5 – develop novel applications of a technological innovation triad in modern food systems to improve sustainability and resilience. This requires harnessing the power of ground-breaking technologies such as biotechnology, nanotechnology, and information technology in sustainable food systems, including resilience. How do we harness agricultural engineering technologies to assure food security during pandemics and natural disasters? How do we harness the power of the Internet of Things (IoT), Data Science, Artificial Intelligence (AI), Machine Learning (ML), and Ubiquitous Knowledge Systems (UKS) for sustainable agri-food systems transformation and modernisation and food intelligence?

Conclusions and Future Prospects

The UN SDGs have provided a master plan for transforming our lives into a sustainable future as encapsulated in Agenda 2030. Preceding the launch of the SDGs, the AU's Agenda 2063 – The Africa We Want provides a set of aspirations for future Africa that is economically developed and at peace with itself and the rest of the world. The integrated nature, inter-linkages, and complexity of the SDGs and aspirations of Agenda 2063 are so important that these must be considered to ensure that the goals are realised. Therefore, partnerships, collaborations, and cooperation at individual, organisation, community, regional, national, and global levels are necessary.

Both Agenda 2030 and Agenda 2063 recognise the pivotal role of agriculture in sustainable development, especially in eradicating hunger by assuring food and nutrition security and eradicating poverty. To achieve this, agricultural and biosystem engineering has critical roles to play through the education and training of an appropriately skilled workforce and in carrying out relevant innovative research to produce new knowledge and technologies for sustainable agri-food systems transformation and modernisation. In response to the question of how the agricultural engineering discipline and profession can achieve global relevance within the context of 21st century SDGs and Agenda 2063, we have identified five critical areas for action to: transform agricultural and biosystems engineering education for sustainable development, transform and modernise agriculture for sustainable agri-food systems, promote and advance improved postharvest systems for supply chain management and the reduction of postharvest losses and food waste, develop new agri-food value chains and promote market access, and develop novel applications of a technological innovation triad in modern food systems to improve sustainability and resilience.

To succeed, we argue that agricultural engineers and the agricultural and biosystems engineering profession must contribute to thought leadership at the individual, professional, institutional, local, regional, and global levels. We call for the reform of agricultural and biosystems engineering education to transform and modernise agriculture and food systems for a sustainable, equitable, and just world. We call for special attention to the future of smallholder agriculture in Africa which is currently characterised by high levels of subsistence, low crop yields, and a high incidence of poverty and food insecurity among farmers. This requires moving smallholder agriculture towards a market-led and knowledge-driven business as part of a continuum of agriculture – including small, medium, and large agribusinesses. To assure its local and global relevance in achieving the SDGs and The Africa We Want, agricultural and biosystems engineering education must address these challenges in the curriculum, teaching, and training of future practitioners and thought leaders.

REFERENCES

Allen, S., & de Brauw, A. (2018). Nutrition sensitive value chains: Theory, progress, and open questions. *Global Food Security*, 16, 22–28. https://doi.org/10.1016/j.gfs.2017.07.002

Anwar, M. R., Liu, D. L., Macadam, I., & Kelly, G. (2013). Adapting agriculture to climate change: A review. *Theoretical and Applied Climatology*, 113, 225–245.

Bello, S. (2012). *Agricultural Engineering Principles & Practice*. Ishiagu: Dominion Publishing.

Blanckenberg, A., Opara, U. L., & Fawole, A. O. (2021). Postharvest losses in quantity and quality of Table Grape (cv. Crimson Seedless) along the supply chain and associated economic, environmental and resource impacts. *Sustainability*, 13, 4450. https://doi.org/10.3390/su13084450

Blanckenberg, A., Fawole, O. A., & Opara, U. L. (2022a). Postharvest losses in quantity and quality of pear (cv. Packham's Triumph) along the supply chain and associated economic, environmental and resource impacts. *Sustainability (IF: 3.889, 15/12/2022)*, 14(2), 603. https://doi.org/10.3390/su14020603

Blanckenberg, A., Opara, U. L., & Fawole, O. A. (2022b). Quantifying the magnitude of postharvest fruit losses: A case study of industry packhouse data in the Western Cape, South Africa. *Acta Horticulturae (IF: 0.257, 15/12/2022)*, 1349, 243–250. https://doi.org/10.17660/ActaHortic.2022.1349.34

Castle, E. N. (1982). Agriculture and natural resource adequacy. *American Journal of Agricultural Economics*, 64(5), 811–820.

Clay, J. (2013). *World Agriculture and the Environment: A Commodity-by-Commodity Guide to Impacts and Practices*. Washington: Island Press.

Denning, G. (2023). Agriculture leads to the MDGs: Rural development in Africa. UN Chronile. https://www.un.org/en/chronicle/article/agriculture-leads-mdgs-rural-development-africa. Accessed 25 April.

Djurfeldt, G., Holmén, H., Jirström, M., & Larsson, R. (2005). *The African Food Crisis: Lessons from the Asian Green Revolution*. Budapest: CABI Publishing.

D'Souza, M. J., & Rodrigues, P. (2015). Extreme pedagogy: An agile teaching-learning methodology for engineering education. *Indian Journal of Science and Technology*, 8(9), 828–833.

Evenson, R. E., & Golin, D. (2003). Assessing the impact of the green revolution, 1960 to 2000. *Science*, 300(5620), 758–762.

FAO. (2019). The state of food and agriculture 2019: Moving forward on food loss and waste reduction. Rome. Licence: CC BY-NC-SA 3.0 IGO.

Ferdinand, D. (2009). *Workforce Education and Development Curriculum Responsiveness to Culturally and Internationally Diverse Graduate Students: A Mixed Methods Study*. Carbondale: Carbondale Southern Illinois University.

Fomunyam, K. G. (2019a). Education and the fourth industrial revolution: Challenges and possibilities for engineering education. *International Journal of Mechanical Engineering and Technology*, 10(8), 271–284.

Fomunyam, K. G. (2019b). Emerging technologies in agricultural engineering. *International Journal of Mechanical Engineering and Technology*, 10(7), 172–178.

Fomunyam, K. G. (2019c). Future of agricultural engineering education in Nigeria. *International Journal of Civil Engineering and Technology*, 10(11), 39–45.

Fomunyam, K. G. (2019d). *Understanding Research Context and Developing Your Niche*. Keynote Delivered at ALLU Lecture Series Mangosuthu University of Technology.

Fomunyam, K. G., & Khoza, S. B. (2020). Theorising open curriculum charges as pathway to responsiveness in South African higher education. In K. G. Fomunyam & S. B. Khoza (Eds.), *Curriculum Theory, Curriculum Theorising, and the Theoriser: The African Theorising Perspective* (pp. 53–69). Amstadam: Brill.

Fomunyam, K. G., & Teferra, D. (2017). Curriculum responsiveness within the context of decolonisation in South African higher education. *Perspectives in Education*, 35(2), 196–207.

Gamble, J. (2003). *Curriculum Responsiveness in FET Colleges*. Pretoria: HSRC Press.

Gay, G. (2010). *Culturally Responsive Teaching: Theory, Research, and Practice*. New York: Teachers College Press.

IFPRI ADB. (2007). *Agriculture and rural development for reducing poverty and hunger in Asia: In pursuit of inclusive and sustainable growth. Asian Development Bank (ADB)*. Manila: International Food Policy Research Institute (IFPRI).

Johnson, D. G. (1993). Role of agriculture in economic development revisited. *Agricultural Economics*, 8(4), 421–434.

Johnston, B. F., & Mellor, J. W. (1961). The role of agriculture in economic development. *The American Economic Review*, 51(4), 566–593.

Kähkönen, E., & Hölttä-Otto, K. (2022). From crossing chromosomes to crossing curricula – A biomimetic analogy for cross-disciplinary engineering curriculum planning. *European Journal of Engineering Education*, 47(3), 516–534. https://doi.org/10.1080/03043797.2021.1953446

Kjeldsen-Kragh, S. (2007). *The Role of Agriculture in Economic Development: The Lessons of History*. Copenhagen: Business School Press DK.

Kolmos, A., Hadgraft, R. G., & Holgaard, J. E. (2016). Response strategies for curriculum change in engineering. *International Journal of Technology and Design Education*, 26(3), 391–411. https://doi.org/10.1007/s10798-015-9319-y

Lichtenberg, E. (2002). Agriculture and the environment. *Handbook of Agricultural Economics*, 2, 1249–1313.

Martinez-Taboada, C., Mera, M. J., Amutio, A., Castañeda, X., Felt, E., & Nicolae, G. (2017). The impact of cultural dissonance and acculturation orientations on immigrant students academic performance. *Universitas Psychologica*, 16, 6–19.

Mathur, S. K., & Rodriguez, K. A. (2022). Cultural responsiveness curriculum for behavior analysts: A meaningful step toward social justice. *Behavior Analysis in Practice*, 15(4), 1023–1031.

McAuliffe, M., Hargreaves, D., Winter, A., & Chadwick, G. (2009). Does pedagogy still rule? *Australasian Journal of Engineering Education*, 15(1), 13–18.

Meijerink, G. W., & Roza, P. (2007). *The Role of Agriculture in Economic Development (No. 4)*. Wageningen: Wageningen UR.
Mitchell, J. E., Nyamapfene, A., Roach, K., & Tilley, E. (2021). Faculty wide curriculum reform: The integrated engineering programme. *European Journal of Engineering Education*, 46(1), 48–66. doi:10.1080/03043797.2019.1593324
Moja, T. (2004). Glocalisation: A challenge for curriculum responsiveness. In H. Griesel (Ed.), *Curriculum Responsiveness: Case Studies in Higher Education* (pp. 21–38). Pretoria: South African Universities Vice-Chancellors Association.
Moll, I. (2004). Curriculum responsiveness: The anatomy of a concept. In H. Griesel (Ed.), *Curriculum Responsiveness: Case Studies in Higher Education* (pp. 1–19). Pretoria: South African Universities Vice-Chancellors Association.
Munhuweyi, K., Sigge, G. O., Opara, U. L. (2016). Postharvest losses of cabbages from retail to consumer and the socio-economic and environmental impacts. *British Food Journal*, 118(2), 286–300. http://dx.doi.org/10.1108/BFJ-08-2014-0280
Nturambirwe, J. F. I., Opara, U. L. (2020). Machine learning applications to non-destructive defect detection in horticultural products. *Biosystems Engineering*, 189, 60–83. https://doi.org/10.1016/j.biosystemseng.2019.11.011
Opara, L. U. (2006a). Editorial: A new era in postharvest technology. *International Journal of Postharvest Technology & Innovation*, 1(1), 1–3.
Opara, L. U. (2006b). Editorial: Postharvest technology for linking production to markets. *International Journal of Postharvest Technology & Innovation*, 1(2), 139–141.
Opara, L. U. (2009). Editorial: Harnessing postharvest technology for development. *International Journal of Postharvest Technology & Innovation*, 1(4), 357–359.
Opara, U. L. (2010). Editorial: High incidence of postharvest food losses is worsening global food and nutrition security. *International Journal of Postharvest Technology & Innovation*, 2(1), 1–3.
Opara, U. L. (2013). Professor Emeritus Adel A. Kader (1941–2012): A golden era in postharvest education, research and practice. *International Journal of Postharvest Technology & Innovation*, 3(1), 1–4.
Opara, U. L. (2015).To feed the world, we must save the harvest. *Resource* (March/April), 7.
Opara, U. L. (2023). The role of the basic sciences in transforming and modernizing smallholder agri-food systems. Invited Presentation at the UNESCO International Conference on the Role of the Basic Sciences in Transforming Africa, Kigali, 13–14 June.
Opara, I. K., Fawole, O. A., Kelly, C., & Opara, U. L. (2021a). Quantification of on-farm pomegranate fruit postharvest losses and waste, and implications on sustainability indicators: South African case study. *Sustainability (IF: 3.889, 15/12/2022)*, 13, 5168. https://doi.org/10.3390/su13095168
Opara, I. K., Fawole, O. A., & Opara, U. L. (2021b). Postharvest losses of pomegranate fruit at the packhouse and implications for sustainability indicators. *Sustainability (IF: 3.889, 15/12/2022)*, 13, 5187. https://doi.org/10.3390/su13095187
Otsuka, K, (2007). Changing sources of household income and poverty reduction in rural Asia, 1984–2004. Paper presented at IFPRI-ADB Policy Forum.
Ogude, N., Nel, H., & Oosthuizen, M. (2005). *The Challenge of Curriculum Responsiveness in South African Higher Education*. Pretoria: Council on Higher Education.
Polyakova, T. (2020). Engineering pedagogy: On the way to "education 4.0". *International Journal of Engineering Pedagogy*, 10(4), 4–7.
Pielke Sr, R. A., Adegoke, J. O., Chase, T. N., Marshall, C. H., Matsui, T., & Niyogi, D. (2007). A new paradigm for assessing the role of agriculture in the climate system and in climate change. *Agricultural and Forest Meteorology*, 142(2–4), 234–254.
Ruel, M. T., Quisumbing, A. R., Balagamwala, M. (2018). Nutrition-sensitive agriculture: What have we learned so far? *Global Food Security*, 17, 128–153, https://doi.org/10.1016/j.gfs.2018.01.002
SDGF. (2023). *The Millenium Development Goals*. Sustainable Development Goals Fund. https://www.sdgfund.org/mdgs-sdgs. Accessed 20 April.
Senzanje, A. (2003). Problems faced and advances made by agricultural engineers in southern and eastern Africa. *Agricultural Engineering International: the CIGR Journal of Scientific Research and Development*, 5(1), 1–14.
Timmer, C. P. (2002). Agriculture and economic development. *Handbook of Agricultural Economics*, 2, 1487–1546.
UN. (2015). Resolution adopted by the General Assembly on 25 September 2015 (A/RES/70/1). United Nations General Assembly. 35 pages.
Wall, G. (2006). Agricultural engineering education in developing and transition countries. *International Journal of Engineering Education*, 22(1), 67–70.
Wedekind, V. (2019). Curriculum responsiveness and student employability: An institutional analysis. In G. Kruss, A. Wildschut, & I. Petersen (Eds.), *Skills for the Future: New Research Perspectives* (pp. 79–90). Pretoria: HRSC Press.
Wright, A. J., Williams, N. J., Starling, T., Reynolds, A., & Garcia-Lavin, B. (2023). Deep-structure curriculum liberation for social responsiveness in graduate health service psychology training. *Training and Education in Professional Psychology*, 17(1), 22–30.

4

Top 100 Questions of Importance to the Future of Agricultural Engineering Education, Research, and Practice in Africa

Umezuruike Linus Opara

Introduction

The African continent's strategic plan for socioeconomic development, titled "Agenda 2063 – The Africa We Want" (African Union, 2022), was approved by the African Union in 2013. Agenda 2063's objectives, aspirations, and focus areas stress the necessity of modernising agriculture for higher output and productivity to create an Africa that is wealthy due to an equitable growth and sustainable development. At the African Union Summit in Malabo, Equatorial Guinea, in June 2014, heads of state and government established a remarkable set of realistic concrete targets for agriculture to advance Agenda 2063. The Malabo Declaration on "Accelerated Agricultural Growth and Transformation for Shared Prosperity and Improved Livelihoods" outlines a fresh set of goals that display a more focused strategy for achieving the agricultural transformation agenda for the African continent, which is shared prosperity and enhanced livelihoods. The Malabo Summit reiterated the significance of prioritising agriculture on the African continent's development agenda. It emphasised that fostering agriculture is a vital policy endeavour for driving economic growth and eliminating poverty in the region. Africa's development partners, private sector, investors, and other stakeholders are expected to key into this policy initiative as the guiding principle for collaboration and cooperation.

It is widely recognised among policy makers, researchers, and development practitioners that agriculture has remained the mainstay of most African economies, accounting for nearly 80% of employment and remaining a major source of livelihood, particularly in rural areas. Accordingly, most pundits argue that transforming and modernising African agriculture is a *sine qua non* to its economic growth and best strategy for widespread poverty reduction. In this context, African leaders reached a consensus on the importance of achieving an annual agricultural GDP growth rate of at least 6% and generating employment opportunities for at least 30% of the youth within agricultural value chains in order to fulfil their "Commitment to Halving Poverty by the year 2025, through Inclusive Agricultural Growth and Transformation ". In a similar vein, African leaders pledged to end hunger in their continent by the year 2025. Additionally, they pledged to support various initiatives, including but not limited to promoting sustainable and dependable production methods, offering relevant knowledge, information, and skills to beneficiaries, improving water management systems with a focus on irrigation, ensuring accessible, dependable, and affordable mechanisation and energy resources, and reducing current postharvest losses by half.

To accomplish the ambitious goals set by African leaders in transforming and modernising the continent's agriculture and food systems for inclusive growth and sustainable development, Africa needs agricultural engineers and other technical experts. No other academic discipline or occupation is better equipped to lead the mechanisation, digitisation, commercialisation, and industrialisation of African agriculture, fostering a knowledge-intensive, profitable, and sustainable sector. To drive Africa's Green Revolution, there is a need for a significant number of exceptional young male and female agricultural engineers, along with their professional associations. To achieve this, it is essential to widen the pool of skilled individuals, reduce the loss of talent to other regions, and actively encourage the return of talented individuals who can contribute their expertise. Harnessing the finest talents from Africa and leveraging access to the most advanced engineering technologies globally is indispensable for the advancement and modernization of African agriculture and food systems.

The agricultural engineering discipline and profession has never been more important in the pursuit of widespread socioeconomic development, sustainable food systems transformation, and as part of efforts to mitigate climate change and environmental degradation. As the human population continues to expand and improve its standard of living, there is an escalating demand for plentiful and dependable access to safe, nutritious food, housing, clothing, fibre, and renewable energy sources. In the same vein, agriculture has important roles to play in addressing the problem of climate change, while preserving bioresources and protecting the environment. Currently, Africa stands as the most food insecure region globally, marked by a surging youth population and a significant poverty rate. Given the pivotal and fundamental role of agriculture in the economies of most Sub-Saharan African nations, addressing these global challenges necessitates a

DOI: 10.1201/9780429150111-5

This chapter has been made available under a CC-BY-NC-ND license.

robust comprehension of engineering applications in agriculture, and the translation of this knowledge into impactful technological and policy solutions.

In this Chapter, we discuss top-ranking questions facing agricultural engineering education and research in Africa in response to growing agricultural engineering for Agenda 2063, with particular relevance to the Agriculture We Want. This effort must be considered as a starting point because both educational and research priorities evolve continuously and unpredictably due to emerging concerns, evolving needs, and the discovery of new knowledge.

Agricultural Engineering Responds to the Grand Challenges and Big Questions Facing Africa

Agricultural Engineering is in a position to tackle the significant challenges confronting Africa. However, it remains uncertain whether the full scope of pressing and crucial issues concerning agricultural engineering education and research in Africa (and indeed globally) have been identified or thoroughly evaluated. What questions should the next generation of agricultural engineers in Africa be addressing to contribute to transforming Africa's agri-food systems into an engine for widespread socioeconomic development, wealth creation, and poverty reduction? What questions should agricultural engineering education in Africa be addressing to ensure that Africa's human capacity and skills are developed for both designing and implementing the needed policies and technological interventions across the value chain? What leading research topics should agricultural engineers in Africa be tackling in order to spearhead the required reduction of current massive yield gaps of crops (plant and animal), reduce losses and waste, and increase total factor productivity through sustainable intensification of agri-food systems?

To crystallise its role on the way to actualising Agenda 2063, the Pan African Society for Agricultural Engineering (PASAE), in partnership with and financial support of the Royal Academy of Engineering (RAEng, UK), implemented a project on "Growing Agricultural Engineering in Africa" (Taylor et al., 2021). To start to answer the questions raised above, we set out to compile a list of the top 100 questions facing agricultural engineering education and research in Africa, based on an extensive review of the literature covering agricultural engineering, agricultural development, and related topics, including policies related to agricultural modernisation in Africa. The literature search resulted in a list of 192 questions which were roughly organised into two main groups: (A1-A5) education, and (B1-B11) research and practice.

Within each group, the questions raised were further organised into sub-groups according to topic. During the process, similar questions were either reframed, merged or one of them discarded, and the process was continued until the top-ranking 100 questions emerged. These questions are expected to inspire the development of more region-specific variations that align closely with the priorities of education, research, and society. Overall, AE education accounted for 24 and AE research and practice accounted for 76 of the top-ranking questions identified.

The Future of Agricultural Engineering Education in Africa

The questions identified in relation to agricultural engineering education were categorised into four broad areas that reflect the breadth and depth of issues gleaned from the literature survey – curriculum, pedagogy, quality assurance, and the low public image of agricultural engineering. In this regard, one question stands out:

"How do we attract the best young minds to AE so that they can be best equipped to address the grand challenges facing Africa and humanity at large such as climate change, food security, water security, renewable energy, environmental degradation, and rising youth unemployment?"

Here, we consider the type and quality of education needed to produce, retain, and grow the pipeline of future agricultural engineers that are better equipped to transform Africa's agri-food systems through policy, practice, entrepreneurship, and thought leadership. Making the best possible progress in modernising and transforming Africa's agriculture will require exceptional people and globally competitive skills and talent. We need to radically change our culture and the way we train the future generation of agricultural engineers so that 'agri cultural/biosystems/biological/bioresource engineer' can join 'doctor', 'pharmacist', 'lawyer', 'computer engineer', 'biomedical engineer', and 'vet' in the list of top professions to which our most talented and capable young people aspire.

A1: What curriculum design and content do we need?

Curriculum design, structure and content are essential for organising knowledge for effective teaching and learning. These allow both the teacher and learner to have a clear understanding of what materials will be covered during the course, including the scope, thereby facilitating co-learning and the preparation for learning sessions by both participants. The curriculum will also address issues such as finding the appropriate balance between theory and practical sessions, basic sciences, engineering, and the humanities, etc.

1. What curriculum will prepare agricultural engineers for life-long learning and future work?
2. What is the balance between technical, soft, and entrepreneurial skills?
3. What are the new and emerging knowledge frontiers/topics relevant to AE education?
4. Agriculture vs Biology vs Engineering: what is the foundation of AE?
5. What are the desirable graduate quality attributes of future agricultural engineers?
6. What are the roles of agricultural experiment stations in AE education and training?

7. What is the relevance of smallholder agricultural practices and related hand tools in modern AE curricula?

A2: Pedagogy and epistemology

Questions related to pedagogy and epistemology concern how we train future agricultural engineers for life-long learning. Pedagogy refers to both the theoretical and practical approach employed by teachers to instruct, encompassing the interaction between culture and various learning methods. On the other hand, epistemology concerns the theory of knowledge, exploring the connection between the mind and reality. It is concerned with the mind's relation to reality. What is it for this relation to be one of knowledge? Do we know things? And if we do, how and when do we know things?

8. To what extent are new and modern teaching and learning methods, including online methods, incorporated in AE education?
9. What are the roles of information and communication technologies in future AE education?
10. How do we harness social media to promote effective learning?
11. How do we make Capstone/Design Project courses relevant to smallholder agriculture?
12. How do we improve and increase the 'practical' component of AE education?
13. Does the students' industrial work experience scheme (SIWES) expose AE students to experiential learning relevant to commercial agribusiness?

A3: How do we assure quality and equivalence in agricultural engineering education?

Similar to any other field, the education sector is regulated and administered through particular policies, practices, and guidelines that define the minimum standards of quality and achievement expected from students, teachers, administrators, parents, and the general public. Quality assurance is necessary for programme accreditation (Opara, 2004; Opara et al., 2004a,b; Opara et al., 2006; Opara & Cuello, 2006a, b; Comparetti et al., 2006; Cuello, & Opara, 2007), and most importantly, it provides the blueprint for quality improvement and benchmarking. Both internal and external quality assurance are necessary to achieve the desired objectives.

14. What are the mechanisms and procedures for accreditation of AE degree programmes?
15. What are the internal and external mechanisms for quality assurance in AE education?
16. What benchmarks are required to standardise AE curricula?
17. How do we achieve a Pan African standard for AE curricula, mobility, and cross border recognition of qualifications?
18. How do we ensure that future researchers have the most suitable background knowledge and skill sets to address the research challenges that they are likely to face?

A4: How do we attract and retain the brightest and best young minds to careers in agricultural engineering?

During the past three decades, one of the major challenges facing the agricultural engineering discipline and related academic programmes in tertiary institutions is low enrolment (Opara, 2012, 2003a,b; Opara et al., 2004a,b). As a result, many programmes have been closed or merged into other related disciplines. Attracting, retaining, and graduating students majoring in agricultural and related bio-inspired engineering programmes remains a challenge and these raise critical questions which need to be addressed to assure the future of the discipline and skills needed to achieve the "Agriculture We Want".

19. What actions (and by whom) are required to improve the low public image and perceptions of AE?
20. What's in a name? – Agricultural, Biosystems, Bioresources, or Biological Engineering?
21. How do we ensure that society appreciates the full importance of AE?
22. How do we make AE attractive to top quality students, their guardians, and advisors?
23. How do we grow and retain the pipeline of AE graduates?
24. What are the current destinations of graduates trained in agricultural engineering in the workforce?

Future of Agricultural Engineering Research and Practice in Africa

Research and innovation are critical for socioeconomic development. Sub-Saharan Africa is still largely an agrarian economy, and the successful industrialisation of Africa, like in other developed economies, will be led by the modernisation and transformation of its predominant smallholders into agribusiness value chains. This transformation will require enhancing agribusiness productivity and growth, developing and advancing new knowledge, and developing novel agricultural engineering solutions to address the persistent technological and structural challenges facing subsistent and smallholder agriculture. These challenges include protecting the environment, sustainable conservation, and utilisation of bioresources, and responding to climate change and other emerging global challenges. Agricultural engineering research must also address the challenges that currently limit both intra-Africa and global trade and market access of agri-food products and services. In particular, the African Continental Free Trade Area (AfCFTA) is one of the 13 flagship projects of Agenda 2063: The Africa We Want, of the African Union. With the potential of creating a single market worth US$6.7 trillion by 2030, the objective of the AfCFTA is to significantly boost intra-Africa trade, particularly trade in value-added production and trade across all sectors of Africa's economy (WEF, 2024). The achieve the goals of creating one Africa market, agricultural engineering education and research has important roles to play in developing the required workforce, conducting research and applying relevant technologies (novel products and processes) which contribute to eliminating barriers to trade in Africa and with the rest of the world.

Regarding the future of agricultural engineering research and practice, the leading questions identified were categorised into eleven broad areas:

(i) sustainable production and intensification;
(ii) banishing the hand-hoe to the museum;
(iii) finding best ways to maintain soil quality and health;
(iv) finding best ways to maintain plant and animal health;
(v) investigating how irrigation can contribute to sustainable agricultural intensification;
(vi) how we feed our children's children;
(vii) ensuring energy security for agriculture and rural development;
(viii) finding cost-effective ways to reduce preharvest and postharvest losses and food waste;
(ix) supporting trade and market access towards agricultural transformation and regional integration;
(x) mitigating the impacts of climate change on agriculture and the environment; and
(xi) Harnessing the power of new and emerging technologies to transform African agriculture and food systems for socioeconomic development.

The limited applications of modern agricultural engineering technologies and other improved inputs and services in African agriculture, which contribute to the deep yield gaps, limited value addition, and lack of competitiveness, can be illustrated by the huge deficits in mechanisation intensity, low levels of irrigation, and high degradation of agricultural soils.

Low Levels of Mechanisation

Sustainable intensification of agriculture, including agricultural mechanisation, is energy dependent (Pimental, 1976). The near-linear positive relationship between the degree of mechanisation (power input and use of improved machinery) and agricultural productivity is well documented in the global scientific literature. Correspondingly, there is overwhelming agreement and a renewed call for action on sustainable agricultural mechanisation as a key input in the modernisation and transformation of the agricultural sector in Africa (Sims et al., 2016). Today, human muscle still accounts for over 80% of the power input across much of Sub-Saharan Africa, while the hand-hoe and cutlass remain the primary tools used in cultivation and other on-farm operations. Similarly, the number and use of modern and improved power sources such as tractors and pumps in Africa is the lowest compared with other regions of the world. Current estimates show that Africa has approximately 5 tractors available for every 1,000 farmers, whereas the USA boasts a significantly higher ratio of 1,600 tractors for every 1,000 farmers (Jones, 2016). The mechanisation of smallholder agriculture remains an intractable problem. Overall, it can be said that Africa lacks the (mechanical) power to modernise and transform its agri-food systems. This deplorable state of modern power and machinery in agriculture has been recognised at the highest levels of leadership in Africa, resulting in the symbolic call to relegate the hand-hoe to the confines of a museum.

Irrigation Deficit and Related Water Harvesting and Water Saving Technologies

Africa's agriculture is thirsty. At the same time, there is rising competition for water between urban demand and agriculture. Approximately 63% of urban regions in sub-Saharan Africa lack access to fundamental water and sanitation services. Although irrigation has the potential to enhance agricultural productivity by at least 50%, the majority of food, feed, and fibre production on the continent relies heavily on rainfall. Currently, the area equipped for irrigation covers just over 13 million hectares, constituting only 6% of the total cultivated area, in stark contrast to Asia's 37%, Latin America's 14%, and the global average of 18%. Numerous governments and development organisations have proposed significant investments in irrigation across Africa. However, the success of these projects relies heavily on geographical, hydrological, agronomic, and economic factors that must be taken into account to assess their long-term viability and sustainability. The lack of sufficient irrigation contributes to Africa's low agricultural productivity. The current very limited application of irrigation in African agriculture underscores the need for research and innovation to optimise crop performance under the difference agro-ecological zones.

Poor Soil Quality and Health

Sub-Saharan Africa (SSA) covers an extensive region of 24.6 million km^2, encompassing a wide variety of soil and land management types. The predominant soil types include Arenosols (21.5%), Cambisols (10.8%), Ferralsols (10.4%), and Leptosols (17.5%). The extent and nature of soil constraints on agricultural productivity vary significantly. Approximately 40% of soils in SSA have low nutrient capital reserves (<10 weatherable minerals), 25% are affected by aluminium toxicity, and 18% have a high leaching potential with low buffering capacity. Soil degradation manifests at various scales, including individual farm fields, farming communities, and landscapes. For instance, gullies present on farms often indicate changes in land use upstream (at the head of the watershed), such as intensive grazing or excessive mechanised agriculture, which can lead to erosion or contamination downstream. Soil degradation, characterised by soil erosion, nutrient depletion, loss of biological diversity, and a decline in water quality and quantity, poses a significant challenge for many countries in sub-Saharan Africa.

Approximately 65% of the agricultural land in SSA suffers from degradation due to inadequate management practices, resulting in a decline in the biological, chemical, and physical quality of the soil. This degradation reduces the soil's ability to support crop production and provide essential ecosystem services. Soil degradation presents a significant challenge for agricultural systems in sub-Saharan Africa. While there is substantial knowledge about the consequences of degraded soils, the evidence base for farm management practices that could prevent and mitigate soil degradation is still developing, and farmers' adoption of such practices remains limited. Despite the low quality and poor health of Africa's agricultural soils, the use of fertilisers in crop production is notably lower compared to other parts of the world. Novel sources of sustainable and environmentally friendly and non-synthetic fertiliser are needed, and agricultural engineers play a crucial role in designing and deploying more effective methods for their application.

B1: How Do We Ensure Sustainable Production and Intensification?

Agricultural production has witnessed significant growth over the last fifty years, fostering economic development in rural and urban areas globally (Pretty et al., 2011). However, the result has been poor in Africa due to a wide range of factors, leading to widespread poverty among farmers, particularly the rural poor. From the early 1960s to the 1970s, Sub-Saharan Africa (SSA) experienced a moderate increase in agricultural production per capita, with intermittent brief periods of growth in the mid-1970s and 1980s. However, in contrast with other developing regions of the world, the overall trend in agricultural productivity growth has been one of decline (Hazell & Wood, 2008). As a result, food insecurity has been a continuing problem in SSA, affecting over 30% of the population (Pfister et al., 2011). In response, many governments in Africa and their external development partners (mainly government-funded donor agencies and philanthropic organisations) have promoted and focused their policies on smallholder/subsistence farming for food security. To date, it is clear that this decades-long approach to sustainable agri-food systems transformation has not yielded the expected results for sustainable economic development and widespread poverty reduction, particularly among the rural population who are responsible for the majority of the food produced and consumed in Africa. Evidence from both developed, developing and emerging regions have shown that sustainable intensification of agriculture can be a major driving force for widespread economic development and poverty reduction. For example, in China, Wang (2013) contended that maintaining agricultural growth over the past few decades played a pivotal role in reducing poverty as a significant portion of the impoverished population resides in rural regions and relies heavily on agriculture for their sustenance.

Many researchers have attributed the low agricultural productivity in Africa to factors intrinsic to the continent and its people, such as climate, soil quality, slavery, and disease. However, Bjornlund et al. (2020) argued that these are not the primary reasons behind the persistently poor performance of African agri-food systems. Instead, the authors traced the historical role of agriculture and asserted that complex agricultural systems once thrived in Africa, supporting food security, manufacturing, and trade before European traders arrived. External interference disrupted these systems in favour of export crops, and even after achieving political independence, the fundamental issues remain unchanged, with resource and wealth extraction continuing to hinder economic development in Africa.

The reported successes of many external development agricultural intensification projects and programmes in Africa have not been sustained nor replicated through domestic agricultural and rural development initiatives. For example, the UK Government's Foresight Global Food and Farming project commissioned analyses on sustainable intensification in 40 projects and programmes across 20 countries during the 1990s–2000s. By early 2010, these initiatives had demonstrated significant benefits for 10.39 million farmers and their families, leading to improvements on approximately 12.75 million hectares of land (Pretty et al., 2011). The research revealed that sustainable intensification had a multiplicative effect on food outputs, as it combined the use of new and improved crop varieties with enhanced agro-ecological management practices, resulting in average crop yields increasing by 2.13-fold. Additionally, sustainable intensification had an additive impact by promoting diversification, leading to the introduction of new crops, livestock, or fish alongside existing staples and vegetables cultivated on the land.

Africa still relies largely on subsistent smallholder agri-food systems, including the slash and burn of grasslands and woodlands for crop and livestock cultivation. Low levels of mechanisation mean that most farmers still rely on human muscle with associated ancient tools (hand-hoe and cutlass) for manual bush clearing, cultivation, weeding, harvesting (Figure 4.1), processing, and transport of food and other agricultural raw materials. Together, these contribute to the very low yield of crops and livestock. The modernisation of agri-food systems using agricultural technologies is crucial to close productivity gaps and reduce the negative impacts of agriculture on the environment.

25. How can AE technologies close the huge yield gaps in Africa?
26. What are the technological bottlenecks facing the adoption of protected cultivation of food, feed, and fibre, especially among smallholders and emerging commercial farmers in Africa?
27. What are the prospects for conservation agriculture among smallholders and emerging commercial farmers in Africa?
28. How do we make precision agriculture work for smallholders and emerging commercial farmers in Africa?
29. Given the increasing popularity and expansion of urban agriculture, including practices like roof-top farming, vertical farming, and basement farming, etc, what are the most effective management strategies to protect human and environmental health from potential risks such as pollution, soil compaction, runoff, erosion, and nutrient losses in urban settings?
30. What role for AE in intensifying the sustainable production, trade, and utilisation of Africa's indigenous plants and livestock for food, feed, fibre, and pharma?

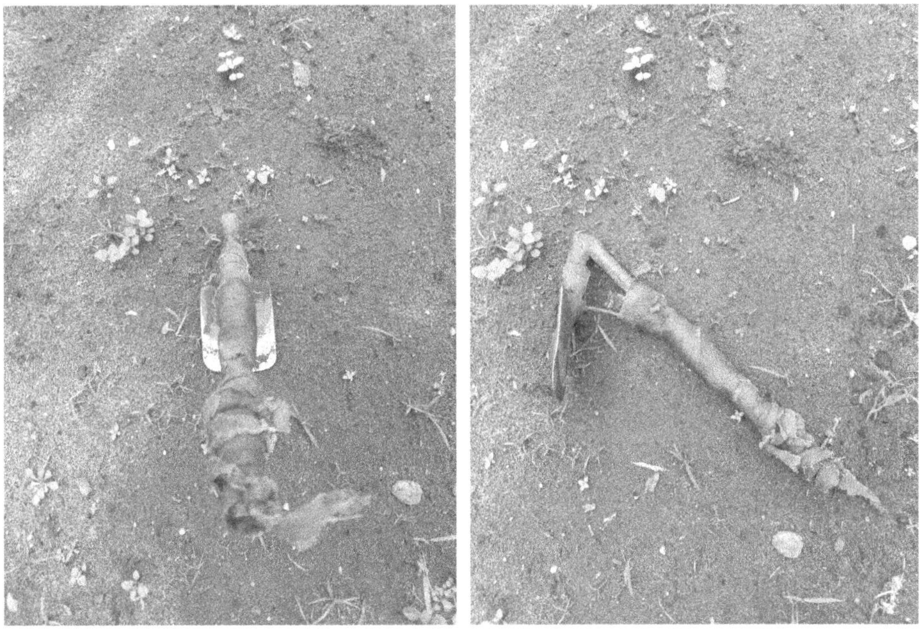

FIGURE 4.1 Front (left) and side (right) views of a typical hand-hoe used for cultivation and harvesting root and tuber crops in Southeastern Nigeria. The handle is traditionally made from trunk and branch of small trees (giving the V shape) but increasing replaced by metal/iron. Notice the wrapping of the handle with discarded cloth materials to reduce bruising of the palms. Photo taken by author in Umubachi Kindred, Umunam Village, Imerienwe, Imo State, in Southeastern Nigeria during the 2023 farming season.

31. What are the technological and socioeconomic models to transit from nomadic herding to ranching and other methods of sustainable livestock intensification?
32. What is the scientific evidence on the impacts of slash-and-burn agriculture on sustainable food systems, habitat loss, environmental degradation, and pollution?

B2: How Do We Banish the Hand-Hoe to the Museum?

The predominance of the hand-hoe and cutlass as primary tools for food production has become a symbol of the primitive status of agriculture in Africa relative to the rest of the world. With the pervasive deployment of modern technologies across other economic sectors in Africa, from communication to transport, health care, fashion, housing, entertainment, etc., it is mind-blowing to imagine that the key technologies used to produce food in most parts of Africa have remained unchanged over millenia. This factor alone is arguably one of the largest contributors to the high level of drudgery and low interest in agriculture among the youth. With the rapidly declining adult population engaged in farming, the widespread use of modern power sources (especially the tractor) and associated implements has become an urgent necessity, including automation and other artificial intelligence enabled technologies.

FIGURE 4.2 Teenage girl on her way home with her peers from farm demonstrating the use of the hand doe. Photo taken by author in Umubachi Kindred, Umunam Village, Imerienwe, Imo State, in Southeastern Nigeria during the 2023 farming season.

Women and girls, who cultivate (Figure 4.2) and process (Figure 4.3) most of the food consumed in Africa, and who suffer the most from the associated ill-health of using the hand-hoe over their lifetime, have rejected it and called for the introduction of modern and productivity-enhancing

FIGURE 4.3 The last steps during cassava processing into garri. Left (pulverising by Mrs Onyewuonyeoma Ogazi) and right (frying by Ms Celicia, nee Ogazi). Photo taken by author in Umubachi Kindred, Umunam Village, Imerienwe, Imo State, in Southeastern Nigeria during the 2023 farming season.

technologies in agriculture. The African Union has called for the banishing of the hand-hoe. How do we achieve this? What is the role of agricultural, biosystems, bioresources, and biological engineering?

33. What is the degree of diversity in hand-hoe designs across the different agro-ecological zones in Africa?
34. To what extent has the ergonomics of past and present hand-hoe designs and their utilisation affected the health of farmers, especially women, in rural Africa?
35. What if agricultural production in Africa relied entirely on the hand-hoe and cutlass?
36. What is the level of agricultural mechanisation intensity across the different agro-ecological zones in Africa?
37. Are there optimal models for selection, cost-effective access and utilisation of improved agricultural mechanisation and automation technologies in Africa?
38. What are the cost-effective pathways for the introduction and utilisation of drones and robots to improve the productivity of smallholder agriculture?
39. How can the African agricultural engineering community better engage and share information with the public and broader scientific communities about the urgent and critical need to mechanise and modernise agriculture?
40. How can renewable energies such as solar accelerate electrification and industrialisation of rural communities in Africa?

B3: What Are the Best Ways to Maintain Soil Quality and Health?

Soils in most parts of Sub-Saharan Africa (SSA) have been reported to be notoriously fragile and degraded (Tully et al., 2015). Researchers have linked soil fertility and overall state of soil health to hunger in Africa (Sanchez, 2002, 2005). Due to the absence of reliable data and only approximate estimates, the true extent and pace of soil degradation in SSA remain largely uncertain. However, it is evident that soil fertility depletion stands as the primary factor leading to the decline in the ability of certain soils to support food production and essential ecosystem services. The main driver behind soil degradation is the expansion and intensification of agriculture aimed at catering to the growing population in the region. To achieve effective solutions to support resilient and sustainable food systems, the integration of agricultural, environmental, and socioeconomic objectives is paramount.

41. What critical soil functional properties and biological diversity are affected by mechanisation operations?

42. To what extent is soil compaction limiting agricultural productivity in Africa, and how can AE reverse this?
43. To what extent is soil erosion limiting agricultural land use and productivity in Africa, and how can AE reverse this?
44. What 'green technologies' and novel production practices can heal and restore damaged soils and ecosystems?
45. What is the economic cost of the soil degradation that may result from mechanisation (e.g., erosion, tillage, and compaction) relative to the value of crop production?
46. Drawing from both historical and recent remotely and closely sensed data, how can geospatial modelling or simulation methods offer enhanced insights into the connections between soil quality, soil management, climate, and agricultural production?
47. How can we enhance soil tillage techniques, residue management, and nutrient applications to foster the sustainability of natural resources, and minimise adverse environmental impacts while ensuring continued food production?
48. How do we raise productivity of marginal lands?
49. In what ways can the worldwide agricultural engineering community improve its communication and interaction with the public and broader scientific communities, encompassing fields like soil and earth sciences, hydrology, and environmental engineering?

B4: What Are the Best Ways to Maintain Plant and Animal Health?

African crops and animals are generally not in good health or highly vulnerable to ill-health due to high incidence of pests and diseases to reduce productivity and contribute to huge preharvest food losses. Due to ineffective cross border policies and practice, the borders of most Africa countries and indeed Africa's borders are highly porous to the movement of goods and people, making them predisposed to the entry of disastrous pests and diseases.

50. What 'green technologies' can be developed from bioresources for better plant and animal health?
51. How can we use knowledge of plants and soil to stop weeds from growing on farms?
52. What traceability systems are required to monitor the disease and contamination of plant and animal foods?
53. What are the prospects of on-farm and postharvest remote sensing of plant diseases and stress?

B5: How Can Irrigation Contribute to Sustainable Agricultural Intensification?

Irrigation is recognised as part of the suite of Green Revolution technologies, in addition to high-yielding crop varieties, fertiliser, the use of agri-chemicals, and mechanisation, which contributed to reported widespread increases in yield and productivity of farmers in Asia and Latin America during the past half century. It is widely acknowledged that the very limited application of these technologies across Africa has contributed to the delayed arrival of Africa's Green Revolution, and it is hoped that the Alliance for Green Revolution in Africa will lead the charge to change this narrative. Questions remain which needs answers to realise this goal.

54. What is the impact of deficit irrigation on crops that feed Africa, especially in water-scarce areas?
55. With the unique climatic conditions in most parts of Sub-Saharan Africa characterised by long periods of hot high temperatures and low relative humidity, what novel techniques can reduce the evaporation of water?
56. What is the evidence that novel water harvesting and storage technologies contribute to improved crop yield and overall agricultural productivity?
57. What are the latest techniques to reduce drainage and enhance water retention of agricultural soils?
58. What is the status of skills capacity, institutions and supply chains for pumps and spare parts necessary for successful irrigation sub-sector?
59. How can we reduce water use in agriculture without reducing yield and crop health?

B6: How Do We Feed Present and Future Generations?

Sub-Saharan Africa remains the hot spot and flash point for food and nutrition insecurity, and very limited resilience among farmers and their households. Both crop yield, productivity, and total production remain low, leading to high food deficits which make smallholder farmers vulnerable to food and nutrition insecurity, malnutrition, and stunting among children. Food deficit is further exacerbated during natural disasters, conflicts, and pandemics such as the recent COVID-19. To address the deficit, many countries rely on food aid donations and imports. Food imports have continued to rise during the past several decades, amounting to over US$ 48 billion per annum. The modernisation of Africa's agriculture and food systems is an imperative to improve productivity and total production through sustainable intensification. To achieve these, some big questions remain while new ones arise.

60. Can Africa feed itself by 2063?
61. How can agricultural engineering improve food availability?
62. How can agricultural engineering enhance access to food?
63. How can agricultural engineering improve food utilisation?
64. Can we use agricultural engineering to prevent malnutrition?
65. In what ways can agricultural engineering promote sustainable consumption?

66. Can we use agricultural engineering to reduce poverty, especially among farmers and rural communities?
67. How can agricultural engineering contribute to the domestication, commercialisation, and sustainable use of wild and under-utilised bioresources for food, feed, fibre, and pharma?
68. What are the best sustainable ways to commercially produce, handle, process, and market alternative new sources of food such as insects, without negatively altering the natural habitats and the environment?
69. Can the local (traditional) crops and animals that feed Africa be transformed into regional and global food value chains?

B7: How Do We Ensure Energy Security for Agriculture and Rural Development?

There is a near-linear relationship between energy input and crop yield. Africa has the lowest level of agricultural mechanisation intensity compared with other regions in the world, as measured by energy input from all sources – human, animals, tractors, and other engine-powered machinery. Human muscle is the largest contributor to agricultural power in Sub-Saharan Africa, which limits productivity and increases labour challenges. Agriculture is recognised as a major contributor to the emission of greenhouse gases (GHGs), contributing to rising temperatures and climate change. While advocating for an energy mix to propel Africa's development, there is tremendous opportunity for renewable energy generation and use in Africa, including autonomous power stations/units for remote rural communities.

70. What roles can agricultural engineering play in developing alternative, clean, and sustainable non-fossil energy sources, such as solar?
71. Which best candidate plants can be commercially grown and processed into biofuels with the least effects on biodiversity, carbon footprints, and food security?
72. How can agricultural engineering enhance the commercial production of "second-generation biofuels" – produced by algae, plant cellulose, etc?
73. Can we develop more efficient and long-term storage systems for solar energy?
74. How can we improve the collection, storage, and conversion of biomaterial wastes into bio-fuel?

B8: What Cost-Effective Ways Exist to Reduce Preharvest (On-Farm) and Postharvest Losses and Food Waste?

The incidence of preharvest and postharvest food losses is very high in Africa, thereby contributing to low crop yields and reduced food availability. While infestation by pests and diseases and inadequate farm management practices contribute to on-farm losses, harvesting and postharvest losses arise from improper harvest management practices, lack of improved postharvest knowledge, and limited application of novel postharvest technologies and agro-processing, among others. While preharvest and postharvest losses contribute to most incidence of food wastage, food waste is increasing becoming a problem among the middle class and in public institutions such as schools and hospitals. Africa's losses amount to over US$40 billion dollars in food produced per annum that does not reach the consumer, excluding the impacts of these losses on the environment, bioresources, and climate change. Investing in education and research efforts for sustainable production as well as sustainable consumption must be part of solution for sustainable agriculture for a sustainable Africa and the world.

75. What are the magnitudes and causes of preharvest (on-farm) losses of food crops and animals at subsistent/smallholder and commercial farms?
76. What the magnitudes and causes of postharvest losses of food crops and livestock products along the value chains of subsistent/smallholders and commercial agribusinesses?
77. Is food waste a problem in Africa?
78. What policies and incentives will promote the reduction of food losses and waste?
79. How can agricultural engineering contribute to reducing preharvest and postharvest food losses?
80. How can agricultural engineering contribute to reducing food waste?
81. Wealth from food waste: what effective processing methods and techniques can valorise food waste into safe food and high-value industrial raw materials?

B9: How Can We Ensure That Trade and Market Access Promote Agricultural Transformation and Regional Integration?

Africa has the lowest levels of intra-regional trade compared to other parts of the world. Trade within Africa has hovered at around 13% during the past decade and has recently increased to about 20% following the adoption of the African Continental Free Commerce Area, compared to about 70% in Europe and Asia. There are technological and policy dimensions to this problem, which can be adequately addressed to promote regional integration, economic development, and improvement in food security through enhanced availability. Developing and setting market standards for trade and exchange is important, including effective monitoring and surveillance systems for movement of people, goods, and services.

82. What are the quality attributes and marketing standards of fresh and processed foods in Africa?
83. How can agricultural engineering research contribute to market access and trade (intra- and global) in agri-food products?
84. What are the best ways to control pests, diseases, and disorders in Africa that are threats to the quality,

safety, marketing, and utilisation of plant and animal food products?

85. What research and new knowledge is required to transform selected local African foods into highly sought after regional and global brands?

B10: How Can We Mitigate the Impacts of Climate Change on Agriculture and the Environment?

Agricultural practices significantly contribute to climate change by negatively affecting the environment, bioresources, greenhouse gas emissions, pollution, and causing temperature rises. Simultaneously, agriculture also faces the consequences of climate change, experiencing shifts in weather patterns, heightened variability, unpredictability, more frequent natural disasters, and increased occurrences of pests and diseases. These climate-related challenges collectively result in reduced crop yields and compromised crop quality. Thus, agriculture is both a contributor and victim of climate change. While we recognise that these impacts of climate change are real in both directions, more evidence-based data is needed for specific agri-food systems and value chain to guide and support policies and investments to mitigate the impacts.

86. What are impacts of climate change on agriculture and food systems in Africa?
87. What are the impacts of climate change on bioresources and environment?
88. What current prediction models best simulate the possible climate change scenarios and impacts on agri-food systems in Africa?
89. Do we know the possible impacts of climate change on incidence of extreme and hazardous events such flooding of agricultural lands, drying of rivers, wildfires, etc?
90. Do we understand the potential impacts of climate change on plant/animal production cycles/seasons vis-à-vis the timeliness of agricultural operations?

B11: How Do We Harness the Power of New and Emerging Technologies to Transform African Agriculture for Socioeconomic Development?

Historically, advancements in agricultural and biosystems engineering education, and related technological advancements for application in agribusiness, have benefited from innovations emanating from other sectors of the economy. While agricultural mechanisation has been recognised as one of the greatest engineering and technological contributions to society during the past century, the benefits have not been realised in many parts of the world due to very low adoption in areas such as in Sub-Saharan Africa. Meanwhile, many new, novel, and emerging technologies have shown remarkable impacts in improving agricultural productivity through sustainable practices. Biotechnology and related technologies such as CRISPR genome editing provide opportunities to dramatically increase crop yield, reduce losses, and incorporate desired quality and safety attributes. Applications of nanotechnology and digitisation in agriculture offer new frontiers for a new agriculture that is more sustainable and delivers novel products and services demanded by consumers. The ongoing explosion in new knowledge and novel technological applications such as artificial intelligence (AI) (Waoo et al., 2023), Machine Learning (ML) (Nturambirwe & Opara, 2020; Nturambirwe et al., 2023), Digital Twins (DTs) (Pylianidis et al., 2021; Shoji et al., 2022), virtual prototyping (Ambaw et al., 2022; Mukama et al., 2020; Mukama et al., 2019), and virtual cold chain (Wu et al., 2018) offer new frontiers for agricultural engineering education and research which can contribute to a sustainable agriculture and a sustainable world.

91. How can agricultural engineers leverage the remarkable advances in biotechnology for sustainable intensification of Africa's agri-food systems?
92. How can agricultural engineers harness the developments in nanotechnology and miniaturisation to radically transform smallholder agriculture into commercial and profitable value chains?
93. Can we utilise the ubiquitous and rising power of information and communication technologies (ICTs) to digitise and connect African agri-food systems into regional and global value chains?
94. How can the integration of data science (DS), artificial intelligence (AI), machine learning (ML), and digital twins (DTs) be mainstreamed to agricultural engineering research?
95. What new scientific approaches will be central to agricultural engineering in the 21st Century?
96. How do we ensure that sound science and best engineering practice inform policy decisions on agricultural development?
97. What is the role of agricultural engineering in preventing and mitigating the impacts of current pandemics and future pandemics on agri-food systems?
98. How do we mainstream agricultural engineering extension services into agricultural development practice?
99. How can we provide effective extension services to farmers and agribusiness operators via smart and non-smart phones?
100. How can we promote and facilitate interdisciplinary collaboration in agricultural engineering research and practice to effectively work towards achieving the AU's Agenda 2063 and the UN's Sustainable Development Goals (SDGs) which address poverty, agricultural modernisation, climate change, the environment, and sustainable development?

Conclusions and Future Prospects

The Green Revolution that has propelled many countries in Asia and Latin America into new trajectories of economic development, through improved food security and widespread poverty reduction, has not had a similar influence in Africa. Sustainable intensification, mechanisation, irrigation, fertiliser application, the use of improved plant and animal genetic resources, soil management, and other recent agricultural technologies have all remained very low in most of Africa. Agricultural engineering education, research, and practice have critical roles to play in modernising and transforming agri-food systems in Africa. Africa needs its own agile and skilled workforce that can embrace and apply modern agricultural engineering technologies across the agri-food value chain, from production to postharvest handling, processing, storage, transport, logistics, and marketing. As the former President of the Federal Republic of Nigeria, His Excellency Olusegun Obasanjo (2016), aptly put it: "But one thing is clear to me as I return to farming: to achieve its potential, African agriculture needs a fresh infusion of innovation and talent".

Future agricultural engineering education must produce lifelong learners who possess the requisite engineering knowledge and understanding of the biological, environmental, and resource dimensions that are necessary for sustainable intensification of the agri-food system. This requires a rethink of agricultural and biosystems engineering education that is aligned with the future of agriculture and food systems that Africa wants – modern, knowledge-driven, technology-enabled, productive, resilient, sustainable, climate-smart, competitive, and profitable. Future agricultural engineering research must contribute to finding sustainable and cost-effective answers to the wide productivity gaps and efficiency challenges facing agriculture (and the environment). Africa's agri-food systems remain largely at the subsistence level, with farm sizes often less than 1 ha and smallholder farmers (mainly women and children) producing over 70% of food consumed. The call to banish the hand-hoe to the Museum in Africa is now urgent more than ever before, not only so as to replace it with tractors and other modern agricultural power sources and types of machinery such as the tractor, but also as a befitting symbol of the modernisation and transformation of African agriculture.

In consideration of these challenges (and opportunities!) facing agricultural engineering education and research in Africa, as well as the complexity of internal and external drivers affecting agriculture such as climate change, the emergence of transformative technologies (e.g., biotechnology, nanotechnology, information and communication technologies, robotics, artificial intelligence, etc.), and the globalisation of education and trade, a set of 100 top-ranking questions of importance to the future of agricultural engineering in Africa have been presented in this chapter. These questions do not address all the challenges facing agricultural and biosystems engineering education, research, and practices in Africa; however, they provide the needed foundation for future efforts to grow agricultural engineering education and research in Africa (and other developing regions) towards achieving both the SDGs and Agenda 2063 – The Africa We Want. Providing answers to these top-ranking questions must begin with Africans – especially agricultural and biosystems engineers – while recognising the importance of partnerships, cooperation, collaboration, and networks. As the former Board Chair of the Alliance for a Green Revolution in Africa put it while announcing the new Africa Food Prize (Masiyawa, 2016), Africa needs its own superheroes to respond to challenges: "But what Africa needs, each and every day, are heroes of a different sort – 'action heroes' with ideas and vision, ready to respond to challenges that could determine the fate of a continent with a land mass larger than the United States, Western Europe, China and India combined. This is a lot of land, with a lot of potential".

Acknowledgement

Prof. UL Opara gratefully acknowledges the Royal Academy of Engineering (RAEng), UK, for funding the project titled "Growing Agricultural Engineering in Africa: Supporting the Operations and Implementation of the Strategic Plan for AfroAgEng" through which the author carried out the study reported in this book chapter on top priority questions facing agricultural engineering education and research in Africa.

REFERENCE LIST

African Union. (2022). Agenda 2063: The Africa we want, overview. https://au.int/agenda2063/overview (Accessed on 10 April 2022).

Ambaw, A., Mukama, M., Fadiji, T. & Opara, U.L. (2022). Fresh fruit packaging design verification through virtual prototyping technique. *Food Packaging and Shelf Life (IF: 8.749, 15/12/2022)*, 32, 100858. https://doi.org/10.1016/j.fpsl.2022.100858

Bjornlund, V., Bjornlund, H. & Van Rooyen, A.F. (2020). Why agricultural production in sub-Saharan Africa remains low compared to the rest of the world – A historical perspective. *International Journal of Water Resources Development*, 36(supl), S20–S53. https://doi.org/10.1080/07900627.2020.1739512

Comparetti, A, Febo, P., Orlando, S. & Mugnozza, G.S. (2006). Tools for quality assurance and assessment of agricultural engineering programmes, tuning subject-area lines and disappearing / rare knowledge in Italy. *Proceedings of Workshop on University Studies of Agricultural Engineering in Europe, Bonn*, September 7–8. https://www.researchgate.net/publication/266137701_Tools_for_quality_assurance_and_assessment_of_Agricultural_Engineering_programmes_TUNING_subject-area_lines_and_disappearing_rare_knowledge_in_Italy (Accessed on 25 April 2023).

Cuello, J. & Opara, L.U. (Guest Editors). (2007). Special issue series: Trends in agricultural, biosystems, and biological engineering education. *International Journal of Engineering Education*, 23(4 Part III). http://www.ijee.ie/contents/c230407.html

Hazell, P. & Wood, S. (2008). Drivers of change in global agriculture. *Philosophical Transactions of the Royal Society*, 363(1491), 495–515. https://doi.org/10.1098/rstb.2007.2166

Jones, R. (2016). Banishing the hoe to the Museum. Impact. Towards Africa's Agricultural Transformation, September–October, pp. 16–18. Nairobi: Alliance for a Green Revolution in Africa.

Masiyiwa, S. (2016). Africa needs its own superheroes to respond to challenges. Impact. Towards Africa's Agricultural Transformation, September–October, pp. 32–33. Nairobi: Alliance for a Green Revolution in Africa.

Mukama, M., Ambaw, A. & Opara, U.L. (2019). A virtual prototyping approach for redesigning the vent-holes of packaging for handling pomegranate fruit. *Journal of Food*

Engineering, 270. https://doi.org/10.1016/j.jfoodeng.2019.109762

Mukama, M., Ambaw, A. & Opara, U.L. (2020). A virtual prototyping approach for redesigning the vent-holes of packaging for handling pomegranate fruit–A short communication. *Journal of Food Engineering*, 270, 109762. https://doi.org/10.1016/j.jfoodeng.2019.109762

Nturambirwe, J.F.I., Hussein, E.A., Vaccari, M., Thron, C., Perold, W.J. & Opara, U.L. (2023). Feature reduction for the classification of bruise damage to apple fruit using a contactless FT-NIR spectroscopy with machine learning. *Foods*, 12(1), 210. https://doi.org/10.3390/foods12010210

Nturambirwe, J.F.I. & Opara, U.L. (2020). Machine learning applications to non-destructive defect detection in horticultural products. *Biosystems Engineering*, 189, 60–83. https://doi.org/10.1016/j.biosystemseng.2019.11.011

Obasanjo, O. (2016). Achieving Africa's agricultural potential. Impact. Towards Africa's Agricultural Transformation, September–October, pp. 28–29. Nairobi: Alliance for a Green Revolution in Africa.

Opara, L.U. (2003a). Image is everything: Undergraduate students' contributions towards better understanding and improvement of public attitudes and perceptions of agricultural engineering. *World Transactions on Engineering and Technology Education*, 2(3), 457±460.

Opara, L.U. (2003b). From farmers to blue-collar professionals: Agricultural Engineering students' perspectives on their career prospects, critical skills, and discipline name change. *World Transactions on Engineering and Technology Education*, 2(3), 399±402.

Opara, L.U. (2004). Outlook on Agricultural Engineering education and research and prospects for developing countries. *Outlook on Agriculture*, 33(2), 101–112.

Opara, U.L. (2012). The need for a professional network of agricultural and biosystems engineers in Africa. *International Journal of Postharvest Technology and Innovation*, 2(4), 426–432.

Opara, L.U., Al-Adawi, S.S. & Al-Shukeili, T.Y. (2004a). Students' perceptions of agricultural engineering education at Sultan Qaboos University in Oman, Part 1: Academic program selection. *World Transactions on Engineering and Technology Education*, 3(1), 139±142.

Opara, L.U., Al-Adawi, S.S. & Al-Shukeili, T.Y. (2004b). Students' perceptions of Agricultural Engineering education at Sultan Qaboos University in Oman, Part 2: Curriculum content and career preferences. *World Transactions on Engineering and Technology Education*, 3(1), 143–146.

Opara, L.U., Perret, J.S. & Ampratwum, D.B. (2006). The evolution of agricultural engineering education at Sultan Qaboos University. *International Journal of Engineering Education*, 22(6), 1139–1148.

Opara, U.L. & Cuello, J. (Guest Editors). (2006a). Special issue series: Trends in agricultural, biosystems, and biological engineering education. *International Journal of Engineering Education*, 22(1 Part I), Agricultural Engineering and the Emergence of Biological Engineering. http://www.ijee.ie/contents/c220106.html

Opara, U.L. & Cuello, J. (Guest Editors). (2006b). Special issue series: Agricultural/ biosystems/ biological engineering education. *International Journal of Engineering Education*, 22(6 Part II). Curriculum Reform, Assessment and Accreditation. http://www.ijee.ie/contents/c220606.html

Pfister, S., Bayer, P., Koehler, A. & Hellweg, S. (2011). Projected water consumption in future global agriculture: Scenarios and related impacts. *Science of the Total Environment*, 409(20), 4206–4216. https://doi.org/10.1016/j.scitotenv.2011.07.019

Pimental, D. (1976). World food crises: Energy and pests. *Bulletin of the Entomological Society of America*, 22(1), 20–26.

Pretty, J., Toulmin, C. & Williams, S. (2011). Sustainable intensification in African agriculture. *International Journal of Agricultural Sustainability*, 9(1), 5–24. https://doi.org/10.3763/ijas.2010.0583

Pylianidis, C., Osinga, S. & Athanasiadis, I.N. (2021). Introducing digital twins to agriculture. *Computers and Electronics in Agriculture*, 184, 105942. https://doi.org/10.1016/j.compag.2020.105942

Sanchez, P.A. (2002). Soil fertility and hunger in Africa. *Science*, 295, 2019–2020.

Sanchez, P.A. & Swaminathan, M.S. (2005).Hunger in Africa: The link between unhealthy people and unhealthy soils. *Lancet*, 365, 442–444. 2.

Shoji, K., Schudel, S., Shrivastava, C., Onwude, D. & Defraeye, T. (2022). Optimizing the postharvest supply chain of imported fresh produce with physics-based digital twins. *Journal of Food Engineering*, 329, 111077. https://doi.org/10.1016/j.jfoodeng.2022.111077

Sims, B., Hilmi, M. & Kienzle, J. (2016). Agricultural mechanization. A key input for sub-Saharan African smallholders. *Integrated Crop Management*, 23. Rome: Food and Agriculture Organization of the United Nations.

Taylor, A.J., Opara, U.L. & Smithers, J.C. (2021). *Agricultural Engineering in Africa: A Key Driver for Transforming Agriculture to Deliver Food Security and to Support Economic Prosperity*. Pretoria: Pan African society for Agricultural Engineering (PASAE/AfroAgEng).

Tully, K., Sullivan, C., Weil, R. & Sanchez, P. (2015). The state of soil degradation in Sub-Saharan Africa: Baselines, trajectories, and solutions. *Sustainability*, 7, 6523–6552. https://doi.org/10.3390/su7066523

Wang, S. (2013). Reducing poverty through agricultural development in China. *IDS Bulletin*, 44(5–6). https://doi.org/10.1111/1759-5436.12057

Waoo, A.A., Pandey, J. & Waoo, A.A. (2023). Chapter 5: Artificial intelligence in agricultural engineering. In A. Ahirwar, P.K. Shukla, M. Shrivastava, P. Maheshwary & B. Gour (Editors), *Innovative Engineering with AI Applications*. Wiley. https://doi.org/10.1002/9781119792161.ch5

WEF, 2024. An Action Plan to Accelerate Global Business and Investment in Africa. White Paper. 36 pages. World Economic Forum, In collaboration withAfrican Continental Free Trade Area Secretariat: Geneva.

Wu, W., Cronjé, P., Nicolai, B., Verboven, P., Opara, U.L. & Defraeye, T. (2018). Virtual cold chain method to model the postharvest temperature history and quality evolution of fresh fruit – A case study for citrus fruit packed in a single carton. Computers and Electronics in Agriculture, 144, 199–208. https://doi.org/10.1016/j.compag.2017.11.034

5

Enhancing the Role of Agricultural and Biological Engineering through Education: Perspective from FAO

Josef Kienzle, Joseph Mpagalile and Karim Houmy

Introduction

While developed countries have made remarkable progress in their agricultural sectors, the same is not the case for developing countries, particularly those in Africa. Yet, this continent is home to a significant proportion of the world's land suitable for expanding sustainable agricultural production, estimated to be around 45% by 2021 (Akiwumi, 2022), offering a promising opportunity for growth and development. This confirms that more effort is needed to increase access to and use of agricultural engineering technologies and the education of human capital in key professions such as agricultural engineers. The application of agricultural engineering and mechanisation technologies has great potential to contribute to current and future agricultural development, particularly in rural areas.

Agricultural engineering is among the major components of the ongoing efforts to modernise agriculture. It remains one of the main pillars of food security for all in achieving the *four betters* of the FAO Strategic Framework 2022–2031: Better Production, Better Nutrition, Better Environment, and Better Life, Leaving No One Behind. Through agricultural mechanisation, countries are able to transform agriculture on many fronts, including crop production, livestock, fisheries, aquaculture, and beekeeping. A recent publication (FAO, 2022) underscored the fact that motorised mechanisation is an important form of automation in agricultural production and a fundamental component of agricultural transformation worldwide. This is echoed in the initiatives of several regions, in which mechanisation has been given high priority. For example, the African Union Agenda 2063: The Africa We Want gives high priority to modernising agriculture for increased productivity and production, using science, technology, innovation, and indigenous knowledge (AUC, 2015). It also underscores that the hand-hoe will be banished by 2025 and the sector will be modern, profitable, and attractive to the continent's youths and women. This will not be realised unless mechanisation is fully embraced.

Agricultural Engineering, when promoted in conjunction with leveraging digitalisation and innovative technologies, can enable farmers and other stakeholders along the food value chain to expedite the enhancement of agricultural productivity. FAO (2020) explained that the use of Information Communication Technology (ICT) tools might help solve challenges such as distortions concerning the availability of tractor hire services that farmers face and which have led to poor access and low utilisation of mechanisation services in developing regions such as Africa. However, it is also evident that a broad perspective is needed to understand the state of mechanisation in the developing world in order to implement the right interventions based on good strategies and appropriate agricultural mechanisation technology for a sustainable future tailored to each region or country (Houmy et al., 2013).

Furthermore, it is worth noting that despite all advantages of agricultural engineering and especially mechanisation, there still exists a clear divide between the level and scope of the utilisation of agricultural engineering and especially mechanisation between developed and developing countries. In most developed countries, agricultural engineering is well-integrated as part of agricultural operations, from land preparation to planting, pest and insect control, harvesting, and postharvest management, including agro-processing and marketing. This has made it possible to carry out such operations in an efficient, precise, and economical way. Unfortunately, the situation in many developing countries seems to be the opposite. The use of agricultural engineering technologies including mechanisation along the food value chain is very low in those countries. Farmers in developing countries still depend on simple traditional tools such as the hand-hoe, which is the opposite of the situation in developed countries where technological advancement has reached a stage of operating semi- and fully-autonomous machines which could automatically undertake field operations without an operator being onboard. Actually, many countries in sub-Saharan Africa are still at the power substitution stage, where efforts are being made to replace manual labour or animal draught power with diesel and electrical powered machinery such as two-wheel drive tractors (2WD) and four-wheel drive tractors (4WD) (FAO, 2018).

DOI: 10.1201/9780429150111-6

Key Definitions

Two terms will form the core of the discussions in this paper, namely agricultural engineering and agricultural mechanisation. These terms are interrelated in the sense that agricultural mechanisation is part of the wider agricultural engineering profession.

Agricultural engineering is part of the engineering profession with the other branches being civil, mechanical, chemical and process, electrical, mining engineering, etc. It is a branch of engineering that deals with the application of engineering technologies in agriculture. According to FAO, agriculture includes crop and livestock production, fishery, aquaculture, and apiculture. Notable areas of study for agricultural engineering include land surveying, levelling, contouring, farm mechanisation, irrigation, postharvest management and agro-processing, economic aspects, environmental control, tools, equipment and machinery used for agricultural operations, ergonomics, standardisation, machinery design, testing, certification, aftersale services (spare parts, repair, and maintenance), quality, safety, etc. Over the years, the definition of agricultural engineering has undergone several changes to include biosystems and biological aspects. These changes reflect a wider application of engineering principles and fundamental concepts of biology to agricultural and biological systems and tools at a scale ranging from the molecular to the ecosystem level.

According to FAO (2018), agricultural mechanisation includes the application of tools, implements and powered machinery and equipment to achieve agricultural production, comprising both crop and livestock production as well as aquaculture and apiculture. Mechanisation covers all levels of farming and processing technologies, from simple and basic hand tools to more sophisticated and motorised equipment. FAO considers agricultural mechanisation as being key in enhancing sustainable food systems. According to FAO (2023a), a sustainable food system is the one that delivers food security and nutrition for all in such a way that the economic, social and environmental bases to generate food security and nutrition for future generation is not compromised. A sustainable food system is therefore profitable throughout, ensuring economic sustainability; it has broad-based benefits for society, securing social sustainability; and it has a positive or neutral impact on the natural resource environment, safeguarding the sustainability of the environment.

The Objective of the Chapter

The objective of this chapter is to discuss the concept of agricultural engineering with a focus on mechanisation and its importance in supporting the ongoing efforts to transform agriculture. In particular, the chapter explores different aspects and branches of the agricultural engineering profession and highlights opportunities where the profession continues to significantly contribute towards enhancing the efficiency of food systems. It also explains different challenges that require urgent attention and need to be addressed. Furthermore, the chapter links the opportunities that the profession provides to the urgent need for technologies and innovations in agriculture, given the many challenges that farmers are facing such as climate change, rapidly increasing populations, dwindling resources available for agriculture, increasing costs, etc.

The Contribution of FAO and Other UN Organisations in Promoting Agricultural Engineering and Mechanisation

From its origins, FAO was designed as a knowledge organisation, and especially so for the classical technical disciplines such as agronomy and agricultural engineering. The organisation of the Agricultural Engineering disciplines within FAO has evolved over the years depending on the needs of the member countries. For example, the organisation split its sections such that the irrigation part now belongs to the Land and Water Division, whereas the livestock part is hosted in the Animal Production Division. The then revamped Rural Infrastructure and Agro-industries Division (AGS) (FAO, 2023b) hosted Agricultural Mechanisation, Food and Agricultural Industries, Postharvest Management, Rural Structures, and Rural Infrastructure themes (FAO, 2011), besides Agricultural Finance and Investment, Agricultural Market Linkages, and Agribusiness Development before it was dissolved at the end of 2015. Following a major internal reform and restructuring in 2015, different Agricultural Engineering themes mentioned above were further reallocated to several divisions. The core part related to agricultural mechanisation was established in the Plant Production and Protection Division (NSP) focusing more on the agricultural engineering during the production phase, i.e., in land preparation, crop establishment, application technologies for pesticide management, harvest and postharvest, and including digital agriculture and agri-robotics. Here a complete re-orientation took place. The Sustainable Agricultural Mechanisation (SAM) sub-team (FAO, 2023c) addresses classical agricultural engineering work and has taken on some existing challenges but also some new ones. Here is a brief summary of the key activities undertaken by our colleagues:

- *Launch of the Framework for Sustainable Agricultural Mechanisation for Africa (F-SAMA):* The Framework highlights ten priority elements of Sustainable Agricultural Mechanisation in Africa (FAO and AUC, 2018) which include:
  - (i) Boosting farm power through appropriate technologies and innovative business models
  - (ii) Promoting innovative financing mechanisms for agricultural mechanisation
  - (iii) Building sustainable systems for manufacture and distribution of agricultural mechanisation inputs
  - (iv) Sustainable mechanisation across agri-food value chains

(v) Innovative systems for sustainable technology development and transfer
(vi) Sustainable transformation of land preparation and crop/animal husbandry practices
(vii) Socio-economic sustainability and the roles of small-scale farmers, women and youth
(viii) Human resources development and capacity building for SAMA
(ix) Need for a long-term vision: policy and strategy issues
(x) Creating sustainable institutions for regional cooperation and networking.

- Support countries mainly in Africa to develop or revise National Agricultural Mechanisation Strategies (NAMS) to align them with the F-SAMA. The recent beneficiaries of such support include Chad, Ghana, Liberia, Zambia, Benin, Togo, and Tanzania. Uganda and Kenya have also revised their NAMS;
- Support countries to develop and promote business models for mechanisation service providers – what works and what type of innovations are coming out? An FAO publication has been realised to provide guidance on this key aspect of mechanisation provision (Houmy, Flores Rojas & Side, 2021);
- Focus on a holistic approach to the promotion of agricultural mechanisation along crop value chains from land preparation to on-farm processing, storage, transport and marketing;
- Linking hardware consideration with agri-business;
- Linking good agronomic practices with scaling tools such as agricultural machinery; and
- Supporting efforts to promote Conservation Agriculture (CA), with its three principles of minimum soil disturbance (no-till), permanent soil cover (mulch, cover crops), and a conducive crop rotation, is a key agronomic concept that when combined with precision planters can make a real difference.
- As indicated above, up-scaling the conservation agriculture (CA) concept requires mechanisation equipment such as precision no-till direct seeders to be available in rural areas and sales points. Therefore the team has been promoting the availability of such equipment based on the actual needs on the ground and on its being part of the mechanisation hire services portfolio of equipment.

Starting from around 2015 the FAO has been seeking partnerships with the private sector agricultural machinery industry and other strategic partners to promote mechanisation. In October 2015, the first Memorandum of Understanding (MoU) was signed with the European Agricultural Machinery Industries Association (CEMA, https://www.cema-agri.org/). Since then, this MoU has been renewed twice in 2018 and 2021. The FAO also, in its effort to promote mechanisation, seeks to collaborate with other partners such as research institutions, mainly the CGIAR Centres such as the International Maize and Wheat Improvement Center (CIMMYT), the International Livestock Research Institute (ILRI), the International Rice Research Institute (IRRI), and AfricaRice and the International Institute of Tropical Agriculture (IITA), always on matters related to sustainable agricultural mechanisation and often combined with CA. On the same subject, FAO is collaborating with relevant NGOs such as the African Conservation Tillage Network. (http://www.act-africa.org/) to promote sustainable agricultural mechanisation (SAM) in Africa.

The United Nations Industrial Development Organisation (UNIDO) is another UN organisation supporting the promotion of mechanisation. UNIDO's focus is on the agricultural machinery production sector, including local production, as well as on business models for mechanisation hire services. FAO and UNIDO have been working together on sustainable mechanisation over the past decades. UNIDO's focus is on areas such as sustainable agricultural machinery supply chains, which the FAO complements by focusing on business models for sustainable agricultural mechanisation with a focus on farmers and rural entrepreneurs (https://www.unido.org). In addition to the above, UNIDO's work has focused on agro-processing and value addition where mechanisation plays a pivotal role.

The Centre for Sustainable Agricultural Mechanisation (CSAM, https://un-csam.org/) based in Beijing is another UN Programme related to sustainable mechanisation. It belongs to the United Nations Economic and Social Commission for Asia and the Pacific (ESCAP). CSAM promotes agricultural mechanisation in Asia and the Pacific region and provides a multi-stakeholder platform for sustainable agricultural mechanisation in Asia and the Pacific. It also supports activities based around the standards, testing guidance, and safety of agricultural tools, equipment, and machinery. CSAM supports the preparation and application of standards in its efforts to enhance mechanisation in the region and to ensure the safety of the operators. In addition, CSAM supports the operations of the Asian and Pacific Network for Testing of Agricultural Machinery (ANTAM). It also contributes to regional, South-South, and Triangular Cooperation to enable the multi-stakeholder dialogue, knowledge sharing, capacity building, and evidence based policy formulation which is vital for the promotion of mechanisation in Asia and the Pacific (Anshuman, 2018).

Opportunities for Agricultural Mechanisation to Support Agricultural Development

In many parts of the developing world, the use of mechanisation is already making a significant contribution to agricultural and rural development. Overall, agricultural mechanisation plays a crucial role in modern-day farming, and its benefits are essential for achieving food security and sustainable agricultural development. Therefore, the agricultural engineer must be able to harness the different benefits of agricultural mechanisation at different levels to ensure:

- *Increased productivity:* Mechanisation must be harnessed to help farmers increase productivity by reducing the time and effort required to perform various tasks such as land clearing, planting, harvesting, postharvest management, and irrigation. Farmers can cover more area in less time with mechanisation, resulting in higher yields;

- *Reduced labour costs:* Mechanisation reduces the need for manual labour, which lowers labour costs. This is particularly important in countries where labour is scarce and expensive. In this context, it also contributes to the creation of new employment opportunities and attracts young people to agriculture;
- *Improved efficiency:* Mechanisation allows farmers to perform tasks more efficiently, with greater precision and consistency. This leads to higher quality products and less waste;
- *Reduced poverty:* Mechanization plays a significant role in achieving food security while also enhancing the livelihoods of people.
- *Better quality of life:* Mechanisation reduces the physical demands of farming, making it easier for farmers to work. This can lead to a better quality of life for farmers and their families. The potential for bringing back the youth to agriculture is huge; and
- *Sustainable agriculture:* Mechanisation can help farmers adopt more sustainable farming practices by reducing chemical inputs, improving soil health, and reducing water consumption, and helps mitigate and adapt to climate change and reduce GHG emissions (no-till/direct seeding through residues, Conservation Agriculture, food loss reduction, drip irrigation, etc.).

However, according to the FAO (2008), one of the key success factors in mechanisation is a sound comprehension of the field situation and the priority operations to mechanise. Location specificity and stakeholder engagement and consultation are therefore essential factors to the efforts to promote successful and sustainable agricultural mechanisation. The agricultural engineering education that is provided, in addition to the technical skills, must enable the trainees to understand these non-technical but highly important factors for the success of any initiative to promote mechanisation.

Ensuring That Agricultural Engineering Education Aims at Addressing Current Needs

Agricultural engineering education must pay special attention to meeting the major challenges faced by farmers and other stakeholders in developing countries. In the 1960s, with the support of many development agencies, countries invested in the creation of many universities by developing facilities and faculty capacity. However, these universities are faced with a multitude of constraints, including that:

- University training and education programmes of the 1970s and 1980s are in decline due to competition with other sectors (ICT, etc.) and because of reduced public funding and a fall in employment opportunities in the public sector (FAO, 2018);
- Training programmes and curricula tend to be relatively static, and need to be revised so that they could be adapted to the needs of the field, given the new requirements of mechanisation; and
- Experience has shown that the education provided is inclined more towards theory than the application of practical knowledge. It also does not seem to adequately capture and reflect the changes emerging in the agricultural engineering technological landscape. Many graduates lack hands-on experience, especially on recent technologies and innovations.

It is therefore important to ensure that agricultural engineering education aims at addressing the current needs in terms of technologies and innovations. This is necessary because agriculture is a rapidly changing field that requires continuous innovation and adaptation to meet the demands of a growing population and the climate change. It is important to ensure that agricultural engineering education addresses the current needs, especially:

- *Keeping up with technological advancements:* Agriculture is becoming increasingly technology-driven with the use of drones, sensors, and artificial intelligence. It is essential to educate future agricultural engineers on these technologies to enable them to adapt to and use them effectively;
- *Meeting the challenges of climate change:* Climate change is a significant threat to agricultural productivity and food security. Agricultural engineers can develop and implement innovative technologies that help to mitigate the impacts of climate change, such as precision irrigation systems and drought-resistant crops;
- *Increasing food production:* The world population is expected to reach 9.7 billion by 2050, and this will require a significant increase in food production. Agricultural engineers can develop new technologies and innovations that increase productivity, reduce waste, and improve the quality of agricultural products; and
- *Promoting sustainable agriculture:* Sustainable agriculture is essential to ensure the long-term viability of the agricultural sector. Agricultural engineers can develop and implement technologies and innovations that reduce the environmental impact of agriculture, such as reducing greenhouse gas emissions and improving soil health.

In the context of developing regions such as Africa, there are many opportunities for agricultural engineers to work in a variety of settings, both in the private and public sectors, and contribute to the development of sustainable and efficient food systems. They can work in:

- *Universities and research institutes:* Responsible for conducting research and teaching courses related to agricultural engineering;

- *Agricultural equipment and machinery manufacturing companies:* Responsible for designing and testing equipment and machinery;
- *Government agencies*: Responsible for developing Sustainable Agricultural Mechanisation (SAM) policies and strategies, as well as food production;
- *Large farms:* Responsible for managing a fleet of equipment and machinery;
- *Agricultural companies:* Responsible for research and development, design and testing, quality control, sales, business development, top management, certification, standardisation, aftersales service provisions, etc.;
- *Consulting firms:* Responsible for providing engineering services to agricultural businesses, government agencies, and other organisations;
- *Non-profit organisations:* Responsible for conducting sustainable agriculture, food security, and rural development projects;
- *International organisations:* Responsible for conducting agriculture and food production projects in developing countries; and
- *High-Level decision-making level:* These are high-level positions where decisions are made to address the current agricultural transformation needs. Agricultural engineers may be required to provide advice based on their technical expertise (as opposed to on the socio-economic and policy level).

Countries, especially in developing regions, still need to comprehend that agricultural mechanisation and the related agricultural machinery supply chains need to be built up and led by the private sector. Public sector procurement of agricultural machinery has proven not to be sustainable. It is here where Governments need to build up more trust in the capabilities of private sector supply chains. Farmers, cooperatives, and groups need to make their own choices as to what type of equipment they intend to get. The financing of such equipment needs to be sorted out between the farmers, the finance institutions and the dealers. There are new leasing schemes that have had great success, also for smaller farmers or women groups (AgLeaseCo, 2019). This shows the right direction of the new thinking, and this will create new needs in the future.

Leveraging Agricultural Engineering in Addressing Future Agricultural Challenges

Several impeding factors have contributed to the need to transform our current food systems. This led to the 2021 United Nations Food Systems Summit (UNFSS), which set the agenda for the globe to engage in conversation around sustainable food systems (FAO, International Centre for Advanced Mediterranean Agronomic Studies & Union for the Mediterranean, 2021). Some of the factors that have led to the need to transform food systems include increasing populations, urbanisation, and climate change. It is estimated that currently 50 percent of the population lives in urban areas, a figure that is expected to reach 70 percent in 2050 (FAO, 2009). This trend is leading to a high demand for increased food production, a demand that is putting pressure on food production, farming systems, and supplies in many countries. Climate change has also led to unpredictability, with the rain seasons in many countries becoming shorter with higher variability within that space and time. This calls for the application of appropriate technologies, including agricultural engineering, to assist farmers in reducing drudgery, enhancing efficiency, and increasing timeliness to cope with these changes.

According to the FAO (2017), many factors require our attention as we strive to transform the way agriculture is practised. These include low productivity, especially in developing countries, mainly due to the current approaches and practices including the low use of improved seeds, fertilisers, and mechanisation. Others include hunger and malnutrition, intensification of natural hazards (floodings, droughts, etc.), lack of a sustainable natural base, and extreme poverty and hunger in some countries, especially in the developing regions.

Based on these scenarios, we can see a clear need to transform the ways that food systems evolve based on the current designs and operations. Given this situation, agricultural engineering promotion has a great role to play in supporting the current efforts to enhance the contribution of agricultural engineering in enhancing crop production. Agricultural engineering can significantly contribute to improving agricultural productivity, irrigation, and rural infrastructure development including farm structures, harvesting, postharvest, and agro-processing to add value.

Agricultural engineering, and mechanisation, in particular, has proven its importance and role in enhancing agricultural productivity, especially when promoted together with other inputs such as improved seeds and fertilisers. The use of agricultural engineering technologies, tools, equipment, and machinery, which is referred to as mechanisation, helps to enhance the efficiency, quality and timeliness of operations, in addition the application of such technologies during harvesting and postharvest management.

Studies conducted by the FAO (FAO, 2006; FAO, 2013) have confirmed that many countries in developing regions are faced with many challenges as they strive to promote agricultural engineering, especially mechanisation. The availability of farm power is still low; many farmers and other stakeholders along the food value chain lack access to mechanisation, and as the result, they do not benefit from it. In particular, women and youths are more affected, and therefore extra effort is required to enhance their participation in the farming, food, and nutrition systems.

It is evident that more effort is still needed to ensure that more appropriate mechanisation business models are selected and applied. Other challenges that have contributed to the low contribution of mechanisation include weak supply chains for equipment and spare parts; and high investment costs.

Furthermore, capacity development, especially for operators and technicians providing repair and maintenance services, is another area that still has weaknesses. We have seen incidences where a tractor, which is an expensive piece of machinery, being given to inexperienced operators or technicians who end up ruining it, leading to high financial losses.

Recommendations

Agricultural engineering education must address current needs in terms of technologies and innovations to equip future agricultural engineers with the knowledge and skills necessary to address the challenges facing the agricultural sector. In this context, the following recommendations, which include relevant parts of the recommendations from FAO and AUC (2018), should be considered:

- Develop programmes adapted to the real need of agricultural engineering and mechanisation. It is recommended to conduct a need assessment at the national level prior to developing such programmes or interventions;
- Reinforce education infrastructure by developing facilities such as workshops, laboratories, and fields for practices, which are necessary for building hands-on capacity in agricultural engineering;
- Revise training curricula for programmes offered by higher education and training institutions. This should include mainstreaming new emerging areas of knowledge such as precision farming, automated systems, robotics, and conservation agriculture;
- Strengthen the awareness of teaching staff of the reality of the field and ensure that they are up to date on new trends in agricultural engineering. This may include refresher courses for lecturers and instructors considering the fast development in innovations including robotics, digitalisation, etc.;
- Create university networks at the national and regional levels to provide a space for dialogue and exchange of experience between teachers, as well as support efforts to develop national and regional societies of agricultural engineering;
- Develop partnerships between universities and various stakeholders, including the private sector, to enable students to acquire practical skills through internships and other activities; and
- Develop research axes in parallel with teaching, enabling universities to contribute to the development of new techniques adapted to the local context.

Conclusion

There is no doubt that agricultural engineering, especially agricultural mechanisation along the food value chain, will continue to play a key role in our effort to modernise farming and food and nutrition systems. As mentioned in this chapter, agricultural engineering as a profession has been around for a long time. Farmers have been using different agricultural engineering equipment such as animal-drawn implements, which have played an important role in agriculture. However, although such equipment is still indispensable in some countries, most of it is being replaced. In reality, traditional technologies cannot cope with the pace of development and its changing demands, needs, and desires. This is especially true if we want to make and keep agriculture as a business attractive to the youth. The current state of Food and Agriculture (FAO, 2022) puts emphasis on the future, where digital automation technologies and the use of modern, precision agriculture and semi-autonomous equipment and in some cases, robotics, are becoming highly significant and appealing, especially for young farmers.

It is therefore important that more effort be made to update agricultural engineering curricula so that these new developments will be included in the preparation of the future agricultural engineers. In addition, it is important that efforts are made to embrace agricultural engineering and mechanisation technologies that will help farmers to address the challenges posed by climate change and increase erratic rainfalls, and reduce drought spells with heat stress. Precision and conservation agriculture technologies and practices will enable farmers and crops to cope with these additional and novel challenges.

Countries must also develop and implement well prepared sustainable agricultural mechanisation strategies (SAMS). Such strategies will provide clear roadmaps on how countries will develop their agricultural mechanisation, including through capacity building in land surveying, soil and water, irrigation, postharvest and agro-processing, and rural structures. It will enable the Directorates of Agricultural Mechanisation in the Ministries of Agriculture to comprehend the complexity of an SAMS, with all its stakeholders from both private and public sectors, as well as from education and research. A SAMS exercise will produce a pool of experts who would assist farmers in accessing and using mechanisation to improve agricultural productivity and provide for more sustainability and resilience in agricultural production.

It is important to implement targeted training programmes, designed to build the capacity of stakeholders such as mechanisation hire service providers, blacksmiths, and agro-mechanics to enhance the local production and repair of implements and machinery. These could be organised in vocational training centres in short or longer-term courses. It would help to comply with the specific needs of the clientele in local areas. In addition, agricultural engineering education should consider gender issues and help to build competency in sustainability (commercial, environmental, and socio-economic) and in agribusiness and finance models. This will help to produce a cadre of highly competent agricultural engineers who will lead decision-making and provide guidance on technical, policy, and strategic matters. Such newly trained engineers with good qualifications will help to propel the agricultural engineering profession by sitting on the desks where future decisions related to the role of agricultural engineering will be made.

Acknowledgements

The authors want thank Dr. Rabe Yayaha, International Maize and Wheat Improvement Center (CIMMYT) for his review, comments, and suggestions to this chapter.

REFERENCE LIST

African Union Commission. (2015). Agenda 2063: The Africa we want. Final Edition, April. Popular Version.

AgLeaseCo. (2019). Mechanizing Zambian farms. At Regional Workshop on experiences sharing on Sustainable Agricultural Mechanization hire services provision, 9–12 December, Kampala Uganda.

Akiwumi, P. (2022, September). Revitalizing African agriculture: Time for bold action. UNCTAD. https://unctad.org/news/blog-revitalizing-african-agriculture-time-bold-action#:~:text=In%202021%2C%2052%25%20of%20employed,are%20found%20within%20the%20continent

Anshuman, V. (2018). Enabling sustainable agricultural mechanization development in Asia and the Pacific through regional coordination and standardization. Centre for Sustainable Agricultural Mechanization (CSAM) United Nations Economic and Social Commission for Asia and the Pacific (UN ESCAP). Unpublished paper presented.

Food and Agriculture Organization. (2006). Farm power and mechanization for small farms in sub-Saharan Africa. Agricultural and Food Engineering Technical Report No. 3.

Food and Agriculture Organization. (2008). Agricultural mechanization in sub-Saharan Africa: Time for a new look. Food and Agriculture Organization, Rome, Italy.

Food and Agriculture Organization. (2009). How to feed the world 2050. High-Level Expert Forum. Rome, 12–13 October. Global Agriculture towards 2050. Food and Agriculture Organization.

Food and Agriculture Organization. (2011). Rural structures in the tropics – Design and development (481pp). Rome, Italy.

Food and Agriculture Organization. (2013). Mechanization for rural development: Issues and patterns in agricultural mechanization – A review. Food and Agriculture Organization, Rome, Italy.

Food and Agriculture Organization. (2017). The future of food and agriculture - Trends and challenges. Food and Agriculture Organization, Rome, Italy.

Food and Agriculture Organization. (2018). Sustainable food systems: Concept and framework. Food and Agriculture Organization, Rome, Italy.

Food and Agriculture Organization. (2020). Use of information and communications technology tools for tractor hire services in Africa. Food and Agriculture Organization, Rome. Italy.

Food and Agriculture Organization. (2022). The state of food and agriculture. leveraging automation in agricultue for transforming agrifood systems. Food and Agriculture Oganization, Rome. Italy.

Food and Agriculture Organization. (2023a). Food systems. https://www.fao.org/food-systems/en/

Food and Agriculture Organization. (2023b). Agriculture and consumer protection department. Rural Infrastructure and Agro-Industries. https://www.fao.org/ag/portal/ags-index/en/?no_cache=1

Food and Agriculture Organization. (2023c). https://www.fao.org/sustainable-agricultural-mechanization/en/

Food and Agriculture Organization & African Union Commission. (2018). Sustainable agricultural mechanization: A framework for Africa. Addis Ababa, p. 127. http://www.fao.org/3/CA1136EN/ca1136en.pdf

Food and Agriculture Organization, International Centre for Advanced Mediterranean Agronomic Studies & Union for the Mediterranean. (2021). food systems transformation – Processes and pathways in the mediterranean: A stocktaking exercise. FAO, Rome. https://doi.org/10.4060/cb7978en

Houmy, K., Clarke, L.J., Ashburner, J.E. & Kienzle, J. (2013). Agricultural mechanization in sub-Saharan Africa – Guidelines for preparing a strategy. *Integrated Crop Management*. FAO, Rome. http://www.fao.org/3/a-i3349e.pdf

Houmy, K., Flores Rojas, M. & Side, C. (2021). Agri-hire in sub-Saharan Africa – Business models for investing in sustainable mechanization. FAO, Rome. https://doi.org/10.4060/cb5071en

Part 2

Curricula Design And Reform

6

Curriculum Design and Application in ABE Education

Ayub N. Gitau and Duncan O. Mbuge

Curriculum Content in ABE Education

ABE Evolution

Traditionally, Agricultural and Biosystems Engineering departments placed a strong emphasis on soil and water resource engineering, precision agriculture, food and process engineering, as well as power and machinery engineering. Graduates in the past were taught how to apply scientific principles when designing machinery and systems to manage the many resources that produce food and fibre with the aim of lowering labour costs, increasing productivity, and elevating the farmer's standard of life (Aguado et al., 2011).

In both developed and developing countries, Agricultural Engineering (AE) currently faces reduced popularity within the industry. In order to avert the diminishing popularity of the course among students and the industry, there was a need to upgrade the curriculum and make it more current and relevant. These changes improved the popularity and gave students the tools they need to pursue lifelong learning and find employment outside of the agricultural industry. The biological engineering component was first incorporated into agricultural engineering curricula in the late 1960s by US engineering colleges. The first US University to start a biological department was Mississippi State University in 1967. Due to the introduction of biological engineering into many other conventional agricultural engineering programmes, department titles had to be changed to reflect the programme changes (Smerdon, 2000).

Since then, a wide range of diverse topics have been incorporated into the traditional Agricultural Engineering due to the introduction of the broader Biosystems Engineering degrees. Biosystems engineering, in its broadest sense, is the use of engineering principles to address issues in biology, food and bioprocess engineering, instrumentation and controls, and environmental engineering. In its broad sense, it may even include some aspects of aquaculture and forestry engineering and biomedical engineering. It is also worth noting that curricula in engineering are often determined by other factors like geographical location, economic activities in the region, as well as the stage of economic development of a country. For instance, universities next to water bodies tend to have units in aquaculture and marine engineering while universities in highly forested areas will have timber and forestry engineering units so as to serve the needs of the local community. Irrigation is normally emphasised in arid and semi-arid areas. It is still common to find curricula that have animal draft power in developing countries since it is still a common and important source of power, while in developed countries, robotics and drones are common and popular tools in ABE. These factors have also influenced emphasis on the development of curricula in ABE (Meredith, 2009).

There are some overarching trends that seem to define the character of ABE curricula, especially emphasis on environmental sustainability and/or the application of renewable energy in most operations. Because of the strong influence of environmental engineering principles, ABE curricula tend to emphasise closed loop systems. For example, food processing will not stop at packaging of food but will go back to look at the water and energy sources and also follow the waste and by-products until they are treated and the by- products applied to some beneficial use. The field of irrigation and water resources engineering also includes watershed management and erosion control and looks forward towards drainage (Payzin, 2010).

Since ABE personnel commonly operate in rural areas, they have tended to take up other engineering operations in those areas. It is not uncommon to find ABE curricula that cover rural roads, buildings, and other infrastructure that can be found in rural areas including domestic water supplies, rainwater management, renewable energy generation, rural electrification, drilling boreholes, food supply logistics, flood control, and sanitation.

ABE Curriculum Content

By creating courses that address current issues, institutions of higher learning can significantly influence/improve productivity. The development and implementation of courses that are attentive to societal socio-economic issues are essential if the ABE profession is to thrive. The engineers at ABE are tasked with enhancing the sustainability of production systems, reducing or eliminating environmental risks, and preserving natural resources. The majority of ABE programmes include agricultural machinery and soil and water resources engineering as fundamental courses in their curricula. The specialisation of mechanisation and machinery traditionally included technologies for proper mechanisation, cultivation, lowering

DOI: 10.1201/9780429150111-8

TABLE 6.1

Categories/Subject classification by the Engineering Accrediting bodies

A	B	C
Mathematics	Engineering Sciences	Complementary Studies
Basic Sciences	Engineering Design and Drawing	

labour costs and improving the quality of farm produce, manufacturing technology, quality and standardisation, testing and evaluation, machinery management, safety and health risks, and human–machine power interaction. With advancements in ABE, the programme has expanded into contemporary domains such as renewable energy application and conservation, biofuels, bioenergy energy–efficient appliances and equipment, and use of Geographic Information Systems (GIS) with controlled traffic and mapping (Arnold et al., 2006).

The technology of postharvest and value-added processing of food, feed, fibre, and industrial crop is a component of the process and food engineering specialty. The focus on soil and water resources engineering also includes technology for land development, hydrology, environment, watershed development, water accumulation and conversation, and preventing soil erosion. Technology for the development of water resources, the lifting, transporting and usage of water, water conservation, controlled precision water application using drip and micro-sprinklers, drainage, and the restoring of flooded terrain using surface and subsurface drainage are also featured.

The core curriculum for all agricultural and biosystem engineering programmes should include foundational courses in physics, chemistry, biology, and pure and applied mathematics and engineering ideas, regardless of the focus on a particular area (Table 6.1.). Curricula need to be updated to reflect this fundamental shift, particularly in the context of introducing agricultural and biosystem engineers to the fundamental concepts of engineering applications in living systems. Biosystems engineering will gain from a shift in curriculum as a field of study and a career in engineering, improving its capacity to fulfil the needs of the also developing chemical, biotechnology, and pharmaceutical industries.

The curriculum should be designed in such a way that basic engineering science courses are taught at years one and two of study. More ABE technical and practical units are then introduced in the second and third years of study. At year four, elective and skill supporting courses, and more design courses in the final (fourth or fifth) year of study (Table 6.2.).

In conclusion, ABE engineers can also work in the developing fields of biomaterials, biofuels, and bio-mechatronics, as well as in the evaluation of food quality, safety, traceability, and the design of sustainable and environmentally friendly systems.

Design/Capstone Projects in ABE Education

Senior year design courses ought to be innovative, interdisciplinary learning opportunities that polish students' academic and professional abilities. While written reports, oral presentations, and team grading improve professional abilities, broad, open-ended challenges promote creativity and provide connections between what at first glance appear to be unrelated engineering scientific fields. The ideal design problem would be one that is both typical of issues resolved by working engineers and pertinent to the students' subject of study (Felder & Brent, 2003), and ultimately addressing a need in society.

Postgraduate students have the option of choosing one of the five de facto thematic areas: Irrigation and Water Resources Engineering, Power and Machinery Engineering, Food and Process Engineering, Environmental Engineering and Structures Engineering, while undergraduates take all thematic areas. Currently, all students generally undertake a two-semester machine component design sequence as their capstone design course (Hughes, 2001).

Given that the courses frequently contain important new technical material, these brief design encounters may not be as thorough as expected and inevitably design work is given less emphasis. The obvious solution to this issue is to add additional capstone courses to the curricula, however this is not feasible since ABE undergraduate curricula are already overloaded and the objective is to decrease rather than expand the program's credit hours.

Internships in ABE Education

A shortage of young professionals who are qualified and engaged and want to work in industry is a global problem. The majority of the time internships have been employed as a practical learning tool, exposing college students to a variety of occupational expectations and abilities in an effort to produce workers who are ready for the job. A deeper understanding of recent interns' views of their lived experiences could contribute to the construction of a more profound understanding of how their perception and the internship experience are interconnected (Anderson, 2015).

For the award of an ABE degree, it is a requirement that students have attended supervised attachments during their course of study. After graduation, graduates should seek to be registered as graduate engineers by the professional bodies. This normally gives engineering trainees a chance to attend internships where they are supervised by registered professional engineers through mentorship programmes. After three years of such supervision, graduate engineers are recommended for registration as professional engineers (K'Akumu, 2022).

In Kenya for instance, students attend three practical attachment sessions in total, of which two are internal (tenable at the University) and one is external (in a private company or government agency). The external attachment is aimed at exposing engineering students to the general operations of organisations and give them a chance to experience being part of the workforce. They participate in all activities including: Routine office work, fieldwork, workshop duties and any other work as may be assigned by the field supervisor. The student is required to complete at least 8 weeks of the attachment. The student is required to maintain a log-book issued by the institution and this forms the basis of the assessment

TABLE 6.2

Model curriculum for bachelor's degree in Agricultural and Biosystems Engineering.

AGRICULTURAL AND BIOSYSTEMS ENGINEERING COURSE DESIGN	CAPSTONE COURSES & DESIGN PROJECT Year V/IV	• Design of ABE tools, machinery, processes, and systems • Modelling and Optimisation • Design projects
	ABE DESIGN COURSES Year IV/V	• Irrigation, Water resources, and Environmental Engineering • Power, machinery, and mechanisation Engineering • Food and Process Engineering • Structures Engineering
	ABE SPECIALISED COURSES Year III	• ABE introductory Courses • Engineering Science Courses • Mathematics Courses • Environmental Courses • GIS Courses
	ABE TECHNICAL COURSES Year II	• ABE Basic Courses • Mathematics Courses • Environmental Courses • GIS and • Entrepreneur/Business Courses
	BASIC ENGINEERING AND SCIENCE COURSES Year I	• Basic Engineering Courses • Mathematics Courses • Basic Science Courses • Computing Courses

when the students report back at the University. The students should be visited and interviewed by University supervisors (Gitau, 2022).

Utilising Internships Effectively

One of the main goals of an internship is to give students a chance to test their knowledge and form a good attitude towards a certain subject or potential future career. Internships are viewed as helpful programmes for everyone engaged and a symbiotic partnership in career preparation. All parties engaged in this arrangement profit. There is no doubt that, in the right programme settings, completing an internship has a direct impact on favourable career indicators given the wealth of literature on the benefits of internships. Research on internships has demonstrated that they can have a positive impact on a person's professional commitment and aspirations, employment satisfaction and career clarity. An internship, according to Harrison and Kennedy (1996), is an arrangement between a student and a business in which the students work for a predetermined amount of time – typically 10 to 16 weeks – shadowing and carrying out duties that are usually advantageous to the company. Additionally, the employer promotes professionalism, education, and skill development.

Internships have grown in popularity within the educational systems over the past 20 years as a means of connecting students as transitioning from academics to the working world. Additionally, internships give students the chance to learn more about a particular field of work without committing fully or taking on the risks and difficulties that come with a regular job. Internships have evolved into a powerful instrument for experiential learning that exposes college students to a wide range of professional expectations and abilities, thereby lowering the percentage of students who leave college without any prior work experience. One of the most valuable attributes that recent graduates can give a potential employer is work experience and internship is one of the best ways for graduates to acquire this skill, making them a desirable candidate for employers (Binder et al., 2015; Gitau, 2022).

Internship has been used to develop knowledge and engage a student in a particular line of work and study. A work experience curriculum in the area of farm management practices was first made available to students in 1855 by Michigan Agricultural College (now Michigan State University). Students attended academic studies at what it would become the first land- grant college in the United States while also taking care of and managing the farm. Despite the fact that internships in the agricultural industry have been practised for more than 150 years, their core principles remain the same. The historical internship model continues to be profitable for the agriculture business due to the benefits it provides to all relevant parties and since it gives students a glimpse into a complex and a diverse profession before they begin their full-time employment opportunities. As the agriculture sector and its employment opportunities develop, so do the value, function, complexity, and obligations for students who choose to take part in internship programmes (Briassoulis & Panagakis, 2003).

Understanding an Internship Experience

It might be possible to strengthen and construct the link between the participant's perception and the internship experiences with better understanding of how recent interns view their experiences. Thus, the importance of structured and supervised internship programmes. Asking an intern for self-reported input on their particular programme is one approach to learn more about the potential advantages and disadvantages of such an experience. Asking students who have completed internships may provide a viable opportunity to study individual perspectives of the internship experience, assisting in the field of research that pulls directly from the thoughts of individual participant. It may be useful to comprehend what an internship experience is like from the perspective of the student in order to conceive a framework that tackles the crucial element of an excellent internship experience (Gitau, 2021). A deeper comprehension of internship programmes is necessary to guarantee that prospective students are provided with an efficient internship program.

The Benefits of Internships

From a university's standpoint, one benefit of internships is that they provide students with a chance to combine academic and experiential learning. Interns are given a chance to test their abilities and preconceived assumptions about a certain career. Students who are making the move to a professional job look for their sense of self and belonging in the workplace, which includes making the right career choice. Knowing oneself at these times of transition can help one feel coherent and grasp one particular stage of life, like college. While simultaneously enrolled as a student, intellectually moving towards a degree and ultimately a vocation, an internship can be a useful way to explore potential career interests. Recent years have seen a change in the focus of internship research, which now emphasises the overall benefits of completing an internship and offers understanding of the procedures, structures, and justifications for its implementation (Bird et al., 2013).

Previous studies have shown that taking part in an internship provides advantages, including advancement in one's career and increased job experience. For instance, doing an internship improves one's ability to construct one's professional identity early in life and increases one's confidence in making career decisions. It is permissible for people to identify their roles and see how their skills might be used in a possible career. Experiences with internships have consistently shown to be a reliable way to offer rich opportunities, enhancing personal progress via experienced learning. Through internships, students can put the ideas, abilities, and knowledge they've acquired during their undergraduate studies to use in practical settings. Students have the opportunity to transition from being a passive learner in the classroom to an active planner who uses professional job-searching strategies by applying for certain opportunities while seeking for an internship. Creating a deliberate approach to expose students to real-world situations and settings where they are able to reflect and learn through experiential opportunities is important, according to the literature on agricultural internships. As more outcomes and discoveries from internship study are finalised, higher education institutions can develop better programmes that will provide students with meaningful experiences that expose them to occupations before they graduate (Blanchard & Fabrycky, 1998).

Facilitation of Internship Programmes

When given the flexibility to choose their own educational needs while using the lecturer as a learning facilitator, the students can build a deeper understanding of knowledge development and application that transfers into their employment. A student's ability to get the most out of their learning experience may be constrained by the absence of a supporting structure by stakeholders in internship programmes. This is since effective programmatic methods and recruiting may be hindered. This issue could further alienate stakeholders during the internship (such as the student and academic advisers or industry mentors). Knowing the key components of the internship experience from the viewpoint of the student can help programming development efforts to strike the correct balance between student ownership of learning and support from the institution and industry (Gitau, 2022). When students transfer from an academic learning environment to an internship experience, internship stakeholders play a crucial role in guiding them to reflect. An authentic learning opportunity is provided by mentoring, which can be official or informal. The student applies existing knowledge of a particular subject to new situations to gauge how well they were taught. The student can discuss how they interpret current events while reviewing the learning process as a result by thinking back on this conversation. In order to help students, actualise the necessary concepts or skills for their internships, stakeholders should set mentorship criteria. By doing so, students will be able to strengthen their interpersonal and teamwork abilities. Stakeholders should work to relate the prior knowledge and skills of intern participants to the situations they encountered during their internship. Participants' processing and learning including how to make choices and deal with issues in the context of unfamiliar situations, become more thoroughly internalised as a result. Effective internships enable transformational coaching and can mimic a real-world work environment without the challenges and complications that can arise from regular employment, so encouraging a student's interest in a particular subject of study (Anderson, 2015).

Two skills that are crucial to students' academic and professional development creative thinking and problem-solving are required to work collaboratively with others and draw from the knowledge and experiences of various groups of people. When transitioning from a formal school to a career, collaboration skills are crucial for success. In a learner-directed programme, the student should have the chance to work with others, engage in active learning and gain fresh insights in a particular area of their line of work. In order to help create these effective and long-lasting learning experiences for students, stakeholders make sure that the required course work is finalised, that tasks and assignments are relevant to the experience and discussions regarding current market trends and needs take place (Beard & Morton, 1999).

Lack of formal information about how to support the learning process and help students move through an internship's phases in order has historically been key obstacles to internship. This challenge arises from absence of a clear and effective tools to support learners in connecting their practical knowledge to the understanding of abstract concepts encountered in the workplace. According to an argument for a more substantial student-centred internship, the student intern would concretely understand the application of taught classroom concepts and directly apply the materials to a task related to their field through the internship experience (Arnold et al., 2006).

It is essential that an internship programme provide benefits to students of all backgrounds. In order to ensure that each learner has a high-impact experience, the structure of the internship should initially inspire the development of a personalised plan, discussing programme goals and learning objectives. Everyone involved gains a firm grasp of the subject matter as a result, which helps to create a purposeful internship experience that includes targeted assessments and beneficial feedback at every turn. Improved communication and comprehension between student participants and academic mentors further benefit the intern's experience and the clarity of their decision to enter the workforce (Bailey et al., 2000).

Internship programmes that offer good mentoring and clear expectations are valued by students. Internship programmes run the danger of not maximising the potentially dynamic and favourably impacting experience for students without strong restrictions and a controlled learning process. Due to this lost chance, one's desire to pursue a career in the Engineering may become less strong (Beard & Wilson, 2002).

The development of thorough, reflective evaluations that allow students to organise their learning and apply it to future planning, such as a career or additional education, is essential for long-term growth. Such learning activities directly involve the learner in applying just-learned material and paradigm shifts outside of the immediate context and help them conceptualise the impact it may have on applications in the future. All stakeholders including the government should support the process of internship for the benefit of the trainees and their country.

Knowledge of Student Background in Curriculum Development

Students normally bring a variety of backgrounds and experiences to college, which they draw on from what they are studying and already know. The variety of their experiences becomes more significant as the student body's diversity grows since they offer fresh perspectives and a variety of prior knowledge to campus (Stolk et al., 2010). Faculties/Schools of engineering at the university level are expanding in various ways, such as the diversity and breadth of their student body's backgrounds and experiences. Biosystems is currently being studied by a growing proportion of students from urban and suburban areas of the Nations. These urban-raised students frequently lack relevant work experience in the field of study when they first enrol in college. This expanding demographic can lead to a rise in the demand for internships in the agricultural sector. Students from diverse backgrounds can strengthen their technical and professional skills through internships, which expose them to these chances (Smerdon, 2000).

Universities throughout the world have put in place programmes that cater for the growing tendency of elderly, returning, or students from urban backgrounds who want to pursue careers in agriculture. For curriculum development, it is of utmost importance to consider the type of curriculum that the learners have experienced in their immediate past level of education. It is vital to check if the learners have had the necessary prerequisites to cope with the expected rigour of the next level of study.

Innovative Teaching and Learning Methodologies

The content of the curriculum is important but the way it is delivered may diminish or amplify the intended effect. For this reason, as we craft ABE curricula, it is important to embrace modern and effective methods of teaching and learning so as to endear the learners to the content. We must be sure that the learner is motivated by our teaching methods to acquire as much knowledge as possible (García-Peñalvo & Colomo-Palacios, 2015). Oftentimes, Universities are not able to afford all the required materials for teaching on-site. But these may be available in partner institutions. It is therefore important to find innovative ways of equipment and material sharing so as to make the most effective and efficient use of limited resources (Briassoulis et al., 2008).

These cutting-edge and contemporary techniques include, for instance, brief lectures, simulations, role-playing, portfolio construction, and Problem-Based Learning (PBL). Remote and virtual laboratories, robotic applications, 3D virtual worlds, advancements in augmented reality, complex data visualisations, and mobile applications are some of the tools that support problem-based learning, cases-based learning, and online or asynchronous learning (Lord et al., 2011). ABE curricula must be alive to the digital transformation made by (Massive Open Online Courses) MOOCs so as to strike a balance between the individualised contact that Agricultural and Biosystems Engineering processes demand and the huge audience of MOOCs. Since mobile devices have become an integral part of our lives, we should maximise on the benefits they may offer learners while mitigating negative influences (Le Deist & Winterton, 2005). Since most of these systems are not previously tested in the education sphere, it is vital to evaluate their effectiveness to enhance the ABE curricula so as to decide on the level of application in the future.

Lifelong Learning (LLL) Competencies

One of the best outcomes of university education is the zeal in learners to continue to seek to be excellent and prestigious in their profession through lifelong learning. This is only likely to happen if the learners were so well motivated during the course of their training that their love for learning drives them to seek knowledge in the field beyond university education and indeed for the rest of their lives. To achieve this ideal, the ABE curricula must bring out clearly the important role that it plays in society and this must be relayed to the students clearly (Guest, 2006). For this reason, curricula must contain courses that promote or include critical thinking, communication, public relations, entrepreneurial, and other complimentary skills. These will enable the graduates to understand and blend into the society they are expected to serve by quickly obtaining employment soon after graduation, finding consultancy assignments and to keep these jobs once they are employed (Koper & Tattersall, 2004).

The professionals in the field should also strive to mentor fresh graduates within professional associations. One of the surest ways to ensure that a graduate is exposed to lifelong learning is through joining professional associations where they can meet mentors to advance them into responsible engineers. The graduates can also benefit from networking with the senior professionals who are likely to help them get jobs and consultancies .

Joining pertinent technical professional associations is critical for a graduate to stay current. Examples of such professional associations in Kenya include the Kenya Society of Environmental, Biological and Agricultural Engineers (KeSEBAE) and the Institution of Engineers of Kenya (IEK). Professional organisations provide several possibilities to network with other professionals, participate in technical and professional development events (such as conferences and seminars), and learn about cutting-edge technology and other technical subjects through tutorials. To present a paper on a project at a conference is a more challenging objective. Engineers can use this to develop communication skills, get feedback, and clearly communicate their project and progress. An extrinsic aim that boosts incentive for ongoing study and effort could be subsequent publication in a journal (Anney, 2014).

Thus, engineers must be committed to their personal growth and adopt habits that will increase their learning efficiency. A successful career is one that "embraces a program of continuous learning and will ensure that you have a career that is filled with many proud successes, a job that gets you respect from your colleagues, a career that is fun and exciting" (Phillips et al., 2010). This commitment for lifelong learning must be ingrained in the ABE curriculum. Professional associations and regulating bodies that manage graduate and professional engineers must also craft relevant curricula to promote lifelong learning. Continuous Professional Development (CPD) is lifelong learning process that applies to all levels of engineers (Moretti et al., 2007).

Quality Assurance in Agricultural and Biosystems Engineering Education

All quality assurance systems subscribe to the maxim that processes and systems are subject to continual improvement, else they will fall short of expectation. Quality Management Systems (QMS) must be put in place as a control tool. For this reason, there is a need to periodically measure the extent of excellence through tools such as learner and employer evaluations. Once this feedback is received, recommendations from it should be implemented in the next curriculum review. Further, there is need to review curricula regularly to incorporate any emerging beneficial trends like Artificial Intelligence (AI), Machine Learning (ML), and the Internet of Things (IoT) as we also deal with large data. Ideally, curricula should be reviewed after every five years' cycle depending on duration of programme. During reviews, consultations must involve all stakeholders in order to capture a wide range of views (Briassoulis & Panagakis, 2006).

Programme Accreditation and Any Other Relevant and Associated Content

Accreditation is an important process that institutions can leverage to show their commitment to maintaining high standards in their programmes and also to set and maintain high standards as envisaged by industry. The beneficiaries of accreditation initiatives are students, the organisation, employers, professions, and society at large in the long run. Accreditation also gives curriculum international recognition, as well as the solace that it promotes best practices to produce learning outcomes and not merely teaching inputs. Accreditation also makes it easy to transfer credits between institutions. For students in particular, accreditation asserts that they have received good quality education that meets the standards of the profession, increases their employability, makes it easy for them to attain professional registration, and confers eligibility to government schemes for loans, grants, and scholarships.

It is for these reasons that ABE programmes must seek accreditation so as to maintain high standards and appeal. ABE curricula must consistently appear in the lists of accredited programmes so as to maintain the prestige of the profession. Accreditation is also important for graduate courses. This is so since it is at this level where increased international mobility is expected and therefore the need for accreditation becomes even more important. There exist five methods of worldwide accreditation as described by Judith Eaton (Eaton, 2007), president of the Council for Higher Education Accreditation (CHEA), in 2007:

a) **International Model**

The International Network for Quality Assurance Agencies in Higher Education (INQAAHE) or UNESCO would be in charge of leading the international accreditation sector in an international paradigm. International recognition and accreditation to the 'Washington Accord' and the body of Systems Engineering (SE) assists students and professionals in recognition across borders.

b) **National Model**

Different national quality standards would be incorporated into the national model agreements, but it is also possible to designate an international coordinator. It would be comparable to, say, the United States, Canada, and Mexico sharing responsibility for its international accreditation space.

c) **Regional Model**

A coordinating body can be established in a regional model as well, although voluntary agreements between agencies for certification or quality assurance accreditation would control the global market. Examples of this paradigm include the EUR-ACE project in Europe and the Bologna process, proposed in the European register.

d) **Institutional Model**

In an institutional approach, like the one used in the US, schools and universities should assume control over and accountability for an area of international accreditation. The means of collaboration should be handled by the institutions, through partnerships, accreditation authorities, and other quality-related organisations like the CHEA or ABET in the United State.

e) **Market Model**

A market model can have trade agreements as the source of responsibility and power for managing international accreditation. It would be comparable to how the General Agreement on Trade in Services (GATS) would have controlled a region of global certification. As a result, decisions made by the GATS regarding the regulation of cross-border flow of people, institutions, programmes, and services could be delegated to accreditation. Competition between institutions of higher learning would improve quality. Low-quality educational institutions would not be able to compete and would be out of pace. This approach would be based on competition, supply, and demand.

Each of these models have benefits and drawbacks. The multinational model, for instance, would offer better quality homogeneity. However, since all schools would be required to adhere to the same accreditation requirements, this model would restrict the capacity to creatively address the problems with international accreditation. The regional and national models are both less centralised and call for some coordination of standards and decision-making in a field of international accreditation. The institutional approach reduces homogeneity while allowing for greater institutional variation in the domain of international accreditation. The market paradigm makes it unclear if the public interest can be adequately served in the international accreditation sector.

Engineering accreditation in Kenya: Case Study

Many engineering programmes were first designed to meet the registration requirements of the United Kingdom's (UK) engineering professional organisations for a variety of historical reasons. The Engineers Registration Act, passed by Kenya's parliament in 1969, six years after the country's independence, created the Engineers Registration Board (ERB). The ERB only accepted engineering degrees from the University of Nairobi and the UK at the time due to the lack of academic institutions. Additionally, the ERB established requirements for board registration. The ERB remained in force until The Engineers Act, 2011, which repealed the ERB Act. In order for engineering degree programmes to be certified for the purposes of registration, the Engineers Board of Kenya (EBK) was created (Gitau, 2021). The EBK established a number of conditions that must be adhered to. Recently, the situation changed and now in addition to the EBK, Commission for University Education (CUE) are also responsible for accreditation. This has significant repercussions (Gola, 2003).

According to the Engineers Act of 2011, the EBK is responsible for establishing and regulating engineering practice in Kenya. The EBK is in charge of regulating engineering standards and developing the capabilities of both individual engineers and engineering organisations. Graduate engineers, professional engineers, consulting engineers, engineering firms, and other entities are all registered with the EBK. The EBK is also anticipated to endorse and accredit Kenyan University engineering programmes. But it might be argued that the EBK's role in accreditation is rather minimal, although being crucial. On its website, accreditation is listed as number 12 out of at least 24 functions and powers (K'Akumu, 2022). In order to be more specific, the EBK is supposed to "Approve and accredit engineering programmes in public and private Universities and other tertiary level educational institutions giving instruction in engineering". The EBK is also expected to create guidelines for engineers in management, marketing, professional ethics, environmental concerns, safety concerns, legal concerns, and any other areas that are pertinent; to create a thorough curriculum for engineers' registration; to hold professional examinations for that purpose; to establish a school of engineering; and to offer tools and opportunities for education, exposure to the world of work, skill acquisition, and ongoing professional development. In other words, the EBK is a professional organisation with several facets that is quite active. However, it has recently engaged with the Commission for University Education (CUE), a government body, regarding accreditation (Memon et al., 2009).

The Commission for University Education (CUE) was created by an Act of Parliament, the Universities Act, No. 42 of 2012, to replace the Commission for Higher Education, which was formed under the Universities Act, Cap. 210B of 1985. In particular, functions 10–12 state the CUE must accredit Universities in Kenya, regulate University education in Kenya and accredit and inspect University programmes in Kenya. According to the CUE website, its mandate is to "promote the objectives of University education, by regulating and accrediting Universities and programs, among other functions". As a result, the CUE has a broad mandate from the Kenyan government to serve all Kenyans by assuring the delivery of high-quality instruction at the University level for everyone. However, similar to the EBK, it has recently expended excessive effort and time on the topic of certification, particularly engineering education accreditation.

It is clear that EBK and CUE both have overlapping accreditation responsibilities: the CUE is in charge of accrediting all University programmes, while the EBK is in charge of accrediting engineering programmes. This is arguably due to conflicting Acts of Parliament that Kenya's MPs approved during the past few years. The difficulty of giving competent Kenyans access to engineering education cannot be reduced to the current standoff between CUE and EBK and conflicting Acts of Parliament; it is crucial to remember this in order to address the accreditation situation. Instead, it results from a multitude of domestic and global structural factors, such as lack of financial resources, an increase in foreign debt, and a constrained capacity for distribution. Additionally, Kenya has made effort to meet this demand for engineering degree programmes, facilities, personnel, and resources (Business Daily, 2013).

Only six of Kenya's twelve engineering Universities now offer degree programmes that are EBK accredited. University of Nairobi, Dedan Kimathi University of Technology, Egerton University, JKUAT, Kenyatta University and Moi University are some of these Universities. In sharp contrast to the EBK, 23 public chartered Universities were fully approved by the CUE as of November 2015. Although it is unclear which programmes are certified and which are not, a prospective engineering student in Kenya might infer from the CUE's list that the engineering programmes at the Technical University of Mombasa or Masinde Muliro University of Science and Technology (MMUST) are completely accredited. However, the EBK contends that they are not. Given this perplexing circumstance, it is crucial that CUE and EBK work together to resolve the issues. This is also arguably a simpler step to take than restructuring, updating, and reconstructing the entire engineering education industry, which cannot be done efficiently, consistently or on schedule without a compromise between CUE and EBK (Natarajan, 2000).

Essentially, it may be said that the CUE's mandate is too broad. It is in charge of regulating and accrediting Kenya's growing number of Universities as well as monitoring all aspects of University education. The CUE disagrees with EBK and the Council of Legal Education (CLE) due to its expansive role. It may be argued that the CUE could rely on EBK's informed advice about curricula and engineering education given their diverse experience. The accreditation of institutional infrastructure, such as engineering laboratories, workshops, and the like, may still fall under the purview of EBK. The EBK would then continue to have control over issues like the credentials of lecturers, curricular standards, and equipment. It is possible that the standards for lecturers, and department leaders and deans will be agreed upon by both parties.

Acknowledgement

We thank Simon M. Thuku, Martin M. Nanje and Robert M. Mathenge for the input in graphics and literature harvest for comparison.

REFERENCES

Aguado, P., Ayuga, F., Briassoulis, D., Panagakis, P., Febo, P., Comparetti, A., & Fehrmann, J. (2011). The transition from agricultural to biosystems engineering University studies in Europe. *Proceedings if the IMSCI*, 1–6.

Anderson, K. P. (2015). Evaluation of undergraduate equine related internship experience by students and employers 1. *NACTA Journal*, 59(3), 234.

Anney, V. N. (2014). Ensuring the quality of the findings of qualitative research: Looking at trustworthiness criteria. *Journal of Emerging Trends in Educational Research and Policy Studies (JETERAPS)*, 5(2), 272–281.

Arnold, S., Warner, W. J., & Osborne, E. W. (2006). Experiential learning in secondary agricultural education classrooms. *Journal of Southern Agricultural Education Research*, 56(1), 30–39.

Bailey, T., Hughes, K., & Barr, T. (2000). Achieving scale and quality in school-to-work internships: Findings from two employer surveys. *Educational Evaluation and Policy Analysis*, 22(1), 41–64.

Beard, C., & Wilson, J. P. (2002). *The Power of Experiential Learning: A Handbook for Trainers and Educators*. Herndon, VA: Stylus Publishing.

Beard, F., & Morton, L. (1999). Effects of Internship Predictors on Successful Field Experience. *Journalism & Mass Communication Educator*, 53(4), 42–53. https://doi.org/10.1177/107769589805300404

Binder, J. F., Baguley, T., Crook, C., & Miller, F. (2015). The academic value of internships: Benefits across disciplines and student backgrounds. *Contemporary Educational Psychology*, 41, 73–82. doi: 10.1016/j.cedpsych.2014.12.001

Bird, W. A., Martin, M. J., Tummons, J. D., & Ball, A. L. (2013). Engaging students in constructive youth-adult relationships: A case study of urban school-based agriculture students and positive adult mentors. *Journal of Agricultural Education*, 54(2), 29–43. doi: 10.5032/jae.2013.02029

Blanchard, B. S., & Fabrycky, W. J. (1998). System Engineering and Analysis (3rd ed.). Virginia: Prentice Hall.

Briassoulis, D., Nikopoulos, E., Ayuga Téllez, F., & Panagakis, P. (2008). The emerging evolution from agricultural engineering to biosystems engineering studies in Europe. Paper presented at AgEng'08, Agricultural and Biosystems Engineering for a Sustainable World, Crete, Greece.

Briassoulis, D., & Panagakis, P. (2003). Current structure of European university studies. *Proceedings of the 1st USAEE Workshop*, Madrid, March 28–29.

Briassoulis, D., & Panagakis, P. (editors) (2006). Restructuring agricultural/biosystems engineering studies and employability in Europe. *Proceedings of the 7th USAEE Workshop*, Vilnius, Lithuania, April 1–2.

Business Daily. (2013). Shortage of engineers pose challenges to Kenya growth. *Business Daily*. http://www.businessdailyafrica.com/Shortage-of-engineers-pose-challenges-to-Kenya-growth/-/539546/1701570/-/43x4y9/-/index.html

Eaton, J. (2007). Un espacio de acreditación internacional. Global University Network for Innovation, La Educación Superior en el mundo 2007: Acreditación para la Garantía de la Calidad, Qué está en juego (pp. 159–164). Barcelona: Mundi Prensa Libros SA.

Felder, R. M., & Brent, R. (2003). Designing and teaching courses to satisfy the ABET engineering criteria. *Journal of Engineering Education*, 92(1), 7–25.

García-Peñalvo, F. J., & Colomo-Palacios, R. (2015). Innovative teaching methods in engineering. *International Journal of Engineering Education (IJEE)*, 31(3), 689–693.

Gitau, A. N. (2021). Inside iconic University of Nairobi school of engineering. *Engineering in Kenya*, pp. 45–47. https://www.iekenya.org/no.004-2021

Gitau, A. N. (2022). Role of academia-industry linkage in engineering the future. *Engineering in Kenya*, pp. 36–37. https://www.iekenya.org/no.009-2022

Gola, M. (2003). Premises to accreditation. A minimum set of accreditation requirements in accreditation models in higher education experiences and perspectives in ENQA. Workshops Reports 3, European Network for Quality Assurance in Higher Education, Helsinki, pp. 25–31.

Guest, G. (2006). Lifelong learning for engineers: A global perspective. *European Journal of Engineering Education*, 31(3), 273–281.

Harrison, R. W., & Kennedy, P. L. (1996). A framework for implementing agribusiness internship programs. *Agribusiness*, 12(6), 561–568.

Hughes, J. L. A. (2001). Incorporating project engineering and professional practice into the major design experience. *Proceedings of 31st ASEE/IEEE Frontiers in Education Conference*, Reno, Nevada, IEEE.

K'Akumu, O. A. (2022). Regulation as struggle for jurisdictional and maximal power: evidence from the reform strategy of Kenya's regulator of engineering practice. *The Theory and Practice of Legislation*, 11(1), 32–51. DOI: 10.1080/20508840.2022.2113201

Koper, R., & Tattersall, C. (2004). New directions for lifelong learning using network technologies. *British Journal of Educational Technology*, 35(6), 689–700.

Le Deist, F. D., & Winterton, J. (2005). What is competence? *Human Resources Development International*, 8, 27–46.

Lord, S. M., Stefanou, C., Prince, M., Chenand, J., & Stolk, J. D. (2011). Student lifelong learning outcomes for diÄerent learning environments. *Proceedings of the ASEE Annual Conference*, Vancouver, British Columbia, Canada, June.

Memon, J. A., Demirdögen, R. E., & Chowdhry, B. S. (2009). Achievements, outcomes and proposal for global accreditation of engineering education in developing countries. *Procedia Social and Behavorial Sciences*, 1, 2557–2561.

Meredith, J. (2009). Launching your career: Lifelong learning—Your key to an enjoyable and rewarding career. The Gold Series: Book 4. IEEE–USA E-Books.

Moretti, B. E., Naessens, V Jr., & Allen, K. S. (2007). Using an engineering design problem to assess attainment of life-long learning. *International Journal of Engineering Education*, 23(1), 131–140.

Natarajan, R. (2000). The role of accreditation in promoting quality assurance of technical education. *International Journal Engineering Education*, 16(2), 85–96.

Payzin, E. (2010). Developments in engineering education in Europe, ASME 2010 international leadership, May 2010 Istanbul.

Phillips, W. M., Peterson, G. D., & Aberle, K. B. (2010). Quality assurance for engineering education in a changing world. *International Journal Engineering Education*, 16(2), 97–103.

Smerdon, E. (2000). An action agenda for engineering curriculum innovation - P.E. presented at the 11th IEEE-USA biennial careers conference, November 2–3, San Jose, CA.

Stolk, J., Martello, R., Somerville, M., & Geddes, J. (2010). Engineering students' definitions of and responses to self-directed learning. *International Journal of Engineering Education*, 26(4), 900–913.

7

Curricula Design and Reform of Agricultural and Biosystems Engineering (ABE) in EU Member and Candidate States

Antonio Comparetti

International Harmonisation of Degree Study Programs in Agricultural and Biosystems Engineering (ABE)

The process of international harmonisation of degree study programs in Agricultural and Biosystems Engineering (ABE) was started by Prof. Giuseppe Pellizzi during the CIGR (Congrès International du Génie Rural/International Commission of Agricultural and Biosystems Engineering) 1989 Conference and was continued by Prof. Pierluigi Febo, both in EU, within EurAgEng (European Society of Agricultural Engineers) SIG RD12 – Education and Communication, and at a global level, within CIGR WG1 (now WG3) – Agricultural Engineering University Curricula Harmonization (Comparetti & Febo, 2018a, b).

Then, four thematic networks were effective:

1) USAEE-TN (University Studies of Agricultural Engineering in Europe – A Thematic Network), comprising 31 HEIs (Higher Education Institutions) from 27 countries, from 2002 to 2006;
2) consortium POMSEBES (Policy Oriented Measures in Support of the Evolving Biosystems Engineering Studies in USA – EU), comprising eight EU and four USA HEIs, from 2006 to 2008;
3) ERABEE-TN (Education and Research in Biosystems Engineering in Europe – A Thematic Network), comprising 35 HEIs from 27 countries, from 2007 to 2010;
4) consortium TABE.NET (Trans-Atlantic Biosystems Engineering Curriculum and Mobility), comprising four EU and two USA HEIs, from 2009 to 2013.

Some examples of significant changes relevant to ABE degree study programs occurred in EU member and candidate states after the end of the ERABEE-TN project: they were the curricula established by the HEIs of 20 EU member and candidate states, i.e., Belgium, Czech Republic, Denmark, Estonia, France, Germany, Ireland, Italy, Latvia, Lithuania, Netherlands, Norway, Poland, Portugal, Romania, Slovak Republic, Slovenia, Sweden, Spain and Turkey (Comparetti & Febo, 2021).

This chapter shows the state of the art on ABE degree study programs in EU member and candidate states 23 years after the last overview, presented during the AgEng2000 Conference, i.e. "The University Structure and Curricula on Agricultural Engineering – An Overview of 36 Countries", Editors Pierluigi Febo and Da-Wen Sun (Febo & Sun, 2000).

Survey on ABE Degree Study Programs Offered by Universities of EU Member and Candidate States

A survey on the ABE degree study programs offered by Universities of EU member and candidate states was carried out in order to focus on the evolution and status of ABE education in these countries in the academic year 2022–2023. This survey was based on the data and information provided upon request by at least one University delegate (ERABEE-TN partner) per each country and on the data and information retrieved from the web-sites of the HEIs of the above 20 EU member and candidate states (Belgium, Czech Republic, Denmark, Estonia, France, Germany, Ireland, Italy, Latvia, Lithuania, Netherlands, Norway, Poland, Portugal, Romania, Slovak Republic, Slovenia, Sweden, Spain and Turkey). Thus, it was possible to measure the study load of the subjects related to Agricultural Engineering and Applied Agricultural Engineering within the BSc and MSc degree study programs offered by the above Universities (Comparetti & Febo, 2021).

International Thematic Networks on Agricultural and Biosystems Engineering Education (ABEE)

Four thematic networks were effective from 2002 to 2013:

1) USAEE-TN;
2) consortium POMSEBES;

DOI: 10.1201/9780429150111-9

3) ERABEE-TN;
4) consortium TABE.NET.

USAEE-TN

USAEE-TN (University Studies of Agricultural Engineering in Europe - A Thematic Network) was funded by the EU Socrates – Erasmus programme, approved and supported by EurAgEng (European Society of Agricultural Engineers), through the SIG (Special Interest Group) RD12 – Education and Communication.

The major objectives of USAEE-TN were to:

- define the core curricula to be used as benchmarks for the degree study programs in Agricultural Engineering in Europe; the on-going development must meet the FEANI (European Federation of National Engineering Associations) criteria for being an Engineering degree study program (this requirement was elaborated in synergy with FEANI) and the "Agricultural Science" core curricula requirements in Agricultural Engineering (this requirement was achieved in synergy with EurAgEng);
- establish a benchmark core curriculum serving the purpose of determining a set of minimum criteria/requirements against which any curriculum can be tested and decided whether or not it meets the criteria for its admission as a particular degree study program in Agricultural Engineering;
- define common accreditation strategies and procedures and establish the bodies/committees to carry out the accreditation of the departments, in order to meet the core curricula requirements.

As the weak part of the Agricultural Engineering degree study programs offered in 2002 in Europe was the inadequate Engineering foundation of the corresponding curricula, the main goals of USAEE-TN were to:

- enhance the Engineering part of the European core curricula in Agricultural/Biosystems Engineering so as to meet the FEANI criteria for being Engineering degree study programs;
- significantly reduce the Agricultural or Biological Sciences part of the European core curricula in Agricultural/Biosystems Engineering, in order to meet the basic knowledge of the corresponding subjects without extending to or competing against or overlapping with the traditional Agronomy degree study programs.

The most paramount outcome of USAEE-TN was the core curriculum of Agricultural/Biosystems Engineering for the first cycle degrees of integrated first and second cycles (i.e. Pivot Point Degree) or for long cycle academic orientation programs of studies. As regards the first cycle BSc degree programs of studies, the proposed core curriculum of Agricultural/Biosystems Engineering allows for 18 ECTS units to be devoted to Applied Agricultural Engineering subjects, while most of the Agricultural Engineering specialisation subjects are offered during the second cycle study program. The first level of studies is considered in the present stage of the core curriculum development of USAEE.

Two different schemes were defined at this level:

A) academic orientation
   - core curriculum for University studies leading to a long cycle degree following an integrated five year program of studies;
   - core curriculum for integrated two-cycle programs of University studies leading to a pivot-point first cycle degree (e.g. BSc) following a three year program of studies (Table 7.1);

B) applications-technological orientation
   - core curriculum for University studies leading to a professional BSc level degree following at most a three year program of studies.

Consortium POMSEBES

POMSEBES (Policy Oriented Measures in Support of the Evolving Biosystems Engineering Studies in USA – EU) was funded by the European Commission, jointly with the US Department of Education, Fund for the Improvement of Post Secondary Education (FIPSE), under the ATLANTIS programme (Actions for Transatlantic Links and Academic Networks for Training and Integrated Studies), in the framework of the 2006–2013 EU–US agreement in higher education and vocational training.

The objectives of the consortium POMSEBES were to:

- define the new (emerging) discipline of Biosystems Engineering;
- identify current/future employers;

TABLE 7.1

Layout of the core curriculum for the first cycle pivot-point degrees of the integrated MSc or long cycle academic orientation study programs in Agricultural/Biosystems Engineering

	Engineering area	Agricultural/Biological Sciences area
General Competences & Electives	Adopted from E4-A1 activity	Adopted from E4-A1 activity
Core Basis	Basic courses from core curriculum	Basic courses from core curriculum
Optional	Optional courses from core curriculum	Optional courses from core curriculum
Course units beyond core curriculum	Depending on institution	Depending on institution

- identify areas of expertise in the discipline (of Biological and Agricultural Engineering);
- identify research needs/opportunities in Biosystems Engineering;
- develop policy measures and accompanying plans in support of the structural development of the new emerging discipline of Biosystems Engineering;
- develop a uniform structure (framework) for compatible programs of the Biosystems Engineering discipline;
- describe the Biological/Biosystems Engineering programs of study in the US and the current trends in the EU, focusing on Education and Research;
- identify the needs of industry (current and future employers);
- develop a pilot accreditation program (in terms of policy/strategy) for the Biosystems Engineering discipline, in conformity to the existing accreditation process conducted by ABET (Accreditation Board for Engineering and Technology) in the US and by EUR-ACE (EURopean-Accredited Engineer, i.e. a decentralised European accreditation system of Engineering degree study programs) in the EU.

ERABEE-TN

ERABEE-TN (Education and Research in Biosystems Engineering in Europe – a Thematic Network) was co-funded by the EU under the LifeLong Learning Programme (LLP).

The aims of ERABEE-TN were to expand the initial USAEE-TN objectives, by developing new activities, and promote the step-by-step critical transition and further structural development of the traditional Agricultural Engineering area to the emerging Biosystems Engineering one. The objectives of ERABEE-TN were to:

- promote synergies between Education and Research, i.e. to promote the structured 3rd cycle programs of studies, also establishing links between education and research in all three cycles;
- support activities in the field of Accreditation and Quality Assurance, i.e. develop advanced European degrees of studies (e.g. European Doctorates) and enhance accreditation and quality assessment and assurance frameworks, integrating the Biosystems Engineering TN activities to the corresponding major developments on the European accreditation and labelling of European Engineering studies (EUR-ACE-Implementation);
- define and upgrade generic and sectorial competences, i.e. implement the TUNING template;
- establish the international attractiveness of European programs of studies in Biosystems Engineering;
- expand Biosystems Engineering programs of studies to include emerging areas such as biofuels, biomaterials, useful waste, etc., as related to the new CAP developments and the European knowledge based bioeconomy framework;
- reinforce the link between Education in the emerging field of Biosystems Engineering and society, introducing the new developments to the non-academic community and the public (Education, Audiovisual & Culture Executive Agency [EACEA], 2010).

Consortium TABE.NET

The consortium TABE.NET (Trans-Atlantic Biosystems Engineering Curriculum and Mobility) was funded by the European Commission, jointly with the US Department of Education, Fund for the Improvement of Post Secondary Education (FIPSE), under the ATLANTIS programme (Actions for Transatlantic Links and Academic Networks for Training and Integrated Studies), in the framework of the 2006–2013 EU–US agreement in higher education and vocational training.

The consortium TABE.NET was targeted to advance in the internationalisation of ABE curricula and develop a global awareness within this discipline. This project is based on and is a follow-up of ERABEE-TN and the data derived from the questionnaire on University-Enterprise cooperation developed by ERABEE-TN was used by the research project coordinated by University of Leon (Spain) entitled "Análisis de los procedimientos y herramientas para la adaptación de los títulos de Ingeniería a las necesidades del mercado laboral nacional e internacional" (Analysis of the procedures and tools for the adaptation of the Engineering studies to the national and international labour market). The objectives of the consortium TABE.NET were to:

- design and build a Biosystems Engineering Digital Library (BEDL) to aid curriculum internationalisation in the discipline;
- widen students' knowledge, as well as teachers' and researchers' information exchange within Biosystems Engineering discipline.

The consortium TABE.NET developed an internet-based tool to facilitate the sharing of teaching examples and learning objects from around the world, in order to enable the internationalisation of ABE degree study programs.

Another output of TABE.NET consortium was a student exchange agreement between University of Illinois at Urbana-Champaign (USA) and University of Bari (Italy), aimed at expanding educational opportunities (four students came from Illinois to Bari, two with a mobility grant and two without it).

EU Member and Candidate States Offering ABE Curricula as ERABEE-TN Project Outcome

With reference to the traditional area of Agricultural Engineering, the new area of Biosystems Engineering developed in order to include research topics and teaching subjects regarding biofuels, biomaterials and quality of products. As far as the impact of ERABEE-TN on teaching and learning, the transition from the traditional area of Agricultural

Engineering to the emerging one of Biosystems Engineering had a direct impact on degree study programs, student learning, information access and curricula advising, as well as contributing towards the harmonisation of higher education in Agricultural and Biosystems Engineering (ABE) in the EU. In fact, when the networking activities of ERABEE-TN started, only two HEIs were offering University degree study programs in Biosystems Engineering, i.e. Catholic University of Leuven (Belgium) and University College Dublin (Ireland).

Instead, during the life-time of the ERABEE-TN project, some remarkable developments were made in the emerging area of Biosystems Engineering, with the establishment of new ABE degree study programs in EU member and candidate states, as an outcome of ERABEE-TN, at the following 10 HEIs: Catholic University of Leuven (Belgium); Aarhus University (Denmark); Estonian University of Life Sciences (Estonia); University College Dublin (Ireland); University of Évora (Portugal); Technical University of Cluj-Napoca (Romania); University of Maribor (Slovenia); Polytechnic University of Cataluña (Spain); University of León (Spain); Uludag University (Turkey) (EACEA, 2010).

According to the survey performed in 2021 (Comparetti & Febo, 2021), as an outcome of ERABEE-TN project, new BSc and MSc ABE curricula were established by the HEIs of 20 EU member and candidate states (Belgium, Czech Republic, Denmark, Estonia, France, Germany, Ireland, Italy, Latvia, Lithuania, Netherlands, Norway, Poland, Portugal, Romania, Slovak Republic, Slovenia, Sweden, Spain and Turkey). The results of this survey, updated to the academic year 2022–2023 and related to 12 EU member and candidate states, are summarised in Table 7.2.

The Universities of at least 12 (Czech Republic, Estonia, Italy, Latvia, Lithuania, Netherlands, Poland, Portugal, Romania, Slovak Republic, Sweden and Turkey) of the above 20 EU member and candidate states offer BSc and MSc curricula in ABE area, including subjects related to Agricultural Engineering and Applied Agricultural Engineering as at least 30% of the total ECTS (European Credit Transfer System) study load.

Instead, the HEIs of other countries (e.g. Denmark, Finland, France and Norway) offer BSc and MSc curricula including subjects related to Agricultural Engineering and Applied Agricultural Engineering as less than 30% of the total ECTS study load.

Moreover, the Universities of 13 EU member and candidate states offering BSc and MSc curricula in ABE area are reported in the following sub-paragraphs.

Czech Republic

The Czech University of Life Sciences Prague (CZU) (https://www.czu.cz/en/) is a public University, with more than 18,000 students (10% are international students), six faculties and one Institute. CZU offers over 170 accredited degree study programs at BSc, MSc and PhD levels (in 9 BSc, 20 MSc and 18 PhD programs the language of instruction is English), is member of the Euroleague for Life Sciences (www.euroleague-study.org) and has implemented the ECTS, which highly facilitates the mobility of students.

Within CZU, the Faculty of Engineering (CZU, n.d.) consists of 10 departments, one Laboratory of Information Technologies, one Technology Certification Centre and one Department of Expert Activities (Expert Institute), as well as being equipped with a research and development mini-brewery, which has especially made a great contribution to the practical instruction of the degree study program in Technological Equipment of Buildings (TZS).

The academic year begins in October and is divided into two semesters.

The Faculty of Engineering offers the following nine 3-year BSc degree study programs and one 2-year follow-up MSc degree study programs in Agricultural/Biosystems Engineering:

1) Agriculture Machinery (ZT);
2) Road and City Automobile Transport (SMAD);
3) Technique and Technology of Waste Treatment (TTZO);
4) Technological Equipment of Buildings (TZS);
5) Trade and Business with Technology (OPT);
6) Information and Control Technology in the Agro-Food Complex (IŘT);
7) Technology and Environmental Engineering (TEE);
8) Maintenance Engineering (IU);
9) Agricultural Engineering (AE) (taught in English and, in the first two years, dealing with a theoretical basis in agricultural technology and, in the third year, focused on the practical application of theoretical knowledge in technical and biological sciences).

After BSc graduation, it is possible to attend a 2-year MSc degree study program, also taught in English. The graduates will achieve theoretical knowledge on the application of exact procedures in agricultural technology, as well as deep knowledge on the theory of technical disciplines that enable them to work in the areas of machine development, technological use and technology maintenance.

Estonia

The University of Tartu is the oldest and largest higher education institution in Estonia and has been the centre of Estonian academic life for almost four centuries. 13,000 students ca. and 3,000 staff members ca. study and work in a wide range of fields at this University. The University of Tartu is ranked in the top 1.2% of the world's Universities, making it one of the leading Universities in Northern Europe.

The Faculty of Science and Technology of the University of Tartu also offers:

- BSc degree study program (240 ECTS) "Technotronics", having the specialisations "Mechanics" and Electronics";
- MSc degree study program (120 ECTS) "Production Engineering", having the specialisations "Automotive Engineering" and "Bioenergetics Engineering".

TABLE 7.2

Survey results, updated to the academic year 2022–2023 and related to 12 EU member and candidate states where ABE curricula were established after the end of the ERABEE-TN (Education and Research in Biosystems Engineering in Europe - A Thematic Network) project

Country	University	Study cycle	Degree study program	Specialisation	Total ECTS	Agricultural Engineering and Applied Agricultural Engineering ECTS (%)
Czech Republic	Prague	BSc	Agricultural Engineering	-	180	38
Czech Republic	Prague	MSc	Agricultural Engineering	-	120	88
Czech Republic	Prague	MSc	Technology and Environmental Engineering	-	120	77
Estonia	Tartu	BSc	Technotronics	Mechanics / Electronics	240	60
Estonia	Tartu	MSc	Production Engineering	Automotive Engineering	120	32
Italy	Bari	BSc	Land and Environmental Science and Technology	-	180	30
Italy	Catania	BSc	Planning and Environmental Sustainability of Territory and Landscape	Sustainable Management of Agro-Forestry Territory	180	47
Italy	Catania	BSc	Planning and Environmental Sustainability of Territory and Landscape	Territory, Environment and Landscape Planning	180	50
Italy	Potenza	BSc	Urban Landscape, Environment and Green	-	180	40
Italy	Perugia	MSc	Sustainable Agriculture	Territory and Landscape	120	37
Italy	Viterbo	BSc	Digital Management of Agriculture and Mountain Territory	Digital Agriculture	180	37
Italy	Viterbo	BSc	Digital Management of Agriculture and Mountain Territory	Digital Management of Mountain Territory	180	32
Latvia	Jelgava	BSc	Agricultural Engineering	Agricultural Machinery	240	43
Latvia	Jelgava	BSc	Biosystems Machinery and Technologies	-	180	33
Latvia	Jelgava	MSc	Agricultural Engineering	Agricultural Technologies and Techniques	120	32
Lithuania	Kaunas	BSc	Agricultural Mechanical Engineering	-	240	56
Lithuania	Kaunas	MSc	Agricultural Mechanical Engineering	-	120	55
Netherlands	Wageningen	BSc	Biosystems Engineering	-	180	45
Poland	Lublin	BEng (Bachelor Engineer)	Agricultural Engineering and Agrotronics	-	210	49
Poland	Lublin	MSc	Agricultural Engineering and Agrotronics	-	90	42
Portugal	Évora	MSc	Precision Agriculture Technologies	-	120	37

(Continued)

TABLE 7.2 (CONTINUED)

Survey results, updated to the academic year 2022–2023 and related to 12 EU member and candidate states where ABE curricula were established after the end of the ERABEE-TN (Education and Research in Biosystems Engineering in Europe - A Thematic Network) project

Country	University	Study cycle	Degree study program	Specialisation	Total ECTS	Agricultural Engineering and Applied Agricultural Engineering ECTS (%)
Romania	Cluj-Napoca	MSc	Biosystems Engineering in Agriculture and Food Industry	-	180	38
Slovak Republic	Nitra	BSc	Management of Machinery Operation	-	180	33
Slovak Republic	Nitra	MSc	Management of Machinery Operation	-	120	35
Sweden	Alnarp	BSc	Agriculture and Rural Management	-	180	42
Turkey	Bursa	BSc	Biosystems Engineering	-	180	77
Turkey	Kahramanmaras Sutcu	BSc	Biosystems Engineering	-	180	71
Turkey	Namik Kemal	BSc	Biosystem Engineering	-	180	77

France

In France, the higher education system includes Universities and other institutions called "Grandes Écoles" or schools of higher education. The "Grandes Écoles" are the most prestigious institutions for higher education in Engineering and Management in France. Most chief executives of major French companies and top managers of French administration are graduates from the "Grandes Écoles".

The "Grande École" Institut Agro Dijon offers the Master degree study program (120 ECTS) "Agricultural, Environmental, Territorial, Landscape and Forestry Sciences" (AEPTF), having the specialisation "Management of Innovative Companies and Technologies for Agricultural Equipment" (GETIA), with a study load in Agricultural Engineering and Applied Agricultural Engineering representing 28% of the total ECTS (L'Institut Agro Dijon, n.d.a).

The Institut Agro Dijon also offers the 3-year Engineering Master ("Diplôme d'Ingénieur") degree study program "Agricultural and Environmental Sciences" (L'Institut Agro Dijon, n.d.b), whose last year is the specialisation "Sciences and Technologies of Agricultural Equipment" (STEA), that is unique in French public higher education (L'Institut Agro Dijon, n.d.c). The students of this program can join the double degree MSc study program "Management of Innovative Companies and Technologies for Agricultural Equipment", offered by the same HEI. The graduates of the "Diplôme d'Ingénieur" can join commercial or marketing departments of farm equipment companies or apply to a PhD program after their Engineering studies and, then, join academia or an industrial R&D department.

The University of Chalon S./Saône offers the Bachelor degree study program "Maintenance and Technology: Polytechnic systems", having a common core curriculum (University Technology Bachelor – BUT) of two years and two optional specialisations of one year, "Technical and Economic Management of Agricultural Equipment" and "Livestock Automation and Robotics" (Institut Universitaire de Technologie [IUT] – Chalon sur Saôn, n.d.).

Italy

In Italy, significant changes relevant to the degree study programs in Agricultural and Biosystems Engineering (ABE) occurred in 2009 and 2010, as a consequence of the new regulations established by the Italian Ministry of Education, University and Research. These changes were aimed at rationalising the degree study programs offered by Italian Universities and were indirectly influenced by stakeholders (i.e., employers), whose needs were taken into account by the Ministry of Education, University and Research, and minimally also by ERABEE-TN activities. The outcomes of these changes were the ABE degree programs, at BSc, MSc and PhD levels, running in the academic year 2009–2010. Moreover, these changes were partially aimed at increasing the attractiveness of the degree study programs, based on the experience of the implementation of 3+2 education system (Bologna process).

In Italy, as an outcome of ERABEE-TN project, the Agricultural Engineering scientific and teaching sector was re-modulated and updated into the new sector "Agricultural, Forestry and Biosystems Engineering", including research topics and teaching subjects regarding the aspects related to Biosystems Engineering, distinguished in the three related Italian traditional sub-sectors: Hydraulics; Mechanics; Structures and Land.

In Italy, the Universities of Bari, Catania, Potenza, Perugia and Viterbo offer BSc and MSc degree study programs with subjects related to Agricultural Engineering and Applied Agricultural Engineering as at least 30% of the total ECTS study load (Table 7.2).

Latvia

The only University in Latvia offering higher education in the area of Agricultural/Biosystems Engineering is Latvia University of Life Sciences and Technologies (https://www.llu.lv/en), which has eight faculties.

The Faculty of Engineering (Latvia University of Life Sciences and Technologies, n.d.a) offers higher education in the area of Agricultural/Biosystems Engineering at all three levels, i.e. Bachelor, Master and PhD.

This faculty offers the 4-year BSc degree study program "Agricultural Engineering", which has two specialisations, "Agricultural Machinery" and "Autotransport", and is taught only in Latvian.

The Faculty of Engineering offers the 3-year BSc degree study program "Biosystems Machinery and Technologies" (Latvia University of Life Sciences and Technologies, n.d.b), which is taught in English, also for foreign students. The graduates of this program can find a job in metal working, mechanical engineering, design, wood processing, agricultural and food industries, trading and other companies related to the design, manufacturing, installation, monitoring, assembly, development and marketing of mechatronic equipment/devices/machines for various biosystems. As another option, the graduates of the above program are encouraged to continue their studies in the MSc degree study program "Agricultural Engineering".

Moreover, the Faculty of Engineering offers the 2-year MSc degree study program "Agricultural Engineering" (Latvia University of Life Sciences and Technologies, n.d.c), which has four specialisations: "Agricultural Technologies and Techniques" and "Machine Designing and Manufacturing" (both taught in Latvian); "Power Engineering" and "Vehicles and Service" (both taught in English).

The graduates from a MSc degree study program can further apply to the 3-year PhD program "Agricultural Engineering" (Latvia University of Life Sciences and Technologies, n.d.d), by choosing one of four specialisations, i.e. "Agricultural Energetics", "Agricultural Machinery", "Transport" and "Machinery Design and Production".

Lithuania

In Lithuania, Vytautas Magnus University (VMU) – Agriculture Academy (AA) is the only state institution of higher education and research awarding diplomas and degrees at BSc, MSc and PhD levels in the areas of Food Sciences, Agriculture, Forestry, Water and Land Resources Management, Bioenergy and Mechanical Engineering, Climate Change and Sustainable Use of Natural Resources.

Within Vytautas Magnus University of Kaunas – Agriculture Academy of Akademjia, the Faculty of Engineering offers BSc and MSc degree study programs in Engineering Sciences (VMU, n.d.).

The Faculty of Engineering offers three 4-year (240 ECTS) BSc degree study programs:

1) Agricultural Mechanical Engineering;
2) Sustainable Engineering;
3) Water and Land Engineering, which has three specialisations, i.e. Aquaculture Engineering, Hydraulic Engineering and Land Use Planning.

Successful Bachelor graduates are offered to choose one of the following 2-year (120 ECTS) MSc degree study programs:

1) Agricultural Mechanical Engineering;
2) Agricultural Engineering and Management;
3) Hydraulic Engineering;
4) Land Use Planning;
5) Sustainable Energy;
6) Transport Machinery Engineering.

There are three doctoral schools in the area of Technology:

1) Environmental Engineering;
2) Mechanical Engineering;
3) Transport Engineering.

Doctoral students can apply for a PhD after 4 years.

The Netherlands

In the Netherlands, higher education is provided by both Universities and Applied Universities (https://info.studielink.nl/en). The academic year begins on the 1st of September and ends in early July, consisting of 40 weeks, each of 42 hours. One ECTS equals a study load of 28 hours, so that one year consists of 60 ECTS. Universities offer higher scientific and artistic education and award 3-year BSc and 2-year MSc degrees, as well as PhD degrees. Universities of Applied Sciences (UAS) provide more practical education, aiming at satisfying the needs of the labour market, as well as awarding UAS BSc degrees and UAS MSc degrees, for a total of four years.

The University of Wageningen has one faculty, 19 BSc degree study programs and 30 MSc degree study programs, e.g. the BSc "Agrotechnology" and the MSc "Biosystems Engineering". The above BSc degree study program "Agrotechnology" is taught in Dutch in the first year, partly in Dutch and partly in English in the second year, and in English in the third year (Wageningen University & Research, n.d.a). The above MSc degree study program "Biosystems Engineering" is taught in English (Wageningen University & Research, n.d.b). This University also offers many BSc 20 week minor degree study programs ("Bachelor's minors"), with 30 ECTS and taught in English (Wageningen University & Research, n.d.c).

Hogere Agrarische School (HAS) University of Applied Sciences offers the following two agricultural related BSc degree study programs:

1) Horticulture and Business Management (in English) (HAS - Green Academy, n.d.a);
2) International Food and Agrobusiness (in English) (HAS - Green Academy, n.d.b).

This University also offers about a dozen interesting 20 week "Bachelor's minors", each with 30 ECTS and taught in English (HAS - Green Academy, n.d.c).

Van Hall Larenstein University of Applied Sciences offers the following two agricultural related degree study programs:

1) Animal Husbandry (4-year BSc in English) (Van Hall Larenstein - University of Applied Sciences, n.d.a);
2) International Agribusiness and Trade (4-year BSc in English) (Van Hall Larenstein - University of Applied Sciences, n.d.b).

AERES University of Applied Sciences offers the following agricultural related degree study programs:

1) several 1 and 2-year degree study programs (AERES University of Applied Sciences, n.d.a);
2) European Food Business (BSc in English) (AERES University of Applied Sciences, n.d.b).

Poland

The Polish higher education system includes Universities, Polytechnics and Higher Schools of Professional (Vocational) Education/Universities of Applied Sciences, both public and private. The 2005 Law on Higher Education distinguishes University HEIs and non-University HEIs (mainly Higher Schools of Professional Education).

A University-type HEI is a Higher Education Institution where at least one organisational unit (e.g. faculty) is authorised to award doctoral degrees (PhD). This institution may provide first cycle degree study programs leading to Bachelor/Engineer degrees and/or second cycle or long cycle degree study programs leading to MSc and PhD degrees.

A non-University HEI is a Higher Education Institution which provides first cycle, second cycle and/or long cycle degree study programs but is not authorised to provide doctoral (third cycle) programs or award PhD degrees. This type of institution includes HEIs referred to as Higher Schools of Professional (Vocational) Education/Universities of Applied Sciences, which are authorised to provide only first cycle degree study programs.

Polish higher education is part of the Bologna three cycle system, so that Bachelor/Engineer degree study programs last for 3 or 3.5 years (6-7 semesters), MSc degrees take 1.5 or 2 years (3-4 semesters) and most PhD programs can be accomplished in four years.

The University of Life Sciences in Lublin offers both the Engineer and MSc degree study programs "Agricultural Engineering and Agrotronics".

Several other Polish Universities also offer degree study programs in Agricultural Engineering, e.g.:

- Bialystok University of Technology;
- Koszalin University of Technology;
- Opole University of Technology;
- Poznań University of Life Sciences;
- University of Agriculture in Krakow;
- University of Warmia and Mazury in Olsztyn;
- Warsaw University of Life Sciences – SGGW;
- Wrocław University of Environmental and Life Sciences.

4.9 Portugal

In Portugal, in 2019, there were 287 HEIs, of which 123 were Universities and 164 were polytechnic education institutions. The academic year begins in September and ends in July, being divided into two semesters. In 2005, following the Bologna process, higher education began to have a new structure of three cycles of studies, leading to the academic degrees of "licenciado" (BSc), "mestre" (MSc) and "doutor" (PhD). This structure was introduced in 2006 and fully implemented from the academic year 2009-2010.

There are 16 Universities and polytechnic education institutions offering BSc degree study programs in the area of Agricultural/Biosystems Engineering, i.e. Agricultural Engineering (4), Food Engineering (4), Agricultural and Animal Engineering (1), Agronomy (6) and Organic Agriculture (1).

Moreover, there are 17 Universities and polytechnic education institutions offering MSc degree study programs similar to Agricultural/Biosystems Engineering, i.e. Agricultural Engineering (7), Animal Production Engineering (1), Mediterranean Agro-Forestry Systems (1), Precision Agriculture Technologies (1), Tropical Agronomy and Sustainable Development (1), Agroecology (1), Agronomy (1), Sustainable Agriculture (2), Organic Farming (1) and Agricultural and Animal Production Engineering (1).

In the MSc degree study program "Agricultural Engineering", sustainable production is a concern and a discipline that is included in the majority of the subjects. The BSc degree study program "Agronomy" offered by the University of Évora is taught in Portuguese, has a 3-year duration and includes 180 ECTS (60 ECTS per year) (University of Évora, n.d.a).

The MSc degree study program "Agricultural Engineering" offered by the University of Évora is taught in Portuguese, has a 2-year duration and includes 120 ECTS (60 ECTS per year, of which 42 ECTS for the thesis) (University of Évora, n.d.b). As an example, the outline of one MSc degree study program offered by the University of Évora, i.e. "Precision Agriculture Technologies" (University of Évora, n.d.c), is shown in Table 7.3.

Romania

The Technical University of Cluj-Napoca, awarded as an "Advanced Research and Education University" by the Order of the Romanian Ministry of National Education no. 5262 of 05/09/2011, is a tertiary Higher Education Institution, having national and international recognition. The Faculty of Automotive, Mechatronics and Mechanical Engineering also

TABLE 7.3

Outline of the MSc degree study program "Precision Agriculture Technologies" offered by the University of Évora (Portugal), School of Sciences and Technology (academic year 2022–2023)

Duration	2 years (4 semesters)
Credits	120 ECTS
Subjects related to Agricultural Engineering	27 ECTS
Remote Sensing and Image Analysis	6 ECTS
Management and Water Quality	6 ECTS
Data Processing Technologies in Precision Agriculture	3 ECTS
Precision Agriculture	6 ECTS
Monitoring and Control Applications in Farm Machinery	6 ECTS
Livestock Facilities Monitoring and Control	6 ECTS
Geographic Information Technologies in Precision Agriculture	6 ECTS
Technologies for the Efficient Use of Irrigated Water	6 ECTS
Subjects related to Agricultural Engineering take up	37% of the hours for the degree (lectures + practical training)

offers the MSc degree study program (180 ECTS) "Biosystems Engineering in Agriculture and Food Industry".

Slovak Republic

In the Slovak Republic, Slovak University of Agriculture in Nitra (www.uniag.sk) is a public University with six faculties and more than 5,000 students. This University offers education in agriculture and related research areas, e.g. Agro-biology, Food Resources, Sustainable Agriculture, Agricultural Production, Biotechnology, Food Technology, Engineering of Agricultural Machinery and Equipment, Computerisation and Automation of Agricultural Equipment, Gardening and Landscape Design, Economics and Management.

The education in the area of Biosystems Engineering is offered by the Faculty of Engineering (former Faculty of Agricultural Engineering), which consists of three institutes:

1) Institute of Agricultural Engineering, Transport and Bioenergetics;
2) Institute of Electrical Engineering, Automation, Informatics and Physics;
3) Institute of Design and Engineering Technologies.

This faculty offers the following BSc degree study programs (3 years -- 180 ECTS):

- Automotive Transport Systems;
- Quality and Safety in Production Technologies;
- Management of Machinery Operation (study program in Biosystems Engineering);
- Control Systems in Production Engineering;
- Production Technologies for Automotive Industry.

This faculty offers the following MSc degree study programs (2 years – 120 ECTS):

- Automotive Transport Systems;
- Quality and Safety in Production Technologies;
- Management of Machinery Operation (study program in Biosystems Engineering);
- Control Systems in Production Engineering.

This faculty offers the following PhD programs (3 years – 180 ECTS):

- Transport Machines and Equipment;
- Control Systems in Production Engineering;
- Machinery and Mechanisation of Agricultural Production (PhD program in Biosystems Engineering);
- Production Technologies.

The education in Biosystems Engineering is strongly connected to the industrial sector. The above faculty provides students with a practical placement at farms, machinery dealers, companies, etc. Students regularly participate in the CLAAS prize competition and others. In the area of Biosystems Engineering, attention is paid to the new machinery developments and technologies used in agriculture, e.g. precision farming. The students who graduate from the Slovak University of Agriculture in Nitra successfully find employment.

Spain

In Spain, there are four professions related to Biosystems Engineering, which are known as "Agricultural Technical Engineer", "Forestry Technical Engineer", "Agricultural Engineer" and "Forestry Engineer". They are charter professions, i.e. certain jobs can be performed only by these professionals. In order to access these professions, one must obtain a certain University degree that complies with the following guidelines established by the national government and the corresponding ministry:

- "Orden" CIN/323/2009 (Agricultural Technical Engineer);
- "Orden" CIN/324/2009 (Forestry Technical Engineer);

- "Orden" CIN/325/2009 (Agricultural Engineer);
- "Orden" CIN/326/2009 (Forestry Engineer).

In order to access the professions of "Agricultural Technical Engineer" and "Forestry Technical Engineer", it is needed to accomplish University BSc studies, having a duration of four years and complying with the aforementioned guidelines. These degree study programs, whose names are slightly variable among Spanish Universities, can be consulted (in Spanish) on the website of the Conference of University Centres offering these curricula (https://agroforestales.com/).

In order to access the profession of "Agricultural Engineer", one must complete University BSc studies meeting the guidelines for agricultural technical engineering and, then, a University MSc degree study program, having a duration of 1.5 or 2 years and meeting the guidelines for agricultural engineers. In order to access the profession of "Forestry Engineer", it is needed to accomplish University BSc studies meeting the guidelines for forestry technical engineering and, then, a University MSc degree study program meeting the guidelines for forestry engineers. The denominations of these 2nd cycle degree study programs are MSc in Agricultural Engineering and MSc in Forestry Engineering, respectively. These degree study programs can be consulted on the website of the Conference of University Centres offering these curricula (https://agroforestales.com/).

In Spain, there are only two study programs having the denomination of Biosystems Engineering, i.e., the following doctoral studies:

- PhD program in Biosystems Engineering from the University of León;
- PhD program in Agri-Food and Biosystems Science and Engineering from the University of Valladolid.

Turkey

In Turkey, higher education is carried out in faculties and vocational schools affiliated with Universities, being directed by the higher education board, i.e. Yüksek Öğretim Kurulu (YÖK). Biosystems Engineering departments emerged in 2009, due to a reconstruction of the departments of Agricultural Machinery and those of Irrigation and Agricultural Structures. The degree study programs offered by the departments of Biosystems Engineering are similar to each other in Turkey.

Biosystems Engineering departments were firstly formed at Bursa Uludağ University (BUU), Kahramanmaras Sutcu Imam University and Gaziosmanpasa University in 2009, followed by the Universities of Namik Kemal, Yuzuncu Yil, Bozok, Mustafa Kemal, Erciyes, Bingol, Adiyaman, Nigde and Bilecik Şeyh Edebali.

Biosystems Engineering education is carried out at undergraduate and postgraduate levels in Turkey. Not all Biosystems Engineering departments offer PhD programs, while most of these departments offer both BSc and MSc degree study programs. An undergraduate (BSc) degree study program is planned for 4 years and 240 ECTS, an MSc degree study program is planned for 2 years and 120 ECTS, and a PhD program is planned for 4 years and 240 ECTS.

The undergraduate degree study program offered by Bursa Uludağ University (BUU) includes the major areas, i.e. Land and Water Resources; Agricultural Structures; Agricultural Machinery Systems; Agricultural Energy Systems. The students are expected to take compulsory and elective subjects, of which at least 25% should have practice sessions, i.e. laboratory and field work, workshops and problem-solving. The students must perform a training program (internship) at the University and in the industry for a sum of 45 working days.

In a graduate program, the students must carry out coursework for two semesters and, then, prepare a thesis in one of the above four major study areas. A combined BSc and MSc is expected to finish in about 6 years, while a combined BSc, MSc and PhD in about 10 years.

Biosystems Engineering Department at BUU has links with commercial farms, land consolidation firms and the machinery industry, to which it provides graduates with high career possibilities.

Challenges and Opportunities in ABEE and Future Perspectives: Past, Present and Future of ABE Degree Study Programs

A harmonisation, a simplification and an increase in the transparency of the degree study programs in Agricultural and Biosystems Engineering (ABE) occurred in EU member and candidate states during the last 23 years. The harmonisation, as an implementation of Bologna process, i.e. the 3 + 2 education system, occurred across the whole EU. Yet, previously the 1st cycle (BSc) was more specialised than the 2nd cycle (MSc) in some countries (e.g. Italy), while currently this situation occurs in other countries (e.g. Spain). Instead, the harmonisation of ABE degree study programs occurred only in some countries, while this was inhibited by the poor or absent relationship between Universities and the national Ministry of Education in the majority of them (e.g. Italy) and, sometimes, also by a weak influence of Agricultural Engineers within each University (e.g. University of Palermo).

The simplification of the degree study programs in this area was obtained through the establishment of new curricula, such as those offered by the above 20 EU member and candidate states as an outcome of ERABEE-TN project. Finally, the increase in transparency was made easier by updating the websites of EU Universities and also by the database on degree study programs in Agricultural/Biosystems Engineering developed under the USAEE-TN and ERABEE-TN projects.

Another result of the ERABEE-TN project was the establishment of new Erasmus+ inter-institutional agreements, e.g. between the University of Palermo (Italy) and Vytautas Magnus University (former Aleksandras Stulginskis University, former Lithuanian University of Agriculture - Lithuania), Agricultural University of Athens (Greece), "Angel Kanchev" University of Ruse (Bulgaria), University of Malta and University of Évora (Portugal). This contributed to increase the student and teaching staff mobility between these Universities, which was one of the objectives of ERABEE-TN project (Comparetti & Febo, 2021).

Some results of the ERABEE-TN project caused changes in the activities and structure of Higher Education in EU: these changes show that the implementation of Biosystems Engineering in Europe is forthcoming. All the following activities were affected by ERABEE-TN project results:

- promotion of the Biosystems Engineering studies accreditation;
- promotion of new degrees and doctorates in Biosystems Engineering;
- synergies with research projects about innovation in education;
- promotion of new teaching innovation groups;
- promotion of new scientific and teaching sectors.

As far as the promotion of the Biosystems Engineering studies accreditation, the active participation of ERABEE-TN members in the EUR-ACE project, aimed at strengthening synergies with the major Engineering Thematic Networks, i.e. EUR-ACE Implementation, EUR-ACE spread and TREE, started to promote the goal of implementing the "EUR-ACE label" accreditation of Biosystems Engineering. The first degree study program in Biosystems Engineering which underwent the EUR-ACE accreditation procedure in the framework of ERABEE-TN was that offered by the University College of Dublin (Ireland), following a site visit by the accreditation agency, i.e. Engineers Ireland on 13–14 December 2010. Along this action it was decided that the EurAgEng recognition process should be encouraged and that volunteer institutions should undergo trial cases. Two Universities, i.e. Polytechnic University of Madrid (Spain) and Latvia University of Agriculture (Latvia), applied for the recognition, so that this process was accomplished.

As far as the promotion of new degrees and doctorates in Biosystems Engineering, the following new degrees were activated by members of ERABEE-TN:

- the MSc degree study program "Biosystems Engineering" ("Master Universitario de Investigacion en Ingenieria de Biosistemas") and the PhD program "Biosystems Engineering" ("Doctorado en Ingeniería de Biosistemas"), offered by the University of León (Spain), were verified by the Spanish Universities Council; previously the proposals of these programs were evaluated by a review panel, consisting of experts who used the protocols of ANECA (National Agency for Quality Assessment and Accreditation); the results of such evaluations were positive; these MSc and PhD programs in Biosystems Engineering were derived from a previous doctorate in Agrarian Technologies ("Tecnologías Agrarias"); the information coming from both ERABEE-TN and POMSEBES consortium projects justified the interest in these new MSc and PhD programs, and it was used to develop their contents; this MSc degree study program was considered strategic by the regional government of "Castilla y León", according to the Regional Strategy for technological development and innovation in this region;
- the MSc degree study program "Biosystems Engineering", offered by the University of Évora (Portugal), was approved by the National Accreditation Agency in 2010 and started in the academic year 2010–2011; this degree study program was constituted by 120 ECTS credits, distributed in four semesters; a common semester was followed by four specialisations, i.e. Bioenergy; Precision Agriculture; Irrigation, Construction and Equipment Engineering; Biophysical Engineering and Ecological Systems;
- the MSc degree study program "Biosystems Engineering in Agriculture and Food Industry", offered by the Technical University of Cluj-Napoca (Romania), started from the academic year 2009–2010 and has a core curriculum that was developed in accordance with ERABEE-TN;
- the MSc degree study program "Technical and Economical Management of Agricultural Engineering", offered by the Higher Educational Agronomic School of Dijon – AgroSup (France), started in 2006 and was positively re-accredited in 2009 for three years;
- the BSc degree study program "Biosystems Engineering", offered by Estonian University of Life Sciences (Estonia), started in 2009, in order to integrate biological and engineering education, thus meeting the expectations and requirements of agricultural producers for such professionals; in the development of this degree study program, a range of stakeholders, i.e. managers of rural companies, farmers, food processors and others closely related to agriculture, were involved; five joint workshops were held with stakeholders and University representatives, so that their conclusions and outcomes could be used to improve the degree study program; the program was a professional Higher Education type, having a duration of four years and 240 ECTS.

As far as the synergies with research projects about innovation in education, the TABE.NET project and ERABEE-TN results were used by a research project implemented by the University of León (Spain).

As far as the promotion of new teaching innovation groups, a new teaching innovation group called BETI (Biosystems Engineering Teaching Innovation) was approved by the Polytechnic University of Madrid (Spain), in order to create educational resources that promote the internationalisation of studies in Biosystems Engineering.

As far as the promotion of new scientific and teaching sectors, in Italy the Agricultural Engineering scientific and teaching sector was re-modulated and updated into the new sector "Agricultural, Forestry and Biosystems Engineering", including research topics and teaching subjects regarding aspects

TABLE 7.4

Updated partnership of the ERABEE (Education and Research in Biosystems Engineering in Europe) Network (2022)

N.	Country	Institution	Faculty/Department/Institute	Partner	E-mail
1	Austria	University of Natural Resources and Life Sciences	Department of Sustainable Agricultural Systems	Elisabeth Quendler	elisabeth.quendler@boku.ac.at
1	Belgium	University of Ghent	Department of Environment	Abdul M. Mouazen	Abdul.Mouazen@ugent.be
1	Bulgaria	University of Ruse "Angel Kanchev"	Department of Agricultural Machinery	Atanas Atanasov	aatanasov@uni-ruse.bg
1	Croatia	University of Zagreb	Faculty of Agriculture	Marija Romic	mromic@agr.hr
1	Cyprus	Cyprus University of Technology	Department of Agricultural Sciences, Biotechnology and Food Science	Michalakis Christoforou	m.christoforou@cut.ac.cy
1	Czech Republic	Czech University of Life Sciences (CULS) Prague	Faculty of Engineering, Department of Technological Equipment of Buildings	Pavel Kic	kic@tf.czu.cz
1	Denmark	Aarhus University	Department of Biological and Chemical Engineering	Morten Dam Rasmussen	mdr@bce.au.dk
2	Denmark	Aarhus University	Department of Electrical and Computer Engineering	Claus Aage Gron Sorensen	claus.soerensen@ece.au.dk
1	Estonia	Estonian University of Life Sciences	Institute of Technology	Timo Kikas	Timo.Kikas@emu.ee
1	Finland	University of Helsinki	Department of Agricultural Sciences	Laura Alakukku	laura.alakukku@helsinki.fi
1	France	Institut Agro Dijon	Department 2A2E	Christelle Gée	christelle.gee@institut-agro.fr
1	Germany	Technical University of Munich	Department of Agricultural Systems Engineering	Heinz Bernhardt	heinz.bernhardt@tum.de
1	Greece	Agricultural University of Athens	School of Environment and Agricultural Engineering, Department of Natural Resources Development & Agricultural Engineering	Demetres Briassoulis	briassou@aua.gr
2	Greece	Agricultural University of Athens	School of Environment and Agricultural Engineering, Department of Natural Resources Development & Agricultural Engineering	Panagiotis Panagakis	ppap@aua.gr
1	Hungary	Hungarian University of Agriculture and Life Sciences (MATE)	Institute of Technology	István Szabo	Szabo.Istvan.prof@uni-mate.hu
1	Ireland	University College Dublin (UCD)	School of Biosystems & Food Engineering	Colm P. O'Donnell	colm.odonnell@ucd.ie
1	Italy	Polytechnic of Bari	Department of Civil Engineering	Giacomo Scarascia Mugnozza	giacomo.scarasciamugnozza@poliba.it
2	Italy	University of Palermo	Department of Agricultural, Food and Forest Sciences	Antonio Comparetti	antonio.comparetti@unipa.it
1	Latvia	Latvia University of Life Sciences and Technologies	Faculty of Engineering	Gints Birzietis	gints.birzietis@llu.lv
1	Lithuania	Vytautas Magnus University	Agriculture Academy, Faculty of Engineering	Kestutis Navickas	kestutis.navickas@vdu.lt
1	Malta	University of Malta	Institute of Earth Systems	George Attard	george.attard@um.edu.mt
1	Netherlands	Wageningen University and Research	Education & Student Affairs	Wilko van Loon	wilko.vanloon@wur.nl
2	Netherlands	Wageningen University and Research	Farm Technology Group	Peter Groot Koerkamp	peter.grootkoerkamp@wur.nl
1	Norway	Norwegian University of Life Sciences	Faculty of Science and Technology	Nils Bjugstad	nils.bjugstad@nmbu.no

(Continued)

TABLE 7.4 (CONTINUED)

Updated partnership of the ERABEE (Education and Research in Biosystems Engineering in Europe) Network (2022)

N.	Country	Institution	Faculty/Department/Institute	Partner	E-mail
1	Poland	University of Life Sciences in Lublin	Faculty of Production Engineering, Department of Machinery Exploitation and Management of Production Processes	Edmund Lorencowicz	edmund.lorencowicz@up.lublin.pl
1	Portugal	University of Evora	School of Science and Technology, Department of Rural Engineering	Jose Rafael Marques da Silva	jmsilva@uevora.pt
1	Romania	Technical University of Cluj-Napoca	Faculty of Automotive, Mechatronics and Mechanical Engineering	Nicolae Filip	Nicolae.Filip@auto.utcluj.ro
1	Russian Federation	Bryansk State Agrarian University	Department of Life Safety and Environmental Engineering	Ivan Petrovich Adylin	vanro1989@mail.ru
1	Serbia	University of Belgrade	Faculty of Agriculture	Goran Topisirovic	gogi@agrif.bg.ac.rs
1	Slovak Republic	Slovak University of Agriculture in Nitra	Faculty of Engineering	Jana Galambošová	Jana.Galambosova@uniag.sk
1	Slovenia	University of Maribor	Faculty of Agriculture and Life Sciences, Department for Biosystems Engineering	Miran Lakota	miran.lakota@um.si
1	Spain	University of Leon	School of Agricultural and Forestry Engineering	Pedro Jose Aguado Rodriguez	pedro.aguado@unileon.es
2	Spain	Polytechnic University of Madrid (UPM)	School of Agricultural, Food and Biosystems Engineering	Francisco Ayuga Tellez	francisco.ayuga@upm.es
1	Sweden	Swedish University of Agricultural Sciences (SLU)	Department of Biosystems and Technology	Knut-Hakan Jeppsson	knut-hakan.jeppsson@slu.se
1	Turkey	Bursa Uludağ University	Biosystems Engineering Department, Agricultural Machinery Systems	Selçuk Arslan	sarslan@uludag.edu.tr
1	Ukraine	Polissia National University	Department of Mechanics and Engineering of Agricultural Ecosystems	Savelii Kukharets	kikharets@gmail.com
1	United Kingdom	Harper Adams University	Department of Engineering	David R. White	drwhite@harper-adams.ac.uk
1	France	INRAE (National Institute of Research for Agriculture, Food and Environment)	Montpellier SupAgro	Jean-Paul Douzals [1]	jean-paul.douzals@irstea.fr
1	Ukraine	Bioenergy Association of Ukraine	-	Semeon V. Drahniev [1]	dragnev@uabio.org

[1] Stakeholders of ERABEE Network.

related to Biosystems Engineering, distinguished in the three related traditional Italian sub-sectors: Hydraulics; Mechanics; Structures and Land; this is a result of the process promoted by the Italian representatives of ERABEE-TN; the new scientific and teaching sector was also approved by all the delegates of the American Society of Agricultural and Biological Engineers (ASABE) (EACEA, 2010).

Even if it is not funded anymore, the ERABEE-TN is continuing to implement the above described objectives of the related project. In fact, during the 6th and last ERABEE-TN workshop, held in Clermont-Ferrand (France) on 9–10 September 2010, the Steering Committee of this TN decided to build up the ERABEE Network, which will work as a non-funded network of Universities/Departments offering degree study programs in the area of Agricultural and Biosystems Engineering. Within this network, the ERABEE-TN partners and new interested ones will stay in touch and support the submission of new proposals, as well as communicate new developments in education and research in Biosystems Engineering.

The project website will also be maintained and updated, for free, by the coordination staff. ERABEE-TN members are going on promoting and disseminating the work carried out within the project's duration, as well as addressing the outcomes and results of the project itself to new target user groups. They also intend to support and promote the synergies and cooperation already established with EurAgEng and EUR-ACE, concerning recognition and accreditation procedures and the organisation of working meetings.

The last accomplished activity of the ERABEE Network was the update of its partnership, which was the task of CIGR WG3 chair, at the beginning of 2022, so that the network is now constituted by 37 partners and 2 stakeholders (Table 7.4).

Acknowledgements

I acknowledge the University delegates (ERABEE-TN partners and collaborators) from each country, who provided me with the data and/or information needed for the survey of the HEIs of 15 EU member and candidate states: Pavel Kic for Czech Republic; Morten Dam Rasmussen for Denmark; Timo Kikas for Estonia; Laura Alakukku and Hanna-Riitta Kymäläinen for Finland; Christelle Gée for France; Gints Birzietis for Latvia; Kestutis Navickas for Lithuania; Wilko van Loon for the Netherlands; Nils Bjugstad for Norway; Edmund Lorencowicz for Poland; Jose Rafael Marques da Silva for Portugal; Nicolae Filip for Romania; Jana Galambošová for Slovak Republic; Pedro Jose Aguado Rodriguez and Francisco Ayuga Tellez for Spain; Knut-Hakan Jeppsson for Sweden; and Selçuk Arslan for Turkey.

REFERENCE LIST

AERES University of Applied Sciences. (n.d.a). *Programmes.* https://www.aeresuas.com/programmes?page=1

AERES University of Applied Sciences. (n.d.b). *European Food Business.* https://www.aeresuas.com/programmes/bachelor-programmes/european-food-business

Comparetti, A. & Febo, P. (2018a). *Harmonisation of Higher Education in Agricultural/Biosystems Engineering.* In EurAgEng 2018 Conference. Wageningen, The Netherlands, 8–11 July.

Comparetti, A. & Febo, P. (2018b). *Harmonization of Higher Education in Agricultural/Biosystems Engineering.* CIGR (International Commission of Agricultural and Biosystems Engineering) Newsletter, December, 115, 46. https://cigr.org/sites/default/files/newsletters/CIGR_NL115.pdf

Comparetti, A. & Febo, P. (2021). *State of the Art on Degree Study Programs in Agricultural/Biosystems Engineering in EU.* In EurAgEng 2021 Conference. Évora, Portugal, 4–8 July.

Czech University of Life Sciences Prague. (n.d.). *Faculty of Engineering.* https://www.tf.czu.cz/en/

Education, Audiovisual & Culture Executive Agency. (2010). *Education and Research in Biosystems Engineering in Europe; A Thematic Network - ERABEE-TN.* Final Report - Public Part. Project number 134306-LLP-1-2007-1-GR-ERASMUS-ENW. Education and Culture DG, Lifelong Learning Programme, pp. 1–24.

Febo, P. & Sun, D.W. (2000). *The University Structure and Curricula on Agricultural Engineering - An Overview of 36 Countries.* FAO, CIGR, EurAgEng, pp. 1–236.

Hogere Agrarische School [HAS] - Green Academy. (n.d.a). *Horticulture & Business Management.* https://www.has.nl/en/bachelors-and-more/bsc-horticulture-business-management/

Hogere Agrarische School [HAS] - Green Academy. (n.d.b). *International Food & Agribusiness.* https://www.has.nl/en/bachelors-and-more/bba-international-food-agribusiness/

Hogere Agrarische School [HAS] - Green Academy. (n.d.c). *Bachelors and More.* https://www.has.nl/en/bachelors-and-more/

Institut Universitaire de Technologie - Chalon sur Saône. (n.d.). *Licence Professionnelle Mention Maintenance et technologie: Systems pluritechniques / Gestion technique et économique des agroéquipements (GTEA) / Automatisme et Robotisation en Elevage (ARE).* https://iutchalon.u-bourgogne.fr/licence-professionnelle-maintenance-des-systemes-pluritechniques-agroequipements

L'Institut Agro Dijon. (n.d.a). *Gestion des entreprises et technologies innovantes pour l'agroéquipement.* https://institut-agro-dijon.fr/formations/masters/getia-gestion-des-entreprises-et-technologies-innovantes-pour-lagroequipement

L'Institut Agro Dijon. (n.d.b). *Agricultural and Environmental Sciences – Engineering Program.* https://institut-agro-dijon.com/programs/international-exchange-students/agricultural-sciences-engineering-program

L'Institut Agro Dijon. (n.d.c). *Sciences et Techniques des Equipements Agricoles.* https://institut-agro-dijon.fr/formations/ingenieur-agronome/sciences-et-techniques-des-equipements-agricoles-stea

Latvia University of Life Sciences and Technologies. (n.d.a). *Faculty of Engineering.* https://www.llu.lv/en/faculty-of-engineering

Latvia University of Life Sciences and Technologies. (n.d.b). *Academic Study Programme - Biosystems Machinery and Technologies.* https://www.llu.lv/en/biosystems_machinery_and_technologies

Latvia University of Life Sciences and Technologies. (n.d.c). *Academic Study Programme - Agricultural Engineering.* https://www.llu.lv/en/agricultural_engineering

Latvia University of Life Sciences and Technologies. (n.d.d). *Doctoral Study Programme Agricultural Engineering.* https://www.llu.lv/en/doctoral-study-programme-agricultural-engineering

University of Évora. (n.d.a). *Agronomy - Bachelor's Degree.* https://www.uevora.pt/en/study/courses/bachelors-and-integrated-master-degrees?cod=9003

University of Évora. (n.d.b). *Agricultural Engineering - Master Degree.* https://www.uevora.pt/en/study/courses/master-degrees?curso=991

University of Évora. (n.d.c). *Precision Agriculture Technologies - Master Degree.* https://www.uevora.pt/estudar/cursos/mestrados?curso=2526

Van Hall Larenstein - University of Applied Sciences. (n.d.a). *Animal Husbandry (BSc).* https://www.vhluniversity.com/study/programmes/bachelor/animal-husbandry

Van Hall Larenstein - University of Applied Sciences. (n.d.b). *International Agribusiness and Trade.* https://www.vhluas.com/study/programmes/bachelor/international-agribusiness-and-trade

Vytautas Magnus University. (n.d.). Agriculture Academy - Faculty of Engineering. *Broader Perspectives - From a Tech Lover to an Inventor.* https://zua.vdu.lt/en/faculties/faculty-of-engineering/

Wageningen University & Research. (n.d.a). https://www.wur.nl/nl/onderwijs-opleidingen/bachelor/bsc-opleidingen/bsc-agrotechnologie.htm

Wageningen University & Research. (n.d.b). https://www.wur.nl/en/education-programmes/master/msc-programmes/msc-biosystems-engineering.htm

Wageningen University & Research. (n.d.c). *List of Bachelor's Minors.* https://www.wur.nl/en/education-programmes/bsc-minors/list-of-bsc-minors.htm

8

Bioresource Engineering Curriculum and Reform at McGill University, Montreal, Canada

O. Grant Clark and Valérie Orsat

Introduction

The current global population of eight billion people is expected to grow to about 10.4 billion by the end of this century (United Nations, 2022). The projected large population, coupled with increasing global affluence, urbanisation, a shrinking agricultural workforce, and a changing climate presents many challenges in sustainable food security, nutritional security, and ecological integrity. These areas lie squarely in the domain of Bioresource Engineering.

Bioresource Engineering is an interdisciplinary programme in engineering and design at McGill University which incorporates foundational knowledge in the biological sciences. As a profession, Bioresource Engineering involves the design, construction, operation, maintenance, remediation, and upgrading of systems that contain biological components, including the technological components of such systems. More simply put, bioresource engineers create technologies to solve problems involving biological materials, living organisms, and ecosystems. These uniquely qualified engineering professionals work to enhance the sustainable use of natural resources. The Bioresource Engineering discipline contributes to sustainable food, energy and ecological systems through strong educational programmes, research activities, and professional and administrative contributions. Similarly, the education, research and service activities of the Department of Bioresource Engineering are continuously evolving in service to the discipline and society at large.

History

McGill University

McGill University is a public, English-language, research university with its main campus at the base of Mount Royal in Montreal, Quebec, Canada (Frost, 1980). The university is named for James McGill, a prominent Scottish merchant whose bequest initiated its formation in 1813. The University of McGill College was formally founded in 1821 through a royal charter granted by King George IV and was renamed in 1885 as McGill University. McGill's mission is to advance learning and disseminate knowledge through excellent education, research, and scholarship that meet the highest international standards and provide a service to society.

Macdonald Campus

Macdonald Campus is thirty kilometres to the west of Montreal, in Sainte-Anne-de-Bellevue. The campus was conceived and bankrolled by Sir William C. Macdonald, an important philanthropist who made his fortune in the tobacco industry (McGill University, n.d.). Construction of the campus began in 1905 and began operation in 1907 as Macdonald College. At its inception, Macdonald College was the largest and most modern agricultural college in Canada. The Rural Adult Education Service was established on the campus in 1938. In 1943, the purchase of 560 ha of adjacent farmland by John W. McConnell expanded the campus to its current 650 ha. In 1965, the college became the Macdonald Campus of McGill University. In 1971, the university accommodated the creation of John Abbott College by leasing a portion of the Macdonald Campus and many historic buildings to the new institution and moving the Faculty of Education to the main campus. Today, Macdonald Campus comprises an academic complex, residences, and sports fields, the Horticulture Research Centre, Macdonald Research Farm, Emile A. Lods Agronomy Research Centre, and Morgan Arboretum.

Administratively, the Macdonald Campus houses the Faculty of Agricultural and Environmental Sciences, School of Human Nutrition, Institute of Parasitology, Farm Management and Technology Programme, and several other research centres and institutes. Within the aforementioned faculty are the Departments of Animal Science, Bioresource Engineering, Food Science and Agricultural Chemistry, Natural Resource Sciences, and Plant Science. The faculty is a founding partner in the Bieler School of Environment. The current enrolment in the faculty exceeds 2,000 undergraduate and graduate students.

Bioresource Engineering Programme

The Bioresource Engineering Programme has its origin in the Manual Training Programme, initiated in 1910 shortly after

DOI: 10.1201/9780429150111-10

the founding of Macdonald College, and has evolved over the subsequent century (Association des ingénieurs en agroalimentaire du Québec, n.d.). In the mid-1960s the undergraduate programme gained its status as a professional Bachelor of Science degree in Agricultural Engineering and the first students graduated in 1971. A postgraduate degree programme was also instituted at that time. The undergraduate curriculum focused on irrigation and drainage, soil and water management, farm structures, and power and machinery. The degree designation was later changed to Bachelor of Engineering and, in 2003, the department was rebranded as Bioresource Engineering in response to the closure of many similar Agricultural Engineering programmes across North America. A decision was made to gradually diversify the programme to include the production of biomass, the processing of commodities into consumer products, and the management of the ecological impacts of those activities. The curriculum continues to adapt to cover emerging methods and technologies.

Programme Scope

Growing societal awareness and concern about issues such as climate change, food security, and the shift away from a linear economy towards circular economic principles are shaping the engineering professions and stimulating the development of new technologies and methods. Engineers must evaluate design options for sustainability, efficiently employing natural resources to produce biologically derived products such as food, fibre, fuels, and high-value chemicals. More stringent equipment standards and continuous advances in areas such as embedded smart technologies, big data, and artificial intelligence improve cost efficiency. Due to the range of potential applications driven, for example, by consumer demand for greater variety of food and products, there is an on-going need to design new technologies. To conform with the demand for sustainability, however, these new technologies must not only increase cost effectiveness, but also elevate social standards for stakeholder communities, and minimise or reverse negative environmental impacts. Accompanying analytical methods such as Life Cycle Assessment and Carbon Footprint Analysis are therefore becoming commonplace in the profession.

McGill's Bioresource Engineering Programme is one of only three similar programmes remaining in Canada, the others being at the University of Manitoba (Winnipeg) and Université Laval (Quebec City). It is necessary, therefore, to maintain not only the customary foundational knowledge related to agricultural, food, and environmental engineering, but to cultivate a broad spectrum of knowledge and skills so that the students trained in this programme can fill the growing demand for qualified engineers in this discipline. This is the approach that was adopted with the rebranding of the programme in 2003. The broader definition of the programme has found a receptive audience with strong enrolment that, before the 2020 coronavirus pandemic, saw annual cohorts of around 60 undergraduate students.

Recent academic hires in the department support the broader programme definition. The research expertise of tenure-track staff now includes the following topics, roughly organised according to the bio-production, bio-process, and bioenvironment engineering streams:

- Biomass production, controlled environments, energy, greenhouse technology
- Sensor systems, remote sensing, geospatial data, precision agriculture
- Sensing technologies, computational methods for smart agriculture
- Advanced, smart, biological, and porous composite materials
- Bio-based polymers in food processing, separation methods for biomaterials production, bioactive compounds, value-addition of agricultural and industrial wastes
- Food process engineering, food properties, heat and mass transfer processes
- Food processing and product quality, extraction of high-value compounds, recovery and valuation of by-products and residues, and production of functional ingredients
- Postharvest technology to ensure the security and safety of food, thermal processing and drying
- Decision support systems for healthy and sustainable food choices based on environmental and nutritional considerations, lifecycle sustainability assessment, and economically circular food systems
- Environmental impacts of cities, bioresource production systems and supply chains connected to cities
- Ecological engineering, circular agricultural economy, organic residuals management, green infrastructure
- Integrated water resources management, participatory systems modelling, the application of artificial intelligence in hydrology
- Beneficial Management Practices to reduce nonpoint source agricultural pollution, green technologies for treating wastewater, flow of water and chemicals in soils
- Water conservation, water pollution, greenhouse gas emissions in agriculture
- Modelling of crops, environment, and field hydrology to predict and manage water quality, engineering of irrigation and drainage systems, and adaptation of those systems to climate change

Programme Structure

Freshman Programme

The department offers a Foundation Year programme to bring students from outside of Quebec to the same academic level as entrants from the provincial system of junior colleges, the *collèges d'enseignement général et professionnel* (CEGEP) (McGill University, 2022a, 2022b).

Bachelor of Engineering – Major Bioresource Engineering

Students in the Major in Bioresource Engineering learn to apply professional engineering skills to create biologically derived products such as food, fibre, fuel, and advanced biomaterials. They are trained to develop systems that sustain and regenerate the health of the ecosystems that provide those services. Graduates of the programme can apply for registration as professional engineers in any Canadian jurisdiction and in some jurisdictions in other countries.

Completion of the Major programme requires seven full-time academic semesters of 13 weeks each; A full-time load is five courses per semester. Embellishments to the programme, such as experiential learning opportunities, Minor programmes, etc., often extend the completion time.

The core course set of the Bioresource Engineering Programme includes: mathematics, mechanical analysis and design, materials, computing, thermodynamics, fluid mechanics, electrical circuits and machines, environmental engineering, sustainability, communications (oral, written, and visual), and instrumentation and control. There is one semester of common courses in the Faculty of Engineering on McGill's main campus which integrates students with their broader peer cohort from other disciplines, encouraging a collaborative, multidisciplinary perspective on the profession. This common semester includes the courses: Responsibilities of the Professional Engineer, Engineering Economy, Engineering Professional Practice, and Design Graphics.

Beyond fundamental engineering principles there are four sets of complementary courses that allow students to adapt the programme to their interests.

- **Complementary Set A** includes one course each of statistics and heat (and mass) transfer.
- **Complementary Set B** (Science and Mathematics) comprises three courses, at least one of which must be related to biological sciences, but the remaining two may be related to other science disciplines.
- **Complementary Set C** (Social Sciences) includes one course about the impact of engineering on society and the environment. It contains another two courses chosen from the social sciences, management studies, humanities, or law.
- **Complementary Set D** (Engineering Design) comprises eleven courses in engineering science and design. Bioresource Engineering is a broad-based discipline, so students may study foundational material of other disciplines depending on their choice of courses. Students have the option of taking up to two courses in Set D from any engineering programme, provided that the chosen courses are taught by a registered engineer and include engineering design and engineering science content.

The engineering curriculum begins with a seminal design course and culminates in a pair of capstone design courses that constitute "a significant design experience conducted under the professional responsibility of a licensed engineer", as required by Engineers Canada (2019). Many courses throughout the Bioresource Engineering Programme include individual and group design work, but three courses are intended explicitly to develop project design skills. The first of these is the cornerstone course Engineering Design I, which is taken at the beginning of the programme. The culmination of the programme is the extensive design project undertaken in the capstone courses Engineering Design II and III, which are taken in sequence during the final two semesters of the programme. In the capstone project, students are expected to perform investigation, research, analysis, and synthesis leading to the design of an original solution to a specific engineering problem that they have defined well and for which they have explored the available possible solutions. The students are mentored by professors in the department to ensure that their project is challenging and involves appropriate use of mathematics, science, design, and the use of engineering tools. Most teams secure real-world clients either on or off-campus who are connected with their project topics, support their efforts, and may in some cases make use of their completed design.

As the Bioresource Engineering Programme changed and broadened, three optional streams were developed to help students focus on their personal interests within the Bioresource Engineering Major. Students may concentrate on of these areas of study by choosing among the complementary courses recommended for the stream of interest.

Bio-Production Engineering Stream

The Bio-Production Engineering stream explores how science and technology can be used in the production of agricultural crops, livestock, and other forms of biomass. Students study the design of machines, robotics, precision agriculture, sensor systems, artificial intelligence and smart production systems, geomatics and geographical information systems, buildings and structures, irrigation and drainage, etc.

Bio-Process Engineering Stream

Students following the Bio-Process Engineering Stream apply engineering principles to the transformation of biological commodities into consumer products such as food, fibre, fuel, and chemicals. This stream includes topics such as: food process engineering; quality, safety, and security of food supply; biological materials and their physical properties; postharvest processing; fermentation; biotechnology; design of bioprocessing machinery; biologically derived chemicals; biologically inspired multifunctional metamaterials; etc.

Bio-Environmental Engineering Stream

The Bio-Environmental Engineering stream helps students to learn how to responsibly manage the impact of their professional activities on ecosystems and natural resources. Some topics emphasised in this stream are: soil and water quality management, organic waste treatment, urban and rural ecology, engineering for sustainability, supply chain management, ecological engineering and nature-based solutions, climate change adaptation, and water engineering.

Bachelor of Engineering – Major Bioresource Engineering – Professional Agrology

Graduate who follow the Professional Agrology Option may apply for registration in the *Ordre des Agronomes du Québec* (Quebec Order of Agronomists), as well as in professional engineering orders and associations. The programme complies with the requirements of two accreditation bodies, making the course selection more prescribed than that of the general major.

Bachelor of Engineering – Honours Bioresource Engineering

The Honours Programme requires that students conduct independent research courses in in consultation with the Programme Director and a supervising professor.

Minor Programmes

While working towards a Bioresource Engineering Major, students can also complete the requirements for an additional minor programme. This allows students to develop an expertise in another area of study. Completion of a minor requires about one additional term of studies. A few popular minors include:

- Agribusiness Entrepreneurship
- Agricultural Production
- Biomedical Engineering
- Biotechnology
- Computer Science
- Construction Engineering and Management
- Environmental Engineering
- Technological Entrepreneurship

Experiential Learning

Workshop Facilities

The department cultivates a hands-on approach to engineering education. The students have access to a large, multi-purpose machine shop with lathes, drill presses, bench grinders, milling machines, automatic saws, and arc and gas welders. Students learn to safely use the machines as part of the laboratory training in several courses. Student groups involved in annual quarter-scale tractor design and robotic design competitions also use the shop regularly. Students can often participate in laboratory research projects and use research equipment as part of their coursework. For instance, in the course Biological Material Properties students use a variety of analytical tools such as an Instron Universal Testing Machine, a Fourier-transform infrared spectrometer, and a rheometer to quantify material properties. In Postharvest Drying and Storage, students learn to use state-of-the art microwave and fluidised-bed drying apparatus.

Field Trips

Field trips give the students the opportunity to see current methods, protocols, practices, and equipment in industry. For example, the Fluid Mechanics class visits consulting engineering firms which develop physical scale models and computational models of large-scale hydroelectric installations. The Organic Waste Management class visits local wastewater treatment and municipal composting facilities. The Agri-Food Buildings class tours agricultural building structures to understand the design and construction methods used in these buildings. In Water Quality Management, students visit agricultural operations that use constructed wetlands to mitigate nonpoint source pollution and best management practices for soil and water conservation.

Internships and Study Abroad

Undergraduate students can access the internship programme of the Faculty of Agricultural and Environmental Sciences to complete one or two 10-week internships prior to graduation. Students receive recognition for the internships as either credited or zero-credit courses on their academic transcripts.

Several field study semesters are administered by McGill, and students that participate in them gain course credits in the same way as if they had completed courses taken on a McGill campus.

Bioresource Engineering students frequently study at universities abroad for one or more semesters as part of an exchange programme.

Student Clubs

The department supports several student clubs and views these activities as a valuable part of the learning experience. The quarter-scale tractor design team, Mutrac (μTrac), has participated every year since 2012 in the annual international competition of the American Society of Agricultural and Biological Engineers (ASABE). The competition allows the students to gain practical experience in machinery design, machine performance, manufacturing, performance analysis, and material properties.

The departmental robotics team competes in the annual ASABE robotics competition, an event in which student teams use robotics to create new solutions to practical problems in agriculture.

Many students use 3D printing technology to complete their projects, reducing the cost of rapid prototyping by up to 90% in some cases. The department has several 3D printers that can produce parts with different levels of design complexity. A design club helps all the design enthusiasts within the department and from other programmes across Faculty of Agricultural and Environmental Sciences.

Professional Accreditation

Canadian Engineering Accreditation Board

Engineers Canada is a national organisation which represents the twelve associations responsible for regulating the licensure and practice of engineering in the provinces and territories (Engineers Canada, 2019). The Canadian Engineering Accreditation Board is a standing committee of that organisation, responsible for the accreditation of all Canadian undergraduate engineering programmes to ensure that each meets acceptable educational standards and that their graduates are therefore eligible to apply for professional registration. Accreditation visits are undertaken at the invitation of a university. A team of experienced engineers is led by a Board member. The university prepares a detailed report for the team before their visit. The team the performs a comprehensive assessment of students, instructors, support staff, curriculum, and facilities. The visiting team conducts a detailed qualitative and quantitative evaluation of the curriculum to ensure it exceeds minimum criteria. The team reports their findings to the Accreditation Board who may grant or renew accreditation of the programme for up to six years.

Accreditation Units

The curriculum content is evaluated to ensure an appropriate academic foundation for every graduate from the programme. This evaluation approach is reflective of the more traditional "teacher-based" model of education and is intended to ensure that all students are exposed to the minimum required educational material. A "minimum path" analysis of the curriculum demonstrates that every graduating student has taken courses that meet all the content and quality criteria, regardless of what allowable combination of courses the student took. This analysis is based on Accreditation Units defined for academic activities (e.g. lectures or labs) based on the hours of contact time between instructors and students. Criteria include a specified minimum number of Accreditation Units in each of these categories: mathematics, natural sciences, engineering science, engineering design, complementary studies, laboratory experience, and safety procedures. Additional rules provide more detailed guidance on the nature of the required content, professional credentials of the instructors, etc.

Graduate Student Attributes

In addition to tracking Academic Units to ensure the quality of the curriculum content, the institution must document competency of the graduating students. This newer approach is rooted in a "learning-based" educational model that requires the department to evaluate how students have improved their capabilities in twelve specific categories. These *graduate student attributes*, defined in more depth by Engineers Canada (2021), include:

- **Knowledge base for engineering**: Mathematics, science, and basic engineering knowledge
- **Problem analysis**: Formulate and solve complex problems
- **Investigation**: Arrive at conclusions through experimentation and interpretation of data
- **Design**: Using a disciplined approach to create devices or systems that meet specific goals
- **Engineering tools**: Develop and apply appropriate methods and technology in their work
- **Individual and teamwork**: Work as effective members of multi-disciplinary teams
- **Communication**: Read, write, speak, listen, and otherwise clearly communicate engineering information
- **Professionalism**: Demonstrate responsible comportment to protect the public and the interests of society
- **Impact of engineering on society and the environment**: Act with consideration of the environmental, economic, health, safety, legal, and cultural consequences of their work
- **Ethics and equity**
- **Economics and project management**: Employ sounds fiscal and management practices
- **Life-long learning**: Intentionally strive to remain competent in their professional capacity

Graduate Attribute Indicators

The department requires an organisational structure to measure the graduate attributes and thereby enable the continuous evaluation and improvement of the programme (Engineers Canada, 2019). Documentation of this structure includes a curriculum map that shows how learning activities (e.g. courses) relate to each attribute, at what level those attributes are evaluated (introduced, developed, or applied), and the point in the programme at which those evaluations are done. There must also be an assessment plan for all attributes. There must be measurable, documented indicators used to measure students' competency in each attribute. These graduate attribute indicators must be documented for each attribute, and collectively summarised for the entire accreditation cycle to describe student learning in all twelve categories. Examples of indicators include assignments, exams, and other graded course evaluation tools. However, indicators may also be assessments or surveys, for example, which do not count towards a course grade. A system of graduate attribute indicators was developed by McGill's Faculty of Engineering and has since been adapted by other universities (Tables 8.1a and b).

The results of this data collection and analysis must clearly demonstrate that graduates of the programme are competent with respect to all attributes. The data should be evaluated by individual instructors, the curriculum committee, and the department to help drive continuous evaluation and improvement at the level of the course, curriculum, and programme, respectively.

Postgraduate Programmes

Students who want a career in research have a variety of postgraduate programmes to choose from within the Department

TABLE 8.1A

Graduate Attribute Indicators. (Adapted with permission from the McGill Faculty of Engineering, n.d.)

Knowledge base for engineering	
KB.1	Recalls and defines information and concepts in mathematics
KB.2	Comprehends information and applies concepts in mathematics
KB.3	Recalls and defines information, first principles and concepts in the natural sciences
KB.4	Comprehends information and applies concepts in the natural sciences
KB.5	Recalls and defines information, first principles and concepts in fundamental engineering science
KB.6	Comprehends information and applies concepts in fundamental engineering science
KB.7	Recalls and defines information, first principles and concepts in specialized engineering science
KB.8	Comprehends information and applies concepts in specialized engineering science
Problem analysis	
PA.1	Identifies and formulates complex engineering problems
PA.2	Develops models from first principles to analyze complex engineering problems
PA.3	Analyzes and solves complex engineering problems
PA.4	Critically evaluates the validity and accuracy of solutions
Investigation	
IN.1	Conducts planned activities (literature review, experiments, measurements, laboratories, etc.) and analyzes data
IN.2	Interprets results and reaches valid conclusions regarding complex engineering problems
IN.3	Formulates hypotheses and designs suitable investigative approaches and/or research methodologies
IN.4	Understands and/or demonstrates appropriate safety protocols
Design	
DE.1	Understands the problem (open-ended complex engineering problem) and defines objectives and constraints
DE.2	Develops a design process considering health and safety risks, applicable standards, economic, environmental, cultural and societal considerations.
DE.3	Researches and develops possible solutions to a complex engineering problem and recommends a final design
DE.4	Implements and evaluates a final design
Use of engineering tools	
ET.1	Selects and uses tools
ET.2	Evaluates tools and identifies their limitations
ET.3	Adapts, integrates and/or creates tools

of Bioresource Engineering, and a broad range of possible topics that mirror the research areas of the academic staff (McGill University, 2023c). Students in these programmes benefit from the varied, multidisciplinary research environment in the Faculty of Agricultural and Environmental Sciences and the close association of the department with the Faculty of Engineering. The different programmes include:

Master of Science in Bioresource Engineering (Thesis)

The Master of Science degree helps students prepare for a research career. Thesis programmes revolve around a large, individual research project resulting in a written thesis and usually published articles in peer-reviewed scientific journals.

Master of Science in Bioresource Engineering (Thesis): Environment

The Environmental Option is offered jointly with the Bieler School of Environment. Students in this programme conduct interdisciplinary research combining engineering and environmental science, politics, economics, and ethics.

Master of Science in Bioresource Engineering–Applied (Non-Thesis)

Non-thesis programmes are course-based. Students usually are already employed in industry or government and desire skills in a specific area to improve their engineering qualifications. Students must be eligible to apply for registration as a professional engineer in Canada.

Master of Science in Bioresource Engineering (Non-Thesis): Integrated Water Resources Management

The Integrated Water Resource Management programme is based on course work and an internship placement, all completed within one academic year. The programme is designed for professionals who seek advanced training about water resource management related to the physical sciences, environment, law, institutions, and socio-economics.

Master of Science in Bioresource Engineering–Applied (Non-Thesis): Environmental Engineering

The programme in Environmental Engineering offers a choice among many courses in collaborating departments. Students

TABLE 8.1B

Graduate Attribute Indicators. (Adapted with permission from the McGill Faculty of Engineering, n.d.)

Individual and teamwork	
IT.1	Participates actively in a team
IT.2	Shares workload
IT.3	Displays good interpersonal skills
IT.4	Develops leadership skills
Communication skills	
CS.1	Understands, interprets and/or assesses oral, written, graphical or visual communications
CS.2	Produces written engineering reports and design documentation
CS.3	Demonstrates competency in the oral communication of complex engineering concepts
CS.4	Demonstrates an ability to give and/or effectively respond to clear instructions
Professionalism	
PR.1	Understands the role of the engineering profession in society
PR.2	Understands the responsibility of professional engineers in protection of the public and its interest
PR.3	Knows pertinent codes, laws and regulations related to the engineering profession
PR.1	Understands the role of the engineering profession in society
Impact of engineering on society and the environment	
IE.1	Understands the social, environmental, economic, health, safety, legal and/or cultural aspects of engineering activities
IE.2	Understands and/or is able to analyze the uncertainties in the prediction of interactions between the different aspects of engineering activities
IE.3	Understands and/or conducts social and/or environmental impact analyses
IE.4	Understands and/or applies the concepts of environmental stewardship, sustainable design and sustainable development
Ethics and equity	
EE.1	Appreciates and articulates issues and dilemmas related to equity
EE.2	Demonstrates knowledge of ethical standards (i.e. Code of Ethics)
EE.3	Understands and/or resolves ethical issues
EE.4	Demonstrates individual accountability
Economics and project management	
EP.1	Understands economic concepts in an engineering context
EP.2	Understands the project management life cycle and its limitations
EP.3	Applies business tools and economics principles in managing engineering projects (or the engineering practice)
Life-long learning	
LL.1	Sets goals
LL.2	Applies appropriate knowledge and skills to learning activities
LL.3	Engages in self-direction and self-evaluation
LL.4	Locates required information
LL.5	Adapts learning strategies to new conditions

are trained as advanced environmental professionals and a university undergraduate degree in engineering is therefore a prerequisite.

Master of Science in Bioresource Engineering–Applied (Non-Thesis): Integrated Food and Bioprocessing

Students learn about the interactions between agriculture and biological products to better design systems for production of food, animal feed, fibre, biofuels, etc. Students learn to assess and improve production systems and to work well in multi-disciplinary teams.

Doctor of Philosophy in Bioresource Engineering

Doctoral degrees are for those who want a career as a research leader in academia, industry, or government.

Doctor of Philosophy in Bioresource Engineering: Environment

The Environment Option, as with the corresponding Master's degree, is offered jointly with the Bieler School of Environment, intended for students interested in environmental issues at the intersection of science, politics, economics, and ethics.

Career Paths

The employment market for graduates of the undergraduate and postgraduate programmes in Bioresource Engineering has been steadily growing. Most students secure good jobs shortly after graduation. Many choose to work in traditional sectors such as production and processing of food, agricultural machinery design, and management of water resources.

However, increasingly strident calls by consumers and regulators for good corporate stewardship of the environment, adoption of nature-based solutions, and use of bio-based materials, for example, have fuelled the creation of employment options in other sectors as well. Graduates find employment in the petrochemical industry, software engineering, construction machinery, railways, automotives, aerospace, electronics, mining, renewable energy, urban infrastructure, waste management, transportation infrastructure, additive manufacturing, and many other fields. They may work for industry, government, non-government organisations, or not-for-profit agencies. Additionally, graduates may choose careers as educators at any level, or translate their skills to other professions including international development, management, medicine, and more.

Current and Future Trends

The Bioresource Engineering Programme must train the engineers needed by society to meet the challenges of today and tomorrow. The department plans, therefore, to maintain key strengths in traditional areas of agricultural, food, and soil and water engineering, while building new capacity in bio-production, bio-process, and bio-environment. Technological trends that are already changing the profession include the rise of remote sensing, precision agriculture, controlled plant production, advanced multifunctional materials, and additive manufacturing.

As we look to the not-too-distant future, we can expect that technology will continue to become more available, less expensive, and increasingly powerful. Dramatic technological changes will include the exponential improvement of artificial intelligence and robotic automation, continued growth of big data, the rise of cloud and edge computing on the back of more powerful and pervasive communications networks, and the eventual advent of operational quantum computing. Changing energy profiles will see the dominance of fossil energy gradually give way to renewable and perhaps to fusion power and force us to confront the terrible existential risks that could accompany a glut of even cheaper and more abundant energy. Expertise in biological technologies such as gene editing and synthetic biology will improve to the point that they are much more routine and effective than they are now.

Social and demographic trends will challenge us with ensuring that our growing, increasingly affluent, resource-hungry population has access to adequate water, food, energy, and bioresources. As the inherited hegemony of European economic colonialism continues to erode, dramatic cultural shifts could fracture global economic systems and allow their reorganisation according to different values systems; this could make room for more diverse modes of production and distribution.

Ecological imperatives already include adaptation to inevitable, accelerating, and uncertain climate change, biodiversity collapse and the accompanying decline in genetic resources, and shrinking stocks of natural capital such as soil, fresh water, forests, and fisheries. Engineers will be increasingly obliged to prioritise climate change adaptation and mitigation, responsible use and regeneration of resources, organic waste management, urban agriculture, and circular economics.

REFERENCES

Association des ingénieurs en agroalimentaire du Québec. (n.d.). *History of Agri-Food Engineering.* https://www.aiaq.qc.ca/en/aiaq-en/agri-food-engineering/history

Engineers Canada. (2019). *Manual of accreditation procedures* (Updated October 2019). https://engineerscanada.ca/sites/default/files/2022-11/Manual_of_Accreditation_Procedures%20%28Accreditation%20Visit%20Cycle%202023-2024%29.pdf

Engineers Canada. (2021). *2021 Accreditation Criteria and Procedures* (Revised October 2021). https://engineerscanada.ca/sites/default/files/2021-11/2021%20Accreditation%20Criteria%20Book%20Word.pdf

Frost, S.B. (1980). *McGill University, Vol. I. For the Advancement of Learning, 1801–1895.* McGill-Queen's University Press.

McGill Faculty of Engineering. (n.d.). *Graduate Attributes and Indicators List.* https://www.mcgill.ca/engineering/files/engineering/graduate_attributes_def_and_indicators_1.pdf

McGill University. (n.d.). Founding Macdonald College. https://www.mcgill.ca/about/history/features/macdonald-college

McGill University. (2022a). *eCalendar–Agricultural & Environmental Sciences Undergraduate Programs.* https://www.mcgill.ca/study/2022-2023/faculties/macdonald/undergraduate/programs

McGill University. (2022b). *eCalendar–Agricultural & Environmental Sciences Graduate Programs.* https://www.mcgill.ca/study/2022-2023/faculties/macdonald/graduate/programs

United Nations. (2022). *World Population Prospects 2022: Ten Key Messages.* United Nations, Department of Economic and Social Affairs, Population Division. https://www.un.org/development/desa/pd/sites/www.un.org.development.desa.pd/files/undesa_pd_2022_wpp_key-messages.pdf

9

Designing Agricultural and Biosystems Engineering Education to Transform Agricultural and Food Systems in sub-Saharan Africa: Insights from the University of Zimbabwe

Emmanuel Manzungu and Brighton M. Mvumi

Introduction

Zimbabwe's economy, like that of many other sub-Saharan African countries, is largely agriculture-dependent. The contribution of the agriculture sector to Zimbabwe's economy has not changed much since independence in 1980. Agriculture contributes 15–18 percent to Gross Domestic Product (GDP), 30 percent to export earnings, 25 percent to total formal employment, accounts for 63 percent of raw materials for agro-based industries, and is the main source of livelihood for 67 percent of the population (Government of Zimbabwe [GoZ], 2018a). There are, however, a number of challenges, which limit the sector from reaching its full potential. The challenges, which are briefly described in Section "Context for Agricultural and Biosystems Engineering Education in Zimbabwe", are a combination of 1) frequent droughts because of a semi-arid climate, 2) a structurally complex agricultural sector due to 90 years of colonialism (1890–1980), a legacy that the post-colonial state has not managed to address adequately (Rukuni, 2006), and 3) poor economic performance that has persisted for half of the country's forty-year post-colonial period.

Since 2000, Zimbabwe's economic performance has been characterised by low growth, resulting in a net decline of 2.8 percent per year on average per capita Gross Domestic Product (GDP) (Sharma et al., 2022). This caused job losses, especially in the formal sector, high informalisation of the economy, poor quality jobs, low incomes, a sharp increase in poverty and emigration of both skilled and non-skilled personnel to regional and international destinations (Sharma et al., 2022). Some estimates put the labour emigration figure at 4 million (Garwe & Thondhlana, 2019).

Agricultural and economic challenges are more pronounced in the smallholder farming sub-sector, which accounts for 80 percent of the total agricultural land, and provides a source of livelihood for over two million households, equivalent to about 10 million people (GoZ, 2018a). This largely subsistence-oriented sub-sector is characterised by low agricultural productivity due to, among other influences, exposure to climate risk, poor soil fertility worsened by low input use, high levels of drudgery because of limited mechanisation, high postharvest losses, and poorly-developed value chains (Rukuni, 2006). These challenges translate to, among others, frequent food and income insecurity for about four million people annually (Zimbabwe Vulnerability Assessment Committee [ZIMVAC] 2019; 2020; 2021). Poverty is more concentrated in the smallholder farming sub-sector as is the case for rural areas in general. More than 90 percent of the poor living in extreme poverty are found in these rural areas (Sharma et al., 2022). The degree to which the challenges facing the smallholder farming sub-sector are successfully addressed as well as those of the entire agriculture sector, is critical to transforming Zimbabwe's agricultural and food systems in particular and the general socio-economic development.

This chapter describes attempts by the University of Zimbabwe (UZ) to address the challenges in the agricultural sector through, among other means, designing relevant and appropriate agricultural and biosystems engineering (ABE) education. UZ is a public university which opened in 1952 as the University College of Rhodesia and Nyasaland making it the oldest university in the country. Initially it was affiliated with the University of London and was later renamed the University of Rhodesia, and became the University of Zimbabwe at independence in 1980, which explains the persistence of some aspects of British education, such as passes in two Advanced (A) Level subjects as entry requirements for university education. The University has eleven faculties (Agriculture, Environment and Food Systems; Arts and Humanities; Business Management Sciences and Economics; Computer Engineering, Informatics, and Communication; Education; Engineering and The Built Environment; Law; Medicine and Health Sciences; Science; Social and Behavioural Sciences; and Veterinary Science) and six specialist research centres and institutes (Centre for Computer Technology Training Development; Centre for Teacher Education and Materials Development; Environment, Climate and Sustainable Development Institute; Institute of Mining Research; Multi-Cultural Language Centre; and Centre for Postgraduate Studies). By virtue of it being the country's oldest university, UZ degree programmes have been directly and indirectly reproduced by other universities in the country. This is also true with respect to the agricultural (and

DOI: 10.1201/9780429150111-11

biosystems) engineering degree programmes, which are the subject of discussion in this chapter.

Agricultural engineering training was introduced at UZ and in the country in 1992, in the form of the Bachelor of Science Honours Degree in Agricultural Engineering. Since then, the curriculum has been reviewed to take account of socio-economic developmental challenges in the Zimbabwean society, and emerging knowledge in the field of agricultural and biosystems engineering. Agricultural (and biosystems) engineering training at UZ has undergone three major phases. The first phase, as already noted, was the introduction of the undergraduate agricultural engineering degree programme. The second phase, coming close to a quarter of the century in 2016, focused on infusing biosystems and environmental engineering aspects into undergraduate and post-graduate degree programmes in addition to agricultural mechanisation, food process engineering, and water and irrigation engineering components. The third phase, introduced in 2019, represented a major shift towards more specialist undergraduate and graduate degree programmes, informed in part, by the quest towards developing a knowledge society as described later. A knowledge society is a society which is economically and culturally characterised by a high degree of dependency on its potential to create scientific and technological knowledge (International Encyclopedia of the Social Sciences, 2022). Universities in Zimbabwe, Africa and world over are regarded as critical actors in creating a knowledge society.

The three phases of ABE education and training at UZ are discussed in this chapter in terms of the objectives, content, and accreditation of the degree programmes. By tracing the evolution of agricultural and biosystems education and training in Zimbabwe, the chapter aims to contribute to an understanding of the extent to which the university training in ABE has served and is not serving a complex agriculture sector that needs to be transformed to contribute to better national development. More specifically, we pose the following questions:

1) To what extent should university curricula be determined by national development objectives and plans?
2) Between comprehensive or special degree programmes, which best serves developing societies like that of Zimbabwe?
3) To what extent should national degree programmes take into account regional and global trends?
4) Should training focus on producing graduates who are employable, self-employed, or who create employment?

The discussion of curriculum changes is preceded by a description of the context in which the changes were and are situated. This is followed by a description of the scope and focus of the Bachelor of Science Honours degree in Agricultural Engineering introduced in 1992. This is followed by a description of the second phase (Section 3) that focused on infusing biosystems aspects into the Bachelor of Science Honours degree in Agricultural and Biosystems Engineering and three Master of Science degree programmes. Thereafter is a description of the third phase that introduced four undergraduate and four (two revised and two new) Masters specialist degree programmes. This is followed by a discussion of the findings in penultimate section and conclusions in last section.

Context for Agricultural and Biosystems Engineering Education in Zimbabwe

Introduction

Any and every degree programme is and should not be designed in a social, political, or economic vacuum. With this understanding, this section provides the context in which ABE education in Zimbabwe has been designed and implemented. The context covers the structure of the agriculture sector, made up of farming sub-sectors with different challenges, and changes in university education in the country and how it has shaped and continues to shape ABE engineering education.

Overview of the Agriculture Sector

At independence in 1980, the agriculture sector in Zimbabwe was made up of four main farming sub-sectors. These include White large scale, Black small scale commercial, Black communal, and the state sub-sector (Table 9.1). As can be seen from Table 9.1, while the White large scale sub-sector accounted for less than 1 percent of the country's population, it accounted for the best agricultural land in terms of production potential (i.e., agro-ecological regions (AER) I and II), per capita land ownership, average farm size, and irrigated area due to good access to water and irrigation infrastructure. The sub-sector was well mechanised. This was in no small part due to favourable land tenure, in the form of title deeds, which allowed farmers to access cheap short, medium, and long term loans. The sub-sector was also heavily subsidised by various colonial administrations and continued to enjoy more or less the same benefits for the greater part, of two decades after independence (GoZ, 2018a). By contrast Black farmers in small scale commercial farming areas and in communal lands faced severe challenges. Agricultural production was conducted on rainfall marginal areas, agriculturally inferior lands, and on small landholdings with limited mechanisation and irrigation infrastructure. Lack of title deeds meant that farmers could not access agricultural finance (Rukuni, 2006).

Since independence in 1980 the structure of the agriculture sector has significantly changed because of the 1980-1998 and fast track (post 2000) land reform programmes aimed at making more and better agricultural land available to the majority Black population (Table 9.2). In addition to black small scale commercial, communal, and state sub-sectors, which remained largely the same, old resettlement areas and A1 and A2 farms emerged out of the land reform process. Old resettlement farms, created between 1980 and 1998, are similar to communal areas and A1 farms, created after the fast-track land reform. Farming is done on comparable land sizes and is accessed through a permit system that provides land for settlement, cropping and grazing. Commercial agriculture is now undertaken on A2 farms, also created after the 2000 fast track land reform programme. As can be seen in

TABLE 9.1

Characteristics of different farm categories in Zimbabwe at independence in 1980. Source: Utete (2003), Moyo (2006), Manzungu et al. (2018).

Land category	Land size		Farm characteristics		Demographic characteristics		Agricultural potential		Irrigation characteristics		
	Ha (million)	% of total land	No.	Average size (ha)	Size (000)	Population density (persons per sq.km)	% in high potential (AER I and II)	% in low potential (AER IV and V)	Size (000ha)	% total irrigated area	Average size (ha)
White large scale	15.5	39.1	4832	2223	1160	10.8	35	43	126	84	40
Black small scale commercial	1.4	3.5	8500	162	166	12.0	19	46	3.6	2	No information
Black communal	16.4	41.4	1000,000	18	5327	32.6	9	74	7.2	5	30*
State	0.3	0.8	55	7644	38	9.0	4	64	13.5	9	4000
National parks and urban land	6.0	15.2	-	-	-	-	-	-	-	-	

*This refers to the average size of the scheme. Average plotholding can be as low as 0.1 ha. AER = agro-ecological region.

TABLE 9.2

Characteristics of farm categories in Zimbabwe before and after the 2000 fast-track land reform programme. Source: Utete (2003), Moyo (2006), Manzungu et al. (2018).

	Before 2000				After 2000			
Land category	Area (million ha)	% total	No of households/ farms	Average farm size (ha)	Area (million ha)	% total	No of households/ farms	Average farm size (ha)
Communal	16.4	42	1, 000, 000	0.6-3.5	16.4	42	1, 000, 000	0.6-3.5
Small scale	1.4	4	8500	125	1.4	4	8500	125
Large scale commercial	11.8	30	4832	2223	3.4	9	N/A	N/A
Old Resettlement	3.7	9	70, 000	12	3.5	9	70, 000	12
A1	-	-	-	-	4.1	11	128, 000	5-6
A2	-	-	-	-	3.5	9	8000	20-120
State and private commercial estates	1.6	1		1500	0.7	2	N/A	4000
Parks/forests and Urban	5.0	15	-	-	5.3	13	-	-
Total	39.6	100	-	-	39.6	100	-	-

N/A Not applicable

Table 9.2, they are much smaller in size compared to the old white commercial farms. How best to mechanise the new smaller commercial farms that do not match the conventional tractors, farm machinery and equipment is a major challenge. The other challenge is that the 2013 Constitution nationalised all agricultural land (GoZ, 2013) such that farmers cannot access private finance. Consequently, practically all farmers in the country now depend on the state for financing agriculture. Apart from the fact that this is a huge cost to the fiscus, there are also challenges relating to efficiency and effectiveness of the financing.

The demographic features in the smallholder farming sub-sectors are both an advantage and disadvantage. Approximately 67 per cent of the country's 16 million total population is under the age of 35 (Zimbabwe National Statistics Agency [ZIMSTAT], 2022). Potentially this provides a lot of labour force. However, the youths are not active in agriculture for a variety of reasons including, among others, drudgery and low incomes. Women constitute about 70 percent of the agricultural labour force (GoZ, 2018a). Unfortunately, they experience a high labour burden due to productive and reproductive roles, because of limited access to adequate and appropriate agricultural mechanisation and irrigation infrastructure.

Agricultural Engineering Issues and Challenges

The discussion of structural challenges facing agricultural production for the different farming categories, needs to be complemented by an appreciation of some of the important ABE engineering issues. Table 9.3 shows the challenges that exist across the board including: Agricultural Mechanisation and Farm Machinery; Soil and Water Conservation; Dams; Irrigation; Postharvest management, agro-processing, and food safety; and Cross-cutting issues. The challenges incorporate technological challenges (in relation to absence or limited), management and managerial issues, and knowledge gaps. These challenges have hardly changed since independence in 1980 and are common to sub-Saharan Africa (Kalibata, 2017).

The agricultural engineering-related challenges are more pronounced in both the old smallholder (communal and old resettlement areas) and the new (A1) farming sub-sectors because of the lack of appropriate machinery and equipment for the land size, affordability challenges, and limited mechanisation extension services.

Overview of University Education in Zimbabwe

Historical Note

University education in Zimbabwe, like other sectors, was affected by its colonial past. The historical account that is given hereunder is drawn from a paper by Garwe and Thondhlana (2019). Before independence, the University of Zimbabwe (then University of Rhodesia) was the only university in the country and was also the only multi-racial educational institution in the country. The Black student population was, however, in the minority. For example, in 1961 Black students accounted for a mere 0.2 percent of the student population. At independence in 1980 student enrolment in university was 2,240, rose up to 9,017 in 1991 and to 49,645 in 2009, and currently practically all students in the country's 24 universities (14 public and 10 private), numbering about 100,000, are Black. The figure, however, represents 8 percent of eligible candidates who pass at least two A level subjects, the minimum entry qualification to university. Other routes to university education other than A level passes, are provided under the Zimbabwe National Qualifications Framework (ZNQF) such as holders of technical and vocational diplomas and mature entry, which values work experience (GoZ, 2018b). Due to high requirements for Mathematics and Physics, only students with A level passes in both are admitted into ABE degree programmes.

TABLE 9.3

Overview of agricultural and biosystems engineering challenges and issues in Zimbabwe.

Focal Area	Challenges/Issues
Agricultural Mechanisation and Farm Machinery	**General** Few manufacturers of agricultural machinery and implements, old and low numbers of tractors, machinery and equipment; Poorly sized tractors and machinery for majority of farm sizes; Limited knowledge to enhance the production and processing capabilities of various farmers with different land holdings and varied enterprises; Non-strategic importation of machinery and equipment with no/little back-up service; Limited and diminishing local manufacture and assembly of tractors, machinery and equipment; Limited or diminished research and development and support to develop machinery and equipment that reflect current farming types; Inadequate testing and certification and local standards for tractors, machinery and equipment; Limited postharvest knowledge and processing ability to increase value addition; Low appreciation of food quality and safety for local consumption and export; Inadequate training of farmers and extension agents **Smallholder farmers** Over-reliance on limited human and animal power; Limited appropriate machinery and equipment; Limited scalable and sustainable alternatives
Soil and Water Conservation	Poor agricultural practices resulting in accelerated run-off leading to soil erosion/land degradation, sedimentation and siltation in rivers and reservoirs worsened by illegal or artisanal mining and reduced groundwater recharge; Limited investment in appropriate soil and water conservation works; Limited/poor operation and maintenance; Limited rainwater harvesting at field and catchment level; Polluted water bodies from agriculture and mining
Dams	**Large dams** Poor operation and maintenance; Lack of adherence to stipulated inspection requirements **Small dams** Poor understanding, design and implementation of community multiple uses of water use; Siltation because of poor land use practices in the catchment; Poor operation and maintenance
Irrigation	Small proportion of irrigated lands (only 5 percent of the arable land is irrigated); High irrigation costs; Limited sustainable energy options for irrigation; Poor water management practices illustrated by low irrigation efficiencies; Inappropriate irrigation technology in particular for smallholder farmers; Irrigation planning, design and management does not take climate change into account; Limited adoption and implementation of climate smart technologies; Limited capacity development for irrigation management
Postharvest Management, Agro-processing and Food Safety	Limited primary, secondary, tertiary processing and value addition; Limited knowledge and education among smallholder farmers and extension officers on postharvest related issues; High levels of postharvest losses across the value chain for both crop and animal produce and products; Limited technologies to reduce postharvest losses; Poor packaging materials and knowledge resulting in poor food quality and compromised food safety standards; Low investment, development, testing and standards development for local consumption and export; Inadequate training of farmers and extension agents
Energy	Inadequate energy at farm and industry level; Limited energy choices; Low use of renewable energy; Poor energy management
Cross-cutting	
Waste management	Low waste-to-energy solutions at farm level; no product development from waste
Decision-support systems	Poor and inappropriate decision-making for the challenges and situations at hand; Low adoption of smart technologies such as GIS, modelling and testing
Gender	No appropriate technologies and measures for women (who are the majority); No access to or ownership of productive assets
Youth	Manual agriculture not attractive; Low income; Slow turnaround of investments; High failure rate and therefore risky; No access to or ownership of productive assets; No access to loans because lack of collateral
Climate change	Poor mainstreaming of climate change into farm activities; Low appreciation of the difference between climate variability and climate change
Occupational health and safety	Exposure to harm and injury; Poor understanding and application of ergonomics principles particularly in the smallholder farming sub-sector; Limited understanding of pest and disease burden due to climate change
Public extension services	Limited mechanisation services to adequately support farmers

To improve the quality of university education in the country, the Government of Zimbabwe established the Zimbabwe Council for Higher Education (ZIMCHE) by an act of parliament (Zimbabwe Council for Higher Education Act [Chapter 25:27]) (GoZ, 2006). Before then universities established their own programmes and curricula. ZIMCHE is mandated to ensure universities and university programmes comply with set standards through appropriate governance, infrastructure as well as teaching and learning facilities, resources, and systems. This is achieved through:

- specifying minimum bodies of knowledge and skills (MBKS) for each study programme within the ZNQF.
- monitoring and evaluating the performance of universities through registration, accreditation, and

academic degree programmes using peer reviews and professional bodies.

- harmonising quality assurance systems and procedures of universities.
- issuing universities with certificates of institutional accreditation and reaccreditation every 5 years.
- conducting assessments for new programmes in line with the university's mandate, adherence to the national critical skills needs and regional and international benchmarking.
- periodic reviews of programmes to keep pace with developments in the various (sub)disciplines.
- identifying external examiners, drawn from local and regional universities, who provide quality assurance of programmes.

Methods of teaching, which include conventional face-to-face methods, methods where students are taught in residence, and block release have been approved. Under the block release system, face-to-face lectures and interaction with the lecturers are spread out over a semester and is restricted to a condensed period called a "block," usually 1 to 2 weeks in length. Lectures are interspaced with assignments and other coursework away from the university. In more recent times, modularisation has also been introduced. Under the system a module is taught face to face over a period of four weeks including all associated learning activities at the end of which is a final examination.

In 2016 ZIMCHE harmonised the credit systems for all degree programmes in the country under the auspices of the Zimbabwe Credit Accumulation and Transfer System (ZIMCATS). Under ZIMCATS, all universities migrated to the notional hour credit system. According to the system, 10 notional hours (made up of lectures, tutorials, seminars, and independent study) are equivalent to one credit. Adoption of ZIMCATS and MBKs is meant to facilitate intra- and inter-learner mobility of students, in alignment with the SADC qualifications framework (SADCQF) (Southern African Development Community [SADC], 2017) which covers higher education and technical and vocational education and training.

The main purpose of the SADCQF includes:

- providing a mechanism for comparability and recognition of qualifications in SADC,
- facilitating mutual recognition of qualifications in all Member States,
- harmonising qualifications wherever possible,
- promoting the transfer of credits within and among Member States and even beyond, and
- creating SADC regional standards where appropriate.

The shift towards enhancing science and technology in university education

The University of Zimbabwe, as the sole university in the country for about a decade after independence in 1980, offered a variety of programmes including those that were science and technology-oriented offered by the faculties of Agriculture, Engineering, Science, Veterinary Science, and Medicine in addition to other programmes. Due to the variety of programmes it offered, the University was referred to as a comprehensive university, a label it shared with Midlands State University, which was established in 1999.

To emphasise the importance of science and technology in society, the Government established the second university in the country in 1991, the National University of Science and Technology (NUST). As the name suggested, its mandate was science and technology. A University was considered in breach of its mandate if 70 percent of the programmes did not reflect its specific mandate.

Ever since the establishment of NUST in 1991, the majority of universities established by the Government were given a science and technology mandate, and invariably included science and technology in their names. The mandates of nine out of the 14 universities established between 1991 and 2019 was science, technology or science and technology. In addition to the National University of Science and Technology established in 1991, these included, Bindura University of Science Education (1996), Chinhoyi University of Technology (2001), Lupane State University (2004), Harare Institute of Technology (2005), Marondera University of Agricultural Sciences and Technology (2015), Gwanda State University (2016), Manicaland State University of Applied Sciences (2016), and Pan African Minerals University of Science and Technology (2016).

As can be expected, these universities offered mostly science and technology programmes, as did UZ and Midlands State University. Other universities with no specific science and technology mandate also offered science and technology programmes. These include Zimbabwe Open University, established in 1998 with a mandate for open and distance learning, and Great Zimbabwe University, established in 2002 with a culture and heritage mandate. Zimbabwe National Defence University was established in 2016 with a mandate for security studies. Some of the 10 private universities also offered science and technology programmes but at a smaller scale.

The push for science and technology in the country's universities received further impetus around 2015 when public universities were required to position STEM (Science, Technology, Engineering, and Mathematics) programmes at the centre of the curriculum. The result was that even those universities with no mandate for science and technology were allowed and encouraged to offer science and technology programmes. A case in point is Great Zimbabwe University, which now offers a full range of science and technology programmes under the Faculties of Science, Agriculture and Medicine. Universities were also encouraged to undertake local manufacturing of science and technology artefacts. For example, the Department of Soil Science and Agricultural Engineering at UZ received funding to manufacture a local tractor and ancillary equipment.

With the advent of the second republic under the leadership of President Emmerson Mnangangwa in 2017, the science and technology thrust received another major boost. Universities were to focus on innovation and industrialisation and ensure that the programmes would result in students with capacity to produce goods and services under Heritage-based Education

5.0 (GoZ, n.d.). Education 5.0 incorporated teaching, research, community service/outreach, innovation, and industrialisation, and was seen as the vehicle for national development (GoZ, 2018b). Innovation hubs were established at all public universities to provide an outlet for the production of goods and services. There was also a wholesale change of programmes focusing on equipping graduates with skills, entrepreneurship and innovation, and industrialisation and employment creation, underpinned by the Education 5.0 philosophy (GoZ, n.d.), Zimbabwe National Qualifications Framework (GoZ, 2018b), and Zimbabwe National Skills Audit (GoZ, 2018c). The country also adopted Vision 2030, which aimed to transform the country into an upper middle income economy by 2030. These national level changes caused a major transformation in ABE training as described later.

Introducing the Bachelor of Science Agricultural Engineering Degree Programme

Process Leading to Introduction of the Programme

As already noted, the Bachelor of Science Honours degree in Agricultural Engineering was introduced in 1992. Its introduction was preceded by consultations coordinated by McGill University in Canada and the Food and Agricultural Organisation. This was against the backdrop of five engineering degree programmes that were offered since the establishment of the Faculty of Engineering in 1970. These included BSc programmes in Civil Engineering, Electrical Engineering, Mechanical Engineering, Mining Engineering, and later a BSc in Land Surveying.

During the discussions, one important question was whether the Agricultural Engineering programme would be administered by the Faculty of Engineering or by the Faculty of Agriculture. For some this was an engineering degree programme that should be placed in the Faculty of Engineering while others thought that it needed to be housed within the Faculty of Agriculture to ensure that the agricultural aspects would not be lost. A decision was taken to house the programme in the Faculty of Agriculture hosted by the then Department of Soil Science and Agricultural Engineering. It was strongly argued that the programme is meant to provide engineering solutions to agricultural practices and processes and therefore should be housed in the Faculty of Agriculture. The plan was to establish a fully-fledged Department of Agricultural Engineering at a later stage.

The programme would maintain very strong links with the other engineering programmes as shown by the structure and in the content of the programme. This was also to make sure that agricultural engineering graduates would not have problems registering with the Zimbabwe Institute of Engineers and practising as engineers.

Programme Structure

Just like other engineering degree programmes in the Faculty Engineering, the duration of the degree programme was set at four years, offered over a minimum of 8 semesters (UZ, n.d.). Each year, referred to as a level, was made up of courses made up of at least 120 course units. The programme was made up of a minimum of 480 course units excluding the communication course. As can be seen from Appendix I, course units for each course ranged between 10 and 15 depending on the Faculty from which the courses were offered.

A course was worth 12 course units and was taught in 6 hours a week with no more than 3 hours being lectures. Each course was made up of course work (made up of practical assignments and in-class tests), which contributed 30 percent to the final mark. The balance of 70 percent was contributed by a three-hour final examination. For courses in the Faculty of Agriculture, a student was not eligible to sit for the final examination if he/she scored less than 40 percent in the course work.

Some courses had pre-requisites, which a student was required to pass before enrolling into a particular course. Failed courses would have to be repeated. There were no supplementary examinations. A student would be allowed to proceed to the next level carrying a maximum of two failed courses with those failing three or more courses required to clear the failed courses before proceeding to the next level. No student was allowed to proceed to level four while carrying a level one course.

The overall mark of the degree classification was based on aggregate marks from Levels 2, 3, and 4 weighted as 15 percent, 25 percent, and 60 percent, respectively. The exclusion of Level 1 from the degree classification was common across the University. The degree would be awarded subject to the candidate obtaining an approved first aid certificate, which was common practice for programmes in the Faculty of Engineering.

Programme Content

Scope and Focus of Courses

For the first two years of the programme, students of the Bachelor of Science Honours in Agricultural Engineering degree programme took courses from the Faculty of Engineering with other engineering students except for Introduction to Agricultural Engineering, which was offered in the second year by the Faculty of Agriculture. As can be seen from Appendix I, Levels 1 and 2 courses were basic courses in mathematics and engineering except for Communication Skills for Engineers, Workshop Practice, and Introduction to Agricultural Engineering. The courses offered in Level III were a mixture of four types of courses. First, were basic engineering courses (Thermodynamics and Heat Transfer and Quantitative Methods).

The second category was made up of foundational courses in agricultural production and agricultural engineering (Soil Science for Engineers; Principles of Animal Science; Hydrology and Hydraulics; Soil Mechanics and Cultivations; Introduction to Agricultural Economics). The third category represented the first set of applied courses (Machinery and Implement Design; Food Engineering; Crop Processing and Storage).

Courses in Level IV were basically applied courses in mechanisation (Agricultural Mechanisation), farm structures (Farm Structures Design; Farm Power), water and irrigation (Soil and Water Conservation Engineering; Irrigation and Drainage Engineering), food engineering (Process Engineering) and business management (Agribusiness Management 1). Quality of teaching was assured through peer and student evaluation of courses using standard tools.

Attachment

Students were required to participate in industrial attachment at a farm (where they would be assigned to an approved entity to gain practical skills or experience) and/or as internship referring to a job-type situation at an approved institution. The Directorate of Practical Agriculture Experience in the Faculty of Agriculture was responsible for approving the entity at which students would be attached. The attachment was done during long vacations at the end of Levels 1, 2, and 3 for a period not less than 24 weeks. Farm attachments were reserved for Level 1. Students on attachment would be visited by Faculty staff for on-site assessments.

During the attachment, the student was required to keep a diary of events and submit a written report at the end of the attachment. Assessment of the attachment consisted of the student diary, the report by the supervisor at the institution to which the student was attached, the mark awarded by the Faculty staff who visited the student, and the student's written report. The attachment was graded as either Pass (to signify that the student could proceed to the next level) or Fail, in which case the student would be required to repeat the attachment.

Research Project

The Research Project, referred to as the Special Project, was a double weighted course undertaken by students in the final year of the programme over two semesters. It accounted for 24 course units. Coursework assessment, which accounted for 30 percent of the final mark, included oral presentation of the proposal and written proposal and presentation of preliminary and final results. The thesis represented the final examination and accounted for 70 percent of the final mark. The thesis could be re-submitted within three months if the student scored between 40 and 49 percent. A student who scored less than 40 percent would repeat the Research Project on a new research topic.

The thesis was marked by two internal examiners excluding the academic supervisor, whose marks should not differ by 10 percentage points. In case of differences, the two markers would be required to discuss to come up with an agreed mark. If there was no agreement the thesis would be referred to the third marker whose mark would be final.

Students who produced the best research projects were eligible for the Student Paperette competition. This was an annual event hosted by the Faculty of Engineering in collaboration with the Zimbabwe Institution of Engineers, to promote research, innovation, and provide engineering solutions, which was open to all final year Engineering students. Judges of projects were drawn from industry. The projects from Agricultural Engineering have been quite competitive. Some of students participated at the annual ZIE National Engineering Schools Award competitions and received top awards.

Accreditation

Since ZIMCHE was not yet established, the programme was accredited by the professional engineering body, the ZIE. Students of the programme were eligible to be associate members of ZIE.

Impact of the Programme

The programme ran for a period of 24 years without any significant change in its curriculum. It produced successful graduates who have been employed or sought further studies in Australia, New Zealand, Italy, and South Africa in such fields as Hydrology, Food Engineering, etc. For example, at the time of writing, one graduate of the programme was elected President of the South African Hydrological Society (SAHS).

It is not only in the academic space where graduates excelled. The Limpopo Department of Agriculture in South Africa found the graduates relevant and versatile and could address all agricultural engineering issues. The comprehensiveness of the programme was attractive to the Department to the extent that it signed a Memorandum of Understanding with UZ to send South African students to follow the programme. The programme started in 2011 and ended after the first cohort because of financial challenges. As holders of Matric qualification, the South African students had to follow a bridging course to cover basic mathematics and physics courses.

Infusing Biosystems Components Into Degree Programmes

Beginning in 2011, a process of reviewing the curriculum of the Bachelor of Science degree in Agricultural Engineering programme was initiated to consider changes in the agricultural industry in Zimbabwe and expansion in knowledge in agricultural engineering. The process took close to five years to complete during which local (including the public and private sectors and past students), regional, and international stakeholders were consulted. The result was a revised undergraduate degree programme and new MSc degree programmes (see below).

Bachelor of Science Honours in Agricultural and Biosystems Engineering

After the internal University processes, the documentation was submitted in 2015 to ZIMCHE for registration and accreditation. ZIMCHE conducted written and physical check assessment. The programme was accredited in 2016.

Learning Outcomes

The name of the revised degree programme was Bachelor of Science Honours in Agricultural and Biosystems Engineering

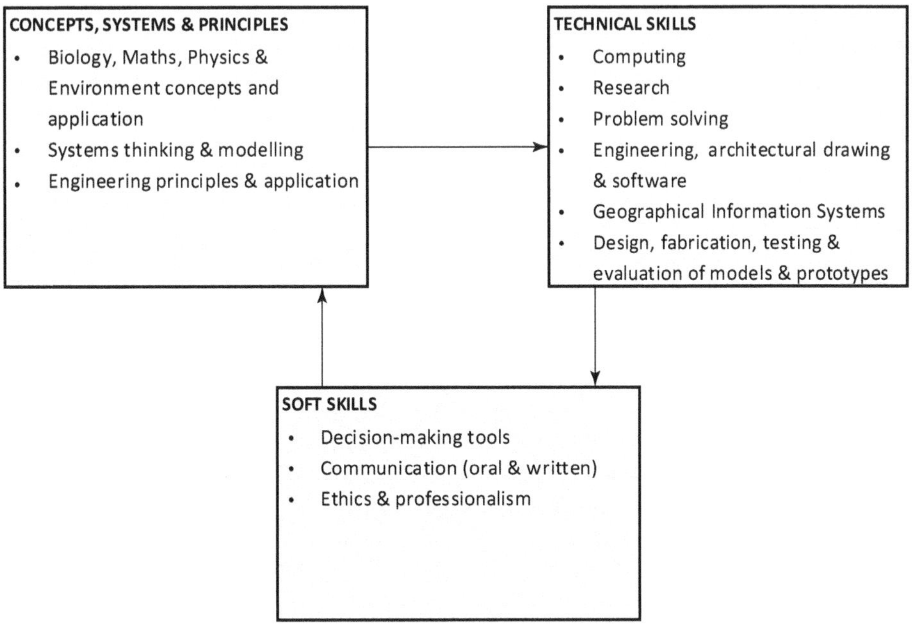

FIGURE 9.1 Skillset of the BSc Honours in Agricultural and Biosystems Engineering.

to indicate inclusion of biosystems and environmental engineering aspects. This was achieved by incorporating biotechnology-related courses in recognition of the fact that agricultural engineering in some cases deals with manipulation of living organisms (Figure 9.1 and Table 9.4). The skillset was crafted around three areas: 1) understanding of Concepts, Systems and Principles; 2) impartation of Technical Skills and 3) appreciation of Soft Skills dealing with decision-making, communication, and Ethics and Professionalism.

Programme Structure

The duration of the programme was changed from four to five years. The minimum number of credits (which replaced course units) for the programme was set at 240 credits, excluding the credits for the Communication and Computer Writing course. The number of credits per course ranged from 4 to 8 depending on the contact hours (Appendix II). For example, a course worth 4 credits was made up of 3 hours of lectures and a 1 hour tutorial per week while a course worth 5 credits was made up of 3 hours of lectures, 1 hour of tutorials, and 3 hours of practical time per week.

TABLE 9.4

Comparative analysis of key components of the Design Project versus the Research Project under the BSc Honours Biosystems and Agricultural Engineering

Dimension	Design Project	Research Project
Objective	Equip students with skills to originate an idea and take it through design and related processes in various aspects of engineering that (may) lead to production of new models and prototypes	Equip students with skills to identify a research problem and develop appropriate design(s) and data collections tools and methods, undertake data collection and analysis, and effectively communicate results through oral and written format
Scope	Concept Note; Design Process including design tools (including relevant software); Product Design and Analysis including product improvement and manufacturing; Feasibility Study including product commercialisation	Research Proposal; Project Document File; Thesis document; PowerPoint presentations of proposal and thesis summary
Contact time/ credits	Equivalent to one course	Equivalent to two courses
Assessment	Coursework (Oral and written Concept Note; Oral and written Feasibility report) and Final Report; Coursework contributes 40% to final mark while drawings of models and prototypes contribute 60% of the final mark	Coursework (Oral and written Research proposal; Presentation of preliminary and final results) and Thesis document; Coursework contributes 30 percent while Final Thesis (about 12,000 words long) contributes 70 percent to final mark
Grading	A candidate who fails the thesis component with a mark below 40 percent fails the course and repeats the course on new topic; Re-submission is allowed for students who attain 40 to 49 percent in the thesis	

Apart from the change from course units to credits, there were also other changes. The attachment period would not be done during vacation as previously but in the fourth year (penultimate) of the programme. It would at least be 30 weeks long (equivalent to two semesters). The attachment period was equivalent to four courses or 20 credits. Another significant change was the introduction of a Design Project in the third year of the study (as was the practice in other engineering programmes) in addition to the Research Project in the final year. The differences between The Design and Research Projects. To ensure integrity of the courses, detailed course outlines were to be kept in the Departmental Office.

Some features of the predecessor programme were retained. This included the notion of pre-requisite courses, repeating failed courses, coursework and examination weighting (and the need for a student to score at least 40 percent coursework to be eligible to sit for the examination), external examiners, and peer and student course evaluation. The aggregate marks from Levels 2, 3, 4, and 5 contributed 15 percent, 25 percent, 10 percent, and 50 percent respectively to the degree classification.

Programme Content

Scope and Focus of Courses

As for the predecessor programme, Levels 1 and 2 (the first two years of the programme), were made up of basic mathematics and engineering courses offered by the Faculty of Engineering. There were, however, some important differences. The Introduction to Agricultural Engineering course was replaced by a course named Agricultural and Industrial Production Systems. The objective of the course was to familiarise students with major production systems in the agricultural, manufacturing, and mining sectors as a foundation for applied courses in the four clusters of courses that constituted the programme, namely 1) Bio-agricultural mechanisation, 2) Postharvest Science and Food Processing, 3) Water and Irrigation Engineering, and 4) Environmental Engineering. Figure 9.2 shows the courses under each cluster.

Effort was made to ensure that each cluster would be underpinned by several basic courses. Thus, Plant Biology and Principles of Biochemistry underpinned Postharvest Science and Food Processing as well as the Bio-agricultural mechanisation cluster. Originally the idea was to have one dedicated biology-related course that addressed engineering properties of biological materials. This, however, was not possible because the two departments, Crop Science and Animal Science, that were supposed to offer the course did not have the human resources to offer such a specialised course. The option was for students to follow similar existing courses. The same challenge related to the incorporation of Statistical Methods and Experimental Design (which, although it appeared on the course list, was not being offered under the predecessor programme), which was identified as critical to enhancing research skills. The end result was that the programme became somewhat cluttered.

To further refocus the programme on biosystems aspects, the Workshop Practice course was changed to a Workshop and

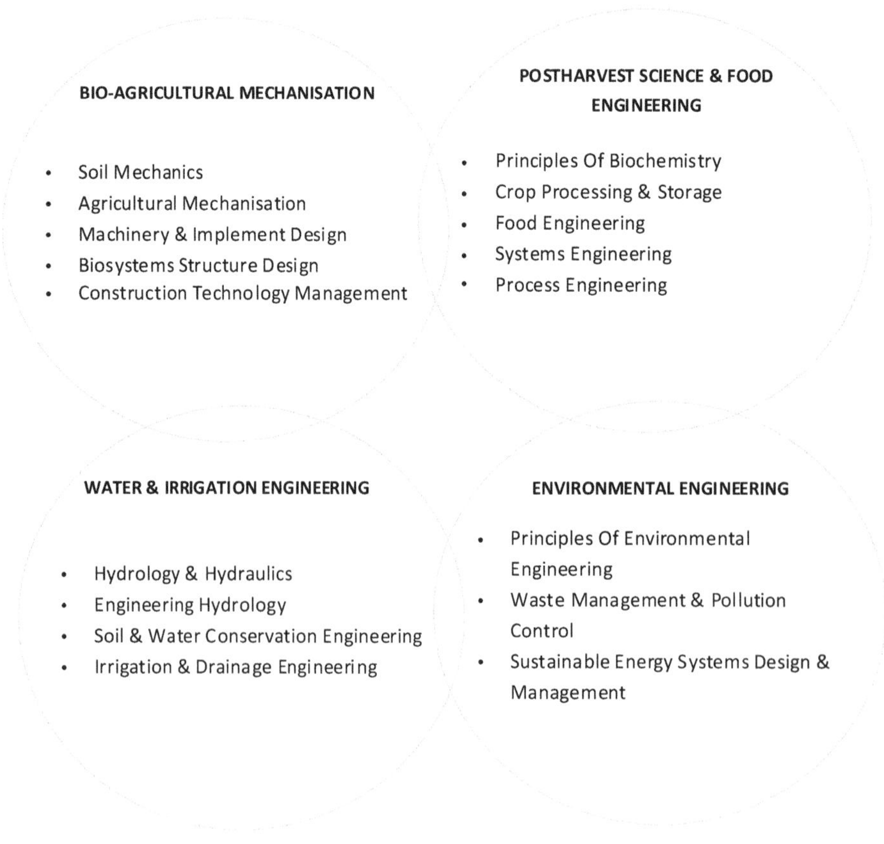

FIGURE 9.2 Clusters of courses of the BSc Honours in Agricultural and Biosystems Engineering.

Laboratory Skills course, which included principles, concepts, and skills related to biological materials in addition to the conventional engineering skills.

Water and Irrigation Engineering were underpinned by Engineering Geology (as well as Bio-agricultural mechanisation) and Hydrology and Hydraulics. The latter was retained from the predecessor programme. Construction Technology and Management was introduced as an applied course in Bio-agricultural mechanisation. Principles of Environmental Engineering underpinned the two applied courses in the Environmental Engineering cluster, i.e., Waste Management and Pollution Control and Sustainable Energy Systems Design and Management.

Geographic Information Systems (GIS) was introduced to ensure students appreciated the principles and application of GIS in engineering. Another new course, Decision Support Systems, was meant to equip students with decision-making tools in complex situations. Last but not least, the objective of not just producing job seekers was achieved by introducing Entrepreneurial Skills Development for Agriculture (across the faculty) in addition to Introduction to Agricultural Economics and Farm Business Management courses.

Industrial Attachment

The aim of the attachment was defined as appreciating the organisation and functioning of entities that are involved in agricultural and biosystems engineering. To this end students would be attached to institutions that deal with various aspects of biosystems and agricultural engineering to promote familiarity with different production processes and systems as well as the associated products and waste, determine the physical, financial, and human resources, and suggest improvements to the production systems and processes and products.

The Department would work with the Faculty Attachment Coordinator to ensure regular interaction with the student through periodic visits. Assessment was made up of the visits by the Department (20 percent), the student's written report (30 percent), an oral (20 percent) at the end of the attachment, and the mark awarded by the institution where the student was attached. This would contribute 20–25 percent of the final mark.

Design and Research Projects

In addition to the Research Project in the final year of the Programme, the Design Project was introduced in the penultimate year of the programme. Table 9.4 shows the components of the two types of projects.

MSc Programmes

Apart from introducing the Bachelor of Science Honours in Agricultural and Biosystems Engineering programme, the curriculum review also resulted in the introduction of four Master of Science (MSc) programmes, namely an MSc in Science Agricultural Mechanisation Systems and Analysis, an MSc in Food Processing Systems and Technology, an MSc in Environmental Systems Engineering, and an MSc in Industrial Ecology and Environmental Management (UZ, 2014a,b, c, d, e). Appendix III shows the constitution of the programmes in terms of learning outcomes, career opportunities, and programme structure and content. The two programmes shared the same structure (Part 1 and 2) as well as the general approach to learning outcomes, career opportunities, and research projects.

To cater for students who were in employment, the Programmes would run for five semesters (30 months) on a part time basis and three semesters (18 months) on a full-time basis. Each semester would normally consist of 13 weeks of teaching followed by 2 weeks of formal examinations. However, this option was never pursued as most students preferred the 30 month-version.

Part 1 was made up of 12 courses while Part II was entirely made up of the Research Project. Each course was made up of 60 contact hours (48 hours of lectures and 12 hours of seminars) and 48 hours of practicals, equivalent to 20 course units. The thesis was equivalent to 80 course units.

The three programmes shared two courses, Research Methods and Data Handling and GIS Applications, which provided scientific and technological skills that apply to research. Due to the strong biological/biosystems components, the MSc Agricultural Mechanisation Systems and Analysis and MSc in Food Processing Systems and Technology had several shared courses such as 1) Engineering Properties of Biological Materials, 2) Process Systems Design and Control, 3) Value Chain Analysis and Business Management, 4) Storage and Packaging.

Specialist Degree Programmes

Background

This section describes the introduction of undergraduate specialist degree programmes in 2019. This represented a significant philosophical departure from the comprehensive degree programmes, which had characterised agricultural engineering training since its inception 30 years ago. There is also a description of new MSc programmes. These changes were driven by a change of philosophy in university education in the country as described above and driven by Vision 2030.

The curriculum changes were informed by the following:

- orientation of degree programmes towards Vision 2030.
- ensuring that degree programmes produce goods and services as captured by Education 5.0 that encompassed innovation and industrialisation in addition to research, teaching, and community service.
- enlarging the scope of the faculty from the Faculty of Agriculture to Faculty of Agriculture, Environment, and Food Systems.
- establishment of a stand-alone Department of Agricultural and Biosystems Engineering.

- adoption of modular teaching where each module was to be taught in 150 notional hours[1] (90 contact and 60 independent study hours) and examined over a period of four weeks (UZ, 2018a, b, c, d, e).
- students were expected to learn a regional language (such as Shona, Ndebele, and Swahili) and an international language (such as German, Portuguese, Spanish, French).

Unlike the introduction of the Bachelors of Science Honours Degree in Agricultural and Biosystems Engineering and the Masters programmes described before, the changes were implemented over a very short period (six months), and did not reflect on the strengths and weakness of the predecessor programmes.

Undergraduate Programmes

In 2019, the Bachelor of Science Honours Degree in Agricultural and Biosystems Engineering was replaced by the following specialist programmes:

- Bachelor of Science Honours Degree in Agricultural Equipment Design and Manufacturing Systems;
- Bachelor of Science Honours Degree in Agricultural Processing, Plant Design, and Systems Engineering;
- Bachelor of Science Honours Degree in Food Engineering; and
- Bachelor of Science Honours Degree in Water Resources and Irrigation Engineering.

The degree programmes more or less coincided with the clusters of the Bachelor of Science Honours degree in Agricultural and Biosystems Engineering degree (Postharvest Science and Food Engineering; Water and Irrigation Engineering; and Bioagricultural Mechanisation) except for Environmental Engineering cluster. Appendix IV shows the modules and credits for each degree programme.

Programme Structure

The duration of the programmes was changed from five years to four years. This was a return to the original Bachelor of Science Honours Degree in Agricultural Engineering, at least in relation to the duration. There were significant changes. These included the adoption of notional credits instead of course units, as was the case with the Bachelor of Science Honours Degree in Agricultural Engineering, and courses worth 5 credits, as was the case with the Bachelors of Science Honours Degree in Agricultural and Biosystems Engineering. Each module (modules having replaced courses) is worth 15 credits. One credit is worth 10 notional study hours and 10 notional hours consist of 6 contact hours and 4 independent study hours. Each module is made up of formal lectures (20 percent), practicals (40 percent), and students' independent study (40 percent).

The Design Project in Level 3 of the predecessor programme was dropped. What was retained was the Research project worth 30 credits and Farm or Industrial attachment worth 60 credits but organised differently as described below.

The minimum number of credits for each programme was set at 480 credits, with each level accounting for 60 credits. Continuous assessment would contribute 50–60 percent of the overall mark. A student is required to pass both the continuous assessment and final examination. Degree classification is based on aggregate marks from Levels 1, 2, 3, and 4 and would contribute, weighted respectively as follows, 10, 20, 30, and 40 percent.

Programme Content

List and Arrangement of Modules

As can be seen from Appendix IV, the programmes are made of eight modules from Level 1 to 3, and seven modules in Level 4.

Out of the eight modules in Level 1 that are common across the programmes, five are engineering-related (Calculus for Engineering; Linear Algebra and Probability for Engineering; Engineering Drawing and Computer Aided Drafting and Design; Electrical Principles; Engineering Properties of Biological Materials), two deal with Technical Skills (Information and Digital Skills; Workshop Practice and Laboratory Skills), while one focuses on Social Skills (Students Development and Life Skills).

The eight modules in Level 2 are also common except the Rheology module for BSc Food Engineering programme. Five modules are engineering-related (Electrical Machines and Energy Conversions; Fluid Mechanics; Strength of Materials; Theory of Structures; Thermo-fluids) one is in mathematics (Advanced Mathematics – Methods for Engineering Ordinary Differential Equations and Statistics for Engineers), one introduces students to Economics (Introduction to Economics), and the last one is concerned with language (Language Acquisition and Intercultural Communication).

In Level 3, common modules across the programmes are Agricultural Machinery Design; Materials Engineering and Manufacturing Processes; Mechatronic Systems; Geographic Information Systems; Systems Automation and Instrumentation; Research and Innovation Skills and Critical Conscientiousness and Life Skills; and Industrial Attachment. Particular modules for the BSc Agricultural Processing, Plant Design, and Systems Engineering are Agricultural Processing Systems Design; Business Management of Agroprocessing Plants; and Soil Mechanics, while for the BSc Water Resources and Irrigation Engineering these include Hydraulics; Hydrology and Hydrogeology; and Principles of Irrigation Design. Peculiar modules for the BSc in Food Engineering are Food and Feed Processing; Food Product and Plant Design; Quantitative Bio-Spectroscopy.

Level 4 of the BSc Agricultural Engineering Design and Manufacturing Systems is made up of the Equipment Manufacturing Systems Design Management and Modelling; Manufacturing Technology; Product Design; Modelling and Simulation of Mechanisation Systems; Sustainable Energy Systems; Entrepreneurship; Professional Ethics Communication and Philosophy; and the Research Project.

The particular modules for the BSc Agricultural Processing, Plant Design, and Systems Engineering include Postharvest Management of Biological Materials; Processing Equipment Selection; Procurement and Plant Design; Processing Plant Environmental Management; and Processing Plant Operations Modelling. Particular modules for the BSc Water Resources and Irrigation Engineering include Sprinkler Irrigation System Design; Micro-Irrigation System Design; Sprinkler Irrigation System Design; Surface Irrigation System Design and Management; and Water Resources Engineering. The peculiar modules for the BSc Food Engineering are Food and Beverage Science and Technology; Food Manufacturing; Food Processing: Machines and Process Design; and one option from Dairy and Meat Processing or Food quality, Packaging and Safety Management.

Attachment

The attachment period was changed from a full year under the BSc Agricultural and Biosystems Engineering to be done over four vacations as was the case with the original BSc Honours in Agricultural Engineering programme. The difference was that the attachment period would accumulate credits in Level III worth 20 credits. The types of attachment (Box 9.1) were somewhat similar to the original BSc Honours in Agricultural Engineering programme, in that they were spread over the vacations. However, there are questions regarding the practicality of such an arrangement. For example, Model B contradicts with the four vacations mentioned above and can disrupt the rest of the learning cycle. It is difficult to see how Model C can be implemented given its open-ended nature. It is also worth noting that that during consultations leading to the introduction of the BSc Honours Agricultural and Biosystems Engineering programme, industry preferred a one-year attachment period or at least nine consecutive months.

BOX 9.1: TYPES OF ATTACHMENT FOR UNDERGRADUATE DEGREE HONOURS PROGRAMMES IN THE FACULTY OF AGRICULTURE ENVIRONMENT AND FOOD SYSTEMS AT THE UNIVERSITY OF ZIMBABWE. SOURCE: UZ (2018A).

MODEL A (ASSOCIATED WITH UNIVERSITY ACTIVITIES)

Attachment can be carried out at the Agro-industrial Park or any university facilities where the University has close control.

MODEL B (OUTSIDE THE UNIVERSITY)

Attachment can be done at an institution, farm, industry, or company which has a capacity to train students. This attachment can take up to one year. In this model, the student can even go to another university in or outside the country with clearly defined learning outcomes or to undertake a full module for specific skills.

MODEL C (CYCLIC ATTACHMENT)

Students get attached to institutions, industries, farms, or companies for short periods from second year up to the fourth year. The objective is to cater for all activities which change according to seasons. The learning is seasonal, and the students can go for a month or two months to learn a particular skill. On completion of a particular period, students can either come back to the university to do other modules or proceed to another institution, industry, company, or farm for a different skill.

Research Project

Unlike the BSc Honours Agricultural and Biosystems Engineering, which has a Design and Research Project, the four programmes included a Research Project, as was the case with the original programme. The difference was that the Research Project should contribute to the production of goods and or services.

The assessment would be based on the research methods used, project report, and contribution/production of goods and services. A student is required to open and maintain an up-to-date project file detailing the research process that is being stipulated. The conditions relating to failing the thesis remained the same. A student who fails the thesis by at least 40% is permitted to re-do the project and resubmit within three months. However, a candidate who fails by less than 40% is required to repeat the research project on a new topic.

Master of Science Programmes

At the Masters level, the MSc Food Processing Systems and Technology and MSc in Environmental Systems Engineering were (slightly) revised content-wise. The MSc Agricultural Mechanisation Systems and Analysis was abandoned. Two new programmes, MSc Agricultural Equipment Manufacturing Systems and MSc Irrigation Systems Design Engineering, were introduced. Appendix V shows the Objective, learning outcomes, and list of modules.

Programme Structure

Full time programmes are scheduled to last at least 24 months (four semesters), while part time programmes last a minimum of 30 months (five semesters). The programmes are cover in two parts, Parts 1 and 2. Part 1 is the taught component of 8 modules while Part 2 compromises a research project. Part 1 is worth 192 credits and Part 2 is worth 180 credits, making a total of 372 credits for the programme.

A module is worth 24 credits made up of 240 notional hours (144 contact hours and 96 independent study) and is completed in four weeks. One credit is worth 6 contact hours and 4 independent study hours. Modules shall comprise of lectures, practicals, fieldwork, research, seminar presentations, community engagements and industrial farm/industrial exposure, and independent learning. Lectures shall account for 20% of the contact and practicals 40%. Continuous assessment shall contribute 50% of the overall mark and the student must pass both.

Programme Content

Arrangement of Modules

The Applied Research Methods is the only module common to all four programmes. Sustainable Energy Systems Design and Management is common for three programmes (MSc Agricultural Equipment Manufacturing Systems; MSc Irrigation Systems Design Engineering; and MSc Environmental Systems Engineering) although in the last programme it is optional. GIS, Remote Sensing, and Modelling is also common to three programmes (MSc Irrigation Systems Design Engineering; MSc in Food Processing Systems and Technology; and MSc Environmental Systems Engineering). The other modules are particular to each of the four programmes.

The difference between the old and new MSc in Environmental Systems Engineering is that Water Supply and Water Treatment Design was combined with Hydrology (resulting in the dropping of Environmental Hydrology and Hydrogeology, Environmental Modelling and Design, and Operation of Landfills and Tailing Dams) in the new programme. Environmental Process Design was changed to Environmental Systems Design, and Environmental Law and Reporting was combined with Environmental Accounting. The other courses remained more or less the same.

It is also important to compare the old and new MSc in Food Processing Systems and Technology. Courses that were dropped from the new programme included Engineering Properties of Biological Materials; Storage and Packaging; Food Preservation, Safety and Hygiene. New courses that were added included Advanced Food and Feed Processing and Manufacturing Technologies; Advanced Livestock Product Processing and Technology as an optional course; and Postharvest Systems and Food Loss Reduction Technologies as a core module. The other four courses remained the same.

Another point of comparison is between the new specialist undergraduate programmes and the new MSc programmes. The BSc Honours in Food Engineering and the MSc Food Processing Systems and Technology share two courses. This is because the undergraduate course is an engineering degree while the MSc is not an engineering programme. This means graduates of the BSc in Food Engineering have no career path in food engineering. The BSc in Water Resources and Irrigation Design Engineering and MSc in Irrigation Engineering share several courses. In the BSc there are such modules as Micro-Irrigation System Design; Sprinkler Irrigation System Design; Surface Irrigation System Design and Management; Sustainable Energy Systems; Geographical Information Systems; and Principles of Irrigation Design, which are comparable to Design and Management of Surface Irrigation Systems; Pressurised Irrigation Systems Design and Management; Sustainable Energy Systems Design and Management; and GIS, Remote Sensing, and Modelling in the MSc programme. This overlap may render the MSc programme less useful for the graduate of Water Resources and Irrigation Design Engineering programme.

The BSc Agricultural Engineering Design and Manufacturing Systems and MSc Agricultural Equipment Manufacturing Systems also share some modules such as Sustainable Energy Systems for the former and Sustainable Energy Systems Design and Management for the latter. There are also similar modules dealing with computer-aided design, manufacturing, modelling, and business management (see Appendix IV and V.) The incremental value of the Masters programme is a point of discussion.

Research Project

The regulations and conditions for the Research Project are the same as for undergraduate programmes in terms of producing goods and services and assessment procedures. The Research Project in Part 2 must lead or contribute to the generation or production of specific goods and/or services.

Discussion

The evolution of agricultural (and biosystems) engineering training at the University Zimbabwe highlights some important issues in terms of learning outcomes, programme structure and content, and type of graduates. The undergraduate programmes that were offered over the three phases were honours degree programmes, the durations of which were at least four years after six years of high school (they were thus one year longer than a comparable degree programme in South Africa). All programmes had a strong engineering tradition, not least because the first degree was established with strong input from the Faculty of Engineering and the Zimbabwe Institution of Engineers. The first two years were devoted to providing students with basic engineering concepts and skills followed by two years of familiarising students with skills in applied engineering fields such as food processing, water resources and irrigation, agricultural mechanisation. 'Soft' skills, industrial attachment, and research projects were also critical elements of the programmes.

There were, however, some important differences which are worth noting. First, the original programme was a comprehensive agricultural engineering programme while the successor degree programme, although comprehensive, was infused with biosystems and environmental engineering aspects. The third set of degree programmes were specialised, focusing on food engineering, equipment design, agricultural manufacturing, and water resources and irrigation engineering. Unlike the first two phases where there were consultations, the introduction of the specialist degree programmes was informed by a philosophical change in university education based on Vision 2020 and Education 5.0. We shall return to the merits of such an approach as well as the merits/demerits of the comprehensive and specialist programmes later on in the context of the discussion of role of degree programmes in society. Suffice it at this point to observe that the introduction of the MSc Food Processing Systems and Technology and the MSc in Environmental Systems Engineering in the second phase proved to be popular because it was the first time such programmes were on offer. Virtually all the students were employed and used the acquired knowledge to contribute to the public and private sector.

The type of graduate coming out of the training was conceptualised differently. The degree programmes under the first and second phases aimed to produce employable graduates that could be hired locally, regionally, and internationally. To this end they were equipped with engineering, communication, and business skills. The programme in the second phase placed more emphasis on entrepreneurship, as an acknowledgement that graduates would not always find jobs. The degree programmes under the third phase, on the other hand, were conceptualised as producing graduates capable of producing goods and services leading to innovation and industrialisation in line with national interests as captured by the national vision (Vision 2030) and a new education philosophy (Education 5.0). While the objective was clear, the structure and content of the programmes did not bear this out. For example, innovation and entrepreneurship were not full modules. The requirement for students to follow a regional and international language could be interpreted as a means to make the graduate more marketable jobwise. It is also not clear how limiting the attachment period to three short vacation periods (doing away with a one year-long attachment which had been suggested during consultations leading to the introduction of the BSc in Agricultural and Biosystems Engineering) would contribute to a graduate with practical skills. The same observation can be made with regards to excluding the Design project, which was more practical-oriented and retaining the Research project, which was more academic. Excluding environmental engineering and the associated courses weakened environmental engineering aspects, which are now critically important in this era of climate change. Granting this, the programme was cluttered with too many courses.

The refocusing of two Masters programmes (MSc Food Processing Systems and Technology and MSc in Environmental Systems Engineering) and the introduction of two new programmes (MSc Agricultural Equipment Manufacturing Systems and MSc Irrigation Systems Design Engineering) was informed by the change in philosophy, as has already been explained. It would have been informative to find out the demand for refocused and new MSc programmes.

The extent to which the degree programmes have served and are serving the national course and society in general is not as straightforward as is commonly assumed. Any degree programme, no matter how well designed and implemented, cannot solve all the agriculture-related and developmental problems in the country. It can only make a contribution. Even so, that contribution may not be so straightforward because of challenges in the economy, as has been the case in Zimbabwe in the last two decades, and also because of the dynamic nature of development and changes in society. This means that the cause and effect relationship between degree programmes and national development is difficult, if not impossible, to establish. Despite these caveats, it is still important to reflect on how well the programmes have served and are likely to serve the Zimbabwean nation and society at large. To try and answer this question, we return to the four questions we posed at the beginning of the chapter.

The fact that degree programmes are designed and implemented in a specific national context to serve specific national interests means that the question of whether degree programmes should serve national objectives is a valid one. The real question is, how do universities identify the national interest which the programmes must take into account? Usually, as was done for the BSc in Agricultural Engineering and BSc in Agricultural and Biosystems Engineering degree programmes, national consultations are held with as many stakeholders as possible –government, semi-government, private, and non-governmental organisations – to determine the perceived stakeholder interests which are then taken to be national interests. Depending on the breadth and depth of the consultation, the extrapolation of stakeholder interests to national interests can be misleading. The other way to determine the national interests, as was done for the four specialist undergraduate degree programmes (BSc Agricultural Engineering Design and Manufacturing Systems, BSc Agricultural Processing, Plant Design and Systems Engineering, BSc Water Resources and Irrigation Engineering and BSc Food Engineering) as well as the four masters programmes (MSc Agricultural Equipment Manufacturing Systems, MSc Food Processing Systems and Technology, MSc Irrigation Systems Design Engineering and MSc Environmental Systems Engineering) is to draw the same from national policies and plans. The programmes make specific reference to Vision 2030 (too broad) and Education 5.0. The danger is that the national policy/plans may represent the view of the state and not society in general, may be political statements more than anything, and may be too narrow and short term. It is also important that universities are more imaginative and futuristic than dependent on national policies and plans. This is not to suggest that they should ignore national development objectives. Long term national development objectives rather than national plans, which are by definition short term, should provide guidance. It is also critical to include current and futuristic technological trends such as reverse engineering, machine learning, and drone technology to achieve precision agriculture. But as pointed out elsewhere in the chapter, how these will benefit smallholder farmers must be carefully considered and not left to chance.

The debate regarding the appropriateness of comprehensive or specialist degree programmes is of critical importance in developing countries like Zimbabwe. Millions of smallholder farmers in rural areas are more likely to benefit from graduates of a comprehensive degree programme rather than graduates of specialist programmes, as attested by the reasoning of the Limpopo Department of Agriculture. While comparable services can be obtained from specialist graduates, this would come at higher cost because of the need to employ specialists with a narrow skillset. Beyond providing services to the public sector, it is debatable whether better value is obtained from graduates of specialist undergraduate or Master's degree programmes. As was pointed out before, this depends on the incremental knowledge and skills the Master's programme provides. In Zimbabwe, the fact that the government dominated the provision of agricultural machinery and equipment means that there was no space for private actors in the agricultural machinery and equipment value chain.

While there should be no debate about whether degree programmes should serve national interests, the debate should be centred around the extent to which they should take into account regional and international trends. Designing degree programmes that are inward-looking assumes that the economy can absorb all the graduates. In Zimbabwe this is clearly not the case, as 20,000 university graduates are added to the job market every year, in addition to the college graduates. A small fraction of the graduates finds jobs, mostly of low quality. Rather than designing 'nationalistic' programmes, a better option is to design programmes that make the graduates competitive regionally and internationally, which has been the case with the BSc Agricultural Engineering. Apart from ensuring the mobility of the graduates, the country stands to benefit from remittances. Zimbabwe gets about one billion US dollars every year. (https://www.pindula.co.zw/Remittances_and_economic_growth_in_Zimbabwe). This view lends support to the approach of benchmarking programmes with regional and international trends.

The change in emphasis from producing employable graduates during the first and second phases of degree programmes to producing graduates who can employ themselves and create employment for others can be a double-edged sword. While in principle there is nothing wrong with such an ambition, there is a need for a close analysis of the situation. As shown by Sharma et al. (2022), the Zimbabwean economy is failing to create good jobs, especially in the agriculture sector. This means that even if the agricultural engineering graduates are self-employed, this is unlikely to be beyond subsistence. In such circumstances the prospects of creating employment for others are even more remote.

As the thrust for a knowledge society gathers in Zimbabwe, it is important to understand the attributes of a knowledge society and chart a way forward with regard to how this can be attained. According to the International Encyclopedia of the Social Sciences (2022), a knowledge society has seven attributes. These include: (1) mass and polycentric production, transmission, and application of knowledge; (2) the price of most commodities is determined by the knowledge needed for their development and sale rather than by the raw material and physical labour needed to produce them; (3) the majority of the population attains higher education; (4) the majority of the population have access to information and communication technologies and to the Internet; (5) a large portion of the labour force is made up of knowledge workers with a high degree of relevant education and experience; (6) individuals and the state invest heavily in education and research and development; and (7) organisations innovate continually. As this chapter has shown, while there is movement in a positive direction with regard to all of these attributes in the country, much ground remains to be covered.

Conclusions

This chapter traced the evolution of agricultural (and biosystems) engineering training at the University Zimbabwe over a period of 30 years. The analysis yielded some important insights, not just in relation to the structure and content of the programmes, but also in terms of the role of the degree programmes in society. These include the extent to which university curricula must be informed by national interests; the appropriateness of comprehensive or specialised degree programmes for a developing country like Zimbabwe; the extent to which national degree programmes consider regional and international trends; and whether training ought to focus on producing graduates that are employable, self-employed, or that create employment. At a more fundamental level, the thrust towards a knowledge society must be based on an understanding of what that means and entails. These are relevant questions that must be posed against a backdrop in which African universities are being called to embrace the Fourth Industrialisation (Ayentimi & Burgess, 2019). The findings presented in this chapter can be applied to many sub-Saharan African countries, which face similar conditions.

Acknowledgements

We want to thank Tinashe Nyabako, a former student of the Bachelor of Science Honours degree in Agricultural Engineering, for engaging in deliberations about the programme as well as drawing all of the illustrations in this chapter. We also wish to extend our sincere gratitude to Mr Phillip Nyamugafata and Dr Elijah Nyakudya, long-serving lecturers on courses in the Bachelor of Science Honours in Agricultural Engineering degree programme, for reviewing the chapter.

NOTE

1. A notional study hour was defined as the unit measuring the amount of work a student undertakes per week for a period of 4 weeks of study. Ten notional hours (made up of lecturers, tutorials, seminars, and independent study) are equivalent to 1 credit.

REFERENCE LIST

Ayentimi, A.T., & Burgess, J. (2019). Is the Fourth Industrial Revolution Relevant to sub-Sahara Africa? *Technology Analysis and Strategic Management*, 31(6), 641–652. https://doi.org/10.1080/09537325.2018.1542129.

Garwe, E.C., & Thondhlana, J. (2019). Higher Education Systems and Institutions, Zimbabwe. In Shin, J.C., & Teixeira, P. (Eds.), *Encyclopedia of International Higher Education Systems and Institutions.* https://doi.org/10.1007/978-94-017-9553-1_479-1.

Government of Zimbabwe [GoZ]. (2006). Zimbabwe Council for Higher Education Act: Chapter 25:27, Harare, Zimbabwe.

GOZ. (2013). Constitution of Zimbabwe Amendment (No. 20) Act 2013, Harare, Zimbabwe.

GoZ. (2018a). National Agriculture Policy Framework (2018–2030), Ministry of Lands, Agriculture and Rural Resettlement, Harare, Zimbabwe.

GoZ. (2018b). Zimbabwe National Qualifications Framework, Ministry of Higher and Tertiary Education, Science and Technology Development, Harare, Zimbabwe.

GoZ. (2018c). Zimbabwe National Skills Audit, Ministry of Higher and Tertiary Education, Science and Technology Development, Harare, Zimbabwe.

GoZ. (n.d.). Community-Led Education 5.0 for Zimbabwe 's Development: Positioning Education 5.0 in the Communities for National Development, Ministry of Higher and Tertiary Education, Science, Innovation and Technology Development, Harare, Zimbabwe.

International Encyclopedia of the Social Sciences. (2022). Knowledge Society. Encyclopedia.com. https://www.encyclopedia.com/social-sciences/applied-and-social-sciences-magazines/knowledge-society (Retrieved 4 January 2023).

Kalibata, A.M. (2017). Preface. In AGRA. Africa Agriculture Status Report: The Business of Smallholder Agriculture in Sub-Saharan Africa. *Alliance for a Green Revolution in Africa (AGRA)*, (5), vi–vii. Nairobi, Kenya.

Manzungu, E., Moyo, S., Boehlert, B., & Cervigni, R. (2018). Potential Impacts of Climate Change and Adaptation Options in Zimbabwe 's Agriculture Sector. Sam Moyo African Institute for Agrarian Studies, Harare.

Moyo, S. (2006) The Evolution of Zimbabwe 's Land Question. In Rukuni, M., Tawonezvi, P., Eicher, C., Munyuki-Hungwe, M., & Matondi, P. (Eds.), *Zimbabwe's Agricultural Revolution Revisited*, University of Zimbabwe Publications, Harare. 144–163.

Rukuni, M. (2006). Evolution of Agricultural Policy: 1890–1990. In Rukuni, M., Tawonezvi, P., Eicher, C., Munyuki-Hungwe, M., & Matondi, P. (Eds.), *Zimbabwe's Agricultural Revolution Revisited*, University of Zimbabwe Publications, Harare. 29–61.

Sharma, D.J., Ananda P.A.E.Z., Rodas, A.P., Hoy, C., Kurasha, F.M.N., Alwang, J., & Chingozha, T. (2022). Reversing the Tide: Zimbabwe Poverty Assessment, World Bank, Washington DC.

Southern African Development Community [SADC]. (2017). The Southern African Development Community Qualifications Framework (SADCQF), Gaborone, Botswana.

University of Zimbabwe [UZ]. (n.d.). Regulations for Undergraduate Degree Programmes, Faculty of Agriculture, Harare, Zimbabwe.

UZ. (2014a). Regulations for Undergraduate Degree Programmes, Faculty of Agriculture, Harare, Zimbabwe.

UZ. (2014b). Regulations Master of Science Degree in Agricultural Mechanisation Systems and Analysis, Faculty of Agriculture, Harare, Zimbabwe.

UZ. (2014c). Regulations for the Master of Science Degree in Food Processing Systems and Technology, Faculty of Agriculture, Harare, Zimbabwe.

UZ. (2014d). Regulations Master of Science Degree in Environmental Systems Engineering, Faculty of Agriculture, Harare, Zimbabwe

UZ. (2014e). Regulations Master of Science Degree in Industrial Ecology and Environmental Management, Faculty of Agriculture, Harare, Zimbabwe.

UZ. (2018a). Regulations for Undergraduate Degree Programmes, Faculty of Agriculture Environment and Food Systems, Harare, Zimbabwe.

UZ. (2018b). Regulations for the Master of Science Degree in Food Processing Systems and Technology, Faculty of Agriculture Environment and Food Systems, Harare, Zimbabwe.

UZ. (2018c). Regulations Master of Science in Agricultural Equipment Manufacturing Systems, Faculty of Agriculture Environment and Food Systems, Harare, Zimbabwe.

UZ. (2018c). Regulations Master of Science Degree in Environmental Systems Engineering, Faculty of Agriculture Environment and Food Systems, Harare, Zimbabwe.

UZ. (2018e). Regulations Master of Science Degree in Irrigation Engineering, Faculty of Agriculture Environment and Food Systems, Harare, Zimbabwe.

Utete, C. (2003). Report of the Presidential Land Review Committee, Volume 1, Harare, Zimbabwe.

Zimbabwe National Statistics Agency [ZIMSTAT]. (2022). Zimbabwe 2022 Population and Housing Census: Preliminary Report on Population Census, Harare, Zimbabwe. https://www.scribd.com/document/584362392/Zimbabwe-Census-2022-Preliminary-Report# (Retrieved 14 March 2023).

Zimbabwe Vulnerability Assessment Committee [ZIMVAC]. (2019). 2019 Rural Livelihood Assessment Report, Harare, Zimbabwe.

ZIMVAC. (2020). 2010 Rural Livelihood Assessment Report, Harare, Zimbabwe.

ZIMVAC. (2021). 2021 Rural Livelihood Assessment Report, Harare, Zimbabwe.

Appendix I. List of courses for the Bachelor of Science Agricultural Engineering degree programme

Course	Course Code	Pre-requisite	Units
LEVEL 1			
Engineering Drawing and Design	ME101		20
Engineering Mathematics I Paper 1	MT104		10
Computer Science for Engineers	EE102		15
Communication Skills for Engineers	CGEN103		10
Engineering Mechanics	ME103		15
Electrical Principles	EE101		15
Engineering Materials	MG103		15
Engineering Mathematics 2 Paper 2	MT105		10
Workshop Practice	WSP180		10
Total course units for Level 1			**120**
LEVEL 2			
Electrical Machines II	EE202	EE101	15
Engineering Mathematics II Paper 1	MT204	MT104	10
Engineering Survey	CE203		15
Theory of Structures	CE201	ME101	15
Strength of Materials	CE205		15
Thermofluids II	ME204	ME103	15
Engineering Mathematics II Paper 2	MT205	MT105	10
Introduction to Agricultural Engineering	AGEN201		12
Fluid Mechanics	CE202		15
Total course units for Level 2			**122**
LEVEL 3			
Food Engineering	AGEN302	ME204/ CE205	12
Soil Mechanics and Cultivations	AGEN304	CE205	12
Introduction to Agricultural Economics	AGEC111		12
Quantitative Methods	AGEN306		12
Hydrology and Hydraulics	AGEN310		12
Thermodynamics and Heat Transfer	ME309	ME204	12
Soil Science for Engineers	SL300		12
Crop Processing and Storage	AGEN303	CE205	12
Machinery and Implement Design	AGEN311		12
Principles of Animal Science	AS208		12
Total course units for Level 3			**120**
LEVEL 4			
Agricultural Mechanisation	AGEN404	AGEN310/ SL300	12
Farm Structures Design	AGEN406	AGEN201/ ME101	12
Agribusiness Management 1	AGEC215		12
Special Project	AGEN460		24
Irrigation and Drainage Engineering	AGEN411	AGEN310	12
Process Engineering	AGEN405	AGEN302	12
Soil and Water Conservation Engineering	AGEN410	AGEN310/ SL300	12
Farm Power	AGEN408	ME309	12
Agricultural Systems Engineering	AGEN407	AGEN306	12
Total course units for Level 4			**120**
Total course units for the programme			482

Appendix II: List of courses and credits for the Bachelor of Science Honours in Agricultural and Biosystems Engineering programme

Course	Course Code	Pre-requisite	Credits
LEVEL 1			
Engineering Drawing and Design	ME101		8
Engineering Mathematics 1 Paper 1	MT104		6
Computer Science for Engineer	EE103		6
Communication Skills for Engineer	CGEN103		4
Plant Biology	CR102		5
Principles of Biochemistry	ASXXX		5
Electrical Principles	EE101		6
Engineering Mathematics 2 Paper 1	MT105		6
Engineering Mechanic	ME103		6
Introduction to Agricultural Economics	AGEC11		5
Materials Science	AGENXX		5
Introduction to Statics for Agriculture	CR110 AS/ CR102		5
Sub-total			67
LEVEL 2			
Workshop and Laboratory Skills	AGENXXX		5
Engineering Mathematics II Paper 1	MT204	MT104	8
Fluid Mechanics	CE202		6
Electrical Machines II	EE202	EE101	6
Engineering Survey	CE203		6
Theory of Structures	CE201	ME101	6
Strength of Materials	CE205		6
Thermofluids II	ME204	ME103	6
Engineering Mathematics II Paper 2	MT205	MT105	8
Professional Communication Skills in Engineering	CSGENXX		4
LEVEL 3			
Quantitative Methods	AGEN306		5
Hydrology and Hydraulics	AGEN310		5
Thermodynamics and Heat Transfer	ME309	ME204	6
Geographical Information Systems	AES208		5
Statistical Methods and Experimental Design I	CR210		5
Environmental Engineering	AGENXXX		5
Food Engineering	AGEN302	ME204/CE205	5
Soil Mechanics	AGENXXX	CE205	5
Crop Processing and Storage	AGEN303		5
Machinery and Implement Design	AGEN311		5
Farm Business Management	AGEC2XX		5
Engineering Geology	GL204		4
Design Project	AGENXXX		5
LEVEL 4			
Industrial Attachment	AGENXX		5
LEVEL 5			
Systems Engineering	AGENXX	AGEN306	5
Agricultural Mechanisation	AGEN404	AGEN310	5
Biosystems Structures Design	AGENXXX	ME101/CE201/ CE205	5
Sustainable Energy Systems Design and Management	AGENXXX		5
Research Project	AGEN460		20
Waste Management and Pollution Control	AESXXX		5
Entrepreneurial Skills Development for Agriculture	AGECXXX		4
Irrigation and Drainage Engineering	AGEN411	AGEN310	5
Process Engineering	AGEN405	AGEN302	5
Soil and Water Conservation Engineering	AGEN410	AGEN310	5
Construction Technology and Management	AGENXX		5
Decision Support Systems	AGEN4XXX		5

Appendix III: Overview of MSc Agricultural Mechanisation Systems and Analysis, MSc in Food Processing Systems and Technology, and MSc Environmental Systems Engineering

Name	MSc Agricultural Mechanisation Systems and Analysis	MSc in Food Processing Systems and Technology	MSc Environmental Systems Engineering
Learning Outcomes	Knowledge of: • key components and concepts that underpin agricultural mechanisation and processes in the agricultural production chain • safety and risk issues in agricultural mechanisation • policies and regulations governing agricultural mechanisation • technology and skills in relevant in planning and management of agricultural mechanisation services • ethics, professionalism and entrepreneurship skills. Capacity to/of: • apply process systems design and operation management systems to agricultural mechanisation • assess technologies and systems in terms of cost effectiveness and environmental suitability • design, supervise, monitor and evaluate mechanisation projects • enhance value addition of food and agricultural products • design systems and operations that enhance labour productivity • research skills in investigating and solving problems.	Knowledge of • key components and concepts that underpin primary, secondary and tertiary food processing • technology and managerial skills in planning and management of primary, secondary and tertiary food processing • policies and regulations governing food processing safety and risk issues in food processing • Ethics, professionalism and entrepreneurship skills. Capacity to/of: • apply process systems design and operation management systems to food/processing; • assess technologies and systems in terms of cost effectiveness and environmental suitability • design, supervise, monitor and evaluate food processing projects; • enhance value addition of food and agricultural products; • design systems and operations that enhance labour productivity; • research skills in investigating and solving problems.	Knowledge of • components, concepts and principles of Environmental Systems Engineering; • policies, laws and statutes related to environmental systems engineering and management • life cycle assessment of technologies and projects; • Ethics and professionalism related to environmental systems engineering. Capacity to/of: • apply systems thinking to identify and solve environmental problems; • design integrated environmental systems that are environmentally sustainable and cost effective; • assess appropriateness of existing designs to important environmental parameters;
Career opportunities	• National level public and semi-public organisations • Regional organisations; international research and development agencies • Private companies and Non-Governmental Organisations • Consultancy and Entrepreneurship • Academic and development-related studies		
Part 1 Courses	**Core:** Engineering Properties of Biological Materials; Process Systems Design and Control; Value Chain Analysis and Business Management; Agricultural Production Systems; Storage and Packaging; Field Mechanisation; Farm Operations Planning and Modelling; Animal Handling and Management; Research Methods; Data Handling and GIS Applications **Optional:** Sustainable Energy Systems Design and Management; Food Processing Machinery and Equipment; Cereal and Oilseeds Processing; Roots, Tuber and Fibre Crop Processing	**Core:** Engineering Properties of Biological Materials; Food Chemistry and Human Nutrition; Process Systems Design and Control; Value Chain Analysis and Business Management; Storage and Packaging; Food Preservation, Safety and Hygiene; Food Processing Machinery and Equipment; Research Methods; Data Handling and GIS Applications	**Core:** Environmental Hydrology and Hydrogeology; Environmental Process Design; Water Supply and Wastewater Treatment Design; Design and Operation of Landfills and Tailing Dams; Sustainable Energy Systems Design and Management; Environmental Modelling; Air Quality and Noise Control; Solid and Hazardous Waste Management; Integrated, Safety Health and Environmental Management; Research Methods; Data Handling and GIS Applications **Optional (at least One):** Environmental Accounting; Environmental Law and Reporting; Green Technologies
Part II (Research Project)	The objective is to equip students with knowledge and skills in the research process from problem identification right up to thesis write up in relation to: • collate, critique, analyse and synthesise information from different sources • conceptualising, designing and implementing a research project; • spatial analysis and graphic presentation of engineering drawings using appropriate ICT; • design and implementation of quality control systems in the management of materials, products and services		

Appendix IV: Common and particular modules for the BSc Agricultural Engineering Design and Manufacturing Systems, BSc Agricultural Processing, Plant Design and Systems Engineering, BSc Water Resources and Irrigation Engineering, and BSC Food Engineering

Programme	Level 1 (8) modules	Level 2 (8) modules	Level 3 (8) modules	Level 4 (7) modules
BSc Agricultural Engineering Design and Manufacturing Systems	**Common modules: Calculus for Engineering; Linear Algebra and Probability for Engineering; Engineering Drawing and Computer Aided Drafting and Design; Electrical Principles; Engineering Properties of Biological Materials; Information and Digital Skills; Workshop Practice and Laboratory Skills; Students Development and Life Skills**	**Common modules: Electrical Machines and Energy Conversions; Fluid Mechanics; Advanced Mathematics (Methods for Engineering Ordinary Differential Equations and Statistics for Engineers); Strength of Materials; Theory of Structures; Thermo-fluids; Introduction to Economics; Language Acquisition and Intercultural Communication**	**Common modules:** Agricultural Machinery Design; Materials Engineering and Manufacturing Processes; Mechatronic Systems; Geographical Information Systems; Systems Automation and Instrumentation Research and Innovation Skills Critical Conscientiousness and Life Skills; Industrial Attachment	**Common modules**: Equipment Manufacturing Systems Design Management and Modelling; Manufacturing Technology; Product Design; Modelling and Simulation of Mechanisation Systems; Sustainable Energy Systems; Entrepreneurship, Professional Ethics Communication and Philosophy; Research Project
BSc Agricultural Processing, Plant Design and Systems Engineering			**Peculiar modules:** Agricultural Processing Systems Design; Business Management of Agroprocessing Plants; Soil Mechanics	**Peculiar Modules**: Postharvest Management of Biological Materials; Processing Equipment Selection; Procurement and Plant Design; Processing Plant Environmental Management; Processing Plant Operations Modelling
BSc Water Resources and Irrigation Engineering			**Peculiar modules:** Hydraulics; Hydrology and Hydrogeology; Principles of Irrigation Design	**Peculiar modules:** Sprinkler Irrigation System Design; Micro-Irrigation System Design; Surface Irrigation System Design and Management; Water Resources Engineering
BSc Food Engineering		**Peculiar module**: Rheology	**Peculiar modules**: Food and Feed Processing; Food Product and Plant Design; Quantitative Bio-Spectroscopy	**Peculiar modules:** Food and Beverage Science and Technology; Food Manufacturing; Food Processing: Machines and Process Design; and one **option** from Dairy and Meat Processing; Food quality, Packaging and Safety Management

Appendix V: Objective, learning outcomes, and modules for the MSc Agricultural Equipment Manufacturing Systems, MSc Food Processing Systems and Technology, MSc Irrigation Systems Design Engineering, and MSc Environmental Systems Engineering.

Name	MSc Agricultural Equipment Manufacturing Systems	MSc Food Processing Systems and Technology	MSc Irrigation Systems Design Engineering	MSc Environmental Systems Engineering
Objective	Develop graduates with advanced capabilities to create automated and optimised manufacturing systems for agricultural equipment covering small medium and large scale enterprises and applying state of the art knowledge platforms	Develop graduates with the capabilities to create, operate and review food processing system, technologies and plants.	Develop graduates with skills in planning, designing, construction, management of irrigation schemes and research and innovation in irrigation technologies to transform and improve rural livelihoods through irrigation development	Produce graduates with skills in the design of appropriate, relevant and cost-effective clean environment-related systems and technologies that meet the stipulated environmental requirements of industry in Zimbabwe and the southern African region, and beyond
Learning Outcomes	Product design agricultural machines and equipment using modelling and simulation software; Design different interfaces in control systems of machines and manufacturing systems: semi and fully automated; Design advanced manufacturing systems for machines and equipment given design specifications, total work content and /design models	Understand and apply properties of agricultural/biological materials pertinent to food processing; Carryout design, modelling and operation food manufacturing systems (mechanical, production and control pathways); Use CAD software in designing food processing plant: Piping and Instrumentation Design; Apply various aspects of industrial automation; Develop lean, agile and sustainable enterprises; Develop food safety management systems; Understand legislation in food safety management and principles of food preservation; Show aptitude for the implementation of construction systems and facilities	Understand the integrated nature of water resources management including the concept of sustainable development in irrigation development and management; Understand principles of water supply and treatment design for application at both small and large irrigation schemes and their surroundings; Competently apply irrigation engineering principles to select and optimise irrigation system design and irrigated crop production to maximise returns; Be competent to select and recommend appropriate energy technologies for existing energy deficit areas as well as principles involved in solar systems design; Participate in the creation of new knowledge and products in irrigation and drainage engineering and management through research and transfer of knowledge to farmers and other stakeholders; Work with communities and other stakeholders in transforming and improving rural livelihoods through irrigation development; Create successful enterprises in irrigation and in value addition of irrigated products	Knowledge of components, key concepts and principles of Environmental Engineering; Capacity to apply systems thinking and environmental modelling to solve environmental problems that take into account biophysical and socio-economic factors as well as safety and health issues; Skills in identification of projects that are innovative, cost-effective and environmentally sustainable; Skills in conducting life cycle assessment of technologies, systems and projects throughout the value chain; Skills in implementation of quality control in the management of materials, products and services; Skills in conceptualising, designing and implementing a research project; Capacity to collate, critique and synthesise information from different sources; Knowledge of policies, laws and statutes related to environmental systems engineering and management; Inculcate ethics and professionalism related to environmental systems engineering and management; Research and entrepreneurial skills to promote own ideas into profitable enterprises in collaboration with relevant industries and support systems.

Name	MSc Agricultural Equipment Manufacturing Systems	MSc Food Processing Systems and Technology	MSc Irrigation Systems Design Engineering	MSc Environmental Systems Engineering
Part 1 Modules	Advanced Agricultural Manufacturing Technology and Systems Design; Agricultural Manufacturing Automation and Robotics; Concurrent and Computer Aided Design; Maintenance Engineering and Management; Management Modelling; Sustainable Energy Systems Design and Management; Logistics, Operations and Business Management; Applied Research Methods	Food Chemistry and Human Nutrition; Food Processing Machinery and Equipment; Postharvest Systems and Food Loss Reduction Technologies; Process Systems Design and Control; Valorisation of Food and Agricultural Products; GIS, Remote Sensing and Modelling; Applied Research Methods plus one optional from Advanced Food and Feed Processing and Manufacturing Technologies; Advanced Livestock Product Processing and Technology	Design and Management of Surface Irrigation Systems; Pressurised Irrigation Systems Design and Management; Water Supply and Wastewater Treatment Design; Integrated Water Resources Management; Sustainable Energy Systems Design and Management; Construction Technology and Management; GIS, Remote Sensing and Modelling; Applied Research Methods	Air Quality and Noise Control; Environmental Systems Design; Hydrology, Water Supply and Wastewater Treatment; Integrated Safety, Health and Environmental Management; Waste Management; GIS, Remote Sensing and Modelling; Applied Research Methods plus one optional from Environmental Law, Accounting and Reporting; Sustainable Energy Systems Design and Management
Part 11 (Research Project)	The objective is to equip students with skills and tools to be competent to conceptualise and implement a scientific and design research project.			

10

Reshaping Agricultural and Biosystem Engineering Degree Programs for Student Success and Retention

Thomas J. Brumm and Steven K. Mickelson

Introduction

The history of the Agricultural and Biosystems Engineering (ABE) department at Iowa State University (ISU) dates back to 1905 when a new department was established to advance an agricultural engineering degree programme. The first major offered focused on applying mechanical science to agricultural problems. The course of study included topics ranging from electrical engineering and machine design to land levelling and farm structures. The ABE Department granted the first agricultural engineering degree in the world in 1910 to Jacob E. Waggoner. The "father" of agricultural engineering, J. Brownlee Davidson, was the first department chair at ISU. He had a vision for combining engineering and agriculture to meet the needs of a world with a growing population needing food, feed, and fiber.

The department has always emphasised a hands-on learning experience to prepare graduates for the workplace. Today, ABE remains committed to supporting ISU students academically and professionally by offering unparalleled degree programs that prepare them for future endeavors beyond college life. With its unwavering dedication towards student success, ABE is considered one of the top departments at ISU and internationally – continuing a legacy of excellence for generations to come.

The department currently has four degree programmes with several options within each major. Majors include agricultural engineering (AE), biological systems engineering (BSE), agricultural systems technology (AST) and industrial technology (ITEC). The engineering majors are in the College of Engineering. The technology majors are in the College of Agriculture and Life Sciences. The AST programme was created in the 1970s to provide graduates with solid math, science, and agriculture backgrounds who could manage systems and solve problems that didn't need an engineering design. Biological Systems Engineering was added in 2008 to meet the needs of the industry for engineers who understood biological processes within engineering design. The ITEC programme was integrated into the department in 2004. It was initially in the ISU College of Education but was found to have a better fit within ABE, complementing the AST programme but with a focus on manufacturing and occupational safety.

ISU has a long history of focusing on student success and retention but made great strides in the 1990s by adding programming to improve graduation numbers. ABE also has had a strong history of providing excellent faculty and staff, student advising and programming to help students succeed. As a relatively small department in the College of Engineering and the College of Agriculture and Life Sciences, our undergraduate degree programs' success is critical to the department's viability.

In the late 1990s, the department was at a critical juncture, with falling student numbers, poor retention rates, and the need for an outcome assessment approach to the new ABET 2000 criteria. In 1997, for instance, our undergraduate enrollment was approximately 200 students, and our graduate enrollment was in the 50s. Of the approximately 40 new agricultural engineering first-year students, we "lost" 30 percent of them after their first semester, i.e., they were no longer enrolled in agricultural engineering. Although we were doing a good job of recruiting students into the department, we weren't doing a very good job of retaining them.

At this time, the authors suggested that the ABE faculty consider adding hands-on laboratory experience to the curriculum to give new students exposure to the exciting experiences they would have later in their degree programme. They also recommended that the faculty consider ABE joining other ISU departments in adding a living/learning community (LLC).

We will outline our approach to programme growth through several critical student success and retention programs we implemented over twenty-plus years. By 2017, we grew to over 900 undergraduate and graduate students. How did this happen? Living-Learning Communities were a significant factor.

Living-Learning Communities

In 1995, ISU took on a new initiative called living and learning communities (LLCs). After attending an LLC conference, ISU leaders determined that ISU could benefit significantly from implementing LLCs across campus. ISU's Learning Community Initiative was a central administrative unit that provided funding for LLC programming, peer mentors,

DOI: 10.1201/9780429150111-12

tutoring, and assessment. ISU provided this office roughly $600K of annual, competitive funding to distribute to LLCs across campus.

LLCs consist of two parts: living communities and academic learning communities. Living communities (LvCs) are pre-arranged living environments (e.g., residence hall floors) where students from a major, department, or focus area (e.g., Agronomy, Agricultural and Biosystems Engineering, Women in Science and Engineering, etc.) live in proximity to each other for the benefit of planned engagement around a common theme. For the ABE LvC, we reserved approximately 40 spaces out of 80 on male and female adjacent dorm floors for students in ABE majors. Our first-year students were not required to live on an ABE living community residence hall floor, but it was a popular student choice because it gave them early preference in choosing their residence hall.

We hired a peer mentor to live on one of the dorm floors to provide programming and mentoring. Programming included tutoring help with math, physics, and chemistry courses, social events with ABE faculty and staff, service-learning projects (e.g., community service with learning outcomes aligned with the program), and informal mentoring and guidance.

Goals for the ABE LvC included (Mickelson et al. 2001):

- Residents in the LvC tutor each other
- Programs in the residence hall focus on issues related to academic success
- An ABE mentor for the LvC helps to create a supportive environment on our floor
- Students in the LvC share with each other what they are learning in class
- Students from the LvC ask each other about their classes
- Students in the LvC encourage each other to attend class
- Students are satisfied with the amount of concern their LvC mentors show towards their academic success
- Students are satisfied with the academic environment that has been created in their LvC
- Students believe the residents in their LvC engage in behaviors that support their academic success
- Students in the LvC form individual study groups
- There is a feeling of positive academic competition in the LvC
- The academic resources mentor (from the Department of Residence) helps to create an academic environment in the LvC
- The resident community advisor helps to create a positive living environment

Learning communities (LCs) are a set of common courses that students in the community take at the same time and place to enhance their learning experience. For our first-year, first-time ABE students, our set of courses included Freshman Orientation (1 credit), English Composition I (3 credits), and Engineering Computer Graphics and Design (3 credits) in the fall semester, and Experiencing ABE (1 credit), English Composition II (3 credits), and Engineering Problem Solving and Programming (3 credits) in the spring semester. Initial course linkages (i.e., scheduling students in the same sections of courses) included math, physics, and chemistry courses, but the feedback from the students was not positive: it was difficult to manage since students coming into the department had varying math and science backgrounds. We instead focused on courses where we had more control of course scheduling and content. The goal was to have some common learning outcomes across all linked courses.

As part of LCs, course linkages provide a platform for individuals to leverage their knowledge base and engage in collaborative learning experiences. LCs allow individuals to explore the various components of their educational development and gain insight into their strengths and weaknesses while engaging with their peers in dynamic problem-solving scenarios (Mickelson 2000; Harms et al. 2001a; Harms et al. 2001b; Mickelson and Brumm 2005; and Mickelson et al. 2007). Through such activities, students build an understanding of the relationships between different subjects, ideas, and concepts by connecting them across disciplines. In addition, course linkages provide students with practical experience that can be applied directly to their academic pursuits. By being exposed to diverse perspectives, students can use these insights from various sources to develop creative solutions to challenging problems.

LCs that use course linkages enable members to learn from each other's experiences. For example, a student struggling with a particular concept may benefit from hearing how another student tackled the same task or issue but used a completely different approach. These linkages allow learners to stretch beyond what they have learned on their own and discover new ways of thinking about topics which can lead to further understanding and development.

Harms et al. (2001a) outlined our ABE LC Objectives*:*

- To build community for entering first-year students within the Agricultural Engineering (AE) and Agricultural Systems Technology (AST) curricula
- To increase the retention of first-year students in the AE and AST programs
- To increase recruitment of students into the ABE curricula, especially underrepresented students (women and minorities)
- To enhance learning and team skills using a collaborative, learning-based educational methodology in the learning community courses
- To improve written communication skills by creating a writing link between the first-year composition courses and other technical courses in the AE and AST curricula

Additionally, we created the following specific objectives, which served as tangible guides for programme planning:

- To build excitement for the fields of engineering and technology

- To increase student involvement within the department of ABE
- To increase student interaction with the ABE faculty
- To increase student interaction with ABE upper-level students
- To have students learn about the differences between the options within the AE and AST curricula
- To develop team skills through the use of collaborative, learning-based assignments
- To introduce students to various problems (areas of interest) within the agricultural engineering and technology field
- To experience hands-on laboratories related to the AE and AST options
- To increase involvement in professional societies and student branches
- To introduce technical writing skills during the first year of study
- To make first-year composition courses more meaningful to students
- To establish career development and job preparation
- To receive academic guidance related to curriculum issues

Student collaborations within LLCs can help foster greater trust between members as they learn more about each other's capabilities while stimulating meaningful dialogue around complex issues that cannot be addressed through traditional classroom activities alone. Also, course linkages within Living and Learning Communities provide an environment where individuals feel comfortable sharing their thoughts without fear of judgment or criticism, which is often lacking in traditional teaching approaches. This open atmosphere encourages members to express themselves openly and freely without worrying about repercussions for taking risks or making mistakes. Such dialogue leads to deeper engagement with the material covered and greater personal growth. Additionally, it allows instructors and facilitators to better tailor instruction according to the group's needs, since everyone's input is critical for developing successful strategies for tackling complex concepts or situations.

Peer mentoring is a critical part of creating and maintaining successful learning communities. The importance of peer-to-peer relationships in these learning environments cannot be overstated. Students often benefit more from their peer mentors than from traditional teacher-student interactions. The peer mentors hired for our LLCs were high-performing, upper-class ABE students interested in mentoring the first-year students. Peer mentors were hired to engage with the first-year LC students for at least one to two hours a week in the classroom and one hour a week outside the classroom. This allowed adequate time for relationships of trust to develop.

The responsibilities of the peer mentors included (Harms et al. 2001a):

- Attending one of the core classes each week
- Participating in group activities in and outside the classroom
- Answering academic and curriculum questions
- Referring students to appropriate academic resources
- Coordinating out-of-class activities
- Providing feedback on classroom assignments
- Providing in-class guidance when appropriate
- Meeting with the core group coordinator once a week

There are several key strategies for successfully implementing peer mentoring within LCs. First is advocating for open communication among participants. These strategies include sharing ideas and knowledge and listening to each other, allowing different perspectives to be heard. Mentoring should be reciprocal. Mentors and mentees can learn from one another and offer their insights into the material being studied or discussed. Second is embracing diversity and using it as an advantage when forming learning communities and engaging in peer mentoring activities. By bringing students with different backgrounds and experiences together, a greater variety of perspectives are available for exploring topics creatively that would not necessarily have been available otherwise. Third, consistent progress evaluations and feedback between mentors and mentees ensure everyone is engaged and benefits from their community involvement. We encouraged mentors to set goals for themselves and their mentees so that everyone involved is motivated to participate fully and make meaningful contributions inside and outside their sessions.

By following these key strategies, LCs can become powerful learning environments where peers can support each other emotionally as well as academically through meaningful dialogue about topics related to their classes or personal lives, fostering a sense of inclusion, collaboration and understanding between members which helps promote deeper engagement with course material through constructive criticism coming from peers with similar interests. Our research has shown that LCs can provide many student benefits, such as increased academic performance, better communication skills and improved problem-solving abilities. In addition, LCs can help foster a sense of belonging and connection between students and build strong relationships and trust within the group.

Hands-on learning courses were integral to the success of our LCs, providing students with meaningful opportunities for personal growth and development. Hands-on learning courses can increase student engagement and promote peer cooperation. Additionally, hands-on classes can help extend students' knowledge beyond the classroom by introducing them to new technologies and allowing them to conduct experiments in real-world settings. The benefits of hands-on learning strategies are not limited purely to academic achievement – they can also go far towards building individual character traits such as confidence, self-esteem, and communication skills. Furthermore, these experiences allow students to gain a sense of ownership over their community while developing leadership capabilities, problem-solving abilities, and organisational skills, all of which serve them well throughout life.

One key component of course design for such an LC would be an appropriate selection of activities designed to foster active learning among all members. Such activities include problem-solving exercises, group discussions, field trips where students have hands-on opportunities to apply what they are learning in real-world contexts, and other projects that allow learners to demonstrate their understanding creatively. An additional important factor is creating a safe and supportive atmosphere where everyone feels comfortable participating and expressing themselves without fear of judgment or criticism - something that can only be achieved when instructors understand the importance of building trust and fostering relationships among all class members.

Furthermore, instructors should ensure their classes are structured to maximise each student's chance for successful completion by incorporating individual assessments and team assignments so that everybody can showcase their particular strengths and areas of expertise. It would also be beneficial if educators were prepared to use teaching techniques like cooperative learning or peer mentoring so that students can continue receiving support from each other even after class has ended.

One key course developed for our first-year experience was "Experiencing ABE". The course was designed during a faculty retreat to address concerns related to student retention. The faculty constructed potential hands-on engineering and technology laboratories from existing junior and senior courses that could be modified to expose first-year students to the kind of laboratories they would experience later in their curriculum. The goal was to increase the excitement for their chosen major and increase retention. This course was first offered in the spring of 1998.

Specific course objectives included (Mickelson 2000):

- Build community for AE first-year students
- Develop team skills
- Introduce students to various agricultural engineering experiences
- Learn of the various options within agricultural engineering
- Experience hands-on laboratories related to the AE options
- Increase student involvement within the ABE department
- Increase involvement in our professional society and student branch
- Increase student interaction with upperclassmen
- Increase student retention in the AE program
- Build excitement for engineering
- Career development and job preparation
- Faculty mentoring in helping make option decisions
- Develop report-writing skills
- Receive academic guidance on registering for classes

Peer mentors participated in this one-credit course to help instructors deliver the laboratories. Each lab was taught by the instructors who, later in the curriculum, would teach the upper-level course that would include a similar laboratory. Example laboratories included tractor power, fluid power, fluid mechanics, electric motor control, and grain aeration.

This approach demonstrated that living-learning communities could benefit students in various ways. Not only do they encourage collaboration, but they also provide important social and psychological support for those involved. Studies have shown that learning communities can lead to improved academic performance, higher motivation levels, better grades, enhanced self-esteem, and reduced stress among students. Figure 10.1 shows the agricultural engineering retention rates from 1997 to 2001 (Mickelson et al. 2007). The data from this study indicate that students who are part of an LC tend to have higher academic success and retention rates than their peers in traditional courses. Specifically, students in learning communities have reported higher satisfaction levels with their college experience, increased self-esteem, more meaningful relationships with peers and faculty members, and improved stress management.

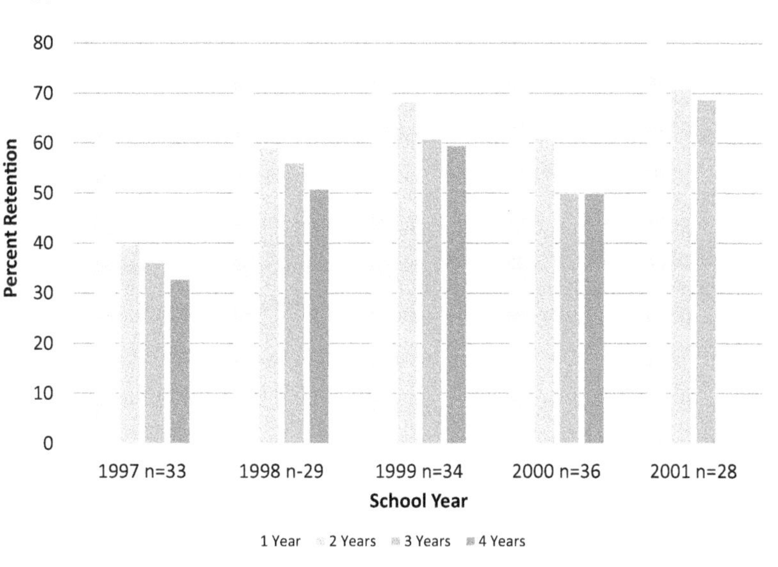

FIGURE 10.1 Agricultural engineering retention rates from 1997 to 2001 (Mickelson et al. 2007).

In addition to these benefits, living-learning communities can help bridge the gap between different cultures and backgrounds and foster greater understanding between students from diverse backgrounds. LLCs can improve cultural awareness and understanding between different groups, helping create a more inclusive learning environment. These communities also often provide access to resources and opportunities that would otherwise not be available to participants. For instance, some of our research has explored how LCs have helped connect students with tutors or mentors who could provide valuable advice or support while pursuing their education goals (Mickelson et al. 2001).

In addition to the positive effects mentioned above, the research also revealed that the department's learning communities helped create a supportive environment for diverse populations of students. In particular, ABE's learning communities provided resources for minority students and first-generation college attendees, which helped them succeed academically as well as socially. LLCs included mentorship by faculty members and upperclassmen peers who could guide them during difficult times.

The success of any given LLC depends upon several factors, such as its structure, size, goals and objectives, leadership style, and overall atmosphere. To ensure that all members have equal opportunity to contribute meaningfully within the group setting, it is vital that every student feels safe expressing their ideas without fear of ridicule or conflict with other members. Additionally, clear expectations should be established so that everyone understands what is expected from them inside and outside the classroom. This philosophy was implemented in all ABE LLC programming

Ultimately, Iowa State University's Department of Agricultural and Biosystems Engineering found that its LLC initiatives significantly improved student retention rates. These results suggest that departments and universities should invest more resources into developing similar programs for their student body to produce successful outcomes for current and future generations of learners.

Workplace Competencies

In 2000, ABET implemented new accreditation criteria. ABET (www.abet.org) is an organisation made up of members of professional societies that accredits engineering, engineering technology, computer science, and applied science degree programs worldwide. The new criteria shifted the focus of accreditation from providing education to a more comprehensive understanding of abilities needed for professional success programme graduates. These changes were designed to ensure that students graduating from an accredited programme would be prepared for their professional careers and have the necessary tools to succeed in their chosen fields.

Additionally, ABET started requiring programs to use outcome-based assessment methods rather than emphasising course-specific requirements. ABET developed 11 student learning outcomes required of all engineering programs (Box 10.1). These outcomes defined what engineering students should be able to do or have at graduation. Programs must measure their students' performance and prove that students achieved these mostly ability-based outcomes.

BOX 10.1: ABET 2000 STUDENT LEARNING OUTCOMES.

(a) an ability to apply knowledge of mathematics, science, and engineering
(b) an ability to design and conduct experiments, analyze, and interpret data
(c) an ability to design a system, component, or process
(d) an ability to function on interdisciplinary teams
(e) an ability to identify, formulate, and solve engineering problems
(f) an understanding of professional and ethical responsibility
(g) an ability to communicate effectively
(h) the broad education necessary to understand the impact of engineering solutions
(i) a recognition of the need for, and the ability to engage in, lifelong learning
(j) a knowledge of contemporary issues
(k) an ability to use the techniques, skills, and modern engineering tools

It is difficult to directly measure an ability as it combines skills, attitudes, and behaviors. The College of Engineering at Iowa State University (Brumm et al. 2006a) engaged with their stakeholders (employers, alumni, faculty, and students) to identify and validate 14 workplace competencies that encompassed the 11 required ABET outcomes (Box 10.2). Only two of these competencies were knowledge-based. The rest are often considered "soft skills". These 14 competencies resonated with employers of Iowa State engineering graduates, often used in their companies for employee evaluation and development. A different set of stakeholders may not identify the same competencies – the competencies in Box 10.2 may be potentially unique to Iowa State's engineering programs and employers.

BOX 10.2: THE 14 IOWA STATE UNIVERSITY WORKPLACE COMPETENCIES.

Analysis and Judgement
Cultural Adaptability
General Knowledge
Integrity
Quality Orientation
Communication
Customer Focus
Initiative
Planning
Teamwork
Continuous Learning
Engineering Knowledge
Innovation
Professional Impact

Observable Key Actions define competencies. Box 10.3 gives the definition and Key Actions of the Communication competency. When a person performs the Key Actions of competency, they are thought to have demonstrated that competency.

BOX 10.3: IOWA STATE UNIVERSITY COMMUNICATION COMPETENCY DEFINITION AND KEY ACTIONS

COMMUNICATION:

Clearly conveying information and ideas through a variety of media to individuals or groups in a manner that engages the audience and helps them understand and retain the message.

KEY ACTIONS

- Organises the communication – Clarifies purpose and importance; stresses major points; follows a logical sequence.
- Maintains audience attention – Keeps the audience engaged through use of techniques such as analogies, illustrations, body language, and voice inflection.
- Adjusts to the audience – Frames message in line with audience experience, background, and expectations; uses terms, examples, and analogies that are meaningful to the audience.
- Ensures understanding – Seeks input from audience; checks understanding; presents message in different ways to enhance understanding.
- Adheres to accepted conventions – Uses syntax, pace, volume, diction, and mechanics appropriate to the media being used.
- Comprehends communication from others – Attends to messages from others; correctly interprets messages and responds appropriately.

The ISU Agricultural and Biosystems Engineering Department integrated workplace competencies into its curriculum in two ways: (1) STARs; and (2) supervisor assessment of student demonstration of workplace competencies in internships.

STARs (Situation/Task, Action, Result) effectively answer behavioral-based questions such as "tell me about a time when you took initiative to accomplish a difficult task". In responding using the STAR format, a person discusses the situation and the required task, the action(s) the person took and the result of their actions. Behavioral-based questions are used by most employers of ISU graduates when interviewing candidates for engineering positions.

Classroom discussions in learning community courses focus on understanding and developing workplace competencies. Assignments ask them to document their demonstration of competencies using the STARs format (Brumm et al. 2006b.). Students create a library of STARs that they use to prepare for interviews. Students often told us that they responded to one or more of the questions they asked during an interview using a STAR they had previously created.

The particular Key Actions associated with a workplace competency can be observed. An external observer can rate a person on demonstration of key actions on a Likert scale according to the following: when given a chance, how often does the person perform the key action? 5 = always or almost always, 4 = usually, 3 = often, 2 = sometimes, and 1 = never or almost never. People can self-assess as well. Thus, numerical ratings demonstrating competency can be determined.

Approximately 80% of the engineering students in the ABE department (AE and BSE) and 100% of our technology students (AST and ITEC) have an internship experience before graduation, a requirement for graduation. To receive academic credit for the internship experience, students self-assess themselves on the 14 workplace competencies and ask their internship supervisor to assess them. Over 80% of supervisors complete the assessment of their student interns.

This feedback allows individual students to compare their self-assessment with their supervisor's assessment. The technology students write reflections that compare the two assessments and propose action plans for improvement.

Summarising the results from hundreds of assessments allowed us to identify the workplace competencies that were the strongest and weakest in our students. While the average for nearly all competencies assessed by supervisors was 4.0 or higher, Table 10.1 is typical of the ranking of the top-five and bottom-five ranked competencies.

There is a surprising degree of agreement between supervisors and students in assessing their demonstration of competencies. Interestingly, our students often rank themselves slightly lower than their supervisors. We were initially surprised that the Innovation competency was often ranked lower than other competencies. However, many students in their first professional employment experience are often not in a position or are reticent to demonstrate innovation.

We identify the weakest competencies and emphasise their improvement by looking at the competency assessments as top-five and bottom-five. Thus, we focus many of our STAR assignments on the lower-ranked competencies. Note that competency weakness or strength judgments should not be based on data from a single term as they may vary depending on the particular set of students. A longer-term view (e.g., averages over two or three years) should be taken.

TABLE 10.1

Typical top- and bottom-ranked workplace competencies of Agricultural and Biosystems Engineering (ABE) students.

Self	Relative Ranking	Supervisor
Integrity	1	Integrity
Professional Impact	2	Cultural Adaptability
Quality Orientation	3	Professional Impact
Teamwork	4	Quality Orientation
Cultural Adaptability	5	Planning
Analysis and Judgment	10	Analysis and Judgment
Communication	11	General Knowledge
Initiative	12	Communication
Customer Focus	13	Customer Focus
Innovation	14	Innovation

Student development of workplace competencies gives them a head-start on their professional careers by preparing them for job interviews and helping them meet employer expectations. Our curricula actively incorporate workplace competencies, aiding them in this development.

Continuous Improvement

With the new ABET 2000 accreditation criteria, the department created a continuous improvement process that incorporated outcomes assessment as illustrated in Figure 10.2 (Brumm et al. 2006c).

The program's mission, goals, and objectives are reviewed (and changed as necessary) every three years in consultation with stakeholders and our industrial advisory board, concurrent with the ABET accreditation cycle. These inform our curricular efforts.

Each course in the curriculum is examined to determine to the extent it addresses each of the outcomes. Not every course addresses every outcome, but the entire curriculum is constructed to adequately address all the outcomes. Once the outcomes are mapped to the individual classes, we determine from the outcomes-competency matrix which competencies are addressed in each class taught in our department.

The primary evidence of student outcome achievement is direct measures of performance: key assignments, workplace competencies evaluations of students on internships by their supervisors, employer evaluation of graduates after five years and the results of the Fundamentals of Engineering (FE) exam. We also used indirect measures: senior exit surveys, student evaluations of instruction (SEI), post-graduate surveys, external programme reviews such as the now defunct CRSESS reviews conducted by the US Department of Agriculture, advisor evaluations, and placement statistics. Indirect measures provide additional information but are not primary in evaluating the attainment of outcomes.

We use workplace competency evaluations in internships to determine the attainment of the outcomes. The achievement of each programme outcome can be determined by how well a student demonstrates a subset of the 14 competencies associated with that outcome. During the validation process with stakeholders (Brumm et al. 2006a), a competency-outcomes map with weighting factors for the competencies for each outcome was determined (Table 10.2). Stakeholders were asked to rate the importance of each competency to each outcome on a 5-point scale. Any competency ranked less than 2.5 was not considered as important for that outcome.

If a perfect rating of 5 for each competency were attained for a given outcome, a 100% score for that outcome would be the sum of the products of 5 (perfect rating) and the weighting factor. Given the rating of a student's demonstration of the workplace competencies (almost always less than 5), a score for that student could be similarly calculated. That score is the percentage achievement of a particular outcome by that student. By aggregating the competency ratings of many students in the degree programme over time, the degree to which the program's students achieved each outcome can be determined.

Figure 10.3 demonstrates the achievement of outcomes of AE students on internships over the years 2006–2011 as determined by the competency-outcomes mapping. We consider 80% to be the successful attainment of an outcome.

An important part of our continuous improvement process is not pictured in Figure 10.3. We needed to understand the role of each course in our curriculum, how the course learning outcomes contributed to the overall student learning outcomes

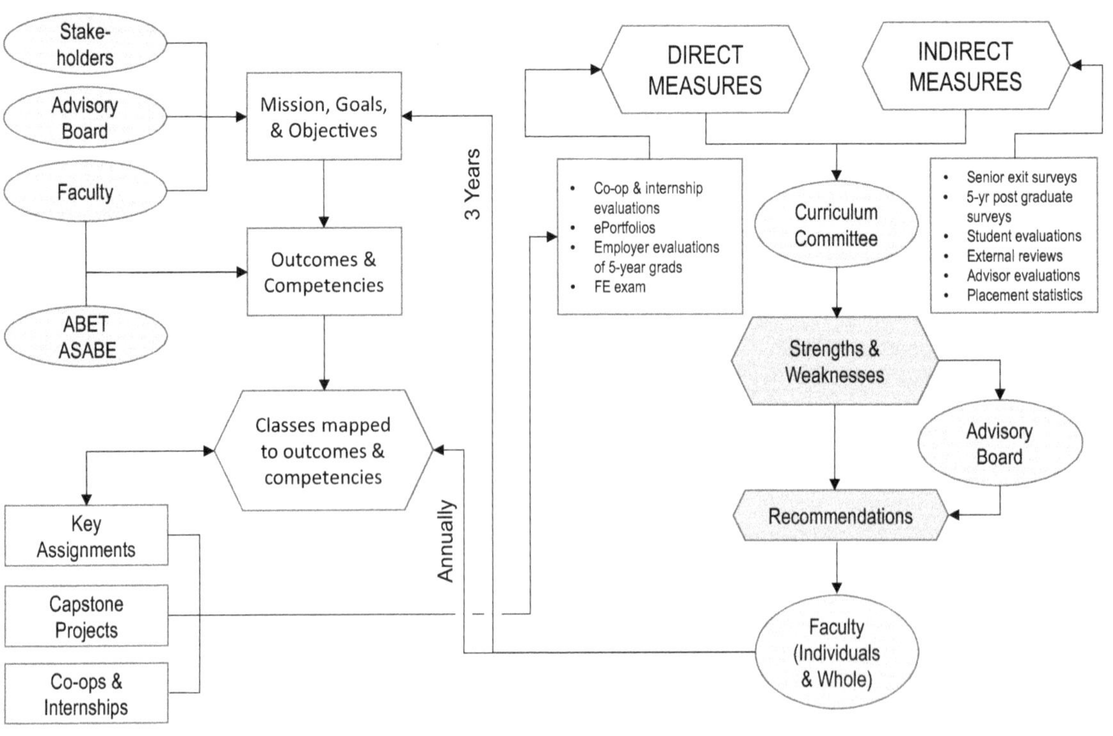

FIGURE 10.2 The Iowa State University Agricultural and Biosystems Engineering (ABE) 2006 continuous improvement plan.

TABLE 10.2

Competency-outcome mapping. Numerical values are the importance of each competency to the outcome on a 5-point scale. (Brumm et al. 2006a)

ABET Student Outcome (a-k)	ISU Workplace Competency													
	Analysis & Judgment	Communication	Continuous Learning	Cultural Adaptability	Customer Focus	Engineering Knowledge	General Knowledge	Initiative	Innovation	Integrity	Planning	Professional Impact	Quality Orientation	Teamwork
a	4.3		3.8			4.8		3.5						
b	4.5		3.6		3.4	4.4		3.7	4.0		4.1		4.3	3.4
c	4.5	4.0	3.8	3.0	4.2	4.4		3.9	4.3		4.2		4.1	3.8
d	3.6	4.7		4.3	3.7			4.0		4.3	3.8	3.9		4.9
e	4.4	3.7	3.8		3.6	4.7		4.1	4.2				3.9	3.6
f	3.5		3.6	3.7			3.8			4.7			3.3	
g		4.9			4.0		3.8	3.7				4.2		
h	3.5		3.9	4.1		3.4	3.9							
i			4.6					4.1						
j	3.1		3.8	3.8			3.7							
k	4.0		4.2	2.6		4.3		3.7					3.6	

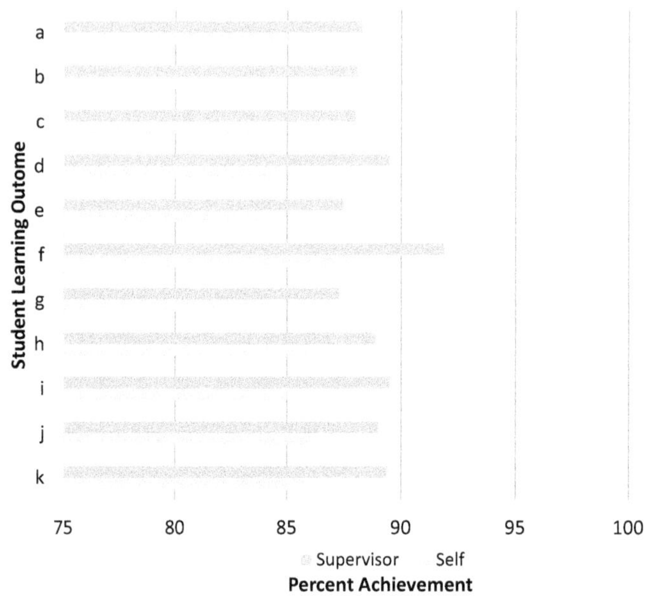

FIGURE 10.3 Achievement of student learning outcomes of agricultural engineering students based on student internship workplace competency assessments for 2006–2011 (supervisor n=114, student n=140).

and how the key assignments in each class demonstrated achievement of course learning outcomes. Thus, we implemented what we have come to call the "Saunders Protocol", a system for reaching this understanding (Saunders et al. 2008).

The curriculum committee conducts a review session with the course instructor for every departmental course in our four curricula. Over six years (our accreditation cycle), each course is evaluated. The evaluation proceeds as follows:

1. The instructor provides the course syllabus, textbook, the instructions for a course key assignment (an assignment where students should demonstrate one or more of the course outcomes) and anonymised examples of student work for that key assignment.
2. The session begins with a verbal introduction to the course and the key assignment by the instructor.
3. The committee members then examine and discuss the students' work, evaluating whether or not the desired outcomes were demonstrated. During this examination, the instructor is not allowed to participate or answer questions.
4. After the examination, the instructor was allowed to participate in further discussions of the class and key assignments.
5. Recommendations for change and improvement (if identified) of the class or assignment were made in collaboration with the instructor.

Instructors universally commented that this review protocol helped them think more deeply about the course and the impact of assignments on student learning. Sometimes it was difficult for the committee to see the demonstration of the knowledge or skills the instructor intended for the assignment. Feedback to the instructors helped them improve the assignments and the overall course. Thus, we saw nearly every course make improvements.

One result of the "Saunders Protocol" was unforeseen. As a whole faculty, we rarely dove deeply into the courses in the curriculum. Our courses were essentially siloed from each other with presumptions about what was being taught and occasional instructor frustration in upper-division courses when students came to the course unprepared. The evaluation sessions became rich discussions of what we were teaching and expectations for what students should learn as they matriculated through the curriculum. We were gaining a new understanding of how courses connected (or should have been) with each other. Changes were made, such as introducing topics earlier in the curriculum or repeating topics at multiple points throughout the curriculum, that improved student learning. We continue to find the Saunders Protocol valuable in our continuous improvement process.

Not every new technique or method that we tried was successful. One example is the use of post-graduate evaluations of our graduates five years after graduation by our graduates' employers (direct measure) and the graduates themselves (indirect measure), as shown in Figure 10.2. Our survey return rate from graduates was usually under ten percent despite numerous attempts and strategies to improve participation. Furthermore, we learned that employers were reluctant to provide evaluations of our graduates as that was a private personnel matter. Thus, we no longer include post-graduate evaluations in our continuous improvement process.

We also implemented electronic portfolios (ePortfolios) as a direct measure of student demonstration of workplace competencies and, thus, achievement of student learning outcomes (Brumm et al. 2003). Students were asked to identify and upload certain examples (papers, reports, projects, and autobiographical information) that demonstrated workplace competencies, reflect on the significance of these examples and explain their selection process for the instructor or audience. All these artifacts and reflections were stored in an electronic database (ePortfolio) that we constructed. When done correctly, the portfolio as an educational artifact shows intellectual growth and gives the assessor of this growth a range of performances that indicate the student's intellectual and technical development. Upon evaluating a student's portfolio, we would have evidence of how well they demonstrated workplace competencies.

Although ePortfolios are powerful tools for learning and reflection, we encountered difficulties after three years of implementation. First, the time it took for faculty to evaluate portfolio artifacts and reflection was burdensome. Second, we used a self-constructed online system for the portfolios. The graduate student who constructed the system graduated and we eventually had no one that could manage and troubleshoot the software. Finally, the portfolios weren't effectively integrated into the curriculum because most classes did not require them to place artifacts into the portfolio. The portfolio was required for a one-credit senior seminar class, but many students had not saved artifacts or had little enthusiasm for creating the portfolio. Thus, we abandoned electronic portfolios for our continuous improvement process and relied instead on collecting representative examples of key assignments from the instructors for evaluation.

We expected that not every innovation we tried in our continuous improvement process would work. Post-graduate evaluations and electronic portfolios were not viable for us. However, living and learning communities, workplace competencies, course evaluations and continuous improvement were significant innovations that changed our curricula and improved student outcomes. We would not have achieved these successes without taking risks in trying new approaches in our curricula.

Creating a Welcoming Environment

The ABE department at ISU is committed to fostering a diverse and equitable climate that supports all students, faculty, and staff. Through various initiatives, the department promotes culturally responsive teaching practices, an inclusive campus environment, and equitable opportunities for success in their academic programs and research activities. Several efforts have been initiated to promote an equitable climate within the department. These efforts include establishing a Diversity, Equity, and Inclusion (DEI) Committee responsible for developing policies and procedures to ensure everyone is treated fairly and with respect, regardless of race or gender. The ABE DEI Committee meets monthly to coordinate departmental-level diversity programming and includes underrepresented and female students in ABE. The committee has also been working to engage our international student population better. The committee's mission is to create a welcoming environment to help increase the representation of women and other underrepresented minorities in ABE's undergraduate and graduate student populations and inform our recruitment and retention efforts.

We have developed a multi-tiered strategy to make ABE a more welcoming department. Our ongoing efforts to make ABE a more welcoming department include training faculty (four times annually: one faculty meeting per semester and at each semi-annual retreat), staff (at fall retreat), peer mentors (beginning of each semester), student club officers (each fall), and teaching assistants (each semester) on topics ranging from inclusiveness to addressing in-classroom issues. We conduct specific programming for our Women in ABE group (speed networking, coffee breaks, and professional development). ABE faculty helped to revise a previously adopted Classroom Code of Conduct (ABE 2023a), which was included in all syllabi starting in Spring 2017. This Classroom Code of Conduct has been physically mounted in all ABE classrooms. We also added diversity-related criteria to annual faculty annual performance reporting requirements.

As part of our department's strategic plan, one of our departmental guiding principles includes diversity and inclusion. The Guiding Principles says, "As the Agricultural and Biosystems Engineering Department works towards achieving our mission, all members of the department are expected to act in accordance with Iowa State University's values and the identified departmental values and guiding principles".

The ABE departmental values and guiding principles are (ABE 2023b):

- Diversity and inclusion of students, staff, faculty, stakeholders, and viewpoints
- Meaningful and collaborative partnerships with domestic and global stakeholders
- Excellence in the scholarship of learning, discovery, and outreach
- A systems approach that increases sustainability
- Culture of lifelong learning

Departmental key performance indicators used to measure that the department has a welcoming learning atmosphere for all interested students include:

- 100% of undergraduate curricula have diversity, equity, and inclusion (DEI) learning outcomes in each year of the four-year plan
- 100% of student club leaders, peer mentors, undergraduate, and graduate teaching assistants receive DEI training every year
- 100% of faculty participate in two or more hours of DEI training every year

The ABE Code of Classroom Conduct developed by ABE faculty strives to ensure an environment conducive to learning, growth, and respect for all. The Code recognises the invaluable contributions of every student and faculty member, regardless of individual backgrounds or beliefs, and encourages constructive dialogue. It promotes mutual respect between students, faculty, and staff through open communication and active listening. Additionally, the Code encourages students to be aware of their biases, strive for excellence in learning, practice professionalism in behavior, and conduct research with integrity.

The department regularly monitors its environment to ensure compliance with its Code of Classroom Conduct policies.

Faculty members are encouraged to remain alert for situations that could lead to conflicts or harm community members. If a problem arises that poses a threat or challenge to the safety or wellbeing of any member(s) within the class atmosphere, then swift action is taken to address this issue appropriately. This action includes holding all parties accountable while ensuring everyone is treated fairly according to due process guidelines outlined in our Code of Classroom Conduct policy. To ensure this standard is upheld, faculty members must be aware of their biases when grading or interacting with students. Furthermore, students are encouraged to report any violations they experience or witness so that prompt action can be taken by the administration or staff members who have been trained in handling sensitive issues involving discrimination or harassment.

The department's Code of Classroom Conduct is invaluable in promoting an inclusive learning environment where everyone can thrive without fear, reflecting all unique perspectives while fostering successful engineering careers within Iowa State University's diverse student population.

In 2017, department course syllabi were updated to include the ABE Code of Classroom Conduct, which also included statements about respect, diversity, harassment, and discrimination. A course syllabus is important for ensuring clear course information and expectations. By providing clear expectations, faculty can ensure that their students have all the information they need to understand what is required from them before they begin studying agricultural and biosystems engineering at Iowa State University.

The department holds a teaching workshop during our faculty retreat each fall that includes directions on what should be included in each course syllabus. These include:

1. The syllabus should present a list of learning objectives that articulate what the students are expected to learn by the end of the course. This list should include theoretical concepts and practical applications, enabling the students to develop skills applicable outside of the academic context.
2. The syllabus should outline all course content, including required readings, lectures, and assignments. A calendar with deadlines for assignments, exams, and other important dates should also be included.
3. The syllabus should indicate what student conduct guidelines must be followed to remain enrolled in the course. It is important to note any attendance policies and rules about cheating or plagiarism for students to remain aware of their responsibilities during their studies.
4. A clear grading policy must be included on the syllabus so that students know how their success will be evaluated throughout the semester. This policy should include descriptions for each grade and any extra credit opportunities available to help them succeed academically.

We created the ABE Students of Color programme to support underrepresented students within our department to promote diversity, equity, and inclusion. This programme focuses on providing support services to underrepresented students so that they can reach their full potential in their studies. ABE Students of Color offers financial assistance to those in need, mentorship opportunities from alums, informal dinners with ABE faculty and staff, and career-building workshops to help students gain valuable skills and knowledge for their future endeavors.

Through this programme, ABE demonstrates its commitment to creating a welcoming and equitable campus environment where everyone can feel respected and valued regardless of their background. The department understands the importance of creating a vibrant, diverse community where everyone is supported. Students from historically underrepresented minority groups feel more welcome in the ABE department, which offers personalised attention and assistance for those seeking academic achievements. Through tailored mentoring programs and workshops designed for students of color, the ABE department provides an essential resource for students who often face cultural barriers when entering higher education settings. Furthermore, these innovative initiatives seek to create an inclusive learning environment that reflects everyone's perspectives while promoting successful engineering careers within the university's diverse student population.

The department has also taken steps to increase its representation of traditionally underrepresented graduate student minorities by providing resources such as fellowships and scholarships to support them as they pursue their graduate degrees.

In addition to providing resources to help ensure student recruiting success, the department has implemented several measures to increase diversity among its faculty. These include advertising faculty job postings using a variety of posting sites and outlets. Search committees must be trained to minimise bias in evaluating and interviewing candidates, including how to write unbiased preferred qualifications in the job description. Search committees are also held accountable by the college administration to demonstrate how they have pursued and obtained a diverse candidate pool.

Overall, the ABE department strives hard to create an accessible and welcoming atmosphere for all members of its community by supporting equity-based policies, increasing access for underrepresented minorities, offering scholarship opportunities, creating culturally sensitive learning environments through workshops, and investing in diversity initiatives geared towards encouraging more women into engineering fields. With these initiatives, there has been a notable increase in the number and diversity of students enrolled in undergraduate and graduate programs within the department. As a result of increasing minority representation, we can incorporate more diverse perspectives into classroom research projects. The impact can also be seen through research projects being conducted within the department. With a greater variety of voices contributing to studies, researchers can explore problems with a broader range of approaches than previously possible, allowing them to better develop innovative solutions that could benefit society.

Faculty and staff members continue to create an inclusive environment for all students by implementing strategies such as mentoring programs, guest lectures from diverse professionals and workshops on how best to create a welcoming space for learners from all backgrounds. These efforts are helping to create an atmosphere where everyone is valued and respected. As a result, more students feel comfortable speaking up and sharing their thoughts in class discussions – something that may not have happened if not for these initiatives.

Summary

The Agricultural and Biosystems Engineering department at Iowa State University has adopted numerous new practices and changes in our curricula to better prepare students for their professional careers, improve student retention and academic performance, and create a welcoming environment for all. We have created a welcoming and productive environment that supports student success through Learning and Living Communities, workplace competencies, continuous improvement and diversity, equity, and inclusion programming. We look for ways to improve so that our department will continue to focus on student success for the next 25 years and beyond.

Acknowledgements

The authors would like to acknowledge the contributions of the Department of Agricultural and Biosystems Engineering faculty, staff, and students related to the student successes gained through the innovative programming described in this chapter. We would also like to thank our spouses for their support over the years of our educational efforts.

REFERENCE LIST

Agricultural and Biosystems Engineering. (2023a). ABE code of classroom conduct. Department of Agricultural and Biosystems Engineering, Iowa State University, Ames, IA. https://www.abe.iastate.edu/abe-code-of-classroom-conduct/

Agricultural and Biosystems Engineering. (2023b). Mission and vision. Department of Agricultural and Biosystems Engineering, Iowa State University, Ames, IA. https://www.abe.iastate.edu/abe-department/mission-and-vision/

Brumm, T.J., Mickelson, S.K. & Ellertson, A.P. (2003). Using ePortfolios to develop and assess ABET-aligned competencies. *Proceedings of the Annual Meeting of the American Society for Engineering Education*, Nashville, TN (June).

Brumm, T.J., Hanneman, L.F. & Mickelson, S.K. (2006a). Assessing and developing program outcomes through workplace competencies. *International Journal of Engineering Education*, 22(1): 123–129.

Brumm, T.J., Mickelson, S.K. & White, P.N. (2006b). Integrating behavioral-based interviewing into the curricula. *National Association of College Teachers in Agriculture (NACTA) Journal*, 50(2): 28–31.

Brumm, T.J., Mickelson, S.K., Steward, B.L. & Kaleita-Forbes, A.L. (2006c). Competency-based outcomes assessment for agricultural engineering programs. *International Journal of Engineering Education*, 22(6): 1163–1172.

Harms, P.C., Mickelson, S.K. & Brumm, T.J. (2001a). Using a first-year learning community to help meet departmental program objectives in agricultural and biosystems engineering. *Proceedings of the American Society for Engineering Education*, Albuquerque, New Mexico (June).

Harms, P.C., Mickelson, S.K. & Brumm, T.J. (2001b). Using learning community course links to bring meaning to the first-year engineering curriculum. *Proceedings of the American Society for Engineering Education*, Albuquerque, New Mexico (June).

Mickelson, S.K. (2000). Retention of freshmen agricultural engineering students through an experiential lab course. *ASEE Annual Conference Proceedings CD-ROM.* Session 2608, pp. 1–13. St. Louis, Mo. (June).

Mickelson, S.K., Harms, P.C. & Brumm, T.J. (2001). Building community for first- and second-year students in the Agricultural and Biosystems Engineering Department at Iowa State University. *Proceedings of the American Society for Engineering Education*, Albuquerque, New Mexico (June).

Mickelson, S.K. & Brumm, T.J. (2005). Measuring the success of learning communities. *Proceedings of the Annual Meeting of the American Society for Engineering Education*, Portland, OR (June). Best Paper Award.

Mickelson, S.K., Harms, P.C. & Brumm, T.J. (2007). Longitudinal study of learning communities in agricultural and biosystems engineering. *International Journal of Engineering Education*, 23(4): 672–682.

Saunders, K., Brumm, T.J., Brooke, C., Mickelson, S.K. & Freeman, S.A. (2008). Assessing student work to support curriculum development: An engineering case study. *Journal of Learning Communities Research* 3(3): 47–62.

11

Essential Contextual Knowledge for Agricultural and Biosystems Engineering Education in Low-income Countries

James P Trevelyan and Andrew L. Guzzomi

Introduction

Since pre-history, people who we would call engineers today have provided artefacts, such as infrastructure and tools, enabling people to be more productive, to do more with less effort, time, energy and material resources. Irrigation systems, roads, vehicles, and communication systems are common examples. Starting with the industrial revolutions of the 18th century, rising agricultural productivity in high-income countries (HICs) has reduced the number of people needed for agricultural production from over 60% of the working population to around 1% today in Europe and the USA (Senzanje, 2003). At the same time, in many HICs, agricultural production not only meets domestic requirements but also provides large surpluses for export. This transformation has also completely restructured societies from small rural communities to large and diverse urban communities with high quality education, social welfare, and healthcare services. Large productivity increases, not only in agriculture, enabled HICs to invest in education, infrastructure, and technological advances, facilitating further productivity improvements.

Agricultural and biosystems engineers played a significant role in this productivity transformation, providing labour-saving machinery such as irrigation, tillage machines, mechanical harvesters, tractors, instruments, electronic ID tags for animals, and post-harvest processing and storage solutions for farmers. In common with food processing, biological, and biomedical engineers, they design for the intrinsic variability of living organisms where large differences in morphology and properties are normal (Opara, 2004). Engineers also created factories to process agricultural products into higher-value packaged food with much longer shelf life than the raw products from farms. Farming itself has become a far more finance- and knowledge-intensive occupation for which post-secondary education is at least highly desirable if not essential. Huge farming enterprises run by corporations have gained easier access to the investment capital to acquire complex modern farming machinery.

In many densely populated low-income countries (LICs), this social, economic, and technical transformation is still in progress. Half the world's food is produced by 1.5 billion small-scale farmers in non-industrialised countries with 80% of food produced by small-scale farmers (Bragdon & Smith, 2015). Data from the World Bank shows agricultural employment ranging from 35 to 45% in LICs like India, Pakistan, Nigeria, and Sudan, accompanied by increasing urbanisation.[1] Agricultural transformations in HICs relied on easily processed mineral deposits, many in colonised countries, and low-cost fossil-fuelled energy that generated pollution and greenhouse emissions. Most LICs with large populations desire a similar transformation but have to work with higher costs and environmental constraints, so different approaches will be needed. Therefore, the future challenge for agricultural and biosystems engineers, particularly in LICs, is to enable large productivity improvements without relying on unsustainable mineral extraction, costly chemicals to control pests, and fossil-fuelled energy. They will need to minimise the need for refrigerated storage and transport while minimising waste in challenging climates and long farm to market transport systems. Engineering schools will need to reshape agricultural engineering curricula to enable graduates to meet these challenges (Opara, 2004). Also, many might question the wisdom of practically eliminating the agricultural labour force as in HICs. Instead, they argue, increasing the value of agricultural products, moving up the 'value chain', might create sufficient resources to meet the need for high quality social welfare, education and health services for large rural populations.

Today, agricultural engineering is a minority engineering discipline in many HICs and agriculture typically forms a smaller part of their economies. However, in LICs, there is still the potential for large productivity gains and possibly greater opportunities for agricultural engineers than in HICs where most agricultural engineers find work today. Frey and Osborne (2017) indicate that, within the US HIC context, some agricultural engineering tasks may be susceptible to job automation with advances in information technology. Significant improvements in global agricultural productivity could be achieved if the majority of the world's food producers who operate small scale farms in LICs were the beneficiaries of agricultural engineering R&D (Moss, Nichols, Foster, Ryan, & Guzzomi, 2021) and this will require considerable human ingenuity. For that reason we have chosen to focus on LICs in this chapter.

Senzanje (2003) has reported that agricultural engineers in LICs today face many challenges gaining relevant employment

DOI: 10.1201/9780429150111-13

This chapter has been made available under a CC-BY-NC-ND license.

in their field with reasonable remuneration. This is a common issue in many engineering disciplines in LICs and arises partly because current engineering school teaching programmes do not enable graduates to meet the needs of local firms (Blom & Saeki, 2011; Tilak & Choudhury, 2021). As Domal (2010) and Trevelyan (2013, 2020, 2022) have observed in India, many engineering graduates today find themselves almost completely unprepared for work, especially in smaller firms and government enterprises. The first author's research interviews in South Asia demonstrated that many engineering graduates find higher paid work as programmers with a couple of months training than as engineers provides additional evidence that four or years of engineering education has not prepared them well enough for local engineering workplaces. Senzanje's (2003) observations help to demonstrate that the Indian experience can be generalised to Africa as well.

Further evidence of education shortcomings comes from observations over several decades showing that engineers see much of their work as 'not real engineering'. This phrase captures the frustrations experienced by many engineers who yearn for more technically challenging tasks in their work (Bailyn & Lynch, 1983; Perlow & Bailyn, 1997) and who perceive socio-technical interactions as "interruptions" (Perlow, 1999). Many educators mistakenly shape the expectations of students by describing engineering in terms of design and technical problem-solving, activities that constitute a very small proportion of the working life for most engineers, especially in LICs (Sheppard, Colby, Macatangay, & Sullivan, 2006).

In this chapter, we draw on new insights from engineering practice research that offer explanations for these difficulties. It should be possible to improve engineering education, enabling graduates to better meet the needs of local firms, thereby improving engineering graduate employment outcomes in LICs (see Figure 11.1).

Today, most engineering educators, referred to in this chapter as 'faculty', have limited if any practice knowledge having been recruited mainly for their research abilities. This is a global issue: Cameron, Reidsema, and Hadgraft (2011) have presented data for Australia, and a similar situation exists in other countries.

It is not unreasonable, therefore, to ask "how can faculty properly educate students without understanding how novice engineers are expected to perform in local workplaces?" For the last two decades at least, most engineering education programmes have adopted professional engineering competency definitions such as those published by the International Engineering Alliance (2021), ABET and others. The assumption implicit in these definitions is that students who demonstrate sufficient competency can practice effectively as engineering novices, anywhere. Although there is still very limited evidence from LICs, there is sufficient to seriously question this assumption.

However, there are several obstacles that prevent faculty from acquiring practice knowledge. Most agricultural engineers gain employment in places far from engineering schools, making it difficult for faculty to experience engineering work practices while teaching. The lack of practice knowledge among faculty also makes it expensive for firms to help faculty acquire work experience though limited term secondments because they would require close supervision and instruction by experienced engineers.

However, we think it is possible for faculty to acquire improved knowledge about engineering practice that could help them provide a more authentic context for students building their knowledge of mathematics and engineering science.

Some faculty would argue that students can gain awareness of engineering practice by working on industry projects and through internship experiences. However, students start with limited notions of engineering practice because their coursework focuses on technical problem-solving and analysis. As a result, most graduates do not see socio-technical interactions as "real engineering", a perception that is strongly reinforced by the professional engineers they meet. This preconceived mental framework hinders their learning about socio-technical interactions with other people that dominate professional practice as explained in section 2.

FIGURE 11.1 Residential construction in Melbourne and Islamabad in November 2022. There are obvious differences in practices.

So far, there have been no detailed ethnographic research studies of agricultural and biosystems engineering practice. Therefore, in this chapter we will draw on published research studies of engineers in other disciplines and draw on our personal agricultural engineering work experiences to provide some insights into agricultural engineering work.

This knowledge can help faculty in several ways. First, it may help faculty understand that engineering work, especially in LICs, is very different from notions of design and technical problem-solving advanced by many engineering schools today. Second, this knowledge may help faculty more appropriately shape the expectations and cultural awareness of students. Third, this knowledge can help faculty design more authentic education experiences that would enable students to understand the context in which they will use technical analysis and design methods learnt in their coursework and improve students' motivation to learn. Fourth, we hope it will stimulate some faculty to research practice in local firms to further build knowledge and understanding about agricultural and biosystems engineering work.

With a focus on LICs, there is inevitably some overlap between agricultural and biosystems engineering and humanitarian engineering that focuses on capacity-building in predominantly rural low-income communities (Gupta, Singh, Sharma, Chatterjee, & Saha, 2019). One distinction between the two is that agricultural and biosystems engineering initiatives, other than research, require economic justification whereas humanitarian engineering tends to be motivated by altruism. That said, it is important to note that altruism can be linked with economic benefits as Paul Polak (2008) showed in his work at the Stanford D-School.

What We Know from Engineering Practice Research

Most of the research on engineering work has come from social scientists in HICs with some contributions from engineering education researchers. Until the 1980s, most researchers were interested in engineers as a social class in HICs and also the dynamics of professions such as engineering (e.g. Bailyn & Lynch, 1983; Layton, 1986; Meiksins & Smith, 1996; Noble, 1979; Perrucci, 1971; Smith & Meiksins, 1995). The rise of Japan's economic power in the 1980s stimulated comparative studies to understand how engineers contributed to that success (Bailyn & Lynch, 1983; Bratton, 1991; Button & Sharrock, 1994; Kilduff, Funk, & Mehra, 1997; Lam, 1997; Lynn, Piehler, & Kieler, 1993). As Japan's economic success receded in the 1990s, research interest shifted to workplace studies on individual technicians and engineers, partly inspired by Zussman's and Bucciarelli's seminal works studying the activities of engineers (Bucciarelli, 1994; Zussman, 1985). In comparison with engineers, many aspects of technicians' work are relatively easy to observe, though it can be hard to identify the cognitive skills that contribute to their performances (Bechky, 2003; Flesher, 1993; Orr, 1996). Many studies were motivated by an interest in understanding engineering design and the thinking behind it, then thought by many to represent the essence of engineering work (Dym, Agogino, Eris, Frey, & Leifer, 2005; Eckert et al., 2004; Ewenstein & Whyte, 2007; Henderson, 1999; Moritz, 1996; Wong & Radcliffe, 2000). Dominique Vinck and his collaborators studied engineers in a wider range of industrial settings with the longer term aim of understanding practice and systematic influences on workplace behaviour (Vinck, 2003). Even so, Barley (2005) concluded that very little was known about the social organisation of engineering work at that time, and called for grounded interpretive studies of engineers at work.

Several researchers have since responded to Barley's call. Even so, engineering practice research, empirical investigations on what engineers actually do in their work, is still in its infancy. A few hundred publications since 2005 make up a small collection alongside technical engineering literature. However, the results already pose significant challenges for engineering faculty. Not the least of these is the finding that engineers spend relatively little time on solitary technical analysis and calculations for which much of their formal education has prepared them, supported by a vast body technical literature. The greater part of engineers' work consists of social interactions with other people (Williams, Figueiredo, & Trevelyan, 2013).

Models of Engineering Work

Following occupational classifications outlined by Howell and Wolf (1991), we can identify three groups of engineering work activities.

Group 1: ***interactions with physical objects,*** often referred to as 'hands-on' activity, requiring manual dexterity, perception-motor coordination, and tacit knowledge accumulated from many years of experience.

Group 2: ***cognitive interaction with abstract objects***, requiring substantial periods of solitary work, often with computers serving as intermediary communication devices. As for group 1 activities, tacit knowledge accumulated through years of experience enables high levels of performance.

Group 1 and group 2 activities also involve collaboration when the artefacts are large or complex and beyond the capacity of one person to perform the required work.

Group 3: ***socio-technical interactions with other people*** enable engineers to plan, organise, collaborate in, and coordinate physical (group 1) and cognitive (group 2) activities by other people. Many of these interactions are synchronous (e.g. face to face dialogues, conferences or meetings, either physically present or by phone or video calling or teleconferencing), relying substantially on oral and written communications supplemented by non-verbal cues and gestures. They may be in formal meetings or informal workplace settings, or socialising outside the office or workplace. Engineers also interact asynchronously through email correspondence, text messages, with boundary objects (Star & Griesemer,

1989; Whyte, Lindkvist, & Jaradat, 2016) such as images, two- and three-dimensional drawings, digital data using workflow and project management systems, customer relationship management systems, word processors, and many other forms of computer-mediated communication. Computer systems have enabled people to work with artefacts so elaborate that the details exceed the memory capacity of people to recall accurately (e.g. Whyte et al., 2016).

It would be a mistake to presume that these interactive performances are non-technical. Trevelyan (2014,Ch7-12), Jesiek, Buswell et al. (2019), Blandin (2012) and others have identified socio-technical performances required for technical collaboration in which technical expertise is a prominent if not pre-eminent factor. Engineering firms have evolved highly structured ways to manage interactive performances as exemplified in project management processes, including the critical element of technical specifications (Trevelyan, 2014,Ch10). Such performances demand high-quality relationships (Korte, 2009; Lutz & Paretti, 2021) with high levels of trust and engagement as most interactions tend to rely more on informal leadership and referent power rather than organisational authority. These workplace performances are not distinct nor exclusive, but rather overlap with and reinforce each other.

Numerous research reports confirm that professional engineering work predominantly consists of Group 3 activity, with varying but less significant involvement in Group 2 activity (Jesiek et al., 2019; Trevelyan, 2014). Few professional engineers participate directly in Group 1 activities (e.g. Korte, 2009; Korte, Brunhaver, & Sheppard, 2015; Lutz & Paretti, 2021; Trevelyan & Tilli, 2008). Figure 11.2 illustrates this finding.

Magarian and Seering (2021) identified responsibilities that distinguish professional engineers and reviewed earlier discussions on what is considered professional engineering work. They argued that the primary distinguishing aspect of professional engineering work is responsibility for the efficacy and safety of products, services, processes, and systems through governance of design. Furthermore, professional engineers often hold responsibility for the financial success of a project along with managing other people involved. Extensive group 3 activities frequently stem from these responsibilities.

Technicians' work mainly requires group 1 activities, with a steadily increasing component of group 2 activities as more tools require programming skills. As supervisors, group 3 performances may also become a significant component of their work.

Technologists mainly include programmers and drafters interacting with computer systems, along with plant and system engineers interacting with power grids, chemical process plants, ships, aircraft, telecommunications networks, and other complex artefacts. They will typically adjust settings in the intermediary computer systems. While their work primarily involves group 2 activities, supplemented with some experiences of group 1, collaborative work also requires group 3 activity.

Where professional engineering predominantly requires Group 3 activity, engineering education typically emphasises solitary Group 2 activity and, in doing so, leaves a significant

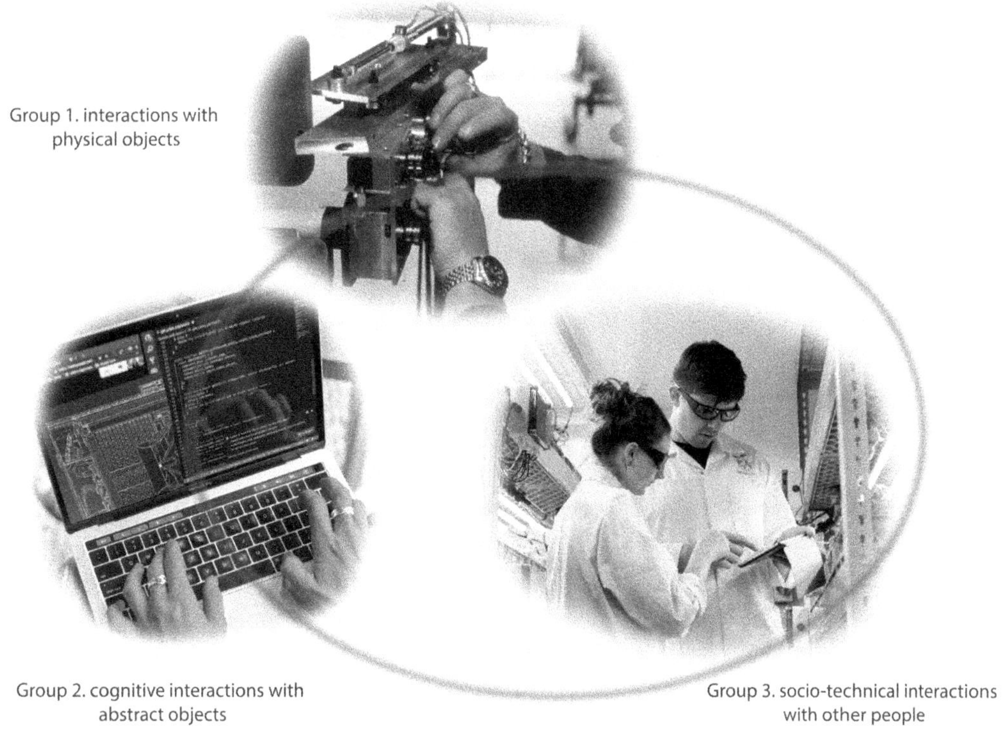

FIGURE 11.2 Research demonstrates that occupational work activities for professional engineers mostly requires group 3 activity, socio-technical interactions with other people, with some hands-on work and some cognitive interactions with abstract objects.

gap between education and workplace practices. Students learn little if anything about Group 3 activity, collaboration methods, and the organisational contexts in which engineering work takes place.

Trevelyan (2014, Ch3) presented a model of professional engineering practice showing project-related activities, reproduced as Figure 11.3. Activities in the upper row result in a detailed technical and commercial case to support the final investment decision. Engineers start by identifying possible engineering solutions that address clients' needs. In the next activity, they conceive solution details, either by drawing on similar solutions developed before, or by designing new solutions. Then, they analyse the likely performance of one or more solutions to determine technical and commercial feasibility. Preparing the solution and investment case represents around 10% of the total expenditure. The main activity in the lower row is delivering the chosen solution, as far as possible within time and resource constraints.

Few if any engineering education programmes address the grey shaded areas of the diagram. Therefore, this representation provides a different view of the education–workplace gap in terms of solution delivery activities that are not addressed to a significant extent in engineering education. As in the earlier model, these activities comprise substantial collaborative work requiring frequent socio-technical interactions with other people.

Research studies provide insights that explain why Group 3 socio-technical interactions require so much time. Most engineers need to rely on expertise beyond their personal knowledge, often arranging for others with the required expertise to contribute skilled collaborative performances (Anderson, Courter, McGlamery, Nathans-Kelly, & Nicometo, 2010; Blandin, 2012; Itabashi-Campbell & Gluesing, 2013; Trevelyan, 2007, 2010). Engineers also need to advocate for resources and particular interpretations of requirements (Mukerji, 2009; Sandberg, 2000).

As explained later, agricultural engineering practice requires a broad range of expertise beyond that capacity of students to learn in a four- or five-year agricultural engineering curriculum. Agricultural engineers, therefore, need to draw on the expertise of people from many other specialist disciplines. This activity has been described as boundary-crossing work (Asplund & Flening, 2021; Jesiek, Mazzurco, Buswell, & Thompson, 2018; Jesiek, Trellinger, & Mazzurco, 2016; Jesiek, Trellinger, & Nittala, 2017). Wilde and Guile (2021) describe "conversations, debates, deliberations and recollections of previous experiences that may occur face-to-face or be facilitated by computer-mediated communication, which inspire professionals with the same or different specialisms to think imaginatively about how to tackle project problems". They refer to this as 'immaterial activity', socio-technical interactions that generate ideas that, at some point in time, emerge as solutions even though they are never explicitly mentioned or costed in project work plans.

As the Figure 11.3 caption explains, engineering education programmes mainly address analysis and performance prediction capabilities and hardly mention the processes needed to deliver working solutions: construction, manufacture, assembly, testing, and commissioning. In many if not most instances, solution delivery activities are performed by different teams. However, relying entirely on document packages passing from one team to another is seldom sufficient for the original technical intentions to be realised faithfully. Design and performance evaluation engineers, therefore, also spend much of their time monitoring implementation activities (Trevelyan, 2014, Ch 9, 10). In particular, they evaluate technical interpretations that mutate in the minds of implementation teams as they work around regulatory, environmental, safety, expertise, logistical, and financial constraints. The effects of these constraints may be unfamiliar for the engineers who conceived the original technical intentions. For example, it is not uncommon for a contractor or technician to re-work many design details to make construction easier and cheaper.

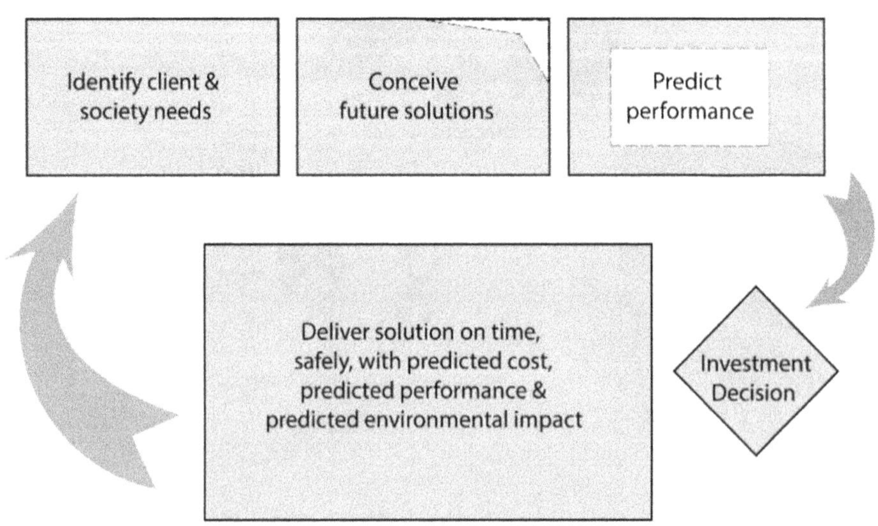

FIGURE 11.3 A sequence of professional engineering activity related to projects, starting with identifying client and society needs (Trevelyan, 2020). The lighter shaded sections with dashed outlines indicate aspects addressed in engineering education programmes. Most aspects are unfamiliar for engineering graduates.

We can see this pattern of re-interpretation in detailed published accounts of events leading up to the collapse of walkways in the atrium of the Kansas City Hyatt Regency hotel in 1978 (Gillum, 2000; Luth, 2000; Moncarz & Taylor, 2000; Pfatteicher, 2000). 114 people died and the small city endured decades of trauma afterwards. Many faculty use this event to illustrate an engineering ethics failure.

However, the detailed published accounts also tell us much about engineering practices. They reveal how the original design was re-interpreted by the construction contractor such that forces in the walkway supports were far higher than in the original design. As a result of mis-communication between the supervising engineers and the construction contractor's drafting team, this mistake remained undetected. Contributing factors included adopting a fast-tracked project plan in which many design details were only considered after construction had started, several other critical issues arising during construction, and a decision by the steel fabrication contractor to outsource completion of design drawings to a subcontractor because they won another large contract. Therefore, the engineers who should have been able to use their technical expertise to identify where interpretations diverged too far from the original intentions were under extreme pressure to complete their work quickly. They did not allow sufficient time to identify the force overload and decide whether to advocate for corrective actions. In among the countless socio-technical interactions every day, in meetings, by email, text messages, phone calls, reports, and other information systems, engineers are called upon to make rapid, consequential technical decisions, often based on imprecise information.

The Hyatt Regency investigations provided a detailed account of structural and construction engineering work practices. Normally, engineers record few details of their complex, interdependent socio-technical activities. It is only because the Hyatt collapse was extensively investigated and studied to prevent a recurrence of a similar disaster that we have such extensive records in this instance. Each of the four cited papers includes different aspects of the events leading to the disaster and, as a collection, provide insightful reading for any engineer.

These accounts also help to explain why tacit knowledge plays a such a significant role in these decisions. Goold and Devitt (2013) provided supporting evidence for this finding, showing that engineers tend to apply their mathematics and science knowledge in rapid technical assessments more as tacit knowledge than by applying slower methods learned in classrooms. Kahneman explained similar ideas in his book, *Thinking Fast and Slow* (2011). Tacit knowledge is, by definition, knowledge of which one is unaware (Polanyi, 1966), and this helps to explain frequent comments by engineers that they hardly apply any of the science and mathematics they learned in engineering schools. They do apply the knowledge, but without necessarily realising it.

Technicians – fabricators and machinists working on the development of agricultural implements and machines – also reinterpret instructions and designs represented by drawings and conversations with agricultural engineers. Here are some examples from personal observations.

- A technician re-interprets welding details specified by the engineer, e.g., making stitch welds instead of fully welding components. The technician may not notice or fully understand welding symbols on the drawing, or may use their own contextual knowledge to select an appropriate welding technique.
- A technician produces parts with dimensions significantly different from those shown on the sketch/drawing, using their contextual knowledge to reinterpret the engineer's requirements.

Experienced engineers have learned technical coordination techniques (Trevelyan, 2007), particularly the need to discuss critical technical requirements with technicians and technologists before work starts, and to regularly monitor work for unacceptable reinterpretations.

These examples reflect contextual knowledge, or experience, suggesting that agricultural implements do not require precision. This may be acceptable for low cost machinery. However, increased agricultural system performance requirements demand more attention to detail in design and fabrication. In HICs, farms require higher performance machines with advanced capabilities and can afford the additional cost. Furthermore, farmers expect higher quality and precision associated with that higher capital cost. Agricultural engineers can add significant value by recognising market drivers that influence machinery designs for better adoption by end users across diverse markets (Cavallo, Ferrari, Bollani, & Coccia, 2014). Some engineers, particularly in HICs, still see agricultural engineering as 'crude' engineering as reflected by phrases such as 'it's a bit agricultural', a derogatory term to incorrectly imply lower standards in agricultural engineering. Recent work by Moss et al. (2021) draws attention to high quality and attention to detail demonstrated by on-farm developed prototype devices, particularly considering limited resources requiring considerable design ingenuity. Other instances demonstrating how increased agricultural performance requirements lead to new technology, including tractor manufacturers developing structural engine gearboxes and dry-sump technology often incorrectly attributed to formula 1 race car developers.

Drawing on these examples, faculty can readily devise education experiences that enable students to develop contextual knowledge based on observations of practice.

Social Culture Influences on Engineering Practices

Uniform accreditation criteria applied in many countries under agreements such as the Washington Accord (International Engineering Alliance, 2021) rely on an implicit assumption that engineering practice is similar everywhere. While Figure 11.1 shows obvious physical differences, there are also significant cultural differences that influence the socio-technical interactions that dominate professional engineering practice.

The term 'social culture' represents habitual ways in which people interact with each other, denoted by *habitus,* the word used by Bourdieu (Nash, 2003). These patterns reproduce themselves in a society through child-rearing, education, and day-to-day social interactions through which people engage with each other in family and public spaces.

Since Group 3 socio-technical interactions with other people form the largest component of professional engineering work, it is to be expected that the social culture of the society hosting an engineering enterprise influences practices that depend on social interactions. Of course, one is so immersed in the social culture of one's home country or region that it is almost impossible to notice how it influences day-to-day life. Temporary immersion in a different culture can make the influences of one's own culture more obvious. Many people experience this when they start working in a new company or institution even within their home country. Similarly, engineers may be exposed to aspects of this when dealing with international companies, suppliers, or manufacturers. Each company or institution will develop a culture that reflects both the host society and the organisation's influences. Work or an extended stay in another country can also make cultural differences more obvious.

In contrast to most HICs, LIC societies in the 'Global South' are often characterised by complex patterns of social behaviour where perceived reputation, socio-economic status, caste, tribal identity, and language strongly mediate power structures and hence collaborative performances (e.g. Waris & Kokab, 2017).

Several studies in LICs have described cultural influences on engineering practice. Coelho (2004) studied water supply engineers in Chennai, a city that recently completely ran out of drinking water supplies for several months ("How Chennai, one of the world's wettest major cities, ran out of water", 2021). Matemba (2020), while researching engineering education in East Africa, also observed significant cultural influences on engineering practice as part of her fieldwork. Kusimo and Sheppard (2019) studied engineering work in a Nigerian factory (2019) where higher skilled workers tended to see lower skilled workers as lazy and lacking an appropriate working attitude. Trevelyan (2022) showed how high-performing engineers create a sub-culture in their own work group to counteract the influence of the host society culture that inhibited knowledge sharing essential to enact distributed expertise (Trevelyan, 2010). The first author's research evidence base includes 7 years working first-hand with engineers in Pakistan, ethnographic interviews, field studies, and extensive visits to India and Pakistan over 25 years (For details, see online appendices, Trevelyan, 2014)

The following excerpt from Trevelyan (2022) illustrates how inhibitions on knowledge sharing arise from social *habitus*.

> A junior South Asian engineer acting as a production supervisor is considered to have high social status relative to production workers. He listens to his manager and later briefs workers on what has to be done. No questions are asked either by the junior engineer or the production workers because that can imply an unwillingness to listen properly. Without being able to ask clarifying questions, the listeners are acutely aware of their own unresolved uncertainties. Workers patiently wait for directions, and expensive machinery lies idle in the meantime. The engineer runs from one worker to the next explaining every small action to each of them in turn. While a casual observer might see this as laziness or 'lack of appropriate work attitude', it may be wiser for the worker to wait for a supervisor or engineer to be present to issue directions and therefore take the blame if an action turns out to be incorrect. Work stops in the absence of visible supervision. Inaction inevitably contributes to low productivity. When his manager asks for a progress report the next day, the junior engineer remains silent instead of reporting production shortfalls.

Domal (2010) and Trevelyan (2013) observed several other factors that contribute to the high costs of engineered services in the LICs they studied. For example, engineering firms lacked the systems and procedures that strengthen technical collaboration in engineering firms in HICs. While engineers in HICs also missed the significance of indirect labour costs (e.g. supervision, training, insurance, transport, accommodation, rest breaks, protective equipment), the proportion of indirect costs in a low-income country is much higher, and can dominate labour costs (Trevelyan, 2014, Ch13). Engineers in LICs missed the significance of productivity and so these two factors combined to create a misperception that labour is much cheaper in a low-income country when, if anything, the reverse is true once indirect costs and productivity are taken into account. Sales engineers representing specialist engineering suppliers are relatively rare in LICs so most novice engineers miss out on the extensive workplace training provided by sales engineers in HICs (Darr, 2000). Few engineers in LICs are entrusted with detailed financial information about their employer or project so it is difficult for engineers to exercise the level of financial accountability that would be normal in a wealthy country, limiting their ability to generate economic value for their employers (Trevelyan & Williams, 2018). Company owners commonly maintain different and contradictory accounts. Taxation accounts tend to understate income. Accounts created to support borrowing tend to overestimate income. It is not uncommon for cash transactions to form a significant proportion of turnover and expenses, often with limited if any documentation.

Local language barriers also play a part. Skilled hands-on production workers often speak a different local language or dialect from the engineers. The engineers are likely to have been educated in a language inherited from European colonial powers, mixing that with the national language and their own mother tongue in casual conversations. They may not be fully proficient and literate in any language. Production supervisors translate engineers' instructions into local dialects for production workers, often losing much in translation, magnifying apparent uncertainties. In the presence of so much apparent uncertainty, a rational response for production workers is to choose inaction unless someone more senior is present to provide guidance and take the blame for misunderstandings.

Engineered service cost differences are startling. For example, the real economic cost of safe drinking water across most of South Asia is 10–50 times the cost in Australia, mostly because one has to carry water rather than rely on pipe networks

(Trevelyan, 2014, Ch 1, 13). Similar data came from a study in Uganda in 1990, with a different cost base (Whittington, Mu, & Roche, 1990). Electricity on demand is typically five times as expensive because of the need for alternative supplies from batteries or generators. Even construction at identical quality standards is 50–60% more expensive than New York. In the absence of understanding how cultural constraints affecting engineers' performances contribute to these costs, many people attribute the differences to corruption, a questionable hypothesis (Trevelyan, 2014, Ch 13).

These studies help us understand the difficulties experienced by engineers seeking employment in LICs (Senzanje, 2003) where the social culture presents significant barriers for the engineering collaboration making it more difficult than in HICs. Engineers require cultural awareness to overcome these barriers and practice effectively, particularly in LICs.

Mapping Agriculture and Biosystems Expertise

Many agricultural engineers face challenges well beyond the intrinsic variability in morphology and properties of biological materials. Animals, as sentient organisms, not only exhibit individual behaviour traits but also require culturally-specific handling by people because of differences in the ways that people consider animals in relation to themselves (e.g. Singer, 1989).

Therefore, in addition to the fundamental mathematics, physical, and biological sciences that support relevant engineering topics, agricultural and biosystem engineers need access to expertise in a far wider range of topics than can feasibly be addressed adequately in a four- or five-year engineering curriculum. Therefore, students need to learn how to access expertise beyond their prescribed studies and work experience, an activity referred to as 'boundary-spanning' (Asplund & Flening, 2021; Jesiek et al., 2018).

Trevelyan (1992) described the development of robots for shearing sheep in Australia between 1974 and 1992, an ambitious and successful agricultural engineering undertaking performed by a team of mechanical, mechatronics, and electronic engineers. The team worked with professional shearers and drew on expertise from many others including animal physiologists, animal behaviour specialists, agricultural economists, occupational health and safety experts, ultrasonic radar specialists, specialist agricultural engineers, shearing contractors, wool marketing experts, wool classers, farmers, sheep breeding experts and many others. In addition, the team's engineers had to work with animal welfare organisations and appear before special commissions to address political concerns arising from the animal-rights movement. Boundary-spanning work was an integral part of the team's daily experiences, both within the team and with outsiders.

One of the most challenging boundary-spanning aspects was translating human shearing skills into explicit software code and electro-mechanical components (Figure 11.4). Team members acquired elementary manual shearing skills early in the project. However, they sought advice from professional

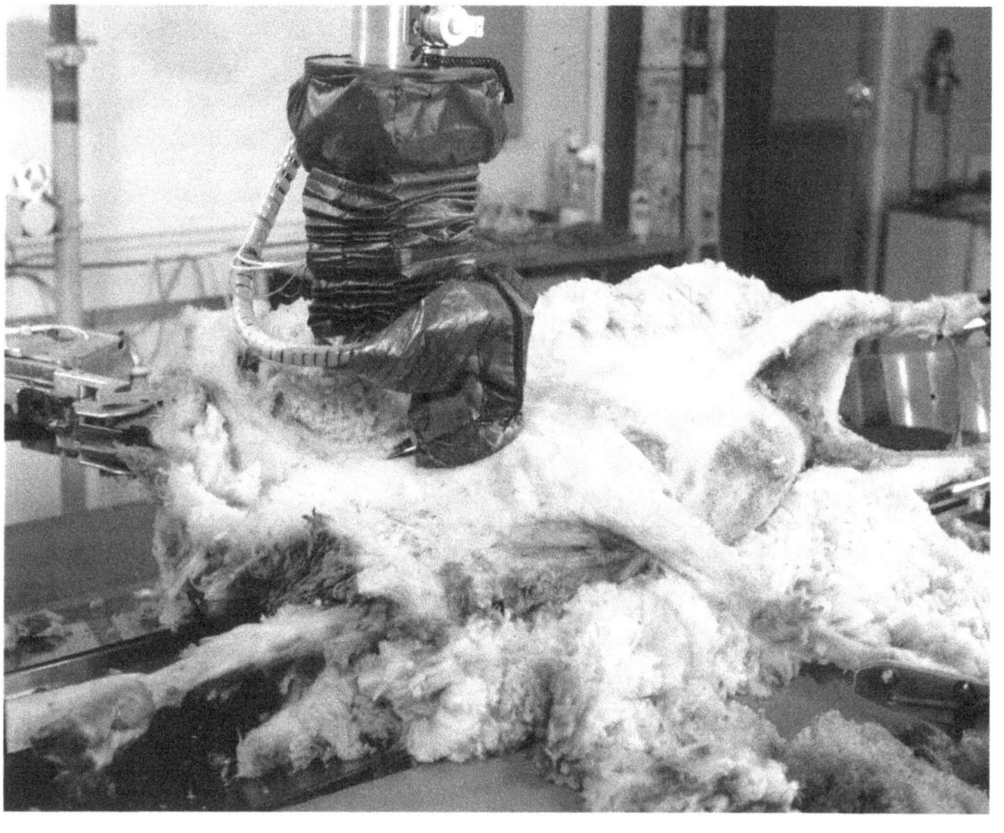

FIGURE 11.4 Robotic sheep shearing, 1989. The team tapped the expertise of human shearers to improve the robot's performance. Video of the robot operating is available at: https://www.youtube.com/watch?reload=9&v=6ZAh2zv7TMM

shearers and later a professional shearing instructor joined the team as a full-time member to help improve the shearing actions of the robot. It was difficult to perceive the shearer's actions because the wool-cutting tool is almost entirely hidden by wool for much of the time. Furthermore, it was not possible to measure the forces applied by the shearer because the cutter's mechanical oscillation motion imparts large periodic forces. However, when team members programmed the robot and prepared the shearing comb and cutter in accordance with the shearer's instructions, the results did not match expectations. Gradually, it became clear that the shearer was instinctively translating his skilled actions to ones that *he thought* were appropriate for the robot. He had prepared a shearing comb with more rounded tips than he used for himself, thinking that this would make it less likely that the robot would cut the skin. This affected the response of the sensing circuit that detected skin contact. The more rounded comb tips required a *larger* force from the robot to penetrate the densely packed wool fibres close to the skin and make electrical contact with the skin. This *increased* skin friction forces, gave rise to wrinkles ahead of the comb tips, making skin cuts more likely. Later, the team prepared comb tips with sharper points and the shearing performance improved to a level similar to a human shearer, with much fewer skin cuts than a normal human shearer.

We can see in this interaction how boundary-spanning work can be challenging, especially when there are different thinking styles and knowledge types. Translating *tacit* sensory-motor knowledge acquired by the shearer over decades of experience into explicit software code and precisely defined shapes for the shearing comb tips was not easy and required many seemingly 'immaterial' conversations, debates, discussions, and humour (Wilde & Guile, 2021), causing considerable concern for project managers insisting on a tight development schedule.

The multidisciplinary nature and boundary spanning aspects typical of agricultural engineering endeavours are also reflected in recent development of alternative site-specific weed control technologies. Low weed density in cropping fields coupled with advances in optical sensing technology is driving demand for site-specific weed control technologies that remove the need for wasteful whole-field herbicide treatments (Walsh et al., 2020). This led to the development of the first broadacre site specific spot tillage device – the Weed Chipper[2] – compatible with conservation cropping systems. Whilst tillage and site-specific weed control is not a novel concept, the team aimed to mechanise the process by developing a rapid response tyne triggered by commercial optical sensing technology (Figure 11.5).

As outlined by Walsh *et al.*, the concept was conceived during meetings in 2012 of leading Western Australian farmer-innovators and with farmers in Queensland and New South Wales experiencing significant herbicide resistant weed growth. With support from the Grains Research Development Corporation (GRDC), the project team of weed scientists, agricultural engineers, machinery manufacturers, and farmer-innovators was formed to develop the targeted tillage system. The team prioritised rapid development and wide user acceptance, and all the stakeholders contributed to the design. Therefore, the engineers, led by the second author, based their design on widely recognised tyne systems to maximise farmer acceptance in terms of familiarity, serviceability, and technology awareness. They modified the commercial hydraulic breakout tyne system on a Shearer Trashworker[3] triggered with WeedIt camera technology.

FIGURE 11.5 The Weed Chipper in operation in a fallow field in NSW. The tynes which are held in a stand-by position above ground are triggered by cameras to chip out weeds.

In work conducted at the Centre for Engineering Innovation: Agriculture & Ecological Restoration, Moss et al. (2021) showed how farmers can often contribute significant innovations. Their research on harvesting technology for subterranean clover (a legume which buries its seeds in pods under the ground) showed how close collaboration between farmers and agricultural engineers can result in commercially useful innovations, even in relatively small industries. A key contextual factor that drove innovation in the industry was the relative isolation of the farmer-inventors. Both team diversity and farmer skill versatility increased inventive outcomes (Singh & Fleming, 2010): isolated farmers have to operate as skilled mechanics, farmers, and business owners (Moss et al., 2021). However, most farmers cannot afford significant research and development (R&D) activity, especially as they are now increasingly affected by climate variability.

Australian farmers have long contributed to research through industry levies, amplified with government funding, helping to ensure that research priorities align with the interests of farmers who can benefit directly from research outcomes (Moss et al., 2021). Mobile telecommunications technology could help with similar initiatives in LICs where government support is often weak or non-existent.

Reliable mobile telecommunications and high-integrity information systems have the potential to transform farming (Opara, 2004; Sigrimis, Hashimoto, Munack, & De Baerdemaeker, 1999), and access to capital (Asongu & Boateng, 2018; Kendall & Voorhies, 2014). Traditionally, farmers in many LICs have relied on finance from powerful actors such as landlords and produce-buyers for seed, fertiliser, fuel, and machinery investments. Mobile phone systems are transforming finance because they provide trustworthy means for financial transfers, a service that formerly relied on banks that served a small minority who could meet their creditworthiness requirements. As a result, a far larger proportion of the population in low-income countries can access electronic banking and access to credit, bypassing traditional 'gatekeepers' such as landlords and corporations. This change also enables farmers to bypass traditional supply chains that favour bulk commodities while providing the credit that farmers needed.

Mobile phone technology might enable even small farmers, in sufficient numbers, to contribute significant finance through producer levies to support industry-specific agricultural and biosystems research and development in LICs. It is also essential for engineers to develop social networks with farmers to encourage them to feel safe enough to share their own innovative ideas so that commercially useful inventions can emerge through collaboration.

Implications for Agricultural and Biosystems Engineering Faculty

Particularly in LICs, faculty work long hours for low pay under time pressure and have to make up for the limitations of local school education. There are limited if any opportunities for research, and very limited access to research literature. Competency statements (International Engineering Alliance, 2021) now frame the educational objectives of most engineering education programmes, even in countries that have not ratified the Washington Accord, emphasising what graduates *can do* rather than what they *know*. Accreditation agencies such as ABET (2021), Engineers Australia (Engineers Australia, 2011) and others issue similar lists of competencies.

Competency Statements Have Limitations

Unfortunately, competency statements are brief and require contextual knowledge for understanding and, in its absence, can cause misunderstandings and inappropriate learning by students (Hager, 2004).

For example the competency "ability to cut a cake" relies on common contextual knowledge that tells us that the cutting instrument is a long enough knife to reach from the centre of the cake to the edge. However contextual knowledge relevant for cutting a cake comprises implicit and tacit knowledge (knowledge that the user is unaware of), and therefore is unwritten. In the absence of contextual knowledge, competencies can easily be misinterpreted by faculty and students alike. As Conrad (2017) recently reported, student writing emulates attributes valued by faculty: long and complex sentences with numerous reference citations. This style of writing is likely to cause confusion in typical engineering workplaces that value short, concise statements with simple sentence structures and clear standardised drawings (which can be hand drawn) where possible. Gonczi (2013), citing Hager's influence, concludes that competencies need to be assessed, at least in part, by experienced practitioners who have relevant contextual knowledge.

Suggestions for Faculty

The aim of this chapter has been to provide agriculture and biosystems engineering faculty with some insights into engineering as *practice researchers* have reported from field studies. We conclude with a series of suggestions that faculty could readily implement within existing curriculum frameworks, enabling students to develop more appropriate knowledge, skills and expectations.

Naturally, there are many on-going improvements in agriculture engineering education that have broad applicability. An emphasis on creativity combined with an appreciation for the economics of small-scale agriculture, including the value of unpaid labour (Whittington et al., 1990), can help graduates identify problems worth solving. A strong focus on freehand sketching to build visual perception skills followed by instruction on manual drafting and then CAD can help designers develop solution ideas that "make sense" as outlined in Guzzomi et al. (2012). Students need to learn that design is an iterative process in which, more often than not, the designer does not have all the necessary information and must start with assumptions or guesses, and that there is no perfect solution with so many unknowns.

The specific suggestions we have described below are particularly aimed at addressing issues that affect graduates in LICs, and which can easily be implemented by individual faculty without significant curriculum changes.

Students graduating from any engineering school enter a wide range of occupations, many in different countries and many embark on non-engineering careers. This reality can lead one to argue that engineering faculty should focus on the science and maths capabilities common to all careers in a particular engineering discipline. Graduates from many LICs obtain employment in HICs. Therefore, education programmes need to address issues that are relevant across diverse employment locales. However, in this paper, we argue that understanding how culture influences engineering practices is likely to empower graduates in any career, particularly agriculture and biosystems engineering careers in LICs.

1. We suggest the most important research finding that faculty can pass on to students is that it is possible to earn a salary in low-income countries (LICs) that is at least as high as in high-income countries (HICs). Typical engineering salaries in LICs are one fifth to one third of salaries in HICs, motivating many LIC engineers to migrate to HICs. Our research exposed engineers who attracted much higher remuneration, as high or higher than for an equivalent position in HICs, largely by their ability to foster effective collaboration, and also an ability to create significant value for their firms (Trevelyan & Williams, 2018). There is no need to migrate to a wealthy country to gain a high salary and income security.

   The first author located networks of LIC engineers earning high salaries by starting with specialist engineering suppliers distributing relatively high-cost components and materials. These companies compete with other distributors selling similar products at much lower prices. The high-cost supplies can be justified by the superior value they provide to the user, for example by providing high quality and reliability, helping to ensure that high-value assets operate reliably without costly breakdowns or frequent unplanned maintenance. In many LIC firms, purchasing departments insist on procurement processes that select the lowest cost components. However, high-cost components often provide much greater value. Engineers who can successfully influence purchasing departments to focus on value rather than purchase cost are likely to be among those earning much higher salaries. These engineers can serve as role models for students to demonstrate how an understanding of value generation can lead to higher remuneration and job satisfaction.
2. Students who learn early on that socio-technical interactions will take most of their time will be less likely to experience a mismatch between expectations and the reality of engineering work as engineers, and also that these interactions are essential for success. Trevelyan (2020) provides guidance to help students rapidly acquire socio-technical workplace skills.

   Faculty can also point out that classroom interactions are socio-technical. Education, after all, is a socio-technical process in which students learn as much from each other as from faculty, and much of the learning occurs through complex social interactions in and out of class.

   This is not an easy task for faculty or students. There are strong beliefs in most engineering communities that social interactions and most collaborative engineering activities such as technical coordination (Trevelyan, 2007) and project management are 'not real engineering'. Students and engineers identify real engineering as the Group 2 mathematical and science-based activities learned in engineering schools. Changing this perception is difficult and will require frequent repetition.

   Collaborative learning methods can provide a setting in which students can appreciate the significance of socio-technical interactions (Smith, Sheppard, Johnson, & Johnson, 2005). Traditional group projects may not satisfy the requirement for collaborative learning because students tend to subdivide the work between them and work separately on each part (Leonardi, Jackson, & Diwan, 2009; Sheppard, Macatangay, Colby, & Sullivan, 2009; Tonso, 2006). Jigsaw puzzle learning is one truly collaborative method that, in many ways, mirrors the kinds of distributed expertise encountered in engineering workplaces (Brown et al., 1993; Trevelyan, 2010). In the first phase, groups of students each learn distinctly different technical topics. Next, after being allocated to teams in which each student acts as a subject-matter expert, they create solutions for problems that can only be solved by students contributing their respective expertise and collaborating together. An important feature of collaborative learning is that more able group members need to help teach the less able members as any of them will be called upon to present the group's work.
3. Faculty can explain the notion of productivity to students, helping them understand why economic development is so challenging in LICs. Many people in LICs imagine that their countries remain poor because of corruption by elites and economic mismanagement by incompetent rulers. While these factors cannot be eliminated from any analysis of productivity, the significant role of engineers in providing the means by which people can be more productive is very often ignored. Indeed, engineers are barely mentioned in UN documents related to the Sustainable Development Goals (e.g. United Nations, 2017a, 2017b; United Nations Development Programme (UNDP), 2017; United Nations Development Programme (UNDP), 2019) and the potential for engineers with regard to the UN's Decade of Ecological Restoration not clearly apparent (Masarei, Erickson, Merritt, Hobbs, & Guzzomi, 2021). To help faculty and students in making this connection, we have advanced a succinct new

definition of engineering, based on our research observations of engineers at work:

> Engineers are people with specialized technical knowledge who conceive, deliver, operate and sustain artificial objects, systems and processes that enable people to do more with less effort, time, materials, energy, uncertainty, health risk, and environmental disturbances. (Trevelyan, 2019, 2020)

4. Faculty can help students learn that the cost of engineering activity in LICs is likely to be higher than in HICs, for similar product quality, reliability, and fitness for purpose. This will seem counter-intuitive for many, even most faculty. Data on the cost of safe drinking water provided in this paper, easily verified, can help students understand this. This is the result of low productivity. This is also an opportunity: engineers who can reduce costs to levels comparable with HICs will be well rewarded by their firms.
5. At the same time, faculty can draw on published material to explain why labour costs are higher in LICs once productivity is factored in. Even in China, where the overall productivity is about 3–4 times lower than in HICs, the intrinsic cost of manufacturing products is higher than in the USA, Europe, and Japan. China manages to provide cheaper products than these countries by a combination of subsidies to manufacturing companies and, in some instances, offering cheaper products with lower levels of quality and performance (Trevelyan, 2020). Chinese firms accept profit margins and investment returns much lower than typical HIC firms. China can also achieve economies of scale with production volume unattainable in most countries.[4]
6. Faculty can encourage students to ask questions to clarify lecture presentations, something that can conflict with local culture. One way to overcome cultural resistance is to ask them to work in groups of two or three students adjacent to each other in the lecture hall. Each group has 3–5 minutes to come up with a question written on a piece of paper which is placed on a pile at the front of the class. The instructor then draws questions at random and has no way to know which student asked the question.
7. Faculty can provide an authentic context for engineering science textbook practice problems to help students appreciate the context in which they will apply the relevant methods. For example, faculty can provide worked solutions to textbook problems, simulating design calculations by engineers, but with authentic features such as missing or extra details, some incorrect assumptions, and mistakes in some of the workings (sparingly applied). Students would have to apply textbook methods to verify the results independently and detect mistakes. Such approaches have been shown to be useful in related contexts (Hesterman, Guzzomi, & Stone, 2007). They would need to find data from reputable sources to substitute for missing information. Some data should be provided with large uncertainty limits so that students learn to identify worst case and likely case scenarios and perform appropriate calculations. At least for some of the time, students could be required to work under very tight time constraints to simulate the realities of practice not only to provide the solutions whether numerical answers or design but also the limitations/caveats associated with the constrained information, time to work on the solution, etc. Asking students to prepare a short presentation on how engineers use mathematical and scientific methods has been demonstrated to be a powerful motivator, significantly improving student learning (Goold, 2015).
8. Faculty can encourage students to perform detailed ethnographic observations on practicing engineers by shadowing an engineer or small-scale farmer for a week, spending about half of each day observing and taking notes, and the other half of each day writing a detailed account of their observations earlier in the day. These observations can then provide case study material to help provide authentic contextual framing for textbook problems in class exercises.
9. Faculty can draw students' attention to the influence of culture on engineering practice by reflecting on student and instructor behaviour in the education setting. For example, the instructor can encourage students to discuss how comfortable they feel in posing questions to other faculty in the presence of their peers. The instructor could prompt a discussion on how comfortable students would feel having a drink with or eating with their faculty, such as a senior professor. Faculty could prompt students to observe how people behave in situations with large power differences, for example a servant speaking with the owner of a house. They could then compare these interactions with their own behaviours with their parents, and behaviours when with their peers. The aim would be to sensitise students to become more observant with respect to social interactions and how people behave with each other in different social situations.
10. Faculty can alert students to interpretation differences. For example, when students perform classroom exercises, their grades often reflect interpretations of the exercise that differed from those of the instructor rather than lack of knowledge or willingness to work hard at studies. Students should be made aware that interpretation differences are a fundamental attribute of human beings. Everyone will have a different interpretation of the same words and drawings or diagrams. Some differences will be minor, but some can be very significant and in engineering these differences can have catastrophic consequences. As engineers, students will encounter

similar interpretation differences. A class exercise to explore interpretation differences could take the following general format.

Stage 1) Students construct a simple model from readily available materials such as wire, cardboard, adhesive tape, etc. It does not have to be a functional model.

Stage 2) Students create drawings of their models and, if necessary, a written set of instructions or a specification. The written materials and drawings are placed in envelopes.

Stage 3) Students draw envelopes at random (excluding their own, of course) and use the drawings and instructions to replicate the original model without being able to see the original model.

Stage 4) Students compare their constructed replica with the original model. Most of the replicas will be different from the original models in some respects because inexperienced students neglect to specify important features and dimensions. The results are often amusing and sometimes surprising. Students can reflect on the interpretation differences to distinguish deviations resulting from inadequate specifications and drawings from those resulting from differences in interpreting text and drawings.

11. Faculty can encourage students to participate in extra-curricular activities, particularly part-time work, and reflect on the social interactions that they take part in to help them become more observant of human behaviour. Students working as volunteers can learn collaboration and coordination techniques that they will find helpful in engineering workplaces. Service learning of this type has been shown to facilitate student learning in the US and has now begun to be implemented in some Indian engineering schools (Dustker, Reddy, Kandakatla, Joshi, & Oakes, 2021).

There are many instructional techniques known to improve the quality of student learning in higher education (Schneider & Preckel, 2017; van Alten, Phielix, Janssen, & Kester, 2019). Many of these methods can be implemented to improve student learning about engineering practices.

One cannot be sure that improving engineering education will influence what happens in engineering workplaces where organisational culture can overwhelm earlier behavioural influences (Buch, 2016). However, by researching engineering practice over time, and observing how education influences workplace practices, it should be possible to improve employment outcomes for agricultural and biosystems engineers in low income countries and significantly increase the wellbeing of rural communities.

Acknowledgements

The authors would like to acknowledge financial and organisational support from The University of Western Australia and other funding agencies that contributed indirectly to this paper. Hundreds of engineers and many engineering employers in several countries participated in research interviews and field observations contributing to this research. Emeritus Prof. Bill Williams of Instituto Politécnico de Setúbal, Portugal provided many helpful editorial comments and suggestions.

Disclosure Statement

No conflict of interest was reported by the authors.

NOTES

1. Data from the World Bank at https://data.worldbank.org/indicator/SL.AGR.EMPL.ZS data obtained 8 December 2022.
2. Video available at https://www.youtube.com/watch?v=9cKo5MWZseI
3. https://johnshearer.com.au/
4. The first author's substantial recent experience working first-hand with engineers in Chinese appliance manufacturing factories has provided these insights, also frequently referred to in business commentary on Chinese manufacturing.

REFERENCES

ABET. (2021). *2021–2022 Criteria for Accrediting Engineering Programs*. Retrieved from https://www.abet.org/wp-content/uploads/2021/02/E001-21-22-EAC-Criteria.pdf

Anderson, K. J. B., Courter, S. S., McGlamery, T., Nathans-Kelly, T. M., & Nicometo, C. G. (2010). Understanding engineering work and identity: A cross-case analysis of engineers within six firms. *Engineering Studies*, *2*(3), 153–174. doi:10.1080/19378629.2010.519772

Asongu, S., & Boateng, A. (2018). Introduction to special issue: Mobile technologies and inclusive development in Africa. *Journal of African Business*, *19*(3), 297–301. doi:10.1080/15228916.2018.1481307

Asplund, F., & Flening, E. (2021). Boundary spanning at work placements: Challenges to overcome, and ways to learn in preparation for early career engineering. *European Journal of Engineering Education*, 1–20. doi:10.1080/03043797.2021.1889467

Bailyn, L., & Lynch, J. T. (1983). Engineering as a life-long career: Its meaning, its satisfactions, its difficulties. *Journal of Occupational Behaviour (pre 1986)*, *4*(4), 263–283.

Barley, S. R. (2005). What we know (and mostly don't know) about technical work. In S. Ackroyd, R. Batt, P. Thompson, & P. S. Tolbert (Eds.), *The Oxford Handbook of Work and Organization* (pp. 376–403). Oxford: Oxford University Press.

Bechky, B. A. (2003). Sharing meaning across occupational communities: The transformation of understanding on a production floor. *Organization Science*, *14*(3), 312–330. doi:10.1287/orsc.14.3.312.15162

Blandin, B. (2012). The competence of an engineer and how it is built through an apprenticeship program: A tentative model. *International Journal of Engineering Education*, *28*(1), 57–71.

Blom, A., & Saeki, H. (2011). Employability and skill set of newly graduated engineers in India. World Bank Policy Research Working Paper 5640. Retrieved from http://papers.ssrn.com/sol3/papers.cfm?abstract_id=1822959#

Bragdon, S. H., & Smith, C. (2015). Small-scale farmer innovation. 22. Retrieved from http://quno.org/areas-of-work/food-sustainability

Bratton, J. (1991). Japanization at work: The case of engineering plants in Leeds. *Work, Employment and Society, 5*(3), 377–395.

Brown, A. L., Ash, D., Rutherford, M., Nakagawa, K., Gordon, A., & Campione, J. C. (1993). Distributed expertise in the classroom. In G. Salomon (Ed.), *Distributed cognitions: Psychological and educational considerations* (pp. 188–228). Cambridge: Cambridge University Press.

Bucciarelli, L. L. (1994). *Designing Engineers*. Cambridge, MA: MIT Press.

Buch, A. (2016). Ideas of holistic engineering meet engineering work practices. *Engineering Studies, 8*(2), 140–161.

Button, G., & Sharrock, W. (1994). Occasioned practices in the work of software engineers. In M. Jirotka & J. A. Goguen (Eds.), *Requirements Engineering: Social and Technical Issues* (pp. 217–240). London: Academic Press Ltd.

Cameron, I., Reidsema, C., & Hadgraft, R. G. (2011, December 5–7). *Australian engineering academe: A snapshot of demographics and attitudes*. Paper presented at the Australasian Conference on Engineering Education, Fremantle.

Cavallo, E., Ferrari, E., Bollani, L., & Coccia, M. (2014). Attitudes and behaviour of adopters of technological innovations in agricultural tractors: A case study in Italian agricultural system. *Agricultural Systems, 130*, 44–54. doi:https://doi.org/10.1016/j.agsy.2014.05.012

Coelho, K. (2004). *Of Engineers, Rationalities and Rule: An Ethnography of Neoliberal Reform in and Urban Water Utility in South India* (PhD). University of Arizona, Tucson. Retrieved from http://hdl.handle.net/10150/280681

Conrad, S. (2017). A comparison of practitioner and student writing in civil engineering. *Journal of Engineering Education, 106*(2), 191–217. doi:https://doi.org/10.1002/jee.20161

Darr, A. (2000). Technical labour in an engineering boutique: Interpretive frameworks of sales and R&D engineers. *Work, Employment and Society, 14*(2), 205–222.

Domal, V. (2010). *Comparing Engineering Practice in South Asia and Australia*. (PhD thesis). The University of Western Australia, Perth. Retrieved from http://research-repository.uwa.edu.au/files/3224816/Domal_Vinay_2010.pdf

Dustker, S. M., Reddy, B. S., Kandakatla, R., Joshi, G. H., & Oakes, W. C. (2021). *Role of Reflection in Service Learning-Based Engineering Programs: A Cross-Cultural Exploratory and Comparative Case Study in India and the USA*. Paper presented at the 2021 ASEE Virtual Annual Conference Content Access.

Dym, C. L., Agogino, A. M., Eris, O., Frey, D., & Leifer, L. J. (2005). Engineering design thinking, teaching, learning. *Journal of Engineering Education, 94*(1), 103–120.

Eckert, C., Blackwell, A., Bucciarelli, L., Clarkson, J., Earl, C., Knight, T., Macmillan, S., Stacey, M., & Whitney, D. (2004, May 18–21). *What Designers Think We Need to Know About Their Processes: Early Results from a Comparative Study*. Paper presented at the International Design Conference - DESIGN 2004, Dubrovnik, Croatia.

Engineers Australia. (2011). Stage 1 competency standard for professional engineers. Retrieved from http://www.engineersaustralia.org.au/

Ewenstein, B., & Whyte, J. (2007). Beyond words: Aesthetic knowledge and knowing in organizations. *Organization Studies, 28*(5), 689–708. doi:10.1177/0170840607078080

Flesher, J. W. (1993). An exploration of technical troubleshooting expertise in design, manufacturing, and repair contexts. *Journal of Industrial Teacher Education, 31*(1), 34–56.

Frey, C. B., & Osborne, M. A. (2017). The future of employment: How susceptible are jobs to computerisation? *Technological Forecasting and Social Change, 114*, 254–280. doi:10.1016/j.techfore.2016.08.019

Gillum, J. D. (2000). The engineer of record and design responsibility. *Journal of Performance of Constructed Facilities (ASCE), 14*(2), 67–70. doi:0887-3828/00/0002-0067–0070

Gonczi, A. (2013). Competency-based approaches: Linking theory and practice in professional education with particular reference to health education. *Educational Philosophy and Theory, 45*(12), 1290–1306.

Goold, E. (2015). Investigating engineering practice is valuable for mathematics learning. *Teaching Mathematics and Its Applications, 34*(1), 3–15. doi:10.1093/teamat/hru026

Goold, E., & Devitt, F. (2013). Mathematics in engineering practice: Tacit trumps tangible. In B. Williams, J. D. Figueiredo, & J. P. Trevelyan (Eds.), *Engineering Practice in a Global Context: Understanding the Technical and the Social* (pp. 245–279). Leiden: CRC/ Balkema.

Gupta, R. K., Singh, D. P., Sharma, M., Chatterjee, S., & Saha, S. K. (2019). *Improved Technologies Under the Rural Technology Action Group (RuTAG) at IIT Delhi*. Paper presented at the Rural Technology Development and Delivery, Singapore.

Guzzomi, A. L., Maraldi, M., & Molari, P. G. (2012). A historical review of the modulus concept and its relevance to mechanical engineering design today. *Mechanism and Machine Theory*, 50(1), 1–14. https://doi.org/10.1016/j.mechmachtheory.2011.11.016

Hager, P. (2004). The competence affair, or why vocational education and training urgently needs a new understanding of learning. *Journal of Vocational Education & Training, 56*(3), 409–433. doi:10.1080/13636820400200262

Henderson, K. (1999). *On Line and On Paper: Visual Representations, Visual Culture and Computer Graphics in Design Engineering*. Cambridge: M.I.T. Press.

Hesterman, D. C., Guzzomi, A. L., & Stone, B. J. (2007). *How to Mark Markers*. Paper presented at the Proceedings of the 2007 AaeE Conference, Melbourne.

How Chennai, one of the world's wettest major cities, ran out of water. (2021, February 4). *The Economic Times*. Retrieved from https://economictimes.indiatimes.com/news/politics-and-nation/how-chennai-one-of-the-worlds-wettest-major-cities-ran-out-of-water/articleshow/80680182.cms

Howell, D. R., & Wolff, E. N. (1991). Trends in the growth and distribution of skills in the U.S. workplace, 1960–1985. *ILR Review*, 44(3), 486–502. https://doi.org/10.1177/001979399104400306

International Engineering Alliance. (2021). *Graduate Attributes and Professional Competences Version 4 : 21 June 2021*(pp. 18). Retrieved from https://www.ieagreements.org/

Itabashi-Campbell, R., & Gluesing, J. (2013). Engineering problem-solving in social contexts: 'Collective wisdom' and 'ba'. In B. Williams, J. D. Figueiredo, & J. P. Trevelyan (Eds.), *Engineering Practice in a Global Context: Understanding the Technical and the Social* (pp. 129–158). Leiden: CRC/ Balkema.

Jesiek, B. K., Buswell, N. T., Mazzurco, A., & Zephirin, T. (2019, June 10–12). *Toward a Typology of the Sociotechnical in Engineering Practice.* Paper presented at the 8th Research in Engineering Education Symposium (REES 2019), Cape Town.

Jesiek, B. K., Mazzurco, A., Buswell, N. T., & Thompson, J. D. (2018). Boundary spanning and engineering: A qualitative systematic review. *Journal of Engineering Education, 107*(3), 380–413.

Jesiek, B. K., Trellinger, N., & Mazzurco, A. (2016). *Becoming Boundary Spanning Engineers: Research Methods and Preliminary Findings.* Paper presented at the American Society for Engineering Education Annual Conference and Exposition.

Jesiek, B. K., Trellinger, N., & Nittala, S. (2017). *Closing the Practice Gap: Studying Boundary Spanning in Engineering Practice to Inform Educational Practice.* Paper presented at the 2017 IEEE Frontiers in Education Conference (FIE).

Kahneman, D. (2011). *Thinking, Fast and Slow.* London: Penguin, Allen Lane.

Kendall, J., & Voorhies, R. (2014). The mobile-finance revolution; How cell phones can spur development. *Foreign Affairs, 93*(March/April), 9. Retrieved from https://www.foreignaffairs.com/print/node/1108438

Kilduff, M., Funk, J. L., & Mehra, A. (1997). Engineering identity in a Japanese factory. *Organization Science, 8*(6), 579–592.

Korte, R. (2009). How newcomers learn the social norms of an organization: A case study of the socialization of newly hired engineers. *Human Resource Development Quarterly, 20*(3), 285–306. doi:10.1002/hrdq.20016

Korte, R., Brunhaver, S. R., & Sheppard, S. D. (2015). (Mis) Interpretations of organizational socialization: The expectations and experiences of newcomers and managers. *Human Resource Development Quarterly, 26*(2), 185–208.

Kusimo, A. C., & Sheppard, S. D. (2019). *Manufacturing Bowler Ability: How Resource Scarcity Hinders Team Coordination During Manufacturing in Nigerian Factories.* Paper presented at the ASME International Design Engineering Technical, Conferences and Computers and Information in Engineering Conference IDETC/CIE2019, Anaheim, CA. Retrieved from https://asmedigitalcollection.asme.org/IDETC-CIE/proceedings-pdf/IDETC-CIE2019/59278/V007T06A028/6454036/v007t06a028-detc2019-97820.pdf

Lam, A. (1997). Embedded firms, embedded knowledge: Problems of collaboration and knowledge transfer in global cooperative ventures. *Organization Studies, 18*(6), 973–996.

Layton Jr, E. T. (1986). *The Revolt of the Engineers. Social Responsibility and the American Engineering Profession.* Baltimore, MD: Johns Hopkins University Press.

Leonardi, P. M., Jackson, M. H., & Diwan, A. (2009). The enactment-externalization dialectic: Rationalization and the persistence of counterproductive technology design practices in student engineering. *Academy of Management Journal, 52*(2), 400–420.

Luth, G. P. (2000). Context and chronology of the Hyatt regency collapse. *Journal of Performance of Constructed Facilities (ASCE), 14*(2), 51–61. doi:0887-3828/00/0002-0051-0061

Lutz, B., & Paretti, M. C. (2021). Exploring the social and cultural dimensions of learning for recent engineering graduates during the school-to-work transition. *Engineering Studies, 13*(2), 132–157. doi:10.1080/19378629.2021.1957901

Lynn, L. H., Piehler, H. R., & Kieler, M. (1993). Engineering careers, job rotation, and gatekeepers in Japan and the United States. *Journal of Engineering and Technology Management, 10*(1 & 2), 53–72.

Magarian, J. N., & Seering, W. P. (2021). Characterizing engineering work in a changing world: Synthesis of a typology for engineering students' occupational outcomes. *Journal of Engineering Education,* 110, 458–500https://doi.org/10.1002/jee.20382.

Masarei, M. I., Erickson, T. E., Merritt, D. J., Hobbs, R. J., & Guzzomi, A. L. (2021). Engineering restoration for the future. *Ecological Engineering, 159,* 106103. doi:https://doi.org/10.1016/j.ecoleng.2020.106103

Matemba, E. E. (2020). *Globalisation of Engineers: African Professional Engineering Education Perspectives* (PhD). Curtin University, Perth. Retrieved from https://espace.curtin.edu.au/bitstream/handle/20.500.11937/82165/Matemba%20E%202020.pdf?sequence=1&isAllowed=y; http://hdl.handle.net/20.500.11937/82165

Meiksins, P., & Smith, C. (1996). *Engineering Labour: Technical Workers in Comparative Perspective.* London: Verso.

Moncarz, P. D., & Taylor, R. K. (2000). Engineering process failure — Hyatt walkway collapse. *Journal of Performance of Constructed Facilities (ASCE), 14*(2), 46–50. doi:0887-3828/00/0002-0046–0050

Moritz, E. (1996). Synthetic, pragmatic, analytic — A comparison of the Japanese, American and German approaches to machine tool design. In L. Rasmussen & F. Rauner (Eds.), *Industrial Culture and Production.* London: Springer-Verlag.

Moss, W., Nichols, P. G. H., Foster, K. J., Ryan, M. H., & Guzzomi, A. L. (2021). Farmer-driven innovation lessons from a case study of subterranean clover seed production. *Prometheus, 37*(4), 353–370 (318 pages). Retrieved from https://www.jstor.org/stable/48676453

Mukerji, C. (2009). *Impossible Engineering: Technology and Territoriality on the Canal du Midi.* Princeton, NJ: Princeton University Press.

Nash, R. (2003). Social explanation and socialization: On Bourdieu and the structure, disposition, practice scheme. *Sociological Review, 51*(1), 43–62. doi:10.1111/1467-954X.00407

Noble, D. F. (1979). *America by Design: Science, Technology, and the Rise of Corporate Capitalism.* New York: Oxford University Press.

Opara, L. U. (2004). Outlook for agricultural engineering education and research and prospects for developing countries. *Outlook on Agriculture, 33*(2), 101–111. doi:10.5367/000000004773973082

Orr, J. (1996). *Talking About Machines: An Ethnography of a Modern Job*. Ithaca, New York: Cornell University Press.

Perlow, L. A. (1999). The time famine: Towards a sociology of work time. *Administrative Science Quarterly*, *44*(1), 57–81. doi:10.2307/2667031

Perlow, L. A., & Bailyn, L. (1997). The senseless submergence of difference: Engineers, their work, and their careers. In S. Barley & J. Orr (Eds.), *Between Craft and Science: Technical Work in US Settings* (pp. 230–243). Ithaca, New York: Cornell University Press.

Perrucci, R. (1971). Engineering: Professional servant of power. In E. Friedson (Ed.), *The Professions and Their Prospects* (pp. 119–133). Beverly Hills, CA: Sage.

Pfatteicher, S. K. A. (2000). "The Hyatt horror": Failure and responsibility oin American engineering. *Journal of Performance of Constructed Facilities (ASCE)*, *14*(2), 62–66. doi:0887-3828/00/0002-0062–0066

Polak, P. (2008). *Design for Affordability*. Paper presented at the American Association for Engineering Education (ASEE), Pittsburgh, PA. Retrieved from https://www.ideglobal.org/story/tribute-to-paul-polak

Polanyi, M. (Ed.) (1966). *The Tacit Dimension*. Garden City: Doubleday.

Sandberg, J. (2000). Understanding human competence at work: An interpretive approach. *Academy of Management Journal*, *43*(1), 9–25.

Schneider, M., & Preckel, F. (2017). Variables associated with achievement in higher education: A systematic review of meta-analyses. *Psychological Bulletin*, *143*(6), 36. doi:10.1037/bul0000098

Senzanje, A. (2003). Problems faced and advances made by agricultural engineers in southern and eastern Africa. *Agricultural Engineering International: The CIGR Journal*, *5*, 14. Retrieved from https://hdl.handle.net/1813/10359

Sheppard, S. D., Colby, A., Macatangay, K., & Sullivan, W. (2006). What is engineering practice? *International Journal of Engineering Education*, *22*(3), 429–438.

Sheppard, S. D., Macatangay, K., Colby, A., & Sullivan, W. (2009). *Educating Engineers*. Stanford, CA: Jossey-Bass (Wiley).

Sigrimis, N., Hashimoto, Y., Munack, A., & De Baerdemaeker, J. (1999). Prospects in agricultural engineering in the information age. *Agricultural Engineering International: The CIGR Journal*, *1*, 21. Retrieved from https://ecommons.cornell.edu/handle/1813/10205

Singer, P. (1989). A utilitarian defense of animal liberation. In L. P. Pojman, P. Pojman, & K. McShane (Eds.), *Environmental Ethics: Readings in Theory and Application* (7th ed., pp. 96–115). Boston, MA: Cengage Learning.

Singh, J., & Fleming, L. (2010). Lone inventors as sources of breakthroughs: Myth or reality? *Management Science*, *56*(1), 41–56. doi:10.1287/mnsc.1090.1072

Smith, C., & Meiksins, P. (1995). System, society and dominance effects in cross-national organisational analysis. *Work, Employment and Society*, *9*(2), 241–267.

Smith, K. A., Sheppard, S. D., Johnson, D. W., & Johnson, R. T. (2005). Pedagogies of engagement: Classroom-based practices. *Journal of Engineering Education*, *94*(1), 87–101.

Star, S. L., & Griesemer, J. R. (1989). Institutional ecology, 'translations' and boundary objects: Amateurs and professionals in Berkeley's museum of vertebrate zoology, 1907–39. *Social Studies of Science*, *19*(3), 387–420. doi:10.1177/030631289019003001

Tilak, J. B. G., & Choudhury, P. K. (2021). Employment and employability of engineering graduates in India. *Journal of Contemporary Educational Research*, *5*(3), 14. doi:10.26689/jcer.v5i3.1825

Tonso, K. L. (2006). Teams that work: Campus culture, engineer identity, and social interactions. *Journal of Engineering Education*, *95*(1), 25–37.

Trevelyan, J. P. (1992). *Robots for Shearing Sheep: Shear Magic*: Oxford University Press.

Trevelyan, J. P. (2007). Technical coordination in engineering practice. *Journal of Engineering Education*, *96*(3), 191–204. doi:10.1002/j.2168-9830.2007.tb00929.x

Trevelyan, J. P. (2010). Reconstructing engineering from practice. *Engineering Studies*, *2*(3), 175–195. doi:10.1080/19378629.2010.520135

Trevelyan, J. P. (2013). Observations of South Asian engineering practice. In B. Williams, J. D. Figueiredo, & J. P. Trevelyan (Eds.), *Engineering Practice in a Global Context: Understanding the Technical and the Social* (pp. 223–244). Leiden: CRC/ Balkema.

Trevelyan, J. P. (2014). *The Making of an Expert Engineer*. London: CRC Press/Balkema - Taylor & Francis.

Trevelyan, J. P. (2019). Transitioning to engineering practice. *European Journal of Engineering Education*, *44*(6), 821–837. doi:10.1080/03043797.2019.1681631

Trevelyan, J. P. (2020). *Learning Engineering Practice*. London: Taylor & Francis.

Trevelyan, J. P. (2022). A social science research agenda for engineering practice. In S. H. Christensen, A. Buch, E. Conlon, C. Didier, C. Mitcham, & M. Murphy (Eds.), *Engineering, Social Sciences, and the Humanities: Have their Conversations Come of Age?* doi:10.1007/978-3-031-11601-8_13

Trevelyan, J. P., & Tilli, S. (2008, June 20–22). *Longitudinal Study of Australian Engineering Graduates: Perceptions of Working Time*. Paper presented at the American Society for Engineering Education Annual Conference, Pittsburgh.

Trevelyan, J. P., & Williams, B. (2018). Value creation in the engineering enterprise: An educational perspective. *European Journal of Engineering Education*. doi:10.1080/03043797.2017.1421905

United Nations. (2017a). The sustainable development goals report 2017. Retrieved from http://sdgactioncampaign.org/wp-content/uploads/2017/07/TheSustainableDevelopmentGoalsReport2017.pdf

United Nations. (2017b). Transforming our world: 2030 agenda for sustainable development. Retrieved from https://sustainabledevelopment.un.org/content/documents/21252030%20Agenda%20for%20Sustainable%20Development%20web.pdf

United Nations Development Programme [UNDP]. (2017). The sustainable development goals report 2017. Retrieved from http://www.un.org/en/ecosoc/docs/pdfs/fina_08-45773.pdf

United Nations Development Programme [UNDP]. (2019). *Human Development Report 2019 Beyond Income, Beyond Averages, Beyond Today: Inequalities in Human Development in the 21st Century*. New York: United Nations Development Programme,.

van Alten, D. C., Phielix, C., Janssen, J., & Kester, L. (2019). Effects of flipping the classroom on learning outcomes and satisfaction: A meta-analysis. *Educational Research Review*, *28*(100281), 18. doi:10.1016/j.edurev.2019.05.003

Vinck, D. (Ed.) (2003). *Everyday Engineering: An Ethnography of Design and Innovation*. Boston, MA: MIT Press.

Walsh, M. J., Squires, C. C., Coleman, G. R. Y., Widderick, M. J., McKiernan, A. B., Chauhan, B. S. ... Guzzomi, A. L. (2020). Tillage based, site-specific weed control for conservation cropping systems. *Weed Technology*, *34*(5), 704–710. doi:10.1017/wet.2020.34

Waris, M., & Kokab, R. U. (2017). Baradarism and politics of the Sarghoda district. *Pakistan Journal of History and Culture*, *38*(1), 59–81. Retrieved from http://www.nihcr.edu.pk/Latest_English_Journal/pjhc%2038-1,2017/4.%20Baradarism%20and%20Politics,%20Waris%20awan,%20Rizwan.pdf

Whittington, D., Mu, X., & Roche, R. (1990). Calculating the value of time spent collecting water: Some estimates for Ukunda, Kenya. *World development*, *18*(2), 269–280. doi:https://doi.org/10.1016/0305-750X(90)90052-Y

Whyte, J., Lindkvist, C., & Jaradat, S. (2016). Passing the baton? Handing over digital data from the project to operations. *Engineering Project Organization Journal*, *6*(1), 2–14. doi: 10.1080/21573727.2015.1115396

Wilde, R. J., & Guile, D. (2021). Client-facing interprofessional project teams: The role of engineers' 'situated judgment'. *Engineering Studies*, *13*(3), 185–204. doi:10.1080/19378629.2021.2005073

Williams, B., Figueiredo, J. D., & Trevelyan, J. P. (Eds.) (2013). *Engineering Practice in a Global Context: Understanding the Technical and the Social*. Leiden, Netherlands: CRC/Balkema.

Wong, W. L. P., & Radcliffe, D. F. (2000). The tacit nature of design knowledge. *Technology Analysis and Strategic Management*, *12*(4), 493–512.

Zussman, R. (1985). *Mechanics of the Middle Class: Work and Politics Among American Engineers*. Berkeley, CA: University of California Press.

12

Novel Models for Delivering Agricultural and Biosystems Engineering Education in Developing Countries: Trends, Challenges and Future Prospects

Brighton M. Mvumi, Peter Rwakatiwana and Tinashe Madzivanzira

Introduction

Of late, the globe has been seized with numerous disruptors in the education delivery systems. These include conflict, or wars, natural disasters such as cyclones, earthquakes, floods, and pandemics such as the Ebola virus and COVID-19 (corona virus disease 2019). However, it is the COVID-19 pandemic which accelerated the unbundling of higher education and triggered the development of innovative and effective approaches to reaching out to learners. This emanated from the restricted movement and lockdowns that were imposed by governments to curb the spread of COVID-19.

On the other hand, as costs of learning are rapidly rising against a general economic stagnation or decline, especially in developing countries, the decrease in enrolment is inevitable. This presents a challenge to the long-term viability of the traditional face-to-face education system.

The other challenge is whether, in a rapidly changing technological environment, driven mainly by the microchip, institutions of higher learning are producing a sufficient labour-force that is fit for purpose based on the conventional four-year degree programme. It goes without saying that the Universities which are able to incorporate digital learning can survive pandemics and environmental and economic disruptors.

During the COVID-19, classes were mainly conducted over the internet with students having to log-on their gadgets and learn from their hostels or from home. Similarly, lecturers and administrators had to facilitate academic business from offices or homes as well to continue to reach out to their learners. In Africa, we have mainly relied on imported technologies, many of which do not fit very well in our context. This challenges University institutions to re-think about an education system that prioritises problem-solving, adaptability, and creativity, which are essential ingredients for sustainable innovation.

The population in sub-Saharan Africa (SSA) is dominated by the youth, with 62% being under the age of 25 (United Nations Department of Economic and Social Affairs [UNDESA], 2017). This age-group provides fertile ground for digital learning as these youths are generally more techno-savvy compared to the relatively older generation. This, therefore, provides a huge opportunity to revolutionise the education system to impact positively, on industry, even well after COVID-19, and have graduates who are prepared for any disruptive change in society. The digital world has also given birth to a number of higher learning institutions offering degree programmes of dubious credibility. Consequently, unsuspecting customers are being fleeced of their hard-earned money, as the programmes are often not accredited by a recognised regulatory authority.

This Chapter reviews available tertiary learning models based on digital technologies in terms of usefulness, trends, challenges, and possible solutions. The Chapter also examines how developments in the digital world apply to the context of facilitating Agricultural and Biosystems Engineering (ABE) education delivery and learning in tertiary institutions such as universities, focusing on developing countries, especially in SSA, and considering their socio-economic environment.

Global Trends in Novel Technologies for Delivering Teaching and Learning in ABE Education

ABE Education With Little or No Human–Machine Interaction

The introduction of computer technology resulted in a fundamental paradigm shift from traditional engineering to the current technology-driven practice (Abumandour, 2022). The main drivers of this change are technology's ability to enhance discovery, innovation, collaboration, and flexibility (Jhurree, 2005; Koretsky & Magana, 2019). In a similar manner, these tectonic shifts in ways of doing business due to technological advancement are not confined to workshops and industries; they are also being experienced at a significant scale in lecture rooms and experimental fields. Novel learning technologies are being adopted at a fast pace in engineering education worldwide because they offer improved instructive teaching, a wider access to education and a flexible learning environment (Ntaba & Jantjies, 2019).

DOI: 10.1201/9780429150111-14

Although the adoption of technologically-oriented pedagogies was evolving at its own pace, it accelerated from 2019 onwards when educators were coerced to switch from traditional to e-learning technologies due to COVID-19 induced lockdowns (Mkwizu, 2022; Abumandour, 2022). The world over educators and researchers are in search for effective pedagogues that can be employed in different disciplines to optimise the use of modern visualisation technologies that can be used to augment current educational experiences for learners in the absence of face-to-face lectures (Ntaba & Jantjies, 2019; Abumandour, 2022). An effective ABE novel learning model must cater for educational content, hands-on activities, lab-based work, design, and simulation-oriented work, which are the main pillars of engineering education. These fundamental learning issues contribute to the whole learning model (Figure 12.1). Most available learning technologies in developing countries do not adequately cover all the components and hence they are found wanting in the process of producing a competent ABE graduate. The new developments in novel learning technologies are focusing on developing pedagogies that fully encompass all the technical elements.

Furthermore, the learning model must also satisfy cognitive learning, which is composed of three interlinked categories of learning theories, i.e., the behaviourist approach, the cognitive approach, and constructivism (Yilmaz, 2011; Çeliköz et al., 2016; Malik, 2021). These cognitive learning domains must be considered in totality because they impact and depend on each other (Çeliköz et al., 2016) (Figure 12.2).

Cognitive learning is an active learning technique that assists one to learn how to maximise the potential of one's brain to acquire new knowledge, link it easily with existing ideas, and retain it (Çeliköz et al., 2016; Malik, 2021). It mainly focuses on the brain's ability to absorb and retain information based on experience, senses, and reflection. All these cognitive learning approaches are key to the development of an effective ABE education. Cognitive learning is important in ABE education because it enhances learning and comprehension, boosts confidence, improves problem-solving skills, help learners to learn new things faster and to think in abstract terms.

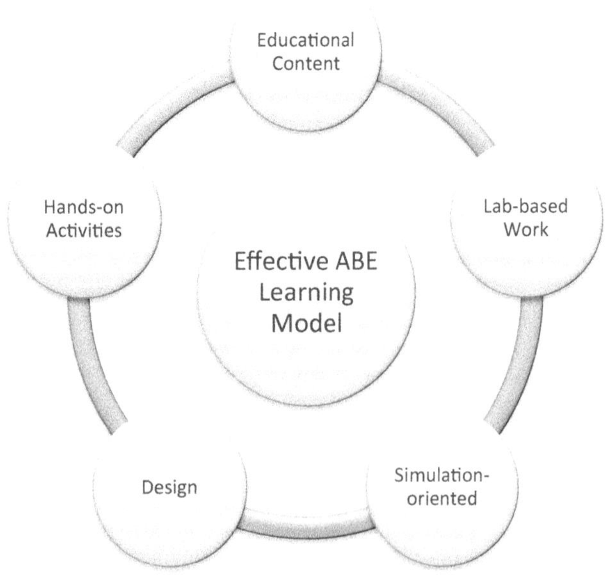

FIGURE 12.1 Technical issues to be covered by an effective Agricultural and Biosystems Engineering (ABE) learning model

A number of blended approaches are being used globally. For example, Wu et al. (2021) tested the learning effectiveness of using radio frequency identification (RFID) together with mixed reality technologies in a higher education institution and they concluded that immersive technology resulted in improved learning performance for knowledge acquisition and skills application. Although RFID is widely used, it is not very effective when used alone, thus Wu et al. (2021) suggest the need for blended methods.

According to Talespin (2022), Pricewaterhouse Coopers (PwC) carried out a study in 2020 to test the immersive learning efficacy of Virtual Reality (VR) soft skills training, their findings were that it outperformed e-learning and classroom learning across metrics (Figure 12.3).

In addition, PwC's VR soft skills training study indicated that the confidence of learners trained by VR was up by 275% which had a 40% and 35% improvement over classroom and e-learning training, respectively. The implication of these results is that if properly applied VR-powered immersive learning can significantly improve learners' knowledge acquisition, retention, and attitude. Therefore, the adoption of new technologies with minimal human–machine interaction in ABE education is important because it equips students with professional knowledge to meet the current and future demands of the digital industry which include artificial intelligence (AI), drone technology, precision agriculture, agro-robotics, Big Data, machine learning, internet of things (IoT), smart agriculture, 3D computing, etc.

Immersive Learning Technologies

Neuroscientific behavioural research findings on how people learn, which showed that the brain treats Virtual Reality (VR) experiences as if they were real, led to the emergence of immersive learning. Immersive learning, therefore, is an experiential training methodology that uses VR to simulate real-world scenarios and train learners in a safe, fascinating, and engaging learning environment. It makes use of a number of cognitive learning principles. The effectiveness and user engagement are facilitated by a combination of advanced learning theory, data science, spatial design, and the existence of the VR. Consequently, the learner's experience under a virtual scenario is more or less similar to actual experiences in reality. Immersive learning technologies have great potential in ABE teaching and learning because they have been proven to be effective in improving the confidence of learners, their preparedness and engagement, as well as their knowledge absorption and acquisition. It also brings with it the much needed benefit of reaching a wider learning community at once, especially if a cloud-based platform is employed by the instructor.

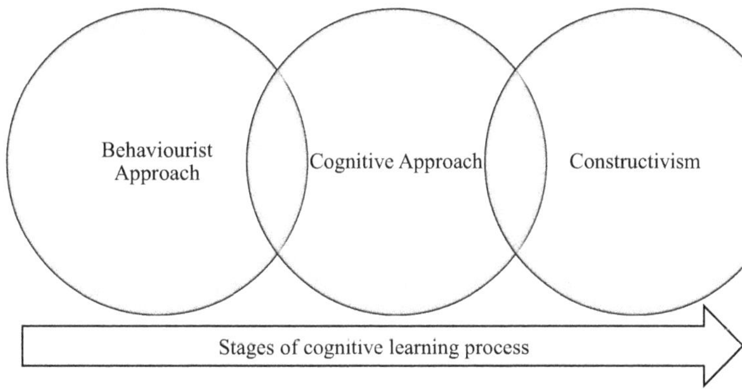

FIGURE 12.2 Interlinkages in cognitive learning approaches required for effective Agricultural and Biosystems Engineering education

FIGURE 12.3 Immersive learning efficacy survey results for PwC Soft skills training in 2020 (Talespin, 2022) (VR = Virtual Reality; PwC = Pricewaterhouse Coopers).

The 'one-size-fits-all' approach used in traditional classroom-based learning has its own well-known inadequacies, because the learner population comprises unique individuals who require different learning styles (Ntaba & Jantjies, 2019). However, making use of immersive learning technologies that ensure engagement and interactive content offers a more inclusive and accessible learning experience, particularly for visual learners. This then presents a psychological challenge for a group of aural leaners that tend to perform better under traditional learning techniques. Nevertheless, immersive learning has been found to be very effective for most learners in honing the knowledge, skills, and attitudes required for the engineering profession (Abumandour, 2022), thus it is very much applicable in ABE education in both the developing and developed worlds. New skills, attitudes, and techniques can be learnt and perfected by learners through artificial, digitally-created content and environments that mimic real-life situations so that the learners become active participants whose decisions and actions have a bearing on the outcomes.

Immersive technologies can be subdivided into four classes, i.e., reality technologies, 360-degree video, simulations, and game-based learning (Wu et al., 2021).

Reality Technologies

Reality technologies which include virtual reality, augmented reality (AR), and mixed reality (MR) are effective in providing learners with spatial, self, and social presence in an environment almost identical to the physical environment (Markowitz et al., 2018; Ntaba & Jantjies, 2019). VR is widely reported in literature as one of the main technologies used for learning in higher education, especially in science, engineering, and medicine (Ding & Li, 2022).

a) Virtual Reality (VR)

VR technology is defined as "using complex technology to form synthetic stimulation to replace real-world sensory information, which means that users enter the virtual scene and use special helmets, data gloves, or input devices such as keyboards and mice to interact with the virtual environment in real-time" (Ding & Li, 2022). Therefore it is an artificial environment that completely immerses learners in alternative digital worlds which are almost identical to the real one. The learner makes use of VR headsets such as a HTC Vive or Oculus Quest that are sometimes used together with headphones to access content. In addition, hand controllers may be used by the learners to actively participate in the virtual environment. According to Ryan (2015) immersion, interactivity, and imagination are the three fundamental characteristics that make VR a unique enabling learning tool in higher education.

VR allows for gaming, modelling, and simulation as well as the provision of virtual laboratories and workshops that are of great value in engineering education. Therefore, this immersive learning technology is very much applicable in ABE education. For instance, in mechanisation, learners can virtually carry out maintenance and repair tasks and get feedback while in the comfort of their homes. On the other hand pipe sizing, pump selection, or irrigation system installation can be done virtually by an irrigation engineering student. There are numerous examples where VR can be applied in ABE education. However, its application in ABE education in developing countries is still limited because of the high financial costs associated with its implementation.

b) Augmented Reality (AR)

AR makes use of digital assets such as a camera with multimedia effects to capture real-world aspects and interlink them with digital content (Ntaba & Jantjies, 2019). The multimedia effects on cameras help content creators to make text, images, video, or three-dimensional models and animations that support real world aspects. Therefore, augmented reality blends the real world with digital content, unlike in VR where the real world is excluded. For instructional information, a 2D form of digital assets can be employed because it is more effective. Learners can make use of their mobile devices, such as smartphones and tablets, to access content. AR learning content is thus easily accessible which makes it more suitable for poorly resourced ABE learners in developing countries. Just like VR, AR can be used for effective instruction in engineering with the added advantage of incorporating real world aspects. Therefore AR presents a number of potential applications in ABE education although it is a relatively new learning technology.

c) Mixed Reality (MR)

A combination of virtual and augmented reality elements resulted in mixed reality. MR is more similar to AR in that it interlinks digital content with real world aspects. However, the content interacts with objects in the real world and is attached to it. Real world objects in MR can obscure digital assets and this distinguishes it from AR reality where there is no obscurity.

360-Degree Videos

A 360-degree video is mostly live-filmed instead of being generated using a computer. The content can be accessed using a VR headset and the learner is fully immersed in an environment. However, the filmmaker's viewpoint is the one that directs the learner. It allows the learners to move their heads to see the world around them without any interaction with their surroundings because they cannot move independently. A 360-degree video is valuable to ABE learners because it can be used as a way of conducting virtual field trips at a low-cost.

Simulation and Game-Based Learning

In the past, realistic simulations were mostly used in high value educational settings such as space training (Ntaba & Jantjies, 2019). However, with the introduction of reality technologies many institutions are now including simulation in their educational environment. Simulation will help ABE learners to practise, acquire the necessary skills, and improve their mental capabilities.

There is an increase in digital game-based learning applications in engineering education due to its ability to motivate, train, and create a sense of reality for the learner (Udeozor *et al.*, 2022; Ntaba & Jantjies, 2019). In addition, digital games are becoming popular because of their ability to enhance the cognitive, affective, and psychomotor skills of the learners. Many studies have shown that students like and enjoy game-based learning (Udeozor *et al.*, 2022).

Advances in technology have enabled the design and development of real-life situational learning games that learners enjoy and are attracted to. With an unprecedented increase in VR- and AR-based simulation games, there is also a marked increase in their adoption in the teaching and training of engineering students. Digital games can be used to assist students to understand difficult concepts in engineering by simply using a laptop, VR, and AR. In addition, games can also be used by learners to comprehend concepts that require dangerous experiments or are expensive and difficult to conduct. Therefore non-digital and digital games have a big potential in ABE education in developing countries because of the relatively young population found in these nations and the enjoyable nature of gaming.

Other Emerging Technologies

The inclusion of novel technologies for delivering teaching and learning in ABE education is mandatory, especially when one considers that most processes in agriculture are being automated (Primicerio et al., 2012; Talaviya et al., 2020). Since agriculture plays a significant role in the economies of most developing countries, it is impossible to ignore the emerging technologies that are bound to boost productivity and precision in this sector, which are mainly based on artificial intelligence (AI). Biology, mathematics, computer science, engineering, psychology, and linguistics, as well as other disciplines, are said to be the foundations on which AI is built (Talaviya et al., 2020). AI mainly focuses on producing technologies that function almost like a human brain and thus its development depends on studying how the human brain thinks, learn, make decisions, and works, while at the same time being in a position to solve problems. AI concepts, therefore, also include novel learning technologies within themselves in order to improve efficiency and effectiveness. Thus AI intelligence software is used to develop intelligent systems which are then used to solve problems in agricultural applications.

Just like humans, the developed software rely on training data fed into them and use intelligent devices to make appropriate decisions and actions. This resulted in the advent of a number of useful disciplines within AI, such as Machine Learning (ML) and Deep Learning (DL) (Amani *et al.*, 2020). Talaviya *et al.* (2020) defines AI as the science of making intelligent machines and programmes while ML is defined as the ability to learn something without openly being programmed. DL on the other hand is said to be concerned with the learning of deep neural networks.

AI technologies in agriculture are transforming production processes by improving crop production, monitoring, security, harvesting, postharvest processing, marketing, and distribution. The use of AI at almost all stages of the value chain in agriculture is becoming increasingly visible, thus there is a need to consider the impact of these technologies and their incorporation in current ABE education pedagogies and beyond. Since the use of automated systems such as agricultural robots and drones is expected to be the mainstay of the future (Primicerio et al., 2012) in agriculture and machinery production, there is a need for ABE learners to understand these systems, systems which will be expected to have a bearing on their performance. Immersive learning technologies such as reality technology and gaming can be helpful for ABE learners to quickly comprehend these sophisticated but useful technologies. The transformation of pedagogues in ABE education in developing countries should be done bearing in mind that farming is increasingly becoming a mixture of technology- and biology-based skills, and so learners need to skilled in these fields in relation to country-specific requirements and the anticipated future dynamics of technological change. For instance, Chatbots (virtual assistance), which are currently being used to assist farmers to solve problems, may also be the in-thing in future ABE education, considering that they have been effective in personalised communication, especially where machine learning is involved. It is, therefore, not surprising that in future, ABE learners will be learning from machines and vice versa, with effective interaction for the benefit of agriculture, production, and humanity as a whole.

ABE Education Technology Application and Associated Challenges in Developing Countries

Which Technologies Are Appropriate to Developing Countries?

ABE education requires that students have both theoretical knowledge and practical skills and experiences. Technologies applicable in ABE can be classified as technologies for teaching, practical experience, supervision of research work, and student assessment. There must be a relationship and balance in terms of the theoretical information delivered and the practical (labs and workshop practices).

The teaching and learning models for developing countries should ensure that the learners have the requisite skills to produce goods and services which entails the requirement for skills and knowledge on innovativeness and industrialisation. The use of technology is critical for student engagement and learning in face-to-face, blended, and online learning forms of learning and teaching (Wanner, 2015). The various technology applications applied in the delivery of ABE education must support engagement and learning. In addition, the development of goods and services requires an in-depth understanding of the subject matter hence the need to focus degree programmes, and not offer overly-wide content programmes. There is also a need to produce students that fit well in the corporate world upon graduation.

Active learning and student engagement are pre-requisites for effective learning and towards achieving both academic and non-academic outcomes, for example, leadership skills, critical thinking skills, and other job-related skills. Student engagement and quality of education in institutions of higher learning are directly linked, as it enhances student experience and learning outcomes (Wanner, 2015). The provision of a learning environment where active learning and student engagement are provided enhances the competitiveness of the learning institution among its competitors.

The Just-in-Time Teaching (JITT) approach, a form of active learning pedagogy, promotes self-learning among students (Madiraju et al., 2020), and can be applied when delivering ABE education. One of the key features of the JITT approach is that students research new ideas before attending lectures making them more alert to confusing concepts (Wanner, 2015). The warm-up material is to be submitted before the lecture commences and provides the lecturer with an insight as to what the students understand, allowing them to distribute focus and effort accordingly (Wanner, 2015). This approach is useful in ABE education as it enhances the development of critical skills, for example, problem-solving and critical thinking, skills key in executing engineering duties such as designing.

Multimedia learning (Figure 12.4) is a learning model that can be utilised in ABE education. In multimedia learning, the learner engages in three cognitive processes which are: a) selecting, where verbal and visual information is selected to yield text and image bases, b) organising, where the text and image bases are converted into verbally-based and visually-based models, and c) integrating, when the learner builds connections between corresponding events occurring in both the verbally-based and visually-based models (Mayer & Moreno, n.d.). Learners that use multimedia learning show better problem-solving capabilities than those that use either verbal or visual information separately (Mayer & Moreno, n.d.).

Traditional Teaching Technologies

Microsoft PowerPoint is a relatively old technology which is widely used for teaching and learning especially in institutions of higher learning (Wanner, 2015). The use of PowerPoint technology is argued to structure student thought and student engagement before and during lectures (Wanner, 2015). PowerPoint technology enhances students' participation during lecturers which improves their understanding of the content being delivered. If PowerPoint slides are shared with the

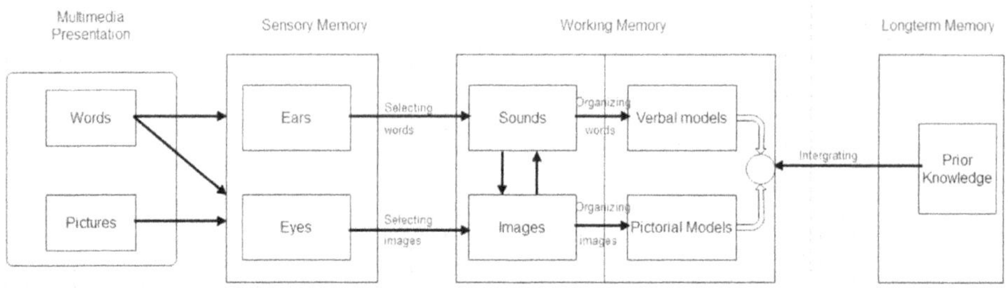

FIGURE 12.4 The cognitive theory of multimedia learning (adapted from Mayer, 2010).

learners before the content is delivered in lecture, they provide students with a guide of what to expect during the lecture, prepare questions or even research on grey areas which improves the capacity of understanding for the learners. During lectures, Microsoft PowerPoint slides also act as a guide for the content deliverer. PowerPoint slides can be used by learners as well, to deliver group technology learning activities when presented with specific tasks.

Social Media Learning Technologies

Combining social media with higher education will bring the institutions and students closer together (Moghavvemi *et al.*, 2018). YouTube can be regarded as an effective E-learning tool if the video is relevant to the subject at hand (Moghavvemi *et al.*, 2018; Mahasneh et al., 2020). Moghavvemi *et al.*, (2018) argues that the effectiveness of YouTube as a learning and teaching tool lags behind other social media tools such as Facebook. However, trends in developing countries for example Zimbabwe show that students at institutions of higher learning use YouTube to view lectures, video tutorials of worked examples, and animations of concepts, while the Facebook tool is usually used for announcements. YouTube on its own is usually not used as a teaching tool in the classroom, but as an out–of-class tool where students get information to reinforce lectures. Like any other tool that requires internet connectivity, challenges affect its functionality. However, the gadgets needed to access YouTube can be accessed easily by majority of the students.

Complementing traditional teaching with online learning improves students' performance and their ability to grasp concepts easily (Moghavvemi *et al.*, 2018). The implementation of YouTube when teaching courses that require practical experience, for example engineering, nursing, and medicine, enhanced student performance. Students who watched YouTube videos performed better than those that did not (Moghavvemi *et al.*, 2018).

E-Learning and Online Learning Technologies

E-learning in general refers to the application of electronic tools, soft and hardware applications, as well as web-based processes to learning (Abbasy & Quesada, 2017). A breakthrough in the field of online learning was made possible through the development of the World Wide Web and the rapid spread of the internet across the globe (Bates, n.d.). Web-based technologies and the internet have become popular platforms for developing and introducing distance learning in the online classroom (Jasim *et al.*, 2021). Electronic teaching and training is widely being implemented all over the world to provide training, technical, and administrative courses, as well as certificates, diplomas, and degrees from colleges, universities, and professional development and industrial training centres (Jasim *et al.*, 2021). Some universities in developing countries formally use electronic Learning Management Systems (eLMS) for learning material sharing between students and lecturers and for student assessments. This model is, however, faced with challenges of poor internet connectivity. Though not formalised at institutions of higher learning, the WhatsApp social media platform is being used for sharing course material among students and lecturers. The majority of students in institutions of higher learning own a smartphone or any other smart device that can support the WhatsApp social media application. Moreover, data required for the platform to be used can be afforded by many students. The creation of student groups on WhatsApp, ease of access (accessibility), and ease of sharing are some of the characteristics of the WhatsApp platform.

To conduct practicals, technicians and lecturers can prepare videos and reading material demonstrating the procedure of the practical as well as industrial applications, to prepare the students for the physical practical session. These videos can be uploaded either on the eLMS platform or on students' social media, for example WhatsApp, as well as through email addresses. This enables the students to conduct the practical with confidence, while remaining motivated as the skills and knowledge that they are acquiring have value and practical applications in industry. There is, however, a need to develop ways to protect the shared content so that it does not fall into the wrong hands, or that other technicians and lecturers in the same field can simply access the content and use it without proper acknowledgement.

Collaborative digital platforms, for example Jamboard, can be used during online learning sessions to elaborate on difficult concepts or even deliver lectures on courses that require calculations, for example engineering mathematics, structural engineering, and solid and fluid mechanics. Internet of Things (IoT) is a model based on the presence of a variety of objects, for example actuators, Radio Frequency identification tags, and sensors, all of which are able to interact with each other (Abbasy & Quesada, 2017). IoT is the use of connected objects and devices to manage data obtained by sensors (Abbasy &

Quesada, 2017). IoT can be best described as the interconnectivity of devices (things) over the internet (Jasim *et al.*, 2021). It allows for the sharing of knowledge, expertise, and experiences in ABE by practitioners, learners, and any other people across the globe in an effective manner. Warm-up work in preparation for the lecture can be shared with students using various IoT, social media, and web-based platforms. This preparatory work can be in the form of quizzes or review of videos or even essays (Wanner, 2015).

IoT platforms can be used for anonymous contributions to the learning curriculum and evaluation of content deliverers by learners, which gives the learners an opportunity to contribute to the improvement of the course without fear (Abbasy & Quesada, 2017). Furthermore, simulations, animations, and use of video tutorials for practicals and industrial visits can be implemented. However, it is important for students to physically participate during practicals and workshop sessions to develop hands-on competencies.

Technologies such as virtualisation and cloud computing give education practitioners access to a large database, of resources (Abbasy & Quesada, 2017). This means that experts from various institutions across the globe can exchange information within their areas of specialisation with ease and without any restrictions from geographical boundaries.

Computer Aided Design (CAD) and simulation is handy in ABE education. The simulations provide the visual model of the design in parts, components, assemblies, and in processing and manufacturing plants. This allows learners to observe the performance of the design and evaluate the effect of various external parameters or the effect of altering any parameters on the performance of the design without physically building and testing the design. The students will still be required to build the model using the parameters attained from the CAD and simulation processes, gaining workshop skills in the process, and to physically test the model. Ending only on simulations will create pseudo engineers who can only design on software but do not have the confidence to develop solutions to problems faced by the world.

Online editing platforms, for example Google Docs, can also be handy in ABE education – for instance, when students are given group assignments, they can contribute remotely. However, the benefit of live group discussions in sharpening the output is missed when using Google Docs, unless the students organise themselves to discuss, using Google Meet, for example.

Time management for research students in ABE, both undergraduate (during the final year research project) and postgraduate, can be difficult. The development of an online project management platform that analyses students' uploaded workplans and progress data could be helpful. The platform could have algorithms that notify the users if the progress is going off track. Both the student and supervisor/advisor must have access to the platform. Such a platform can assist in writing progress reports.

How Can These Technologies Be Best Applied in Developing Countries?

There is a need for governments to facilitate the provision of gadgets that allow information dissemination by controlling prices, providing grants, facilitating learners' and instructors' acquisition of the gadgets in instalments, and promoting development of internet infrastructure to improve coverage and to regulate the price of accessing the internet. Chimbunde (2022) suggested that for internet access to improve among online learning users, the governments of developing countries must liberalise the internet market to de-monopolise the provision of internet services. To improve learners' access to online learning material, institutions offering ABE education can enter into agreements with Internet Service Providers (ISVs) and have the databases and websites containing the learning material accessible free of charge for actively registered students.

The British Open University introduced a fully-online programme where students were completely off-campus (Bates, n.d.). Students were, however, capacitated with a computer, modem, and printer, all of which they had learnt to use. There is a need to capacitate students with the necessary tools and gadgets (both software and hardware) and provide training and information on how to use the technology (Bates, n.d.).

Embracing New Teaching and Learning Models in Developing Countries

Why New Approaches?

Just like other online applications, immersive learning technologies encourage lifelong learning, flexible learning, and are not limited to the learner's geographic location or gender (Jhurree, 2005; Ntaba & Jantjies, 2019). Thus it is an effective ABE pedagogy that can be employed during the case of pandemics associated with continuous lockdowns such as COVID-19 (Abumandour, 2022). Implementation of VR-powered immersive learning technologies will unlock opportunities for the female child who is normally disadvantaged in most developing countries (Mkwizu, 2022). Furthermore, institutions and countries will benefit from long-term investment in the associated technology infrastructure. Literature survey findings by Ntaba and Jantjies (2019) indicated that adoption of AR and VR in higher education contributed to learner engagement and enhanced learning outcomes. Furthermore, they concluded that the existing ICT are mainly used as tools that facilitate educator and learner interaction, but are inadequate in providing the necessary learner experience. Therefore, ICT tools alone are not effective in ABE education that requires a full learner experience for deeper understanding, skills acquisition, and memory retention. A panacea to this problem is to employ immersive learning technologies.

ABE-offering institutions and learners must adopt VR-powered immersive learning technology because students will have a full-body experience that results in improved engagement, faster understanding, and a sense of presence to the learner. In addition VR-powered immersive learning technology permits learners to have an unrestricted number of practices until they have fully understood the concepts and acquired the necessary skills (Ding & Li, 2022), mainly because operating virtual laboratories is less costly than running real laboratories (Koretsky & Magana, 2019). VR-powered immersive learning technology results in emotional experiences for the learners that are of paramount importance for cognitive learning and remembrance. There is also some distinctive interactive learning that can be achieved under VR-powered immersive learning technology. Moreover, it allows learners to get useful real-time feedback on their actions and decisions. In ABE education real-time feedback is particularly important when dealing with practical situations.

Therefore, VR is a fantastic learning tool that can be employed by ABE educators because it completely immerses the learners to levels where they feel like being in an environment identical to the real-life situation. According to Wu et al. (2021), immersive technologies enhance teaching quality and learning performance in engineering and other technical subjects. Immersive learning in VR allows learners to continuously practice both the theoretical concepts and practical skills required to become a fully-fledged Agricultural and Biosystems engineer by making use of an environment that mimics reality.

VR content can either be in the form of 360-degree Video or Computer-Generated Images (CGI). The 360-degree Video is used for replicating the actual environment because it is filmed on location and thus it is handier in ABE education when dealing with practicals and field visits. CGI are also important for theory delivery and simulation. VR-powered immersive learning technologies are also important in ABE education because by their nature they are designed to permit interaction with the virtual environment to tune the learner's brain into the right mood for learning and knowledge acquisition, which is essential for hard and diversified sciences like ABE.

Institutions can make use of valuable VR data to assess the performance of learners and instructors to their advantage. The findings from the analysis of VR data can be used for continuous improvement. Useful VR data can be in various forms which, *interalia*, include usage, performance, learner attention and engagement, feedback quality, as well as forecasted data that can be used for developing machine learning based models. ABE educators and learners in developing countries stand to benefit from properly designed VR-powered immersive learning technology because it creates the spatial, self, and social presence for the students that is key for knowledge acquisition and retention.

The adoption of immersive learning technologies in ABE education is advantageous in that complicated engineering and science concepts can be simulated to enhance skills absorption and knowledge retention. Game based learning, reality technologies, and 360-degree videos can be employed to improve learning effectiveness. Unlike other fields, ABE learners' thrust and interest is on acquiring hands-on experience with practical applications rather than attending lectures (Wu *et al.*, 2021). Hands-on experience is of great importance to developing nations due to the need to develop adequate infrastructure and equipment in these countries. However, strategies in developing countries should be slightly different from those of developed because of a myriad of technical challenges experienced in these countries. As such, previous studies on engineering education highlighted the need for adopting different blended learning approaches for effective transfer of both theoretical and practical knowledge to the learner.

What Is Required for Effective Implementation?

Government Policy

In a world of competition, developing nations should embrace technology at all costs for them to be competitive globally. There is need for government policy-makers in these countries to move with technological trends in global education reform. Policy-makers must understand and appreciate the importance of introducing novel learning models in ABE education to improve the skills base required for the creation of sustainable food systems that can render a country food secure. Even though many developing countries in Africa have technology-related policies (Mkwizu, 2022), there is need to develop sector specific digital policies and for a continuous scanning of the environment to take advantage of technology applications in those areas. This will help governments predict future trends that are bound to affect the normal way of doing business and assess their state of preparedness to deal with the alternative scenarios.

To facilitate the adoption of novel learning models in ABE education, governments must create an enabling environment for technology acquisition, support training of trainers in the field, avail funds for uninterrupted network access at ABE-offering institutions and provide the means of establishing virtual laboratories at a minimal cost. A case in point is that of removing import duty on all immersive learning technologies equipment and offering grants, subsidies, or reduced taxes to ABE institutions that are adopting immersive learning approaches.

Developing countries must develop a strategy to wean themselves from being perpetual consumers of learning technologies from elsewhere through capacitating their research and development units involved in this area. Countries must focus on how to leverage emerging novel learning technologies to improve learner experiences and skills acquisition in ABE education for the benefit of their economies. Research should not be limited to adoption and integration but must span across all areas, including new learning technology development where possible, to ensure sustainability. Quick adaptation to low-cost proven technologies that were developed in the developed world should be promoted to enable institutions to improve their ABE learning approaches which in turn will have a positive impact on the associated economies.

For successful adoption of these technologies, policy-makers must properly plan for how current pedagogies in ABE education can be integrated with new novel learning approaches to effectively equip the graduate for the digital era and beyond. In resource-limiting situations, a gradual approach of blended technology application is recommended until the institutions have been capacitated enough to opt for a complete switch.

An inclusive policy development strategy should be adopted to get the buy-in of all key stakeholders, i.e., the government, academic institutions, academics, parents or guardians, and learners. This is necessary to avoid challenges associated with plan implementation in the future. An inclusive policy development is also important for changing public and professional opinions on immersive learning technologies and other novel technology-oriented pedagogues that can be applied in ABE education.

According to Jhurree (2005) technologies can be manipulated for use in illicit and unethical behaviour. Therefore, to curb this, the government must also put in place laws that regulate the usage of these novel technologies so that users do not infringe on other people's rights.

Technical and Managerial Issues

Each of the novel learning technologies has different technical requirements for its successful implementation in developing countries such as Zimbabwe. For example, VR-powered immersive learning technology requires the close collaboration of experts from a variety of disciplines. These disciplines, among others, include user experience (UX) design, user interface design (UI), data science, instructional design, filmmaking and sound design. Immersive content specialists are also required to create an environment that fully engages the learner for an effective learning experience. In addition the means of VR delivery to learners must be adequate and uninterrupted to create a truly immersive training scenario. Therefore, there is need for adequate infrastructure and training to support the technology to be employed. Infrastructure may also include hardware, software, video games, networks, and internet access. The most widely used components of VR are a computer and a headset and these must be readily available to learners.

One approach for effective implementation is to adopt the 'Will, Skill, Tool' (WST) model suggested by Knezek & Christensen (2016), a novel pedagogy for intensive technology-oriented professions. In this case, the instructors' wills and skills are said to be embedded into sophisticated technological tools to facilitate the learning process. However, to encourage quick adoption of immersive technologies, there is a need to assess and examine students using the new technology-based learning environment that was used as a medium for exploration, knowledge and skills acquisition (Jhurree, 2005).

Although immersive learning technology is a promising pedagogy for ABE education, its successful development and implementation can be challenging especially for developing countries such as Zimbabwe. Figure 12.5 shows the factors that must be considered to develop and implement a successful immersive learning technology as well as the process that must be followed.

There is need to treat the transformation as a project so that all stakeholders are consulted beforehand (Jhurree, 2005). According to Levine (1998) as cited by Jhurree (2005), the implementation plan for transformation from traditional to technologically-oriented approaches should encompass the following:

- Formulating a planning team
- Collecting and analysing data
- Formulating the visions, goals, and objectives
- Exploring available technology
- Determining training and staffing needs
- Determining a budget and funding sources
- Developing an action plan
- Implementing the plan
- Evaluation

These recommended implementation strategies can be fused into suggestions in Figure 12.5. The proven three-phased approach of strategic planning, management control, and operational control should also be employed during the implementation of technology-driven pedagogues in ABE education, bearing in mind that most ABE programmes tend to be multidisciplinary in nature. Hence, different disciplines might require a number of approaches depending on their nature. Emphasis should be placed on improving the robustness of each stage by following the Plan-Do-Check-Act repeated cycle. During operation the systems should always be checked for continuous improvement that enriches the learner experience and knowledge acquisition.

Any ABE-offering institution in developing countries that is willing to make a switch to these novel learning approaches must do a self-assessment of its capabilities and be honest to itself before making a choice of the possible approaches, because most of them are inadequately resourced (Jhurree, 2005). In some cases, a phased implementation approach should be followed for effective performance. In addition, it is recommended that ABE institutions should develop a priority list to help them in implementing the novel learning technologies.

Training on how to use the technology is important to both the learner and the educator for successful implementation of these technologies. It is important to ensure that the training offered is in synchrony with the available infrastructure so that there is quick practice and implementation of learnt skills. There is need for provision of updated and relevant continuous professional development programmes on the hands-on instructional use of novel learning technologies to all ABE educators in developing countries so that their effectiveness can match or exceed the global trends. Cooperation,

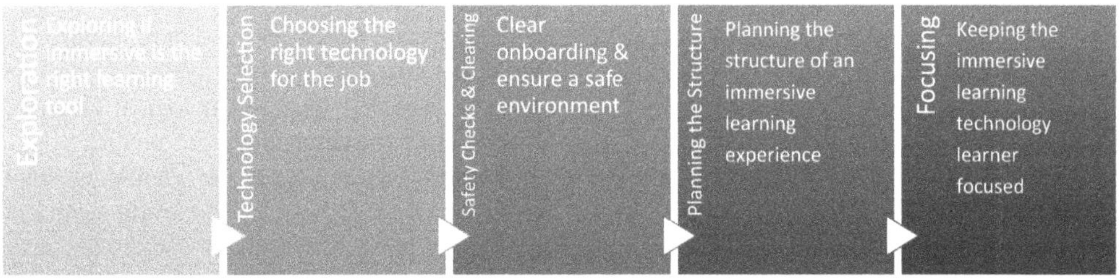

FIGURE 12.5 Steps to be considered for successful implementation of immersive learning technology (adapted from Future Visual, Brighton, United Kingdom).

collaboration, and twinning and exchange programmes among ABE institutions of different capabilities across the globe can assist academics and students in developing countries to embrace and appreciate the importance of these technologies at a faster rate.

Limiting and Enabling Factors

A number of innovative learning models have been proposed (Križani´c, 2020; Tiernan, 2010; Tang *et al.*, 2020). However, it is still difficult for most engineering educational institutions to do away with lectures in the modules because they have proven to be an effective means for knowledge dissemination and acquisition (Wu et al., 2021). Nevertheless conventional teaching approaches, such as face-to-face, are vulnerable to human distraction, hence the need for immersive learning technologies.

Most institutions of higher learning in developing countries do not have enough expertise and infrastructure to create a truly immersive learning environment. Where the experts exist, they are not adequately trained to cater for the new VR-powered immersive learning technologies. On the other hand, learners also face a barrage of challenges relating to information and communication technologies (ICT) both technically and economically. The technical challenges include the absence of ICT infrastructure in some remote areas and poor networks. Costly data bundles and receiving gadgets are some of the major economic challenges faced by learners. Ding and Li (2022) opine that the current devices used for VR-powered immersive learning technologies in higher education, i.e., mostly computers and headsets, are not readily portable. Lack of guidance and training in the use of VR equipment before class is also highlighted as one of the impediments to its effective application.

As a result of these challenges experienced in developing nations, some scholars are advocating for using immersive learning as a complement to traditional learning rather than a complete overhaul (Ntaba & Jantjies, 2019). However, it is suggested that this process be done gradually on a case by case basis depending on institutional and country policy and capacity until traditional learning approaches are almost completely replaced by immersive learning technologies and other novel learning models.

Another limitation for some of the new learning technologies is that in some cases learners are vulnerable to psychological challenges. According to Ntaba and Jantjies (2019) some scholars opine that learners in virtual environments tend to exhibit feelings of confusion, isolation, frustration, anxiety, constant tiredness, pressure, marginalisation, and demotivation. However, this psychological challenge can be overcome if varied interesting learning methods such as gaming and simulation are employed within the virtual environment. In addition, due to VR-powered immersive learning technology's flexibility, learners can adjust the pace of learning to their own liking to avoid cases of exhaustion. On the part of educators, the challenge is how to infuse engineering modules that are grounded in theories for effective learning experience under immersive learning technologies (Koretsky & Magana, 2019).

ABE institutions in developing nations have the potential to fully utilise the novel learning technologies because most of these countries already have an ICT policy that can act as a foundation for development. In addition the population in most developing countries, e.g., in SSA, is mainly composed of youths who can quickly adapt to these learning technologies. The will and zeal of learners in these countries after COVID-19 lockdown experiences is also one of the enabling factors. Social and economic challenges in most of these countries negatively affect learners in such a way that they would be happy to free up more time to hustle as a result of using immersive technologies. High literacy rates in some of these countries can be taken as a key enabler to the adoption of novel technologies. Lack of real laboratory and workshop infrastructure at most ABE-offering institutions will encourage learners to switch to immersive technologies that have virtual laboratories and workshops for a real experience.

Case Study: Lessons Learnt in Technology Application at the University of Zimbabwe.

Education 5.0 and the Demand for Goods and Services

In the past, tertiary education was based on Model 3.0, focusing on teaching, research, and community service. In 2019 the Government of Zimbabwe launched Model 5.0 to incorporate two additional pillars: innovation and industrialisation; a deliberate recognition of the two as socio-economic drivers for poverty alleviation and prosperity (Murwira, 2019). The underlying principle of Education 5.0 as a scientific and developmental philosophy is that higher and tertiary education which does not produce goods and services is not relevant (Government of Zimbabwe [GoZ], n.d.).

The Government of Zimbabwe's vision through Education 5.0 is that graduates from Tertiary Education Institutions should be equipped with skills that empower them to become innovative towards societal development through the application of transformative science and technology knowledge that delivers goods and services (GoZ, n.d.)

At University of Zimbabwe (UZ) level, the philosophical change was guided by the Strategic Plan (2019–2025) which pioneered the implementation of Education 5.0. In the UZ Strategic Plan (University of Zimbabwe [UZ], n.d), there is clear articulation of the desire to embrace new ways of conducting cutting-edge research, driven by a clear understanding of the needs, wants, and circumstances of society, knowing what the University should cause (e.g. a new generation of companies and industries), and most importantly, how to come up with innovative models, methodologies, tools, strategies, and policies that trigger technological advancements for modernisation of the Zimbabwean economy.

Furthermore, the Strategic Plan puts emphasis on critical thinking, creativity, and innovation, and problem-solving among graduates and researchers/scholars as the cornerstones for the generation of quality knowledge and a robust evidence-base upon which decisions can be made towards modernisation of the Zimbabwean economy.

Experiences With ABE Education Delivery

High cost of data bundles is prevalent among students in developing countries, thus they face challenges in accessing online classes (Chimbunde, 2022), even learning material uploaded on e-learning platforms. ISVs created internet access data packages tailor-made for students undertaking online lessons but the existing infrastructure failed to handle the increase in data traffic leading to internet access challenges. To operationalise online learning in developing countries, governments in developing countries must support ISVs to improve internet infrastructure and network coverage. About 82% of students in sub-Saharan Africa do not have access to the internet due to various limiting factors which include expensive data bundles, poor network coverage, and weak network signals (Chimbunde, 2022).

Unavailability of well-trained personnel, poor digital infrastructure, and poor cyber security are some of the hindrances to online learning in developing countries (Chimbunde, 2022). Most learning technologies being used in developing countries are foreign products, and users, both students and lecturers, need to be trained to use the technologies. For example, there was lack of knowledge by lecturers on ways to convert hardcopy material into softcopy material which can be updated online (Chimbunde, 2022). The issue of cyber security is one of the largest threats to online learning but it is hardly talked about.

Students from developing countries, for example Zimbabwe, are not financially capacitated and cannot afford to pay for the licenced software that is used in various ABE education, for example CAD and simulation software. Software providers, for example Autodesk, have student versions of their software which students can access and download for free on their website. To enhance access, platforms similar to the Access to Global Online Agricultural Research (AGORA) can be developed specifically for students in developing countries to access licensed software free of charge.

There exists a positive correlation between lecture attendance and overall student performance (Wanner, 2015). However, it can be difficult to track student attendance and conduct assessments, for example tests and examinations, when using online learning tools and platforms. Students have a tendency to just log onto the platform and do other things that are not learning-related. The development of artificially intelligent IoT systems that keep track and monitor student attendance during online learning is key. Systems that allow remote student assessment for theoretical examinations making use of Artificial Intelligence (AI) and Artificial General Intelligence (AGI) can revolutionise online learning technology. It is also difficult to ascertain that the tasks completed by students on e-learning platforms for assessment were done by them or another person was tasked to complete them on their behalf.

With regards to creating video tutorials, the supporting technologies, for example high resolution cameras, may not be available due to a lack of funding to purchase them. Students require smart gadgets, for example laptops, smartphones, and tablets that connect to the internet, to access the information provided online.

When online learning became mandatory due to the COVID-19 pandemic movement restrictions, students and lecturers encountered challenges in the areas of accessing the internet (poor network coverage and bandwidth), owning user-friendly tools, and the high cost of the data bundles (Chimbunde, 2022). The most challenging aspect of delivering ABE education virtually is how to dispense practical skills. ABE is a very practical discipline which relies on hands-on development of skills.

Future Perspectives/Outlooks on ABE Education

The anecdotal evidence suggests that higher education institutions offering ABE have to transform or become obsolete in future if they are to continue being relevant in serving the agricultural industry in particular. As the population continues to increase, there will always be a demand for food, but innovative ways of incorporating a good understanding of the biosystems application is required if this demand for the food is to be met.

It is necessary to develop ABE education programmes that can deliver economic and societal needs and so a rapid transition is required towards embracing more disruptive change through digital technology. However, it is critical that ABE education programmes meet certain minimum quality requirements, e.g., in alignment with the Southern African Development Community (SADC) qualifications framework (SADC, 2017) which covers higher education and technical and vocational education and training in southern Africa.

The ABE profession also needs to develop and have a professional accreditation body that can set the minimum standards to avoid unscrupulous or bogus higher learning institutions dishing out degrees over the net. ABE is very much an applied discipline. It is therefore imperative that hands-on skills are developed somehow in order for the candidates to be able to produce goods and services. So, Hybrid models of face-to-face learning by doing will continue to be relevant to the sector with self-learning as learners are encouraged to tinker with different software in solving problems on their own.

Immersive learning technologies are proving to be effective for ABE. However, one cannot completely rule out the use of traditional pedagogies in developing countries because of the myriad of technical challenges that these nations face. ABE institutions in developing countries are at different paces of development. It is recommended that they select a gradual educational transformation that best suits their environment. VR and Gaming that have been proven to be effective in ABE in developed countries should be adapted and blended with traditional approaches to achieve the learning objectives with the hope of finally having more of the immersive technologies and other emerging technologies as economies permit. As it stands, it is inevitable that the future of ABE education lies in immersive technologies and artificial intelligence with a minimum blending of traditional pedagogies as human involvement is declining in the process. ABE institutions in developing countries must gradually address their technical and economic challenges in order to keep pace with the changing world of engineering education and remain relevant now and in the near future. Thus government policy is key in these digital transformations and will largely determine the pace of the transition.

REFERENCES

Abbasy, M. B., & Quesada, E. V. (2017). *Predictable Influence of IoT (Internet of Things) in the Higher Education*, *7*(12), 914–920. https://doi.org/10.18178/ijiet.2017.7.12.995

Abumandour, E. T. (2022). Applying e-learning system for engineering education –challenges and obstacles. *Journal of Research in Innovative Teaching & Learning*, *15*(2), 150–169 doi: 10.1108/JRIT-06-2021-0048

Anami, B. S., Malvade, N. N., & Palaiah, S. (2020). Deep learning approach for recognition and classification of yield affecting paddy crop stresses using field images. *Artificial Intelligence in Agriculture*, *4*, 12–20.

Bates, T. (n.d.). Online learning tools and technologies. In *Tony Bates Associates Litd* (pp. 1–17). https://www.tonybates.ca/wp-content/uploads/2008/07/online-learning-tools

Çeliköz, N., Erişen, Y., & Şahin, M. (2016). Chapter 3: Cognitive learning theories. In Z. Kaya & A.S. Akdemir (Eds.), *Learning and Teaching: Theories, Approaches and Models* (1st ed., pp. 31–47). Ankara: Çözüm Eğitim Yayıncılık.

Chimbunde, P. (2022). Funding the online teaching and learning in developing countries: Insights from Zimbabwe. *Educational Technology Research and Development*. doi: 10.1007/s11423-022-10163-3

Ding, X., & Li, Z. (2022). A review of the application of virtual reality technology in higher education based on web of science literature data as an example. *Frontiers in Education*, *7*, 1048816. doi: 10.3389/feduc.2022.1048816

Future Visual. (n.d.). Immersive learning – What it is, benefits and how to implement. Retrieved on 3/02/2023 from: https://www.futurevisual.com/blog/immersive-learning/

Government of Zimbabwe (GoZ). (n.d.). *Doctrine Education 5.0, Heritage – Innovation - Industrialisation*. Harare: Ministry of Higher and Tertiary Education, Science and Technological Development.

Jasim, N. A., Salim AlRikabi, H. T., & Farhan, M. S. (2021). Internet of Things (IoT) application in the assessment of learning process. *IOP Conference Series: Materials Science and Engineering*, *1184*(1), 012002. https://doi.org/10.1088/1757-899x/1184/1/012002

Jhurree, V. (2005). Technology integration in education in developing countries: Guidelines to policy makers. *International Education Journal*, *6*(4), 467–483.

Knezek, G., & Christensen, R. (2016). Extending the will, skill, tool model of technology integration: Adding pedagogy as a new model construct. *Journal of Computing in Higher Education*, *28*, 307–325. doi: 10.1007/s12528-016-9120-2

Koretsky, M. D., & Magana, A. J. (2019). Using technology to enhance learning and engagement in engineering. *Advances in Engineering Education*, Spring, 1–53.

Križani´c, S. (2020). Educational data mining using cluster analysis and decision tree technique: A case study. *International Journal of Engineering Business Management*, *12*, 1847979020908675. doi: 10.1177/1847979020908675

Levine, J. (1998). Planning strategically for technology integration. [Online]. http://www.coe.uh.edu/insite/elec_pub/HTML1998/el_levi.htm [Accessed 11 July 2001].

Madiraju, C., Tellez-Corrales, E., Hua, H., Stec, J., Nauli, A. M., & Brown, D. M. (2020). Analysis of student perceptions of just-in-time teaching pedagogy in PharmD microbiology and immunology courses. *Frontiers in Immunology*, *11*, 1–12. doi: 10.3389/fimmu.2020.00351

Mahasneh, D., Shoqirat, N., Singh, C., & Hawks, M. (2020). From the classroom to Dr. YouTube: nursing students' experiences of learning and teaching styles in Jordan. *Teaching and Learning in Nursing*. doi: 10.1016/j.teln.2020.09.008

Malik, S. (2021). Learning theory of cognitivism and its implications on students learning. *World Wide Journal of Multidisciplinary Research and Development*, *7*(5), 67–71.

Markowitz, D. M., Laha, R., Perone, B. P., Pea, R. D., & Bailenson, J. N. (2018). Immersive virtual reality field trips facilitate learning about climate change. *Frontiers in Psychology*, *9*, 2364. doi: 10.3389/fpsyg.2018.02364

Mayer, R. E. (2010). Applying the science of learning to medical education. *Medical Education*, *44*(6), 543–549. doi: 10.1111/j.1365-2923.2010.03624.x

Mayer, R. E., & Moreno, R. (n.d.). *A Cognitive Theory of Multimedia Learning: Implications for Design Principles* (pp. 1–10). https://citeseerx.ist.psu.edu/document?repid=rep1&type=pdf&doi=7b18026a2534894bc6698a2a87757b0c2d3940f5

Mkwizu, K. H. (2022). Virtual reality and open schooling: challenges and opportunities. *International Journal of Open Schooling*, *1*(1), 96–108.

Moghavvemi, S., Sulaiman, A., Jaafar, N. I., & Kasem, N. (2018). Social media as a complementary learning tool for teaching and learning: The case of youtube. *International Journal of Management Education*, *16*(1), 37–42. doi: org/10.1016/j.ijme.2017.12.001

Murwira, A. (2019). *Towards Revitalizing the Roles of Universities in Development (Zimbabwe)*. Harare: Ministry of Higher and Tertiary Education, Science and Technological Development.

Ntaba, A., & Jantjies, M. (2019). Open distance learning and immersive technologies: A literature analysis. Conference: International Conference Cognition and Exploratory Learning in Digital Age 2019, 51–60. doi: 10.33965/celda2019_201911L007

Primicerio, J., Di Gennaro, S. F., Fiorillo, E., Genesio, L., Lugato, E., Matese, A., & Vaccari, F. P. (2012). A flexible unmanned aerial vehicle for precision agriculture. *Precision Agriculture*, *13*(4), 517–523.

Ryan, M. L. (2015). Narrative as virtual reality 2: Revisiting immersion and interactivity in *literature and electronic media*. Baltimore: JHU Press. Retrieved on 3/02/2023 from: https://www.academia.edu/16494139/Narrative_as_Virtual_Reality_II_Revisiting_Immersion_and_Interactivity

Southern African Development Community (SADC). (2017). The Southern African Development Community Qualifications Framework (SADCQF). Gaborone, Botswana.

Talaviya, T., Shah, D., Patel, N., Yagnik, H., & Shah, M. (2020). Implementation of artificial intelligence in agriculture for optimisation of irrigation and application of pesticides and herbicides. *Artificial Intelligence in Agriculture*, *4*, 58–73.

Talespin. (2022). What is immersive learning? – An introduction to learning in the metaverse Retrieved on 3/02/2023 from: https://www.talespin.com/blog-post/what-is-immersive-learning-an-introduction-to-learning-in-the-metaverse

Tang, Y. M., Au, K. M., Lau, H. C., Ho, G. T., & Wu, C. H. (2020). Evaluating the effectiveness of learning design with mixed reality (MR) in higher education. *Virtual Reality*, *24*, 797–807. doi: 10.1007/s10055-020-00427-9

Tiernan, P. (2010). Enhancing the learning experience of undergraduate technology students with LabVIEW™ software. *Computers & Education*, *55*(4), 1579–1588. doi: 10.1016/j.compedu.2010.07.001

Udeozor, C., Toyoda, R., Abegão, F., & Glassey, J. (2022). Digital games in engineering education: Systematic review and future trends. *European Journal of Engineering Education*, doi: 10.1080/03043797.2022.2093168

United Nations Department of Economic and Social Affairs [UNDESA]. (2017). World population prospects 2017. https://www.un.org/development/desa/pd/sites/www.un.org.development.desa.pd/files/files/documents/2020/Jan/un_2017_world_population_prospects-2017_revision_databooklet.pdf

University of Zimbabwe. (n.d.). *University of Zimbabwe Strategic Plan 2019–2025*. Harare.

Wanner, T. (2015). *Enhancing Student Engagement and Active Learning through Just-in-Time Teaching and the Use of PowerPoint*, *27*(1), 154–163.

Wu, C. H., Tang, Y. M., Tsang, Y. P., & Chau, K. Y. (2021). Immersive learning design for technology education: A soft systems methodology. *Frontiers in Psychology*, *12*, 745295. doi: 10.3389/fpsyg.2021.745295

Yilmaz, K. (2011). The cognitive perspective on learning: its theoretical underpinnings and implications for classroom practices. *The Clearing House*, *84*, 204–212.

13

Current and Future Trends in Agricultural Engineering Education in Nigeria

Ogbonnaya Chukwu

Introduction

Agricultural engineering education pertains to the socio-economic and technological advancement of a nation by addressing pertinent issues. It involves combining creativity and practicality with a scientific approach. When the scientific foundation encompasses living beings, their by-products, and/or natural resources, it gives rise to a diverse range of engineering branches, such as Agricultural, Biological, Biosystems, Bioresources, and Bioenvironmental Engineering (Chukwu, 1999). With the progression of industrialisation and the enhancement of the economic status of farmers and processors, the quality and quantity of agricultural inputs, management techniques, farm produce quality, and value addition methods will continue to undergo changes (Singh, 2000). To meet the evolving requirements of agriculture, especially in the agro-industrial sector, it is crucial to update the agricultural engineering curriculum periodically. Typically, it takes around 5 to 10 years for newly trained engineers to make effective and efficient contributions to research, policy-making, and decision-making. To achieve significant results, the planning of education must be based on the future needs for at least a decade (Singh, 2000; Department of Agricultural and Bioresources Engineering [DABE], 2021).

Agricultural Engineering involves designing, producing, utilising, and managing technical methods and processes for the production, storage, treatment, and processing of agricultural goods. This field applies the principles of Engineering and Technology to solve problems in the biological realm, including plants and animals. The profession aims to develop effective and eco-friendly techniques for managing and producing food, fibre, timber, and renewable energy for the global population (Chukwu, 1998). Lately, the profession has been referred to by various names, including Bioresources Engineering, Agricultural and Biological Engineering, and Agricultural and Environmental Engineering.

The definition of Agricultural Engineering has been revised to encompass the application of engineering principles and fundamental concepts of biology to agricultural and biological systems and tools. These systems and tools can range in scale from molecular to ecosystem level and are utilised for the safe, efficient, and environmentally-sensitive production, processing, and management of agricultural, biological, food, and natural resource systems. It draws upon Animal Biology, Plant Biology, as well as Mechanical, Civil, Electrical, and Chemical Engineering principles, making it a broader scope engineering discipline than any other. Agricultural Engineering presents a diverse range of challenges, as engineering solutions are required for the efficient production of food, feed, fur, fuel, and fibre under various environmental conditions and production practices. This field encompasses various areas of practice, including (Igbeka, 1991, 2002):

- Utilisation and processing of agricultural products for food, fodder, as well as for alternative energy sources such as solar and biomass, or bio-renewable products.
- Development of technology for treatment, upgrading, and recycling of biomaterials to be used as soil improving or fertilising materials.
- Planning, implementation, and maintenance of rural infrastructure (roads, water supply, transport, electricity, cottage industries) to enhance rural living standards.
- Ensuring the availability of safe and abundant food, clean water, renewable fuel and energy, safe working conditions, and a healthy environment for the world's population.
- Implementation of soil and water conservation techniques such as erosion control, tillage, irrigation, drainage structures, landscape maintenance, biological waste management, and public services such as clearing ditches and cutting hedges.
- Development of automated computer-aided control systems and robotics for field and postharvest operations, food processing, and animal production.
- Designing agricultural machinery, equipment, structures, processes, and systems.
- Management and processing of poultry, dairy, fish, and forestry products.

DOI: 10.1201/9780429150111-15

The Evolution of Agricultural Engineering as a Profession

The application of engineering principles and techniques to the field of agriculture has been there long before the term agricultural engineering was used. In most cases, agriculture adapted its engineering from the disciplines of Civil, Mechanical, Electrical, Chemical, and Architectural engineering. The first time that agricultural engineering was listed as a course in any University or equivalent institution was in 1896, in the catalogue of the University of Nebraska (Stewart, 1979). A conference was called late in 1906 at the University of Illinois which was attended by three men, F.R. Crane of the host university, J.B. Davidson of Iowa State College, and C.A. Ocock of the University of Wisconsin. These gentlemen met to discuss instructional methods and the conference was so successful that it was agreed to make it an annual event.

During the second meeting, which was held December 27 and 28, 1907, in the agricultural engineering building at the University of Wisconsin in Madison, the American Society of Agricultural Engineers was formally founded and initiated as a national engineering society. In 1910, Iowa State College had the distinction of conferring the first degree of Bachelor of Science in Agricultural Engineering and by 1925 the number of institutions conferring such degrees was 10, reaching 41 in 1950. This was how a profession which was to change the agricultural development of the USA started.

On the European side of the Atlantic, developments were also taking place (Gibb, 1988). Generally, however, in Britain, farming went on at a leisurely pace. With its 'Golden Age' in the Nineteenth century now long past, and with the ready availability of cheap imported staple foodstuffs from the colonies, there was little incentive for the majority of British farmers or engineers to emulate their American cousins. This state of laissez-faire was to be broken by the German sea blockade of Britain in the World War I of 1914–1919. With the ending of the war and the resumption of international trading in the 1920s the first incursion of mechanical power into British farming slowed down, almost to a standstill. But the seeds had been sown, and in 1924 the Oxford Institute of Research in Agricultural Engineering was established by the University of Oxford.

Also, even in the depressed years of the early 1930s were sporadic efforts by a handful of pioneers (farmers and engineers) which resulted in the introduction to British farms of the first combine harvesters, balers, and other machines, on whose successors future farm mechanisation in Britain was to become dependent. From this background there arose in 1938 the impetus and initiative to set up the Institution of British Agricultural Engineers, which not until eleven years later altered their name to the Institution of Agricultural Engineers.

The evolution of Agricultural Engineering as a profession started, as we can see, in the United States and later spread to Europe and Great Britain. The people who started it all were engineers with basic training in either mechanical, civil, chemical, or electrical engineering. They were very dedicated to the idea of having a profession which applied engineering technology to solving agricultural problems. These people have been referred to at different instances as agricultural engineering converts but I would like to call them 'born-agains'; they were very aggressive, charismatic, and enthusiastic about their new-found profession. They faced new challenges and they accepted these challenges with determination, hard work, and dedication. It was this attitude that propelled agricultural engineering into recognition and helped achieve the goals the founding fathers set in the United States (Igbeka, 2002).

The Evolution of Agricultural Engineering Education in Nigeria

Agricultural Engineering as a profession in Nigeria, according to Aboaba (1988), started sometime in 1959 through the conversion process. Agricultural and Engineering graduates were sent to a two-year postgraduate training in the University of Durham, Britain, leading to a National Diploma or Master of Science in Agricultural Engineering. The first degree in Agricultural Engineering awarded by a Nigerian University was in 1967 by the University of Nigeria, Nsukka. There were two graduates, and I will beg your indulgence to mention their names: Messrs U.P.C. Okudo and E. Nwalo. However, in 1966, nine Nigerians with Bachelor of Science in Agricultural Engineering degrees who were trained in Israel, under the UNESCO/FAO/Israel fellowship, returned to take up positions in the country's Ministries of Agriculture and Universities. Six of them were from the then Eastern Region while three were from the West. It can be said, therefore, that the first set of Nigerian Agricultural Engineers with basic degrees in Agricultural Engineering that worked in the country was made in 1966 in Technion, the Israel Institute of Technology, Haifa. The practice of agricultural engineering as a profession in Nigeria, as we can see, dates back to only in the sixties. At present, there are 28 universities (Table 13.1; National Universities Commission, 2022) awarding at least Bachelor's degree in Agricultural Engineering in Nigeria; and 22 Polytechnics and 15 Colleges (Table 13.2a and Table 13.2b, respectively; National Board for Technical Education, 2022) awarding at least National Diplomas in Agricultural Engineering.

Initiatives to Enhance Agricultural Engineering Education in Nigeria

The Nigerian Society of Agricultural Engineers founded in 1967, which later became The Nigerian Institution of Agricultural Engineers, is a regulatory organisation that establishes standards for the practice of agricultural engineering in Nigeria. It operates under the umbrella of the Nigerian Society of Engineers, which oversees the regulation of all engineers in the country. In conjunction with the National Universities Commission and Council for Regulation of Engineering in Nigeria (COREN), it has contributed immensely to development

TABLE 13.1

Universities awarding a Bachelor's Degree in Agricultural Engineering in Nigeria

S/No	University	Year Established
1	University of Ibadan	1948
2	University of Nigeria, Nsukka	1960
3	Ahmadu Bello University, Zaria	1962
4	Obafemi Awolowo University, Ile-Ife	1962
5	University of Ilorin	1975
6	University of Maiduguri	1975
7	Bayero University, Kano	1975
8	Rivers State University of Science and Technology, Port Harcourt	1979
9	Federal University of Technology, Owerri	1980
10	Federal University of Technology, Akure	1981
11	Enugu State University of Science and Technology, Enugu	1981
12	Federal University of Technology, Minna	1982
13	Olabisi Onabanjo University, Ago-Iwoye, Ogun State	1982
14	Federal University of Agriculture, Makurdi	1988
15	Abubakar Tafawa Balewa University, Bauchi	1988
16	Modibbo Adama University of Technology, Yola	1988
17	Federal University of Agriculture, Abeokuta	1988
18	Ladoke Akintola University of Technology, Ogbomosho	1990
19	University of Uyo	1991
20	Imo State University, Owerri	1992
21	Michael Okpara University of Agriculture, Umudike	1992
22	Nnamdi Azikiwe University, Awka	1992
23	Niger Delta University, Yenagoa	2000
24	Akwa Ibom State University, Mkpat-Enin	2000
25	Taraba State University, Jalingo	2008
26	Kwara State University, Malete, Ilorin	2009
27	Adeleke University, Ede	2011
28	Federal University, Oye-Ekiti	2011

of Agricultural Engineering education in Nigeria. The Senate or Academic Board of each institution sees that academic standards are maintained. As part of the policy of maintaining high academic standards, each institution stipulates the conditions that prospective students must have attained and obtained certain grades in specific subjects before they can be admitted into the Department of Agricultural Engineering at the 100/200/300 levels.

TABLE 13.2A

Polytechnics awarding a National Diploma/Higher National Diploma in Agricultural Engineering in Nigeria

S/No	Polytechnic	Year Established
1	Kaduna Polytechnic, Kaduna	1956
2	Institute of Management and Technology, Enugu	1965
3	The Oke-Ogun Polytechnic, Saki	1970
4	Auchi Polytechnic, Auchi	1973
5	Kwara State Polytechnic, Ilorin	1973
6	Ramat Polytechnic, Maiduguri	1976
7	Waziri Umaru Federal Polytechnic, Birnin Kebbi	1976
8	Federal Polytechnic, Ado-Ekiti	1977
9	Federal Polytechnic, Bida	1977
10	Federal Polytechnic, Nekede, Owerri	1977
11	Lagos State Polytechnic, Ikorodu	1977
12	Lagos State Polytechnic, Ikorodu	1977
13	Plateau State Polytechnic, Barkin-Ladi	1978
14	Imo State Polytechnic, Umuagwo, Owerri	1978
15	Federal Polytechnic, Bauchi	1979
16	Federal Polytechnic, Mubi	1979
17	Rufus Giwa Polytechnic, Owo	1979
18	Federal Polytechnic, Oko	1979
19	Hassan Usman Katsina Polytechnic, Katsina	1983
20	Nuhu Bamalli Polytechnic, Zaria	1989
21	Federal Polytechnic, Nasarawa	1993
22	Delta State Polytechnic, Ozoro	2003

TABLE 13.2B

Colleges awarding a National Diploma/Higher National Diploma in Agricultural Engineering in Nigeria

S/No	College	Year Established
23	Federal College of Agriculture, Ibadan	1921
24	Akperan Orshi College of Agriculture, Yandev	1926
25	Samaru College of Agriculture, Zaria	1931
26	Federal College of Agriculture, Ishiagu	1955
27	Federal College of Agriculture, Akure	1957
28	College of Agriculture, Zuru	1976
29	Federal College of Forestry Mechanisation, Afaka Kaduna	1976
30	College of Agriculture, Lafia	1977
31	Mohamet Lawan College of Agriculture, Maiduguri	1977
32	College of Agriculture, Jalingo	1979
33	Plateau State College of Agriculture, Garkawa	1998
34	Federal College of Land Resources Technology, Kuru, Jos	2000
35	Federal College of Land Resources Technology, Owerri	2000
36	Audu Bako College of Agriculture, Danbatta	2002
37	Oyo State College of Agriculture and Technology, Igboora	2006

Agricultural Engineering Education in Nigeria: a Case Study of the Department of Agricultural and Bioresources Engineering, Federal University of Technology (FUT) Minna

Vision

To be a world class centre of excellence committed to the development and improvement of agricultural resources through engineering for economic, social, and cultural wellbeing.

Mission

The Department is committed to the advancement of engineering knowledge in agriculture through excellent teaching, learning, and research and to empower our graduates to be self-reliant.

Philosophy

Agricultural and Bioresources Engineering integrates engineering science and design and applied biological sciences for the solution of problems involving plants, animals, and natural environment. It deals with patterns of relationships among organisms and their environments, and engineering design to develop processes, machines, and systems that influence, control, or utilise biological materials and organisms for the benefit of the society.

Objectives of the Programme

(a) To educate engineers who can effectively apply engineering principles to various aspects of agriculture, such as agricultural mechanisation, crop production, processing and storage, farm and business management, irrigation, rural electrification, farm structures, drainage, and erosion control.

(b) To educate engineers who can economically select, efficiently use, maintain, and repair labour-saving equipment utilised in agriculture to maximise the benefits derived from them.

History of the Programme

The Department of Agricultural and Bioresources Engineering was established in 1983 as the Department of Agricultural Engineering. The degree of Bachelor of Engineering (Agric. Engineering) was approved for the Department in 1983. In 2008, the name of the Department was changed to Agricultural and Bioresources Engineering. The first set of graduates passed out in 1989. The Department currently provides undergraduate and postgraduate degree programmes up to the PhD level, which have received official accreditation. The staff strength rose from three in 1987 to thirty-five (twenty-three academic staff and twelve non-academic staff) in 2022. The Department places a strong emphasis on research and provides opportunities for training to its staff. Additionally, practical exposure is prioritised for students.

Staff Development/Research

Staff development is highly encouraged in the Department. The University has laid down rules on study leave/fellowship and conference/workshop attendance (physical and virtual), and tries to encourage staff as far as financial allocation permits. Currently, the Department has one academic staff member studying for a PhD degree in Europe and two studying for PhD degrees in Nigeria. Also, three staff members in the department (two teaching staff and one non-teaching staff) are studying for Master's degrees in Agricultural and Bioresources Engineering in Nigeria.

Current Students Population at the Undergraduate Level

100 Level–137	200 Level–151	300 Level–147
400 Level–103	500 Level–102	

Student Welfare

Every student is assigned an academic staff member as an adviser, and they can directly interact with any of the Department staff or the Head of Department, in addition to the University's Students Affairs Officer, to address any issues or concerns. In the event of unsatisfactory attention, students may escalate their grievances to the Head of Department, who is responsible for addressing the matter at departmental meetings. If necessary, the Head of Department may refer the issue to the Dean of School or Board of Studies, who may further escalate it to the Senate for academic matters or the Students' Affairs Officer for other matters.

Examination

The setting, conduct, evaluation schemes, moderation schemes (internal and external for degree examinations), and issuance of results are as in the "Approved Senate Regulations governing the conduct of examination" and the "General Academic Regulations for degree courses". The relevant details are set out below:

General Academic Regulations for Degree Courses

(a) The unit of course shall be the semester, one semester units being when a class meets one hour every week for the semester or three or four hours every week in the laboratory for the semester or the equivalent in workshop or field work time. The size of course shall, as much as possible, be a maximum of three units and its duration shall be one semester except for projects and design courses which may carry more than three units and may last more than one semester.

(b) A core course is one which must be registered for and passed by a student to get his/her degree and is counted towards the classification of his/her degree.

(c) An elective course is either compulsory or optional. All elective courses taken are used in final computation for the degree classification.
(d) A mandatory course is a course that must be completed and passed before a student can obtain their degree. It also counts towards the degree classification.
(e) An audited course is attended by a student without writing the exam for it.
(f) A prerequisite course is a course that must be passed before a student can enrol in a higher-level course.
(g) A co-requisite course is a course that can be taken concurrently with another specified course.

Student Work Load

(a) A student is allowed to register for up to 24 units per semester, while the minimum number of units they can register for is 16 units per semester.
(b) Grading System

The grading system is presented in Table 13.3 while the final assessment and class of degree are presented in Table 13.4.

To classify the degree level, the CGPA will take into account courses ranging from the 100 to the 500 level, or from the 200/300 level to the 500 level courses for Direct Entry students. A student with a National Diploma is admitted directly into the programme at the 200 level, while his or her counterpart with a Higher National Diploma is admitted at the 300 level.

TABLE 13.3

Grading system for the Bachelor of Engineering programme at FUT Minna

Marks	Letter Grade	Grade Point
70 – 100	A	5
60 – 69	B	4
50 – 59	C	3
45 – 49	D	2
40 – 44	E	1
0 – 39	F	0

TABLE 13.4

Final assessment and class of degree at FUT Minna

Class	Cumulative Grade Point Average (CGPA)
First Class	4.50 – 5.00
2nd Class Upper Division	3.50 – 4.49
2nd Class Lower Division	2.40 – 3.49
3rd Class	1.50 – 2.39
Pass	1.00 – 1.49

Duration of Semester

Typically, a semester consists of fifteen weeks allocated for teaching, with an additional two weeks set aside for exams, totalling seventeen weeks, excluding government-declared public holidays and mid-semester breaks.

Continuous Assessment

A student's progress is assessed regularly throughout the course, and they receive periodic updates on their performance. The final exam for each course usually takes place at the end of the semester when the course is offered. The overall grade for the course is determined by combining the student's score in various assessments, including quizzes, tests, assignments, and attendance, as well as the final exam. The continuous assessment, which includes all the assessments except the final exam, counts for a maximum of 40% of the total grade.

In addition, the student must have a CGPA of 2.0 and above to remain in good academic standing. Also, students with a session CGPA greater than 4.5 at each level of study are usually recommended for the University Scholarship during the subsequent session.

Excursions to industrial establishments and film shows are also arranged from time to time. Students are encouraged to join professional bodies such as the Nigerian Institution of Agricultural Engineers and Nigerian Society of Engineers, as student members.

Admission Requirements

The minimum requirements for admission into the Bachelor of Engineering in Agricultural Engineering are as follows:

(a) **University Tertiary Matriculation Examination (UTME) Entry Requirements**
(i)*GCE Ordinary Level/SSCE Qualification*

Five credits at General Certificate of Education (GCE) Ordinary Level or Senior School Certificate Examination (SSCE), which must include Mathematics, Physics, Chemistry, English Language, and Biology/Agricultural Science.

(ii) *UTME Subjects*

Candidates are required to sit for the UTME in English Language, Mathematics, Chemistry, and Physics. The admission of a candidate through the UTME depends on his/her performances at the matriculation examination and subsequent post UTME screening (which is university-based).

(b) **Direct Entry Requirements**

GCE 'A' Level Interim Joint Matriculation Board (IJMB) or Higher School Certificate (HSC) passes in Mathematics and Physics at Credit Level and passes in English Language and any other subject at GCE 'O' Level or SSCE are required.

(c) **Waivers and Special Considerations**

(i) Candidates holding the National Diploma (ND) with a minimum of Lower Credit or its approved equivalents obtained in Agricultural Engineering or cognate Departments from recognised Polytechnics or Schools/Colleges of Technology may be admitted into the 200 level.

(ii) Candidates holding the Higher National Diploma (HND) with minimum of lower credit or its approved equivalents obtained in Agricultural Engineering or cognate Departments from recognised Polytechnics or Schools/Colleges of Technology may be admitted into the 300 level.

(iii) The course(s) to be taken from lower levels by candidates admitted through Direct Entry (ND and HND or their equivalents), will depend on the overall assessment of the candidate's academic transcript.

Duration of the Programme

The programme duration is five years, with each year divided into two semesters (see Tables 13.5A-I). There are a total of ten semesters. However, students stay only for one semester at 400 level; the second semester at 400 level is used for the compulsory Student's Industrial Work Experience Scheme (SIWES) programme. A student that cannot graduate within the stipulated period of five years is given an opportunity to spill over for a maximum duration of two and a half years (or 5 semesters), after which his/her studentship is terminated.

Minimum Academic Standard to Proceed from the 100 level to the 200 Level

All students are expected to pass the following minimum credit hours taken at the 100 level in Mathematics, Chemistry and Physics before proceeding to the 200 level.

10 Credits of Mathematics 8 Credits of Chemistry
8 Credits of Physics

Industrial Training

As part of the five-year degree programme, students are required to complete one year of industrial training, which consists of different components. Firstly, there is a ten-week Student's Work Experience Programme (SWEP) which takes place internally at the end of the 200 level. Secondly, there is a twelve-week Student's Industrial Work Experience Scheme I (SIWES I) which is carried out in an industry setting at the end of the 300 level. Finally, there is a twenty-four-week SIWES II which is also completed in an industry setting and starts at the end of the first semester in the 400 level. Together, these three components make up the one-year industrial training requirement.

Graduation Requirements

For a student to graduate, it is required that the student passes a minimum of 26 credit units in basic sciences (10 units Mathematics, 8 units chemistry and 8 units Physics). Also, he/she must pass 130 credit units in major Engineering courses that will include four credit units (elective) within the department and 2 credit units (elective) outside the department. A minimum of 5 credit units is required from social sciences and 6 credit units from general studies. The student's work experience programme and industrial work experience scheme are mandatory for graduation and are worth 10 and 38 credit units, respectively. However, these programmes are not factored into a student's final grade calculation. Therefore, a minimum of 221 credit units are required for a student to graduate. The courses except SWEP and SIWES are graded (see Table 13.6).

100 LEVEL: SEMESTER I

TABLE 13.5A

Semester 1 of Year 1 programme

Course Code	Course Title	General Studies	Basic Sciences
MAT 111	Algebra and Trigonometry	-	3
MAT 112	Vector, Geometry and Dynamics	-	3
CPT 111	Elementary Computer	-	2
PHY 113	Mechanics	-	3
CHEM 111	General Chemistry I	-	3
CHEM 112	Inorganic Chemistry	-	2
GST 104	Introductory Economics	2	-
GST 110	Grammar and Writing	3	-
GST 211	Logic and Philosophy	2	-
	Total Units	**7**	**16**

100 LEVEL: SEMESTER II

TABLE 13.5B

Semester 2 of Year 1 programme

Course Code	Course Title	General Studies	Basic Sciences
MAT 121	Differential and Integral Calculus	-	3
STA 127	Probability and Statistics	-	3
PHY 123	Properties of Matter	-	2
PHY 126	Electricity and Magnetism	-	3
CHEM 121	General Chemistry II	-	3
CHEM 191	Practical Chemistry	-	2
PHY 100	Practical Physics	-	2
GST 103	Nigerian Peoples and Culture	2	-
GST 121	Reading, Comprehension and Oral English	2	-
EHG 121	History and Philosophy of Science	1	
	Total Units	**5**	**18**

200 LEVEL: SEMESTER I

TABLE 13.5C

Semester 1 of Year 2 programme

Course Code	Course Title	Major Eng'g Courses
EMG 211	Engineering Mathematics I	3
MEE 212	Applied Mechanics	3
MEE 216	Material Science	2
ABE 211	Introduction to Agric. & Bioresources Engineering	2
ABE 213	Engineering Drawing I	2
CIE 214	Fluid Mechanics I	2
CIE 215	Strength of Materials	2
ELG 217	General Engineering Laboratory I	3
ESG 218	Engineers in Society	1
EEE 219	Basic Electrical Engineering I	2
	Total Units	**22**

200 LEVEL: SEMESTER II

TABLE 13.5D

Semester 2 of Year 2 programme

Course Code	Course Title	Major Eng'g Courses
EMG 221	Engineering Mathematics II	3
CHE 222	Engineering Thermodynamics	2
CHE 228	Engineering Chemistry	2
MEE 223	Engineering Drawing II	2
MEE 225	Workshop Practice in Manufacturing Technology	2
ELG 227	General Engineering Laboratory II	3
EEE 224	Computer Programming	3
EEE 226	Information Technology in Engineering	2
EEE 229	Basic Electrical Engineering II	3
	Total Units	**22**

300 LEVEL: SEMESTER I

TABLE 13.5E

Semester 1 of Year 3 programme

Course Code	Course Title	Major Eng. Courses	Social Sciences	Electives
EMG 311	Engineering Mathematics III	3	-	-
MEE 311	Theory of Machines	2	-	-
MEE 312	Engineering Economics	-	2	-
MEE 313	Machine Drawing and Design	2	-	-
ABE 312	Applied Hydraulics in Agric. & Biores. Eng.	2	-	-
CIE 311	Engineering Survey and Photogrammetry	2	-	-
CIE 314	Engineering Geology	2	-	-
APT 212	Principles of Animal Production	-	-	2
CRP 211	Principles of Crop Production	-	-	2
SSD 211	Basic Soil Science	-	-	2
GST 311	Entrepreneurial Studies	-	2	-
	Total Units	**13**	**4**	**6**

300 LEVEL: SEMESTER II

TABLE 13.5F

Semester 2 of Year 3 programme

Course Code	Course Title	Major Eng. Courses	Electives
EMG 321	Engineering Mathematics IV	3	-
ABE 320	Bio-systems Laboratory Practical II	2	-
ABE 321	Workshop Practice	2	-
ABE 322	Farm Structures and Environments	2	-
ABE 323	Mechanics of Soil and Machine Interactions	3	-
ABE 324	Tractor Driving and Farm Operation	2	-
ABE 325	Engineering Properties of Biomaterials	2	-
ABE 326	Hydrology	3	-
MEE 324	Computer Programming & Engineering Application	-	2
MEE 326	Engineering Metallurgy	-	2
	Total Units	**19**	**4**

LONG VACATION–SIWES (3 MONTHS)
400 LEVEL: SEMESTER I

TABLE 13.5G

Semester 1 of Year 4 programme

Course Code	Course Title	Major Eng. Courses
ABE 411	Farm Power	3
ABE 412	Engineering Communication	1
ABE 413	Farm Structures and Environmental Control	3
ABE 414	Plant and Animal Production Machinery Systems	2
ABE 415	Processing and Storage of Biomaterials	2
ABE 416	Bio-systems Laboratory Practical II	2
ABE 417	Irrigation and Drainage Engineering	3
ABE 418	Engineering Mathematics V	3
	Total Units	**19**

400 LEVEL: SECOND SEMESTER–SIWES (6 MONTHS)
500 LEVEL: SEMESTER I

TABLE 13.5H

Semester 1 of Year 5 programme

Course Code	Course Title	Major Eng. Courses	Social Sciences	Electives	
				Within	Outside
ABE 511	Farm Electrification	3	-	-	-
ABE 512	Soil & Water Conservation Engineering	3	-	-	-
ABE 514	Bio-systems Laboratory Practical III	2	-	-	-
ABE 515	Design of Agricultural Machines	2	-	-	-
ABE 519	Computer Simulation & Design of Agricultural Bio-systems	2	-	-	-
AET 515	Diffusion of Innovation	-	-	-	2
MEE 515	Engineering Management & Law	-	2	-	-
	Electives	2	-	2	-
	Total Units	**14**	**2**	**2**	**2**

500 LEVEL: SEMESTER II

TABLE 13.5I

Semester 2 of Year 5 programme

Course Code	Course Title	Major Eng. Courses	Social Sciences	Electives Within
ABE 521	Industrial Studies & Entrepreneurship	-	2	-
ABE 522	Biomaterials Handling & Farm Transportation Equipment	2	-	-
ABE 523	Final Year Project	6	-	-
ABE 524	Bio-systems Laboratory Practical IV	2	-	-
ABE 526	Land Clearing & Development	2	-	-
	Electives	-	-	4
	Total Units	**12**	**2**	**4**

TABLE 13.6
Graduation requirements

	Mode of Entry		
Courses Required	**100 Level**	**200 Level**	**300 Level**
Basic Sciences	26	-	-
Major Eng. Courses	116	116	86
Electives within	6	4	4
Electives outside	12	10	10
Social Sciences	8	8	8
General Studies	10	5	5
SWIES	-	-	-
SWEP	-	-	-
Total	**178**	**143**	**113**

Postgraduate Specialisation in Agricultural Engineering

The current and future needs in the agricultural sector for self-sufficiency call for continuous training of highly qualified experts to work in the manufacturing, design, and construction sectors, as well as highly trained specialists to work in Higher Institutions and Research Centres. Agricultural Engineering in the tropics, especially in developing countries, needs not only a more appropriate technology to meet the demand of great number of farmers in the production of agricultural crops, but also the maintenance of acquired technology to store, process, and preserve agricultural products and to maintain a balance with regard to soil–plant–machine interrelationships.

In order to find solutions to these complex problems, the Department of Agricultural and Bioresources Engineering, FUT Minna has mounted postgraduate programmes, namely a Postgraduate Diploma (PGD) in Agricultural Engineering, a Masters in Engineering (MEng) in Agricultural Engineering, and a Doctor of Philosophy (PhD). The curriculum for each postgraduate programme has been designed, upgraded and is constantly being reviewed to train engineers for these higher degrees who can take up these challenges. This is by solving practical engineering and technological problems based on solid theoretical and research work. In order to attain modern standards in engineering design and construction and data analysis, computer application is highly emphasised and applied in running the postgraduate programmes. The different options for the award of the above degrees/diploma are presented below.

Farm Power and Machinery

This specialisation includes numerical analysis and computation, experimental design and analysis, advanced farm machinery design and analysis, dynamics of agricultural vehicles, advanced studies in traction and transport mechanisms, soil–plant–machine systems, instrumentation and measurement, ergonomics for agricultural engineers, and advanced tractor design.

Crop Processing and Storage

This specialisation should include numerical analysis and computation, experimental design and analysis, engineering properties of biomaterials, drying of agricultural products, environmental engineering for plants and animals, design of agricultural structures, advanced studies in agricultural product storage, and advanced processing of agricultural products.

Soil and Water Engineering

This specialisation includes numerical analysis and computation, experimental design and analysis, advanced soil and water conservation, agricultural land drainage engineering, irrigation engineering, water conservation engineering, soil conservation, agricultural waste management, surface water hydrology, advanced hydraulics and water engineering, engineering hydrology, and advanced groundwater resources development.

Food Engineering

This specialisation should include numerical analysis and computation, experimental design and analysis, food biotechnology, food microbiology, food plant design, and unit operations in food processing.

Farm Structures and Environmental Control

This specialisation includes numerical analysis and computation, experimental design and analysis, farm site investigation and introduction to geophysical methods, foundation engineering, concrete technology, design of steel structures, environmental health engineering, remote sensing applications, wood structures, and structural engineering concept and design.

Current Population of Students at the Postgraduate Level

Master of Engineering –52 PhD–25 Postgraduate Diploma in Agricultural Engineering–4

Future Trends in Agricultural Engineering Education in Nigeria

To shape the future of agricultural engineering education in Nigeria, it is essential to consider both societal and industrial demands. With the global population projected to hit 9 billion by 2050 and agricultural production expected to increase by over 70% by then, it is evident that a greater number of agricultural engineers will be necessary to address the world's increasing demand (Kaliyan, 2022).

If you are an agricultural engineer, your role is to help manage and optimise the production and utilisation of agricultural crops. You could help develop a new crop that has a higher water efficiency or help fix sustainability issues by developing a new crop that is more resilient to environmental changes. Your work may also focus on improving the quality of food by designing a new type of food processing unit. All these call for a redirection in agricultural engineering education.

The future of agricultural engineering is bright and exciting. The discipline is evolving to meet the needs of our changing world. As advancements in technology make it easier to grow food in different regions, and as agricultural engineering is increasingly utilised in the fields of food security, food processing, and biosecurity, the importance of education in this area is becoming indispensable to keep up with the new techniques and approaches., the future is looking more and more promising.

Top Agricultural Engineering Education and Technology

Recent agricultural engineering education and technology are keeping up with the ever-changing needs of the industry. Some of the more popular technologies include precision farming, remote sensing, and robotics. These help us to get a better understanding of your crops, leading to the development of more efficient crops. More recent developments to make farming more efficient are the drones and automated planting machines. There is also a big focus on making farming easier and safer for the farmer.

With the rapid advancement of technology education, agricultural engineering is becoming more important to tackle challenges in the global food market. The future of agricultural engineering is to integrate technology with biology and the social aspects of agriculture to create sustainable, resilient, and adaptable landscapes.

Conclusions

i. Agricultural engineering is a significant discipline in Nigeria that plays a crucial role in increasing agricultural productivity. Its benefits include improving input efficiency, conserving resources, reducing postharvest losses, and adding value to agricultural produce.
ii. For agricultural engineers to expand their scope, they need specialised academic training in precision irrigation/farming, watershed management, postharvest technology, aquaculture, food engineering, value addition to farm produce, and agricultural waste management and utilisation.
iii. To achieve this, the agricultural engineering curriculum should be redesigned to allow undergraduate students to choose their specialisation with guidance and counselling from academic advisers and appropriate elective selection.
iv. Nigerian universities should expand their postgraduate specialisations in agricultural engineering to include options such as energy and power in agriculture, dairy, and aquacultural engineering.

REFERENCES

Aboaba, F.O. (1988). Agricultural Engineering: A Potent Force in Food/Fibre Production in Nigeria: Past, Present and Future. University Lecture, University of Ibadan, Ibadan.

Chukwu, O. (1998). Technological Self-Reliance & Industrial Development in Relation to Nigerian Educational System: The Perspective of an Engineer. Paper Presented at the 3rd National Conference of Nigerian Teachers Association, FCE Kontagora, September 21–25, 1998.

Chukwu, O. (1999). Agricultural Mechanisation as a Strategy for Rural Development of Niger State, Nigeria. *Proceedings of 1st National Conference of Nigerian Society of Engineers (NSE), Minna Branch*, Vol. 1, pp. 36–46.

Department of Agricultural and Bioresources Engineering. (2021). Updated NUC Self-study Form 2021. Federal University of Technology Minna, Nigeria.

Gibb, J.A.C. (1988). Agricultural Engineering Perspective: An Account of the First Fifty Years of the Institution of Agricultural Engineers (1938–1988). The Institution of Agricultural Engineers, Silsoe.

Igbeka, J.C. (1991). The Agricultural Engineer, Agricultural Mechanisation and Appropriate Technology. Inaugural Lecture Delivered on Behalf of Faculty of Technology. University of Ibadan, Ibadan, May 30, 1991.

Igbeka, J.C. (2002). Agricultural Engineering and the Society. Paper Delivered at the South West Regional Chapter Meeting, Institute of Agricultural Research and Training, Moor Plantation, Ibadan, March 7, 2002.

Kaliyan, N. (2022). The Future of Agricultural Engineering. *Researcher in Agricultural Engineering Field; Passionate About Grain, Biomass, and Sustainability!* January 12, 2022. https://www.linkedin.com/pulse/future-agricultural-engineering-article-written-ai-writing-kaliyan?trk=pulse-article

National Board for Technical Education (NBTE). (2022). Accreditation Status of Programmes Offered in Polytechnics and Similar Tertiary Institutions in Nigeria. NBTE, Kaduna, Nigeria.

National Universities Commission (NUC). (2022). List of Nigerian Universities That Offer Degree Courses in Agricultural Engineering and Years Founded. NUC Abuja, Nigeria: Annual Publications.

Singh, G. (2000). Agricultural Engineering Education in India. *Agricultural Engineering: the CIGR Journal of Scientific Research and Development*, Vol. II, pp. 1–16.

Stewart, R.E. (1979). 7 Decades that Changed America: A History on the American Society of Agricultural Engineers 1907–1977. ASAE Publication.

14

The Need for an Equipment Engineering Course in ABE Education

John K. Schueller

Introduction

Throughout most of history, the greatest concern for most people has been acquiring enough food to feed themselves. It still remains a vital concern to provide enough food, feed, fibrer, and fuel for what is now over eight billion people and to do so in an economically, environmentally, and socially sustainable way. But now, due in a large part to the advances in agricultural equipment, only a small fraction of the population in developed countries needs to work in agriculture, thereby freeing the rest of the population to devote their time and energies to building the rest of modern society (Schueller, 2000). Those advances in agricultural equipment, both historically and now, are often due to innovations by farmers and those in supporting industries. But many advances, and the practical realisations of many of the other advances, often depend upon the efforts and achievements of engineers, especially agricultural engineers. As technologies and societies have changed, it is important to examine how equipment engineering instruction should be conducted to meet contemporary needs.

Engineering is the use of scientific and mathematical principles to design and build structures, machines, and other items and systems. Although humans have long built structures and tools, the use of the term "engineering" and instruction in it is a relatively recent phenomenon. Although initially applied to those designing and building military structures and machines, it was extended to "civil" tasks and different branches of engineering were developed. In the United States of America (USA), the "founder societies" covered the branches of chemical, civil, electrical, mechanical, and mining engineering. Later, there were branches aligned with particular industries, such as automotive, aerospace, nuclear, petroleum, and agricultural engineering. As the author of this chapter is most familiar with the USA engineering context, the history and situation in the USA will be the basis of this chapter. Although there are differences in the situations in other countries – for example, Germany has the overarching Verein Deutscher Ingenieure (VDI) which covers all engineers – casual observations indicate somewhat similar developments in other developed countries which may make these observations applicable to them.

Through much of the 20th century, agricultural engineering and its instruction at universities was relatively uniform in developed countries. A major part was instruction in "power and machinery" which covered farm power and field machinery. In the latter half of the century, common textbooks were used in developed countries. Since tractors were the most important piece of farm power and machinery, there were often separate courses on that subject, using textbooks such as *Tractors and Their Power Units* (Liljedahl et al., 1989). The other agricultural field machinery would be taught in a course based upon an appropriate textbook based upon the country, such as *Principles of Farm Machinery* (Kepner et al., 1982) or *Farm Machinery and Equipment* (Smith and Wilkes, 1976) in the U.S.A., *Farm Machinery* (Culpin, 1976) in the United Kingdom, or *Sel'skokhozyaistvennye Mashiny (Agricultural Machines)* (Klein et al., 1970) in the Soviet Union. The agricultural engineering graduates with the resulting knowledge found many jobs with equipment manufacturers and with governmental or educational organisations. Agricultural engineering graduates working in other non-equipment fields, such as drainage, irrigation, soil conservation, crop processing, structures, and food engineering also learned this subject in the common agricultural engineering curricula.

However, agricultural engineering university educational programmes eventually suffered both supply and demand problems. Historically, most of the supply of students of agricultural engineering were farm children. As the percent of the population who were farmers drastically declined, the percent of potential university students with farm backgrounds proportionately declined. And many farm children saw expanding opportunities in nonagricultural subjects.

Agricultural engineering university instruction in developed countries is typically at public universities supported by governmental bodies, in the USA's case, usually by state governments at flagship land-grant universities. A common problem in populous states such as California, Texas, Florida, and Illinois is that the farm children have to compete for flagship university admission with large populations of urban and suburban applicants with often-superior pre-university educations. Consequentially, farm children may not obtain admission at the appropriate university if they want to study agricultural engineering.

Eventually the USA achieved, except for some specialty crops such as fresh market fruits and vegetables, full mechanisation and the farm equipment industry reached maturity.

DOI: 10.1201/9780429150111-16

Engineering advancements such as automated testing, finite element techniques, and computer-aided design made agricultural equipment engineers more productive. The combination of a mature industry and more productive engineers reduced the demand for agricultural engineers instructed in traditional equipment engineering topics. Accordingly, the reduction in potential textbook purchasers caused the private publishers, with some exceptions such as *Fundamentals of Tractor Design* (Renius, 2020), to abandon the agricultural engineering textbook market. To fill the gap, professional societies published textbooks. For example, the American Society of Agricultural and Biological Engineers produced *Engine and Tractor Power* (Goering and Hansen, 2004) and *Engineering Principles of Agricultural Machines* (Srivastava et al., 2006) and the Society of Automotive Engineers published *Machine Design for Mobile and Industrial Applications* (Krutz et al., 1999). Although these textbooks provided strong support, offerings and enrollments in agricultural field equipment courses declined.

This reduction in supply of students and demand for agricultural equipment engineering graduates caused USA agricultural engineering departments to refocus. The overwhelming majority of such departments no longer give first degrees in agricultural engineering. If one examines the "Educational Programs in Agricultural and Biological Engineering and Related Fields" in ASABE's *2022-2023 Member Roster* (American Society of Agricultural and Biological Engineers, 2022), it can be seen that both "biological engineering" and "biosystems engineering" bachelor's degrees are offered at more universities than "agricultural engineering" degrees and that there is a diversity of degree titles, including "bioengineering", "bioenvironmental engineering", "bioproducts and biosystems engineering", etc.

The rationale for changes varied between universities and may have had multiple causes. In many cases, the change was spurred due to declining enrollments with the hope that the new degree titles would appeal to a larger number of students. It was also facilitated by the observation that the combination of biological and engineering knowledge was a useful preparation for future work in many industries and professions.

As space at those flagship institutions is limited, there is often strong political pressure on the institutions to maximise throughput from admissions through graduation, thereby forcing the institutions to limit or reduce hours to degree or to accept pre-university work towards degree requirements. These limitations and reductions constrain curriculum coverages at the same time there is a need to expand the topics of instruction. The refocusing of the ABE degrees has resulted in, and continues to demand more, instruction in biological subjects. In addition, there needs to be more instruction in electrical/electronic topics such as sensors and controls and in computer topics such as artificial intelligence.

The above-discussed issues greatly affect the ability to provide much instruction in agricultural equipment engineering in university programmes. Yet such instruction is still needed to provide graduates who can design and improve the equipment necessary to produce food, feed, fibrer, and fuel crops in an economically, environmentally, and socially sustainable manner. Consequently, this chapter presents a course design believed to be of use in developed countries to produce such needed graduates. As the concept of "course" can have widely different meanings, here it refers to a post-secondary study involving a lecture, maybe with a supporting practicum or laboratory experience, conducted one to three times per week for a semester lasting three to four months.

Contemporary Needs

The diversification of degree programmes and curricula is partly due to the variations in backgrounds, needs, and employment opportunities in particular states and nearby states. Despite a high degree of cultural homogeneity in the USA, students very often desire to have careers near where they are raised. Accordingly, more instruction on agricultural equipment engineering is appropriate at universities located near agricultural equipment manufacturers.

However, despite what is written above, it would seem that some instruction would be appropriate for all agricultural/biological/biosystems engineers (hereafter, just referred to as "ABE") on the topic of equipment engineering. Even ABE engineers who work in irrigation, food processing, controlled environments, bioreactors, etc., will work with such items as geared systems, prime movers, pumps, etc. This chapter therefore proposes a rough outline of a lecture course that will meet those needs.

Course Goals

This chapter proposes a single semester 3rd or 4th year course for engineers to meet the basic equipment engineering needs of agricultural, biosystems, or biological engineers. It will prepare the graduates to be productive engineers in the agricultural equipment industry. It will also prepare the graduates who are not in equipment engineering specialties to understand machinery basics and have skills that are applicable to their specialties.

Given the size of many ABE departments compared to other larger engineering departments and the wide range of topics which the students must be instructed in, this may be the only course on equipment engineering that is offered. However, larger ABE departments or those who have many graduates go to the agricultural equipment industry may have subsequent follow-on courses with more focused concentration on particulars of agricultural equipment. Conversely, some of the topics discussed below may be covered in other courses at some institutions and may therefore be neglected in this proposed equipment engineering course.

Characteristics of a Modern Equipment Engineering Course

A modern agricultural equipment engineering course should build upon the preparation that the university provides in the areas of engineering mechanics. These should be taken by the students before enrolling in this course. Knowledge of statics would be necessary. It also is desirable that the student have knowledge of the basics of both dynamics and the calculation of deflections and stresses, commonly known as strength of materials or mechanics of materials. Fluid mechanics would be helpful, although it may not be able to be a prerequisite depending upon the program's course schedule.

Historically, the equipment engineering course in agricultural engineering programmes focused on a particular set of crop production machines, including such items as tractors, ploughs, and planters. In the contemporary situation, such a focus is not so defendable. The course must be applicable to a vast array of machines, including those which are not used in crop production. This can be partially achieved by concentrating on the components of machines, such as shafts, pumps, and gearing, rather than on complete machines.

As a consequence of the diverse employment opportunities of ABE graduates, the course must be such that students in the various ABE specialisations and a wide range of career paths will benefit from it. For example, the knowledge gained on pumps should apply to graduates who are working on tractors, irrigation systems, bioreactors, dairy processing, and other fields. Accordingly, the instruction should be mainly on fundamentals that are widely applicable to many types of systems. But, of course, there should be some practical examples to illustrate the application of the fundamentals. These examples should not only include crop production equipment engineering, but also examples from other specialisations to maintain interest in the course from students with other ABE specialisations. For example, selecting a bearing can be applied to a shaft used in many different specialisations.

Given the diverse and complex systems ABE engineers work on, they need to be exposed to using many different information sources to aid them in their analyses and designs. Of course, textbooks are useful. But ABE specialisations are such that textbooks alone are not sufficient in engineering practice. So the students need to learn how to use standards and technical papers, as well as learning contemporary advances and techniques from the course instructors. The modern student also should learn how to assess the reliability of online information and when it can be used.

Common Equipment System Concerns

System and Component Goals

There are common goals which apply to systems and components. Performance is one of them. Evaluations of measurements and constraint conformance are necessary to quantify how the machine performs. Typical criteria and how to evaluate against the criteria should be taught to the students. Static evaluations are intuitive to the students and dimensioned, dimensionless, and percentage of absolute and relative errors should be easily grasped. Dynamic performance is also important for some components and systems. Relevant topics at this point in the students' studies include response time, settling time, and percent overshoot.

Similarly, the concepts of efficiency should be mastered. This should include understanding why efficiency is important and how to calculate efficiency, including an overall system efficiency when components have various individual efficiencies. Typical efficiencies of a variety of components and systems should be presented.

Reliability and durability should be part of the course. Calculating system reliability from component reliabilities is one of the most important aspects. How factors of safety are calculated and used should also be part of this. Given that agricultural systems often have large unknowns and variabilities, the need for factors of safety needs to be understood by the students.

Safety is another important topic which is widely applicable. Identifying modes of failure and their consequences should be emphasised. This is an appropriate time to introduce codes and standards. Besides individual operator and bystander safety, there should also be some discussion of environmental impacts, drawing upon issues of health and efficiency as well as general environmental impact.

The need for economic considerations should be emphasised. However, given the dynamic nature of costs, it is difficult to include material in this course which would be relevant. But students can be taught to do pricing online for homework and design problems assigned as part of their work in this course. An important concept for the agricultural equipment industry is the effect of volumes on pricing as agricultural equipment can range from very low volume entities, such as harvesters for certain specialty crops, to high volume parts, such as the ground engagement parts used in large numbers each on a range of tillage implements.

Common Calculations

Before discussing the engineering of particular components, the students need to know how to do some common calculations. The most important of these are to calculate energy, power, and efficiency. In the USA this must be done with both traditional and SI units. In other countries, SI along may be sufficient, although some non-SI units, such as bar and tonne, may be necessary.

Calculations need to be made for both translational and rotational mechanical systems. For electrical, the calculations need to be made for both DC and AC systems. And calculations need to be made for fluid and hydraulic systems. Of course, the calculations also need to be made for combination systems. Calculating the energy consumed or supplied, the power consumed or supplied, and the efficiency of any part of the system is a common calculation that students must be able to do. Again, some of the goals and calculations of this section, and some of the following sections, may be neglected if they are covered in other courses.

Selected Topics for Coverage

Materials

In order to design appropriate structures and machines, ABE students must have some knowledge of materials used in such structures and machines. This knowledge should include the designations by which they are known in industry, their common uses, and such characteristics as relative expense, density, surface characteristics, corrosion resistance, and manufacturability. The performance characteristics which should be approximately known including yield strength, tensile strength, ductility, and modulus of elasticity.

Although lightweight materials have greatly increased in importance relative to steel and iron alloys in many fields, such as in the aerospace, biomedical, and automotive industries, steel and iron alloys retain much of their importance to ABE students since having light weights, while important for some applications, is not as dominant in many agricultural and biosystems applications as other fields. Because steels are widely used for agricultural equipment and other agricultural applications, ABE students should learn the appropriate characteristics and advantages of the common alloys of carbon, alloy, and stainless steels. This should include manufacturability, strength, brittleness, and corrosion resistance. The common lack of understanding that increasing material strength does not substantially increase modulus of elasticity, and hence deflection resistance, needs to counteracted. Agricultural equipment is also a relatively large user of cast irons, so the students should also learn a little about grey, white, malleable, nodular, and CGI cast irons. Given their importance in handling fertilisers and agricultural chemicals, as well as their great significance to those students in processing and food specialisations, stainless steels such as 304 and 316 should be covered.

Aluminium alloys are the class of nonferrous materials most important for agricultural applications as they have increased in importance due to their lower density. Cast and wrought aluminium alloys and their characteristics should be a part of this course. Although it would be nice to include them, other common engineering metals, such as copper (bronze and brass), magnesium, and titanium alloys, are not common enough in agricultural applications to warrant more than minimum discussion.

Plastics and composites are becoming more important in agricultural applications. Some fundamentals should therefore be taught. However, they do not need to be covered to the levels of ferrous and aluminium alloys as their use is more integrative and cosmetic in agricultural applications and structural failure is often less impactful.

Manufacturing

For a design to be realised, it needs to be manufactured. Some ABE students have backgrounds from farm shops in which they have some manufacturing experience. But many do not and manufacturing knowledge should not be assumed. Manufacturing often makes big differences in component or system cost and performance. Therefore, ABE students should understand common manufacturing processes and their characteristics. They should also understand how component design details, required tolerances, production volumes, and performance requirements have a big impact on optimum manufacturing processes and conduct. This design for manufacturing knowledge is typically absent in fresh engineering graduates, but having some of it would position the graduates well to be successful.

Casting and molding is used extensively in manufacturing components for agricultural applications. Expendable mold (sand) casting's use for cast irons should be known, along with the resulting differences in casting compressive and tensile strengths. The tradeoffs between expendable mold and permanent mold (diecasting) of aluminium alloys should become known. A short discussion of injection molding is probably sufficient for plastics manufacturing. The students should learn to design to avoid sharp corners and other stress concentrations for objects to be manufactured by casting and molding. Instruction should also promote fillets, uniform wall thickness, good draft angles, and other good design principles, as well as favourable mold filling and casting cooling temperature gradients.

Additive manufacturing, commonly also known as 3D printing or rapid prototyping, has recently been greatly advancing in capabilities and in commercial adoption. Additive manufacturing uses computer controlled machines to build a solid part from particles or liquids which conforms to the solid model of the desired component. Although the additive manufacturing processes are slow during the actual manufacturing process, the parts can generally be produced within a shorter time period because there is no need to acquire tooling, fixturing, or specialised machines. Generally, the additive processes are less competitive when production volumes are increased. Additive manufacturing allows more complexity in manufacturing geometries, but surface conditions and strengths can be more problematic. Additive manufacturing has not made a significant impact in agricultural applications at this point due to often-higher per-component costs, but that may change in the future and ABE graduates should be prepared.

Material strength and ductility are very important for material deformation processes. Bulk material deformation processes such as forging, rolling, bulk bending, and extruding are important methods for producing shaped materials for subsequent processes. However, time limitations for this course would probably preclude anything more than description of the processes and including some guidelines for designing with extrusions. Sheet metal deformation processes studied should include shearing, bending, and deep drawing. Guidelines on bending radii and drawing ratios should be discussed.

Instruction on material cutting and removal processes, also now known as subtractive manufacturing, should concentrate on the traditional processes of milling, turning, and drilling. They are important to understand since such processes consume the most manufacturing labour time and cost. Design considerations for maximising producibility and minimising costs should be presented to the students. Examples include ample corner radii when turning and perpendicular surfaces for drilling. Other common practices such as broaching and tapping also deserve mention and discussion.

Since laser and waterjet cutting are now popular, they should also be included. The ability to form complex shapes of holes and edges with no tooling compared to other sheet metal processes should be discussed. The potential for aiding inventory control and assembly through kitting also merits attention.

Parts and components produced by the above processes eventually need to be combined to produce subassemblies which need to be joined together to produce assemblies which need to be joined together to produce machines. A design decision first needs to be made whether there should be subsequent easy disjoining for repurposing, replacement, repair, or recycling. Welding and related processes need to be learned. The same applies for fasteners and assembly. The instruction on fasteners needs to include standard thread sizes and forms. There should be discussion of the current trend towards more monolithic components.

Structures and Frames

Most ABE students will have previously taken a course titled mechanics of materials or strength of materials in which they have analyzed deflections and stresses of structures. Some may have also learned some finite element analysis in which they became capable of analyzing more complex structures. But some review is usually necessary of calculating beam deflections and stresses according to failure theories in objects. Generally, the previous courses should be supplemented by some discussion of the advantages of fillets, bracing, gradual transitions, closed cross-sections, and other design choices in promoting reliability for the typically uncertain agricultural environments and loadings.

Prime Movers

Agricultural field equipment is a major power consumer. As are equipment in other agricultural applications such as environmental control, crop processing, and irrigation. So ABE graduates need to understand the various alternatives for power supply and their advantages and disadvantages.

As agriculture is being electrified, ABE students should learn about the various electrical power sources. They should learn about AC mains power sources, including the differences in voltages, frequencies, and connectors in different countries, and how to calculate amperage and power. They should also learn about batteries currently used and planned for the future for agricultural equipment, including power capacities, energy storage capacities, and charging requirements and times. Fuel cells should get minor coverage, as also should the topics of rectifiers, inverters, cables, and connectors.

Many ABE graduates will have to select electrical actuators for machinery and equipment. Most of these will be AC and DC electric motors. An understanding of the most important of the many types of motors and their characteristics is necessary. This would definitely include AC induction and brushed and brushless DC motors. Speed-torque relationships should be emphasised. Thermal management and safety (including potential for use in explosive, flammable, or wet environments) should be included. Methods of controlling speed and torque should be discussed including variable frequency and pulse-width modulation techniques. As electric motors tend to be larger and heavier than other power sources, discussions should be conducted on which types of applications have a large size or heavy weight as a potential a serious detriment. Besides motors, there should be some lesser instruction in other electric actuators, including solenoids and torque motors.

Despite the increasing electrification in agriculture, internal combustion engines will remain important in agriculture. ABE students will still need to learn about them, although instruction to the level of detail of previous generations may not be possible within the course time constraints. Mobile agricultural equipment historically moved from gasoline to diesel. But it appears that future equipment might have a diversity of fuels. So the ABE graduate should know some about engines which use diesel, gasoline, biofuels, methane, and hydrogen. They should know about the engine cycles for both spark and compression ignition. Students should learn about environmental effects including pollutants such as hydrocarbons, particulate matter, and NO_x, as well as greenhouse gases. Characteristics, typical performance, and relevant equations should be learned for the various engines and fuels.

Power Transmission

Except for some fans, pumps, and other applications, the motors and engines discussed above usually run at speeds that are much faster than many agricultural applications need to operate at. Even if the prime mover can operate at the needed speed, weight and size restrictions often require the prime mover to operate at the higher speeds in order to have acceptable power-to-weight or power-to-size ratios. In addition, agricultural equipment design often requires that the power be consumed at a different location or many different locations than the location of the prime mover. Power transmission is therefore very important. Given that prime movers are usually selected from existing options designed by electrical or mechanical engineers but power transmission for agricultural equipment is often designed by ABE graduates, power transmission should be given greater emphasis than prime movers.

Agricultural applications are typically very demanding of power transmission systems. They are often high power and/or high torque. There often are shocks or impacts, sometimes these are even repetitive. Power transmission systems have to be designed for both impacts and fatigue. Of course, they should also be designed for efficiency. Given that loadings are often changing and unknown, designing such systems is challenging. But ABE students should be given proper instruction on characteristics and analysis methods.

The typical and appropriate starting point is to teach students how to design shafts. This includes performing analyses for designing for torque, bending, and combined loads. Since shafts are usually designed rather than selected from existing designs for a specific application by ABE graduates, the instruction should be complete in teaching analyses of shaft stress and deflection. Methods of mounting (shoulders, retaining rings, etc.) and driving (keyways, splines, etc.) the shafts should be introduced along with an understanding of their stress concentration effects.

Where there is relative movement, such as supporting shafts, there needs to be bearings or bushings. The materials used for solid bearings should be described and their characteristics understood, including their P, V, and PV limits. The various types of rolling element, both ball and roller and both radial and thrust, bearings should be discussed with their characteristics and advantages. For example, roller bearings should include cylindrical, tapered, spherical, and needle types. This would include mounting methods and statistical life calculations.

The required speed and torque changes in most agricultural applications are done with gearing. The ABE student should learn the various types and characteristics of involute gearing, including spur, helical, bevel, and worm. The student should learn the various parameters of gearing, such as the pitch and pressure angle, and the equations for torque, speed, and forces. Therefore, the ABE graduate should be able to specify the proper gears needed for an application.

The students should also learn some about common conventional geartrains, included countershaft and reverted. Of course, they should be able to calculate the speeds and torques in a conventional geartrain. The students should also know of the existence and some of the advantages of planetary gearsets. They should also know a little about power-split and power-recirculating transmissions. As leadscrews and ballscrews are becoming more popular, especially as they are increasingly connected to electric motors to replace hydraulic linear actuators, there should also be an introduction to them in this course.

Agricultural applications also frequently use flexible power transmission elements. Belts, particularly V-belts, are commonly used. Characteristics, standard sizes, length calculations, and power calculations should be learned by the students. Similar learning should also be insured for roller chains. The question of when to use gears, V-belts, or roller chains should be explored by the students.

Fluid Components

Many agricultural machines and equipment either use fluids to transmit power or have to move fluids. Hence, components to handle fluids have long been an important part of agricultural engineering. It is therefore important that agricultural engineers be able to understand and design fluid components and systems.

Hydraulics for transmitting power, often known as fluid power, is a topic that is taught as a separate subject in some ABE degree programmes. But if it is not taught at a particular university, it would be appropriate in this course to provide an introduction to the topic. Students should become capable of converting between the mechanical speed and torque or force to and from the hydraulic flow and pressure for pumps, motors, and linear actuations. The students, of course, should be able to calculate mechanical and hydraulic powers. In doing so, they should have an understanding of volumetric, mechanical, and overall efficiencies and how to apply them in the conversions and calculations.

Students should learn the characteristics of hydraulic pumps and motors of various types, including external gear, internal gear and gerotor, balanced and unbalanced vane, and axial and radial piston designs. They should similarly learn about hydraulic cylinders. The students should learn the characteristics of hoses and tubing and how to select diameters for appropriate fluid velocities.

ABE graduates should have some beginning understanding of hydraulic circuits and how to select valving, including directional and relief valves, to achieve acceptable performance and safety. They should learn the functions of reservoirs and how to size and select appropriately. They should at least be aware of pressure-compensated and load-sensing controls.

The students should also learn a little about other fluid systems and their components. The most important would be to learn about centrifugal pumps and pressure-flow pump curves. The characteristics of centrifugal pumps in comparison to positive-displacement pumps should be taught. Similarly, fans should also be discussed, including the characteristics of different types.

Integrated Assemblies

The previous section of this chapter discussed individual components. This is most appropriate for a modern machinery engineering course in that many of the students will end up pursuing careers which do not involve field crop production equipment engineering but will use these components in their other applications. In addition, this will be useful for graduates who go into equipment engineering as initial job assignments often start at the component selection or design level.

It is a desired and important experience for students to learn how to integrate components into a good system. The open-ended nature of such a project prepares students for employment where the assignments are not finely structured and detailed. However, the assignment of a sophisticated and complicated machine task, such as "design a tractor", is too complex, time-consuming, and abstract to demonstrate and teach the meticulous care and attention to detail that is required to be a good engineer. A project with a smaller scope is better.

An example project appropriate for a final project for this course would be the design of a simple geared speed reducer. The student would be supplied with the input and output speeds and torques. The student would then have to select appropriate gears, design a shaft with proper connection to the gears and means of maintaining gear and shaft alignment, select appropriate bearings, and design the gearbox which encloses the components. Such an assignment will exercise the knowledge gained and allow the student to build confidence.

If the student had access to centrifugal pump analysis software, another potential project would be to design a centrifugal pump to meet assigned pressure-flow characteristics and pump input speed. After the vanes had been iteratively designed with the software to meet the requirements, the student would then have to design the shaft and bearings. The design of the casing would follow with the student having to design the pump's mounting points as part of the casing.

Related Courses

The above course was formulated to be the first equipment engineering course for ABE undergraduate students. It was designed to provide a minimum basis for agricultural equipment engineers and useful information for engineers in other ABE fields. Depending upon the local situation, there would be other courses that would be supporting, complementary, or subsequent to this course. A supporting laboratory experience would be a helpful addition. For example, the disassembly of a Ford 8N tractor and its redesign to separate the clutching of the PTO and from the hydraulic system power supply proved to be a very successful "hands-on" experience for students at the University of Florida (Carroll et al., 2017).

Given the increasingly dominant reach of digitisation in society, including agriculture, it is necessary that ABE students have expertise in that area. So it is expected that those students will take a complementary course in which they will learn about sensors and controls. Given the complexity of agricultural and biological systems and environments with a combination of physical, chemical, and biological variables, it is vital that ABE graduates be able to understand sensing and control of such systems. It would be advantageous if they also had some knowledge of networking, especially ISO 11783, known as ISOBUS, and artificial intelligence. The course could also cover some of the topics inherent in spatially-variable crop production, now generally known as precision agriculture. Although sensing and control could be included in the course described here, the amount of information and the level of knowledge required of ABE graduates will demand a separate course on the subject.

It might be advantageous to have a separate follow-on course on tractors at universities located where a substantial number of graduates will work for tractor manufacturers or will need deeper tractor knowledge. Such a course could utilise the specialised 2020 book *Fundamentals of Tractor Design* by Prof. Karl Renius as a textbook. Of course, the course should also have some updating with additional materials accounting for the progress in digitisation and electrification of tractors. Such a course has been very successful at the University of Florida, as it shares a target audience with mechanical engineers and achieves an annual enrollment of 70 to 80 students.

Another follow-on course might be a course in contemporary crop production machinery. Using the knowledge from this course, the students could learn about the various implements and vehicles used for tillage, crop establishment (planting), crop promotion (fertilisation), crop protection (weed, pest, and fungus control), harvesting, and material transport. This course would again be slanted towards local agricultural characteristics and employment opportunities and likely only successful where there is a strong regional equipment industry.

Moving Forward

The need for an appropriate contemporary equipment engineering course has been building for years as neither the alternative of no course in equipment engineering nor the alternative of a historically traditional course fulfil the contemporary needs of undergraduate programmes offered by ABE departments in universities in developed countries. The course content described in this chapter may help fulfil those needs. A competent faculty member in an ABE department could develop such a course and experimentally determine the efficacy and efficiently by offering it. The generation of a corresponding textbook or other teaching materials could then facilitate the widespread adoption of such a course.

REFERENCES

American Society of Agricultural and Biological Engineers. 2022. *2022–2023 Member Roster.* American Society of Agricultural and Biological Engineers, St. Joseph, MI.

Carroll, B.F., J.D. Abbitt, and J.K. Schueller. 2017. Tractor disassembly and modification for experiential learning. ASABE Paper No. 1700011. American Society of Agricultural and Biological Engineers Annual International Meeting, Spokane, Washington, DC.

Culpin, C. 1976. *Farm Machinery.* 9th ed. Granada Publishing, London.

Goering, C.E., and A.C. Hansen. 2004. *Engine and Tractor Power.* 4th ed. American Society of Agricultural and Biological Engineers, St. Joseph, MI.

Kepner, R.A., R. Bainer, and E.I. Barger. 1982. *Principles of Farm Machinery.* 3rd ed. AVI Publishing, Westport, CT.

Klein, N.I., I.F. Popov, and V.A. Sakun. 1970. *Sel'skokhozyaistvennye Mashiny. (Agricultural Machines).* Kolos Publishers, Moscow.

Krutz, G.W., J.K. Schueller, and P. Claar III. 1999. *Machine Design for Mobile and Industrial Applications.* 2nd ed. Society of Automotive Engineers, Warrendale, PA.

Liljedahl, J.B., P. K. Turnquist, D. W. Smith, and M. Hoki. 1989. *Tractors and Their Power Units.* 4th ed. Van Nostrand Reinhold, New York.

Renius, K.T. 2020. *Fundamentals of Tractor Design.* Springer Nature, Cham.

Schueller, J.K. 2000. In the service of abundance: Agricultural mechanization provided the nourishment for the 20th century's extraordinary growth. *Mechanical Engineering* 122(8):58–65

Smith, H.P., and L.H. Wilkes. 1976. *Farm Machinery and Equipment.* 6th ed. McGraw-Hill, New York.

Srivastava, A.K., C.E. Goering, R.P. Rohrbach, and D.R. Buckmaster. 2006. *Engineering Principles of Agricultural Machines.* 2nd ed. American Society of Agricultural and Biological Engineers, St. Joseph, MI.

15

Accreditation Program for Engineering Education in Japan Applicable to Agricultural and Rural Engineering Education

Yoshiyuki Shinogi

Introduction of JABEE

The Japanese Accreditation Board for Engineering Education, or JABEE, is a non-governmental organisation that evaluates and accredits educational programmes that foster professionals from the standpoints of the "knowledge and abilities expected for professionals" and a "benchmark of social requirements". JABEE-accredited education is provided by a department, or a course within the Departments of Engineering, Agriculture, or Science. This includes all educational degree programmes in Agricultural and Rural Engineering.

Established in 1999, JABEE began accreditation in 2001. "Accreditation is valuable for higher education institutions, as well as students". A student who has completed a JABEE-accredited program is on the path to becoming a "Professional Engineer", an international license for engineering professionals. This license signifies that the recipient has received an internationally recognised engineering education. In most signatory countries and regions of the Washington Accord, "only graduates who have completed engineering education accredited by a signatory of the accord are eligible to be Professional Engineers".

This chapter is a synthesis relative to the subject matter in Japan as reported by the JABEE (https://jabee.org/en/) and the Faculty of Agriculture at Kyushu University (Kyushu University, 2023).

The Washington Accord and Professional Engineers

The Washington Accord, established in 1989, is an international accord of bodies that accredit bachelor-level engineering programmes. "Privileges granted to graduates of programs accredited by their country's accreditation bodies or by Washington Accord signatory organisations vary from signatory to signatory". The path to becoming a "Professional Engineer" in any country is exclusive to those who have completed an accredited programme, JABEE became a signatory in 2005.

In countries that are signatories to the Washington Accord, most engineering education programs are accredited by a national accreditation body. However, this is not the case in Japan; not all engineering programs are accredited by JABEE, leading to unfortunate cases where graduates of non-JABEE programs could not become 'Professional Engineers' in their home countries upon their return. To avoid such cases, the Malaysian government only grants government scholarships to those admitted to JABEE-accredited programs.

In Japan, the "Professional Engineer" title is licensed by the Japanese government, specifically the Ministry of Education, Culture, Sport, Science, and Technology (MEXT). From a worldwide perspective, to be called a "Professional Engineer", an individual is required to complete an accredited engineering education programme provided by a higher education institution such as a "Professional Engineers, Japan (IPEJ)", "International Professional Engineers Agreement (IPEA)", or "Asia-Pacific Economic Cooperation (APEC)", "which are under international and regional frameworks for the mobility of professional engineers". "Under the International Engineering Alliance (IEA), the Washington Accord, IPEA, and APEC Engineer have been discussing consistency from the perspective of quality assurance of the international engineers" (JABEE, 2023).

Accreditation Programme

> JABEE accredits education provided by a department or a course within the department of engineering, agriculture, or science. JABEE accreditation criteria adhere to the principles of the Washington Accord which is an international framework for the accreditation program and education of internationally recognised equivalent programs
>
> **(JABEE, 2023).**

JABEE supports self-improvement in the quality of education of the programmes through a peer-reviewed evaluation process by teams formed from the same specialisations. Accreditation criteria check whether the program provides students with a proper education that instils attributes such as a specialised

DOI: 10.1201/9780429150111-17

knowledge of science and technology, engineering design ability, communication skills, the ability to work in a team, and expected professional ethics.

The concept of "engineering design" has been defined as follows:

"Engineering design" describes the concept of creating a repeatable and open-minded process to develop a system, element, or component under suitable conditions, taking into consideration the economy, environment, society, ethics, health and safety, possible manufacturing, and sustainability, which have all stemmed from the accumulation of science and technologies, such as mathematics, basic science, engineering, and arts and social sciences (JABEE, 2023).

"Graduates from JABEE-accredited programs are exempted from the first-step examination of the 'Professional Engineer', a nationally qualified license". To foster international professionals and contribute to societal and industrial development, educational programming in the Engineering, Agricultural and Science departments of many higher education institutions has been accredited.

"JABEE's accreditation is a third-party accreditation. In third-party accreditation, the education institution is required to publicise learning outcomes and ensure societal accountability" (JABEE, 2023). Additionally, the responsibility of education must rely upon organisations such as the department, and not individual faculty members.

The second advantage is that education is improved through accreditation. The evaluation focuses on reform towards "outcomes-based" education, educational improvement by the "Plan-Do-Check-Act-PDCA-cycle", "Engineering design" education, and teamwork education, which are all particularly important for professional education.

JABEE accreditation criteria was developed by considering the concept of the Washington Accord. "JABEE-accredited programs are recognised as internationally equivalent; this is the third advantage. International recognition is an effective support for JABEE program graduates to continue studying or to work abroad. Foreign students who wish to study engineering in Japan are gradually choosing JABEE-accredited programs, since if a program is unaccredited by JABEE" (JABEE, 2023), they would be ineligible for a professional engineer qualification in their countries. "The admission of foreign students is important for the development and internationalisation of the university" (JABEE, 2023).

Non-JABEE-accredited program graduates are required to take the first-step evaluation of the national professional engineer evaluation, whereas "JABEE graduates are exempted" (JABEE, 2023). This requirement is a significant burden to prepare for years after graduation while working on the first step of evaluation, which is composed of academic elements, making an exemption an advantage.

JABEE Accreditation Evaluation

JABEE evaluation and accreditation confirms that the engineering educational program continuously meets JABEE requirements, and that the quality of the engineering education program is based on a reliable and transparent evidence. "Comprehensive review visits are normally conducted at six-year intervals. Programs whose accreditation validity has expired are listed separately. Discipline" (JABEE, 2023) shows which "Highly Recommended Items by Discipline" in "Criteria for Accrediting Engineering Education Programs" (JABEE, 2023) are applied to evaluate the programme,

"Recognition of the substantial equivalency of the accredited engineering education programs by other signatories of the "Washington Accord" has been effective since 2005" (JABEE, 2023). JABEE accreditation for the Rural Engineering section began in 2002 and has been increasingly developed and elaborated upon through the years since. By 2019, significant changes have been made to accreditation systems for evaluations.

JABEE confirms the system and rules to ensure the appropriate future operation of each educational programme, as well as the evidence to prove operational histories and possibilities which can be determined for effective and appropriate operation. Terms of accreditation validity are granted based on the applicability of the criteria evaluated at a maximum of six years. However, accreditation terms may be shortened depending on application progress and elaboration.

Evaluation takes approximately one year after a request for evaluation, usually beginning around late April and ending in May of the next year; Japanese semesters and the fiscal year begin at the start of April. After the receipt of the request from a university, an evaluation team is created (Figure 15.1).

The team utilises a "self-review report" and "on-site evaluation" for their purposes. An evaluation report is submitted to the field evaluation committee of each academic society and checked. If there are any merged fields, some coordination among the fields is carried out. After that, this is submitted to JABEE's "Evaluation and Accreditation Coordination Committee" and overall coordination is done. The final evaluation for the report is approved by the "JABEE Accreditation Commission", and finally approved by the "JABEE Board of Direction".

A "Self-review report" is initially prepared by the programme, and it is delivered to an evaluation team. The evaluation team assesses this report with additional and related supporting documents. The evaluation team then proceed to "on-site evaluation". The team creates "Program Review" and "Executive Summary" reports to evaluate the programme, The program's university can review these reports and make an opposition improvement report in response. The evaluation team subsequently creates a second evaluation report, and finally issues a "Judgement Result Notification" to both the university and also to the "Evaluation Committee of the Related Engineering Fields", thus completing the evaluation.

Evaluation judgment was previously done in the following categories: Accept (A), Concern (C), Weakness (W), and Deficiencies (D).

1) Accept (A): Applicable evaluation item meets accreditation criteria.
2) Concern (C): Applicable evaluation item currently meets accreditation criteria. However, improvement is needed. Measures for continuous and complete

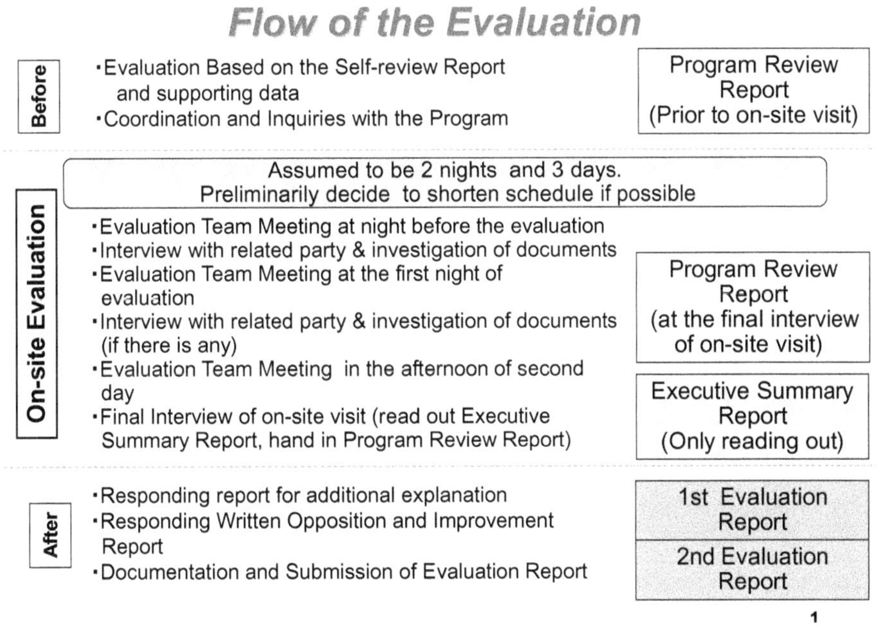

FIGURE 15.1 JABEE evaluation workflow (JABEE, 2023).

accordance with accreditation criteria must be taken.

3) Weakness (W): Applicable evaluation item almost meets current accreditation criteria. However, the degree of its accordance is low and requires improvement. Measures to increase the degree of applicability are required.
4) Deficiencies (D): Applicable evaluation item fails to meet accreditation criteria, and the program may not continue its current education system (JABEE, 2023).

This evaluation scheme has changed since 2018; there are now only three categories for evaluation, namely: "Satisfy (S)", "Weakness (W)", and "Deficiencies (D)". Similarly, under the "D" evaluation, the program cannot continue the education scheme. The workflow of the evaluation team is shown in Figure 15.2.

Primarily due to the COVID-19 pandemic, elaborations have been added for online evaluations. Almost all processes and procedures are currently online, and the evaluation team no longer visits the relevant university; almost all evaluations are now carried out remotely, such as through Microsoft Excel files and online interviews and meetings. To alleviate overburdens for both the program and the evaluating team, almost all processes have been vastly simplified.

JABEE Common Criteria for Accreditation

"JABEE common criteria is provided to accredit education programs in higher education institutions that establish, implement, evaluate and improve education. A program seeking accreditation must satisfy all four criteria as listed below and demonstrate its conformity with appropriate evidence" (JABEE, 2023).

The four criteria for evaluation are: 1) Learning Outcomes, 2) Educational Methods, 3) Achievement of Learning Outcomes, and 4) Educational Improvement.

Learning Outcomes

The program defines and publishes the profiles of autonomous professionals for the benefit of the faculty and students. The program takes account of traditions, resources, and the fields of graduates to define the profile of the professionals, while also considering the requirements of society and the demands of the students.

The program defines its learning outcomes, of which all the graduates are required to acquire by the completion of the programme, These outcomes are published and publicised to the faculty and students. The learning outcomes are the milestones for autonomous professionals (criterion 5.1) and include knowledge and abilities as defined below. These criteria also consider the items related to knowledge and abilities, as defined in the category- and discipline-specific criteria:

1. Ability to conduct multi-dimensional thinking with knowledge from a global perspective.
2. Ability to understand the effects and impacts of professional activities on society as well as the nature of professionals' social contributions and responsibilities.
3. Knowledge of and ability to apply mathematics, natural science, and information technology.
4. Knowledge of and ability to apply related professional fields.

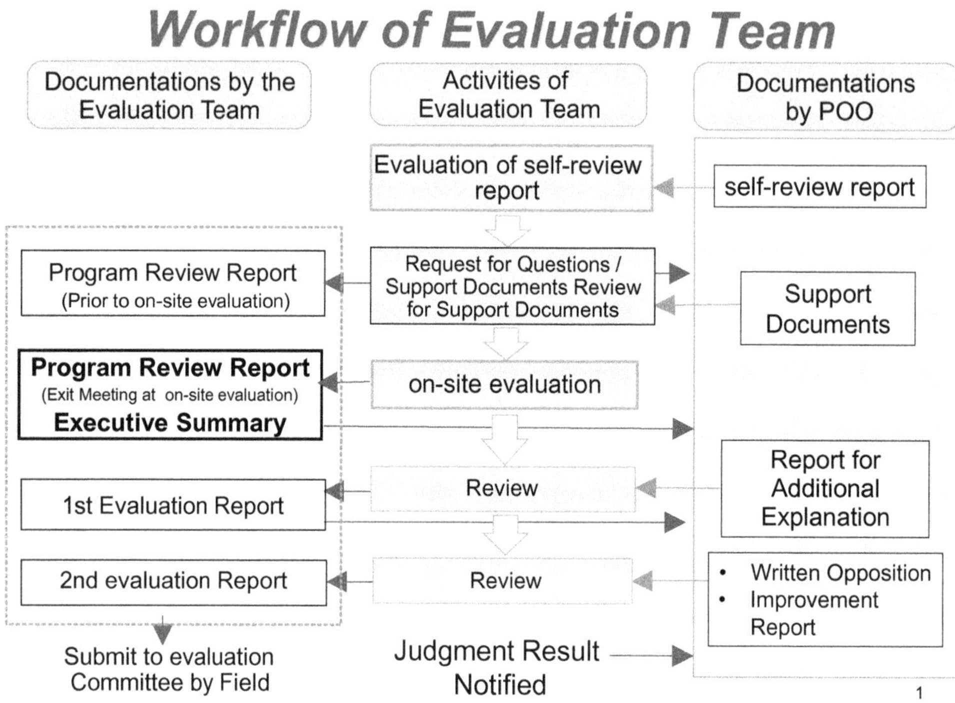

FIGURE 15.2 Workflow of Evaluation Team (JABEE, 2023).

5. Ability to design according to the requirements of society by utilising various sciences, technologies, and information.
6. Communication skills, including logical writing, presentation, and debating.
7. Ability to learn independently and continuously.
8. Ability to manage and accomplish tasks in a planned way within constraints.
9. Ability to work in a team. (JABEE, 2023)

Educational Methods

Within the curriculum based on published curriculum policies, "the program defines criteria and methods for assessing the degree of achievement for each learning outcome" (JABEE, 2023). "The program should define the correspondence of each course vis-à-vis learning outcomes, learning and educational contents and outcomes to be achieved, as well as the criteria and methods for assessment" (JABEE, 2023). These are publicised to the faculty and the students through the syllabus. "Required items related to the educational contents are defined in category-specific criteria; the syllabus is an important piece of information for both teachers and students" (JABEE, 2023).

Implementation of Education based on Syllabus and Encouragement for Independent (active) Learning

The program should implement education based on the syllabus. The programmes should encourage student independence and active learning, including continual self-reflection for achieving learning outcomes.

Faculty and Supporting System

> The program provides the faculty and the support system to appropriately implement education based on the curriculum as defined in Criterion 5.2.1. The support system includes a mechanism to smoothly implement education based on the curriculum by coordinating courses and improving qualitative progress based on evaluating educational activities by the faculty. Additionally, the support system is publicised to the faculty. Highly recommended items related to the faculty and the support system are defined in the discipline-specific criteria. (JABEE, 2023)

Admission Policies

The program publishes its admission policies for admitting students, who have the appropriate knowledge and abilities necessary for education as based on the curriculum.

Educational Environment and Student Support

"The program, or the higher education institution to which the program belongs, is equipped with facilities, resources, and structures necessary to implement the education and learning support for students and also dedicates necessary efforts to maintain, operate and update them" (JABEE, 2023). The faculty, supporting staff, and students are informed of such efforts.

In 2019, some minor changes have been conducted, with the main points as following:

For Educational Methodology, there are five criteria:

> 2.1 Educational process subject to the curriculum policy, 2.2 Promotion of independent learning and achievement of education based on the syllabus, 2.3 Achievement of the teachers' group and supporting systems, 2.4 Acceptance of students based on admission policy, 2.5 Operation of educational environments and its publicisation (disclosure). For 3; Learning and Educational Achievement Goals, there are two categories: 3.1 Achievement of learning and education goals, 3.2 Checking of knowledge and ability achievements. For 4; Educational Improvement, there are two categories: 4.1 Achievement of internal quality certification systems and publicisation, and 4.2 Sequential improvement systems.
>
> **(JABEE, 2023)**

Achievement of Learning Outcomes

Within each course, "the program evaluates the degree of achievement of the learning outcomes while following the assessment criteria and methods as defined in the syllabus". The program also reviews and confirms that all program graduates have achieved all the program learning outcomes upon completion of said programme. The evaluation of the degree of achievement also includes recognition of the credits earned by students from other programmes (e.g., other departments and/or other higher education institutions).

Review of the Degree of Achievement of the Graduates in Terms of Knowledge and Abilities

The program reviews and confirms that all program graduates have acquired knowledge and abilities from Criteria 1 to 9 as embodied in the learning outcomes.

Educational Improvement

"The program, or the higher education institution to which the program belongs, implements internal quality assurance to confirm that educational activities are provided following to Criteria 1 to 3" (JABEE, 2023). The contents of these internal quality assurance activities are relayed to the faculty. The mechanism of internal quality assurance considers both societal and student requirements and demands and includes a self-checking function.

Continuous Improvement

"The program has a mechanism to continuously improve educational activities based on review results" (JABEE,2023).

Highly Recommended Items by Discipline

"Additional items highly recommended by the discipline are defined in the discipline-specific criteria" (JABEE, 2023).

Educational Improvement of Kyushu University – A Case Study

Educational improvement is widely diverted. An example of our Kyushu university (JABEE Agroproduction and Environmental Engineering program) is introduced below (Kyushu University, 2023).

Curriculum Policy

"Classes are arranged that cover nature, social issues, science for creating, regenerating, and protecting soil, water, and meteorology in rural regions for agriculture" (JABEE, 2023). Students are educated according to these topics and clear educational goals are set; the evaluation results are determined through an alliance of research fields and laboratories.

Purpose of Education (JABEE Program)

"This course aims at creating an international professional and specialist in Rural Engineering", who combines broad expertise and techniques with problem-solving capabilities and leadership skills" (JABEE, 2023). The student would have their knowledge of technologies and their ability to understand and develop new technologies enriched, and their ability to understand cultural diversity and modern societal problems would be honed, as would their skill in identifying problems, finding solutions, and communicating and creating presentations in these areas.

Diploma Policy (JABEE Program)

By the end of the course, students should have achieved:

- A: Acquired basics of mathematics, natural science, and the ability to understand and develop new technologies.
- B: Acquired an objective understanding of the diversity of different cultures and the problems which are found in modern society.
- C: Acquired knowledge and techniques in Rural Engineering required by a specialist.
- C1: A deep knowledge of soil, water, substructure, and environment which are the basis of Rural Engineering.
- C2: A deep understanding of multi-disciplinary knowledge, primarily in agronomy and environmental sciences to acquire wide expertise and techniques.
- C3: The ability to create and execute an experimental plan, data analysis, and report writing.
- D: Acquired the required knowledge and skills for a social and professional person.
- D1: Developed an understanding of engineering ethics which is necessary for a professional engineer.
- D2: Acquired the capability to work in a team sharing responsibility and respecting the rights of the team.
- D3: Acquired the capability to communicate in a language, such as English.

D4: Acquired the techniques of information processing.
E: Ability to identify problems, solve them, and make a presentation.
E1: Acquired the ability to set an agenda and make and execute a research plan.
E2: Acquired the capability to resolve spontaneous issues and work independently until it is resolved.
E3: Acquired the capability to conduct presentations of results.
F: Acquired with global society, the ability as professional specialists to identify the role of agricultural development not only in Kyushu, Japan but also in the wider Asia-monsoonal regions.

(Kyushu University, 2023)

Education improvement systems

A checking system for educational modification is present; the outline is shown in Figure 15.3. Each month, a "Consideration of section meeting" is held. Teachers of relevant sections should attend this meeting, where various issues regarding education, research, and operational issues are always discussed. There are many sub-committees: "Education contribution reward selection", "Educational environmental checking", "Documentation arrangement", "Program completion evaluation", "Educational improvement", "External educational improvement", and finally "Internal educational committees". The "Program external evaluation" committee make a final decision on the programme, All committees have individual and strictly followed rules.

The discussion of each committee is reported to the "Consideration of section" meeting. One of the most significant sub-committees is the "External educational improvement and Program external evaluation" committee. Five to seven non-affiliated and reputable peoples are invited from outside to evaluate present education systems and reflect on these systems periodically. Every year, these committee meetings are held to introduce yearly activities and outcomes to all team members. The results and discussion are regularly shared to all teacher members. In addition, undergraduate students are invited to evaluate all lectures and classes. Any potential issues are further discussed at the "Consideration of section" meeting. Teachers that have made significant contributions to education are also rewarded in these meetings on a biannual basis. This committee is additionally comprised of several students and invited external persons. Alumni are also often invited, based on PDCA cycles.

Figure 15.4 shows the numbers of promotion, JABEE-completion students, and taking ratio (%). At Kyushu University, after a one-and-a-half-year fundamental education period, students pass on to our two-and-half year professional education course. Around 20 students are usually passed on, or promoted, to the Agro-production and environmental engineering education course. A small fluctuation exists, but a large majority students take this JABEE course. Only JABEE programmes have been offered recently to students.

JABEE Accreditation outcomes

The accumulative number of JABEE-accredited programmes in Fiscal Year (FY) 2016 is 501, and the total of JABEE-accredited program graduates has reached approximately 260,000 by the same year.

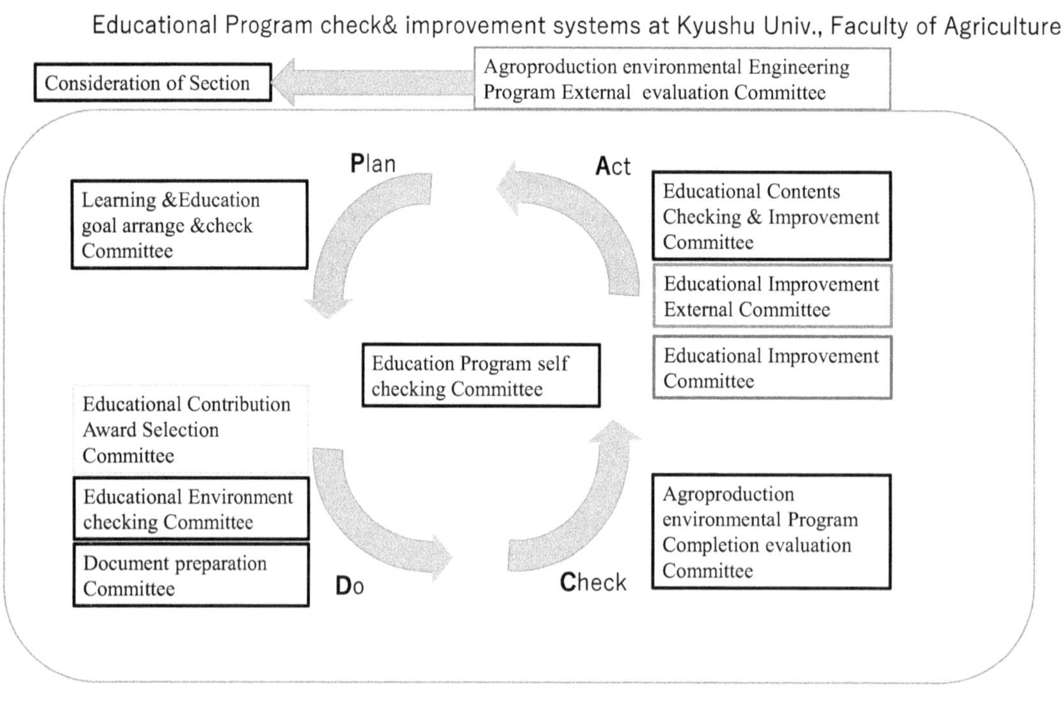

FIGURE 15.3 Education program check and improvement systems at Kyushu University, Fukuoka, Japan.

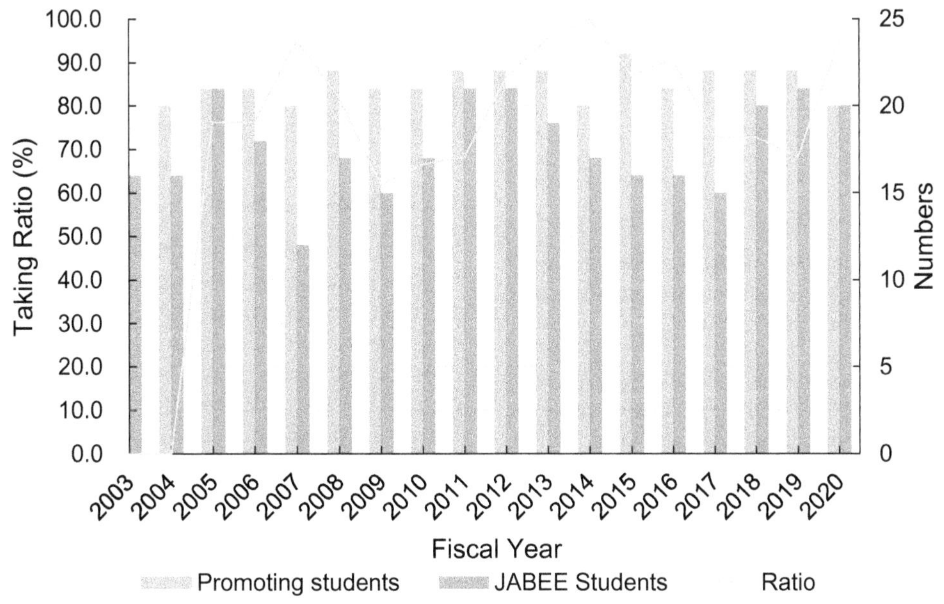

FIGURE 15.4 Numbers of promoting and completing students at Kyushu University from 2006 to 2023.

The 20-year anniversary of JABEE for rural engineering programmes was celebrated in September 2022 at Kanazawa City, as well as an anniversary symposium at the annual technical conference of the Japanese Society of Irrigation, Drainage and Land-Reclamation Engineering (JSIDRE).

Regarding the main outcomes there were 29,828 (4,848) examinees for the "Professional Engineers" exam in 2021 (parenthesis numbers are JABEE-completion examinees). This license is one of the most significant outcomes and reputedly one of the most difficult ones in Japan.

The construction section has the largest number of examinees at 17,625, with the general technologies control section as the runner-up with 3,473. 1,500 (147) examinees were for agriculture. There were 1,819 female candidates in total, with 524 completing JABEE programmes. There were 2,659 (438) students who successfully passed, with 243 (55) for agriculture. The average success was 11.6% (11.4 %). There are four categories regarding agriculture: agriculture, forestry, fishery, and biological engineering. There were 150 (12) successors, with the ratio being between 10.3 to 18.5% (0 to 21.1%), and the average ratio was 13.0%.

There are five categories for agriculture such as livestock, agronomy and foods, rural engineering, rural planning, and plant protection. Rural engineering has the largest number of both examinees and successful examinees, but statistically has the lowest success ratio (10.1%) compared with the average and other categories (10.3–18.5 %).

According to the examinee analysis data, the largest age range for examinees and successful examinees is 40 years, followed by a 30-year range. The highest successful ratio is the 30-year range (13.5%), with the 40-year range being second. The average successful age is 43.9 years old, and the average age is 42.5 for both examinees and successful examinees. There are various examinees from government officers and private companies. The highest success ratio is found among government officers (20.1%), and local government officers are slightly behind at 18.1%. The final education is over Master's (13.9%) and undergraduate at 10.6%, on average.

The latest successful examinees of the "Professional Engineers" examination of all the engineering section were announced in March 2022. Table 15.1 shows the numbers of examinees, final successful candidates, JABEE program completers, and agricultural programmes completers in these eight years.

Regarding the "Agricultural Technologies Sections", there were 80 successful examinees, with five completing JABEE programmes. As of April 2022, there are 14,363 "Individual Professional Engineers" registered for JABEE, and 67,448 for the Agricultural Engineering section.

Regarding "Rural Engineering section", numbers of examinees and successful JABEE program examinees were the largest in 2017; the number of final examinees peaked in 2015 with a subsequent decline afterwards. In 2017, 19 out of 36

TABLE 15.1

Professional Engineer Successors during 20142021 (JABEE, 2023).

Year	2014	2015	2016	2017	2018	2019	2020	2021
Final Examinees	23,207	24,878	25,032	26,253	25,914	24,326	20,365	22,903
Final Successors	3,498	3,649	3,648	3,501	2,365	2,819	2,423	2,659
JABEE Program Completers	106	158	249	274	274	310	379	438
Agricultural Program Completers.	1	4	7	1	1	5	4	5

universities carried out JABEE programmes. As of 2022, only 13 programmes still remain active. There were 438 successful examinees, which was 16.0% of the entrance exam candidates, numbering at 2,659.

Miscellaneous

In addition to supporting the improvement of professional education programmes in higher education institutions through the accreditation of education programmes,

> JABEE implements promotional activities by organising some symposia and workshops. JABEE supports the reform to outcomes-based education, as well as the improvement of education through the PDCA cycle, engineering design education, and teamwork education, all of which are particularly important for professional education. Accreditation bodies of professional education have an internationally consensus that educational improvement is made through the accreditation of education programs. International Engineering Alliance (IEA), of which the Washington Accord is a member (and JABEE is a signatory of the Washington Accord) continuously discusses important issues of professional education, such as the definition of the profiles of professionals. JABEE introduces these international trends to Japanese education institutions"
>
> **(JABEE, 2023).**

As previously mentioned, accreditation systems have been simplified to reduce burdens on both program and evaluation parties. In the post-COVID era, almost all accreditation is now being conducted online, with the majority of accreditation files and contents utilising Microsoft Excel. Communication among team members and program side members is also primarily conducted online, vastly simplifying the processes.

> JABEE was recognised on March 31, 2010, by the Japanese MEXT as an evaluation and certification agency for the "Professional Graduate Schools based on the provisions of Article 110 of the School Education Act". JABEE's scope of evaluation and certification covers professional graduate schools, which intend to foster advanced professionals in the industrial fields of Information Technology, Innovation for Design and Engineering, Embedded Technology, or Nuclear Technology.
>
> **(JABEE, 2023).**

JABEE issues a certificate to schools that have been appropriately evaluated. JABEE submits self-review reports and evaluation and certification reports to the MEXT; self-review reports are made by professional graduate schools. These reports are subsequently published on the JABEE homepage.

Current issues

There are some issues in JABEE systems, especially for rural engineering section fields. They are as following:

Accreditation schemes

The accreditation scheme is rigid and concrete, but this attribute is occasionally overburdening both for the program and the evaluation sides. Despite the evaluation process being significantly simplified, however, there are still large burdens for both sides. Many improvements have been implemented, and successes are already beginning to be observed; the quality of education is not expected to worsen. If education quality does not become worse easily, six-year periods may seem to be too short and further simplified evaluation systems can be expected, especially in cases of continuous evaluation.

Accreditation systems

Accreditation systems have been carried out primarily by technical associations like the rural engineering association, namely the Japanese Society of Irrigation, Drainage, and Land-Reclamation Engineering (JSIDRE) secretariat. Evaluation team members must take a special training course to be certified by JABEE, as widely-diverging evaluation criteria has been reported. This course may be a big burden for evaluation teachers, but uniform evaluation must be provided. To resolve this issue, further discussion must be conducted over potential solutions.

Accreditation fee

Reliable accreditation is costly. The program university must prepare around 10 thousand dollars every six-year evaluation period, which is a massive amount for many universities and other higher education institutions. With the simplification of the accreditation scheme, as well as many systems shifting online, it may be possible to reduce costs. Alternatively, payment in installments is another potential method that may be used as required.

Demonstration and identification

The "Professional Engineer" designation has yet to become widespread or easily recognised in Japan. Because of this factor, it is more difficult to encourage student attendance of JABEE programmes. JABEE and relevant parties must emphasise their mission on a larger scale within Japan. In Japan, despite the presence of manufacturing, the professional engineers are still not always recognised. The concept of Professional Engineers and JABEE may be further publicised through online media. In addition, accreditation schemes must be transparent; faculty, students, and the public must all be able to access information regarding the accreditation system, accreditation programmes, and standards.

One of significant issue is the reduction of Human Resource budgets at universities. As a result, the individual teacher has become busier with education and research, lacking the time to provide proper engineering education. Additionally, in some universities, programmes cannot be operated due to a lack of teachers or rehabilitation of the faculties and education systems; these issues are increasing in severity in some cases.

Perspectives

JABEE's impacts and effects are widespread, and society must understand its necessity. The government, and any private company, should also consider its necessity and impacts. For example, cost-sharing with private companies is expected. Since almost all universities have been facing budget reductions, some donations can be provided from some private companies with a corresponding recruitment privilege of graduate students.

In Japan, education systems are widely different in senior high school and university. Passive-type education for the entrance exam of the university is present for senior high school but active learning type education is expected at university. These gaps are widely understood.

To ensure a proper engineering education system can be provided for future students, many of these issues must be shared and discussed with many stakeholders.

REFERENCE LIST

JABEE. 2023. Japan Accreditation Board for Engineering Education ,. https://jabee.org/en/. Accessed April 17, 2023.

Kyushu University. 2023 Faculty of Agriculture. (, Policy. https://www.agr.kyushu-u.ac.jp/english/policy/. Accessed April 17, 2023

16

Process for Improvement and Evaluation of the Agricultural Engineering Curriculum at the Instituto Tecnológico de Costa Rica.

Natalia Gómez-Calderón, Karolina Villagra-Mendoza, Isabel Guzmán-Arias, Milton Solórzano-Quintana, Adrián-Enrique Chavarría-Vidal and Andrea Soto-Grant

Curricular Redesign 2007: Deepening the Disciplinary Foundations

The process of redesigning the curriculum of the undergraduate programme in Agricultural Engineering at the Instituto Tecnológico de Costa Rica was the most detailed intervention of the course in its thirty years of existence. In 2007, when the programme reached its maturity, it was detected that new demands for the development of this discipline has arisen both nationally and internationally, as well as in the social and university context, so rethinking the programme was necessary. The main objective of this process was to update the curriculum integrally and to have a higher quality of training in the field.

Redesign Context

One of the goals of the Instituto Tecnológico de Costa Rica (ITCR) is to train professionals in the technological field who can combine mastery of their discipline with a clear awareness of the socio-economic, cultural, and environmental context in which technology is generated, transferred, and applied, allowing them to participate critically and creatively in national productive activities (ITCR, 2005).

The institution has ensured that this principle is respected and, to this end, since the late 1990's, has established and adopted a series of policies and decisions aimed at continuous improvement and accrediting the quality of its academic programmes (Fëdorov et al., 2003). As a result of the implementation of these decisions, between 2002 and 2007, about one third of the undergraduate programmes offered by the institution received the seal of approval of the Sistema Nacional de Acreditación de la Educación Superior (National System for the Accreditation of Higher Education, SINAES) and the Canadian Engineering Accreditation Board (CEAB).

It is also noteworthy that specifically the Agricultural Engineering programmes offered by the ITCR (Agricultural Engineering and Administration, Agronomy Engineering and Agricultural Engineering), with the advice of the Academic Development Center (CEDA), have been placed at the forefront of the self-improvement process, knowing that it is a complex and sensitive issue of national interest to renew and strengthen agriculture in Costa Rica (Antillón et al., 2005). It is very important that the careers directly related to this field generate deep and comprehensive processes to revise, update, and innovate their curricula.

As an example of the evolution of ITCR's academic offerings, we can cite the self-evaluation and comprehensive redesign of the curriculum of the Agricultural Engineering degree programme, which was completed on its 30th anniversary in 2006. This fact has allowed it to make a new commitment to the students and to the country, in addition to offering an educational programme based not only on several decades of tradition, but also on the consideration and recognition of emerging socio-economic, disciplinary, curricular, and pedagogical trends. Below are some of the key features of the redesign undertaken in 2007: its background, the methodology used, and the main outcomes achieved.

Background

The regional and global development situation in 2007 led to the imminent opening of markets in all productive sectors and aggressive competition for market placement, which implied a demand for the best production conditions. In the agricultural sector, the dependence on multiple natural variables underscores the need for high efficiency in production.

On the other hand, the economic and social development of our country has been linked to agricultural production as an activity that generates sources of employment, wealth, and foreign exchange (Arias, 2005). Consequently, pressure was exerted to significantly improve aspects directly related to the activity, such as respecting workers' rights, agricultural certification, environmental management and optimising resources. Not only do these elements add value to the products, but they also allow us to take measures to preserve natural resources and to contribute to the development of society.

In this complex context, the role of Agricultural Engineering was evaluated in 2007, based on an objective that consisted in generating rural projects related to sustainable production,

DOI: 10.1201/9780429150111-18

from their conception, design, construction, selection, and planning to their management and maintenance. All this to improve the agro-socio-economic conditions and to reduce the possibility of adverse scenarios that nature may present. Also, to facilitate the work of the related population and to increase production for the benefit of society in general.

At that time, graduates of this career worked in state institutions (such as local governments, the Instituto Costarricense de Electricidad (Costa Rican Institute of Electricity, ICE), the Ministerio de Agricultura y Ganadería (Ministry of Agriculture and Livestock, MAG), the Servicio Nacional de Riego y Avenamiento (National Irrigation and Drainage Service, SENARA), the Instituto Nacional de Aprendizaje (National Institute of Learning, INA), and the Instituto Costarricense de Acueductos y Alcantarillados (Costa Rican Institute of Potable Water and Sewers, AyA)), for the national banking system, in professional technical schools, universities, production companies (such as companies that produce intensive crops for export and specialise in non-traditional products), and companies that sold specialised equipment to solve technical problems in production farms (such as irrigation systems and agricultural machinery).

The school of Agricultural Engineering seeks to permanently strengthen its response to the context by redesigning the profession, thanks to a curriculum that allows the training of qualified professionals to create, adapt and integrate the technology necessary for the use of resources and productive forces, thus contributing to the development of the country and the transformation of Costa Rican society.

In 2007, the continuous search for quality covered all the elements involved in the training of professionals in this discipline. The school of Agricultural Engineering, aware of the need to expand its horizons with a vision of leadership towards the use of the latest technologies, e.g., the design, construction, and management of structures for crop protection, the use of geographic information systems, the development of specialised software, the acquisition of new communication skills in English, the stimulation of leadership and management skills, the strengthening of research and motivation for entrepreneurial activity, etc.

In order to properly integrate all these variables, the career redesign implies following an organisation and a congruent methodology.

Methodology

The methodology for the redesign of the Agricultural Engineering curriculum is characterised by its planning, legality, balance, prospectivity, and driven character, in addition to its participatory and consensual nature.

Planned and Legal Character

Redefining the programme to focus on development and innovation was a complex but an achievable challenge in the medium term. In this case, the work began in July 2005 and ended in November 2006, with the official launch in 2007. From the beginning, there was a roadmap that guided all efforts during this period, as this work was part of a broader process of career self-evaluation and self-improvement. The conceptual map, which represented the route of the curricular redesign of the ITCR undergraduate degree in Agricultural Engineering, consisted of a SWOT analysis (Strengths, Weaknesses, Opportunities, and Threats), the preparation of the study plan (SP), and the elaboration of the professional academic profile (PAP). The existence of a roadmap gave the redesign a systematic character, placing each element in relation to the whole and allowing each stage to be planned and adapted in a congruent manner.

The map is based on an earlier analysis of a broad but selected range of academic-administrative sources that guide and standardise curriculum development at the institution. For this purpose, the following sources were used: the *Manual for the Design of Study Plans and Curricular Updates* (ITCR, 2000), the *Academic Model of the Technological Institute of Costa Rica* (ITCR, 2003), the *Methodological Guide for Curricular Design in Study Plans* (Quesada Solano et al., 2001), the *Flow for the Creation of New Careers or the Modification of Existing Careers* (Consejo Nacional de Rectores. Oficina de Planificación de la Educación Superior [CONARE], 2009), the *Indicators for Measuring the Incorporation of the Environmental Dimension in the Curricula of the State Universities of Costa Rica* (CONARE, 2002), the *Guide for the Elaboration and Self-Evaluation of the Course Program* (Fëdorov, 2004), and the *Curricular Guidelines for Engineering Degrees in Ibero-America* (Asociacion Iberoamericana de Instituciones de Enseñanza de Ingenieria [ASIBEI], 2005), among others.

In this section, it is important to highlight the incorporation of the program's curriculum development and innovation process within a major objective: quality improvement and international accreditation. Therefore, the elements of the map, especially those related to the curricular structure and the skills of the graduates, have been designed considering the references provided by the national and international organisations that accredit programmes in the field of engineering. Among these indicators and standards are those provided by the Costa Rican Agency for the Accreditation of Engineering and Architecture Programs (CFIA, 2004), the National System for the Accreditation of Higher Education (SINAES, 2000), the Council for the Accreditation of Engineering Education (CACEI, n.d.) of Mexico, ABET (ABET, 2004) in the United States, the Canadian Engineering Programs Accreditation Board (CEAB, 2004), the Japan Accreditation Board for Engineering Education (JABEE, 2002), and the Accreditation Agency for the Accreditation of Engineering and Informatics Studies (ASIIN, n.d.) in Germany. The compliance of the redesign with these guidelines and regulations gave the process the appropriate legal profile and, in addition, made the renewed programme eligible for international certification of its quality.

Balanced and Prospective Character

It is stated that the renewal of the curriculum is a balanced and prospective process, since it takes as a basis all the good traditions and advantages that the programme has acquired in the course of its 30 years of existence. For example, the educational strengths that already existed in the areas of

Basic Science, Engineering Sciences, and Soil and Water Resources Management and Conservation were respected when the curricular structure of ITCR's undergraduate degree in Agricultural Engineering was redesigned.

However, due to the evolving and innovative nature of the process, the best practices of the past were viewed through the prism of current political, socio-economic, disciplinary, curricular, and pedagogical trends. Thus, the new curriculum of 2007 considers the implications of diversifying, growing, modernising, and opening up the agricultural sector in Costa Rica, both in its subject matter and in the teaching and learning methodology. For this reason, the training of a new generation of agricultural engineers has focused on strengthening aspects related to entrepreneurship, social and environmental responsibility, research, management, regulations, communication, administration, maintenance of machinery, and development of production technologies.

Of course, working in the field related to the agricultural sector and engineering education was very helpful when it came to the respective analyses. The sources published by the United Nations Educational, Scientific and Cultural Organisation (UNESCO, 1998), the International Labor Organisation (OIT, n.d.) the Ibero-American Association of Engineering Education Institutions (ASIBEI, 2005), the Central American Higher University Council (Fëdorov & Guzmán-Arias, 2007), the Tuning project (González & Wagenaar, 2003), the 6 x 4 UEALC project, as well as the writings of some national and Latin American authors (Arredondo, n.d.; Gómez Campo & Celis Giraldo, 2005; Lechuga, 2002; Márquez & Marquima, n.d.; Pereyra, 2000; Posada, n.d.; Pulido, n.d.; Soto Perdomo, 1993; Tünnermann, n.d.) and so forth, have served as valuable inputs.

Likewise, similar (or rather, complementary) national curricular offerings were considered, such as the programme of the same name at University of Costa Rica (UCR & Centro de Evaluación Académica [CEA], 1998). This engineering design course emphasises aspects of postharvest handling, while the ITCR course has traditionally focused on the production process.

Driven character

On the other hand, it is stated that the redefinition of the programme is the result of a process led by a Coordinating Committee, which was composed of six people: the school's director, two professors, an administrator, a student, and an advisor from the Academic Development Center (CEDA). One of the professors served as coordinator of the commission. During the final year of the process, she was assigned 20 hours per week as part of her workload to perform this function. Another professor was assigned 8 hours per week to perform work related to the process.

In addition, and as in almost all similar processes carried out by ITCR, the Coordinating Commission received support from a CEDA official, who was assigned 8 hours a week to assume this task. The curricular and pedagogical advice given by this department, designed to support the ITCR faculty, implies total involvement and permanent accompaniment. However, this is not a directive role, since the decision making, and the responsibility derived from it are fully assumed by the Commission in charge of the redesign and by the aforementioned School Council. Therefore, the role of the consultant was based on the systematisation of the theory and quality experiences accumulated at CEDA and consisted in mapping the paths to be followed, showing the possible alternatives, their implications, scope, advantages, and disadvantages, providing curricular intelligence and methodological support to the Commission and the school undertaking the self-evaluation and redesign process.

To successfully reach the culmination of this process, it is necessary to generate and modulate cohesion forces projected towards the internal and external sectors of the academic unit and the institution. This required the very active participation of the management and a strong leadership capacity by the Coordinating Commission.

Participatory and Consensual Nature

Another quality of the process was its participatory and consensual nature. The redesign methodology validated and combined the information obtained from bibliographical sources and the contributions generated during the workshops and work meetings held with the participation of academics from the faculty of Agricultural Engineering, representatives of academic service departments, students, graduates, businessmen, and employers in the agricultural sector.

For instance, considering the help from the different sectors mentioned before and as part of the redesign process, an assessment was done regarding the national and international context in which graduates are introduced to the curriculum.

This analysis revealed the need to reinforce aspects of the programme related to social and environmental responsibility, the promotion of development in rural areas, continuous updating and facing the revitalisation and internationalisation of markets, among other things. These conclusions led to the reinforcement of curricular and pedagogical elements that contribute to the development of new skills in the areas of humanistic training, communication, entrepreneurship, management, regulations, mechanisation, and development and adaptation of production technologies.

The involvement of graduates and employers is an example of the representative nature of the process. The interest and contributions of the delegates from companies and institutions dedicated to irrigation until the year of evaluation (Durman Esquivel, Irrigación de Costa Rica, Amanco, Riegos Modernos), management and maintenance of agricultural machinery (Instituto Nacional de Aprendizaje (INA), Macori S.A., Hacienda Juan Viñas), and development and adaptation of technologies for production (Linda Vista, Melones de Costa Rica), among others, from a practical point of view, were very important to redefine the professional academic profile and innovate within the structure of the curriculum of the programme, strengthening the curricular elements related to administration, research, English level, and the development of technologies for production and their adaptation.

The redesign route was marked by a series of key moments, which brought together different interested sectors and allowed to outline the path to follow in a participatory and consensual manner. This was the case of the workshop entitled "*Towards an inspiring, precise and shared goal*", which began the review of the vision and mission of the course at that time, with the participation of all the faculty and administrative staff and with the representation of the students, graduates, and employers of the career. In the same way, the career objectives were elaborated based on a workshop called "*Where are our actions headed?*" The curricular principles were redefined through a workshop called "*Building on solid foundations – Fundamental ideas governing the curriculum*", followed by several working sessions over a period of about three weeks. The redesign of the professional academic profile of the graduate in Agricultural Engineering required a significant investment of time and work, and the goal of the redesign was achieved in the workshop "*Graduates of the ITCR's undergraduate degree in Agricultural Engineering: a competent, competitive and committed professional*".

In order to confirm the legal and consensual nature of the process, each of the intermediate products of the career redesign (the vision and mission of the programme, the object of study, the substantiation and justification, the curricular principles, the professional academic profile, the objectives, the curricular structure, the methodological and evaluation principles, course programmes, the transition plan, equivalency system, dissemination, and attraction plan) were gradually approved by the respective school councils.

Once the process was completed, the full report was submitted to the Academic Development Center for technical curricular analysis, which gave a positive opinion. Finally, in November 2006, the curriculum redesign of the ITCR's undergraduate programme in Agricultural Engineering was presented and approved by the Faculty Council, which consists of representatives from all academic departments of the institution.

The methodology used to carry out the curricular redesign of the programme, supported by the necessary resources, made it possible to integrate and utilise the contributions of various sources and social, professional, and academic actors, who, under the guidance of the Coordinating Commission, the advice of CEDA and the consent of the School Council, were able to produce a renewed, relevant, legal, and consensual training proposal in a prudent period of time.

Redesign Achievements

The features described in the preceding section highlight the quality of the implementation of a series of changes, the result of a highly reflexive process, in the context of the reality of the career environment and the institution.

The first product obtained was the update of the vision and mission of the career. The specific areas of the curriculum have been redefined and a commitment has been made to maintain the level of performance required to strengthen teaching, research, and extension. Therefore, these two elements of redesign guided the revision, conception, and maturation of the other fruits obtained.

Within the organisation of the inputs that led to the following stages, an analysis of strengths, weaknesses, and threats was conducted to identify opportunities for improving the programme, giving rise to a solid justification of the process. These steps allowed a clear redefinition of the object of study, the specification of the areas and sub-areas of the discipline and the incorporation of entrepreneurship. On the other hand, it is expected that the object of study will be further nurtured by research, the results of which will be efficiently transferred to students and producers (ITCR & Escuela de Ingeniería Agrícola [IA], 2006).

Subsequently, ten curricular principles were officially established to define the most operative elements of the renewed curriculum. These principles are: vertical sequence, horizontal coherence, continuity, disciplinary integration (in terms of multidisciplinary and interdisciplinarity), flexibility, identity, social and academic linkage and relevance, universality, internationalisation, and transversality. This last principle encompasses the topics of quality, entrepreneurship, leadership, citizenship, critical thinking, adaptability, research, information management, teamwork, communication, an environmental dimension, and sustainability, safety, and integral health.

Another outstanding achievement was the redesign of the professional academic profile, using a model that that combines competencies with fundamental learning, integrating the three main areas of training: personal-citizen, academic-engineering, and professional-labour.

In addition, new career objectives have been set that will allow maintaining a horizon and make constant progress towards the state defined in the guiding elements of the curriculum. These objectives include, for instance to maintain a constant analysis of the surroundings in order to be able to respond to the demands of the environment, in terms of updating personnel and the curriculum, new technologies, and actions to support the productive sector (ITCR & IA, 2006).

On the other hand, there has been a redefinition of the curriculum structure in terms of the eight major fields of the curriculum: Basic Sciences, Social and Economic Sciences, Computer Sciences, Engineering Sciences, Management and Conservation of Water and Soil Resources, Administration and Maintenance of Agricultural Machinery, Development and Adaptation of Production Technologies, and the Final Graduation Project. In the redesign of this structure, greater congruence with other elements of the curriculum was sought. The academic weight of each area was redistributed, and the curriculum diagram was restructured by eliminating, creating, and modifying relevant courses (Figure 16.1).

Figure 16.1 shows that the main changes affect four of the eight fields that make up the curriculum of the programme, based on the analysis of the trends and perspectives of the development of the sector and the discipline at the national and international levels. For Engineering Sciences (ES) and Management and Conservation of Water and Soil Resources (MCWSR), their relative presence tends to decrease. While the formative weight of Social and Economic Sciences (SES) and Development and Adaptation of Production Technologies (DAPT) increases. There has also been an increase in the number of credits corresponding to the Final Graduation Project (FGP).

Another result of redesigning the curriculum was to provide more flexibility. The redesigned curriculum includes three elective courses, and students can choose from a broad list of

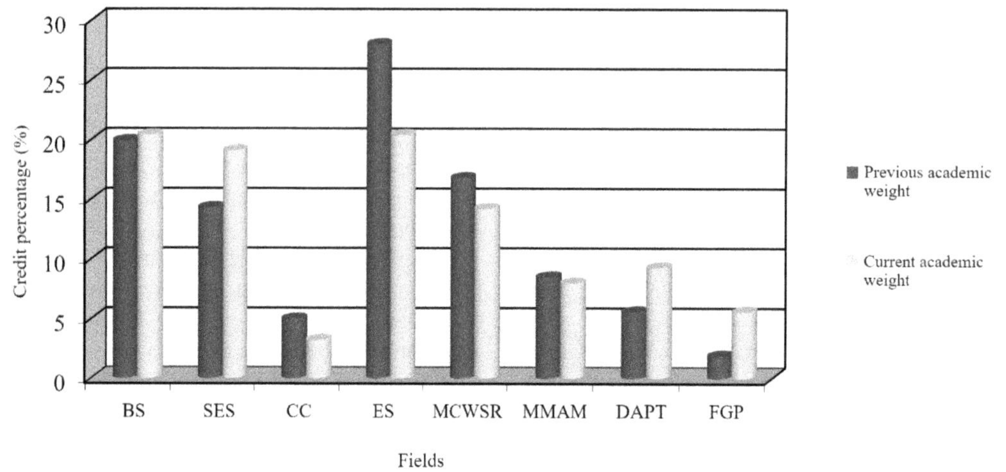

FIGURE 16.1 Variation in the percentage of credits corresponding to each course field. The following acronyms are used in the graph: BS – Basic Sciences; SES – Social and Economic Sciences; CC – Computer Sciences; ES – Engineering Sciences; MCWSR – Management and Conservation of Water and Soil Resources; MMAM – Management and Maintenance of Agricultural Machinery; DAPT – Development and Adaptation of Production Technologies; and FGP – Final Graduation Project (adapted from Fëdorov and Guzmán-Arias, 2007).

courses in the area that appeals to them the most. For example, in the field of Development and Adaptation of Production Technologies there are courses such as Environmental Protection and Wastewater Treatment, Greenhouse Crop Management, Agricultural Certification, or Geographic Information Systems.

In the end, all the guiding elements, including the methodological and evaluative principles, are applied through the programmes of each of the 63 subjects of the curriculum, and ready for implementation into practice.

The implementation of all these changes and their continuous evaluation have been considered to establish a career management model aimed at achieving the success of the rethinking of Agricultural Engineering. Therefore, it was necessary to create a transition programme for regular students. This would allow for the gradual implementation of the redesigned curriculum. It was also crucial to create an outreach plan to introduce the new approach of the curricular offerings to the target population of this important discipline.

Conclusions of the 2007 Curriculum Redesign

In conclusion, it was possible to redesign a comprehensive curriculum for the Agricultural Engineering career in a timely manner, thanks to a participatory, coordinated, and consensual work effort, supported by a methodology that combines tasks, responsibilities, controls, evaluations, and individual and group decisions, taking into account current curricular, pedagogical, and disciplinary trends.

This modified curriculum, with a total of 157 credits distributed over 10 academic periods, achieves a greater coherence of the elements that make it up. It includes 62 semester courses, humanistic training centres, and a final graduation project of one year's duration. The renewed Agricultural Engineering Program of ITCR opened its doors in the first semester of 2007, aware of the importance of a critical and constant analysis of the academic and disciplinary environment.

Curricula Redesign 2021: Towards Digital Agriculture

The curriculum created in 2007 (called 1007 Plan) has evolved into the current programme, called the 1008 Plan. The 1008 Plan of the Agricultural Engineering programme of the Instituto Tecnológico de Costa Rica has been accredited by the Canadian Engineering Accreditation Board (CEAB) since 2014 and has been reaccredited by the Accreditation Agency for Engineering and Architecture Programs (AAPIA) until the year 2026. In 2021, a new process of analysis, evaluation, and transformation of the Agricultural Engineering career began, in order to generate a new plan whose axes take into account the insertion of soft skills through the evaluation of the attributes of the graduates, in addition to the consideration of the Sustainable Development Goals of the United Nations (UN), which the Instituto Tecnológico de Costa Rica has incorporated into its policies.

Objectives of the Curriculum Redesign Initiated in 2021

The overall objective of this process is to conduct a comprehensive redesign of the curriculum of the undergraduate degree in Agricultural Engineering to incorporate emerging technologies in the field of performance of the degree, as well as other disciplines and related sciences. The work commission of the school is composed of three professors and researchers and a consultant from the Academic Development Center (CEDA) of the Instituto Tecnológico de Costa Rica. The following are the specific objectives of the redesign:

a. Define the needs in the national and international environment to establish the professional performance profile according to the needs found that should cover the curriculum of the undergraduate degree in Agricultural Engineering at the Instituto Tecnológico de Costa Rica.

b. Analyze and evaluate the sciences, technologies, and tools in the field of Agricultural Engineering, natural sciences, and other related disciplines that guarantee the relevance of the education, research, and extension work of the undergraduate degree in Agricultural Engineering at the Instituto Tecnológico de Costa Rica.

c. To establish a curriculum according to the defined needs, in order to satisfy the professional performance profile of the graduates of the Agricultural Engineering career at the Instituto Tecnológico de Costa Rica

Rationale for the Agricultural Engineering Curriculum

The Development of the Discipline at the Global, Regional, and National Levels

The International Commission of Agricultural and Biosystems Engineering (CIGR) defines Agricultural Engineering as the discipline that applies the principles of engineering, science, and technology to ensure food security and the sustainable use of natural resources (Zazueta Ranahan, 2017).

According to the United Nations (UN), the world's population will reach 10 billion by 2050. That's 32% more than was estimated in 2015 (Food and Agriculture Organization [FAO], 1996). This puts intense pressure on the agricultural production sector to produce more food with fewer resources and with minimal environmental impact, in addition to climate change and the closing of the agricultural frontier (FAO & Intergovernmental Technical Panel on Soils [ITPS], 2015). As markets open in all sectors, regional and global development trends are intensifying competition for product placement, and those who do not have the capacity to adapt their production systems to more efficient models will be at a distinct disadvantage (FAO, 1996).

This will require transforming current production systems to increase productivity and yields, while considering ecological concerns, ongoing challenges of infrastructure for drought and flood management, access to credit, insurance, technology, and knowledge (FAO, 1996).

Agricultural Engineering has become the main bastion for the reform of the productive apparatus towards a better use of all resources, due to the tendency to reduce the area sown, the number of machines used to make work more efficient, the climatic variations that affect crops and their influence on the availability of water and, in general, on the productive infrastructure. Likewise, considering the reduction of arable land due to the growth of urbanised areas, Agricultural Engineering has all the necessary tools to transform traditional agriculture into a highly productive activity, allowing fields to produce more quickly and efficiently with less available resources (FAO, 1996).

In Costa Rica, Agricultural Engineering is able to keep arable land in production throughout the year by designing hydraulic structures to convey water, precision irrigation systems, and applying drainage systems that allow the recovery of land that would otherwise be unproductive. On the other hand, the application of new or improved methods, technologies, tools, and machines to the work of cultivation in all its stages of development and postharvest, improves its efficiency by allowing them to take place in less time, making possible more harvests during the year, and by improving the conditions for the crop, thereby increasing the yield (Gómez-Calderón, 2016). In addition, choosing the right equipment for the right soil conditions and applying it at the right time saves on inputs and protects the environment (Liu & Li, 2013).

For this reason, the agrifood system requires the intervention of Agricultural Engineering to increase production efficiency, deal with climate change, ensure overall sustainability and resilience, strengthen rural areas, provide incentives for sustainable and competitive support policies for the industry along the value chain, and develop bioproducts and processes (circular economy). This must be approached through sustainable intensification, including the adoption of smart agricultural technologies and practices. Digitalisation, technologies, and tools have the potential to transform agriculture into a smart and agile sector, while at the same time addressing the major challenges it faces (CIGR, n.d.).

The "*Status of the World's Soils Resources*" identified 15 major degradation problems in Latin America, the first two of which, erosion and compaction, pose challenges for integrated resource management, as both conditions also involve water management (FAO & ITPS, 2015). Agricultural Engineering is the discipline that brings together the knowledge of hydrology, soil management, and controlled traffic farming, which allows for the quantification of, soil losses through field methods, statistical data modelling, or image analysis, which proposes systems for the conservation of resources through infrastructure and management with adapted, designed, emerging, or improved agricultural machinery, to propose systems for the conservation of resources through infrastructure and management for the correct preparation of soils according to the needs of the productive systems and the analysis of the physical, relief, and micro-relief restrictive conditions of the land and to propose, as far as possible, measures to overcome the aforementioned restrictions for the proper development of crops within the different production systems and the various conditions and locations. This refers to the management and use of land and water in a way that enables sustainable development and conservation of natural resources.

By optimising the use of soil, water, inputs, and mechanisation, and by applying technologies and tools in a wide range of environments and production systems, Agricultural Engineering is the ideal tool for technical and technological development through the application of the principles of precision agriculture.

According to the results of the XIII Latin American Congress of Agricultural Engineering (Villagra-Mendoza, 2018) held in San José, Costa Rica, Agricultural Engineering is the discipline called on to capitalise on multidisciplinary opportunities to achieve orderly progress, in the Latin American region, towards agriculture 4.0, since agriculture is increasingly under pressure to use technology as a facilitator in a contemporary environment where agriculture is a social stabilising factor that has to do with food, energy, degradation, climate change, pollution, and international markets, for which it is required to increase efficiency and innovation in all processes and equip the next generations with comprehensive skills, precision agriculture being one of the ways to achieve smarter production (Villagra-Mendoza et al., 2018) as shown in Figure 16.2.

Socio-economic Needs and Political Context

The minister of Agriculture, Fishing, and Rural Sector of the government of Alvarado Quesada (2018–2022), pointed out in the management report from May 2018 to April 2019 that the political guidelines of the sector are the frame of reference for the actions to be carried out by the institutions that constitute the area, whose objective is to contribute to the social and economic wellbeing of the population linked to agriculture in a manner that is increasingly resilient, under the principle of teamwork, and in favour of the greatest number of people possible, especially the most vulnerable. It should also be noted that these guidelines aim to achieve a technologised, competitive, inclusive, and sustainable agriculture, driven by public institutions that are agile, modern, and coordinated with the productive sector (Secretaría Ejecutiva de Planificación Sectorial Agropecuaria [SEPSA], 2019).

However, the macroeconomic figures show that economic activity (as measured by the real growth of the Gross Domestic Product, GDP) grew by 2.7% in 2018, compared to a growth of 3.4% in the previous year, and in terms of Agricultural Value Added (AVA), there was a slowdown, with growth reaching only 2.4% (compared to a growth of 3. 7% in 2017), due to the increase in pineapple exports, driven by the application of better practices for pest and disease control, which led to an improvement in productivity per hectare and fruit quality (SEPSA, 2019). This proves the absence of public policies to support national farmers, an omission that has been maintained in recent years and at the same time penetrates various issues of concern to the sector.

Despite this condition, the agricultural sector absorbed 11.9% of the national workforce in 2018, showing a growth trend, as it registered an increase of 2.9% compared to the same period in 2017. The primary sector ranks second in the country as a generator of employment, with 12.3% of the total national workforce. Nevertheless, in 2018, 25.1% of rural households lived below the poverty line, and 8.3% were in extreme poverty; the average monthly income in rural areas was 41.5% lower than in urban areas (an amount of 486,766 colones less) (SEPSA, 2019).

This condition of the rural sector is also in contrast to the fact that agricultural products accounted for 43.7% of total national exports during the year. These placements grew by 0.8% (more than $40 million) compared to 2017. A total of 890 agricultural products were exported to 131 foreign markets, with bananas, pineapples, syrups and concentrates, and palm oil accounting for 57.2% of these exports. The agricultural trade balance is in surplus, meaning the country is a net agricultural exporter (SEPSA, 2019).

For the first quarter of 2020, agriculture was noted as one of the main sources of employment in the country, employing 270,000 people. However, this was already 9000 jobs less than in the same period of 2019. It was also reported that financing,

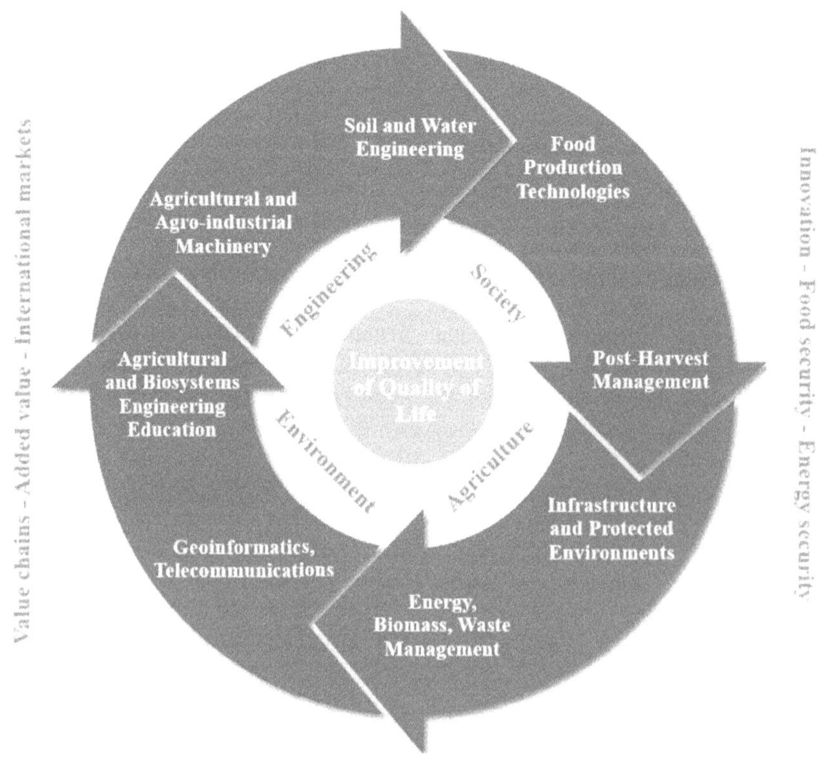

FIGURE 16.2 Challenges of Agricultural Engineering towards an agriculture 4.0 derived from the XIII CLIA 2018 (adapted from Villagra-Mendoza et al., 2018).

climate change, the difficult leap to precision agriculture, the lag in agrochemical uses, and intermediaries are the main challenges currently facing the activity, according to several stakeholders in the sector (Casa Presidencial, 2020)

Summa Magazine (Revista Summa, 2019) identifies other challenges, such as the new tax regulation of Costa Rica's basic food basket, which excludes from the exemption many agricultural products, as well as raw materials of great importance to agribusiness, putting at risk the hundreds of thousands of jobs provided by the agricultural sector, in addition to the lack of application of trade defence instruments and regulations to protect phytosanitary and animal health heritage, which results in the absence of support for the trade defence panel on issues related to the restriction of imports of products that compete with domestic producers. On top of this, the lack of access for agricultural credit, which accounts for only 3% of all credit granted by the national financial system, and the excessive bureaucracy imposed on the vast majority of operators, are further obstacles.

More than a decade ago, Arias (2005) highlighted the need to promote policies that create the appropriate environment to maintain the competitiveness of the agro-export sector, including programmes for the modernisation of companies and productive reconversion, which would include adding value to exports, commercial intelligence, market research, business management, along with research and technology transfer, which are fundamental elements for companies to become competitive, especially the small ones.

To remain competitive in a globalised market, innovation and the generation of ideas are essential. A system of research and technology transfer, integrated with the university system, must be strengthened so that producers can benefit from innovation for productivity improvement, process mechanisation, and production under controlled conditions. Moreover, Costa Rica has a certain disadvantage compared to other countries producing the same products due to its high labour costs (without considering the other challenges currently added), which makes strengthening the research and development component essential (Arias, 2005).

Summa Magazine (Revista Summa, 2019) stated that international markets are becoming increasingly competitive, and given this reality, Costa Rican farmers need technological tools that will allow them to compete on a level playing field.

By 2019, Costa Rica had 480,000 hectares planted with agricultural crops and more than one million hectares of pasture for livestock production. This is in addition to 93,000 farms dedicated to agricultural production, a source of work and income for more than 500,000 inhabitants of coastal and rural areas (Revista Summa, 2019).

Agriculture, fisheries, aquaculture, and agroindustry are fundamental pillars for Costa Rica's development, which urgently requires high investments in knowledge and technology development, so the updated curriculum must take this situation into account.

Similar International and Domestic Offers

Table 16.1 summarises the careers offered in Costa Rica with which it shares some area of knowledge. The similarities and differences are described with the purpose of identifying the elements that can improve the Agricultural Engineering programme at the Instituto Tecnológico de Costa Rica (ITCR).

Outside of Costa Rica, descriptions of agricultural and biosystems engineering courses are very similar to the Agricultural Engineering course offered by the Instituto Tecnológico de Costa Rica. In general terms, most schools define it as the branch of engineering that deals with the design of agricultural machinery, soil management, land development, livestock mechanisation, and automation, and the efficient planting, harvesting, storage, and processing of agricultural products.

Table 16.2 summarises the international Agricultural Engineering programmes offered in America. The large number of engineering programmes offered around the world demonstrates the importance and impact they have on production and resource management systems.

Internal and Administrative Dimension of the Instituto Tecnológico de Costa Rica

Article 51 of the Organic Statutes of the Instituto Tecnológico de Costa Rica establishes that, in order to carry out its activities, the Institute is organised into units, the basic unit of which is the Department. It also states that the academic departments are those responsible for teaching, research, and extension, and adds that the generic name of the academic departments is "school" (ITCR, 2005).

The Agricultural Engineering career is supported by the work of the school of the same name, to which the Instituto Tecnológico de Costa Rica has assigned resources in infrastructure, support materials for teaching, research, and extension activities, and teaching and administrative staff, so that through its proper management it can fulfil the responsibilities assigned by the institution.

The response of the Agricultural Engineering Career to the mission of the Instituto Tecnológico de Costa Rica has been through a curriculum that allows the training of quality professionals who have generated, adapted, and integrated the necessary technology to utilise resources and productive forces, contributing to developing the country and transforming Costa Rican society through the improvement of the quality of life.

By using the latest technologies and high-quality human resources in teaching, engineering, science and technology, and extension and research, a better academic preparation has been obtained in support of the training of the graduates in agricultural technology that the country needs.

The Agricultural Engineering programme meets the country's needs in water and soil management, mechanisation, and machinery, and developing and adapting production technologies.

Agricultural Engineering School

By putting theoretical concepts into practice and constantly updating knowledge, the professional identity of the professor and the student has been developed. In view of this constant training and updating of the professionals that are part

TABLE 16.1

Course offerings in Costa Rica with similar areas of knowledge.

Course	University	Credits / No. of courses	Similar field	Differences with ITCR Agricultural Engineering
Agricultural and Biosystems Engineering	*Public Universities:* • Universidad de Costa Rica (UCR)	178 / 65	Conceptually, it focuses on applied physics to agricultural production biosystems.	Agricultural Engineering at ITCR focuses on the management of natural resources for production and supply, such as soil conservation systems and water management, conveyance, and consumption through efficient production techniques and technologies.
Environmental Engineering	*Public Universities:* • Instituto Tecnológico de Costa Rica (ITCR)	Not specified/ 66	One of its five competencies is to locate and select viable water sources.	Environmental Engineering focuses on the application of hydrology for water treatment systems, sanitation, and soil remediation, while Agricultural Engineering focuses on the application of hydrology for soil management and conservation, flood risk assessment, urban and agricultural drainage systems, aqueduct systems, and water availability for irrigation.
Bachelor's Degree in Water Resources Engineering	*Public Universities:* • Universidad Técnica Nacional (UTN)	180 / 62	According to the description of the bachelor's degree in Water Resources Engineering, it seeks to train engineers with the necessary knowledge, skills and attitudes in the design, implementation, and management of hydrological systems, capable of carrying out scientific research processes for the evaluation, control and development of innovative systematic strategies that contribute to the mitigation of water problems of natural and anthropogenic origin. It also seeks to train professionals with the capacity to design water catchment, conduction, storage and treatment systems, as well as modelling of watercourses for their use, that comprehensively evaluate the different problems for the generation of proposals that guarantee a safer society and an adequate use of water resources.	It focuses only on the hydrological component in the field of Agricultural Engineering. Agricultural Engineering focuses on the formation of design abilities of pressurised systems or water conveyance and design of hydraulic structures.
Agronomy Engineering	*Public Universities:* • Universidad de Costa Rica (UCR) • Universidad Nacional (UNA) • Instituto Tecnológico de Costa Rica (ITCR) • Universidad Nacional a Distancia (UNED) • Universidad Técnica Nacional (UTN) *Private Universities:* • EARTH University	Between 144 – 160/ 49 or more	Focuses on the management of phytosanitary production systems. Some offers in the country include specialisation in irrigation and drainage.	Agricultural Engineering does not focus on agricultural science, but on engineering and technology applied to agricultural systems. The difference with engineering careers in agronomy, which include specialties in irrigation and drainage, is that agronomy indicates the water requirements of crops and Agricultural Engineering specialises in the mechanical and hydraulic design of the system.

TABLE 16.1 (CONTINUED)

Course offerings in Costa Rica with similar areas of knowledge.

Course	University	Credits / No. of courses	Similar field	Differences with ITCR Agricultural Engineering
Civil engineering	*Public Universities:* • Universidad de Costa Rica (UCR) *Private Universities:* • Universidad Autónoma de Centro América (UACA) • Universidad Latina de Costa Rica • Universidad Fidélitas	173 / 61 plus electives Private universities do not specify the number of credits and the number of courses.	Hydraulic and river engineering structures	Agricultural Engineering applies the knowledge of civil engineering and construction as a basis for the field of water management, design of hydraulic structures for irrigation or drainage water conveyance, dikes for flood control in agricultural areas.
Mechanical engineering	*Public Universities:* • Universidad de Costa Rica (UCR) *Private Universities:* • Universidad Autónoma de Centro América (UACA) • Universidad Latina de Costa Rica • Universidad Fidélitas	156 / 55 plus electives Private universities do not specify the number of credits and the number of courses.	Mechanisms and Machine Elements	Agricultural Engineering applies the knowledge of mechanical engineering to the elements and mechanisms of agricultural machinery, to the behaviour of the mechanical properties of the soil and its interaction with the machinery.
Hydrologic Engineering	*Public Universities:* • Universidad Nacional (UNA)	The number of credits and the number of courses is not specified.	Hydrological and hydraulic systems modelling	Hydrologic engineering focuses only on the hydrologic components of a system. Agricultural Engineering focuses on the formation of design abilities of pressurised systems or water conveyance and design of hydraulic structures.

of the Agricultural Engineering career, it can be affirmed that there is the maturity and academic rigour necessary to train professionals with a undergraduate degree or higher, capable of adapting and incorporating the necessary technology to achieve an efficient use of production resources and the improvement of production systems, while protecting natural resources, contributing to national sustainable development.

Recently, the researchers of the school have managed to maintain a constant flow of projects (both research and extension) in the three defined lines of research, making important contributions to the productive sector, the public sector, and the general population. In order to widen the scope of the projects and to attract external resources to ensure the successful completion of each project and the continuity of the lines of extension and research, several connections have been established directly with small and medium producers, government institutions, private companies, and international organisations. This has allowed the generation of a broad range of didactic material and technology (the latter means any technical information, adaptation, software, etc., which is of interest and applicability for the areas of Agricultural Engineering), which has also been strengthened in its transfer and production as equipment to work.

It is expected to have greater feedback from the world environment as part of the globalisation process in which the career is immersed. For this purpose, liaisons are maintained with universities that have similar disciplines worldwide and with networks such as the Latin American Association of Agricultural Engineering (ALIA) and the CIGR.

The course contributes to the development of the discipline, through constant contact with employers to know what their real needs are in the labour and technological field, through a diagnostic system for proper technological management that allows the school to remain at the forefront in the use of technologies of a leading school, within a leading university, thus the career will keep its professionals in constant training and updating.

Agricultural Engineering Course Vision

The course will have the leadership in the training of agricultural engineers with a undergraduate degree. Furthermore, it will lead research and extension projects in water and soil resource management and conservation, agricultural machinery management and maintenance, and production technology development and adaptation, thus contributing to the development of society.

Mission of the School of Agricultural Engineering

Using the latest technologies, guided by a constant analysis of the environment in which we operate and supported by highly

TABLE 16.2

International offer of Agricultural Engineering courses in America.

Zone	Country	Universities
North America	United States of America	Auburn University
		Cal Poly San Luis Obispo
		Clemson University
		Cornell University
	Canada	Dalhousie University
		Florida A&M University
		Iowa State University
		Kansas State University
		Louisiana State University
	Canada	McGill University
		Michigan State University
		Mississippi State University
		North Carolina State University
		North Carolina Agricultural & Technical State University
		North Dakota State University
		Ohio State University
		Oklahoma State University
		Oregon State University
		Penn State University
		Purdue University
		South Dakota State University
		Texas A&M University
		University of Arizona
		University of Arkansas
		University of California, Davis
		University of Florida
		University of Georgia
		University of Illinois
		University of Kentucky
		University of Nebraska-Lincoln
	Canada	University of Manitoba
		University of Minnesota
		University of Missouri
	Canada	University of Saskatchewan
		University of Tennessee
		University of Wisconsin
	Canada	Université Laval
		Utah State University
		Virginia Polytechnic University
	Mexico	Universidad de Chapingo
		Universidad Autónoma Agraria Antonio Narro, Mexico
Central America	Nicaragua	Universidad Nacional de Ingeniería
	Costa Rica	Instituto Tecnológico de Costa Rica
		Universidad de Costa Rica
South America	Colombia	Universidad Nacional de Colombia
	Brazil	Universidade Federal do Ceará
		Universidade do São Paulo
		Universidade Estadual de Campinas
		Universidade Federal de Viçosa
		Universidade Federal de Pelotas
		Universidade Federal de Lavras
		Universidade Federal da Paraíba
	Venezuela	Universidad de Los Andes
	Ecuador	Universidad de Manobí
	Perú	Universidad Nacional Agraria La Molina
		Universidad Nacional Pedro Ruiz Gallo de Lambayeque
	Chile	Universidad de Chile
	Bolivia	Universidad Mayor de San Simón
		Universidad Autónoma Gabriel René Moreno (U.A.G.R.M.)

qualified and constantly updated human resources, we integrate teaching with research and extension to prepare competent, competitive, and enterprising Agricultural Engineering graduates, committed to ethics and the integral development of the country.

Objectives of the ITCR Undergraduate Degree in Agricultural Engineering

- Achieve technological leadership by integrating research with teaching and extension in soil and water management and conservation, agricultural machinery management and maintenance, and production technology development and adaptation.
- Define and develop, within the study plan, coures, areas, axes, and methodologies that allow students to acquire the competencies defined in the professional academic profile.
- Analyse, evaluate, adapt, and integrate emerging technologies in agricultural, natural, and other engineering fields to ensure excellence in teaching, research, and extension.
- Maintain a constant analysis of the field to respond to the demands of the environment in terms of personnel and curriculum updating, new technologies, and actions to support the productive sector.
- Develop an academic and administrative platform that allows it to have high-level professors, well-trained administrative staff, modern infrastructure and equipment, and an appropriate environment for developing teaching, research, and extension activities.
- Coordinate the implementation of suitable measures within the current curriculum for the enrolment of students whose backgrounds make them ineligible according to the established procedure, with a selection modality that guarantees the incorporation and success of students with potential for the study of Agricultural Engineering.
- Perform continuous self-evaluation and self-regulation through collective efforts in order to obtain and maintain accreditation as a means of improving professional quality.

Center for Research and Extension in Agricultural Technology and Engineering (CETIA)

In response to the technological needs of the country's agricultural environment, ensuring food security and strengthening the work that has been done, and to ensure the sustainability of research and extension projects, as well as the maintenance of current linkages and the generation of others in the short term, the need for the Center for Research and Extension in Agricultural Technology and Engineering (CETIA) was raised to support the position of the ITCR in issues related to the application of engineering and technology in our tropical agriculture. The Institutional Council of the ITCR has approved the operation of CETIA in January 2023.

CETIA will have the following areas of focus: soil management and conservation, agricultural and agro-industrial mechanisation and automation, precision agriculture, digital agriculture, water management for conveyance and distribution, fluvial flood risks, environmental flows, hydrology and hydraulics, agricultural and urban drainage, soil physics and mechanics, irrigation systems, normalisation and standardisation in Agricultural Engineering, data management for modelling and simulation scenarios and marketing of agricultural products. These areas are among the strategic technical assistance and research lines of Agricultural Engineering school at the ITCR.

Soil Management and Conservation

Area focused on groundwork decisions or above ground infrastructure, e.g., terracing, land levelling, and other soil conservation works, site preparation, subsurface drainage systems, or surface water runoff control. International certification and environmental management systems for agro-export companies consider erosion quantification. The research centre can measure and make recommendations to minimise soil erosion through direct field measurements, remote sensing, or modelling of conditions, as well as recommend control measures and controlled agricultural machinery traffic systems and determine the degree of compaction and its effect on soil behaviour.

Agricultural and Agroindustrial Mechanisation and Automation

This area is related to the design or selection of agricultural mechanisation solutions for any stage of the crop, asset management at all stages, including machinery management, maintenance workshop, supply inventories, precision farming equipment in agricultural machinery, whether propelled, self-propelled, or stationary. The increase in the fleet of agricultural machinery has created a great demand for professional managers of this type of resource, and a great need for training and professional updates in the evaluation and assessment of machinery, data management and operational indicators for fleet monitoring, economic analysis for asset replacement, failure analysis of agricultural elements and mechanisms, power calculations required for each type of soil and work, etc., to reduce the impact of machine use on company budgets and the environment (soil and air).

Precision Agriculture

Precision agriculture consists of applying different resources in the time, space, and quantity necessary for each management zone defined based on crop yield maps (Molin, 2017). Each management zone has different systems for soil and water conservation, drainage, irrigation, farm traffic control, and application of inputs at all stages of production, according to its climate, topography, soil and ground conditions, water availability, required crop care at different stages, and postharvest management. This can include production tools in protected and controlled environments, geographic information systems, remote sensors, image capture and processing, statistical data management software, weighing systems built into harvesters or conveyors, and more. Since precision agriculture has been introduced in Costa Rica in extensive crops, there is a need to create or adapt tools for crops grown in small or medium areas (Gomez-Calderón & Solorzano-Quintana, 2017). according to the classification of the MAG Decree N° 37911. The entry of CETIA in this field represents the positioning of ITCR in the technological management of agriculture.

Digital agriculture

The application of Industry 4.0 to agriculture consists of the management of all production processes through the Internet of Things, mechatronics, and cloud computing (environmental sensing, remote sensing, artificial intelligence, and big data management), allowing control and automation, while storing information that allows decision-making based on real-time data transmission (AGRICULTURERS, 2017). It complements precision agriculture, especially in irrigation systems, fertigation, and greenhouses with controlled environments according to the conditions of the moment. The school of Agricultural Engineering has worked hand in hand with the mechatronics engineering career, with excellent results in research on controlled environments and different production substrates to reduce pressure on the soil resource. Scientific products have been published and patents are in the process of being obtained. This is an evolving issue around the world, so there are great opportunities for innovation, especially in tropical agriculture, because state-of-the-art technology tends to be developed in the temperate zones of the world, and there are great opportunities to develop solutions for conditions in the tropics.

Water Management for Conveyance and Distribution

Water management, its conveyance, and its subsequent distribution consists of systems that allow access to water resources in an efficient, equitable, timely and quality manner. The approach to this type of problem includes hydraulic design, operation and maintenance manuals, economic and environmental feasibility studies, and the preparation of management plans from initial planning to start-up. The School of Agricultural Engineering has worked on topics related to irrigation projects, agricultural and urban drainage, evaluation, and design of aqueducts, whether municipal, of with the Instituto Costarricense de Acueductos y Alcantarillados (Costa Rican Institute of Potable Water and Sewers, AyA) or (Local Management Associations for Potable Water and Sewers (ASADAS), through research and extension projects and theses to productive areas or for human consumption.

Fluvial Flood Risks

Flood dynamics studies analyze the impact of extraordinary river floods on populated areas, infrastructure, and agricultural production areas by determining parameters such as the area affected, the magnitude of hydraulic variables, and the duration of flooding. This is done through hydraulic and hydrological studies using photogrammetric techniques, mathematical modelling, programming, and geographic information systems. Since 2015, the school of Agricultural Engineering has been cooperating with the Comisión Nacional de Prevención de Riesgos y Atención de Emergencias (National Commission for Risk Prevention and Emergency Care, CNE) in the generation of flood spots. Research projects have been carried out on the subject, in collaboration with both public institutions and the private sector, technical training has been made with international experts and faculty of our school, in addition to the consulting services in this field.

Determination of Environmental Flow

Determining environmental flows is a fundamental issue in integrated water resources management because it represents the precise quantity and quality of water needed to sustain freshwater and estuarine ecosystems, as well as the livelihoods and wellbeing of people who depend on the ecosystem. The school of Agricultural Engineering, in collaboration with the Ministerio de Ambiente y Energía (Ministry for Environment and Energy, MINAE), has developed research on the adaptation and validation of different calculation methodologies for the determination of the environmental flow regime. This issue is a national priority on which the school and the ITCR have been protagonists in the progress made in the last five years. This linkage has been a source of external funding for research projects in the field, in addition to directly contributing to water resource management in Costa Rica.

Hydrology and Hydraulics

Hydrology is the study of phenomena related to the spatio-temporal distribution and properties of groundwater and surface water, including the study of precipitation, soil moisture, runoff, watershed streamflow, and evapotranspiration. This allows the study of the watershed response under deficit, normal, and surplus conditions, providing the basis for drainage design, irrigation, water resource management planning, emergency response, risk management, and many other applications. Hydraulics is the tool that allows for the design and the operation of the civil engineering works that make possible the attenuation, the conveyance, and the use of the water resources. This subject is important in the development of Agricultural Engineering, since it is a common axis of application in many of the fields developed in this document.

Agricultural and Urban Drainage

A drainage system must be designed to allow the evacuation of runoff in such a way that it does not cause inconveniences in agricultural or urbanised areas, using different techniques according to the requirements of the crop or the characteristics of the urban centre. In this area, the school has participated in the quantification of floods in urban and agricultural areas through research and final graduation projects. CETIA's potential contribution would be to assist communities with their regulatory plans.

Soil Physics and Mechanics

Soil physics and mechanics covers the analysis of the behaviour of soil and its responses to different phenomena of natural or anthropic origin; in other words, it is a branch of applied mechanics that deals with the response of the soil to stress states or loads present in its physical environment. In addition, the analysis of multiple variables allows the establishment of relationships with the spatio-temporal behaviour and crop yields in plantation agriculture, which is fundamental for the development of precision agriculture techniques and soil conservation and management. CETIA has the capacity to perform soil laboratory analysis and soil engineering tests in the field and is able to provide sales and consulting services that reflect the improvement of agricultural practices in the country.

Irrigation Systems

This field includes the design and implementation of an irrigation system using any technology available on the market. It also includes the evaluation of installed systems in order to find the most suitable solutions for each productive operation, based on a scientific analysis of the variables involved. The school of Agricultural Engineering has always been a national academic reference for best practices in this field. According to the College of Engineering Technologists (CITEC), currently more than 60% of those who are professionals in Agricultural Engineering work in activities related to agricultural, landscape, and sports irrigation, so an Irrigation Code is being developed in the CFIA that will regulate the practice of this discipline. This represents an opportunity for the Instituto Tecnológico de Costa Rica, through the school, to become a certifying body for crop irrigation equipment and processes, which means providing a service capable of generating economic resources for the management of CETIA.

Normalisation and Standardisation in Agricultural Engineering

Engineering practice involves the use of a variety of technical standards designed to ensure proper operation and durability of systems and equipment under ethical and professional criteria. International standards are a reference that in many cases allow the construction of national regulations, where the school is called to become a certifying body for equipment and processes, representing an enormous potential for attracting external resources, as in the case of irrigation. The agro-export sector is conditioned to be certified with environmental standards of process quality and safety. CETIA can be a service provider, including services for the State, such as the registration, verification, and calibration of machinery for the application of crop protection products required by the State Phytosanitary Service (SFE-MAG).

Data Management for Modelling and Simulation Scenarios

Modelling the variables involved is part of the analytical processes that precede engineering designs and natural resource management actions. Access to and management of data is critical to the success of modelling and simulating possible scenarios. In this field, the school has gained a lot of experience through research projects in hydrological and hydraulic modelling. Nonetheless, modelling and information management are used in all areas of research and can become another service offered to other institutions and companies, which has been done on several occasions with the National Banana Corporation (CORBANA) and the Comisión Nacional de Prevención del Riesgo y Atención de Emergencias (CNE). The school of Agricultural Engineering has a Modelling and Data Analysis Laboratory (LAMOD), which is in the process of being equipped with hardware and software.

Marketing of Agricultural Products

For several years, the school of Agricultural Engineering has been conducting research on various crops under greenhouse conditions. This has led to the use of production to obtain resources that contribute in some way to the school's research. Our goal is to continue to explore commercialisation modalities and other alternatives, including scheduled deliveries, in order to generate steady revenue to support this and other research efforts. This initiative can also generate training activities and model production stations.

Rationale for the New Curriculum 2024

The rationale for the redesign of the undergraduate programme in Agricultural Engineering started in 2021, being described as a step on the part of the course towards digital agriculture, managed to continue the foundations of the redesign made in 2007, based on the study of the development of the discipline at the global, regional, and national levels, the study of the socio-economic needs and the political context, similar international and national offers, and the internal and administrative dimension of the Instituto Tecnológico de Costa Rica. The following aspects are the basis for the definition of the operational elements of the programme (curriculum, curricular structure, and transition plan).

Object of Study

The object of study was approved by the Agricultural Engineering School Council on March 22, 2021. It was proposed by the curriculum committee in charge of the redesign and improved by working groups in which the faculty of the school was involved. The final text is as follows: "*The purpose of the Agricultural Engineering degree is to study the efficient management of resources for production and related data to improve the quality of life in terms of food, energy, climate change, pollution and access to international markets through the design and evaluation of engineering systems for soil, water, structures and environment, machinery and technologies for the management of natural resources and production, applying quality control in the systems*".

Disciplinary Areas

On June 21, 2021, the Agricultural Engineering School Council approved the disciplinary areas on which the career redesign is based. The CEDA advisor led the workshops. The workshops covered the following areas:

General Education

Is the ability to critically analyse the environment through the enrichment of social, economic, political, cultural, and environmental sensitivity and the improvement of communication skills and logical thinking.

Exact and Natural Sciences

Is the ability to analyse the knowledge of the laws that govern the universe and thus the foundations of engineering, composed of the disciplines of mathematics, physics, chemistry, and biology.

Engineering Sciences and Technologies

Is ability to apply the base knowledge to the solution of specific problems or needs of each area of Agricultural Engineering with scientific and technological support. It is an intermediate discipline between exact and natural sciences and Agricultural Engineering knowledge.

Soil and Water Technology and Engineering

Is the ability to design and implement systems that guarantee optimal hydraulic, hydrological, hydric, and edaphic conditions for production, ensuring the appropriate use, management, and conservation of soil and water resources to achieve the development of urban and rural societies.

Structural Engineering and Technology

Is the ability to design and evaluate microenvironments and works for the management of natural resources and production through the application of engineering knowledge to make them functional and economical, ensuring sustainability, health, and human welfare.

Technologies, Equipment, and Energy

Is the ability to manage equipment and technologies (from the selection, acquisition, maintenance, management of supplies and spare parts warehouses, and final disposal), ensuring their adequate useful life, productivity, efficiency, sustainability, and quality of agricultural systems and products at all stages (land preparation, planting, protection, harvesting, and transportation of agricultural products, postharvest handling of products, cooling networks, and agro-food processing), through

TABLE 16.3

Summary of the Agricultural Engineering Course Profile (Barquero-Alfaro & Chinchilla-Brenes, 2020).

Description	Occupational tasks	Related fields	Interests
Engineering design, management, preparation, maintenance, and improvement of adequate and efficient agricultural processes, such as irrigation, drainage and energy. It focuses on the use, management and conservation of soil and water resources, as well as the design and management of machinery and resources to meet agricultural needs.	• Improve production with precision agriculture technology, automation, and digitisation trends. • Define production management zones for irrigation, drainage, and soil conservation. • Evaluate soil characteristics. • Design and evaluate irrigation systems for agricultural production and green space maintenance. • Design drainage systems for land prone to flooding. • Design and evaluate farm machinery. • Develop maintenance plans for agricultural equipment.	• Surveying Engineering • Civil Engineering • Geography • Environmental Engineering • Integrated Water Resources Management • Agricultural Engineering • Forest Engineering	• Agricultural Production, Irrigation and Drainage Systems • Water Supply • Soil Conservation • Agricultural Machinery • Skills • Mathematics, Physics and Chemistry • Data Synthesis and Analysis • Anticipating risks • Teamwork • Field work

technologies and engineering of components, machines and systems, making use of energy sources that provide rational and economical applications following environmental protection aspects.

Management, Ergonomics, and Processes

Is the optimisation of operational planning, coordination, and processes engineering by applying current legislation and regulations (laws, regulations, standards, codes, certifications). It seeks to improve the efficiency of operations, economy, health, ergonomics, safety, and the occupational hygiene of employees. It allows the traceability of the management of natural resources or production, food safety, labour, social, environmental, and quality aspects of the processes.

Professional Profile of the Agricultural Engineering Programme

The Department of Orientation and Psychology (DOP) of the Vice Rectorate of Student Life (VIDA) of the Instituto Tecnológico de Costa Rica, conducted the professional profile of Agricultural Engineering. Barquero-Alfaro & Chinchilla-Brenes (2020) designed the tests and carried out the review and validation process of the profile obtained. As a result of the questionnaires used and the review and validation process with experts in the first phase of the research, the professional profile of the career was elaborated. A comprehensive document was obtained, which is presented in Table 16.3 as an executive summary for the purposes of illustration.

What to Expect in the Future of Agricultural Engineering

The new curriculum is a step towards digital agriculture, taking into account the data revolution that is taking place all over the world as a result of the processes of automation and digitalisation of systems. Although the fields of action of the career remain the origin of the discipline, it's necessary to acquire transdisciplinary and interdisciplinary knowledge. This will allow a better management of agricultural production from an engineering point of view, using the best tools.

On the other hand, the demands of international quality have led the Instituto Tecnológico de Costa Rica to adopt the quality standards of the Washington Accord, to follow the policies of the United Nations Sustainable Development Goals, and to insert itself in the world with the relevance and updating that it requires.

REFERENCE LIST

ABET. (2004). *Criteria for Accrediting Engineering Programs Effective for Evaluations during the 2005–2006 Accreditation Cycle.* Engineering Accreditation Commission. ABET Inc.

Agencia Registrada de Acreditación de Estudios de Ingeniería e Informática [ASIIN]. (n.d.). *Información dirigida al mundo universitario. Requisitos y principios para la realización de estudios de ingeniería e informática a nivel de Bachelor y Máster.* Accreditation Agency for Study Programmes in Engineering Informatics, Natural Sciences and Mathematics [ASIIN]. Retrieved December 12, 2005, from https://www.asiin.de/en/about-asiin.html

Agriculturers. (2017). *Agricultura de precisión y trabajo compartido: Los grandes desafíos.* AGRICULTURERS. https://agriculturers.com/

Antillón, A., Chang, F., Gámez, R., Macaya, G., & Villalta, G. (2005). *Informe final de carácter general no vinculante al Presidente de la República sobre el Tratado de Libre Comercio entre Centroamérica, República Dominicana y Estados Unidos de América.* Editorial Tecnológica de Costa Rica.

Arias, O. (2005). Retos para la agricultura en Costa Rica. *Agronomía Costarricense, 29*(2), 157–166. http://www.mag.go.cr/rev_agr/v29n02_157.pdf

Arredondo, V. (n.d.). *Políticas de flexibilidad curricular en educación superior.* Universidad Autónoma de Tamaulipas. Retrieved January 22, 2006, from https://www.uat.edu.mx/

Asociacion Iberoamericana de Instituciones de Enseñanza de Ingenieria [ASIBEI]. (2005). *Directrices curriculares para carreras de ingenieria en iberoamerica.* ARFO Ediciones e Impresiones Ltda.

Barquero-Alfaro, A., & Chinchilla-Brenes, S. (2020). *Perfil de tareas, intereses y habilidades vocacionales de los estudiantes de la carrera de Enseñanza de la Matemática con Entornos Tecnológicos y las Ingenierías Agrícola y en Computadores del Instituto Tecnológico de Costa Rica.* Instituto Tecnológico de Costa Rica [ITCR].

Campo, V. M. G., & Celis Giraldo, J. E. (2005). Factores de innovación curricular y académica en la educación superior. *Revista Iberoamericana de Educación, 5*(1), 71–81. https://doi.org/10.17533/udea.unipluri.12205

Canadian Engineering Accreditation Board [CEAB]. (2004). *Accreditation Criteria and Procedures.* Canadian Engineering Accreditation Board [CEAB]. https://engineerscanada.ca/

Casa Presidencial. (2020). *Con una inversión de ₡4.267 millones, Gobierno impulsa programa de alta tecnología para sector agroproductivo.* Presidencia. https://www.presidencia.go.cr/sobre-casa-presidencial/

Colegio Federado de Ingenieros y de Arquitectos de Costa Rica [CFIA]. (2004). *Criterios y procedimientos de acreditación: Programas de ingeniería, 2004.* CFIA, Departamento de Formación Profesional.

Consejo de Acreditación de la Enseñanza de la Ingeniería [CACEI]. (n.d.). *Manual del CACEI.* Consejo de Acreditación de La Enseñanza de La Ingeniería [CACEI]; CACEI. Retrieved December 12, 2005, from https://www.cacei.org/

Consejo Nacional de Rectores. Oficina de Planificación de la Educación Superior [CONARE]. (2002). *Indicadores para medir la incorporación de la dimensión ambiental en los currícula de las Universidades Estatales de Costa Rica.* CONARE, OPES. https://repositorio.conare.ac.cr/handle/20.500.12337/2475

Consejo Nacional de Rectores. Oficina de Planificación de la Educación Superior [CONARE]. (2009). *Fluxograma para la creación de nuevas carreras o la modificación de carreras ya existentes.* CONARE. https://repositorio.conare.ac.cr/handle/20.500.12337/2366

Fëdorov, A. (2004). *¿Cómo programar un curso? Guía para evaluar y autoevaluar el programa del curso.* Editorial Tecnológica de Costa Rica.

Fëdorov, A., & Guzmán-Arias, I. (2007). Desarrollo e innovación en el curriculum del programa de la licenciatura en ingeniería agrícola del Instituto Tecnológico de Costa Rica. *Revista Eduación, 31*(1), 77–89.

Fëdorov, A., Hernández, F., Lira, R., Masís, C., & Sánchez, R. (2003). *Diagnóstico institucional sobre el estado de la gestión de los procesos de autoevaluación con propósitos de mejoramiento continuo y de acreditación, de las carreras del ITCR, a partir de 1998.* Proyecto CAMINA 2003, ITCR, Centro de Desarrollo Académico, material inédito disponible en el archivo del centro.

Food and Agriculture Organization [FAO]. (1996). *Food for All.* FAO. https://www.fao.org/3/x0262e/x0262e00.htm#TopOfPage

Food and Agriculture Organization [FAO], & Intergovernmental Technical Panel on Soils [ITPS]. (2015). *Status of the World's Soil Resources (SWSR) – Main Report.* FAO. https://www.fao.org/3/i5199e/i5199e.pdf

Gómez-Calderón, N. (2016). *Fundamentos de maquinaria agrícola.* Editorial Universidad Estatal a Distancia.

Gomez-Calderón, N., & Solorzano-Quintana, M. (2017). Agricultura precisa y exacta en tiempo, lugar y cantidad de insumos utilizados. *InvestigaTEC, 29*, 20–21. https://revistas.tec.ac.cr/index.php/investiga_tec/article/view/3144

González, J., & Wagenaar, R. (2003). Tuning educational structures in Europe. Informe final. Fase 1. *Estudios Sobre Educación, 5*, 239. https://doi.org/10.15581/004.5.26536

Instituto Tecnológico de Costa Rica [ITCR]. (2000). *Manual para el diseño de planes de estudio y actualizaciones curriculares.* ITCR, Vicerrectoría de Docencia and Centro de Desarrollo Académico (CEDA).

Instituto Tecnológico de Costa Rica [ITCR]. (2003). *Tercer Congreso Institucional: Modelo académico del ITCR.* ITCR, Taller de publicaciones.

Instituto Tecnológico de Costa Rica [ITCR]. (2005). *Estatuto Orgánico del Instituto Tecnológico de Costa Rica.* Editorial Tecnológica de Costa Rica.

Instituto Tecnológico de Costa Rica [ITCR], & Escuela de Ingeniería Agrícola [IA]. (2006). *Rediseño de la carrera de Licenciatura en Ingeniería Agrícola*, Costa Rica.

International Commission of Agricultural and Biosystems Engineering [CIGR]. (n.d.). *Who we are.* International Commission of Agricultural and Biosystems Engineering [CIGR]. https://www.cigr.org/

Japan Accreditation Board for Engineering Education [JABEE]. (2002). *Japan Accreditation Board for Engineering Education, JABEE.* Japan Accreditation Board for Engineering Education [JABEE]. https://jabee.org/en/

Lechuga, C. (2002). *Evaluación y Acreditación de la Enseñanza de la Ingeniería en Centroamérica: Estudio de Diagnóstico de la Demanda y Oferta Académicas en los Mercados Laborales en el Sector de ingeniería.* SP-CSUCA-REDICA.

Liu, X., & Li, X. (2013). Application of information technology for Precision agriculture. ASABE Annual International Meeting, 7004:1–7004:6. https://doi.org/10.13031/aim.20131620239

Márquez, A. D., & Marquima, M. (n.d.). *Evaluación, acreditación, reconocimiento de títulos, habilitación: enfoque comparado.* CONEAU. https://eco.mdp.edu.ar/cendocu/repositorio/00214.pdf

Molin, J. P. (2017). *Curso: Agricultura de precisión, enfoque en la ingeniería.* Escuela de Ingeniería Agrícola, ITCR.

Organización de las Naciones Unidas para la Educación la Ciencia y la Cultura [UNESCO]. (1998). Declaración mundial sobre la educación superior en el siglo XXI: visión y acción. World Conference on Higher Education in the Twenty-First Century: Vision and Action. https://unesdoc.unesco.org/ark:/48223/pf0000113878_spa

Organización Internacional de Trabajo [OIT]. (n.d.). *Competencia laboral.* Organización Internacional de Trabajo [OIT]. Retrieved October 25, 2005, from https://www.oitcinterfor.org/

Pereyra, A. (2000). *Integración regional y reformulación curricular: Una mirada a la enseñanza de las ciencias sociales en el MERCOSUR.* ANEP, Consejo de educación secundaria.

Posada, R. (n.d.). *Formación superior basada en competencias, interdisciplinariedad y trabajo autónomo del estudiante.* Facultad de Educación, Universidad del Atlántico.

Pulido, H. (n.d.). Flexibilidad Curricular. *Seminario Fortalecimiento de La Capacidad Académica de Las Instituciones de Educación Superior.*

Quesada Solano, M. E., Cedeño Suárez, M. A., & Zamora Calvo, J. M. (2001). *El diseño curricular en los planes de estudio: aspectos teóricos y guía metodológica.* EUNA.

Revista Summa. (2019, May). Costa Rica: Sector agropecuario se encuentra en desaceleración económica. *Revista Summa.* https://revistasumma.com/costa-rica-sector-agropecuario-se-encuentra-en-desaceleracion-economica/

Secretaría Ejecutiva de Planificación Sectorial Agropecuaria [SEPSA]. (2019). *Informe de gestión del Sector Agropecuario, Pesquero y Rural: mayo 2018 – abril 2019.* SEPSA. http://www.sepsa.go.cr/docs/2019-006-Informe_Gestion_SectorAgro_2018-2019.pdf

Sistema Nacional de Acreditación de la Educación Superior [SINAES]. (2000). *Manual de acreditación: Convocatoria del año 2000.* SINAES - CONARE.

Soto Perdomo, R. (1993). Propuesta para un modelo curricular flexible. *Revista de La Educación Superior, 85*, 103–108. http://publicaciones.anuies.mx/pdfs/revista/Revista85_S2A4ES.pdf

Tünnermann, C. (n.d.). Una Nueva Política para la Educación Superior en América Latina y Caribe. *IV Encuentro de Estudios Prospectivos Región Andina: Sociedad, Educación y Desarrollo.*

Universidad de Costa Rica [UCR], & Centro de Evaluación Académica [CEA]. (1998). *Ingeniería Agrícola: Plan de estudios - Bachillerato y Licenciatura.* Universidad de Costa Rica [UCR], Centro de Evaluación Académica [CEA]. https://www.cea.ucr.ac.cr/index.php/es/

Villagra-Mendoza, K. (2018). Memorias del XIII Congreso Latinoamericano y del Caribe de Ingeniería Agrícola. *XIII CONGRESO LATINOAMERICANO Y DEL CARIBE DE INGENIERÍA AGRÍCOLA.* https://doi.org/10.18845/mct.2018.459

Villagra-Mendoza, K., Gómez-Calderón, N., & Solórzano-Quintana, M. (2018). La ingeniería agrícola une a Latinoamérica y el Caribe. *InvestigaTEC, 33*, 22–23. https://revistas.tec.ac.cr/index.php/investiga_tec/article/view/3876

Zazueta Ranahan, F. S. (2017). The International Academy of Agricultural and Biosystems Engineering: A new instrument for recognizing the profession. *Resource Magazine, 24*(2), 31. https://elibrary.asabe.org/abstract.asp?aid=47677

17

Agricultural Engineering Education at Sultan Qaboos University, Sultanate of Oman – Historical Evolution, Curriculum Design and Reform, Students' Perceptions, and Future Prospects

Umezuruike Linus Opara and Yaseen Al-Mulla

Introduction

Even though oil and gas play a major role in the current national economy of Oman, a sizeable part of those who live in the villages still rely on the agriculture and fishing industries for a living. The Omani culture and food system significantly relies on agriculture and other land-based businesses. In the Sultanate of Oman, agricultural engineering (AE) education plays a significant part in the country's overall growth. When Sultan Qaboos University was founded over three decades ago, one of the first undergraduate degree programmes offered in the College of Agriculture and Marine Sciences was Agricultural Mechanization (later renamed Agricultural Engineering).

Due to internal and external pressures influencing universities and agriculture-related programmes, the field of agricultural engineering has seen significant changes in focus during the past ten years (Opara et al., 2004a, b; Opara et al., 2006a, b). The College of Agriculture and Marine Sciences at Sultan Qaboos University (SQU), the main institution in the Sultanate of Oman, houses the AE academic programme. Developing adequate and creative solutions for the sustainable exploitation and management of agricultural and marine bioresources, and to offset the effects of human interventions on the environment, is a challenge for AE in Oman and similar arid nations. The government's 2020 Vision and the 5-year national development plan both stress the need to boost the contributions of the agricultural and marine sectors in continued attempts to diversify the economy, despite the fact that the oil and gas industries dominate.

As the top academic institution in the nation, Sultan Qaboos University is expected to take the lead in bringing the government's development strategy for the agricultural and marine industries to fruition. Expanding perspectives on the use of contemporary engineering science and technology to promote agriculture through an AE curriculum (Opara, 2004) was one of the goals for engineering education set at the university's founding in 1986. The first alumni of this programme, which began in 1986, graduated in 1990. The idea, curriculum, and nomenclature of the AE programme have all experienced considerable modifications since it first started.

In the Sultanate of Oman, AE programmes are offered only by Sultan Qaboos University (SQU), located in Muscat. When the College of Agriculture was founded in the mid-1980s, the programme was one of the pioneers. Ten years later, Agricultural Mechanization was replaced with Bioresource and Agricultural Engineering (BAE) as the department's name (and focus). Issues facing both the College and the Bioresource and Agricultural Engineering programme include the extremely low student interest in agriculture and associated fields of study. Although students are admitted into the College, getting them to stay has proven to be quite difficult. Understanding students' attitudes towards the college and the major, as well as their perceptions of the image/profile of the AE profession, the curriculum content, and their future employment, is crucial to developing an effective intervention strategy to increase the appeal of AE among students.

This chapter's goals are to outline the development of the AE undergraduate and postgraduate programmes at Sultan Qaboos University and to suggest future paths for its improvement in response to the agricultural, food and environmental challenges facing the Sultanate and the world in the twenty-first century.

Development of the Agricultural Mechanization BSc Degree Programme

When Sultan Qaboos University's College of Agriculture was established, the Department of Agricultural Mechanization (and its academic programme) was one of the pioneering departments. A 4-year Bachelor of Science (BSc) Agricultural Mechanization programme was the academic major offered, and it concentrated on the technical hands-on aspects of farm power, machinery, and irrigation systems as closely connected to tillage, crop planting and protection, harvesting and processing, irrigation, and management of water conservation structures. Courses in applied mechanics, engineering science, surveying, electrical systems, the design, usage, and administration of machinery and power units, as well as computer applications, were offered to students. Fieldwork, lab work, workshop activities, and lectures

DOI: 10.1201/9780429150111-19

were all incorporated into the educational programme. A typical curriculum for the agricultural mechanisation programme taught between 1985 and 1995 is shown in Table 17.1 (DABE, 2006). Although the overall number of credit hours needed was close to that of regular engineering programmes, the total number of credits for the Mathematics & Basic Sciences courses was around 50% less than that needed to be accredited as a professional engineering programme. As a result, Sultan Qaboos University's agricultural mechanisation programme was specifically created to provide engineering technologists with the technical skill necessary to collaborate with professional engineers.

The goal of the agricultural mechanisation programme was to train graduate experts who would offer technical and applied engineering services to resolve operational issues associated with mechanised agriculture and the exploitation of fishery resources. Graduates were required to hone their skills in the production and processing of field crops, the choice, use, maintenance, and repair of equipment, as well as the extension and consulting services to farmers regarding mechanisation. The first group of seven students to complete the Agricultural Mechanization degree programme – out of the original ten enrolled – graduated in 1990.

TABLE 17.1

The 1986 Degree Plan for BSc in Agricultural Mechanization, College of Agriculture, Sultan Qaboos University (DABE, 2006); Cr = Credits

Course Code	Course Description	Cr	Course Code	Course Description	Cr
University requirements			Departmental requirements		
ARAB 1001	Arabic	3	AGRN 2502	Introduction to Agronomy	3
HIST 1010	Oman & Islamic Civilization	2	AMEC 1201	Introduction to Agricultural	
LANC ****	As defined by College	6		Mechanization	3
SOCI 3320	Contemporary Omani Society	1	AMEC 2301	Mechanics I	3
Electives Requirement			AMEC 2302	Workshop Practice I	3
********	The student chooses 6 credits from any University electives	6	AMEC 2303	Materials	3
			AMEC 2304	Electrical Systems	3
			AMEC 2306	Land Surveying	3
	Total credits defined	18	AMEC 2311	Mechanics II	3
			AMEC 2312	Workshop Practice II	3
College requirements			AMEC 2313	Engineering Science	3
BIOL 1009	Introductory Biology	3	AMEC 3401	Mechanics III	3
BOTN 2001	Introductory Plant Biology	3	AMEC 3402	Power Units	3
CHEM 1009	Introduction to Chemistry	3	AMEC 3404	Machinery Applications	3
CHEM 2201	General Chemistry I	4	AMEC 4403	Machine Design	3
COMP 2001	Microcomputers in Agriculture	3	AMEC 4405	Automatic Controls	3
LANC 2040	English for Agriculture I	3	AMEC 4406	Farm Machinery Management	3
LANC 2041	English for Agriculture II	3	AMEC 4412	Soil-Machine Relations in Tillage	
MACS 1010	Foundation Mathematics	6		& Traction	3
PHYS 1009	General Physics	3	AMEC 4413	Hydraulic & Pneumatic Systems	3
PHYS 2207	Physics for Agriculture	4	AMEC 4414	Special Problems	3
			ECON 4500	Agricultural Extension Methods &	
	Total credits defined	35		Techniques	3
			MATH 2296	Mathematics for Agriculture	3
Elective requirement - choose 3 credits from			SWAT 2201	Introduction to Soil and Water	3
ECON 3001	Introduction to Agricultural		SWAT 3301	Soil Physics	3
	Economics	3	SWAT 3402	Irrigation Principles	3
ECON 4001	Farm Management	3	SWAT 4406	Irrigation Systems	3
ERSC 2901	Earth Sciences for Agriculture	3			
HORT 2515	Introduction to Horticulture	3		Total credits defined	75
PLNT 3005	Statistical Methods &				
	Experimental Design	3			
SWAT 3303	Hydrology	3	Total credits defined for the degree		131
SWAT 3305	Agroclimatology	3	Total credits required for the degree		130

Evolution of the Bioresource and Agricultural Engineering BSc Degree Programme

The Philosophy

To replace the Agricultural Mechanization programme with a new one that is more focused on engineering, the programme underwent significant internal assessments. Initial plans were for a new major in bioresource engineering to take the place of the mechanisation programme (Opara, 2004), but subsequently the nomenclature bioresource and agricultural engineering (BAE) was chosen. The department and/or its academic major are still frequently referred to by the phrase "Bioresource" by many academics and students. This may help to explain the discrepancy in official documents on the precise name of the major. For example, the departmental record in the 1997 College Annual Report refers to the new major chosen by students as "Bioresource and Agricultural Engineering", however, the Annual Reports for 1998 and 1999 just mentioned "Bioresource Engineering" (DABE, 2006).

The new programme name was thought to be a condensed version of biological resources engineering, a word that was quickly becoming one of the titles adopted (wholly or in part) by certain AE schools at North American colleges (Opara, 2004; Opara & Cuello, 2006a, b). The term "bioresource engineering" in the Omani context was also presumed to imply engineering related to tasks like soil and water conservation, forestry and range production and management, fisheries habitat enhancement, wild lands management, and other tasks geared towards the preservation of biological resources like food and agricultural products (Opara, 2004). The Bioresource Engineering programme was intended to produce professional engineers.

Name Change: Incorporating the Biological Dimension

The University Council approved the creation of a new department with the name Bioresource and Agricultural Engineering (BAE) in 1996, taking the place of the mechanisation programme. The goal of the newly established department was to produce professional engineering graduates in the fields of bioresource and agricultural engineering, conduct research with a focus on local issues, and offer services that improve quality of life. The BAE programme was intended to last four years (excluding the period of one or more semesters of intensive English language courses at the university). The bulk of key engineering courses, including drawing and computer graphics, thermodynamics, statics, dynamics, and fluid mechanics, are completed by participants in the programme alongside their peers at the College of Engineering (Table 17.2).

The goal of the programme is to produce graduates who can use their scientific knowledge, alongside engineering science, and practical problem-solving skills in agriculture, the environment, and food industry. The guiding principle behind the restructuring was to create and put into effect an educational programme that is thorough on the engineering aspects of production agriculture, but also broad enough to equip students to find career paths in several other sectors that are not primarily connected to production agriculture. The strategy demanded a curriculum that combines lectures with hands-on laboratory exercises, practical design experience, mechanical workshop practice, and field manufacturing activities. Power and machinery engineering, soil and water engineering, and food and process engineering are the three primary areas that the programme's array of departmental electives allows graduates to finish their studies.

Pedagogy of Teaching and Learning

For our graduates to be "trouble shooters" and "system integrators" who can function as members of multi-disciplinary technology teams, the teaching and learning methodology offered throughout all BAE courses focuses strong emphasis on exposing students to sophisticated problem-solving methodologies. In BAE classes, class numbers seldom surpass 20 students, allowing teachers to connect directly with students and offer them individualised learning experiences. Students typically complete the 120-credit BAE undergraduate programme over the course of eight semesters. Table 17.2 presents the degree programme that was authorised and put into effect in 1996 (Opara et al., 2006a). The courses are broken down into three primary categories: departmental requirements (64 credits), college prerequisites (38 credits), and university prerequisites (18 credits).

To ascertain their degree of placement in the Intensive English Language programme (IELP), students who are admitted into the university undertake an English language test. This programme, which is subdivided into six stages, aims to improve the students' level of "general" English to better prepare them for their academic courses. Students participating in the IELP focus on English and are not permitted to take any other courses concurrently. Individuals with outstanding placement test results may enter the Credit English Language Programme right away (CELP). After two semesters, most students leave the intense programme. Students must complete two additional English language courses after finishing the IELP, each of which is worth three credits and has six hours of tutoring each week. Students who are enrolled in credit English courses typically pursue additional university-required courses and/or fundamental scientific courses concurrently. Together with six credits of electives of the student's choosing, the institution requires that all students take courses on Islamic civilisation, Omani society, and related social and historical issues. Students within the College of Agriculture and Marine Sciences must choose an area of specialisation after completing two fundamental science courses. Only students with excellent results in math, physics, and chemistry courses are permitted to register for the Bioresource and Agricultural Engineering major.

Final-Year Design Project

The 6-credit engineering design project course in the final year is a crucial part of the undergraduate programme. The purpose of the engineering design project is to give the students the opportunity to hone their integrative skills in innovation, problem-solving, teamwork, communication, and use of

TABLE 17.2

The 1996 degree plan for BSc in Bioresource and Agricultural Engineering

Course Code	Course Description	Cr
University requirements		
ARAB 1001	Arabic	3
HIST 1010	Oman & Islamic Civilization	2
LANC 2040	English for Agriculture I	3
LANC 2041	English for Agriculture II	3
SOCI 3320	Contemporary Omani Society	1
Electives Requirement		
********	General Electives	6
	Total credits defined	18
College Requirement		
COMP 2001	Microcomputers in Agriculture	3
ECON 4500	Aric Extension Methods & Techniques	3
LANC 0990	Intensive English I	0
MATH 1192	Mathematics for Agriculture I	4
MATH 2193	Mathematics for Agriculture II	3
Elective requirement - choose 13 credits from		
BIOL 1091	Plant Biology for Agriculture	3
BIOL 1092	Animal Biology for Agriculture	3
CHEM 1091	General Chemistry for Agriculture	4
CHEM 2091	Organic Chemistry for Agriculture	3
PHYS 1091	Intro to Physics for Agriculture	4
Elective requirement - choose 13 credits from		
ANSC 2001	Principles of Animal Science	3
BIOR 2001	Introduction to Bioresource Engineering	3
BIOR 2201	Introduction to Agricultural Machinery	3
BIOT 2201	Introduction to Agricultural Biotechnology	3
CROP 2510	Introduction to Crop Production	3
ECON 2003	Introduction to Agricultural Economics	3
FISH 2001	Introduction to Marine Science & Fisheries	3
FOOD 2071	Principles of Food Science	3
NUTR 2301	Introduction to Human Diet	3
SOCI 3821	Principles of Rural Sociology	3
SOIL 2201	Introduction to Soil & Water	3
	Total credits defined	38

Course Code	Course Description	Cr
Departmental requirements		
BIOR 4301	Design Project	6
BIOR 4800	Bioresource Engineering Internship	3
MATH 2108	Calculus II	3
PHYS 2001	General Physics II	3
MECH 2101	Basic Mechanics I (Statics)	4
MECH 3152	Thermodynamics I	3
MECH 3202	Dynamics	3
MECH 5540	Unit Operations	3
Department core electives-choose 36 credits from the following courses or any other approved by HoD		
BIOR 3003	Soil Water Concepts in Irrigation Engineering	3
BIOR 3004	Agricultural Watershed Hydrology	3
BIOR 3005	Land Surveying	3
BIOR 3101	Properties of Foods & Agricultural Materials	3
BIOR 3102	Elements of Food Engineering	3
BIOR 3201	Power Units	3
BIOR 4003	Soil Engineering for Agriculture	3
BIOR 4004	Design of Irrigation Systems	3
BIOR 4005	Specialized Applications in Soil & Water Engineering	3
BIOR 4900	Bioresource Eng Special Problems	3
CIVL 2000	Remote Sensing & Geographic Information Systems	3
CIVL 4046	Fluid Mechanics	3
ECON 3010	Production Economics & Business Management	3
FOOD 3076	Food Processing I	3
FOOD 4076	Food Processing II	3
FOOD 4077	Food Product Development	3
MATH 3304	Ordinary Differential Equations	3
MECH 3027	Fluid Mechanics	3
MECH 3146	Engineering Drawing & Computer Graphics	3
MECH 4160	Numerical Methods for Engineers	3
MECH 4171	Control Engineering	3
SOIL 3311	Soil Physics	3
STAT 3001	Statistical Methods in Agriculture	3
WATR 3402	Irrigation Principles	3
Total credits defined		64
Total Credits defined for the Degree		120

information and computer technology. During two semesters, students work in groups of 2–4 students on a real-world engineering challenge in one of the department's key focus areas (power and machinery engineering, food and process engineering, and soil and water engineering), under the guidance of faculty. The way the design project course is administered has undergone several adjustments.

The Fall semester was originally devoted to a teaching component on engineering design projects and preparation of design proposals, which were then put into practice and evaluated during the Spring semester. The credit hours have recently been formally divided into 3 credits per semester. The coordinating faculty gave introductory lectures to final-year students on engineering design and teamwork at the start of

TABLE 17.3

Comparison of the curriculum in Bioresource and Agricultural Engineering (2000 Degree Plan) with ABET

	ABET Requirements		BAE Curriculum	
Requirements Subject area (Component)	**Credits**	**(Year)**	**Credits**	**(Year)**
Mathematics and basic sciences	32	(1.0)	16	(0.5)
Basic engineering	48	(1.5)	50	(1.6)
General education	16-48	(0.5-1.5)	57	(1.8)
Intensive English	-	-	0	(0.5-1.0)
Total	96-128	(3-4)	123	(4.4-4.9)*

* Include at least one summer session for Internship.

the fall semester (these were typically given over the course of 3 weeks). Following this, the teams of students were assigned to different faculty members. Each team of students collaborates over the course of the final 12 weeks of the fall semester to create and present a proposal for a design project. At the conclusion of each semester, a comprehensive course written report and an oral presentation are required.

Beginning with the academic year 2003–2004, the department returned to the old concept, which called for an engineering design project to be taught in one semester and to be implemented and tested in a subsequent semester. Traditionally, only faculty evaluated courses, but in more recent years, both faculty and students have taken part in assessing oral presentations and each faculty member has given a grade for the report that his or her project team produced. Each design project team receives the required instruction and materials to support the implementation of their design project during the Spring Semester based on the accepted design proposal. Students initially registered in one design project course (BIOR 4301; 6 credits) for the duration of the academic year; however, the course was later divided into two (BIOR 4300 and BIOR 4301, for Fall and Spring, respectively). Normally, each design project team receives the same final mark or grade at the end of each semester, but in some cases, students have received different grades after faculty provided evidence to the Department Board that a particular student's contribution was either spectacular or fell short of the minimum expectation. To provide them hands-on experience in a relevant corporate sector or government organisations, final-year students are also obliged to complete an internship course (3 credits).

Benchmarking Against ABET Accreditation

Early BAE degree programmes included a total of credits that met the ABET standard, but there were not enough credits in math and fundamental science. A comparison of the degree programmes for the 1999 and 2000 cohorts with the typical ABET requirement for professional engineering education serves as one illustration (Table 17.3). It should be mentioned that numerous US universities offer 135-credit programmes in agricultural (and related) engineering at the same time. Even though in 1996, the programme had an engineering-based curriculum, the departmental programme shows through ongoing review and evaluation that fundamental engineering and scientific components needed to be reinforced. A new curriculum was recommended in 1999 after a thorough examination of the degree plan and the courses offered. The department committed to adapt and enhance the curriculum by adding additional mathematics and engineering-based courses to fulfil ABET standards after holding a workshop to meet the accreditation criteria for ABET (DABE, 2001). The updated BAE programme (135 credits) was created and put into practice (Table 17.3). It includes sufficient mathematics and fundamental sciences (35 credits), a basic engineering course (51 credits), and general courses (49 credits).

The new BAE curriculum (Figure 17.1) and a generalised framework for comparable programme approved by ABET (Figure 17.2) follow the same arrangement of courses, with the exception that the BAE curriculum had fewer (6%) mathematics courses. The physical sciences and engineering specialty's additional courses help to explain this disparity. The intention to pursue ABET accreditation of the AE programme at Sultan Qaboos University remained. However, a complete application for ABET accreditation could not be developed immediately until the BAE programme curriculum was further reviewed and improved later following a strategic review. It was considered that a revised curriculum that is more closely aligned with ABET standard and thus associated with the College of Engineering could give higher chances for success in terms of prospective students and programme accreditation given

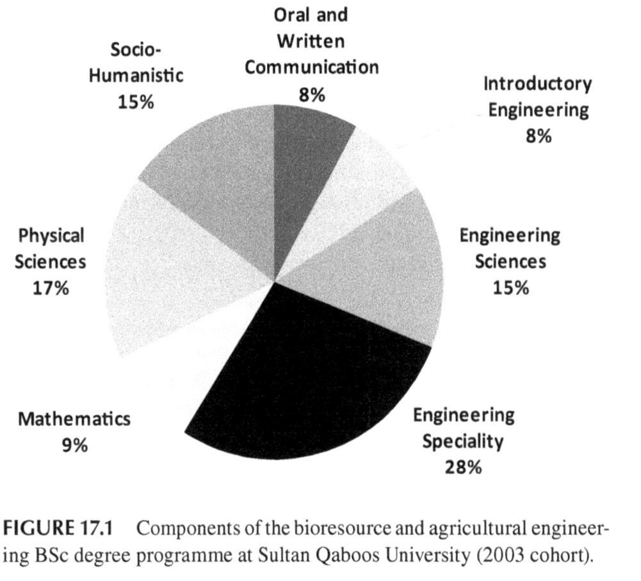

FIGURE 17.1 Components of the bioresource and agricultural engineering BSc degree programme at Sultan Qaboos University (2003 cohort).

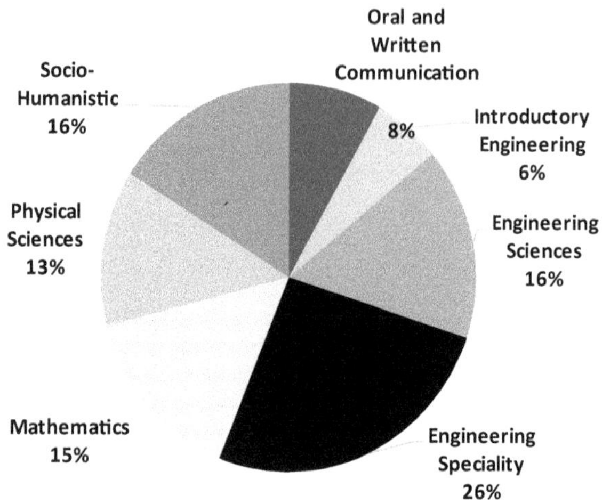

FIGURE 17.2 Components of an ABET-certified engineering degree programme (Eide et al., 2002).

the extremely low attractiveness of courses in the College of Agriculture and Marine Sciences among Omani students.

The Emergence of the Agricultural Engineering BSc Degree Programme (2004)

A proposal to integrate the departments of soil and water sciences and bioresource and agricultural engineering was approved by the university council in the summer of 2004. The merger thereafter became official on September 1, 2004, thanks to a Qarar (Order/Decree) issued by the university president. This involves renaming the department "Soils, Water, and Agricultural Engineering" and changing the name of the major from "Bioresource and Agricultural Engineering" to "Agricultural Engineering". A group of external assessors first suggested the merger of the College's small departments, which have an average faculty size of six, in 2001. Also, the merger was explored to combine similar fields, eliminate overlap, and lessen administrative and service obligations within the College. The three majors previously provided by the two previous departments – Water Technology, Soil and Water Sciences, and Agricultural Engineering – are maintained under the new department. There was some concern, especially among AE faculty and staff, that the programme may gradually lose its visibility and attractiveness unless coordinated efforts are made to market and promote it among students and industry, given the higher number of faculty and related teaching and research activities in the soil and water areas.

Student Perceptions and Attitudes Towards Agricultural Engineering Education And Discipline

Poor public perception (Opara, 2003a) and a rapidly-dwindling traditional rural base for student recruitment (Opara, 2003b) are issues AE education is facing. These challenges have often led to low enrolment of students relative to similar degree programmes in Engineering or Agriculture. In extreme cases, this has resulted in the closure of departments of AE and the academic programme, such as Massey University in New Zealand and Stellenbosch University in South Africa. In the 21st century, when "smart" industries and "knowledge" economies are anticipated to be the norm (Opara, 2001a,b; Opara, 2004), these issues are predicted to get worse'

As key stakeholders, the perspectives, and attitudes of students towards AE education, including the names of the discipline, department, and academic programme, curriculum content and employment prospects, are important. Studies by Opara et al. (2006a) explored attitudes towards academic programme selection among AE students at Sultan Qaboos University. In a separate study, students discussed how they felt about the public image and the curriculum content (Opara et al., 2006b). Preferences for a career, a desired discipline name, and a degree title. These results demonstrated that more than 78% of students questioned did not initially favour the College of Agriculture and Marine Sciences (which offered the Agricultural Engineering major). Yet, out of the 10 academic programme majors available, 72% of applicants who were accepted to study for a degree in the College chose to major in Bioresource and Agricultural Engineering as their first option (Opara et al., 2006a). However, most students who registered in the AE programme did not like the fundamental sciences and engineering courses, including the management courses (Opara et al., 2006a).

A large majority of students claimed that the public awareness of AE and grasp of its place in society were very poor. Agricultural engineering's negative reputation was mostly attributable to its relationship with agriculture (74%) rather than engineering (26%). The inclusion of the word "Engineering" and/or "Technology" was preferred by most students over "Agriculture" and/or "Science", The students' poor interest in names that were associated with "Biology" was linked to the great regard that society has for engineers as well as the students' strong associations with "machines" and math/physics rather than engineering and biology, respectively. Students also recommended doable actions to improve the reputation, attractiveness, and visibility of AE among students through focused advertising campaigns and neighbourhood outreach initiatives.

Students also suggested a number of doable strategies to improve the public's perception of, and profile for, AE, such as (i) targeted advertising, publicity, and promotion of related events and accomplishments, (ii) extension and community outreach programme, (iii) connecting students with industry and potential employers, (iv) expanding employment opportunities for agricultural engineers outside of the agricultural sector, and (v) changing the names of College and degree programmes to make them more appealing to the students and their families, (vi). Increase the amount of 'practical' fieldwork and laboratory by revising the curriculum.

Assessing the Impacts of Curriculum Reform and Name Changes

The low number of graduates generated during the mid- to late 1990s indicates that the number of students majoring in the agricultural mechanisation programme seldom surpassed five

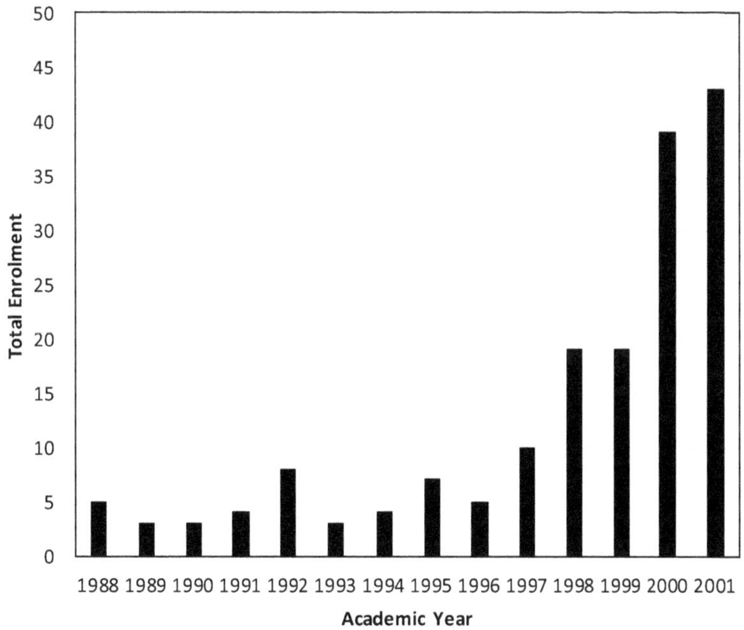

FIGURE 17.3 Total enrolment in the department (Years 2–4 only). Note that the new curriculum in BSc Bioresource and Agricultural Engineering was introduced in 1996.

(Figure 17.3). The number of students enrolling in the new major steadily increased with the inception of the new Bioresource and Agricultural Engineering programme in 1996. The inclusion of the term "engineering" in the department name and programme title was considered as a key component in recruiting more students, according to studies by Opara (2003a, b) and Opara et al. (2004a, b) on the views and preferences of senior undergraduate students in the department. The trend may be characterised by students' preferences among the 10 majors available at the institution in 1998. The general choice for Bioresource and Agricultural Engineering among the 143 students who completed the standard college questionnaire to express their desire for a Major was as follows: First choice scored 16.78%, second choice 13.29%, and third choice 14.68%. To thoroughly analyse the long-term effects of the curriculum restructuring on student retention and employment, more thorough data collection, monitoring, and assessment are needed.

Transition to the Agricultural Engineering Degree Programmes, BSc and BEng

After merging the Department of Bioresource and Agricultural Engineering and the Department of Soil and Water Sciences, forming a new Department of Soils, Water and Agricultural Engineering, which took place on 1 September 2004, the BSc degree in "Bioresource and Agricultural Engineering" (BSc BAE) was changed to a BSc in "Agricultural Engineering" (BSc AE). The main reason for removing "Bioresource" was due to the translation of this word into the Arabic language, which reflects the biology aspect of the name, caused confusion over what the real nature of this programme was, especially in relation to engineering.

The structure of the BSc BAE degree plan for the 2003 cohort, before the merging process, consisted of 18 credits for University Requirements, 43 credits for College Requirements, and 59 credits for Departmental Requirements with a total of 120 credits as detailed in Table 17.4. However, after the merging process the degree plan of BSc AE has been restructured to be with a total of 129 credits (Table 17.5). While the number of College Electives reduced by 3 credits, the major requirements increased by 6 credits after adding "Differential Equations for Engineers" and "Basic Mechanics". Major Electives increased by 6 credits as well after adding two courses to the list.

Despite the removal of "Bioresource" from the degree, the term 'Engineering' in the degree title did appeal to many of the students admitted to the newly formed department. This is reflected in the number of enrolments to the AE programme in 2005, a year after the merging took place (Figure 17.4).

Nevertheless, the number of complaints from the programme's graduates, after the merging process, increased. There were based mainly on what they were facing in terms of not being accepted as engineers by stakeholders and potential employers. As per the common perception in Arabic countries, every student who graduates from the college of Agriculture is designated as an "Agricultural Engineer", including those from non-academically engineering-based programmes. To mitigate this problem, the Department and College came up with a solution of providing an official letter, signed by the AE programme coordinator and the Head of the Department for each graduating AE student certifying that the AE programme at Sultan Qaboos University is an Engineering programme. Although this step has contributed to some extent in finding more and related jobs for the programme's graduates, it did not convince the newly admitted students, as is shown in Figure 17.4, as student enrolment decreased from 20 students in 2005 to 8 students in 2007.

In 2006, the University administration formed an ad-hoc committee to explore forming a joint AE programme between the College of Agricultural and Marine Sciences, the college

TABLE 17.4

Structure of the BSc Bioresource and Agricultural Engineering degree plan for the 2003 cohort (120 credits) prior to the merger of the Department of Bioresource and Agricultural Engineering and the Department of Soil and Water Science, forming a new Department of Soils, Water and Agricultural Engineering which took place on 1 September 2004.

Course Code	Description	Cr	Course Code	Description	Cr
University Requirements			Departmental Requirements		
ARAB 1001	Arabic	3	BIOR3003	Soil & Water Concepts in	
HIST 1010	Oman & Islamic Civilization	2		Irrigation Engineering	3
LANC 0970	Intensive English Language Course	0	BIOR3102	Elements of Food Engineering	3
LANC 2040	English for Agriculture I	3	BIOR3202	Principles of Machinery	3
LANC 2041	English for Agriculture II	3	BIOR3203	Postharvest Technology	3
SOCI 3320	Modern Omani Society	1	BIOR3306	Computer Programming	3
XXXX 0000	University Elective	3	BIOR4300	Design Project I	3
XXXX 0000	University Elective	3	ECON 4500	Agricultural Extension	
	Subtotal	18		Methods & Techniques	3
			BIOR4800	Bioresource Internship	3
College Requirements			MATH2107	Calculus I	4
BIOL 2101	General Biology I	4	MATH2108	Calculus II	3
CHEM 2101	General Chemistry I	4	MEIE3146	Drawing & Computer Graphics	3
MATH 1106	Pre-calculus Math	4	MEIE3152	Thermodynamics	3
PHYS 2101	General Physics I	4	MEIE3201	Statics	3
CAMS 2000	Introduction to Agricultural &		PHYS2102	General Physics II	4
	Marine Sciences	1	SWAT4310	Fundamentals of Fluid	
COMP 2001	Computer Applications I	3		Mechanics	3
CAMSxxxx	Computer Applications II	3	Elective requirement - choose at least 12 credits		
CAMS 4001	Management & Business Skills	3	BIOR2300	Workshop Practice	3
CAMS 4002	Seminar & Presentation Skills	2	BIOR 3005	Land Surveying	3
STAT 3001	Statistics in Agriculture	3	BIOR3101	Properties of Food &	
ECON 2003	Introduction to Agricultural	3		Agricultural Materials	3
	Economics		BIOR3201	Power Units	3
			BIOR3308	Instrumentation & Control	
Elective requirement - choose at least 9 credits from				Systems	3
ANSC 2001	Principles of Animal Science	3	BIOR4006	Arid Zone Hydrology	3
BIOR 2001	Introduction to Bioresource &		BIOR4007	Applied Irrigation Engineering	3
	Agricultural Engineering	3	BIOR4008	Agricultural & Marine Waste	
BIOR 2201	Introduction to Agricultural			Management	3
	Machinery	3	BIOR4100	Food & Process Engineering	3
BIOT 2201	Introduction to Agricultural		BIOR4103	Heat and Mass Transfer	
	Biotechnology	3		Tillage & Soil-Machine	3
CROP 2510	Introduction to Crop Production	3	BIOR4202	Relationships	3
FISH 2001	Introduction to Marine Science &		BIOR4304	Modelling & Analysis of	
	Fisheries	3		Biophysical Systems	3
FOOD 2071	Principles of Food Science	3	BIOR4900	Bioresource & Agricultural	
NUTR 2301	Introduction to Human Diet	3		Engineering Special Problems	3
SOIL2201	Introduction to Soils & Water	3	CIVL2000	Remote Sensing & GIS	3
MASFxxxx	Marine Environment & Production	3	FOOD3075	Sanitation & Quality Control	3
	Subtotal	43	FOOD4171	Food Packaging	3
			MATH4172	Differential Equations for	
				Engineers	3
			WATR3303	Elements of Hydrology	3
			WATR4315	Water Quality	3
			WATR4400	Groundwater Hydrology	3
				Subtotal	59

that hosts the AE programme, and the College of Engineering, a college that shares the engineering nature of the AE programme. Since the College of Engineering is an ABET-accredited college, the committee invited an ABET consultant to review the AE programme and to recommend the way to cope the programme's study plan with ABET requirements. The outcome of the ad-hoc committee was in favour of establishing the AE programme between the two colleges due its

TABLE 17.5

Structure of the BSc Agricultural Engineering Degree Plan (129 credits) after restructuring of the curriculum and the merger of the Department of Bioresource and Agricultural Engineering and Department of Soil and Water Science, forming a new Department of Soils, Water and Agricultural Engineering.

Course Code	Description	Cr	Course Code	Description	Cr
University Requirements			Departmental Requirements		
ARAB 1001	Arabic	3	SWAE 3201	Power Units	3
HIST 1010	Oman & Islamic Civilization	2	SWAE 3203	Postharvest Technology	3
LANC 0970	Intensive English Language Course	0	SWAE 3310	Fundamentals of Fluid Mechanics	3
			SWAE 3306	Computer Programming	3
LANC 2040	English for Agriculture I	3	SWAE 3308	Instrumentation & Control Systems	3
LANC 2041	English for Agriculture II	3			
SOCI 3320	Modern Omani Society	1	SWAE 4302	Irrigation Principles	3
XXXX 0000	University Elective	3	SWAE 4300	Agricultural Engineering Design Project	3
XXXX 0000	University Elective	3			
	Subtotal	18	SWAE 4800	SWAE Internship	3
			FSHN 3102	Elements of Food Engineering	3
CAMS Requirements			AGEC 4500	Agricultural Extension Methods & Techniques	3
BIOL 2101	General Biology I	4			
CAMS 2000	Introduction to Agricultural & Marine Sciences	1	MATH 2107	Calculus I	4
			MATH 2108	Calculus II	3
CAMS 3002	Computer Applications II	3	MATH 4172	Differential Equations for Engineers	3
CAMS 4001	Management & Business Skills	3			
CAMS 4002	Seminar and Presentation Skills	2	MEIE 2129	Basic Mechanics	3
CAMS 2001	Computer Applications I	3	MEIE 3104	Engineering Drawing & Computer Graphics	3
CHEM 2101	General Chemistry I	4			
AGEC 2003	Introduction to Agricultural Economics	3	MEIE 3141	Thermodynamics	3
			PHYS 2102	General Physics II	4
MATH 1106	Pre-calculus Math	4	Elective requirement – choose 18 credits from		
PHYS 2101	General Physics I	4	SWAE 2300	Workshop Practice	3
CROP 3001	Statistics in Agriculture	3	SWAE 3005	Land Surveying	3
			SWAE 3003	GIS for Environmentalists	3
Elective requirement – choose at least 6 credits			SWAE 3303	Elements of Hydrology	3
ANSC 2001	Principles of Animal Science	3	SWAE 3305	Agro-climatology	3
BIOT 2201	Introduction to Agricultural Biotechnology	3	SWAE 4006	Arid Zone Hydrology	3
			SWAE 4202	Tillage & Soil-Machine Relationships	3
CROP 2501	Introduction to Crop Production	3			
FISH 2001	Introduction to Marine Science & Fisheries	3	SWAE 4203	Principles of Machinery	3
			SWAE 4204	Advances in Postharvest Technology	3
FOOD 2071	Principles of Food Science	3	SWAE 4301	Agricultural Engineering Design Project II	3
NUTR 2301	Introduction to Human Diet	3			
SWAE 2001	Introduction to Agricultural Engineering	3	SWAE 4304	Modelling & Analysis of Biophysical Systems	3
SWAE 2002	Introduction to Agricultural Machinery	3	SWAE 4402	Design & Management of Irrigation Systems	3
SWAE 2201	Introduction to Soils and Water	3	SWAE 4400	Groundwater Hydrology	3
	Subtotal	40	SWAE 4900	SWAE Special Problems	3
			FSHN 3075	Sanitation & Quality Control	3
			FSHN 3101	Properties of Food & Agricultural Materials	3
			FSHN 4100	Food & Process Engineering	3
			FSHN 4171	Food Packaging	3
				Subtotal	71

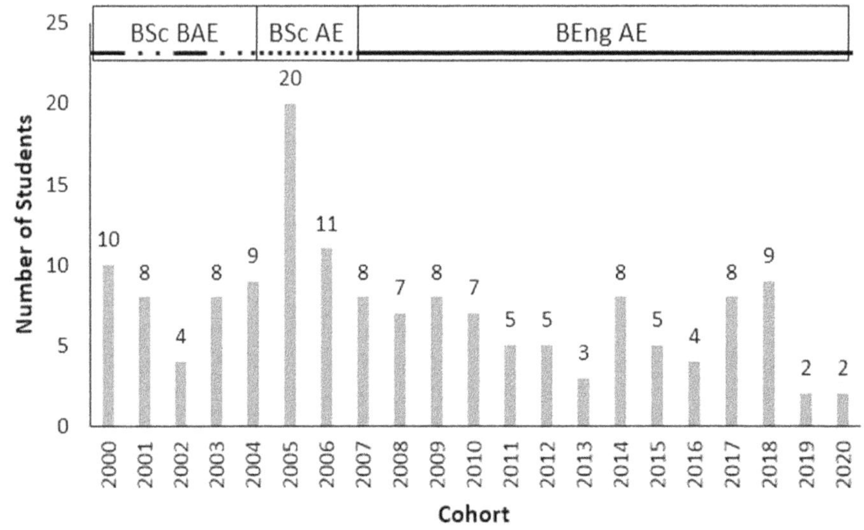

FIGURE 17.4 Student enrolments in the undergraduate Agricultural Engineering degree programme during the three periods of degree name changes.

unique role, as an engineering-based agricultural discipline, in meeting the current national, regional, and world challenges of meeting both food and water securities. In addition to the positive evaluation of the ABET consultant, one of the encouraging factors of establishing the joint programme was that the consultant found that the existing resources of the programme were good enough for such an establishment. Thus, the new joint B.Eng programme in Agricultural Engineering was established in June 2007 and accepted its first batch of students in September 2007.

The B. Eng. in AE degree offered under this joint programme had the following foreseen advantages:

1. It will strengthen the programme by attaching it to an ABET-accredited College
2. It will be the only programme to confer a B.Eng outside the College of Engineering
3. It will increase the reputation of the programme
4. The number of enrolments will be increased as the programme will be open to students of both colleges
5. Graduates will be better qualified with a degree benchmarked with ABET
6. Higher chance of employment

However, the following requirements were needed to be implemented for the joint programme:

1. Restructure the study plan to cope with the ABET requirements.
2. Run a job market exploration through workshops and surveys to explore the potential employers' assessment of the new joint programme.
3. Follow new regulations for admitting new students to the joint programme
4. Build a strategy on how the joint programme is administered by both colleges
5. Form a steering committee from both colleges to oversee the management, performance, and progress of the joint programme

The study plant was restructured by increasing the total number of credits from 129 to 136 credits with an emphasis on: 1) adding Linear Algebra & Multi-variant Calculus courses and Electrical Engineering fundamentals courses; 2) substituting the general physics courses with physics for engineering; and 3) making the internship equivalent to 0 credits and extend it to 10 weeks. Table 17.6 details the latest B.Eng. AE study plan for the cohort of 2021. The B. Eng. AE study plan was benchmarked with the ABET requirements as shown in Figure 17.5. Thus, the number and type of AE courses do form course categories that match all the eight required categories of ABET, with regard to which ABET requires 8% of Oral and Written Skills and 16% of Engineering Science courses. The same percentages also apply to the AE courses. ABET requires 3% General, 15% Mathematics, 13% Physical Sciences, 26% Engineering Specialisation, and 13% Socio Humanistic courses. AE courses in these categories do match the ABET percentages by a difference of +/- 1–2% due to one to two credits' difference in each of these categories. The only category that differs more than 2% is the Introductory Engineering. ABET requires 6% of this category's courses while AE Introductory Engineering courses forms 11% of total courses due to the nature of the programme which requires additional introductory courses in the fields of soils and crop production.

Two workshops which included surveys on the joint programme were conducted, in 2007 and 2013, targeting representatives from different major stakeholders and potential employers in the country from both public and private sectors including the Ministry of Agriculture & Fisheries, Muna Noor Manufacturing & Trading LLC, the Ministry of Civil Service, Occidental Oman Inc., Ministry of Regional Municipality, Environment & Water Resources, Royal Court Affairs, Muscat

TABLE 17.6

Degree Plan of the Bachelor of Engineering (BEng) in Agricultural Engineering for the 2021 cohort as a joint programme with College of Engineering (136 credits).

Course Code	Course Description	Cr	Course Code	Course Description	Cr
University Requirements (6 Credits)			Major Requirements (80 Credits)		
	Arabic	2	BIOL 2101	General Biology	4
	Contemporary Omani State &		CAMS 2003	Introduction to Food & Resource	
	People	2		Economics	3
	Oman & Islamic Civilization or		CAMS 3000	Seminar & Presentation Skills	2
	Islamic Culture	2	CAMS 3001	Biometry & Experimental Design	
University Electives (6 Credits)		6		in AMS	3
			CAMS 4001	Management & Business Skills	3
College Requirements (35 Credits)			PLNT 2515	Introduction to Plant Sciences	3
CAMS2000	Introduction to Agricultural &		ECCE 3015	Electrical Engineering	
	Marine Sciences	0		Fundamentals	3
CHEM 2101	General Chemistry I	4	FSHN 3101	Properties of Food and Agricultural	
LANC 2145	Communication in Agricultural			Materials	3
	Sciences	3	FSHN 3102	Elements of Food Engineering	3
LANC 2146	Academic Writing in Science	3	MEIE 2129	Basic Mechanics	3
MATH 2107	Calculus I	4	MEIE 3104	Engineering Drawing & Computer	
MATH 2108	Calculus II	3		Graphics	3
MATH 3171	Linear Algebra & MV Calculus	3	MEIE 3141	Thermodynamics I	3
MATH 4174	Differential Equations for	3	SWAE 2001	Introduction to Agricultural	
	Engineers			Engineering	3
PHYS 2107	Physics for Engineering I	4	SWAE 2201	Introduction to Soil & Water	3
PHYS 2108	Physics for Engineering II	4	SWAE 2302	Fundamentals of Agricultural	
SWAE 2307	Workshop Practice I	1		Machinery	3
SWAE 3306	Computer Programming	3	SWAE 3001	GIS for Environmentalists	3
SWAE 4801	Agricultural Engineering Internship	0	SWAE 3005	Land Surveying	3
			SWAE 3201	Power Units	3
Major Electives (9 Credits)			SWAE 3203	Postharvest Technology & Quality	
MEIE 3102	Solid Mechanics	3		Management	3
FSHN 3107	Food Sanitation & Quality Control	3	SWAE 3303	Elements of Hydrology	3
MEIE 3262	Material Science & Engineering	3	SWAE 3307	Workshop Practice II	2
SWAE 3325	Desalination and Membrane		SWAE 3308	Instrumentation & Control Systems	3
	Technology	3	SWAE 3310	Fundamentals of Fluid Dynamics	3
SWAE 4001	Remote Sensing of the		SWAE 3402	Irrigation Principles	3
	Environment	3	SWAE 4300	Agricultural Engineering Design I	3
FSHN 4101	Food & Process Engineering	3	SWAE 4301	Agricultural Engineering Design II	3
SWAE 4202	Tillage & Soil Machine Relations	3	SWAE 4304	Modelling & Analysis of Bio-phyisical	
SWAE 4203	Principles of Machinery	3		Systems	3
SWAE 4204	Advances in Postharvest				
	Technology	3			
SWAE 4302	Agriculture in Controlled				
	Environment	3			
SWAE 4400	Groundwater Hydrology	3			
SWAE 4402	Designs & Management of Irrigation				
	Systems	3			

Municipality, and Al Ansari Trading Enterprise LLC, in addition to representatives from SQU related colleges, departments, and units including the College of Engineering, Career Relations, and the Agriculture Experiment Station.

Major outcomes of the workshops were:

- 94% of the participants showed their support to the proposed joint programme.
- 89% of the participants agreed that the joint programme's courses can produce graduates equipped with needed knowledge and skills in the field of Agricultural Engineering.
- 89% of the participants recommended to expose the students with more practical training.
- 100% of the participants showed their interest in providing internship to the students as well as

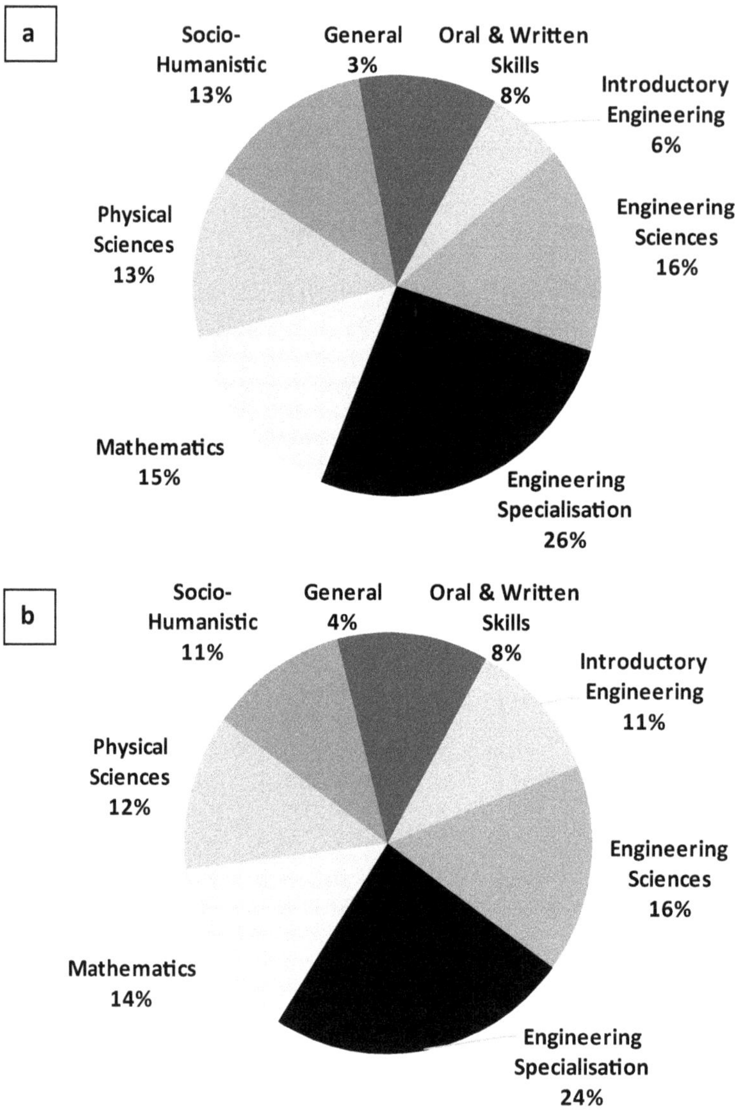

FIGURE 17.5 Comparison of the undergraduate curriculum structure and content in (a) the recommended ABET and (b) the Bachelor of Engineering (BEng) in Agricultural Engineering at Sultan Qaboos University.

encouraging their organisations in employing the programme graduates.

- A request to build up a database containing statistics on the programme's students from the number of the enrolment to their job status after graduation.
- To avoid the misconception of the Arabic translation of "Agricultural Engineering", which refers to anyone who graduated from college of Agriculture, a request was made to come up with a clear job description of the programme for potential employers as well as targeting the media in this regard.
- A recommendation of changing the programme name to cope with the worldwide direction of this discipline. The suggested name was "Biosystems Engineering"

The new criteria regulating the admission of new students to the joint programme included: 1) students' high school grades should meet the College of Engineering entrance requirements; 2) students must get at least C+ in Math, Chemistry, and Physics; and 3) the cumulative GPA should be no less than 2.50.

Following the recommendation of the ABET consultant, the new BEng programme was domiciled in the Soils, Water, and Agricultural Engineering department in the College of Agricultural and Marine Sciences. However, the academic affairs were managed by a Programme Steering Committee representing both colleges. To ensure the dual management and supervision, the following three procedures to implement were emphasised:

1. Signatures of both deans should appear on the AE programme certificate
2. The phrase "Joint programme" and "B. Eng." Should appear on the on the transcript
3. Students' records be accessible to the supervisor no matter from which of the colleges the student is admitted to the joint programme

An advisory board consisting of related stakeholders and potential employers was also established to provide a guidance body on the enhancement and improvement of the joint programme to meet the up to date job market requirements. Moreover, the programme underwent 8 academic evaluations in years 2009, 2011, 2013, 2014, 2016, 2018, and 2021, through the invitation of 6 ABET related consultants, one national body reviewer, and one international body reviewer.

Despite the fact that the academic curriculum of the joint programme has successfully satisfied all above mentioned evaluation processes throughout the years since its establishment, the joint programme did not contribute to an increase in enrolment. As can be noted from Figure 17.4, the number of enrolments to the programmes decreased from 7 students in 2008 to 2 in 2020. Three main disadvantages of implementing this programme have surfaced since its establishment:

1. The new admission requirements are difficult for students to meet
2. The major's name is still not appealing to students as it is still mixed up with the designation given to all graduates from the College of Agriculture
3. Difficult dual management of the programme

Overview of Postgraduate Programmes in Agricultural Engineering

In 1997 Sultan Qaboos University started offering a postgraduate programme in Agricultural Engineering, Master of Science in Bioresource Engineering. In 2006 the degree name changed to Master of Science in Agricultural Engineering. It is a combined course and thesis programme with a total of 30 credits consisting of 12 credits core courses, 12 credits elective courses, and 6 credits for thesis. Table 17.7 details the degree plan of the MSc in AE which is offered with the following research themes:

1. Mechanisation and Automation
2. Controlled Environment Agriculture
3. Precision Agriculture
4. Soil/land and Water Engineering
5. Postharvest and Process Engineering
6. Power and Energy in Agriculture

Different mechanisms are used to evaluate the MSc students during their studies, including: the ability to write research proposals, theses, and publishable papers, the ability to run lab-based and field-based analyses, and their presentation skills. The admission process to the MSc programme is summarised in Figure 17.6. Figure 17.7 shows the enrolment to the programme since 2013.

Similar to other postgraduate programmes at Sultan Qaboos University, the MSc in AE programme is managed as shown in Figure 17.8.

During the curriculum design and improvement, the curriculum was benchmarked against similar well-known regional (King Saud University) and international (Washington State University) programmes and aligned with the Oman National Qualification Framework using 35 indicators as listed in Table 17.8. The benchmarking resulted in a good comparability among the three universities.

TABLE 17.7
Degree Plan of MSc in Agricultural Engineering (30 credits)

Course Code	Course Title	Cr
Core Courses		
CAMS 6001	Advanced Statistics & Experimental Design	3
CAMS 6002	Research Methodology	3
SWAE 6004	CAD Applications in Irrigation Engineering	3
SWAE 6301	Advanced Controlled Environment Agriculture	3
	Total	12
Elective Courses*		
SWAE 6009	Advanced Topics in SWAE	3
SWAE 6203	Non-destructive Testing Techniques for Biological Products	3
SWAE 6302	Precision Agriculture	3
SWAE 6303	Advanced Remote Sensing Techniques for Environmental Applications	3
SWAE 6311	Multiphase Flow & Chemical Transport	3
SWAE 6312	Soil & Groundwater Contamination & Clean-Up	3
SWAE 6405	Integrated Water Resources Management	3
FSHN 6008	Novel Methods in Food Preservation	3
MEIE 6013	Renewable Energy	3
	Total	12
SWAE 7002	Thesis	6
	Total	6

*Students are allowed to take two (2) higher level (4000 and above) undergraduate courses and any postgraduate courses outside the degree plan subject to the approval of the Head of Department and the Supervisor. CAD: Computer Aided Design, CAMS: College of Agricultural and Marine Sciences, SWAE: Soils, Water and Agricultural Engineering, FSHN: Food Science and Nutrition, MEIE: Mechanical and Industrial Engineering

Summary and Conclusions

At the inception of Sultan Qaboos University in 1986, the Department of Agricultural Mechanization, which offered an undergraduate degree in Agricultural Mechanization, was among the founding departments in the College of Agriculture and Marine Sciences (CAMS). During the past 38 years, the names of the academic department, degree programme, and degree structure have changed several times in response to low interest among prospective students and, thus, declining enrolments. The first major reform resulted in the establishment of a BSc degree programme in Bioresource and

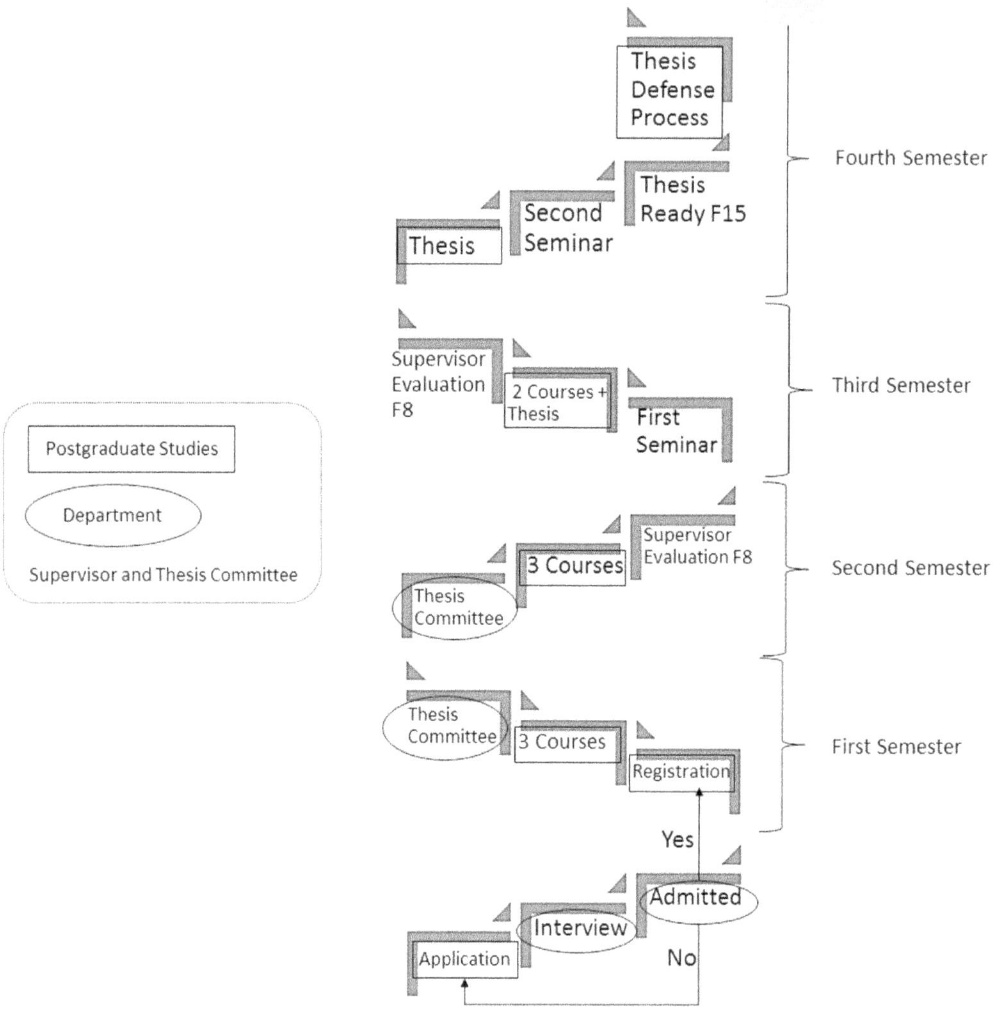

FIGURE 17.6 Steps towards admission to the MSc in Agricultural Engineering programme. (Source: adapted from 2022 Self-Study Report of the academic programme review on MSc in Agricultural Engineering programme)

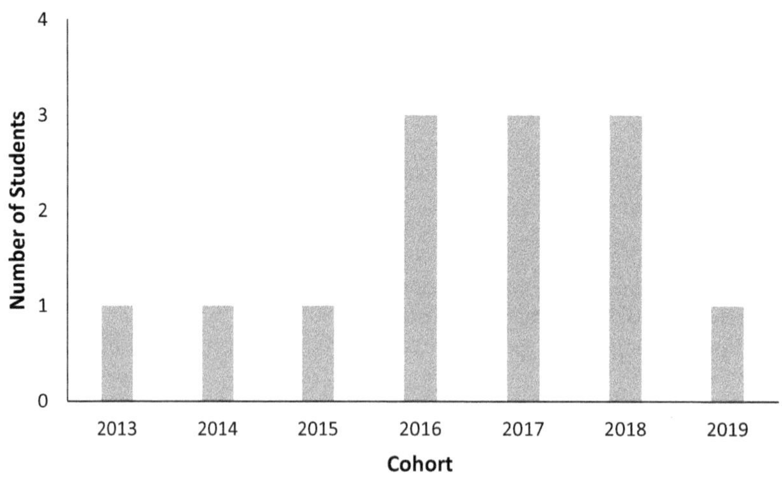

FIGURE 17.7 Students enrolments in the MSc AE postgraduate programme.

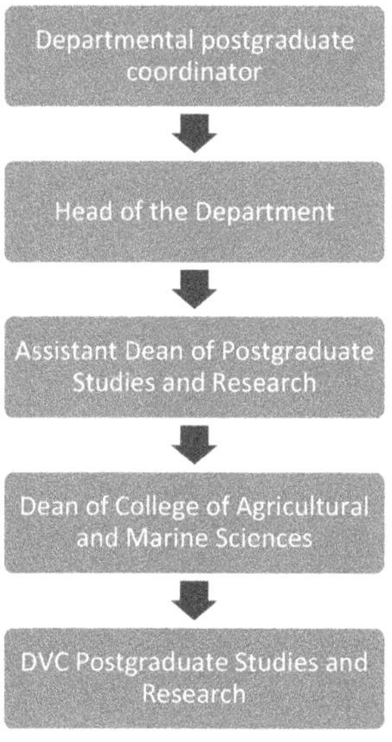

FIGURE 17.8 Management of MSc in AE programme.

Agricultural Engineering in the Department of Bioresource and Agriculture Engineering with 120 credits, following by the offering of the degree of BSc in Agricultural Engineering under the Soils, Water and Agricultural Engineering Department with 129 credits. However, as an engineering programme hosted at a non-engineering college, the necessity of official academic recognition of its engineering aspect has been raised for two reasons. First, the graduates faced difficulties in getting employment as they were not recognised as engineers due to a common perception in Oman stemming from the university's calling all graduates from the College of Agriculture "agricultural engineers". Second, the above reason has contributed to discouraging prospective students to enrol in the programme, especially as they are required to take extra courses in which they often have academic difficulties. Students' perceptions, attitudes, and feedback on the undergraduate degree programme suggested their preference for a more engineering- rather than agriculture-oriented programme design and name. Furthermore, the low public image and profile of AE was identified as a major contributor to low preference among students.

In response to these challenges, a new joint programme was established between the College of Agricultural and Marine Sciences (programme host) and the College of Engineering. The new programme offers a Bachelor of Engineering (BEng) degree in Agricultural Engineering which has been benchmarked with the Accreditation Board for Engineering and Technology (ABET). Based on this reform, the curriculum was restructured with high admission criteria and increased total credits from 129 to 136 credits. The joint (inter-college) degree offering was envisioned to overcome the problems faced by the programme and its graduates, including poor image and low profile among students and the general public, and low enrolment. However, unseen pitfalls emerged, including the difficulty of new students in meeting the new admission requirements for an engineering programme, and the technical issue of facilitating the bimodal administration of the joint programme under two colleges. The College of Agricultural and Marine Sciences also offers postgraduate degrees (MSc and PhD) in Agricultural Engineering, which have been benchmarked alongside similar programmes in the region and North America in terms of curriculum contents and graduate attributes.

TABLE 17.8

Indicators used for Benchmarking MSc AE at SQU with similar programmes at King Saud University, Saudi Arabia (KSU) and Washington State University, USA (WSU). AES: Agricultural Experiment Station.

No	Indicators	No	Indicators
1	Entry requirements	18	Status
2	Credit load	19	Research Output
3	Degree structure	20	Total Faculty
4	Study duration	21	Domestic Faculty
5	Teaching evaluations & assessments	22	International Faculty
6	Supervision evaluation & assessments	23	Total Students
7	Tracks/Streams	24	International Students
8	Scholarships/Assistantships	25	Student/faculty Ratio
9	Research funding	26	Scholarships
10	Research collaborations	27	Academic Reputation
11	Career path support for graduates	28	Citations per Faculty
12	Access to library resources	29	Employer Reputation
13	Support in research logistics & infrastructure e.g. AES	30	Faculty Student Ratio
14	Availability of key lab instruments	31	International Students Ratio
15	Cost of living for the graduate student & dependents	32	Courses Composition (College Required)
16	World Universities Ranking QS	33	Courses Composition (Major Required)
17	Size	34	Courses Composition (Major Electives)
		35	Research Tracks (research areas)

Future Prospects

The reforms made in curriculum content and names of the academic programme and department, and the transition to a joint BEng degree programme resulted in immediate rises in student enrolments, these impacts were not sustained, thereby raising questions about the long-term viability of AE education in Oman. To assure a future for the AE programme to increase student enrolment and retention, further revision of the admission criteria will be required without jeopardising the standards for an accredited engineering programme. It also requires finding the best mechanism for the bimodal administration of the programme so as to reduce administrative bottlenecks. There is a need to strengthen the link between the academic programme and industry, thereby making the programme more responsive to the needs of society and the country at large. Given the critical importance of engineering and related technologies in ensuring sustainable agriculture and food systems, a multi-stakeholder approach is necessary to assure a future for AE education and research in Oman, and indeed globally.

REFERENCES

DABE. (2001). Handbook, Department of Agricultural and Bioresources Engineering, College of Agricultural and Marine Sciences, Sultan Qaboos University, Sultanate of Oman.

DABE. (2006). Handbook, Department of Agricultural and Bioresources Engineering, College of Agricultural and Marine Sciences, Sultan Qaboos University, Sultanate of Oman.

Eide, A.R., Jenison, R.D., Mashaw, L.H., & Northup, L.L. (2002). *Engineering Fundamentals and Problem Solving*, 4th edn, McGraw-Hill, New York.

Opara, L.U. (2001a). Historical evolution and tasks for agricultural engineering in the new millennium, in S. Kosutic (ed.), *Proceedings of the 29th International Symposium on Agricultural Engineering*, University of Zagreb, Zagreb, Croatia, 1–19.

Opara, L.U. (2001b). Agricultural engineering education and research in knowledge-based economy, in S. Kosutic (ed.), *Proceedings of the 30th International Symposium on Agricultural Engineering*, University of Zagreb, Zagreb, Croatia, 33–36.

Opara, L.U. (2003a). Image is everything: Undergraduate students' contributions towards better understanding and improvement of public attitudes and perceptions of agricultural engineering, *World Transactions on Engineering and Technology Education* 2(3), 457–460.

Opara, L.U. (2003b). From farmers to blue-collar professionals: Agricultural Engineering students' perspections on their career prospects, critical skills, and discipline name change, *World Transactions on Engineering and Technology Education* 2(3), 399–402.

Opara, L.U. (2004). Outlook on agricultural engineering education and research and prospects for developing countries, *Outlook on Agriculture* 33(2), 101–112.

Opara, L.U., Al-Adawi, S.S., & Al-Shukeili, T.Y. (2004a). Students' perceptions of agricultural engineering education at Sultan Qaboos University in Oman, Part 1: Academic program selection, *World Transactions on Engineering and Technology Education* 3(1), 139–142.

Opara, L.U, Al-Adawi, S.S., & Al-Shukeili, T.Y. (2004b). Students' perceptions of agricultural engineering education at Sultan Qaboos University in Oman, Part 2: Curriculum content and career preferences, *World Transactions on Engineering and Technology Education* 3(1), 143–146.

Opara, L.U., Perret, J.S., & Ampratwum, D.B. (2006a). The evolution of agricultural engineering education at Sultan Qaboos University, *International Journal of Engineering Education* 22(6), 1139–1148.

Opara, L.U., Al-Adawi, S.S., & Al-Shukeili, T.Y. (2006b). Student perceptions of the public image of agricultural engineering and their preferred name for the discipline and title degree, *International Journal of Engineering Education* 22(1), 59–66.

Opara, U.L., & Cuello, J. (Guest Editors). (2006a). Special issue series: Trends in agricultural, biosystems, and biological engineering education, *International Journal of Engineering Education*, 22(1) Part I: Agricultural Engineering and the Emergence of Biological Engineering http://www.ijee.ie/contents/c220106.html

Opara, UL & Cuello, J. (Guest Editors). (2006b). Special issue series: Agricultural/biosystems/biological engineering education, *International Journal of Engineering Education*, 22(6) Part II: Curriculum Reform, Assessment and Accreditation http://www.ijee.ie/contents/c220606.html

18

Agricultural Engineering Education at Michael Okpara University of Agriculture, Umudike, Nigeria

Joseph C. Adama, Stephen O. Babalola and Chioma V. Agu

Introduction

Evolution of Agricultural Engineering Education in Nigeria

After independence, the Eastern and Western governments in the first republic saw the need for agricultural engineers to drive the agricultural development programmes of these governments. In 1966, nine Nigerians were granted scholarships to be trained in Israel under the UNESCO/FAO/Israel fellowship programme (Igbeka, 2002). Some of these Nigerians included Emmanuel Odigboh, Okezie Abariukwu, Felix Azi, Steve Akparanta, Silas Izugbokwe, N.O.A. Ajai, and V.O. Osunsaya (Abariukwu, 2023; Igbeka, 2023; Akinbolade, 2023).

Before the return of the first set of graduate agricultural engineers, a course in agricultural engineering was offered by the University of Nigeria, Nsukka. The first engineering faculty in Nigeria is the Faculty of Engineering, University of Nigeria, Nsukka. It was established as the College of Engineering in 1961. In 1963, the Department of Agricultural Engineering was established in the university as Nigeria's first Department of Agricultural Engineering. According to Igbeka (2002), "the first agricultural engineering degree was awarded in 1967 by the University of Nigeria, Nsukka". Two graduates were in the set, namely, U. P. C. Okudo and E. Nwalo. Their convocation was hurriedly organised on the eve of the declaration of civil war.

Now, many public and private universities in Nigeria offer Agricultural Engineering at degree level. An estimated 50 universities are offering agricultural engineering at degree level in Nigeria. Table 18.1 lists some of these universities.

On average, the admission quota to the universities by the Joint Admissions and Matriculation Board for agricultural engineering is 80 students per session per university. That means 4000 students are admitted to study agricultural engineering in Nigeria per session. This is the estimated current situation.

The application of technology in improving agricultural activities brought the necessity of agricultural engineering in solving engineering problems on the farm through mechanisation. Until 1960, most of the problems encountered in farm practices were resolved with the consultation of civil engineers and agriculturists. The service rendered by these consultants from other fields and their interest in agriculture made some of them broaden their training and education, giving room to what we call agricultural engineering today.

Agricultural engineering can be classified either as a profession or as training. As a profession, the agricultural engineer has a background in agriculture and specialises in planning, designing, developing, and managing systems required in agricultural production, processing, distribution, marketing, and consumption. As training or education, it is the branch of engineering involved in the design and production of farm power and machinery, the planning and construction of farm structures, environmental control, drainage systems, soil and water conservation, bioprocessing, postharvest operations, and farm electrification.

Agricultural engineering is the driver and agent of agricultural mechanisation in almost all the agricultural sectors in Nigeria. Intensification on research in agriculture through mechanisation was extended to Umudike as one of the centres for research because of its conducive environment for agricultural engineering and research activities.

The Department of Agricultural Engineering

The Michael Okpara University of Agriculture, Umudike (MOUAU) (formerly called the Federal University of Agriculture) was established in May 1993 with a core mandate to impart agricultural education scientifically and practically, undertaking applied research and such extension services as would assist the achievement of national self-sufficiency in food production and catalysing as well as sustaining rural development (MOUAU, 2021).

The Department of Agricultural Engineering is the first engineering department in the College of Engineering and Engineering Technology of the university. The department was established in the 2001/2002 academic session along with the College of Engineering and Engineering Technology. The Department runs 5-year degree programmes leading to the award of the Bachelor of Engineering (B. Eng) Degree.

DOI: 10.1201/9780429150111-20

TABLE 18.1

Some Universities in Nigeria and the Names of the Departments

S/N	Name of University	Name of the Department
1	University of Nigeria, Nsukka Enugu State	Agricultural and Bioresources Engineering
2	University of Ibadan, Ibadan Oyo State	Agricultural and Environmental Engineering
3	Obafemi Awolowo University, Ile Ife Osun State	Agriculturaland Environmental Engineering
4	Ahmadu Bello University, Zaria Kaduna State	Agricultural and Bioresources Engineering
5	Bayero University, Kano Kano State	Agricultural and Bioresources Engineering,
6	University of Maiduguri, Borno State	Agricultural and Environmental Engineering
7	University of Ilorin Kwara State	Agricultural and Biosystems Engineering
8	Abubakar Tafawa Balewa University Bauchi, Bauchi State	Agricultural Engineering
9	Federal University of Technology, Owerri, Imo State	Agricultural Engineering
10	Federal University of Technology, Akure, Ondo State	Agricultural Engineering
11	Federal University of Technology, Minna Niger State	Agricultural and Bioresources Engineering
12	Modibo Adama University of Technology, Yola Adamawa State	Agricultural and Environmental Engineering
13	University of Uyo, Akwa Ibom State	Agricultural and Food Engineering,
14	Nnamdi Azikiwe University, Awka Anambra State	Agricultural and Bioresources Engineering
15	Federal University of Agriculture, Abeokuta, Ogun State	Agricultural Engineering
16	University of Agriculture, Markurdi Benue State	Agricultural and Environmental Engineering
17	Michael Okpara University of Agriculture, Umudike Abia State	Agricultural and Bioresources Engineering
18	Enugu State University of Science and Technology, Enugu Enugu State	Agricultural and Bioresources Engineering
19	River State University of Science and Technology, Port Harcourt River State	Agricultural and Environmental Engineering
20	Niger Delta University, Yenagoa Bayelsa State	Agricultural and Bioresources Engineering
21	Niger Delta University of Science and Technology, Yenagoa Bayelsa Akwa Ibom State	Agricultural and Environmental Engineering
22	Landmark University, Omu-Aran	Agricultural and Bioresources Engineering
23	Federal University Oye, Ekiti State	Agricultural and Bioresources Engineering
24	University of Benin, Benin City, Edo State	Agricultural Engineering
25	Imo State University, Owerri Imo State	Agricultural Engineering
26	Kano State University of Science and Technology, Wudil, Kano State	Agricultural Engineering
27	Kwara State University of Science and Technology. Molete Kwara State	Food and Agricultural Engineering
28	Olabisi Onabanjo University	Agricultural Engineering
29	Oyo State Technical University, Ibadan Oyo State	Agricultural Engineering
30	Afe Babalola University, Ado Ekiti, Ekiti State	Agricultural Engineering
31	Osun State University, Osun State	Agricultural Engineering
32	Adeleke University, Ede Osun State	Agricultural Engineering
33	Ladoke Akintola University of Science and Technology, Ogbomosho Oyo State	Agricultural Engineering

(Adama, 2017; Samphina Academy, 2023; O.J. Olaoye, personal communication, March 31, 2023; W.A. Lamidi, personal communication, March, 30,2023; M. Enebaifo, personal communication, March 28, 2023).

The programmes

The department runs two programmes. These are undergraduate and postgraduate. The undergraduate programme is a five-year programme leading to the award of a Bachelor of Engineering (B.Eng.) Honours Degree in Agricultural Engineering. The postgraduate programme leads to the award of a Master's or Ph.D. degree in a specialised option in the department. There are four options, especially at the postgraduate level: Bioprocessing/Storage and Food Engineering, Farm Power and Machinery Engineering, Soil and Water Resources Engineering, and Structures and Environment. The agricultural engineering programme in the university combines all these options to produce broadly trained graduates that are versatile for professional jobs in both the public and private sectors. This study will focus on the undergraduate programme.

Entry Requirements

The entry requirements for agricultural engineering programmes in Nigeria are the same as in other engineering programmes. These requirements are as contained in the brochure of the Joint Admission and Matriculation Board (JAMB, 2022) are presented below:

(i) The admission requirement for UME candidates into the course is a minimum of 5 (ordinary) credit passes in West African Senior Secondary School Certificate (WASC/SCE) or General Certificate of Education (GCE) or National Examination Council Ordinary level (NECO) in not more than two sittings. The credits must be obtained in English language, Mathematics, Physics, and Chemistry. The remaining one is preferably Biology or Agricultural Science. A pass in JAMB that meets the departmental cut-off mark determined by the university. The candidate must not be below 16 years of age.

(ii) Direct entry candidates: Direct entry admission is based on a combination of O'Level results with the following qualifications:

  (a.) General Certificate of Education or Higher School Certificate (A Level two papers)

  (b.) Ordinary National Diploma (Upper credit)

  (c.) Higher National Diploma (Upper credit).

Period of Formal Studies in the University

The university runs a two-semester calendar in one session with about 15 weeks. The departmental course of study is structured in such a way that a minimum of five years is required. The first year is devoted to preliminary university courses. The second and third-year courses expose the student to some related courses and basic foundation college courses designed to broaden his/her knowledge and appreciate engineering. The remaining two years are devoted to specialised courses in the field of agricultural engineering. Year 2 and 3 students go for 3 months of industrial training at the end of the session. Year 4 students undertake a 6-month industrial training from the beginning of Year 4 second semester in places related to their area of specialisation. An excursion is taken during each academic year as part of the requirements for the award of B.Eng.

Registration of Courses

All students register for their courses at the beginning of each semester according to levels in line with the course curriculum. The normal registration period is five days from the beginning of the semester. For fresh students, late registration begins after matriculation and ends after the specified period normally indicated by the Senate-approved university calendar.

Every student should meet his or her academic adviser, who will guide him on the mode of course registration. Students above the 100 level must ensure that they reregister for their failed courses before registering for courses at their current level. Any course registered earns the student an F grade if the examination is not taken by him.

Existing Grading Systems

The examination of students on the programme is carried out on continuous assessment in addition to examination at the end of each semester. Quizzes, tutorials, take-home assignments, attendance of lectures, tests, etc., comprise the continuous assessment, and it carries 30% of the mark. The objective of introducing continuous assessment is to compel the students to:

- Attend classes
- Pay attention in class as lectures are in progress
- Participate in group assignments
- Study after the lectures

The written examination carries 70% of the mark. A student must have attended at least 75% of lectures to be eligible to write the semester final examination. In addition, he/she must have done class assignments and tests for continuous assessment to a final grade on the subject. Examination done without a continuous assessment will earn a student a grade of Failure (F) regardless of a score over 70%. Table 18.2 presents the grading system. From Table 18.2, 70–100% scores are given grade 'A', the highest grade. The lowest grade is 'F', obtained by a candidate who scores 39 or below.

There is a proposal for a review of the grading system, which is seen to be outdated. The new grading system is a modification of the existing system in Table 18.3. This new system is shown in Table 18.3. In the new system, the 'E' grade is removed so that a score of 44 and below means failure. This new grade system is yet to be implemented in the university.

Programme design

In pursuance of the mission and mandates of the university, specific features are incorporated into the agricultural engineering programme to equip students for the challenges faced after graduation. These include:

- Common foundation year courses at 100 and 200 levels for all engineering students

TABLE 18.2

Existing grading system (ABE, 2019b).

Score (%)	Grade	Grade Point (GP)	Description
70-100	A	5	Excellent
60-69	B	4	Very good
50-59	C	3	Good
45-49	D	2	Fair
40-44	E	1	Pass
39 and Below	F	0	Fail

TABLE 18.3

New grading system yet to be implemented (ABE, 2019a).

Score (%)	Grade	Grade Point (GP)	Description
70-100	A	5	Excellent
60-69	B	4	Very good
50-59	C	3	Good
45-49	D	2	Fair
44-Below	F	0	Fail

- Workshop practice, laboratory work, tutorials, seminars on specific topics, farm machinery, and field operations.
- Broad-based engineering and interaction between students and professionals
- Adequate knowledge in the areas of engineering management, economics, and law.
- Design projects with a bias towards local applications.
- Supervised Students' Work Experience Programme SWEP. This is offered in the second semesters of 200 and 300 levels of study. The programme is done in-house and aimed at preparing the students for industrial and workshop practices.
- Supervised Students' Industrial Work Experience Scheme, SIWES, offered after the 400-level first-semester examination. It is industry based and offered as the only course in the second semester of the 400-level.
- Final year projects in which students work alone under supervision in specific options of Agricultural Engineering.
- General studies courses
- Entrepreneurial courses

The academic programme has been placed to offer challenges and to encourage the development of ingenuity and originality in students.

The Name and Course Curriculum

The department was established as the "Department of Agricultural Engineering (EAE)". The curriculum of the department has undergone a number of reviews. The name changed to Agricultural and Bioresources Engineering (MOUAU, 2010). This name change and the curriculum reviews were necessitated by the need to align with emerging technologies globally, attract students into the department, reduce the exodus of the admitted students to sister departments after the first year, and increase employment opportunities for the products. Since establishment, the curriculum of the department has been reviewed three times. Now, the National Universities Commission (NUC) has changed both the name and its course structure. The new name is "Department of Agricultural and Biosystem Engineering (ABE)".

The primary duties of the academic and technical staff of the university are teaching, mentoring, research and development, and then service to society if the opportunity comes. Students are the most important people in the university system. Naturally, agricultural engineering is a programme that requires knowledge from other areas of study. A graduate of agricultural engineering should remain more versatile than graduates of other engineering professions. Apart from the basic engineering courses, the students are offered separate courses in Mechanical, Civil, and Electrical Engineering. This makes agricultural engineering graduates of the department unique. After going through these courses, referred to as "borrowed" courses, the students come "home" in the department to do classroom work, laboratory exercises, workshop practices, and field operations. As contained in the course curriculum and discussed earlier, this works comprises 70% examination and 30% continuous exercise. The impacts of the curriculum on the students are discussed.

Existing Course Curriculum

The curriculum presented here is currently in use in the department and was developed in line with the special training needs and the minimum academic standard of the programme. The courses offered in the department for the undergraduate programme are presented in Tables 18.4 to 18.17. It contains the course code, course title and credit load.

The 100-level courses (Tables 18.4 and 18.5) are mainly foundation courses drawn from the colleges of physical sciences and general studies, and are mostly physics, chemistry, and mathematics courses. The general studies courses include French, German, Use of English, Use of Library, Social Science, and Nigerian History. Two courses, Introduction to Engineering and Computer Applications and Information

TABLE 18.4

100 Level courses first semester (ABE, 2019b).

S/N	Course Code	Course Title	Credit Units
1	ENG 111	Introduction to Engineering	1
2	MTH 112	Elementary Mathematics 1	3
3	PHY 112	Elementary Physics I	2
4	PHY 117	General Physics Lab I	1
5	CHM 113	General Chemistry I	3
6	CHM 114	Practical Chemistry I	1
7	GSS 111	Use of English I	1
8	GSS 112	Nigerian History	2
9	GSS 114	Elementary French I	1
10	GSS 115	Basic German I	1
11	GSS 116	Use of Library	1
		Total units	18

TABLE 18.5

100 Level courses second semester (ABE, 2019b).

S/N	Course Code	Course Title	Credit units
1	MTH 122	Elementary Mathematics II	3
2	MTH 123	Introduction to Vectors	3
3	PHY 121	General Physics II	2
4	PHY 122	Elementary Physics II	2
5	PHY 127	Physics Lab II	1
6	CHM 121	General Chemistry II	3
7	CHM 124	Practical Chemistry II	1
8	GSS 121	Use of English II	2
9	GSS 124	Elementary French II	1
10	GSS 125	Basic German II	1
11	GSS 126	Social Science	2
12	ENG 121	Computer Applications and Information Technology	2
		Total units	23

TABLE 18.6

200 Level courses first semester (ABE, 2019b).

S/N	Course Code	Course Title	Credit Units
1	ENG 211	Engineering Thermodynamics I	3
2	ENG 212	Workshop Technology/Practice I	2
3	ENG 213	Basic Electrical Engineering, Workshop Technology/Practice I	3
4	ENG 214	Engineering Drawing I	2
5	ENG 215	Engineering Mechanics	3
6	MTH 211	Mathcmatical Methods I	3
7	MTH 214	Linear Algebra I	2
8	GSS 212	Peace & Conflict Resolution Studies	2
9	GSS 217	Philosophy and Logic	2
		Total Units	22

TABLE 18.7

200 Level courses second semester (ABE, 2019b).

S/N	Course Code	Course Title	Credit Units
1	ENG 221	Strength of Materials I	2
2	ENG 222	Engineering Drawing II	2
3	ENG 223	Computer Programming	3
4	ENG 224	Materials Science	2
5	ENG 225	Fluid Mechanics I	3
6	ENG 226	Engineer in Society	1
7	MTH 221	Mathematical Methods II	3
8	STA 224	Statistics for Physical Science & Engineering	3
9	GNT 221	Entrepreneurship Studies	2
10	ABE 221	Agricultural and BioResources Engineering Concepts	1
11	ENG 200	Student Work Experience Program (SWEP I)	1
		Total Units	23

Technology, are the only engineering courses offered at 100-level.

At 200-level courses (Tables 18.6 and 18.7), there are gradual shifts from physical sciences and general studies to general engineering courses. These engineering courses are from mechanical, civil, computer, and electrical engineering departments. The course Engineer in Society is a college course done at 200-level. The objectives of the courses are to enable the students to understand the environment and integrate it with engineering knowledge. These will prepare students for the challenges facing the engineering profession in the world in general and Nigeria in particular. A course in agricultural engineering is being done by the students at the two hundred level. The course is called Introduction to Agricultural Engineering. The objective of the course is to enable the students to have a sense of homecoming and to appreciate the course they have come to study.

The 300-level courses offered are shown in Tables 18.8 and 18.9. At this level, courses from agricultural colleges are introduced. These courses include Introduction to Soil Science, Agricultural Biotechnology, Introduction to Agricultural Extension and Rural Sociology, Crop Production for Engineers, and Animal Production for Engineers. At this level, students are offered second-tier courses on entrepreneurship. The course in agricultural economics was dropped, and engineering economics was introduced in the second review of the curriculum.

These entrepreneurship courses were introduced by the National Universities Commission across all the universities in the country. The courses are meant to make the students self-reliant after graduation in the midst of no employment opportunities.

The 400-level courses (Table 18.10) are only offered in the first semester. The students spend the first semester in the school, and after the semester examination, they proceed on industrial training called SIWES. The SIWES (Table 18.11) carries 15 credit units. The students are paid a stipend to

TABLE 18.8

300 Level courses first semester (ABE, 2019b).

S/N	Course Code	Course Title	Credit Units
1	ENG 311	Engineering Economics	2
2	ENG 313	Engineering Analysis	3
3	ABE 311	Engineering Hydraulics I	2
4	ABE 312	Soil Mechanics II	2
5	ABE 315	Automotive Services and Maintenance	2
6	SSA 311	Introduction to Soil Science & Soil Tillage	2
7	EME 311	Mechanics of Machine I	2
8	EME 315	Strength of Materials II	3
9	CCS 311	Agricultural Biotechnology	1
10	GNT 311	Business Development and Management	2
11	AEX 311	Introduction to Agric. Extension & Rural Sociology	2
		Total Units	23

TABLE 18.9

300 Level courses second semester (ABE, 2019b).

S/N	Course Code	Course Title	Credit Units
1	ABE 321	Machine Component Design and Presentation	2
2	ABE 322	Agricultural Land Surveying	3
3	ABE 323	Engineering Hydrology	3
4	ABE 324	Geology for Agricultural Engineering	3
5	CPP 324	Crop Production for Engineers	3
6	CAS 321	Animal Production for Engineers	3
7	EME 321	Engineering Thermodynamics II	3
8	EME 324	Engineering Metallurgy	2
9	ENG 326	Technical Report Writing and Presentation	1
10	ENG 300	Student Work Experience Program (SWEP II)	1
		Total Units	24

TABLE 18.10

400 Level courses first semester (ABE, 2019b).

S/N	Course Code	First Semester Course Title	Credit Units
1	ENG 418	Computational Methods in Engineering	3
2	ABE 412	Agro-Resources Structures and Environmental Control	3
3	ABE 413	Directed Studies and Seminar	1
4	ABE 414	Engineering Hydraulics II	3
5	ABE 415	Engineering Properties and Handling of Biomaterials	3
6	ABE 416	Irrigation and Drainage Principles	3
7	ABE 417	Field Operations and Management of Farm Power & Machinery	3
8	ABE 418	Soil and Water Conservation Engineering	3
9	GNT 411	Practicum	2
		Total Units	24

TABLE 18.11

400 Level course second semester (ABE, 2019b).

S/N	Course Code	Course Title	Credit Units
1	ENG 400	Industrial Training (Students Industrial Work Experience Scheme)	15
		Total Units	15

support what their parents give them. The industries give the students some stipends too. The emphasis is not on the stipend but on what they learned, the challenges they faced in the industry, and how these challenges were handled. At the end of the training, the students write a report and make an oral presentation.

The 500-level courses were designed in line with the three major areas of specialisation in agricultural engineering practice. The students are offered core courses for the first (Table 18.12) and second (Table 18.13) semesters. These courses are Agricultural Land Development, Agro-Resources Power and Machinery Systems Management, Principles of Agro-Industrial Management and Business Law for the first semester. And then, Electrification and Alternative Energy Sources and the final project, Agricultural and Bioresources Mechanization, Design of Agro Resources Machinery I, Agro-Resources Transportation and Ergonomics, Environmental Engineering, and Foundation Engineering for the first semester.

Other courses offered are according to the option the candidate intends to specialise. These are in Table 18.14 for Processing and Storage Engineering, Table 18.15 for Soil and Water Engineering, and Table 18.16 for Farm Power and Machinery Engineering.

Some Courses fall under electives, as shown in Table 18.17. These are Mechanics of Deformable Bodies, Agro Resources Machinery, Packaging and Containerization Engineering, and Agricultural Land Drainage.

TABLE 18.12

500 Level courses first semester core courses (ABE, 2019b).

S/N	Course Code	Course Title	Credit Units
2	ABE 511	Bush Clearing and Land Development	2
3	ABE 512	Agro-Resources Power and Machinery Systems Management	2
4	ABE 513	Principles of Agro-Industrial Management and Business Law	3
5	ABE 500	Final Year Project	3
6		Total	10

TABLE 18.13

500 Level first semester: Bioprocess and Food Engineering option (ABE, 2019b).

S/N	Course Code	Course Title	Credit Units
1	BFE 511	Food Engineering	3
2	BFE 512	Biomass Engineering	2
3	BFE 513	Solar Energy Applications in Bioprocess Engineering	2
4	BFE 514	Processing and Storage of Agro-Resources Products	3
		Total	10

TABLE 18.14

500 Level first semester: Soil and Water Engineering option (ABE, 2019b).

S/N	Course Code	Course Title	Credit Units
1	SAE 511	Design of Irrigation Systems	3
2	SAE 512	Rural Water Supply and Sanitation	3
3	SAE 513	Design of Irrigation and Soil Conservation Structures	3
4	SAE 514	Aquacultural Engineering	2
		Total	11

TABLE 18.15

500 Level first semester: Agricultural and Bioresources Power and Machinery Engineering option (ABE, 2019b).

S/N	Course Code	Course Title	Credit Units
1	APM 511	Agricultural and Bioresources Power	2
2	APM 512	Agro-Resources Machinery I	1
3	BFE 512	Biomass Engineering	2
4	BFE 513	Solar Energy Applications in Bioprocess Engineering	2
5	BFE 514	Processing and Storage of Agricultural Products	3
		Total	10

TABLE 18.16

500 Level courses second semester core courses (ABE, 2019b).

S/N	Course Code	Course Title	Credit Units
1	ABE 521	Design of Agro-Resources Machinery	2
2	ABE 522	Electrification and Alternative Energy Sources	3
3	ABE 523	Agro-Resources Transportation and Ergonomics	2
4	ABE 524	Agricultural and Bioresources Mechanization	2
5	ABE 525	Environmental Engineering	3
6	ABE 526	Foundation Engineering	3
7	ABE 500	Final Year Project	3
		Total	18

TABLE 18.17

500 Level courses second semester elective courses (ABE, 2019b).

S/N	Course Code	Course Title	Credit Units
1	PPF 521	Mechanics of Deformable Bodies	2
	SAE 521	Agricultural Land Drainage	2
2	APM 521	Agro-Resources Machinery II	2
3	BFE 522	Packaging and Containerization Engineering	2
		Total	8

The New Academic Curriculum

In 2018, the National Universities Commission (NUC) reproduced a draft of a curriculum of courses called Benchmarks Minimum Academic Standard (BMAS) in Nigerian universities and circulated same to all the universities and other stakeholders for study and input. In early 2021, a committee was set up to study and harmonise the inputs from the universities and other stakeholders and develop a course curriculum for courses in Nigerian universities. The committee has completed the assignment. The committee's report was first sent to professional bodies to give input and then to the universities. The universities, in turn, made the report available to departments for further input. The review at the university level is being completed.

The courses, referred toas the Global Course Structure, are categorised into:

- Basic Sciences
- General Studies Courses
- Management & Humanities
- Basic Agriculture
- Basic Engineering
- Core Courses

In the NUC Report (2021), the new programme has "courses that support emerging engineering practices in agriculture such as the use of drones and robots, the overriding intervention of renewable energy in agriculture, the engineering of livestock and fisheries, need to deepen the design thinking and creative skills of students and the gradual shift from agricultural engineering to biosystems engineering as well as reflecting the impact of climate change on agricultural technology applications".

Some of these courses are:

- ABE 102: Introduction to Agricultural and Biosystems Engineering (2 credit units)
- GET 102: Engineering Graphics and Solid Modelling (2 credit units)
- ABE 202: Machine Drawing and Advanced Solid Modelling(2 credit units)
- GET 306: Renewable Energy Systems and Technology (3 credit units)
- GET 307: Introduction to Artificial Intelligence, Machine Learning and Convergent Technologies (3 credit units)
- ABE 307: Biosystems Engineering(2 credit units)
- ABE 401:InstrumentationandMeasurementinAgriculturalandBiosystems Engineering(3 credit units)
- ABE 502: Aquaculture and Agroponics Engineering (2 credit units)
- ABE 503: Livestock Production Engineering (2 credit units)
- ABE 505: Drone and Robot Technology in Agriculture(2 credit units)

Apart from ABE 102 Introduction to Agricultural and Biosystems Engineering, other courses in the curriculum are new. These courses would constitute 70% of the entire course structure. Departments are mandated to introduce courses peculiar to the programme. The new courses are being developed by the department and could not be presented here. Find the reproduced courses in Tables 18.18 to 18.23, as recommended by the committee.

A Comparative Analysis of the Existing and New Course Curricula

The comparison discussed here is between the existing curriculum and the new curriculum called the Core Curriculum Minimum Academic Standard (CCMAS), which is 70% of the total course contents. Recall that departments have been mandated to add 30% of courses in areas specific to them, especially at higher levels. "The CCMAS documents are uniquely structured to provide for 70% of core courses for each programme while allowing universities to utilise the remaining 30% for other innovative courses in their specific focus areas. In addition to the overall Learning Outcomes for each discipline, there are also Learning Outcomes for each programme and course. In general, programmes are typically structured such that a student does not carry less than 30 credit units

TABLE 18.18

Global course structure of the new course structure (NUC, 2021).

Level	GST	ENT	Basic Sciences	Faculty(GET)	Programme(ABE)	SIWES	Total
100	4	0	16	3	2	0	25
200	2	2	0	23	0	3	30
300	2	2	0	12	20	4	24
400	-	0	0	0	0	8	8
500	-	0	0	5	14	0	18
Total	8	4	16	43	34	15	105

GST= General Studies; ENT = Entrepreneurship; GET = General Engineering and Technology; SIWES = Students Work Experience Scheme

TABLE 18.19

100 Level courses of the new course structure (NUC, 2021).

Course Code	Course Title	Units	Status	LH	PH
GST 111	Communication in English	2	C	15	45
GST 112	Nigerian People and Culture	2	C	30	-
CHM101	General Chemistry I	2	C	30	-
CHM102	General Chemistry II	2	C	30	-
CHM107	General Practical Chemistry I	1	C	-	45
CHM108	General Practical Chemistry II	1	C	-	45
MTH 101	Elementary Mathematics I	2	C	30	-
MTH 102	Elementary Mathematics II	2	C	30	-
PHY 101	General Physics I	2	C	30	-
PHY 102	General Physics II	2	C	30	-
PHY 107	General Practical Physics I	1	C	-	45
PHY 108	General Practical Physics II	1	C	-	45
GET 101	Engineering Society	1	C	15	-
GET 102	Engineering Graphics and Solid Modelling I	2	C	15	45
ABE102	Introduction to Agricultural and Biosystem Engineering	2	C	30	-
Total		25			

Key LH = Lecture Hours, PH = Practical Hours, C = Core course

TABLE 18.20

200 Level courses of the new course structure (NUC, 2021).

Course Code	Course Title	Units	Status	LH	PH
GST 212	Philosophy, Logic and Human Existence	2	C	30	-
ENT 211	Entrepreneurship and Innovation	2	C	30	-
GET 201	Applied Electricity I	3	C	45	-
GET 202	Engineering Materials	3	C	45	-
GET 204	Students Workshop Practice	2	C	15	45
GET 205	Fundamentals of Fluid Mechanics	3	C	45	-
GET 206	Fundamentals of Thermodynamics	3	C	45	-
GET 209	Engineering Mathematics I	3	C	45	-
GET 210	Engineering Mathematics II	3	C	45	-
GET 211	Computing and Software Engineering	3	C	45	-
GET 299	SIWESI	3	C	9Weeks	
Total		30			

Key LH = Lecture Hours, PH = Practical Hours, C = Core course

TABLE 18.21

300 Level courses of the new course structure (NUC, 2021).

Course Code	Course Title	Units	Status	LH	PH
ENT 312	Venture Creation	2	C	15	45
GST 312	Peace and Conflict Resolution	2	C	30	-
GET 304	Technical Writing and Communication	3	C	45	-
GET 305	Engineering Statistics and Data Analysis	3	C	45	-
GET 306	Renewable Energy Systems and Technologies	3	C	30	45
GET 307	Introduction to Artificial Intelligence, Machine Learning and Convergent Technologies	3	C	45	-
ABE301	Design of Machine and Structural Elements	2	C	15	45
ABE302	Animal Production	2	E	30	-
ABE303	Crop Production	2	E	30	-
ABE304	Farm Management, Rural Sociology and Agricultural Extension	2	E	30	-
ABE305	Soil Science	2	E	30	-
ABE306	Land Surveying and Geographic Information System	2	E	15	45
ABE307	Biosystems Engineering	2	C	30	-
ABE308	Rural Infrastructural Engineering	2	E	30	-
GET 399	SIWESII	4	C	12weeks	
Total		24			

Key LH = Lecture Hours, PH = Practical Hours, C = Core course

TABLE 18.22

400 Level courses of the new course structure (NUC, 2021).

Course Code	Course Title	Units	Status	LH	PH
ABE 401	Instrumentation and Measurement in Agricultural Biosystems Engineering	3	E	30	45
GET 499	SIWES III	8	C	-	24 weeks
Total		8			
Student Industrial Work Experience (SIWES)					
GET 299	SIWES I	3	C	-	9 weeks
GET 399	SIWES II	4	C	-	12 weeks
GET 499	SIWES III	8	C	-	24 weeks
Total		15			

Key LH = Lecture Hours, PH = Practical Hours, C = Core course

TABLE 18.23

500 Level courses of the new course structure (NUC, 2021).

Course Code	Course Title	Units	Status	LH	PH
GET 501	Engineering Project Management	3	C	45	-
GET 502	Engineering Law	2	C	30	-
ABE501	Environmental and Social Impact Analysis	2	C	30	-
ABE502	Aquaculture and Agroponics Engineering	2	C	30	-
ABE503	Livestock Production Engineering	2	C	30	-
ABE504	Greenhouse Technology	2	C	30	-
ABE505	Drone and Robot Technology in Agriculture	2	C	15	45
ABE599	Final Year Project	3	C	15	90
Total		18			

Key LH = Lecture Hours, PH = Practical Hours, C = Core course

TABLE 18.24

Course structure of the new and old courses compared

	GST		ENT		Basic Sciences		Faculty (GET)		Programme (ABE)		SIWES		Total	
Level	New	Old	New	Old	New	Old	New	Old	New	Old	New	Old	New	Old
100	4	12	0	0	16	25	3	3	2	0	0	0	25	40
200	2	4	2	2	0	11	23	26	0	2	3	1	30	46
300	2		2	2	0	10	12	17	20	17	4	1	24	47
400	-	-	0	2	0	-	0	3	0	19	8	15	8	39
500	-	-	0	-	0	-	5	-	14	28	0	-	18	28
Total	8	16	4	6	16	46	43	49	34	66	15	17	105	200

or more than 48 credit units per session" (NUC,2021). This comparison does not include the 30% for innovative courses, which are still under review.

The course structures for the existing and new curricula are presented in Table 18.24, showing a reasonable decrease in the total credit load for the courses at all levels. For instance, the GST courses offered by the students are 8 units in the new curriculum but 16 in the old curriculum. Basic sciences fell from 46 credit units to 16 credit units. GET was reduced from 49 credit units by 6 units to 43 in the new curriculum. Agricultural Engineering (ABE) courses have a total of 34 credit units compared to the existing curricula (66 credit units) calculated from core/compulsory courses in the old curriculum. All the SIWES courses have a total of 15 credit units in the new curriculum but 17 credit units in the old curriculum.

The total credit load for the new curriculum is 105 without the expected 30%, while the total credit load for the existing curriculum is 200. With 30% of 105 (60 credit loads) coming from departments and colleges, the total credit load for the new curriculum will be 165.

A number of courses are dropped or merged. For instance, at 100-level (First and Second Semesters), the following general studies courses are dropped: Use of Library, Basic German I, Elementary French I, and Farm Practice for the first semester; and Use of English II, Introduction to Vectors, Elementary French II, Basic German II, and Social Science for the second semester. The following courses are added: Philosophy, Logic, and Human Existence; Nigerian People and Culture; and Communication in English. The codes for courses changed as follows; GSS to GST, ENG to GET, SWEP I & II to SIWES I & II, and GNT to ENT. The courses Engineering Graphics and Solid Modelling I, as a GET course, and Introduction to Agricultural and Biosystem Engineering, as an ABE course, were introduced at 100. While Introduction to Agricultural and Biosystem Engineering existed at a 200-level as Concepts of Agricultural and Biosystem Engineering, Engineering Graphics and Solid Modelling I was new. The basic science courses retained their names. However, the number of credit units was slashed to a maximum of 2 against a maximum of 3.

At the 200-level, only one general studies course with 2 credit units was retained as against 2 courses with a total of 4 credit units. All the basic science courses, such as Linear Algebra I, and Mathematical Methods I and II gave way to faculty (GET) courses named Engineering Mathematics I and II. The course Entrepreneurship Studies GNT 221 was renamed Entrepreneurship and Innovation ENT 211 without changing the credit hours. In addition, the following courses were dropped at this level: Engineering Drawing, Engineering Mechanics, Peace & Conflict Resolution Studies (GSS 212), and Philosophy & Logic (GSS 217). The courses GSS 212 and 217 gave way to Philosophy, Logic and Human Existence, GST 212.

For the 300-level courses, the courses in both the old and new curricula are mostly departmental and agriculture-based. The following faculty courses were dropped from the level; Engineering Economics (ENG 311), Engineering Analysis (ENG 312), Engineering Thermodynamics II (EME 321), and Business Development and Management (GNT 311). Other courses in the old curriculum that are not in the new curriculum are Strength of Materials, Automotive Services and Maintenance, Soil Mechanics, Mechanics of Machines, Agricultural Biotechnology, Engineering Hydraulics, Machine Components Design and Presentation, Geology for Agricultural Engineers, and Engineering Metallurgy. New courses were brought in which were not in the existing curriculum: Venture Creation, Peace and Conflict Resolution (moved from 200-level in the old curriculum), Renewable Energy Systems and Technologies, Introduction to Artificial Intelligence, Machine Learning and Convergent Technologies, Design of Machine and Structural Elements (existed as Machine Components Design and Presentation with increased contents), and Rural Infrastructural Engineering. The courses from agriculture are retained, but there were changes in the course codes, numbers, and content. The content now has more engineering components than the old curriculum.

The 400-level courses are more for industrial training called the Students Industrial Work Experience Scheme (SIWES). As in the old curriculum, coursework is in the first semester. The second semester is devoted to the supervised SIWES. In the first semester of the session, one course, Instrumentation and Measurement in Agricultural and Biosystems Engineering, was listed in the new curriculum; other courses are to be filled by the departments and faculties.

New courses were brought in at the 500-level. These courses are common to all the engineering disciplines in all the universities in the country: Engineering Project Management, Engineering Law, Environmental and Social Impact Analysis, Aquaculture and Agroponics Engineering, Livestock Production Engineering, Greenhouse Technology, and then Drone and Robot Technology in Agriculture. From the design, departments are to develop the curriculum for the final courses as it suits them using the procedure/guidelines and according to the options (if any). The courses will indicate the core courses and electives.

Contents of Basic Sciences, Humanities, and Other Courses on the Agricultural Engineering Degree Programme

The coursework offered to agricultural engineering students at the Michael Okpara University of Agriculture, Umudike, cuts across the fields of basic sciences (Chemistry, Physics, Mathematics, Computer Science, Statistics), Humanities, Entrepreneurship, Agriculture, and other engineering courses. Further analysis was done on the data presented in Table 18.24 to determine the percentage contributions of these courses to the department's degree programmes. Results show that the general studies courses contribute 16 credit loads which are 8% of the course content. The entrepreneurship courses have 6 credit units which are 3%. Basic sciences have 46 credit units (23%), and other engineering courses have 49 credit loads (24.5%) to contribute. Overall, these constitute 58.5% of the course content apart from SIWES.

Some Supervised Final Year Projects in the Department

In the final year, students undertake a project under the supervision of an academic staff member. The process for implementing the supervised final project in the department starts from the 400-level. In the first semester of the 400-level, students present an oral presentation in a course called Seminar. The course has two components, written and oral presentation. The written presentation is assessed and scored by the supervisor. The written component is graded over 70%. The presentation is done in the presence of the entire academic staff and graded over 30%. At this stage, students are trained on how to write a technical report and present their work to the staff. At the presentation, the students are advised to search for a possible area to carry out their projects as they go for industrial training.

In the first semester of the final year, the students are assigned supervisors for the final project. It is the duty of the head of department to assign students to academic staff for project supervision. This is followed by a proposal where the academic staff members will agree on the project topic, objectives, problem statement, proposed methodology, etc. The student's project work is usually defended in the presence of an external examiner who is invited from another institution.

In selecting a project topic, the following factors are taken into consideration. These factors are:

- Need of the society
- Availability of fund
- Possibility of completing or making appreciable progress within the time limit
- Availability of materials(instruments, consumables, equipment, etc.)
- Global changes

Students undertake their final year project by choosing a specific option (Farm Power and Machinery, Bioprocessing and Storage, or Soil and Water). The following are some of the student projects in the different options.

Project titles on Farm Power and Machinery

1. Development and performance of washer-integrated cassava peeling machine
2. Machinery selection for tillage operations in the rainforest agro-ecological zone
3. Development of a tractor-mounted fertiliser drop spreader
4. Development of a tractor-drawn ginger harvester
5. Effect of operational speed, depth of cut, and moisture content on fuel consumption and field efficiency of disc plough in sandy-loam soil
6. Effects of tillage methods on the growth and yield of "ugu" (*Telfairiaoccidentalis*) and okra (*Abelmoschusesculentus*)
7. Design and development of a mechanical rotary power inter-flow crop weeder
8. Effect of tillage method on soil properties in Umudike environment
9. Design and fabrication of a Zambezi Almond nut cracking machine;

Project titles on Bioprocessing and Storage

1. Effect of cabinet drying and oven drying on the physical properties of periwinkle types
2. The effect of cabinet and oven drying on the chemical properties of periwinkle types
3. Design and construction of a low-cost subsurface drip irrigation system
4. Assessment of biomechanical properties of different native plans in Abia state in relation to slope stabilisation
5. Innovative bioshield: use of biogeotextiles for erosion control
6. Prediction of soil losses in MOUAU through erodibility index
7. Design and evaluation of solar water distiller
8. Effects of moisture content on the physical properties of melon seed
9. Effect of fermentation on the quality characteristics of cassava flour
10. Effect of different enzyme substrates for the performance of broiler fed peeled cassava root meal based diet
11. Absorption characteristics of *Irvingia* and breadfruit seeds.
12. Flowability and compressibility of flamboyant pod (rubber bean)
13. Comparison of storage stability of two varieties of *Irvingia* seed (Ogbono) in five different storage materials

14. Effect of drying conditions on the viability and vigour of okra seeds.
15. The effect of softeners on energy consumption, cooking time, and quality of black eye beans
16. Effect of drying on the viability and vigour of fluted pumpkin seeds
17. Development of a portable evaporative cooler suitable for indoor application using polythene foam pads.
18. Effect of food softeners on cooking time, energy consumption, and quality of breadfruit.
19. Effect of moisture content on physical and thermal properties of alligator pepper pods
20. Energy consumption and drying process of cocoyam cylinder in a convective dryer
21. Evaluation and management of a selt watering raised soil bed with grated rice husk as an agro-fibre
22. Evaluation of non-conventional based feed for catfish
23. Processing and evaluation of feeds from selected non-conventional materials and for sustainable poultry production
24. Evaluation of storage strategies in order to curb postharvest losses of breadfruit seeds in Nigeria (*Treculia Africana*)
25. Drying of turmeric rhizome (*Curcuma longa*) Using a passive solar dryer
26. Determination of selected engineering properties of soursop seeds
27. Comparative evaluation of the physiochemical properties of pig dung-based tilapia feed and other tilapia feeds
28. Determination of bioenergy potentials in some biomass materials
29. Effects of smoking and oven drying on the proximate composition and sensory properties of aba knife fish (*Gymnarchusniloticus*)
30. Evaluation of storage strategies in order to curb postharvest losses of breadfruit seeds (*Treculia Africana*) in Nigeria
31. Design, construction, and testing of a passive evaporative cooler for the short-term preservation of fresh tomatoes
32. Determination of the physical and mechanical properties of cocoyam, corns, and cormels in relation to postharvest processing
33. Determination of some engineering properties of African oil bean kernel
34. Moisture-dependent physical properties of soybean and wheat grains
35. Determination of functional and aerodynamic properties of maize flour
36. Energy requirement for drying of different sample geometries of some selected fruit vegetables
37. Design, fabrication, and testing of a hand-operated poultry deep litter waste removal machine
38. Evaluation of some selected engineering properties of African star apple fruits and seed
39. Influence of split angle on the process parameter of turmeric rhizome (*Curcuma* lo*nga* l.)
40. Evaluation of the effects of different drying methods on the nutrient content of unripe plantain
41. The effects of drying methods on the quality characteristics of fresh turmeric rhizomes

Project titles on Soil and Water

1. Comparative evaluation of the performance of the sprinkler and check basin system using supplementing irrigation for tomato production
2. Performance evaluation and energy usage of an irrigation pump
3. Determination of water requirement of an indigenous crop (scent leave) on loamy soil for optimum production
4. Assessment of gully erosion sites in Ekwulobia/Oko axis, Anambra state
5. Construction of bitter leaf washing and shredding machine
6. Modification of a forage chopping machine
7. Modification and performance evaluation of a cereal grinding machine
8. The effects of landscape on soil loss from erosion sites
9. Development and performance evaluation of a triple air-flow solar drier
10. Evaluation of surface draining system, which, when installed, collects excess water/water runoff from roads
11. Development and performance evaluation of a double air-flow solar dryer
12. Chemical analysis of wastewater from the cassava peeling process
13. Development and a low-cost solar-powered water pumping system
14. Design and fabrication of solar water heater
15. Evaluation of the impact of untreated sewage effluent on farmlands
16. The effect of moisture content on the mechanical and aerodynamic properties of melon seeds
17. Land suitability evaluation for soybean farming using GIS and AHP techniques
18. Effect of oil spillage on soil physical and chemical properties in river states
19. Development and evaluation of solar power automated sprinkle-on irrigation system
20. Comparative analysis of cans, baskets, and perforated bags as storage structures for okra crops
21. Comparative evaluation of the effectiveness of underground pits as a storage structure for cassava, potatoes, and cabbages

22. Comparative analysis of concrete using rice husk ash and saw dust ash as a partial replacement for cement
23. Organic amendment of soil physiochemical properties using cow dung
24. Scientific application of nanotechnology for water purification
25. The role of soil in purifying wastewater effluence: a study of fishpond water
26. A study of the parameters needed in the design of a motorised yam harvester
27. Evaluation of the effect of anthropogenic activities and domestic waste on surface water quality
28. Development of a digital cone penetrometer
29. Determination of the physical and mechanical properties of saw dust from tropical wood species
30. Design and construction of a portable rainfall simulator
31. Evaluation of the impact of untreated sewage effluent on farmland
32. Design of soil erosion control structures for Ohuhu Nsulu erosion site at Isiala Ngwa North LGA, Abia state
33. Evaluation of waste and waste management in the Umuahia metropolis
34. Selected properties of Bambara Groundnut (*Vigna subterranea* [L.] Verdc.) as influenced moisture content
35. Design of water supply pipeline network
36. Development and evaluation of a prototype solid waste incinerator
37. Design and development of a portable water quality monitoring device
38. Evaluation of waste management techniques in Nigerian Breweries Aba, Abia state.
39. Assessment of the impact of sediment deposits in a water reservoir (a study of NRCRI dam, Umudike, Abia state).
40. Design of erosion control structure for Mgbom Community Primary School, Afikpo North L G A, Ebonyi state.
41. Effect of catchment hydrology on soil erosion at Ndoru
42. Effects of wastewater from fishpond on some selected engineering properties of soil (case study of MOUAU fish farm)
43. Spatial and temporal quality evaluation of the Idaw river for irrigation purposes
44. Determination of soil erodibility indices in Olokoro farmland
45. Soil water status studies of MOUAU farmland
46. Design and fabrication of plastic water tank washing machine
47. Effect of long storage of rainwater and bore-hole water for portable use
48. Physical characterisation of periwinkle shells and rice husks as coarse aggregate for livestock housing

Academic and Technical Staff Disposition

a). Academic

The department of agricultural engineering at MOUAU has twenty-four academic staff made up of six professors, two associate professors, and four senior lecturers. Twelve are lecturer I cadre. On gender, two of the twenty-four academic staff are females, and twenty-two are males. Eleven staff members have Ph.D.s, and twenty-three have master's degrees. This is presented in Table 18.25.

b). Technical Staff

The department has seven technical staff (Table 18.26) who are experienced in the field. The technologists comprise three Principal Technologists I, two Engineers I, one Technologist I, and one Engineer II. One of the technical staff members is female, and six are males.

Impacts of some reforms on Agricultural Engineering Education at Michael Okpara University of Agriculture, Umudike

The major reforms in Agricultural Engineering Education at the Michael Okpara University of Agriculture, Umudike, include a name change, curriculum reviews, course grading systems, examination system, results computation, and reconciliation, etc. Our discussion will focus on the name change and curriculum review because they have the highest effects on the system.

Several internal and external factors make the need for the name change and curriculum review imperative. Some of these factors are listed without the option of priority:

- The enrolment into the programme is declining, static at a point and/or with no appreciable increase
- A number of students who are offered admission to take the course apply to change to other departments after their first year. They use agricultural engineering to gain admission (springboard) only to transfer to other departments, especially when they attain a good grade point
- The students who come for the programme are offered many courses in other departments. Some of them who are not strong enough tend to be scared away, discouraged, and would look for a way out.
- There are emerging technologies which are making the conventional methods of carrying out operations in agriculture obsolete
- Need for entrepreneurship and self-employment in the absence of government jobs
- The high cost of operations and materials calls for alternatives such as solar, biotechnology, etc.

TABLE 18.25

Academic staff disposition of the Department. M = Male F = Female (ABE, 2019a)

S/N	Name and Gender	Rank	Qualifications	Option
1	Prof. V.I.O. Ndirika (M)	Professor	Ph.D.; M.Sc.; B.Tech. (Agric Eng)	Postharvest Engineering
2	Prof. K.J. Simonyan (M)	Professor	Ph.D.; M.Sc.; B.Eng. (Agric Eng)	Postharvest Engineering
3	Professor I.E. Ahaneku (M)	Professor	Ph.D.; M.Sc.; B.Sc. (Agric Eng)	Soil and Water Engineering
4	Prof. A.B. Eke (M)	Professor	Ph.D.; M.Sc.; B.Eng. (Agric Eng)	Postharvest Engineering
5	Prof. U.N. Onwuka (M)	Professor	Ph.D.; M.Sc. Food Process Eng	Food Engineering
6	Prof. J.C. Adama (M)	Professor	Ph.D.; M.Eng.; B.Eng. (Agric Eng)	Farm Power and Machinery Engineering
7	Dr. U.J. Etoamaihe (M)	Professor	Ph.D.; M.Eng.; B.Eng. (Agric Eng)	Food Engineering
8	Dr. M.C. Ndukwu (M)	Professor	Ph.D.; M.Eng.; B.Eng. (Agric Eng)	Crop Processing and Storage Engineering
9	Dr. Oduma Okechukwu (M)	Associate Professor	Ph.D.; M.Eng.; B.Eng. (Agric Eng)	Farm Power and Machinery Engineering
10	Dr. Augustine Onyekachi Igbozulike (M)	Senior Lecturer	Ph.D.; M.Sc.; B.Eng. (Agric Eng)	Processing Engineering
11	Dr. James Ehiem (M)	Senior Lecturer	Ph.D.; M.Eng.; PGD; HND (Agric Eng)	Processing and Storage Engineering
12	Akani O. Akani (M)	Lecturer I	M.Sc.; B.Sc. (Agric Eng)	Farm Power and Machinery Engineering
13	C.U. Orji (M)	Lecturer I	M.Eng.; B.Eng. (Agric Eng)	Postharvest Technology
14	C.C. Emeka Chris (F)	Lecturer I	M.Eng.; B. Eng (Agric Eng)	Soil and Water Engineering
15	Olawale Pearse (M)	Lecturer I	M.Eng.; B.Eng. (Agric Eng)	Soil and Water Engineering
16	Dr. Elijah Ugwu (M)	Lecturer I	Ph.D.; M.Sc.; B.Eng. (Agric Eng)	Soil and Water Engineering
17	Onu Olughu Onu (M)	Lecturer I	M.Eng.; B.Eng. (Agric Eng)	Postharvest Engineering
18	Chukwuka Austine Onyenwoke (M)	Lecturer I	M.Eng.; B.Eng.; HND (Agric Eng)	Postharvest Engineering
19	Stephen Oluwaseun Babalola (M)	Lecturer I	M.Sc.; B.Tech. (Agric Eng)	Farm Structures and Environ. Control
20	Francis Orji (M)	Lecturer I	M.Sc.; B.Eng. (Agric Eng)	Soil and Water Engineering
21	Precious Ehiomogue	Lecturer I	M.Sc.; B.Eng. (Agric Eng)	Soil and Water Engineering
22	Goziechukwu Inekwe (M)	Lecturer I	M.Sc.; B.Eng. (Agric Eng	Power and Machinery Engineering
23	Chidinma Ikechukwu Ede (F)	Lecturer I	M.Eng.; B.Eng. (Agric Eng)	Soil and Water Engineering
24	Okosa Ikechukwu (M)	Lecturer I	M.Eng.; B.Eng. (Agric Eng)	Postharvest Engineering

TABLE 18.26

Technical staff disposition of the Department (ABE, 2019a)

S/N	Name of Staff	Rank/Designation	Qualifications
1	Francis Okechukwu Nkwazema (M)	Principal Technologist	M. Eng. PGD; HND, ND
2	Okafor-Yadi Osemedua (M)	Principal Technologist	PGD; HND
3	Nancy Nkechi Amanze (F)	Principal Technologist	PGD; HND; ND
4	Cyprian Dirioha (M)	Engineer I	B. Eng.
5	Paul Tosin (M)	Engineer I	M.Eng., B.Eng.
6	Joseph Antigha Edet (M)	Technologist II	M.Eng., B.Eng.
7	Joseph Ogbonnaya Unachukwu (M)	Engineer II	B.Eng.

In view of these points, it became evident that a name change and curriculum review were imperative and would have to be a continuous exercise.

Name Change

The first attempt to change the name from agricultural engineering to something else started in 1999. At the Annual General Meeting of the Nigerian Society of Agricultural Engineers (now Nigerian Institution of Agricultural Engineers), which took place at Bauchi, a report of a three-member committee set up to consider a uniform name for agricultural engineering departments in universities and polytechnics in Nigeria was presented. The committee was chaired by Prof. M. O. Faborode.

The committee came up with some recommendations:

TABLE 18.27

Changes in the names of some courses to reflect the new name of the Department (ABE, 2019b)

S/N	New Name	Former Name
1	Agricultural and Bioresources Engineering Concept	Agricultural Engineering Concept
2	Agro-Resources Power and Machinery Systems Management	Power and Machinery Systems Management
3	Agro-Resources Structures and Environmental Control	Farm Structures and Environmental Control
4	Solar Energy Applications in Bioprocess Engineering	Solar Energy Applications in Agricultural Engineering
5	Processing and Storage of Agro-Resources Products	Processing and Storage of Agricultural Products
6	Agricultural and Bioresources Power	Farm Power
7	Agro-Resources Machinery I,	Farm Machinery I,
8	Agro-Resources Machinery II	Farm Machinery II
9	Biomass Engineering.	Biomass Engineering
10	Agro-Resources Transportation and Ergonomics	Farm Transportation and Ergonomics
11	Agricultural and Bioresources Mechanization	Agricultural Mechanization

- Agricultural and Environmental Engineering,
- Agricultural and Biological Engineering,
- AgroMechanical and Environmental Engineering

These recommendations were not implemented. However, the need for a name change continued to arise. At a global level, professional societies have changed or modified their names. For instance, the American Society of Agricultural Engineers changed its name to the American Society of Agricultural and Biological Engineers; the International Commission for Agricultural Engineering changed its name to International Commission for Agricultural and Biological Engineering. The same happened to the professional bodies in India, Canada, etc.

Many departments of agricultural engineering in Nigeria have changed its name to one of the following:

- Agricultural and Environmental Engineering,
- Agricultural and Bioresources Engineering,
- Agricultural and Biosystems Engineering

As stated in sub-section "The Name and Course Curriculum", the department was established with the name "Department of Agricultural Engineering (EAE)". In 2010, the department changed its name to the Department of Agricultural and Bioresources Engineering (ABE). However, this name has not received approval from the National Universities Commission.

Now the National Universities Commission has set up a committee to, among other things, consider a uniform name for all the departments in Nigerian universities. For the department of agricultural engineering, the name "Agricultural and Biosystem Engineering" was recommended and adopted. Universities have been communicated to regarding this name change.

Course Curriculum Review

From its inception to now, the department has reviewed its course curriculum twice. These reviews were done in-house. The last review was done in 2010 when the name was changed to Agricultural and Bioresources Engineering. Some courses changed their names and synopses to reflect the departmental name change (Table 18.27).

The Impacts

Student Enrolment

Figure 18.1 shows the trend of students enrolling in the department for 14 academic sessions (2001/2002 and 2002/2003 and then 2009/2010–2020/2021). As shown in Figure 18.1, 7 students were admitted in the 2001/2002 session. In 2002/2003, the number increased to 18. From there, the number of students kept increasing up to the 2017/2018 session when the number was 155. From the 2018/2019 session, a gradual decrease in enrolment was noticed. The decline is not proportional. For instance, the number of students enrolled declined from 155 in 2017/2018 to 142 in 2018/2019 (a difference of 13 students). While in the 2019/2020 session, the number was 140, a difference of 2 students. A report by the departmental examination officer (Orji, 2023) indicates that even though registration is still ongoing, the number of new students who have registered so far at present 100-level is 112.

Students graduation

The number of students that graduated from the department for a selected number of academic sessions is shown in Figure 18.2, where 7 students graduated in 2005/2006 and 18 graduated in 2006/2007. The highest number of 102 was recorded in 2017/2018 followed by the 2019/2020 set, which was 100.

Graduated Versus Enrolled

A comparative study of Figures 18.1 and 18.2 show that there have been fluctuations in the number of admitted and graduated students over a number of sessions. Apart from the first and second sets of admissions, the number that graduated has been lower than the number that enrolled. In all, there are remarkable improvements in the number of enrolments and graduates compared to what was obtainable in discipline many years back. In the department, there is a reasonable increase in both students that are admitted/enrolled into the department and those that graduated from the department over the years.

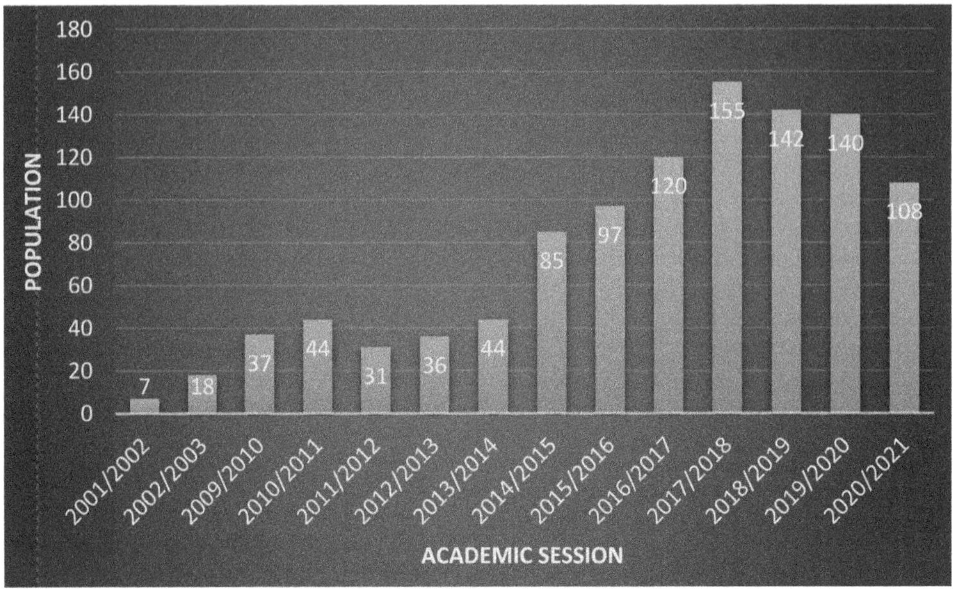

FIGURE 18.1 Trends in Students' Enrolment in the Department over a given period.

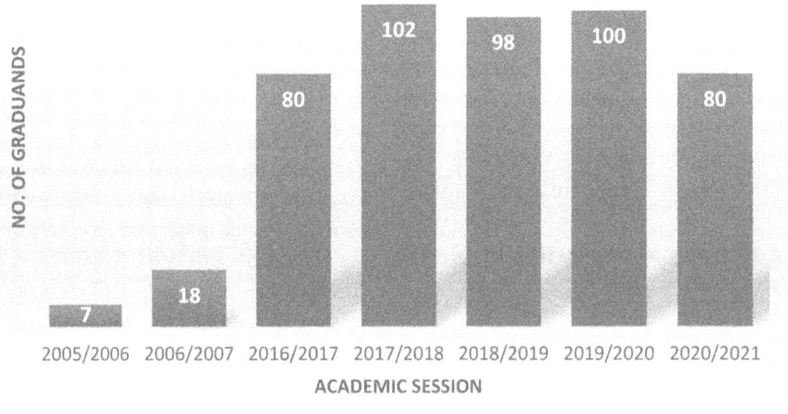

FIGURE 18.2 Trends in Students' Graduation in the Department over a given period.

It is difficult to state categorically that these improvements in the status of enrolled and graduated students are due to the name changes or curriculum reviews only, although they must have contributed. This is because other external factors influence the choice of study course, enrolment, and final graduation in the country. Some of these include government policy, advocacy by professional bodies, job availability, peer group, scholarship openings and global changes.

From experience, it is difficult to have all the students enrolled to graduate simultaneously. There are also internal and external forces that are unconnected with the name or curriculum including:

- Financial constraints
- Health issues
- Lifestyle
- Interference (family, friends, sports, etc.)
- Politics

Students' final year projects

The change in the name of the department and previous reviews of the course curriculum have not seriously impacted the projects carried out by the students in the department. This can be seen in the projects developed in the department. From the project titles, there is no noticeable departure from the existing project works, which have been conventional. Some of the projects undertaken by the students have been presented in the subsection "Some Supervised Final Year Projects in the Department."

Feedback from Industry

As the department prepares for accreditation, a report on the graduates from the employers is required. The department had gone through a number of accreditation programmes/visits. These are those of the National Universities Commission (three times) and the Council for the Regulation of Engineering in Nigeria (two times). To fulfil this requirement, the department requested and received reports on our graduates' performance from their employers. Some establishments where our graduates work include Independent Electoral Commission, Abia State, United Bank for Africa, Enugu State Polytechnic, Michael Okpara University of Agriculture, Umudike Alex Ekwueme Federal University, Ndufu Alike, World Bank assisted International Fund for Agricultural Development(IFAD) Enugu, the National Assembly, etc. All the reports from the employers show that the employees are doing very well in the duties assigned to them.

For those students who remained in the department to graduation, they are always more versatile. They compete with civil, mechanical, chemical, and other engineering graduates in job interviews and in the field of practice. Hence, products of the department are offered appointments in mechanical, civil, electrical, biomedical, mechatronic, and other engineering fields. This is in addition to jobs in areas such as tillage, management of agricultural machinery, erosion control, irrigation, products processing, etc., reserved for agricultural engineering graduates. Also, the products of the department are found working in areas of biotechnology, waste recycling, environmental management, artificial intelligence, etc.

Industrial Training

Students of the department undergo three stages of industrial training: Students Work Experience Programme, SWEP I (Table 18.7), Students Work Experience Programme, SWEP II (Table 18.9), and the Students Industrial Work Experience Scheme, SIWES (Table 18.11). These training stages give our students an edge over students from other institutions. Sometimes, the experience acquired by the students in this programme guides them in life even after graduation. This is because, at the place of training, the students should pass through different departments/units of the establishment. Some notable and relevant establishments where our students frequently undertake the SIWES programme are the National Centre for Agricultural Mechanization Ilorin, National Root Crops Research Institute, Umudike, River Basin Development Authority, Agricultural Development Programme, etc. The students are supervised by both the university and the Industrial Training Fund. The SIWES programme is a good source of information to the Department, the College, and the University on the students' performance in the industry. The Industrial Training programmes of the department have impacted positively on the students. For instance, many students get their final project direction from experience gathered at the place of the SIWES.

Change of course

Experience shows that since the change and curriculum review, the number of students that apply to transfer from the department to other departments after the first or second year has reduced drastically. For the past 7 years, except for this session when two students applied to the Senate for a change of course, no exodus of students to other departments was recorded.

Again, one cannot attribute this reduction in the number of students that change from agricultural engineering to name change and curriculum review only as there are other factors, as listed above, that could contribute to the reduction.

Conclusion and Future Prospects

Conclusion

Agricultural engineering education at the Michael Okpara University of Agriculture Umudike is examined from the point of view of the programme run by the Department of Agricultural Engineering in the university, which are reforms (name changes, curriculum review, mode study), the impacts of the reforms on student enrolment, graduation, research outputs, employment, and employer ratings.

There are internal and external factors which make these reforms imperative: declining enrolment by students into the programme, the rate at which the students change to other departments after their years of study, the inability of some students to cope with heavy course loads, emerging technologies which are making the conventional methods of carrying out operations in agriculture obsolete, the need for entrepreneurship and self-employment in the absence of government jobs, and the high cost of operations and materials which calls for alternatives such as solar, biotechnology, etc. All the reforms are to address these issues.

The curriculum reviews introduced new courses to reflect the changing trends in society. In each case, there are shifts from the basic science, agriculture, and humanities content of the courses to the engineering content of the courses. The new curriculum approved by the National Universities Commission emphasised a radical shift from all other courses, including core engineering courses, to the agricultural engineering content of the course.

Findings show a reasonable increase in both students who are admitted/enrolled into the department and those who graduated from the department over the years. Also, employment of graduates in different parts of the world and in many sectors has been encouraging. However, it is difficult to state categorically that these improvements in the status of enrolled and graduated students are due to the name change or curriculum reviews. This is because there are other external factors that influence the choice of course of study, enrolment, and final graduation in the country. Some of these include government policy, advocacy by professional bodies, job availability, peer group, scholarship openings,

Prospects

The new curriculum developed by the NUC will give agricultural engineering a strong base in the university as it captures many areas such as biotechnology, digitalisation, artificial intelligence, and precision agriculture, which are components of the Nigeria Digital Agriculture Strategy. The curriculum will lead to staff training and the engagement of new staff to handle the new courses in the department.

The Federal government's resolve to diversify the economy through agriculture, the increase in the population requiring more people to feed, and the drudgery associated with hand-powered agriculture requiring mechanisation, are some of the factors that make the prospects of agricultural engineering in the university bright.

There are proposals in the pipeline to mount a separate college/faculty for the agricultural engineering programme within the university.

Recommendation

Reforms for the agricultural engineering programme in the university should be an ongoing project. All the stakeholders should be involved in advocacy nationwide to create awareness of the role of agricultural engineering in society.

Acknowledgements

The authors wish to acknowledge O. I. Abarikwu, J.O Olaoye, M. Enabifo, Wasiu Limidi and F. N. Orji for giving us information via virtual communication. We are also grateful to the Head, Department of Agricultural Engineering Michael Okpara University of Agriculture, Umudike, for giving us access to the documents in the office where we obtained necessary data and information.

REFERENCE LIST

Abariukwu, O.I. (2023). Personal communication with the authors via phone call. March 21.

Adama, J.C. (2017). Unpublished Field Survey on List of Universities Offering Agricultural Engineering in Nigeria. Submitted to the Editor in Chief, Agricultural Mechanization in Africa, Asia and Latin America(AMA).

Akinbolade, J. (2023). Personal communication with authors. April 22,

Igbeka, J.C. (2002). Agricultural Engineering and the Society. *NSAE Quarterly*, Vol. 3, No. 2. June–April. Published by the Nigerian Society of Agricultural Engineers.

Igbeka, J.C. (2023). Personal communication with authors. April 21.

Orji, F.(2023). Communication with the Authors via Phone Call. March 17.

Michael Okpara University of Agriculture Umudike. (2010). Department of Agricultural and Bioresources Engineering Michael Okpara University of Agriculture, Umudike. Students' Hand Book. Published by PATES 07068022608.

ABE. (2019a). Department of Agricultural and Bioresources Engineering, Michael Okpara University of Agriculture, Umudike. Self Study Form (Unpublished).

ABE. (2019b). Department of Agricultural and Bioresources Engineering, Michael Okpara University of Agriculture, Umudike. Students' Hand Book. Published by FUTURE PRINTS: Umudike

Michael Okpara University of Agriculture Umudike. (2021). Historical Background of Michael Okpara University of Agriculture, Umudike. Retrieved 2021-06-03. www.mouau .edu.ng

Joint Admissions and Matriculation Board. (2022). Guidelines for Admission to First Degree, National Diploma (ND), and Nigerian Certificate in Education (NCE). http://www .jambng.org,ng

National Universities Commission. (2021). Core Curriculum and Minimum Academic, Standards for Nigerian Universities System (CCMAS): Engineering/Technology, Abuja, Nigeria. ..

Samphina Academy. (2023). Universities in Nigeria that Offer Agricultural Engineering. https://samphina.com.ng/

19

Agricultural Biosystems Engineering Education in Indonesia – A Journey

Hadi K Purwadaria

History of Agricultural Biosystems Engineering (ABE) in Indonesia

The Development of Agricultural Mechanisation to Agricultural Engineering (AE) and Agricultural Biosystems Engineering (ABE) Study Programmes

Indonesia is the largest archipelago in the world, spreading latitudinally from 6° North to 11° South and longitudinally from 95° West to 141° East, or can be summarised as 5° S and 120° E (Wikipedia, 2022a). The total area is 1,919,440 km²; however, since the land is only 24% and the water is 76%, it has a long coastline of 54,720 km. That constitutes an Exclusive Economic Zone (EEZ) of 6,159,032 km². Compared to Indonesia, the USA, for example, has a latitude and longitude of 37° N and 95° W, an area of about two times Indonesia at 9,826,675 km², but a coastline of about 1/3 of Indonesia's 19,920 km. Indonesia is flanked by two large continents, Asia and Australia, and by two large oceans, the Indian and Pacific Oceans.

The Indonesian population was 276.2m in 2022, ranking Indonesia as the 4th major populated country in the world, next to the USA, which has a population of 338.9m (World Population Review, 2022). Since the number of islands in Indonesia is very large – either 13,466 as surveyed by the National Coordinating Agency for Survey and Mapping (Bakosurtanal), Indonesia, 18,703 by the National Institute of Aeronautics and Space (LAPAN), or 17,508 by the CIA World Factbook – Indonesia faces a big challenge in delivering a qualified education equally to all people occupying many of the islands. Fortunately, most people are flocking to the five bigger islands of the country. Starting with the largest, the major islands in consecutive order are Papua (New Guinea), Kalimantan (Borneo), Sulawesi (Celebes), Sumatra, and Java. Though, the most populated islands are in reverse with their sizes: Java has 56.1%, Sumatra 21.7%, Sulawesi 7.4%, Kalimantan 6.2%, and Papua 3.2% of the total Indonesian population.

With Indonesia's geographical and population background, the development of ABE education across the country can be better appreciated. As may be common in other nations, ABE education started with Agricultural Mechanization then later transformed into AE, and presently some are transforming into ABE. The first study programme in Agricultural Mechanization was founded in the Faculty of Agriculture at Andalas University, city of Padang, province of West Sumatra, in 1957. Andalas University was the fourth university established by the government, following the Gadjah Mada University at Yogyakarta, DIY, the University of Indonesia at Jakarta, DKI Jakarta, and the Airlangga University at Surabaya, East Java.

The Agricultural Mechanization study programme in the Faculty of Agriculture spread from Andalas University to Gadjah Mada University in 1963 and IPB University (IPB) at Bogor, West Java in 1964. IPB was the former campus of the University of Indonesia, focusing on agricultural education and was located 60 km south of Jakarta, separate from the main campus in Jakarta. It received the Presidential Decree to stand out as an independent university in 1963. It should be noted that the first campuses of IPB were built in cooperation with the US government under the Kentucky Contract Team in 1955. It is important to mark that the Agricultural Mechanization study programme at IPB was established in the new Faculty of Agricultural Technology and Mechanization, which later transformed into the Faculty of Agricultural Technology (Mandang et al., 2018), not in the Faculty of Agriculture as in the two previous universities. This initiative would be followed later on by other similar universities in Indonesia. This first step taken by Agricultural Mechanization at IPB University will be followed up with a series of other initiatives characterising the study programme as the trendsetter of AE/ABE development in Indonesia until now.

Many of the faculty staff in the Agricultural Mechanization study programme at IPB had been fortunate to get the opportunity to pursue their Master's and PhD degrees in the USA under the Kentucky Contract Team Programme in 1957–1966, and the Midwestern University Consortium for International Assistance – United States Agency for International Development (MUCIA-USAID) programme in 1970–1981 which extended to 1988 (USAID, 1988). In the USA, the IPB faculty staff realised that they were studying AE, which had significant differences from Agricultural Mechanization in the curriculum course contents and degree recognition.

DOI: 10.1201/9780429150111-21
This chapter has been made available under a CC-BY-NC-SA license.

In returning home, six US and one Japanese PhD alumni observed that they could restructure the existing Agricultural Mechanization curriculum into the AE curriculum and they could deliver AE engineering courses, so they decided to transform Agricultural Mechanization into an AE study programme, also changing the name of the faculty to the Faculty of Agricultural Technology in 1981.

It took about one generation of education (10–15 years) for other universities to follow the step since they were required to prepare a sufficient number of PhD faculty staff to manage the AE study programme by sending them either to the AE graduate programme at IPB University that had been initiated in 1979 or to AE graduate programmes abroad. Table 19.1 illustrates the transformation process at IPB and the other 16 universities offering AE/ABE in Indonesia. In the other universities, the changes from Agricultural Mechanization to AE study programmes happened in 1989 (STIPER Agricultural Institute), 1991 (Hasanuddin University), 1994 (Brawijaya University), 1995 (Sriwijaya University), 1996 (Andalas University, Gadjah Mada University, Padjadjaran University, and Sam Ratulangi University), and 1998 (Jember University).

When other universities established the AE study programme at a later stage (1995–2014), they did not start with Agricultural Mechanization but directly constructed the AE curriculum. An observation from Table 19.1 indicated that eight universities did exactly that: Udayana University and Lampung University in 1995, North Sumatra University in 1996, Syiah Kuala University in 1997, Soedirman University in 2000, Mataram University in 2005, and Jambi University in 2014.

By sending many of the faculty staff for AE PhD degrees abroad as early as 1974 who returned home in 1979–1980, the AE study programme of IPB University created AE graduates who worked at many universities in the provinces. After getting their PhD degrees either abroad or back at their alma mater at the AE graduate studies of IPB University, they set up the AE study programme at their universities. Out of the 16 other AE study programmes in Indonesia, the alumni of IPB played significant roles and leadership at about eleven universities.

The transformation from the AE into the ABE study programme at IPB University was unique because it took a stepping stone in the Mechanical and Biosystems Engineering study programme. In other universities, the transformation of AE into an ABE study programme proceeded directly. The process of transformation is still going on in Indonesia. In 2022, only six universities were transformed into the ABE study programme; another is under process (Table 19.1), while others have not yet fallen into a similar step. One of the reasons could be the limited number of PhDs they have, so they are waiting to complete their staff education abroad or at IPB University, Gadjah Mada University, or other national universities that have now offered graduate study.

In an infographic map, the current situation can be presented in Figure 19.1, illustrating the spread of the AE and ABE study programmes across provinces in Indonesia. The number marked on the map is the same number listed in Table 19.1 so the viewing of Figure 19.1 should be accompanied by looking at Table 19.1.

The map showed one of the big challenges faced by Indonesia in establishing quality education equally from island to island, from the western to the eastern part of the country. Indonesia consists of 34 provinces distinguished by different shades on the map (Figure 19.1). The government recently added 4 more provinces in Papua, the largest island (not yet appearing on the map), even though the population is the lowest among the five biggest islands in the nation. AE and ABE study programmes exist in 17 universities in 13 provinces located on the three bigger islands of Java, Sumatra, and Sulawesi, and the two smaller islands of Bali and West Nusa Tenggara. Agricultural engineers in Indonesia still have to work hard to extend the AE or ABE higher education to other universities in the provinces located in the other bigger islands of Kalimantan and Papua, and the smaller islands of East Nusa Tenggara and Maluku as well. Interestingly, only one private university – Stiper Agricultural Institute, Yogyakarta, DIY (Stiper, 2022) offers AE/ABE, while the 16 others are state universities.

Indonesia's National Accreditation Board for Higher Education accredits all university study programmes. Sixty-five per cent of Indonesia's 17 AE/ABE study programmes are accredited A, while others received accreditation B, and none got C. In 2014, the Mechanical and Biosystems Engineering Study Programme at IPB University took a further step to being accredited by JABEE (Japan Accredited Board for Engineering Education), a signatory member of the Washington Accord. The duration of accreditation is valid until 2019. After transforming into Agricultural and Biosystems Engineering, it obtained IABEE (Indonesia Accreditation Board Engineering Education) accreditation from 2020–2022. IABEE was accepted as a provisional signatory member of the Washington Accord in 2019 and as a full signatory member in 2022.

Definition of Agricultural Mechanization and ABE

Agricultural Mechanization in Indonesia was defined as a discipline of science that explores natural resources and energy for the development of human creativity in agriculture for the prosperity of human beings. The definition was formulated at the First Symposium of the Indonesian Association of Agricultural Mechanization Professions at Ciawi, Bogor, in 1967 (Setiawan et al., 2006, p. 1). Later, the association was transformed into the Indonesian Association of Agricultural Engineers (PERTETA), up to the present.

The definition of ABE is more varied among the ABE study programmes across the country. For example, IPB University defines ABE as a science in engineering and its applications for machinery and process modelling related to sustainable biological systems in food and agriculture (Teknik Pertanian & Biosistem IPB, 2022). Gadjah Mada University emphasises that ABE focuses on using engineering and management in biomass production to fulfil people's needs and maintain environmental sustainability (Departemen Teknik Pertanian dan Biosistem UGM, 2022). Furthermore, Brawijaya University describes ABE as the applications of production technology using agricultural products, materials, and natural power, emphasising the environment and the agricultural production and processing system (Teknik Pertanian dan Biosistem UB, 2022).

TABLE 19.1

The transformation of Agricultural Mechanisation to AE and ABE study programmes in universities in Indonesia.

No	Founded Study Programme, University, Year	Transformation 1, Year	Transformation 2, Year	Transformation 3, Year
1	Agricultural Mechanization, Faculty of Agriculture, Andalas University, Padang, West Sumatera, 1957	AE, Faculty of Agriculture, 1996	ABE. Faculty of Agricultural Technology, 2020	
2	Agricultural Mechanization, Faculty of Agriculture, Gadjah Mada University, Yogyakarta, DIY, 1963	Agricultural Mechanization, Faculty of Agricultural Technology, 1964	AE, 1996	ABE, 2020
3	Agricultural Mechanization, Faculty of Agricultural Mechanization and Technology, IPB University, Bogor, West Java, 1964	AE, Faculty of Agricultural Technology, 1981	ABE/Mechanical and Biosystems Engineering, 2010 *Accredited by JABEE from 2014-2019*	ABE, 2017 *Accredited by IABEE from 2020-2022*
4	Agricultural Mechanization, Faculty of Agriculture, Brawijaya University, Malang, East Java, 1974	AE, Faculty of Agriculture, 1994	AE, Faculty of Agricultural Technology, 1998	ABE, 2021
5	Agricultural Mechanization, Faculty of Agriculture and Forestry, Hasanuddin University, Makassar, South Sulawesi, 1980	AE, Faculty of Agriculture and Forestry, 1991	AE, Faculty of Agriculture, 2008	
6	Agricultural Mechanization, Faculty of Agriculture, Padjadjaran University, Bandung, West Java, 1983	AE, Faculty of Agriculture, 1996	AE, Faculty of Agro-Industrial Technology, 2005	
7	Agricultural Technology, Faculty of Agriculture, Jember University, Jember, East Java, 1983	AE, Faculty of Agricultural Technology, 1998		
8	Agricultural Mechanization, Faculty of Agriculture, Sriwijaya University, Palembang, South Sumatera, 1984	AE, Faculty of Agriculture, 1995		
9	Agricultural Mechanization, Faculty of Agriculture, Sam Ratulangi University, Manado, North Sulawesi, 1984	AE, Faculty of Agriculture, 1996		
10	Agricultural Mechanization, Faculty of Agriculture, Stiper Agricultural Institute, Yogyakarta, DIY, 1985	AE, Faculty of Agricultural Technology, 1989		
11	AE, Rector Office, Udayana University, Denpasar, Bali, 1995	AE, Faculty of Agricultural Technology, 2004	ABE, 2020	
12	AE, Faculty of Agriculture, Lampung University, Bandar Lampung, Lampung, 1995			
13	AE, Faculty of Agriculture, North Sumatera University, Medan, North Sumatera, 1996	ABE, Faculty of Agriculture, 2021		
14	AE, Faculty of Agriculture, Syiah Kuala University, Banda Aceh, DI Aceh, 1997			
15	AE, Faculty of Agriculture, Soedirman University, Purwokerto, Central Java, 2000			
16	AE, Faculty of Agriculture, Mataram University, Mataram, West Nusa Tenggara, 2005	AE, Faculty of Food Technology and Agroindustry, 2012	ABE, 2022 (under process)	
17	AE, Faculty of Agricultural Technology, Jambi University, Jambi, Jambi, 2014			

In comparison, the American Society of Agricultural and Biological Engineers (ASABE) mentions that ABE refers to the application of engineering science and designs to the processes and systems involved in the sustainable production, post-production, and processing of safe food, feed, fibre, timber, and other agricultural and biological materials, and the efficient (ASABE, 2022). Discussing the evolution of AE education into Biosystems Engineering education in Europe, Briassoulis et al. (2008) defined Biosystems Engineering as a science-based engineering discipline that integrates engineering science and design with applied biological, environmental and agricultural sciences, broadening in this way the area of application of engineering sciences not strictly to agricultural sciences, but to the biological sciences in general, including the agricultural sciences. This was repeated by Holden et al. (2020, p. vii) in an ASABE-published textbook stating that

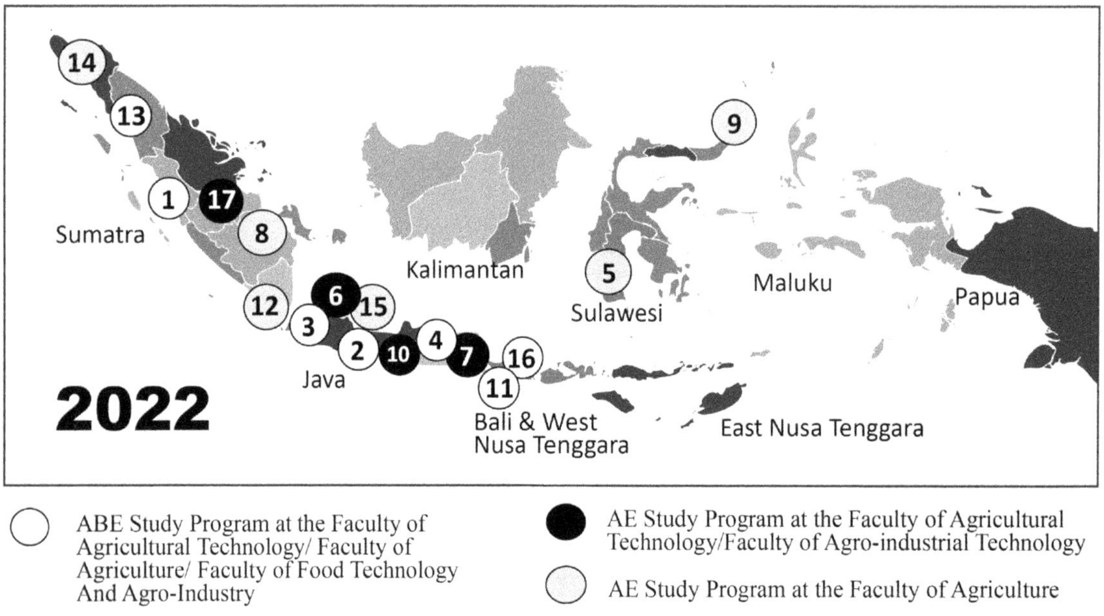

FIGURE 19.1 Map of universities offering Agricultural Engineering/Agricultural Biosystems Engineering (AE/ABE) Study Programmes in Indonesia in 2022.

Biosystems Engineering integrates engineering science and design with applied biological, environmental, and agricultural sciences.

The Development of Divisions in Agricultural Mechanization and AE/ABE Study Programmes

The development of divisions under Agricultural Mechanization and AE/ ABE study programmes in Indonesia is closely influenced by the visions of the faculty staff, the resources owned by the related university, the participation of the faculty staff in either the national or international cooperation programmes in agricultural development, the graduate education experienced by the faculty staff, and the development of AE/ABE in the world. The weight of the influential factors may not be divided equally, but they are connected. For example, the vision of the faculty staff will motivate them to find the resources needed to deliver good teaching and research and to find and participate in national and multinational programmes enhancing the agricultural programmes in Indonesia. Their graduate education experiences and the network they established while studying abroad or domestically support their motivation.

In 1978–1990, the JICA (Japan International Cooperation Agency) Programme significantly improved the education quality of the AE study programme at IPB by strengthening the resources and capacity building of the Faculty of Agricultural Technology, IPB University, including the AE study programme. The programme provided 10,000 m^2 of faculty buildings, laboratories, equipment, a pilot plant, and a massive graduate scholarship. As an alumnus of a Japanese university, the AE faculty staff, Dr Kamaruddin Abdullah – later a full professor – had been playing a key role in obtaining the JICA grant. It was also one of the major factors that motivated the decision to change from the Agricultural Mechanization study programme to the AE study programme at IPB in 1981 (Table 19.2), and the action helped support the AE-IPB as the trendsetter in the AE/ABE education development in Indonesia. This case was a model of how a visionary faculty staff was highly motivated to leverage his network into a significant multinational cooperation that fully benefited all.

National and multinational cooperation programmes in postharvest technology to develop grain and horticultural product postharvest machinery in Indonesia occurred a lot in the decade of 1980–1990. Early in the 1980s, the UNFPA project provided a programme to establish small soybean processing enterprises for homemakers in Java, Bali, and West Nusa Tenggara villages. An AE staff was involved in manufacturing tofu and tempeh processing machinery and helped the small plants operate as business entities. From the mid-1980s until the mid-1990s, AGPP (ASEAN Grain Postharvest Programme), later on broadened to ACPHP (ASEAN Crops Postharvest Programme) to also cover other crop commodities, GTZ, IDRC, and IRRI handed over several programmes to distribute out paddy threshers, maize shellers, soybean threshers, horticultural and tuber dryers, and paddy combine harvesters to provinces in Java, Sumatra, and Sulawesi. The projects involved not only AE faculty staff from IPB but also from Gadjah Mada University and Andalas University as well. The trend in the national development encouraged the IPB faculty staff of Food Technology and AE to form and open a new Master's degree in Postharvest Technology in 1983, led by the AE faculty staff as the Chair, which later on was put permanently under the AE study programme. The interest in this field was also indicated by several AE study programmes creating divisions of either Postharvest Engineering or Postharvest Technology at Andalas University, Udayana University, and most recently, Jambi University (Table 19.2).

The ABE study programme at IPB does not contain the Division of Soil and Water Engineering like in the other universities (Table 19.2) because the division moved to form a new

TABLE 19.2

Development of divisions in Agricultural Mechanisation, AE, and ABE study programmes.

University	Divisions of the Agricultural Mechanization Study Programme	Divisions of the AE Study Programme	Divisions of the ABE Study Programme
IPB	1964* 1. Farm Machinery 2. Agricultural Product Processing Machinery 3. Food Processing Machinery 4. Soil and Water Engineering 5. Farm Structure 6. Farm Electrification	1981 1. Farm Machinery Engineering 2. Soil and Water Engineering 3. Agricultural Mechanization System and Management 4. Food and Agricultural Product Processing Engineering 5. Energy and Electrification 6. Farm Structure and Environment 7. Agricultural Ergonomics and Electronics	2010 1. Mechanical Engineering and Automation 2. Biosystems Engineering 3. Renewable Energy Engineering 4. Bioinformatics Engineering
Other Universities	After 1964 1. Similar to IPB but selected divisions only 2. Modification of some selected divisions from Agricultural Mechanization IPB: 2.1. Farm Machinery and Equipment 2.2. Agricultural Product Technology 3. Unique divisions created by University: 3.1. Postharvest Engineering by Andalas University in 1974	After 1981 1. Similar to IPB but selected divisions only 2. Modification of some selected divisions from AE/ABE-IPB: 2.1. Farm Machinery and Equipment Design; and Farm Power and Machinery 2.2. Land and Water Resources Engineering 2.3. Management for Farm Machinery and Equipment; Management and Information System for Farm Machinery and Equipment 2.4. Agricultural Product, Food and Postharvest Engineering 2.5. Postharvest Technology 2.6. Thermal System and Renewable Energy Engineering 2.7. Bioinformatics and Information Engineering 2.8. Oil Palm Mechanization 2.9. Agronomy and Mechanization Technology 2.10. Agro-Informatics Technology 3. Unique Divisions created by Universities: 3.1. Instrumentation by Jember University in 1998; and Instrumentation and Control by Hasanuddin University in 2008 3.2. Bioprocess Engineering by Mataram University in 2005 3.3. Pre-Harvest Engineering by Udayana University in 2005 3.4. Environmental Engineering/Control and Conservation by Mataram University in 2005 and Jember University in 2011 3.5. Agricultural Energy, Automation, and Informatics by Jember University in 2011 3.6. Bio-Environmental Management and Control Engineering, and 3.7. Information and Agro-Industrial Engineering by Soedirman University in 2016 3.8. Farm Workshop by Syiah Kuala University in 2016, and Mechanical and Farm Workshop Engineering in 2020 3.9. Oil Palm Industrial Engineering, 3.10. Mechatronics and Agricultural Automation, and 3.11. Hortismart Technology by Stiper Agricultural Institute in 1985	After 2010 1. Maintain previous AE divisions 2. Modification of some selected divisions from AE-IPB: 2.1. Soil and Water Resources Engineering 2.2. Equipment Design and Ergonomics 2.3. Agricultural Engineering Management System 3. Unique Divisions created by Universities: 3.1. Biophysics by Gadjah Mada University in 2020 3.2. Natural Resources Management by Udayana University in 2020 3.3. Biosystems Mechatronics by Brawijaya University in 2021

* Andalas University, founded in 1957 with no division
Gadjah Mada University, founded in 1963 with two divisions: Farm and Agricultural Machinery; and Soil and Water Preservation Engineering

study programme, in Civil and Environmental Engineering, that was still under the Faculty of Agricultural Technology in the same time with the transformation of AE into ABE study programme in 2010.

Other Issues Addressed by AE-IPB University as the Trendsetter in the Education System

Higher education in Indonesia in the era after independence, around 1945, was still following the conventional system. Until 1980, undergraduate students experienced annual evaluations that would hand them the decision to pass or not pass to the next level. If the students were granted not pass, they had to repeat all courses at the same level they took previously regardless of the grade they gained for the courses. All the courses with grades A, B, C, D, to F were to be repeated, giving them a boring and frustrating load. This system caused many students to stay as recidivists as they were called for more than two years at one level, thus stacking up the number of students at every level, producing many dropouts, and lengthening the study duration unnecessarily. The ratio of output over input at the universities became small, and the cost of education was high and had to be shouldered by the government since, at the time, students in state universities were free from tuition.

In 1981, the fresh PhD graduates returning from the USA to the AE study programme at IPB University observed that the lagging situation could not be further tolerated if the Indonesian higher education system wanted to emerge in an equal quality with other global universities. By chance, two of them were appointed as Dean and Vice Dean for Academic Affairs of the Faculty of Agricultural Technology, so they introduced the credit unit system evaluation every semester instead of an annual evaluation system. The credit unit system allowed the students to take new courses offered for the next semester, accepting the validity of A-, B-, and C-graded courses of the previous semester not to be repeated. However, students had to repeat the D- and F-graded courses from the previous semester in addition to the new courses. The new system also cut the internship from two times to one time, trusting in one high-quality experience rather than in quantitative numbers. The internship moved to the end of the 6th instead of the 7th semester, avoiding conflict with the research manuscript preparation, starting early in the 8th semester.

Many, including the IPB rectorate office, were opposed to the initiatives, thinking that they required a good computer system which was still scarce at the time, but it was proven successful even though it operated manually. Surprising to many people, some of the first batch students exposed to this system could finish their BS-AE degree in seven semesters or 3.5 years, another advantage of this kind of system. When IPB University, as a whole, adopted the credit unit system, it was broadly followed by other universities in Indonesia, improved, and much later computerised. Before long, as everybody realises, the credit unit system, called SKS, is now adopted as a natural part of the Indonesian higher education system as in other universities in the world.

The Curricula of the AE/ABE Study Programme

The AE/ABE study programme for undergraduates in Indonesia is 4 years of study consisting of 8 semesters. Febo and Comparetti (2013, pp. 2–3) and Comparetti and Febo (2018) note that in Europe, the length of study for the ABE programme is planned to be 3 years, called pivot-point. This means the European ABE BS system will have a one-year advantage. It further impacts the MSc education system for ABE. In Europe, an MSs in ABE can be completed in 3 + 2 years, while in Indonesia it takes 4 + 2 years. No information is available on whether the US education system will shorten the 4-year BS degree. However, the difference with the Indonesian education system is that the US system does not include research in the BS study programme.

The structure of the AE/ABE curricula, in general, can be presented as listed in Table 19.3 taking an example of ABE curricula at IPB University (Widodo & Supriyanto, 2020). The terms and the total credits for the undergraduate programme may be slightly different from one university to another university, however, the principles are similar.

TABLE 19.3

The structure of ABE curricula at IPB University (Widodo & Supriyanto, 2020).

No	Element	SKS	Percentages
1	Common and Fundamental Courses	36	24.5
2	Core (Engineering) Courses	36	24.5
3	In-Depth/Competencies Courses	30-31	21.0
4	Enrichment Courses	22	30.0
5	Final Projects	22	30.0
	Total	146-147	100.0

The core engineering courses included i. Engineering Drawing, ii. Engineering Mechanics, iii. Strength of Materials, iv. Measurement and Instrumentation, v. Fluid Mechanics, vi. Engineering Thermodynamics, vii. Heat Transfer, viii. Electricity and Electronics. This has similarities with the core engineering courses listed by Briassoulis et al. (2013, p. 310), i.e., i. Engineering Graphics and Design (CAD), ii. Mechanics (Statics), iii. Strength of Materials, iv. Mechanics (Dynamics), v. Fluid Mechanics, vi. Applied Thermodynamics, vii. Heat and Mass Transfer, viii. Electricity and Electronics, and ix. System Dynamics.

The Final Projects may include several courses and projects allocated in Semesters 7 and 8. The courses are usually the capstone courses such as Engineering Design and Agricultural Machinery and Equipment Design. The projects may cover Engineering Design Project, Research Manuscript, which is an undergraduate research report, Seminar on Research Proposal, Seminar on Research Results, and Scientific Publication.

Two curricula of the ABE Study Programmes are selected to represent AE/ABE education in Indonesia in general. First, from the IPB University, Bogor, West Java, as the trendsetter of AE and ABE development in the nation, and the second is from Udayana University, Den Pasar, Bali, which is relatively newly founded and developed but already has an ABE Study Programme. The curricula of the ABE Study Programme for undergraduates at IPB University are presented in Table 19.4 (Widodo & Supriyanto, 2020). The curricula of the ABE Study Programme for undergraduates at Udayana University, Denpasar, Bali are presented in Table 19.5 (Wirawan et al., 2022).

Current Research in AE/ABE Integrated with Emerging Technology

Research is carried out by the faculty staff and intertwined with undergraduate education. In the universities offering graduate courses, research is part of the education system from undergraduate to graduate. In 2022, out of the 17 universities with AE/ABE study programmes, three offer MSc and PhD graduate study: IPB University at Bogor, West Java, Gadjah Mada University at Yogyakarta, DIY, and Padjadjaran University at Bandung, West Java. Three other universities offer only MSc graduate study: Andalas University at Padang, West Sumatera; Brawijaya University at Malang, East Java; and Hasanuddin University at Makassar, South Sulawesi.

TABLE 19.4

Curriculum of the ABE Study Programme at IPB University, Bogor, West Java (Widodo & Supriyanto, 2020).

No	Courses	SKS*	No	Courses	SKS
Semester I			**Semester II**		
1	Pancasila (Indonesian Way of Life)	2	1	Religion	3
2	English	3	2	Indonesian Language	2
3	Introduction to Mathematics	3	3	Introduction to Agriculture	2
4	Biology	3	4	Calculus	3
5	Physics	3	5	Chemistry	3
6	General Sociology	3	6	General Economics	3
7	Introduction to Entrepreneurship	1	7	Introduction to Agricultural Technology	2
	Total	**18**		**Total**	**18**
No	**Courses**	**SKS**	**No**	**Courses**	**SKS**
Semester III			**Semester IV**		
1	Principles of Agronomics	3	1	Surveying	2
2	Calculus II	3	2	Fluid Mechanics	3
3	Engineering Mechanics	3	3	Engineering Mathematics	3
4	Engineering Materials	2	4	Engineering Drawing	3
5	Workshop	3	5	Strength of Materials	2
6	Engineering Characteristics of Agricultural Materials	2	6	Engineering Thermodynamics	3
7	Engineering Programming	3	7	Automatic Control	3
8	Measurement and Instrumentation	3	8	Engineering Statistics	3
	Total	**22**		**Total**	**22**
No	**Courses**	**SKS**	**No**	**Courses**	**SKS**
Semester V			**Semester VI**		
1	Engineering Hydrology	3	1	Engineering Entrepreneurship	1
2	Machine Elements	3	2	Farm Machinery and Equipment	3
3	Integrated Laboratory Practices: Engineering Mechanics and Materials	1	3	Irrigation and Drainage Engineering	2
4	Engineering Economics	3	4	Agricultural Product Processing Engineering	2
5	Heat Transfer	2	5	Agroindustry Management	3
6	Engine and Driving Power	3	6	Energy and Farm Electrification	3
7	Farm Structure and Environment	2	7	Scientific Writing and Presentation Methods	1
8	Agricultural Mechanization System and Management	3	8	Greenhouse and Hydroponics Technology	2
			9	Elective 1	2-3
	Total	**20**		**Total**	**19-20**
No	**Courses**	**SKS**	**No**	**Elective Courses**	**SKS**
Semester VII			**Semester VI**		
1	Food Process Engineering	3	1	Principles of Biosystems Process Engineering	2
2	Engineering Design	2	2	Ergonomics and Work Safety	3
3	Agricultural Machinery and Equipment Design	3	3	Agricultural Informatics Engineering	2
4	Integrated Laboratory Practices: Biosystems Environmental Control Engineering	1	4	Refrigerated Engineering	3
5	Postharvest Engineering	2	**Semester VII**		
6	Thematic Community Services	3	1	Robotics Engineering	3
7	Seminar on Research Proposal	1	2	Relationship of Soil and Farm Machinery and Equipment	3
8	Elective 2	3	3	Renewable Energy Conversion Engineering	3
9	Elective 3	3	4	Audit Energy	3
	Total	**21**	5	Supporting Course from Other Study Programmes	3
Semester VIII					
1	Research Manuscript	6			
	Total	**6**			
	Grand Total 146-147 SKS				

*SKS: *Satuan Kredit Semester* (Semester Credit Unit)

TABLE 19.5

Curricula of the ABE Study Programme at Udayana University, Bali, Denpasar (Wirawan et al., 2022).

No	Courses	SKS*	No	Courses	SKS
Semester I			**Semester II**		
1	Indonesian Language	2	1	Soil Biophysics	3
2	Social Science and Cultural Principles	2	2	Thermodynamics	2
3	Introduction to Agricultural Technology	2	3	Engineering Materials	2
4	English	2	4	Fluid Mechanics	3
5	Pancasila (Indonesian Way of Life)	2	5	Statistics	2
6	Mathematics	2	6	Heat Transfer	2
7	Physics	2	7	Engineering Mechanics	2
8	Climatology	3	8	Agricultural and Biosystems Science	3
9	Biology	2	9	Calculus	2
10	Chemistry	2			
	Total	**21**		**Total**	**21**
No	**Courses**	**SKS**	**No**	**Courses**	**SKS**
Semester III			**Semester IV**		
1	Engineering Drawing	2	1	Principles of Programming	2
2	Workshop	3	2	Engineering Economics	2
3	Electronics	3	3	Experimental Design	2
4	Strength of Materials	2	4	*Subak*** Irrigation Engineering	2
5	Hydrology	2	5	Land Surveying	2
6	Measurement and Instrumentation	3	6	Energy and Electrification	3
7	Machinery and Implements	2	7	Food Process Engineering	3
8	Irrigation and Drainage Engineering	2	8	Engineering Design	2
9	Engineering Mathematics	2	9	Power and Combustion Engine	2
			10	Integrated Irrigation Laboratory Practices	2
	Total	**21**		**Total**	**22**
No	**Courses**	**SKS**	**No**	**Courses**	**SKS**
Semester V			**Semester VI**		
1	Scientific Method	2	1	Smart Farming	3
2	Research Operations	2	2	System Dynamics	3
3	Mechatronics	3	3	Entrepreneurship	3
4	Supply Chain Management	2	4	Internship	3
5	Quality Management System	3	5	Elective 1	3
6	Geographical Information System	2	6	Elective 2	3
7	Greenhouse and Fertigation Engineering	3	7	Elective 3	3
8	Physiology and Postharvest Engineering in Horticulture	4			
	Total	**21**		**Total**	**21**
Elective Courses for Semester VI					
No	**Elective Courses**	**SKS**	**No**	**Elective Courses**	**SKS**
1	Remote Sensing	3	10	Refrigerated and Freezing Engineering	3
2	Multivariate Analysis	3	11	Drying Engineering	3
3	Farm Land Conservation	3	12	Packaging Engineering	3
4	Soil, Water, and Crops Relationship	3	13	Estate Crops Processing Engineering	3
5	Ergonomics and Work Safety and Health	3	14	Fishery Product Processing Engineering	3
6	Farm Environment and Biosystems	3	15	Agricultural Product Processing Engineering	3
7	Energy Conversion Engineering	3	16	Fresh Product Retail	3
8	Energy Audit	3	17	Supply Chain Analysis	3
9	Waste Processing Engineering	3			
No	**Courses**	**SKS**	**No**	**Courses**	**SKS**
Semester VII			**Semester VIII**		
1	Religion	2	1	Seminar on Research Results	1
2	Civics	2	2	Scientific Publication	1
3	Research Proposal	1	3	Research Manuscript	4
4	Community Services	3	4	Research Manuscript Exam	1
5	Major Design Project	5			
	Total	**13**		**Total**	**7**
			Grand Total 147 SKS		

*SKS: *Satuan Kredit Semester* (Semester Credit Unit)

***Subak* is a traditional Balinese irrigation system attached to the community culture and life

Emerging technologies such as AI, IoT, robotics, control and automation have been applied and integrated into Indonesia's current AE/ABE research. With the emerging technologies, the AE/ABE study programmes have taken steps to improve the laboratory facilities and equipment to conduct qualified research. Each university has its research roadmap conforming to national and provincial agricultural and economic development. In general, the coverage of research will be influenced by the interest of the faculty staff, taking into consideration their opportunity: i) to get the research funding from the Ministry of Education and Research, ii) to build cooperation with other parties such as regional government, industries, farmer or other user communities, and international institutions, and iii) to obtain access to appropriate instruments and equipment for running the experiments.

To observe the ongoing research implemented in AE/ABE education in Indonesia, two examples are provided: the ABE-IPB and Gadjah Mada Universities, both offer MS and PhD degrees added with Brawijaya University, which offers an MS degree only. ABE Study Programme at IPB University has four divisions as follows. The Division of Mechanical Engineering carries out research in technology applications to increase plant productivity through the development of biosensing, system monitoring, and variable rate control. Research themes, among others, are 1) the development of portable sensing instruments for non-destructive quality evaluation of agricultural products, 2) the development of machinery, robotics, and control systems for plant factories, 2) the development of precision machinery for the mechanisation of oil palm plantations, and 3) the development of machinery for food crops and terramechanics.

The Division of Biosystems Engineering works in a broad spectrum of research topics starting with 1) non-destructive quality evaluation for horticultural products, and estate crops using process imaging, Near Infrared Reflectance, and Ultrasonics, 2) design of packaging material and system for the distribution of perishable fresh agricultural products, 3) smart greenhouse and its control system, 4) the improvement of the parboiled rice process, 5) physical characteristics and process of Nano fibril soy protein, 6) microencapsulation of essential oil, 7) valorisation of whey protein, 8) development of fishing gear, 9) application of fine bubble technology for agri-aqua farming, and 10) plasma fine bubble technology for postharvest handling of agricultural products.

Division of Bioinformatics Engineering focuses on developing and improving PRECIPALM (Precision Agriculture Platform for Oil Palm) to predict the soil's content of N, P, K, and Mg. Data of the oil palm plantation area was extracted from the images captured using remote sensing Satellite Sentinel 1 and 2 with a multispectral camera and then inputted to the AI model producing real-time nutritional distribution in a colourful regional gradation spatial map indicating respected nutrition content. The map is useful for identifying the type of fertiliser and its total needs calculation in planting oil palm trees in the region. PRECIPALM has been developed in cooperation with East Kalimantan Fertilizer Factory and has gained copyright for 50 years from the Ministry of Law and Human Rights. Further, the division is looking for the application of AI and precision agriculture to predict the Basal disease attacking the oil palm and to estimate the production of the oil palm plantation.

The Division of Renewable Energy Engineering concentrates on the following fields: 1) thermo-chemical processing of biomass focusing on macroalgae and microalgae, 2) biofuel, biomass, and biohydrogen technology by exploring the system design and identifying the raw materials to produce the biofuel and biohydrogen, 3) thermal system simulation, modelling, and applications of heat pumps, solar flat plate collectors, biodiesel reactors, and biomass grate furnaces, 4) development of energy efficient solar drying, fluidised and rotary bed drying systems, and 5) energy and exergy analysis for the process of palm oil waste products into energy.

Research at the ABE Study Programme at Gadjah Mada University (*Departemen Teknik Pertanian dan Biosistem UGM*, 2022) is classified into four divisions. The Division of Land and Water Resources Engineering explores research in the fields of 1) the development of food production areas based on precision farming, 2) the modernisation of land and water resources management to support the sustainable growth of agriculture, and 3) the development of CSA-Climate Smart Agriculture to mitigate the climate change and to increase the agricultural productivity.

The Division of Biosystems Mechanical Engineering focuses on 1) the development of mechanisation system 4.0 with the orientation for humanised technology to create smart food estates, 2) the design of automated farm machinery that is based on the application of precision agricultural technology, and 3) the development of biomass energy as the source for the renewable energy to support the sustainable agricultural growth.

The Division of Food, Postharvest, and Bioprocess Engineering does research in 1) controlled postharvest handling technology to improve the quality and minimise losses of the agricultural products, 2) the implementation of the automated system for the machinery and equipment to support the food processing industry, and 3) the application of bioprocess technology through the chemical and biochemical design in food industry planning and management.

The Division of Biological Environment Control Engineering observes in-depth the following topics: 1) the design and utilisation of smart greenhouse technology to support the growth of environmentally friendly agriculture, 2) the development of controlled farm structure and environment to improve the quality and yield of the agricultural products, and 3) biomass agricultural waste handling technology to increase the value added of the agricultural system and to support the sustainable agriculture.

Research and development at the ABE Study Programme, Brawijaya University, Malang, East Java, also cover four divisions. The Division of Farm Power and Machinery develops agricultural machinery design and performance testing, biodiesel processing systems with ultrasonic, biodiesel processing using the supercritical reactor and continuous supercritical reactor, the purification of bioethanol to achieve fuel grade quality, the production of hydrogen energy, and low-temperature seed drying systems.

The Division of Food and Agricultural Product Processing Engineering explores a broad spectrum from the drying kinetic and physical characteristics of specific agricultural products such as the flower of *Clitoria spec* L. for herbs and the fruits of *Pangium edule* Reinw. ex Blume for foods, and the seeds of *Sapindus rarak* for traditional detergent; the design of solar and solar hybrid dryers; new inventions in vacuum frying systems; hypobaric storage of fruits at low temperatures; the application of electric field pulses for milk and fruit juice pasteurisation; non-destructive quality evaluation of oil palm fruits and horticultural products; the biopolymer and bioceramic based membrane synthesis for food processing and production of new biomaterials from waste products; to bioplastic processing from root crops.

The Division of Biosystems Mechatronics consists of bioinstrumentation, control, and system engineering in a plant factory, a fully controlled closed bioproduction system using plant response-based sensing, and the application of IoT and AI in optimisation modelling in agricultural engineering problems. Some topics, for example, are the use of computer visions and Artificial Neural Networks to control plant water content and plant nutrition based on the Speaking Plant Approach; smart greenhouse design and control; the use of photonics engineering, chemometrics, and machine learning for plant diseases detection, quality evaluation, and fruit grading and sortation; and the development of bio-sensing instrumentation for non-destructive quality evaluation.

Division of Natural Resources and Environmental Engineering focuses on irrigation engineering such as sprinkle irrigation, sub-surface irrigation, drip, and trickle irrigation; and land and water conservation, as well as environmental conservation at the watershed area.

In comparison with the ABE divisions in the three universities in Indonesia, it will be worthwhile to look at the ABE study programme in Europe and US. Briassoulis et al. (2013, p. 310) indicated that there are six specialisations incorporated as the mid-level competencies in a Biosystems Engineering study programme in Europe: Bioprocess Engineering; Bioenergy Systems; Bio-based Materials; Biosystems Informatics and Analysis; Structural Systems, Materials, and Environment for Biological Systems; and Water Resources Engineering.

Ting (2010, pp. 1–2) analysed the core competencies in Biological/Biosystems Engineering in the US from four major areas: Automation, Culture, Environment, and Systems. When these four major areas intersect, they will breed emerging ABE disciplinary relevance and impact areas. Many are similar to the previously mentioned areas, and more, such as Biomass and Renewable Energy; Precision and Information Agriculture; Agricultural and Biosystems Management; Food Quality and Safety; Spatially Distributed Systems; Bio-sensors; Bionanotechnology; and Intelligent Machinery Systems.

Unique Initiatives Applied in AE/ABE Education

Substituting Internships for Real Experience in Farming and Postharvest Processing

The internship has been interwoven with the Agricultural Mechanization curriculum since it was first offered as a study programme in 1957. It went along with the transformation of Agricultural Mechanization to AE and ABE curricula. Accounted to 3 SKS, a student will go for an internship in food and agricultural industries such as beverages plants, food canning factories, and agricultural machinery manufacturers; in plantations of estate crops such as oil palm, rubber, tea, coffee, cocoa, and tobacco; and in watershed management authority institutions for 1–3 months. During this period, the students are obliged to do the internship under the company's direction and write reports afterwards approved by the company and their academic advisers.

Challenges happen when some industries, plantations, or offices that traditionally are willing to accept internship students decline to do so at one time for any reasons that maybe occur in their internal management. With an average of 100 students per batch per university, the faculty staff is pressured to help all students find a good internship programme simultaneously at a representatively sized company or institution. It should be noted that the time for internships is the same for all universities, i.e., at the end of the even semester every year.

Hence an alternative idea: the notion of offering the students to practice farming and postharvest processing on campus land instead of going to industries was considered interesting. The idea was implemented in an even semester of the year 1986. Students were divided into ten people per group, and each group determined the seasonal crop they wanted to plant, like maize, soybeans, mungbeans, and groundnuts. They developed a plan from planting to postharvest processing until marketing, implementing their knowledge using the agricultural machinery provided by the university for land tillage, seed planting, water irrigation system, threshing, and drying. Observing how each group solved the problems they faced during the practices was worthwhile for the students and the faculty staff. This interesting idea would be adopted, except the students did not have enough time because the internship should be completed over three months before the next semester began. Since the crop harvest was done only after three months, the students needed longer to do postharvest processing and marketing. They, unfortunately, had to do that overlaid with the new courses in the next semester. Thus, this interesting and beneficial alternative was dropped and never applied anymore.

Intensive Homework and Quizzes for Challenging Courses

Undoubtedly, the engineering curriculum contains mathematics and courses loaded heavily with mathematics. Students need to practice problem-solving in these kinds of courses. To facilitate the students, these courses provide intensive homework given at every class meeting, evaluated by the faculty staff immediately, returned to the students, and discussed at the next class meeting. To maintain students' interest and concentration, the delivery time is limited to two hours each 50 minutes, so courses with 4 credits will be divided into two class meetings in one week, each in two hours. Instead of providing the student with a two-time evaluation at the middle and the end of the semesters, the courses conduct 5 times of quizzes, breaking the course materials into 5 equal-load sections and evaluating each time after each section.

The initiative proved to be successful in leveraging the students' knowledge, indicated by the higher grades gained by the students. The courses are removed from the students' "killer courses" list, and they feel a lot easier when taking them. The challenge is finding the faculty staff willing to do extra time to evaluate the homework consisting of 5 problems twice a week for an average of 100 students, plus the additional 3 times more quizzes equal to mid-term and final semester exams. It is unfortunate that after being implemented for about 30 years, the initiative has been removed since the study programme could not find any more faculty staff willing to spend extra time after the retirement of the initiator.

Encouraging Students to Initiate a Startup

As far back as June 1997, when the world monetary crisis happened and hit Indonesia very hard, not only in the national economy but also in socio-politics, the opportunity for university graduates in Indonesia to find a job steeply decreased. It was intensified because the IDR conversion to USD depleted to one-eighth in a short time, leaving a traumatic shock to everybody with increasing flaring inflation.

Fortunately, just before the monetary crisis, the AE study programme had started offering the Entrepreneurship course in semesters V, VI, or VII. It was also doubly lucky that the Ministry of Education and Culture provided a national programme encouraging groups of students to initiate a startup, awarding them little seed money but sufficient, and competitively based on a selection of their business proposals.

The Entrepreneurship course might replace the mid-term and final semester evaluation with the business proposal developed by the students per group of four. The business proposals were then submitted to the Ministry of Education and Culture to be selected for a seed money grant. Out of an average of 25 proposals submitted annually, the AE/ABE students at IPB University successfully obtained 16 startups to be funded. Once the students gained the seed money, they started the business. Some startups were successful, and some failed. However, the important thing was that students had experience in doing commercial business, which could become another alternative for a career after graduation. The initiative established at the end of the 1990s has been adopted continuously until now.

Providing an Alternative To the Research Manuscript

It has been narrated in the previous sub-chapter that Indonesian BS curricula covers 8 semesters in 4 years and includes research in the Final Project in Semester VIII, called the Research Manuscript (Table 19.4). It has been compared to the recent European education offering a 3-year bachelor programme, and the US system with a 4-year bachelor without research. Some universities in the Philippines have taken the same move as in Europe. Japanese undergraduate programmes still adopt research as the final assignment for the students. In Australia, students who do not take BS for Honor do not need to complete research. Undoubtedly, the BS for Honor is prepared for students who want to continue into advanced studies. Overall, it could be concluded that in the Asian continent, there are many universities applying research as part of the BS curriculum.

The controversial issue is that research for the BS study programme only benefits the graduates who plan to continue for their advanced degree and work in research and research-related institutions like consultant offices, NGOs, and Research & Development units in the industries. However, several attempts by visionary faculty staff in Indonesia to cut research as a compulsory project for a BS degree have not been fruitful, and neither is the cut of a 4-year study into a 3-year study.

In light of the situation, initiatives have been applied to provide alternatives to the Research Manuscript. One is to allow students to intern in a startup company for three months, advised by the academic adviser and the business incubator management in the university where the startup company is taken as the resident incubatee. The first trial in 1989 at IPB University proved to be successful. Ten students from semester VIII who originated from several different study programmes were registered under this programme, interned at the startups, and wrote a manuscript about the commercial business operation of the startups as the final assignment. Even though the alternative is still provided, not many students enrol in this one. The in-depth observation indicated that most faculty staff encourage undergraduate students to do research.

The reason is that the faculty staff needs research to fulfil the requirement necessary for their annual performances evaluated by the university and the Ministry of Education, Culture and Research Technology. The accumulative points of performance that include research, scientific publications, writing books, teaching, and community services are the means for faculty staff to climb to a higher level of career until reaching the peak as a professor. Since the number of graduate students is far too low compared to the number of undergraduates, the convenient way out is to involve the undergraduate students as well in doing research.

New Initiative by the Ministry of Education, Culture and Research Technology

In 2020, the New Minister of Education, Culture, and Research Technology, who supervised both state and private universities, introduced the Independent Learning-Independent Campus (MBKM) system, meaning every university student is free to choose the eight education channels offered in the programme based on the student's passion and interest for either one or two semesters given quite a generous credits of maximum 20 SKS per one semester and 40 SKS per two semesters (*Direktorat Jenderal Pendidikan Tinggi*, 2020). Except for an education channel called the Students Exchange channel, the system further means the student may replace 7-8 core courses per semester in the student's curricula with one of the other seven channels in the programme. The eight channels are as follows.

1. Student Exchange: Student Exchange allows students to enrol for one semester in other universities taking similar courses to the original courses in their study programme's curricula or to enrol in other study programmes in the same university taking different courses from their study programme's curricula.

2. Internship: A channel opens for students to intern in industries for one full semester. There are two alternatives called free-form and structured form. In the free-form, the host industries evaluate the students for various components agreed upon by both the industry and the students before the internship. In the structured form, the lecturer from the core courses replaced by the internship will grade the students' performances during the internship.
3. Teaching Assistantship: The Teaching Assistantship channel allows students to become assistants in teaching at the following levels: elementary school, junior, and senior high school. This channel is specifically provided for study programmes in education, resulting in the graduates' profession as a teacher in elementary and high schools.
4. Research Assistantship: Research Assistantship is a channel provided for students who have a passion for research and plan to have a career as a researcher. The duration can be taken for one or two semesters. Research assistantships can be carried out in research institutions and study centres.
5. Humanity Project: This channel allowed students to join national or international humanity projects as foot soldiers such as pioneering natural disaster relief and refugee handling projects in one or two semesters.
6. Entrepreneurship Programme: The university offering this channel has made it compulsory to develop an overall entrepreneurship programme of 20 SKS with detailed structures constituting an integrated programme that may cover the micro-credential courses and the project setting up a startup. For this specific programme, the Ministry identifies and selects partner institutions that are trusted to have more practical skills in delivering the entrepreneurship programme. Students could select the partners identified by the Ministry or other plans set up by their university.
7. Independent Study/Project: Independent Studies/Projects are specific projects based on innovative ideas by a group of students from interdisciplinary studies participating in specific international competitions such as in science, engineering, law, or business. The other alternative is the independent project created by the study programme that has an innovative idea to support the student's education.
8. Villages Development/Thematic Community Services: This channel extends thematic community services in many university curricula for all study programmes. Students stay in the villages, living in the middle of the community, identifying the community problems, finding a solution, and implementing it. The usual 3 SKS thematic community services for 3 months now can be extended into 20 SKS for one semester or 40 SKS for two semesters. The Ministry encouraged the university to form cooperation with the Ministry of Villages, Disadvantaged Regions, and Transmigration for this specific channel.

The Ministry has socialised and promoted the new initiative starting in 2020. It grants higher KPI (Key Performances Index) for universities with more students involved in the new initiative. High incentives are given for students joining the eight channels in the programme initiation in 2020. Unfortunately, the incentives are decreasing in subsequent years as it becomes more likely to be integrated into the curricula by all universities compared to the initial year in 2020.

Some challenging issues that may be arisen during the full implementation of the new initiative in years ahead are the following.

1. Since most universities execute the timing of the new initiative programme in semesters V, VI, and VII, it means that a student joining the new initiative will replace either basic or core expertise courses that characterise the study programme commonly allocated in those semesters with the choice of the student's interests. The student may gain good skills from one of the eight channels but also loses part of the expertise foundation required to build the competencies of the related study programme.
2. The readiness of universities and lecturers varies in accepting students from other universities to attend the same courses they are interested in on the Student Exchange channel. Universities with a large student body will face a tough challenge in coordinating the time, classes, and lecturers when additional students are coming in a massive number. As a result, they may decline the option to host the Student Exchange.
3. Big universities in Indonesia may accept 100–300 students for each study programme annually. The critical factor is the capability of each study programme in managing this number of students to select different kinds of the eight programme channels. This may result in the limitation of offering only one or two of the eight related channel options.
4. The decreasing incentives for the students, after a good one in the initiation period of 2020, may discourage many from joining the new initiative afterwards.
5. Students taking the Student Exchange and Entrepreneurship channels by attending courses in excellent far-away universities or partner institutions did not have problems from 2020 to the middle year of 2022 because the classes are conducted online. After the pandemic subsided, students will face difficulties in supporting their living and travel far away to the host universities or partner institutions to attend conventional classes, except online classes remain offered by the hosts.
6. The principles of the Ministerial initiative are to provide freedom for university students to choose how to build their competency according to their passion and interests. However, the Ministerial system of granting higher KPI (Key Performances Index) to the universities with higher numbers of students joining the initiative programmes may contradict the

initiative's independent philosophy because it may motivate the universities to pressure students to join the programme.

The Contribution of ABE Graduates to Indonesia and the World

From 1957 until now, the total number of bachelor graduates in Agricultural Mechanization and AE/ABE study programmes from around Indonesia is estimated at over 25,000. Many of the graduates worked at government institutions such as the Ministries of Agriculture; Forestry; Marine Affairs and Fisheries; Education and Culture; Cooperatives and SMEs; Industry; Economics; Internal Affairs; Foreign Affairs; Law and Human Rights; the Investment Coordinating Board, the National Research and Innovation Agency (BRIN); and Regional Governments. Some are elected to legislative seats as members of the House of Representatives. In 65 years, some took the highest-level position, like the Minister of Agriculture (refer to Box 19.1), a province governor, or a career that looked as unlikely in the ABE profession as the Director General and Inspector General of the Min. Law and Human Rights, however, climbing from the Directorate of IPR (Intellectual Property Rights) was at least an indirect connection with the agricultural machinery design.

BOX 19.1: DR H. ANDI AMRAN SULAIMAN

Dr. IR. H. Andi Amran Sulaiman, M.P., the Minister of Agriculture, Republic of Indonesia in the cabinet period of 2014–2019, graduated as an Agricultural Engineer in 1993 from the Hasanuddin University, Makassar, South Sulawesi. He continued his studies to gain a Master's degree in 2003 and a Ph.D. in 2012 (Wikipedia, 2022b).

In his early career, he worked at the plantation state company of PT Perkebunan Nusantara XIV. He then quit the job and built his own company that expanded fast. His business became a holding company, Tiran Group, that covers 11 business enterprises such as rodenticide production, sugar cane, and oil palm plantations, sugar cane factory, distributors of food and other products, and mining in gold and nickel. As a creative businessman, he patented a rodent fogger with Patent No. ID S000001297 that later became commercially produced.

He has been very keen on leveraging the development of agricultural engineering professionals, and he assumed the Chair of the Association of Agricultural Engineers for South Sulawesi Province, Indonesia from 2006–2014.

He was appointed and served as a dedicative Minister of Agriculture under the Working Cabinet of President Joko Widodo, Republic of Indonesia in 2014–2019. Currently, on the 23rd October 2023 he was appointed again as the Minister of Agriculture until 2024.

In the early development, when the universities in Indonesia were still growing and expanding from the central government to the provincial capitals, many alumni of agricultural engineering filled up the needs of faculty staff in the newly established Agricultural Mechanization and AE/ABE study programmes across the country. The Department of Agricultural Mechanization IPB and Gadjah Mada Universities, as patriarchs, bred first generations of faculty staff assisting the foundation of many AE/ABE study programmes at the state universities in the provinces. When the postgraduate studies came to the education systems in agricultural engineering, the majority of the alumni working as faculty staff in the state universities went back to their alma maters (IPB and Gadjah Mada Universities) and pursued advanced degrees at the MSc and PhD levels. Some of the fortunate alumni followed the models of their professors to apply to various outstanding universities abroad to get their Master's and PhD degrees.

The continuing education in AE/ABE undergraduate and graduate programmes from generation to generation has been breeding distinguished full professors who produced successful graduates who later on gained full professors for themselves, and these later generations of professors have also educated their graduates who now have achieved full professors as well. Thus, three consecutive generations of full professors are now teaching AE/ABE in the universities in 13 out of 38 provinces in Indonesia. One of them is Prof. Dr Bayu Taruna Widjaya Putra, ABE Study Programme at Jember University, Jember, East Java, who was ordained as a full professor at 38 (Box 19.2). So far, he has been the youngest ABE professor in Indonesia.

BOX 19.2: PROF BAYU TARUNA WIDJAJA PUTRA

Prof Dr. Bayu Taruna Widjaja Putra is the youngest ABE Professor in Indonesia so far. He achieved his full professorship at 38 years of age in December 2022. He has published over 16 scientific articles, seven of them at Q1 Scopus level journals in the last five years. He is deeply committed to teaching and research. He wrote four books in Bahasa Indonesia (Indonesian language) on the ABE emerging technology to facilitate student learning.

His research focuses on photogrammetry and remote sensing, precision agriculture, AI, and deep and machine learning. As the leader, he makes a lot of improvements in the Laboratory of Precision Agriculture and Geoinformatics, ABE Department at Jember University, Jember, East Java.

Bayu wrote a book chapter, "Smart food sensing and IoT technology", in a book entitled *Bio- and Nano-Sensing Technology for Food Processing and Packaging*, published by the Royal Society of Chemistry in 2022. He has received international recognition from the KWEF (Kurita Water Environment Foundation), Japan as the best researcher in 2019. He also obtained a credential as a Deep Learning Professional from PyImageSearch Raspberry Pi for Computer Vision in 2021.

He is an active fellow member of several international professional organisations such as IEEE and ISPRS. He is nominated as the country representative of Indonesia in ISPA (International Society of Precision Agriculture).

He graduated from AE Study Program at Jember University, and received his M.Eng. and Ph.D. degrees from the Agricultural System and Engineering Department, AIT, Bangkok, Thailand.

Some significant portion of graduates ventured into the private sector and state companies, spreading out a broad interest from plantations, agricultural machinery manufacturers, food and beverage industries, various types of industries outside agriculture, banks, retail, and restaurants, on to as far as the automotive industry. Some built their dreams of opening enterprises, being licensed by multinational corporates, or becoming inventors of their startups. They proliferated in the big cities but were also rooted in small villages doing farming and managing village workshops for agricultural machinery. They produced controlled sensors for smart greenhouses, closed-house systems for chickens, and food processing machinery.

AE/ABE inventors have founded their own companies, or started from a small one and grown bigger. Some received a patent for their invention. They successfully transform their inventions into innovations. Dr Anang Lastriyanto from ABE Study Programme, Brawijaya University, Malang, East Java, is among them; his profile is presented in Box 19.3. The multiplier effect of his invention, the vacuum frying machine, could be illustrated as follows. Take for an assumption that 1600 out of over 2000 machines he manufactured (Box 19.3) were sold to small and medium-scale food industries. The industries used the machines for the production of any kind of fruit chips and fish-dried chips. If it says each industry would hire 5–10 employees, the impact of the innovative machines improved the living quality of 8,000–16,000 people, not counting the employees in the machinery manufacturing plant, the farmers and fishermen who supplied the raw materials for the food industries, the packaging industries, and the distribution chains. Further, the food chips produced by the vacuum frying machine have lower oil content than the deep-fried chips. This, too, contributes to the healthier food product for the consumer.

BOX 19.3: DR ANANG LASTRIYANTO

An Agricultural Engineering faculty staff member who has been successful in developing a manufacturing company for food processing machinery, Dr. Anang Lastriyanto is an innovator of vacuum frying machines.

He designed, tested, and modified his invention as early as in the year 1995. He later established a small workshop that manufactured his machines and sold them to food industries making fruit chips. He claimed that the machine will reduce the oil content in the chip products as much as 15–20% compared to deep frying oil products.

He kept doing research and development to improve his machines. As the results, vacuum fried products in the market not only covered a broad spectrum of fruit and root crop varieties but also fish-dried chips and meat products. He generously served buyers with several days training at his places for free. He further built a simple boarding house for the operators who needed cheap accommodation during the training.

By 2022, over 2000 units of his machines at different capacities have been sold to small and medium-scale food industries, and also to education institutions around Indonesia. He patented his invention under the Patent Right No. ID P0026372B. In the last years, he founded a legal company to ensure the sustainability of his enterprise, now turned to a bigger plant. His new machine ready for commercialisation is an integrated honey processing with pasteurisation, evaporation, vacuum cooling and filling, thus keeping the original quality of honey.

With all his career in business, he is a dedicated faculty staff at AE/ABE Study Programme, Brawijaya University, Malang, East Java. He pursued his MSc degree at IPB University, and his PhD at the Faculty of Engineering at his home university.

An AE faculty staff member plays a key role in the foundation and management of a business incubator in the university and its development towards the Science Techno Park, as marked by the development of the business incubator at IPB University, Bogor, West Java. The business incubator of IPB University is among the first four business incubators established in Indonesia in 1994. Later, the same AE faculty staff member also contributed significantly to the national programme to build and strengthen business incubators around Indonesia. To this day, over 180 business incubators are claimed as members by the AIBI – Association of Indonesian Business Incubators (AIBI, 2022), not to mention many business incubators that are not members of the association, such as for-profit incubators owned by conglomerate companies.

With this brief observation, it could be concluded that agricultural engineers/agricultural and biosystems engineers in Indonesia have participated in national economic development and left their signatures in the community welfare and education, which has undoubtedly improved in the last decades.

Agricultural engineers/ agricultural and biosystems engineers in Indonesia have also contributed to the world. Some became principal investigators in research in other countries, like Dr Nanik Purwanti, faculty staff at the ABE Study

Programme, IPB University, Bogor, West Java in Ireland in 2019–2021 (Box 19.4). Others actively published their scientific articles in Q1-level Scopus scientific journals, like Prof. Dr Budi Indra Setiawan, faculty staff at the Civil, and Environmental Engineering faculty, IPB University, Bogor, West Java (Box 19.5). They might be invited as a visiting scholar (Box 19.5), as editors in the proceedings of international conferences or seminars (Purwadaria et al. 2002, and Purwadaria et al. 2013), and as writers of chapters in books together with worldwide recognised experts (Purwadaria et al. 2006, Purwadaria et al. 2016, Zhang et al. 2020a, and Zhang et al. 2020b). Many lead research and development projects in Indonesia, cooperating with regional and international institutions.

BOX 19.4: DR NANIK PURWANTI

A woman agricultural engineer who has a reputation for getting international research funding, Dr. Nanik Purwanti was a singular principal investigator of Increasing Processability and Overall Valorisation of Irish Acid Whey in Teagasc Food Research Centre, Moorepark, Ireland for three years in 2019–2021. The research aimed to characterise the effect of seasonal and processing variation on acid whey stream from casein manufacture. This research explored the effects of pretreatments of acid whey on the manufacture of whey protein concentrate and permeate powders. The research was in close collaboration with Arrabawn Co-operative Ltd Ireland, a major manufacturer of acid whey powder with interest in increasing volumes of acid whey products. The research was co-funded by the European Union, under the Horizon 2020 Marie Skłodowska-Curie Actions in partnership with Enterprise Ireland.

Her work was recognised by L'Oreal-UNESCO for Women in Science (FWIS) National Fellowship Programme in 2014, from which she was granted a research award about Whey Protein Stabilised Emulsion for Microcapsules Materials in 2014–2018. She extended the research by exploring soy protein isolates (SPI) from Indonesian soybean varieties and SPI fibrils to replace the functionality of whey protein for microencapsulation. She was also nominated by the L'Oreal–UNESCO FWIS National Fellowship to represent Indonesia in the International Rising Talents Programme 2015.

Nanik has published over 30 scientific articles in several international and national journals. She is a dedicated faculty staff member at the ABE Study Programme, IPB University, Bogor, West Java where she graduated as Bachelor of AE. She obtained her MSc and PhD degrees at Wageningen University & Research, the Netherlands, in the years 2007 and 2012, respectively.

BOX 19.5: PROF BUDI INDRA SETIAWAN

An agricultural engineering alumnus who is most active in scientific publications, Prof Budi Indra Setiawan also published more than 50 other articles in other scientific journals both international and domestic at the same five years. Not to count his articles in Proceedings. He is the author of a book chapter in *Supply Chain Resilience: Reducing Vulnerability to Economic Shocks, Financial Crises, and Natural Disasters*, published by ERIA-Springer in 2020.

His researches covered cooperation both with national and international institutions, and linked to national corporate industries as well. He has made significant contributions to national policy in land and water management. In 2014–2019 he was nominated as a member of the Ministerial Expert Team working for the Minister of Agriculture, Republic of Indonesia.

Gaining international reputation, he has been invited as visiting scholars to several international universities and institutions prior to these last five years. He has been serving as one of the editors of the PAWEE – Paddy and Water Environment Engineering Journal since 2003, and eventually received the International Award from PAWEE in 2010. He served as a member of Technical Section 1: Land and Water Engineering, CIGR since 2007.

He was a dedicated faculty staff member at the AE Study Programme previous to moving, in 2010, into the Civil and Environmental Engineering Study Programme that he helped found at IPB University, Bogor, West Java.

Budi Indra obtained his AE bachelor degree at IPB University, then pursued his MSc and PhD at the University of Tokyo in Japan. He achieved his full professor in the field of Soil Physics and Hydrology in 2004.

Shaping for the Future

Entrepreneurship Programme as an Alternative to Research in the Final Project

Looking at the interests of the undergraduate students with the creation of startups that conforms to the growing interests of Gen Y, it will be most appropriate if the AE/ABE curriculum fully integrates the entrepreneurship courses and practices into the Final Project. Many startups founded by AE/ABE alumni reached success in their businesses. For example, the following two startups have already grown into legal companies. One is from Gadjah Mada University producing an IoT farming solution platform combined with IoT multi-sensor equipment for oil palm plantations (*PT Merapi Tani Instrument,* 2022), and another one is from IPB University producing Agritech Data Processing 4.0 and sensor technology equipment for the smart farming system, for either indoor or outdoor farming (Box 19.6).

BOX 19.6: BUNG DAKA PUTERA

A biosystem engineering graduate, Bung Daka Putera built his startup while he was still in his junior year in 2019. His idea became both the 1st Winner of Indonesian Acceleration for students' startups, and the 1st Winner of Agri-Maritim technology in 2019.

Eventually, right after his graduation from ABE-IPB University in 2020, he and some of his friends built their company PT Kharisma Agri Inovasi, manufacturing agritech data processing 4.0 and sensor technology equipment for smart farming system either indoor or outdoor farming.

In 2021, the startup was successful in getting a seed money grant from the Ministry of Research and Technology/BRIN (National Agency for Research and Innovation). In the same year the startup was listed as among the Best 15 in the ASEAN Startup Challenge.

So far, the company has been successful in selling over 175 units of monitoring and controlling devices for 48 + greenhouses and a 156 ha ginger farm.

His enterprise can be found at http://kharismacrop.com

It will only be fair that the education system prepares and supports the interests of undergraduate students and their expectations for their careers after graduation. Rather than a top-down initiative taken by the Ministry of Education, Culture, and Research Technology at Channel 6 of Independent Learning-Independent Campus with a total of 20 SKS entrepreneurship programmes as discussed in Sub-Chapter 1.4., it would be more sustainable if the universities take the initiative in a bottom-up way with the education system substituting the entrepreneurship programme for the 6 SKS of a research manuscript.

At many universities in Indonesia, the entrepreneurship skill of the graduates is strengthened not only by the curriculum but also by the institutions called business incubators. Fifteen out of 17 universities offering AE/ABE Study Programmes have a business incubator at the university level (AIBI, 2022). Two business incubators have developed into Science Techno Parks, one focusing on agribusiness and agroindustry at IPB University Science Techno Park at Bogor, West Java, and the other on the biomedical industry at Gadjah Mada University Science Techno Park at Yogyakarta, DIY. It is also interesting to note that the founder of the business incubator at IPB University in 1994 is an AE faculty staff member, as well as the current director of the IPB Science Techno Park, while the manager of the Gadjah Mada University business incubator is an AE alumnus. Using the business incubators' services, the entrepreneurship programme, as an alternative for research in the Final Project, will ensure more diversions for the graduate career. One of the successful startups incubated by a business incubator has been illustrated in Box 19.6, an incubatee in the IPB business incubator.

Towards Secure Food Availability, Human Wellness and a Better Environment

Summarising the evolution of Agricultural Mechanization education into AE/ABE in Indonesia and visualising the future, a roadmap for AE/ABE education could be structured as illustrated in Figure 19.2. Since the transformation of Agricultural Mechanization into AE/ABE has been narrated, the following describes the last part of the roadmap which is the AE/ABE education beyond 2023.

Nowadays, AE/ABE scientists and students discuss climate change because of greenhouse gas emissions, sustainable agriculture, food security, precision farming, and Industry 4.0. They are researching in these areas, making presentations in scientific seminars and conferences, and publishing articles in scientific journals. However, looking at the curricula in Tables 19.3 and 19.4, few courses support the mission to implement sustainable agriculture and food security through precision farming and Industry 4.0. There are some, like Robotics Engineering in the ABE curricula of IPB University; Smart Farming, and Remote Sensing in the ABE curricula of Udayana University. In other universities like Jember University, there are GIS, IoT, Machine Learning and Deep Learning, and Precision Agriculture courses.

Some additional core courses in emerging technology that could be considered to achieve the necessary needs for the ABE undergraduate curriculum are Plant Factory/Inside-Door Farming, Smart Greenhouse/Smart Farming, Mobile Application Based Control/Bio-Sensing/Bio-Process Control, Image Processing, Non-Destructive Quality Evaluation of Agricultural Products, and Mitigation for Climate Change in Farming.

People in the world, as well as in Indonesia, are facing an alarming increase in the human population, and thus an increasing demand for food supply. There is also a trend of acute interest in human health and wellness. Meanwhile, in activities to satisfy the necessary foods and wellness, people create environmental problems, such as deforestation and production of gas emissions, causing the build-up greenhouse effect, climate change, and natural disasters. A sustainable agricultural system must be developed by the intensive multidisciplinary approach of ABE with other fields such as agronomy, food technology, mechanical and electrical engineering, and biomedical engineering. In finding a solution to minimise the impact of climate change, the agricultural system will best integrate local wisdom traditionally proven successful into emerging technology.

So far, plant factory is applied to common vegetables that could be sold at high prices. In the future, other tropical vegetables and herbs must be grown in the same system at a reduced cost.

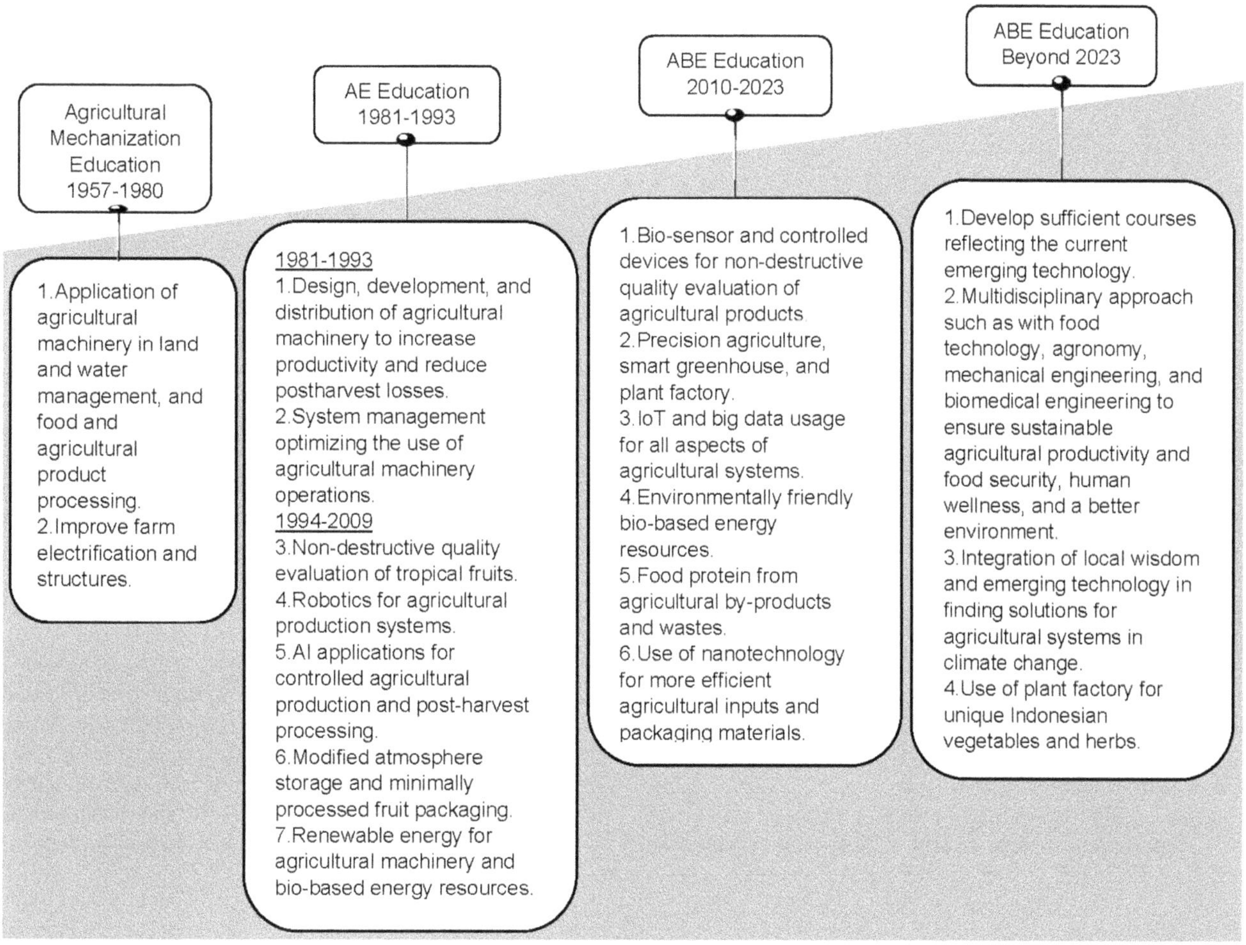

FIGURE 19.2 Roadmap of Agricultural Engineering/Agricultural Biosystems Engineering (AE/ABE) education in Indonesia.

REFERENCE LIST

American Society of Agricultural and Biological Engineers. (2022). https://www.asabe.org/

Association of Indonesian Business Incubators. (2022). https://aibinetwork.com/

Briassoulis, D., Panagakis, P., Nikopoulos, E., & Ayuga, F. (2008). *The Emerging Evolution from Agricultural Engineering to Biosystems Engineering Studies in Europe*. Research Gate. https://www.reserchgate.net/publication/ 233853648

Briassoulis, D., Gallego, E., Pantaleo, A.M., Holden, N.M., Owende, P., Ting, K.C., & Mallikarjunan, K. (2013). The "threads" of biosystems engineering. *Transactions of ASABE*, 57(1):307–330. https://doi.org/10.13031/trans.56.10185.

Comparetti, A., & Febo, P. (2018). *Harmonization of Higher Education in Agricultural/Biosystems Engineering*. Research Gate. https://www.researchgate.net/publication/337891787

Febo, P., & Comparetti, A. (2013). Biosystems engineering curricula in Europe. *Journal of Agricultural Science and Technology*, 3(1): 1–9. https://www.reserchgate.net/publication/280882439

Holden, N.M., Wolfe, M.L., Ogejo, J.A., & Cummins, E.J. (Eds.). (2020). *Introduction to Biosystems Engineering*. ASABE and Virginia Tech Publishing.

Indonesia. (2022). Departemen Teknik Pertanian dan Biosistem UGM. Department of Agricultural and Biosystems Engineering UGM. Gadjah Mada University. https://tpb.tp.ugm.ac.id/id/

Indonesia. Direktorat Jendral Pendidikan Tinggi. (2020). *Buku panduan Merdeka Belajar-Kamus Merdeka (Guide Book on Independent Learning-Independent Campus)*. Kementerian Pendidikan dan Kebudayaan. https://dikti.kemdikbud.go.id/wp-content/uploads/2020/05/Buku-Panduan-Merdeka-Belajar-Kampus-Merdeka-2020-1.pdf

Mandang, T., Suryaatmadja, S.L., Mangunwidjaja, D., Soekarto, S.T., & Pandjaitan, N.H. (2018). *Sejarah dan perkembangan FATETA IPB (History and Development of the Faculty of Agricultural Technology, IPB University)*, Suparno, O., & Sugiyono (Ed.). IPB Press.

PT Merapi Tani Instrumen. (2022). http://www.mertani.co.id

Purwadaria, H.K., Budiastra, IW., Suroso, A.I.W., & Heldman D.R. (2006). Low IT in developing countries. In *CIGR Agricultural Engineering Handbook Volume VI* (pp.501–523). ASABE Publisher.

Purwadaria, H.K., Fardiaz, D., Kardono, L.B.S., & McElhatton, A. (2016). Tempe from traditional to modern practices. In McElhatton, A., & El Idrissi, M.M. (Eds.), *Modernization of Traditional Food Processes and Products* (pp.145–160). ISEKI Food Series. Springer. https://doi.org/10.1007/978-1-4899-7671-0_13

Purwadaria, H.K., Seminar, K.B., Suroso, Tjokronegoro, H.A., & Widodo, R.J. (Eds.). (2002). *Proceedings of the 2nd IFAC-CIGR Intelligent Control for Agricultural Applications 2001.* Pergamon Press.

Purwadaria, H.K., Srzednicki, G., & Kanlayanarat, S. (Eds). (2013). *Proceedings of the 2nd Asia Pacific Symposium on Postharvest Research, Education and Extension 2012.* Acta Horticulturae 1011. ISHS.

Setiawan, B.I., Tambunan, A.H., Hermawan, H., Desrial, & Gardjito. (2006). Agricultural engineering education in Indonesia. *Agricultural Engineering International: The CIGR Ejournal*, VIII. Invited Overview Paper No. 1, January. https://www.researchgate.net/publication/268424346

Stiper Agricultural Institute. (2022). https://home.instiperjogja.ac.id/

Teknik Pertanian & Biosistem IPB. (2022). *Agricultural & Biosystems Engineering IPB.* IPB University. https://tmb.ipb.ac.id/id/

Teknik Pertanian dan Biosistem UB. (2022). *Agricultural and Biosystems Engineering UB.* Brawijaya University. https://tep.ub.ac.id/s1-tep/

Ting, K.C. (2010). Development and perspectives of agricultural engineering towards biological/biosystems engineering. *Journal of Agricultural Engineering*, 41(1): 1–5. https://doi.org/10.4081/jae.2010.1.1

USAID. (1988). *Development Assistance and Institution Building the IPB/US University Experience.* R.13. IPB MUCIA. https://pdf.usaid.gov/pdf_docs/pnabe735.pdf

Widodo, S., & Supriyanto. (2020). *Panduan programme sarjana mayor Teknik Pertanian dan Biosistem (Undergraduate Programme Guide for ABE).* Department of Mechanical and Biosystems Engineering, Faculty of Agricultural Technology, IPB University.

Wikipedia. (2022a). *Geography of Indonesia.* https://en.wikipedia.org/wiki/Geography_of_Indonesia

Wikipedia. (2022b). *Andi Amran Sulaiman.* https://id.wikipedia.org/wiki/Amran_Sulaiman

Wirawan, IP.S., Triani, IG.A.L., Gunadnya, I.B.P., Arnata, IW. Puspawati G.A.K.D. Wiadnyani, A.A.I.S, Arihantana, N.M.I.H., Anggreni, A.A.M.D., & Madrini, I.A.G.B. (2022). *Pedoman akademik programme sarjana (Academic Guide for Undergraduate Programme).* Faculty of Agricultural Technology, Udayana University.

World Population Review. (2022). *Total Population by Country 2022.* https://worldpopulationreview.com/countries

Zhang, S.L., Jiang, X.K., & Purwadaria, H.K. (2020a). Field production of Konjac. In Srzednicki, G., & Borompichaichartkul, C. (Eds.), *Konjac Glucomannan* (pp.115–160). CRC Press.

Zhang, S.L., Purwadaria, H.K., Borompichaichartkul, C., and Tripetch, P. (2020b). Konjac industry in major producing countries. In Srzednicki, G., & Borompichaichartkul, C. (Eds.), *Konjac Glucomannan* (pp.223–254). CRC Press.

20

Agricultural and Food Engineering Education in Hungary

László Baranyai and Lien Le Phuong Nguyen

Introduction

The agriculture and food industries are traditionally important sectors of the Hungarian economy. The geography of the country is in favour of agriculture and forestry. The production volume of agriculture, forestry and fishing has gradually increased by 15.86% from 1998 to 2021, but the share of the sector in gross value added continuously decreased to 57.14% in the same period (Hungarian Central Statistical Office, 2022a). The current share of gross value added of the sector is 4%. According to the land usage of Hungary, the largest areas are arable land (81.93%), followed by grassland (15.18%), orchard (1.65%), and vineyard (1.18%). Regarding livestock production for slaughter, poultry is in first place with 962,901 tonnes, followed by pigs (578,685t) and cattle (114,799t). The tendency shows that poultry has increased to almost double since 1998, while pigs have decreased by 18.48% and fell to the second place. The significant change is result of strategy because cardiovascular illnesses were the primary causes of death (35.95% in 1998, and 27.44% in 2021) in Hungary (Hungarian Central Statistical Office, 2022b). Cardiovascular illnesses and mortality firmly depend on diet and poor nutrition is known to increase risk (Bhandari et al., 2022). Consumer awareness in nutrition was rising. The available food per capita was 683.7kg in 2020, summarised from vegetables and fruits (189.9kg), milk and dairy products (170.1kg), flour and rice (94.3kg), and meat (72.5kg) at the top of the list, while fish (6.4kg) was at the end. In comparison with 1998, one can see that the structure of available food did not change, but the quantities did. The amount of fish increased dramatically by 3.6kg (128.57%), while vegetables and fruits increased by 28kg (17.29%) and milk and dairy products increased by 20.5kg (13.7%). On the other hand, sugar and honey decreased by 9.1kg (21.62%). Unfortunately sugar and honey are together in these statistics, but honey production shows an improving tendency until 2017 and sharp decline after that. Presently its production in Hungary is below the level of 1998. According to the changed volumes in food consumption, the Hungarian population pays more attention to healthy nutrition now than it did 25 years ago. The education of farmers and food producers had an essential role in this transition.

Digital Agriculture Strategy

The Government of Hungary (2019) announced the Digital Agriculture Strategy in the decree 1470/2019 (VIII. 1). It was predicted that food production in Hungary had 60% more potential. A comparative study of French and Hungarian agriculture confirmed that dairy, cereal, oilseed, and protein crops production in French farms was more efficient than the Hungarian one (Latruffe et al., 2012). However, it was also observed that Hungarian technology was more productive. Digital technology in agriculture became important factor in competition as it can improve effectiveness and profitability. On the other hand, the EU common agriculture politics focuses on protection of the environment, which requires precision agriculture. Large companies with strong financial backgrounds had already started digitalisation, but most of the Hungarian agriculture – small farms and small and medium enterprises (SME) – need help. The main purpose of the Digital Agriculture Strategy decree is to contribute to the increasing profitability of agriculture by means of the collection and utilisation of information, the automation of technology, and the deployment of robotics. In order to reach the main goals, three strategic goals and eight horizontal goals were identified:

- Strategic goals:
    - production: increase the application of precision agriculture (S1)
    - farms: application of enterprise management and decision support software (S2)
    - products: development of product tracking systems and online business (S3)
- Horizontal goals:
    - human resource:
        - development of digital competencies of agriculture stakeholders (H1.1)
        - make digital agriculture counselling available for producers (H1.2)
    - research, development, and innovation:
        - development of digital agriculture innovation environment (H2.1)

DOI: 10.1201/9780429150111-22

  - development of startup ecosystem for digital agriculture (H2.2)
- public administration and services:
  - decrease the cost of access to public services and data (H3.1)
  - improved legislation to facilitate the utilisation of digital technology (H3.2)
  - development of collection and processing of data of the sector (H3.3)
- politics and funding:
  - support the spread of precision agriculture (H4.1)

The above-mentioned strategic goals require new training programmes so that stakeholders can gain their new digital skills. Development, usage, and maintenance of the proposed systems need professionals with the appropriate digital skills. The horizontal goals of human resource development, and research, development, and innovation show the tasks of the education system, especially higher education, and research institutes. Based on surveys conducted in Hungary, it is mainly farmers below the age of 40, qualified in higher education, working on at least 300 hectares, who are using precision farming and digital technology. It was confirmed in a recent survey that people in the age group of 31–40 had the best digital competences in terms of information processing, communication, content creation, safety, problem solving, and precision agriculture (Gaál et al., 2021). It was also found that approximately 15% of the farmers would seek advice from advisory services, most of them prefer using information shared in the community (Fieldsend et al., 2019). This latter study observed that many farmers actively acquire and share knowledge in the absence of formal services.

The Digital Agriculture Strategy evaluated the current situation (in 2019) and identified the challenges for agriculture, livestock, horticulture, viticulture and oenology, forestry, and aquaculture. Some challenges raise global questions that apply to all sectors.

Agriculture used precision farming early as soon as global positioning systems (GPS) became available. It was built into the technology of tillage, plant protection, irrigation, and harvesting. The development of information technology and sensors help data acquisition, data evaluation, decision making and intervention. Although big data technology and cloud computing are available, farmers have no skills to manage those systems. There are integrator companies who deliver ready to use solutions to farmers, but those companies typically represent one developer. Their software and data are specific to the selected developer. Farmers can benefit from whole systems rather than single elements. The drawback of these complex systems is the knowledge required to use them effectively.

Precision livestock farming provides personalised technology depending on the needs of the individual animal. Initially, digital technology was used only to identify the animals. The opportunities are far greater, such as use in behaviour and health monitoring, optimisation of feed, and environmental optimisation.

Horticulture uses similar precision farming solutions to field agriculture. Global positioning systems allow for the collection of plant-specific information. More complex and sophisticated systems can be used in greenhouses and hydroponics, such as environment control including illumination colour. Both agriculture and horticulture can utilise drones and robots to perform measurements, distribute plant protection chemicals, remove weeds, harvest produces, etc.

Viticulture is ahead of other sectors of horticulture in terms of the application of precision farming. The cutting-edge technologies, such as satellite imaging, zone mapping of vegetation indices, ripeness, etc., are already used in the orchard. The digital technology with appropriate models can help in the identification of optimal harvest date, actions against infections and insects, make irrigation plans, etc.

Forestry used precision farming to replace manual data collection at the beginning. Due to the combination of digital maps, aerial photography, and satellite images, forestry was pioneer in the development of geographic information systems. Precision farming technology can help in the registration of forests, monitoring their health and the number of plants, controlling self-driving tools.

Aquaculture is using precision farming to monitor lakes, control feeding and water quality, and monitor aquaculture facilities for normal operations (water, feeding, electricity), water quality, temperature, Oxygen level, etc. The Ministry of Agriculture and the Institute of Agricultural Economics (Hungary) announced the Aquaculture Information System (HALir) online service in 2017 to provide up-to-date information with maps and easy to use reports.

Dombi (2018) evaluated statistical models on two sectors of the Hungarian economy, including agriculture. The study concluded that there are available models to predict production of agriculture and their analysis revealed that drought and machinery are the most significant factors. Machinery positively affects capital stocks. Kemény et al. (2017) concluded in their investigations that investment to precision agriculture pays back within 7 years, independently from the farm size.

Human Resources

The Digital Agriculture Strategy evaluated the behaviour of agriculture enterprises in Hungary according to their behaviour and likely actions in digital development. The evaluation is based on the survey of the Institute of Agricultural Economics, Hungary (2016). In their answer, the majority of the companies identified the lack of qualified employees as the most significant barrier to long-term development. The employee's knowledge shall improve together with the production technology. Medium or large enterprises, where raw materials are processed and leaders have high level professional qualifications, are more likely to send employees to training. Besides general knowledge, practical skills are of great interest as well. The feedback from the companies helps schools, colleges, and universities to improve their curricula. On the other hand, they also indicate the opportunity of short-term training programmes and adult education and validate the life-long learning concept. Agriculture enterprises in Hungary increased the number of their dual interns, who participate in joint programmes and study at the university while working at the company. According to the feedback, students in the dual programmes are more active and loyal to the companies. The agriculture sector is continuously looking for qualified workforce. The survey in 2016 shown that there are open positions for all levels of qualification (50% skilled worker, 20% high school, 10% university). The study published by Nótári et al. (2013) revealed that average qualifications in agriculture are very low, below high school completion. This needs to be addressed to improve the efficiency, quality, and safety of food and agricultural products.

Human resource development is clearly assigned to the education system. Horváth and Szabó (2019) investigated how top executives interpret the concept of industry 4.0 and its driving forces and barriers. The biggest challenge was found to be the lack of training programmes to develop competences in the rapidly changing technological environment. Especially at rural areas the digitalisation can be challenging. A recent study revealed that the main barriers are socio-cultural, drivers are mostly economic, and impacts are balanced between social and economic aspects (Ferrari et al., 2022). The pure knowledge transfer is not sufficient in the industry. Practical skills are required, preferably without internal or in-house training. Curricula shall be improved to include them, especially the management of the modern digital technologies and complex systems. The dual training programmes are popular due to the combination of the strong theoretical background of the school (university) and the practical training opportunity directly in the factory. The dual type students study together with others but work at the company during recess. Students also prefer it since they have jobs and salaries already from the first day at school.

Innovation

A white book was published in Hungary about research, development, and innovation in the agriculture sector (Biro, 2014). The book has evaluated the innovation environment in Hungary and concluded that research and development in agriculture is very fragmented with low budgets and low intentions of cooperation. One of the specific aims is the application of information technology as an interdisciplinary field. In response to the suggestion, a Higher Education and Industry Cooperation Centre of Agriculture Informatics (AgIT FIEK) was established at Szent István University in Gödöllő (presently called Hungarian University of Agriculture and Life Sciences). The main goal of the centre is to facilitate and drive IT development in agricultural production. At the same time, the European Innovation Partnership (EIP-AGRI) has been launched to connect stakeholders in agriculture and forestry. The European programme is to facilitate research on real world problems and development solutions to practical challenges in the sector.

In 2019, a co-fund project called ICT-AGRI-FOOD (ICT Technologies for Sustainable Food Systems) began. It provides a framework for co-funded projects towards the "use of smart digital technology to make European food systems more sustainable, resilient and secure by the insertion of digital technology into the agri-food sector" (Federal Office for Agriculture and Food, 2019). As it is stated on the website "Protecting the health of humans, animals and plants at every stage of the food production process is a key public health and economic priority for the EU. Smart digital technology can help safeguard health along the whole agro-food chain". The ICT-AGRI-FOOD facilitates research on precision farming and robotics in agriculture to provide information (environmental impact, origin, nutritional value, safety, and integrity) to stakeholders, including consumers.

Higher Education in Agriculture in Hungary

The higher education system in Hungary offers training programmes for agriculture in all levels (Table 20.1). Vocational training is available at universities for those who did not join bachelor (BSc) programmes but already finished high school and completed its final test. These vocational trainings are 2 years long, and universities issue a certificate about successful completion. This certificate is not equivalent to a bachelor's degree. The bachelor programmes (BSc) in agriculture are typically 7 semesters long (3.5 years), consisting of one year training of basic knowledge followed by one year transition to

TABLE 20.1

Agriculture training programmes in Hungarian universities

University	VT*	BSc	MSc	PhD
Eszterházy Károly Catholic University (n.d.)	X	X		
Hungarian University of Agriculture and Life Sciences (n.d.)	X	X	X	X
John von Neumann University (n.d.)	X	X	X	
Óbuda University (n.d.)		X		
Széchenyi István University (n.d.)	X	X	X	X
University of Debrecen (n.d.)	X	X	X	X
University of Nyíregyháza (n.d.)	X	X		
University of Pécs (n.d.)	X	X		
University of Sopron (n.d.)		X	X	X
University of Szeged (n.d.)	X	X	X	
University of Tokaj (n.d.)	X	X		
University of Veterinary Medicine (n.d.)		X	X	X

* VT = vocational training in higher education.

professional courses. Students have to choose their specialisation at the end of the second year. The third year is filled with specific courses. The last semester is purely practical training, what students spend at companies. During that semester, they are employed at a company and prepare report about their activities at the end. Master's programmes (MSc) are two years long, and not all universities offer Master's training in agriculture. The doctoral programme (PhD) is available at 5 universities in the field of agriculture. The PhD programmes are 4 years long, divided into 2 parts. The first part includes courses and finishes with a complex exam. In this complex exam, students have to pass an oral exam in two topics (or subjects) approved by the doctoral school, and also have to present the progress of their research work. The second part of the PhD programme requires a successful complex exam. The second part consists only scientific research and students have to fulfil formal requirements (such as number of publications) to be able to make a public defence and achieve their PhD title.

There are vocational advanced studies training programmes offered for engineers who already received their bachelor (BSc) degree. These professional training programmes are usually 1.5 or 2 years long, designed with corresponding style timetables (weekend classes) so that participants can do their job parallel to the training. Despite the corresponding style and limited number of lessons, these programmes are focusing on practical skills and include several industrial case studies. Such vocational advanced studies are available at all universities. Many of those programmes are available since they focus on specific topics, such as "fruit and vegetable processing" or "palinka production".

As a result of the comparison of available training programmes (Table 20.1), three universities offer education in agriculture at all levels: Hungarian University of Agriculture and Life Sciences, Széchenyi István University, and University of Debrecen.

Hungarian University of Agriculture and Life Sciences

The university started working in its current form in February 2021. Universities of five campuses (Budapest, Gödöllő, Gyöngyös, Kaposvár, Keszthely) and agricultural research institutes joined to establish the Hungarian University of Agriculture and Life Sciences (n.d.). The oldest regular higher education campus of agriculture in Europe started at Keszthely in 1797. This belongs to the Hungarian University of Agriculture and Life Sciences presently. The higher education institute of horticulture in Budapest was established in 1853. The university is working on the field of agriculture from the beginning of its history. The university organised its education portfolio into six groups:

- agriculture and food science
- economy and social sciences
- environment protection and environmental sciences
- engineering and information technology
- arts
- pedagogy and liberal arts

Four of the above-mentioned groups offer training programmes related directly to agriculture and food science. This university has the largest capacity of state scholarship students in the field of agriculture.

Széchenyi István University

The Széchenyi István University is located in Győr (North-West of Hungary) and focuses on the development of the city and the regions nearby. The university was established in 1718 as Jesuit academy of theology. The academy opened towards other scientific fields and non-theology students from 1745. Presently it cooperates with international companies, such as Audi, NOKIA, and Philips, to provide valuable practical skills on the labour market. The university organised its education portfolio into nine groups following the structure of faculties:

- agriculture and food sciences
- architecture, civil engineering, and transport sciences
- automotive engineering (in cooperation with AUDI Hungary)
- business and economics
- health and sport sciences
- humanities, education, and social sciences
- law and governance
- mechanical engineering, informatics, and electrical engineering
- performing arts

Only one group is related directly to agriculture, however other groups offer education in logistics, enterprise management, mechanical engineering, etc. what can be valuable for agriculture business as well. The capacity of state scholarship students at the university in the field of agriculture is approximately 12% of that of the Hungarian University of Agriculture and Life Sciences.

University of Debrecen

The University of Debrecen was established in 1538 to offer education in theology. The university opened its faculty of human medicine in 1918 and medicine became the strongest education programme in its portfolio. The university organised its education portfolio into thirteen groups following the structure of faculties:

- law
- human medicine
- humanities
- health
- dentistry
- economics and business
- child and special needs education
- pharmacy
- informatics

- agricultural and food sciences, and environmental management
- engineering
- science and technology
- music

Only one group is related directly to agriculture but the faculties of health, economics and business, informatics, and engineering also offer valuable skills for the sector. The capacity of state scholarship students at the university in the field of agriculture is approximately 20% of that of the Hungarian University of Agriculture and Life Sciences.

Hungarian University of Agriculture and Life Sciences

The Hungarian University of Agriculture and Life Sciences was formed by the Hungarian Government in 2021 to make the largest multidisciplinary university with a focus on agriculture. The institution has 9795 students, 1550 international students and 1123 staff (Quacqarelli Symonds, 2023). The education programmes are available in Hungarian and English as well. Besides the campuses, education takes place in Kisvárda, Nyírbátor and Szarvas (Hungary), Berehove (Ukraine), Komarno (Slovak Republic), Miercurea Ciuc and Odorheiu Secuiesc (Romania), and Senta (Republic of Serbia). The university organised the education into nineteen institutes (Hungarian University of Agriculture and Life Sciences, n.d.):

- Szent István Campus (Gödöllő)
  - Institute of Agricultural and Food Economics
  - Institute of Agronomy
  - Institute of Aquaculture and Environmental Safety
  - Institute of Environmental Sciences
  - Institute of Genetics and Biotechnology
  - Institute of Mathematics and Basic Science
  - Institute for Physical Education and Sports
  - Institute of Rural Development and Sustainable Economy
  - Institute of Technology
  - Institute for Wildlife Management and Nature Conservation
- Buda Campus (Budapest)
  - Institute of Food Science and Technology
  - Institute of Horticultural Science
  - Institute of Landscape Architecture, Urban Planning and Garden Art
  - Institute of Plant Protection
  - Institute of Viticulture and Oenology
- Kaposvár Campus
  - Institute of Animal Science
  - Institute of Animal Nutrition
  - Institute of Pedagogy
  - Institute of Arts

The above listed campuses provide the headquarters of the institutes, but education programmes are running at all other locations as well. There are twelve bachelor (BSc) training programmes in the field of agriculture, four vocational training programmes in higher education, sixteen master (MSc) programmes, twenty-eight vocational advanced studies programmes. The university has twelve doctoral schools and nine of them are directly related to agriculture:

- Doctoral School in Animal Science (animal husbandry)
- Doctoral School of Animal Biotechnology and Animal Science (animal husbandry)
- Doctoral School of Biological Sciences (biology)
- Doctoral School of Economic and Regional Sciences (business and management, regional sciences)
- Doctoral School of Environmental Sciences (environmental sciences)
- Doctoral School of Food Sciences (food sciences)
- Doctoral School of Horticultural Sciences (plant breeding and botany)
- Doctoral School of Landscape Architecture and Landscape Ecology (agricultural engineering)
- Doctoral School of Management and Organisational Science (business and management)
- Doctoral School of Mechanical Engineering (agricultural engineering)
- Doctoral School of Plant Science (plant breeding and botany)
- Festetics Doctoral School (animal husbandry, environmental sciences, plant breeding and botany)

The practical skills are critical in agriculture and food industry and are expected by companies when they search for new employee. Education programmes strengthen practical training. All students have to participate in company training and make a written report and oral presentation about their activity. The practical training can be organised in Hungary and in abroad as well.

According to the higher education legislation (No. CCIV in 2011) and the 230/2012., VIII. 28 decree of the government (Government of Hungary, 2012), all bachelor (BSc) training programmes have to include a continuous and one semester long practice. It is mandatory, not optional, and typically takes place in the last (7th) semester. Regular students have a 12–15 week long practice, and corresponding students have a shorter, 4–6 week long practice. Master's (MSc) students prepare their 4 weeks practical training in spring and complete it during the summer. Agricultural engineers in 5-year training programmes (MSc) have 80 hours (2 weeks) of production practice, and 160 hours (4 weeks) management practice, both are organised in summer and two times during their education.

Improvement of curricula is a natural procedure and absorption of modern technology is visible in the description of subjects. This process is facilitated by the university as it joined the 5G Coalition and Artificial Intelligence Coalition in Hungary. These technology platforms are able to provide real time data collection and evaluation in big quantity (big data). Agriculture can benefit from modern information technology solutions and the knowledge already appears in curricula. The Hungarian University of Agriculture and Life Sciences (n.d.) has eleven bachelor (BSc) training programmes in the field of agriculture:

- agricultural and food industry mechanical engineering
- agricultural engineering
- agricultural engineering on rural development
- animal breeding
- environmental protection engineering
- food engineering
- horse breeding and equestrian sport organiser
- horticultural engineering
- landscape architecture and garden art
- viticulture and oenology engineering
- wildlife management

In all education programmes, students meet the Institute of Mathematics and Basic Science, who offer optional courses on artificial intelligence, machine learning, statistics, data mining, advanced sensor technology. The cutting-edge technology also appears in the specialised subjects, especially the application of information and communication technology (ICT). Authors of this chapter believe that biosystems engineering (or biochemical engineering), what is classified as engineering science, belongs to agriculture, as well. Biochemical engineering students learn environmental protection, enzyme technology, probiotics, what fits to the scope of agriculture. There are short non-degree programmes on

- agriculture
- herbs and spices
- stud farm management
- viticulture and oenology

There are sixteen master (MSc) training programmes in the field of agriculture:

- agricultural and food industry mechanical engineering
- agricultural biotechnology
- agricultural water management
- animal breeding
- environmental management
- environmental protection
- food safety and quality engineering
- food engineering
- foraging and forage safety engineering
- horticultural engineering
- landscape architecture
- plant medicine
- plant breeding
- rural development
- viticulture and oenology
- wildlife management

As is visible in the education portfolio, topics cover all major fields of agriculture. University professors do their research on these fields and build their research results into the education programmes. The Quacqarelli Symonds university ranking shows that the Hungarian University of Agriculture and Life Sciences obtained the rank of #201-250 in the subject of agriculture and forestry in the year of 2022 (Quacqarelli Symonds Limited, 2023). The best ranking criteria are the citations per paper, and the employer reputation. This is promising result for a relatively young brand dated back to 2021, the establishment of the university in its present form.

The relationship between the university and the agriculture sector is more visible through the twenty-eight vocational advanced study programmes. These have strong focus on a narrow topic and are offered to students who have already obtained a higher education degree. In the field of agriculture, the following programmes are available (Hungarian University of Agriculture and Life Sciences, n.d.):

- agricultural and food industry product designer
- apiary
- beer master
- biological soil management
- cattle breeding
- chocolate, tea, coffee making
- cynology
- ecological farming
- ecological green land maintenance
- environmental protection
- fishery and aquaculture
- food supply chain quality control
- food supply chain quality
- fruit and vegetable processing
- herb recognition and utilisation
- historical garden management
- irrigation
- palinka master
- plant genetics and breeding
- plant protection
- precision agriculture
- soil engineering
- sowing management
- special translator on agriculture and natural sciences
- viticulture and oenology

- waste management and recycling
- wildlife management
- wood examiner and nurser

The vocational advanced studies are practice oriented and both individuals and enterprises invest into these training programmes to gain knowledge on the latest developments.

International Cooperation and Development of Education

The Hungarian University of Agriculture and Life Sciences, and its predecessors in title, have strong international activity. Besides research projects, there are initiatives to improve education and try new training models. Such project is the E[3]UDRES[2] European University Alliance, where the E[3]UDRES[2] means "Engaged and Entrepreneurial European University as Driver for European Smart and Sustainable Regions". This is a network of higher education institutions in Europe, who work together and collaborate in the fields of research, teaching, innovation, and much more (St. Pölten University of Applied Sciences, 2023). The project is co-funded by the Erasmus+ Programme of the European Union. Nine institutions form the core of the network:

- St. Pölten University of Applied Sciences, Austria (lead)
- Polytechnic Institute of Setúbal, Portugal
- Hungarian University of Agriculture and Life Sciences, Hungary
- UC Leuven-Limburg University of Applied Sciences, Belgium
- Politehnica University of Timisoara, Romania
- Vidzeme University of Applied Sciences, Latvia
- Saxion University of Applied Sciences, The Netherlands
- Fulda University of Applied Sciences, Germany
- Jyväskylä University of Applied Sciences, Finland

The activity of the network is organised in three levels:

- The executive board is composed of the university management members from all institutions, and responsible for strategic development, vision, mission, and culture.
- The board of coordinators is composed of all site coordinators at each institution and work package leads, and responsible for implementation and planning, also promotes exchange of knowledge and experiences.
- The board of student representatives is composed of students of all member institutions and advocates for the interests of students.

The activities of the network focus on smart and sustainable regions, and the future university. Especially this latter is interesting from the point of view of education development, as the network builds a multi-university campus across Europe. Three new education formats were introduced successfully to students and teachers as well: hackathons, bootcamps and I Living Labs.

Hackathons

The hackathons are real-life challenges, which are introduced to international teams to be solved within a short time. Hackathons are organised once per year. Local businesses and institutions provide the problems for the event. Teams are working continuously on the solution for 36–48 hours, meanwhile mentors and experts help them with input and refine ideas. Jury is voting for the solutions they support the most. The challenges announced for Hackathon 2022 were (St. Pölten University of Applied Sciences, 2023):

- developing sustainable food products
- design of a booking platform for care services
- digital technologies and advanced manufacturing
- add value in musculoskeletal healthcare
- sustainable tourism for a mountain valley
- development of a startup village

The challenges were submitted by companies and member universities. Solutions require creative thinking in multidisciplinary team.

Bootcamps

The bootcamp events are 5 days long, where students come together and work on regional challenges. These events are organised once per year. During the bootcamp, students work in small teams to a smarter and more sustainable society. Participants gain knowledge on design thinking, business models, and soft skills like international collaboration. Students are awarded with 3 ECTS (European Credit Transfer System) points for participating the bootcamp.

I Living Labs

The I Living Labs are university courses in which students are set to work on complex problems (challenges). The cooperation of international students with vastly different background is called trans-disciplinary work by the project. There is no conventional teaching activity, teachers play coach role in these courses and provide support to the teams. In this aspect, university course is only a framework that provides the opportunity of regular team meetings and discussions. Coaches guide students and stimulate their skill development. The I Living Labs strengthen analytical thinking and problem solving, build so called future skills. The I Living Labs finish with showdown event, a presentation of solutions of the teams.

Stipendium Hungaricum Scholarship

The Stipendium Hungaricum scholarship programme was founded by the Hungarian Government in 2013. It is supervised by the Ministry of Foreign Affairs and Trade and managed by the Tempus Public Foundation (2023a). The scholarship is based on bilateral agreements between governments of Hungary and the partner country. The mission of the programme is "to promote the internationalisation of the Hungarian higher education and to attract top foreign students from all around the world who can establish personal and professional attachments to Hungary while enjoying high quality education in the heart of Europe" (Tempus Public Foundation, 2023a). The scholarship covers the following provisions:

- tuition fee
- monthly stipend of approximately 103 EUR for students of non-degree, bachelor and master programmes, approximately 331 EUR for doctoral students in the first 2 years, and approximately 425 EUR for doctoral students in the second 2 years.
- Accommodation contribution of approximately 95 EUR per month
- medical insurance and support to health care services approximately 153 EUR per month

Around one thousand students participated in the Stipendium Hungaricum at the beginning and eleven thousand students won scholarship for the academic year of 2021/2022 (Tempus Public Foundation, 2023b). According to the number of students, bachelor (BSc) programmes are the most popular (38%), followed by master (MSc) training programmes (34%) and doctoral research (PhD) (18%). When compare the submitted applications of 2021/2022 academic year to the available quota, doctoral research achieved the highest ratio with 1220%, followed by the bachelor (BSc) training programmes with 321% and master (MSc) programmes with 157%.

Presently around twelve thousand Stipendium Hungaricum students are studying in 28 higher education institutions in Hungary. More than 600 full degree and non-degree programmes are running, primarily in English. Students came to Hungary from nearly 90 countries. The applications for the academic year of 2022/2023 shows that students from Asia (55.4%) and Africa (39.5%) cover already 95% of the interest. Central and South America is represented only with 2.8% and Europe with 2.2%. The top five most popular fields of study for 2022/2023 academic year were:

- computer science and information technology
- economic science
- engineering science
- medical and health science
- natural science

Regarding the institutions, 51.5% of applications were submitted to the universities offering education in the field of agriculture. The largest number of applications targeted the universities with human medicine education programmes as University of Debrecen (20.4%) and University of Pécs (8.9%). The Hungarian University of Agriculture and Life Sciences obtained the third place (8.5%) and it is the top university with primarily agricultural profile.

REFERENCES

Bhandari, B., Liu, Z., Lin, S., Macniven, R., Akombi-Inyang B., Hall, J., Feng, X., Schutte, A.E., Xu, X. (2022). Long-term consumption of 10 food groups and cardiovascular mortality: A systematic review and dose response meta-analysis of prospective cohort studies. *Advances in Nutrition*, 14(1): 55–63, DOI: 10.1016/j.advnut.2022.10.010

Biró, S. ed. (2014). *[In Hungarian] Innováció a magyar agrár- és vidékfejlesztésben (In English: Innovation in the Hungarian Agriculture and Rural Development).* Institute of Agricultural Economics.

Dombi, M. (2018). Modeling the material stock of manufactured capital with production function. *Resources, Conservation & Recycling*, 138: 207–214. DOI: 10.1016/j.resconrec.2018.07.015

Eszterházy Károly Catholic University. (n.d.). https://uni-eszterhazy.hu/en

Federal Office for Agriculture and Food. (2019). ERA-NET cofund ICT-AGRI-FOOD. https://www.ictagrifood.eu

Ferrari, A., Bacco, M., Gaber, K., Jedlitschka, A., Hess, S., Kaipainen, J., Koltsida, P., Toli, E., Brunori, G. (2022). Drivers, barriers and impacts of digitalisation in rural areas from the viewpoint of experts. *Information and Software Technology*, 145: 106816. DOI: 10.1016/j.infsof.2021.106816

Fieldsend, A.F., Voitovska, Y., Toirov, F., Markov, R., Alexandrova, N. (2019). A sustainable approach to fostering agricultural knowledge sharing in conflict-affected areas of Eastern Ukraine. *NJAS - Wageningen Journal of Life Sciences*, 89: 100293. DOI: 10.1016/j.njas.2019.02.004

Gaál, M., Molnár, A., Illés, I., Kiss, A., Lámfalusi, I., Kemény, G. (2021). Chapter 11 - Where do we stand with digitalization? An assessment of digital transformation in Hungarian agriculture. In Bochtis, D., Achillas, C., Banias, G., Lampridi, M. (Eds.), *Bio-Economy and Agri-production*, Academic Press, pp. 195–206. DOI: 10.1016/B978-0-12-819774-5.00011-4

Government of Hungary. (2012). 230/2012 (VIII. 28) decree about vocational training programs in higher education and practical trainings in higher education.

Government of Hungary. (2019). 1470/2019 (VIII. 1) decree about digital agriculture strategy of hungary.

Horváth, D., Szabó, R.Z. (2019). Driving forces and barriers of industry 4.0: Do multinational and small and medium-sized companies have equal opportunities? *Technological Forecasting & Social Change*, 146: 119–132. DOI: 10.1016/j.techfore.2019.05.021

Hungarian Central Statistical Office. (2022a, January 31). 19Agriculture. STADAT database. https://www.ksh.hu/stadat_eng?lang=en&theme=mez

Hungarian Central Statistical Office. (2022b, January 31). 22.1.1.10. Mortality by frequent causes of death and by sex. STADAT database. https://www.ksh.hu/stadat_files/nep/en/nep0010.html

Hungarian University of Agriculture and Life Sciences. (n.d.). https://en.uni-mate.hu

Institute of Agricultural Economics. (2016). [In Hungarian] Agrármunkaerőpiaci barométer felmérés (in English: Survey on agriculture labour market), Research report, Budapest.

John von Neumann University. (n.d.). https://nje.hu/en

Kemény, G., Lámfalusi, I., Molnár, A. (Eds.) (2017). [In Hungarian] A Precíziós Szántóföldi Növénytermesztés Összehasonlító Vizsgálata (in English: Comparative study of precision arable crop production). AKI, Budapest, p. 170. DOI: 10.7896/ak1703

Latruffe, L., Fogarasi, J., Desjeux, Y. (2012). Efficiency, productivity and technology comparison for farms in Central and Western Europe: The case of field crop and dairy farming in Hungary and France. *Economic Systems*, 36: 264–278. DOI: 10.1016/j.ecosys.2011.07.002

Nótári, M., Berde, C., Ferencz, Á. (2013). Human resources management and education in Hungarian agriculture. *Procedia - Social and Behavioral Sciences*, 81: 632–637. DOI: 10.1016/j.sbspro.2013.06.488

Óbuda University. (n.d.). https://uni-obuda.hu/main-page/

Quacquarelli Symonds Limited. (2023, January 31). QS world university rankings. https://www.topuniversities.com

Széchenyi István University. (n.d.). https://admissions.sze.hu/welcome

St. Pölten University of Applied Sciences. (2023, January 31). Engaged and entrepreneurial European university as driver for European smart and sustainable regions. https://eudres.eu

Tempus Public Foundation. (2023a, January 31). Stipendium Hungaricum. https://stipendiumhungaricum.hu/about

Tempus Public Foundation. (2023b, January 31). Stipendium Hungaricum program statistics. https://tka.hu/palyazatok/7619/statisztikak

University of Debrecen. (n.d.). https://www.edu.unideb.hu

University of Nyíregyháza. (n.d.). https://english.nye.hu

University of Pécs. (n.d.). https://international.pte.hu

University of Sopron. (n.d.). http://international.uni-sopron.hu/home

University of Szeged. (n.d.). https://u-szeged.hu/english

University of Tokaj. (n.d.). https://unithe.hu/homepage/

University of Veterinary Medicine. (n.d.). https://univet.hu/en/

21

Food Engineering Education in Nigeria

Akindele Folarin Alonge, John Olajide and Patrick O. Ngoddy

Introduction

Food Engineering is a relatively new engineering discipline in Nigeria. It is an offshoot of Agricultural Engineering having been the latest specialised area in the USA where it all started in the mid-sixties. Food Engineering also evolved from other branches of engineering which include Chemical Engineering and Mechanical Engineering. Food Engineering evolved from other disciplines, though it was housed in Agricultural departments in some Universities, while in others it was housed in Chemical Engineering (Niranjan, 2017).

Agricultural Engineering has had its share of non-recognition, misrepresentation, and marginalisation in the country over the years. Even among the elites and colleagues in engineering field, agricultural engineering is no less than tractorisation. To many, agricultural engineering is synonymous with tractor and farm machinery. Farm Power and Machinery is a specialised area that has to do with various sources of power in the farm and the machineries used from land clearing to the harvesting of crops. While the tractor is a prime mover and tractorisation is a major thing in farming, it noteworthy that it covers less than twenty-five percent of the Farm Power and Machinery Option (Onwualu et al., 2006).

Here comes Food Engineering with challenges similar to that of Agricultural Engineering. Many people also confuse this with Food Technology or Food Science. The misunderstanding has been based basically on what Food Engineering is all about, that is, the difference between Food Engineering, Food Technology, and Food Science. For Food Science, the misunderstanding has been on the difference between Food Science and Food Technology, Human Nutrition, Home Economics, and Hotel and Catering Management.

Each of them is vital and interrelated but are not the same. This paper attempts to provide information on Food Engineering as a relatively new discipline and what it offers.

Definitions of Terms

Food Engineering

Food engineering is the application of science and engineering to the evaluation, utilisation, processing, packaging, distribution and storage of food. Food engineering is a relatively new field of study but is based on concepts that have evolved over a significant period of time. For example, fish and meat preservation and the dairy industry have been worked on for hundreds of years. Food canning and edible oil processing at factory level dates back for only about one hundred and seventy years. Food engineering is the latest branch of Agricultural Engineering that is highly developed in the United States. Food Engineering evolved from other disciplines though it was housed in Agricultural departments in some Universities while in others it was housed in Chemical Engineering.

Food Engineering studies are designed to give students a solid background in Food Chemistry, Microbiology, Quality Control, Food Processing, Principles of Packaging, Principles of Storage, Factory Operation, Administration, Sanitation, Food Laws, and Unit Processes. The programme also includes, courses in General Biochemistry, Analytical Chemistry, Human Nutrition, and Statistics, and selected basic engineering courses such as fluid mechanics, engineering thermodynamics, engineering materials and engineering mechanics, workshop technology, metallurgy and manufacturing practices, machine theory and practice, technical report writing, computers and their applications, strength of materials, heat transfer, etc. It also includes general crop production, general animal husbandry, and general livestock production. Food Engineering is generally involved in the design of food machinery, storage systems, and the design of processes and preservation.

The course of studies presented here aims at providing adequately trained and qualified staff (i.e. skilled and competent professionals) in various fields of food, production, fields, industries, food commodity research stations, food canning and edible oil processing factories, beverage industries, and general areas related to the aforementioned activities.

Food Technology

This is a science that deals with the techniques and principles involved in processing and preservation of food substances. The application of food science helps in manufacturing safe, wholesome, and nutritious food products. The study of food technology is to develop new methods, systems for keeping food products safe and resistant from natural harms such as bacteria and other micro-organisms. Food processing helps in preservation, enhances the flavour, and reduces the toxins

DOI: 10.1201/9780429150111-23

in the food product which results in better distributional efficiency and easy marketing of the food product.

Food technology involves all aspects of food science including production, processing, and distribution. Professionals in this field work to improve manufacturing methods through preservation, storage, and new product development. Food technology is the application of food science to the preservation, processing, and packaging of food, and the development of new food products. Food technologists have a strong background in the sciences. They use their knowledge to create and improve food products. They also apply food science research to improve food safety, quality, storage, preservation, packaging, processing, and distribution. They work with Food Scientists to enhance the nutritional value of foods.

Food Science

Food Science is an exciting field of study that applies sciences such as Chemistry, Biochemistry and Microbiology to discover ways to improve the taste, nutrition, and quality of the world's food supply. It deals with the processing of raw agricultural produce into food products that are acceptable for human consumption. A Food Scientist possesses the skills necessary to convert these raw materials into the safe, attractive food products and beverages available on our supermarket shelves. Food Scientists constantly seek to improve existing techniques by which new products are developed, processed, packaged, stored, and transported.

Food Science is the discipline in which the engineering, biological, and physical sciences are used to study the nature of foods, the causes of deterioration, the principles underlying food processing, and the improvement of foods for the consuming public. It is one of the life sciences and usually considered distinct from the field of nutrition. It is a highly interdisciplinary applied science and incorporates concepts from different fields such as microbiology, biochemistry, chemical engineering, and many others. In order to apply food science and put it to practical use food technology comes in to play. Food science is the study of the physical, biological, and chemical makeup of food and the concepts underlying food processing methods.

Food and Human Nutrition

It is the application of the science of nutrition to the prevention and treatment of disease and promotion of health. It is the process by which nutrients in food are converted into body tissues and supply energy to sustain human life and good health (Kent-Jones et al., 2023). This is quite different from Food Engineering. There are seven major classes of nutrients: carbohydrates, fats, fibre, minerals, protein, vitamins, and water. A healthy diet throughout life promotes healthy pregnancy outcomes, supports normal growth, development, and ageing. It helps to maintain a healthy body weight, and reduces the risk of chronic disease leading to overall health and wellbeing.

Home Economics

This is arts and science that deals with home management. It is more all-embracing as it includes cooking, nutrition, purchase, preparation, service of food, interior design, clothing and textiles, child development, family relationships, home budgeting, and household economics.

Food Engineering in Nigeria

Food Engineering started in the United States of America around the sixties and from there spread around the world. The first B.Sc. programme in Food Engineering started at the University of Massachusetts in 1974 under the leadership of Professor J.T. Clayton. Prior to that time, there were earlier efforts in England and Europe to develop the discipline in the nature of so called "Taught M.Sc." courses in Food Engineering. B.Sc. courses in Food Engineering commenced later in Israel, Great Britain, France, Germany, The Netherlands, Czechoslovakia, Brazil, and Chile.

The Food Engineering degree programmes at LAUTECH and University of Uyo (UNI-UYO) were the first of their kind not only in Nigeria but also in the whole of Africa. The programmes in Nigerian Universities date from the mid-1990s. The first set of Food Engineers was produced from LAUTECH in 1997. Those from UNI-UYO graduated in 2003. The Council for Regulation of Engineering in Nigeria (COREN) accredited (albeit, partially) the LAUTECH programme in 2003. In Nigeria, it started in late nineties with Ladoke Akintola University of Technology. Ogbomoso, University of Uyo as pioneering Universities. The father of Food Engineering in Nigeria is Prof Patrick Ngoddy. He was one of the earliest people that had a higher degree in the USA (Michigan State University) in the late sixties.

Those who are currently Food Engineers are first of all Agricultural Engineers who specialised in Crop Processing and Storage or Farm Machinery. Others include Mechanical Engineers, Chemical Engineers and Food Technologists who had higher degrees and training in agricultural engineering. There are over six universities with Food Engineering programmes; Ladoke Akintola University of Technology, Ogbomoso; University of Uyo, Uyo, Obafemi Awolowo University, Ile-Ife, Federal University of Agriculture, Abeokuta for postgraduate studies only; University of Ilorin, Ilorin, Federal University of Technology, Minna (at postgraduate level). Additional four universities will soon start food engineering. Some of the universities like Obafemi Awolowo University housed Food Engineering in the Department of Food Technology. Table 21.1 shows the List of Universities offering Food Engineering in Nigeria.

Food Engineering versus Food Science and Technology

Food Scientists and Technologists use their foundational skills in food science (i.e., knowledge of the bio-chemo-physical nature of foods) in addressing the challenges (i.e., problem solving) in food manufacturing, preservation, and product development. Food Science and Technology seeks to train them in complex and complementary subjects of food quality,

TABLE 21.1

List of universities with Food Engineering programmes in Nigeria

S/n	Name	Geo-political location	Year Founded	Availability of UG & PG programme	Faculty/Department where programme is domiciled
1	Ladoke Akintola University of Technology, Ogbomoso, Oyo State	South West	1990	Both UG and PG	Faculty of Engineering
2	University of Uyo, Uyo, Akwa Ibom State	South-South	1996	Both UG and PG	Faculty of Engineering
3	Obafemi Awolowo University, Ile-Ife, Osun State	South West	1997	Both UG and PG	Dept of Food Technology
4	Federal University of Agriculture, Abeokuta, Ogun State	South West	2001	PG	Dept of Food Science and Technology
5	University of Ilorin, Ilorin, Kwara State	North Central	2014	UG	Faculty of Engineering
6	Federal University of Technology, Minna, Niger State	North Central	2010	PG	Dept of Agricultural and Bioresources Engineering
7	Kwara State University, Malete, Ilorin. Kwara State	North Central	2020	UG	College of Engineering
	Proposed				
8	Michael Okpara University of Agriculture, Umudike, Abia State	South East		UG	Not known yet
9	Federal University of Technology, Owerri, Imo State	South East		UG	Not known yet
10	University of Ibadan, Ibadan, Oyo State	South West		UG	Not known yet
	UG = Undergraduate PG = Postgraduate				

food safety, and food laws that govern the domestic and internal marketing, trade, distribution, and consumption of food (Ngoddy, 2021).

Those in Food Engineering are trained as engineers who could bring skills of Engineering Science and Technology to problem solving on challenges in the design and making-operational food processes, food machinery and equipment, and food plants for food manufacturing. To perform these triangular functions of process design, machine design and plant design. Food Engineers must be trained to work hand in hand with Food Technologists with whom they must mutually share a sufficient base of understanding of the scientific nature of Food (its bio-thermo-physical nature) and its technological behaviour during handling, manufacturing, storage, marketing, distribution, and consumption.

Ngoddy (2021) further stated that the logic of the base of mutual understanding between the Food Technologists and Food Engineers is that they could only effectively work together if, and only if, they are equipped to understand enough of the nuances of Food Engineering, and in turn, Food Engineers must understand enough of the nuances of Food Technology for them to work creatively together so that food processes, food machines and food plants designed and installed by the Engineers can be run efficiently and profitably by Food Technologists.

For the food manufacturing and distribution industries and profession, engineering design assumes the triangular configuration of three complementary and interlocking components, namely;

1. **Food Process Design**: which strategises unit processes into the sequence of treatment steps, leading from an aggregation of raw materials to finished end products, packaged as desired for the intended market;
2. **Food Machine Design**: that creates machine-technology systems as units of complementary components, which define each desired unit-operation to accomplish specified tasks in the process step, linking *seriatum* from step-to-step in the manufacturing system.
3. **Food Plant Design**: which translates the design process sequence (of process design) and the complementary machine-technology units) of machine design) into a harmonised physical reality in space, on a factory floor inside a factory building, which is located within an integrated complex of buildings, with facilities that make up the integral entity called a Food Plant. It is a self-contained campus of synergistic parts which today must pass the acid test of Environmental Impact Assessment (EIA) approval to operate in an ecosystem, the environmental and socio-economic quality of life which it must compromise.

Training in Food Engineering

Like other engineering fields, food engineering training is designed to meet national needs and aspirations vis-à-vis industrial development and technological emancipation. The minimum standards specified by Nigerian Universities Commission (NUC) and COREN BMAS are put in place to meet these needs and produce graduates in Food Engineering with sufficient academic backgrounds coupled with sufficient practical experience. These graduates are to rise up to the challenges of a developing economy. Such graduates are resourceful and creative, and must be able to perform the following functions;

i. to design engineering projects and supervise their construction.
ii. To design and make components, machines, equipment, and systems.
iii. to design and develop new products and production techniques in food industries.
iv. to install and maintain complex engineering systems so that they can perform optimally in our environment.
v. to adapt and adopt exogenous technology in order to solve local engineering problems.

The general philosophy therefore is to produce technically skilled graduates with high academic standard; to provide services for the food industries and allied institutions at all levels in food process design, food equipment design and fabrication, maintenance and evaluation of food processing equipment and plants.

The primary objectives of the Food Engineering programme are to train Food Engineers specifically for the following purposes;

(i) to design and fabricate machinery and develop processes for food processing and preservation
(ii) to develop systems for storage and packaging of food materials
(iii) to develop new techniques for utilising presently unused food resources and by-products of food manufacturing for direct human consumption
(iv) to develop manpower that can work in higher educational institutions, research institutes, and national and international agencies

Prospects for Food Engineering Graduates

Job opportunities are available in food industries. Professional food technologists are employed as Plant Managers, Supervisors, Sales Personnel, and research Personnel in the Food industries. They are also involved in any of the areas of Food Technology such as milling of cereals; canning of meats, fish, fruits, fruit juices, vegetables, milk, etc; confections – jams, jellys, candies, etc.; dehydration and freeze drying of meats, fish, eggs, milk, root crops, etc.; dairy products; curing and smoking of meats and fish; cooking oils and fats. They are also involved in development of new food products and/or improved food products and equipment.

They can also work in Government research establishments. They are employed by the government research establishments that are devoted to the development of new attractive and nutritious products from indigenous foodstuffs as well as by those establishments devoted to the solution of fundamental problems affecting the food supply of the country. Moreover, the Federal and State Governments require the services of Food Technologists as field supervisory and inspection personnel, and to protect the consumer through the development and application of effective analytical food testing methods needed to back up the food and drug decree.

They can also work in private enterprises. The training that will be provided will be such that individual who wishes to establish his own food industry would have acquired the necessary skills and competence. There are many struggling food industries in Nigeria that can become successful food industries given management and supervision/personnel with technological "know-how".

They can also be engaged in teaching and research. Food Engineers and Technologists who complete postgraduate studies will be prepared for opportunities in teaching and research at the University or Polytechnic level. They can also be engaged in consultancy services; food technologists with postgraduate qualifications are in demand to undertake feasibility studies for newly planned food industries and to provide expect advice regarding problems within existing ones.

Admission Requirements

A. Entry Requirements

The entry requirements into the department is as follows:

1. UME Entry requirements
   i) Candidates should have five Ordinary level credits in science subjects including English Language, Mathematics, Physics, Chemistry, and a credit in Biology, Agricultural Science, Health Science, or Food Nutrition.
   ii) Candidates should offer the following science subjects in UME.

English Language
Physics
Chemistry
Mathematics

2. Direct Entry Requirements
   i) Candidate should possess two Advanced level passes or equivalent in the relevant science subjects.
   ii) Candidates with Ordinary National Diploma (OND) in Food Technology with minimum of Upper Credit.

iii) Candidates with Higher National Diploma (HND) in Food Technology with Upper Credit may be accepted into the 300 level.

3. Other requirements are as specified in the JAMB Brochure.

Minimum Duration

The programme will run on a full-time basis for a minimum of 10 semesters for UME students and minimum of 8 semesters for direct students. The students will be expected to go for industrial training for a semester in the 400 level. A maximum of 14 semesters for UME and 12 semesters is allowed for direct entry students

Requirements for a Degree

1. According to NUC standard, to be eligible for a degree, a candidate must satisfactorily complete a minimum of 155 units including:
   (a) 10 units of general Studies courses
   (b) 6 units of management courses

Professional Bodies Relevant to Food Engineering

The professional bodies relevant to Food Engineering include the Nigerian Society of Engineers (NSE), Council for Regulation of Engineering in Nigeria (COREN), the Nigerian Institution of Agricultural Engineers (NIAE), and the Nigerian Institution of Food Science and Technologists (NIFST).

The Council for the Regulation of Engineering in Nigeria is a regulatory body. It was established by Decree 55 of 1970 and 27 of 1992 (amendment), now the Engineers (Registration, etc.) Act CAP E 11, 2004 to regulate and control the practice of Engineering in Nigeria in all its aspects and ramifications (COREN, 2017c).

The mandate of COREN therefore includes:

- Accreditation of Engineering Programmes;
- Registration of Engineering Personnel;
- Regulation and Control of Engineering Practice.

The law empowers COREN to carry out Accreditation of Engineering Programmes in Universities, Polytechnics and Technical Colleges.

Section (1) (b) of Decree 55 of 1970 and 27 of 1992 (amendment), now the Engineers (Registration, etc) Act CAP E 11, 2004, empowers COREN to:

> Determine what standards of knowledge and skill are to be attained by persons seeking to become registered as engineering personnel and to raise those standards from time to time as circumstances may permit. For this purpose, the Act provides that COREN shall conduct visitation to Engineering Institutions in Nigeria or elsewhere for the purpose of:
>
> (a) Accrediting their courses (sections 7 and 8 of the Decrees);
> (b) Withdrawing any previous approvals, if so warranted and as prescribed in sub-sections (3) – (5) of section 7 (COREN, 2017b).

Similarly, Section 9 Sub-section 6 of the Act provides (COREN, 2017a):

> An educational institution for the training of persons in the Engineering profession shall submit syllabus of its programme, content and minimum facilities to the Council for approval before a course approved by the National Universities Commission or the National Board for Technical Education is commenced.

In Nigeria, NSE serves as the overall governing body for all engineering specialisations. It is essential that all members register with and be a part of this organisation. A group of Nigerian graduate engineers and students in the UK created the organisation on February 16, 1958. The Society was officially opened at the Nigerian House in London. The NSE is listed as a company limited by guarantee with the Corporate Affairs Commission. It was established in 1967 as a company limited by guarantee without a share capital. On December 1, 1998, the original Memorandum and Articles of Association were modified. Similar to some of its equivalent professional bodies in other professions, the NSE has distinguished itself via forward-thinking and creative programmes to become a platform for the members' professional development and the technological advancement of the nation (NSE, 2022).

COREN is the body that registers all engineers, technologists, technicians, artisans, and craftsmen. It is compulsory for all those who wishes to practice engineering in Nigeria register with this council. COREN was established by decree 55 of 1970, amended by decree 27 of 1992, and by the Engineers (Registration, etc.) Act, CAP E11 of 2004 Law of the Federal Republic of Nigeria. The Act establishes COREN as a statutory body of the Federal Government empowered to regulate the Practice of Engineering in all aspects and ramifications in Nigeria. COREN has a mandate through the Decrees empowered the Council to regulate and control the training and practice of engineering in Nigeria and to ensure and enforce the registration of all engineering personnel (i.e. Engineers, Engineering Technologists, Engineering Technicians, and Engineering Craftsmen) and consulting firms wishing to practice or engage in the practice of engineering.

In 1975, the NIAE began as the Nigerian Society of Agricultural Engineers, NSAE, with the following objectives; "to promote the science and art of engineering in Agriculture; to encourage original research; to foster and promote agricultural engineering education; to advance in every possible way the standards of agricultural engineering; to promote healthy interaction between members and allied professionals; and to encourage the enhancement of professional competence of its members" (NIAE, 2021). In 1999, a merger of the NSAE with the Nigerian Society of Engineers, NSE occurred, and thus the NSAE transformed into the Nigerian Institution of Agricultural Engineers, NIAE, a division of the NSE. NIAE currently houses Food Engineering, being an offshoot of Agricultural Engineering. Agricultural Engineers examine and help in interviewing all those who are registering with both COREN and NSE.

Food Engineers cannot function well without Food Scientists and Technologists. The Nigerian Institute of Food Science and Technology is the only legally recognised non-profit organisation in Nigeria that speaks for Food Professionals who work in government, universities, industry, and research institutes. The mission of NIFST is a manifestation of the outstanding human resources of NIFST. Founded in 1976, the Institute has its membership spread across ten chapters within Nigeria and beyond the shores of Nigeria. The Governing Council currently pilots the affairs of NIFST (NIFST, 2023).

The only regulatory body for the training and practice of Food Engineering like other engineering and degree programmes is the National Universities Commission (NUC). Both COREN and NUC over the years have developed and continue to review and upgrading the Benchmark Minimum Academic Standards (BMAS) for Undergraduate and PGD Programmes in Engineering. In 1962, the NUC was founded as a consultative agency within the Cabinet Office; however, in 1974, it became a statutory body. The Federal Ministry of Education (FME) oversees the parastatal NUC. There is a Governing Council for the Commission. Over the years, the Commission has grown from a small office in the cabinet office to fill a significant branch of the government's development and management of Nigeria's higher education system. The Commission's primary responsibilities include "granting approval for all academic programmes run in Nigerian universities; granting approval for the establishment of all higher educational institutions offering degree programmes in Nigerian universities; ensure quality assurance of all academic programmes offered in Nigerian universities; and channel for all external support to the Nigerian Universities" (NUC, n.d.).

Both COREN and NUC are involved are involved in relations. COREN over the years has developed and continue to review and upgrade the Benchmark Minimum Academic Standards (BMAS) for Undergraduate and PGD programmes in engineering.

Challenges Confronting Food Engineering Stakeholders: Emerging Issues

The NSE have been clamouring for the commencement of a Food Engineering Division. We want to rush into these so that the new course will not be stillborn. In other climes like those of the American Society of Agricultural and Biological Engineers (ASABE) and Canadian Society of BioEngineering (CSBE), all the agricultural related programmes are under the same umbrella. That is to say that Agricultural Engineering, Food Engineering, Environmental Engineering, Wood Products and Forestry Engineering, and Ecological Engineering all belong to ASABE. Even the general umbrella society for agricultural engineering in the world, known as CIGR, accommodates all agricultural engineering related programmes. CIGR is International Commission of Agricultural and Biological Engineers (Alonge, 2019a, b).

Currently in Nigeria, Food Engineering, Wood Products Engineering and Agricultural Engineering, Environmental Engineering all belong to Nigerian Institution of Agricultural Engineers (NIAE). I have advocated in the past for a change of name from NIAE to the Nigerian Institution of Agricultural and Biological or Biosystems Engineers (NIABE). It was felt that we are not grown enough to do that apart from the fact that there was a challenge of registration and accounts with the Corporate Affairs Commission (CAC) and banks respectively. However, in October 2018, Prof Ngoddy mandated Prof John Olajide and Prof Folarin Alonge to start a Food Engineering Forum through which we sought to have members to interact and find a common front to discuss issues that has to do with the future of the profession. The inaugural meeting was held in International Institute of Tropical Agriculture (IITA) on October 22, 2018 during the CIGR Section VI Conference. Plans are on the way to get a constitution and other things in place. This may eventually lead to a full blown society.

Another vital issue has to do with the current COVID-19 pandemic's challenges and how they relate to food engineering, which need to be attended to. Other challenges are discussed below.

Challenges of Food Engineering in Nigeria

Non-availability of a Benchmark Minimum Standard (BMAS) for the Food Engineering Programme.

The Food Engineering programme, being new, had the initial problem of accreditation with the relevant regulatory bodies, namely: the National Universities Commission (NUC) and the Council for the Regulation of Engineering in Nigeria (COREN) due to lack of the academic template required for the exercise. This problem was solved with the development of academic template in LAUTECH in 2002.

Provision of Required Engineering Personnel for the Food Industry

There has been a great need to provide expertise that can serve the Nigerian Food Industry adequately in the area of designing of food processes and food equipment. It was in mid-1990s that LAUTECH produced the first set of Food Engineers that were trained specifically for this purpose because they have adequate knowledge about nature and the properties of food raw materials and food processes, which are prerequisites for the effective design of food processes and machinery.

Lack of Functional and Very Strong Machine Building Industry

Lack of a functional and very strong machine building industry that is required to translate functional equipment and machinery prototypes developed by Food Engineers into commercialis-able processing machinery that is affordable.

A Typical Food Engineering Programme – University of Uyo

A typical Food Engineering Programme, as seen in University of Uyo (2018), is as presented below. This is currently in use. However, this is undergoing review in line with the NUC Core Curriculum Minimum Academic Standard (CCMAS). The CCMAS will likely come into effect from 2023.

Programme Structure

(i) **Duration of Study in the University:**

The period of formal study in the University comprises nine semesters of course work involving lectures, tutorials, practicals, fieldwork in industry, seminar presentations, the timely completion of assigned design and/or research projects under staff supervision; and one semester of industrial attachment. The minimum duration for the B.Eng. Food Engineering Degree Programme is five years through UME; four years with GCE (A/L) or OND; and three years with HND or B.Eng. Degree provided a candidate satisfies all other University and Faculty requirements as stipulated. The maximum duration for the programme is seven and a half years, after which as student is no longer qualified for graduation with the exception of some special cases, for instance suspension of studies.

(ii) **Academic Load**

The course requires full-time studies in the University. A full-time student is required to register for a minimum of 15 course units and a maximum 24 course units in a given semester as stipulated by the University Senate. No student in the Food Engineering Programme shall carry more than a maximum load of twenty-four (24) credit hours per semester. The permission to register for less than the minimum or more than the maximum load may be granted by the Senate upon recommendations of the faculty. The stated objectives are realised through the courses available in the departmental curricula summarised in the programme outline which is presented in Table 21.2

(iii) **Planned Field/Industrial Visits**

One such visit at least per year in years 2, 3 and 5.

(iv) **Graduating Requirements**

The final award and the class of the degree shall be based on the cumulative grade point average (CGPA) obtained by each candidate in all prescribed courses and approved by the Senate of the University. The final cumulative grade point average shall be calculated on the basis of the total number of all the course units registered for during the course of the undergraduate programme.

TABLE 21.2

Requirements for a Bachelor of Engineering

Course name	Credit units
General Studies Courses	10 units
Seminar	1 unit
Physical Sciences Courses	35 units
Engineering Mathematics	12 units
Computer Programming	4 units
Student Industrial Training (SIWES)	4 units
Student Work Experience Programme (SWEP)	2 units
Civil Engineering Courses	4 units
Mechanical Engineering Courses	17 units
Electrical Engineering Courses	6 units
Food Engineering Courses	43 units
Chemical Engineering Courses	12 units
Final Year Project	6 units
Faculty of Engineering Courses	12 units
Microbiology/Biochemistry Courses	9 units
Agricultural Engineering Courses	5 units
Basic Agriculture Courses	5 units
Electives	x units
Total	**195 – 232 units**

Note: To be eligible for the award of the Bachelor of Engineering Degree (B.Eng.) in Food Engineering, a student must satisfactorily complete a minimum of 195 and not exceeding a maximum of 223 credit units.

Class of Degree: The class of degree shall be awarded as follows:

Level	CGPA
First Class (Hons)	4.50 – 5.00
Second class (Upper Division, Hons)	3.50 – 4.49
Second class (Lower Division, Hons)	2.40 – 3.49
Third Class (Hons)	1.50 – 2.39
Fail	< 1.49

Course Structure

The Department operates the course credit system with evaluation based on continuous assessment and the end-of-semester examination. Teaching is done through lectures, seminars, tutorial, laboratory practicals, and fieldwork, and each course lasts for one semester. Examination is conducted in every course at the end of the Semester that the course is taken. In addition to the end-of-semester examination, undergraduate's performance is continuously assessed throughout the semester by means of tests, assignments and term papers. Performance by course examination carries 70% while continuous assessment carries 30%. Courses are assigned weights called credit units. One credit unit shall mean one hour of lecture or tutorial per week per semester, or the following equivalent may apply:

(a) Two hours of Seminar,
(b) 3Three hours of Laboratory or field work

Course Code: Course code carries a three letter index preceding the numerical codes. The following letter codes are applicable in structuring the courses offered in the programme.

AGE = Agricultural Engineering Courses
ANS = Animal Science Courses
BIO = Biological Science Courses
CHM = Chemistry Science Courses
CHE = Chemical Engineering Courses
COS = Crop Production Courses
CPE = Computer Programming Courses
CVE = Civil Engineering Courses
ELE = Electrical Engineering Courses
ENG = English Language Courses
FDE = Food Engineering Courses
GRE = General Engineering (faculty based) Courses
GST = General Studies Courses
MEE = Mechanical Engineering Courses
MCB = Microbiology Courses
MTH = Mathematics Courses
PHY = Physics Courses
SOS = Soil Science Courses
ZOO = Zoological Courses

Numerical Codes: This also is a 3 number digit code following the letter code, constituted as the year of study, the semester and the serial number.

Example:

111 = 100 level, first semester, course serial number 1
212 = 200 level, first semester, course serial number 2
123 = 100 level, second semester, course serial number 3
428 = 400 level, second semester, course serial number 8

Therefore FDE 417 = Food Engineering course at 400 level in first semester with serial number 7.

Continuous Assessment and Examination

The standard of tests and examinations include setting, conducting, evaluation schemes, moderation schemes – internal and external – for degree examinations, and the issuance of results. The head of department is the Chief Examiner for the department. There are two examination officers for the department, one per programme, for computation of results and presentation of same to the Business Committee of Senate. The Department conducts its examination in line with the University Senate's guidelines. This is carried out semester by semester according to the University's calendar.

Continuous Assessment: Students that register for a course are to be continuously assessed by means of tests, quizzes, tutorial assignments, and reports. Other means may be adopted as appropriate and consistent with the objectives and conduct of the course as determined by the department.

Industrial Training: Industrial training is an important component of the programme and is broken down as follows:

(i) SWEP-1: Long Vacation of year 2 - 3 months
(ii) SWEP-2: Long Vacation of year 3 - 3 months
(iii) SIWES: Second Semester of year-4
(iv) SIWES: Long Vacation of year-4
→ 6 months

Due to its importance, grading of the six month industrial attachment is based on the following criteria:

a. A logbook prepared according to specifications by the faculty
b. A final report prepared by the student based on the experience
c. An evaluation by the training officer of the company where the student had worked
d. Periodic visit by the University (faculty staff)

Examination: Examination is administered at the end of each semester for all course offered during the semester including courses that are purely practical. The duration of each examination shall be determined by the department or faculty (for faculty courses). The course unit will be the major determining factor.

Final Mark: The total mark for all courses shall be 100%, where 30% shall normally be allocated to continuous assessment for all course except courses that are purely practical which may carry a continuous assessment mark of not more than 80%, the pass mark in any course shall be 45% beginning from the 2013/2014 session. Students admitted prior to this session have a pass mark of 40%.

The level of achievement/grade of the performance of a student in examination, guided by the National Universities Commission Core Curriculum and Minimum Accreditation Standards (NUC-CCMAS, 2022, December 27), is presented in Table 21.3.

TABLE 21.3

Grading system for students' examination.

SCORE (%)	LETTER GRADE	GRADE POINT (GP)
70 - 100	A	5.0
60 - 69	B	4.0
50 - 59	C	3.0
45 - 49	D	2.0
0 - 44	F	0.0

B.Eng Degree in Food Engineering Course Syllabus

The curriculum of the Food Engineering (B.Eng.) programme is designed for five years, comprising classroom, laboratory, and workshop exercises, Student Workshop Experience Programme, and Industrial training. It is tabulated per semester and by year as follows:

Year 1: Formal University lectures, practicals, and examination (Table 21.4)

Year 2: Formal University lectures, practicals, and examination followed by a three month industrial attachment period (SWEP I) during the long vacation. (Table 21.5)

Year 3: Formal University lectures, practicals, and examination followed by a three month industrial attachment period (SWEP I) during the long vacation. (Table 21.6)

Year 4: Formal University lectures, practicals, and examination during the first semester and a 6-month industrial attachment period (SIWES) in the second semester. (Table 21.7)

Year 5: Formal University lectures, practicals, seminar, projects, and examination. (Table 21.8)

Emerging Issues in Food Engineering Education

One of the good things that have been to Food Engineering and indeed all Engineering programmes is the review of the curriculum. According to NUC-CCMAS (2022, December 27), the CCMAS are designed for the education and training of undergraduate students wishing to obtain first degrees in the different areas of engineering and technology in the Nigerian university system. The CCMAS document is expected to guide institutions in the design of curricula for their engineering and technology programmes by stipulating the minimum requirements. Being such, institutions are encouraged to take due cognisance of the CCMAS while bringing necessary innovation into the content and delivery of their programmes towards achieving the overall goals of engineering education and training in the country.

TABLE 21.4

Year 1 (100 level courses) of the Food Engineering (B.Eng.) programme at the University of Uyo, Nigeria.

YEAR 1 – 100 LEVEL COURSES					
FIRST SEMESTER					
		UNITS			
COURSE CODE	COURSE TITLE	L	T	P	U
PHY 111	General Physics	2	1	0	**3**
PHY 112	General Physics Laboratory I	0	0	1	**1**
MTH 111	General Mathematics I	3	1	0	**4**
CHM 111	General Chemistry I	3	0	0	**3**
CHM 112	General Chemistry Laboratory I	0	0	1	**1**
GST 111	Use of English I	2	0	0	**2**
GST 112	Philosophy and Human Existence	2	0	0	**2**
BIO 112	General Biology I	3	0	0	**3**
BIO 113	General Biology Laboratory I	0	0	1	**1**
GRE 111	Engineering Drawing I	0	0	1	**1**
	TOTAL	**15**	**2**	**4**	**21**
SECOND SEMESTER					
COURSE CODE	**COURSE TITLE**		**UNITS**		
		L	**T**	**P**	**U**
PHY 121	General Physics II	2	1	0	**3**
PHY 122	General Physics Lab. II	0	0	1	**1**
CHM 121	General Chemistry II	3	0	0	**3**
CHM 122	General Chemistry Laboratory II	0	0	1	**1**
MTH 121	General Mathematics II	3	1	0	**4**
GRE 121	Engineering Drawing II	0	0	1	**1**
GST 121	Use of English II	2	0	0	**2**
GST 122	Science and Basic Computer Technology	2	0	0	**2**
BIO 122	General Biology II	3	0	0	**3**
BIO 123	General Biology Laboratory II	0	0	1	**1**
GST 123	Citizenship and Peace Studies	2	0	0	**2**
	TOTAL	**17**	**2**	**4**	**23**

TABLE 21.5

Year 2 (200 level courses) of the Food Engineering (B.Eng.) programme at the University of Uyo, Nigeria.

YEAR 2 – 200 LEVEL COURSES					
FIRST SEMESTER					
		UNITS			
COURSE CODE	**COURSETITLE**	**L**	**T**	**P**	**U**
GRE 211	Engineering Mathematics I	2	1	0	**3**
GRE 212	Engineering Society	1	0	0	**1**
GRE 213	Engineering Drawing III	0	0	1	**1**
GRE 214	Workshop Practice I	0	0	1	**1**
GST 211	Entrepreneurial Studies I	2	0	0	**2**
CVE 211	Strength of Materials I	1	1	0	**2**
ELE 211	Electrical Engineering I	1	1	0	**2**
MEE 211	Engineering Mechanics I	1	0	1	**2**
MEE 212	Engineering Materials I	1	1	0	**2**
CPE 211	Computer Programming & Application	1	0	1	**2**
FDE 211	Introduction to Food Engineering	2	0	0	**2**
FDE 212	Introduction to Solar Energy	1	0	0	**1**
MCB 211	General Microbiology I	2	0	1	**3**
	TOTAL	**15**	**4**	**5**	**24**
SECOND SEMESTER					
COURSE CODE	**COURSE TITLE**		**UNITS**		
		L	**T**	**P**	**U**
GRE 221	Engineering Mathematics II	2	1	0	**3**
GRE 222	Engineering Drawing IV	0	0	1	**1**
GRE 223	Workshop Practice II	0	0	1	**1**
GRE 224	Student Work Experience Programme (SWEP I)	0	0	1	**1**
GST 221	Entrepreneurial Studies II	2	0	0	**2**
MEE 223	Engineering Mechanics II	1	1	0	**2**
MEE 221	Thermodynamics & Heat Transfer	1	1	1	**3**
CVE 221	Strength of Material II	1	1	0	**2**
MEE 222	Fluid Mechanics I	1	1	0	**2**
ELE 221	Electrical Engineering II	1	0	1	**2**
CPE 221	Computer Programming Applications II	1	0	1	**2**
FDE 221	Food Engineering Unit and Dimension	1	0	0	**1**
	TOTAL	**11**	**5**	**6**	**22**

- The CCMAS has ten unique features. They are listed as follows;

1. Courses are designed to produce engineers who are imbued with critical thinking, analytical and modern design skills, and who are able to initiate entrepreneurial business and start-up manufacturing ventures.
2. Courses have well-defined learning outcomes, aligned with current COREN BMAS accreditation requirements, and reflecting the standards in top global institutions.
3. Emphasis on the 4th Industrial revolution (4IR) technical and digital skills, rooted in hands-on practicals and experiential industry-learning relevant to the national socio-economic context.
4. SIWES now 15 units, all mandatory and credited for the second semester 6-month supervised industry immersion.
5. Introductory courses to get all fresh students, including those by direct entry, acquainted with programme fundamentals, to "catch and initiate them young"
6. Introductoryartificialintelligence,machinelearning and convergent technologies, renewable energy systems, and technology introduced and mandatory.
7. Critical industry-standard design concepts introduced early to get students acquainted with new additive and subtractive technologies.
8. APIComputingandSoftwareEngineering,dataanalytics, and engineering communication strengthen programmes with critical concepts and skills.
9. Students are able to continue deepening their knowledge in creative designs and start-ups of businesses even during forced breaks.
10. Curriculum is student-centred, with smart semester load that gives ample time to reflect on the studies and sharpen their creative capabilities.

TABLE 21.6

Year 3 (300 level courses) of the Food Engineering (B.Eng.) programme at the University of Uyo, Nigeria.

YEAR 3 – 300 LEVEL COURSES					
FIRST SEMESTER					
		UNITS			
COURSE CODE	**COURSE TITLE**	**L**	**T**	**P**	**U**
AGE 311	Engineering Statistics	2	0	0	**2**
ANS 211	Principles of Animal Production	2	0	1	**3**
BCH 211	General Biochemistry I	2	0	1	**3**
CHE 314	Chemical Engineering Process Calculations	1	1	0	**2**
CHM 213	Organic Chemistry II	2	0	0	**2**
CHM 217	Organic Chemistry Laboratory II	0	0	1	**1**
FDE 311	Food Chemistry	2	0	1	**3**
FDE 313	Unit Operations in Food Processing	2	0	0	**2**
FDE 314	Unit Operations in Food Processing Laboratory I	0	0	1	**1**
GRE 311	Engineering Mathematics III	2	1	0	**3**
MEE 311	Theory of Machines	1	1	0	**2**
	TOTAL	**16**	**3**	**5**	**24**
SECOND SEMESTER					
COURSE CODE	**COURSE TITLE**	**UNITS**			
		L	**T**	**P**	**U**
GRE 321	Engineering Mathematics IV	2	1	0	3
GRE 322	Technical Report Writing	1	0	0	1
GRE 323	Student Work Experience Programme II (SWEP II)	0	0	1	1
FDE 324	Engineering Thermodynamics	1	1	0	2
CHE 323	Mass Transfer (Separation Process)	2	1	0	3
MEE 323	Machine Design	1	1	0	2
COS 221	Principles of Crop Production	2	0	1	3
FDE 325	Process Instrumentation	2	0	0	2
FDE 321	Animal Food Product technology 1	2	0	0	2
	TOTAL	**13**	**4**	**2**	**19**

TABLE 21.7

Year 4 (400 level courses) of the Food Engineering (B.Eng.) programme at the University of Uyo, Nigeria.

YEAR 4 – 400 LEVEL COURSES					
FIRST SEMESTER					
		UNITS			
COURSE CODE	**COURSE TITLE**	**L**	**T**	**P**	**U**
MEE 413	Mechanics of Machines	2	1	0	**3**
GRE 412	Engineering Economics	2	0	0	**2**
FDE 411	Unit Operations in Food Processing II	2	0	0	**2**
FDE 412	Unit Operations in Food Processing Laboratory II	0	0	1	**1**
FDE 413	Plant Food Products Technology I	2	0	0	**2**
FDE 414	Principles of Human Nutrition	2	0	0	**2**
FDE 415	Food Process Engineering	2	0	0	**2**
FDE 416	Unit Operations in Food Processing Laboratory III	0	0	1	**1**
FDE 417	Food Engineering I	2	0	0	**2**
FDE 418	Heat Transfer	2	0	0	**2**
AGE 415	Design of Agricultural and Food Machines	2	0	0	**2**
MCB 316	Food Microbiology	2	0	1	3
	TOTAL	**20**	**1**	**3**	**24**
SECOND SEMESTER					
COURSE CODE	**COURSE TITLE**		**UNITS**		
		L	**T**	**P**	**U**
GRE 421	Industrial Training [6 Months]	0	0	4	**4**

TABLE 21.8

Year 5 (level 500 courses) of the Food Engineering (B.Eng.) programme at the University of Uyo, Nigeria.

YEAR 5 – 500 LEVEL COURSES					
FIRST SEMESTER					
		UNITS			
COURSE CODE	**COURSE TITLE**	**L**	**T**	**P**	**U**
GRE 511	Engineering Management and Law	2	0	0	**2**
FDE 517	Food Biotechnology	3	0	0	**3**
FDE 512	Food Engineering II	2	0	0	**2**
FDE 518	Seminar	0	0	1	**1**
FDE 514	Food Process Design	2	0	0	**2**
FDE 513	Food Engineering Lab	0	0	1	**1**
FDE 516	Plant Food Products Technology II	3	0	0	**3**
FDE 515	Food Machinery and Equipment Design	1	0	1	**2**
	Electives	4	0	0	**4**
	Total	**17**	**0**	**3**	**20**
COURSE CODE	**ELECTIVE**	**UNITS**			
		L	**T**	**P**	**U**
CHE 512	Process Dynamics and Control I	2	0	0	2
CHE 515	Process Optimization	3	0	0	3
CHE 413	Separation Process II	2	1	0	3
SECOND SEMESTER					
COURSE CODE	**COURSE TITLE**	**UNITS**			
		L	**T**	**P**	**U**
FDE 545	Project	0	0	6	**6**
FDE 521	Food Packaging	3	0	0	**3**
FDE 523	Animal Food Product Technology II	3	0	0	**3**
FDE 525	Food Standards and Quality Control	2	0	0	**2**
FDE 526	Pollution Control and Waste Management in Food Industry	2	0	0	**2**
FDE 527	Food Plant Design	2	0	0	**2**
	Elective	2	0	0	**2**
	Total	**14**	**0**	**6**	**20**
COURSE CODE	**ELECTIVE**	**NO OF UNITS**			
		L	**T**	**P**	**U**
CHE 521	Process Dynamics and Control II	2	0	0	2
FDE 544	Solar Energy Engineering & Control II	2	0	0	2

The Food Engineering curriculum like other engineering disciplines has the features. It is hoped that this will benefit all the stakeholders.

Conclusion and Recommendations

This difference between Food Engineering and Food Technology, Home Economics, and Food Nutrition has been highlighted in this paper. Although Food Engineering is relatively new in Nigeria, it has a lot of contributions to make towards industrial manufacture of high quality, nutritious, safe, and moderately affordable food products. The Food Engineering curriculum in Nigeria should incorporate development of new scientific and technological concepts/principles and deployment of digital facilities. This is imperative so as to make indispensable contributions towards overcoming the global challenges of food insecurity and sustainable food manufacturing systems.

REFERENCES

Alonge, A. F. (2019a, November). Innovations in agriculture and technology for sustainable food sufficiency. *The Nigerian Engineer Magazine*, pp. 5–9. Published by the Nigerian Society of Engineers (NSE), Abuja.

Alonge, A. F. (2019b). The role of agricultural and food engineers in agric business value chain. Paper presented at the MAY 2019 Edition of the Nigerian Society Engineers' (NSE) Uyo Branch Technical Seminar at Akwa Ibom State Water Corporation Hall, Uyo, on Sunday May 19.

Council for the Regulation of Engineering in Nigeria. (2017a). *Accreditation*. https://www.coren.gov.ng/about-coren/23-council/73-accreditation

Council for the Regulation of Engineering in Nigeria. (2017b). *Education & Training*. https://www.coren.gov.ng/about-coren/council/council-committees/education-training

Council for the Regulation of Engineering in Nigeria. (2017c). *COREN History*. https://www.coren.gov.ng/about-coren/coren-history

Kent-Jones, D. W., Truswell, A. S., Weininger, J., & Carpenter, K. (2023, April 26). Human nutrition. Encyclopedia Britannica. Retrieved May 8, 2023, from https://www.britannica.com/science/human-nutrition .

Ngoddy, P. O. (2021). Festschrift in Honour of Engr. Professor Patrick Obi Ngoddy @80. Chapter 3. Published by Nigerian Institution of Food Science and Technologists (NIFST). Printed by University of Nigeria, Nsukka Press, Nigeria. ISBN 979-978-985- 480-6

National Universities Commission. (n.d.). *About Us.* Retrieved April 25, 2023, from https://www.nuc.edu.ng/about-us/

National Universities Commission Core Curriculum and Minimum Accreditation Standards. (2022, December 27). *Engineering and Technology.* https://nuc-ccmas.ng/engineering-and-technology/

Nigerian Institution of Agricultural Engineers. (2021). *Our History.* https://niae.net/1/history/

Nigerian Institute of Food Science and Technology. (2023). *About Us – Brief History.* https://nifst.org/about-us

Nigerian Society of Engineers. (2022). *History of NSE.* https://nse.org.ng/about-nse/history-of-nse

Niranjan, K. (2017). Reengineering food engineering. *The Chemical Engineer*, 909. p. 3033. Available at http://centaur.reading.ac.uk/70389/

Onwualu, A. P., Akubuo, C. O., & Ahaneku, I. E. (2006). *Fundamentals of Engineering in Agriculture*, First Edition. Immaculate Publications Limited, Enugu.

University of Uyo. (2018). Food engineering programme handbook. Published by the Department of Agricultural and Food Engineering, Faculty of Engineering, University of Uyo, Uyo, Akwa Ibom.

22

Agricultural Engineering Higher Education in Turkey

Can Ertekin and Ahmet Colak

Introduction

Agricultural Education During the Ottoman Empire

Although Turkey is an agricultural country, establishing schools in this field happened quite late. The first attempts to establish a school providing education in agriculture can only be traced back to the Tanzimat period. The first agricultural education started in the Ayamama Farm in Istanbul-Yeşilköy in 1846 when the Agricultural School was established by order of the grand vizier of the time, Buyuk Resit Pasa (Eris et al., 2005). In the establishment of this school, the Grignon High Agricultural School in France, with its 4-year education period, was taken as an example (Eris, 2004).

This school, which was attached to the Ministry of Public Works in 1850, also carries the pioneering role of being the first vocational and technical education institution that started education in our country. In addition, some modern agricultural tools and machines were brought from Europe, and their use was taught to the students. Thus, the school became the first educational institution in our country where modern agricultural tools were used. These machines were also introduced to the public, and farmers were encouraged to use modern agricultural tools. The interest in this multi-purpose agricultural school, which was established with great hopes, waned due to reasons such as an inability to attract students from the provinces, the inability of students to study full time, not benefiting enough from foreign experts, and insufficient teaching materials and resources, especially books; on September 27, 1851, the school was destroyed and closed (Kadioglu, 2008). Later, one of the two schools that provided education at a level considered higher education today was in Bursa in 1891; the other was opened in 1893 in Istanbul-Halkalı.

Halkalı Agricultural School Âlisi

The attempt to open a new agricultural school came to the fore only in 1878–79 when Ahmed Cevdet Pasha was the Minister of Trade and Agriculture. The establishment works continued until 1891 for reasons such as buying a place in Halkalı and having the building built.

The school, which was named Halkalı Ziraat Mekteb-iI Âlisi, had to be closed for a short time during the First World War, and when it was reopened, a Farm Machinist Branch was opened under the school in order to expand and increase the use of new agricultural tools among farmers. The school, which had to be closed again during the Armistice of Mudros, was reorganised during the Republic period. Immediately after the Republic, many young people who graduated from Halkalı Agricultural School were sent to Germany to receive further education and learn modern research methods. In July 1927, law number 1109, "Law on the Establishment of Agriculture and Veterinary Institutes and Ali Schools and the Improvement of Agricultural Education", was enacted. With the enactment of this law, the establishment of Ankara Higher Agricultural Institute, which was established to provide higher education in agriculture, veterinary and forestry, started and Halkalı Agricultural School was closed in 1928 (Askin, 2017; Cesme, 2014).

Regulation of Education of Agriculture (Tedrisat-ı Ziraiye Nizamnâmesi)

The Regulation of Education of Agriculture, issued on February 10, 1912, signed by Sultan Mehmed Reşad, Grand Vizier Said Pasha, and Aristidi, Minister of Commerce and Agriculture, classified agricultural schools under four headings as Labour Schools, Farm Schools, Surgery Schools, and District Agricultural Schools. It is understood that a four-stage education system was planned in the field of agriculture in the Ottoman Empire. It aimed to train Farmer-Sharecroppers (partner), who could read and write in Turkish and manage their own field by teaching new agricultural techniques practically to farming children with Farm Schools in the second stage. The third tier, Agricultural Surgery Schools, is programmed to train farm butlers. At the top level, there are Agricultural District Schools. They trained technical staff to manage large farms and teachers for agriculture schools where theory and practical application were taught. Boarding students were admitted to all the schools in these four levels, and the student's expenses and stipends were covered. Although the provisions brought and the targets set by this regulation, which invalidated all previous regulations regarding agricultural schools, could not be fully implemented due to the conditions brought about by the First World War, which emerged immediately after its publication, this regulation was the first attempt to deal with and organise agricultural education as a whole (Kadioglu, 2005).

DOI: 10.1201/9780429150111-24

Agricultural Education in the Republican Period

Bursa Agricultural School and Halkalı Agricultural School, two schools that provide education at a level that can be considered higher education, continued their activities until the Republic's first years. Many of the graduates of Halkalı Agricultural Mekteb-i Âlisi were sent abroad, especially to Germany, starting in 1923; these young people were given the opportunity to receive further education there and to get acquainted with modern research methods.

The opening of the Higher Institute of Agriculture is a second milestone in the development of Turkish universities, as well as the re-establishment of Istanbul University. Both organisations indicate a determined orientation towards modern scientific understanding and methods unique to Europe. In preparation for this breakthrough, under the direction of consultant Oldenburg, a German expert committee came to Ankara upon invitation in the spring of 1928. The "Sixth Monthly Establishment Studies in Turkey Report" detailed the work of 11 delegation members between 11.04.1928–30.09.1928 and was presented by the top government adviser Dr. Oldenburg (duplicated notes, unpublished) (Ciftci, 2008).

With the implementation of the Law on the Establishment of Agriculture and Veterinary Institutes and Ali Schools and the Improvement of Agricultural Education, most of the 12 Secondary Agricultural Schools were closed. In 1928, Halkalı Ziraat Mekteb-i Âlisi, as it was then known, was also closed. His students were transferred to the Istanbul High Forest School. The High Veterinary School in Istanbul, which left the Ministry of Economics, continued its education by being attached to the Ministry of Education (Ciftci, 2008).

By establishing institutes and laboratories in Etlik, Ankara, the Oldenburg delegation laid the foundations of a temporary establishment, namely the Higher Agricultural School, which was opened in 1930. From 1930 to 1933, four Germans (Prof. Eckstein, Prof. Kotte, Prof. Jessen and Prof. Christiansen-Weniger) and a Luxembourgian (Prof. Lucius) worked, conducted education and training in this school, apart from Turkish academic staff (Widmann, 2000) In connection with the law, Ankara Higher Agricultural School was opened in 1930 in the old Agricultural School building on the Keçiören road, under the Ministry of Agriculture and as a pioneer of the Higher Agricultural Institute (Yazıcıoglu, 1958). This high school, which was opened for four years of education, was a boarding school. The students started higher education in the second year after interning at Atatürk Forest Farm in the first year. While Ankara Higher Agricultural School continued to operate in this way, the foundation of the Higher Agricultural Institute was laid in 1928 by the then Deputy of Agriculture Sabri Toprak and the construction of the buildings was started.

Prof. Dr. Falke, who later took office as the Rector at the Higher Agricultural Institute, arrived in Ankara during the autumn of 1932. He returned to Germany to prepare for the needs of all laboratories, to ensure that these vehicles arrived on time and to determine if the German Professors teaching at the Ankara Higher Agricultural School also returned to their hometown before YZE was opened because, according to the YZE Law, scientists who will teach at the Institute must be professors in their hometowns (Akman, 1978).

On the other hand, in 1932, Geneva University pedagogy professor and former rector Albert Malche was invited to Turkey by Mustafa Kemal ATATÜRK and was asked to prepare a report that would form the basis of the University Reform. Upon this invitation, Malche came to Istanbul on January 16, 1932, and quickly began to prepare his report by visiting faculties and clinics, examining learning environments such as laboratories, classrooms, and libraries, and talking with lecturers, students, and administrative officers. He evaluated the results he obtained from the questions he submitted to the Darülfünun Emini in writing on February 18, 1932, and submitted his report to the government on May 29, 1932 (Widmann, 2000; Ozata, 2007).

Report on Istanbul Darülfünûn prepared by Malche, is a detailed education evaluation study of 3 parts. In this report, Malche first explained the aims and methods of his research, then conveyed the critical analysis of Darülfünûn and presented detailed solution proposals in the last section (Aras et al., 2007).

The main points highlighted in this report are as follows (Ozata, 2007; Kalayciogullari, 2009);

- Darülfünûn does not seem very loyal and supportive of the laws on scientific autonomy and legal personality granted to it.
- Faculty members do not conduct any serious research that can be considered scientific. Seminar, laboratory, and application studies are insufficient.
- Turkish books are very few. Libraries do not have enough books. The closure of the Darülfünûn library at 16:00 and the fact that no books are given out hinder the students' work.
- There is no scientific cooperation between the faculties of Darülfünûn.
- Western languages are not given enough importance; students do not know foreign languages.
- Although there are many and large classrooms in Darülfünûn, they are not used enough. Some of the places reserved for laboratories, seminaries, clinics, and institutes remain old and primitive.
- Curriculums are both overloaded and encyclopaedic; lecture notes have been taught for many years without any changes or improvements.
- A training method is used, which does not include personal work and research, thinking and speaking, and consists of repeating the present.
- The teaching staff is very crowded. The low wages of professors oblige them to take side positions. Faculty members place their external work first and their main duties second.
- There is no consensus of opinion and ideal among faculty members. Because of the grouping between them, academic management is in the hands of the groups.

Unfortunately, most of the findings in this 95-page report by Malche, such as "lack of scientific cooperation", "lack of a wider research environment", "weakness of the faculty members' belonging to the institution", and "the view of education disconnected from society and introverted" are unfortunately, as a result of today's Turkey, also valid in universities.

In the recommendations section of the report, Malche emphasised that it is not enough to establish new chairs, especially in line with the Western example, but that educators who will create a new scientific culture and spirit are necessary. According to Malche, only in this way can a new generation of thinkers, questions, and researchers be created with a real science and research environment (Widmann, 2000). It is known that our leader Mustafa Kemal read this report with great care and took notes.

In the light of these developments, Ankara Higher Agricultural School, which was established in Ankara in 1930 and adopted the same education system as the German Agricultural Faculties, joined the Higher Agricultural Institute in the 10th year of the Republic with the directive of Atatürk, 3 years after graduating. It was transferred to the Higher Agricultural Institute, which was opened on October 30, 1933. The Faculty of Agriculture, one of the faculties of the Higher Institute of Agriculture, was ready as an excellent facility with its buildings, classrooms, laboratories, dormitories, sports facilities, and all other outbuildings, as the first higher education institution of Turkey in a modern sense, the day after the 10th-anniversary celebrations of the Republic. It was opened by Chief Deputy İsmet İnönü on October 30, 1933 (Ozsan et al., 1995; Eris et al., 2005). Atatürk, in his speech at the opening of the 4th Term of the 3rd Legislative Year of the Turkish Grand National Assembly on November 1, 1933, said, "The necessity of establishing our economic order becomes more inevitable and urgent every day. It is true, however, that our progress on the kind and quantity of our agricultural produce has put our nation in a rather comfortable and presumably secure position. There is no doubt that the high engineers to be trained by the Higher Agricultural Institute, which was opened with all its faculties, will be beneficial to the country in the future" (TBMM, 2023).

In the opening years, 23 highly respected German scientists with the title of Ordinarius Professor were appointed to the directorships of the Institutes affiliated with the aforementioned faculties. Some of them have also undertaken the duties of Rector and Dean. About 40 young Turkish scientists, 5 of whom were associate professors, were sent to Germany from 1927 onwards to complete their doctorates and return to their homeland; they also started to work in their faculties and related institutes in 1933 (Ozsan et al., 1995). In addition to this strong academic structure at the opening date of the Higher Agricultural Institute, it has truly magnificent buildings necessary for contemporary education; laboratories equipped with the most modern tools, equipment and equipment; lecture halls and classrooms; a central library and printing house; 60 hectares of land for research and applications; dormitories and cafeterias necessary for the social life of students; indoor and outdoor sports facilities suitable for all kinds of sports imaginable (including fencing, equestrian sports, ice hockey, and tennis), as well as a strong administrative and servant staff (Ozsan et al., 1995).

Within the High Institute of Agriculture, Faculties of Agriculture, Veterinary, Natural Sciences (Faculty of Science) and Agricultural Arts took form. The education and training system of this institute is based on 4-year general agriculture education (Abaci et al., 2020). In 1934, Istanbul High School of Forestry was attached to the Institute under the name of Faculty of Forestry. In this way and with its self-governing autonomous structure, the Institute is a fully modern university structure with its Senate, Faculty Members Board, Institute Grand Board, Faculty Boards, and the authority to grant doctorates and academic titles (Ozsan et al., 1995).

This law, which was prepared in full, provided the birth of an institution with a university character. In this way, the mentality of accepting universities only as educational institutions was demolished, and 2 new tasks were given to the university besides teaching. Research, publication and teaching form the basis of today's modern universities.

To develop this institution, designed according to the model of Western Universities, only German professors were appointed to the chairs established in the Faculty at the beginning, and the courses were given in German. Assistants were selected from young people who graduated from Halkalı Higher Agricultural School and specialised in Germany. These young people finished their specialisation education in Germany, returned to the country and were appointed assistants to German Professors. German Professors, who stayed at YZE from 1931 until 1942, played an important role in the establishment and development of our Faculty. The tradition of academic study they established continues to this day.

The Higher Institute of Agriculture (YZE) has opened a new era in the field of scientific research and teaching in the Republic of Turkey and has brought innovations in the modern sense, namely:

- For the first time, he established the tradition of doing research in the western sense,
- It has made it compulsory for young people who will become faculty members to complete a doctorate,
- A gradual development is envisaged for faculty members and assistantships. These are, respectively, candidate assistant, second and first-year assistant, Chief assistant (on PhD), Laboratory chief, second and first-year associate professor (on the habilitation thesis), second and first-year professors (on the commission report), second and first class Ordinary professorships. are the stages,
- The full-time working principle has been applied at YZE,
- The principle of election of the Rectors and Deans has been applied,
- Article 14 of the Law states, "The Grand Council of the Institute is composed of Ordinarius and professors and associate professors, branch chiefs, assistants, teachers and civil servants and three executive directors (members) to be chosen by each faculty student from among themselves. This assembly is called for a meeting either by the rector himself, or by the Divan, or upon the request of a total of twenty-five

> persons from associate professors, branch chiefs and assistants". As can be seen, the Great Council of the Higher Agricultural Institute consists of not only faculty members but also assistants, lecturers, civil servants, and three members to be chosen from among students from each faculty. This is an indication that there is pluralism and polyphony in the administration.

With the Higher Institute of Agriculture Law (No. 2291), the principles and traditions formed in this science centre set an example for the Universities Law (No. 4936) enacted in 1946.

After the Universities Law No. 4936 was enacted in 1946, there were tendencies to include the Institute as a university within the framework of this law; both the government and the parliament have taken many initiatives. However, the Higher Institute of Agriculture could not maintain its integrity under "İnönü Ziraat University" or "Ankara Agricultural University". With the "Additional Law to the Universities Law" enacted in 1948, the Faculty of Agriculture and Veterinary Medicine, including Natural Sciences and Agricultural Arts, was affiliated with Ankara University, established in 1946, and the Faculty of Forestry was affiliated with Istanbul University. Later, the Natural Sciences staff in the Faculty of Agriculture was transferred to the Faculty of Science of Ankara University, and Ankara University Faculty of Agriculture returned to its current structure (Ozsan et al., 1995).

Until 1955, when the Ege University Faculty of Agriculture was opened in İzmir, Ankara University Faculty of Agriculture remained the only Higher Education Institution in the field of agriculture in Turkey. Ege University Faculty of Agriculture was founded in 1955, and Atatürk University Faculty of Agriculture in 1957; Ankara University Adana Faculty of Agriculture was founded in 1967–1970 and merged with Çukurova University in 1973. Agricultural Faculties of Harran and Ondokuz Mayıs Universities in 1976, and Uludag University Faculty of Agriculture in 1980 (Ozsan et al., 1995). Until the YÖK system was established after 1980, the number of Faculty of Agriculture established in the 57-year Republican period was 7. After the Higher Education Law No. 2547 and the establishment of the Higher Education Council, the number of Agricultural Faculties increased rapidly in our country. There were 4 units in 1982, 1 unit in 1983, 2 units in 1987, 8 units in 1992 and 1 unit in 1995, a total of 23. Today, new Faculties of Agriculture are opened with the decision of the Council of Ministers, and these faculties start to demand students within 1–2 years (Ciftci, 2012; Eris et al., 2005). As of January 2023, the Faculty of Agriculture or the faculties that train Agricultural/Biosystem Engineers across Turkey (Figure 22.1) are listed below (YOK, 2023);

- Adıyaman University Faculty of Agriculture, Adıyaman
- Akdeniz University Faculty of Agriculture, Antalya
- Alanya Alaaddin Keykubat University Rafet Kayisli Faculty of Engineering, Antalya
- Ankara University Faculty of Agriculture, Ankara
- Ataturk University Faculty of Agriculture, Erzurum
- Aydın Adnan Menderes University Faculty of Agriculture, Aydın
- Bilecik Seyh Edebali University Faculty of Agriculture and Natural Sciences, Bilecik
- Bingol University Faculty of Agriculture, Bingol
- Bolu Abant Izzet Baysal University Faculty of Agriculture, Bolu
- Bursa Uludag University Faculty of Agriculture, Bursa

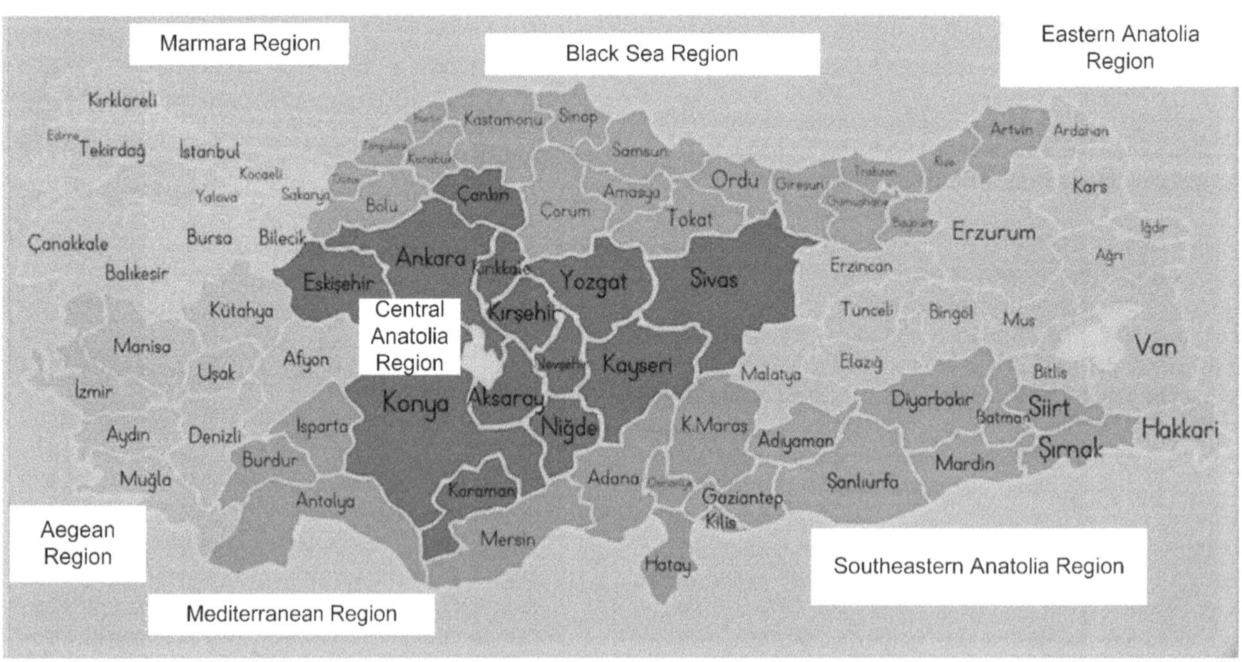

FIGURE 22.1 Provinces and regions of Turkey (Anonymous, 2023a).

- Canakkale Onsekiz Mart University Faculty of Agriculture, Canakkale
- Cukurova University Faculty of Agriculture, Adana
- Dicle University Faculty of Agriculture, Diyarbakir
- Düzce University Faculty of Agriculture, Düzce
- Ege University Faculty of Agriculture, Izmir
- Erciyes University Faculty of Agriculture, Kayseri
- Eskişehir Osmangazi University Faculty of Agriculture, Eskişehir
- Harran University Faculty of Agriculture, Sanliurfa
- Hatay Mustafa Kemal University Faculty of Agriculture, Antakya
- Igdir University Faculty of Agriculture, Iğdır
- Isparta University of Applied Sciences, Faculty of Agriculture, Isparta
- Kahramanmaras Sutcu Imam University Faculty of Agriculture, Kahramanmaras
- Kilis 7 Aralik University Faculty of Agriculture, Kilis
- Kirsehir Ahi Evran University Faculty of Agriculture, Kırsehir
- Kocaeli University Faculty of Agriculture, Kocaeli
- Konya Food and Agriculture University, Faculty of Agriculture and Natural Sciences, Konya
- Konya Technical University, Faculty of Agricultural Sciences and Technologies, Konya
- Malatya Turgut Ozal University, Faculty of Agriculture, Malatya
- Mugla Sitki Kocman University, Fethiye Faculty of Agriculture, Mugla
- Muş Alparslan University Faculty of Applied Sciences, Muş
- Necmettin Erbakan University Faculty of Agriculture, Konya
- Nigde Omer Halisdemir University, Faculty of Agricultural Sciences and Technologies, Nigde
- Ondokuz Mayıs University Faculty of Agriculture, Samsun
- Ordu University, Faculty of Agriculture, Ordu
- Pamukkale University Faculty of Agriculture, Denizli
- Recep Tayyip Erdogan University Faculty of Agriculture, Rize
- Sakarya University of Applied Sciences, Faculty of Agriculture, Sakarya
- Selcuk University Faculty of Agriculture, Konya
- Siirt University Faculty of Agriculture, Siirt
- Sirnak University Faculty of Agriculture, Sirnak
- Sivas Science and Technology University, Faculty of Agricultural Sciences and Technologies, Sivas
- Tekirdag Namık Kemal University, Faculty of Agriculture, Tekirdag
- Tokat Gaziosmanpaşa University Faculty of Agriculture, Tokat
- Usak University Faculty of Agriculture, Usak
- Van Yüzüncü Yıl University Faculty of Agriculture, Van
- Yaşar University Faculty of Agricultural Sciences and Technologies, İzmir
- Yozgat Bozok University Faculty of Agriculture, Yozgat.

Before it came to the present situation, in parallel with the management and order changes that occurred at various times in Turkey, there were serious changes in the laws related to Higher Education and, naturally, in the agricultural epithet. For example, in the Faculties of Agriculture, the education period, which was 4 years at first, became 5 years for a period, then turned to 4 years, the "lecture" system was abolished, and a structuring emerged within the concepts of "Department", "Department" and "Science". The Department was abolished, some were newly established, and changes were made to academic titles. The general purpose of these changes was expressed as increasing the quality of education (Eris, 2004).

Education and Training Systems Applied in Agricultural Faculties

Description of Education Systems Implemented from 1930

Since 1930, seven different education and training systems have been applied in universities. The years of their implementation are shown in Table 22.1.

TABLE 22.1

Education and training systems applied in agricultural faculties

Implementation Years	Teaching duration	System	Given Title
1930–1953	4 years (1-year Internship + 3 years)	General Education	Agricultural Engineer
1953–1960	4 years	Partition System	Agricultural Engineer
1960–1967	4 years	Partition System	Agricultural Engineer
1967–1977	5 years	3.5 + 1.5 Years System	Agricultural Engineer
1977–1999	4 years	Partition System	Agricultural Engineer
1999–2002	4 years	Undergraduate Programs System	Agricultural Engineer
2002–2009	4 years	3+1 System	Agricultural Engineer
2009 and Continuing	4 years	Partition System	Agricultural Engineer*

*Food Engineering, Landscape Engineering and Fisheries are awarded diplomas directly under their own names

Education System Implemented Between 1930 and 1953

Students enrolled in the faculty do an internship on the farms for 11 months (beginning of November–end of September), and students who are successful at the end of the internship are entitled to continue their second year, while students who fail the internship exam have to repeat the internship. Theoretical lessons were given to students in grades 2–4, and the applications of these lessons were made. Students who graduate generally graduate with no section specified on their diplomas.

Education System Implemented Between 1953 and 1960

No academic year was allocated for practical studies; practical studies were distributed during the holidays of the 4-year teaching period. Students entering the faculty indicate which of the eight departments they will enter in their petition, and when they graduate, it is written which department they graduated from on their diplomas.

Education System Implemented Between 1960–1967

Students make their department preferences while registering with the faculty. Preparatory and supplementary courses (required and elective) are 135 hours, with 45 credit hours from required departmental courses, totalling 180 credit hours with at least 22.5 hours per semester (Ciftci, 2004).

Education System Implemented Between 1967 and 1977

Students take common general courses in the first 7 semesters and the department courses they prefer in the last 3 semesters. Required credit hours from general courses are 164 hours, with 49 from departmental courses, totalling 213 credit hours with 21.3 credit hours per semester.

Education System Implemented Between 1977 and 1999

Students make their department choices during their OSYM applications. Required credit hours are 73, with 77 from departmental and non-departmental courses, totalling 150 credit hours with 18.75 hours per semester (Ciftci, 2004).

Education System Implemented Between 1999 and 2002

According to this regulation, students choose the undergraduate programme in their application to OSYM, take the common courses of the undergraduate programme in the first 6 semesters and the courses they prefer in the last 2 semesters. Required credit hours are 103 to 109 in the first 6 semesters, and 39 + 6 = 45 departmental courses, totalling 150 credit hours, which is 18.75 per semester (Ciftci, 2004).

Education System Implemented Between 2002 and 2009

According to this regulation, students choose their courses according to their OSYM application. They are enrolled in Agricultural Engineering or Landscape Architecture programmes and take common general courses in the first 6 semesters and the departmental courses they prefer in the last 2 semesters. Required credit hours include 111 from general courses and 36 from departmental courses, totalling 147 credit hours, which is 18.38 credit hours per semester (Ciftci, 2004).

Education System Implemented 2009–Ongoing

The Department System, which was implemented between 1953–1960, 1960–1967 and 1977–1999, was reinstated after 2 different education systems were implemented in a short period of 10 years. Students make their department preferences in the OSYM placement programme and are placed directly into departments according to their scores (Ciftci et al., 2010; Ortas & Ciftci, 2015).

General Structure of the Turkish Education System

In 1981, Turkey's higher education administration was comprehensively restructured under the new Higher Education Law (No. 2547). The system became centralised, with all higher education institutions tied to the Council of Higher Education (YÖK). After this restructuring, all institutions of higher education were designed as universities. Expansion of higher education throughout the country was achieved, application to higher education was centralised, and a central university exam and placement were introduced. In addition to public universities, the first non-profit foundation university in Turkey started to provide education for students in 1986.

Since 2012, compulsory education in Turkey has been 12 years long and is divided into three stages (primary, elementary, and secondary education) (YOK, 2023). The education system in Turkey continues, as shown in Figure 22.2. Agricultural/Biosystem Engineering titles are given in the relevant Faculties. A Faculty (College) is defined as a division conducting higher education, scholarly research, and publication. Various departments and programmes may be connected to it. A Bachelor's degree is awarded after completing a four-year course and 240 ECTS of study. The study duration for dentistry, veterinary medicine and pharmacy is five years, and for medicine, six years. The qualifications in these four fields are considered equivalent to a Master's degree. The undergraduate education period in Agricultural/Biosystem Engineering is 4 years. A Master of Science (Science specialisation, master, higher engineering) is a higher education that requires at least two semesters of 12-credit education based on undergraduate education and a thesis that will reveal the results of the research in the field. On the other hand, a Ph.D. is at least six years based on a bachelor's or a master's degree). It is a higher education that covers at least four semesters and aims to reveal the results of original research.

Agricultural Engineers' Authority, general working areas and duties were defined with the enforcement of the "Regulation on

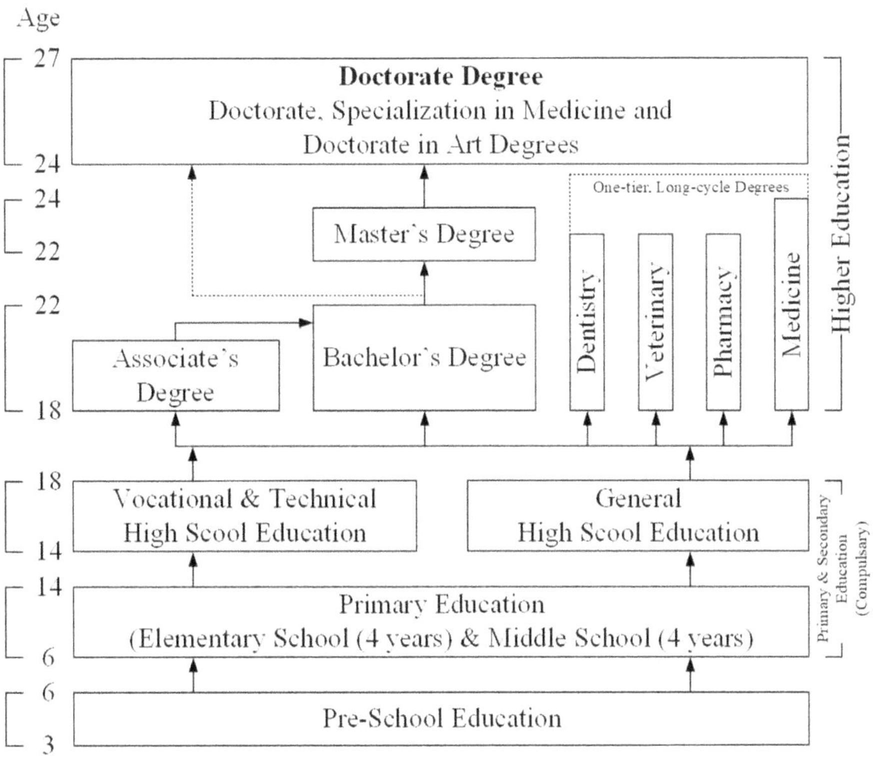

FIGURE 22.2 General structure of the Turkish education system (YOK, 2023).

the Duties and Powers of Agricultural Engineers". According to the 6th article of the Law on Agricultural High Engineering dated 6/5/1960 and numbered 7472, it was decided by the Council of Ministers on 18/12/1991, and the Order of the Code was published on the Official Gazette Date: 24/1/1992, No: 21121: 5, Volume: 31, p. 2537. The scope of the regulation follows: The duties and authorities of agricultural graduate engineers, agricultural engineers who graduated from agricultural faculties and agricultural graduates and agricultural engineers who graduated from agricultural faculties or agricultural higher schools in foreign countries and whose equivalence has been accepted by the Council of Higher Education are shown (Anonymous, 1992).

Today, undergraduate education is given in the following departments in the Faculties of Agriculture or similar names;

- Horticulture
- Plant Protection
- Herbal Production
- Herbal Production and Technologies
- Biosystems Engineering
- Genetics and Life Sciences
- Food Engineering
- Poultry Breeding
- Molecular Biology and Genetics
- Landscape Architecture
- Fisheries Engineering
- Dairy Technology
- Agricultural Economy
- Farm Machinery and Technologies Engineering
- Agricultural Biotechnology
- Agricultural Genetic Engineering
- Agricultural Structures and Irrigation
- Seed Science and Technology
- Soil Science and Plant Nutrition
- Animal Science

In these departments, a 4-year education is given. When it comes to Master's and Doctorate programmes, there are different programmes in Science Institutes or Graduate Education Institutes at different universities;

- Sustainable Agriculture
- Sustainable Agriculture and Food Systems
- Dairy Technology
- Dairy and Products Technology
- Agricultural Sciences
- Agricultural Sciences and Technologies
- Agricultural Economy
- Agricultural Management
- Farm Machinery and Technologies Engineering
- Agricultural Policy and Extension
- Agricultural Biotechnology
- Agricultural Genetic Engineering
- Field Plants
- Agricultural Structures and Irrigation
- Medicinal and Aromatic Plants
- Seed Science and Technology

- Soil Science
- Soil Science and Plant Nutrition
- Animal Science
- Bee and Bee Products
- Horticulture
- Horticultural Cultivation and Breeding
- Plant Nutrition
- Plant Breeding
- Plant Breeding Genetics and Biotechnology
- Plant Breeding and Genetics
- Plant Protection
- Herbal Biotechnology
- Herbal Production and Technologies
- Biosystems Engineering
- Industrial Plants
- Industrial Plants and Technologies
- Entomology
- Phytopathology
- Food Engineering
- Food Technology
- Animal Breeding
- Animal Biotechnology
- Animal Production and Technologies
- Poultry Breeding
- Fruit Cultivation and Breeding
- Organic Farming
- Landscape Architecture
- Aquaculture

The number of instructors working in the Faculties of Agriculture in Turkey is given in Tables 22.2–22.9 for 2022. As can be seen, there are 1199 Professors, 495 Associate Professors, 653 Assistant Professors, 41 Lecturers, and 494 Research Assistants, a total of 2875, of which 33.7% were female. In addition, the number of students studying in undergraduate programmes and enrolling in the 2021–22 academic year is also given in Table 22.10–22.16. The total number of registered and newly enrolled students is 21074 male and 11032 (34.4%) female, with a total of 32106 students. When it comes to MSc and PhD, there are 7011 (43.7% Female) and 2917 (47.2% Female) students, respectively, in the 2021–22 academic year (YOK, 2023).

Agricultural Faculties Education Programs Evaluation and Accreditation Association (ZIDEK)

The Agricultural Faculties Education Programs Evaluation and Accreditation Association (ZIDEK) is an independent institution that contributes to improving the quality of Agricultural Engineering education by carrying out accreditation, evaluation, and information studies for Agricultural Faculties education programmes in various disciplines. ZIDEK was established with the application made on March 03, 2011, with 7 founding members to accredit different undergraduate programmes within the faculties of Agricultural Engineering in Turkey and the Turkish Republic of Northern Cyprus. The Ankara Governorship Provincial Directorate of Associations approved the Association's charter on 17 June 2011 with the Agricultural Faculties Education Programs Evaluation and Accreditation Association (ZIDEK)/06-100-100 Register Number. As of 31 December 2021, the ZIDEK General Assembly, the most authoritative board of the Association for the Evaluation and Accreditation of Education Programs of Agriculture Faculties, comprises 46 members.

As a National External Evaluation and Accreditation Agency, ZIDEK was given a "Quality Evaluation and Registration Certificate" by the Higher Education Quality Board (YÖKAK) on 08 July 2020. Evaluation and accreditation studies for the Faculty of Agriculture Education Programs started in 2021.

According to ZIDEK, students who have reached the graduation stage must prove that they have achieved the following programme outcomes;

a) Adequate knowledge of mathematics, science and related engineering (Agricultural Engineering, Forestry Engineering, Fisheries Engineering, etc.) disciplines; ability to use theoretical and applied knowledge in these fields in complex engineering problems.
b) Ability to define, formulate and solve complex problems in related Engineering fields and to select and apply appropriate analysis and modelling methods for this purpose.
c) The ability to design a complex system, process, device, or product related to the engineering field, under realistic constraints and conditions, in other words, to meet certain requirements, taking into account the available possibilities and the current state of the field in question, and the ability to apply modern design methods for this purpose.
d) Ability to select and use modern tools necessary for analysing and solving complex problems encountered in related Engineering applications and the ability to use information technologies effectively.
e) Ability to design, conduct experiments, collect data, and analyse and interpret results to investigate complex problems or discipline-specific research issues encountered in the relevant Engineering field.
f) Ability to work effectively in disciplinary and multidisciplinary teams; individual study skills
g) Ability to write effective reports in the field, to understand written reports, to prepare design and production reports, to make effective presentations, to take and give clear and understandable instructions.
h) Awareness of the necessity of lifelong learning; the ability to access information, follow developments in science and technology, and constantly renew oneself.

TABLE 22.2

Number of academics (Male-M/Female-F) in related faculties in different universities in Mediterranean Region of Turkey in 2022.

UNIVERSITY/ FACULTY	TYPE	Professor			Assoc. Prof. Dr.			Assist. Prof. Dr.			Lecturer			Res. Assist.			Total		
		M	F	T	M	F	T	M	F	T	M	F	T	M	F	T	M	F	T
Akdeniz Univ./Agriculture	State	45	6	51	21	5	26	9	12	21	0	1	1	3	8	11	78	32	110
Alanya Alaaddin Keykubat University/ Enginnering	State	1	1	2	0	0	0	0	0	0	0	0	0	0	0	0	1	1	2
Cukurova Univ./ Agriculture	State	49	28	77	13	3	16	10	5	15	0	0	0	20	10	30	92	46	138
Hatay Mustafa Kemal Univ./Agriculture	State	40	9	49	12	4	16	18	9	27	0	0	0	6	5	11	76	27	103
Isparta Applied Sciences Univ./Agriculture	State	42	14	56	12	7	19	8	9	17	1	0	1	14	7	21	77	37	114
Kahramanmaras Sutcu Imam Univ./ Agriculture	State	36	6	42	7	3	10	16	2	18	0	1	1	4	6	10	63	18	81
Mediterranean Region		**213**	**64**	**277**	**65**	**22**	**87**	**61**	**37**	**98**	**1**	**2**	**3**	**47**	**36**	**83**	**387**	**161**	**548**

TABLE 22.3

Number of academics (Male-M/Female-F) in related faculties in different universities in Aegean Region of Turkey in 2022.

UNIVERSITY / FACULTY	TYPE	Professor			Assoc.Prof.Dr.			Assist.Prof.Dr.			Lecturer			Res.Assist.			Total		
		M	F	T	M	F	T	M	F	T	M	F	T	M	F	T	M	F	T
Aydin Adnan Menderes Univ./ Agriculture	State	47	9	56	10	9	19	18	15	33	2	0	2	5	7	12	82	40	122
Ege Univ./Agriculture	State	61	38	99	10	17	27	12	6	18	0	3	3	16	20	36	99	84	183
Mugla Sitki Koçman Univ./Agriculture	State	1	0	1	0	0	0	0	0	0	0	0	0	0	0	0	1	0	1
Pamukkale Univ./ Agriculture	State	1	0	1	0	0	0	2	2	4	0	0	0	0	0	0	3	2	5
Yasar Univ./ Agricultural Sciences And Technologies	Foundation	1	0	1	0	3	3	2	1	3	0	0	0	0	2	2	3	6	9
Usak Univ./ Agriculture	State	4	1	5	1	1	2	3	7	10	0	0	0	1	1	2	9	10	19
Aegean Region		**115**	**48**	**163**	**21**	**30**	**51**	**37**	**31**	**68**	**2**	**3**	**5**	**22**	**30**	**52**	**197**	**142**	**339**

TABLE 22.4

Number of academics (Male-M/Female-F) in related faculties in different universities in Central Anatolia Region of Turkey in 2022.

UNIVERSITY/ FACULTY	TYPE	Professor			Assoc.Prof.Dr.			Assist.Prof.Dr.			Lecturer			Res.Assist.			Total		
		M	F	T	M	F	T	M	F	T	M	F	T	M	F	T	M	F	T
Ankara Univ./ Agriculture	State	72	44	116	6	7	13	10	13	23	5	4	9	17	15	32	110	83	193
Kayseri Erciyes Univ./ Agriculture	State	16	3	19	11	3	14	9	5	14	1	0	1	8	3	11	45	14	59
Konya Food And Agriculture University	Foundation	2	0	2	2	2	4	0	4	4	0	0	0	2	2	4	6	8	14
Eskisehir Osmangazi Univ./ Agriculture /	State	11	6	17	9	5	14	11	10	21	0	0	0	7	4	11	38	25	63
Nigde Omer Halisdemir Univ./ Agronomics And Agricultural Technologies	State	6	3	9	4	3	7	3	4	7	0	0	0	10	5	15	23	15	38
Necmettin Erbakan Univ./Agriculture	State	0	0	0	1	0	1	4	1	5	0	0	0	1	0	1	6	1	7
Kirsehir Ahi Evran Univ./Agriculture	State	9	2	11	14	6	20	15	13	28	0	0	0	3	3	6	41	24	65
Selcuk Univ./ Agriculture	State	55	5	60	13	5	18	14	22	36	5	1	6	6	11	17	93	44	137
Sivas Science And Technolgy Univ./ Agricultural Sciences And Technologies	State	3	0	3	3	0	3	5	2	7	0	0	0	3	4	7	14	6	20
Yozgat Bozok Univ./ Agriculture	State	3	1	4	6	5	11	7	6	13	1	0	1	2	8	10	19	20	39
Central Anatolia Region		**177**	**64**	**241**	**69**	**36**	**105**	**78**	**84**	**162**	**12**	**5**	**17**	**59**	**55**	**114**	**395**	**240**	**635**

TABLE 22.5

Number of academics (Male-M/Female-F) in related faculties in different universities in Eastern Anatolia Region of Turkey in 2022.

UNIVERSITY/		Professor			Assoc.Prof.Dr.			Assist.Prof.Dr.			Lecturer			Res.Assist.			Total		
FACULTY	TYPE	M	F	T	M	F	T	M	F	T	M	F	T	M	F	T	M	F	T
Ataturk Univ./ Agriculture	State	64	15	79	12	4	16	9	13	22	0	0	0	13	15	28	98	47	145
Bingol Univ./ Agriculture	State	9	0	9	5	0	5	10	3	13	0	0	0	6	3	9	30	6	36
Igdir Univ./ Agriculture	State	9	0	9	8	2	10	14	8	22	0	1	1	8	6	14	39	17	56
Malatya Turgut Ozal Univ./Agriculture	State	11	0	11	2	1	3	7	4	11	0	0	0	2	1	3	22	6	28
Van Yuzuncu Yil Univ./Agriculture	State	16	10	26	11	15	26	15	5	20	1	1	2	14	4	18	57	35	92
Mus Alparslan University/Applied Sciences	State	2	0	2	3	0	3	6	2	8	4	2	6	4	3	7	19	7	26
Eastern Anatolia Region		**111**	**25**	**136**	**41**	**22**	**63**	**61**	**35**	**96**	**5**	**4**	**9**	**47**	**32**	**79**	**265**	**118**	**383**

TABLE 22.6

Number of academics (Male-M/Female-F) in related faculties in different universities in Marmara Region of Turkey in 2022.

UNIVERSITY /		Professor			Assoc.Prof.Dr.			Assist.Prof.Dr.			Lecturer			Res.Assist.			Total		
FACULTY	TYPE	M	F	T	M	F	T	M	F	T	M	F	T	M	F	T	M	F	T
Bursa Uludag Univ./ Agriculture	State	48	19	67	9	21	30	4	4	8	1	0	1	7	7	14	69	51	120
Bilecik Seyh Edebali University	State	3	1	4	2	2	4	7	4	11	0	0	0	1	3	4	13	10	23
Canakkale Onsekiz Mart Univ./ Agriculture	State	33	10	43	9	3	12	18	6	24	0	0	0	6	6	12	66	25	91
Duzce University/ Agric. & Natural Sciences	State	2	2	4	2	4	6	5	3	8	0	0	0	3	2	5	12	11	23
Kocaeli Univ./ Agriculture	State	4	1	5	1	3	4	0	3	3	0	0	0	1	1	2	6	8	14
Sakarya Applied Sciences Univ./ Agriculture	State	3	0	3	4	0	4	7	1	8	0	0	0	2	1	3	16	2	18
Tekirdag Namik Kemal Univ./ Agriculture	State	47	14	61	11	9	20	18	11	29	1	0	1	6	2	8	83	36	119
Marmara Region		**143**	**47**	**187**	**38**	**42**	**80**	**59**	**32**	**91**	**2**	**0**	**2**	**26**	**22**	**48**	**265**	**143**	**408**

TABLE 22.7

Number of academics (Male-M/Female-F) in related faculties in different universities in Black Sea Region of Turkey in 2022.

UNIVERSITY / FACULTY	TYPE	Professor			Assoc.Prof.Dr.			Assist.Prof.Dr.			Lecturer			Res.Assist.			Total		
		M	F	T	M	F	T	M	F	T	M	F	T	M	F	T	M	F	T
Bolu Abant Izzet Baysal Univ./Agriculture	State	11	2	13	9	3	12	9	1	10	0	0	0	7	0	7	36	6	42
Duzce Univ./Agriculture	State	2	2	4	2	4	6	5	3	8	0	0	0	3	2	5	12	11	23
Ondokuz Mayis Univ./ Agriculture	State	46	12	58	16	4	20	12	5	17	0	1	1	18	13	31	92	35	127
Ordu Univ./Agriculture	State	16	2	18	7	3	10	6	5	11	1	1	2	11	7	18	41	18	59
Recep Tayyip Erdogan Univ./Agriculture	State	3	0	3	0	1	1	1	3	4	0	0	0	3	5	8	7	9	16
Tokat Gaziosmanpasa Univ./Agriculture	State	31	7	38	9	3	12	10	15	25	1	1	2	7	8	15	58	34	92
Black Sea Region		**107**	**23**	**130**	**41**	**14**	**55**	**38**	**29**	**67**	**2**	**3**	**5**	**46**	**33**	**79**	**234**	**102**	**336**

TABLE 22.8

Number of academics (Male-M/Female-F) in related faculties in different universities in Southeastern Anatolia Region of Turkey in 2022.

UNIVERSITY/ FACULTY	TYPE	Professor			Assoc.Prof.Dr.			Assist.Prof.Dr.			Lecturer			Res.Assist.			Total		
		M	**F**	**T**	**M**	**F**	**T**	**M**	**F**	**T**	**M**	**F**	**T**	**M**	**F**	**T**	**M**	**F**	**T**
Adiyaman Univ./ Agriculture	State	3	0	3	5	0	5	2	1	3	0	0	0	1	0	1	11	1	12
Dicle Univ./ Agriculture	State	17	4	21	7	3	10	10	1	11	0	0	0	3	0	3	37	8	45
Harran Univ./ Agriculture	State	27	2	29	13	4	17	8	6	14	0	0	0	8	1	9	56	13	69
Kilis 7 Aralik Univ./ Agriculture	State	0	0	0	1	1	2	2	2	4	0	0	0	0	1	1	3	4	7
Siirt Univ./ Agriculture	State	5	1	6	13	5	18	14	13	27	0	0	0	10	9	19	42	28	70
Sirnak Univ./ Agriculture	State	2	1	3	1	1	2	9	3	12	0	0	0	1	5	6	13	10	23
Southeastern Anatolia Region		**54**	**8**	**62**	**40**	**14**	**54**	**45**	**26**	**71**	**0**	**0**	**0**	**23**	**16**	**39**	**162**	**64**	**226**

TABLE 22.9
Number of academics (Male-M/Female-F) in related faculties in different regions across Turkey in 2022.

	Professor			Assoc.Prof.Dr.			Assist.Prof.Dr.			Lecturer			Res.Assist.			TOTAL		
REGION	**M**	**F**	**T**	**M**	**F**	**T**	**M**	**F**	**T**	**M**	**F**	**T**	**M**	**F**	**T**	**M**	**F**	**T**
Mediterranean	213	64	277	65	22	87	61	37	98	1	2	3	47	36	83	387	161	548
Aegean	115	48	163	21	30	51	37	31	68	2	3	5	22	30	52	197	142	339
Central Anatolia	177	64	241	69	36	105	78	84	162	12	5	17	59	55	114	395	240	635
Eastern Anatolia	111	25	136	41	22	63	61	35	96	5	4	9	47	32	79	265	118	383
Marmara	143	47	190	38	42	80	59	32	91	2	0	2	26	22	48	265	143	408
Black Sea	107	23	130	41	14	55	38	29	67	2	3	5	46	33	79	234	102	336
Southeastern Anatolia	54	8	62	40	14	54	45	26	71	0	0	0	23	16	39	162	64	226
Total	**920**	**279**	**1199**	**315**	**180**	**495**	**379**	**274**	**653**	**24**	**17**	**41**	**270**	**224**	**494**	**1905**	**970**	**2875**

TABLE 22.10

Number of undergraduate students (Male-M/Female-F) in various universities in Mediterranean and Aegean Regions of Turkey, 2021–2022

			NEW ADMISSIONS			TOTAL NUMBER OF STUDENTS		
UNIVERSITY	**FACULTY**	**PROVINCE**	**M**	**F**	**TOTAL**	**M**	**F**	**TOTAL**
Akdeniz University	Agriculture	Antalya	262	104	366	1271	494	1765
Alanya Alaaddin Keykubat University	Engineering	Antalya	15	16	31	20	20	40
Hatay Mustafa Kemal University	Agriculture	Hatay	79	35	114	576	244	820
Isparta University Of Applied Sciences	Agriculture	Isparta	97	40	137	511	214	725
Cukurova University	Agriculture	Adana	225	95	320	1500	681	2181
Kahramanmaras Sutcu Imam University	Agriculture	Kahramanmaras	85	46	131	524	283	807
Mediterranean Region			**763**	**336**	**1099**	**4402**	**1936**	**6338**
Aydin Adnan Menderes University	Agriculture	Aydin	175	114	289	1030	598	1628
Usak University	Agriculture	Uşak	50	31	81	289	131	420
Yasar University	Agronomics And Agric. Technologies	İzmír	19	11	30	19	11	30
Ege University	Agriculture	İzmír	278	171	449	1584	830	2414
Aegean Region			**522**	**327**	**849**	**2922**	**1570**	**4492**

TABLE 22.11

Number of undergraduate students (Male-M/Female-F) in various universities in Central Anatolia Region of Turkey, 2021–2022

			NEW ADMISSIONS			TOTAL NUMBER OF STUDENTS		
UNIVERSITY	**FACULTY**	**PROVINCE**	**M**	**F**	**TOTAL**	**M**	**F**	**TOTAL**
Ankara University	Agriculture	Ankara	318	233	551	1907	1284	3191
Kayseri Erciyes University	Agriculture	Kayserí	86	42	128	598	289	887
Eskisehir Osmangazi University	Agriculture	Eskíşehír	122	66	188	607	273	880
Yozgat Bozok University	Agriculture	Yozgat	44	16	60	172	60	232
Necmettin Erbakan University	Agriculture	Konya	8	1	9	45	36	81
Konya Food And Agriculture University	Agriculture And Natural Sciences	Konya	10	38	48	41	134	175
Nigde Omer Halisdemir University	Agronomics And Agricultural Technologies	Niğde	51	29	80	260	179	439
Selcuk University	Agriculture	Konya	185	91	276	1064	528	1592
Kirsehir Ahi Evran University	Agriculture	Kirşehír83	Kirşehír83	41	124	403	167	570
Central Anatolia Region			**907**	**557**	**1464**	**5097**	**2950**	**8047**

TABLE 22.12

Number of undergraduate students (Male-M/Female-F) in various universities in Eastern Anatolia Region of Turkey, 2021–2022

			NEW ADMISSIONS			TOTAL NUMBER OF STUDENTS		
UNIVERSITY	FACULTY	PROVINCE	M	F	TOTAL	M	F	TOTAL
Ataturk University	Agriculture	Erzurum	77	47	124	520	274	794
Bingol University	Agriculture	Bingol	19	5	24	106	54	160
Igdir University	Agriculture	Igdir	29	7	36	109	43	152
Van Yuzuncu Yil University	Agriculture	Van	60	31	91	366	137	503
Mus Alparslan University	Faculty Of Applied Sciences	Mus	8	4	12	54	16	70
Malatya Turgut Ozal University	Agriculture	Malatya	47	25	72	232	130	362
Eastern Anatolia Region			240	119	359	1387	654	2041

TABLE 22.13

Number of undergraduate students (Male-M/Female-F) in various universities in Marmara Region in Turkey, 2021–2022

			NEW ADMISSIONS			TOTAL NUMBER OF STUDENTS		
UNIVERSITY	FACULTY	PROVINCE	M	F	TOTAL	M	F	TOTAL
Bilecik Seyh Edebali University	Agriculture And Natural Sciences	Bílecík	50	19	69	174	102	276
Bursa Uludag University	Agriculture	Bursa	223	214	437	1209	981	2190
Canakkale Onsekiz Mart University	Agriculture	Çanakkale	134	57	191	937	395	1332
Duzce University	Agriculture/	Düzce	54	17	71	174	83	257
Kocaeli University	Agriculture	Kocaelí	29	13	42	29	13	42
Tekirdag Namik Kemal University	Agriculture	Tekírdağ	111	65	176	812	376	1188
Sakarya University Of Applied Sciences	Agriculture	Sakarya	60	44	104	79	53	132
Marmara Region			**661**	**429**	**1090**	**3414**	**2003**	**5417**

TABLE 22.14

Number of undergraduate students (Male-M/Female-F) in various universities in Black Sea Region of Turkey, 2021–2022

			NEW ADMISSIONS			TOTAL NUMBER OF STUDENTS		
UNIVERSITY	FACULTY	PROVINCE	M	F	TOTAL	M	F	TOTAL
Bolu Abant Izzet Baysal University	Agriculture	Bolu	49	22	71	321	149	470
Ondokuz Mayis University	Agriculture	Samsun	150	66	216	989	446	1435
Ordu University	Agriculture	Ordu	39	28	67	220	182	402
Recep Tayyip Erdogan University	Agriculture	Rize	9	1	10	68	25	93
Tokat Gaziosmanpasa University	Agriculture	Tokat	108	36	144	523	212	735
Black Sea Region			**355**	**153**	**508**	**2121**	**1014**	**3135**

TABLE 22.15

Number of undergraduate students (Male-M/Female-F) in various universities in Southeastern Anatolia Region of Turkey, 2021–2022

			NEW ADMISSIONS			TOTAL NUMBER OF STUDENTS		
UNIVERSITY	FACULTY	PROVINCE	M	F	TOTAL	M	F	TOTAL
Harran University	Agriculture	Sanliurfa	124	60	184	692	423	1115
Kilis 7 Aralik University	Agriculture	Kilis	11	3	14	32	12	44
Siirt University	Agriculture	Siirt	72	32	104	290	140	430
Sirnak University	Agriculture	Sirnak	9	4	13	59	18	77
Adiyaman University	Agriculture	Adiyaman	19	4	23	83	39	122
Dicle University	Agriculture	Diyarbakir	111	43	154	575	273	848
Southeastern Anatolia Region			**346**	**146**	**492**	**1731**	**905**	**2636**

TABLE 22.16

The total number of undergraduate students (Male-M/Female-F) in different regions across Turkey, 2021–2022

	NEW ADMISSIONS			TOTAL NUMBER OF STUDENTS		
REGION	M	F	TOTAL	M	F	TOTAL
Mediterranean	763	336	1099	4402	1936	6338
Aegean	522	327	849	2922	1570	4492
Central Anatolia	907	557	1464	5097	2950	8047
Eastern Anatolia	240	119	359	1387	654	2041
Marmara	661	429	1090	3414	2003	5417
Black Sea	355	153	508	2121	1014	3135
Southeastern Anatolia	346	146	492	1731	905	2636
Total	**3794**	**2067**	**5861**	**21074**	**11032**	**32106**

TABLE 22.17

Programmes that have been granted accreditation and/or monitoring as of 2022

UNIVERSITY/FACULTY/PROGRAM NAME	STARTING DATE	FINISHING DATE
Ankara University Faculty of Agriculture Department Of Animal Science	28.03.2022	30.09.2024
Bursa Uludag University Faculty of Agriculture Department of Biosystems Engineering/Department of Animal Science	28.03.2023	30.09.2025
Cukurova University Faculty of Agriculture Department of Horticulture/ Department of Plant Protection/Department of Farm Machineries And Technologies Engineering/ Department of Soil Science And Plant Nutrition	28.03.2022	30.09.2024
Cukurova University Faculty of Agriculture Field Crops	28.03.2023	30.09.2025
Ege University Faculty of Agriculture Department Of Dairy Technology/Department Of Farm Machineries And Technologies Engineering/ Department Of Farm Structures And Irrigation/Department Of Field Crops/Department Of Animal Science	28.03.2022	30.09.2024
Ege University Faculty Of Agriculture Department Of Holticulture/Department Of Plant Protection/Department Of Agricultural Economics/ Department Of Soil Science And Plant Nutrition/	28.03.2023	30.09.2025
Tekirdag Namik Kemal University Faculty Of Agriculture Department Of Horticulture/Department Of Plant Protection/Department Of Agricultural Biotechnology/Department Of Animal Science	28.03.2023	30.09.2025

i) Knowledge of ethical principles, professional and ethical responsibility, and standards used in engineering practices.
j) Knowledge of business practices in the field of engineering, such as project management, risk management and change management; awareness of entrepreneurship, innovation; information about sustainable development.
k) Information about the effects of engineering practices on health, environment and safety in universal and social dimensions and the problems of the age reflected in the field of engineering; awareness of the legal consequences of engineering solutions.

The training plan should also include the following components;

a) Mathematics and basic science education for at least one year or 32 credits, or at least 60 ECTS (Labour-based Credit system) credits. Basic science education should be in accordance with the relevant programme and should be supported by experimental studies.
b) At least one and a half years or at least 48 credits or at least 90 ECTS (Labour-based Credit system) credits in basic engineering sciences and vocational engineering education in accordance with the relevant programme.
c) General education that complements the technical content of the education programme and is in line with the programme objectives.
d) Having at least 25% of applied courses.
e) At least 240 ECTS for graduation from four-year undergraduate programmes.

Programmes that have been granted accreditation and/or monitoring as of 2023 are given in Table 22.17 (Anonymous, 2023b).

Discussion and Conclusions

The new paradigms developing in the world of teaching have started to validate a system in engineering teaching, which is student-oriented with innovative teaching and a why-should-I-know-focused curriculum; all unnecessary information is removed, and the creativity of the human brain is highlighted. This practice of lifelong learning is widespread in developed universities and should be made more effective in higher education institutions in our country.

An important point to consider is that the engineer must be sure he has received adequate training and will make the necessary evaluations properly. Developing technology makes it necessary to renew oneself and to be a lifelong learner, especially in engineering. Students must acquire educational competencies of the twenty-first century in all areas, since agriculture is becoming increasingly more coordinated with industry and incorporating modern automation and communication technologies to solve problems, these new competencies and skills are sought in Agricultural Engineers. In summary, the new generation of Agricultural Engineers should know the geography and natural resources of the country well, and should be prone to field studies. However, they should have a strong solution-oriented practice that will recognise and solve problems. In addition, they should have sufficient skills in agricultural techniques and technologies, have the qualifications to establish good relations with farmers, know the sector, and have strong communication. They should be willing to learn continuously, write well, and have high speaking and persuasion skills. In addition, they should be open to new ideas, be forward-thinking, and value teamwork and interdisciplinary collaboration. They should have knowledge of engineering and agricultural economics, have adopted an environment and ecosystem-friendly understanding, developed managerial and leadership skills, and know the rules and scope of economics. They should include those who are technology literate and able to adapt to digital technologies, have developed the ability to use technology, media and social media with informatics methods and applications, and have learned how to access information. Finally, Agricultural Engineers who have a universal understanding of culture, can write and speak a foreign language at a level that can reach international sources, and have internalised professional ethics should be trained.

REFERENCE LIST

Abaci N.I., Koksal O., Atsan T., Demiryurek K., Ozcatalbas O., Cukur T., Bostan D.B. (2020). Tarimsal Ogretim ve Yayim Hizmetlerinde Mevcut Durum ve Gelecek. Turkiye Ziraat Muhendisligi 9. Teknik Kongresi, s. 629–637, Ankara, Turkey (in Turkish).

Akman A. (1978). Turkiye'de Ziraat Yuksek Ogretim Reformunun Anatomisi Ankara Yuksek Ziraat Enstitusu (YZE). Ankara Universitesi Ziraat Fakultesi Yayinlari: 739, Derlemeler: 430, 93 s., Ankara (in Turkish).

Anonymous. (1992). Ziraat Muhendislerinin Gorev ve Yetkilerine Iliskin Tuzuk. *Resmi Gazete*, 24/1/1992, No: 21121, 31, s. 2537 (in Turkish).

Anonymous. (2023a). Türkiye Haritası, Siyasi, Sehirler, Bolgeler, Ankara. https://www.hurriyet.com.tr/egitim/turkiye-haritasi-siyasi-sehirler-bolgeler-goller-ve-turkiyenin-komsulari-isimleri-ile-listesi-buyuk-renkli-41889121

Anonymous. (2023b). ZIDEK Tematik Analiz Raporu. Ziraat Fakulteleri Egitim Programlari Degerlendirme ve Akreditasyon Dernegi, Ankara (in Turkish).

Aras N.K., Dolen E., Bahadir O. (ed.), (2007). Turkiye'de Universite Anlayisinin Gelisimi. TUBA Yayinlari, Ankara (in Turkish).

Askin U. (2017). Ankara Yuksek Ziraat Enstitusu: Tam Bagimsizlik Yolunda Bir Anit Kurum. *Turk Dunyası Arastirmalari*, 117, 85–114 (in Turkish).

Cesme V. (2014). Halkali Ziraat Mektebi: Eğitimi, Eğitimci Kadrosu, Örnek Çiftliği Ve Yayinlari. *Osmanlı Bilimi Araştırmaları*, XVI/1, 73–99 (in Turkish).

Çiftçi C.Y. (2004). Geçmişten Gunumuze Turkiye'de Ziraat Muhendisligi Egitiminin Degişimi. 50. Yıl Kutlama ve Tarım Haftası Etkinlikleri. TZMO Tarim ve Muhendislik. Uluslararası Sempozyum 12–16 Ocak, Ankara, s. 304–332 (in Turkish).

Ciftci C.Y. (2012). 165. Yılında Tarımsal Yuksekogretim. *Ziraat Muhendisligi*, 359 (in Turkish).

Ciftci C.Y. (2008). Kurulusunun 75. Yılında Yuksek Ziraat Enstitusu (1933–1948). Ankara Univ. Ziraat Fakultesi Yayınları, Yayin No:1496, 728 s., Ankara (in Turkish).

Ciftci C.Y., Ortas I., Celik S., Kendir H., Saglam S. (2010). Tarimsal Yükseköğretim Programinda Degisimler ve Sorunlar. VII. Türkiye Ziraat Mühendisliği Teknik Kongresi, 11–15 Ocak 2010, Ankara (in Turkish).

Eris A. (2004). Turkiye'de ve Dunya'da Universiter Tarim Egitimi. I. Ulusal Ziraat Fakültesi Ogrencileri Kongresi, 28 Ekim, Ege Universitesi, Ziraat Fakultesi, Izmir (in Turkish).

Eris A., Ciftci C.Y., Ismailcelebioglu N., Direk M. (2005). Tarimsal Yuksekogrenim. Turkiye Ziraat Muhendisligi VI. Teknik Kongresi, 03–07 Ocak, Ankara (in Turkish).

Kadioglu S. (2005). Osmanli Doneminde Turkiye'de Ziraat Okullari Uzerine Notlar: Tedrisat-ı Ziraiye Nizamnamesi. *Kutadgubilig: Felsefe Bilim Arastirmalari*, Sayi: 8, s. 239–257 (in Turkish).

Kadioglu, S. (2008). Ankara Yuksek Ziraat Enstitusu'nde Multeci Bilim Adamlari. *Osmanli Bilimi Arastirmalari*, Cilt: 9, Sayı: 1–2, s. 183–197 (in Turkish).

Kalaycioguullari, I. (2009). Darulfunun'dan Universite'ye "Gelenek"ten Araştırma"ya. Ankara Universitesi, Dil ve Tarih-Cografya Fakultesi Dergisi, Cilt: 48, Sayi: 1, s. 43–59, Ankara (in Turkish).

Ortas, I., Ciftci C.Y. (2015). Tarimsal Yuksekogretimde Degisimler ve Yeni Arayislar. Türkiye Ziraat Muhendisligi VIII. Teknik Kongresi, 12–16 Ocak, Ankara (in Turkish).

Ozata M. (2007). Atatürk Bilim ve Universite, TÜBİTAK Populer Bilim Kitaplari, Ankara (in Turkish).

Ozsan M., Ergene A., Dinc U., Berkman A., Koycus C., Ozkaya T., Tatlidil F., Akturk D. (1995). Ziraat Muhendisligi ve Lisansustu Ogrenimi. Türkiye Ziraat Muhendisligi IV. Teknik Kongresi, 9–13 Ocak, Ankara (in Turkish).

TBMM. (2023). Mustafa Kemal Atatürk'un Turkiye Buyuk Millet Meclisi Acilis Konusmalari. Kütüphane ve Arsiv Hizmetleri Baskanligi, Ankara. https://www5.tbmm.gov.tr /kutuphane/meclis_acilis.html (in Turkish).

Widmann H. (2000). Atatürk ve Universite Reformu. Y. Tukel, Z. Guler, Kabalci Yayinevi, 533 s., Istanbul (in Turkish).

Yazicioglu T. (1958). Ankara Universitesi Ziraat Fakultesi ve Memleketimizde Yuksek Ziraat Ogretiminin Tarihcesi. Türk Yuksek Ziraat Muhendisleri Birligi Nesriyati, 182 s. Ankara (in Turkish).

Yuksek Ogretim Kurumu. (2023). Ankara. http://yok.gov.tr

23

Agricultural and Biosystems Engineering Education in Ghana

Emmanuel Yaovi Hunnuor Bobobee

Introduction

Agricultural and Biosystems Engineering (ABE) is a field of engineering that applies the principles of engineering, mathematics, soil science, plant biology, animal biology and economics to the development, planning, design, operation, and management of sustainable agricultural production systems, the processing of products, and the distribution of those products in local, regional, and global markets (Yang et al., 2020). This description is similar to the one given by Field and Solie (2008). In U.S. institutions, biosystems engineering is considered a broader term that includes biological engineering and agricultural engineering (Johnson, 2012). According to Ting (2009), the overarching mission of biosystems engineering that encompasses agricultural, food, and biological engineering, is to integrate life and engineering for enhancement of complex living systems (Ulloa, 2021). Agricultural and biosystems engineering education in Ghana is offered at the undergraduate and postgraduate levels in several traditional and technical universities (Bobobee et al., 2021). The programme is designed to provide students with the knowledge and skills necessary to solve problems in the field of agricultural and biosystems engineering and the agricultural food value chain. The programme is structured to provide a broad-based education in the basic sciences, engineering sciences and agricultural sciences (Meinke et al., 2009).

An important goal of higher education in Agricultural Engineering in Ghana, in addition to the specialised study and professional training, is to endow students with a wide range of knowledge to meet the demands of agriculture, industry, and rural development in Ghana and a rapidly changing world (Diao et al., 2019). Consequently, the professional agricultural engineering programmes include courses from the basic engineering sciences, agriculture, the humanities, and entrepreneurship (Babatope et al., 2020). The agricultural and biosystems engineering training involves the specialised areas of Farm Power and Machinery, Soil and Water Engineering, Food and Postharvest, Storage and Processing Technology, Agro-environmental Control, and Bioenergy Engineering.

Students studying Agricultural and Biosystems Engineering in Ghana are trained in areas such as crop and soil science, agricultural mechanisation, agricultural structures and environmental control, irrigation and drainage engineering, postharvest and food processing, bioenergy, and agro-industrial management. Graduates from these programmes work in various industries, including commercial farms, agricultural machinery and equipment manufacturing, sale and service industries, irrigation and drainage systems, food and agro-processing, the Ghana government Ministry of Food and Agriculture, development agencies, and Non-Governmental Organisations. The traditional universities offer bachelor of science (BSc) degrees and the technical universities offer bachelor of technology (BTech) degrees and higher national diplomas (HND), which are competency based (Table 23.1).

The visions of the Ghana Agricultural and Biosystems Departments focus on developing into national and international centres of academic excellence to provide leadership for study, research, and extension work in agricultural and Biosystems engineering appropriate to tropical and Ghanaian conditions (Clark, 2002). This is to contribute effectively to, and sustain, the overall socio-economic development of Ghana and Africa. The various departments do this through: (i) the continued development and motivation of existing staff and future staff from recruitment, retraining, and retention, (ii) collaboration with engineers, scientists and industrialists within and outside the country, (iii) joint research efforts and linkages with institutions inside and outside the country, (iv) strengthening linkages between the Departments and national institutions for the extension of suitable and innovative technologies for food production, handling, processing and storage, rural agro-industrial development, and for the good management and sustainability of the rural and agricultural environment.

Historical Background

Agricultural Engineering education in Ghana has a long and rich history, dating back to the colonial era. During the colonial period, the emphasis was on the training of agricultural technicians and farm managers to support the colonial agricultural economy. In 1948, the University College of the Gold Coast, which later became the University of Ghana, was established to provide higher education in various disciplines, including agriculture. As Ghana gained independence in 1957, the government's focus shifted to agricultural development as a means of achieving economic growth and development.

DOI: 10.1201/9780429150111-25

TABLE 23.1

Ghanaian universities offering Agricultural Engineering programmes at different levels.

UNIVERSITIES	PROGRAMMES					
	HND	BTech Agric. Eng.	BSc Agric. Eng.	BSc Agric. Mech./Tech	MPhil/ MEng/ MSc	PhD
Kwame Nkrumah University of Science & Technology (KNUST), Kumasi			X	X	X	X
University of Ghana (UG) Legon Accra			X		X	X
University of Energy and Natural Resources (UENR), Sunyani			X			
University of Cape Coast, Cape Coast*					X	X
University for Development Studies, Tamale			X	X		
Cape Coast Technical University, Cape Coast		X				
Ho Technical University, Ho**	X	X				
Tamale Technical University, Tamale	X	X				
Bolgatanga Technical University, Bolgatanga	X	X				
Wa Technical University, Wa	X	X				

(Source: Bobobee et al. (2021)

*University of Cape Coast offers agricultural engineering at the post graduate levels in Irrigation, Mechanisation and Food & Postharvest presently

** Ho Technical University runs a 2-year top up BTech programme for HND graduates

This renewed focus on agriculture led to the expansion of Agricultural Engineering education in Ghana. However, it was not until the 1960s that Agricultural Engineering began to be taught as a separate discipline. The first programme in Agricultural Engineering was established at the Kwame Nkrumah University of Science and Technology (KNUST) in Kumasi in 1960. The programme was initially part of the Department of Agricultural Science, and its focus was on providing technical training in farm machinery and equipment operation and maintenance to graduates in agriculture.

Prior to the 1970s, agricultural engineering was taught as part of the agricultural science programmes. However, in the 1970s, the KNUST began to offer a dedicated Agricultural Engineering programme. In 1974, the University of Ghana in Legon, Accra, also introduced a Bachelor of Science degree programme in Agricultural Engineering. The programme was later restructured and renamed as Agricultural and Biosystems Engineering in the 1990s. Since the 1970s, several other universities and polytechnics (later technical universities) have also introduced Agricultural Engineering programmes. The technical universities are; Cape Coast Technical University, Ho Technical University, Tamale Technical University, Bolgatanga Technical University, and Wa Technical University.

In the 1970s and 1980s, the emphasis on Agricultural Engineering education in Ghana focused on agricultural mechanisation and technology development. The goal was to increase agricultural productivity through the introduction of modern technology and equipment. The curriculum included courses in farm machinery design and maintenance, irrigation and drainage systems, and postharvest processing.

In the 1990s and 2000s, the programme was later restructured and renamed as Agricultural and Biosystems Engineering at KNUST to reflect the evolving nature of the field. The focus of the programme also expanded to include environmental control and management, bioenergy and bioprocessing, and sustainable agricultural systems. Today, Agricultural and Biosystems Engineering education in Ghana continues to evolve to meet the needs of a rapidly changing agricultural sector. The curriculum has expanded to include courses in precision agriculture, biotechnology, bioenergy, and agroforestry. The focus is on developing innovative solutions to the challenges facing Ghana's agriculture, including climate change, food security, and rural development.

Organisation, Teaching Programmes, and Activities

Majority Agricultural and Biosystems Engineering Departments in the traditional universities were established as component departments of the Faculties of Agriculture to offer courses in agricultural mechanisation and related disciplines to students at the undergraduate and diploma programmes. The KNUST Department started the 4-year undergraduate degree programme to train professional agricultural engineers in 1971. Cape Coast University will soon start a BSc programme in the School of Engineering, but presently runs MSc, MPhil, and PhD programmes in agricultural engineering. The technical universities, which were initially Polytechnics, run a uniform BTech programme. They all started with HND and later introduced the BTech and MTech degrees. Ho technical university runs a two-year top up BTech programme for HND graduates.

For the award of the B.Sc. engineering degree, a student requires a minimum of 120 and a maximum of 168 credits. In the traditional universities, the programme of studies for the bachelor degree as shown in Figure 23.1, requires between 59–72 credit hours in professional agricultural engineering courses, 47–71 credits in basic engineering sciences, 12–22 credits in mathematics, and 0–12 credits in languages or social sciences.

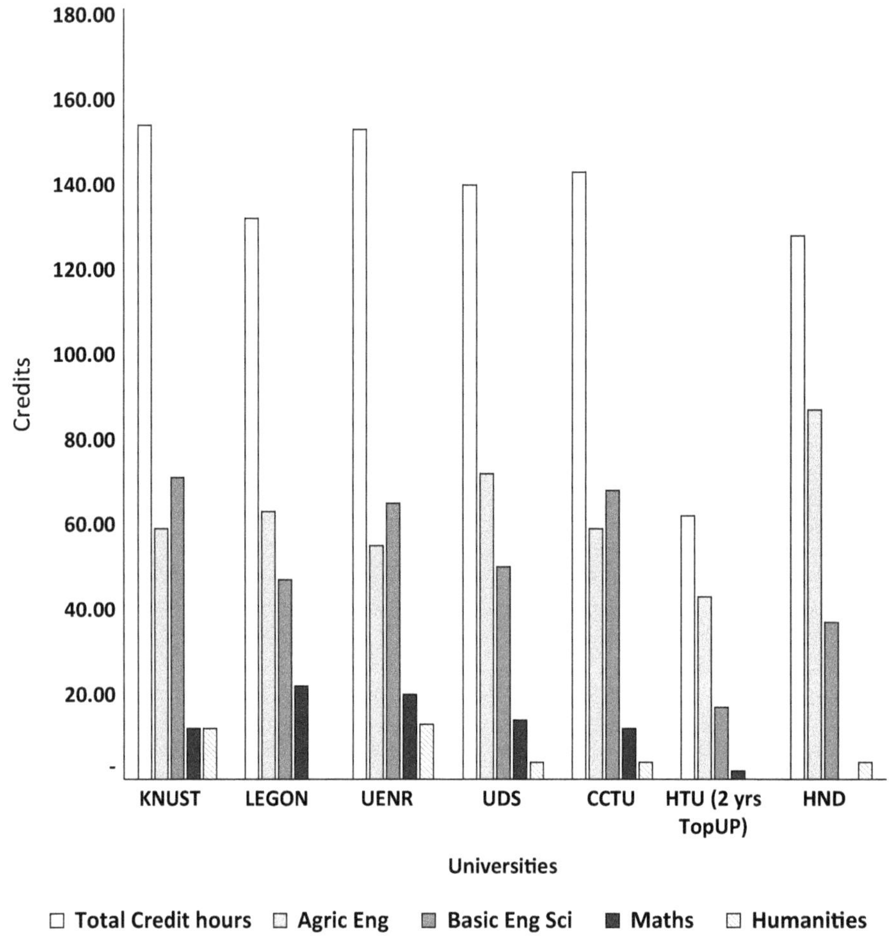

FIGURE 23.1 Components of undergraduate Agricultural Engineering programmes in Ghanaian universities.

In the technical universities, the BTech degree requires 143 credit hours made up of 85 credits in core agricultural engineering subjects, 98 credits in Basic engineering subjects, 18 credits in Mathematics, and 6 credits in humanities. The two-year top up BTech programme for HND graduates at Ho require less credits as shown in Figure 23.1.

Curriculum

The curriculum for Agricultural and Biosystems Engineering in Ghana covers a broad range of topics. Students are trained in areas such as crop and soil science, agricultural mechanisation, agricultural structures and environmental control, irrigation and drainage engineering, postharvest and food processing, bioenergy and agro-industrial management. The curriculum also includes practical training in laboratory work and fieldwork.

The programme is designed to equip students with the necessary knowledge and skills to design and develop agricultural systems that are sustainable, efficient, and productive. The curriculum is continually updated to reflect the latest advances in Agricultural and Biosystems Engineering, ensuring that graduates are well-prepared for the job market in the emerging fourth industrial revolution (Industry 4.0)

The curriculum for Agricultural and Biosystems Engineering in Ghana covers a wide range of topics to equip students with the knowledge and skills required to design and develop sustainable, efficient, and productive agricultural systems.

The curriculum typically consists of both core courses and elective courses, and is designed to provide a balance of theoretical and practical training. The courses offered vary depending on the specific institution, but the following are some of the common core courses in the curriculum:

- Introduction to Agricultural and Biosystems Engineering
- Crop Science and Soil Science
- Agricultural Mechanization and Farm Power
- Agricultural Structures and Environmental Control
- Irrigation and Drainage Engineering
- Postharvest and Food Processing Engineering
- Agro-industrial Management
- Bioprocess Engineering
- Renewable Energy and Bioenergy

- Agricultural Policy and Development
- Rural Sociology and Extension Education
- Engineering Mathematics and Statistics

The curriculum includes laboratory work and fieldwork to provide students with hands-on experience and practical skills. Students undertake projects in their final year, which often involves working on real-life problems faced by the agricultural industry. This project provides students with the opportunity to apply the knowledge and skills they have acquired throughout the programme.

The curriculum is continuously reviewed and updated to keep with the latest advances in Agricultural and Biosystems Engineering. As a result, the programme prepares graduates for the ever-changing needs of the agricultural industry. Graduates have the necessary skills to work in various sectors such as manufacturing, research, government agencies, and non-governmental organisations.

Challenges

The education of Agricultural and Biosystems Engineering graduates in Ghana faces several challenges, including:

Inadequate infrastructure: Most academic institutions offering Agricultural and Biosystems Engineering education in Ghana lack the necessary infrastructure, including research facilities, laboratories, and modern equipment, to support effective teaching and research in the field.

Insufficient funding: Agricultural and Biosystems Engineering education in Ghana is not adequately funded, which limits the ability of academic institutions to invest in infrastructure, research, and innovation.

Shortage of qualified faculty: There is a shortage of qualified and experienced faculty members in Agricultural and Biosystems Engineering in Ghana, which limits the quality of teaching, research, and innovation in the field.

Limited industry collaboration: There is need for increased collaboration between academic institutions and industry players in the agricultural sector to enrich the curriculum to align with the current and future needs of the industry.

Limited research and innovation: There is a limited amount of funding for research and innovation in Agricultural and Biosystems Engineering in Ghana, which affects the development of new technologies and techniques to improve the efficiency, productivity, and sustainability of agriculture in Ghana.

Lack of awareness: There is a lack of awareness among students and the general public about the importance of Agricultural and Biosystems Engineering and the opportunities available in the field.

Brain drain: Many of the skilled professionals in Agricultural and Biosystems Engineering in Ghana migrate to other countries for better opportunities, which limits the pool of skilled professionals available in the country.

Career Opportunities

Agricultural and Biosystems Engineering education in Ghana offers numerous career opportunities for graduates. Graduates from Agricultural and Biosystems Engineering programmes in Ghana like their counterparts anywhere in the world have a wide range of career opportunities. They can work in various industries, including agricultural machinery and equipment manufacturing, irrigation and drainage systems, food processing, agro-processing, the public and private sectors, or start their own businesses. They can also work in research institutions, government agencies, and non-governmental organisations, providing technical support and expertise in Agricultural and Biosystems Engineering. Some of the career opportunities available include:

Agricultural Engineer: As an agricultural engineer, one can work in the agricultural industry, where there will be responsibility for designing and developing systems that improve the efficiency and productivity of agricultural processes. This may include designing irrigation systems, farm machinery, farm structures and other agricultural equipment.

Environmental Engineer: With a degree in Agricultural and Biosystems Engineering, one can work as an environmental engineer, where one will be responsible for developing and implementing strategies to manage and mitigate the environmental impact of agricultural activities.

Research Scientist: Opportunities exist to work as a research scientist in the agricultural sector, to be responsible for developing and testing new technologies and techniques to improve agricultural productivity and sustainability.

Technical Sales Engineer: Engineers can work in the private sector, where they will be responsible for selling and marketing agricultural equipment and technologies. This may include farm machinery, irrigation systems, and other agricultural products.

Entrepreneur: Engineers can start their own businesses to provide services or products to the agricultural industry. For example, one can start a company that designs and manufactures farm machinery or provides agricultural consultancy services.

Government Agencies: Engineers can work in government agencies such as the Ministry of Food and Agriculture, where they will be responsible for implementing agricultural policies and programmes.

NGOs: Engineers can also work in non-governmental organisations in the agricultural sector, where they will be responsible for providing technical support to farmers and rural communities.

Role

Agricultural and Biosystems Engineering education plays a crucial role in the development of Ghana's agricultural sector and the country's overall development. Here are some of the ways in which Agricultural and Biosystems Engineering education contributes to development in Ghana:

Improving agricultural productivity: Agricultural and Biosystems Engineering education equips students with the knowledge and skills to design, develop, and implement new technologies and techniques to increase agricultural productivity, such as irrigation systems, improved crop production, and mechanisation.

Enhancing food security: Agricultural and Biosystems Engineering education contributes to the development of innovative and sustainable food production systems that can increase the quantity and quality of food produced in the country. This helps to enhance food security and reduce hunger.

Reducing postharvest losses: Agricultural and Biosystems Engineering education enables students to develop and implement innovative technologies and techniques to reduce postharvest losses, such as improved drying, storage facilities and transportation systems.

Enhancing environmental sustainability: Agricultural and Biosystems Engineering education equips students with the knowledge and skills to develop and implement sustainable agricultural practices that minimise the negative impact of agriculture on the environment, such as soil degradation, erosion and water pollution.

Creating employment opportunities: Agricultural and Biosystems Engineering education provides students with the knowledge and skills to work in a wide range of roles in the agricultural sector, such as engineers, researchers, consultants, and entrepreneurs. This helps to create employment opportunities and contribute to economic development.

Fostering innovation: Agricultural and Biosystems Engineering education contributes to the development of a culture of innovation and research in the agricultural sector in Ghana. This helps to foster the development of new technologies, techniques, and solutions to the challenges facing the sector.

Future outlook

The future of Agricultural and Biosystems Engineering education in Ghana is promising and is expected to continue to grow in importance. The agricultural sector is currently facing a range of challenges, including climate change, soil degradation, and water scarcity, among others. These challenges present opportunities for the development and application of new technologies and innovations in agriculture, which will require the expertise of Agricultural and Biosystems Engineers.

The following are some of the key areas that are likely to shape the future of Agricultural and Biosystems Engineering education in Ghana:

Sustainable agriculture: There is a growing demand for sustainable agricultural practices that can help mitigate the adverse environmental impact of agricultural activities. Agricultural and Biosystems Engineers will be instrumental in the development of technologies and techniques that promote sustainable agriculture, such as precision farming and conservation agriculture using frontier technologies of the fourth industrial revolution.

Climate-smart agriculture: Climate change is expected to continue affecting agriculture, with more frequent and severe droughts, floods, and extreme weather events. Agricultural and Biosystems Engineers will play a crucial role in the development of climate-smart agriculture practices, such as the mechanised production of drought-resistant crop varieties and water-efficient irrigation systems.

Digital agriculture: There is a growing trend towards the use of digital technologies in agriculture, such as drones for precision farming and the Internet of Things (IoT). Agricultural and Biosystems Engineers will be at the forefront of the development and application of these technologies, which will help to improve agricultural productivity and efficiency.

Food processing and value addition: Agricultural and Biosystems Engineers will also be involved in the development of technologies and techniques for food processing and value addition. This will help to reduce food waste and increase the value of agricultural products.

Public-private partnerships: The future of Agricultural and Biosystems Engineering education in Ghana will be shaped by strong public-private partnerships. Such collaborations will be essential in driving innovation and research, and in ensuring that graduates of the programme are well-prepared to meet the needs of the industry and society.

To advance the education of Agricultural and Biosystems Engineering in Ghana, the following measures can be taken:

Investment in infrastructure: There is a need for increased investment in infrastructure, including research facilities, laboratories, and modern equipment, to enhance the quality of education and research in the field.

Collaboration with industry: Collaboration between academic institutions and industry players is crucial to ensure that the curriculum of the programme aligns with the current and future needs of the industry. This will ensure that graduates are well-prepared to meet the demands of the industry and contribute to the growth and development of the agricultural sector.

Faculty development: There is a need for continuous training and development of faculty members to keep up with the latest trends and innovations in the field. This can help to enhance the quality of teaching, research, and innovation in the field of Agricultural and Biosystems Engineering.

Research and Innovation: There is a need for increased funding for research and innovation in the field of Agricultural and Biosystems Engineering to develop new technologies and techniques that can improve the efficiency, productivity, and sustainability of agriculture in Ghana.

Collaboration with other countries: There is a need for collaboration with other countries to exchange knowledge and expertise in the field of Agricultural and Biosystems Engineering. This can help to enhance the quality of education and research and promote the development of new technologies and techniques in the field.

Promotion of the program: There is a need to promote the programme to attract more students and encourage them to pursue a career in the field of Agricultural and Biosystems Engineering. This can help to increase the pool of skilled professionals in the industry and contribute to the growth and development of the agricultural sector in Ghana. There are several tools that can be used to enhance the education of Agricultural and Biosystems Engineering in Ghana, including:

Computer-aided design (CAD) software: CAD software can be used to design and simulate agricultural systems, such as irrigation systems, soil erosion control structures, and other agricultural equipment. This tool can help students to visualise and understand the complex systems and processes involved in agricultural engineering.

Geographic Information System (GIS) software: GIS software can be used to map and analyze the spatial distribution of crops, soils, and other agricultural resources. This tool can help students to understand the spatial aspects of agricultural systems and develop strategies for optimising the use of resources.

Simulation software: Simulation software can be used to simulate the behaviour of agricultural systems and processes under different conditions. This tool can help students to test different scenarios and understand the impact of various factors on the performance of agricultural systems.

Laboratory equipment: Modern laboratory equipment, such as soil analysis kits, plant growth chambers, crops and soil health analyses tools, heart rate monitors for drudgery analyses, and water quality meters, can be used to conduct experiments and analyze samples. This tool can help students to develop hands-on skills and understand the physical and chemical properties of agricultural materials.

Mobile applications: Mobile applications can be used to provide access to relevant information, such as weather forecasts, crop management practices, and market prices. This tool can help students to stay up to date with the latest developments in the industry and develop problem-solving skills.

Online resources: Online resources, such as digital libraries, scientific journals, and online courses, can be used to provide students with access to a vast amount of information and knowledge. This tool can help students to deepen their understanding of the subject matter and develop critical thinking skills.

Conclusion

Agricultural and Biosystems Engineering education in Ghana at the tertiary level has come a long way since the 1970s. The introduction of dedicated degree programmes and the expansion of these programmes to other universities and technical universities in Ghana have provided students with a wide range of career opportunities. The curriculum is continually updated to reflect the latest advances in the field, ensuring that graduates are well-prepared for the job market. The future of Agricultural and Biosystems Engineering education in Ghana looks bright, and it will continue to play a critical role in the country's agricultural development.

Presently in Ghana, agricultural engineering university education is provided exclusively by public institutions supervised and funded by the nation. Four public traditional universities and five technical universities provide tertiary education and research in agricultural engineering.

At the Departments of Agricultural and Biosystems Engineering in Ghana, to qualify for the award of the B.Sc. degree in engineering, a student requires a minimum of 120 and a maximum of 168 credit hours after studying the programme with the following components: Humanities, <9%, Mathematics, 8–17%, Basic Engineering Sciences, 36–46% and Agricultural Engineering 39–52%. Similarly in the technical universities, to graduate BTech students must obtain 143 credit hours made up of 85 credits in core agricultural engineering subjects, 98 credits in Basic engineering subjects, 18 credits in Mathematics, and 6 credits in humanities

Interest in the programme has been growing among prospective applicants since the inception of the course with one student in 1971 at KNUST, Kumasi.

All agricultural engineering and mechanisation graduates belong to the Ghana Society of Agricultural Engineers (GSAE). Those who graduate from engineering faculties with strong engineering fundamentals become registered professional engineers during their senior years with the Ghana Institution of Engineering as part of their continuing education and professional development programmes.

In addition, some of the Agricultural Engineers are admitted in other disciplines of higher degree programmes. It is therefore recommended that to strike a balance of Agricultural Engineers, from their production to requirements, the engineers should be brought from other disciplines. Alternatively, the entry seats should be enhanced or more of the institutions be open-ended.

The education of Agricultural and Biosystems Engineering in Ghana faces several challenges, including inadequate infrastructure, insufficient funding, shortage of qualified faculty, limited industry collaboration, limited research and innovation, lack of awareness, and brain drain. Addressing these challenges will require concerted efforts from government, academic institutions, industry players, and other stakeholders to invest in infrastructure, increase funding, develop faculty members, foster industry collaboration, promote research and innovation, raise awareness, and create better opportunities for skilled professionals in the field.

Agricultural and Biosystems Engineering education in Ghana provides graduates with diverse and exciting career opportunities in the agricultural industry. With the skills and knowledge gained from this programme, one can make significant contributions to improving the efficiency, productivity, and sustainability of the agricultural sector in Ghana.

Agricultural and Biosystems Engineering education plays a crucial role in the development of Ghana's agricultural sector and the country's overall development. By equipping students with the knowledge and skills to design, develop, and implement new technologies and techniques, Agricultural and Biosystems Engineering education helps to increase agricultural productivity, enhance food security, reduce postharvest losses, enhance environmental sustainability, create employment opportunities, and foster innovation in the agricultural sector in Ghana.

Agricultural and Biosystems Engineering education in Ghana has a bright future, with many opportunities for growth and development. The sector is essential in ensuring food security and sustainability, and graduates of the programme will play a vital role in shaping the future of agriculture in Ghana.

The education of Agricultural and Biosystems Engineering in Ghana can be advanced by investing in infrastructure, collaborating with industry, developing faculty members, funding research and innovation, collaborating with other countries, and promoting the programme to attract more students. These measures will help to enhance the quality of education and research and promote the growth and development of the agricultural sector in Ghana.

The education of Agricultural and Biosystems Engineering in Ghana can be enhanced using various tools, including computer-aided design software, GIS software, simulation software, laboratory equipment, mobile applications, and online resources. These tools can help to provide students with a comprehensive and practical understanding of agricultural engineering and prepare them for the challenges and opportunities of the future.

REFERENCES

Babatope, A. A., Samuel, T. M., Ajewole, P. I., & Anyanwu, O. M. (2020). Competence-driven engineering education: A case for T-shaped engineers and teachers. *International Journal of Evaluation and Research in Education*, *9*(1), 32–38.

Bobobee, E.Y.H., Addo, A., Asante, E.A., & Bessah, E. (2021). Agricultural engineering education in Ghana. In *Agricultural Engineering in Africa – A Key Driver for Transforming Agriculture to Deliver Food Security and to Support Economic Prosperity* (p. 79). PASAE Publication.

Clark, N. (2002). Innovation systems, institutional change and the new knowledge market: Implications for third world agricultural development. *Economics of Innovation and New Technology*, *11*(4–5), 353–368.

Diao, X., Hazell, P., Kolavalli, S., & Resnick, D. (2019). *Ghana's Economic and Agricultural Transformation: Past Performance and Future Prospects* (p. 320). New York, NY: Oxford University Press.

Field, H. L., & Solie, J. B. (2008). *Introduction to Agricultural Engineering Technology: A Problem-Solving Approach* (3rd ed.). New York: Springer.

Johnson, T. (2012). Biological engineering: What it means to me. *Resource*, *19*(4), 14–15.

Meinke, H., Howden, S. M., Struik, P. C., Nelson, R., Rodriguez, D., & Chapman, S. C. (2009). Adaptation science for agriculture and natural resource management—Urgency and theoretical basis. *Current Opinion in Environmental Sustainability*, *1*(1), 69–76.

Ting, K. C. (2009). Development and perspectives of agricultural engineering towards biological/biosystems engineering. In *Proceedings 2009 AIIA Conference* (pp. 12–16). Ischia Porto: Italian Society of Agricultural Engineering (AIIA).

Ulloa, J. G. (2021). *Applied Biomedical Engineering Using Artificial Intelligence and Cognitive Models*. New York: Elsevier.

Yang, X., Medford, J. I., Markel, K., Shih, P. M., De Paoli, H. C., Trinh, C. T., … Tuskan, G. A. (2020). Plant biosystems design research roadmap 1.0. *BioDesign Research*, *2020*, 8051764. DOI: 10.34133/2020/8051764.

24

ABE@Illinois

K.C. Ting and Ronaldo Maghirang

Agriculture and Food – A Large-Scale Bio-Based Economic Engine

The agriculture and food sector is a large-scale economic engine that can be characterised as systems of systems. Its many complex constituent systems are locally operated and globally connected. A top-level abstraction of agriculture and food systems is shown in Figure 24.1. This diagram depicts the linearity of resource use and waste disposal throughout the value chain, as commonly seen in today's commercial activities. Agriculture-based materials are moved through the value-chain stages of production, harvest/storage, processing/packaging, sales, and consumption, with additional resource inputs going into and unused discarded wastes coming out of each value-chain stage in a relatively independent fashion. In contrast, a circular agriculture and food systems diagram is shown in Figure 24.2. There is a renewed interest among the practitioners to increase the circularity to make agriculture and food systems more sustainable. Potential opportunities for circularity improvement include minimising system inputs, effectively transforming resources into usable products at every value-chain stage, enhancing loop-closing through waste management and resource recovery, intensifying reuse of materials and energy within each stage and the entirety of the value-chain, minimising net system outputs, and regenerating the environment external to the system. In this sense, the agriculture and food systems of systems need to be broadly defined to also include energy, water, environment, climate change, economics, society, etc. (American Society of Agricultural and Biological Engineers [ASABE], 2021; Ellen Macarthur Foundation, 2019; National Research Council, 2015; National Academies of Sciences, Engineering, and Medicine [NASEM], 2019).

Our society is facing grand challenges in terms of food security/safety, energy sufficiency, environmental sustainability, and family/community resiliency. These challenges are the causes to many multi-faceted and multi-scale problems. The key challenges surrounding the improvement of the

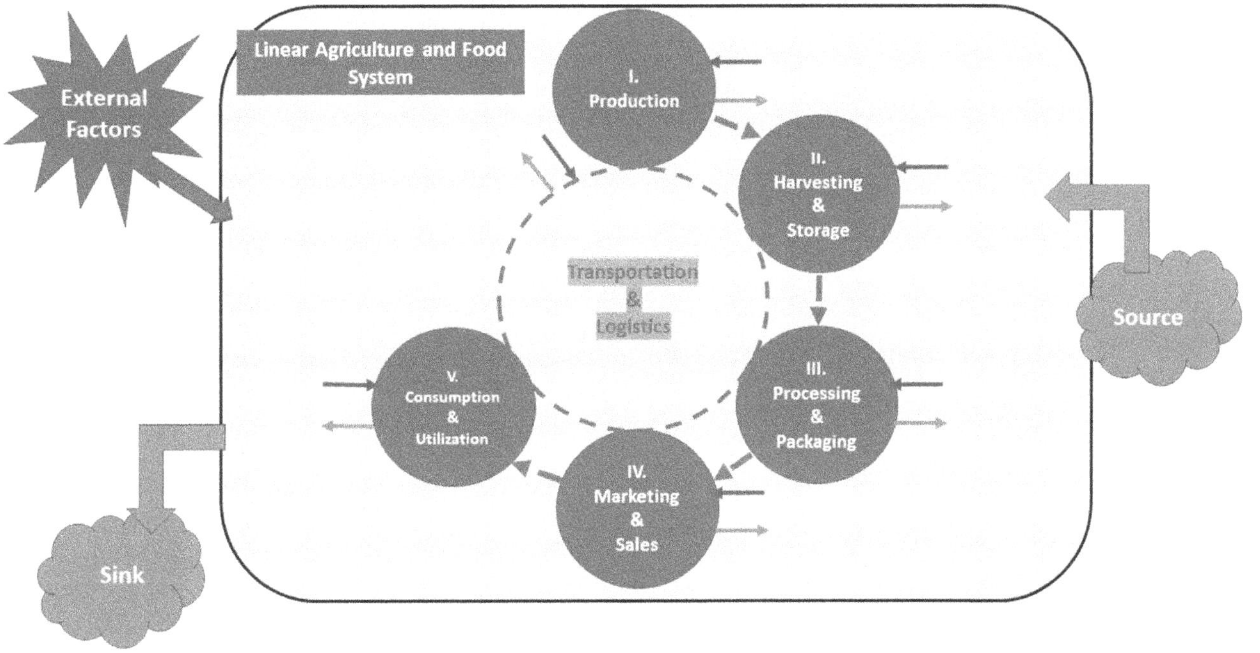

FIGURE 24.1 Concept diagram of linear agriculture and food systems.

DOI: 10.1201/9780429150111-26

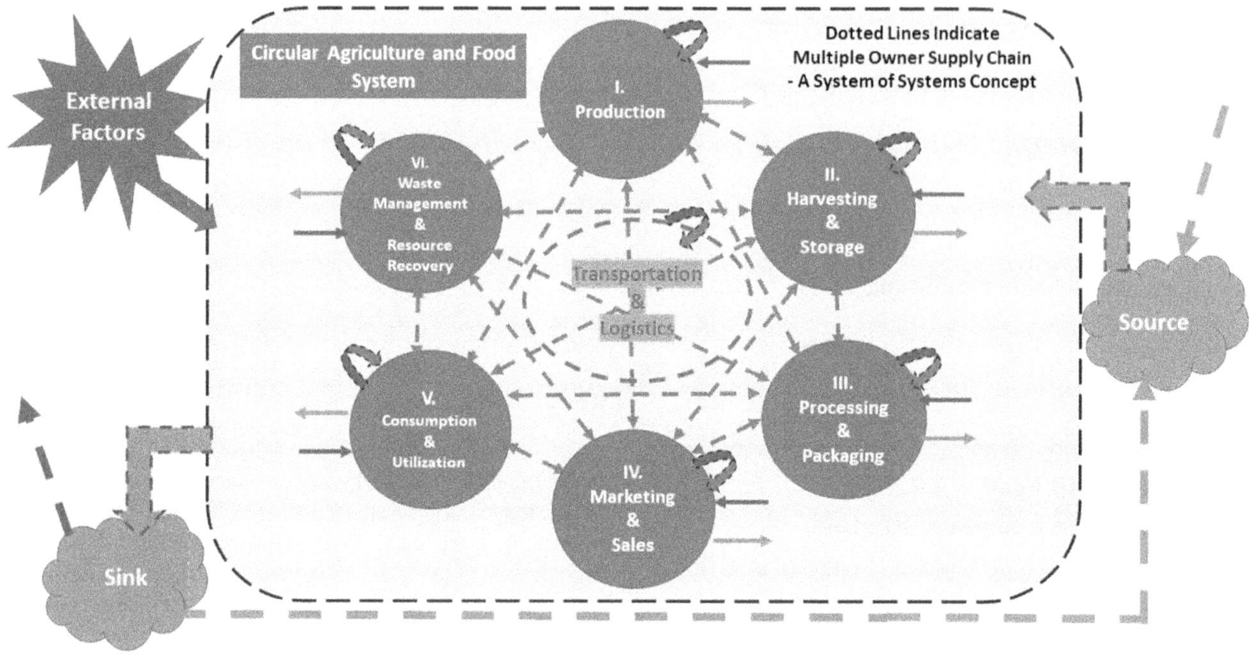

FIGURE 24.2 Concept diagram of circular agriculture and food systems.

effectiveness and competitiveness of large-scale agriculture and food systems of systems are to:

- Minimise environmental impact
- Optimise economic return
- Provide management capabilities
- Monitor and ensure intelligent resource use
- Understand governing constraints
- Enable creative productivity
- Interface with other economic sectors
- Identify value-added opportunities
- Create wealth and generate jobs

The U.S. and many other parts of the world have made great strides in addressing domestic and global agriculture and food production and provision by advancing knowledge and related educational activities. The work will need to continue and go beyond addressing isolated needs to integrate knowledge to maximise its impact. This integration of knowledge can be achieved by deploying a science-based, grand-scale systems-of-systems approach to provide a roadmap for solution-based decision support accessible to multi-sector collaborators across the globe. Agricultural and biological engineering, and closely related disciplines, will be major contributors to this locally operated and globally connected effort.

Changes in "Agri-xxx and/or Bio-yyy Engineering", Commonly Referred to as ABE

The "Agricultural Engineering" discipline has changed rapidly in the past 30 years. The change has been viewed as inevitable and necessary by the leaders and practitioners in this disciplinary area. Most changes have occurred in academic units on university campuses and in professional societies, including the American Society of Agricultural and Biological Engineers (ASABE). At this time, no academic department in the U.S. carries the simple name of "Agricultural Engineering." Almost all the academic departments contain the term "biological" or the prefix "bio." The changes, especially in the subjects of research activities and contents of educational programs, are still on-going and in large scale. The process of change has brought about both new opportunities and expected challenges.

In the past, the agricultural engineering discipline was best described by four distinct sub-areas: Power & Machinery, Electrical Power & Processing, Soil and Water, and Structures & Environment. The recent emphases are biology, environment, information, food, resources, automation, systems, and nanotechnology. While these new emphases will continue, visionary leaders in the discipline have started addressing the needs of flexibility, globalisation, customer-centered approach and satisfaction, multidisciplinary integration, life-long learning, team building/collaboration, and leadership development.

The factors and environments that influence the change have been:

- Clientele and stakeholders, including students of diverse backgrounds and with broad interests.
- Marketplace and industries
- Land-Grant missions (in the U.S.) and government policies
- Accreditation requirements for undergraduate engineering programs
- Professional engineering licensing

- University and college strategic plans
- Accountability and reward systems
- Resource availability and funding decisions
- Centralised decision-making versus grassroot empowerment
- Curriculum and faculty development

The examples of changes that have occurred are as follows:

- Academic department name
- Curriculum
- Faculty specialty
- Student recruitment
- Professional affiliation
- Research and professional interests
- Funding model
- Collaboration and partnership

Although the academic department names have many variations that include some forms of "Agri-xxx and/or Bio-yyy Engineering," they are commonly referred to as "ABE" or "BAE" in an aggregated fashion. The Department of Agricultural and Biological Engineering at University of Illinois at Urbana-Champaign (ABE@Illinois) has experienced all the changes shown above within the past years. In 2005, the membership of American Society of Agricultural Engineers (ASAE) voted to change its name to American Society of Agricultural and Biological Engineers (ASABE). Coincidently in the same year, ASAE P-210 (currently ASABE EOPD-210), a committee consisting of academic programme administrators, held a special meeting during the society's annual international meeting to assess the impact of proliferation of academic department and degree names. Six key issues were identified: (1) degree programme curricula and their accreditation, (2) student recruitment, (3) industry recognition and placement of graduates, (4) differentiation from bioengineering, (5) consistency between the degree programme names and the ABET accreditation criteria for "Agricultural Engineering" and "Biological Engineering," and (6) influence of local (campus level) politics. A recommendation was made at the conclusion of the meeting to name department and degree programme names as "Agricultural Engineering" or "Biological Engineering" according to the criteria used for their ABET accreditations.

Recent ASABE Initiatives

At the turn of the 21st century, the U.S. National Academy of Engineering identified the greatest engineering achievements of the 20th century as (Wulf, 2000)

1. Electrification
2. Automobile
3. Airplane
4. Water supply and distribution
5. Electronics
6. Radio and television
7. Agricultural mechanisation
8. Computers
9. Telephone
10. Air conditioning and refrigeration
11. Highways
12. Spacecraft
13. Internet
14. Imaging
15. Household appliances
16. Health technologies
17. Petroleum and petrochemical technologies
18. Laser and fiber Optics
19. Nuclear technologies
20. High-performance materials

Agricultural and biological engineers have been contributing in the development and implementation of many of the above engineering achievements, especially as they apply to agriculture, food, environment, water, and energy domains.

At that same time, ASABE also selected 15 top achievements, closely related to its disciplinary areas, from the 20th Century, such as:

1. Agricultural tractor
2. Rural electrification
3. Self-propelled combine
4. Milking machine
5. Cotton picker
6. Conservation tillage
7. Agricultural electronics
8. Center pivot irrigation
9. Precision farming (GPS)
10. Farm safety
11. Seeding technology
12. Biofuels feedstock
13. Drip & trickle irrigation
14. ASABE standards
15. Refrigerated on-farm milk

ASABE Global Initiative

In 2014, ASABE embarked on a Global Initiative Visioning process. The outcome of the initiative has been summarised in a white paper entitled "Global Partnerships for Global Solutions: An Agricultural and Biological Engineering Global Initiative." The key areas, all in the context of climate change and sustainability, for global collaboration were identified as

1. Improve food productivity
2. Reduce food losses and waste

3. Enhance energy conservation and efficiency
4. Develop adaptable renewable energy systems
5. Improve water availability, conservation, and efficient use
6. Provide clean water for multiple uses (human consumption, agriculture, recreation, ecosystems services, biodiversity, etc.)

As part of the actions of this initiative, three global conferences have been held on the themes of food *("Engineering and Technology Innovation for Global Food Security," October 24-27, 2016, Stellenbosch, South Africa)*, water *("Global Water Security Conference for Agriculture and Natural Resources," October 3-6, 2018, Hyderabad, Telangana, India)*, and energy *("Sustainable Energy for a Sustainable Future," October 24-27, 2022, San Jose, Costa Rica)*. All these activities have strong influence in the current and future development of ABE, locally and globally.

ASABE Circular Bioeconomy Systems Initiative

In 2020, ASABE initiated a Roundtable 2020 on "Transforming toward Circular Food and Agriculture Systems." Subsequently, a Task Force was appointed in September 2021 to address Circular Bioeconomy Systems (CBS) and to guide the development of ASABE's circular economy concepts applied to food, fiber, environment, and energy systems. The Task Force was charged with 1) increasing awareness and engagement in CBS across ASABE communities, 2) establishing strong connections with other engineering and science disciplines and stakeholders, and 3) recommending a structure in ASABE for the long-term engagement of our society in applying circular economy concepts. The CBS Initiative has generated highly impactful outcomes and the activities continue to evolve. Detailed information on the CBS Initiative may be found at www.asabe.org.

Domains and Core Competencies of the ABE Academic/Professional Discipline

Like many other university academic units, ASABE@Illinois stakeholders, including faculty, staff, students, external advisors, and partners, have deliberated over the years the department's sustainability and competitiveness. The overarching value proposition of ABE has been to "Make agriculture and food systems work well, better, together, smarter, and wiser to accomplish their purposes" as depicted in Figure 24.3. Accordingly, the vision story of ABE@Illinois is "Creating a world with abundant food, energy, water, and a healthy environment."

ABE focuses on the enhancement of complex living systems in agriculture, food, environment, and energy. The ABE@Illinois undergraduate degree programs focus on the biological interfaces of plants, animals, soils, microorganisms, and humans with the design and management of environments, machines, mechanisms, processes, structures, and computer models. The core competencies (i.e., the fundamental and essential disciplinary knowledge and skills of graduates) consist of automation, culture (i.e., biological sciences), environment, and systems (Figure 24.4). Included in the core competencies are soft skills, understanding of sustainable business models, development of human capital, and life-long learning. As of the planning horizon of 2015–2020, the application domains (i.e., the subjects of disciplinary relevance and impact) include Agricultural and Biological Systems and Technology (*including precision and information agriculture, plant and animal production, sustainable intensification, and health and safety*); Food and Bioproducts (*including processing and manufacturing, and security and safety*); Energy (*including renewable energy and energy efficiency*); Water (*including land and water resources, and water quality and use*); Environment (*including air, soil, and water quality and built environment*); and Biological Engineering (*including biotechnology, biosensing, big data, informatics, and analytics*).

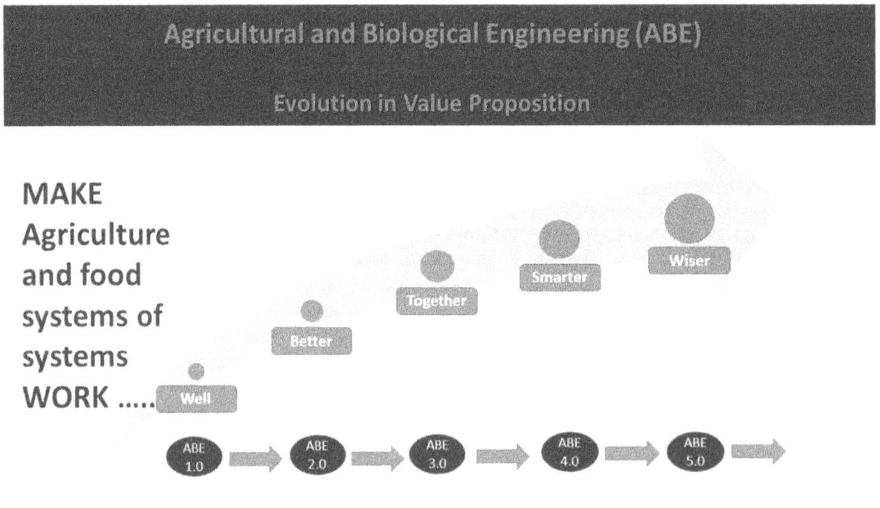

FIGURE 24.3 Evolution of Agricultural and Biological Engineering value proposition.

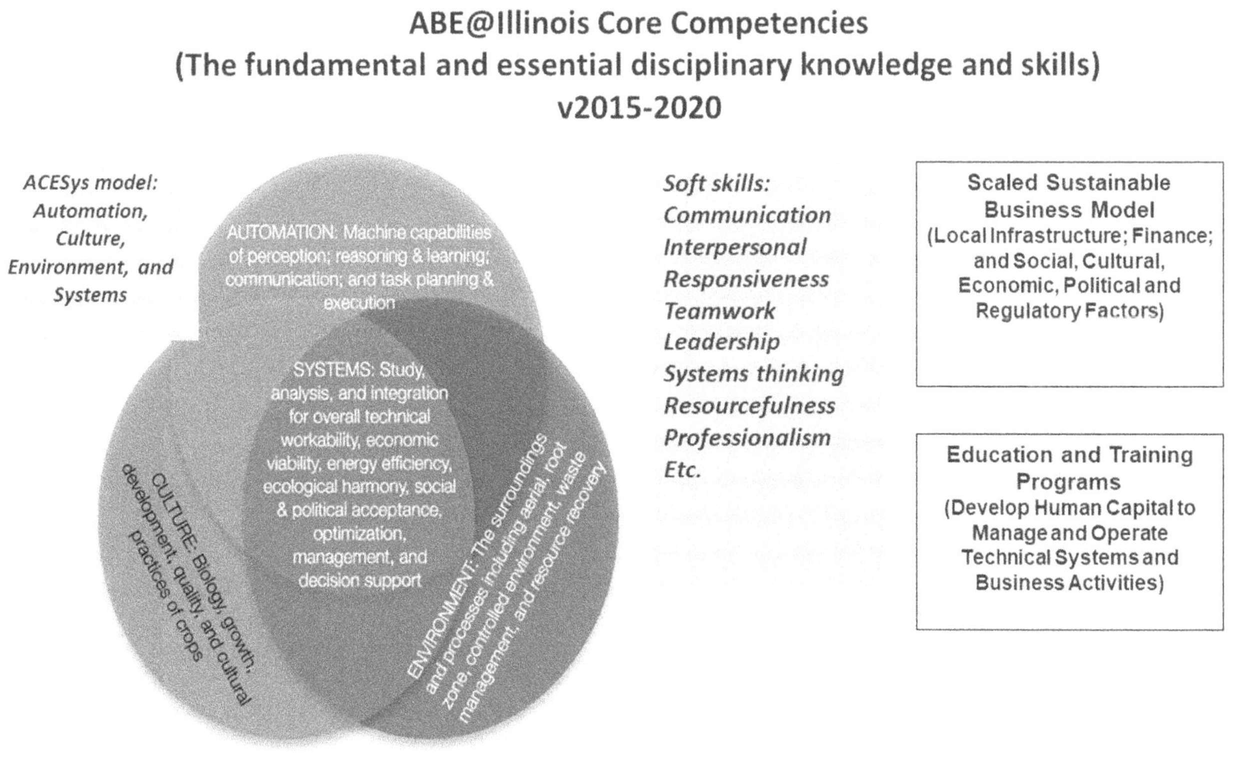

FIGURE 24.4 ABE@Illinois core technical competencies and soft skills.

Agricultural mechanisation was among the 20th century's greatest engineering achievements ranked by the National Academy of Engineering. It is among the long-term emphases of agricultural engineering, together with structures and environment, soil and water, and electric power and processing. The work in these subject areas has made impressive impacts to the agriculture and food sectors' productivity and environmental stewardship in many parts of the world. Based on the achievements in these traditional areas, ABE@Illinois has been evolving towards bio-based, intelligence empowered, and systems-scoped engineering to guide our development into the foreseeable future.

ABE discipline's contribution to the ever evolving agriculture and food systems of systems needs to be more forward thinking and take a more holistic approach. ABE@Illinois has continuously been in the process of strategic visioning. Being an integral part of a Land-Grant campus, our faculty continue to contribute to the university's teaching, research, and extension functions, as well as professional and public service (including economic development) responsibilities. Engineering emphasises design processes to deliver operational effects that satisfy required functions and purposes. Traditionally, ABE discipline has been to design and implement operations that involve biological processes; i.e. the concept of "bringing engineering to life." Biological and life sciences have, therefore, been viewed as requirements (aka constraints) in the design processes. The new vision of ABE is also to proactively incorporate life sciences into engineering in designing bio-based systems; i.e. using life sciences as resources for engineering work (that is, therefore, the concept of "bringing life to engineering"). It is important to recognise that ABE's overarching goal is to enhance lives including human, plants, animals, and microorganisms. Contributing to the ABE domains requires knowledge and skills in the subject areas of automation, culture, environment, and systems (the ACESys paradigm shown in Figure 24.4). The core competencies of ACESys are described as follows (Ting, 2013):

> "Automation" deals with information processing and task execution related to a system's operation. The purpose of automation is to equip engineering systems with human-like capabilities of perception, reasoning/learning, communication, and task planning/execution. Commonly seen automation topics are instrumentation, control, computerisation, mechanisation, modeling, machine vision, robotics, artificial intelligence, etc.
>
> "Culture" includes the factors and practices that can directly describe and/or modify the growth and development of biological objects. The cultural factors, such as morphological and physiological conditions as well as genetic expressions are important in monitoring growth, development, and functions of biological objects. The cultural practices may include operations which directly alter biological states and activities.
>
> "Environment" encompasses the surroundings and processes of biological objects, which consist of climatic and nutritional, as well as structural/mechanical conditions. Understanding and control of environmental factors has been perceived as a major engineering challenge in agricultural production and bio-processing.

"Systems" analysis and integration is a methodology that starts with the definition of a system and its goals, and leads to the conclusion regarding the system's workability, productivity, reliability, and other performance indicators for decision support purposes. The success of systems analysis relies on the effective use of intelligence, including data, information, knowledge, and wisdom. Two key resources in systems analysis are: (1) intelligence about individual components and/or constituent systems, as well as their interrelationships and (2) methods of information gathering and processing for creating value-added intelligence.

ABE@Illinois Educational Programs (As of 2023)

ABE@Illinois delivers outstanding curricular and co-curricular educational opportunities that reflect and respect individual and programmatic needs and perspectives. The department offers undergraduate and graduate programs in ABE and Engineering Technology and Management for Agricultural Systems (ETMAS). The ABE undergraduate programme equips graduates in addressing grand challenges related to food, water, energy, and the environment. The programme is ABET-accredited under both the agricultural engineering and biological engineering programme criteria. The curriculum is flexible, allowing students to tailor their education to their interests in developing, designing, manufacturing, testing, and managing agricultural and biological engineering systems in the domain areas of Bioprocess Engineering and Industrial Biotechnology, Soil and Water Resources Engineering, Off-highway Equipment Engineering, Renewable Energy Systems Engineering, and Sustainable Ecosystem and Environment Engineering.

In response to the changing and emerging needs of industry, society, and students, the curriculum is currently being transformed to include a strong grounding in computing and data science.

The ETMAS undergraduate programme prepares graduates for careers in the management and application of engineering technologies for agricultural, construction, renewable energy, food, and environmental systems. The curriculum is also flexible, allowing students to select from several specialisations, including the following:

- Agricultural production and processing: Apply engineering and management principles to agricultural production and food and feed processing systems.
- Construction management: Plan and manage construction projects for residential, agricultural, commercial, or industrial sectors.
- Digital and precision agriculture: Apply and manage advanced engineering technologies in production agriculture.
- Energy and the environment: Develop and manage projects involving the application of renewable energy systems or environmental sustainability.

The department provides ample opportunities for high-impact experiential learning experiences for both programs. For example, the department offers a senior capstone course for each programme that allows students to partner with a client to solve real-world problems. Additionally, many students complete an internship or co-op and many participate in hands-on undergraduate research.

The department also places a high priority on graduate education on the application of engineering principles to solve problems of agricultural production, utilisation, environmental control, and biological systems and to improve the quality of life. The department offers the following programs:

- Master of Science (MS) and Doctor of Philosophy (PhD) in ABE
- MS in ETMAS and Professional Science Master's (PSM) in ETMAS
- PSM in Bioprocessing and Bioenergy
- Master of Engineering in Digital Agriculture (online)

The MS programme in ABE is research-based and is designed to prepare students for advanced positions in the agricultural and biological engineering industry, consulting, and government, as well as for further graduate studies. The PhD degree in ABE is research-oriented programme designed for students who wish to pursue a career in academia or research.

The graduate degree programme in ETMAS combines the knowledge of agricultural, biological, and physical sciences with managerial and technical skills. Students focus on applying engineering principles, studying the technologies used in agriculture, and integrating concepts in the food and agricultural industries. The ETMAS graduate programme has two paths: (1) MS in ETMAS and (2) MS with PSM in ETMAS. The PSM programme is full-time, does not require a thesis, and is generally completed in 18 months.

The department through the Integrated Bioprocessing Research Laboratory (IBRL) also offers the PSM in Bioprocessing and Bioenergy. It is a business-integrated programme designed to prepare learners for lucrative careers in the bioenergy life cycle - from feedstock supply to biofuel transportation and distribution.

Starting Spring 2023, the department offers a Master of Engineering with concentration in digital agriculture. It is an interdisciplinary programme from the Center of Digital Agriculture and the Departments of Agricultural and Biological Engineering, Crop Sciences, and Computer Science. Students take classes from all three departments, and they will complete a capstone course working with industry partners on real-world problems. The programme is fully online and can be completed in 12 months with full-time study or taken part time.

ABE@Illinois Research Programs (As of 2023)

ABE@Illinois conducts mission-driven research to address challenges related to food, agriculture, and the environment. Current major research areas include artificial intelligence

(AI) for agricultural systems, sustainable bioprocessing and industrial biotechnology, hydrothermal liquefaction (HTL) of wet biomass and biowaste into crude oil, photosynthesis engineering, agricultural and industrial safety and health, off-highway equipment engineering, and soil and water resources engineering. Brief descriptions of these areas are below. Major state-of-the-art facilities include the Integrated Bioprocessing Research Laboratory (https://ibrl.aces.illinois.edu/), Illinois Autonomous Farm (https://digitalag.illinois.edu/autonomous-farm/), Distributed Autonomous Systems Laboratory (http://daslab.illinois.edu/), and Hydrothermal Liquefaction Laboratory.

- AI for food and agricultural systems: Researchers in the department are working on creating new technologies for robotics and AI to address critical labor challenges facing agriculture today. Recent successes include the creation of phenotyping robots, and ongoing projects span from soft robotics to health and safety aspects of robotics. These efforts include the creation of the Illinois Autonomous Farm as a collaboration between ABE and the Center for Digital Agriculture.
- Sustainable bioprocessing and industrial biotechnology: Researchers are discovering and optimising products for use as food, feed, and energy, involving biofuels production bioprocessing technologies, quality and value of bioproducts, recovery of nutrient and nutraceutical.
- Hydrothermal liquefaction (HTL) of biowaste to crude oil: Wet biomass is converted into biocrude oil, using thermochemical conversion technology, at elevated temperatures and pressures. The focuses of the HTL lab are on the conversion of animal manure, algae, food waste, and crop residues followed by the subsequent upgrading involving distillation, catalytic hydrotreating, and esterification to yield high-value oil product including bio-kerosene as a jet fuel.
- Agricultural and industrial safety and health: Researchers work on the protection of agricultural workforce's safety and health by exploring the association between stress and mental health and identifying risk and protective factors for mental illness.
- Off-highway equipment engineering: Researchers are working to innovate and create equipment, control systems and machinery for the agricultural, construction, and mining industries with specific focuses on machine vision systems, data acquisition systems, unmanned aerial systems, risk management, and precision agriculture.
- Soil and water resources engineering: Food security, economic growth, and ecosystem sustainability are interwoven to the use and conservation of soil and water resources. Current research focuses are on water drainage and erosion control, conservation best management practices, ecosystem services, global changes and their impact on water resources and ecosystem dynamics, nutrients, sediment and pathogen transport, and cyber-informatics infrastructure and computational tools.
- Photosynthesis engineering: Research focuses are on engineering plants with improved photosynthesis by integrating genomics, proteomics, and synthetic biology. Specifically, natural diversity in photosynthetic enzymes from non-model and emerging model species is investigated to source materials for manipulating energy harvesting and dissipation in crops.

ABE@Illinois Extension/Outreach Programs (As of 2023)

The department is committed to improving food and energy supplies on a global scale. Locally, we facilitate this effort through collaboration with the Illinois Extension outreach. Our Extension faculty and specialists work with farmers to improve livestock waste management systems, air-quality control in livestock facilities, farm safety, drainage, and chemical applications. One of the major areas is the Agricultural Safety and Health programme, which seeks to provide research-based information and resources that assist individuals and organisations in the understanding and prevention of injury and illness while working in and around Illinois agriculture. Additionally, agricultural work includes a number of unique stressors, such as impacts of weather conditions and events, financial concerns about commodity prices, loan interest rates, and managing the farm business, long work hours, working with heavy machinery, and at times, working closely with family. Stress among people in agriculture is important because it can contribute to negative mental health outcomes, and people in agriculture have higher rates of depression, anxiety, suicidal ideation, and deaths by suicide compared to the general population. Faculty in ABE offer several programs to help people who work in agriculture to learn more about stress and stress management, as well as programs for agricultural community members to learn more about farm stress, mental health, and how to help someone who is struggling or in crisis.

ABE@Illinois Economic Development Engagements (As of 2023)

Research in the department drives innovation in various areas, including agricultural operations, industrial biotechnology, energy from biowaste, etc., that spurs economic growth. For example, the Integrated Bioprocessing Research Laboratory (IBRL) in the department is a bioprocessing product research and development facility that bridges the gap from basic discovery to commercialisation of bioproducts. The facility serves as an invaluable asset to bioenergy and bioprocessing companies by partnering them with the human and equipment capabilities of the university. Since IBRL's opening in 2018, it has profoundly impacted industrial bioprocessing by researching and derisking more than 600 projects with more than 80 industrial partners to develop new starches, low caloric sugars, food coloring pigments, alternative food proteins, cosmetics, and biofuels, as well as optimising unit operations related to milling, dewatering, separations, and fermentation technologies. The work at IBRL has directly translated to hundreds of millions of dollars in economic activity in the US.

Visioning and Strategic Framing Processes

Strategic vision and planning are important processes in charting future directions and identifying critical actions in advancing the goals of an organisation. Typical and effective visioning processes and outcomes in developing strategic plans are illustrated in this section.

The Processes

1. Prepare for a brainstorming session
    - Determine when and where
    - Select participants (facilitator, support staff, integration leaders, discussion leaders, discussants)
    - Perform SWOT (Strengths, Weaknesses, Opportunities, and Threats) analysis
    - Distribute the result of SWOT analysis
    - Formulate and distribute brainstorming meeting agenda
    - Arrange logistics (meeting rooms, audio/visual equipment, name tags, refreshments, lunches, etc.)
2. Conduct brainstorming meeting
3. Summarise information gathered during the brainstorming meeting
4. Prepare a draft report
5. Review and revise the draft report with an executive team
6. Distribute the draft report to the SWOT analysis and brainstorming meeting participants and request input
7. Finalise the draft report into a strategic plan by taking the input into consideration
8. Prepare and distribute various versions of the strategic plan for internal and external communications
9. Implement the strategic actions
10. Monitor, evaluate, and modify the strategic intent

Example SWOT Analysis Questions

1. What are the organisation's major (internal) strengths?
2. What are the organisation's major (internal) weaknesses?
3. What are the organisation's major (external) opportunities?
4. What are the organisation's major (external) threats?
5. What are the organisation's clienteles and/or markets now served? Are there additional clienteles and/or markets that should be served? If so, which and why?
6. What must the organisation do to improve its productivity, sustainability, and competitiveness?

Example SWOT Analysis Results

Strengths:

- High quality people
- Participating in two top-ranked colleges
- Quality of teaching and care of the students
- International presence and reputation
- Good research and teaching capabilities
- Systems approach to solving complex problems
- Ability to solve difficult problems
- Ability to lead and participate in multidisciplinary activities
- Many promising "early" career faculty members

Weaknesses:

- Biological engineering components
- Practical understanding of industry needs
- Competitive grants and visibility at Federal level
- Funding in general (especially endowments)
- Gender diversity among faculty
- Student recruitment
- Lack of super-stars
- Lack of quality laboratory space
- Our organisational structure

Opportunities:

- Expanded biological aspect in our discipline
- Geographical location close to major related industries
- Funding for agriculture/biology/health/environment
- Encouragement for technology commercialisation and entrepreneurship
- Alumni relationship and support
- Multi-disciplinary, multi-institutional projects
- Faculty excellence hiring initiative
- International activities
- Distance learning and continuing education
- Demand for M.S. programs
- Recruitment of students from urban areas

Threats:

- Eroding completeness of core competencies due to a "shrinking" faculty size
- Lake of recognition for the importance of agriculture
- Lack of government support
- Low undergraduate student enrollment

- Lack of a clear guidance in balancing research/teaching/outreach responsibilities
- Ability (or inability) to sustain a high ranking

Example Brainstorming Meeting Agenda

1. Arrival – Refreshments
2. Opening Remarks and Charge to the Attendees
3. Strengths, Weaknesses, Opportunities, and Threats (SWOT) Analysis and current unit Values (What we believe in), Vision (What we aspire to be), Mission (What we do best for whom), and Domains (The subjects of disciplinary relevance and impact)
4. Discussing SWOT and updating unit Values, Vision, Mission, and Domains
   - discuss the SWOT analysis and to identify what is most apparent in each section, i.e. "What jumps out?"
   - discuss validity of Values, Vision, Mission and Domains – noting any significant reasons to change any of them
5. Goals (Where we are going to succeed) – breakout groups, report back, and prioritisation
   - breakout groups to each identify a certain number of primary goals
   - reports from breakout groups to consolidate a common set of goals and prioritise a certain number of top goals
6. Objectives (Sub-goals and Milestones) – breakout groups, report back, and prioritisation
   - breakout groups to each identify a certain number of objectives with specific targets/milestones (what and when) for each of the common goals
   - reports from breakout groups to identify a common set of objectives and prioritise a certain number of objectives for each common goal
7. Indicators of Success (How we are going to measure success) – breakout groups, report back, and prioritisation
   - breakout groups to each identify essential performance indicators for the identified goals and objectives (e.g., one per objective)
   - reports from breakout groups to define a common set of performance indicators and prioritise the key performance indicators
8. Actions (What things we need to make happen and get done) – breakout groups, report back, and prioritisation
   - breakout groups to each identify the essential actions to initiate, make progress, or accomplish objectives, (e.g., one per objective)
   - reports from breakout groups to define a common set of actions and prioritise actions
9. Revisit Values, Vision, Mission, Domains, Goals, Objectives, Indicators of Success, and Actions
   - review and modifications
10. Adjourn

Example Strategic Plans

Figure 24.5 shows an abbreviated version of ABE@Illinois Vision 2020 Strategic Plan published on September 21, 2015. A strategic plan is a living document that needs to be revisited and likely revised periodically.

Example Strategic Initiatives

In a comprehensive strategic plan, strategic initiatives are frequently stated in more detail. As an illustration, some example strategic initiatives are shown here:

The ABE department will continue to be globally preeminent and locally relevant in its research, teaching, outreach/extension, and economic development programs. The department has been recognised for constantly providing leadership in exploring new technical areas relevant to its discipline. It has also been valued as contributing members on interdisciplinary teams. In the next five years, the department will direct its effort towards the following technical areas. Each focus area incorporates systems management and safety/health dimensions.

Initiative 1: Agricultural Automation

Farming and related activities will continue to rely on intelligent machines to be productive and competitive. Automated systems are equipped with human-like intelligence and capabilities. Information, electrical/electronic, and mechanical technologies need to be integrated to create automated machines. Fundamental understanding and enabling technologies that are necessary for advancing the functions of agricultural automation are as follows:

- Perception: concept of measurement, sensors and sensing techniques, data collection devices, data processing methods/software, and analog and digital instrumentation.
- Reasoning and learning: mathematical, statistical, logical, and heuristic analysis of information for decision support, as well as data mining, modeling, simulation, intelligent inference and optimisation, and experiential learning by machines.
- Communication: real-time data sharing, wired and wireless transmission of data, networking, and cyberspace.
- Task planning and execution: process control, environmental control, materials handling, actuators, intelligent machines, and robotics.

There are challenges and opportunities for agricultural automation. The challenges are commonly known as to (1) make economic returns acceptable or even attractive to the investors, (2) decide the appropriate level of automation and the balance between fixed automation and flexible automation,

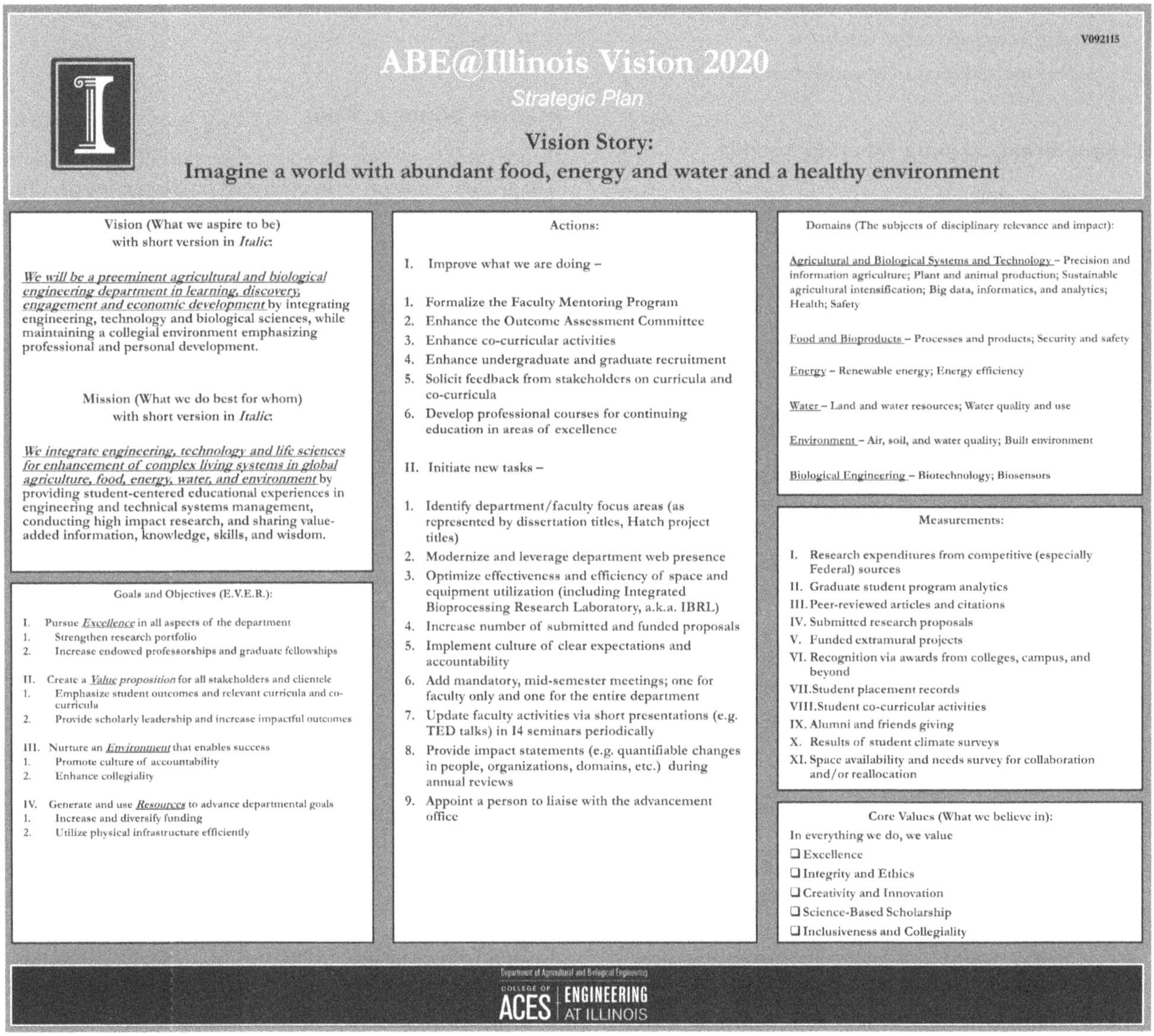
V092115

ABE@Illinois Vision 2020

Strategic Plan

Vision Story:

Imagine a world with abundant food, energy and water and a healthy environment

Vision (What we aspire to be)
with short version in *Italic*:

We will be a preeminent agricultural and biological engineering department in learning, discovery, engagement and economic development by integrating engineering, technology and biological sciences, while maintaining a collegial environment emphasizing professional and personal development.

Mission (What we do best for whom)
with short version in *Italic*:

We integrate engineering, technology and life sciences for enhancement of complex living systems in global agriculture, food, energy, water, and environment by providing student-centered educational experiences in engineering and technical systems management, conducting high impact research, and sharing value-added information, knowledge, skills, and wisdom.

Goals and Objectives (E.V.E.R.):

I. Pursue *Excellence* in all aspects of the department
1. Strengthen research portfolio
2. Increase endowed professorships and graduate fellowships

II. Create a *Value proposition* for all stakeholders and clientele
1. Emphasize student outcomes and relevant curricula and co-curricula
2. Provide scholarly leadership and increase impactful outcomes

III. Nurture an *Environment* that enables success
1. Promote culture of accountability
2. Enhance collegiality

IV. Generate and use *Resources* to advance departmental goals
1. Increase and diversify funding
2. Utilize physical infrastructure efficiently

Actions:

I. Improve what we are doing –

1. Formalize the Faculty Mentoring Program
2. Enhance the Outcome Assessment Committee
3. Enhance co-curricular activities
4. Enhance undergraduate and graduate recruitment
5. Solicit feedback from stakeholders on curricula and co-curricula
6. Develop professional courses for continuing education in areas of excellence

II. Initiate new tasks –

1. Identify department/faculty focus areas (as represented by dissertation titles, Hatch project titles)
2. Modernize and leverage department web presence
3. Optimize effectiveness and efficiency of space and equipment utilization (including Integrated Bioprocessing Research Laboratory, a.k.a. IBRL)
4. Increase number of submitted and funded proposals
5. Implement culture of clear expectations and accountability
6. Add mandatory, mid-semester meetings; one for faculty only and one for the entire department
7. Update faculty activities via short presentations (e.g. TED talks) in I4 seminars periodically
8. Provide impact statements (e.g. quantifiable changes in people, organizations, domains, etc.) during annual reviews
9. Appoint a person to liaise with the advancement office

Domains (The subjects of disciplinary relevance and impact):

Agricultural and Biological Systems and Technology – Precision and information agriculture; Plant and animal production; Sustainable agricultural intensification; Big data, informatics, and analytics; Health; Safety

Food and Bioproducts – Processes and products; Security and safety

Energy – Renewable energy; Energy efficiency

Water – Land and water resources; Water quality and use

Environment – Air, soil, and water quality; Built environment

Biological Engineering – Biotechnology; Biosensors

Measurements:

I. Research expenditures from competitive (especially Federal) sources
II. Graduate student program analytics
III. Peer-reviewed articles and citations
IV. Submitted research proposals
V. Funded extramural projects
VI. Recognition via awards from colleges, campus, and beyond
VII. Student placement records
VIII. Student co-curricular activities
IX. Alumni and friends giving
X. Results of student climate surveys
XI. Space availability and needs survey for collaboration and/or reallocation

Core Values (What we believe in):
In everything we do, we value
❑ Excellence
❑ Integrity and Ethics
❑ Creativity and Innovation
❑ Science-Based Scholarship
❑ Inclusiveness and Collegiality

Department of Agricultural and Biological Engineering
COLLEGE OF ACES | ENGINEERING AT ILLINOIS

FIGURE 24.5 Abbreviated ABE@Illinois vision 2020 strategic plan, version September 21, 2015.

(3) integrate automation with crop growth/development and cultural tasks, (4) consider the possibility of multiple use of the systems that enable automation, (5) increase and/or create market demand for the agricultural products, and (6) continuously improve research/development and education/training capabilities. The opportunities that could be explored to resolve these challenges are that (1) the technology readiness for agricultural automation is high, (2) the past success in agricultural mechanisation is impressive, (3) advances in other economic sectors, such as manufacturing, business, etc., are transferable to agriculture, (4) economic viability of automated systems is improving, (5) market demands for high and uniform quality agricultural products are increasing, and (6) the capacities to adapt/develop emerging technologies and create spin-off technologies are very much available. This is the opportune and exciting time to further the impact of automation to agriculture.

Initiative 2: Bio-Energy and Bio-Products

Bio-energy (and related bio-products) is one of the strategic areas of the campus. Our department has been a major leader/contributor in bio-energy research including ethanol production from corn, conversion of biomass to crude oil, and characteristics of biodiesel and their effects on engine performance.

Renewable energy produced from crops such as corn, soybeans and biomass is an extremely important and timely research need for our state and the U.S. Research is needed to modify the dry-grind corn process to increase efficiencies of conversions, to create distillers dried grain with solubles (DDGS) with higher protein quality and with lower fiber content, and to develop other co-products, such as phytosterols and corn fiber oil. Additional work is needed to reduce the required energy inputs in converting biological materials into ethanol and high valued co-products. Research efforts on bioprocessing should focus on (1) process design, (2) evaluation for process efficiency, and (3) evaluation of co-products for industrial and human and animal food uses.

Conversion of soybean oil and other renewable vegetable oils into biodiesel fuel is also emerging as a very important research area. The diesel engine is the primary power source for the transportation sector and, therefore, diesel fuel is vital to the economy. As much as alternatives are being sought for gasoline with ethanol being the forerunner, alternatives to petrodiesel must also be developed. Even though blends of biodiesel are already commercially available, research is needed to evaluate the compatibility of this renewable fuel with the engine. Opportunities exist to exploit the unique properties of these fuels by modifying the engine or optimising performance and reducing emissions when consuming such fuels. In addition, a wide array of renewable source materials are suitable for the production of biodiesel, which presents some challenges in terms of fuel quality, but also opportunities for genetically modifying the source materials to arrive at a renewable fuel source that has optimised properties for combustion, emissions reduction, stability, and operability.

Our Department is uniquely positioned to lead the development of technologies to economically produce bio-fuels and value-added byproducts from livestock manure and crop residues. Manure transportation, treatment, and use cost billions of dollars every year in the U.S. On the other hand, the need for bio-fuels is growing. Thermochemical conversion (TCC) is a mother nature's method for the formation of petroleum from deceased hydrocarbons under high temperature and pressure. The method can be engineered to provide solutions to the livestock problems.

Initiative 3: Sustainable Environment

Maintaining, and potentially improving, environmental quality and sustainability is a critical need for the future. While much is known about individual processes that are occurring and are influenced by our resource use and lifestyle decisions, less is known about interactions and system-wide processes. Our ability to assess the impact of resource use decisions, whether related to land, water, plant, animal or people, has greatly improved over the past few decades but, again, interactions, unforeseen impacts, and system-wide response are not nearly as well understood. Those shortcomings may lead us to make policy decisions that are appropriate at the individual process scale, but counterproductive when broader scale impact is evaluated. In essence, we need to enhance our abilities to understand and base decisions on processes and impact at the ecosystem scale.

- Information: One of the most important thrusts in the environmental area is the collection of data that can be used for planning and decision-making. Our ability to quickly assess impact and/or make environmentally beneficial decisions will depend upon the quantity and quality of information we have and the tools we have to consider those data separately and in combination. Resource data varies both spatially and temporally, so geospatial location of data and tools to utilise that information is a necessary tool for students, staff and faculty.
- Processes: The continual refinement of our understanding of the processes that influence environmental quality and sustainability is critical to the overall goal of an improved environment. As processes are identified and examined in one portion of the environment, they must be considered again in others to see if there are additional influences in different contexts. Additionally, the matter of scale again comes into play, as processes that are critical at a finer scale may have little observable impact at a broader scale. The "one-size-fits-all" approach is not likely to be applicable.
- Simulation: Given the potentially vast spatial and temporal nature of environmental processes, computer simulation models will continue to be important tools for prediction and assessment. However, each model requires extensive testing and refinement for the situation it was intended to represent. Unfortunately, models have been, and continue to be, used outside their originally intended application and context. We must continue to develop, refine, and verify simulation models that enhance our understanding.
- Socioeconomic considerations: As one might expect, resource use decisions are influenced by the knowledge, attitudes, values, and abilities of those making the decisions. It is critical that environmental process assessments that involve people making decisions consider socioeconomic characteristics and influences of the decision-makers.

Targeted areas of thrust in environmental sustainability research include improving and protecting environmental quality pertaining to surface and groundwater, air, and soil; developing technologies to provide productive environments for animals, workers, plants, and desirable microorganisms; and reducing the adverse environmental impact from biological production facilities.

Initiative 4: Biological Engineering

Traditionally, the field of agricultural engineering, like many other engineering disciplines, has been heavily based on physical and chemical sciences in its activities. However, unlike many other engineering disciplines, agricultural engineering has mainly been targeting its applications to bio-processing and bio-production. Biological science has been included in agricultural engineering work mostly as a "constraint." It is now an opportune time to proactively consider biological science as an integral part of our discipline and utilise it as a resource in our work. The integration of biological, physical, and chemical sciences with engineering and technology provides a powerful platform for addressing systems level issues relevant to agricultural and food production. The biological engineering programme in our department emphasises bioenvironmental, bioprocessing, bioresources, and biosystems issues as related to systems involving agriculture, food, environment, and energy.

The intersection between the life sciences and engineering sciences has created unprecedented opportunities and challenges in what is now considered Biological Engineering. There is much to be gained from applying engineering, to understand and control biological processes as well as to develop new engineering paradigms and technologies inspired by biological processes such as self-assembly and self-healing. We also recognise that the interface between engineering and biological systems is very broad and ranges from engineering at the nanometer scale (biotechnology and nanobiotechnology) to ecological and life support systems scales. Research and educational programs will be created in the following focus areas:

- Biological nanotechnology: Using nanoscale biological components such as proteins and nucleic acids to create devices or systems of devices that provide functions not natively seen in nature (e.g., biosensors, nanotherapeutics with functional biological nanocomponents etc).
- Programmable biotechnology: Using tools from computer sciences and molecular genetics to implement intelligent functionality and reasoning in living organisms at the molecular or genetic level. This field is also sometimes known as synthetic biology (e.g., whole-cell biosensors, metabolic engineering, tools for molecular biosciences etc.)
- Biological device design: Using living, biological components in an engineering design, be it at the nano scale or at the ecosystem scale, is accompanied with very specific challenges. We recognise that engineering with biological components requires the development of specifications on the allowable environmental states for the biological components. We will strive to elucidate and codify the principles behind engineering using biological components. This also involves the development of mathematical and information-based tools on the analysis and simulation of biological systems.

Initiative 5: Systems Informatics and Analysis

Accessing information has become less of a challenge today compared to the past. The available quantity of information is overwhelming; however, much more will continue to be needed and created. The quality of information is equally important. Furthermore, how the information is generated and used has strong impacts to its applicability and usefulness. Therefore, a real challenge is how to create, gather, and synthesise information for beneficial use.

Systems analysis is a creative and critical thinking approach that facilities scientific information generation, synthesis, and application for problem solving. The purpose is to understand the holistic performance of a system consisting of interrelated components united to operate in specified ways and/or achieve certain goals.

The overall objective of this initiative is to create necessary infrastructure to allow the analysis, optimisation, and control to occur for emerging biological systems. Thus, for the next five years the emphases will be:

- *Complex design techniques* will be developed to understand the interactions in biological systems that make them inherently variable, though resilient in nature.
- *Decision support* will be provided to bring the results of modeling and analysis research on systems ready for delivery to the user community in a means easily tractable and transparent enough for easy implementation.
- *Early reliability measurement techniques* will be made available for a wide array of applications where the feasibility of life testing may not be readily available due to cost related, ethical, or time constraints.
- *Holistic agro-ecosystem Perspectives* will be investigated to enhance the study of such systems at the ecosystem scale, especially as they become more complex.
- *Multi-scale modeling* will be conducted to provide the fluid motion of data and information from models describing activity across a wide array of detail spanning from the nano- to the ecosystem-scales.
- *Sustainable development* will be bolstered by the consideration of systems from a holistic perspective and the use of innovative control and decision-making techniques, especially in areas where constraints are tight.

Example Teaching Resource Retreat Action Items

Retreats conducted to identify opportunities for improving academic degree programs are important planning activities. Some examples of goals and action items identified during departmental undergraduate teaching resource retreats are:

Goal 1: Continue our student recruiting and retention efforts

- Better internal communications among advisory groups by creating an advisors' outreach component to the department's Recruiting Committee with the guidance of the undergraduate programme coordinator.
- Enhance departmental staff/faculty participation and involvement in both recruiting & retention activities, including faculty extracurricular involvement and student clubs.
- Continue existing recruitment plan and work on methods to enhance such as home-town recruiting & career-based approaches.

Goal 2: Recruit students in other majors to take our existing courses

- Identify the means by which the department's courses can be listed as technical electives in other degree programs

- Review existing minors and identify opportunities for new minors

Goal 3: Add more sections to the existing courses that have attracted high college and campus wide interests

- Restructure our curricula, including eliminating courses with low enrolment.

Goal 4: Create new courses that will attract wide student interest

- Create broader appeal

Goal 5: Adjust existing courses to meet general education requirements

- Review/survey/analyze how our students currently satisfy their general education requirements and tie these results to curricula streamlining

Goal 6: Publicise, aggressively, potentially popular courses

- Obtain email lists of advisors in other degree programs, and use them to disseminate appropriate course information
- Consider means of rewarding or recognising faculty involved in large instructional unit courses
- Create and/or improve our web presence for each course in our programs
- Develop department's teaching themes message

ABE in the Global Environment - Local Relevance and Global Preeminence

How to Learn and Succeed Globally

There are many signs that we are living in a global environment, now and in the future. We are immersed in many surroundings and activities of global nature in our daily life, including media, entertainment, education, trade, commercial products, restaurants and supermarkets, community, customer service, transportation, travel, research, science/engineering/technology, political/social/economical issues, diplomacy, ecology, etc. It is, therefore, important to provide holistic educational experience to students including international engagement and activities. The benefits are many, including experiencing world class education, appreciating diverse cultures, understanding issues of global impact, learning and practicing foreign languages, enjoying new and fun ways of learning, integrating information and knowledge, establishing global partnership, becoming a systems thinker, and gaining personal and professional confidence. On the other hand, there are potential disadvantages from not having international engagement, such as lack of international experience in personal and professional life, not having the advantage of accessing global resources, potentially having narrow view on issues, missing a life-changing experience, not getting a complete educational and professional experience while studying and working in university, may not be able to operate in a global environment, and missing opportunities to be "at the table" to actively address global issues. Some valuable personal and professional skills that are essential for functioning in a global environment are interpersonal and group communication, cultural understanding, diplomacy, negotiation, etiquette, team building and leading, networking, problem solving, and traveling, living, and working in true international settings.

Global engagements of ABE

ABE discipline and academic units have been in existence globally for well over 100 years. ABE activities are truly locally operated and globally connected. Many ABE educational and research activities have reached the level of local relevance and global pre-eminence. It is, therefore, enlightening to look at the global engagements of ABE, especially in the past 20 years.

There have been many individual-to-individual collaborative research and education activities between ABE@Illinois faculty members and their colleagues in many parts of the world in the past 20 years. Formal partnerships in student mobility have also occurred including visiting/exchange scholars, summer apprenticeship programmes, summer study abroad, international team competitions, 3+2 degree completion programs, corporate sponsored international scholarships, etc.

It is well-known that students, faculty, and academic units may face various challenges and opportunities in carrying out global engagement activities. For example, students need to integrate activities into their degree curriculum, seek financial support, demonstrate value to their future employers and in their professional practice, and incorporate activities as service-learning experiences. Faculty members will likely need to have institutional support and incentives in their performance evaluations, promotion/tenure considerations, management of resources (such as time and funding), and student recruitment. On the other hand, departments and colleges will need to provide incentives for both students and faculty, as well as organisational and financial resources. Commonly seen institutional support in promoting global engagements are administrative support, formal exchange agreements, special tuition rates, dedicated scholarships, study abroad information and workshops, and pre-travel orientations.

Concluding Remarks – Where Do We Go From Here?

All academic departments and degree programs are pursuing excellence. ABE discipline is no exception. Almost all units have some quantitative definitions and measurements for being "great" and "the best." Examples, on a per faculty full-time equivalent (FTE) basis, include student majors, instructional units, peer-review (impact factor weighted?) journal publications, citations, extramural (competitive?) funds, and extension and outreach deliverables and products. Less publicised, but equally important, qualitative considerations are doing good science and generating long-term sustained impact, locally relevant and globally preeminent, successful implementation of diversity with a solid core, students competing favourably in the market place (including faculty and leadership positions), high calibre people interested in joining or being associated with the unit, strong industry support and clientele satisfaction, unique programs that are highly valued, leadership in teaching, research, extension/outreach, service, and economic development, excellent national and international reputation/ranking, and effective integration of multi-disciplinary, multi-functional, multi-regional, and international programs and activities. Some departments have been foregrounding specific actions in their effort to continue advancing their goals, such as participation in and leading large centre-like grants/programs, establishing thematic laboratories in strategic areas, and increasinng outcomes of development and advancement activities (including creation of named and endowed professorships). Looking at a bigger picture, ABE, being an academic unit, will need to continue to align with campuses, colleges, universities, and beyond, as well as periodically conduct market study, SWOT analysis, strategic planning, resource generation and deployment, internal and external communications, advancement and development effort, outcome assessment, and continuous improvement.

Looking into the future, it seems that ABE is converging more and more towards a non-compound name of "agricultural biosystems engineering," with no "and" in the name. First of all, broadly defined agriculture (which is ABE's major domain) will continue to be important in providing food, feed, fiber, fuel, and in furnishing and managing safety/health, environment, and sustainability issues. Secondly, engineering is one of our core competencies. Thirdly, bio-based systems are important deliverables of ABE's work. Many organisations, including governments, national academies, the United Nations, professional societies, non-governmental organisations, and corporations have started to increasingly emphasise systems thinking and systems approach. In addition, the name "agricultural biosystems engineering" keeps the ABE as its acronym and, for some departments, is inclusive in representing our programs in agricultural biological engineering and systems management. One possible description of agricultural biosystems engineering is "Enhance sustainable productivity of food and agriculture systems of systems through innovative biosystems that are empowered by intelligent engineering analysis/solutions, machines, and systems management." In fact, a college level unit at a university in South Asia has just changed its name to "Faculty of Agricultural Biosystem Engineering." This is by no means a suggestion to change any ABE department name at this moment. This is just to point out that a potential direction that ABE is heading toward. In any case, ABE will continue to empower with knowledge and wisdom for life to enable people to learn and succeed globally. ABE's science, engineering, technology, and management-based mission will continue to be amazing and exciting, and its impacts will continue to be long lasting.

Acknowledgements

The faculty, staff, students, advisors, and partners of ABE@ Illinois have contributed, in the past 20 years, to the information presented in this article. Their active participations in providing inputs in both verbal and written forms are most enlightening and deeply appreciated.

REFERENCE LIST

American Society of Agricultural and Biological Engineers. (2021). *Resource*, March/April. American Society of Agricultural and Biological Engineers.

Ellen Macarthur Foundation. (2019). *Cities and circular economy for food*. Ellen Macarthur Foundation.

National Academies of Sciences, Engineering, and Medicine. (2019). Innovations in the food system: Exploring the future of food. In *Proceedings of a workshop*. Nation Academies Press.

National Research Council. (2015). *A framework for assessing effects of the food system*. National Academies Press.

Ting, K.C. (2013). A systems concept for controlled environment plant production. *Resource*, March/April, p. 6. ASABE.

Wulf, W.A. (2000). Great achievements and grant challenges. *The Bridge*, vol. 30, no. 3&4, Fall/Winter, pp. 5–10. National Academy of Engineering.

25

A Profile of Agricultural and Biosystems Engineering Education in Indonesia

Y. Yuwana

Institutional Establishment

Agricultural and Biosystems Engineering Education (ABEE) in Indonesia was early established in the 1960s under the name of agricultural mechanisation. In the 1980s the name of agricultural mechanisation changed to agricultural engineering, as the name of agricultural mechanisation was associated with the exploitation of large-scale machinery and equipment in the agricultural sector which might cause unemployment. In advance of scientific and technological developments in related fields, in the 2010s biosystem terminology was added to complement the name of agricultural engineering (History of Agricultural Engineering GMU, 2009; Department of Mechanical Engineering and Biosystem IPB, 2022; Agricultural Engineering and Biosystems UNAND, 2022). At the time of its establishment, ABEE was hosted in agricultural technology institutions, which were considered as part of the wider domain of educational institutions. In progress, ABEE is considered as software and referred to as a study programme, while its associated hardware is a department. In Indonesia, the department is the smallest academic administrative unit under the faculty, while the faculty is the administrative unit under the university.

In general, the agricultural technology educational structure with which the AB study programme is affiliated can be in the form of a department or faculty. A developed agricultural technology faculty usually comprises three departments or study programmes, namely Agricultural and Biosystems Engineering, Agricultural Industrial Technology, and Agricultural Products/Food Processing. The existence of departments is a requirement to establish faculties, in which at least two departments can apply to establish a faculty (History of Agricultural Engineering GMU, 2009; Department of Mechanical Engineering and Biosystems IPB, 2022; Agricultural Engineering and Biosystems UNAND, 2022; Agricultural Engineering and Biosystems S1 UB, 2022).

ABEE is clustered into a vocational and academic programme, in which the vocational programme consists of a Diploma D3 while the academic programme includes 3 strata, namely the Bachelor (S1), Master (S2), and Doctoral (S3) strata. The vocational programme is usually hosted at a higher school polytechnic college or institute whereas the academic programme resides in a university. Now in Indonesia, there are about 36 ABEEs spread across various government higher education institutions which include 8 D3, 22 S1, 4 S2, and 2 S3. ABEE's name varies Agricultural Engineering, Agricultural Mechanisation, Agricultural Mechanisation Technology, and Processing Engineering Mechanisation for the D3; Agricultural Engineering, Agricultural Machinery Engineering and Biosystem, Agricultural mechanisation, and Environmental Engineering and Bioprocess Technology for the S1; Agricultural Engineering, Agricultural Engineering, and Biosystem, and Postharvest Technology for the S2; and Agricultural Engineering Science for the S3 (PDDikti, 2020).

The fields of study in ABEE comprise Biosystem Mechanical Engineering, Land and Water Resources Engineering, Food Engineering, Postharvest and Bioprocess, Biological Environment Control Engineering, Natural Resources Engineering, Energy and Agricultural Machinery, Agricultural Biophysics, Agricultural Environmental, Building Engineering, Mechanical, and Automation Engineering, Renewable Energy Engineering, Bioinformatics Engineering, and Ergonomics and Agricultural Electronics.

Curricula Status in Indonesia Regulation

ABEE is established as a manifestation of the application of the Three Higher Education Mandates of the Indonesian Education System (Putri, 2003), which states that institution of higher education is constrained to conduct education, research, and community service. It is also stated that education is a constructive effort to provide an atmospheric learning process where students can potentially grow with noble character, intelligence, personality, self-control, and religious and spiritual strength, in addition to appropriate skills for themselves, society, and national building (Yudhoyono, 2012a).

Higher education covers diploma, bachelor, master, and doctoral programmes, as professional, and specialist programmes, carried out by tertiary institutions. This education level functions to: a. provide capability, character, and civilisation in the frame of national life; b. develop academics with skilled, creative, competitive, responsive, innovative, and cooperative characters by implementing the Mandate of Higher Education;

DOI: 10.1201/9780429150111-27

and d. develop scientific knowledge by taking into account and implementing humanity values. The objectives of Higher Education are to a. develop religious human beings with a noble character as well as cultured, skilled, competent, creative, capable, knowledgeable, independent, and healthy beneficial for the nation; b. provide scientific and technological graduates to meet national interests and competitiveness; c. conduct research producing science and technology beneficial to national progress, civilisation advancement, and humankind welfare; and d. perform community service based on the resulting research works to educate people and promote human welfare.

It is further explained that the research is carried out by employing scientific principles and, when community service is conducted, by implementing reproducible science and technology to enhance people's social life and welfare.

The higher education curriculum comprises propositions and configurations respecting teaching purposes, content, material, and methods to guide the activity of learning. In general, a study programme curriculum in Indonesian tertiary institutions is structured based on the national education system, implemented based on higher education national standards to achieve national qualifications. The curricula are proposed by tertiary institutions in a flexible, relevant, and contextual manner.

Curriculum development has to be built with strong philosophy, sociology, psychology, history, and argumentation (Junaidi et al., 2020). Philosophical foundations provide philosophical guidance for the design, implementation, and improvement of education quality (Ornstein & Hunkins, 2014), how science and technology are mastered by the students to understand and improve their individual and social life (Zais, 1976). The curriculum development should provide facilities and an environment relevant to the learners to study and experience individually and socially (Ornstein & Hunkins, 2014). The curriculum should prompt students' motivation and curiosity. The curriculum should be able to facilitate learning all the time through the glory of the nation's past, preserving it until now to look at a better life in the future. The curriculum should have a strong constitutional foundation starting from the time of design, development, implementation, evaluation, and quality assurance to guarantee the achievement of its goals.

Indonesian Higher Education Standard

The Indonesian Standard of Higher Education comprises education, research, and community service standards which are the minimum requirements that must be met by all tertiary institutions in Indonesia. These standards consist of: a. competence of graduate standard; b. content of learning; c. process of learning; d. assessment of educational learning; e. lecturer and personnel; f. facilities and infrastructure; g. management of learning; and h. finance. These standards must be used as the basis for formulating the Indonesia Qualification Framework (Makarim, 2020a).

Indonesia Qualification Framework

Adapted to the European Qualification Framework and the Asian Qualification Reference Framework, each of which consists of 8 levels, and the Australian Qualification Framework which consists of 10 levels, the Indonesian Qualification Framework comprises 9 levels of qualification, level 1 as the lowest level up to level 9 as the highest level. The Indonesian Qualification Framework includes: a. levels 1 to 3 are qualified for the position of operators; b. levels 4 to 6 are dedicated to position technicians or analysts; c. levels 7 to 9 are qualified for the position of experts. Based on the learning outcome's equalisation, the Indonesian Qualification Framework equalises a. Graduates of basic education as level 1; b. Graduates of the lowest secondary education as level 2; c. Graduates of Diploma 1 as at least level 3; d. Graduates of Diploma 2 as at least level 4; e. Graduates of Diploma 3 as at least level 5; f. Graduates of Diploma 4 as level 6; Graduates of Masters as level 8; h. Graduates of Doctoral Degrees as level 9; i. Graduates of professional education as level 7 or 8; j. Graduates of specialist education as 8 or 9 (Yudhoyono, 2012b).

The general description of the Indonesian Qualification Framework includes a. belief in God. b. completing their duties with personality, ethics, and morals; c. acting as a good citizen; d. being good at working together and socially and being environmentally sensitive; e. having an open mind; f. prioritising the interests of the people and the nation.

The Indonesian Qualification Framework for D3 is equivalent to qualification level 5 as described: a. having the ability to complete various jobs with the right method and measurable performance both qualitatively and quantitatively; b. mastering the theory of certain fields of science and solving problems procedurally; c. being able to organise work groups and make written reports appropriately; d. being responsible for the results of his or her work and the work of his or her group.

The Indonesian Qualification Framework for S1 is equivalent to qualification level 6 which is described as: a. able to implement their expertise to solve various problems; b. mastering the theoretical concepts of a particular field of knowledge in depth to formulate problem-solving; c. able to make the right decisions by choosing alternative solutions independently and in groups; d. responsible for their work and their organisation.

The Indonesian Qualification Framework for S2 is equivalent to level 8 qualification including a. being able to conduct research in their field for developing knowledge, technology, and/or art, to create innovative tested works; b. being able to employ inter or multi-disciplinary approaches to solving problems in their scientific field; c. being able to perform research and development to obtain recognition at national and international levels.

The Indonesian Qualification Framework S3 is equivalent to level 9 which is described as: a. being able to perform research in their field to produce new knowledge, technology, and/or creative, original, and tested works; b. being able to utilise inter-, multi-, and trans-disciplinary approaches to solving problems in their field; c. being able to accomplish research and development beneficial for mankind and gaining recognition at national and international levels.

Relationship between Qualification Framework and Higher Education Standard

To meet the Qualification Framework, the graduate attitude, general skill, special skill, and mastery are formulated. The formulation of attitude includes: 1. being pious; 2. upholding religion, morals, and ethics as human values in duties; 3. contributing to the quality of life improvement based on "Pancasila" (Indonesian Fundamental Ideology) in society, nation, state, and civilisation; 4. acting as noble citizens characterised with nationalist glory; 5. respecting religion, belief, culture, findings, opinion, and viewing diversity; 6. cooperating and being concerned for society and the environment; 7. obeying the law in the nation and social life; 8. upholding norms, ethics, and academic values; 9. demonstrating their expertise towards work with a responsible attitude independently; and 10. upholding independence, struggle, and entrepreneurship spirits.

General skills of graduates for the D3 incorporate: a. being able to accomplish a wide scope of work and analyze information with an assortment of standardized and non-standardized strategies; b. being able to configure qualitative and quantitative performance; c. being able to solve the problems encountered using their expertise in effective and innovative methods; d. being able to construct working reports properly as their professional responsibility; e. being able to work cooperatively, communicatively, and innovatively; f. being able to accomplish job under their responsibility; g. being able to perform self-evaluation and develop their competency independently; h. being able to manage scientific documents to guarantee their security as well as preventing plagiarism.

The general skill of S1 graduates incorporates 1. being able to utilize consistent, basic, orderly, and inventive consideration within the setting of the advancement of science and innovation that pays consideration to and applies the values of humanity in agreement with their ranges of ability; 2. being able to configure autonomous performance qualitative and quantitatively; 3. being able to ponder the suggestion of the advancement knowledge that takes into consideration and applies humanity values according to their competency based on logical standards, strategies, and morals to create arrangements, thoughts, and plans to compile a scientific portrayal; 4. being able to research to construct a thesis and upload it on the college publication media; 5. being able to make the right decision to solve the problem according to their expertise based on analyzed data; 6. being able to manage institutional networking internally and externally; 7. being able to accomplish individual and group achievement under their responsibility; 8. being able to perform self-evaluation of a workgroup of their responsibility and manage independent learning; 9. being able to manage scientific documents to guarantee their security as well as prevent plagiarism.

General skills of graduates for the S2 include: 1. able to think logically, critically, systematically, and creatively with regard to employing research, designs, or artistic works of art in their educational domain with respect for the values of humanity beneath their fields of ability, define logical conceptions and comes about of considers based on logical principles, procedures, and morals in the form of a thesis or other identical work transferred on the scientific media of university, as well as papers published in accredited national journals or international reputable journals; 2. being able to carry out scholastic approval or considers concurring to their field of mastery in understanding issues in pertinent communities or businesses through the advancement of their information and skill; 3. being able to define thoughts, comes about of thought, and logical contentions in a capable and based on scholastic morals, and communicate through the media to the scholarly community and the more extensive community; 4. being able to distinguish research domains and then putting them in a roadmap of research using inter- or multi-disciplinary approaches; 5. being able to create choices within the setting of problems within the improvement of science and innovation that consider and apply the humanity values involved in scientific work; 6. being able to oversee, create, and keep up systems with colleagues and peers inside the organisation and the more extensive scientific community; 7. being able to extend the capacity of learning autonomously; 8. being able to record, store, secure, and verify scientific information to guarantee legitimacy and anticipate literary theft.

General skills of graduates for the S3 include: a. being able to create findings based on research employing scientific method; b. being able to integrate holistic scientific works and report them in the form of dissertations and scientific papers qualified for reputable international journals; c. being able to select and conduct up-to-date and relevant research topics employing a holistic approach as well as using every single available resource; d. being able to construct a road map of research employing a holistic approach based on a main target advancing to massive targets; e. being able to propose argumentation and problem solving respecting scientific methods and moral ethics and deliver them to the public and media; f. being able to manifest strong leadership in their institutional responsibility; g. being able to conduct scientific management affairs of their obligation; and h. being able to establish and sustain academic networking.

ABEE Competencies

Graduate competencies equalised by the Indonesian Higher Education Standards are minimal standards. Specific skills are formulated by study programme associations or by each tertiary institution under the vision and mission and excellence set by the concerned tertiary institution as the basis for establishing its curriculum. Graduate competence is a part that can be performed by graduates with relevant expertise after their academic accomplishment. Graduate profiles can be formulated based on the achievement of studies in response to jobs prescribed by governmental and industrial sectors as well as the demand for scientific and technological development. For ABEE at the D3 level, the competency formulation is not as complex as for the S1, S2, and S3.

The competencies of ABEE graduates compiled from several leading universities in Indonesia are as follows. The competencies of ABEE's graduate for the D3 (D3 Graduate Competency, 2021; D3 Agricultural Mechanisation Technology Politek Payakumbuh, 2022) include: 1. being able to complete work within the scope of agricultural mechanisation by evaluating various data employing proper methods; 2. being able to exhibit performance within the scope of agricultural mechanisation with qualitative and quantitative measurement; 3. being able to solve problems for workshop work, operation, and maintenance of agricultural machine tools with characteristics and contexts appropriate for agricultural mechanisation, based on creative thinking, carried out job independently for the targeted results; 4. being able to accurately write reports on the results of assigned job operations, agricultural workshops, agricultural machinery services, and maintenance of agricultural machinery tools for the clients; 5. being able to manage their work; 6. being able to be responsible for the working group's achievement; 7. being able to conduct supervision and evaluation of the work assigned to workers under their responsibility; 8. being able to conduct self-evaluation of working groups in the field of agricultural mechanisation and develop their knowledge and skill independently; 9. being able to manage scientific documents as well as avoid plagiarism; 10. being able to operate and carry out maintenance of agricultural tools and machines according to Standard Operating Procedures; 11. being able to construct, manufacture, and perform simple modifications of agricultural tools and machinery according to the requirements of the technical drawings or models requested; 12. being able to operate, maintain, and repair minor damage to tools and machines for processing agricultural products (postharvest, food, plantations such as palm oil, tea, rubber, cacao, coffee, and fruit processing); 13. being able to test and assess the performance of agricultural tools and machines according to Indonesian National Standard; 14. being able to manage agricultural workshop, greenhouse, or screen house work according to Standard Operating Procedure and able to manage machinery service management business; 15. being able to supervise activities related to agricultural tools and machinery under predetermined quality standards; 16. being able to make business proposals and reports in the field of agricultural workshops and agricultural tools and accountable machine services; 17. being able to work together in work groups and make comprehensive written reports per their competence; 18. being able to manage businesses in the field of agricultural machine tools on a small and medium scale; and 19. being able to become a service entrepreneur and market agricultural machinery tools;

The competencies of ABEE graduates for the S1 (S1 Programme, 2011; S1 Graduate Profile IPB, 2020; S1 Graduate Profile UNPAD, 2021) comprise attitudes, knowledge, and general and special skills.

The attitudes include the ability to work in a professional group, to understand professional ethics techniques, to generate professional ideas and concepts based on knowledge of agricultural engineering, and to carry out the management of agricultural and biosystems resources.

The knowledge competence includes 1. ability to master the properties of agricultural products, the management principles in monitoring and utilizing agricultural and biosystems resources, and the innovative development of agricultural techniques, and to identify and select the proper standard for mechanical elements according to engineering principles; 2. ability to understand the basic science, the basics of engineering, the basics of agricultural science, freeze drying technology, the mechanisation of agricultural technology (on the farm and off-farm), the research methods and development of agricultural techniques, the entrepreneurial characteristics of agribusiness and agroindustry, and to set design goals and parameters according to expected product technology.

The general skills include 1. ability to select proper engineering materials standards and to assess the strength of materials precisely; 2. ability to calculate product effectiveness and efficiency; 3. ability to employ scientific methodology for the advancement of the agricultural engineering domain; 4. ability to manage agricultural natural resources, equipment, and processing in agricultural systems; 5. ability to have a professional attitude, have an entrepreneurial spirit, be responsive to agricultural engineering developments, conduct effective communication, and obey engineering ethics; 6. ability to perform strong professional ethics; 7. ability to demonstrate strong professional leadership; 8. ability to provide problem-solving and entrepreneurial skills in the agricultural engineering domain; and 9. ability to utilise relevant findings in the field of agricultural engineering.

The specific skills include: 1. ability to make design drawings, design demands, and design simple equipment, components, or processes required in agricultural engineering operations conforming to ISO standards; 2. ability to provide problem-solving including design, manufacture, management, agricultural machinery, building biological systems (agriculture in a broad sense), processing agricultural and food products, renewable energy, and systems in agriculture; 3. ability to perform engineering drawings; 4. ability to operate and control instrumentation systems; 5. ability to employ vacuum frying and evaporation techniques; 6. ability to construct appropriate agricultural machinery and equipment; and 7. ability to analyze the impact of engineering solutions on the environment and society using a multi-disciplinary approach.

The competencies of ABEE's S2 graduates include 1. being able to compile the agricultural engineering and biosystems principles to identify, understand, explain, evaluate/analyze problems using a scientific approach and method; as well as solving problems through design, construction, and testing of equipment as well as managing natural resources and production processes with an inter- or multi-disciplinary approach; 2. being able to perform inventory, identification, analysis/evaluation, and processing for biomaterial and natural resources in an environmentally sound manner; 3. being able to perform research and publish the results in national and international scientific forums and journals; 4. being able to master scientific methods, so that they can act as professionals, researchers, and scientists in the agricultural and biosystems engineering domain; 5. being able to perform professionally and creatively in their work and careers for comprehensive scientific development per engineering ethics and community life norms, as well as communicate scientifically effectively and be responsive to science and technology developments; 6. being able to act as a leader and policy maker based on engineering principles and management systems in the field of agricultural and biosystems engineering; and 7. mastering and applying engineering design principles and management systems as a basis for decision-making and developing competitive technology based on raw materials, local wisdom, and technology in tropical industrial agricultural systems in a sustainable and environmentally friendly manner.

The competencies of ABEE's S3 graduates include: 1. mastering philosophy of science, and research methodology; 2. identifying, explaining, evaluating, and analyzing problems employing a philosophical approach and scientific method; 3. developing science, technology, and art by researching to produce genuine and useful findings for society; 4. solving problems through design, construction, and testing of equipment as well as management of natural resources and production processes with holistic approaches; 5. leading advance research and disseminating the results in national and international scientific forums and journals; 6. performing and behaving professionally and innovatively in their work and careers for comprehensive scientific development under engineering ethics and the norms of society's life; and 7. communicating scientifically effectively and being responsive to science and technology developments.

Curriculum Reform

ABEE's curriculum is formed and reformed based on graduate competencies. The curriculum is developed and determined by higher education institutions regarding the Indonesian Higher Education Standards. The competencies referred to are at least fulfilling the following curriculum elements: a. personality foundation; b. science, technology, arts, and/or sports mastery; c. skill ability building to work; d. character building to work according to mastered skills and expertise; e. social life rules mastery by the expertise choice in the work.

The curriculum is the lifeblood of a learning programme so its existence is dynamically designed, implemented, and evaluated periodically according to the science, technology, and arts competencies needed by the community and college graduates users. The turbulent development of science and technology in the 21st century follows a logarithmic pattern, leading the Indonesian Higher Education Standards to also follow these changes. In six years, the Indonesian Higher Education Standards have undergone three changes (Nuh, 2014; Nasir, 2015; Makarim, 2020a) and now finally meet the Ministry of Education and Culture's policy on Independent Learning for Independent Campus. This change is often mistakenly perceived as a necessity that every time the education minister changes, the education curriculum also changes. However, changing the education curriculum is a necessity as long as it does not conflict with the educational philosophy and applicable regulations (Junaidi et al., 2020).

The stages of the curriculum reform begin with Graduate Learning Outcomes based on graduate competencies. Graduate Learning Outcomes formulation is also based on the curriculum evaluation results of the ongoing Graduate Learning Outcomes curriculum, tracer studies, and feedback from the relevant stakeholders. The evaluation of the curriculum also engaged the scientific and technological advancement of the related fields, the change in the labour market, as well as vision and mission set up by the institution. The curriculum is further developed by identifying and defining study materials and courses that are structured each semester during the study period. The preparation of curriculum documents includes 3 stages, namely: curriculum establishment, learning processes, and learning programme evaluation.

In detail, curriculum reform begins with graduate profiles analysis relevant to the market demands and learning contents as well as teaching materials provided by the higher education institution under their disciplines (scientific vision). This stage is followed by formulating courses and their credit weights, and organising the courses in a simple matrix form comprising a. graduate profile determination and Graduate Learning Outcomes formulation; b. learning materials prescription and course formulation; c. course matrices and curriculum mapping compilation. Learning Outcomes are prescribed respecting the results of graduates tracking, stakeholder feedback, associations' suggestion, consortia recommendation, scientific and academic trends, and the curriculum evaluation results. The Graduate Learning Outcomes formula should accommodate knowledge, science, and technology relevant to the 4.0 industrial and the future era as well as human collaboration, artificial intelligent systems, the Internet of Things (IoT), cybernetics, intelligent machines, and environmental synergistics (Suherman et al., 2020). In the end, the Graduate Learning Outcomes formulation must refer to Indonesian Higher Education Standards and Qualification Framework descriptors according to their level of education. The Graduate Learning Outcomes have also reflected the vision, mission, the uniqueness of tertiary education institutions.

After the Graduate Learning Outcomes have been formulated, they are followed by the determination of learning materials and the formation of courses. Some Graduate Learning Outcomes items that are suitable as the basis for the formation of courses containing knowledge, skills, and attitudes are selected. The learning material can serve several Graduate Learning Outcomes. Each Graduate Learning Outcomes item contains learning material that will be used to form courses. The learning material may contain one or more branches of knowledge assigned by similar higher education forums. The next learning material is described in more detail into learning material. The level of learning materials refers to the Indonesian Higher Education Standards. Learning materials can be updated or developed according to the knowledge, science, and technology advancement and the strategy of higher education institutions. The determination of learning materials needs to involve scientific groups/laboratories in higher education. The formation of a course based on the selected learning material is continued by creating a matrix bridging attitudes, general skills, specific skills, and knowledge.

The next stage is the determination of courses. The determination of courses for the ongoing curriculum is carried out by evaluating each course based on the evaluated learning outcomes beforehand. The evaluation is carried out by examining how far each course is related to the formulated learning outcomes of existing courses. The formation of new courses is conducted by using several learning outcome points assigned to them. This is then followed by the determination of the credit weight. The amount of credits for a course is assigned according to the time spent by students to be able to have the abilities formulated in that course. The determining elements for the estimated weight of credits regarding the Indonesian Higher Education Standards are: a. the ability level that must be reached; b. the learning material level that must be mastered; and c. learning methods/strategies chosen to achieve these abilities. Furthermore, the weight of the credit for each determined course reveals a. the ability level that must be achieved in the learning outcomes assigned to the course which is represented in the course learning outcomes; b. the learning material level that is equated with the time required for learning activities to achieve each learning outcomes item assigned to the course; c. the chosen learning method.

After the courses and the credit weights have been determined, the course organisation is arranged in the structure of the curriculum. The stages of the curriculum structure formulation with the appearance of a matrix of course organisation per semester take into account the following matters: 1. the planned course learning stages to obtain learning outcomes; 2. the accuracy of the course status adjusted to the ability alignment level and course integration both vertically and horizontally; 3. the fact that the student study load is normally between 8–10 hours per day per week, equivalent to 17–21 credits per semester, and 4. the fact that the drafting process involves all study programme lecturers and is then determined by the study programme. The course organisation in the structure of the curriculum needs to be performed precisely to guarantee that the student learning stages are appropriate, ensuring that learning is carried out efficiently and effectively to obtain learning outcomes. The course organisation in the structure of the curriculum consists of horizontal and vertical organisations (Ornstein & Hunkins, 2014). The organisation of horizontal courses in semesters is intended to broaden students' discourse and skills in a wider context. For example, in the same semester, students learn about science and humanities in the context of achieving abilities according to one of the Graduate Learning Outcomes points on the assigned skills. Meanwhile, the organisation of courses vertically at the semester level is intended to provide students with mastery of abilities relevant to the learning difficulty level to reach the learning outcomes predetermined by the higher education institution.

The government gives authority to higher education institutions to develop and reform curricula according to the vision, mission, and excellence to be achieved by the higher education institution with which ABEE is affiliated. The curriculum reform comprises analyzing, designing, developing, implementing, evaluating, and following up on improvement stages conducted by the higher education institution (Ornstein & Hunkins, 2014). Each stage in the curriculum reform is performed utilising the Indonesian Higher Education Standards which consist of eight standards, namely the Standard of Graduate Competency, Standard of Learning Content, Standard of Learning Process, Standard of Learning Assessment, Standards of Lecturer and Education Personnel, Standards of Learning Facilities and Infrastructure, Standard of Management, and Standards of Learning Finance.

Curricula Content and Running

Curricula Content

In general, the ABEE curricula content and structure consist of 1. basic Engineering and national courses; 2. Agricultural and Biosystem Engineering Technical Courses; 3. Agricultural and Biosystem Engineering Specialisation Courses; 4. skill supporting and elective courses; and 5. capstone course and thesis. The curricula structure must contain national mandate courses including Religion, "Pancasila", Citizenship, and Indonesian Language. The curricula are diverse among ABEE institutions. Figure 25.1 presents the structure of ABEE curricula which are grouped into General and Basic Courses, Programme Core Courses, In-depth Courses, Enrichment Courses, and Project Courses for the D3 (D3 Graduate Competency, 2021; Agricultural Mechanisation Technology, 2019) and S1 (DTPBSFTPUGM, 2021; Satria, 2020; Education Guide 2020/2021 FTP UB, 2020) levels; Compulsory, Elective, and Field Concentration Courses for the S2 (Master Program Curriculum, 2021; Curriculum K2020, 2021; Education Guide 2020/2021 FTP UB, 2020) level; Compulsory and Elective Courses for the S3 (Curriculum of Doctoral Programme, 2021; Curriculum K2020, 2022) level, compiled from several leading ABEE institutions in Indonesia. Table 25.1 shows the names of the course under the groups. Table 25.2 shows an

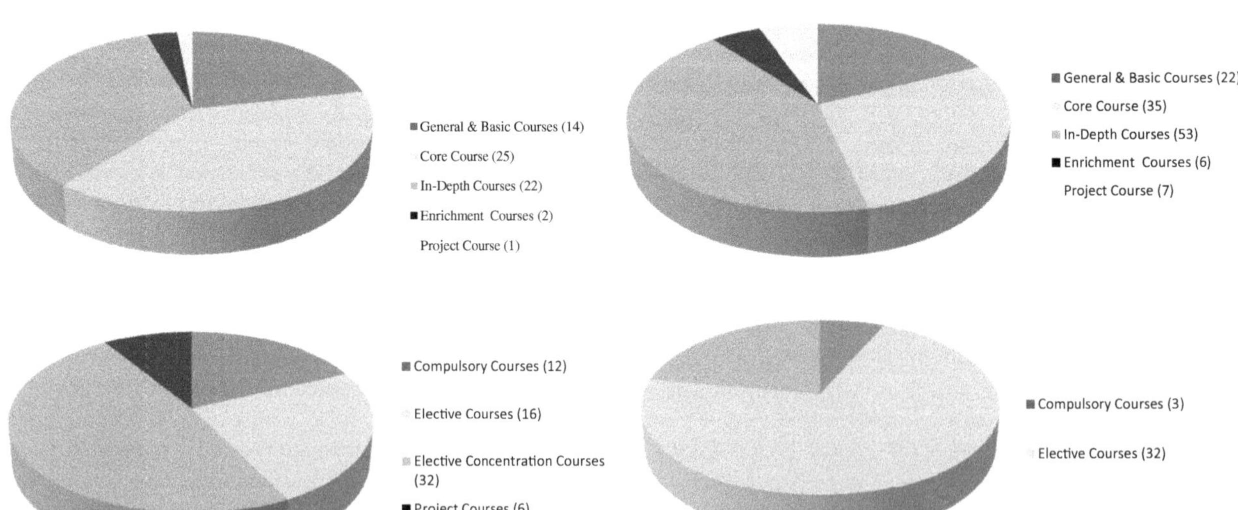

FIGURE 25.1 Composition and number of courses of Agricultural and Biosystems Engineering Education D3, S1, S2, and S3.

TABLE 25.1

Agricultural and Biosystems Engineering Education (ABEE) competencies for the D3, S1, S2, and S3 programmes

The curriculum of D3 (Diploma program)

General and Basic Courses

No.	Subject	Credits
1	Religion	(1-1)
2	Pancasila Education	(2-0)
3	Civic education	(2-0)
4	Indonesian	(1-1)
5	English	(1-1)
6	Applied mathematics	(1-1)
7	Agronomy	(1-1)
8	Rural Sociology	(1-1)
9	Food Crop Production Technology	(1-2)
10	Computer application	(1-2)
11	Applied Physics	(1-2)
12	Material Strength	(1-2)
13	Introduction to Plant Cultivation	(1-1)
14	Statistics	(1-1)

Core Courses

No.	Subject	Credits
1	Occupational Safety, Health, and Safety	(1-1)
2	Modern Agricultural Technology	(1-1)
3	Heat Transfer	(1-2)
4	Soil and Water Conservation Engineering	(1-1)
5	Drive Motor	(1-2)
6	Control and Lubrication System	(1-1)
7	Scientific Writing Techniques	(1-1)
8	Machine Elements	(1-1)
9	Machine Design	(1-1)
10	Agricultural Tractor	(1-2)
11	Vegetable Production Technology	(1-2)
12	Engineering Materials Knowledge	(2-1)
13	Mapping	(1-1)
14	Engineering drawings	(1-2)
15	Agricultural Energy and Electricity	(1-1)
16	Basic Electronics	(1-1)
17	Territorial Measurement	(1-2)
18	Organic Farming System	(1-1)
19	Occupational Health and Safety	(1-1)
20	Agricultural Climatology	(1-1)
21	Scientific Paper Writing Techniques	(1-1)
22	Agricultural Cultivation Machine Tools	(1-1)
23	Instrumentation	(1-1)
24	Mechanical Education	(1-1)
25	Agricultural Communications	(1-1)

In-depth Courses

No.	Subject	Credits
1	Irrigation and Fertigation Engineering	(1-1)
2	Hydroponic Engineering	(1-1)
3	Soil Processing Engineering	(1-1)
4	Agricultural Cultivation Machines and Tools	(1-2)
5	Harvesting Technique	(1-1)
6	Agricultural Products Handling Techniques	(1-2)
7	Engineering Economics	(1-1)
8	Spatial Information Technology	(1-1)
9	Green House	(1-2)
10	Agricultural Products processing techniques	(1-1)
11	Agricultural Machinery and Equipment Management	(1-1)
12	Agricultural Machinery and Equipment Testing	(1-1)
13	Agricultural Workshop	(1-2)
14	Mechanical Technology	(1-1)
15	Farm Building	(2-2)
16	Workshop Management	(1-1)
17	Alsin Maintenance and Care	(1-1)
18	Professional ethics	(1-1)
19	Independent Project	(0-3)
20	Postharvest Techniques	(2-1)
21	Alsintan Design	(1-2)
22	Sustainable Natural Resources Management	(1-1)

Enrichment Courses

No.	Subject	Credits
1	Entrepreneurship	(1-1)
2	Field Work Practice (PKL)	(0-4)

Project Courses

No.	Subject	Credits
1	Final Project (TA)	(0-4)

The curriculum of S1 (Undergraduate Programme)

General and Basic Courses

No.	Subject	Credits
1	Religion	(2-1)
2	Pancasila Education	(1-0)
3	Civic education	(1-0)
4	Indonesian	(1-1)
5	English	(1-0)
6	Basic Biology	(2-1)
7	Physics	(2-1)
8	Chemical	(2-1)
9	Economy	(2-0)
10	Innovative Agriculture	(2-1)
11	Mathematics	(3-0)
12	Statistics and Data Analysis	(3-0)
13	Computational Thinking	(2-0)
14	Sociology	(2-0)
15	Sport	(0-1)
16	Calculus I	(1-0)
17	Introduction to Agricultural Technology	(2-0)
18	Introduction to Agronomy	(2-0)
19	Introduction to Information systems	(2-0)
20	Microbiology	(2-0)
21	Introduction to Agricultural Engineering and Biosystems	(2-0)
22	Computer Application	(1-1)

Core Courses

No.	Subject	Credits
1	Electricity and Electronics	(2-0)
2	Engineering Mechanics	(2-1)
3	Engineering drawings	(1-1)
4	Engineering Programming	(2-1)
5	Engineering Thermodynamics	(2-1)
6	Engineering Materials Knowledge	(2-0)
7	Hydroclimatology for Agriculture and Biosystems	(2-0)
8	Relations of Soil, Water, and Plants	(2-0)
9	Technical Characteristics of Agricultural Materials	(2-0)
10	Fluid Mechanics	(2-1)
11	Heat and Mass Transfer	(1-1)
12	Instrumentation and Control	(2-1)

(Continued)

TABLE 25.1 (CONTINUED)

Agricultural and Biosystems Engineering Education (ABEE) competencies for the D3, S1, S2, and S3 programmes

13	Material Strength	(2-0)
14	Manufacturing Technology	(1-1)
15	Agricultural Material Knowledge	(2-0)
16	Ergonomics	(2-0)
17	Research Methods and Scientific Presentation	(2-0)
18	System Analysis and Modeling	(2-0)
19	Climatology	(2-1)
20	Properties of Agricultural Materials	(2-1)
21	Chemical environment	(2-1)
22	Environmental Physics	(2-1)
23	Food Materials Knowledge	(2-0)
24	Life Physics	(2-1)
25	Properties of Engineering Materials	(2-1)
26	Engineering Profession	(1-0)
27	Mapping	(2-1)
28	Fundamentals of Soil Physics	(2-1)
29	Agricultural Environmental Engineering	(2-1)
30	Power in Agriculture	(2-0)
31	Agricultural Cultivation	(2-0)
32	Experimental design	(2-0)
33	Ergonomics, Occupational Health, and Safety	(2-0)
34	Machine and Soil Dynamics	(2-0)
35	Air Pollution Control and Management	(2-0)
In-depth Courses		
No.	Subject	Credits
1	Agricultural Mechanization Systems and Management	(3-0)
2	Engineering Economics	(2-0)
3	Engineering Mathematics	(2-1)
4	Food Processing Techniques and Agricultural Products	(3-0)
5	Agricultural Cultivation Machines and Tools	(2-1)
6	Ergonomics1)	(1-1)
7	Motor Fuel and Agricultural Power1)	(2-1)
8	Bioconversion Principle2)	(2-0)
9	Energy Conversion and Conservation2)	(2-1)
10	Engineering Design	(2-0)
11	Agroindustry Management	(2-0)
12	Irrigation and Drainage Engineering1)	(2-1)
13	Robotic Technology1)	(2-1)
14	Postharvest Techniques2)	(3-0)
15	Biosystem Process Engineering2)	(2-1)
16	Measurement and Instrumentation Techniques	(2-1)
17	Mass Biomass Cultivation Technique	(2-0)
18	Agroclimatology	(2-0)
19	Industrial Sanitation Engineering	(2-0)
20	Storage and Packaging Techniques	(2-0)
21	Operations Unit	(2-0)
22	Biosystem Machine Planning	(2-0)
23	Non-destructive Evaluation Technology	(2-0)
24	Principles of Irrigation Engineering	(2-1)
25	Soil and Water Conservation Engineering	(2-1)
26	Watershed Engineering and Management	(2-0)
27	Disaster Risk Reduction in the Agricultural Sector	(2-0)
28	Non-destructive Evaluation Technology	(2-0)
29	Introduction to Control Systems	(2-1)
30	Agricultural Environmental Engineering	(2-1)
31	Waste Treatment Process Engineering	(2-1)
32	Beverage Process Engineering and Confectionery	(2-1)
33	Biosystem Machine Construction	(2-1)
34	Bioenergy Engineering	(2-1)
35	Geographic Information System on Agriculture	(2-1)
36	HVAC Facility Design	(2-0)
37	Environmental Management and Waste Management	(2-0)
38	Bioprocess Engineering	(2-0)
39	Food Quality Evaluation	(2-0)
40	Food Industry Process and Equipment Design	(2-0)
41	Geographic Information System on Agriculture	(2-0)
42	Environmental Measurement	(2-0)
43	Farm Building	(2-0)
44	Biomaterial Process Engineering	(2-1)
45	Food Process Engineering	(2-1)
46	Drying Technique	(2-0)
47	Refrigeration and Freezing Techniques	(2-0)
48	Modeling and Simulation Techniques	(2-1)
49	Territorial Measurement	(2-0)
50	Factory Design	(3-0)
51	Biosystem Robotics	(2-1)
52	Postharvest Handling Technology	(2-0)
53	Design of Agricultural Machinery and Tools	(2-0)
Enrichment Courses		
No.	Subject	Credits
1	Fundamentals of Agronomy	(2-1)
2	Calculus II	(3-0)
3	Enrichment activities outside the Study Program (ILIC)	
4	Entrepreneurship	(2-0)
5	Community service	(0-4)
6	Field Work Practice (Internship)	(0-3)
Project Courses		
No.	Subject	Credits
1	Capstone Course 1: Precision Agriculture Technology	(2-2)
2	Essay	(3-0)
3	Capstone Course 2: Plant Production Techniques in Controlled Environments	(2-2)
4	Thematic Real Work Lectures	(1-3)
5	Capstone Project: Design Project	(0-3)
6	Seminar	(1-0)
7	Thesis	(6-0)
The curriculum of S2 (Master Programme)		
Compulsory Courses		
No.	Subject	Credits
1	Research Methodology	(2-0)
2	Engineering Analysis Method	(2-0)
3	Advanced Engineering Mathematics	(3-0)
4	Modeling and Simulation Techniques	(3-0)
5	Industrial Farm Management System	(3-0)
6	Biosystem Engineering	(3-0)
7	Process Design and Manufacturing Systems	(2-0)
8	Special Topics	(2-0)
9	English	(2-0)
10	Design Engineering	(3-0)
11	Research methods	(2-0)
12	Economic Techno	(3-0)

TABLE 25.1 (CONTINUED)

Elective Courses

No.	Subject	Credits
1	Biosystems Environmental Engineering	(2-1)
2	Design and Testing	(2-0)
3	Analysis and Evaluation of Agricultural Engineering and Biosystems Projects	(2-0)
4	Renewable Energy Conversion Analysis	(2-0)
5	Physical Properties of Materials and Agricultural Products	(2-1)
6	Mechatronics of Agricultural Machinery and Tools	(3-1)
7	Energy Conversion Technique	(2-0)
8	Instrumentation and Testing	(2-0)
9	Advanced Environmental Conservation Techniques	(2-0)
10	Advanced Water Supply and Irrigation	(2-0)
11	Waste Management and Treatment	(2-0)
12	Advanced Drainage	(2-0)
13	Agricultural Mechanization Project Management	(2-1)
14	Sustainable Agricultural Engineering	(2-0)
15	Agricultural Engineering Information System	(2-1)
16	Agricultural Product Marketing Management	(3-0)

Land and Water Resources Engineering Concentration Elective Courses

No.	Subject	Credits
1	Geographic Information System Application	(2-0)
2	Soil Physics	(2-0)
3	Engineering and Management of Irrigation Systems	(2-0)
4	Development and Management of Water Resources	(2-0)
5	Engineering Hydrology	(2-1)
6	Spatial Technology	(2-1)
7	Natural Resource Management Engineering	(2-1)
8	Environmental Climatology and Hydrology	(2-0)

Biosystem Mechanical Engineering Concentration Elective Course

No.	Subject	Credits
1	Energy Conversion & Conservation Engineering	(2-1)
2	Management of Mechanized Farming Systems	(2-0)
3	Soil Mechanics	(2-0)
4	Agricultural Machinery Design and Testing	(2-0)
5	Control and Sensor Systems	(2-0)
6	Agricultural Production Engineering and Biosystems	(2-0)
7	Process Engineering for Agriculture and Biosystems	(2-0)
8	Agro Biosystem Machine Design	(2-1)
9	Renewable Alternative Energy	(2-1)
10	Bioproduct Process Technology	(2-1)
11	Mathematical Modeling and Simulation	(2-1)

Food Engineering, Postharvest, and Bioprocess Engineering Concentration Elective Courses

No.	Subject	Credits
1	Agricultural Product Process Engineering	(2-1)
2	Advanced Packaging and Storage Techniques	(2-0)
3	Advanced Heat and Mass Transfer	(2-0)
4	Drying and Cooling Techniques	(2-0)
5	Postharvest Physiology	(2-0)

Agricultural Mechanization Management Elective Course

No.	Subject	Credits
1	Decision-Making Techniques	(2-1)
2	Analysis of Agricultural Mechanization Systems	(2-1)
3	Management of Labor and Agricultural Machinery	(3-0)

Bioenvironmental Engineering Concentration Elective Course

No.	Subject	Credits
1	Advanced Agricultural Building Environmental Engineering	(2-1)
2	Advanced Agricultural Land Environmental Engineering	(2-0)
3	Advanced Air Conditioning Techniques	(2-0)
4	Advanced Agricultural Waste Treatment Process Techniques	(2-0)
5	Advanced Biopolymer Process Engineering	(2-0)

Project Courses

No.	Subject	Credits
1	Colloquium	(0-1)
2	Research proposal	(0-2)
3	Thesis	(0-6)
4	Thesis Seminar	(0-1)
5	National Scientific Publications	(0-2)
6	Thesis Exam	(0-2)

The curriculum of S3 (Doctoral Programme)

Compulsory Courses

No.	Subject	Credits
1	Philosophy of Science and Dissertation Studies	(2-0)
2	Agricultural Engineering Comprehensive Study	(2-0)
3	Agricultural Engineering Science	(3-0)

Elective courses

No.	Subject	Credits
1	Mathematical Analysis	(2-0)
2	Communication of research results	(2-0)
3	Bioenergy Engineering	(2-0)
4	Bioproduction Mechanical Engineering	(2-0)
5	Bioinformatics Management System	(2-0)
6	Biosensing and Robotics Engineering	(2-0)
7	Advanced Irrigation	(2-0)
8	Soil Dynamics	(2-0)
9	Advanced Hydroclimatology	(2-0)
10	Soil and Water Conservation Techniques for Sustainable Agriculture	(2-0)
11	Food Product Engineering	(2-0)
12	Transport Phenomena	(2-0)
13	Food Material Preservation Techniques	(2-0)
14	Advanced Bioprocess Engineering	(2-0)
15	Agricultural environmental management and policy	(2-0)
16	Agricultural Pollution Control and Environmental Restoration	(2-0)
17	Analysis and Arrangement of Agricultural Areas	(2-0)
18	Digital Data Processing Techniques	(3-0)
19	Numerical Analysis for Engineering	(3-0)
20	Agricultural Equipment and Machinery Control Engineering	(3-0)
21	Mechanical Engineering and Automation for Precision Agriculture	(3-0)
22	Agricultural and Food Machinery Design	(2-1)
23	Food Process Engineering	(3-0)
24	Biological Process Engineering	(3-0)
25	Agricultural Product Distribution Techniques	(3-0)
26	Greenhouse Environmental Control Engineering	(3-0)
27	Optimization Methods in Engineering	(3-0)
28	Solar Energy Conversion Technique	(3-0)
29	Special Topics	(3-0)

(*Continued*)

TABLE 25.1 (CONTINUED)

30	Teramechanics	(3-0)
31	Agricultural Mechanization Strategy	(3-0)
32	Energy Analysis for Agricultural Engineering	(3-0)
Compulsory S3 Courses (Doctoral By-Research Programme)		
No.	Subject	Credits
1	Writing Qualification Examination	(2-0)
2	Oral Qualification Examination	(2-0)
3	Colloquium	(1-0)
4	Research proposal	(2-0)
5	Dissertation Seminars	(1-0)
6	National Scientific Publications	(2-0)
7	International Scientific Publications 1	(3-0)
8	International Scientific Publications 2	(3-0)
9	Closed Exam	(2-0)
10	Research and Dissertation	(12-0)

Notes: (Class-Laboratory),(2-1), 2 credits class and 1 credit laboratory practical work (practice)

example of a curriculum structure in the form of course offerings each semester (Graduate Curriculum, 2021).

Running Curricula

According to the Indonesian Higher Education Standards, the learning process comprises characterising, planning, implementing, and student learning. The learning process is conducted interactively, holistically, integrally, scientifically, contextually, thematically, effectively, collaboratively, and student-centrally.

Planning of the learning process prepared for each course is presented in the Plan of Semester Learning containing at least the study programme name, the course name and code, semester placing, credits set up, the supporting lecturer name, graduate learning course achievements, the final planned graduates, relevant learning material, learning method, learning time, student learning experience, learning assessment, and learning references utilised. The Plan of Semester Learning must be improved regularly. The student research and community services relating to learning process must be based on the research and community standards.

The learning prepared through curricular exercises must utilise successful learning strategies and methods taking after the characteristics of the course to realise certain abilities specified within the course arrangement of satisfying graduate learning results. The learning strategies alluded to can be chosen for the usage of learning in courses counting gather talks, reenactments, case ponders, collaborative learning, agreeable learning, project-based learning, problem-based learning, or other learning strategies, which can successfully encourage the fulfillment of graduate learning results. Each course can utilise one or a combination of a few learning strategies and be accommodated in a frame of learning. The execution of learning can be within the frame of addresses, reactions, and instructional exercises, classes, practica, studio hone, workshop hone, field hone, work hone, investigations, plans, or improvements, military preparations, student exchanges, apprenticeships, business work; and/or other forms of community benefit.

The learning application can be done inside and outside the higher education institution. The learning usage exterior of the higher education institution could be a learning prepare that comprises learning in other higher education institutions at the same tertiary institution; learning within the same higher education institution at diverse tertiary education institutions; learning in other higher education institutions at a diverse tertiary institution; and learning at a non-higher education institution. Learning outside of the higher education institution is conducted based on a participation assertion among higher education institutions or other related partners and the teaching outputs are acknowledged through a credit exchange component.

The application of learning exercises incorporates student trades, work, educating help in educative units, research, helpful ventures, entrepreneurial exercises, free studies, and building village building exercises/genuine practical work (Dikti GuideBook, 2020).

The student learning load is expressed in the number of credits, called the units of semester credit. A semester is an effective time unit of the learning process consisting of a minimum of 14 weeks, out of mid and final exams. Two semesters comprise one academic year. A minimum of 16 face-to-face classroom meetings, including two exams, also applies for the intermediate semester lectures.

The maximum period and minimum study load for the fulfillment of an educational programme are summarised as follows: 5 academic years and 108 credits for the D3, 7 academic years and 144 credits for the S1, 4 academic years after completing the S1 and 36 credits for the S2, and 7 academic years after completing the S2 and 42 credits for the S3. Institution of higher education is authorised to propose the period of educational programmes (D3, S1, S2, and S3) less than the maximum limit.

One credit of the learning process may account for 50 minutes of class meeting/week/semester, 60 minutes of regular laboratory or practical work/week/semester, 100 minutes of seminar/week/semester, 70 minutes of independent activity, and 170 minutes of the other learning activities/week/semester.

The student learning process and outcome assessment are performed according to assessment standards comprising principles, techniques, instruments, mechanisms, procedures, implementation, and reports.

Report of assessment for course taking success qualifications are stated: letter A equals to number 4 for the category of very good, letter B equals number 3 for the category of good, letter C equals number 2 for the category of sufficient, letter D equals number 1 for the category of less, and the letter E equals the number 0 for the category of very poor. Universities can propose sub-letters and sub-numbers to grade the interval of 0 to 4.

The graduate outcome assessments per semester are assigned for the Semester Achievement Index. The results of an evaluation of learning outcomes at the final Study Programme are revealed by the Cumulative Grade Point Average. The Semester Grade Point Average is expressed in terms of the amount determined by summing up the multiplication of the

TABLE 25.2

Example of S1 Curriculum in semester arrangement

Semester 1

		Credits		Status	
No.	Name of Course	Class	Laboratory	C	E
1	Mathematics I	2		√	
2	Physics I	2	1	√	
3	Chemistry I	2	1	√	
4	Cell Biology	2	1	√	
5	Statistics I	2		√	
6	Introduction to Agronomy	2	1	√	
7	Pancasila	2		√	
8	*Introduction to Agricultural Technology*	2		√	
9	Religion	2		√	
	Subtotal	18	4		

Semester 2

		Credits		Status	
No.	Name of Course	Class	Laboratory	C	E
1	Mathematics II	2		√	
2	Physics II	2	1	√	
3	Chemistry II	2		√	
4	Statistics II	2		√	
5	*Introduction to Information Systems*	2		√	
6	*Microbiology*	2		√	
7	*Engineering Drawing*	1	2	√	
8	*English*	2		√	
9	*Citizenship Education*	2		√	
	Subtotal	19	3		

Semester 3

		Credits		Status	
No.	Name of Course	Class	Laboratory	C	E
1	*Engineering Mathematics*	2		√	
2	*Thermodynamics*	2		√	
3	*Electricity and Electronics*	2	1	√	
4	*System Analysis and Modelling*	2		√	
5	*Climatology*	2	1	√	
6	*Programming and Computer Applications*	2	1	√	
7	*Engineering Economics*	2		√	
8	*Agricultural Material Properties*	2	1	√	
	Subtotal	17	4	√	

Semester 4

		Credits		Status	
No.	Name of Course	Class	Laboratory	C	E
1	*Engineering Mechanics*	3		√	
2	*Fluid Mechanics*	2		√	
3	*Heat Transfer*	3		√	
4	*Mass Transfer*	2		√	
5	*Hydrology*	2		√	
6	*Engineering Material Properties*	2		√	
7	*Measurement and Instrumentation Engineering*	2		√	
8	*Engineering Ethics*	1			√
9	*Farm Power*	2	1		√
10	*Ergonomics*	2			√
11	*Surveying*	2	1		√
12	*Land Resources Engineering and Management in Tropics*	2			√
13	*Biomass Cultivation Engineering*	2			√
14	*Agroclimatology*	2			√
15	*Industrial Sanitation Engineering*	2			√
16	*Storage and Packaging Engineering*	2			√
17	*Special Topics I*	2			√
	Subtotal	22	2		

TABLE 25.2 (CONTINUED)

Semester 5

		Credits		Status	
No.	Name of Course	Class	Laboratory	C	E
1	*Agricultural Production Machinery*	2	1	√	
2	*Fundamental to Soil Physics*	2	1	√	
3	*Agricultural Environmental Engineering*	2	1	√	
4	*Unit Operations*	2	1	√	
5	*Entrepreneurship*	2		√	
6	*Agriculture and Biosystem 4.0*	2		√	
7	*Biosystem Machinery Design*	2			√
8	*Mechatronics and Robotic Biosystems*	2			√
9	*Pneumatic and Hydraulic Systems*	2			√
10	*Non-destructive Evaluation Technology*	2			√
11	*Principles of Irrigation Engineering*	2			√
12	*Soil and Water Conservation Engineering*	2			√
13	*Watershed Engineering and Management*	2			√
14	*Disaster Risk Reduction in Agriculture*	2			√
15	*Biomaterial Process Engineering*	2			√
16	*Environmental Chemistry*	2			√
17	*Environmental Physics*	2			√
18	*Food Process Engineering*	2			√
19	*Drying Engineering*	2			√
20	*Refrigeration and Freezing Engineering*	2			√
21	*Food Material Science*	2			√
22	*Special Topics II*	2			√
	Subtotal	20	4		

Semester 6

		Credits		Status	
No.	Name of Course	Class	Laboratory	C	E
1	*Introduction to Control System*	2	1	√	
2	*Biophysics*	2	1	√	
3	*Agricultural Waste Treatment Process Engineering*	2	1	√	
4	*Postharvest Engineering*	2	1	√	
5	*Internship*	2		√	
6	*Biosystem Machinery Constructions*	2	1		√
7	*Bioenergy Engineering*	2	1		√
8	*Irrigation Management*	2			√
9	*Geographic Information System in Agriculture*	2	1		√
10	*Heating, Ventilation, dan Air Conditioning Facilities Design*	2			√
11	*Air Pollution Control and Management*	2	1		√
12	*Environment and Waste Management*	2			√
13	*Bioprocess Engineering*	2			√
14	*Food Material Quality Evaluation*	2			√
15	*Design Process and Equipment of Food Industry*	2			√
16	*Beverage and Confectionery Engineering*	2			√
17	*Special Topics III*	2			√
	Subtotal	15	4		

Semester 7

		Credits		Status	
No.	Name of Course	Class	Laboratory	C	E
1	*Thesis*	3		√	
2	*Community Services*	3		√	
	Subtotal	6	0		

Semester 8

		Credits		Status	
No.	Name of Course	Class	Laboratory	C	E
1	*Seminar*	2		√	
2	*Final Design Project*	4		√	
	Subtotal	6			
	Total Credits	**123**	**21**		

Notes: Lab = laboratory practical work, C = Compulsory, E = Elective letter grades for each course taken together with its credit divided by the number of credits of courses taken in a single semester. The Grade Point Average (GPA) for the study programme accomplishment is assessed using the same method.

Students of D3 and S1 are declared to pass their studies after completing all prescribed courses with a GPA equivalent to or greater than 2.00. Graduated students from the D3 and the S1 will be awarded the predicate of satisfactory, very satisfactory, or cum laude according to the next category: GPA of 2.76 to 3.00 for a satisfactory predicate, GPA of 3.01 to 3.50 for a very satisfactory predicate, and GPA of more than 3.50 for a cum laude predicate.

Graduated students from the S2 and S3 can be given the title of satisfactory, very satisfying, and cum laude according to the next category: GPA of 3.00 to 3.50 for a satisfactory predicate, GPA of 3.51 to 3.75 for a very satisfactory predicate, and GPA of more than 3.75 for a cum laude predicate.

For example, the graduation requirement for the D3 programme is to have taken 111–112 credits of courses. The S1 graduation requirements (S1) include the number of credits that must be taken: 144–148 credits consisting of 126 credits of compulsory courses and 8–22 credits of elective courses, cumulative grade point average ≥ 2.0, length of study 4 years (4–7 years); the number of credits with a D score is a maximum of 25% of the total credits, and there are no courses with an E score. For the S2, the number of credits that must be taken is 40–50 credits consisting of 32 credits of compulsory courses (20 credits lecture, 12 credits for thesis) and interest courses 8–18 credits, grade point average ≥ 3.0, length of study 2 years (2-4 years). The graduation requirements for the Gadjah Mada University S3 include: having taken ≥ 40 credits consisting of 4 credits for compulsory courses, 8 credits for interest/elective courses, and 28 credits for a dissertation; cumulative grade point index > 3.25; passing a comprehensive exam; pass the closed exam; and have published a minimum of 3 papers in accredited national scientific journals or 1 paper in accredited international scientific journals, or 1 paper in accredited national scientific journals, and open exams.

Graduate Attributes

Students who are declared to have graduated from the D3, S1, S2, and S3 are entitled to obtain diplomas, degrees, and certificates accompanying diplomas. Graduates of academic or vocational degrees have the right to use academic titles or vocational degrees or titles. Degrees for graduates of academic education consist of: a. Scholar, written behind the name of the person entitled to include the letter S. followed by the initials of the higher education or field of study; b. Master, which is written behind the name of the person entitled to include the letter M. followed by the initials of the higher education or field of study; and c. Doctor, written before the name of the person entitled to include the abbreviation Dr. The educational title for the D3 graduates, which is written behind the name of the person entitled, includes the abbreviation A.Md. and is followed by the initials of the higher education or area of expertise. For example, the degrees and titles for ABEE graduates for the D3, S1, S2, and S3 are A.Md.T, S.T., M.T., and Dr.T. (or Dr.) respectively (Nuh, 2014).

Graduates' Careers

ABEE alumni are spread across various sectors such as well-known agribusiness companies, especially plantation commodities, the banking sector, especially large government and privately-owned banks, the government itself, especially several ministries such as the Ministry of Agriculture, the Ministry of Industry, the Ministry of Research, Technology, and Higher Education, the Ministry of Environment and Forestry, the Ministry of Public Works and Public Housing, the Ministry of Manpower and Transmigration, the Ministry of Home Affairs, Indonesian research institutes, logistics affairs agencies, drug and food control agencies, regional planning agencies, private companies in various fields, and entrepreneurs. Alumni are considered by users to have a very good reputation and credibility. In recent years, the demand for ABEE education alumni from several well-known sectors, both private and government, has increased. The fields of work for ABEE alumni include production managers or supervisors in agricultural equipment/machinery companies and agricultural and food product processing factories; designers of agricultural tools/machinery or production systems/processing of agricultural products and agricultural buildings; managers at a plantation company, clearing and preparing land for agriculture and transmigration; systems analysts or programmers in an Information Technology company; irrigation and drainage network planners; entrepreneurs in the field of agricultural and food product processing, farming, and others; project evaluators in banks and consulting bureaus; executive staff in government Departments and Agencies; researchers at various research institutions, as well as teaching staff at various tertiary institutions or education and training centers. The forms of industry that most ABEE alumni are involved in include the food industry; the manufacturing industry; the processing, handling and storage of agricultural products; the application and development of control systems; tool and machine design; agroindustry entrepreneurship; and analysis-based agroindustry.

D3 graduates usually fill jobs as site managers in agroindustry companies, especially food crop agriculture, plantations, or vegetable-based products companies, engineers of simple parts/elements of agricultural machine tools in agroindustry companies or agricultural workshops, the mechanic at an agricultural workshop, operators of agricultural machine tools in agroindustry companies, especially food crops, plantations, or vegetable based products companies, agricultural machinery service business managers, and agricultural machinery entrepreneur training, rental, maintenance, and repair and marketing services.

Following the Indonesian Agricultural Engineering Association, S1 graduates work as designers, managers, system analysts, science and technology developers, and entrepreneurs, as well as ABEE D3 teachers and vocational school teachers relevant to ABEE.

Most graduates of the S2 and S3 work as planners, planning and development formulators, and policymakers in companies and government institutions, as researchers, and in various research institutes, and lecturers in tertiary institutions.

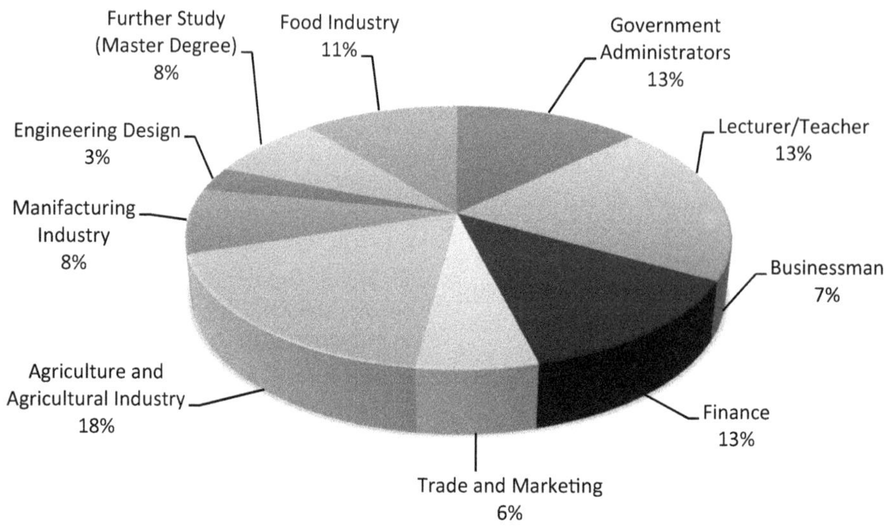

FIGURE 25.2 Agricultural and Biosystems Engineering Education graduate occupational fields (TEP Career Opportunities, 2022).

Figure 25.2 shows the percentage distribution of occupational fields or careers of ABEE graduates as a result of a tracer study conducted by Brawijaya University (TEP Career Opportunities, 2022).

From Figure 25.2, it can be seen that most ABEE alumni work in the agricultural sector and the agricultural industry, followed by 3 sectors, namely government employees, educators, and the financial sector, then the food industry sector and other fields of work.

Programme and Institution Accreditation

Accreditation is an assessment activity under predetermined criteria based on Indonesian Higher Education standards. Accreditation is carried out to determine the eligibility of study programmes and higher education institutions based on criteria that refer to the Indonesian Higher Education Standards. The government established the Indonesian Accreditation Board for Higher Education to develop an accreditation system. Study programme accreditation as a form of public accountability is carried out by an independent accreditation institution. An independent accreditation institution is an independent institution formed by the government or an independent institution formed by a community that is recognised by the government on the recommendation of the Indonesian Accreditation Board for Higher Education (Yudhoyono, 2012a).

Accreditation is an external quality assurance system as part of the higher education quality assurance system. Accreditation aims to: a. determine the eligibility of study programmes and higher education based on criteria that refer to the Indonesian Higher Education Standards; and b. externally guarantee the quality of study programmes and higher education both in the academic and non-academic fields to protect the interests of students and the community (Makarim, 2020b).

Accreditation for study programmes is carried out by an Independent Accreditation Board. If the IAB has not been formed, accreditation for the study programme will be granted by the Indonesian Accreditation Board for Higher Education. The Independent Accreditation Board and Indonesian Accreditation Board for Higher Education prepare accreditation instruments under their respective authorities and are determined under statutory provisions. according to Regulation-Indonesian Accreditation Board for Higher Educa tion-Number-1-2022-regarding-Accreditation-Mechanisms. The stages of accreditation consist of: a. evaluation of data and information; b. Accreditation rating determination; and c. monitoring and evaluation of accreditation rating (Buchori, 2022).

The data and information evaluation stage includes: a. higher education leaders submit applications to the Independent Accreditation Board for study programme accreditation and/or the Indonesian Accreditation Board for Higher Education Accreditation; and b. the Independent Accreditation Board and/or the Indonesian Accreditation Board for Higher Education evaluates the adequacy of data and information on study programmes and/or higher education, using data and information from higher education database. Evaluation of the adequacy of data and information is carried out by assessors.

The accreditation ranking stage includes: a. the Independent Accreditation Board and/or the Indonesian Accreditation Board for Higher Education processes and analyzes data and information from higher education applicants for accreditation, to determine the accreditation rating of study programmes and/or higher education; and b. the Independent Accreditation Board and/or the Indonesian Accreditation Board for Higher Education announce the accreditation rating of study programmes and/or higher education under their respective authorities.

Accreditation Criteria

The accreditation rating is determined using the accreditation criteria. The accreditation criteria are accreditation benchmarks that refer to national higher education standards. The Indonesian Accreditation Board for Higher Education determines the focus of the assessment into criteria that include the commitment of tertiary institutions through the study programme management unit to the capacity and effectiveness of education which consists of 9 (nine) criteria as follows, 1. vision, mission, objectives, and strategy; 2. caretaker, governance, and cooperation; 3. students; 4. human resources; 5. finance, facilities, and infrastructure; 6. education; 7. research; 8. community service; and 9. outcomes and the Three Higher Education Mandate achievements.

The assessment of vision, mission, objectives, and strategy is focused on clarity of direction and commitment, and consistency of the study programme development conducted by the study programme management unit to achieve performance and quality targets based on planned, effective, and directed mission and programme steps in the context of realisation of university and study programme visions.

The caretaker, governance, and cooperation assessments are focused on the leadership performance and effectiveness, the effectiveness of governance, the system of resource management, systems of quality assurance, the system of communication and information technology, objectives of programmes and activities in respect to the realisation of the vision and mission, as well as the establishment and implementation of cooperation and partnership strategies in terms of academic and non-academic sustainable manner nationally, regionally and internationally to foster the study program's competitiveness.

The assessment of students is focused on the system effectiveness of fair and objective new admission, the properness of students to lecturers, and the educational staff's ratio properness to support the effective and efficient implementation of learning programmes, as well as to foster programmes and student involvement in interests, talents, and professionalism.

The assessment of human resources is focused on the effectiveness of the system of recruitment, resource availability with respect for quantity, qualifications and competencies, programme development, rewards, punishments, and employment termination, for lecturers and education staff to perform teaching, research, and community service activities.

In finance, facilities, and infrastructure, the financial assessments are focused on adequacy, effectiveness, efficiency, and accountability, as well as the sustainability of financing to support the implementation of education, research, and community service. The facility and infrastructure assessments are focused on their fulfillment in terms of availability, accessibility, and utility as well as security, safety, and health for academic affairs and community services.

The education assessment is focused on the curriculum policy, curriculum development, and curriculum suitability within the field of study programme along with the strengths and advantages of the curriculum, academic culture, learning process, assessment system, and quality assurance system to sustain learning outcomes achievement with respect for the realisation of vision and mission of the higher education institution.

The assessment research is focused on the commitment to research excellence and suitability according to the higher education institution's vision and mission.

The assessment of community service is focused on the commitment to perform excellent community services in quantity, type, advantages, and suitability, as well as the coverage of the service area.

The assessment of the Three Higher Education Mandate outcome and achievement is focused on achieving graduate qualifications and competencies in the form of a clear picture of the profile and learning outcomes of graduates from study programmes, graduate tracing, feedback from graduate users, and public perception of graduates undergraduate learning outcomes/competences stipulated by study programmes and tertiary institutions regarding the IQF, the number and excellence of scientific publications, the number of citations, the number of intellectual property rights, and the benefits/impact of research results on the realisation of the vision and implementation of missions, as well as the contribution of community service to social development and empowerment, economy, and social welfare.

Accreditation Rating

Rating of accreditation of the study programme and higher education consists of: a. Good; b. Very Good; and c. Superior. A superior predicate is obtained if the accreditation score is more than or equal to 361, and meets the requirements for a superior rating. A very good predicate is obtained if the accreditation score is more than or equal to 361, but does not meet the requirements for a superior rating. A very good predicate can also be obtained if the accreditation score is more than or equal to 301, and less than 361, and meets the requirements for a very good rating. A good predicate is obtained if the accreditation score is more than or equal to 301, and less than 361, but does not meet the requirements for a very good rating. A good predicate can also be obtained if the accreditation score is more than or equal to 200, and less than 301, without requiring a rating. A score of more than or equal to 200, but which does not meet the requirements it needs to be accredited, and either meets or does not meet the requirements for a superior or very good rating, is still not accredited. Values below 200, either meeting or not meeting the requirements need to be accredited, the status is still not accredited.

Here is the meaning of accreditation values based on the Study Programme Accreditation Instrument (Sasongko, 2019). The achievements of the current ABEE accreditation ratings are still partially using the current and previous Study Programme Accreditation Instrument, in which the previous Study Programme Accreditation Instrument consisted of 7 criteria, namely 1.vision, objectives, and achievement strategies; 2. governance, leadership, management systems, and quality assurance; 3. students and graduates; 4. resources and human resources; 5. curriculum, learning, and academic atmosphere; 6. funding, facilities, and infrastructure and information systems; 7. research, service, or community service, and 8. cooperation.

The followings are the values and meanings of the accreditation values using the 7 criteria: A with a score of 361–400, B with a score of 301–360, C with a score of 200–300, and No accreditation with a score of less than 200.

The current ABEE study programme accreditation ratings in Indonesia are as follows. Out of 20 undergraduate study programmes, 11 are accredited A and 9 are accredited B. From the 8 D3 study programmes, 1 is accredited A, 2 are accredited B, 3 are accredited Good and 2 have not been accredited while out of 4 study programmes S2, 3 are accredited A and 1 is accredited B, and of the 2 S3 study programmes, all of them were accredited A or superior (PDDikti, 2020)

Admission of New Students

Admission of new students for the D3, S2, and S3 programmes is regulated and carried out by the respective tertiary institutions while the admission of new S1 programme students at Public Higher Education is carried out according to the ministerial regulation (Nasir, 2017). The admission of new students for the Public Higher Education S1 is carried out on the following principles: a. fair, namely not discriminating between religion, ethnicity, race, gender, age, social position, physical condition, and level of economic ability of prospective students, while still taking into account the potential and academic achievements of prospective students and the specifics of the study programme at the Public Higher Education concerned; b. accountable, namely carried out under clear procedures and criteria; and c. transparent, namely the implementation of acceptance is carried out openly and the implementation results can be accessed easily.

The admission of new students is done as follows: a. national selection to enter the Public Higher Education S1 is carried out based on the results of tracking prospective student academic achievements, called the Public Higher Education National Selection Admission; and b. the joint selection to enter the Public Higher Education S1 is carried out based on the results of a written exam in printed form and using a computer, called Public Higher Education Joint Selection Admission. In addition to these methods, higher education institutions can carry out independent selections which are regulated and determined by each higher education institution.

The implementation of the Public Higher Education National Selection Admission can be done before the implementation of the final school exam. The Public Higher Education Joint Selection Admission implementation can be done before and after the implementation of the final school exam. The independent selection was carried out after the announcement of the Public Higher Education Joint Selection Admission. results. The number of students (quota) accepted through the Public Higher Education National Selection Admission and Public Higher Education Joint Selection Admission pathways is at least 30% of the capacity of the study programme concerned. If the Public Higher Education National Selection Admission does not comply, it can be transferred to the Public Higher Education National Selection Admission and if it is strong, the Public Higher Education Joint Selection Admission quota cannot be changed, it can be transferred to the self-selection quota to meet the predetermined capacity. The Public Higher Education Joint Selection Admission quota that can be transferred to this independent quota is a maximum of 10%. Public higher education is obliged to find and recruit prospective students who have high academic potential but are economically disadvantaged and prospective students from the frontier, outermost, and disadvantaged areas to be accepted by at least 20% of all new students accepted and spread across all study programmes, which screening is also carried out through the Public Higher Education National Selection Admission, the Public Higher Education Joint Selection Admission, and the independent selection (Nasir, 2017).

Requirements for participants to take part in the Public Higher Education National Selection Admission: a. prospective participants are in the last class of secondary education who will graduate in the current year; b. prospective participants have consistently good academic achievements; c. prospective participants enter the best ranking quota in schools determined based on school accreditation; and d. prospective participants meet other requirements determined by each public higher education. Academic achievement is consistently good and prospective participants enter the best ranking quota in schools which is determined based on school accreditation assessed by the public higher education. Requirements for participants to take part in the Public Higher Education Joint Selection Admission: a. participants already have a diploma or certificate of completion of secondary education; and b. graduates of secondary education in the last three years. Provisions regarding the requirements, methods, procedures, and criteria for independent selection carried out by each public higher education are regulated and determined by the public higher education under statutory provisions.

Requirements to be accepted as a new public higher education student include: a. participate in and be declared to have passed the selection of new student admissions; b. already has an original diploma in secondary education; and c. meet the requirements set by the public higher education.

In general, the ABEE new student capacity per higher education institution ranges from 40–50 students per year. Nationally, the ABEES1 study programme acceptance index, the ratio of the enrolled students to the accepted students, meaning that the higher number is the lower the competitiveness level, is at a moderate level. The profile of the acceptance indices of the ABEE S1 programmes in the last 5 years is presented in Figure 25.3 (PDDikti, 2020).

Challenges, Opportunities, and Prospects

ABEE Challenges

ABEE enrollment data shows that the competitiveness of incoming students sharply decreased in the last 5 years. This problem must be overcome so as not to cause a decrease in quality which will lead to a decrease in the performance of the learning process. This downward trend also occurs in other agricultural-related fields of education. In Indonesia, the farming profession is seen as a disadvantaged profession. This sector is often used as a place to accommodate workers who cannot find work in other sectors. It is not surprising that the number of farmers in Indonesia has now reached 38.77 million. A phenomenon that has been happening for a long time in Indonesia is that everything related to the word agriculture is perceived as marginal by society. The name of agricultural engineering is also marginalised. The community does not yet understand precisely this term and the related field involved.

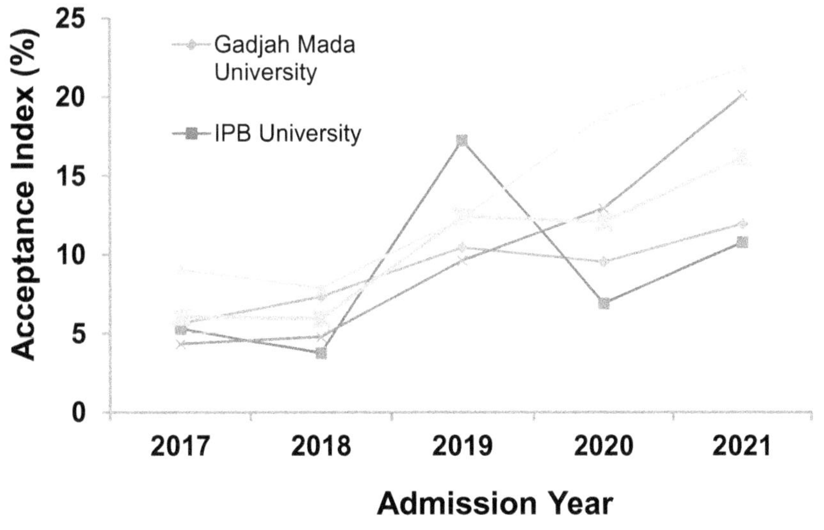

FIGURE 25.3 Agricultural and Biosystems Engineering Education incoming student acceptance competitiveness.

The agricultural term is more dominantly understood by the people than the engineering term. Some even say that agricultural engineering is engineering that is not original, moreover the fast progress in the field of engineering in the RI 4.0 era has made ABEE's name even more obscure. ABEE is considered not to employ technology as sophisticated as other engineering education. Thus ABEE students also feel that they are not as cool as other engineering education students, even though ABEE graduates already employ the engineering term as a graduate title. Such an image is also exacerbated by the fact that the majority of ABEE students come from people with lower to middle incomes whose financial capacity in education is also limited.

This limited financial capacity has an impact on the income of ABEE institutions because the student education costs are adjusted to the income of their parents, where the students with high-income parents pay high rates and the students with low-income parents pay low rates. Due to there being no financial compensation from the government respecting that problem, the small income received by ABEE educational institutions will affect their ability to provide resources for education delivery which in turn will affect the quality of the learning process. The majority of ABEE institutions also face the inability to provide proper laboratory instruments for serving a good learning process. ABEE's laboratory instruments are often less sophisticated than those available to research institutions and companies, leading to an effect on the readiness level of ABEE alumni in the workplace, which also contributes to the quality of AB. Some alumni consider that ABEE's knowledge is too broad, and the courses offered are too numerous but not in-depth which reinforces their unpreparedness in the workplace. The ABEE institution running with these typical courses configuration and content is at risk of unsustainability considering emerging media applications such as "ChatGPT". Facing this problem, in the future, it is necessary to bring up more choices of field concentrations within ABEE by offering courses with in-depth studies. Some alumni stated that many users, both government institutions and private companies, did not fully understand the competence of ABEE graduates, so they were hesitant to employ ABEE graduates. This may also be one of the reasons for the decline in public interest in studying at ABEE. Other alumni said that they are forced to work outside their fields.

The declining quality of students and the declining ability of educational institutions to provide educational resources are likely to be the cause of the unsatisfactory ABEE in Indonesia. This is reflected in the number of ABEE S1 programmes, around half of which are still rated B accreditation.

ABEE Opportunities

These above facts indicate that revitalising ABEE by adding biosystem terminology to ABEE has not been able to improve ABEE's image in society. This suggests that ABEE branding needs to be pursued by generating big ideas and creating great works in future related fields.

ABEE branding needs to be done step by step. Limitations of learning facilities are overcome by utilising the facilities available at companies and research institutions, through networking. Collaborative learning methods and cooperative learning can be used for this. Companies can also be used as partners as a vehicle for learning by mutually reinforcing cooperation. Universities can find solutions to real problems faced by companies, as well as become learning media. In this case, collaborative-problem-based-learning may be proposed. The ABEE institution together with companies can build joint projects that generate added value by implementing project-based learning.

ABEE learning must also be in touch with high technology perceived by young people today such as ICT, IoT, AI, drones, independent machines, and so on. This effort should be done while increasing learning resources. By doing this, it can align with other engineering educations. If the ABEE brand image increases and the ABEE public interest increases tremendously, then the number of ABEE institutions can be expanded to other tertiary institutions since in Indonesia there are 2990 private tertiary institutions and 125 state tertiary institutions. Indonesia also has 24.33 million school students.

ABEE Prospects

The ABEE branding is advanced with ABEE's science and technology contents oriented to utilise the potential of human resources and local natural resources to create added values having an impact on income or economic performance while contributing to finding solutions for the world problems, such as food crises, energy crises, agricultural land crises and so on, to encourage progress and prosperity of nations and human welfare.

As an illustration, Indonesia's territory is 5.2 million km2 (37% land and 63% sea) 5,110 km long and 1,887 km wide consisting of 17,508 islands with the second largest biodiversity in the world after Brazil, which is inhabited by 275.77 million people comprising 1340 tribes and 715 languages. The total area of Indonesia's wetlands is 7,463,948 ha. The area of oil palm plantations reaches 16.38 million ha, the largest in the world. The area of the coffee plantation is 1,230,001 ha. The area of coconut plantations reaches 3,654,478 Ha, meanwhile, the area of rubber plantations reaches 3.69 million ha. Indonesia is estimated to have up to 33,393,570 ha of swampland. The Indonesian peatland is about 1,343 million ha. Indonesia reaches 81,290 km, placing Indonesia in second position as a country having the longest coastline in the world after Canada. The number of Indonesian people working in the agricultural sector is 38.77 million. Through agricultural mechanisation, ABEE should be able to contribute to improving the management and utilisation of these natural resources. The ABEE Curricula should contain science and technology on wetland mechanisation both on-farm and off-farm to improve the efficiency of rice production and postharvest processing to ensure adequate food needs in Indonesia since rice is an Indonesian staple food. Agricultural mechanisation also needs to be brought into the plantation sector, especially to streamline land preparation, maintenance of crops, and evacuation of harvested products from the fields. To do this the challenge is not simple because paddy fields and plantation areas in Indonesia are hilly, which makes it difficult to be accessed by agricultural machinery and equipment. Thus the mechanisation of agriculture requires the design of large-scale drones which is both a challenge as well as an opportunity for ABEE. Agricultural mechanisation can also be employed for agricultural cultivation in swamp areas. For this purpose, it is necessary to prepare amphibious system agricultural tractors to actuate agricultural equipment. Specially designed machines and equipment are also required for agriculture in Indonesia's coastal sand areas. This is also a challenge and an opportunity for ABEE. The success in cultivating hostile areas for agriculture will help the world to overcome the food crisis.

Agricultural mechanisation requires enormous energy to operate the machines and equipment in one hand. But on other hand, the world is overshadowed by an energy shortage, especially fossil energy. So it is necessary to design special proposed machines, especially those operating on farms, to be driven by renewable energy, such as solar, wind, hydro, and ocean wave energies. The energy of biomass can also be explored for utilisation. The biomass resource potential in Indonesia is about 50,000 MW coming mainly from oil palm, sugar cane, and coconut plantations and waste forest products. The current latest report indicates that the potential of world biomass energy is about 42 EJ–350 EJ in 2100 (Exa = 10^{18}). This effort can be dedicated to ABEE's contribution to dealing with the world energy crisis.

Agricultural mechanisation needs to be operated massively in the postharvest stage to accelerate the preparation of raw materials for the agricultural product processing industry to enhance the downstream process and diversification of the end products of this industry. The down-streaming of the agricultural industry, in addition to increasing the added value of products, will also create thousands of industries/factories bringing up massive jobs. These new jobs can be used as a way to migrate workers from the on-farm sector to the off-farm sector. If this is done on a national scale, the agroindustry can be used as a locomotive for the transformation of agrarian countries into industrial countries which will increase people's incomes leading to human prosperity.

Emerging technology in ABEE can be explored to facilitate and enable farming on lands that are difficult even those that currently cannot be cultivated, such as swamps, peatlands, and sandy soils, even deserts. ABEE institutions around the world can become pioneers to set up research projects involving other educational institutions to green the desert. Solar energy can be harvested and conversed to electricity here. Giant–scale power banks can be built to supply the electricity to operate engines for various farm machinery and equipment. Employing these devices seawater can be processed into fresh water and streamed as the water supply for desert agriculture to produce food. By doing this successfully, ABEE will contribute favours to the world in overcoming the food crisis, which currently has more than 350 million population suffering from an acute food crisis. This success will later become the ultimate of ABEE's service to the world.

REFERENCE LIST

Agricultural Engineering and Biosystem S1 UB. (2022). Faculty of Agricultural Technology, Brawijaya University. https://tep.ub.ac.id/pendidikan/program-studi/sarjana-teknik-pertanian-dan-biosistem/ (Accessed on December 20, 2022).

Agricultural Engineering and Biosystem UNAND. (2022). Department of Agricultural Engineering and Biosystem, Faculty of Agricultural Technology, Andalas University: http://tpb.fateta.unand.ac.id/profil/ (Accessed on December 20, 2022).

Agricultural Mechanisation Technology. (2019). Public Agriculture Polytechnic Paykumbuh. https://tmp.ppnp.ac.id/kurikulum/ (Accessed on December 15, 2022).

Buchori, I. (2022). Regulation of the National Accreditation Board of Higher Education Number 1 of 2022 Concerning Accreditation Mechanism for the Accreditation That Was Performed by the National Accreditation Board of Higher Education, 10 p.

Curriculum K2020. (2021). Master Programme of Agricultural Engineering and Biosystem, Faculty of Agricultural Technology, IPB University. https://fateta.ipb.ac.id/id/master-of-agricultural-and-biosystem-engineering/ (Accessed on March 10, 2023).

Curriculum K2020. (2022). Doctoral Programme of Agricultural Engineering Science, Faculty of Agricultural Technology, IPB University. https://pasca.ipb.ac.id/wp-content/uploads/2022/08/K2020-TEP-S3-1.pdf (Accessed on December 17, 2022).

The Curriculum of the Doctoral Programme. (2021). Doctoral Programme of Agricultural Engineering Science, Faculty of Agricultural Technology, Gadjah Mada University. https://tpb.tp.ugm.ac.id/id/pendidikan/program-s3/kurikulum-program-doktoral (Accessed on December 17, 2022).

D3 Agricultural Mechanisation Technology Poliek Payakumbuh. (2022). https://pmb.ppnp.ac.id/program-studi-detail/ (Accessed on December 11, 2022).

D3 Graduate Competency. (2021). Diplome D3 Agricultural Mechanisation Technology POLBAN BOGOR (2022). https://polbangtan-bogor.ac.id/page/teknologi-mekanisasi-pertanian / (Accessed on December 24, 2022).

Department of Mechanical Engineering and Biosystem IPB. (2022). Faculty of Agricultural Technology, IPB University. https://ipb.ac.id/faculty/index/faculty-of-agricultural-technology/department-of-mechanical-and-biosystem-engineering (Accessed on March 10, 2023).

Dikti GuideBook. (2020). Guidebook of Independent Learning for Independent Campus. Directorate General of Higher Education Ministry of Education and Culture, 1st Edition, 34 p.

DTPBSFTPUGM. (2021). Operating Procedure 2021 Curriculum and Transitional Rules Study Programme of Bachelor Agricultural Engineering, 32 p.

Education Guide 2020/2021 FTP UB. (2020). Education Guide Faculty of Agricultural Technology, Brawijaya University, 305 p.

Graduate Curriculum. (2021). Department of Agricultural Engineering and Biosystems, Faculty of Agricultural Technology, Gadjah Mada University. https://tpb.tp.ugm.ac.id/id/pendidikan/program-s1/kurikulum-program-sarjana (Accessed on December 17, 2022).

History of Agricultural Engineering GMU. (2009). Department of Agricultural Engineering and Biosystems, Faculty of Agricultural Technology, Gadjah Mada University. https://tpb.tp.ugm.ac.id/id/sejarah-teknik-pertanian-ugm (Accessed on December 18, 2022).

Junaidi, A., Wulandari, D., Arifin, S., Soetanto, H., Kusumawardani, S.S., Wastutiningsih, S.P., Utama, M.S., Cahyono, E., Hertoto, G.F., Syam, N.M., Jumaipa, H.W.Y., Putra, P.H., Wijayanti, C., & Jobih. (2020). Preparation of Higher Education Curriculum in the Era of Industry 4.0 to Support the Independent Learning-Independent Campus, DJPTTKPK Press, 4th Edition, 140 p.

Makarim, N.A. (2020a). Regulation of the Minister of Education, and Culture of the Republic of Indonesia Number 3 of 2020 Concerning National Standards for Higher Education, 74 p.

Makarim, N.A. (2020b). Regulation of the Minister of Education and Culture Republic of Indonesia Number 5 of 2020 Concerning Accreditation of Study Programmes and Higher Education, 30 p.

Master Programme Curriculum. (2021). Department of Agricultural Engineering and Biosystems, Faculty of Agricultural Technology, Gadjah Mada University.

Nasir, M. (2015). Regulation of the Minister of Research, Technology and Higher Education Republic of Indonesia Number 44 of 2015 Concerning National Standards for Higher Education, 71 p.

Nasir, M. (2017). Regulation of the Minister of Research, Technology, and Higher Education Republic of Indonesia Number 90 of 2017 Concerning Admission of New Graduate Programme at State Higher Education, 11 p.

Nuh, M. (2014). Regulation of the Minister of Education and Culture Republic of Indonesia Number 154 of 2014 Concerning Science and Technology Cluster and Graduate degrees College, 64 p.

Nuh, M. (2014). Regulation of the Minister of Education and Culture Republic of Indonesia Number 49 of 2014 Concerning National Standards for Higher Education, 53 p.

Ornstein, A.C., & Hunkins, F.P. (2014). *Curriculum: Foundations, Principles, and Issues*. Pearson Education Ltd, Edinburgh Gate, Harlow, Essex CM20 2 JE, England. Printed and bound in Vivar, Malaysia. ISBN13:978- 1-978-292-16207-2.

PDDikti. (2020). Higher Education Database, Ministry of Education and Culture. https://pddikti.kemdikbud.go.id/ (Accessed on December 27, 2022).

Putri, M.S. (2003). The President Republic of Indonesia, Laws of the Republic Indonesia Number 20 of 2003 Concerning National Education System, 37 p.

S1 Graduate Profile IPB. (2020). Department of Mechanical Engineering and Biosystem Faculty of Agricultural Technology, IPB University. https://tep.ub.ac.id/s1-tep/pendidikan/profil-lulusan/ (Accessed on December 29, 2022).

S1 Graduate Profile UNPAD. (2021). Study Programme of Agricultural Technology, Faculty of Agricultural Industrial Technology, Padjadjaran University. https://ftip.unpad.ac.id/tep-profil-lulusan/ (Accessed on December 30, 2022).

S1 Programme. (2011). Department of Agricultural Engineering and Biosystem, Faculty of Agricultural Technology, Gadjah Mada University. https://tpb.tp.ugm.ac.id/id/pendidikan/program-s1 (Accessed on December 26, 2022).

Sasongko, D. (2019). Regulation of Higher Education Accreditation Board Number 5 of 2019 Concerning Study Programmes Accreditation Instrument, 4 p.

Satria, A. (2020). Decree of Rector of Bogor Agriculture University Institute Concerning Graduate Education Programme Guide Bogor Agricultural Institute Year 2020 Edition, 679 p.

Suherman, M., Wijoyo, H., & Indrawan, I. (2020). Industry 4.0 vs Society 5.0. Pena Persada, Banyumas, 1st Edition, 109 p.

TEP Career Opportunities. (2022). Career Opportunities of Agricultural Engineering and Biosystem, Department of Biosystem Engineering, Brawijaya University. https://tep.ub.ac.id/s1-tep/kemahasiswaan-alumni/peluang-karir/ (Accessed on December 17, 2022).

Yudhoyono, S. B. (2012a). Laws of the Republic Indonesia Number 12 of 2012 Concerning Higher Education, 65 p.

Yudhoyono, S.B. (2012b). Regulation of the President of the Republic Indonesia Number 8 of 2012 Concerning Indonesia National Qualification Framework, 8 p.

Zais, R. S. (1976). *Curriculum: Principle and Foundations*. New York: Harper & Row.

26

Evolution and Status of Agricultural and Bioresources Engineering Education at the University of Nigeria

Chinenye C. Anyadike and Joel N. Nwakaire

Introduction

Agricultural and Bioresources Engineering is a discipline that deals with the design, construction, and management of agricultural systems, as well as the use of biological resources for energy, food, and other human needs. The goal of this field is to find sustainable and environmentally friendly ways of producing food, fibre, and other products from the land, and to make use of renewable energy sources. The vital roles Agricultural and Bioresources Engineering plays in providing solutions to the world's pressing problems, such as food security, environmental degradation, and energy production cannot be overemphasised. This rapidly evolving field that plays a critical role in addressing the challenges facing agriculture and the environment, encompasses a wide range of activities, including soil and water conservation, crop production, waste management, renewable energy, and rural infrastructure. Solving these complex issues in agriculture and environment needs the integration of interdisciplinary approaches, and this requires collaboration between engineers, scientists, and stakeholders from different fields and sectors, as well as an appreciation for the social, economic, and cultural context in which agricultural and bioresources systems operate. Hence, the curricula of Agricultural and Bioresources Engineering programmes typically include courses in core engineering disciplines, such as mathematics, physics, and computer science, as well as specialised courses in areas like irrigation, soil mechanics, and animal production. Many programmes also include hands-on laboratory and field experiences, which provide students with practical skills and knowledge.

In recent years, there has been a growing recognition of the need for reforms in Agricultural and Bioresources Engineering education. This is driven by changes in the world's population, food systems, and the environment, which are putting increasing demands on natural resources and creating new challenges for agriculture. Furthermore, the incorporation of new technologies and innovation such as advances in precision agriculture and digital technologies are revolutionising the way that agriculture is managed, and it is essential that engineers are trained in these technologies so that they can contribute to their development and implementation. Reforms in Agricultural and Bioresources Engineering curricula also aim to promote an awareness of the ethical and social responsibilities of engineers. This includes an understanding of the impacts of engineering solutions on different communities, as well as the importance of considering social and environmental issues in the design and development of new technologies.

The need to keep pace with the changing needs of society and the environment has always necessitated reforms in Agricultural and Bioresources Engineering curricula. This is to ensure that graduates are equipped with the knowledge, skills, and attitudes they need to meet the changing trends and challenges of the 21st century and contribute to sustainable development (Adepoju & Nwulu, 2020). The reforms that have been implemented in recent years have helped to ensure that ABE graduates have the knowledge and skills they need to meet the challenges of the 21st century. Globally the discipline of Agricultural Engineering has undergone considerable changes in orientation during the past decade in response to both external and internal pressure affecting universities and agriculture related projects (Opara, 2004).

Historical Perspectives on the Evolution of ABE at the University of Nigeria

Established in 1962, the University of Nigeria (UNN) Department of Agricultural Engineering (as it was known then) is the oldest in Nigeria. Started and developed by academic personnel from Michigan State University (M.S.U), the Department benefited from the best of the land-grant philosophy of many of the bigger and better known American Universities, a philosophy that found optimum expression in their Faculties of Agriculture. So, at the beginning under the Headship of Professor J.O. Boyed, the Department together with the Department of Agricultural Mechanisation was based in the Faculty of Agriculture for administrative purposes while, at the same time, it was based in the Faculty of Engineering for its academic regulation and administration. As a result of this arrangement, the Department was very well equipped as it had free use of, and direct responsibility for maintenance and repair of, all the agricultural machinery and equipment in the Farm Operations Unit of the Faculty of Agriculture. To facilitate the

DOI: 10.1201/9780429150111-28

discharge of that responsibility the Department was provided with a very well equipped Fabrication Workshop, a Carpentry/Farm Structures Workshop, and a Farm Power and Machinery Workshop in the Farm Operations Centre. Similarly, with the first two years of the five-year programme being the same for all the Departments of the Faculty of Engineering, students of Agricultural Engineering Department took the foundation engineering courses and the workshop technology courses in well-furnished mechanical engineering workshops, civil engineering workshops/laboratories and electrical engineering workshops/laboratories (Agricultural and Bioresources Engineering, 2021)

The Department, therefore, had a very solid foundation and, until 1966, enjoyed steady growth and development under the combined efforts of the MSU team and the Dutch-Technical-Aid academic personnel, headed from 1964 by Dr. B.F. Cargill (Late) of MSU. The first set of graduates literally graduated on the eve of the Nigerian civil war in 1967. After the ruins and devastation of the three years of civil war the Department struggled to re-establish itself and succeeded admirably in doing so.

The interruption caused by the civil war notwithstanding, the Department of Agricultural and Bioresources Engineering (DABE), has for years continued to play a pioneering and leadership role in the development of the discipline of Engineering in Nigeria. DABE has done so by contributing and helping to establish most of the other Agricultural Engineering Departments in the country, through assistance with programme/curriculum development, external examination, staff assessment/recruitment, and training of academic staff. Since 1975, staff members of the Department have served the national professional society, the Nigerian Institution of Agricultural Engineers (NIAE), which is a division of the Nigerian Society of Engineers, in various capacities. Staff members of the Department have also served as active members of the Advisory Committee and resource persons for the National Centre for Agricultural Mechanisation. In 2004, the Department pioneered the name change to Agricultural and Bioresources Engineering in line with global developments, to strengthen the discipline for the production of food, feed, and fibre, and to incorporate the relevant specialties of Aquacultural and Forest and Biomass Engineering into its curriculum. This development has since become a model to other Departments in Nigerian universities

The revised curriculum provided four options of minor Specialisation in the final year of study, namely (1) Power Systems, Machinery, and Mechatronics Engineering (2) Soil, Water Resources and Aquacultural Engineering, (3) Food and Bioprocess Engineering, and, (4) Structures and Environmental Engineering. New courses like Aquacultural Engineering, Forest and Biomass Engineering, Precision Agriculture, Decision Support and Data Acquisition Systems, Renewable Energy, etc., were introduced to reflect current developments in the discipline. The aim of adding Bioresources Engineering to the curriculum was to embrace the full range of its capabilities in training and providing the needed personnel for the development of national self-sufficiency in food, fibre, and timber products. This development has since become a model to other Department of Agricultural Engineering in Nigerian universities (Mbajiorgu, 2017). In the first two years of the programme, study concentrates on the engineering sciences common to all engineering disciplines. From the third year onwards, discipline specialisation courses are introduced. This finally leads to students, in the final year of the programme, taking courses with emphasis on one of the aforementioned options: However, there is still need for curriculum revision after more than a decade of the last review, considering the emerging technological trend and the 4.0 industrial revolution. The Nigerian University Commission (NUC) and Council of Regulation of Engineering Practices (COREN) are current driving the curriculum change in universities (COREN, 2020).

Perspectives on Future Dimensions of ABE Curriculum Development.

Agricultural and Bioresources Engineering is a rapidly evolving field that is crucial to addressing the challenges of sustainable development and food security. As such, the development of the Agricultural and Bioresources Engineering curriculum must keep pace with these changes to ensure that students are equipped with the knowledge and skills they need to succeed in this field. A future-oriented perspective on Agricultural and Bioresources Engineering curriculum development considers the latest trends, innovations, and technologies in the industry, as well as the evolving needs of society and the environment. By adopting a forward-thinking approach to curriculum development, we can ensure that Agricultural and Bioresources Engineering graduates are well-prepared to tackle the challenges of the future and contribute to a more sustainable and equitable world (Dada & Oyetunde, 2020).

Agricultural and Bioresources Engineering Education: The Need for Reforms

The world is facing numerous challenges related to food security, environmental sustainability, and the efficient use of resources. The global population is expected to reach 9.7 billion by 2050, and food production must increase by 70% to meet the increasing demand (Tripathi et al., 2019), As stated earlier, Agricultural and Bioresources Engineering plays a critical role in addressing these challenges by providing innovative solutions to improve agricultural productivity, reduce waste, and conserve resources.

However, the field of Agricultural and Bioresources Engineering has evolved significantly in recent years, with new technologies, equipment, and processes being developed and implemented. The rapid pace of change in the industry highlights the need for curricula reforms to ensure that students receive education and training that is up-to-date and relevant to the needs of the industry. The curricula of Agricultural and Bioresources Engineering programmes must keep pace with these changes to provide students with the knowledge, skills, and abilities required to succeed in the industry. This necessitated the ongoing curriculum review in Nigerian universities.

Agricultural and Bioresources Engineering (ABE) curricula reform is of utmost importance in today's world as it plays a crucial role in addressing the challenges faced by the agriculture and bioresources industries. The ABE field is an interdisciplinary field that combines engineering principles and practices with biological and agricultural sciences to create sustainable and efficient agricultural and bioresources systems.

The demand for food production is increasing globally, and the agriculture and bioresources industries are facing several challenges, such as climate change, declining soil fertility, and water scarcity, to mention a few. ABE curriculum reform is necessary to equip students with the necessary skills and knowledge to tackle these challenges and provide sustainable solutions for the agriculture and bioresources industries.

ABE curricula reforms should focus on incorporating the latest technologies and practices in agriculture and bioresources engineering, such as precision agriculture, renewable energy, and waste management. These technologies are crucial in improving the efficiency and sustainability of agriculture and bioresources systems, and ABE students should be well-versed in these areas.

Moreover, ABE curriculum reform should also focus on interdisciplinary education, as the ABE field encompasses several disciplines, including engineering, biology, and agriculture. This interdisciplinary education is crucial in providing students with a well-rounded education, which is essential in solving complex problems in the agriculture and bioresources industries.

ABE curricula reform should also emphasise hands-on learning experiences. This can be achieved by incorporating practical projects, internships, and co-op programmes into the curriculum. These hands-on learning experiences provide students with a deeper understanding of the practical applications of ABE concepts, which is crucial in preparing them for careers in the agriculture and bioresources industries.

ABE curriculum reform is essential in ensuring that students are equipped with the necessary skills and knowledge to address the challenges faced by the agriculture and bioresources industries. The reforms should focus on incorporating the latest technologies and practices in ABE, interdisciplinary education, and hands-on learning experiences. These reforms will play a crucial role in creating a sustainable future for the agriculture and bioresources industries.

The Importance of Agricultural and Bioresources Engineering Curricula Reforms

Agricultural and Bioresources Engineering curricula reforms are critical for ensuring that engineers are equipped to address the complex challenges facing agriculture and the environment. By integrating interdisciplinary approaches, incorporating new technologies and innovations, and promoting an awareness of ethical and social responsibilities, Agricultural and Bioresources Engineering education can contribute to the development of sustainable solutions for the future. Other benefits of curricula reforms are as discussed below.

Alignment with Industry Needs

One of the key benefits of curriculum reforms in Agricultural and Bioresources Engineering is the alignment of education and training with the needs of the industry. The reformed curricula must reflect the current and emerging trends, challenges, and opportunities in the industry to ensure that students receive the education and training they need to be successful in their careers. The reforms should focus on the integration of new technologies, equipment, and processes into the curriculum to provide students with hands-on experience and practical skills relevant to the industry.

Agricultural and Bioresources Engineering curricula play a crucial role in preparing students for careers in the agriculture and bioresources industries. It is essential that these curricula align with industry needs to ensure that graduates are equipped with the knowledge and skills necessary to succeed in their chosen careers.

To align with industry needs, curricula should cover topics that are relevant to the current and future demands of the agriculture and bioresources industries. For example, courses in precision agriculture, sustainable agriculture, and renewable energy are becoming increasingly important as the industry shifts towards more sustainable and efficient practices. Additionally, courses in food safety, bioprocessing, and waste management are essential for students who wish to work in the food and bioproducts industries.

In addition to covering relevant topics, curricula should also provide students with hands-on experience through internships, co-op programmes, and other experiential learning opportunities. These experiences give students the opportunity to apply what they have learned in the classroom to real-world situations, and they also provide students with valuable connections to industry professionals.

Another important aspect of alignment with industry needs is ensuring that curricula are flexible enough to adapt to changing industry demands. This can be achieved through continuous review and revision of course offerings, as well as through partnerships with industry organisations and associations. These partnerships can provide valuable insights into the skills and knowledge that are most in demand, and they can help to ensure that curricula are constantly evolving to meet the changing needs of the industry.

Finally, it is important to provide students with access to the latest tools and technologies used in the industry. This can be achieved through partnerships with industry organisations and through the inclusion of hands-on laboratory experiences in the curricula. In this way, students can gain hands-on experience with the latest technologies and techniques, and they can be better prepared to succeed in their careers.

In conclusion, alignment with industry needs is critical to the success of Agricultural and Bioresources Engineering curricula. By covering relevant topics, providing hands-on experience, being flexible and adaptive, and providing access to the latest tools and technologies, these curricula can ensure that graduates are well-prepared to succeed in their careers.

Preparing Students for the Future

Agricultural and Bioresources Engineering curricula reforms must also focus on preparing students for the future of the industry. The industry is evolving rapidly, and students must be equipped with the knowledge, skills, and abilities required to adapt to changing trends and challenges. Reforms should include the integration of interdisciplinary subjects such as data analytics, artificial intelligence, and renewable energy to ensure that students are prepared for the future of the industry.

The field of Agricultural and Bioresources Engineering is crucial for sustainable development and food security in the future. It requires engineers who are equipped with the latest knowledge and skills to tackle the challenges facing the agricultural and bioresources sector. As such, preparing students for the future in Agricultural and Bioresources Engineering Curricula should focus on providing a comprehensive education that equips them with the technical, practical, and soft skills needed to succeed in this field (Aliu, 2021).

Technical skills are the foundation of an agricultural and bioresources engineer's education. Students must be trained in the latest technologies, methods, and practices used in the industry. This includes topics such as precision agriculture, sustainable resource management, and bioprocessing. The curriculum should provide hands-on training and real-world experience in these areas to ensure students are ready for the challenges they will face in their careers.

Practical skills are also crucial for success in this field. Agricultural and bioresources engineers must be able to design, build, and operate systems that meet the needs of the agricultural sector. This requires a strong understanding of engineering principles and an ability to apply these principles in real-world situations. The curriculum should include laboratory and field-based experiences that allow students to apply what they have learned and develop the practical skills needed to succeed in their careers.

In addition to technical and practical skills, agricultural and bioresources engineers must also possess soft skills such as communication, teamwork, and leadership. These skills are critical for success in this field as engineers must work with stakeholders from a variety of backgrounds, including farmers, government agencies, and industry partners. The curriculum should include opportunities for students to develop these skills through group projects, presentations, and leadership activities.

Finally, preparing students for the future in Agricultural and Bioresources Engineering Curricula must also include a focus on sustainability and ethical considerations. Agricultural and bioresources engineers play a critical role in balancing the needs of the agricultural sector with the needs of the environment. The curriculum should include courses that address sustainability and ethical considerations in engineering practices, such as sustainable resource management, environmentally responsible design, and responsible technology use.

In conclusion, preparing students for the future in Agricultural and Bioresources Engineering Curricula requires a comprehensive education that equips them with the technical, practical, and soft skills needed to succeed in this field. This includes a focus on the latest technologies and practices, hands-on training and real-world experience, soft skills development, and a commitment to sustainability and ethical considerations. By providing students with this education, we can ensure that they are well equipped to tackle the challenges facing the agricultural and bioresources sector and help drive sustainable development and food security in the future.

Promoting Sustainability

Another important aspect of Agricultural and Bioresources Engineering curricula reforms is the promotion of sustainability. The field of Agricultural and Bioresources Engineering plays a critical role in promoting sustainable agriculture, reducing waste, and conserving resources. Reforms should focus on incorporating sustainable practices into the curriculum, including the use of renewable energy sources, waste management, and conservation of natural resources. This will help prepare students for careers in the industry that are focused on sustainability and environmental protection.

Promoting sustainability in Agricultural and Bioresources Engineering Curricula is crucial for the future of our planet. Sustainability involves balancing economic, social, and environmental factors to meet the needs of current and future generations. Agricultural and Bioresources Engineering is a unique field that provides solutions to improve the efficiency of the agricultural and bioresources industries, while also protecting and preserving the natural resources that support them.

To promote sustainability in the curricula, there are several key areas that need to be addressed. Firstly, students should be taught about the importance of sustainable practices in agriculture and the impact of traditional farming practices on the environment. This includes topics such as soil conservation, water management, and waste reduction.

Secondly, the curriculum should include courses on renewable energy sources, such as wind and solar power, and how they can be used in the agricultural and bioresources industries. The use of technology such as precision agriculture, remote sensing, and geographic information systems, can help farmers make more informed decisions about their land and resources, leading to more sustainable practices.

Thirdly, the curriculum should include courses on sustainable design, which focuses on designing systems and processes that minimise waste and reduce environmental impacts. This could include topics such as life cycle analysis, waste reduction and management, and environmental impact assessments.

Finally, students should be encouraged to take an interdisciplinary approach to sustainability by collaborating with students and faculty from other fields, such as environmental science, biology, and economics. This will provide students with a well-rounded understanding of sustainability and the ability to tackle complex sustainability challenges.

In conclusion, promoting sustainability in Agricultural and Bioresources Engineering Curricula is essential to prepare future engineers to address the complex challenges of sustainable agriculture and bioresources. The inclusion of sustainability in the curriculum will help students develop the skills and knowledge necessary to create a more sustainable future for our planet.

Encouraging Innovation and Creativity

Innovation and creativity are crucial components in the field of Agricultural and Bioresources Engineering. To foster these qualities in students, it is necessary to encourage them in the curriculum. The following are some of the ways in which this can be achieved:

1. Problem-Based Learning (PBL) – This teaching method allows students to work on real-world problems and develop innovative solutions. This approach helps students to apply their technical knowledge and think critically, which is essential for creating innovative solutions.
2. Interdisciplinary Approach – Agricultural and Bioresources Engineering draws from many disciplines including biology, chemistry, physics, and mathematics. Encouraging students to take courses in these areas will broaden their perspectives and lead to creative thinking.
3. Incorporating Entrepreneurship – By teaching students how to turn their ideas into business ventures, they can learn how to be innovative and creative while also considering the economic viability of their solutions.
4. Encouraging Students to Participate in Research – Allowing students to work on research projects and presenting their findings at conferences can provide a platform for students to showcase their innovative ideas.
5. Incorporating Industry Visits and Workshops – Visiting companies and attending workshops to understand industry needs can provide students with a better understanding of real-world problems and the importance of innovation and creativity.
6. Cross-cultural Collaboration – Encouraging students to collaborate with their peers from different cultures can foster new perspectives and lead to more creative solutions.

In summary, incorporating innovative and creative thinking in Agricultural and Bioresources Engineering curricula is essential to prepare students for the challenges they will face in the future. By incorporating problem-based learning, interdisciplinary approaches, entrepreneurship, research opportunities, industry visits, and cross-cultural collaboration, universities can ensure that their students are well equipped to tackle the challenges of the future.

Integrating Interdisciplinary Approaches in ABE Curriculum Reforms

Integrating interdisciplinary approaches in agricultural and bioresources engineering curricula reforms is essential for providing students with a comprehensive and practical education in this field. Agricultural and bioresources engineering involves the application of engineering principles and technology to agriculture, forestry, and other natural resource-based industries. As a result, students in this field require a broad understanding of subjects ranging from biology, ecology, and agronomy to mechanics, thermodynamics, and electrical engineering.

In recent years, there has been a growing recognition of the need for interdisciplinary approaches in education to address the complex and interconnected challenges facing the agricultural and bioresources sector. For example, the use of precision agriculture technologies and the development of sustainable agriculture systems require a deep understanding of both engineering principles and agronomic practices. Similarly, the design and implementation of sustainable forestry systems and the management of renewable energy resources require a blend of engineering and ecological knowledge.

To effectively integrate interdisciplinary approaches in agricultural and bioresources engineering curricula, it is important to involve experts from a range of fields in the curriculum development process. This includes agricultural scientists, agronomists, ecologists, and engineers. Collaborating with these experts will ensure that the curriculum is grounded in real-world challenges and that it prepares students for careers in the industry.

The curriculum should also incorporate hands-on learning opportunities that allow students to apply the interdisciplinary knowledge they have gained in real-world settings. For example, students could participate in interdisciplinary projects that involve designing and implementing sustainable agriculture systems, developing precision agriculture technologies, or managing renewable energy resources. These projects could be carried out in collaboration with industry partners and could provide students with valuable experience and exposure to the realities of the field (Ismail & Mohammed, 2015).

Another way to integrate interdisciplinary approaches in agricultural and bioresources engineering curricula is to incorporate cross-disciplinary courses. For example, a course in sustainable agriculture could be taught by an agronomist and an engineer, with each bringing their unique perspectives and expertise to the course. This approach allows students to learn from experts in both fields and to understand the interconnections between engineering principles and agronomic practices.

It is also important to incorporate interdisciplinary approaches into research projects in the field. This could involve collaboration between agricultural scientists, engineers, and other experts to address real-world challenges facing the agricultural and bioresources sector. For example, a research project could focus on developing sustainable agriculture systems that incorporate precision agriculture technologies and that are designed to reduce environmental impact. This type of research project would require a blend of expertise from multiple fields, including agronomy, ecology, and engineering.

Integrating interdisciplinary approaches in agricultural and bioresources engineering curricula reforms is essential for providing students with a comprehensive and practical education in this field (Rickus, 2006). This can be achieved through involving experts from a range of fields in the curriculum development process, incorporating hands-on learning opportunities and cross-disciplinary courses, and incorporating interdisciplinary approaches into research projects in the field. By doing so, students will gain a deep understanding of the complex and interconnected challenges facing the agricultural and bioresources sector, and they will be well-prepared for careers in this dynamic and growing field.

Components of Agricultural and Bioresources Engineering Curriculum Reforms

Agricultural and Bioresources Engineering curricula reforms typically consist of several key components. These components include:

Interdisciplinary education: Agricultural and Bioresources Engineering is a multidisciplinary field that requires knowledge and skills from multiple fields, including agriculture, biology, and engineering. Agricultural and Bioresources Engineering curriculum reforms aim to provide students with a comprehensive interdisciplinary education that prepares them for careers in the field. This interdisciplinary education helps students to understand the complex relationships between the different components of Agricultural and Bioresources Engineering and how they interact to provide sustainable and innovative solutions.

Hands-on experience: Agricultural and Bioresources Engineering is a practical field that requires hands-on experience in order to develop the necessary skills and abilities. Agricultural and Bioresources Engineering curricula reforms often include hands-on experiences such as laboratory work, field work, and design projects. These experiences help students to develop practical skills that are essential for careers in the field and support their professional development.

Focus on sustainability: Agricultural and Bioresources Engineering curricula reforms place a strong emphasis on sustainability, including the use of renewable energy sources, environmental protection, and sustainable agriculture practices. This focus helps to equip graduates with the skills and knowledge they need to develop sustainable solutions for the world's most pressing problems.

Integration of technology: Agricultural and Bioresources Engineering curricula reforms aim to integrate the latest technological advancements in the field into the education process. This includes the use of computer simulations, virtual reality, and other technological tools. The integration of technology helps to provide students with a more hands-on, interactive education and prepares them for careers in a rapidly evolving field.

Collaborative learning: Agricultural and Bioresources Engineering is a field that requires collaboration and teamwork. Agricultural and Bioresources Engineering curricula reforms often incorporate collaborative learning opportunities, such as group projects and team-based design challenges. These experiences help students to develop important teamwork and communication skills that are essential for careers in the field.

Career development opportunities: Agricultural and Bioresources Engineering curricula reforms aim to provide students with the opportunity to develop their careers through internships, co-op programmes, and other professional development opportunities. These opportunities allow students to gain hands-on experience in the field, make professional connections, and prepare for careers after graduation (Fernando, 2006).

Preparing Students For the Future in Agricultural and Bioresources Engineering Curricula

The field of Agricultural and Bioresources Engineering is constantly evolving, with new technologies and innovations being developed every day. These advancements have the potential to revolutionise the way food is grown, processed, and distributed (Babatope, 2020). As such, it is crucial that the curricula for Agricultural and Bioresources Engineering programmes be updated regularly to ensure that students are equipped with the most current knowledge and skills. Some of the incorporation of new technologies and innovations in Agricultural and Bioresources Engineering curricula reforms will be discussed.

1. Precision Agriculture: Precision Agriculture is an innovative technology that enables farmers to optimise the use of resources and improve crop yields. This technology utilises GPS, remote sensing, and other technologies to gather data on soil conditions, weather patterns, and crop growth, allowing farmers to make informed decisions about planting, fertilising, and harvesting. To prepare students for careers in Precision Agriculture, Agricultural and Bioresources Engineering curricula should include courses on data analysis, geospatial technologies, and precision farming practices.
2. Biotechnology: Biotechnology is a rapidly growing field that has the potential to revolutionise the food industry. Advances in genetic engineering and biochemistry have led to the development of new crop varieties that are resistant to pests and diseases, have improved nutrient content, and are better suited for changing environmental conditions. Agricultural and Bioresources Engineering programmes should include courses in molecular biology, biochemistry, and genetic engineering to ensure that students are equipped with the knowledge and skills necessary to work in this field.
3. Renewable Energy: The use of renewable energy sources, such as wind and solar power, is becoming

increasingly important as the world moves towards a more sustainable future. Agricultural and Bioresources Engineering programmes should include courses in renewable energy technologies and their applications in agriculture, such as the use of wind turbines and solar panels on farms. This will ensure that students are equipped with the knowledge and skills necessary to design and implement renewable energy systems for agricultural operations.

4. Robotics and Automation: Robotics and automation are increasingly being used in agriculture to improve efficiency and productivity. From autonomous tractors and drones to robotic harvesting systems, these technologies have the potential to transform the way food is grown and processed. Agricultural and Bioresources Engineering curricula should include courses in robotics and automation, covering topics such as control systems, sensor technology, and programming.

The incorporation of new technologies and innovations in Agricultural and Bioresources Engineering curricula reforms is essential to ensure that students are equipped with the knowledge and skills necessary to succeed in this rapidly evolving field. From Precision Agriculture and Biotechnology to Renewable Energy and Robotics and Automation, the incorporation of these advancements will enable students to be at the forefront of innovation and make a positive impact on the world (Akeel et al., 2019) By keeping curricula up-to-date and relevant, Agricultural and Bioresources Engineering programmes will continue to produce graduates who are well-prepared to meet the challenges of the future.

Conclusions and Recommendation

The department of Agricultural and Bioresources Engineering, University of Nigeria has played a significant role in the development of agricultural mechanisation in Nigeria's agricultural landscape. It has effectively contributed to the development and continuous operation of many other departments of agricultural engineering in Nigeria. The department needs to be repositioned to effectively train graduates that have relevant industrial skills to fit into the job market and contribute effectively to the sustainable development goals of decent work, clean environment, no hunger, and poverty reduction. It is imperative to design and implement a robust reformed curriculum that meets the needs of the evolving agrarian society considering the 4th industrial revolution. In that light, earlier curriculum reforms do not aid the department to equip the student for the future. Agricultural and Bioresources Engineering curricula reforms play a critical role in producing graduates who are equipped with the necessary skills and knowledge to meet the challenges of the 21st century. These reforms aim to provide students with a comprehensive education in the field of Agricultural and Bioresources Engineering, including its principles, techniques, and methods. The key components of Agricultural and Bioresources Engineering curricula reforms, including interdisciplinary education, hands-on experience, focus on sustainability, integration of technology, collaborative learning, and career development opportunities, are crucial in preparing graduates for careers in the field and supporting their professional development. In conclusion, Agricultural and Bioresources Engineering curricula reforms are essential for the development of sustainable and innovative solutions to the world's most pressing problems.

REFERENCE LIST

Adepoju, O. O., & Nwulu, N. I. (2020). Engineering students innovation competence: A comparative analysis of Nigeria and South Africa. *International Journal of Engineering Pedagogy*, 10(6), 147–155. https://doi.org/10.3991/ijep.v10i6.14695

Akeel, U. U., Bell, S. J., & Mitchell, J. I. (2019). Assessing the sustainability content of the Nigerian engineering curriculum. *International Journal of Sustainability in Higher Education*. 20(4), 590–613. https://doi.org/10.1108/IJSHE-11-2018-0217.

Aliu, O. (2021, October 14). Council adopts outcome based education approach to train engineers. Vanguard. https://www.vanguardngr.com/2021/10/council-adopts-outcome-based-education-approach-to-train-engineers/ 2021.

Agricultural and Bioresources Engineering. (2021). Self study document. University of Nigeria, Nsukka.

Babatope, A, A., Samuel, T. M., Ajewole, P. I., & Anyanwu, O. M. (2020). Competence-driven engineering education: A case for T-shaped engineers and teachers. *International Journal of Evaluation and Research in Education*, 9(1), 32–38.

Council of Regulation of Engineering Practices. (2020). Outline of Outcome-Based Education (OBE) approach for engineering programmes in Nigerian Universities, 1st Ed. Council for the Regulation of Engineering in Nigeria.

Dada, L. T., & Oyetunde, S. M. (2020). Quality and unemployment issues in curriculum and graduate outputs of Nigerian Universities. *Journal of Positive Psychology and Counselling*, 6, 134–145.

Fernando, S., Murali, N., & Bhushan, S. (2006). The need to reform agricultural engineering curricula in developing countries. *International Journal of Engineering Education*, 22(1), 79.

Ismail, S., & Mohammed, D. S. (2015). Employability Skills in TVET curriculum in Nigeria federal universities of technology. *Procedia—Social and Behavioral Sciences*, 204, 73–80. https://doi.org/10.1016/j.sbspro.2015.08.111.

Mbajiorgu, C. C. (2017). Revised five-year standard undergraduate academic programme in agricultural & bioresources engineering. University of Nigeria.

Opara, L. U. (2004). Outlook for agricultural engineering education and research and prospects for developing countries. *Outlook on Agriculture*, 33(2), 101–112.

Rickus, J. L. (2006). Building an integrated undergraduate biological engineering program in an agricultural and biological engineering department: Incorporating the student perspective. *International Journal of Engineering Education*, 22(1), 53.

Tripathi, A. O., Mishra, R., Maurya, K. K., Singh, R. B., & Wilson, D. W. (2019). Estimate for world population and global food availability for global health. In Ram B. Singh, Ronald Ross Watson and Toru Takahashi (Eds.) *The role of Functional Food Security in Global Health* (pp. 3–24). Academic Press.

27

Development of the Agricultural and Biosystems Engineering Programme at the Botswana University of Agriculture and Natural Resources (BUAN)

Rejoice Tsheko

Historical Perspective

The beginnings of the Botswana University of Agriculture and Natural Recourses (BUAN) can be traced back to 1959 in Mahalapye where, prior to independence, agricultural extension and any investigational work was carried out as early as the 1930s at a town located 200km north of Gaborone, (International Service for National Agricultural Research [ISNAR], 1990).

The Botswana Agricultural College (BAC) was later established at Sebele where the agricultural training school was relocated from Mahalapye, to just north of the city of Gaborone in 1967. The BAC offered certificate programmes in agronomy and animal husbandry (Botswana College of Agriculture, 2001). During this period, Botswana was one of the poorest countries in the world with a population of about 500,000 and the agricultural sector contributed about 40% of the GDP with over 95% of the population residing in rural areas (Central Statistics Office [CSO], 1982). In 1991, BAC was transformed into the Botswana College of Agriculture (BCA).

The small number of trained scientists was a major factor limiting the capacity for Botswana to address the various problems in agricultural production and natural resources management (International Service for National Agricultural Research, 1990). During this period, there were less than five locals trained as agricultural engineers. The college became an associate college of the University of Botswana and started offering the BSc Agriculture programme, and the first degree students were admitted in 1988. This programme included some basic agricultural engineering courses and, further, a Diploma in Agricultural Engineering programme was also introduced at the college (Botswana College of Agriculture, 2000–2001). The engineering courses were developed and offered by the Department of Agricultural Engineering and Land Planning. Most of the staff teaching these courses had MSc level qualifications only, with their first degrees not necessarily in Engineering (Botswana College of Agriculture, 2000–2001). This limited the impact of these programmes as there was a lack of a strong grounding in engineering science and design. During this period, there was a concerted effort to recruit new staff with BSc Engineering degree qualification and to train existing staff to a PhD level.

In 2018, the Botswana College of Agriculture was finally transformed into the Botswana University of Agriculture and Natural Resources (BUAN). It also ceased to be an associate college of the University of Botswana (Botswana University of Agriculture and Natural Resources Act 12 of 2015). With the university status, the Department of Agricultural Engineering and Land Planning changed its name to the Department of Agricultural and Biosystems Engineering and embarked on developing a fully fledged Biosystems Engineering programme following the Washington Accord model and Engineering Council of South Africa (ACSA) guidelines, and sought accreditation from the Engineering Registration Board (ERB) and the Botswana Qualifications Authority (BQA).

Status of Agricultural and Biosystems Engineering Education in Botswana (ABEN)

The global recognition of the importance of sustainable management natural resources such as air, water, land, food, and other biological resources and the impacts of global environmental change and non-sustainable use of the earth's resources has obligated the Agricultural Engineering profession to metamorphosise into Agricultural and Biosystems Engineering and hence produce engineers competent to lead our collective utilisation of finite resources (American Society for Agricultural and Biological Engineers, 2011).

Across the world, there are different variations in the offering of ABE education, mostly dependent on the needs of the country. These include programmes such as Biosystems Engineering, where graduates receive multidisciplinary training in areas such as microbiology, biochemistry, molecular biology, bioinstrumentation, in vitro culture, and bioreactor operation, as well as techniques for bioenergy and biomaterial production, environmental bioremediation, and aquatic organism production (Universitat Politècnica de Catalunya BarcelonaTech, 2022), And Biological Engineering

DOI: 10.1201/9780429150111-29

programmes, which entail the application of engineering principles to protect natural resources and to produce food, biofuels, pharmaceuticals, and other biobased products (University of Florida, 2023), and produces engineers who can create and adapt engineering knowledge and technologies for the efficient and effective production, processing, storage, handling, and distribution of food, feed, fibre, and other biological products, while at the same time providing for a quality environment and preserving and protecting natural resources (Oklahoma State University, 2022). Some programmes break down the Biosystems Engineering into Bioprocess Engineering, Ecological Engineering, and Forest Engineering (Auburn University, 2023.).

Botswana is a middle-income country aspiring to gain a high-income status through a knowledge based and diversified economy (Government of Botswana, 2016). Vision2036 has identified the agricultural sector as having the potential to impact on the livelihoods of many people through provision of food and employment. The agricultural sector's contribution to GDP has been shrinking from about 44% at independence in 1966 to less than 2% in 2022 due to the growth of other sectors of the economy and previous policies such as food security (Ministry of Agriculture, 2009) (Figure 27.1).

Agricultural production in Botswana has always been low with yields of less one tonne per hectare for grains and this has been attributed to low and erratic rainfall, poor soils (low fertility and organic matter), a lack of adequate and competent extension staff, lack of draft power for timely operations, lack of evidence-based policies, and politically driven social programmes (Ministry of Agriculture, 2009).

The agricultural sector contributed about 12 percent to formal employment in 1972, compared with only 1.7 percent in 1998 (Botswana Institute for Development Policy Analysis, 2005), which is a huge decline, showing the less important role played by the sector in formal employment creation (International Labour Organization, 2007). Additionally, agricultural wages have consistently been the lowest, and the incidence of under-employment reported as widespread. To fulfil the conflicting needs such as the need to conserve the environment and also increase agricultural production, the ABE curriculum developed is an attempt to strike a delicate balance between production and conserving the environment while also availing the graduates with increased opportunities for employment in all the sectors of the economy. The contribution of the agriculture sector to employment is heavily contested depending on the objectives and whether it is formal employment or subsistence; the mid-term review of the National Development Plan 11 indicated that the sector employs about 80,000 adults, which is 12% of total employment (Ministry of Finance and Economic Development, 2020). Temoso, Villano, and Hadley (2015) concluded that in order to improve agricultural growth in Botswana there is a need for institutional strengthening and capacity building, infrastructure development, technology development and transfer, and institutional policy development and reforms.

Curriculum Development

From the literature, the following speciality areas of ABEN are common namely; Biological Engineering, which addresses problems and opportunities presented by living things and the natural environment; Natural Resources, which deals with understanding the complex mechanics of the resources, so that they can be used both efficiently and sustainably; Power Systems and Machinery Design, which focuses on designing advanced equipment; Structures and Environment, which deals with the design of humane animal housing, efficient and sustainable

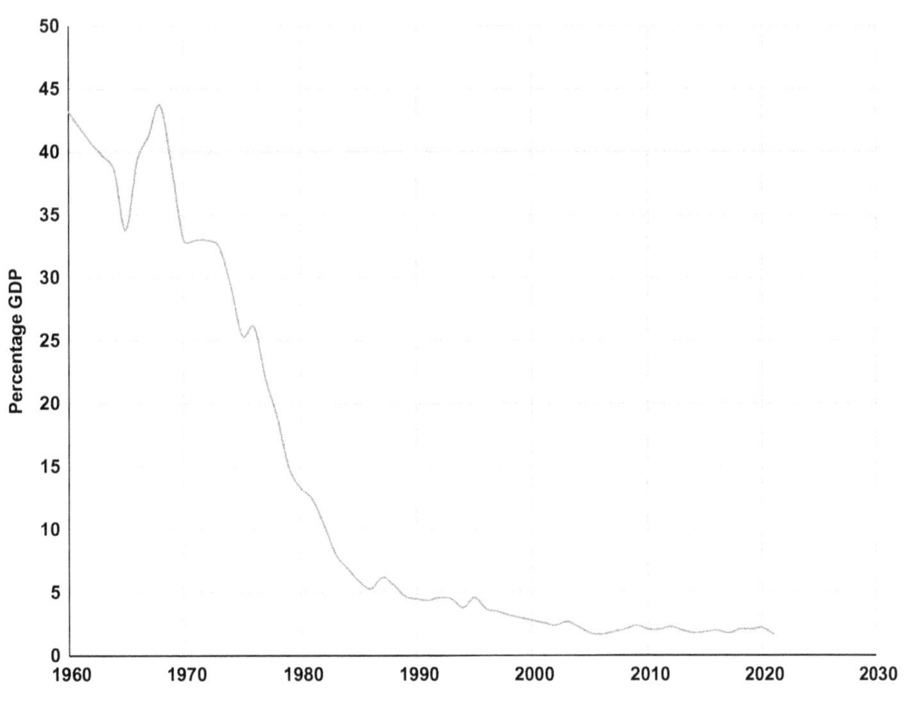

FIGURE 27.1 Contributions of the Agriculture Sector to the GDP from 1960 to 2021 (Statistics Botswana, n.d.).

storage structures, and greenhouses; Food and Bioprocess, which applies the understanding of microbiological processes to develop useful products, to treat municipal, industrial, and agricultural wastes, and to improve food safety; Information and Electrical Technologies which covers geographic information systems, global positioning systems, machine instrumentation and controls, robotics, remote sensing, electromagnetics, and machine vision, sensors, and spectroscopy; Forest Engineering, which applies engineering to solve natural resource and environment problems in forest production systems and related manufacturing industries; and Energy Systems, which deals with the identification and development of viable energy sources (cleaner and more efficient) such as biomass, methane, and vegetable oil (Aguado et al., 2011).

In developing the ABEN programme at BUAN, the key guiding principles in selecting emphasis (speciality areas) were based on the international trends and the needs and constraints playing in the Botswana and Southern Africa space. These include limited teaching laboratories and especially equipment and most importantly the level of development of the agricultural sector. The level of industrialisation is still in its infancy; moreover, most farmers are not at the level where they can adequately provide guidance to the students during the industrial attachment period, hence the need for the university to have well equipped laboratories for teaching and research.

Past Approach

The previous programmes curricula – namely the BSc Agriculture, Diploma in Agricultural Engineering, Higher Diploma in Agricultural Engineering, BSc Agricultural Mechanisation, and BSc Soil and Water Conservation Engineering – were more focussed on agricultural production. Curriculum development activities included all the stakeholders and the consultation took place in a variety of formats, including face-to-face meetings, staff/faculty mobility, and questionnaires. These programmes were neither accredited by an engineering body nor by the qualifications authority; that is they were developed at the department level then endorsed by the faculty, the Senate, and were approved by the University Council. The programmes were occasionally subjected to peer reviews from other universities on a consultancy basis and this was an expensive exercise with limited impact. The initial programmes – namely the BSc Agriculture and Diploma in Agricultural Engineering – were not semester programmes; that is, students wrote exams at the end of the year of study and emphasis was placed on the examination marks. The Higher Diploma in Agricultural Engineering, BSc Soil and Water Conservation Engineering, and BSc Agricultural Mechanisation are semester programmes, where the final grade is based on 50% continuous mark performance and 50% on the final examination. This was meant to improve the students' competencies through continuous learning and to avoid memorisation being used at the end of the year to pass end year examinations without being competent in the subject.

The BSc Agriculture Curriculum

The BSc Agriculture was a fulltime programme lasting a minimum of four years or eight semesters, each of sixteen weeks duration. The students enrolled in this programme were required to undertake a total of sixteen weeks of supervised Field Practical Training (FPT) in the agricultural industry and carry out a special/research project during study. The objectives of this programme were not well articulated in the prospectus, but it appears that the designers of the programme wanted to produce a graduate with some knowledge in all the fields of agriculture including agricultural engineering but not necessarily competent in any one of these. In Table 27.1 only the agricultural engineering courses in the BSc Agriculture programme are listed. This programme was overloaded and lacked many attributes, especially engineering science and design aspects, as it was presented more as a technology course. Students joining this programme in year two would have spent one year doing basic sciences at other universities (BAC was an Associate College of the University of Botswana during this period).

lists the ABEN courses in the BSc Agriculture programme. The programme covered crop production, animal production, agricultural economics, and agricultural extension courses.

TABLE 27.1

Course structure of the BSc Agric Programme showing only the Engineering courses (Botswana College of Agriculture, 2001).

Year 2 (semesters 3 & 4)		
AEL210	Agricultural Engineering I	Ael211 introduction to agricultural engineering Ael212 land surveying Ael213 land use planning
AEL220	Agricultural Engineering II	Ael221 Introduction to agricultural Mechanization Ael222 workshop practice
Year 3 (semesters 5 & 6)		
AEL310	Agricultural Engineering III	Ael311 hydrology & climatology Ael312 Soil & water conservation Ael313 land evaluation and management
AEL320	Agricultural Engineering IV	Ael321 Soil/Plant/Water Relationship AEL322 Irrigation Technology
Year 4 (semesters 7 & 8)		
AEL410	Agricultural Engineering V	AEL411 Agricultural Mechanization AEL412 Farm Structures AEL413 Agricultural Processing

This was like trying to fit four degrees in one programme; it is understandable why this approach was taken since, according to the International Service for National Agricultural Research (1990), there were only twenty-seven professional staff at the department of agricultural research (DAR), of whom only two were agricultural engineers. It was also estimated that about sixty-five professional agricultural science staff and two hundred technicians were needed at DAR to create a critical mass. These numbers were even worse at the Ministry of Agriculture, hence the need for a rapid human resource development strategy. When the BSc Agriculture programme was introduced, there were very few (less than 5) Batswana graduates of the BSc Agricultural Engineering, and thus most of the BSc Agriculture programme ended up being employed as Agricultural Engineers (Irrigation, Soil and Water Conservation, Water Engineering, Mechanisation, Agricultural Processing and Structures, and Post-Harvest) in the Ministry of Agriculture. These graduates were not competent as they lacked engineering science, design, and other competencies one would expect in an engineer.

The Diploma in Agricultural Engineering Curriculum

The Diploma in Agricultural Engineering was a fulltime programme lasting a minimum of three years or six semesters each of sixteen weeks duration as described in Table 27.2. The students enrolled in this programme were required to undertake a total of fifteen weeks of supervised Field Practical Training (FPT) in the agricultural industry. The objectives of this programme were not fully articulated, and it incorporated almost all areas of the traditional agricultural engineering programme though not in details and it could have served as a preparatory programme for student entering a BSc Agricultural Engineering programme. Most graduates of this programme joined the public sector because the private sector was very limited. These graduates also had a limited design competence.

The Higher Diploma in Agricultural Engineering Curriculum

The objective of the programme was to produce agricultural engineering technicians by collaborating with other departments and other external stakeholders especially the agricultural industry either public or private sector. This programme was a normalisation and update of the diploma in agricultural engineering which was supposed to be two years or four semesters according to the diploma programmes regulations but was instead six semesters. It also incorporated the introduction of additional semester of basic science courses, core, optional, general education, and elective courses in the course structure for flexibility and enhancement of science fundamentals. Some courses were dual coded to BSc level (especially the technology ones) to facilitate articulation when a student

TABLE 27.2

Course structure of the DAEng Programme (Botswana College of Agriculture, 2001).

Semester 1		
BS010	Mathematics	
BS011	Biology	
BS012	Chemistry	
BS013	Physics	
BS014	Communication Skills	
Semesters 2 & 3		
BS017	Mathematics and Statistics	Mathematics, statistics
ASP017	Livestock Production I	Animal nutrition, animal health
CSP018	Crop Production I	Soil science, crop science
AEE010	Socio-Economic Sciences I	Principles of economics, rural sociology, farm records & accounts
AEL011	Technical Drawing & Computing	Technical drawing, computing
AEL012	Engineering Mechanics & Material	Mechanics, engineering materials
AEL013	Workshop Technology	Metal workshop, wood workshop
AEL014	Surveying	Land surveying, soil surveys & land use planning
Semesters 4 & 5		
CSP023	Crop Production II	Horticulture, crop protection
AEE020	Socio-Economic II	Agricultural extension, supervision and administration
AEL022	Soil & Water Conservation	Soil mechanics, soil conservation, water harvesting & conservation
AEL023	Hydrology	Fluid mechanics, surface water hydrology, ground water hydrology
AEL024	Farm Power & Energy	Electricity, farm power
AEL025	Principles of Irrigation	Hydraulic structures, design of irrigation systems
AEL026	Agricultural Machinery and Mechanization	Agricultural Machinery and Mechanization
AG020	Project	
Semester 6		
AEL030	Irrigation Management	Project planning & evaluation, irrigation systems and management
AEL031	Farm Structures	Farm Structures
AEL032	Post-Harvest Technology	Post-Harvest Technology
AG030	Project	

decided upgrade to pursue a technology degree programme. So, the programme duration was six semesters long and included twelve weeks of field practical training in the industry. Examining this programme closely, it had the same flaws found in the diploma in agricultural engineering programme since the graduates were lacking in technician competency and were also not competent in design, being more suited as technologists. The higher diploma in agricultural engineering programme outlined in Table 27.3 was just a stop-gap measure while a fully fledged engineering programme was being developed. The Department of Agricultural Engineering and Land Planning made a lot of industrial collaboration. The scope of where the students went for their Field Practical Training was expanded from predominately farms to other engineering industries. This change was based on the feedback from students who complained that the farms were limited in engineering equipment and other areas of engineering such as food processing, earthworks engineering, and mechanical engineering, such as tractor and motor mechanics. The students were now sent to milling industries, brewery, meat processing, grain silos, soil engineering labs and many more industries. This intervention prepared the graduates of the programme to be market ready on graduation and further contacts were established for future employment, and the also the programme was indirectly marketed.

Figure 27.2 shows that when the higher diploma in agricultural engineering programme was introduced the number of students enrolled in the diploma in agricultural engineering programme dwindled as students now preferred a higher diploma programme instead of an Ordinary diploma programme, as it gave them an opportunity to articulate to degree programmes. One of the improvements in the course structure from a diploma to a higher diploma was an additional semester of basic science courses to strengthen the science background of the students and the dual listing of some courses (diploma/BSc); this made it easier for the students to either transfer into a degree programme during study or later complete a degree programme in a shorter time due to the award of credits for dual listed courses. This was only possible for applied courses but not engineering science courses. The students were still required to do all the engineering science courses even when they held a higher diploma in agricultural engineering. One of the challenges in this programme structure was the difficulty of teaching similar courses at different levels that is either pitching a diploma level course to a degree level or vice versa.

TABLE 27.3

Course structure of the HDAEng Programme (Botswana College of Agriculture, 2008).

Semester 1	
BSB110	Mathematics I
BSB111	General and Inorganic Chemistry
BSB112	Physics I
BSB113	Biology of Cells
GEC111	Communication and Study Skills I
GEC121	Computing and Information Skills Fundamentals I
Semester 2	
BSB120	Mathematics II
BSB121	Physical and Organic Chemistry
BSB122	Physics II
BSB123	Biodiversity
GEC112	Communication and Study Skills II
GEC122	Computing and Information Skills Fundamentals II
Semesters 3	
AEB212	Introduction to Agricultural Economics
ELD120	Soil Mechanics
ASB210	Introduction to Animal Science
ELD215	Engineering Drawing
CSB211	Principles of Crop Production
Elective	
Semesters 4	
ELD227	Fluid Mechanics and Hydraulics
ELB320	Hydrology and Climatology
ELD214	Mechanics
AEB322	Farm Business Management
	Options
ELB 211	Soil and Water Conservation
ELD223	Soil Survey and Land Use Planning
ELD210	Irrigation Practices
Elective	
Semester 5	
ELB220	Farm Workshop Practice
ELD213	Computer Applications in Engineering
ELD224	Engineering Materials
ELD310	Fundamentals of Electricity
ELD321	Farm Power
Elective	
Semester 6	
ELB321	Tractor and Implements Technology
ELB310	Farm Structures
ELB410	Agricultural Mechanization
ELB411	Agricultural Processing
ELB322	Design of Irrigation Systems

The BSc Agricultural Mechanization Curriculum (BSc AGM)

The objective of this technology programme was to improve sustainable agricultural production in Botswana using mechanisation. This is a technology programme that was developed in line with the Botswana Government National Master Plan for Arable Agriculture and Dairy Development programme (National Master Plan for Arable Agriculture & Dairy Development program, 2002). The policy objectives for the agricultural sector, which guided this programme were:

- Improvement in food security at both household and national levels,
- Diversification of the agricultural production base,
- Increased agricultural output and productivity,
- Increased employment opportunities for the fast-growing labour force,
- Provision of a secure and productive environment for agricultural producers,

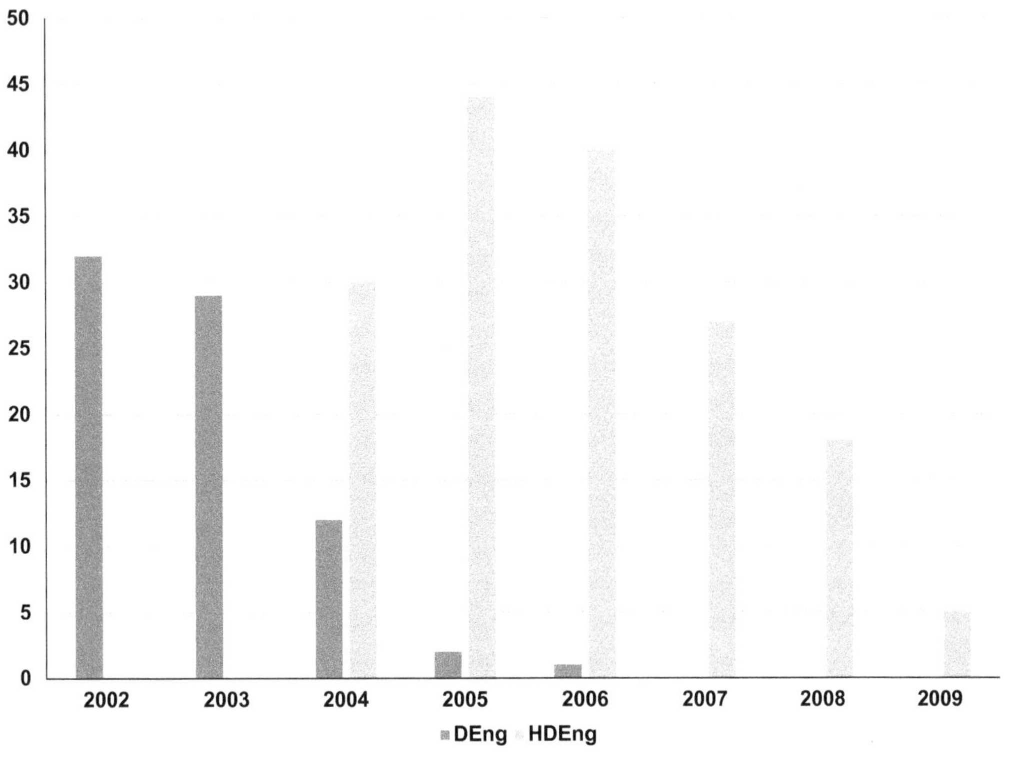

FIGURE 27.2 Student enrolment in the DAEng and HDAEng. Programmes.

- Conservation of scarce agricultural and land resources for future generations.

The programme duration is a minimum of eight semesters and include two six-week blocks of Field Practical Training (FPT), totalling six credits to the programme. The enrolment averages 88 students per year, of whom 76% are males and 24% females, and the enrolment peaked to 148 in 2012. The course structure of the BSc AGM programme is shown in Table 27.4.

The BSc Soil and Water Conservation Engineering Curriculum (BSc SWCE)

The objective of this technology programme was to improve sustainable agricultural production in Botswana by appropriate land and water management strategies. This is a technology programme that was developed in line with the Botswana Government National Master Plan for Arable Agriculture & Dairy Development programme (National Master Plan for Arable Agriculture & Dairy Development program, 2002). The policy objectives for the agricultural sector, which guided this programme were:

- Improvement in food security at both household and national levels,
- Diversification of the agricultural production base,
- Increased agricultural output and productivity,
- Increased employment opportunities for the fast-growing labour force,
- Provision of a secure and productive environment for agricultural producers,
- Conservation of scarce agricultural and land resources for future generations.

The programme duration is a minimum of eight semesters and include two six-week blocks of Field Practical training, totalling six credits to the programme. The enrolment averages 148 students per year, of whom 65.5% are males and 34.5% females, and the enrolment peaked to 180 and above students in the years 2012, 2018, 2019, and 2020. The course structure of the BSc SWCE programme is shown in Table 27.5.

Figure 27.3 shows the student enrolment for the BSc AGM and BSc SWCE from 2007 to 2022, the BSc SWCE has a higher enrolment and for both programmes, the number of enrolled students is decreasing year on year. This decrease might be due to market saturation, as almost all the initial graduates got employment, but with time, a smaller number of students were employed on graduation.

For both programmes, the number of students enrolling in the programme is gradually decreasing, although the BSc SWCE continues to attract more students with a better gender distribution. This programme is performing better in terms of attracting students because the graduates of this programme are able to get employment in other sectors of the economy and can also register with the Engineering Registration Board while the BSc AGM graduates can only get employment opportunities from the agricultural sector. The name of a programme plays a role into how the graduates are perceived by the market;

TABLE 27.4

Course structure of the BSc AGM Programme (Botswana College of Agriculture, 2015).

Semester 1	
BSB110	Mathematics I
BSB111	General and Inorganic Chemistry
BSB112	Physics I
BSB113	Biology of Cells
GEC111	Communication and Study Skills I
GEC121	Computing and Information Skills Fundamentals I
Semester 2	
BSB120	Mathematics II
BSB121	Physical and Organic Chemistry
BSB122	Physics II
BSB123	Biodiversity
GEC112	Communication and Study Skills II
GEC122	Computing and Information Skills Fundamentals II
Semesters 3	
AEB212	Introduction to Agricultural Economics
ELB120	Soil Mechanics & Foundations
BSB221	Biometry I
ELB220	Farm Workshop Practice
	Options
CSB211	Principles of Crop Production
ASB210	Introduction to Animal Science
GEC (not prescribed)	
Elective	
Semesters 4	
ELB213	Computer Applications in Engineering
ELB310	Farm Structures
ELB340	Farm Implements Technology
ELB341	Measurement Techniques in Agro-Biological Sciences
	Options
ELB214	Statics
ELB215	Engineering Design
GEC222	Problem Solving with Spreadsheets
Semester 5	
ELB325	Environmental Impact Assessment
ELB331	Renewable Energy
ELB322	Tractor Power
ELB333	Electrical Power
	Options
ELB210	Land Surveying & Evaluation
ELB420	Land Drainage
Semester 6	
ELB230	Fluid Mechanics
ELB311	Heating, Ventilating, and Air Conditioning
AEB322	Farm Business Management
	Options
ELB342	Crop Harvesting Technology
ELB450	Food Preservation and Storage
GEC (not prescribed)	
Semester 7	
BSB321	Biometry II
APB410	Project
ELB423	Current Issues in Agricultural Engineering
ELB443	Agricultural Machinery Testing
GECs (not prescribed)	
Electives	
Semester 8	
ELB410	Agricultural Mechanization Management
APB420	Project
AEB422	Project Appraisal and Evaluation
	Options
ELB216	Fluid Power
ELB411	Agricultural Processing
ELB330	Introduction to Remote Sensing
ELB440	Geographical Information Systems
GEC (not prescribed)	

TABLE 27.5

Course structure of the BSc SWCE Programme (Botswana College of Agriculture, 2017).

Semester 1	
BSB110	Mathematics I
BSB111	General and Inorganic Chemistry
BSB112	Physics I
BSB113	Biology of Cells
GEC111	Communication and Study Skills I
GEC121	Computing and Information Skills Fundamentals I
Semester 2	
BSB120	Mathematics II
BSB121	Physical and Organic Chemistry
BSB122	Physics II
BSB123	Biodiversity
GEC112	Communication and Study Skills II
GEC122	Computing and Information Skills Fundamentals II
Semesters 3	
AEB212	Introduction to Agricultural Economics
ELB120	Soil Mechanics & Foundations
BSB221	Biometry I
ELB210	Land Surveying and Evaluation
	Options
CSB211	Principles of Crop Production
CSB311	Crop Physiology
CSB310	Vegetable Production
GEC	
Semesters 4	
ELB245	Irrigation Technology
ELB230	Fluid Mechanics
ELB221	Soil/Plant/Water Relations
	Options
ELB214	Statics
ELB215	Engineering Design
GEC222	Problem Solving with Spreadsheets
ELB213	Computer Applications in Engineering
GEC (not prescribed)	
Semester 5	
ELB322	Design of Irrigation Systems
BSB321	Biometry II
ELB211	Soil & Water Conservation

(Continued)

TABLE 27.5 (CONTINUED)

Course structure of the BSc SWCE Programme (Botswana College of Agriculture, 2017).

	Options
ELB310	Farm Structures
ELB220	Farm Workshop Practice
ELB333	Electrical Power
GEC (not prescribed)	
Semester 6	
ELB360	Irrigation Water Supply and Conveyance
ELB365	Irrigation and Water Quality
	Options
ELB325	Crop Harvesting Technology
ELB330	Introduction to Remote Sensing
ELB440	Geographical Information Systems
Elective	
GEC (not prescribed)	
Semester 7	
ELB320	Hydrology & Climatology
ELB420	Land Drainage
APB410	Project
ELB423	Current Issues in Agricultural Engineering
Electives	
Semester 8	
ELB421	Watershed Management
AEB322	Farm Business Management
APB420	Project
AEB422	Project Appraisal and Evaluation
ELB424	Ground Water Hydrology
	Options
ELB426	Wind Erosion
ELB411	Agricultural Processing

for example the BSc AGM graduates are trained in mechanisation including agricultural but the non-agricultural industries like earth moving companies do not easily see past the name of the programme. As discussed previously, the department of Agricultural and Biosystems Engineering aggressively pursued an industry partnership strategy. These efforts resulted in students undertaking Field Practical Training (FPT) with many non-agricultural industries, and thus the industry appreciated the graduates' competencies, and, furthermore, the students were market ready on graduation and the programmes were indirectly marketed to the industry.

Present Approach

The ABE courses and programmes from 1991 to 2022 were mostly addressing agricultural production issues except the BSc SWCE programme which had a wider scope; that is, it also addresses soil and water conservation beyond agricultural production. Two issues were identified as limitations to the programmes. Namely that the graduates of the BSc AGM programme can only get employment in the agriculture sector (which contributes about 2% to the GDP) and further graduates of the two programmes are more technology oriented, since they have a weak background in basic science and engineering science and, further, they lack engineering professional competencies such as design. The BSc AGM's problems are further worsened by the fact that although they could be employed in earthmoving or even mining industries, the degree name tends to limit their desirability in the sectors beyond agriculture. A programme name such as BSc Mechanization could have made a difference in the employability of the graduates.

The new BEng. Biosystems Engineering programme is focussed on both the agricultural production but with more emphasis on Biosystems Engineering issues – that is, understanding the complex mechanics of natural resources and how to use the efficiently and in a sustainable way. Biosystems Engineering (BE) integrates engineering science and design with applied biological, environmental, chemical, and agricultural sciences to develop and apply engineering solutions to problems in biological systems (Wolfe et al., 2011). The curriculum development activities involved all stakeholders including the Engineering Registration Board and the Botswana Qualification Authority for accreditation. "Accreditation is the voluntary, peer review process utilised by higher education institutions and industry practitioners to evaluate academic degree programmes. The accreditation process is used by universities, colleges and specific degree programmes to clarify instructional goals and objectives, enhance programme content and improve programme delivery. It is vital to the continued success of our profession" (American Society of Agricultural and Biological Engineers, 2011). The engagement strategy or needs assessment took place in a variety of formats, including face-to-face meetings, staff/faculty mobility, and virtual networking. The new programme is being accredited by both the Engineering Registration Board (ERB) and Botswana Qualification Authority (BQA). It is planned that in future the new programme will be put through the American Society of Agricultural and Biological Engineers (ASABE) for accreditation, which is one of 33 professional bodies that comprise ABET. ABET is an ISO 9001 Quality Management Systems (QMS) certified non-profit making accrediting agency for engineering, computing, engineering technology, and applied science degree programmes, primarily in the United States, but is now gradually inclusive of other parts of the world.

One of the challenges which tend to dilute the rigour of Agricultural Engineering or Biosystems Engineering is that these programmes are usually housed in the Faculty of Agriculture, not the Faculty of Engineering. For the Biosystems Engineering programme, the department opted to register for the Manufacturing, Engineering, and Technology as the field of study and Agriculture and Nature Conservation as the sub-field of study in order for the programme to follow the Washington Accord model and hence be accredited by the Botswana Engineering Registration Board.

Bachelor of Engineering in Biosystems Engineering (BEng. BE)

From the historical observation at BUAN, especially from the BSc SWCE, a programme which does not only focus on agricultural production tends to attract more students and improves the male to female ratio of the enrolled students. This is so

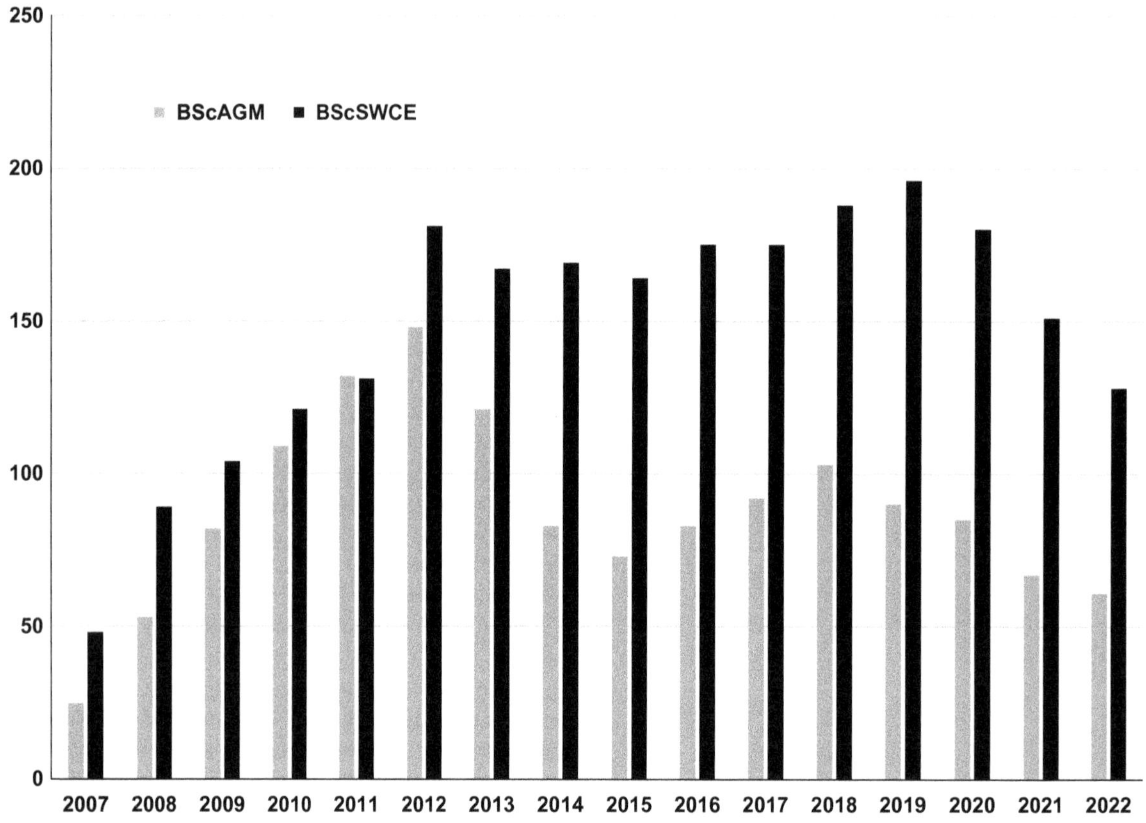

FIGURE 27.3 Student enrolment in the BSc AGM and BSc SWCE Programmes.

because there are more opportunities for employment. This observation resonates with the transformation which started in the USA then later in Europe, Australia, New Zealand, Asia, and lately in some parts of Africa, where Biosystems Engineering programmes replaced traditional Agricultural Engineering programmes. The ASABE, formerly the American Society of Agricultural Engineers (ASAE), was established in 1907 and went through several transformations and debates, such as the concept of Engineering of Biology proposed in 1937 by C.O. Reed of Ohio State University and later in 1962 by E.H. Kidder of Michigan State University; these attempts were meant to broaden the scope of the profession. The North Carolina State University was the first to change its department name from Agricultural to Biological and Agricultural (ASABE, 2007). With time, several economic conditions affecting the agricultural industry, such as the loss of some traditional areas of employment, the emerging new areas of employment, and the Society membership, finally forced the ASAE to transform into ASABE in 2005, and this also introduced transformations to agricultural and biosystems engineering education (ASABE, 2007). As part of the transformation into the future, Mary Leigh Wolfe was the first woman elected an ASABE Fellow.

The proposed programme is a five year (ten semester) programme at the National Credit Qualification Framework (NCQF) level 8 with 632 credits which would lead to the award of a Bachelor of Engineering degree in Agricultural and Biosystems with specialisation in Mechanisation and Structures; Soil and Water Engineering; Agro and Bio Processing; Sustainable Energy; and Bio and Agro Waste. The emphasis is on modern technologies applied to both agricultural production and non-agricultural applications. The programme rational is to produce high calibre market ready graduates for the agricultural and natural resources sectors graduates for the local and international market.

The programme is driven by The Vision 2036 (Government of Botswana, 2016). Pillar No.1 envisages that by 2036 Botswana will have graduated from a middle income to a high-income status, with a knowledge based and diversified economy. The National Development Plan 11 (Ministry of Finance Development Planning, 2017) provides a medium-term plan towards achieving Vision 2036. The National Policy on Agricultural Development (Ministry of Agriculture, 2014) has identified poor infrastructure, overdependence on rainfed agriculture, weak agro industries, and the effects of climate as some of the barriers to development of the agricultural sector. The planned programme will address skills development in the areas of irrigation, agro processing, and climate-smart agriculture, as well as the provision of sustainable (renewable) energy as an input in agricultural production. The National Energy Policy objectives seek to promote the sustainable production of bioenergy and utilisation of waste for energy production (Ministry of Mineral Resources, Green Technology and Energy Security, 2021). The agricultural sector is key in that there is tremendous potential to generate energy from agro waste such as cow dung and waste from abattoirs and agro processing industries. Production of bioenergy sources such as biodiesel and ethanol relies on agricultural feedstocks. The Botswana Education and Training Sector Strategic Plan for 2015–2020 emphasises improving the quality and relevance of education (Education & Training Sector Strategic Plan, 2015).

The Human Resource Development Council (HRDC) has a mandate to publish the professions and competencies that are required by the job market on an annual basis (Human Resource Development Council, 2016). The Department of Agricultural and Biosystems Engineering conducted a national needs assessment survey in 2018 on the areas of specialisation that are needed by the agriculture–based business industry in Botswana (Department of Agricultural and Biosystems Engineering, 2018). The 120 respondents were asked to prioritise specialisations and areas that were highly ranked include Soil and Water Engineering, Soil and Water Management, Irrigation Technology, and Irrigation and Drainage. Medium ranking areas include Agricultural Mechanisation, Agricultural Machinery, and Geoinformatics (Department of Agricultural and Biological Engineering, 2018). This survey was biased towards agricultural production, but this was augmented by government policy objectives, especially in agro processing, renewable energy, climate-smart agriculture, and 4IR.

Based on these findings, the Department of Agricultural and Biosystems Engineering has now developed a BEng in Biosystems Engineering programme which we believe will meet both the nation's requirements and the international standards set by the Washington Accord for graduate attributes and accreditation of engineering programmes (International Engineering Alliance, 2014), and the requirements of the Engineering Council of South Africa (ECSA). This is to ensure that graduate engineers produced by BUAN meet international standards. At local level the programme has been developed in consultation with the Botswana Engineers Registration Board to ensure that graduate engineers can be registered as professionals by the Board and the programme has been developed in line with the BQA mission of ensuring that all learners acquire quality assured awards.

BEng Programme Objectives

The main objective of the BEng programme is to train graduates who can use their knowledge of mathematics, biological and physical sciences, and engineering principles to solve problems relating to the production, handling, and processing of biological materials, in an environmentally sustainable manner.

Specific objectives are:

1. To train competent and ethical professionals capable of addressing challenges associated with the sustainable production, management, and use of biological materials,
2. To produce graduates who are innovative and can create technologies,
3. To train engineering professionals who can communicate designed engineering solutions effectively and clearly,
4. To train graduates who are lifelong learners and can work in diverse, multidisciplinary, and multicultural environments. This aligns with the BEng core graduate competencies (exit level outcomes and alignment to graduate attributes required for accreditation of engineering programmes).

In line with the objectives of the BEng programme, its graduates will be able to demonstrate competencies in the following areas: technical skills, practical skills, entrepreneurship, climate-smart adaptability, management, communication, innovation, analytical skills, teamwork and interpersonal relation skills, and information and communication technology skills.

Technical and Practical Skills

Graduates of the BEng programme will demonstrate competence in the application of first principles in mathematics, basic sciences, and engineering sciences to solve problems relating to the production, handling, and processing of biological materials, in an environmentally sustainable manner Such technical and practical skills will be evidenced by their ability to:

1. Apply mathematical, numerical analysis and statistical knowledge and methods to solve engineering problems,
2. Utilise physical laws and knowledge of the physical world as a foundation for the engineering sciences and the solution of engineering problems,
3. Use the techniques, principles, and laws of engineering science at a fundamental level and more specifically in the agricultural and biosystems engineering area to:
   a. Identify and solve open-ended engineering problems,
   b. Identify and pursue engineering applications,
   c. Work across engineering disciplinary boundaries through cross disciplinary literacy and shared fundamental knowledge,
4. Select, use, and maintain modern engineering tools and technology, including adaptive tools and new technology, to facilitate work activity.

Innovation and Analytical Thinking Skills

BEng graduates will have competencies to enable them to employ unique analyses, generate innovative ideas, and to visualise "the big picture". They will also display competence in the use of logical thought processes to analyse information and draw conclusions.

Entrepreneurship and Management

The BEng graduate will be able to use information on basic business principles, trends, and economics to successfully integrate into and make positive contributions in the operation and management of existing business entities where they would be engaged and also those that they would initiate and develop for themselves through self-employment.

Climate-Smart Adaptability

The BEng graduate will display the capability to apply their knowledge and skills to adapt to new, different, or changing requirements brought about by climate change in the relevant context.

Teamwork and Interpersonal Relation Skills

The BEng graduate will display readiness to take on the role of team member, and to work effectively and cooperatively with others from diverse backgrounds.

Information and communication technology (ICT) or 4IR

With respect to ICT, BEng graduates will demonstrate competence to:

1. Use information technology and related applications, including adaptive devices and software, to convey and retrieve information,
2. Use appropriate ICT–based engineering methods, skills, and tools, which is evidenced by their ability to:
   a. Select and assess the applicability and limitations of ICT–based methods, tools, and skills,
   b. Properly apply the method, tool, or skill,
   c. Critically test and assess the end results produced by the method, tool, or skill,
   d. Create computer applications as required.

Communication Skills

The BEng graduate will demonstrate competence to communicate effectively, both orally and in writing, with engineering audiences and the community at large, as evidenced by their ability to:

1. Use appropriate structure, style and language for purpose and audience,
2. Use effective visual and/or graphical support,
3. Apply methods of providing information for use by others involved in engineering activities,
4. Deliver fluently,
5. Meet the requirements of the intended audience.

The BEng. Biosystem Engineering Programme Structure

Table 27.6 shows the course structure of the BEng programme, with the following details given for each of the 68 courses within the programme: course code, course name, category (mandatory or not), the number of credits as defined by BQA, notional number of learning hours, classification (i.e. mathematics, basic science, computing and IT, engineering science, design and synthesis, as well as "complimentary" or "discretionary") and finally the type of the course (core, optional, or GEC).

The total number of credits accrued by an individual learner would be 632 from a total of 53 courses taken, selected from the 68 courses offered. Out of the 632 credits earned by a BUAN ABE BEng graduate, 432 would be from 36 core courses (68%), 156 would be from 13 optional courses (25%), and 44 would be from 4 GEC courses (7%). The curriculum of the programme should guarantee the graduates of the programme to have prerequisite competencies to enter the engineering profession and practice successfully. The structure of the programme includes two winter sessions where the students will do hands on training in the industry to complement the practical sessions offered during the course of study, this should enhance the practical skills hence the graduates should be market ready at graduation.

TABLE 27.6

The Biosystems Engineering Programme Structure (Department of Agricultural and Biological Engineering, 2018).

Level 100 – Semester 1					
Course Name	**Category**	**Credits**	**Notional hours**	**ECSA classification**	**Type**
General Mathematics	Mandatory	12	120	Mathematics	Core
General Biology	Mandatory	12	120	Basic Science	Core
General Physics	Mandatory	12	120	Basic Science	Core
General Chemistry	Mandatory	12	120	Basic Science	Core
Computing & Information Technology	Mandatory	8	80	Computing & IT	GEC
Communication & Study Skills	Mandatory	12	120	Complimentary	GEC
Level 100 – Semester 2					
Course Name	**Category**	**Credits**	**Notional hours**	**ECSA classification**	**Type**
Calculus & Analytical Geometry	Mandatory	12	120	Mathematics	Core
Chemistry 2	Mandatory	12	120	Basic Science	Core
Biology 2	Mandatory	12	120	Basic Science	Core
Statics	Mandatory	12	120	Engineering Science	Core
Engineering Statistics (Biometry)	Mandatory	12	120	Complimentary	GEC
Level 200 – Semester 3					
Course Name	**Category**	**Credits**	**Notional hours**	**ECSA classification**	**Type**
Differential Equations – PE	Mandatory	12	120	Mathematics	Core
Environmental Biophysics	Mandatory	12	120	Engineering Science	Core
Fluid Mechanics	Mandatory	12	120	Engineering Science	Core
Dynamics	Mandatory	12	120	Engineering Science	Core
Material Science	Mandatory	12	120	Engineering Science	Core

(Continued)

TABLE 27.6 (CONTINUED)

The Biosystems Engineering Programme Structure(Department of Agricultural and Biological Engineering, 2018).

Level 200 – Semester 4					
Course Name	**Category**	**Credits**	**Notional hours**	**ECSA classification**	**Type**
Numerical Analysis	Mandatory	12	120	Mathematics	Core
Engineering Design 1	Mandatory	12	120	Design & Synthesis	Core
Thermodynamics	Mandatory	12	120	Engineering Science	Core
Soil Mechanics & Foundations	Mandatory	12	120	Engineering Science	Core
Structural Mechanics (Strength of Materials)	Mandatory	12	120	Engineering Science	Core
Winter Session I					
Course Name	**Category**	**Credits**	**Notional hours**	**ECSA classification**	**Type**
Industrial Placement 1	Mandatory	12	120		Core
Level 300 – Semester 5					
Course Name	**Category**	**Credits**	**Notional hours**	**ECSA classification**	**Type**
Geomatics (Engineering Surveying)	Mandatory	12	120	Engineering Science	Core
Heat & Mass Transfer	Mandatory	12	120	Engineering Science	Core
Instrumentation & Control	Mandatory	12	120	Engineering Science	Core
Irrigation Agronomy / Solute Transport	Not Mandatory	12	120	Discretionary	Option
Power systems / Food Microbiology	Not Mandatory	12	120	Discretionary	Option
Level 300 – Semester 6					
Course Name	**Category**	**Credits**	**Notional hours**	**ECSA classification**	**Type**
Renewable Energy	Mandatory	12	120	Discretionary	Core
Bio-environmental Engineering	Mandatory	12	120	Engineering Science	Core
Functional Analysis & Design of Agricultural Field Machinery / BioEnergy Systems / Hydraulics /	Not Mandatory	12	120	Engineering Science	Option
Hydrology & Water Resources / Waste Treatment / Fluid Power	Not mandatory	12	120	Discretionary	Option
Engineering Economics	Mandatory	12	120	Complimentary	GEC
Winter Session II					
Course Name	**Category**	**Credits**	**Notional hours**	**ECSA classification**	**Type**
Industrial Placement 2	Mandatory	12	120		Core

(Continued)

The number of credits in the BEng Biosystems engineering shall be 632, as shown in Table 27.7, and warrants the degree to be classified at NCQF level 8.

As shown in Table 27.8, the programme has 68% biosystems engineering courses at core, 25% courses as optional, and 7% courses as general education to produce a well-rounded graduate.

Some of the equipment and facilities for the engineering programme at BUAN are listed in Figures 27.4–27.6. There are also other agricultural production equipment and facilities available for teaching and research.

Challenges and Opportunities in ABE Education

The challenges in the agricultural and biosystem engineering education in Botswana are threefold, namely: 1) limited resources such as a) human resources (there is lack of critical mass of instructors and lecturers with Agricultural or Biosystems Engineering degrees), and b) teaching resources (shortage of teaching facilities and teaching equipment – since its establishment, BUAN has not been resourced to a satisfactory level; however recently, some teaching and research equipment has been procured from Germany), 2) Agriculture Sector limitations (the sector is very small and contributes less than 2% to the GDP and, further, the sector is not developed enough to provide adequate on the job training to the students during the course, and 3) government policies (most of the policies address the social rather than commercial aspects of agriculture because most people earn a living from subsistence agriculture). These constraints tend to hinder the agricultural sector from growing and increasing agricultural productivity, and also increasing the number of real jobs which could be created if the sector were to become highly commercial.

Since the Botswana University of Agriculture and Natural Resources is the main institution that provides training in ABE at tertiary level, it is imperative to collaborate with technical colleges and oversee the training of technicians and craftsmen; high schools which provide the students who enrol in the ABE programme. BUAN has to influence the Science, Technology, Engineering, and Math (STEM) curriculum throughout the value chain (pre-schools, primary schools, and high school). BUAN should link with vocational institutions who should provide vocational training so that BUAN can concentrate on professional training, which is mainly problem solving in order to increase agricultural production sustainably and improve the sustainable extraction of wealth from abundant natural resources.

TABLE 27.6 (CONTINUED)

The Biosystems Engineering Programme Structure (Department of Agricultural and Biological Engineering, 2018).

Level 400 - Semester 7					
Course Name	**Category**	**Credits**	**Notional hours**	**ECSA classification**	**Type**
Computer Programming	Mandatory	12	120	Computing & IT	Core
Irrigation Systems Design	Mandatory	12	120	Discretionary	Core
Agribusiness	Mandatory	12	120	Complimentary	Core
Bio & Agro Processing / Electrical & Electronic Principles	Not mandatory	12	120	Discretionary	Option
GIS / Unit Operations in Agro Processing	Not mandatory	12	120	Discretionary	Option
Level 400 - Semester 8					
Course Name	**Category**	**Credits**	**Notional hours**	**ECSA classification**	**Type**
Climate-Smart Agriculture	Mandatory	12	120	Complimentary	Core
Professional Ethics	Mandatory	12	120	Complimentary	Core
Entrepreneurship	Mandatory	12	120	Complimentary	Core
Engineering Project Management	Mandatory	12	120	Discretionary	Core
Organic Waste Processing/ Machine Design	Not mandatory	12	120	Discretionary	Option
Level 500 - Semester 9					
Course Name	**Category**	**Credits**	**Notional hours**	**ECSA classification**	**Type**
Design Project 1	Mandatory	12	120	Design & Synthesis	Core
Engineering Design 2	Mandatory	12	120	Complimentary	Core
Mechatronics & Robotics / Air Pollution	Not mandatory	12	120	Complimentary	Option
Solar Thermal Energy / Engineering Properties of Biomaterials	Not mandatory	12	120	Complimentary	Option
Applied Soil and Crop Dynamics for Off–road Machinery / Agro Meteorology	Not mandatory	12	120	Discretionary	Option
Level 500 - Semester 10					
Course Name	**Category**	**Credits**	**Notional hours**	**ECSA classification**	**Type**
Design Project 2	Mandatory	12	120	Design & Synthesis	Core
Biosystems Product Processing	Mandatory	12	120	Complimentary	Core
PV Systems / Remote Sensing	Not mandatory	12	120	Complimentary	Option
Land Remediation / Land Drainage	Not mandatory	12	120	Complimentary	Option
Environmental Risk Assessment / Water Quality	Not mandatory	12	120	Discretionary	Option

TABLE 27.7

Credits Summary Table (Department of Agricultural and Biological Engineering, 2018).

Semester	Credits	Notional hours
Semester 1	68	680
Semester 2	60	600
Semester 3	60	600
Semester 4	60	600
Winter Session I	12	120
Semester 5	60	600
Semester 6	60	600
Winter Session II	12	120
Semester 7	60	600
Semester 8	60	600
Semester 9	60	600
Semester 10	60	600
Grand Total	**632**	**6320**

TABLE 27.8

Course Structure Statistics (Department of Agricultural and Biological Engineering, 2018).

Course by Type	Number of Courses	Proportion (% of credits)
Core (total 36 courses)	36 (total 432 credits)	68
Optional (total 28 courses)	13 (total 156 credits)	25
GEC (total 4 courses)	4 (total 44 credits)	7
Total (68 courses)	**53**	**100**

Finally, there are new areas and technologies which can continue to enhance the scope of the ABEN Education and profession. These include geoinformation, earth observation, robotics and controls, machine vision, artificial intelligence, nano technology, precision agriculture and resources management, communications, and digital agriculture.

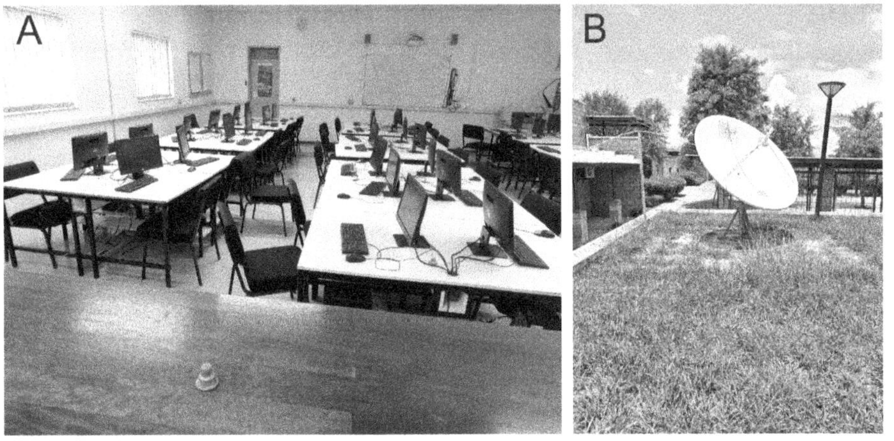

FIGURE 27.4 Geoinformation lab and a EUMETSAT receiver dish.

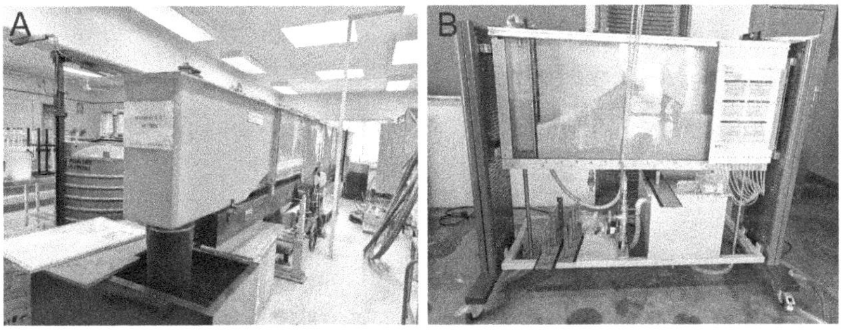

FIGURE 27.5 Hydraulics lab equipment.

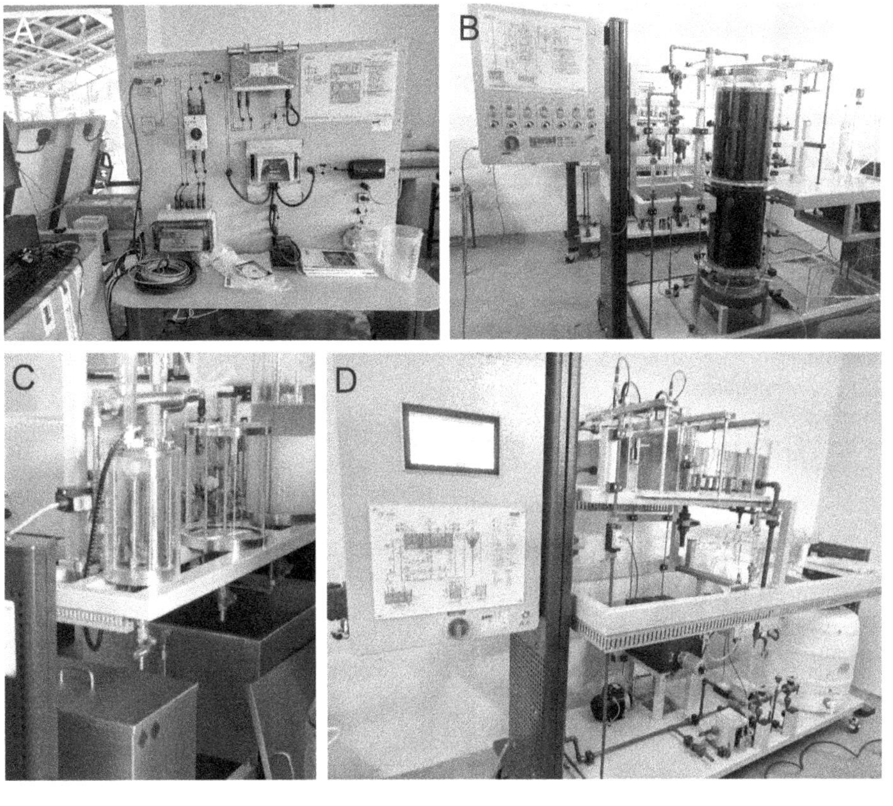

FIGURE 27.6 Solar energy network bench, mini Biodiesel plant, and two activated sludge water treatment plants.

REFERENCE LIST

Aguado, P., Ayuga, F., Briassoulis, D., Panagakis, P., FEBO, P. Comparetti, A., Scarascia-Mugnozza, G., O´Donnell, C., Navickas, K., and Fehrmann, J. (2011). The Transition from Agricultural to Biosystems Engineering University Studies in Europe. https://www.iiis.org/CDs2011/CD2011SCI/EISTA_2011/papersPdf/EA444OM.pdf

American Society of Agricultural and Biological Engineers. (2007). Engineering for a Sustainable Tomorrow - 1907–2007. https://www.asabe.org/Portals/0/AboutUs/100yr-Part1.pdf

American Society of Agricultural and Biological Engineers. (2011). American Society of Agricultural and Biological Engineers. Retrieved January 5, 2023, from http://www.asabe.org/

Auburn University. (2023). Retrieved January 5, 2023, from https://bulletin.auburn.edu/undergraduate/samuelginncollegeofengineering/departmentofbiosystemsengineering/

Botswana College of Agriculture. (2001). Prospectus 2000–2001. Botswana College of Agriculture, Botswana, Gaborone.

Botswana College of Agriculture. (2008). 2008–2009 Prospectus. Botswana; Gaborone.

Botswana College of Agriculture (2015). 2015–2016 Undergraduate Prospectus. Botswana, Gaborone.

Botswana College of Agriculture. (2017). 2017–2018 Undergraduate Prospectus. Botswana University of Agriculture and Natural Resources, Botswana; Gaborone.

Botswana Institute for Development Policy Analysis. (2005). Challenges for Agricultural Diversity in Botswana Under the Proposed SADC-EU Economic Partnership Agreement (EPA). BIDPA Working Paper Number 27. https://core.ac.uk/download/pdf/19919273.pdf

Botswana Qualifications Authority. (2021). Manual for Development, Validation and Registration of Qualifications on the National Credit and Qualifications Framework April 2021 Version 2, Botswana Qualifications Authority, Formal Sector Employment Stats Brief March 2018, Statistics Botswana. https://www.bqa.org.bw/framework/

Botswana University of Agriculture and Natural Resources Act, Cap 12, Botswana Government, 2015.

Central Statistics Office. (1982). Agricultural Census 1982. Botswana. http://catalog.ihsn.org/catalog/2059#metadata-metadata_production

Department of Agricultural and Biological Engineering. (2018). Proposal for New Bachelor of Engineering in Biosystems Engineering (B.Eng.) Programme.

Education and Training Sector Strategic Plan. (2015). Botswana Government. https://www.gov.bw/sites/default/files/2020-03/ETSSP%20Final%20Document.pdf

Government of Botswana. (2016). Achieving Prosperity for All. https://statsbots.org.bw/sites/default/files/documents/Vision%202036.pdf

Human Resource Development Council. (2016). Agriculture Sector HRDC Plan. Botswana. https://www.hrdc.org.bw/sites/default/files/Agriculture%20Sector%20HRD%20Plan.pdf

International Engineering Alliance. (2014). 25 Years, Washington Accord, 1989–2014 Celebrating International Engineering Education Standards and Recognition; International Engineering Alliance. https://www.ieagreements.org/accords/washington/

International Labour Organization. (2007). https://www.ilo.org/global/about-the-ilo/newsroom/news/WCMS_082596/lang--en/index.htm

International Service for National Agricultural Research. (1990). Review of Botswana's Agricultural Research System (ISNAR Report No. R51). ISNAR, The Hague. https://pdf.usaid.gov/pdf_docs/PNABG593.pdf

Ministry of Agriculture. (2009). Botswana Agriculture Sector Review. Government Printer, Botswana.

Ministry of Agriculture. (2014). Policy on Agricultural Development. Government Printer, Gaborone.

Ministry of Finance Development Planning. (2017). National Development Plan Eleven (NDP 11). Government Printer, Gaborone. https://www.finance.gov.bw/images/NDP_11_2017-2023.pdf

Ministry of Finance and Economic Development. (2020). Mid-Term Review of National Development Plan 11 Draft. https://www.gobotswana.com/sites/default/files/ndp_11_mtr_draft_final_july_ndp_11_draft_mfed_31_july_2020.pdf

Ministry of Mineral Resources, Green Technologies and Energy Security. (2021). National Energy Policy. Government Printer, Gaborone. https://www.bera.co.bw/downloads/National%20Energy%20Policy%20Final%20April%202021.pdf

National Master Plan for the Arable Agriculture and Dairy Development. (2002). NAMPAADD Guidelines. https://www.africaportal.org/documents/5144/NAMPAADD_GUIDELINES.pdf

Oklahoma State University. (2022). Retrieved January 23, 2023, from http://catalog.okstate.edu/engineering-architecture-technology/biosystems-agricultural-engineering/

Statistics Botswana. (n.d.). 1960–2021. Retrieved January 23, 2023, from https://www.statsbots.org.bw/

Temoso, O., Villano, R.A., and Hadley, D. (2015). Agricultural Productivity, Efficiency and Growth in a Semi-Arid Country: A Case Study of Botswana. *African Journal of Agricultural and Resource Economics*, 10(3), 192–206. https://afjare.org/media/articles/2-Temoso-et-al.pdf

University of Florida. (2023). https://catalog.ufl.edu/UGRD/colleges-schools/UGENG/BE_BSBE/

Universitat Politècnica de Catalunya BarcelonaTech. (2022). https://eeabb.upc.edu/en/Studies/bachelors-degree/biosystems-engineering

Wolfe, M., Holden, N., Briassoulis, D., Ayuga, F., and Mugnozza, G. (2011). Trans-Atlantic Biosystems Engineering Curriculum and Mobility (TABE.NET). 22.1540.1-22.1540.8. 10.18260/1-2--18353. https://www.researchgate.net/publication/344534550_Trans-Atlantic_Biosystems_Engineering_Curriculum_and_Mobility_TABENET

Part 3

Pedagogy, Epistemology and Emerging Trends

28

Developing Faculty Competency for the Delivery of Active Learning Methods

Fedro S. Zazueta, Patrizia Busato and Remigio Berruto

Some Guiding Principles

A successful online course is one in which at the end of the course the student has acquired the subject matter knowledge at the desired level, retains this knowledge over time, and is able to use this knowledge as a scaffold for the acquisition of deeper knowledge. In addition, a well-designed course results in gaining "soft" skills that are essential in today's workforce, such as communication, critical and complex thinking.

The actors in creating an online learning environment include at a minimum one or more subject matter experts, an instructional designer, learning asset developers such as media production technicians knowledgeable of the online delivery tools such as a course management system. Key to developing a high-quality online course (or module) is:

1) Knowledge of the student's current level of competency and motivators that lead to student engagement.
2) A clear statement of the high-level goals of the course.
3) A clear statement of the specific behaviours and competencies resulting from the course.
4) A clear statement of how the student will be evaluated, that is, how will the student demonstrate gain of the intended knowledge. Note that this applies to the overall goal of the course as well as to specific behaviours and competencies.
5) Access to technology and technology tools.
6) The capacity to create or have access to quality learning assets that when experienced by the student lead to acquiring the desired knowledge. Examples of these include videos, simulations, readings, projects, and many others.

Knowledge Basics

This section is not intended to be a review of knowledge and learning theory. However, it addresses some basic notions that are useful in the development of online courses. It is important to recognise that knowledge of a subject can vary in breadth and depth. To illustrate this consider asking the asking the same question to three individuals and comparing the answers. If the question *What is gravity?* is asked of a 6th grader the answer might be: "The force that pulled the apple that fell on Newton's head". If the same question was asked to a student of engineering level physics the answer might be: "A force between two objects that is directly proportional to the product of the mass of each object and inversely proportional to the square of the distance separating their centres of gravity". Finally, if the same question was asked of a Ph.D. in physics, the answer might be: "Gravity is the result of the curvature by massive objects of the space and time around them. The strength of the gravitational 'field' at any point in space or time is just the degree to which that coordinate in space-time is invisibly curved". There are levels of knowledge involved as deeper learning and mastery of a subject matter by any individual takes place.

To organise higher forms of thinking and learning in education a taxonomy is useful (Bloom et al., 1956) in three domains of educational activity:

1) Cognitive: Mental skills (knowledge).
2) Affective: Growth in feelings or emotional areas (attitude or self).
3) Psychomotor: Manual or physical skills (skills) used.

Over time, Bloom's knowledge taxonomy was revised into its current form and is often represented as shown in Figure 28.1 (Forehand, 2005). From the lower to a higher level in the knowledge domain:

1) *Remembering:* Recall or retrieve previous learned information. (The student defines, describes, identifies, knows, labels, lists, matches, names, outlines, recalls, recognises, reproduces, selects, states).
2) *Understanding:* Comprehending the meaning, translation, interpolation, and interpretation of instructions and problems. State a problem in one's own words. (The student comprehends, converts, defends, distinguishes, estimates, explains, extends, generalises, gives an example, infers, interprets, paraphrases, predicts, rewrites, summarises, translates).

DOI: 10.1201/9780429150111-31

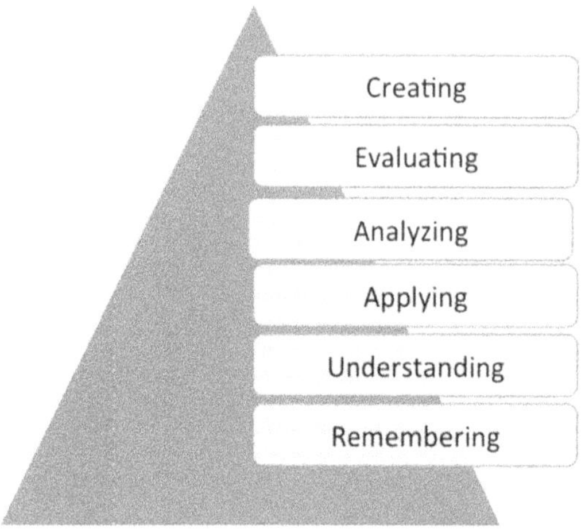

FIGURE 28.1 Revised Bloom's Taxonomy in the Cognitive Domain

3) *Applying:* Use a concept in a new situation or unprompted use of an abstraction. Applies what was learned in the classroom into novel situations in the workplace. (The student applies, changes, computes, constructs, demonstrates, discovers, manipulates, modifies, operates, predicts, prepares, produces, relates, shows, solves, uses).
4) *Analyzing:* Separates material or concepts into component parts so that its organisational structure may be understood. Distinguishes between facts and inferences. (The student analyses, breaks down, compares, contrasts, diagrams, deconstructs, differentiates, discriminates, distinguishes, identifies, illustrates, infers, outlines, relates, selects, separates).
5) *Evaluating:* Make judgments about the value of ideas or materials. (The student appraises, compares, concludes, contrasts, criticises, critiques, defends, describes, discriminates, evaluates, explains, interprets, justifies, relates, summarises, supports).
6) *Creating:* Builds a structure or pattern from diverse elements. Put parts together to form a whole, with emphasis on creating a new meaning or structure. (The student categorises, combines, compiles, composes, creates, devises, designs, explains, generates, modifies, organises, plans, rearranges, reconstructs, relates, reorganises, revises, rewrites, summarises, tells, writes).

In addition to the knowledge dimension there is a cognition dimension to knowledge from a lower to a higher level (Krathwohl, 2002):

1) *Facts:* A specific and unique data or instance.
2) *Concepts:* A class of items, words, or ideas that are known by a common name, includes multiple specific examples, shares common features. There are two types of concepts: concrete and abstract. It includes knowledge of terminology and of specific details and elements.
3) *Processes:* A flow of events or activities that describe how things work rather than how to do things. There are normally two types: business processes that describe workflows and technical processes that describe how things work in equipment or nature. They may be thought of as the big picture, of how something works. It includes knowledge of classifications and categories, principles and generalisations, theories, models, and structures.
4) *Procedures:* A series of step-by-step actions and decisions that result in the achievement of a task. There are two types of actions: linear and branched. It includes knowledge of subject-specific skills and algorithms, techniques and methods, and the criteria for determining when to use appropriate procedures.
5) *Principles:* Guidelines, rules, and parameters that govern. It includes not only what should be done, but also what should not be done. Principles allow one to make predictions and draw implications. Given an effect, one can infer the cause of a phenomena. Principles are the basic building blocks of causal models or theoretical models (theories).
6) *Metacognition:* Includes strategic knowledge, knowledge about cognitive tasks including appropriate contextual and conditional knowledge, self-knowledge.

Note that knowledge of a subject can vary widely within Bloom's Taxonomy. On any subject an individual may be performing at the lowest level, in which an individual retrieves relevant knowledge from memory to recall basic elements of the subject (remember, factual). This can be compared to cognition awareness and knowledge on the same subject to create new knowledge (create, metacognitive), which is a much more complex level of behaviour in the taxonomy.

Thus, it is very important that a clear articulation of learning behaviours (objectives) be made to ensure that the student can perform at the desired cognitive and knowledge level. The traditional "the material was covered in the class" provides no clear indication of the level of competence of the student. In addition, these objectives should be aligned with the desired outcomes.

Learning Theory

Several learning theories have been proposed and used in the design of learning environments (Wang, 2008). Some of the most relevant are behaviourism, cognitivism, and constructivism.

Behaviourism is a worldview that operates on a principle of "stimulus-response". It assumes that a learner is essentially passive and responds to external stimuli. The learner starts as a "tabula rasa" and behaviour is shaped through positive or negative reinforcement. Learning is defined as a change of behaviour in the learner.

Cognitivism is a paradigm where the learner is viewed as an information processor. Knowledge is seen as a schema, or symbolic mental construction. Learning is a change in in a

TABLE 28.1

Example behaviours related to the level and dimension of knowledge.

	Levels of Knowledge					
Knowledge Dimension	**Remember**	**Understand**	**Apply**	**Analyze**	**Evaluate**	**Create**
Facts	List	Summarise	Respond	Select	Check	Generate
Concepts	Recognise	Classify	Provide	Differentiate	Determine	Assemble
Processes and Procedures	Recall	Clarify	Carry out	Diagnose	Judge	Design
Principles	Identify	Explain	Troubleshoot	Rank	Justify	Predict
Metacognition	Distinguish	Predict	Use	Deconstruct	Reflect	Create

learner's schemata. Cognitivism responds to behaviourism by recognising that people require active participation to learn and changes in behaviour are an indication of what occurs within the learner's brain.

Constructivism postulates that learning is an active and constructive process in which the learner is the information constructor. (An individual constructs his/her own subjective reality linked to prior knowledge). It views learning as an active and contextualised process in which knowledge is constructed (as opposed to acquired). This construction is based on the learner's personal experience and hypothesis about the environment, bringing past experiences and cultural factors into a learning situation.

The practical importance of these learning theories, as they relate to pedagogy, is that behaviourism helps in understanding and articulating learning expectations in terms of behaviour. Cognitivism and constructivism, on the other hand, help understand how higher levels of learning can be achieved through the learner's experience and social interaction.

Thus, with a clear understanding of the learner's desired level of knowledge and dimension, there are three direct implications for successful development and delivery of an environment that leads to successful learning:

1) A clear articulation of the knowledge to be acquired by the learner can be made in terms of the learner's expected behaviour.
2) Assessment of the learner's acquired knowledge can be made by observing the learner's behaviour.
3) The design of a learning environment and experience for the learner must focus on acquiring the desired behaviour through the learner's experience, built upon existing knowledge, and social interaction.

Table 28.1 shows examples of different behaviours in the knowledge domain (Bloom's Taxonomy) and dimension. It is important to note that the context in which the behaviour is used is important.

Stages in Developing a Learning Environment (Course)

To create a course, it is useful to think about three main stages (Figure 28.2):

1) What the student will learn.
2) How the student will be assessed, and
3) How the course will be constructed to create a learning experience for the student that leads to success.

It is important to note that developing course content follows the clear definition of goals, objectives, and assessment.

Learning Goals

A challenge in the development of a degree programme or competency is to design courses in such a way that the sequence of learning activity leads to discipline specific learning goals, workplace skills, soft skills, and accreditation competencies. These are developed progressively by the learner over the course of the programme and respond to the mission of the university. Accreditation and regulatory and professional bodies each define their own learning goal sets. Aligning educational goals is a complex undertaking as illustrated in Figure 28.3. To this chapter, a course has a set of goals that are high level and integrated into the larger curricula.

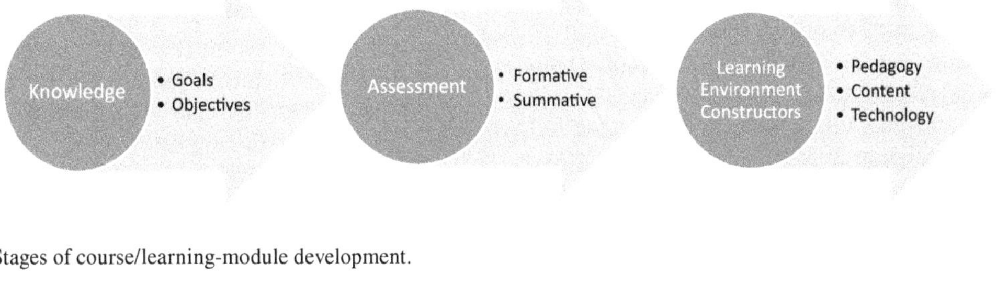

FIGURE 28.2 Stages of course/learning-module development.

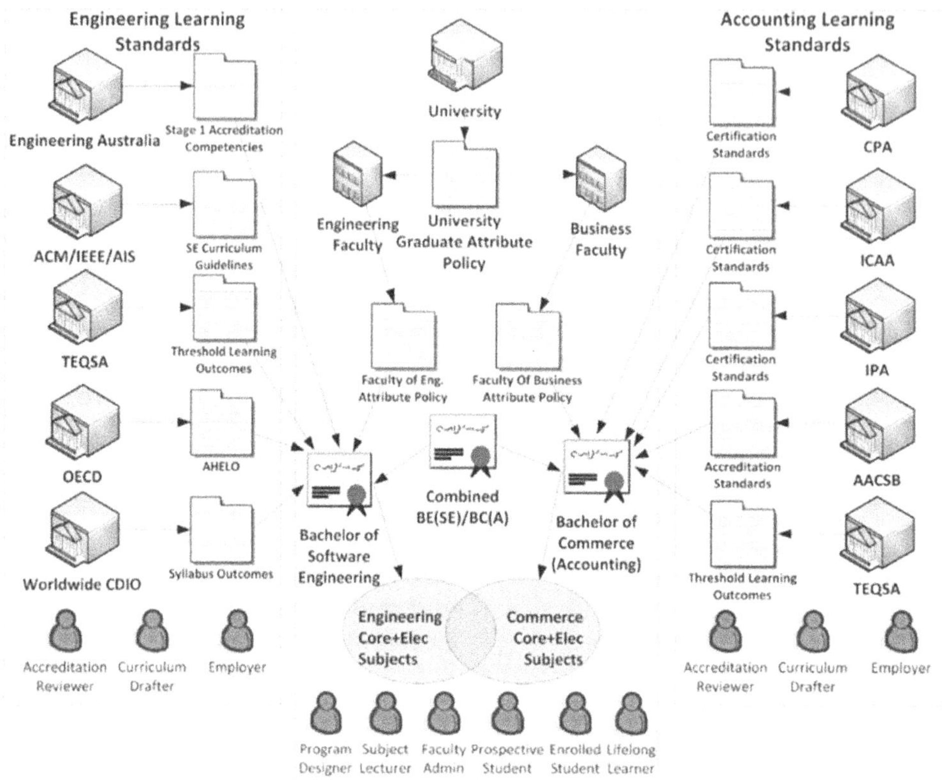

FIGURE 28.3 Illustration of complexity of learning goal requirements for two-degree programmes. (Gluga et al., 2013)

Learning Objectives and Learning Objects

Given a set of learning goals the definition of learning objectives follows. A learning objective is the target behaviour of the student that is to be acquired through the learning exercise. It is a clear statement describing what a student is expected to learn from a lesson. It provides details of what the student can do after completing the instruction process. A learning objective is composed of the following:

1) An action verb that identifies a measurable behaviour of the student. The verb is explicitly associated with a knowledge level in Bloom's Taxonomy.
2) A description of the condition under which the behaviour is to be performed.
3) The criteria or standard defining acceptable performance by the student. This criterion is the basis for assessment.

The importance of learning objectives goes well beyond indicating to the learners what they will know and be able to do after the successful completion of some learning activity. Well-crafted learning objectives guide the rest of the course development process. Course materials, assignments or activities, and assessments should all be selected to align with the learning objectives. A question to consider when building a course from learning objectives is: How any element of the course (video, discussion, project, etc.) relates back to one or more of the learning objectives? Learners should not be asked to read or review material that is not relevant to one of the objectives. Nor should they be assessed on skills or knowledge which is not in one or more of the objectives. When writing a learning objective, it is important for the instructor and the instructional designer to have a clear understanding of what level of performance is required from the student in the cognitive and knowledge domains. Table 28.2 shows examples of a well written and a poorly written learning objective.

A learning object encapsulates all necessary elements and resources to ensure that a learning objective can be attained and assessed. It must contain the following types of elements:

1) A single learning objective. As stated earlier, these clearly specify measurable behaviours that the student must exhibit upon completion of the learning process.
2) Knowledge domain resources. These consist of the content of the course and its structure. They are the materials used by the student to achieve the learning objectives. It is important that these resources be closely associated to the learning objectives, covering the breadth and depth of the subject matter.
3) Instructional information. This contains information relevant to the presentation of the content and its sequence. The rhythm of the presentation is based on the abilities of the student.
4) Assessment. These are the resources needed to determine if the student successfully attained the learning objectives. It is important that the assessment

TABLE 28.2

Examples of learning objectives.

Learning objective examples	
Well Written Learning Objective	**Comment**
Given the geometry, material and steady flow rate in a single pipe system, the student will be able to calculate the pressure distribution with less than 1% error using Bernoulli's modified equation.	Has one observable behaviour, conditions or context, and expected level of performance and proficiency criteria.
Poorly Written Learning Objective	
Students will understand the use of Bernoulli's modified equation and apply it to flow rate problems.	Understand is not an observable behaviour (internal). Has more than one verb, understand and apply. No context or proficiency criteria

resources be focused specifically on the learning objectives at the appropriate cognitive level.

5) Metadata. These are a collection of standard data that describe the LO and are used primarily for classification and search purposes. It also includes information about the pedagogical use of the material, authoring, versioning, and other type of data. There are various metadata standards, including IEEE's WG12 and Dublin Core.

A learning object is associated with a single learning objective, where the learning objective is the highest resolution of knowledge. The student learning environment is built focusing on a scaffolded sequence of learning objectives. A learning objective and its associated assessment items as well as materials related to the learning activities that the student will undertake constitutes a learning object.

It is important to consider, in the ontology of the LO, that students have learning style preferences. Thus, it is good to provide assets related to the objectives of the LO that address different learning styles. A useful tool for visualising the scope of the assets from the point of view of learning styles is to construct a matrix of learning objectives organised by the cognitive levels of Bloom's Taxonomy and learning style preference. During the design process described below, the result of this stage of the LO is a catalogue of assets (those currently available and those not available but needed) and a diagram that explains the relationships between the resources and the corresponding objectives.

Assessment Basics

The purpose of assessment is to evaluate, measure and document knowledge acquisition by the student. In education two forms of assessment are often used, formative and summative assessment.

Formative assessments are conducted throughout the learning process and are diagnostic as well as evaluative of specific skills. An accepted formal definition of formative assessment is (Council of Chief State School Officers [CCSSO], 2023):

> Formative assessment is a planned, ongoing process used by all students and teachers during learning and teaching to elicit and use evidence of student learning to improve student understanding of intended disciplinary learning outcomes and support students to become self-directed learners. Effective use of the formative assessment process requires students and teachers to integrate and embed the following practices in a collaborative and respectful classroom environment:
>
> - Clarifying learning goals and success criteria within a broader progression of learning;
> - Eliciting and analysing evidence of student thinking;
> - Engaging in self-assessment and peer feedback;
> - Providing actionable feedback; and
> - Using evidence and feedback to move learning forwards by adjusting learning strategies, goals, or next instructional steps.

Note that formative assessment is not only a tool to evaluate the performance of the student, but also a tool to assist the student and the instructor in the learning process. Formative assessment evaluates the learner as engagement as the learning process occurs and is generally implemented using benchmarks closely associated to learning objectives.

Summative assessment is used to provide assessment of learning (AoL) and assessment as learning (AaL). That is, it is used to determine how well knowledge was acquired after an instructional period is completed, and to provide the student evidence of learning, self-regulation, and reflection on the acquired knowledge. In conducting this type of assessment, it is important that:

1) The purpose or goals of the instructional period are clearly articulated,
2) Objective benchmarks are used, and
3) Often, to include a high stakes component.

Summative assessment is typically conducted at the end of a module, unit, project, course and is closely related to a learning goal. Figure 28.4 illustrates the types of assessment and how they are used (Rao & Banerjee, 2023). In this figure the criteria referenced are methods that use an established standard to assess student learning, such as the rubrics discussed below. Ipsative (of-the-self) formative assessment conducted by the student is related to reflection and self-assessment of

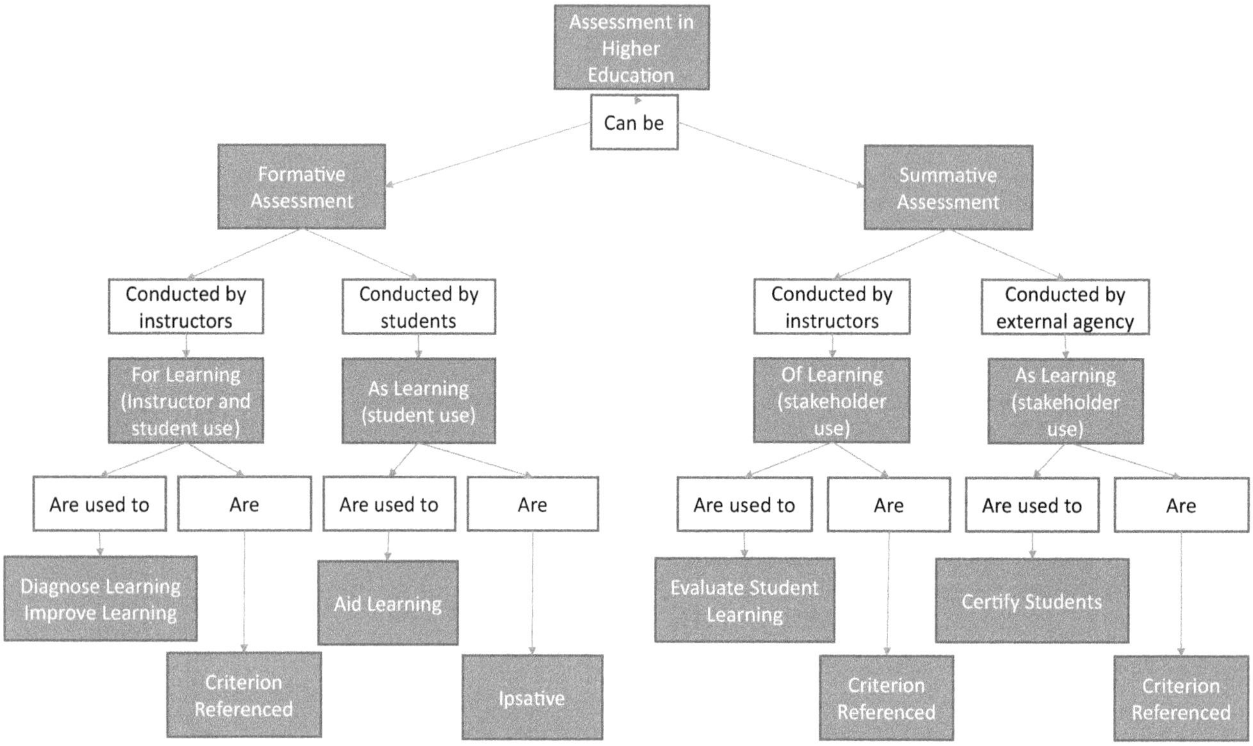

FIGURE 28.4 Assessment types and their use (adapted from Rao & Banerjee, 2023).

learning, connecting previous performance to current performance leading to metacognition (Education Endowment Foundation, 2017).

During the delivery of the course, assessments provide feedback that is timely, frequent, constructive (what and how), and help the students reflect on their knowledge (why). The alignment of goals, learning objectives, learning activities, and summative assessments is critical to student success. It is important that assessments match the cognitive levels of the course learning objectives.

Finally, the use of rubrics is highly recommended for assessment at any level, particularly summative (Rao & Banerjee, 2023). A rubric provides an objective criterion by which an assessment will take place. Take for example the rubric shown in Figure 28.5, to assess the student's capacity to "work effectively in teams". The rubric specifies the dimensions of the item being assessed, it provides a scale for the performance of the learner, and it establishes clear criteria to apply the scale to the different dimensions of the assessment. Also, rubrics ensure that evaluations are uniform and fair.

Active Learning Pedagogy (Flipped Classroom)

The learning environment to be created for the student is largely affected by the pedagogy to be used and the available technology. Advances in technology and learning theory and practice have created new directions and opportunities for pedagogy in engineering education. An active learning pedagogy is the flipped classroom. The flipped classroom is unique in its combination of active, problem-based learning constructivist ideas and direct instruction methods based on behaviourist principles. This pedagogical approach is enabled by technological advances that permit the transmission and duplication of information at very low cost and various means, and the trend in education to make learning student centred. A simple description of a flipped or inverted classroom is one where activities that traditionally take place in the classroom, take place outside the classroom in a flipped classroom, and vice versa, as shown in a simplified manner in Figure 28.6.

A definition of flipped classroom that accommodates theoretical frameworks may describe it not in terms of what is done in the traditional classroom, but in terms of human interaction. Thus, a flipped classroom is one in which learning activities not requiring human interaction take place outside the classroom (enabled by technology) and learning activities requiring human interaction take place in the classroom (virtual or physical). Figure 28.6 illustrates this definition of the flipped classroom (Busato et al., 2016b). Note that by this definition of a flipped classroom, activities requiring human interaction may occur face-to-face or virtually and in synchronous and asynchronous manners.

The focus of activities not requiring human interaction is for the student to understand and apply basic concepts related to the subject matter of the course in preparation for activities requiring human interaction that focus on higher levels of learning in Bloom's taxonomy. Some activities that do not require human interaction are readings, closed problem solving, self-assessments, and extensive use of video.

Early studies show that quality video lectures that are short and focused on a single topic outperform traditional lectures. Also, online homework is equally effective as paper and pencil. These, coupled with quizzes for self-evaluation provide a basis for the student to engage in activities requiring human

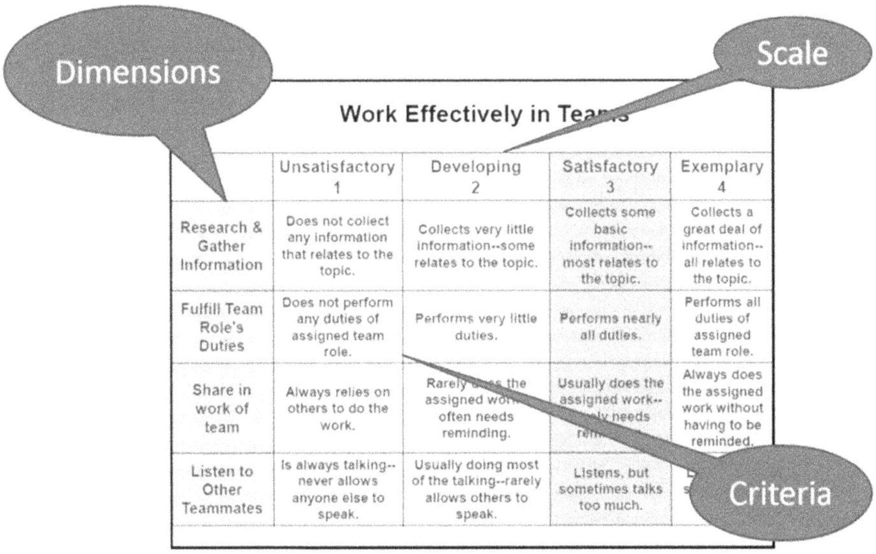

FIGURE 28.5 Elements of a rubric.

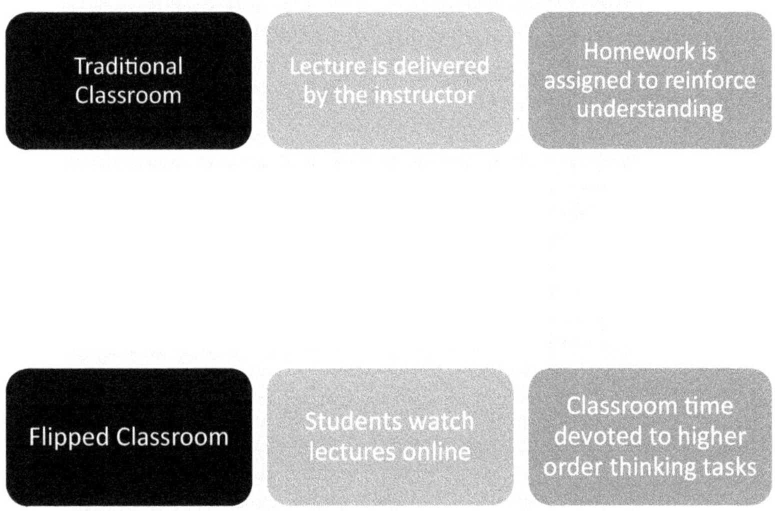

FIGURE 28.6 The flipped classroom.

interaction focused on higher level skills such as communicate effectively; identify, formulate, solve engineering problems, and work in teams. Specific activities requiring human interaction include the use of face-to-face and online discussion boards used to post and answer questions (students and faculty alike) and carefully crafted open-ended problems. This approach provides an opportunity to develop activities for active learning, such as cooperative learning, peer-assisted learning, and problem-based learning. It is important to note that activities are not limited to those shown in Table 28.3. The number and type of activities can be diverse provided they focus on efficiently achieving a learning outcome and the learning style of the students.

Actors in Online Course Development

The actors in the development of a course may include one or more subject matter specialists, an instructional designer, an instructional technologist, media specialist, web developer, programmer, librarian, and others depending on the complexity of the materials used in the course. An individual may have

TABLE 28.3

Learning theory and behaviours in a flipped classroom environment.

Flipped Classroom	
Does not require Human Interaction	**Requires Human Interaction**
Behavioural Learning Theory	**Constructivist Learning Theory**
Example activities: • Video lectures • Practice problems • Quizzes	Example activities: • Question/answer • Discussion • Collaborative open-ended problem solving

more than one of the roles for these actors. For example, it is typical that the instructional designer and technologist may be one person.

Development of high-quality, reusable content requires a substantial investment in competent professionals and technology services. Table 28.4 shows different actors involved in the development of a learning object, a module, or a course. It is important to note that not having this ideal team does not preclude a faculty member from development and online delivery as some of these roles can be carried out by a single individual. An experienced instructional designer is a highly valuable resource who can also act as a project manager and instructional technologist in the development of an LO.

The actors for the delivery of an online course or module include the instructor and depending on the number of students, teaching assistants. Also, technical support staff, usually a help desk. The primary role of the instructor and teaching assistants is to ensure at an individual level that students are successful. In addition, to moderate activities that involve social interaction as well as individual student performance in these. Much of the success of an online course lies in the ability of the instructor and teaching assistants to keep students engaged. To achieve this, constant attention and opportune intervention is essential during the delivery of the course.

The most important actor is the learner. To be successful a learner requires motivation, an appropriate level of prerequired knowledge, and access to technology appropriate to the course. These must be clearly understood and articulated when the high-level goals of the learning object, module or course are designed. Also, to make opportune progress during the course the most important factor is time. Other than personal matters that may arise, a learner that fails to advance in the course is often lacking in one or more of these.

Technology

The technology options that can be used in design of learning environments are too numerous to list here. However, a fundamental in the implementation and delivery of a learning

TABLE 28.4

Actors in the development of online learning environments.

	Composition of a high-level team for developing a learning object (LO)
Actor	**Role**
Subject matter specialist	This expert (typically, one or more faculty members) defines the learning objectives and is an expert in the knowledge domain of the LO. The subject matter expert also provides ideas for structuring the content and content sequence of the object.
Instructional designer	Ensures pedagogical integrity, making sure that learning objectives are at the appropriate level and that educational resources and tasks, including assessment, are correctly implemented at the right level, sequence, and depth while attending to the overall quality of the LO.
Instructional technologist	Contributes to the selection of the technology(es) that best support student learning. Ensures that that the technology used works as expected and provides training for faculty and students.
Project manager	Responsible for making sure that all members of the team understand their role in the project, is responsible for setting deadlines, assigns resources and ensures timely communication among team members. The project manager's main charge is to make certain that the project is done on budget and on time.
Librarian	Responsible for aiding in the selection of course materials, references and copyright issues.
Graphic designer/product designer	Responsible for the overall usability of the project, including accessibility compliance. This actor also ensures the aesthetic quality of the product and more importantly, of the quality of the interaction that the user experiences. In addition, the graphic designer creates all visual components of the LO, including graphics, text, animation, and other components.
Project integrator	Responsible for the correct assembly of the assets that conform the LO as well as all aspects related to content management. In addition, the project integrator ensures that interaction among components occurs correctly and at the right time. This person is also responsible for maintaining the metatags that describe the LO.
Videographer/Photographer	Responsible for on-site and studio acquisition of graphic material as well as post processing of the materials to deliver an LO component according to specifications.
Programmer	Develop the necessary software assets that are components of the LO. This may include scripts, macros, and software, as well as integration or connections to external software such as databases, simulations, decision support systems, statistical and artificial intelligence tools, and other applications requiring specialised code.
Help desk manager	Prepares held desk staff to assist faculty and staff during the delivery of a course or module to ensure that issues related to technology do not impede the learning process and the student focuses on learning.

object, module, or course is a Learning/Course Management System (LMS/CMS).

LMS and CMS have become a vital part in the delivery of online instruction and to supplement the traditional classroom (Krouska et al., 2017). These two terms are often used interchangeably although there is a difference between the two. The primary function of a CMS is to provide a platform for online student access to learning materials, tracking student performance, storing student submissions, and mediating communications between students and instructors. On the other hand an LMS has a more holistic set of functions that allows instructional content management, provides tools for assessment of the student and organisational goals including progress towards reaching these goals. An LMS provides data that allows supervision of the learning process for the organisation. There are numerous LMS and CMS systems available. Some of the major commercial LMS include Blackboard Learn, Canvas, and Desire2Learn. Open sources systems such as Moodle and Sakai are also available.

It is important to note that, although the interfaces for LMS/CMS are well designed from a usability perspective and their functionality and use in learning is well understood, it is important that students and faculty be trained and supported in their use. In addition, some of these systems may need a large resource investment, and as a result require institutional support.

Figure 28.7 shows a typical arrangement of a module for instruction related to logistics. Note that there is an introduction to the course followed by learning objectives, activities, and formative assessment.

ADDIE Methodology

Quality online course development is well understood, and instructional design methodologies are well tested and commonly used. ADDIE is a common methodology for instructional design that addresses the stages of course development (Branch & Kopcha, 2014. This methodology is well tested and is composed of the phases shown in Figure 28.8. The major activities that take place during each of the phases of ADDIE are:

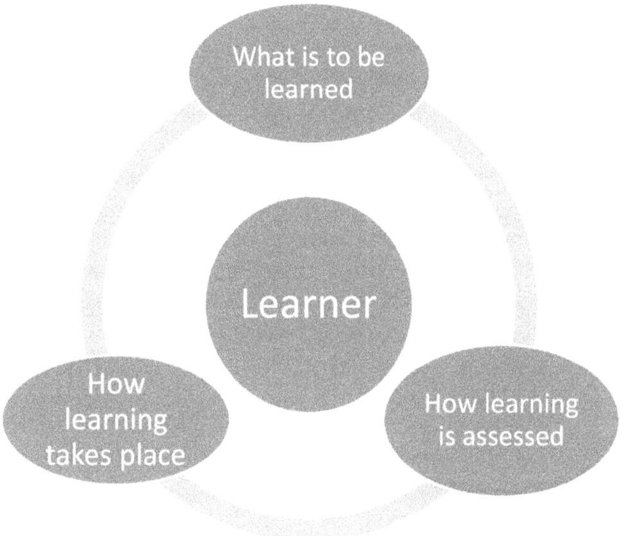

FIGURE 28.8 The ADDIE methodology.

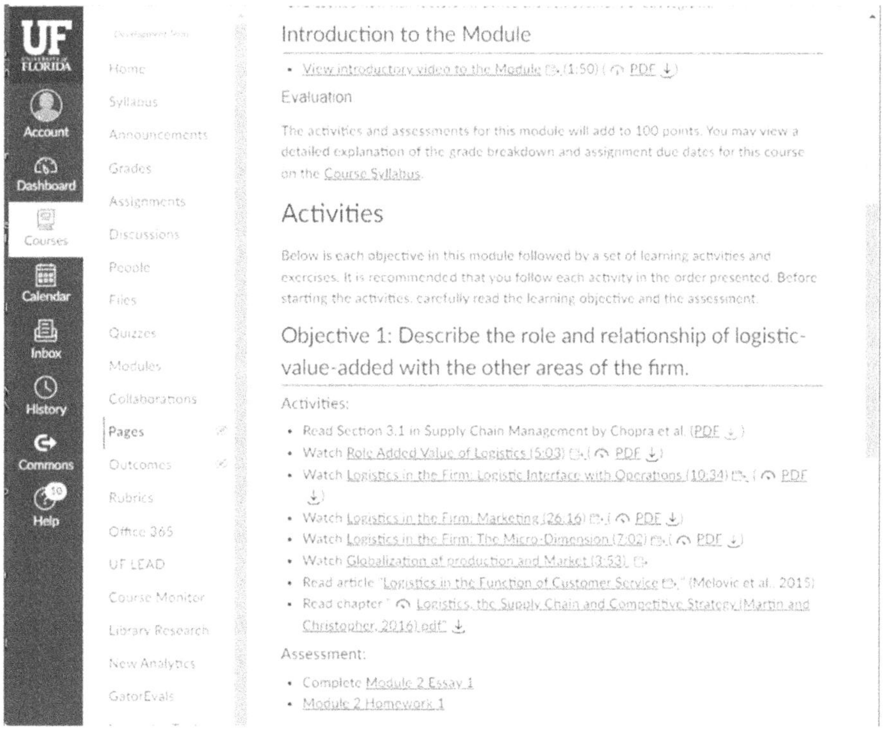

FIGURE 28.7 Screen Capture from a course using an LMS.

1) Analysis. The primary purpose is to articulate clearly what the instructional problem is. Instructional goals and objectives are established at this level, as well as identifying learner knowledge and skills.
2) Design. This stage focuses on developing learning objectives, assessment items, and learning assets. It follows a logical and orderly method for identifying, developing, and evaluating strategies to attain the course's goals. This stage requires great attention to detail.
3) Development. In this stage the instructional designer works with the staff to create and assemble the learning assets that were designed in the previous phase. Includes testing and debugging. In general, this process will move forwards quickly if the design phase is executed carefully.
4) Implementation. During the implementation phase all functional components of the course are assembled. Also, training for the instructor is provided. It is a good practice to develop a manual that covers course curriculum, learning outcomes, methods of delivery, and student assessment procedures. It may also be required to conduct training for the learners.
5) Evaluation. Performance methods are used to measure how well the objectives were achieved. That is, the level of success the learner reaches in retaining and demonstrating acquired skills and understanding. As a general guideline, the evaluation focus can be on understanding of the material, long term retention, and critical thinking skills. Also important at this stage, is to measure how well the course materials facilitate effective learning by the student.

Note that evaluation of each stage is ongoing.

Learning To Create Online Materials for Instruction

Over the course of the past decade the authors have conducted a workshop during professional meetings at international venues that focuses on the concepts discussed above. The workshop is a combination of tutorials and hands-on activities to familiarise interested individuals with online learning and flipped classroom pedagogy. The workshop is conducted as a combination of presentations and hands-on exercises as described in the agenda below. The presentations focus on introductory concepts related to:

1) Online learning. A review is conducted of online teaching/learning experiences by diverse institutions with a special focus on improving quality or learning and reducing the cost of instruction. In addition, the ADDIE methodology is presented to familiarise teaching faculty with the development process of an online course. Barriers and conditions for success are reviewed, including required services and infrastructure, faculty and student competencies and skills, and required operational support.

TABLE 28.5

Components of a faculty training workshop on the flipped classroom

Flipped Classroom Workshop
Welcome and introduction
Flipped classroom: What is it? Why it works!
Service and infrastructure requirements
Online instructional design concepts
Goals and Learning Objectives
Hands On: Writing a learning objective and assessment.
Best Practices for Teaching online
Delivering a flipped classroom course: A real example
Hands on: Creating an engaging flipped classroom activity
Using online discussions
Questions and answers

2) Flipped classroom. The flipped classroom methodology is presented, including learning theory, research findings, and the development and implementation of course modules for use with this innovative pedagogy. The hands-on group exercises focus on writing a learning objective and assessment and creating realistic and engaging active learning activities.

The agenda is shown in Table 28.5 and is conducted as a half-day, full day, or two day event, as time allows. Longer time allows for more detailed hands-on exercises.

The details of the materials for the workshop are too extensive for the space allowed here but can be downloaded in the link provided below by Busato et al. (2016a) and Center for Instructional Technology and Training (2017). A second workshop that focuses on training instructors in the flipped classroom, using a flipped classroom format, is available at Busato and Zazueta (2017).

Summary

Development of learning environments require a clear definition of what is to be learned expressed in terms of observable student behaviour, how the learner will demonstrate that they have acquired the behaviour that demonstrates they have learned, and a learning environment where the student can acquire the target knowledge. The capacity of faculty to achieve this requires a basic understanding of learning theory, instructional design, and skills in the use of associated technologies.

REFERENCE LIST

Bloom, B.S. & Krathwohl, D.R. (1956). *Taxonomy of educational objectives: The classification of educational goals, by a committee of college and university examiners. Handbook 1: Cognitive domain.* New York: Longmans.

Branch, R.M., & Kopcha, T.J. (2014). Instructional Design Models. In J. Spector, M. Merrill, J. Elen, & M. Bishop (Eds.), *Handbook of Research on Educational Communications and Technology.* Springer, New York. https://doi.org/10.1007/978-1-4614-3185-5_7

Busato, P., Berruto, R., & Zazueta, F.S. (2016a). Flipped Classroom Workshop: A Workbook for Active Learning Pedagogy and Andragogy. ASABE International Meeting, Orlando, FL, 13p.

Busato, P., Berruto, R., Zazueta, F., & Silva-Lugo, J. (2016b). Student Performance in Conventional and Flipped Classroom Learning Environments. *Applied Engineering in Agriculture*, vol. 32, no. 5, pp. 1–10. SSN 0883-8542 https://doi.org/10.13031/aea.32.11298

Busato, P., & Zazueta, F.S. (2017). Introducción a los sistemas de Control Computarizados. https://fsz.ifas.ufl.edu/control/

Center for Instructional Technology and Training. (2017). Online Learning Workshop. https://fsz.ifas.ufl.edu/OnLineLearningWorkshop.htm

Council of Chief State School Officers. (2023). Revising the Definition of Formative Assessment. https://ccsso.org/resource-library/revising-definition-formative-assessment

Education Endowment Foundation (2017). Meta-cognition and Self-Regulation. https://educationendowmentfoundation.org.uk/evidence-summaries/teaching-learning-toolkit/meta-cognition-and-self-regulation/

Forehand, M. (2005). Bloom's Taxonomy: Original and Revised. In M. Orey (Ed.), *Emerging Perspectives on Learning, Teaching, and Technology*. http://www.coe.uga.edu/epltt/bloom.htm

Gluga, R., Kay, J., & Lever, T. (2013). Foundations for Modelling University Curricula in Terms of Multiple Learning Goal Sets. *IEEE Transactions on Learning Technologies*, vol. 6, no. 1, pp. 25. https://doi.org/10.1109/TLT.2012.17

Krathwohl, D.R. (2002). A Revision of Bloom's Taxonomy: An Overview. *Theory into Practice*, vol. 41, no. 4, pp. 212–218. https://doi.org/10.1207/s15430421tip4104_2

Krouska, A., Troussas, C., & Virvou, M. (2017). Comparing LMS and CMS Platforms Supporting Social e-Learning in Higher Education. 8th International Conference on Information, Intelligence, Systems & Applications (IISA), Larnaca, Cyprus, pp. 1–6. https://doi.org/10.1109/IISA.2017.8316408.

Rao, N.J., & Banerjee, S. (2023). Classroom Assessment in Higher Education. *Higher Education for the Future*. https://doi.org/10.1177/23476311221143231

Wang, V. (2008). Fundamentals of Learning Theories. In L. Tomei (Ed.), *Encyclopedia of Information Technology Curriculum Integration* (pp. 347–352). IGI Global. https://doi.org/10.4018/978-1-59904-881-9.ch057

29

Educating the Compleat Biological Engineer

Arthur T. Johnson

Introduction

Agricultural engineering is an applications-based engineering discipline. That is, engineering situations in agricultural engineering are all related in some way to their use in agriculture. Some agricultural engineering problems may require the use of methods from civil engineering, mechanical engineering, electrical engineering, or chemical engineering, but the ultimate use is all in agriculture.

Biological Engineering, on the other hand, is a science-based engineering discipline. Biological Engineering practitioners incorporate the science of biology as an essential part of their discipline. The range of applications of Biological Engineering techniques is as broad as the science of biology. Agriculture is just one of the many applications areas possible with Biological Engineering approaches. Thus, agricultural engineering and Biological Engineering are, by definition, somewhat orthogonal to each other; some, but not all, of agricultural engineering can fit into the purview of Biological Engineering, and agriculture is but one use for Biological Engineering.

The other foundational engineering disciplines are mostly based on specific areas of the physical sciences; civil engineering and mechanical engineering are based on the science of mechanics; electrical engineering, as its name suggests, is based on electrophysics. All engineering disciplines are fundamentally based on physics, but chemical engineering incorporates essential elements from the science of chemistry in its domain. In that respect, Biological Engineering is somewhat analogous to chemical engineering by including an additional science into its area of expertise.

The foundational engineering disciplines were established as it became necessary to solve problems that arose in newly emerging fields of technology. Civil engineering is the most ancient of these, dating back at least to Roman times. The other disciplines followed when it became clear that applications and associated problems were becoming widespread. It has only been relatively recently that knowledge in biology has expanded greatly, and that it has become clear that biosystems can be of use to solve problems of importance to society. Thus, until relatively recently, there was no need for a field of Biological Engineering to emerge. However, we are now at the point where biological applications are becoming important for other engineering disciplines as well.

The concept of a discipline of Biological Engineering was proposed during the late 1960s (Arthur, 1970), but the real impetus towards formation of the new field came during the late 1980s and early 1990s when undergraduate enrollment in agricultural engineering academic programs plummeted, causing deep concern among academicians that existence of their programs could no longer be justified (Johnson, 2006). Attention turned to the one science unique (at that time) to agriculture, biology, with the hope that a broader, science-based engineering discipline would attract students who were not particularly interested in working in an agricultural setting (Johnson & Davis, 1990; Johnson & Phillips, 1995). Soon after, interest in developing a Biological Engineering curriculum also developed within the ranks of chemical engineering, although their concept of Biological Engineering centered specifically on biochemicals. The agricultural engineering approach is much broader, just as the issues in agricultural engineering are broad, incorporating mechanics, electricity, transport processes, and even genetics in those curricula.

Basic Curriculum

The basic blueprint for a Biological Engineering curriculum was laid out in a final report for a US Department of Agriculture (USDA) project authored by Garrett et al. (1992). Biological Engineering undergraduate students need to take the basic set of engineering subjects, beginning with physics, chemistry, and the introductory engineering sciences, such as statics, dynamics, strength of materials, transport processes (fluid flow, heat transfer, and mass transfer), electrical circuits, and, perhaps, instrumentation, machine design, and materials properties.

There should be a significant element of design in the upper-level undergraduate curriculum, culminating in a capstone design course in the senior year. However, incorporating design projects in as many upper-level engineering science courses as possible helps to promote the sound engineering judgment necessary for most to become successful Biological Engineering practitioners. There are many limitations and regulations that must be considered for many practical Biological Engineering problems. It is only the rare undergraduate who makes the correct decisions the first time a practical engineering problem is

DOI: 10.1201/9780429150111-32

encountered. It is much better for the student to err in the relatively benign confines of the university classroom rather than later in the more unforgiving world of professional practice.

General education courses, such as English, Technical Writing, History, and other courses necessary to produce a well-rounded graduate should not be minimized. These courses are important to negotiate the nontechnical aspects of professional life.

Biology Requirement

The distinguishing feature of a Biological Engineering undergraduate educational curriculum is a strong emphasis on biology, and it is unlikely that standard biology courses given by biologists for biologists meet the needs of an engineering discipline dedicated to solving problems either using biological entities or applied to biological systems. A Biology for Engineers course should be developed by engineers and taught as an engineering science course (Johnson, 2019; Johnson, 2021). The course needs to emphasize principles of biology that can be useful in design processes. Engineers need to know about biological principles in the same ways that they know about transport processes, mechanical principles, electrical principles, chemical principles, and materials principles. These principles form the basis of engineering applications. When dealing with any system, including biological systems, it is best to know the likely outcomes of an idea before spending much time or effort developing it.

Studying principles and generalities exhibited by biological entities gives an understanding of how that entity is supposed to work, and such a concept is important to have before creative approaches to the solution of biological problems can occur. Rather than defying the basic laws of chemistry and physics, biological responses are constrained by them, just as are responses of inanimate objects. The Biology for Engineers course should demonstrate limitations imposed by chemical and physical environments in which the biological entity is located. There are also constraints imposed by other biological entities in the surrounding environment. This concept of physical, chemical, and biological environments and their effects on expectations of biological responses is an important one for Biological Engineers to understand.

One subject that will not usually be found in a traditional biology course is an introduction to control systems principles. But, biological reactions are often used to control actions and responses, so it is important that the Biological Engineering student has at least some knowledge of sensors, actuators, and control mechanisms. Feedback, feed forward, and open-loop systems need to be included, all with examples drawn from the field of biology.

Also necessary for inclusion in the Biology for Engineers course are principles of toxicity and communications. Each of these is important to the understanding of biological responses, and may not be taught in traditional biology courses. Principles of optimization can also be enlightening when studying biological conditions.

Cycles of a physical, chemical, or biological nature should also be included. Whereas introduction of the cyclical nature of substances, such as water, nitrogen, or phosphorous, is usually taught in an environmental context, cycles are also important for the effectiveness and toxicity of medicines taken internally. So, biological cycles need to be taught to all students, regardless of specific interests.

As with all other engineering science courses, quantitative approaches to problem solving are expected. But, first, basic concepts of how things work must be presented and understood. After that, then predictive equations are appropriate. In the field of biology, there are many characteristics that scale with some physical measurement, such as body mass. This field is called allometry. Allometric equations can be useful for predicting responses of a biological system when actual data is unknown or unavailable.

And don't forget psychology. With increasing evidence of aspects of sentience (self-awareness) present in animals at many hierarchical levels (Johnson, 2020a), it becomes important to understand how these animals process sensory inputs and modify expected responses. Designs involving human operators should also take into account human factors effects, the part of psychology that may greatly affect product acceptability.

It is important that all of these courses maintain the broadest possible scope. Whether the graduate intends to practise in environment or biomedicine, which represent two potential popular application areas, a thorough knowledge of all areas of biology is important because living things are not isolated, and the physical, chemical, and biological environments in which they find themselves must be completely understood in order to make sound engineering judgments for potential problem solutions. Knowledge in one potential application area is usually applicable to many other application areas, as well.

The broad, general nature of the Biology for Engineers course should extend to all hierarchical levels, from the individual gene to the largest ecosystem. As it turns out, responses at these different levels can be very similar in a general sense, depending on the surrounding physical, chemical, and biological environments. It is necessary to teach such responses in a comprehensive way so that the student develops general prediction concepts that can be applied with confidence to problems with unfamiliar parameters. Principles rather than details need to be the aim of such a course.

The ability to see connections across biological levels and realms can be the key to new solutions for Biological Engineering problems. Examples of these are:

1. Common functionality between human kidneys and environmental wetlands.
2. Surface coatings to repel mussels from ocean vessels can also be used to keep microbes from sticking to medical instruments.
3. Principles used to preserve wildlife species are opposite to ones needed to eliminate microbial infections.
4. Information legacies take many forms, depending on hierarchical level, but are necessary for all species.
5. Sensing and reaction are important at all biological hierarchical levels.

At the completion of the upper-level engineering science courses, including Biology for Engineers, students should have at least nascent abilities of:

1. **Knowledge of biological principles and generalisations in order to design products and products involving living things.**
2. **Ability to transfer knowledge from a known biological system to a different, unfamiliar biological system.**
3. **Ability to avoid the unintended consequences involved when dealing with any biological component.**

Competencies Expected of Biological Engineers (from Johnson, 2010, used with permission)

Biological Engineering courses and curricula can result in graduates who vary greatly in their abilities to deal with biological systems. Whereas there has been general agreement about a better understanding of biology and biological applications for all Biological Engineers, specific recommendations of competencies have been formulated (Johnson, 2010) with the idea that a minimal list of principles and basic concepts expected of all Biological Engineering graduates would standardize commonalities no matter what educational institution was attended.

Before there can be any confidence in the uniformity of Biological Engineering education, there must be an understanding of the necessary knowledge to be possessed by all Biological Engineers. Of course, uniformity will only come when there is consensus about the essentials that must be known to be successful Biological Engineers. The set of concepts and principles set forth here is intended as the minimum amount of knowledge that all Biological Engineers should possess to represent the field adequately. Basic science, engineering, mathematical, and liberal arts competencies have been assumed to be in addition to this list.

Each of these concepts and principles is listed, in no particular order except for the first, and some explanation is given for each. Explanations, however, are not meant to be exhaustive or complete, but rather as a short justification of the importance of that particular item to the listing.

1. Survival and Reproduction of the Genes. This is the first principle of biology, and the dominating theme of all biological actions. Living things will compete, control, modify, adjust, kill, sneak, cover up, and scrap in order to survive and reproduce. Death is so catastrophic to any living thing that no measure is spared in order to live.

Survival and reproduction are important at many hierarchical levels. The obvious level for which survival and reproduction are important is at the organism level, but it is the genes that are driven to survive and reproduce. They do this in cooperation with other genes by surrounding themselves with survival mechanisms built into cells, tissues, organs, and organisms. There is competition among genes, causing some genes to take advantage of others. So-called cheating genes duplicate themselves at the expense of other genes. *Transposons* are genes that move from one genetic location to another in order to duplicate themselves within the genome. Killer genes cause toxins to be produced that eliminate competitive genes. The ultimate competition in biology is among the genes.

Survival and reproduction extend to groups of individuals as much as to individuals themselves. Forming herds, schools, or human families gives mutual cooperative support for survival mechanisms. Members of these groups assist each other to fend off threats poised by predators and to locate sufficient food and other resources to meet the group demands. If the survival advantage of group membership is great enough, then there may be a large sacrifice in order to belong to the group. Examples of this are the ceding of breeding rights to dominant males in a herd or physical mutilation for some human groups.

Even among plants, there are group survival mechanisms in play. *Masting* in oak trees occurs when all oaks in an area form copious quantities of acorns in certain years and not others. This increases the probability that some acorns will survive consumption by hungry animals. Seventeen-year locusts synchronize emergence for the same reasons. Plants also communicate attacks by insect pests through airborne chemicals that induce other plants to buttress defences.

2. Form is Related to Function. Related to modularity is the realization that form and function are closely aligned. More acute hearing requires larger ears. Night vision requires pupils that allow more light to be captured. Heat maintenance requires heavy insulation. Efficient mass transfer requires a large surface area. Each of these can be predicted from the laws of physics or chemistry. From form, one can infer function; from function, one can predict form.

Realizing that organisms have been perfected over many generations to deal with environmental stresses is the key to using biological solutions to solve engineering problems of a similar nature. Copying biological solutions is called *biomimetics*, and this is one of the easiest and surest ways to formulate a problem solution.

3. Modularity and Incremental Change. Biology and technology operate similarly. Each begins with a new prototype and proceeds from there towards improvements by making incremental changes. Successful parts are maintained; others are improved. This similarity between biological evolution and technological evolution is the paradigm for all technological progress. It is useful for Biological Engineers to be familiar with this concept. Not only does it supply context for engineering efforts, but it also can be a vehicle for greater understanding of biological organisms and why they appear as they do.

4. Environment Influences Outcomes. Actions and structures of living beings are strongly dependent on physical, chemical, and biological environmental factors (Johnson, 2019). Biological actions and responses are not made without accounting for environmental factors. These actions can take the form of organismal cooperation or competition (biological environment), initiating heat-loss mechanisms (physical environment), or detoxification of cyanide (chemical environment). But environment has more profound effects at the genetic level. Environmental factors can determine the expression of certain genes. The ultimate manifestation of this is the formation of *epigenetic markers* that determine genetic expression for generations at a time.

Biological engineers must be aware constantly of the role that environment plays in biological responses. Environment can determine not only the degree of action, but also whether it occurs at all.

5. Physical Limits Cannot Be Exceeded. Living things do not abrogate the laws of physics; they reaffirm them. Mechanical strengths of materials do not change because they are included in living creatures. Exceed these limits, and they break.

Likewise, diffusion limits do not change because materials move inside living cells. The maximum distance that oxygen can diffuse passively at a rate sufficient to sustain life is equal to the thickness of two cells. Any farther than this, and other mechanisms (such as circulation) must be used to deliver oxygen. To accommodate physical limitations, biological things can manipulate shapes, sizes, and compositions of materials. The means that biology has developed to circumvent physical limitations are often good models for engineering designs, but, in the final analysis, physical limitations must be recognized.

6. Relationships Among Energy, Order, Entropy, and Information. Passing information from one biological unit to another requires communication channels. These channels may involve chemical, visual, auditory, touch, taste, magnetic, electrical, or neural mechanisms. Many communications in nature are chemical. Bacteria communicate their presence to others within a biofilm with chemicals called *autoinducers*. Insects indicate the presence of mating adults by airborne chemicals called *pheromones*. Many mammals mark their territories with chemicals sprayed on prominent physical features.

Communications are important for organisms not only to know the presence of others, but to anticipate the intentions of others. Thus, signals of aggression, acceptance, threats, or love elicit emotional responses that are meant to enhance survival.

Communications result in information that leads to order. This, in turn, reduces entropy, at least locally. In order to maintain order, a constant supply of energy is necessary. Stop the flow of energy, and death ensues, the organism decays, and entropy again increases. Shannon's relationship between information and the probability of the occurrence of an event, where information content is higher for rarer events, can also be used to determine the entropy of the order maintained by an abundant supply of energy (Shannon & Weaver, 1949).

7. Reversible Chemical Reactions. Chemical reactions that are extremely vigorous, generate too many products, or that have extreme behaviour in one way or another are not usually useful biologically. Life requires agility, both physically and chemically. Chemical reactions that can easily be reversed are the useful kind. For instance, adenosine triphosphate (ATP) is a very useful compound for storing and releasing chemical energy. ATP is converted into adenosine diphosphate (ADP) through hydrolysis, with the liberation of energy to drive other chemical

reactions. ATP is used almost universally in living things as an energy-storage chemical. ATP is neither a very energetic compound nor a very unenergetic compound; it is intermediate in energy. That way, it can store and release energy when needed.

Maintaining the human blood alkalinity level at about a pH of 7.4 is necessary for maintenance of internal equilibrium. A bicarbonate buffering system is used to convert circulating carbon dioxide into carbonic acid or bicarbonate ion. These processes either soak up or release hydrogen ions. Thus, human blood pH can be maintained constantly, even during severe exercise when metabolic acids are released into the blood.

Proteins are used that have different forms and purposes with little change in energy levels. Thus, very little energy is needed to control protein configurations and functions.

Not all of these reactions are strictly reversible chemically, but they do represent

reactions that move easily between reactants and products, so the energy differences between chemicals on both sides of the equation are small.

8. Molecular Shapes and Chemical Mass Action. Much of the chemical action attributed to enzymes and proteins is the result of molecular shape. The complex folding of proteins from simple strings of amino acids into molecules resembling knotted ropes results in parts of the molecule that fit well with other molecules. When two or more such molecules physically locate next to each other they are usually close enough to react readily.

The chemical law of mass action states that chemical reactions are much more likely to progress when concentrations are high. Enzymes, by bringing reactants close together, increase the local concentrations manyfold. As an added benefit, because enzymatic action is based on physical shape, the enzyme does not take part directly in the chemical reaction and is not consumed.

Chemistry inside a cell depends very strongly on the physical shapes of molecules. In addition to the enzymes described above, genetic duplication depends upon the complementary shapes of the *pyrimidines* cytosine and thymine fitting physically with the *purines* adenine and guanine. Also, the physical integrity of the cell structure depends on the physical gel resulting from the way that polar water molecules line up with one another.

9. Osmosis. Osmotic pressure is a powerful motivator. For instance, tall trees do not have any physical means to pull water from the ground to the very top of the tree if the tree is taller than the height of water that can be supported by atmospheric pressure. This height is about 30 metres. However, water that evaporates from leaves at the treetop reduces the concentration of water, so this diffusion gradient pulls water as far up as it needs to go.

10. Benefits of Redundancy and Alternative Pathways. The success of biological entities is largely a result of dependency on redundancy and alternative pathways. This gives organisms the ability to adjust to new environments and new challenges.

Redundancy is at the heart of the immune system. There are many ways to deal with foreign invaders, and, if one or several of these is overwhelmed, there are other lines of defence to be overcome. The organism is thus usually successful at maintaining self-integrity.

There are alternate metabolic pathways that allow the organism to substitute one type of metabolism for another. Thus, the muscles normally act with aerobic metabolism, but are capable of anaerobic metabolism when oxygen is in short supply. Microbes in a bioreactor can be induced to produce certain biochemical products when given reactants different from those that are most efficiently used or preferred. In this way, alternative pathways can be exploited for economic benefit.

One place where redundancy has not usually been consciously recognized is in the information legacy to pass information from one generation to the next. This function has usually been recognized as belonging to the genome. It is the genes, after all, that are uniquely posed to supply the basic information source for the chemical mechanisms of the cell. Nevertheless, there are at least three other transgenerational information sources that have not been as widely recognized. The first of these is *memes*, or cultural information taught by certain individuals to

others. Through this cultural information, survival is enhanced among animals, such as meerkats that teach their young how to catch and eat dangerous prey, or sheep and goats that teach their young about the differences between poisonous and nonpoisonous plants, or humans who school their young for an extended childhood.

The second additional information legacy is the microbes passed from one individual to another. These microbes have a profound effect on health and wellbeing. Probiotic microbes are passed from mothers to their young to help them digest food, calibrate their immune systems, and prevent dangerous diseases.

The third, and perhaps not the last, information legacy is in the form of misfolded proteins, called *prions*, that have been found to be able to be passed from one generation to another, and have also been found to be able to replicate without a genetic DNA or RNA model. Many prions are associated with dangerous diseases, such as *transmissible spongiform encephalopathy*, or *Alzheimer's disease*. Other prions, however, help yeast cells cope with radically different environments, or assist in animal memory retention by altering frequently used neural synapses (Saltus, 2010).

The principle of redundancy has been taken very seriously when failures cannot

be tolerated. NASA depends upon redundant components for its space ventures, for instance. Automobile safety systems are also highly redundant. The Biological Engineer attempting to control a dangerous microbe, as an example, should incorporate redundant controls to assure that the objectives of the process are met.

11. Adaptation Requires Energy and Resources. Life has developed as a means to survive and reproduce, with the emphasis on reproduction. Energy and resources are required to form new offspring, and energy and resources needed to adapt to environmental demands uses energy and resources otherwise able to be used to reproduce. Thus, environments that demand adaptation in order to survive reduce the ability to reproduce.

If maximum reproduction is desired, then optimum environments need to be supplied to the organism. Such would be the case for breeding animals or producing a field of grain.

If reproduction is not the goal, but something else is desired, then reproduction will suffer. Such is the case in a bioreactor to produce an economically important chemical from genetically modified organisms. Modification of the genome to improve the yield of a desired metabolite reduces the ability of the organism to reproduce. Thus, there may be more of the desired product, but it may take longer for the population to reach the critical level.

12. There is Competition for Limited Resources. It can be assumed that every niche that can support life, however limited, will contain living things to the maximum extent possible. This makes all biological systems limited by resource availability.

Competition among species and among individuals of the same species keeps them from unchecked growth. *Staphylococcus aureus*, a dangerous microbe, is present on the skin of nearly all humans, yet normally causes little or no trouble because other microbes compete successfully with *Staph* and limit its influence. Human beings, as a matter of fact, contain about 10^{14} cells, only about ten percent of them of human origin. The remainder are bacteria, small mites, and other microbes that compete for space and resources on the skin or in the digestive system.

It would be wrong to ignore competition among microbes. *Competitive inhibition*

keeps most microbes in check and does not allow them to become a threat to the host. Trying to disinfect all microbes, therefore, removes many benign competitors of the more dangerous types. The result can be worse than if no disinfection was attempted.

Competitive inhibition can be used to control diseases of plants as well as of humans. A biological problem can often be solved in a biological way. To control one organism, it is often best to look for a natural enemy.

13. Antagonistic Mechanisms Give Precise Control. When it is necessary to control a process to a very close tolerance, antagonistic mechanisms are used. The idea is to attempt to control in two directions simultaneously. Then, the resulting movement is the difference between the two efforts. Fine movements of the fingers are achieved with antagonistic muscle control. Exquisite control of glucose level in the blood is achieved by means of two hormones, insulin and

glucagon, that have opposite effects.

An alternative to antagonistic control is to use active control in one direction and use stored energy to restore the system to its resting position. Such is the case with human respiration at rest. Inhalation is active, and passive exhalation is achieved when stored gas in the lungs is released into the atmosphere. The restoring force is produced by gas compression and elastic stretching of the tissues. Active exhalation is initiated during exercise, when it becomes important to remove gases quickly from the lungs.

Biological systems, even up to the ecological level, use a lot of antagonistic control, sacrificing energy efficiency for rapidity and precision.

14. Optimization Saves an Incremental Amount of Energy. Life is energy intensive. Thus, it may not appear to make much sense to save additional energy through optimization. Nevertheless, the biological environment is competitive, so, in a place where all the competition uses as much energy as is available, a small difference in energy requirements may have a large reproductive effect. Especially subject to optimization are overhead processes, such as blood circulation and digestion, or survival processes, such as locomotion. It appears that heart pumping and respiration, especially during heavy exercise, are optimized to reduce the rate of energy consumption, as is the energy used by walking and running. There are probably optimized biochemical processes, and nutrient recovery and utilization in whole ecosystems appear to be optimized.

Optima may either be narrow and highly precise, or broad with little penalty for deviation from the optimum value. Biological optima usually conform to the latter (Johnson, 2013). Thus, we find that the rates of respiration around the optimum value to minimize energy expenditure during exercise vary both above and below the best rate, but center on the optimum. The same can be seen to be true both for heart rates and walking speeds.

One intriguing question remaining in biology is why there are so many genetic variants still present in the genomes of different organisms. Selective survival should have extirpated the less-favorable genes many generations ago. Looking at the genome as an optimal solution, however, infers that there will remain genetic variability in a manner similar to other biological optima. A slightly smaller energy demand in an organism could pay off in a huge difference in reproductive potential.

15. Directed Evolution. Relatively recently, it has been realized that evolutionary principles can be used as an engineering design tool. These principles are simple: set a series of design goals, start the design process with a prototype, make some change by some process that may include randomness, test the resulting new prototype against the design goals. If the new prototype is better than the old, even incrementally, keep it and make another change. Test again and change again. Continue to cycle until the design goals are met or until no further improvement can be made.

Directed evolution has been used to design improved microwave antennas, buildings, and new enzymes that do not exist in nature.

16. Analogical Thinking. The ability to see connections among different physical and biological systems can be valuable for the biological engineer (Johnson, 1999; Schreuders & Johnson, 1999). If, for instance, all biological units, whether they are at the cellular, tissue, organ, organism, group, population, or ecosystems levels have certain traits in common. Some of these have been described above. If these commonalities can be understood as concepts, then they can be useful in forecasting expected biological behaviors at all levels and help to achieve the three Biological Engineering goals given in the previous section on Biology for Engineers.

Analogical thinking begins with characterizing physical laws obeyed by biological entities. Classifying all variables into either effort or flow and understanding the relationships between these two can be a start (Johnson, 1999; Schreuders & Johnson, 1999). Analogical concepts in chemistry and the engineering sciences follow. When these are all brought together, biology is no longer a science consisting of a long list of facts, but is, instead, a coherent set of concepts and fundamentals that can be of value in formulating designs for new and improved products and processes.

17. Looping and Successive Approximation. Although it may appear as if there are elegant control algorithms embedded in the central nervous systems of animals, the real control mechanisms more likely involve trial-and-error procedures.

Sensors feeding information directly to the brain can invoke motor responses that result in changes of sensor outputs. These, in turn, would invoke other changes. If these other changes bring the control system closer to the set point, then further changes in the same direction are attempted. This continues until the set point is passed, and then the invoked change is in the opposite direction. This successive approximation procedure results in hunting around the desired set point. The exact set point is not always met, but deviations are usually small.

18. Reliability Curves. These are so universal in shape that they apply to automobiles, computers, and tools, as well as to human beings. A product just off the assembly line has a relatively high probability of failure. This is a result of faulty components or improper assembly. After this initial high rate of failure, called the *burn-in* period, comes a relatively long period of satisfactory performance with low rate of failure. This is the normal working period. When ageing begins, the rate of failure again increases and continues to climb. For very, very old products, the failure rate seems to level off at a high rate.

This progression of failures, high when new, decreasing to a low value for most of its working life, and then increasing again with age has been attributed to redundant faulty components. Failure of one or two components can be tolerated as long as other parts can assume the functions of those that no longer function. When the last part fails, the whole product fails.

Humans and other organisms are assembled from redundant, imperfect components. They follow the classical reliability curve, just as human-made products do (Gavrilov & Gavrilova, 2004). All engineers should be aware of the product reliability cycle and plan accordingly.

19. Immediate Availability with Circulatory System. Instant hot water in your hotel room? With a hot water circulation system, hot water is transported in a cycle from source to far end and back again. Tapping into this circulation avoids the wait while pipes are purged of noncirculating cold water.

Circulatory systems in higher-level animals have common design characteristics that engineers are well to be aware of. First, to deliver oxygen to the tissues, the smallest vessels must be in intimate contact with the tissues. Small vessels have large resistances that dissipate large amounts of power. To overcome this, flow rates in these small vessels must be kept low, so there must be a large number of them in parallel. Also, to minimize energy requirements, there must be a small number of very large vessels to feed the smaller ones and collect from the smaller vessels. Thus, there must be a branching pattern from large to small and from small to large vessels.

Living things use circulatory systems to deliver oxygen, glucose, and hormones to various locations in the body whenever and wherever they are needed, without the delay that would be present without a circulatory system. This mechanism works so well that it can be used for applications other than in living beings.

20. Proportional Plus Derivative Sensors. It is common that biological receptors produce outputs (usually neural in animals) that are related not only to the level of the stimulus but also to the rate of change of the stimulus. This incorporates a derivative element in the downstream control. Derivatives anticipate future stimulus positions, and so improve stability of the control system.

For stimuli that are chronically present, the control system that includes sensors and controllers often adapts to the continuous presence of the stimulus. This has the effect of saving attention for those elements of the environment that have meaning.

21. Weber-Fetchner Law. Biological sensing is inherently nonlinear, giving rise to the Weber-Fetchner Law, and its variants, Weber's Law and Stevens's Power Law. Each of these is somewhat different quantitatively but expresses the fact that the just-noticeable difference in a stimulus depends on the level of that stimulus already present. Thus, the ability to see a light depends on the background light level; the ability to hear an additional sound depends on the background sound level, and so on.

The effective additional level of a stimulus appears to be so universal in biology that the same principle should be expected for enzymes, neural activity, and running speed, to name a few. To expect the ability to recognize an additional component to be influenced by the amount already there can influence Biological Engineering designs of all kinds. Perhaps the most important part of understanding the concept behind the Weber-Fetchner relationship is that it clearly identifies limits to discernment sensitivity.

22. Young's Principle. Very selective, narrowband sensors require a lot of energy to operate. In addition, there are only a few instances in biology where such selectivity is warranted. One of these is the detection of pheromones to indicate reproductive opportunities.

Biological sensors are nearly all broadband, with responses that spread over a wide range of input qualities. Signals from these sensors are not very informative, because they could represent a wide range of possible stimuli. Yet, organisms are very much aware of their specific surroundings. How is this possible?

The answer is a number of broadband sensors with center stimuli that differ from each other by finite amounts. Their outputs overlap. When contributions from each of these sensors are compared, the contribution of each sensor, when added to the others, can determine exactly what type of stimulus has been received.

Examples in biology abound. One of the best examples is the colour sensors in the retina of the eye. There are three types of receptors, each sensitive to a different range of light frequencies. They are broadband, and their outputs overlap. A specific colour is determined by the relative contributions of each of these receptors. Thus, perception of a myriad of colours is possible from the outputs of only three types of receptors. A similar mechanism allows determination of the physical position of touching without having to have touch sensors at each location on the skin.

23. Hagen-Poiseuille Formula. This formula relates vessel resistance to fluid viscosity, vessel length, and vessel diameter. What it tells us is that resistance is extremely sensitive to diameter, being inversely proportional to diameter to the fourth power. A small difference in diameter can have a large effect on resistance. This equation was developed for laminar flow in a uniform, straight conduit. These conditions rarely apply in biology. Nevertheless, the Hagen-Poiseuille formula is a simple, quantitative explanation of the effects of changing vessel size, and, as such, is a valuable rough guide even when flow is turbulent and the vessel is not straight or uniform.

In the bodies of animals, there are many sphincter muscles that encircle fluid-flow passageways. The Hagen-Poiseuille formula tells us that it is not necessary for the sphincter muscles to completely close these passageways in order to control fluid flowing in them. A small contraction has a large effect. When exposed to nitric oxide, arteriolar sphincter muscles relax and blood rushes in. It doesn't take much.

24. The Law of Laplace. The Law of Laplace results from a force balance on the shell of a sphere or cylinder and relates pressure inside the object to tensile stress in the wall, wall thickness, and radius of the object. It gives a clear reason why bacteria are rounded, flimsy capillaries can contain blood at high pressure, and why lung surfactant is necessary to keep respiratory alveoli from collapsing into one huge sac with a surface area too small to sustain life. The implications of the Law of Laplace are counterintuitive, because it states that smaller vessels can contain higher internal pressures without bursting. Architects have used the Law of Laplace as the basis of building shell designs that withstand high loads without heavy thick roofs and walls. Biological Engineers should understand this principle in order to avoid intuitive errors in judgment.

25. Class 3 Levers. First-class levers come to mind whenever a mechanical advantage is sought. First-class levers place the fulcrum between the applied force and the load such that a smaller force can be applied to lift a heavier load by lengthening the moment arm. However, the arms and legs don't work that way.

Muscles produce high forces, but cannot produce them very quickly or with large displacements. Thus, they are matched well with third-class levers, where the force is applied between the load and the fulcrum. There is a mechanical disadvantage of a third-class lever; the applied force must be higher than the load force. However, the third-class lever multiplies both the speed and the amplitude of the force displacement, so it allows more rapid motion and larger excursions than muscles are capable of generating by themselves.

Additional Concepts. Other concepts taught to engineering students are, of course, also important for competent Biological Engineers to know. These include force, mass, and energy balances, Bernoulli's equation, Fick's equation, Ohm's law, thermodynamic laws, Newton's laws, and others. Maxwell's equations, Kirchhoff's laws, and the Navier-Stokes equations are just as important for specialists within Biological Engineering. These have not been included in the above list because they are not as unique to Biological Engineering as are many of the above. Understanding them does not necessarily assist in understanding biological nature. The same is true for concepts related to mathematical modelling, chemical equilibrium, pharmacokinetics, chemical kinetics, and a whole host of other information important for some Biological Engineers. What were listed, however, are concepts and principles suggested to be expected of all Biological Engineers, as the core of what distinguishes Biological Engineering from other disciplines.

Becoming Ready for an Engineering Career

Using the word "compleat" in the title of this chapter was not a misspelling. The word "compleat" means having all necessary or desired elements or skills. Up until this point, the discussion was all about technical skills to be taught to undergraduate Biological Engineering students. But, the world of professional engineering practice is much richer than just technology.

Students take engineering courses to learn techniques for solving problems. Thus, most engineering courses taken by undergraduate students are highly technical in nature. But, there are many additional techniques and skills that can be learned along the way. Other types of knowledge can also be incorporated into engineering science courses without diminishing the value of the engineering techniques being taught (Johnson, 2020b, used with permission). These other skills, of an ancillary nature, can improve the value of a course for future engineers.

The objective of good teaching should be to facilitate student learning, and to help them learn as much as possible, including many "soft" skills that would be very valuable later on in their careers. There is much more to engineering practice than just technical knowledge, and it is these other aspects of professional life that can also be taught in class. I call this "collateral learning". These collateral skills can be taught in upper-level engineering science classes without detracting from the technical focus of each class. In fact, these collateral learning skills actually enhance the value of these classes to the students. The following are some of these collateral skills that the students can be given a chance to learn.

1. Learning to Work in Teams. Working engineers are no longer lone rangers on their jobs; they often work in teams composed of others with multiple talents, the better to find new and improved solutions to assigned tasks. So, it makes sense to give students the opportunity to learn all about working in groups while still in an academic environment where learning is the objective of educational programs. These same skills would have to be learned eventually on the job, where the environment for error is not as lenient as it is in the academic classroom.

Initially, students in my classes were just assigned to groups of three to five and were then free to work out the details of how they would work together. However, students soon made it known that working with others in this way was not necessarily part of their skill sets and was a new mode for performing in class. I had given them no information about how to make their groups operate successfully. These complaints were heard, and we responded by introducing to them the Myers-Briggs personality test. We required all students in the course to take the test, and we spent one whole recitation session explaining what the test measured and how to interpret the results. Whereas we did not directly answer how each student should interface with others in his or her group, we did, at least, open up for them some understanding about introverts and extroverts, thinkers and feelers, judges and perceivers that most had not before had to deal with on a practical level. With the revelation that Biological Engineering student personality types can fill a very broad range (some were enrolled because of their interest in biology and others because they liked math, for instance), students became much more aware of differences among group members and a tolerance developed for those members who were not motivated or skilled in the same ways as others. After that, student groups worked extremely well as a means to produce quality designs and reports.

Incidentally, it soon became clear that groups composed of three members sometimes resulted in isolation of one member by the two others. Thereafter, all groups had either four or five members.

There can be times when some students have difficulty getting along with certain other students. Although there may be some others whom we all would prefer not to be in contact with, but who may have to be dealt with anyway, this is a skill that some students need to learn. Students are capable of learning this ability if necessary.

Fair evaluation of student participation in groups is a challenge that must be faced whenever grades are determined largely by group efforts. Some students may not spend as much time or effort as other members, whereas other students may contribute more than their fair share. To evaluate individual student contributions, each student can be required to evaluate contributions of each student, including themselves, several times during the course. Instructors can reserve the right to increase or reduce each student's final letter grade by one letter based on these evaluations. If the composition of the groups changes during the course, the scores can be based on a broad consensus of many students, and thus, can be a fair assessment, even if one person did not do well in one group. But, if a student was consistently slacking, then that would be easy to tell, and that student's final grade could be reduced by one letter. As it turns out, I raised more grades than I reduced.

2. Learning to Approach Problems of Current Interest. Design problems that are assigned should usually be based upon situations current in engineering practice. These can be inspired by relevant trade publications, and, when a problem of current interest appears that could be modified somewhat to be able to be given to students, a design problem can be developed. So, don't throw away these publications as junk mail when they appear in your inbox.

These problems can be made to conform to technical topics presented in class. Giving design problems of this type to students can make them more interesting to students, and allow them to learn another set of skills, as well.

These problems usually require more information to solve than is given in class. After all, class topics are meant to teach general problem-solving techniques. Many of these problems require outside reading and research to find details related to the problem background and to typical approaches to the solution of similar problems. In this way, students learn to supplement their class work with extra details and to find prior work related to their problems, similarly to how a working engineer would have to operate given a design task by his or her employer. In this way, their design solutions would have a good deal of realism included, thus helping to develop engineering judgment. So, students have to develop information-gathering skills.

At first, my students would bug other faculty members who they thought might have quick answers to their detailed questions. After a while, though, they learned to use the library, and some even called industry people whom they had not previously met to ask questions. I have been told that some students represented themselves as practising engineers to these industry representatives in order to gain access to them and their knowledge.

3. Learning to Solve Vaguely Defined Problems. Design problems assigned in class can be deliberately given with some key pieces of information either missing or given in a nonprecise way. This should be done to hone their abilities to deal with engineering problems not completely specified. The instructor and students can play roles during class sessions after design problem assignment, in which the students are free to ask questions of the instructor, playing the role of a nonengineer client, to help to define what was needed for the design solution. Students have to learn to think about the project, figure out what was not known to them, but was important for them to solve the problem, and to ask pertinent questions in ways that could be understood by a lay person without including words of engineering jargon. For instance, the word "viscosity" would not normally used by a nonengineer, but the word "thickness", or some equivalent might be.

4. Learning About Regulations and Practical Limitations. As with all practical problems engineers must deal with, there are limitations that must be known. For instance, anything involving food is subject to health department rules, and these can vary from location to location. Biomedical instrumentation is subject to Food and Drug Administration rules. Environmental regulations are relevant to wastes and discharges. There are rules and regulations that must be known to limit design options. Students should be made to investigate these before they formulate a final solution to the assigned problem.

5. Learning About Technical Writing. Our students were required to take a technical writing class in their undergraduate curriculum. Unfortunately, this course was taught in the English department and, as such, students were taught writing skills appropriate to compositional writing, but not necessarily for technical writing. As a case in point, students were highly encouraged in their technical writing class to write from the viewpoint of the first person, the author of the writing. Verbs are supposed to be in the active voice.

But good technical writing is hardly ever done that way, as opposed to what they were told in their required technical writing course. When a paper is published in a scientific or technical journal, it seems boastful and self-serving if it is written in the first person. When writing operating instructions, the first-person pronoun "I" never appears. Other personal pronouns also do not appear. Why? The reason for these characteristics is that, in technical writing, the attention should be placed on the action taken rather than on the actor. A good design should be a good design no matter who did the analysis; good data should be good data no matter who conducted the experiment. For these reasons, good technical writing exclusively uses the third person in technical writing, and verbs are used in passive voice. Students should be taught about this distinction and the reason behind it, so that they understand how to write correctly in a professional context.

The mantra of writing in plain language so that a lay person can understand it was repeated many times in class. "Write so that your mother can understand what you are saying". Whereas some technical details cannot be written this simply, students must learn to realize that the merit of their works later in their careers will likely be judged by some nontechnical personnel, perhaps business managers, financiers, or product users. Technical details should be separated from nontechnical descriptions so that both technical and nontechnical readers can be satisfied.

If an engineer produces the best design in the world, but cannot communicate it effectively, then he or she might as well not have produced the design in the first place. This was the motivation given to the class to produce an effective design report as the culmination of their design efforts. Their final product at the end of their assignment was a design report with a particular format given to them as a template. Elements of the report were:

Summary. This is given in two paragraphs: (1) a short, but adequate, description of the problem, and (2) a summary of the design solution. It is this part of the report that will be read by the boss, before he hands it to someone else to evaluate, so the summary should be effective and located right up front.

Table of Contents. This is a formality to help the reader find things in the report.

Introduction. This should include a detailed description of the problem, as well as any background necessary to the problem solution, including, perhaps, current approaches taken for similar problems.

Design Concept. This is a verbal description of the means to solve the problem with enough detail that it can be understood by technical personnel. This section should provide the context to understand what is being calculated in the next section. Any design conclusions need to be included here.

Detailed Calculations. These are the calculations to supplement the approach given in the previous section. Any assumptions made should be given explicitly. Only representative calculations should appear; repetitive calculations need to be moved to the appendix.

Bibliography. Reference materials should be given in a form that makes them easily accessible should the reader wish to look at them. Internet references should include the access date because these references can be changed at any time.

Appendix. This includes supplemental materials and repetitive calculations.

Written descriptions should be supplemented, wherever possible with good, well-described drawings and illustrations. Illustrations taken from other original sources must be properly cited.

6. Learning Time Management. There are other collateral skills that should be learned by individual students. One involves the short time deadline given for submission of the final design report. I usually gave the groups two weeks before the reports were to be turned in. This was done for several reasons. First of all, two weeks is usually the amount of time that students would work on a problem like this anyway, no matter when it was assigned and when it was due. Second, working engineers rarely, if ever, have unlimited amounts of time to complete their assigned tasks. Instead, they must learn to do the best that they can in the time available and to be satisfied with the result. Deadlines are a part of professional life. This requires students to get to work quickly on the problem, and to produce the best possible solution and report in the time available. Engineering solutions, they would learn, were not perfect; there could always be improvements to be made, but engineering solutions had to be the best that could be done in the time available.

Highly motivated college students often strive for perfection, but that is not always possible in a short amount of time. Having students work together in groups on design projects helps in that there can be specialized efforts by different students to contribute to the final product: one student may do library research, another may interview company representatives, and another may make the necessary technical drawings, for instance, but they are all responsible for the final report.

7. Learning to Evaluate Others' Reports. When it comes to reports, some students are better at them than others. Most of the design reports in my courses were written by groups of students working together on the project. There are several ways to deal with this, and to improve the report quality of all students in the end. Student groups submitted multiple copies of their reports, all on the same day, and these were distributed so that other groups would each receive several reports from groups other than their own. Then, each person in each group can be expected to peer-review another report and to submit an anonymous evaluation based upon a template that can be supplied. Working engineers are often expected to evaluate proposals submitted to their companies. By requiring peer evaluations, students are given a chance to learn how this is done in the relatively friendly confines of the classroom, rather than under pressure at work. Some students did amazingly good jobs at peer evaluation, including substantial editing of written materials in the report. These evaluations were also graded by me, for a minimal five percent of the report grade, and then returned back to the group originating the design report, so that they could use the extensive peer feedback to improve their next design report effort. If multiple design problems are assigned in each class, then students have the opportunity to improve their engineering approaches and report writing abilities based upon this feedback from a sampling of their peers. The quality of the reports can become noticeably better as the semester goes on.

One additional advantage of peer evaluation of design reports is that students receive feedback from more than just the instructor. If a peer cannot understand what a particular group has written, then it can have more of an effect than just the teacher saying it. The responsibility for evaluation is spread over more than just the teacher's opinion.

8. Learning to Write Letters. The generation of students towards the end of my teaching career was not familiar with writing letters. They knew how to text among themselves and their friends, but had not had occasion to write formal letters to anybody. Design reports required a letter of transmittal as a courtesy to the customer who commissioned the report. I gave the students a simple template for a letter of transmittal: one sentence that says, "here it is", and perhaps a second short paragraph summarizing the major results or aspects of the report. These might be expanded slightly, but keep the letter short and to the point, I told the students, because the letter is only a formality. After writing several of these, one for each design project report submitted, students became confident that they could write letters of transmittal.

9. Learning Engineering Judgment. Engineering judgment is essential for a successful engineering career, whether it is used to estimate the rate of chemical into a mammalian cell or the cost of production of a new product proposed to management. There is nothing better to help develop judgment than to confront engineering designs based upon real scenarios. Thus, it was soon decided in our department to make all of our upper-level engineering courses require design projects as a major part of the final grade evaluation. I required design projects in my courses. My understanding was that practising engineers were not usually required to take academic tests during their careers, so prelim and midterm tests were eliminated from my courses. However, I found that, without any academic means to prod students to keep up with their studies, students did not understand the technical materials in the course well enough to do well on their design projects. Consequently, students were made to take weekly short quizzes that, in total, contributed a nominal ten percent of their final grade. I even made it easier for a student to miss a quiz or to have an off-day or two by only counting the best ten quiz grades. Nonetheless, students were so motivated that they worked hard to do well on all the quizzes. Given the option of voluntarily missing a few quizzes if that student had high grades on all the other ten, few students exercised their opportunities to miss quizzes and still maximize their quiz average.

When, during one semester, it became obvious that students were preprogramming their calculators to compute numerical answers to expected quiz questions, we decided to eliminate the use of calculators for quizzes. Instead, to help students develop numerical judgment, we graded as correct any estimates that came within ten percent of the correct answer. Engineers should at least be able to estimate to this degree of accuracy.

To add to our students' appreciation of current techniques and practices, several lectures in our capstone design course were devoted to mundane topics of practical hardware and software. One or two class sessions were given by outsiders who lectured about the patenting process and operations of selected government agencies. The idea for these was to prepare students for the constraints imposed upon them once they had graduated and were employed in the outside world. And the capstone design project had to include the fabrication and testing of a real prototype, assisted by our excellent shop personnel. Design projects in other courses only required reports as the final stage.

10. Professional Practice. Two groups of students inadvertently provided lessons in professional practice that I presented to all succeeding classes. The project that they were to design, in this case, was a ventilation system for a government building containing biosecurity laboratories. There were several labs in the hypothetical building, and each one had to be protected against contamination from outside air as well as to prevent contaminants from inside each lab from leaking out.

All groups submitted their reports, as required. All designs included HEPA (High Efficiency Particulate Air) filters to remove contaminant particles from the air stream. Students had checked industrial specifications for pressure drops and particulate efficiencies. The normal efficiency specification was 97%, and all groups but one chose to use this spec for their designs. One group, however, found a single company that specified its filters as 99% efficient. This group chose to specify filters from this one company as the ones that they incorporated into their design. What was wrong with this choice, from a professional standpoint, was that this choice tied them to filters from one particular source, making the choice vulnerable to pricing and availability only from this one company. It is not good engineering design practice to choose to use outlier products if normal specifications would suffice. I used this example to instruct subsequent classes about choices of suppliers.

A second group submitted their report for the same project that included an extra protected bathroom in the building. I'm sure that they expected some praise, if not extra credit, for including this additional work. Instead, I docked them for the extra design. My rationale was this: first, this project had been played as a government contract, specified only to submit designs for the biosecurity laboratories. Extra work submitted as part of this contract would have called into question how much funding was expended for the extra nonessential work, and this may have caused ensuing legal proceedings to recover those funds.

Worse yet, as I told students, a practising engineer who was able to supply the ventilation system for a bathroom in addition to the labs, should propose to the funding agency that they could provide the design, and for a certain amount of money. By providing the agency with the extra design, as this one group did, the engineers were undercutting their own abilities to set professional fees. It was not good to cheapen one's own technical abilities by providing them below market value, unless they were provided to a charitable or public service cause.

I used these two illustrations in subsequent classes to illustrate to the students that practising engineering was different from engineering college instruction. When they graduated, and were responsible for supporting themselves as engineers, there were practical matters to attend to.

11. Learning About Professional Responsibility. As often happened when teams worked together on their designs and reports, they often assigned specific tasks to individual group members to avoid duplication of efforts and work as efficiently as possible to meet the deadline for submission. In one case, one student plagiarized background material from the internet and included this material within the introduction section of the group's report. I recognised that this material had been copied without attribution to the original author and brought the issue to the attention of the University of Maryland Honor Board, which I was obligated to do based on the university honor code. Although only one member of the group had been guilty of plagiarism, all members of the group had signed the report, as required. I had indicated to the students at the beginning of the course that, by signing the report (and also each homework submission), they were attesting to the fact that they had contributed to the report, were each familiar with all details of the report, and that they attested to the fact that all other members of the group also knew the details of the report (this last stipulation spread the responsibility for student learning from the instructor, me, to other members of the group, as well).

There was a lesson to be taught to the students about the meaning of their signatures. Although I conceded that only one student had plagiarized, they all had taken responsibility by signing the report. Before this episode, the other members of the group had not thought much about the meaning of their signature, but they conceded that they were somehow also responsible for the report contents. I eventually advised the Honor Board that I did not want any harsh punishment to conclude this case, but the point was made to this group, and the remainder of the class, about the meaning of a personal signature in a professional context. Thereafter, I used this example in subsequent years to make the point about personal signatures, for another lesson in professionalism.

12. Learning About Ethics. At the end of the semester, there may be class periods left after the last of the technical material has been taught and before the due date for the last of the class assignments. One of these class periods can be used to talk about ethics and professionalism. In my class, I used, as a guide, a prior version of an Institute of Electrical and Electronics Engineers (IEEE) code of ethics, which was simple and unembellished, as some engineering ethics codes are. We talked about individual responsibilities and public responsibilities. We talked about kickbacks taken by Spiro Agnew while he was still Baltimore County Executive in Maryland, and why engineering firms might want to contribute to him to gain favour when contracts were awarded. We talked about the space shuttle Challenger disaster and why it was important to speak up when one's engineering judgment seems to be important. We talked about the responsibilities that engineers assume to foster the development of their subordinates. We talked about engineering as a meritocratic profession where there is no place for discrimination due to race, creed, national origin, or anything other than professional performance (in classes with students with much diversity, this may have been the only time that they heard this message from one of their professors). We talked about collusion about professional fees and how students owed professional service to the people of the state for helping to underwrite the cost of their education. We talked about all of these examples and others to illustrate what would be expected of them to be the best that they could be in a professional career. And, last, but certainly not least, they heard the message that I, and others, expected the very best from them upon graduation. They had proven themselves to me in this course, and now they were prepared to achieve great things in their lives.

The Bottom Line

What educators should try to do for their engineering students is to expose them to a full range of issues that they would need to know once they enter the workforce. Engineers, contrary to popular perception, can be good communicators, socially aware, and interesting (Johnson & Rehkugler, 1990). The students who graduate having already learned these skills can expect to advance in their careers faster than those who need to learn these skills once on the job.

Even if Biological Engineering students do not intend to pursue careers as practising engineers, these skills are still valuable to navigate whatever situations in which they find themselves. For those wishing to attend graduate school, conduct research, go to medical school, or work at the Patent Office, for instance, these skills make them better prepared than many of their peers. They should do well.

REFERENCES

Arthur, R.M. (1970). What Is Biological Engineering, *Agricultural Engineering*, 51: 691.

Garrett, R.E., Davis, D.C., Edwards, D.M., Johnson, A.T., & Rehkugler, G.E. (1992). Improvement of Competencies of Agricultural and Related Biological Engineers, Report of CSRS USDA Agreement No. 90-38411-5734, University of California, Davis, CA.

Gavrilov, L., & Gavrilova, N. (2004). Why We Fall Apart, *IEEE Spectrum*, 41(9): 30–35 (September).

Johnson, A.T. (1999). *Biological Process Engineering: An Analogical Approach to Fluid Flow, Heat Transfer, and Mass Transfer Applied to Biological Systems*, John Wiley and Sons, New York.

Johnson, A.T. (2006). The Making of a New Discipline, *International Journal of Engineering Education*, 22: 3–8.

Johnson, A.T. (2010). Essential Concepts for Biological Engineers, *Biological Engineering*, 3(1): 3–15.

Johnson, A.T. (2012). Biological Engineering: What It Means to Me, *Resource*, 19(4): 14–15 (July/August).

Johnson, A.T. (2013). Teaching the Principle of Biological Optimization, *Journal of Biological Engineering*, 7: 6–12.

Johnson, A.T. (2019). *Biology for Engineers*, 2nd ed., Taylor and Francis, Boca Raton, FL.

Johnson, A.T. (2020a). Consciousness in Animals, *IEEE Pulse*, 11(2): 29–30 (March–April), DOI: 10.1109/MPULS.2020.2984306; https://www.embs.org/pulse/articles/consciousness-in-animals/

Johnson, A.T. (2020b). Collateral Learning, *IEEE Pulse*, 11(4): 37–43 (July–August), DOI: 10.1109/MPULS.2020.3008452 https://www.embs.org/pulse/articles/collateral-learning/ © [2010] IEEE. Reprinted, with permission.

Johnson, A.T. (2021). We Need to Teach Biology to Engineers, *ASABE Resource*, 28(3): 10–13 (May/June).

Johnson, A.T., & Davis, D.C. (1990). Biological Engineering: A Discipline Whose Time Has Come, *Engineering Education*, 80(1): 15–18.

Johnson, A.T., & Phillips, W.M. (1995). Philosophical Foundations of Biological Engineering, *Journal of Engineering Education*, 84: 311–318.

Johnson, A.T., & Rehkugler, G.E. (1990). Career Opportunities in Biological Engineering, *Careers and the Engineer* (Spring): 30–34.

Saltus, R. (2010). A Silver Lining, *HHMI Bulletin*, 23(2): 22–27.

Shannon, C., & Weaver, W. (1949). *The Mathematical Theory of Communication*, University of Illinois Press, Urbana, IL.

Schreuders, P.D., & Johnson, A.T. (1999). A Systems Approach for Bioengineering, *International Journal of Engineering Education*, 15: 243–248.

Arthur T. Johnson received his PhD in 1969 from Cornell University and retired from the University of Maryland in 2009. He served with the US Army in Vietnam, and was awarded the Army Commendation and Bronze Star Medals. At Maryland, he taught Biology for Engineers, Transport Process Design, Basic Electronic Design, and others. He is a registered Professional Engineer, has received numerous teaching awards, held leadership roles in regional, national, and international organisations, and is a Fellow of seven professional organisations. He has written five books, and holds six patents. He operates a 49-acre farm producing fruit for the farmers' market.

30

Innovation in Bioengineering Education

Pilar Barreiro Elorza, Belén Diezma Iglesias and Eva Cristina Correa Hernando

Teaching through Innovation as Underlying Framework

Let us start with the following statement taken from *The Civilised Engineer* (Florman, 2014): according to *Webster's Dictionary*, "civilisation" consists of "progress in education, refinement of taste, and the arts that constitutes culture". If we take this definition literally, the average engineer today is simply not civilised.

Bearing this idea in mind, it is worth reviewing some of the key features of the Spanish universities that may be related to the statement; special emphasis will be placed on the technical studies, before entering the pure-innovation perspective.

The book of Reques Velasco (2009) brings together a multitude of statistical analyses regarding higher education in Spain, mainly based on the yearbook of university statistics and in the publications of the national statistical institute. The study classifies universities at the European level into three main typologies: those determined by demand (demand-led), a model determined by costs (expenditure driven), and an intermediate or mixed model. The first (Spain, France, Italy, and Greece) comes from the Napoleonic university model, which understands education as a constitutional right, as an instrument of social justice and equal opportunities, and therefore practises freedom of access to higher education. In the cost-determined model (USA, Great Britain, Denmark, the Netherlands, and Sweden), there is a strong control over access, either public or by the institution itself; in the mixed model are countries such as Germany, Ireland, and Belgium.

The study defines university activity as typically higher-tertiary level, with an important role in socio-economic and cultural development. Its essential functions of training and basic research are extended to include applied research, local and regional development, and the cultural and scientific dynamics of the territories. The relationship between the university and the territory has been modified, as has the university-society relationship, since both training and research are carried out both inside and outside the universities. The university is highlighted as a structural and compensatory policy instrument for the dynamic rebalancing of any territory and for the reduction of territorial disparities. This means recognising the university's social and economic function, with a prominent role in economic growth by encouraging structural changes in companies towards research, development, and innovation (RDI) activities. The latest data on the number of Spanish universities, according to the Spanish Ministry of Universities, count a total of 83 institutions, 50 public and 33 private, among which only four correspond to purely technical universities (also called UP4.)

Another relevant study is provided by the Global University Network for Innovation (GUNi), an international network that currently gathers 250 members from 80 countries, including higher education institutions, United Nations Educational, Scientific, and Cultural Organisation (UNESCO) chairs, and research centres. GUNi was created in 1999 after the first UNESCO World Conference on Higher Education. On the other hand, United Nations University is an international community of experts dedicated to research work, permanently based in Tokyo, Japan since 1975, and aims at postgraduate education and the dissemination of knowledge to promote the goals and principles of the United Nations Charter.

The GUNi reports (GUNi, 2009, 2013, 2017, 2019, 2022) recall that the purpose of education is to transform rather than to transmit, to offer the opportunity to "light a fire" rather than to "fill a vessel", reflections coined by Yeats to inspire, provoke, and motivate. Therefore, the central educational purpose of higher education institutions should be to explicitly facilitate progressive, reflective, critical, and transformative learning that leads to a better understanding of the need and expression of responsible paradigms for living, for "being", and for "becoming", as individuals and collectively as communities. According to the GUNi, this holistic perspective of educational purpose is the main challenge in the current context.

The report by the GUNi in *New Visions for Higher Education towards 2030: A stepping stone for the GUNi International Call for Action (2022–2025)* has as its motto: "Rethinking Higher Education Institutions (HEIs) for Sustainable and Inclusive Societies". The report states that there are major transformations and crises under development, which include climate change and environmental degradation, demographic pressures, forced migrations, rising inequalities, political pressures, and the transformation of the labour market. In this framework, the GUNi points out that some of these transformations might have a devastating effect on our societies and our planet, and in some ways may become irreversible if clear action is not taken urgently. In any case, they have crucial

DOI: 10.1201/9780429150111-33

implications for HEIs and their role in society and it is of utmost importance to address them. Moreover, the COVID-19 pandemic has exposed pressing issues in higher education and society, and has revealed the requirement for rapid, undeniable transformations such as digitalisation.

The GUNi report from 2022 reflects on three main guiding questions: 1) If we were to create a HEI from scratch today, what would it be like? 2) If we were to reform HEIs, what changes should we put in place and most importantly how would we implement them? 3) What should HEIs look like soon?

The contents that will be covered in this chapter, that are inspired by the pillars for the change and innovation in management for the HEIs in 2030 horizon, are outlined in Figure 30.1 and include:

a) Robotics in agro-environmental engineering curriculum, the AGROTECH challenge as a student competition icon, and the Technical University of Madrid (UPM) blog on agro-industrial robotics.
b) The concept of content and language integrated learning (CLIL), covering its application to its patent analysis, as well as the study of English for specific purposes (academic and vocational studies, research, and innovation, among others).
c) The ATHENS programme as an example of European inter-university collaboration, and other relevant examples of international partnership, including tutoring and mentoring students across their curricula.
d) Some specifics on teaching innovation, technological surveillance, and the European Engineering Learning Innovation and Science Alliance (EELISA) communities (mission-driven working groups that bring together students, teachers, and researchers).
e) The iconic concept of *travailleurs-paysannes* (peasant-workers), and its relationship with the acknowledgement of the depth of knowledge (and error propagation without it), together with the application of reverse engineering as a learning strategy for recent disruptive technologies.
f) The old unwritten laws of engineering (1944) and their relationship with the social skills of young engineers at professional level, with special emphasis on the importance of writing across the curricula, as a training process, for technical reporting activity in the command chain at the workplace.

Robotics in Agro-Environmental Engineering Curriculum; the AGROTECH Challenge as a Student Competition Icon, and the UPM Blog on Agro-industrial Robotics

Agricultural mechanisation and automation are, and have been, very challenging processes for humankind: while the first refers to the substitution of human and animal power with mechanical power, the later considers the substitution of decision making, from a fully human process to a completely automated situation (robotisation).

In the EU, and particularly in Spain, there is a wide range of rural diversity due to regional conditions, so high as to have led to 20 recognised areas in the Common Agricultural Policy in Spain; recently renewed for a four-year period (until 2027). Therefore, the specifications and requirements for a fruitful mechanisation and automation process must be considered locally, within a need for large adaptive optimisation abilities to enhance climate change mitigation and resilience.

As agricultural robotics is a hot topic that greatly engages students and young engineers, it has been our commitment to include this emerging area of research into the degree study plans from the very beginning (2000) of the transition towards the Bologna requirements; some of the advantages of this conception are: objectively quantifying skills improvement, adjusting the mind to open-ended problems, and attracting teachers towards the inclusion of innovation in higher education curricula.

AGROTECH/Robots in Curricula

In 2008, a team of professors from different disciplines (applied physics, applied mathematics and computer science, and rural engineering) felt the need to design a transversal training itinerary that would include optional subjects related to precision agriculture, mechatronics, automated control, and information and communication technologies, so that students

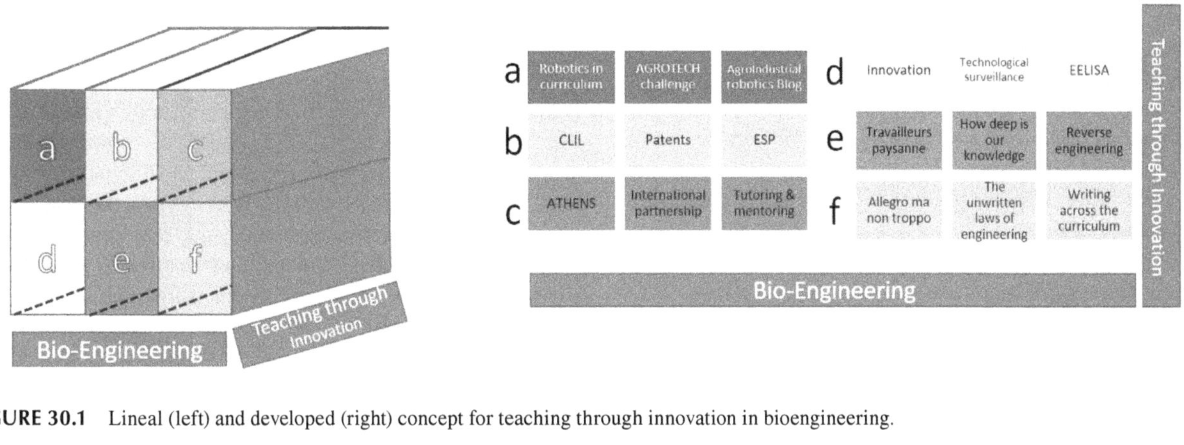

FIGURE 30.1 Lineal (left) and developed (right) concept for teaching through innovation in bioengineering.

would become familiar with a set of knowledge fields that are common to most technical studies but that are not the main objective of agronomy engineering. In this context, the idea arose of creating a new subject, robotics applied to agriculture, which would try to integrate and check to what extent the students had assimilated the concepts worked on in this itinerary.

To complement the experience of our students in the subject of robotics applied to agriculture, by enabling them to demonstrate their knowledge and skills in a multicultural and multidisciplinary context, an international competition was organised.

The proper environment for calling the participants was achieved through the students themselves. In this sense, the Board of European Students of Technology (BEST) proved to be the best networking solution, with a team of enthusiastic and well-prepared people. The Madrid delegation of this student organisation promoted the proposal to its International Steering Committee, finally launching the first edition of the AGROTECH competition in November 2008 (Barreiro Elorza, Recio Aguado, et al., 2009; Barreiro Elorza et al., 2010). Thus, BEST Madrid was in charge of the call and selection of the participants as well as their management during the whole event, while the professors, together with the students, focused on the approach to the problem, in addition to being in charge of arranging adequate facilities for the event.

The challenge: the students (a pool of 40 candidates from 16 European countries selected among 120 proposals) were welcomed by BEST, and team-building took place, also led by the student organisation, considering the participants' fields of specialty. In the first edition of AGROTECH, the professors wanted to address an issue which would be a true agricultural challenge though attainable within a three days' work by untrained participants; for that, Lego Mindstorm NXT was chosen as the platform to work with, which fulfilled several conditions, such as ready-to-use and flexible material. Fruit harvesting, a major task in agriculture, was quickly chosen as the topic for the event, but to prevent problems from its seasonal nature, we decided to build an artificial dwarf tree forest where the students could test the feasibility of selective harvesting. The problem statement was finally ready with the approval of all parties: "Picture your robot, a fully autonomous, sensitive, and sensible harvesting machine, working in a dwarf tree forest on a moon-like surface (white). The season for jujube fruits (a ping-pong ball shaped food) is just about to start, and thus this nice paradise fruit is whitening from orange (immature) to fully white (mature). Look at the nice aromatic fruits on the top branches that need to be detached from the tree, loaded in a bin, and transported along the forest... Would your robot be able to take the challenge?"

The teams had to integrate picking tool and vehicle components for harvesting as much fruit as possible. At the end of the day, eight of ten teams were able to perform the task, with a wide variety of success.

Focusing on the students' feedback, we can summarise some of the comments received, which points to team development as the key to success:

"It was a very interesting competition. It was the first real team cooperation, because at school we already know our team. Here, it was real cooperation, because we also had to know the minds of our teammates and find out how to communicate in the right way. I liked the competition mood, our collaborative feeling. I also liked the serious way of evaluating the robots, the stressful lack of time. Good job! I will go home with a good impression and more knowledge about work cooperation".

Whenever enough multidisciplinary knowledge was available, tasks were clearly distributed among participants. In the words of the students:

"I was glad I was with them (the team), because we each of us had kind of a specialty, so we were able to share the work and be efficient"; "The best issue in the event was team cooperation and task assignment. I trusted the software engineer and he trusted me".

Figure 30.2 gives an overview of the starting point and of the results of the event, which was evaluated as clearly successful by both the participants and the organisers.

We repeated, with two more editions – AGROTECH 2.0 (The Energy Harvester) and AGROTECH Reloaded (Let's Feed the Watermelons).

The formulation has always been a challenge, combining detailed and open-ended work with amusement and inspiration – at least with that intention were the concepts designed and written:

"Harvest at last! Harvest is the most anticipated and desired activity for farmers, but... What if energy and not crops was the target? Are you prepared for an intense energy-demanding world? Picture yourself as a lonely robot in search for renewable energy supplies...and willing to react".

Robots Evolving into Cyber-Physical Systems (CPS)

In 2020, Richard Duro and Pilar Barreiro, decided to integrate some points of view regarding the interest of robotics in agroforestry and the environment, especially regarding the approach of an outsider to either of the domains: cyber-physical systems (perception, artificial intelligence, and robotics), or rural systems (biodiversity, human-adapted ecosystems, and adaption and/or resilience to climate change).

We decided to write a liberal perspective of such a domain (Barreiro Elorza and Duro, 2020) as a common ground for a European joint project (National European Worldwide Applied General Equilibrium) (NEWAGE, Figure 30.3), which, sadly, at the end, was not supported; our NEWAGE concept is still under development, but hopefully will offspring at some time.

In this article, we started by defining the term "environmental robotics" as a part of the tech-science on robots dealing with unsupervised acting/reacting units introduced in an ecological system with or without a specific ecological function. We followed by explaining the concept of ecological robots (ecobots), which adheres to the previous statement by incorporating the capability to harvest energy from their surroundings to make them fully autonomous and potentially prepared for prolonged periods of isolated and unattended life, eventually beyond human existence. One can imagine that such an approach would result in a stimulating starting point, not just for young engineers of varied backgrounds, but also for professionals from the humanities world (Barreiro Elorza, Diezma Iglesias, et al., 2009).

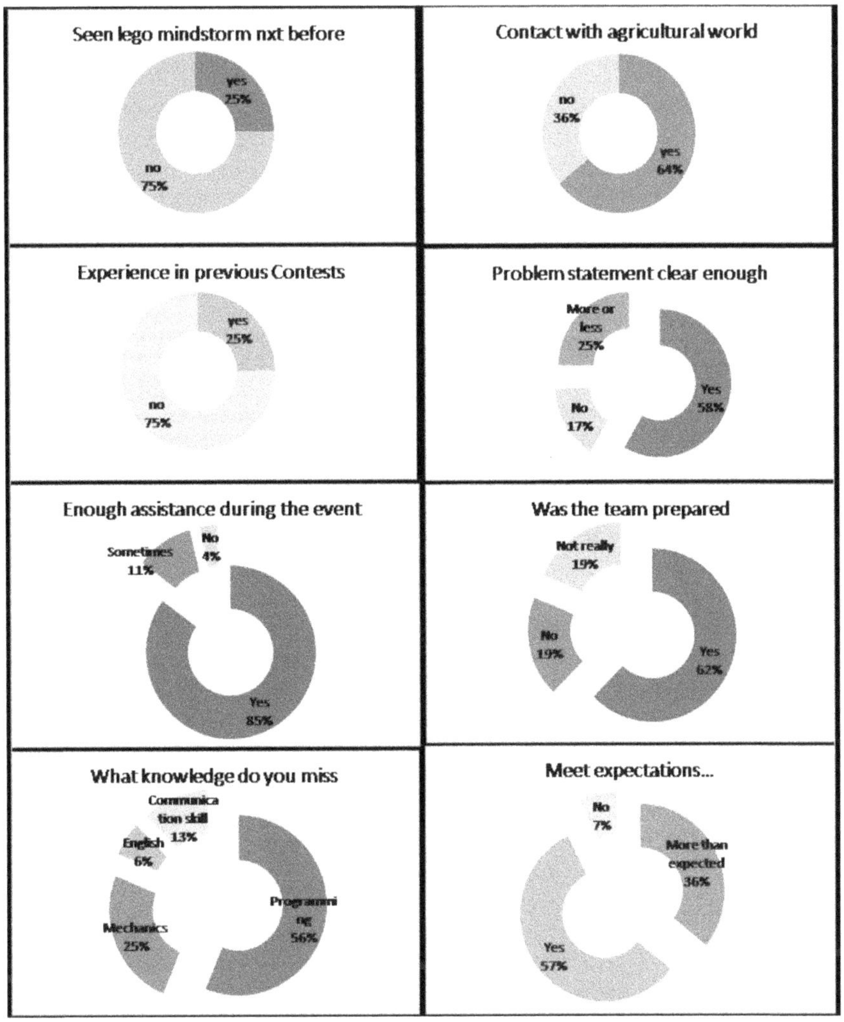

FIGURE 30.2 Summary of quantitative features of the AGROTECH event (Barreiro Elorza, Recio Aguado, et al., 2009).

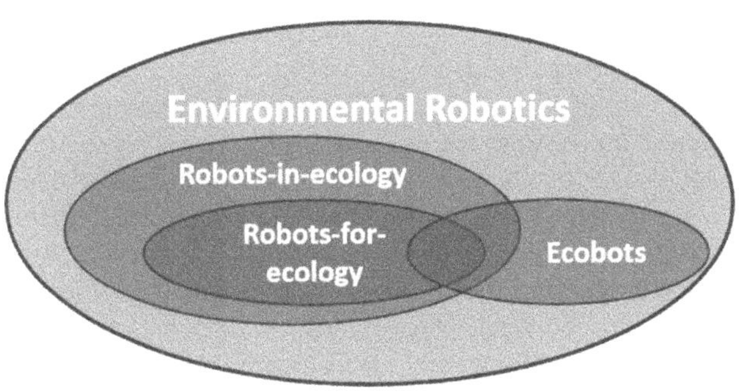

FIGURE 30.3 European Project NEWAGE concept.

The Concept of CLIL, Covering Its Application to Patent Analysis, as Well as the Study of English for Specific Purposes (Academic and Vocational Studies, and Research and Innovation, Among Others)

In this section, we include a variety of reflections and experiences related to the synergy between humanities and engineering, specifically linked to the need for a proper and well-balanced relationship between the communication skills (as considered by modern language faculties) and the professional life of young engineers. More precisely, we focus on CLIL, as well as understanding and using engaging English as de facto language in the workplace or when promoting oneself through personal branding skills.

CLIL/Patents

The origin of english for specific purposes (ESP) goes back to 1988, centred in English for Science and Technology (EST). At that point, the studies tended to be descriptive and focused mainly on statistical grammar counts (the number of words and small structures). One of the most extensive softwares for the quantitative analysis of grammar counts is AntConc, a quick word counter and detailed linguistic analysis tool.

Regarding the concept of CLIL, also in the 1980s, major interest was placed on analysing the relationship between EST grammar or lexicon and the author's rhetorical purposes, while another interesting focus in ESP in the early times was the analysis of the syntactic voice in research articles; an interesting finding in the last decade of the 20th century was that

25% of the verbs in research articles were passive, which in a way influences rhetorical decisions: the way the author leads the reader to jump to the expected conclusions.

In our particular experience, we introduced the analysis of patent-article pairs using CLIL in 2016, in the framework of a workshop entitled "Patent Analysis: Understanding the Inventor's Mind", which was in English and lasted 7.5 hours, with the main objective of making participants aware of the skills required by patent technology elucidation (Barreiro Elorza et al., 2014). The patent is an ideal text to exercise the CLIL methodology in its "model-attached" mode. With this methodology, we contribute to the internationalisation of our university, encourage cultural diversity in the classroom, and bring teaching together through interdisciplinarity. The patent requires close collaboration between content and linguistics, which are interdependent because what is conveyed and how it is communicated affect intellectual property claims and have serious legal and financial implications (Barreiro Elorza & Sancho Guinda, 2015). This exercise (Figure 30.4) allows playing with different levels of complexity adapted to the learners and encourages thinking beyond the obvious in an interesting exercise. We have repeated this seminar in at least four academic seasons with great success.

Following the idea of analysing the interaction between language and content (Figure 30.2), it is worth noting that, while the first decade in the life of ESP relied on EST, between 1981 and 1990, some researchers placed their interest on English for occupational purposes, also referred to as vocational English as a second language, as the study had become increasingly empirical, triangulated, and complex, and required glossaries and rhetorical skills.

Later, with the broad extension and difussion of personal computers, new knowledge and information tools became ubiquitous – posters, slides, etc. – in what was to be renamed as computer-mediated instruction and communication; nowadays there is an extended use of the terms visual and verbal branding, which in the early days of journalism would be called "libro de estilo". This knowledge is of utmost importance for young engineers when incorporated into technical enterprise and the entrepreunership world (Barreiro Elorza et al., 2015).

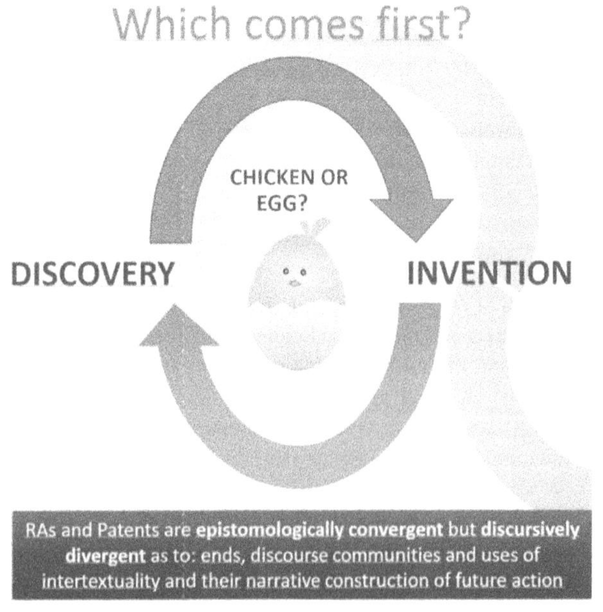

FIGURE 30.4 Features for comparing a research article (RA) and its corresponding patent (P), as twin examples.

At this time, many linguistic studies helping engineering education also focused on the concept of technical communication via computational abstracts, the basis of executive reporting in engineering projects.

Understanding English in the Workplace

This area of study began in 1993, when the World Trade Organisation (the origin of globalisation) began to function, and English became the de facto language. Also at that time, it became clear that higher education would have to evolve from a national basis to an international global understanding; as English became the de facto language in most workplaces (whether offices or workshops), there has been a tendency to analyse how people use the language to negotiate their working lives.

As a result, we understand that such potential for cross-fertilisation between work and language should be considered in engineering studies as a transversal competence that fosters critical thinking and broadens minds in a changing world.

Understanding English in the workplace can be used to achieve task- and/or people-oriented goals – either to successfully achieve a task within a project, or to ensure workers satisfaction. These are general competences of major interest that are not generally considered when designing study plans.

In this context, acknowledging that there are multiple readings of any professional text is to be considered an understanding of the social reality generated by the people involved in the workplace. In a way, this approach may be considered as a kind of exegesis of professional texts, bringing to consciousness tacit wisdom from technical enterprises and technology-based small and medium-sized enterprises (SMEs), of which the student should be aware.

Another important focus when trying to understand the social interactions in the workplace is to analyse the particularities of multicultural, highly hierarchical, pluri-generational characteristics of today's factories, where many engineering students expect to be involved. In this sense, we feel that anticipating those difficulties is also a transversal skill to be (somehow) taught, or at least highlighted during higher education studies. In fact, setting joint practices through technical challenges that bring together people involved in lifelong learning (students or workers of any age), is to be considered a goal: a skill/competence of major demand throughout the working life of technical/engineering professionals that facilitates self-confidence and success for young professionals.

Considering the large number of changes and adaptions to study plans since 2000, one of our innovations in bioengineering education (in 2010) consisted of facilitating meetings and information exchanges between master students on agro-engineering, with farm vocational studies from rural areas in Spain. The feasibility of reuniting such distant backgrounds in a common think tank (co-led by professors from agricultural sociology) led to the following conclusions:

From engineers and future agrorobot builders: pre-design evaluation requires understanding who the potential end-user will be; the feasibility of adjusting the design specifications to achieve user-friendly maintenance; the establishment of new ownership alternatives (purchase versus rental or shared use); and creation of new infrastructures associated with the rural environment, among others.

From future users, the main feedback can be summarised as: the countryside always adapts to all changes; robots will not be viable for small farmers, although they will be viable for cooperatives and large farms; they will bring convenience and reduce physically demanding tasks; they will require specific training; and they confirm the need to create new infrastructures associated with the rural environment.

The activity led to a joint communication in the annual student congress at the School of Agronomy, Food, and Biosystems Engineering (ETSIAAB-UPM) and was considered a great success by all parties. Thus, in this chapter, we take the opportunity to highlight the power of building a common ground from scratch, especially for focus groups that are distant in background and interest but aligned in the end-use perspective.

PROYÉCTATE, an Innovative Experience for Promoting Writing Skills

PROYÉCTATE is an example of how we (as lecturers) have faced the need of educating students in writing under the scope of professional communication skills; it is a project in teaching innovation for engineering degrees, titled accordingly.

PROYÉCTATE integrated a consortium of professors from two higher education institutions: Autonomous University of Madrid (UAM) and UPM, the former related to humanities, while ours is centred in technical studies. More precisely, the team from UAM focused on verbal branding as a hot topic in the framework of linguistics in modern languages.

In the case of the UPM team, an eclectic profile of educators was set, ranging from agroforestry engineering and industrial design to English teaching staff, with which there had a been previous long and fruitful interactions.

In the first phase of this project, linguistics students provided rapid advice to students participating in the ActúaUPM project, a UPM initiative which focuses on the creation of technology-based companies in the university environment (spin-offs). The main objective in PROYÉCTATE was to help students from the different engineering fields covered by UPM to improve the presentation of their projects so that they could materialise; this activity took up just one weekend, in the final hours available between the finalisation of the proposals and the deadline for the submission of proposals.

The second phase turned out to be longer, as it involved several activities. Firstly, a linguistics intern at ETSIAAB-UPM was asked to create a short presentation for a collaborative blog between the two institutions, UAM and UPM, with the aim of bringing the two universities together so that people from the humanities field would understand the ethics behind engineering, and people from the sciences field would understand the importance of knowing how to communicate. To do this, a piece of writing and a cartoon were written to liven up the tone of the blog, so that the students would find it fun as well as useful.

Subsequently, the grant holder (linguistics profile) in charge of coordinating the students was asked to contact the students of the ActúaUPM project and to elaborate a corpus with the most common mistakes. In her words, "contacting the students took some time and, unfortunately, only two responded. However, their assessment of the activity turned out to be very positive and enormously rewarding". Although at first glance the activity may seem simple, the students had to receive specific training to perform all these tasks. Specifically, the UPM provided video tutorials on how to create blogs within the framework of this institution, while the UAM held several sessions with modern languages lecturers, who guided the group of linguistics students in charge of the consultancy. An ad hoc training session was organised to familiarise the linguistics students with the use of the UAM CorpusTool programme, with which a representative corpus of texts was compiled. For their part, the UPM lecturers introduced the PROYÉCTATE coordinator grant holder to the teaching staff, as well as to every one of the students and assistants who collaborated in engineering teaching in the agroforestry engineering department of the ETSIAAB-UPM, an essential task to create a genuine integration of the multidisciplinary team.

PROYÉCTATE was a great success, and the jury indicated an improvement in the quality of the proposal. Consequently, in a second edition, the students participating in PROYÉCTATE II were taught how to promote their proposals under the typical elevator pitch example.

Promoting Personal Branding: The Elevator Pitch

This activity took part within the previous innovation project, PROYÉCTATE. It was intended to provide a linguistic consultancy service and support on personal branding to those teams of students that reached the final step in the ActúaUPM award. It consisted of promoting the entrepreneurial idea by means of what is called an elevator pitch – a slang term used to describe a brief speech that outlines an idea for a product, service, or project; the name comes from the notion that the speech should be delivered in the short time period of an elevator ride (investopedia.com).

To this end, the professors from UAM (humanities) gave a very inspiring seminar entitled "How language can help you (or not) to get your project financed". It covered some mandatory stages in the design of the video: 1) What do I do? (What experience do I have? What do I do now? Do I provide a solution to a problem?); 2) Why is it worth investing in me (What do I offer? Why is my proposal valuable? What resources do I have?); and 3) the video itself (added value and differentiation; reach; consistency between form and content; originality and quality).

The seminar also provided a rubric on four dimensions: a) added value and differentiation; b) scope; c) consistency in form and content; and d) originality and quality. Even though a detailed description is beyond the scope of this chapter, it is worth noting that three of the initiatives reached the final stage, with the videos available on YouTube (Estrada et al., 2015; Muñoz, 2015; Sánchez et al., 2015).

The Advanced Technology Higher Education Network/SOCRATES (ATHENS) Programme as an Example of European Inter-University Collaboration, among Other Relevant Examples of International Partnership, Including Tutoring and Mentoring Students across Their Curricula

In this section, we summarise our point of view concerning the most engaging features and results, regarding the relevance of international partnership and the mobility of students in the higher education domain based on the Advanced Technology Higher Education Network / SOCRATES (ATHENS) programme .

ATHENS Courses UPM107 and UPM88

The Erasmus programme, set in 1989, has been for many years the most important vehicle for the exchange of students among European countries (the EU being founded in 1993); many of us consider it the main tool enhancing the feeling of being a free citizen, and that it reflects a common ancient European culture.

More recently, the ATHENS programme has come into action. The ATHENS network is made up of a total of 15 European HEIs (one per country), with the UPM being the Spanish university member of the network. In the framework of the ATHENS programme, the participating universities organise one-week courses (twice a year, in March and November) in which students from all the universities of the network participate.

The ATHENS courses consist of 30 hours (lectures, laboratory, technical visits, internships, etc.) spread over one week of classes. The subject of the course is chosen by the teacher, as well as its level of difficulty (general or specialised), the language in which it is taught (preferably English), the timetable, and the prerequisites to be met by the participating students. The course may be taught by one or more teachers. The ATHENS week was established with the support of the European Communities SOCRATES programme, receiving an annual subsidy of €50,000 from 1997–2001. Today, the programme is mainly founded by contributions from the member institutions, among which the UPM is a very active member. In our case, we have participated since 2014, mostly without interruption.

Alternative International Exchange, Tutoring, and Mentoring

As may be recalled, convergence towards a common European area began in 1987 with the establishment of the Erasmus programme for the exchange of students and teachers between European universities. Nowadays, we find not only Erasmus but Erasmus+. Erasmus is for fun, but Erasmus+ is for sport, in the sense that the latter is intended for more mature students, at a master's or PhD level, with a high level of commitment (auto-exigence).

Another interesting tool concerns joint degrees; although they are not allowed in some countries, more and more national regulations seem to be adapting, partly due to the positive influence of complementary programmes such as Erasmus Mundus. The main problem in such cases is always the need for accreditation of transnational programmes.

At the ETSIAAB-UPM, more than a decade ago, we (among others) proposed a master's degree called AgroTIC, with the possibility of carrying out national and foreign stays in the private sector, given the technical nature of the courses. The philosophy of these activities was intended to physically bring together students, teachers, and equipment to demonstrate and use the different technologies that cannot be demonstrated in the classroom. The proposal considered providing funding for this internship, as attracted from companies and manufacturers of such equipment, interested in its promotion; in addition, public funds would be attracted from mobility calls, such as Erasmus. Though this master's degree was finally disregarded, the proposed structure is more and more incorporated in the successive study plans at ETSIAAB-UPM.

Curricular and Non-Curricular Internships

External academic placements are a training activity whose objective is to contribute to the integral training of the student, complementing theoretical and practical learning. This activity favours the development of technical, contextual, and personal skills in a professional setting, and provides practical experience that facilitates initial employment and future employability; internships promote the values of innovation, creativity, and entrepreneurship.

On the other hand, internships are positive for companies, which receive the motivation and talent of young engineers, as well as for universities, which thus have additional ways to demonstrate the quality of the training plan.

Our syllabuses include two types: a) curricular, configured as academic activities that are part of each specific syllabus (with an assigned number of ECTS (European Credit Transfer and Accumulation System) between six and 15); and b) extracurricular, which are those that students carry out on a voluntary basis, without being configured in any particular syllabus, although they are included in the European Diploma Supplement.

300 curricular internships and 100 extracurricular internships are carried out annually at ETSIAAB-UPM. The internships mobilise 270 professional tutors and 130 academic tutors, who collaborate to ensure that the activities and the acquisition of the student's learning results are adequate; the tutors also evaluate the competences acquired by the student during their stay in the company.

External placements are formalised by signing agreements, 40% with companies, 37.8% with public institutions outside the UPM, or with the Ministry of Agriculture or the Food and Agriculture Organization; 21.6% of placements are carried out in institutions attached to the UPM. Concisely, depending on the degree, between 50% and 100% of graduates carry out internships in companies.

The University-Business Foundation Network (REDFUE) recognises the importance of internships and, together with the Spanish Engineering Institute, organises the annual *Concurso de Distinciones a la Excelencia en las Prácticas de Alumnado de Ingeniería* (Awards Competition for Excellence in Engineering Student Internships). As a result, internships in companies have become one of the fundamental tools of the university learning ecosystem guaranteeing the interaction between public academia and the private sector (Barreiro Elorza, 2010).

Massive Open Online Courses

The purpose of the UPM, as a public university, is to promote free, online, and accessible knowledge for the whole society. For this reason, since 2016, the UPM has been launching calls for educational innovation hrants for massive open online courses (MOOCs) through the Miríadax platform.

MOOCs are courses developed primarily by teachers, published openly, and universally accessible via the internet. They allow, among other things, the creation of learning communities and access to knowledge and expertise to millions of people, asynchronously and free of charge, with predictive and adaptive learning models.

Miríadax was started in 2013 by the Santander Bank and the Spanish telecom company Telefónica, through the Red Universia (a network of universities in Latin America) and Telefónica Digital Education. It is based on the open-source platform WEMOOC. Miríadax offers courses from 45 universities based in nine countries (European Commission et al., 2017).

MOOCs, being online courses (allowing users to follow them flexibly) and free of charge, attract thousands of people eager to obtain a degree from a prestigious university in a few weeks or months, without spending a single euro. In fact, they have become one of the most widely used teaching tools in Spain. The way they work is simple: the student only has to access the website, choose the course he/she likes the most, and register for it. Although it is true that only a few make it to the end – in this sense, they are reminiscent of what happens with collections by instalments – they have made it possible for people with few resources to obtain degrees from Harvard, Yale, Princeton, or MIT (Arrieta, 2014). For example, the European Union considers MOOCs as an opportunity for free digital learning for migrants and refugees (Colucci et al., 2017).

The UPM currently has the largest number of MOOCs published nationally. Since 2016, and through different calls, the UPM has delivered 85 MOOC courses, with 295 editions, on the Miríadax platform. More than 1,000 teachers are involved. The topics are as varied as the interests of the users: programming applied to science and engineering, applied educational innovation, free software and open knowledge, social networks applied to teaching, and climate change and sustainability are some of the courses that the UPM has launched.

In our case, LPF_TAGRALIA as the coordinating research group, or with the specific participation of some of its members, such as the authors of this chapter, has participated in several MOOCs and corresponding editions, specifically the courses entitled "Frutura Technologies: Fruit Quality" (Innovación Educativa, n.d.) and "How to manage agricultural and environmental risks?" (Innovación Educativa, 2019).

Specifics of Teaching Innovation and the EELISA Communities: Mission-Driven Working Groups That Bring Together Students, Teachers, and Researchers

Given the main assumption that we live in a changing world, and that most of conventional workplaces will undergo a radical change soon, it seems wise not only to teach content, but to familiarise the students with the idea of needing to cultivate a sense of innovation, and more specifically with the concept of technological surveillance.

Also in this section, we introduce the European Engineering Learning Innovation and Science Alliance (EELISA) communities: mission-driven working groups that bring together students, teachers, and researchers from all partner universities with prestigious professionals, grassroots organisations, citizens, private companies, and public institutions to find innovative solutions to real-world challenges. EELISA launched the EELISA Days on 10 November 2022, the date chosen in commemoration of the birthday of Elisa Leonida Zamfirescu, the woman who inspires the acronym of the alliance, as well as the World Science Day for Peace and Development.

The EELISA communities offer students the possibility to gain EELISA Credentials, a passport in which students collect the evidence or "badges" for their contributions to social challenge-solving processes.

We have selected both the ATHENS network and the EELISA communities as icons for the inclusion of innovation and technological surveillance in our students' curricula (Barreiro Elorza et al., 2015).

Learning Community (Technology Watch)

A learning community is a partnership in which the participants collaborate in a balanced way, on an equal footing, and which is focused on continuous reflection and questioning. There is constant learning with and from others. It is a model of professional development in which learning becomes a way of life (Figure 30.5).

Teachers, like students, construct our knowledge by integrating recent learning with prior knowledge, putting our ideas into practice, and reflecting on the results. When teachers collaborate with experts on real problems in a way that forces us to justify our actions, deep understanding is fostered and false or outdated shared beliefs are dismantled (Figure 30.6).

The Iconic Concept of *Travailleurs-Paysannes* and Its Relationship with the Acknowledgement of the Depth of Knowledge (and Error Propagation Without It), Together with the Application of Reverse Engineering as a Learning Strategy for Recent Disruptive Technologies

In this section, we want to highlight the particularities of farmers, rural areas, and people thereof, since they hold

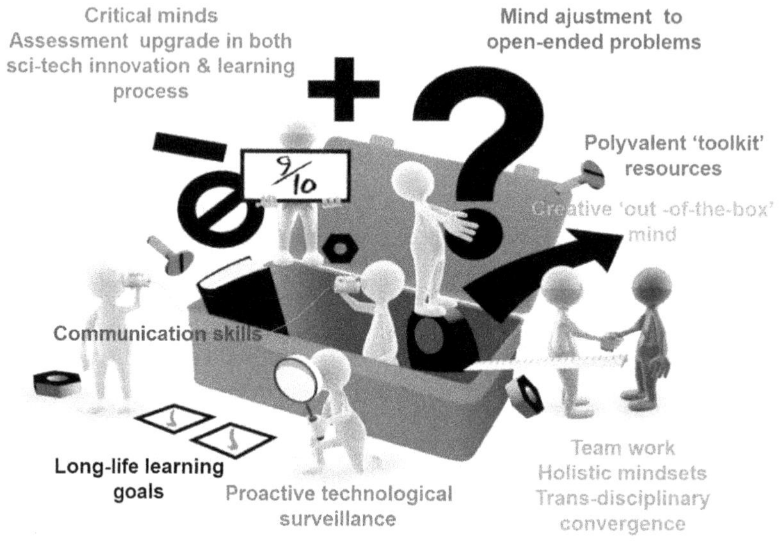

FIGURE 30.5 Toolbox containing all linguistic skills that facilitate and improve engineering communication across the curricula, and lifelong learning.

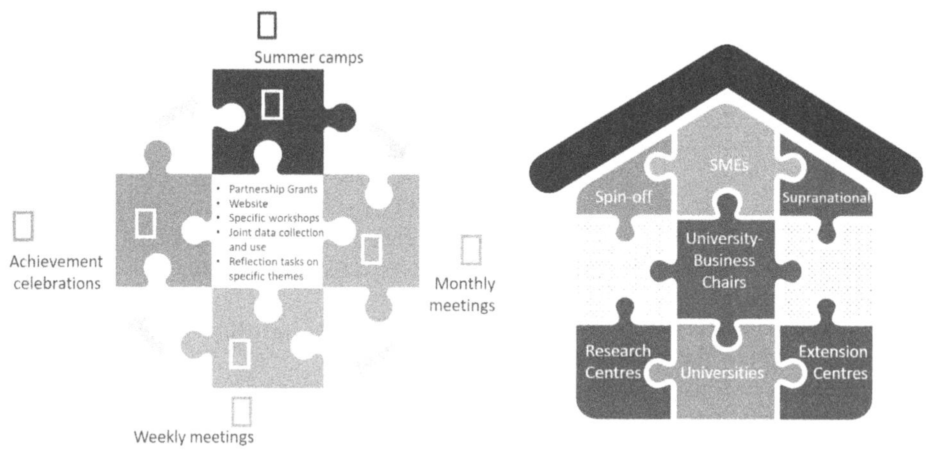

FIGURE 30.6 Linking of learning communities (left) and stakeholders and their interactions in learning communities (right).

fundamental tacit knowledge which is seldom available in formal academic studies; they are depositaries of the context of crop growing, livestock health and welfare, and history which remains unrecorded and, thus, to date neither unfolded nor understood in the digital environment. In this sense, we find relevant that bioengineering students approach the *travailleurs-paysannes* with a sense of understanding of the qualitative nature of the knowledge related to natural environments and the biodiversity they hold and protect (Barreiro Elorza and Gil Sierra, 2018). When trying to unfold and register the wisdom that underlies traditional agronomic cultural practices, the students must reverse the process of understanding, in what is sometimes denominated as reverse engineering process.

On the other hand, we would also like to highlight the importance of including green skills in vocational and higher education studies; green skills referring to the modification of products, services, or operations according to climate change adjustments, requirements, or regulations. Such new competencies will help the SMEs by incorporating a broader point of view, while improving their resilience to disruptive technologies and processes.

Reverse Engineering

According to Chikofsky (Chikofsky & Cross, 1990), the term "reverse engineering" has its origin in hardware analysis, where the practice of deciphering designs from finished products is commonplace. Reverse engineering is regularly applied to improve one's products, as well as to analyse a competitor's products or those of an adversary in military or national security situations. When transferring those concepts to software systems, many of these approaches can be seen to apply to gaining a basic understanding of a system and its structure. However, while the aim in the case of hardware was usually to duplicate the system, in the case of software, it is mostly to gain a sufficient understanding of the design level to facilitate maintenance, strengthen enhancement, or support replacement. According to the same authors, reverse engineering involves extracting design artefacts and building or synthesising abstractions that are less dependent on implementation. Reverse engineering can be started at any level of abstraction or at any stage of the life cycle.

Five years later, Müller and co-workers (Müller et al., 1995) declared that scant engineering research had so far been focused on software maintenance and evolution compared to software construction, and so a dramatic imbalance was already arising in software education engineering, with 50% to 90% of software evolution work being devoted to programme comprehension.

In 2011, efforts were being invested in reverse engineering for large-scale software projects. The challenge in this case is the construction of a feature diagram that reduces the selection of a parent for each feature, in an inherently interactive hierarchical construction process.

Some recent pedagogical studies indicate that imitation and cooperation are among the most important features of humans compared to other primate species, and that they may form the basis of the ability of humans to colonise new niches and ensure continued survival. The educated and evolutionary way to proceed is always to seek both self-welfare and community improvement (Barr et al., 2000).

Incorporating emerging fields of knowledge into higher education curricula implies a fundamental change in the role of teachers who move from the centre to the periphery of expertise. New contents and materials are continuously being developed with the assistance of advanced students who dedicate a great deal of time to research in multidisciplinary blogs and wikis, while the academic staff place the results in the context of well-established core knowledge in electronics and robotics.

Small Farmers' Union /Travailleurs-Paysannes/How Deep Is Your Knowledge

Another field where we have applied our experimental pedagogical points of view (limited though they may be) deals with the interaction between higher education, vocational studies, and experienced professionals, which in the case of agriculture holds and preserves a great deal of tacit and contextual wisdom.

Aligned with this concept, in Spain, since 2018, a new tool called operational groups has been set. Operational groups are a set of agents with different profiles linked to the rural environment who come together to develop and execute an innovative project aimed at finding a solution to a shared problem or to test and put into practice an innovative idea related to a sector specific to the rural environment (agriculture, livestock, forestry, etc.).

The operational groups are made up of farmers, stockbreeders, foresters, agri-food or forestry industries, RDI centres, training and advisory centres, and non-profit institutions, while their size and composition depend on the project; operational groups and innovative projects are funded through the rural development programmes and are used by the European Innovation Partnership for Agricultural Productivity and Sustainability (EIP-AGRI) to connect the different actors in rural areas. This action is part of the EIP-AGRI's objective of promoting and accelerating innovation in rural areas.

An example of an operational group in which LPF_TAGRALIA has participated, is the Small Farmers' Union Madrid Ecoviña study of the benefits of generating a dust cloud in the soil for the prevention of vine diseases (redruralnacional.es). This joint project between farmers, extension centres, and academics aims to obtain a higher quality grape which is a more economical option for the customer and that conserves and improves the characteristics of the environment. The main activity of the project has focused on the evaluation of the technique (and the machine), originally designed by a traditional farmer, further adapted by us for this project, based on different test farms, using different parameters related to plant health and production quality, as well as economic costs and environmental impacts. The results have been disseminated through open days and the machine has also been patented.

The Old Unwritten Laws of Engineering and Their Relationship with the Required Social Skills for Young Engineers at Professional Level, with Special Emphasis on Its Importance for Reporting the Technical Activity in the Command Chain at the Workplace

The Unwritten Laws of Engineering was first published in 1944 by the American Society of Mechanical Engineers: a concise and precious set of advice by W.J. King, a General Electric engineer who became a professor of engineering at the University of California, at Los Angeles (King, 1944). This book is of great interest as it highlights many circumstances where the social skills of young engineers may be crucial, especially in a multicultural and multidisciplinary environment.

We take the opportunity to summarise some of those statements that we consider relevant skills and competencies: a) however menial and trivial your early assignments may be, give them the best effort; b) demonstrate the ability to get things done; c) in carrying a project do not wait passively for anyone; d) confirm your instructions and the other person's commitments in writing; when sent out on a business trip, prepare yourself for it; e) develop a "let's go and see" attitude.

Redesigning Leadership

This section will introduce the audience to the concept of STEAM (science, technology, engineering, art, and maths), as formulated by John Maeda and Becky Bermont in the book *Redesigning Leadership* (Maeda, 2011). It is fascinating the way Becky Bermont explains the synergy of their two very different characters (his, extremely creative and visual, compared to hers of a steadier and more conservative disposition). In fact, each chapter in the book commences with a different haiku selected from among John Maeda's 1,200 statements, at that time, on Twitter (140 characters) (Maeda, n.d.).

Coming back to the concept of STEAM, Angeles Quesada, a philosopher and entrepreneur, explains the STEAM approach as helping people to use the scientific method, ranking ideas in systematic way, and looking for hidden meanings (Equanima,

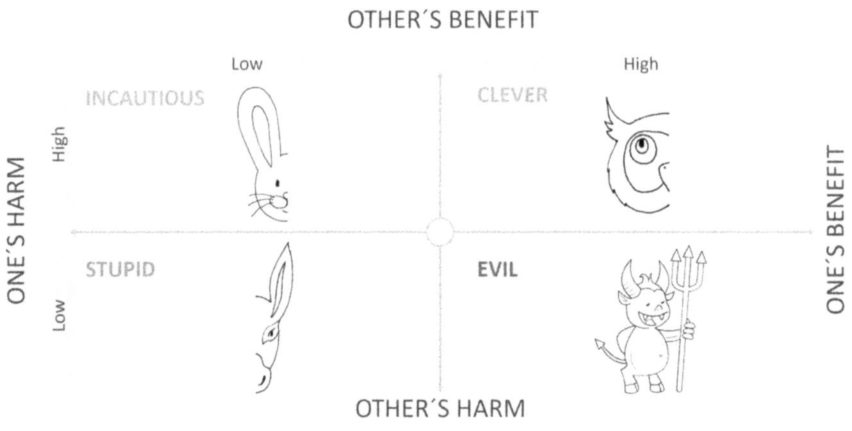

FIGURE 30.7 Cartesian coordinate system for the classification of members of a community (adapted from Cipolla, 1992).

n.d.). Also approaching the STEAM concept, we find Carmen Torrijos, an expert in artificial intelligence and computational linguistics, who indicates the need in STEAM of cultivating patience when people from very different domains approach you, and also having patience with yourself in the opposite case. The STEAM concept represents the knowledge required for near-future professionals in a challenging world.

To summarise, the book *Redesigning Leadership* (Maeda & Bermont, 2011) defines the following underlying idea: The pyramid versus the plum tree is proposed; it is worth noting that the term plum was set in the 18th century as "something desirable". Their philosophy can be summarised as: 1) be a work-in-progress and keep learning; 2) avoid rumours as tumours; 3) be grateful; 4) self-regulate destructive forces; 5) get yourself to smile; and 6) be techno-critical.

Allegro Ma Non Tropo (Cheerful but Not Too Much)

In our years of experience (since 1997), we have detected that the management of human conflicts that arise between students in the development of collaborative work is a fundamental aspect in the face of which we teachers feel helpless. We do not know how to act, we defer our intervention, we try to ignore conflicts, and when we do take them on, we make drastic decisions that often have dramatic consequences, such as the suppression of a working group: the dog is dead and the rabies is over. We look for resolution procedures, usually based on the oral transmission of personal experiences, and we are accustomed to using the universal method known as trial and error. Perhaps it is the text *Teaching with Your Mouth Shut*, by Donald Finkel, which has offered us the most clues in this process of voluntary search by amateur pedagogues with reckless tendencies (Finkel, 2000).

In a particular study presented in 2011, we started from the classification coined in 1975 by Carlo Maria Cipolla (Cipolla, 1992), according to which intelligent people are those who achieve good for themselves and for others; the incautious are those who obtain benefits for others while accumulating losses for themselves; and the wicked are those who obtain benefits for themselves and losses for others (we discard the last category, the stupid, in the study). We do however contemplate the possibility of the existence of dupes with traits of stupidity who make their own losses greater than the gains of others, and with whom a group loses; as with the wicked with traits of stupidity, whose gains are less than the losses of the group. According to the book's hypothesis, those who contribute to the welfare of a group, in addition to the intelligent, are both the unwary and the wicked endowed with traits of intelligence.

According to Professor Cipolla, an efficient group is one that has an unusually high percentage of intelligent individuals who, at the same time, produce enough profit for themselves and for the other members of the community to make progress a fact. Whereas, in a decadent group, one observes, above all, an alarming proliferation of the wicked, with a high percentage of stupidity, and an equally alarming growth in the number of the unwary with traits of stupidity.

Cipolla proposes a Cartesian coordinate system for the identification of the four categories, defined by one's own benefit-detriment (abscissa axis) and by the benefit-detriment of others (ordinate axis) (Figure 30.7).

When applied in the bioengineering domain to our students, we found that the Cipolla approach leads to a structured reflection on the attitudes of the participants in collaborative work, quantifying their effects on the individual and the group. We found that such awareness helps in conflict resolution and supports the establishment of a process of constructive criticism within the groups, and in the general class prospective (Barreiro Elorza et al., 2011).

Concluding Remarks

Our 25-year experience and expertise as senior academics in bioengineering, demonstrates that the number of tools and activities for the improvement and/or reformulation of technical curricula is endless. Therefore, there is a need for a joint action of (dedicated) lecturers, motivated students, and stakeholders, if success is to be attained.

Nowadays, global crises, with notable environmental and geopolitical implications, are increasing the complexity of society's needs. The search for answers to these needs is broadening the scope of engineering in general, and bioengineering specifically, to include teamwork skills, leadership, multicultural and interdisciplinary skills, communication, lifelong learning, and dispositions for sustainability, globalisation, innovation, and entrepreneurship, all with critical thinking. These transformations are broadening the competencies necessary for engineers, requiring the development of innovative curricula. The experiences presented in this paper seek to contribute to this objective, to the extent that the rigidity of most of the current curricula allows it.

Given the difficulty of the mission, we demand an ambitious reflection on the concept of engineering pedagogy, as well as on the preparation of postgraduates in the field of engineering education, as elements that could accelerate the introduction of innovation methodologies, such as those presented in this paper, in formal engineering education.

Let's put an end to an old wish:

> To educate human beings capable of creating a better world is the problem of cultivating human beings who possess a free intelligence with a cheerful disposition.
>
> ...
>
> The ability to set oneself a goal and to direct one's energies towards it cannot be developed under rigid discipline or in absolute freedom.
>
> *Bertrand Russell*

REFERENCE LIST

Arrieta, E. (2014). España se contagia de la fiebre de los cursos MOOC. *Expansión*. Retrieved March 23, 2023, from https://www.expansion.com/2014/08/26/emprendedores-empleo/desarrollo-de-carrera/1409044094.html

Barr, R. E., Schmidt, P. S., Krueger, T. J., & Twu, C.-Y. (2000). An introduction to engineering through an integrated reverse engineering and design graphics project. *Journal of Engineering Education*, *89*(4), 413–418. https://doi.org/10.1002/j.2168-9830.2000.tb00545.x

Barreiro Elorza, P. (2010). *Relación entre la Universidad y las empresas en los aspectos docentes y formativos* VIII Jornadas sobre la actividad docente e investigadora en Ingeniería Agroforestal., Sevilla, España. https://oa.upm.es/8175/

Barreiro Elorza, P., Diezma Iglesias, B., Moya Gonzalez, A., Sancho Guinda, C., & Mendes Fuentes, V. (2015). *Vigilancia tecnológica: un aprendizaje adaptativo a través del pensamiento divergente Technological surveillance: An Adaptive Learning Through Divergent Thinking* CINAIC 2015: Congreso internacional sobre aprendizaje, innovación y competitividad, Madrid. https://oa.upm.es/38524/

Barreiro Elorza, P., Diezma Iglesias, B., Recio Aguado, B., Fuentes Mendez, V., Martinez Ramirez, E., & Morato Izquierdo, M. d. C. (2009). *Fitting into their shoes: How robots help multidisciplinary approaches under cooperative learning* Taller Internacional RED-U sobre ABP y EBL, 2010, Madrid, España. https://oa.upm.es/9725/

Barreiro Elorza, P., & Duro, R. (2020). From environmental robots to cyber creatures: A liberal prospection on the future agro-ecological systems. *Advances in Agriculture, Horticulture and Entomology* (02), 1–12. https://oa.upm.es/58075/

Barreiro Elorza, P., & Gil Sierra, J. (2018). Utopus, la propuesta de un agricultor mallorquín que llega a la NASA. *Vida Rural*, *447*, 50–53. https://oa.upm.es/65923/

Barreiro Elorza, P., Moya Gonzalez, A., & Diezma Iglesias, B. (2011). *Allegro ma non troppo: las dimensiones ocultas en la coordinación de estudiantes* JIMCUE´11 – VI Jornadas Internacionales Mentoring & Coaching:Universidad – Empresa (22, 23 y 24 noviembre, Madrid), Madrid. https://oa.upm.es/32952/

Barreiro Elorza, P., Recio Aguado, B., Diezma Iglesias, B., Mendez Fuentes, V., Morato Izquierdo, M. d. C., & Martinez, E. (2009). *AGROTECH09: Building agricultural robots with lego mindstorm. A multidisciplinary and multicultural approach* EDULEARN '09 Conference. International Conference on Education and New Learning Technologies, Barcelona, España. https://oa.upm.es/5698/

Barreiro Elorza, P., & Sancho Guinda, C. (2015). *Understanding the inventor's mind through patent analysis: A clil team-teaching experience at the Technical University of Madrid* XXXIII Congreso Internacional AESLA: "La Comunicación Multimodal En El Siglo XXI: Retos Académicos y Profesionales". Madrid. https://oa.upm.es/38523/

Barreiro Elorza, P., Sancho Guinda, C., Diezma Iglesias, B., Moya Gonzalez, A., & Gil Sierra, J. (2014). *Patent elucidation for CLIL and vice versa understanding the inventor´s mind* 7th International Conference of Education, Research and Innovation, Sevilla, España. https://oa.upm.es/32814/

Barreiro Elorza, P., Valero Ubierna, C., & Diezma Iglesias, B. (2010). *Aplicación de técnicas de aprendizaje colaborativo en el grado de ingeniero agrónomo (agricultura de precisión) y máster en agroingeniería (robótica en la agricultura)* VIII Jornadas sobre la actividad docente e investigadora en ingeniería agroforestal, Sevilla, España. https://oa.upm.es/8191/

Chikofsky, E. J., & Cross, J. H. (1990, January). Reverse engineering and design recovery – A taxonomy [Article]. *IEEE Software*, *7*(1), 13–17. https://doi.org/10.1109/52.43044

Cipolla, C. M. (1992). *Allegro ma non tropo* (2nd ed.). Crítica.

Colucci, E., Smidt, H., Devaux, A., Vrasidas, C., Safarjalani, M., & Castaño Muñoz, J. (2017). *Free Digital Learning Opportunities for Migrants and Refugees: An Analysis of Current Initiatives and Recommendations for their Further Use* JRC Science for Policy Report, Issue.

Equanima. (n.d.). Pensamiento crítico. Retrieved March 23, 2023, from https://www.equanima.org/pensamiento-critico

Estrada, J. J., Luna, I., Solleiro, J., & Ivorra, B. (2015). *Leasing Design* [Video]. YouTube. https://youtu.be/G8nDolSlJyA

European_Commission, Joint Research, C., Inamorato dos Santos, A., Punie, Y., Conole, G., Tannhäuser, A., Childs, M., Nkuyubwatsi, B., & Witthaus, G. (2017). *Validation of Non-Formal MOOC-Based Learning: An Analysis of Assessment and Recognition Practices in Europe (OpenCred).* Publications Office. https://doi.org/doi/10.2791/809371

Finkel, D. (2000). *Teaching with Your Mouth Shut* (Heinemann, Ed.) Portsmouth, NH: Heinemann Boynton/Cook.

Florman, S.C. (2014). *The Civilized Engineer.* New York: St. Martin's Griffin.

GUNi. (2009). *Synthesis of the GUNi Higher Education in the World Reports 1-3. Higher Education at a Time of Transformation: New Dynamics for Social Responsibility* (1–3). (Higher Education in the World, Issue 1–3).

GUNi. (2013). *Knowledge, Engagement and Higher Education: Contributing to Social Change (5).* Higher Education in the World, Issue 5.

GUNi. (2017). *Towards a Socially Responsible University: Balancing the Global with the Local (6).* Higher Education in the World, Issue 6.

GUNi. (2019). *Humanities and Higher Education: Synergies between Science, Technology and Humanities (7).* Higher Education in the World, Issue 7.

GUNi. (2022). *Higher Education in the World 8: 'Special Issue'. New Visions for Higher Education Towards 2030 (8).* Higher Education in the World, Issue 8.

Innovación Educativa, U. P. M. (n.d.). *FRUTURA Technologies: Quality of Fruits.* Retrieved April 3, from https://innovacioneducativa.upm.es/mooc/informacion-mooc?idmooc=2089

Innovación Educativa, U. P. M. (2019). *¿Cómo gestionar los riesgos agrarios y ambientales?* Retrieved April 3, from https://innovacioneducativa.upm.es/sites/default/files/moocs/catalogo-2019-septiembre.pdf

King, W. J. (1944). *The Unwritten Laws of Engineering.* ASME Press.

Maeda, J. [@johnmaeda]. (n.d.).Twitter. Retrieved April 3, 2023, from https://twitter.com/johnmaeda?lang=es

Maeda, J. (2011). *Redesigning Leadership.* MIT Press.

Maeda, J. & Bermont, R.J. (2011). Massachussets: The MIT Press.

Müller, H. A., Wong, K., & Tilley, S. R. (1995). Understanding software systems using reverse engineering technology. *Object-Oriented Technology for Database and Software Systems*, 240–252. https://doi.org/https://doi.org/10.1142/9789812831163_0016

Muñoz, B. (2015). *This Is Aiding* [Video]. YouTube. https://youtu.be/3eCt_vBKYj0

Reques Velasco, P. (2009). *Universidad, sociedad y territorio.* Universidad de Cantabria.

Sánchez, A., Romera, P., Laschuetza, L., Docavo, P., & Serrán, R. (2015). *GREENERGYM: Elevator Pitch* [Video]. YouTube. https://youtu.be/f0P3LMI5e6Y

31

Nomograph-based Models for Introductory Undergraduate Teaching and Research in Selecting Agricultural Power and Machinery Ownership Systems in Developing Countries

Umezuruike Linus Opara

Introduction

The productivity of agriculture in developing countries, particularly in Sub-Saharan Africa, remains the lowest globally. At the same time, these countries remain the most food-insecure, despite the place of agriculture as the mainstay of the economy and means of livelihood of most of the population who also live in rural areas. The transformation and modernisation of smallholder agriculture in developing countries that is necessary to increase productivity requires the introduction of appropriate mechanisation strategies (Taylor et al., 2021). At local levels, such strategies require the introduction of improved power sources and related machinery in agriculture, from farm to processing and consumption. The benefits that accrue from any sustainable agricultural mechanisation have been presented in numerous literature (Opara, 2023, in this book, CAM-SAM; Kienzle et al., 2023, in this book; Sims et al., 2016; Giles, 1973; Harris et al., 1973). In addition to eliminating the drudgery usually associated with hand-powered cultivation practices, more food and better standard of living can be achieved by the farmer and his family.

Despite a high "willingness index" and favourable level of growing awareness among farmers and other stakeholders of the need to mechanise agriculture, a low "good management index" was identified as a major limiting factor (Anazodo, 1982). Farm power and machinery selection as an important aspect of farm power and machinery management is, thus, a vital factor to consider in discussing the introduction and use of modern technological inputs aimed at enhancing agricultural mechanisation. The predominant traditional smallholder farming systems and associated low rate of work often results in a low cultivated area per season (Odigboh, 1976). In some instances, farmlands have been left fallow much longer (often exceeding four to five years in parts of southeastern Nigeria), while in others, farmers rent out their portion of communal annual land allocation. Others choose to enter into agreement with willing farmers who share half of the planted area during harvest time or deliver half of the harvest to the landowner. In all these cases, farm power is the major limiting factor. Therefore, there is a need to develop a systematic basis for the selection of appropriate alternative power and machinery systems required to enhance the production volume per farmer both at the rural level and the level of large-scale commercial farms.

Given the increasing willingness to adopt new and modern engineering technologies for agricultural production among farmers, coupled with the fact that successful agricultural mechanisation involves the introduction and use of a given set of improved modern machinery, at least to a certain degree, the need for an efficient and systematic method for selecting such machinery becomes an important aspect of overall farm management. Perhaps the most important decision regarding machinery is whether to own specific equipment. The decision may be between owning or hiring; between different sizes, capacities, and ratings; between new and used equipment; between owning one or more than one of a specific machine.

To varying extents, farmers and managers of farm machinery units in many developing countries adopt various methods in planning the introduction and use of modern agricultural machinery on farms. Many, especially the policy makers in government, rely on trade fairs, exhibitions, agricultural shows, and overseas tours to well-established mechanised farms. Others rely on what is readily available in the market, whether appropriate to specific local need or not, while many are guided by experience and history of the use of the machinery model or make in particular areas. Inasmuch as these trade fairs, exhibitions and tours serve several useful purposes for the introduction and use of farm machinery by providing opportunities to know what is already available in the market and to know about recent improvements, the overall seemingly haphazard and non-systematic approach to farm power and machinery selection in many situations has led to a lot of attendant problems in the use of this machinery. These problems range from unsuitability in use and low field efficiencies to repair and maintenance problems, among others. For instance, an historical example in Nigeria showed that about 700 agricultural tractors and related farm implements worth over N23 million purchased by a defunct civilian government of Benue State were reported to lie idle for years because of a wide range of reasons, including the claim by most farmers

DOI: 10.1201/9780429150111-34

that these machines were not suitable for their local needs, both in complexity and flexibility (Anazodo et al., 1989). During the same period, there were widespread reports (Daily Times, 1986) that agricultural equipment worth more than two million naira was still lying idle at a store in the Uzo-Uwani local government area of Anambra State nearly ten years after they were bought. Various reasons advanced for the non-utilisation of this imported equipment ranged from a spare-parts problem to the complexity of the machines.

There is, thus, the need to match equipment with the work available and the environment in which the equipment with the work available and the environment in which the equipment is to be used, to ensure the numerous benefits of successful agricultural mechanisation (Giles, 1973; Harris et al., 1973; Anazodo, 1980). The application of an appropriate and sustainable level of agricultural mechanisation will facilitate the uptake of improved engineering technologies that are urgently needed to transform and modernise agri-food systems and promote rural development. This includes deciding whether to own or how to own specific farm machinery, and this approach combines the attributes and performance characteristics of the machinery system to be selected with the characteristics of the farmer and the farming system into which the machinery will be introduced. This approach also affords a systematic basis for continuous assessment of the decision taken and the response of the system selected, given a set of changing values of the input variables that affect the objective function. The aim of this chapter is to present a simple model, using nomographs (decision charts) to assist in farm power and a machinery-ownership system based on the requirement of timeliness and net economic benefit to crop production.

The model development involves five key steps:

(a) Assemble literature data on agricultural mechanisation (farm power and machinery performance) in selected farming system;
(b) Measure and document any missing data on agricultural mechanisation;
(c) Apply these data to evaluate the cost-benefit relationships aimed at determining the economics of using modern farm power and machinery for crop production systems of interest (maize, rice, cassava, cotton, etc.) under the different machinery-ownership and management systems;
(d) Develop nomographs or alignment charts for each crop production model under the different machinery management systems; and
(e) Develop a flow chart of the model proposed that may aid a computer-based solution.

Farm Power and Machinery Selection – An Overview

Farm power and machinery selection is one of the three major aspects of farm power and machinery (Hunt, 1973), and its deals primarily with the decision processes involved in the choice of a set of machinery for the requirement of farm operations and for a particular system of farm power and machinery ownership and use. The other equally important components of farm machinery management are performance evaluation as well as repair and maintenance. The selection of appropriate size of farm tractor and implements and/or an appropriate machinery management system for the introduction and use of modern farm machinery is thus one of the important decision parameters in the overall agricultural machinery management programme. This becomes even more important when one considers the cost of owning and operating a typical tractor and implement unit, which has increased up to 100% over the years (Mayfield, 1981; Khan et al., 1984). Apart from costs, there are many other factors that affect the selection and use of farm tractor and implements, such as agriculture conditions, farming requirements of soil and crops, management scale, economic considerations, etc.

Proper selection of farm machinery for farm operations is one of the most difficult decisions facing farmers and one serious mistake can be costly, which is usually manifested in a high cost per hectare or per hour and the inefficient execution of field operations. Thus, according to Bowers (1968), before one invests money in a specific machine, there are several questions that must be considered:

(i) Will the machine perform the desired function so as to return maximum profit to the farm enterprise?
(ii) Is custom work readily available in such quantity and quality that it represents the lowest-cost method for getting the job done?
(iii) Is the machine large enough to complete critical operations without severe timeliness losses?
(iv) Is the machine's initial cost excessively high and must it be owned so long before minimum cost is reached that it will be completely obsolete?
(v) Would the money you spend on the machine be better spent elsewhere in the business?

There are several methods already available to assist in selecting individual machines or systems of machines. The more sophisticated programmes involve mathematics; for example, see Opara (2023, in this book, CAM-SAM) on a computer-aided model for selection of agricultural power and machinery for sustainable mechanisation development.

Factors Affecting Farm Power and Machinery Selection

The economic selection of farm tractors and machinery is often an involved problem for any successful mechanisation programme and is usually complex because of the variations in farm size and differences in ecological conditions. Therefore, the efficient selection of field machinery for farm operation calls for an understanding of those factors that measure the performance characteristics of the field machinery, the factors that affect such characteristics, as well as cost considerations, timeliness, and a proper analysis of those factors that characterise the system for which the machinery is being selected. Most of the factors discussed below have been presented in detail by numerous authors (Bowers, 1968; Hunt, 1973; Anazodo, 1975; Kepner et al., 1980; Renoll, 1981).

Machinery Performance

Machinery performance is one of the three aspects of economic performance of field machinery (Hunt, 1973). The other two are power performance and labour performance. Optimum farm machinery management occurs when the economic performance of the total machinery system has been maximised. Good machinery management, therefore, requires that individual operations in a machine system must be adjusted and combined in such a manner that their overall performance returns the greatest profit to the farm business.

Agricultural machinery performance is measured as the rate and quality with which the operations are accomplished. Rate is an important measure because it affects the sensitivity of operations to the seasons and bad weather (this consideration leads to the concept of timeliness, which is a critical consideration in agricultural operations which are highly weather-dependent). The rate of machine performance is expressed in terms of quantity per time. Most agricultural field machine performances are reported as hectares per hour (ha/hr). Harvesting machine performance is sometimes expressed as tonnes per hour, or in the case of balers, bales per hour. Some of the factors that affect the performance of agricultural machinery (Anazodo, 1975) include the system of machinery ownership and use, management, man-machine suitability, cultural practices, ecological factors, technical factors, as well as sociopolitical factors. The influence of these factors varies from place to place.

The performance of agricultural machinery can be assessed qualitatively or evaluated quantitatively. Qualitative assessment may be carried out by way of field demonstrations, and such evaluation may be accomplished by observing how good a quality of job the equipment does under the prevailing circumstances; how easily the local operator handles it while working in his field; and how suitable it is to the ecological conditions and cultural practices in question (Anazodo, 1975). On the other hand, less subjective valuable expressions have been developed for quantitatively evaluating the performance of most agricultural machinery. The following machine performance factors are relevant for the choice of equipment, and they have been extensively discussed elsewhere in the literature (Hunt, 1973; Anazodo, 1975; Renoll, 1981).

Theoretical Field Capacity (TFC)

This is the maximum possible rate of field coverage that would be obtained if the farm machinery were performing its functions at the rated forward speed and always covered 100% of its rated width.

$$\text{TFC} = \text{optimum speed} \times \text{rated width of implement} \quad (1)$$

Even though theoretical capacity cannot be sustained for long periods of time because of the interruptions associated with field operations (such as turns, breakdowns, obstructions in the field, etc.), it is yet valuable information in farm machinery management since it provides a reference base for assessing the actual performance of machines and their operators.

Effective Field Capacity (EFC)

Usually, it is not possible to always operate machines continuously at their rated width of action or without time loss during field operations. Therefore, their effective or actual capacities will be substantially less than their theoretical capacities. EFC, therefore, is a measure of the actual rate of performance of farm machinery and it thus gives the actual capacity of the machinery to accomplish a field operation.

$$\text{EFC} = \text{actual average speed} \times \text{effective width of machine} \quad (2)$$

Field Efficiency

The field efficiency (FE) of farm machinery may be defined as the ratio of the EFC to the TFC, expressed as a percentage:

$$\text{FE} = \frac{\text{EFC}}{\text{TFC}} \times 100\% \quad (3)$$

Measurement of the FE evaluates the ability of a piece of equipment to achieve its maximum possible capacity (theoretical capacity) when operated in a given field. Considering the theoretical field capacity as the ideal field coverage that could be obtained if the machine were performing its functions 100% of the time at the rated forward speed and always using 100% of its rated width, field efficiency, as a percentage, can be expressed as:

$$\text{FE} = 100\left(\frac{\text{To}}{\text{Te} + \text{Th} + \text{Ta}}\right) \quad (4)$$

Where:

To = theoretical operating time per hectare when the equipment is, at all times, operating
with the rated speed (v) and rated width (b).

$\text{To} = \frac{600}{\text{vb}}$ (v in kilometres per hour [km/hr], b in metres, To in minutes per hectare [min/ha]).

Te = effective field operating time per hectare.

Theoretical field capacity (TFC) =
$= \frac{100\ \text{To}}{\text{K}}$

K = percentage of implement width actually utilised.

Th = time lost per hectare due to interruptions that are not proportional to area, such
as filling the tank of a spray machine.

Ta = time lost per hectare due to interruptions that tend to be proportional to area, such as
time lost in turning at the ends of a field.

With the FE established based on time, the EFC can then be determined as the product of the TFC and the field efficiency:

$$\text{EFC} = \text{TFC} \times \text{FE} = \frac{\text{V} \times \text{bFE}}{10}(\text{ha/hr}) \quad (5)$$

Where:

V = ground speed, km/hr

b = machine width, metres

FE = field efficiency, decimal

According to Renoll (1981), equation 5 above has several disadvantages. The value of FE contains several non-productive items, including such things as breakdowns, adjustment, turning at field ends, adding seed and fertiliser, and field delays. Since FE is the sum of the many different factors, it is difficult to obtain reliable values for it. The author, therefore, developed a mathematical expression which is useful in predicting farm machinery performance rates using 14 specific input coefficients including items relating to the actual machine and its use, field size, shape, physical conditions, and machine management. This model can be used to predict the machine's capacity for individual field and farm conditions rather than for average field condition, as is frequently done with other capacity-calculating methods. Renoll's equation is expressed in two parts as follows:

$$T = A + B\left(\text{hours per hectare } [\text{hrs/ha}]\right) \quad (6)$$

$$C = \frac{1}{T}(\text{ha/hr}) \quad (7)$$

Where:

A = time spent actually performing the specific operation, hrs/ha.

B = time used for supporting activities, row-end turnings, and other delays, hr/ha.

T = total time, hrs/ha.

C = performance rate, ha/hr.

Field machine index (FMI)

FMI is used to indicate how well adapted a specific field is for the use of machinery. It indicates the relationship between row length, turning time, and machine effectiveness. It is given by the expression:

$$FMI = \left(\frac{Te - Th - Ta}{Te - Th}\right) \times 100\% \quad (8)$$

Where:

Te = total time for productive field operation.

Th = total non-productive time excluding turning (e.g. time for adjustment in the field).

Ta = total turning time (i.e. time spent in turning at row ends).

FMI is very vital in machinery management, especially in the choice of machines for particular field conditions, because it measures the suitability of the field to the machine. The higher the value of FMI, the more suited a field is to the machine.

Material Efficiency (ME)

The material efficiency of a machine describes the completeness with which it handles the material and the resulting condition of the processed material. This efficiency index is very important because most farm machines handle materials as part of their performance and the quality of the machine's performance is described by the efficiency with which it handles material. Material efficiency has no application with equipment for land clearing and seedbed preparation but is a vital parameter for estimating the quality of work of a harvesting machine. It can be measured using the expression:

$$ME = \frac{A - B}{A} \times 100\% \quad (9)$$

Where:

A = the monetary value or weight of material before being handled by the machine,

such as the estimated yield of a corn field.

B = the actual value or weight of material after being handled by the machine,

such as the actual weight of corn collected by a harvesting machine.

Timeliness of Operation

Timeliness has been defined as the ability to perform an activity at such a time that quality and quantity of product are optimised. Timeliness of operation as a decision criterion in farm power and machinery selection finds great application in deciding between larger or smaller units of equipment, or between one or more than one machine. It may also be important in deciding between ownership and a hiring service. The probability that unfavourable weather may make smaller equipment inadequate cannot be calculated accurately. For this reason, a farmer deciding what equipment to buy must balance the extra cost of the larger (or additional) equipment against the estimate of the risk involved in owning smaller and less-costly equipment. According to Eradat (1983), the extra investment in bigger machinery should not exceed the amount of the reduction in timeliness penalties. Thus, the timeliness of a field operation can have an economic value.

Timeliness coefficient is a factor representing proportionate reduction in return per unit of time spent performing a given activity because of lack of timeliness. This reduction occurs because agricultural operations cannot be performed at arbitrary times throughout the year. Almost every operation required for successful crop production must be timely. Untimely completion of any operation will cause substantial loss of yield and quality, which ultimately will affect the farmer's income. This loss is termed "timeliness penalty" or "timeliness cost". Edwards and Boehlje (1980) described timeliness penalties as the indirect cost of lower crop yields that occurs because farm operations such as planting and harvesting are not completed during the optimal time period. This is because all "time" potential during the growth season cannot be utilised to complete the agricultural operations. The timeliness penalty can thus be considered as an hourly charge against the implement due to its inability to complete an operation within an optimum time. The amount of this charge will depend on the total value of the crop and the timeliness coefficient, or factor K. This factor K is considered the decimal reduction in the value of the crop for each hour required by the operation.

It thus represents the economic significance of delays in the performance of agricultural field operations. In other words, it measures the crop value for each hour or day the completion of a given operation is delayed beyond the operation time for which the maximum crop value will occur.

From the practical assessment of timeliness penalties for machinery selection purposes, Eradat (1983) observed that timeliness penalty constitutes a major part of the whole economy of farm machinery selection and, unless its magnitude is properly evaluated and incorporated in the structure of a machinery selection programme, the reliability of such a programme will be questionable. In the determination of the timeliness coefficients for rice and the factors affecting their value, Chancellor and Cervinka (1974) emphasised that the simplest method for incorporating timeliness penalties in a machinery selection programme is to define a season for every individual operation and consider the machinery system which will complete the work during that season as well as maximise the farm income. The size of the machine required, in terms of its capacity in hectares per hour, can be determined by dividing the number of hectares to be cropped by the number of working hours available for timely field operation. The problem, however, is not that simple, as one may not or perhaps cannot have a single figure for the number of working hours available for timely field operations, since agricultural operations cannot be performed at arbitrary times.

Cost Analysis

The total cost of performing a field operation includes the charges for the implement or machine, for power utilised, and for labour. Machine costs include depreciation, interest on investment, taxes, insurance, shelter, repair and maintenance, lubrication, and fuel (if the machine has an engine). Researchers have for a long time investigated the ongoing cost of owning and operating farm machinery (Bukhari, 1982). Cost consideration is an important factor in farm power and machinery selection because the buyer or farm machinery manager must bear in mind that not only the initial purchase price of the equipment must be considered. Additionally, he should be familiar with the estimate of the operating cost of the equipment. The availability of the information on the operating cost will aid the overall farm machinery planning programme. The total cost of operating field machinery is usually expressed on a per-hectare or hourly basis, or on a production-unit basis (e.g., \$/ha or \$/kg). This cost can be delivered into two groups: fixed or owning costs and variable costs.

Fixed Costs

These costs are related to machine ownership, and occur whether or not the machine is operated. They are dependent on calendar-year time and are independent of use. Fixed-cost items include the cost of interest on the machinery investment, annual depreciation, taxes, housing, and insurance.

Depreciation is the decline in the value of the machinery per year. It is often the largest cost of owning and operating farm machinery. Among the various methods employed in calculating depreciation are the straight-line, constant percentage, estimated value, and the compound interest or sinking-fund methods. Hunt (1973) reported that costs relating to farm machinery operation can only be estimated and, therefore, time should not be spent in using complex depreciation methods. Hunt suggested that the straight-line depreciation method is most reliable for cost approximations. Thus, annual fixed costs including depreciation, interest, taxes, shelter, and insurance may be considered to form a single annual fixed cost as a percentage of purchase price for any estimations.

Expressed mathematically, the depreciation of a farm machinery can be calculated by the straight-line method using equation 10:

$$D = \frac{P - S}{L} \tag{10}$$

Where:

D = depreciation, \$/year or \$/hour.
S = salvage or resale value of equipment (if any).
L = estimated economic length of life in years or hours of the equipment.
P = purchase price of the equipment.

Fixed costs due to taxes, insurance, shelter, and interest on investment vary considerably with time and location (Bowers, 1968).

Variable Costs

These are the cost items which vary in proportion to the amount of use of the implement. They include the costs of maintenance, repair, power and labour, oil, lubricant, and fuel. These are costs clearly associated with use and time. With machines such as combines, balers, and field choppers that may be equipped with auxiliary engines, the cost of fuel and oil for the engine must be included in the total machine charge. For implements such as tractor-pulled ploughs, etc., the cost for fuel and supplementary labour is usually neglected in estimating their cost since these costs may have been levied on the tractor.

Power Requirement and Availability

Tractors are the major source of power in modern agricultural field operation (Bukhari, 1982). The other power source includes human muscles and animal power. The selection of the appropriate farm tractor capacity rating for farm equipment is a problem encountered often in machinery management and thus requires careful consideration, because the cost of power is very significant in the calculation of the overall cost of farm operation. This becomes even more important when one realises that mechanical power is the readily-available substitute to compensate for the farm labour being lost to urban occupations, and that all analyses geared towards mechanisation of agriculture carry the implicit assumption that the energy necessary for such mechanisation programmes will be available (Harris et al., 1973). While reviewing the cost of owning and operating tractors, Bukhari (1982) rightly observed that tractors are the most important source of farm power for the mechanisation of farms, requiring a high initial capital investment, and that the selection of the right tractor means the difference between profit and loss on the investment.

Thus, the selection of farm machinery will depend greatly on the type and size of tractor available or previously chosen. Experience and practical trials provide the best basis for choosing suitable tractors because of the combined influences of agricultural conditions, farming requirements of the soil and crops, management scale, and economic considerations, etc. As the choice of a particular tractor size mainly depends on soil type and the nature of the operations to be carried out for various crops, it is reasonable to make such a choice based upon economic performance. The following factors are relevant to consider in the final selection of tractor for farm operation:

Tractor Size

The size or, rather, capacity of the power unit, usually expressed in horsepower or kilowatts, must be matched with the amount of work to be done as indicated by timeliness and cost. The tractor size must be matched adequately with the field implement to obtain good results and minimise repair and maintenance problems.

Tractor Type and Device for Mounted Implements

For most situations, the tractor to be selected will need to be an all-purpose machine that can be applied to almost all kinds of farm work, including ploughing, harrowing, sowing, weeding, harvesting, and transport. On small farms generally, one tractor is expected to do almost all jobs and there is not much flexibility in the choice of a particular size of tractor. For large farms, one can select two or more tractors of different types and sizes, within the range of the minimum power requirement, to suit various farm operations.

For whichever case the selection is made, an important consideration is to choose tractors that are equipped with good hydraulic control systems in order to ensure flexibility in use and to make it possible for tractor-mounted implements to be used.

Labour Considerations

An important factor in using farming machinery is the labour required to operate it. On the other hand, the availability of labour and required skills also affect the successful introduction and use of any farm machinery in a particular farming system. Labour consideration is a very vital factor during the selection of farm machinery for its introduction and use because improper selection of machinery can lead to considerable labour displacement, with the attendant employment and related social problems (Oram et al., 1979). In most developing countries where unemployment is currently very high, and a high percentage of the population is involved in the farm business at the subsistence level, careful consideration must be given in selecting machines that will elevate labour without displacing it, to the detriment of the rural community. On the other hand, the skill and training of operators as well as their adequate supervision are important factors for efficient machinery use.

Good incentives and recognition of the importance of the tractor operators and their need for training, recognition, and experience will improve the efficiency of machinery use and lead to increased yields (Anazodo, 1975). The effects of these factors on successful agricultural mechanisation have been verified by Giles (1973), who showed that an improperly operated drill can cause a reduction in crop yield. This applies to the other major farm inputs. For example, chemical fertiliser applied in excessive amount or non-uniformly may cause lodging and crop yield loss.

Miscellaneous Factors

An important social factor which has bearing on the choice of equipment, but which cannot be measured quantitatively is the effect of mechanisation on the nature of the life and work of farm workers, including farmers themselves. For example, a machine which achieves timeliness of operation and reduces the drudgery of farm work by improving the dignity of the farmer may have achieved something more important than overcoming a temporary labour shortage. Other social and organisational factors affecting the selection and use of farm power and machinery include the system of machinery ownership and use, the farming system, the size of the farm, the types of crops growing, the cultural practices of the farmers, and the type of soil. The combined influences of these factors are interrelated and further highlight the need for their individual consideration. About a half-century ago, Giles (1973) observed that although mechanisation must be planned nationally and regionally, specific machines and methods must be selected in the initial analysis on an individual farm basis.

Anazodo (1982) summarised the influence of prevailing social-economic factors for machinery introduction and use in any area of interest, that is, sound knowledge and appreciation of the social-economic resources based in an area is a fundamental requirement for the successful application of a new technological input to the agricultural production function. He maintained that the effectiveness of an agricultural modernisation strategy depends to a large extent on the degree of integration of the modern techniques being proposed with the prevailing social-economic factors in the relevant agricultural area.

Farm Power and Machinery-Ownership and Use Systems – An Overview

The system of farm power and machinery ownership and use is the organisational structure into which the selected machinery will be introduced and used. It represents the conditions under which the machinery will be managed. Since it is essential that the farm machinery is suitable for the work it has to do so that the goals of achieving increased food production and better living standards can be realised, it is necessary, therefore, to study and understand the characteristics of the system of machinery's ownership and use, which influences the result of any mechanisation programme.

Numerous farm power and machinery management systems in Nigeria have been identified and studied, but for the purpose of this thesis, four major systems that provide services to farmers who engage in agricultural production have been chosen for discussion and analysis. These are the hand-powered traditional system (HPTS), tractor and equipment hire system (TEHS), sole-ownership and use system (SOUS), and the farmers' cooperative ownership and use system.

Hand-Powered Traditional System (HPTS)

This management system represents the class of farmers in rural agricultural communities who use human muscle as the major and sometimes only source of farm power. The system is characterised by fragmented holdings, which may be uneconomical for machinery application unless land resources are amalgamated. Mixed cropping coupled with varying fallow periods is usually practised. In general, traditional practices are very slow and must be modernised to increase cropping intensity and productivity.

Tractor and Equipment Hire System

This is the farm machinery-ownership and use system whereby farm machinery and implements bought and owned by the government are hired out to farmers for cultivating, harvesting, and other farming services. They could be hired out to farmers on a minimum-deposit basis to provide them with operators who are basically familiar with general farm machinery technology. Thus, this hire service system provides a form of subsidy to the farmers, thereby encouraging the introduction of mechanised practices.

Sole-Ownership and Use System

In this system, the individual farmers only acquire the machines required to carry out farm operations. The farmer owns the equipment and machines, operates them, and is this able to perform many different types of work at the desired and suitable time and manner. Given the same parameters of comparison, such as the percentage of tractors in unserviceable condition, the number of tractor breakdowns per unit, the number of operational tractors, the value of tractorisation intensity, the level of farm tractor utilisation, and the cost of operation, the private ownership and use system has been found to be the most attractive system of farm machinery management.

Farmers' Cooperative Ownership and Use System (COUS)

The COUS approach to agricultural development and access to mechanisation technologies entails the joint association of farmers and other agro-entrepreneurs for mutual benefit including access to agricultural machinery and other services. In many instances, they are designed to attract government assistance for the cooperative rather than individual farmers. A tractor and equipment hiring service unit provides hiring services to the participating farmers, who in turn contribute money to a central purse for the overall running of the association. The organisational structure of this system is such that the government acquires land, develops it, and then leases it out to the association which, thereafter, parcels the land to its members. The society also depends on commercial banks or other financial agencies for loan acquisition through the assistance of the government.

Theory and Principle of Nomographs

The principle of nomographing as a scientific operations research tool (Wilson, 1952) can be applied to finding solutions to the problem of selecting an appropriate farm power and machinery-ownership system, particularly in smallholder agriculture, where the economy of scale may not permit ownership. The basic idea is to generate sufficient statistical data on the decision variables that affect the selection of the farm power and machinery level to be adopted, and then design nomographs which would be quite convenient for reading the numerical values of these decision variables given the changing values of the input variables.

A nomograph, or alignment chart, is a numeric scientific research tool that presents the relationships between two or more variables on a chart. Thus, while graphs have the advantage of showing at a glance the form of the function, the nomograph is more convenient for reading numerical values. Because of this requirement, one basic first step in the design and development of a nomograph is to establish the relationships between the decision variables and/or generate statistically sufficient data on the decision variables that are to be presented on the nomograph. For the first two variables, say A and B, there is a scale for A (not necessarily either linear or straight) and another scale for B. A straight edge is laid across the A and B scale, intersecting them at the desired values. The intersection of this straight edge with the third scale for C is the value of C, which is a function of A and B. Thus, the principle of the nomograph is that a line joining a chosen point on a scale with a chosen point on another scale will intersect a third scale at the value of the roots of the variable for the third scale if they are real and lie within the range specified for the analysis.

Model Development

Statement of the Modelling Problem

The problem addressed in this chapter is the development of a systematic basis that can be used to determine the appropriate level of power and machinery required to increase the level of agricultural mechanisation, which is considered a necessary input for the modernisation and transformation of agri-food systems, particularly in developing countries. Four major systems of farm machinery ownership and management were identified and chosen to represent different levels of agricultural mechanisation. The problem environment is comprised of the individual farmer's crop production environment (soil, crop, production practices, etc.), and the machinery-ownership/management system available.

The Conceptual Model

The model is a decision model that utilises the characteristics of any farmer in terms of farm size, power sources, and cropping pattern, and a cost-benefit appraisal of his farming enterprise, to determine the appropriate level of mechanisation that will be required to achieve and sustain timely farm operations and afford the farm good economic benefits, presented as monetary benefits. The appropriate level of mechanisation for the purpose of this project would be the level of machinery introduction and use through the selection of any one of the machinery management systems reported here, or a suitable combination among them. Thus, a suitable combination for one farmer may be to use the service of a TEHS for all land clearing operations and thereafter to own all the machinery units required for tillage, planting, and harvesting operations. The modelling system involves a two-stage approach in the selection of an appropriate power and machinery system for the successful mechanisation of crop production operations: (a) timeliness determination and (b) cost-benefit determination.

Timeliness Determination

Several investigators have conducted research to study the effect of timeliness on crop yield (Chancellor & Cervinka, 1974). Numerous others have also worked on the practical assessment of timeliness penalties for machinery selection purposes (Burrows & Siemens, 1974; Edwards & Boehlje, 1980; Huan-Wen & Hunt, 1985). Most of the literature cited agree that the simplest method for incorporating timeliness penalty in a machinery system programme is to define a season for every individual operation and consider the machinery system that will complete the work during that season (Sturrock et al., 1977; Eradat, 1983).

Methodology Adopted

a) The first step is to obtain data on the calendar of operations for the crop desired and the environment concerned. Based on rainfall pattern and characteristics and the agronomic-cultural practices required for efficient crop performance, a schedule is developed that would ensure timely field operation. Under a chosen design probability or efficiency level, the operation schedule is used to estimate the available man-days (AMD) for optimum crop performance and completion of critical field operations. The optimum AMD for each critical field operations such as sowing, weeding, harvesting, etc., is also specified.
b) Using the data on the rate of work (ha/hr or ha/man-day) for the execution of each critical farm operation, the required man-days (RMD) for the completion of the entire critical operations is determined for each machinery management system. Assuming an eight-hour standard man-day, for each critical field operations, the RMD can be calculated from the equations:

$$\text{RMD} = \text{Inverse Rate of Work } (\text{ha/man-day}) \times \text{farm size } (\text{ha}) \quad (11)$$

$$\text{RMD} = \text{Inverse Rate of Work } (\text{ha/hr}) \times \text{farm size } (\text{ha}) \;/\; 8 \quad (12)$$

Analysis of Results

Any machinery management system which satisfies the requirement that RMD < AMD provides a possible solution and could also be selected to produce the crop whose calendar of operation was used, assuming that the timeliness of operations were the only decision criteria. If more than one management system satisfies the requirement, the one with the lowest RMD value is ranked first in the list of possible solutions.

Cost-Benefit Determination

The second stage involves the evaluation of the cost-benefit relationship for each of the possible solutions that were determined when the timeliness criterion was analysed. This stage in the selection of an appropriate machinery management system for efficient crop production is equally important because, apart from achieving timely field operations that enhance optimum crop performance, the machinery system adopted should also ensure optimum economic results for the farmer at minimal cost of operations. The improved economic welfare for the farmer has long been advanced as one of the numerous advantages of successful agricultural mechanisation (Esmay & Hall, 1973; Harris et al., 1973; Anazodo, 1980). In reviewing the mechanisation of small farms in tropical Africa, Kyembeh (1986) rightly pointed out that appropriate mechanisation for small farms must, among other criteria, provide incentives to small farmers for its adaptation. Farmers, like other entrepreneurs, are usually responsive to new technology if they are convinced of its benefits.

Methodology Adopted

For the purpose of incorporating the economic benefits criterion in the selection of an appropriate machinery management system for efficient and timely crop production, nomographs were developed for the crop production models studied (maize and cassava) under the farm power and machinery management systems. The decision variables in the nomographs are farm size (ha), cost of all operations, and the net economic benefit. The average rate of work (ha/hr) that can be attained under the machinery management system was incorporated as a constraint limiting the size of farm that can be effectively cultivated for timely farm operation.

For all the management systems that offered possible solutions under the timeliness consideration, the one that brings the highest economic return to the farmer whose characteristics were used in the analysis is the optimum solution and is thus recommended as the appropriate level of machinery to be adopted for crop production operations. However, it is to be noted that the acceptance of this optimal solution by the farmer depends on his capacity to generate enough capital either from personal savings, borrowing from relations, or loan from commercial banks, to defray the total cost of production. Therefore, the optimal solution would not always be the feasible solution. The feasible solution will depend on the farmer's capital outlay, which is usually introduced into the model as a constraint or limiting factor.

Size of Power and Implements

For situations where complete ownership of the machinery system for crop production is the optimal solution, the characteristics of the farmer and farming environment should be used in determining the set of optimum-sized minimum-cost power and implement size and number (for elaboration on models for selecting farm power and machinery size, see Opara [2023, in this book]).

Model Constituents

Model Components: These are the elements of subsystems that describe the overall system. For our system study, the components are the farmers and the machinery management system as a unit. The subsystems of the machinery management system are the four different systems of farm power and machinery management: hand-powered traditional system, tractor and equipment hire system, sole-ownership and use system, and farmers' cooperative ownership and use system.

Variables: These are the characteristics that define the system and are usually paid into the system to generate results or are generated by the system. They include the variable yield of the crop, machine performance characteristics (variable rate of work: operating hours per day, variable lost machine time, speed of field operations), the cost of the various farm inputs (seeds, fertilisers, herbicides, insecticides), labour charges, and machinery costs for field operations.

Constraints: These are the limitations to the operation of the system, such as weather constraints (including tillage and harvest time limits) and capital availability.

Functional relationships: These show the behaviour among variables and parameters. Most of the equations presented in this chapter represent the system of functional relationships.

Parameters: These are the variables in which we can assign numerical values. For example, the standard man-hours used per day for a field operation.

The objective functions are introduced into the model to:

(i) Minimise the level of machinery introduction that is required to achieve and ensure timely farm operation subject to the cost of the investment in machinery and capital availability to the farmers.
(ii) Minimise the cost of the farmer's operations subject to the timeliness requirement of the field operation.
(iii) Maximise the economic benefit expressed as profits.

Figure 31.1 presents a detailed flow chart of the nomograph model for education on selective and sustainable agricultural mechanisation management. In brief, the following steps are incorporated:

i. Characterise the farmer. This includes determining the power and machinery management systems available, farm size, crops grown, and available days to complete critical field operations.
ii. Compute the cost of field inputs and miscellaneous expenditures.
iii. Quantify rate of work, cost of operations, and manual labour charges.
iv. Estimate the working days required for critical operations under each combination of power and machinery management systems.
v. Develop nomographs and use them to determine the total cost and net economic benefit of each cropping system.
vi. Select the appropriate power and machinery system.

Relationships in the Model

The relationships between the various constituents of the model are presented in Figure 31.2. As can be seen, the variables and the constraints are directly related to the two components of the model – the farmer and the machinery management systems. Field inputs depend on the farmer's environment and crop choice, whereas the fieldwork data and the cost of field operations depends on the machinery management system.

A Special Case for the Nomograph Model

Two major features appear peculiar in the nomograph reported here and are worthy of emphasis. First, an attempt has been made to incorporate a limiting factor among the three major decision variables. The result is that, to read off the second and third variables (total cost and economic benefit), the first variable (farm size) must be related to the limiting factor or constraint (the average attainable rate of work). A second feature is that it was possible to incorporate all four decision variables into a single nomograph where the rate of work has been included, because the value depends on the machinery management system considered.

Summary of the Procedure Applied to Designing and Developing a Nomograph

1) Establish the form of interrelationship that existed between the input and output variables in the system under study (that is, the farmer and farming systems, and the farmer power and machinery-ownership and management system).
2) Identify and choose the major decision variables that affect the selection of any farm power and machinery-ownership and management system. These variables will be incorporated into the nomographs.

The following variables were identified and used to develop the nomographs:

(a) Farm size, ha.
(b) Rate of work, ha/hr.
(c) Total cost of operations.
(d) Net economic benefits.
(e) The timeliness requirement.

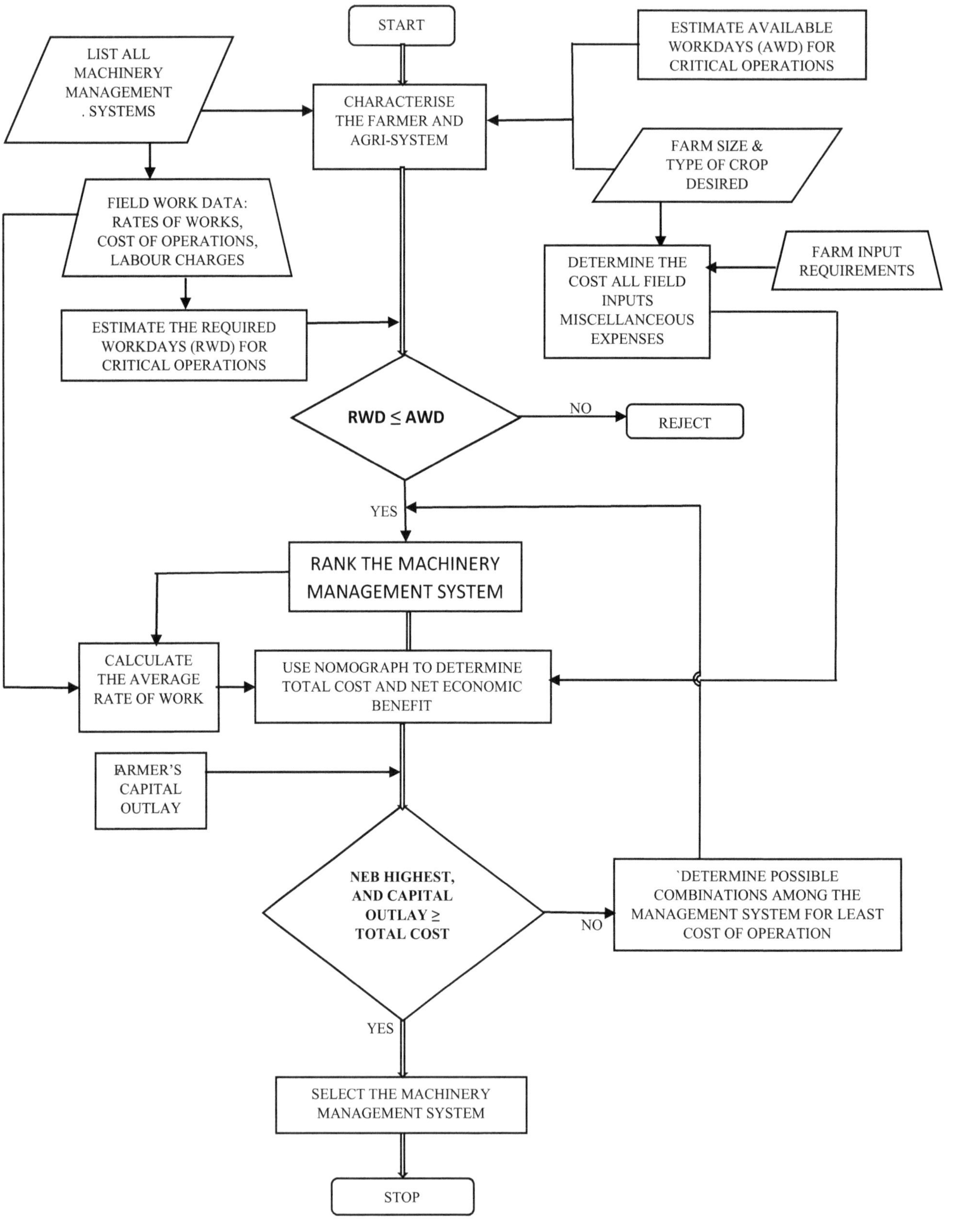

FIGURE 31.1 Flow chart of the nomograph model for education on selective and sustainable agricultural mechanisation management.

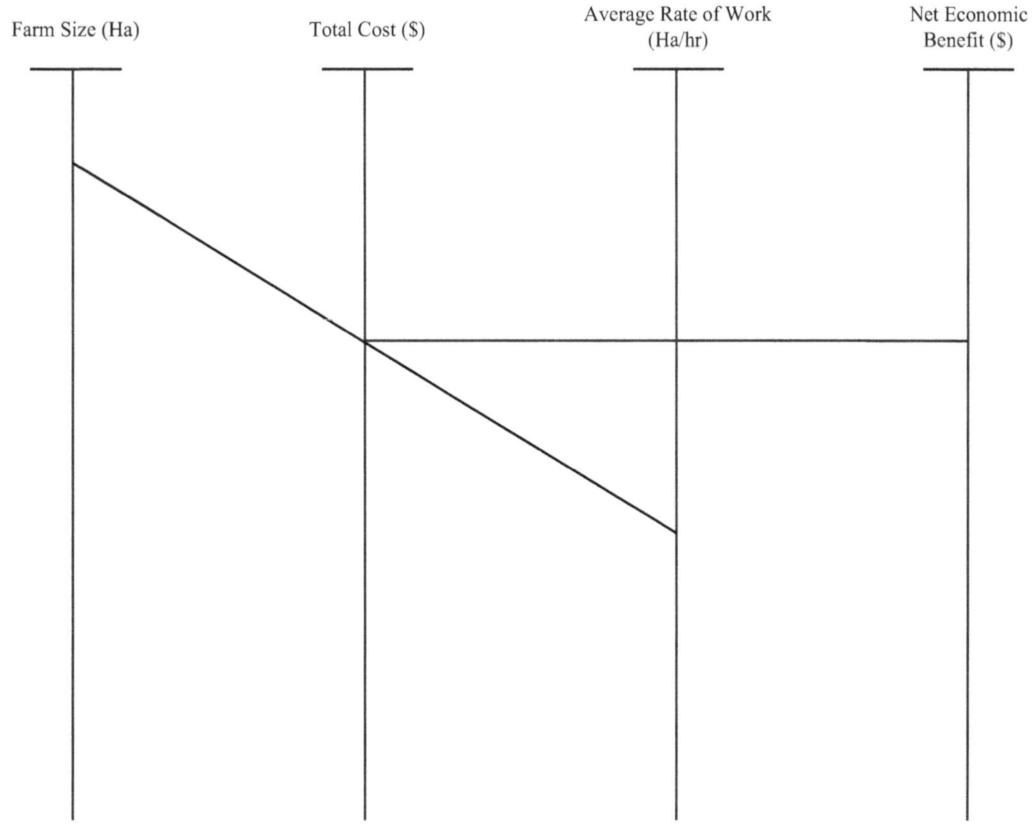

FIGURE 31.2 General format of a nomograph model. For each machinery set under an ownership and management system, the total cost of operations is determined by knowing the farm size and rate of work. Based on the total cost and market prices, the net economic benefit to the farmer is then determined.

Format of the Nomographs

Figure 31.3 shows the general format of a typical nomograph for a specific machinery-ownership and management system. In using these nomographs for the selection of machinery management systems for crop production, the rate of work is treated as a variable, since it can assume different values depending on the management system selected, which is represented by the capacity (work rate, ha/hr) of the machinery. The placement of the variables on the nomograph is predicated on their interdependence and the objective function of net economic benefit to the farmer while achieving timeliness of the operations. The interdependence of the variables are as follows:

a) Net economic benefit = *f*(total cost)
b) Farm size = *f*(rate of work)
c) Total cost = *f*(rate of work and farm size)

The nomographs were developed by plotting the real values into the corresponding columns on the general format displayed. Different scales were chosen for each decision variable. Cost-benefit analyses were carried out to produce each desired crop under each farm power and machinery-ownership and management system. The following cost-benefit appraisal technique was tested:

a) Break yield, breakeven price.
b) Benefit-cost ratio.
c) Financial efficiency.
d) Net economic benefit.

By varying the farm size, data were generated (based on the results of the cost-benefit analysis) for corresponding values of total cost, net cost, and net economic benefits.

Model Assumptions and Limitations

For the model developed to find great applicability, it is assumed that the machinery management system used in the analysis is available and provides adequate services to farmers in different localities. It is also assumed that the price of farm inputs used in developing the nomographs remains approximately stable and uniform over time and location. For a particular farming system and location, a major limitation in the model could be the non-availability of data relating to fieldwork and cost of operations in the mechanisation of certain field operations under the different machinery management systems. These include, for instance, information on crop harvesting using mechanical means such as corn-pickers or combines in rural areas where these are not available for testing. If the rate of work for such operations using mechanical means are incorporated into the model, the optimum farm size

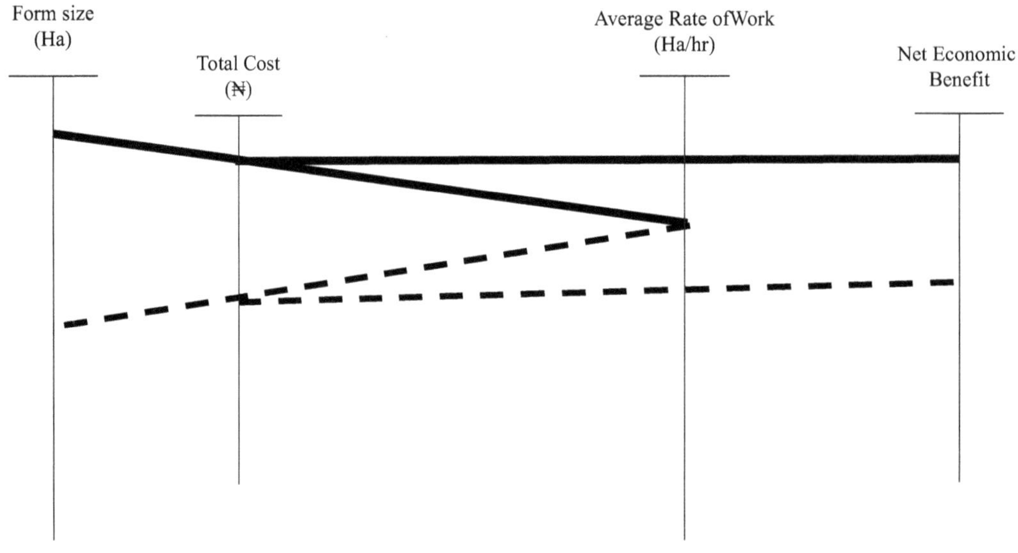

FIGURE 31.3 General format of a typical nomograph showing estimates of the total production cost and net economic benefit for a given rate of work (ha/hr) under two different farm sizes (bold and dashed lines, respectively). The placement of the variables on the nomograph is predicated on their interdependence and the objective function of the net economic benefit to the farmer while achieving timeliness of the operations.

TABLE 31.1

Summary of Results from the Application of the Nomograph Model in Maize and Cassava Production Under Different Mechanisation Systems in Southeast Nigeria.

Machinery-ownership and management system	Crop type	Required working days per hectare (ha)	Maximum farm size attainable, ha
Hand-powered smallholder traditional system (HPTS)	Maize	129	0.89
	Cassava	230	1.45
Tractor and equipment hire system (TEHS)	Maize	23	5.00
	Cassava	25	13.36
Sole-ownership and use system (SOUS)	Maize	23	5.00
	Cassava	7	47.71
Farmers' cooperative ownership and use system (COUS)	Maize	23	5.00
	Cassava	53	6.30

that can be cultivated using modern machinery will increase substantially, except for the hand-powered traditional system, where human muscle is the major and sometimes the only source of power.

Model Testing

The nomograph modelling approach was applied for cassava and maize production in the humid tropics using historical machinery performance data and the economics of crop production and prices from the southeast of Nigeria. The results summarised in Table 31.1 showed that for farm sizes $\geq$ 0.9 ha and 1.5 ha for maize and cassava, respectively, using the services of a tractor-implement system assured timely and profitable agribusiness for the farmer.

Conclusions and Future Prospects

The deployment of modern and cost-effective power and machinery remains a critical limiting factor towards sustainable mechanisation of agriculture and food systems, especially in developing countries. One of the common challenges faced by farmers is whether to own, lease, or buy, and this situation is even more challenging for smallholder farmers, who traditionally rely solely on human muscle and rudimentary implements such as the hand-hoe. Declining availability and high cost of youth labour in rural areas and the rapidly rising age of farmers calls for concerted efforts to introduce modern agricultural engineering technologies in agricultural business practice as well as to sustain them in educational curricula and research. This chapter has introduced a simple but effective nomograph modelling methodology for determining optimum power and machinery systems applicable to different farming systems. It also represents a useful tool for both education and research on sustainable agricultural mechanisation development and warrants incorporation in agricultural and biosystems engineering education programmes and related vocational courses. This is particularly relevant in developing economies, where smallholder farmers and other agribusinesses still dominate and where there is an urgent need to transform the agri-food systems using appropriate, modern, and productivity-enhancing technological innovations.

REFERENCES

Anazodo, U.G.N. (1975). Performance of Agricultural Machinery Under Tropical Conditions. Paper Presented at The National Irrigation and Agricultural Engineers' Seminar, Bagauda, Kano State, August 25–30.

Anazodo, U.G.N. (1980). Agricultural Mechanization as A Catalyst for Rural Development. Paper Presented at the National Seminar on Achieving Event Development in Nigeria. October 18th to 20th November, 1980 at Hotel Presidential Enugu. Organised by the Economic Development Institution (EDI), University of Nigeria, Enugu Campus.

Anazodo, U.G.N. (1982). Mechanization of Family Farm in Nigeria: A Quantitative Analysis of the Problem and Prospects. *AMA*, 13(3 Summer): 59–63.

Anazodo, U.G.N., Opara, L.U. & Abimbola, T.O. (1989). Agricultural Mechanization Study Report. Perspective Plan for Agricultural Development in Nigeria (1989–2004). Federal Agricultural Coordinating Unit (FACU), Ibadan, Nigeria, 118 pages.

Bowers, W. (1968). *Modern Concept of Farm Machinery Management*. Stipes Publishing Company, Champaign, IL.

Bukhari, S.B. (1982). Cost of Owning and Operating Tractors: A Review. *AMA* (Autumn): 61–67, 70.

Burrows, W.C. & Siemens, J.D. (1974). Determination of Optimum Machinery for Corn-Soya Bean Farm. *Transactions of the American Society of Agricultural Engineers*, 17: 1130–1135.

Chancellor, W.J. & Cervinka, V. (1974). Timeliness Coefficients for Rice and Factors Affecting Their Value. ASAE Paper No. 74-1021.

Daily Times. (1986). ₦. 6y Timeshttp://example.com?id=nLie Idle in Uzo-Uwani L. G. A. January 14.

Edwards, S.W. & Boehlje, C. (1980). Machinery Selection Considering Timeliness Losses. *Transactions of the ASAE*, 23: 810–815, 821.

Eradat, K.O. (1983). The Practical Assessment of Timeliness Penalties for Machinery Selection Purposes. *The Agricultural Engineer*, 38(4 Winter): 111–115.

Esmay, M.L. & Hall, C.W. (ed.) (1973). *Agricultural Mechanization in Developing Countries*. Tokyo: Shin-Norinsha Co., Ltd.

Giles, G.W. (1973). Selective Mechanization- An Aid to Increased Yields. ASAE Paper No. 73-512. ASAE St Joseph, Michigan 49085.

Harris, W.L., Bender, F.E. & Esmay, M.L. (1973). Agricultural Mechanization As Related Yields and Production. ASAE paper No. 73-513. ASAE St. Joseph, Michigan 49085.

Huan-Wen, G. & Hunt, D.R. (1985). Optimum Fleet Selection with Power-Based Models. *Transactions of the ASAE*, 28(2): 364–368.

Hunt, D.R. (1973). *Farm Power and Machinery Management*, 6th ed. Iowa State University Press, Iowa, IA, 324 pages.

Kepner, R.A., Bainer, R. & Barger, E.L. (1980). *Principles of Farm Machinery*, 3rd ed. AVI. Pub. Co., Inc., Westport, CT.

Khan, S., Chaudhry, M.S. & Sheriff, S.M. (1984). A Computer Programme of Agricultural Machinery Management: Selection of a Particular Tractor Size. *AMA*, 15(3): 11–13.

Kienzle, J., Mpagalile, J. & Houmy, K. (2024). Chapter 5: Enhancing the Role of Agricultural and Biological Engineering through Education: Perspective from FAO. In: Opara, U.L. (ed.). *Agricultural, Biosystems, and Biological Engineering Education – Global Perspectives and Current Practice*. CRC Press/Taylor & Francis Group.

Kyembeh, N.G. (1986). Mechanization of Small Farms in Tropical Africa. *AMA*, 17(3), 1–5.

Mayfield, W. (1981). A New Method Estimating Farm Machinery Costs. *Transactions of the ASAE*, 24(6): 1446–1448.

Odigboh, E.U. (1976). Engineering the Mechanization of Nigeria Agriculture: Sociological, Economical, and Technological Implications. In *Symposium Proceedings on Appropriate Approach for Accelerated Food Production*. DAE., University of Nigeria, Nsukka, September 10–11, pp. 120–130.

Opara, U.L. (2024). Chapter 36: CAM-SAM: A Computer-Aided Tool for Teaching and Research on Sustainable and Selective Agricultural Mechanization Development. In: Opara, U.L. (ed.). *Agricultural, Biosystems, and Biological Engineering Education – Global Perspectives and Current Practice*. CRC Press/Taylor & Francis Group.

Oram, P., Zapata, J., Alibaruho, G. & Roy, S. (1979). Investment and Input Requirement for Accelerating Food Production in Low-income Countries by 1990. IFPRI Research Report No. 10, September.

Renoll, E.S. (1981). Predicting Machine Field Capacity for Specific Field and Operating Conditions. ASAE Paper No.79-1029. ASAE St. Joseph, Michigan 49085.

Sims, B., Hilmi, M. and Kienzle, J. (2016). Agricultural Mechanization: A Key Input for Sub-Saharan African Smallholders. *Integrated Crop Management*, Vol. 23. Food and Agriculture Organization of the United Nations, Rome.

Taylor, A.J., Opara, U.L. & Smithers, J.C. (2021). Agricultural Engineering in Africa: A Key Driver for Transforming Agriculture to Deliver Food Security and to Support Economic Prosperity. Pretoria, Pan African Society for Agricultural Engineering (PASAE/AfroAgEng).

Sturrock, F.G. Cathie, J., & Payne, T.A. (1977). Economies of Scale in Farm Mechanization: A Study of Cost on Large and Small Farms. Occasional Paper No. 22. Agricultural Economics Unit, Department of Land Economy, Cambridge University.

Wilson, E.B. (1952). *An Introduction to Scientific Research*. McGraw Hill Book Company, Inc., London.

32

The Role of Inclusion, Diversity, Equity, and Access (IDEA) in Agricultural and Biological Engineering

Deepak Keshwani, Jennifer Keshwani and Marybeth Lima

Introduction

The goal of this chapter is to explore the importance of inclusion, diversity, equity, and accessibility (IDEA) in agricultural and biological engineering education and how IDEA principles can be integrated in our educational efforts. We want to acknowledge that the ideas and suggestions presented in this chapter are situated in a North American context. We are equal contributors to this chapter; thus, the order of authors is alphabetical.

What Is IDEA?

As a starting point, we would like to discuss definitions and context of the various words in the IDEA acronym. Lima and Keshwani (2022) in their definitions state the importance of inclusion in our current social context and note its importance for engineering because there are groups who have been on the margins whose needs have not been met in design practices. Zallio and Clarkson (2021), who reviewed IDEA in the design of the built environment, state: "When it comes to definitions, it is helpful to underline that accessibility is about designing and building a solution usable by as many people as possible. Diversity is about guaranteeing that everybody is recognised as a unique human being: it's about empowering people by respecting and appreciating what makes them different, in terms of age, gender, ethnicity, religion, disability, sexual orientation, education, and national origin. Equality is about ensuring everybody has an equal opportunity and is not treated differently or discriminated against because of their characteristics. Equity is about ensuring that everyone has access to the same opportunities. Inclusion concerns the opportunities that everyone should use the same facilities, take part in the same activities, and enjoy the same experiences, including people who have a disability or other disadvantage". Chugh (2018), in her insightful book entitled *The Person You Mean to Be: How Good People Fight Bias*, writes about how diversity is about gateways (getting people in), and how inclusion is about pathways (getting people through). Each of these definitions has useful dimensions in connection to engineering design and practice.

We were unable to find literature involving IDEA in agricultural and biological engineering (ABE), other than the recent piece in *Resource Magazine* (Lima & Keshwani, 2022). Though there is literature involving IDEA in engineering and engineering design (Clarkson et al., 2003; Smith-Doerr et al., 2017; *U.S. Access Board – U.S. Access Board – Home*, n.d.; Walden et al., 2018), there is a dearth of literature with respect to consideration of these factors simultaneously. Two areas of engineering and design that apply IDEA concepts include the digital environment and the built environment. With respect to the former, Microsoft has an extensive online resource with regard to the digital environment here: https://www.microsoft.com/design/inclusive/.

Regarding built environment, (Zallio & Clarkson, 2021) reviewed the use of IDEA in civil engineering and architectural design as related to assessment tools "to highlight the state of the art of assessment tools that consider inclusion, diversity, equity and accessibility (IDEA) with a people-first approach". They found that design work for the built environment focused more on accessibility than the other three areas; that the number of published articles centred on these considerations has increased, especially after 2015; and that holistic, mixed-method assessment tools have "the potential to guarantee accessible, equitable, diverse and inclusive spaces for everyone". These authors also state that "inclusion has the potential to become a form of sustainability as a proclamation about equity and diversity".

We believe that long-standing priorities in ABE, including systems thinking, sustainability, and a focus on people, make members of our discipline particularly well-suited to integrating IDEA into engineering education and practice.

We also recognise that there is an inherently personal dimension to this work, and that at times it can feel uncomfortable, perhaps from a lack of background (most of us have no formal education in these topics), because it feels incompatible or at odds with our training, because it can lead to difficult conversations and personal truths, or for other reasons. Because of this, we share parts of our personal stories with respect to IDEA in ABE in the hopes that readers will better understand and feel more comfortable "taking the plunge" in navigating this dimension of IDEA practice in engineering.

DOI: 10.1201/9780429150111-35

Deepak's story: *In my first year teaching a capstone design course, I developed a reflective assignment where students had to describe global, cultural, and social factors relevant to their design project and associated ethical issues that they should consider. If a factor was not relevant to their design project, students had to justify it. A student in the class approached me about the assignment and felt that their project didn't need to consider any global, cultural, or social factors since their team was designing a specific diagnostic process and tool for use by healthcare providers and their patients. After a rather lengthy discussion, we brainstormed how there could be a myriad of relevant factors including patient demographics, language, accessibility, etc., and how those could prompt some ethical issues. I also intentionally discussed the Accreditation Board for Engineering and Technology (ABET) outcomes related to design and ethics that highlight the importance of careful consideration of social aspects in design. One comment from the student stood out to me. He said that if it was so important, how come this is the first time they've been asked to do this in an engineering course. That comment was really convicting. It made me realise just how much our curriculum perpetuates the technological-social divide in engineering. I found myself wondering if my assignment was even useful as an assessment tool if we are not equipping students to engage with the social context.*

Jenny's story: *Monitoring teamwork in a college engineering classroom is not a passive task. One semester, as I wandered my classroom of sophomore engineering students, my attention fell on one team. Three students. Two agricultural engineering majors and one biological engineering major. Two male. One female. Two white. One Asian American. I was sad, but not surprised, to notice the one female, Asian American, biological engineering student was being completely shut out of the conversation by her two male, white, agricultural engineering major teammates.* Despite all my reminders in class that we are a community, we learn best from each other, we each have strengths and talents to offer, and including others is just the right thing to do; her thoughts, ideas, and abilities were ignored. *As any intentional professor would do, I asked the two white, male, agricultural engineering students to schedule a meeting with me in my office. Which they did. They politely came and listened to me explain the interactions I had observed and why this is unacceptable in our classroom. They explained they were just reserved and more comfortable talking with each other. I assured them I would be watching their team interactions in the future. I did not witness these two students ignore a teammate during the remaining few weeks of the semester. However, I'm not convinced their underlying motivations were impacted by the experience.*

Marybeth's story: *I've plunged increasingly further into IDEA concepts in design by working with communities (often elementary schools) to design playgrounds that are specifically geared toward the community and that place children at the centre of the design process* (Lima, 2013, 2014). *In this context, I was working with a school to replace a pre-kindergarten playground; their current playground consisted of a bucketball, a small geodesic dome climber, and a rusty red gate that my students and I nicknamed "the gate to nowhere"* (Figure *32.1 a). We were motivated to upgrade the playground because the 60 children who played on it had almost nothing to do. They could look through a chain link fence (topped with barbed wire) at a backyard playground that had a tree house, swings, slides, and climbers. We had been collaborating with the school for six months when I asked the principal about the story behind the gate: how did it come to be there? She told me that it had been the entrance to some old play equipment that had to be removed for safety reasons, but whoever had removed the equipment had left the gate behind. "It's the kids' favourite piece of equipment on the playground", she then said. When she saw my shock, she said, "During recess, the teachers line up the kids in front of that gate. They ask the children, 'Where do you want to go today, anywhere in the universe?' Once the child in line has thought of a location, they run through the gate, and yell the location as they do. They take turns and it's their favourite thing to do". What my students and I had nicknamed the gate to nowhere was actually the portal to anywhere.*

FIGURE 32.1 Playground that was the focus of Marybeth's community-based design project.

Even while paying attention to inclusive design and systems thinking, my students and I made so many mistakes in this situation. Our stakeholder analysis and teamwork efforts had not uncovered critical information about the role of the portal; we made bad assumptions that minimised the perspective of the community with whom we were collaborating. For me, this moment was a wakeup call, to remember to ask the right questions at the outset; expressing the soul of the community is critical in community-based design, and in this case, we almost missed it. While we eventually added a new playground *and kept the portal (*Figure *32.1b), the thorny boundary that separates the school from the surrounding community, and that reinforces the "keep out" message to the children, is still there. Designing with IDEA in mind is a journey and an evolution, and I am still actively learning and practising these principles today.*

Engineering Connections to IDEA

Engineering is inherently a service profession with an explicit commitment to improve the human condition for all people and the broader ecosystems with which people interact. Agricultural and biological engineers apply their engineering skills to solve problems related to food, energy, water, agriculture, and related systems. The work we do and the engineering innovations we design directly impact people every day. We need look no further than the preamble of the National Society of Professional Engineers code of ethics to remind ourselves of that commitment (*Code of Ethics | National Society of Professional Engineers*, n.d.). This commitment also shows up implicitly in ABET student outcomes that expect students to consider global, cultural, and social factors in their work as engineers (*Criteria for Accrediting Engineering Programs, 2022 – 2023 | ABET*, n.d.).

As educators, we are tasked with preparing future agricultural and biological engineers for this commitment to improve the human condition for *all people and the broader ecosystems with which people interact.* Embedding IDEA principles intentionally and strategically into our curriculum is one potential strategy. When we think about IDEA in an engineering context, we may primarily associate it with efforts to attract a diverse group to the profession, with the assumption that diverse people will bring diverse perspectives and doing so will help enhance our profession. After all, research indicates that increased diversity is positively correlated with enhanced decision making (Rock & Grant, 2016), financial outperformance (*How Diversity, Equity, and Inclusion (DE&I) Matter | McKinsey*, n.d.), and increased innovation (Phillips, 2014).

While we need to continue our efforts to increase diversity to enhance our profession, we contend that it is not enough. Increasing participation of engineers from diverse backgrounds places the burden of change on the underrepresented groups entering our profession. We believe that embedding IDEA concepts into the practice of engineering helps make this our collective responsibility and that requires intentional integration of IDEA into engineering curricula.

Previously, the authors introduced an inquiry-based framework that imbeds IDEA into the engineering design process (Lima & Keshwani, 2022). In this chapter, we will revisit that framework, explore the necessary knowledge, skills, and abilities that engineers need to develop to utilise the IDEA framework in practice, and discuss curricular implications and resources.

IDEA Framework for Design

A framework for IDEA-informed design was shared in the American Society of Agricultural and Biological Engineers (ASABE) publication, *Resource Magazine* (Lima & Keshwani, 2022). The framework describes specific considerations during the design process that should be addressed to ensure designs and the design process are inclusive, represent diverse perspectives, and are equitable and accessible to all. IDEA principles must be considered with respect to the people who participate in the design process and will be impacted by the people who participate in the design process, as well as the design process itself, the designed artefact, and within the broader context in which we design (see Figure 32.2). A salient feature of this framework is that it fosters an inquiry-based approach that intentionally bridges the technological and social aspects of practising engineering design.

The framework originated after discussions at the 2022 Annual International Meeting of ASABE at two sessions focused on IDEA. During both sessions, it became apparent that there was a gap between our engineering and engineering education practices, and the principles of IDEA. However, this also presented an opportunity for us to adopt a systems perspective when it comes to engineering design. This is a living framework that will and should continue to evolve and we welcome readers' feedback and suggestions to enhance it.

Essential Elements to Support the Framework

Preparing engineering students with the knowledge, skills, and abilities to create inclusive designs is crucial and requires understanding the diverse needs and perspectives of all users. By learning how to design with inclusivity in mind, we can create products and systems that are accessible and beneficial to a wider range of individuals. In this section, we will describe various avenues and strategies for equipping engineering students to implement the IDEA design principles.

Systems Thinking

Systems thinking skills provide the necessary tools to understand, create, and influence systems of interconnected sets of elements coherently organised to achieve something. (Meadows, 2008). Equipping ABE students with systems-thinking skills is not only vital for their professional goals, but also provides the awareness and skill to navigate complex IDEA challenges in their personal and professional lives. Developing systems thinking is crucial for the success of any engineering project. Systems thinking is the ability to understand complex interconnections between different components of a system and how they interact and affect one another. Through this ability, engineers can identify potential weaknesses and strengths of a system, allowing better decisions when designing and constructing a project. Systems thinking also helps engineers develop an understanding of how a system can be improved, and how to make it more efficient and cost effective. Systems thinking is an important skill for future engineers to possess, as it enables them to design and construct more efficient and reliable systems that meet the needs of their communities. The skills and abilities we impart on our students may feel small, but small things add up to create bigger, societal impacts, as reflected in a core principle of emergent strategy and broader systems thinking (Brown, 2017).

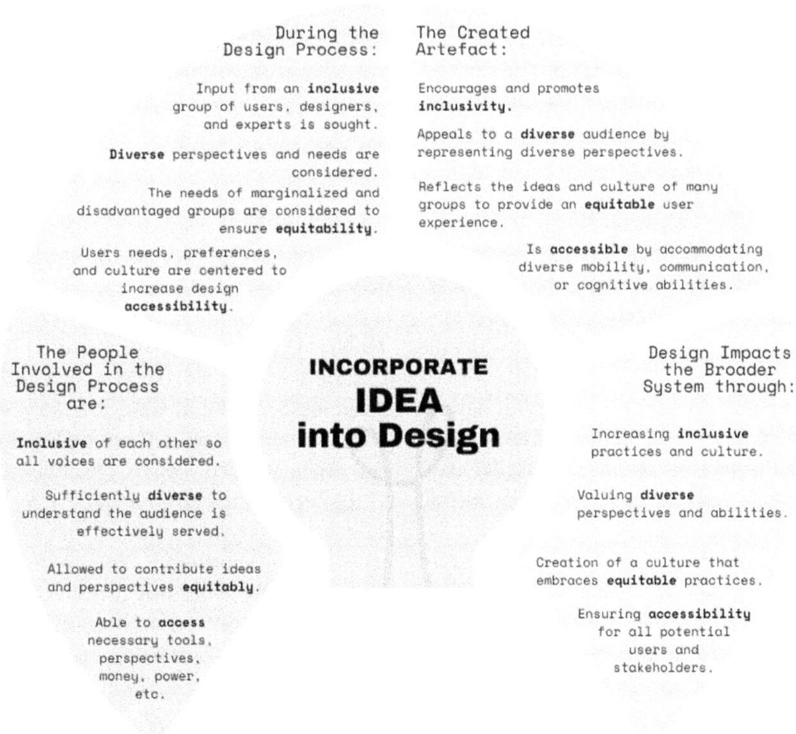

FIGURE 32.2 IDEA Framework for Design (Source: authors' compilation adapted from Lima and Keshwani (2022)).

ABE students and professionals are always working within systems. Corn fields, the human body, and the food production chain are all examples of complex systems ABE works within. ABE programmes provide knowledge of individual components of relevant biological systems through advanced courses in biology, agronomy, and engineering sciences to equip ABE students. However, to prepare students to address core problems instead of simply symptoms of those problems, future engineers must be empowered to view, analyse, and manipulate systems. Systems thinking provides the tools to assess current systems and identify leverage points to alter systems to achieve the desired outcomes. Waters Center for Systems Thinking has identified 14 Habits of a Systems Thinker (Waters Center for Systems Thinking, 2020).

The Habits of a Systems Thinker include aspects of conceptualising a system ("Makes meaningful connections within and between systems"), analysing systems ("Changes perspective to increase understanding"), and taking informed action ("Uses understanding of system to identify possible leverage actions"). Providing opportunities for students to develop these habits with respect to the ABE systems they will impact during their careers is vital. Systems-thinking habits also support student implementation of the four components of the IDEA framework for design, as detailed below.

System Inclusivity

An inclusive system will recognise that there are several groups in the margins whose needs may be overlooked during the design process. Using the systems-thinking habit of "change[ing] perspective to increase understanding" allows engineers to view a design from perspectives other than their own to give space for other voices to enter the design process (Waters Center for Systems Thinking, 2020). Ensuring design inclusivity requires identifying all participants in a system and then seeking feedback and perspectives from all participants.

Communication skills are a good first step to equip students to consider other's perspectives. The ability to interview stakeholders from diverse user groups and truly listen to responses enables engineers to consider many perspectives (Hess & Fila, 2016). Competence in changing perspectives also requires acknowledging personal mental models that frame individual experiences of the world. Spending time with students to work through the culture, beliefs, and attitudes they find "normal" is helpful preparation to provide space to consider alternative perspectives. Also, understanding the history of a system is helpful to identify how the current system function was designed and maintained. Acknowledging that our current cultural system is the result of all the cultures that came before us is a useful experience for students to learn to consider perspectives other than their own.

Value Diversity Within the System

As we highlighted in our personal stories, the habits and character our students develop serves as the foundation of their life-long practice as engineering professionals. The way our students treat their classmates impacts how our graduates value colleagues and supervisors. The perspectives our students consider in their senior design projects guide the ways they consider the user experience of their designs. Reflecting on

Jenny's story above, while the male students acted in response to Jenny's directions, they didn't think to enact a solution themselves – they needed to be prompted. While the students appeared to change the way they treated their classmates, Jenny was left with the thought that they complied out of external pressures (grades, respecting authority), rather than truly understanding the principle. The overall impact likely had not changed the safety of students of colour on my predominantly white campus (Suza, 2023). A system that values diversity will ensure that many perspectives participate in the design process and implementation. One relevant systems-thinking habit is considering "how mental models affect current reality and the future" (Waters Center for Systems Thinking, 2020). This is the ability to consider how the attitudes and beliefs of designers and users impact system function. Creating a diversity-sensitive design requires ABE professionals to acknowledge their personal mental models and seek to understand the mental models or those impacted by their designs.

Similarly, developing skills in perspective-taking equips students to consider how mental models affect current reality and the future. Starting with a baseline of understanding that my personal mental model may be different than others, this systems-thinking habit considers the impacts of different mental models on choices and design reception. Providing examples of designs that failed to fully consider the mental models of the user audiences can be helpful. For example, the development process for American Sign Language gloves (Erard, 2017) was overly dependent on the designer's mental models of how people who are deaf or hard of hearing would like to communicate with people who hear. Seeking to first understand user needs and attitudes toward the problem to be solved would have led the design team to a much different solution.

Improving System Equity

An equitable system will treat everyone fairly and without bias. A relevant systems-thinking habit is that ABE engineers should be able to "recognize that a system's structure generates its behaviour" (Waters Center for Systems Thinking, 2020). This awareness and ability to acknowledge that the systems we interact with daily have an intended function provides ABE engineers with opportunities to improve the equity of those systems. A historical understanding of society and how our culture developed helps equip students to recognise that a system's structure generates its behaviour. The systems we live in are not broken. They are behaving exactly as they were designed to function (Stroh, 2015). The problem is that the original designs were not equitable. While all engineering students participate in humanities courses (history, sociology) which provide a structure to assess the current structure of the systems within which we design, they receive little insight into how to use what they learn in humanities courses in an engineering context. It is incumbent upon engineering educators to make the connections between humanities and professional engineering explicit. This background structure empowers ABE professionals to create solutions that rise above the current limitations of inequitable systems. Additionally, critiquing examples of quick fixes that fail (Stroh, 2015) describes the short-term responses to mitigate systemic issues that lead to ongoing problems – sometimes even worsening problems – provides students skills in identifying when short-sighted solutions negatively impact system equity.

System Accessibility

An accessible system can be used, approached, or understood by everyone. A relevant systems-thinking habit for designing system accessibility is the ability to "consider short-term, long-term and unintended consequences of actions". Providing a short-term solution to a problem that does not consider the broader systemic issues may lead to limited accessibility to a design. For example, developing a cost-prohibitive solution to reducing agricultural water use may work in wealthier countries, but the design will not be feasible in communities with fewer resources to support agricultural technology.

Requiring students to include a discussion of consequences in design analysis is a good first step in equipping students to consider short-term, long-term, and unintended consequences of their designs (Cech, 2014; Lima & Oakes, 2014; Walsh et al., 2019). However, this method does not provide students with an opportunity to personally experience design consequences. Real-world, implementable design experiences through "service-learning" or community engagement exercises have potential to allow students to discover the consequences of their designs. Ensuring that the design experience does not end with the delivery of the design to the community, but instead continues through implementation and collection of user feedback, provides optimal learning opportunities for understanding the impact of short-term, long-term, and unintended consequences.

Engineering Ethics

In the introduction of the chapter, we alluded to the National Society of Professional Engineers Code of Ethics. In alignment with the existing literature base on engineering ethics education (Diduch et al., 2012; Hess et al., 2021; Rottmann & Reeve, 2020), we contend that there needs to be a stronger connection between engineering ethics and IDEA principles. Engineering ethics is one of the primary means to assess and regulate the impact of engineering decisions on the lives of people. Therefore, engineering ethics is crucial to realising IDEA principles.

Historically, both the practice of engineering ethics and the teaching of engineering ethics has used a Rules and Code approach based on contractual obligations. Given our tendency to prioritise technical knowledge over social considerations, this has led to a utilitarian approach to addressing ethical problems in engineering (Bowen, 2009). This tendency to prioritise technical knowledge over social considerations has been identified as one potential explanation for a historic decoupling of ethics and equity in engineering (Rottmann & Reeve, 2020). A contractual approach also implies that the motivation for ethical behaviour is based on social agreements that are enforced or managed by regulations (Bowen, 2009). This approach does then prompt a question about who was at the table when those social agreements and regulations were articulated (formally or informally). This consideration is particularly relevant to ABE, given the past homogeneity in the demographics of our profession and research that suggests that norms of dominant groups can restrict underrepresented voices and views, for example those of women engineers (Faulkner, 2000) or those who are not white (McGee & Martin, 2011). Currently, there is little research on the experiences of underrepresented voices in ABE. Two notable exceptions are studies examining the experiences of female faculty in ABE (Abadie et al., 2009; Cauble et al., 2000). As a profession, we are in the early stages of understanding our past and present as it relates to IDEA.

A common approach to teaching and professional development related to engineering ethics has been using case studies based on two dimensions: an inductive/deductive dimension and a micro/macro dimension, and often some combination of these (Rottmann & Reeve, 2020). An inductive approach will use a specific situation to identify ethical lessons. For example, what can we learn from examining a bridge collapse. A deductive approach establishes a specific moral framework or ethical concern, and then applies that to the analysis of an event, for example, application of care ethics to a specific medical technology, or asking how deciding to build a dam in a particular location leads to inequity in terms of benefits to various groups of people.

A micro approach typically involves an individual or group of engineers and their clients or employers, while a macro approach focuses on collective responsibility in a broader social and political context (Herkert, 2005). For example, a case study looking at the consequences of an employee not following standard machinery testing protocols would be a micro approach. An example of a macro approach would be examining long-term consequences of agricultural machinery automation on rural unemployment.

Case studies offer the opportunity to discuss ethical scenarios for practical situations that are often based on past events or likely-to-happen events. In an ABE context, some of the challenges include a lack of case studies that intentionally apply deductive strategies to situations and a lack of macro-level case studies. It is in the macro dimension where socio-political consequences and the broader impacts of engineering decisions can be explored. It is in these broader impacts that we often see systemic issues arising that impact IDEA. Case studies also need to balance negative behaviours with moral exemplars, and balance shock/disaster situations with more common everyday mundane situations.

From an implementation perspective, for case studies to be effective, they need to be accompanied by an ethical theory or framework that can be applied in a situation for the learning to be internalised and transferable to other situations. There is a range of theories that can be applied in an ABE context. Some examples include virtue ethics (Hooft, 2014), moral foundations theory (Graham et al., 2013), and ethics of care (Held, 2006).

Another barrier to connecting ethics to IDEA is the compartmentalisation of ethics within a specific-time point (i.e. senior design or a senior seminar course) and lack of disciplinary context if taught broadly for multiple majors. The limited research on engineering ethics in the context of ABE that has been done has already recommended as best practice integration across the curriculum and specificity to disciplinary contexts (Rottmann & Reeve, 2020).

A final consideration is curricular flexibility for non-technical course content. Ethical decision making in an engineering context that integrates IDEA principles requires our students to be comfortable discussing and evaluating social and cultural contexts. How do we equip students in this regard? One approach is to lean into the teaching expertise at our institutions from humanities, arts, and social sciences through required general education courses, and build on that initial exposure by intentionally integrating social and cultural contexts into our disciplinary courses throughout the curriculum. Doing so will normalise social considerations in engineering and will counter the over-reliance on the utilitarian approach to engineering ethics.

Stakeholder Analysis

Embedding IDEA principles into engineering practice and design in particular requires an intentional consideration of the impacts of engineering decisions and solutions on people. Stakeholder analysis is one strategy that can be used to centre our engineering practice around people. This approach is often used in project management to effectively assess project requirements and engage with individuals on resource management, decision making, and communication (PMI, 2013). Design-thinking also utilises a stakeholder analysis framework as a means to evaluate desirable product features and develop empathy for the end-users (Köppen & Meinel, 2015; Plank et al., 2021).

We define stakeholders as any individual or group of individuals involved, influenced, or impacted by the design process. Stakeholders would include individuals directly involved with the engineering design process, such as the engineers, employers, clients, but also those not directly involved, such as potential end-users of a product or process, suppliers of resources that might be needed, individuals or groups that might be in proximity (spatially or temporally) to an implemented solution, regulators, policy makers, etc. Embedding stakeholder analysis early in the engineering design process has merit, as it may help identify potential criteria and constraints beyond obvious functional attributes of a design. In the context of IDEA, stakeholder analysis can be used to identify those whose voices and perspectives are hidden or could be diminished in the design process.

We propose a four-step stakeholder analysis that can be summarised as follows:

- **Step 1:** Team-based brainstorming to identify the broadest possible set of stakeholders and clearly define their interest on the outcome from the engineering design process.
- **Step 2:** The team then categorises their interest on a qualitative scale as high, medium, or low.
- **Step 3:** For each stakeholder, the team then defines their influence over the process as high, medium, or low.
- **Step 4:** Create an interest-power chart (Figure 32.3) that maps out the various stakeholders based on their level of interest (on x-axis) and power (on y-axis).

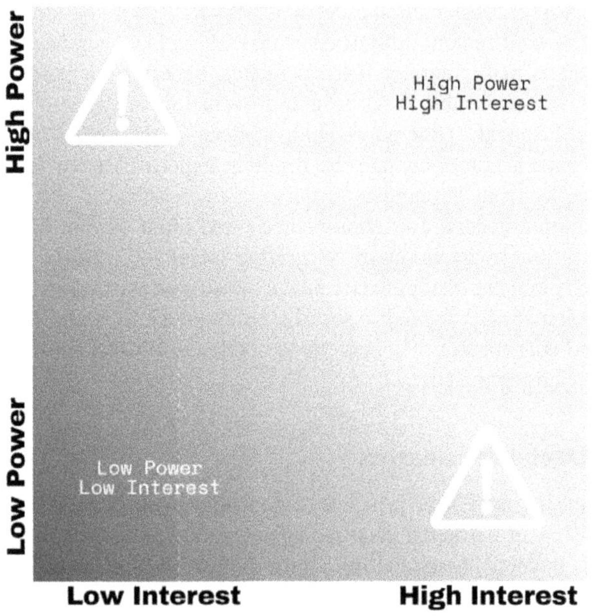

FIGURE 32.3 Power-interest chart to foster stakeholder analysis.

The stakeholders on the bottom-right part of the interest-power chart should be a primary consideration of the design team, as this is the group that is likely to be most disadvantaged from the outcome of the design process. The design team also needs to consider the impact of undue influence from stakeholders in the top left corner of the chart. These individuals and groups have significant power over the system but may not recognise the implications of their power due to their low interest in the system, i.e. the proverbial bull in a china shop. While the analysis can be carried out as an individual, a team-based approach with an external facilitator can minimise individual subconscious bias and groupthink.

Teamwork

Teamwork is widely recognised as a critical area in engineering, since much of engineering practice and design is conducted in a team setting; additionally, this aspect of engineering is explicitly mentioned in accreditation, specifically Student Outcome 5: an ability to function effectively on a team whose members together provide leadership, create a collaborative and inclusive environment, establish goals, plan tasks, and meet objectives.

In the classroom and in practice, teamwork involves more than putting people together and expecting them to function as a team. Being intentional about teamwork can enhance IDEA in engineering practice and design. Instruction on teams and teamwork is important. All team members should understand team dynamics and the best practices of teams (Katzenbach & Smith, 1993; Tuckman, 1965), as well as the ability to take on task roles (recorder, facilitator, gatekeeper, etc.) and behavioural roles (giving input, encouraging, evaluating, etc.). Providing teams with methods to resolve conflict, and a working knowledge of "problem characters" and strategies to avoid them are also important (Jalajas & Sutton, 1984). Finally, creating written documents such as team constitutions, at the outset of a collaboration, encourages teams to stick to the intentions and principles that they set and declare, and can provide clarity at difficult junctures.

We recognise that teamwork is a skill that employers desire in graduates and an essential learning skill, given the extent to which we expect students to work in teams. Therefore, it is crucial that we formalise the act of teaching and assessing teamwork in our curriculum. A good example of a research-based system that can be used for this purpose is the Comprehensive Assessment of Team Member Effectiveness (CATME), which provides detailed terminology about teaming, the ability to intentionally form teams, and fostering constructive peer-evaluation (Loughry et al., 2014).

Teaming fundamentals alone, while critical for the strong functioning of teams, are not in and of themselves sufficient to ensure that IDEA is practised in design. Toward this end, individuals can get a better sense of their implicit bias (https://implicit.harvard.edu/implicit/takeatest.html) and privilege (Chugh, 2018), and as a team, can become familiar with and use liberating structures (https://www.liberatingstructures.com/), which are methods intended to equitably involve everyone in teaming activities.

Role of Community

The community can (and we argue should) play a key role in engineering design. Historically, the design process has not been transparent to those outside of the process. There have been some efforts to engage the community (often defined as "the users") in the design process (Dieter & Schmidt, 2013), but these approaches tend more toward having potential users of a design surveyed for their opinions at the outset of the process, or testing prototypes of a design concept for feedback and subsequent iteration. Though these approaches increase the role of the community in the design, they still leave the design process itself solely to the engineers. In fact, some engineers have the attitude that they need to "dumb down" their elegant creations for the consumer.

A human-centred design approach (Lima & Oakes, 2014) appreciates and respects those who will use the design, but still falls short of the ideal, which is engaging people who will use the design throughout the design process. This approach doesn't mean that engineers need community members to understand the specifics of all design processes, any more than community members need engineers to understand everything about their expertise; the idea is to use our complementary knowledge and skills at each step of the process to co-create an artefact. Regardless of whether or not community members have an engineering background, they can co-design with creative ideas, making and examining the veracity of assumptions, and providing insights that engineers do not have. Bergeron et al. (2019) present a set of principles for community-based design.

Community-based participatory research (CBPR) approaches are well-defined in literature and practice, particularly in health fields, where researchers sought to achieve better health outcomes for patients (Brush et al., 2020; Minkler, 2005). The recognition that "one size does not fit all" and the importance of context and community drove adoption of these approaches. We propose to extend CBPR approaches to engineering, i.e. to practice community-based participatory design, which would, at best, involve the community equitably in terms of power sharing and decision making, and in collaboration across every step of the process (Israel et al., 2001). We have not seen this term previously in the literature, though we believe that there are ample examples of such approaches in the literature: see issues of *The International Journal for Service Learning in Engineering, Humanitarian Engineering, and Social Entrepreneurship* for examples generally. Some specific articles of interest include Leidig and Oakes (2021) and Wixom et al. (2022). We believe that this area of research is new and critical for future development.

Social Justice

The work done in social justice in engineering is a useful lens through which to incorporate IDEA into engineering (Leydens et al., 2022). Leydens and Lucena (2018) present six criteria for engineering for social justice (p. 21). These criteria are further detailed below.

- **Listening Contextually.** This means practising active listening in which hearing the words is only the first step. It is also understanding point of view, context, and perspective with an open mind and heart. Sometimes listening to what is NOT said is an important part of contextual listening. One quote that illustrates the pitfalls of not listening contextually is this: Engineers often walk into the community with the answers when they don't even know the questions.
- **Identifying Structural Conditions.** Identifying those systems, practices, or factors that limit the possibilities of some – for example, wealth inequality, racism, lack of accessibility to health care or clean air, etc. Designing with these conditions in mind is one way to practise engineering with an eye toward IDEA (and a more just world).
- **Acknowledging Political Agency/Mobilising Power.** Political agency and power are factors that are present in all engineering contexts. Engineers need to acknowledge these factors to realise imbedded assumptions and ways of operating that may keep others out (for example, historically underrepresented people being told, "You are not a real engineer"), or may drive engineering practice in particular directions that are antithetical to IDEA and may go unnoticed and thus unaddressed (technology taking precedence over people, profit over sustainability, short-term payoff over long-term wellness, etc.).
- **Increasing Opportunities and Resources.** Looking at ways that engineering can increase opportunities for people (though enhanced access to goods, services, or use of design) and enhance their resources (to better conserve the planet's resources).
- **Reducing Imposed Risks and Harms.** Ensuring that engineering designs have minimal risks for those who use them is a concept that engineers are already familiar with due to codes of ethics. This concept can be broadened by looking at reduced risks and harms through the IDEA lens. Harms are not just physical; they can also be emotional or psychological.
- **Enhancing Human Capabilities**. Leydens and Lucena cite ten capabilities to consider during design (p. 29) as follows:
- Life (of a normal length)
- Bodily health
- Bodily integrity (freedom from assault and the ability to move about freely, etc.)
- Senses, imagination, and thought (which are critical to being fully human)
- Emotions (love, grief, longing, gratitude, and more)
- Practical reason (for critical thinking, freedom of conscience, etc.)
- Affiliation (including protecting institutions that advance compassion and ensuring the social preconditions for self-respect and non-humiliation regardless of sex, ethnicity, sexual orientation, etc.)
- Other species (how we manifest respect for plants, animals, and nature in general)
- Play (recreation, laughter)
- Control over one's political and material environment

Resilience and Learning from Failure

Failure analysis is core to engineering sciences. Testing materials to failure is often the goal of engineering testing. When a material fails, the properties of the material, such as mechanical strength, can be determined. Similarly, in weightlifting, athletes test their muscle strength by lifting the maximum weight they can handle. This approach allows the weightlifters to measure their overall strength. Developing skills to design from an IDEA perspective will push students to test their strength in many ways. Providing examples of using failure as an opportunity to grow – both personally and professionally – may help students shift their perspectives of failure in other areas of life to provide opportunities to learn and improve. Failure in this sense develops resilience.

Developing resilience and the ability to learn from failure in future engineers is an important skill to cultivate so that students can effectively respond to the complex challenges of their field. Fear of failure, even in the context of engineering design, is present beginning in elementary students (Lottero-Perdue & Parry, 2017). As engineers work to solve problems and achieve goals, they must be able to recognise and identify potential obstacles to develop solutions or effect positive change. This skill requires an ability to think critically and persist in the face of adversity. Developing resistance is essential for engineers to successfully tackle difficult engineering problems and navigate complex situations. Additionally, developing resilience can help engineers be more creative and innovative in their work, as they must be able to push beyond their comfort zones and explore new and creative solutions. Resilience can also help engineers remain motivated and focused on the task at hand, even when faced with numerous challenges.

Approaching design from an IDEA perspective will likely be a new experience for our students (and ourselves), which may result in failed attempts and attempting to move forward on wobbly footing. Failure may be a new and uncomfortable experience for engineering students, who often experience academic success. Supporting students in their experience of failure through resilience is vital to preparing IDEA-equipped engineers. The core concept of Dolly Chugh's book, *The Person You Mean to Be* (Chugh, 2018), is that striving toward IDEA excellence requires moving from being a "good person" to constantly striving to be a better person, with the idea that there is no such thing as perfection. This growth mindset develops humility and a willingness to consider other perspectives and life experiences.

Where Do We Go from Here?

Positionality Statements

We started this chapter with stories from our experiences as engineers and educators. We choose to include personal stories to illuminate the inherently personal component of an IDEA approach and to provide tangible examples of how we see IDEA principles play out in our professional lives. These experiences undoubtedly impact how we engage with this topic and our students. To this end, we provide our individual positionality statements (Secules et al., 2021) that provide a glimpse into our identities and experiences that shape our perspectives related to IDEA. Each of us are underrepresented in the engineering profession, and as such, negotiating issues around IDEA in engineering has been and continues to be a reality of our existence in this discipline, as well as an ideal that we endeavour to enhance. We believe awareness of our positionalities is a critical part of moving forward with work on integrating IDEA principles into engineering. We invite you to join us in this self-reflection.

Dr Deepak Keshwani: *I am a tenured Associate Professor and Director of Undergraduate Programmes at the University of Nebraska-Lincoln, an institution with a predominantly white student and faculty body. As a grandchild of refugees, and being an immigrant of colour myself, I identify as an outsider in the North American Higher Educational system while at the same time benefiting from my privilege as a cis-gendered male in a STEM field. I approach my teaching, scholarship, and service related through these salient identities. My motivation for IDEA-related scholarly endeavours stems from a concern about the impact of hyper-polarisation on the education and professional growth of students. This concern stems from my teaching and advising responsibilities that have provided me with extensive interactions with engineering and technology students. These interactions have led me to being involved in student success initiatives that connect to IDEA. For instance, I am currently involved in an institutional-level effort to scale up an ecology of validation model to support the academic success of at-promise students (first-generation, minority, low socio-economic status etc.). Being involved in this effort has forced me to confront my own biases and challenge assumptions I make about my students. My personal call to action is to examine my own pedagogical practices from an IDEA perspective and reimagine engineering curriculum to balance the technological and social aspects of our profession.*

Dr Jenny Keshwani: *I am a tenured Associate Professor and Science Literacy Specialist through Nebraska Extension. I grew up on a sugar beet farm in North Dakota. I studied biological systems engineering and biomedical engineering to eventually earn an interdisciplinary PhD in engineering and oral biology. For the past ten years, I have served as a faculty member in a Midwest R1 Engineering Department with a large extension appointment focused on K12 science and engineering education. I was a first-generation college student from out of state. I was a diligent student but was unprepared to study at a college level and lacked the confidence to ask questions or visit my professors during office hours. I was a solid B student. As a result, I don't value perfect grades and often tell my students that grades don't really reflect their learning or abilities. I encourage my students to push themselves beyond their comfort zone without considering impacts on their GPA and often remind my classes that learning is the goal – not accumulating points. I find identity in my strength in empathy and drive to be an "includer", which led me to build community within my classes and colleagues. My "includer-ness" bolsters my interest in IDEA work. As a result, I lead the Cultivate ACCESS (Agriculture Career Communities to Empower Students in STEM) virtual mentoring programme to empower high school students historically excluded from STEM-related agricultural careers. I also intentionally create community within my classes, teams, and colleagues through team-based activities and by focusing on the whole person during interactions. My work as a science literacy specialist allows me to broaden the participation of educators and students in STEM areas by making science and engineering accessible and meaningful through cultural awareness and personally relevant problem solving. I regularly meet with a community of "Paradigm Shifters" representing various aspects of the education and agroecosystem sectors to practise systems-thinking skills and identify ways to have a positive impact through the small and large aspects of our influence and privilege. I'm thankful for our conversations and how we challenge each other to continue on the journey of becoming better versions of ourselves.*

Dr Marybeth Lima: *I am a tenured professor in biological and agricultural engineering at Louisiana State University, and for the past three years, I have served as Department Chair. I became interested in the Chair position as the result of volunteer experiences I had during COVID. I sought to combat my feeling of uselessness during the summer of 2020 by volunteering at an elementary school I had previously collaborated with on a playground project. The school was serving as a food distribution centre for area children in need, and I was tasked with passing out milk each Wednesday morning during June and July. I became friendly with other volunteers, including members of an LSU men's sports team, who were volunteering to give boxes of non-perishables that were donated by a local company. In this capacity, I had a series of experiences in which (1) I did not speak up at a critical juncture when something* *culturally inappropriate was said (I returned the next week to talk to this person and he had been transferred to another location. I never saw him again); (2) I spoke up at a subsequent critical juncture and, in so doing, opened up a space in which other volunteers were supported; (3) I witnessed the sports team volunteers giving too much food to people in line – they ignored the asked-for amounts and overloaded people (against their will) in an effort to finish early – I confronted the student in charge of the group (he was in charge only for this week, as the company employee who had supervised these students each week was on vacation), who happened to be an engineering student. He told me it was none of my business, and that I couldn't control what he and his group did. They continued piling on food, over the loud complaining of the people in line. The group finished two hours early and people in line after that had no food, only the milk my station was distributing; and (4) being exposed to COVID in the ensuing week, and being unable to return to the site to tell the person who had been on vacation what had happened while he was gone. Through this series of experiences, I was reminded of the critical role of IDEA in engineering and in community, and I sought a leadership role in the hopes that, at least in my department, no engineering student would prioritise their interest in finishing early over children going hungry. I have established a committee on diversity, equity, and inclusion, comprised of all departmental constituents, but mostly students; together, we are working toward creating an environment where all feel welcomed, valued, respected, and engaged.*

Call to Action

In this chapter, we described IDEA's role in ABE and highlighted why these skills are important for future engineers to develop. We also provided strategies to implement in engineering education programmes that can equip students with the ability to consider the diverse needs of all users and stakeholders in the design process. As educators in ABE programmes, it is our responsibility to create opportunities for ourselves and our students to practise IDEA principles. We must foster a culture within our departments and university communities that values IDEA-grounded practices. Our students learn from the educational experiences we provide in the classroom and from their extracurricular activities and observations. How we interact with our colleagues and treat others, the awards and accomplishments we celebrate, and the leaders we choose and follow all shape their understanding of what it means to be a professional engineer. Therefore, it is crucial that we model and promote inclusive behaviour and practices in all these areas. There are several actions we can take collectively and individually to make progress toward imbedding IDEA into the field of agricultural and biological engineering. Here are some of our recommendations for next steps.

Actions for the Profession

- A concerted effort by our professional society is needed to develop case studies and other curricular supports focused on IDEA.
- Increased participation in/with organisations such as the American Society for Engineering Education, where critical dialogue and scholarship related to IDEA in an engineering context is occurring.
- Examination of systemic barriers to the participation and success of underrepresented groups in our profession, and execution of subsequent plans to eliminate these barriers.
- Establish an IDEA-focused student design competition through a professional society, such as ASABE.
- Break down the silos that divide technical content areas. We do our professionals, students, and stakeholders a disservice by suggesting that the decisions made in one sector do not have systemic impacts.
- Break down the wall between social and technical considerations. Engineering is a sociotechnical practice (Cech, 2014); we should explicate the role of humanities in engineering, as we do with math, science, and engineering fundamentals.

Actions for Academic Programmes

- Integrate engineering ethics across the curriculum using a formalised theory or framework to ensure repeated exposure.
- Integrate stakeholder analysis in the curriculum for engineering design classes.
- Increase student awareness of concepts such unconscious bias and groupthink.
- Incorporate activities to develop and practise systems-thinking habits in coursework.
- Integrate principles of humanities and social sciences into core engineering courses. Model to students that a variety of skills are essential to become an ABE professional – not just science and engineering knowledge.
- Include activities and opportunities across the curriculum that require students to use IDEA principles and practices.

Actions for Individuals

- Participate in professional development to gain knowledge of ethical theories and frameworks and their application to engineering situations, and the ability to integrate social and cultural contexts into disciplinary courses.
- Pursue IDEA-related self-development through assessments such as the Intercultural Development Inventory, implicit association tests, etc.
- Practise systems-thinking habits in daily interactions and projects. Read systems thinking books and discuss with a small group of colleagues. A couple of our favourites are cited in the systems thinking section.

We believe that engineering can be a force for good in the world, as we simultaneously agree with Donna Riley's concerns: "We accept too readily the facile self-aggrandising pronouncements of members of the profession that *engineers help society.* To truly answer the question of what engineering has to do with justice, we must also be willing to examine closely and carefully what engineering has to do with injustice" (Leydens & Lucena, 2018, p. xvii).

In order for engineering to be a force toward a more just world, we believe that it is critical to centre IDEA principles in the practice of engineering, including the people (engineers and non-engineers), the designed artefact, the design process, and the larger systems of which we are a part. This work isn't easy, but it's important and we believe it must be done.

REFERENCE LIST

PMI. (2013). *A Guide to the Project Management Body of Knowledge (PMBOK Guide)* (5th ed.). (2013) Project Management Institute: Pennsylvia..

Abadie, A., Christy, A., Jones, J., Wang, J., & Lima, M. (2009). Longitudinal Survey of Female Faculty in Biological and Agricultural Engineering. *Transactions of the ASABE, 52*(4), 1397–1405. https://doi.org/10.13031/2013.27763

Bergeron, C. A., Hargrove, A., Tramontana, B., Steyer, J., Emily, A., Davison, D., Deras, A., Flynn, A., Porte, K. G. L., Garcia, B., Gonzalez, G., Lilly, L., Patterson, E., Vallery, S., Williams, A., Zerkus, J., Ruston, M., & Lima, M. (2019). Design for the Soul of the Community: Using Community-based Design Principles. *International Journal for Service Learning in Engineering, Humanitarian Engineering and Social Entrepreneurship, 14*(1), Article 1. https://doi.org/10.24908/ijsle.v14i1.12648

Bowen, W. R. (2009). *Engineering Ethics Outline of an Aspirational Approach* (1st ed.). Springer London. https://doi.org/10.1007/978-1-84882-224-5

Brown, A. M. (2017). *Emergent Strategy: Shaping Change, Changing Worlds.* AK Press. https://www.are.na/block/14640659

Brush, B. L., Mentz, G., Jensen, M., Jacobs, B., Saylor, K. M., Rowe, Z., Israel, B. A., & Lachance, L. (2020). Success in Long-Standing Community-Based Participatory Research (CBPR) Partnerships: A Scoping Literature Review. *Health Education & Behavior: The Official Publication of the Society for Public Health Education, 47*(4), 556–568. https://doi.org/10.1177/1090198119882989

Cauble, S. M., Christy, A., & Lima, M. (2000). Towards Plugging the Leaky Pipeline: Biological and Agricultural Engineering Faculty in The United States And Canada. *Journal of Women and Minorities in Science and Engineering, 6*(3), 229–249. DOI: 10.1615/JWomenMinorScienEng.v6.i3.40

Cech, E. A. (2014). Education: Embed Social Awareness In Science Curricula. *Nature, 505*(7484), Article 7484. https://doi.org/10.1038/505477a

Chugh, D. (2018). *The Person You Mean to Be.* Blackstone Audio.

Clarkson, J., Keates, S., Coleman, R., & Lebbon, C. (Eds.). (2003). *Inclusive Design.* Springer. https://doi.org/10.1007/978-1-4471-0001-0

Code of Ethics | National Society of Professional Engineers. (n.d.). Retrieved December 22, 2022, from https://www.nspe.org/resources/ethics/code-ethics

Criteria for Accrediting Engineering Programs, 2022–2023 | ABET. (n.d.). Retrieved December 22, 2022, from https://www.abet.org/accreditation/accreditation-criteria/criteria-for-accrediting-engineering-programs-2022-2023/#GC2

Diduch, C., Macisaac, D., Haralampides, K., & Wilson, B. (2012). Engineering and Social Justice through an Accreditation Lens: Expectations and Learning Opportunities for Ethics and Equity. *Proceedings of the Canadian Engineering Education Association (CEEA).* https://doi.org/10.24908/pceea.v0i0.4658

Dieter, G. E., & Schmidt, L. C. (2013). *Engineering Design* (5th ed.). McGraw Hill.

Erard, M. (2017, November 9). Why Sign-Language Gloves Don't Help Deaf People. *The Atlantic.* https://www.theatlantic.com/technology/archive/2017/11/why-sign-language-gloves-dont-help-deaf-people/545441/

Faulkner, W. (2000). Dualisms, Hierarchies and Gender in Engineering. *Social Studies of Science, 30*(5), 759–792.

Graham, J., Haidt, J., Koleva, S., Motyl, M., Iyer, R., Wojcik, S. P., & Ditto, P. H. (2013). Chapter Two - Moral Foundations Theory: The Pragmatic Validity of Moral Pluralism. In P. Devine & A. Plant (Eds.), *Advances in Experimental Social Psychology* (Vol. 47, pp. 55–130). Academic Press. https://doi.org/10.1016/B978-0-12-407236-7.00002-4

Held, V. (2006). *The Ethics of Care: Personal, Political, and Global.* Oxford University Press.

Herkert, J. R. (2005). Ways of Thinking About and Teaching Ethical Problem Solving: Microethics and Macroethics in Engineering. *Science & Engineering Ethics, 11*(3), 373–385. Academic Search Premier.

Hess, J. L., & Fila, N. D. (2016). *The Development and Growth of Empathy Among Engineering Students.* American Society for Engineering Education.

Hess, J. L., Whitehead, A., Jesiek, B. K., Katz, A., & Riley, D. (2021). WIP: Intersections Between Diversity, Equity, and Inclusion (DEI) and Ethics in Engineering. *2021 IEEE Frontiers in Education Conference (FIE),* 1–5. https://doi.org/10.1109/FIE49875.2021.9637059

Hooft, S. van. (2014). *Understanding Virtue Ethics.* Routledge.

How Diversity, Equity, and Inclusion (DE&I) Matter | McKinsey. (n.d.). Retrieved December 22, 2022, from https://www.mckinsey.com/featured-insights/diversity-and-inclusion/diversity-wins-how-inclusion-matters

Israel, B., Schulz, A., Parker, E., & Becker, A. (2001). Community-Based Participatory Research: Policy Recommendations for Promoting a Partnership Approach in Health Research. *Educ Health, 14,* 182–197.

Jalajas, D. S., & Sutton, R. I. (1984). Feuds in Student Groups: Coping with Whiners, Martyrs, Saboteurs, Bullies, and Deadbeats. *Organizational Behavior Teaching Review, 9*(4), 94–102. https://doi.org/10.1177/105256298400900413

Katzenbach, J. R., & Smith, D. K. (1993). The Rules for Managing Cross-Functional Reengineering Teams. *Planning Review, 21*(2), 12–13. https://doi.org/10.1108/eb054404

Köppen, E., & Meinel, C. (2015). Empathy via Design Thinking: Creation of Sense and Knowledge. In H. Plattner, C. Meinel, & L. Leifer (Eds.), *Design Thinking Research: Building Innovators* (pp. 15–28). Springer International Publishing. https://doi.org/10.1007/978-3-319-06823-7_2

Leidig, P., & Oakes, W. (2021). Model for Project-Based Community Engagement. *International Journal for Service Learning in Engineering, Humanitarian Engineering and Social Entrepreneurship, 16*(2), Article 2. https://doi.org/10.24908/ijsle.v16i2.14809

Leydens, J. A., & Lucena, J. C. (2018). *Engineering Justice: Transforming Engineering Education and Practice*. John Wiley & Sons.

Leydens, J. A., Lucena, J. C., & Riley, D. M. (2022, September 15). *Engineering Education and Social Justice*. Oxford Research Encyclopedia of Education. https://doi.org/10.1093/acrefore/9780190264093.013.1772

Lima, M. (2013). *Building Playgrounds, Engaging Communities: Creating Safe and Happy Places for Children*. LSU Press.

Lima, M. (2014). The LSU Community Playground Project: Reflections on 16 Years of an Engineering Service-Learning Program. *International Journal for Service Learning in Engineering, Humanitarian Engineering and Social Entrepreneurship*, 492–508. https://doi.org/10.24908/ijsle.v0i0.5565

Lima, M., & Keshwani, D. (2022, November). Incorporating IDEA into Engineering Design. *Resource Magazine*, 10–13.

Lima, M., & Oakes, W. (2014). *Service-Learning: Engineering in Your Community* (2nd ed.). Oxford University Press.

Lottero-Perdue, P., & Parry, E. (2017). Elementary Teachers' Reflections on Design Failures and Use of Fail Words after Teaching Engineering for Two Years. *Journal of Pre-College Engineering Education Research (J-PEER)*, *7*(1). https://doi.org/10.7771/2157-9288.1160

Loughry, M. L., Ohland, M. W., & Woehr, D. J. (2014). Assessing Teamwork Skills for Assurance of Learning Using CATME Team Tools. *Journal of Marketing Education*, *36*(1), 5–19. https://doi.org/10.1177/0273475313499023

McGee, E. O., & Martin, D. B. (2011). "You Would Not Believe What I Have to Go Through to Prove My Intellectual Value!" Stereotype Management Among Academically Successful Black Mathematics and Engineering Students. *American Educational Research Journal*, *48*(6), 1347–1389. https://doi.org/10.3102/0002831211423972

Meadows, D. H. (2008). *Thinking in Systems: A Primer*. Chelsea Green Publishing.

Minkler, M. (2005). Community-Based Research Partnerships: Challenges and Opportunities. *Journal of Urban Health: Bulletin of the New York Academy of Medicine*, *82*(Suppl 2), ii3–ii12. https://doi.org/10.1093/jurban/jti034

Phillips, K. W. (2014). How Diversity Works. *Scientific American*, *311*(4), 42–47. https://doi.org/10.1038/scientificamerican1014-42

Plank, I. S., von Thienen, J. P. A., & Meinel, C. (2021). The Neuroscience of Empathy: Research-Overview and Implications for Human-Centred Design. In C. Meinel & L. Leifer (Eds.), *Design Thinking Research* (pp. 89–124). Springer International Publishing. https://doi.org/10.1007/978-3-030-76324-4_6

Rock, D., & Grant, H. (2016, November 4). Why Diverse Teams Are Smarter. *Harvard Business Review*. https://hbr.org/2016/11/why-diverse-teams-are-smarter

Rottmann, C., & Reeve, D. (2020). Equity as Rebar: Bridging the Micro/Macro Divide in Engineering Ethics Education. *Canadian Journal of Science, Mathematics & Technology Education*, *20*(1), 146–165. Academic Search Premier.

Secules, S., McCall, C., Mejia, J. A., Beebe, C., Masters, A. S., L. Sánchez-Peña, & Svyantek, M. (2021). Positionality Practices and Dimensions of Impact on Equity Research: A Collaborative Inquiry and Call to the Community. *Journal of Engineering Education*, *110*(1), 19–43. https://doi.org/10.1002/jee.20377

Smith-Doerr, L., Alegria, S. N., & Sacco, T. (2017). How Diversity Matters in the US Science and Engineering Workforce: A Critical Review Considering Integration in Teams, Fields, and Organizational Contexts. *Engaging Science, Technology, and Society*, *3*, 139–153. https://doi.org/10.17351/ests2017.142

Stroh, D. P. (2015). *Systems Thinking For Social Change: A Practical Guide to Solving Complex Problems, Avoiding Unintended Consequences, and Achieving Lasting Results*. Chelsea Green Publishing.

Suza, W. P. (2023). Make Predominantly White Campuses Safe for Black Students. *Nature Human Behaviour*, 1–2. https://doi.org/10.1038/s41562-023-01525-9

Tuckman, B. W. (1965). Developmental Sequence in Small Groups. *Psychological Bulletin*, *63*, 384–399. https://doi.org/10.1037/h0022100

U.S. Access Board—U.S. Access Board—Home. (n.d.). Retrieved March 12, 2023, from https://www.access-board.gov/

Walden, S. E., Trytten, D. A., & Shehab, R. L. (2018). Research-Based Recommendations for Creating an inclusive Culture for Diversity and Equity in Engineering Education. *2018 IEEE Global Engineering Education Conference (EDUCON)*, 1591–1597. https://doi.org/10.1109/EDUCON.2018.8363425

Walsh, H., Dong, A., & Tumer, I. (2019). Towards a Theory for Unintended Consequences in Engineering Design. *Proceedings of the Design Society: International Conference on Engineering Design*, *1*(1): 3411–3420. doi: 10.1017/dsi.2019.348.

Waters Center for Systems Thinking. (2020). *Habits of a Systems Thinker*. https://ttsfilestore.blob.core.windows.net/ttsfiles/habits-single-page-2020.pdf#page=1&view=FitH

Wixom, J. H., Dahlin, E., Child, C., & Mattson, C. A. (2022). Logics of Collaboration: An Ethnography of Engineering Co-design in the Brazilian Amazon. *International Journal for Service Learning in Engineering, Humanitarian Engineering and Social Entrepreneurship*, *17*(2), Article 2. https://doi.org/10.24908/ijsle.v17i2.15709

Zallio, M., & Clarkson, P. J. (2021). On Inclusion, Diversity, Equity, and Accessibility in Civil Engineering and Architectural Design: A Review of Assessment Tools. *Proceedings of the Design Society*, *1*, 2297–2306. https://doi.org/10.1017/pds.2021.491

33

AgTecCollection – Repository Approach to Rescue Analogue Agricultural Engineering Knowledge

Hermann Auernhammer

Agricultural Engineering in the Past

From the development of the first tools at the beginning of field cultivation to autonomous field robotics in today's yield-oriented land management, agricultural engineering has achieved and accomplished tremendous outcomes. Research created the basics for design and manufacturing. Handicraft and industry produced the necessary equipment and machinery. Education, training, and advice enabled the best possible application in practice.

All of this was presented, documented, and disseminated in images, documents, drawings, films, and videos using the technology available at the time. But, until the end of the last millennium, all of this highly valuable material was created almost exclusively in analogue format. To ensure that this documented and indispensable knowledge can also be used by everyone in the digital world of today and tomorrow, its digitisation is essential.

The Long Way

Most of the analogue material mentioned above is found in scientific institutions around the world. There, it is mostly preserved and archived in very good quality, but it increasingly falls into oblivion. In addition, there are unpredictable changes at the respective institutions. Work areas are closed, the responsible persons leave, or spatial changes lead to disposal measures at short notice. Very often, everything available is then disposed of in the trash and thus irrevocably lost forever.

Such a situation arose at the Chair of Agricultural System Technology at the Technical University of Munich (TUM) in 2000. The transition from overhead transparencies to PowerPoint for presentations in lectures and exercises had become necessary. In addition, the department was facing a reorganisation. A possible closure of the Chair was also discussed. And, in addition, there were considerations to put the Chair building, which was built in 1955, to another use or to demolish it. What then to do with all the scientific documents accumulated in the building?

Preliminary Work

Suddenly, a decision that could not be postponed was called for. Finally, it was no longer a matter of just waiting, but of saving what could be saved. Therefore, it was decided to start digital archiving as soon as possible and, due to the need for teaching, to implement a first pilot project with images.

With a single-slide scanner from the time of use for the integration of slides in overhead transparencies, 1,000 slides were digitised. These were then inserted into an Excel spreadsheet with 1,000 rows. Additional columns were used to enter available metadata for each digitised object, including a distinctive identifier, author(s), year, keyword, and available references. The indexing was done without restriction per object and a free choice of keywords.

However, it quickly became apparent that this made a targeted search for keywords impossible because, for example, a tractor had up to five different country-specific and locally used designations (tractor, tower, trekker, towing machine, bulldog). Thus, this solution was unsatisfactory, especially since the amount of work for the single image digitisation was much too high. However, this action yielded 1,000 digitised images with comprehensive metadata, because all the information present on the slide frames had also been documented in the notes.

System Development

Independently of this project, the TUM library had submitted a parallel research proposal to the German Research Foundation for the development and setup of a central media server "mediaTUM®" for the university (Leiss et al., 2010). This was financially supported from 2003 onwards and thus the opportunity for a very fruitful collaboration arose. On the part of the Chair of Agricultural Systems Technology, the experience in dealing with the digitisation of slides, the acquisition and preparation of metadata, and the meaningful design of keywords could be brought into the research project. At the same time, the Chair was able to use the media server to be created and thus realise the development and construction of an agricultural technology archive "AgTecCollection" in three further time periods (Table 33.1).

DOI: 10.1201/9780429150111-36

TABLE 33.1

System Development, Digitisation, and Publication in Continuous Cooperation Between the Library and the Chair of Agricultural Systems Technology at the Technical University of Munich.

Period	Library of the Technical University of Munich	Chair of Agricultural Systems Technology (AST)
2000–2003	Project definition and application to German Research Foundation	Digitisation of 1,000 slides, EXCEL indexing
2003–2006	Development of mediaTUM test system with digitised AST slides	Slide preparation, metadata acquisition, digitisation
2006–2009	System enhancements, definition of AgTec classification, upload of digitised images with metadata	Finishing digitisation, metadata revision, preparation for slide archiving, slide archiving
2009–today	Development of specific tools and system enhancements	Creation of the AgTecCollection with indexing and publication

The main focus was on the digitisation of images in the existing slide collections from teaching and publication activities. Accompanying this, the existing writings of doctoral and post-doctoral theses were to be included in the archiving.

Images

While the open-source programme mediaTUM was being developed in the TUM library by a specially created work team (Team mediaTUM), the acquisition and digitisation of the entire slide inventory could be tackled at the Chair of Agricultural Systems Technology. For this, there was no information about the number of existing objects and which contents were contained in the slides. There was also no information about their use and storage, about their condition, and generally about which materials were used for framing.

From Slide to Digital Copy

Starting from this situation, a process of successive subprocesses emerged, each of which had to cope with many unknowns in addition to recognisable requirements and specifications.

Capture Existing Collections

In consultation with the users of the individual offices in the building, the sequence of the upcoming digitisation action was therefore worked out and agreed upon. According to this, the existing structure of the existing collection was to be adopted unchanged in each case. Derived from this, the respective origin data of the individual objects with owner as author and the field of work could be assigned to the metadata. In addition, from the storage in the form of special slide cabinets (Figure 33.1), hanging folders, folders, and other containers or mechanisms, the assignment could be divided to subareas. This resulted in a basic metadata set related to the individual office.

Finally, it was also determined that, after removal and processing, the objects had to be returned to the removal point and their classification. In this way, it was also possible to obtain any additional information that might be required at a later date.

After completion of a first inspection, a rough estimate showed that a total of about 40,000 individual slides had to be processed. These slides were made exclusively of glass, according to the available technology at the time of production. Some of them had glued frames, and frames made of sheet metal were also found. In the case of more recent objects, the frames were predominantly made of plastic with or without a glass cover for the slide. Finally, many glassless slides with frames made of cardboard were found, which were mostly provided with appropriate imprints as teaching and demonstration material by the agricultural machinery industry.

At the same time, the wide range of slide contents had become visible. The overwhelming majority depicted in black and white or in colour as a photograph the technology, its construction, and its use. Other slides contained texts, overviews, or drawings and mappings. These, in turn, were predominantly in black and white or monochrome. And, finally, it had become apparent that, regardless of the respective content, almost every single slide required more or less intensive cleaning in order to achieve the best possible quality from it as a digital copy.

Planning the Workflow

With the information thus obtained, the planning of the slide digitisation could be undertaken. The focus was on a realistic implementation in terms of time, the acquisition of all available metadata, and a largely automatic and trouble-free digitisation in the highest possible quality as goals. Under these conditions, two required workplaces were set up.

In preparation for digitisation, the individual slide went through the following work cycle after an Excel worksheet was provided for it with the associated origin data and further columns for the metadata to be captured:

1. Manual cleaning with dry microfiber cloth.
2. Applying a consecutive number with a permanent pen and entering it into the column provided for this purpose in the opened Excel entry sheet.
3. Insertion of the slide into a conventional single-slide projector with hand-push operation, which also enabled simple and fast subsequent cleaning.
4. Capturing of the information available in the slide by:
   a. Type (photograph, overview, schematic, diagram, mapping, other).
   b. Orientation (portrait or landscape).
   c. Colour (colour, black and white, monochrome).

FIGURE 33.1 Slide cabinet with subject classification of associated slides.

FIGURE 33.2 Digitisation equipment (single-slide scanner in the rear).

d. Additionally, for non-photographs, available data for: author(s), title, drawing number, illustrator.

5. Inputting all this information into the appropriate columns of the Excel data sheet.
6. Taking out the slide, taking over the information on the slide frame, and adding it to the Excel table, then placing it in a prepared circular magazine.

Independently of these work steps, the prepared slides then had to be fed into the digitisation process. A slide scanner – DigiDia 3600, Reflecta GmbH (Rottenburg, Germany) – with the software CyberView X TWAIN Data Source (copyright manufacturer, 2002) under Windows XP was used as the system (Figure 33.2). After the system start and the introduction of the round magazine into the scanner with the magazine assignment to slide position one, only the number noted on the first slide had to be entered into the software. This was followed by independent digitisation at a defined resolution of 2000 dots per inch (dpi) and storage in tiff format with a time requirement of eight hours per magazine. Finally, the circular magazine had to be removed and the slides contained therein returned to their original storage.

Digitisation

Due to the available human resources at the Chair, auxiliary staff was used to carry out the digitisation. Student assistants agreed to take over the entire process on their own and to complete the task as quickly as possible, regardless of the study situation. They succeeded in organising the necessary preparation process in advance in such a way that the digitisation could be carried out in three shifts per day over seven days a week, largely unaffected by this.

In addition, the entire slide collection of the Chair of Agricultural Engineering at TUM was included in the digitisation process. This Chair had been closed in the fall of

2003 and transferred to the Chair of Automotive Technology. Fortunately, the slide collections were preserved. However, the previously existing classification in the existing departments with their teaching units and fields of work was lost, because space-saving storage in cardboard boxes had taken place.

In total, more than 52,000 slides with their metadata were captured and digitised in only 17 months between November 2003 and March 2005 (Auernhammer et al., 2013). Thus, for the first time, an inventory of existing slide material had been created, the number of which was far above the figures that had been circulating until then. With that, the development of agricultural technology from its origins up to the time of project processing had been made accessible in digital form.

In addition, the now-existing digital stock was supplemented by a further slide stock of more than 5,000 individual objects, because in the meantime the Chair of Agricultural Engineering at the Humboldt University in Berlin had also been closed. And there, too, responsible persons were able to preserve the existing slide stock. This consisted exclusively of glass slides in the format of 90 × 90 mm. They were made exclusively in black and white and they mainly documented the agricultural technology of the former German Democratic Republic. Their contents also consisted of photographs and graphic representations directly comparable to the previously mentioned holdings and – a special situation – for each slide, there was a file card with comprehensive metadata on origin, content, and with keywords according to Chair definition. Understandably, the digitisation of this collection had to be done in a different way:

1. Digitisation was done with the transmitting light scanner UMAX U0104-HBL0

   PowerLook (Techville Inc., Dallas, TX, USA) (Figure 33.3). For each pass, six slides were digitised simultaneously at a resolution of 2,000 dpi. Before saving in tiff format, each digital copy then had to be assigned the label present on the associated slide, thus ensuring the link to the index cards. All this took place in the drawing office accompanying daily activity, with an expenditure of about 15 minutes per six slides to run each edition and return the slides into the slide cabinets handed over from Berlin.
2. On the other hand, the metadata acquisition was again carried out by student assistants. It was done directly comparable to the digitisation of the Chair's own holdings in Excel tables, with an identical layout and one additional column for the existing keywords. The information from the existing index cards was then transferred into these tables.

Thus, after completion of this action, digital copies of almost 56,000 slides were available. For the first time, all existing metadata for all digital copies was also available in digital form.

Securing Slide Collections

Independently of digitisation, importance was attached to permanent archiving of the original slide collections. According to the origin, two different archives had to be used for this purpose while preserving the property rights.

For the slides of the two Chairs from TUM, a direct transfer and acceptance into the university archive could be realised. Under the condition of the clear marking and supply of existing information to each individual object, the archive provided long-term stable cardboard boxes for this purpose. Each box held a total of 200 slides in subunits of ten each, allowing easy access to a single object within the container.

After all existing and digitised slides were classified, each individual box received an insert sheet containing the available data from the Excel spreadsheets created. In addition, an overall printout of the Excel tables was created and a copy was made on a USB disc. The overall package prepared in this way was then handed over to the archive of the Technical University.

FIGURE 33.3 Black-and-white slides on the transmitting light scanner.

In contrast, archiving the slide collection of the Chair of Agricultural Engineering at Humboldt University of Berlin proved to be more difficult. After all available metadata had been digitised and recorded at the Chair of Agricultural Systems Technology in Weihenstephan, property rights were suddenly asserted by Humboldt University Berlin, despite prior agreement. In the process, the legal admissibility of the digitisation carried out was called into question and extended to a demand for deletion of the digital material created. Only after protracted negotiations was it finally possible to find a solution acceptable to all parties:

1. The Chair of Agricultural Systems Technology and the TUM library were granted the right to use the digitised material and its metadata and to incorporate them into the planned agricultural engineering archive. There, however, care was to be taken to ensure that all material remained visible under a designation with reference to Humboldt University.
2. Because it was not possible to realise an orderly and guaranteed return and archiving of the original material to Berlin, it was transferred to the German Agricultural Museum of the University of Hohenheim on the basis of a permanent loan agreement. For this purpose, the entire holdings were transferred to Hohenheim in the slide cabinets, with slides (Figure 33.4) on the one hand and index cards on the other.

FIGURE 33.4 Slide cabinet from Humboldt University.

Metadata Revision

The described process of digitisation with the acquisition of available metadata was largely trouble-free. At the same time, it led to a previously unanticipated high number of available digital copies. However, unavoidable errors arose due to the capture of metadata by different people with different skills and experience.

1. Errors occurred, for example, in the recording of image content in the form of texts and or drawings due to spelling and typing errors.
2. The correct recording of the information on the slide frames was not easy.
3. Also, the origin assignment after the given spatial registration led to errors.
4. Finally, as digitisation progressed, changes and additions had resulted in the Excel capture tables.

After the digitisation was completed, therefore, a thorough revision of all existing metadata had to be carried out. For example, it was very easy, even when time-consuming, to detect and eliminate typing errors. It was also relatively easy to identify and correct discrepancies in author names. In addition, the adaptation of the data entry tables with merging into an overall table with an identical structure was also the subject of this action.

Classification

Parallel to digitisation, the development of the mediaTUM media server at the TUM library had progressed to the point of use (Auernhammer, 2019). It was thus available for an initial indexing (complete description of the objects with metadata) and publication (classification in the archiving system) of available digitised material. For this, however, a corresponding classification structure had to be developed in advance.

The Excel tables created during metadata collection with the subject areas and fields of work recorded there served as the basis for this. However, these did not yet reveal any clear systematics. Therefore, with the support of the specialist staff from the TUM library, an attempt was made to create an agricultural engineering classification from them. This was the only way to ensure fast and largely error-free classification of the digitised material during indexing and publication. This led to the following definitions:

1. The digitised material created is exclusively images. For this purpose, a "collection of images" is defined.
2. The images contained in the digitised objects can be assigned to 19 subject areas without major overlaps. These thus represent the first level of directories in the collection of images.
3. The fields of work contained therein are to be classified into a maximum of three subdirectories. An additional sub-level should be the exception.
4. The designations of the subject areas and work areas describe the image contents and are therefore to be

used as keywords. A keyword index can be derived from this. This limits the number of keywords and prevents spelling and typing errors during indexing.

5. For the collection of images and for all directories contained therein, the synonymous designation in German and English is to be used for their designations from the beginning.
6. A special agrarian schema "document-type images-LT" is to be created to record the metadata for each image. This takes into account not only the fields specified by the library (standard data fields in mediaTUM), but also the information specific to agriculture in the Excel tables. For each image, it is obligatory to enter information on type, keywords, colour, and origin. In the further 23 metadata fields, additional available data can be entered and possible links can be created.

Following these specifications, the mediaTUM team created the required archiving structure in mediaTUM.

Test Indexing

A test indexing and test publication served to get to know the system. At the same time, the metadata schema created was to prove its suitability. The test was done with the first nearly-3,000 images of the slide digitising campaign and contained mostly photographs. Schematic drawings and diagrams were also included. Overviews and mapping played a subordinate role. The test set also covered the required colour assignments, whereby coloured images were in the majority due to the high proportion of photographs.

The keywording was carried out exclusively in German using the previously defined keyword index. Because there was no restriction on the number of keywords per object, entries were made with a maximum of eight keywords. However, four keywords per object covered almost 90% of the keyworded images.

In summary, the first integration of digitised objects into a digital archive was realised with the test indexing and test publication. After the relatively quick familiarisation, the required effort per image was kept within limits. This was accommodated by the fact that the associated metadata was automatically integrated into the already largely advanced metadata-capture scheme when uploaded. This minimised the effort required for additions.

It was also possible to derive conclusions for the indexing and publication of images that did not originate from the Chair.

- Because the origin data for the images of the Chair of Agricultural Engineering were missing at the time of the inventory transfer, special features of this collection cannot be taken into account in the upcoming indexing and publication.
- On the other hand, the available metadata acquisition scheme allows the acquisition of the metadata of the Chair of Agricultural Engineering of Humboldt University Berlin. For this purpose, a special keyword field was included in the scheme, to which the already-available keywords can be automatically assigned during upload. An additional keywording can be omitted, because the collection is to be classified contractually in its own subarea of the archive to be created.

Slide Integration in MediaTUM

Following on from the successful indexing and publication test, the entire digitised holdings could now be uploaded to the media server for indexing and publication. This was done on the basis of the sequential numbers assigned to the digitised items and led to storage in individual directories with 100 objects each. Because the objects were provided with consecutive IDs from the media server, this process can be permanently documented and chronologically classified in mediaTUM on the basis of the ID. This can be illustrated very beautifully at the collection definition with the ID=11274 and at the present time mediaTUM shows more than 1.6 million IDs to date.

During the upload, the metadata available for each object from the Excel collection file were directly assigned to the corresponding fields in the metadata schema "document-type images-LT". The documents created in this way were stored in a buffer and later transferred to the author's working directory as needed during indexing and publication.

Related to this final action for the images remained the request to the responsible collection owner, in this case the author, to clarify the legal issues regarding the copyright of the digital copies before the upcoming publication.

Obtaining Copyrights Prior to Publication

A form prepared by the TUM legal department was available for the uniform treatment of copyrights. In this form, the owner, legal successor, or legal representative of the digital copies of his collection had to agree to their storage, indexing, and use for private and/or scientific purposes and certify this by signature. The respective documents were then to be handed over to the library. In addition, copies were stored in an internal working directory of the collection, making them a permanent part of the collection.

Fortunately, no resistance or restrictions were encountered in obtaining the transfer of rights. Thus, no obstacles stood in the way of publication. Unfortunately, this was no longer generally the case a short time later, because, in the case of scanned dissertations, for example, several authors no longer agreed to their publication and downloading, for more or less incomprehensible reasons.

In all this, with the activities carried out, the commitment of those involved and the very fruitful cooperation with mediaTUM, great things had been achieved by the beginning of 2009.

- In the overall inventory, all agricultural technology slides of the Chair of Agricultural Systems Technology and the Chair of Agricultural Engineering of TUM and the Chair of Agricultural

Engineering of Humboldt University of Berlin have been secured.

- Digitised copies of all slides were created and the metadata available for them were digitally recorded.
- All originals were transferred to archives with documents and tables of contents for long-term preservation.
- The fact that this also made a significant contribution to the development of the mediaTUM media server of the TUM library is an additional important and positive effect of this project.

Indexing and Publication

With the integration of the digitised images and the available transfer of rights, the actual indexing and publication of the images could begin. This was done in the home office using a powerful internet connection by the author in close coordination with a supervisor from the mediaTUM team responsible for the collection. This made it possible to discuss open questions quickly and to implement new wishes and requirements in a system-compatible manner.

As a result of the intensive work on digitisation, it was quickly recognised that, in addition to the slide collections, further extensive analogue image material was available in the form of drawings and photos. This led to several publication cycles, each with additional digitisation and scanning activities.

Slides

In June 2009, the indexing and publication of the scanned slides began. Due to the intensive metadata collection during digitisation, the existing information in the assigned fields had to be checked and the keywording had to be carried out during their indexing. The process benefited from the sequential transfer of the existing slides from the respective subject areas and work areas at the Chair. Consecutive images could thus often be assigned with identical keywords. In addition, their classification could be carried out in the same directory.

In contrast, the images from the Chair of Agricultural Engineering at TUM required much more effort in the absence of information on the subject area and field of work. In contrast to these, the effort required for the images from the Chair of Agricultural Engineering at Humboldt University of Berlin was very low, because there, the required keywords were already integrated into the corresponding metadata field.

The most frequently required rework and additions concerned incorrect colour information, missing names of authors, and missing numbers for drawings. In addition, changes were made to the languages, because only German ("DE") or English ("EN") was provided in the metadata language index, although bilingual as well as other language drawings were available in the image holdings.

In addition, by request, an additional field was integrated into the metadata record. With copy/paste, the available source information of the object with author(s), title, year, origin, and ID was automatically generated for the user and thus directly retrievable without errors.

And finally, at the end of the indexing of the collection in November 2016, another metadata field with the synonymous English keywords was added to the metadata schema. Because their generation was not planned during the indexing in mediaTUM and has not been realised to date, a script-based supplement was realised for this purpose. For this, all German keywords previously available in the index were transferred to an Excel table, where they were supplemented with the synonymous English keywords (to be taken almost completely from the directory names). The mediaTUM supervisor then used a script to make the necessary additions to all published objects in accordance with the German keywords inserted in each object.

In addition, the creation and definition of an internal publication option should not go unmentioned. Without the possibility of access for the public, lists were created there for digital copies that were not worthy of publication, that contained errors, or that contained personal information. Identifiable duplicates were also stored there, as well as objects with missing digital copies. Thus, this digital material is preserved for internal use by the Chair and is protected from oblivion.

Large-format Drawings

Although the slide collections available at the Chair of Agricultural Systems Technology were very extensive, they did not include all the drawings ever made. Rather, slides were made primarily for teaching purposes. In addition, there were those for lectures and presentations of a practical and scientific nature. In this respect, in addition to photographs, only the drawings which showed required correlations in overviews or schematic drawings or which represented selected results in diagrams or mappings were integrated.

Another limiting factor was that slide production in the 1950s and 1960s represented a not-inconsiderable cost factor. The more the drawings deviated from the standard A4 format, the higher the costs.

A thorough search revealed more than 2,000 large-format drawings (<A3) in the Chair building from the time of the Chair's founding until 1974 (when cataloguing of all newly produced drawings began), some of which deviated greatly from the standard dimensions and were over two metres long at most. For their digitisation, it was possible to resort to a large-scale digitiser – Gruse Digital Imaging Equipment, Typ CS 220 StE, (CRUSE Technologies GmbH, Wachtberg, Germany) – in a second digitisation action in the TUM Architecture Museum in 2016 (Figure 33.5.)

Compared to single sheet scanners, this created digitised images with up to six individual drawings on the support surface. They had to be individually isolated, aligned, cropped, and then saved for upload using a standard image processing system. In addition, they had to be assigned drawing numbers for chronological classification according to the format YYYY-B-nnn (year-building number-consecutive number in the year) defined in 1974, so that the entire stock of images could be recorded using this criterion. The subsequent indexing and publication were carried out in the same way as described for the digital slides.

However, this inventory also included many construction and conversion plans of tethered barns for dairy cattle into

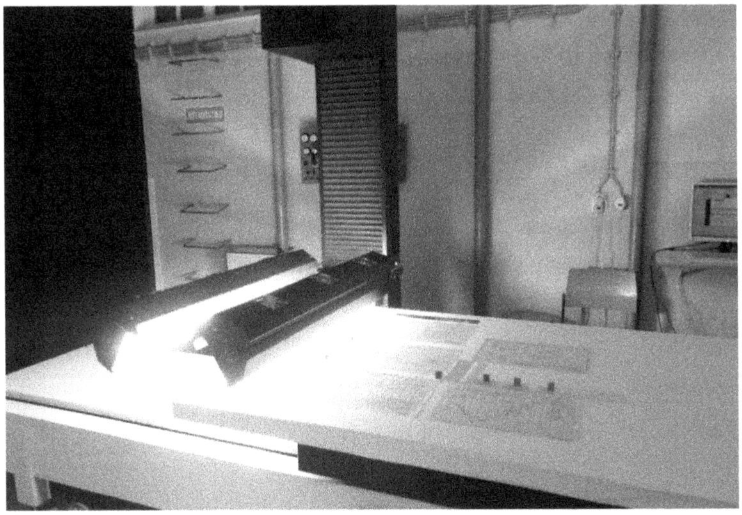

FIGURE 33.5 Large-scale scanner.

loose housing systems. These plans almost always included the names of the owners with their exact addresses. The publication of these plans would therefore have had to be omitted for data protection reasons, unless it had been possible to transfer the rights of the owners or legal successors in painstaking detail. In order to preserve the original drawings without the additional effort of searching for ownership, and to enable their publication, a copy of the digital copy was made in each case. In this copy, the personal data was deleted and the drawing number was supplemented by a "z" following the consecutive number. This meant that nothing stood in the way of publication, while the original digitised copy was saved in the internal area of the archive in a subdirectory for drawings sorted by year.

All the originals of the drawings digitised and prepared in this way were also handed over to the TUM archive. They document outstanding scientific achievements of the 1950s and 1960s, with fundamentals for a more animal-friendly husbandry of dairy cattle. They also have a high-librarian value because all of them were made by hand.

Total Stock of Drawings in A4 and A3 Format

The situation with drawings in standard A4 and A3 formats from 1974 onwards was similar to that with large-format drawings. Of these, too, only a comparatively small number had been transferred to slides. In addition, the slides created could not always reproduce the quality of the original drawings, or a deterioration in quality had remained despite cleaning the slide before digitisation.

For this reason, a third digitisation campaign was launched in 2018. A student assistant digitised all 13,222 available drawings on a printer-scanner. The drawing number contained in the drawings was used to name the digitised files. Their indexing and publication took place in different forms:

1. Drawings published in slides: For these objects, the images of the scanned original drawings were uploaded into the system. Using a specific script, the mediaTUM team then replaced the saved image, so that a high-quality image is now available for download.
2. Drawings not yet published: During the upload, the information contained in the drawing books about author(s), title, draftsman, drawing number, creation date, and references to previous versions are also transferred via Excel files and inserted into the metadata schema. The subsequent indexing and publication were carried out in the usual way of slide digitisation.
3. Drawings with personalised data not yet published: As listed for the large-format drawings, neutralised duplicates with extended drawing numbers had to be created first. For this, the upload with the drawing book data and indexing and publication subsequently took place. On the other hand, the digital copies of the originals were placed in the internal corresponding year indexes.

In agreement with the TUM archive, only the 8,058 drawings made by hand up to 1990 were finally transferred to the archive because, in contrast to the drawings made digitally from 1991, only these have a library value.

Black-and-White Photographs

Following the digitisation of the drawings from 1974, more than 2,000 black-and-white photographs from the founding years of the Chair building, from 1955 to 1968, were scanned in this third digitisation phase.

They were each on index cards with identifiers and information on the back of the card. There, the subject, the content of the photo, the author, the date of the photo, and additional notes were included.

With the transfer of these data into an Excel file and their upload, indexing and publication could be completed in spring 2022.

TABLE 33.2

Published Images by Number, Colour, and Language (as of December 2022).

Type	Number	Share	Colour	B/W	Monocolour	DE	EN
Photograph	31,765	55	26,089	5,666	10	738	89
Overview	6,087	11	372	5,099	616	5,559	347
Schema	9,930	17	754	8,598	578	8,074	451
Diagram	9,094	16	294	7,828	972	8,520	415
Map	558	1	259	295	4	360	139
Total	57,434	100	27,768	27,486	2,180	23,251	1,441

Total Stock of Available Images

The decision to digitise the entire slide material of the Chair of Agricultural Systems Technology and to transfer it into a digital archive has become a mammoth task. In it, however, the slides owned by the Chair represented only a part of the available drawings. It was therefore the right decision to supplement the inventory with all available drawings. The still-available photographic material also had to be included, because this, like photographs in slides, documents technical solutions and applications.

The fact that neighbouring agricultural engineering institutes were closed rather coincidentally during the period of slide digitisation led to valuable additions to the library's own holdings. The resulting considerable number of duplicates of individual objects had to be accepted. Including all subcollections, an archive with a total of 57,434 digital images and associated metadata has thus been created by the end of December 2022 (Table 33.2).

Documents

In contrast to the digitisation of images as slides, drawings, and photographs, the digitisation of writings was carried out according to specific areas of work of the Chair. In the foreground were research results in dissertations and post-doctoral theses. They were followed by writings of research topics in significant research projects. In addition, all student research papers were to be recorded. And finally, additional attention was paid to the recording and integration of all the publications of the professors working at the Chair so far.

Dissertations and Post-doctoral Theses

Since all previous professors of agricultural engineering worked in the same building, all dissertations and post-doctoral theses were either in the original version or in an identical duplicate in the department library. This made it possible to remove one copy at a time, open it at the spine, and digitise it in the batch feeder of a printer-scanner largely automatically. In general, the document was then processed using optical character recognition (OCR) to extract the text and convert the scanned document into a searchable pdf.

Because initial publication tests were already being carried out in parallel with mediaTUM's system development, the mediaTUM team first had to develop an agricultural engineering-specific metadata-capture scheme, "dissertation-LT". This then allowed the upload to be carried out via the workflow provided by mediaTUM:

1. In the user's working directory, the "upload" function is available.
2. The desired object(s) are copied there and the upload is started.
3. The object type must be selected before the internal object creation. Possible assignments to the corresponding metadata schema are book, book contribution, booklet, congress volume, congress contribution, journal editor, journal contribution, report, dissertation, post-doc thesis, student research project, miscellaneous.
4. After that, the object is provided in the working directory for indexing and publication.

Digitisation as well as indexing and publication of all dissertations and post-doctoral theses took place in parallel with the indexing and publication of the slide images. In contrast to these, all document metadata schemata contain an abstract in addition to document-specific entries. Because this was missing in many older dissertations, the indexing of the dissertations was much more time-consuming than that of the images.

In addition, the respective assignment of copyrights during publication was important, because some of the former doctoral students, as owners of the dissertation, had not permitted its free use. In these cases, no thumbnail is displayed in the AgTecCollection and, of course, the otherwise-available link to download the document is missing.

Subject-Specific Research Reports

In the past, the research work of the Chair has always focused on comprehensive research topics in addition to current individual projects. In a specific digitisation project in 2016–2017, an attempt was therefore made to capture all available reports and publications and make them available in the AgTecCollection.

Schlüter Publication Series

A highlight can be seen in the documentation of Schlüter events, which are only available in full in a few places. As a tractor manufacturer on site in Freising, Schlüter had visionary foresight as a businessman and engineer to initiate and drive the development towards large tractors in Europe. All this was done in close coordination with the Chair of Agricultural Systems Technology and resulted in two series of publications:

1. The respective latest products in connection with adapted technology were presented at annual large events with up to 20,000 visitors. The Schlüter booklets contain the documentation of the integrated presentation event.
2. Customers of Schlüter also had the opportunity to participate in the annual training seminars. The lectures and discussions of these events are documented in the Schlüter seminars.

Due to the close cooperation, the Chair received the copyright for both series of publications after the sale of the tractor plant. These have become an indispensable part of the AgTecCollection.

Agricultural Labour Science

Regardless of the technical orientation, work science has always been a focal point of the research work at the Chair. In addition to many technical publications, this led to the development of the first deterministic simulation system based on standardised work elements.

This gradually covered almost all work in the field of indoor and outdoor farming. The outstanding work in cooperation with colleagues from neighbouring Switzerland is also included. It is understandable that these pioneering works with their publications had to be included in the AgTecCollection (see also the "Research Data" section).

Collaborative Research Centre "Production Techniques in Cattle Husbandry"

Already at the end of the 1950s, the first considerations for dairy cattle husbandry in loose housing were carried out at the Chair, in order to replace the usual tethering in modified housing systems and to grant the animal the freedom it deserves (Eichhorn, 1965). They led to a 12-year special research project for production techniques in cattle farming (Wenner et al., 1987).

Without any doubt, the results obtained can be seen as the birth of today's intensively discussed demand for "animal-friendly husbandry systems". At the same time, the first electronic control systems were developed in this project. Accordingly, precision livestock farming also had its birth in this research project (Auernhammer et al., 1985). Many of the documents from this area were preserved and are also part of the AgTecCollection.

Electronics

In continuation of the use of electronics in animal husbandry, the first approaches to its use in field work were realised at the end of the 1980s. The pioneering work on the standardisation of an agricultural BUS system finally led to the ISOBUS for tractor-implement communication with integration into farm management, which is recognised and used worldwide today (Oksanen & Auernhammer, 2021).

In addition, the use of GPS and yield measurement sensors in the combine harvester laid the foundations for precision farming (Auernhammer et al., 1991). The documents of the two areas are an indispensable component for an objective assessment of these technologies in the AgTecCollection.

Publications of the Chair Professors

Again, in the course of many years of the exclusive building use by the Chair of Agricultural Systems Technology, it was possible to deal with the publications of the professors active so far. Based on the available publication directories, it was possible to find almost all publications, to digitise them, and to process them with OCR.

Also, with this indexing, much attention had to be paid to the integration of abstracts. Incomplete information in the publication indexes as well as in the existing documents made indexing difficult. Publishers' copyrights also had to be observed, which is why the decision was made in favour of using "not downloadable" in cases of doubt. An overview of the publications made is given in Table 33.3.

Research Papers

Finally, all existing student research project papers were scanned and processed with OCR. Their indexing and

TABLE 33.3

Publications of Professors Working at the Chair of Agricultural Systems Technology from 1946 to 2007 with Numbers, Language, and Numbers of Reviewed Publications.

Author *	Period	Numbers	DE	EN	Others	Reviewed
Prof. Dr -Ing. M. Hupfauer	1946–1966	115	109	3	3	–
Prof. Dr -Ing. W. Brenner	1955–1969	67	67	–	–	–
Prof. Dr H.-L. Wenner	1969–1989	116	116	–	–	–
Prof. Dr M. Estler	1980–1996	402	393	8	1	2
Prof. Dr -Ing. G. Englert	1988–1998	55	51	4	–	–
Prof. Dr H. Schön	1990–2000	44	35	9	–	–
Prof. Dr H. Auernhammer	1996–2007	257	139	109	9	39

* See publication listings at https://mediatum.ub.tum.de/?item=1304963_0.html

publication took place exclusively under the copyright "not downloadable". However, according to custom at the Chair, a copy will be made available upon request if confirmation of exclusively scientific use is given with a corresponding mention in the bibliography.

With additional publications of conference papers as pdf documents and selected brochures, a total of 2,408 documents were digitised and published in the AgTecCollection. After completion of all activities mentioned above, the following document types are included in the collection according to type and publication (see Table 33.4).

These can be assigned to two publication areas:

1. Rows one through three in Table 33.4 list those which show the research work externally with a share of 54%. Even if the German language dominates, the 21% of publications in English must not be overlooked. In terms of their publication, these were more recent, and thus reflect the change in publication behaviour.
2. These publications are contrasted by the internal publications (in rows four to six), with almost equal relative shares. These publications are almost exclusively in German, although other languages are still used in isolated cases.

System Expansion and Documentation

During the indexing and publication of images and documents, the media server mediaTUM had been extended to also archive moving images in films and videos. Later, research data were added. For these areas, too, publications were made in collections defined for this purpose.

Films and Videos

In addition to documents and images, the moving image, i.e. film and later video, also had a firm place in scientific work. They were predominantly used in two different forms:

1. Films were an integral part of experiments in order to document the technology used and the course of the experiment, in addition to recording purely measured values. In retrospect, situations that could not be traced back from the measured values could be traced back to the circumstances in which they were recorded. On the other hand, high-speed cameras, for example, made it possible at a very early stage to see more than meets the eye and to record and later evaluate these situations. In the agricultural sector, this was especially true for crop movements in closed units. The recording of animal behaviour under changing conditions was also a typical field of application for moving image technology.
2. In addition, films and videos supplemented the presentations in teaching, lectures, and other presentations in order to show the use of the technology even more realistically. In contrast to the experiment, this application describes the implementation in practice.

Due to the different recording techniques and the analogue material created with it, its digitisation requires much more effort than that of images and documents. Therefore, so far, only the first copies available in digital form have been published in the collection created for this purpose. More than 50 additional copies are in preparation.

Research Data

With increasing international integration in research and further development in data processing, measurement data, once determined, are becoming more and more important. The demand for their doubtless documentation is therefore justified not only in the economic sense. Rather, this opens up many new research approaches and much more reliable results and result classifications for follow-up investigations.

Related to the research activities at the Chair with many pioneering approaches, the first data were therefore also made accessible and published in the collection created for this purpose. Especially mentioned are:

1. Research data from agricultural labour science. They comprise working-time elements determined in Germany and Switzerland for indoor and outdoor farming with all their statistical parameters. Their linkage to work processes in different aggregation leads over several levels up to total work sequences of production processes (Auernhammer, 1990). The required programmes are assigned to these data in Fortran source code and as executable DOS

TABLE 33.4

Published Documents by Number, Share, and Language (as of December 2022).

Type	Numbers	Share	DE	EN	Others
Books and book contributions	193	8	168	20	5
Conference books and papers	376	16	201	168	6
Journal articles	734	30	699	31	4
Research reports	457	19	438	14	3
Dissertations and post-doc theses	141	6	139	2	0
Term papers, others	507	21	490	12	5
Total	2,408	100	2,135	247	23

programmes. With the stored procedures, these can be used interactively directly after the download.

2. Research data from the beginnings of GPS use in local yield measurement in the combine harvester. To date, all yield maps created over the entire duration of the research project on the Scheyern experimental farm from 1990 onwards have been published (Auernhammer et al., 1994). They describe the local yield conditions in different resolutions. Trace maps are available from the underlying GPS data. The underlying measurement data have been processed and are ready for publication soon.
3. Measurement data from animal observation on alpine pastures (Braunreiter et al., 2008). The individual values obtained with GPS collars in conjunction with additional sensors on the test animals allow analysis of animal behaviour over the time of the grazing period, with the associated weather data providing additional information.

Activity Documentation

Accompanying all work on digitisation, indexing, and publication, the respective activities were entered in so-called activity listings with the required naming of the date and published. In this way, all work carried out to date can be traced back to year of its first beginning. The links contained therein also allow direct access to newly available objects and to changes and additions. It also contains links to publications on and about the AgTecCollection.

Further links refer to presentations with which the AgTecCollection has been presented in Germany and worldwide so far. For the first time, a link to a first digital twin of the AgTecCollection at the German Digital Library is included (see row four in Table 33.5, righthand column).

At the same time, the second row on the right, after the provisional completion of indexing and publication, also contains the total publication status at the end of 2022. Its development from 2009 to 2022 is shown in Figure 33.6.

Images and writings will only be added in small numbers if such material is found in the successing Chairs. On the other hand, the outstanding films/videos as well as the research data will be added.

Accesses and Downloads

An archive lives on its use and at the same time the use confirms to the archivist the meaningfulness of his work and contributes thereby to his motivation. Usage data can also provide information on the demand for the archived material and thus on the user-friendliness of its presentation with the available search algorithms. Access analyses and download analyses are therefore an indispensable tool.

In mediaTUM, from the beginning of the publication in 2009, the monthly top-50 accesses and downloads could be retrieved with a specially integrated tool until 2018. However, the restriction to the top 50 only covered objects with an average of more than 18 accesses and more than eight downloads per month. The total number of objects accessed and downloaded per month is therefore many times higher. It should also be noted that the access figures by object include the directories accessed in addition to images and documents. In contrast, the top-50 accesses and downloads by country include all countries, because the maximum number of countries per month was not exceeded until 2018.

This tool was disabled in 2018 and replaced by the open-source programme Matomo in 2019. Therefore, directly comparing time series with the available data over the entire period for indexing and publication is not possible.

Accesses and Downloads of Images and Documents

Despite the contents in the access and download statistics not being directly comparable, both showed clear differences in numbers. On the one hand, these are due to the integration of English keywords in the objects of the archive. On the other hand, they also showed the changed indexing activities, with a focus on images until 2016 and on documents thereafter (see Figure 33.6).

In order to make a reliable comparison, two limitations must be made. This concerns the initial year of indexing (2009), with only a few months available. It also concerns 2018, with an unexplained accumulation of accesses in some months.

Therefore, the analysis is done for two timespans – 2010 to 2015 and 2016 to 2017 – each with complete annual records.

TABLE 33.5

Activity Listing for the Year 2022 (https://mediatum.ub.tum.de/?item=11274_22.html).

Time	Activity
31-12-2022	At the end of the year, the collection contains 54,035 images, 2,609 scriptures, six films/videos, and 14 objects in the research data for public access. In addition, 4,184 images, 130 documents, and seven research data are stored in the protected internal area. The total workload for the AgTecCollection work this year was 480 hours.
08-12-2022	In a keynote speech at the XX CIGR World Congress 2022 in Kyoto on 8 December, 2022, the AgTecCollection was presented under the title "Repository AgTecCollection – An Attempt to Rescue Analog Agricultural Engineering Knowledge".
13-07-2022	After extensive rework (removing duplicates, correcting and adding comments, inserting available dxf-files for the years 1986 to 1990 into the attachments of the drawings), the publication of a total of 53,315 images is completed. This section of the AgTecCollection will now be integrated as a copy into the "German Digital Library" and can also be used there by anyone.
30-03-2022	Team mediaTUM supplements the English keywords with a list of synonyms for all German keywords.
07-02-2022	All 2,159 photos in black and white of the work of Prof. Brenner and co-workers from the period 1956 to 1968 are recorded and published.

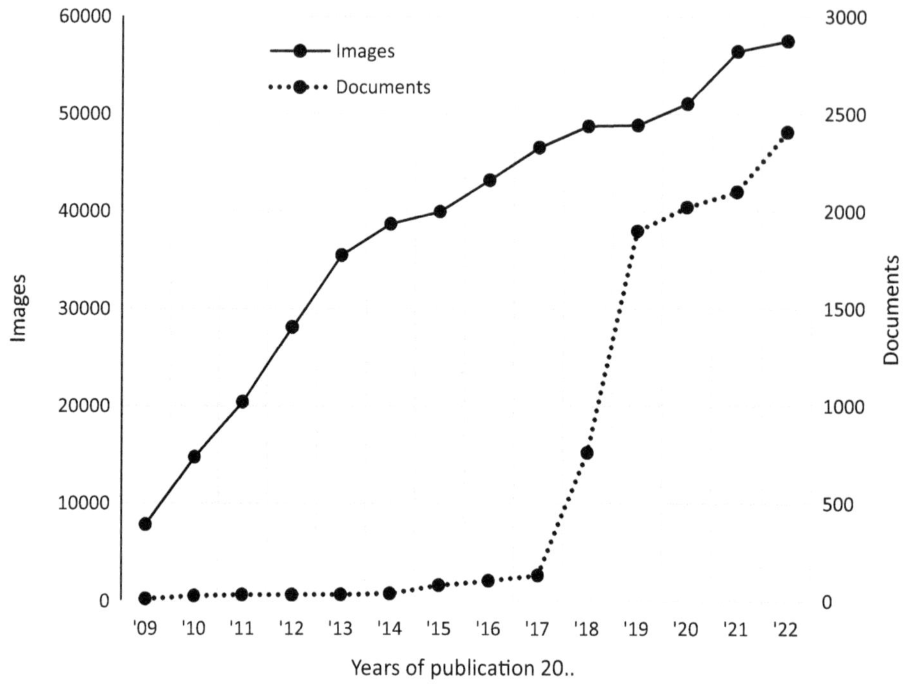

FIGURE 33.6 Indexing and publication for images and documents from 2009 to 2022.

TABLE 33.6

Top-50 Accesses Per Year and Per Object to Directories, Images, and Fonts in the AgTecCollection from 2010 to 2017.

	Top-50 Accesses/Year		Top-50 Accesses/Object/Year	
Object Group	**2010 to 2015**	**2016 to 2017**	**2010 to 2015**	**2016 to 2017**
Directories	12,283	12,558	69.0	49.5
Images	2,283	6,351	10.0	23.5
Documents	876	3,734	16.5	40.0
Total	15,442	22,642	95.0	113.5

TABLE 33.7

Top-50 Downloads Per Year of Images and Documents from the AgTecCollection from 2010 to 2017.

	Top-50 Downloads/Year		Top-50 Downloads/Object/Year	
Object Group	**2010 to 2015**	**2016 to 2017**	**2010 to 2015**	**2016 to 2017**
Images	2,002	5,065	2	4
Documents	0	26,087	0	278
Total	2,002	31,152	2	282

For the accesses, the parameters created for this purpose are shown in Table 33.6.

With almost identical accesses per year to the directories in the two time periods, very high increases become visible for images and documents from 2016 to 2017, with a tripling and a quadrupling, respectively. In relation to the respective objects per year, a further change is also visible. After that, the accesses are concentrated on fewer directories. For images and documents, on the other hand, they increase almost identically by 150% in each case.

Because directories are not downloadable, the data for images and documents in the top-50 download data cannot be compared directly with that of the accesses. Rather, these contain many more objects and thus much more strongly represent the total published objects and the changing download behaviour of users. These differences are very clearly visible in Table 33.7.

Unmistakably, the downloads from 2010 to 2015 are very much concentrated on the increasing number of images. Compared to them, the documents completely fade into the background. This changes into the opposite in 2016 to 2017. Now, the downloads of the now-published documents dominate. It should be noted that these were almost exclusively documents in German. In terms of downloads per object, this leads to two downloads for images and nearly 300 for documents. The surprisingly high value for the documents is mainly caused by the available textbooks, the writings of the Schlüter major events, and the Schlüter seminars.

Compared to the annual analysis, the available download data also allow an analysis in terms of monthly events, with a differentiability into four activity periods (Table 33.8).

After that, the download behaviour changes a lot after 2015. Until then, the differentiated time cycles were almost on par with their downloads per month. After that, November becomes the main download month, followed by June. The other months, on the other hand, are far behind. The reasons for this are certainly manifold, with the scientific use of downloads with the start of the semester (November) and the end of the semester (June) presumably exerting a major influence.

TABLE 33.8

Seasonal Download Activity for 2010 to 2015 and 2016 to 2017.

Time Span	Top-50 Downloads/Month		Top-50 Downloads/Object/Month	
	2010 to 2015	**2016 to 2017**	**2010 to 2015**	**2016 to 2017**
December to May	1,114	3,815	4	44
June	778	8,610	3	100
July to October	948	3,463	4	40
November	958	17,993	4	233
Weighted average	1,028	5,279	4	62

Accesses and Downloads by Countries

In contrast to the accesses and downloads by object, those for countries are comprehensive and thus describe the worldwide demand behaviour for the created collection. Starting from nearly 40 countries with accesses, these increase to over 80 countries by 2012 with an annual increase. Thereafter, too, there is a steady, but diminished, increase, reaching the total number of 118 countries as early as 2018. This is reflected in the download Figure 33.7.

If, on the other hand, attention is given to the number of downloads by country, the picture is very different. Accordingly, the annual increase is much lower. In relative terms, around 36% or one-third of downloads occurred in 2009. This share rises steadily to 66% by 2016 and, after a very sharp increase, reaches more than 93% in 2018. This increase can clearly be attributed to the integration of English keywords that took place at the end of 2016. In other words, the more the keywords are adapted to the user languages, the higher the download figures.

This effect is expressed via the English keywords in the increasingly international use of the collection. It becomes very clear when the newly added countries are considered in the downloads (Table 33.9).

Another important piece of information is obtained by analysing the number of downloads for the countries in question, looking at the relative proportions and also the frequency by download over the months (Table 33.10)

First, it can be seen from this that with the available analysis tool, about one-third of the accessing countries could not be clearly assigned. This makes it understandable that this tool was deactivated at the end of 2018.

Secondly, it shows a dominance in the downloads by Germany, the US, and Austria, together comprising about 80%. This is surprising for the German-speaking countries Germany and Austria, but not for the USA. In addition, a look at the number of months with downloads confirms that these countries can also be seen as "loyal users" because downloads were made from them almost every month.

From these figures, however, variations in behaviour and demand can also be seen. For example, the total downloads from Oman occurred in only four months, which suggests that a very specific use took place, whether the downloads were used for presentations or found expression in a written work.

Another important piece of information is included in this table: without exception, the top-20 download countries, with almost 99% of all downloads, belong to the northern hemisphere. On the one hand, this is understandable, because roughly the same technologies are used there. But on the other hand, it is also incomprehensible, because the documented bases for design, construction, and use of the documented techniques are the same worldwide. This again raises the question

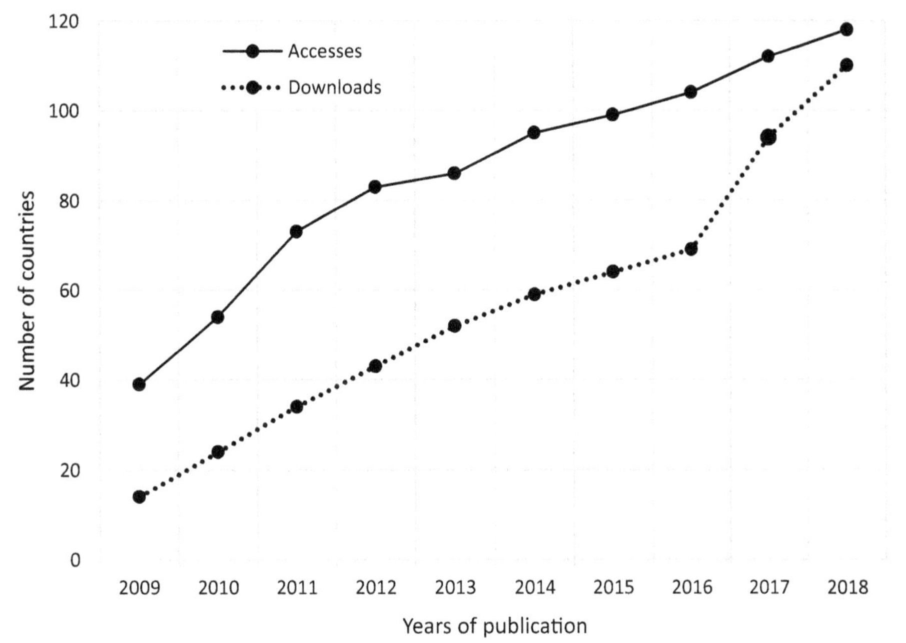

FIGURE 33.7 Accesses and downloads by country from 2009 to 2018.

TABLE 33.9

Newly Added Countries with Downloads from 2009 to 2018.

Year	Added Countries with Downloads
2009	Austria, Belgium, Switzerland, Germany, France, Italy, Netherlands, Romania, Luxembourg, Latvia, Poland, Thailand, Bosnia and Herzegovina
2010	Denmark, Norway, Iran, Spain, United States, Canada, China, Czech Republic, Serbia, Slovak Republic
2011	Croatia, Slovenia, Estonia, Finland, Great Britain, Japan, Sweden, Turkey, Hungary, Bulgaria
2012	Satellite Privider, Syria, Liechtenstein, Oman, Lithuania, Russia, Brazil, Armenia, Israel
2013	Greece, Portugal, India, Lebanon, Colombia, Ukraine, Mexico, European Union, South Africa, Taiwan
2014	Ireland, Singapore, Turkmenistan, Egypt, New Zealand, Kuwait, Iceland
2015	Australia, Azerbaijan, Korea (South), Philippines
2016	Argentina, Aland Islands, Chile, Macedonia, Malaysia
2017*	Indonesia, United Arab Emirates, Morocco, Hong Kong, Kenya, Suriname, Viet Nam, Cambodia, Moldova, Mongolia, Algeria, Uganda, Cameroon, Peru, Pakistan, Venezuela, Bangladesh, Belarus, Papua New Guinea, Dominican Republic, Ethiopia, Guatemala, Kazakhstan, Bolivia, Jamaica
2018	Ivory Coast, Paraguay, Tanzania, Uruguay, Uzbekistan, Rwanda, Namibia, Saudi Arabia, El Salvador, Yemen, Albania, Kyrgyzstan, Guinea, Nepal, Panama, Iraq

* Newly added countries with downloads after integration of English keywords

TABLE 33.10

Top-20 Countries with Downloads by Number, Share, and with Downloads Made by Month with Relative Share (as of 12/31/2018).

Rank	Country	Downloads	Share	Months	Share
1	Not recognised	161,946	38. 29	110	99. 1
2	Germany	156,251	36. 95	111	100
3	USA	61,849	14. 62	83	74. 8
4	Austria	15,975	3. 78	108	97. 3
5	China	3,981	0. 94	34	30. 6
6	Switzerland	3,462	0. 82	95	85. 6
7	Netherlands	3,004	0. 71	98	88. 3
8	Polen	2,068	0. 49	89	80. 2
9	Great Britain	1,299	0. 31	55	49. 5
10	Oman	1,223	0. 29	4	3. 6
11	Belgium	1,218	0. 29	62	55. 9
12	Ukraine	997	0. 24	28	25. 2
13	Italy	994	0. 24	78	70. 3
14	Greece	992	0. 23	29	26. 1
15	Canada	976	0. 23	65	58. 6
16	France	933	0. 22	59	53. 2
17	Rumania	604	0. 14	54	48. 6
18	Luxemburg	519	0. 12	65	58. 6
19	Hungary	518	0. 12	48	43. 2
20	Russia	425	0. 10	43	38. 7
...	...	...	...	...	...
110	Iraq	1	0	1	0

of whether the integration of English keywords into the documents alone can cover the real worldwide requirements.S

User Interface

As in all retrieval programmes, the simple search in mediaTUM is the first step for the user, with a comprehensive search of all available metadata and also of the stored documents. However, this usually results in finding and identifying a very large amount of content and, above all, too little differentiated content. Therefore, search options that are very strongly subject-oriented and focused on subject areas are more helpful.

In the AgTecCollection, the "advanced search", with up to ten criteria to be considered, can be used for this purpose. The options listed in parentheses are available for selection:

- Document type (Photograph, Overview, Scheme, Diagram, Map).
- Authors (Index).
- Brand (Index).
- Colour (coloured, monochrome, black and white).
- Origin (Index).
- Language (Index).
- Year (from–to).
- Keywords (Index).

Targeted searches can thus be specified very strongly and usually lead to the desired goal in a short time. Advanced search thus becomes a very powerful tool. Just like the "simple search", it presents the results in different display formats, allowing the user to adapt the result display to their own wishes on the interface been using.

Because, from the user's point of view, very often unique searches in a certain subject or work area are required, this was taken into account with specific search options. For this purpose, so-called selection tables were created via a subject-specific entry point (Table 33.11).

This allows the search to be targeted to forms of representation in a subject area. If, for example, the user clicks on the "List" option for diagrams of harvesting technology, all objects available in this area are listed. Their display can be based on two criteria. Usually, the year of publication and the author(s) in ascending or descending order are the most

TABLE 33.11

Example of a Bilingual Selection (German and English) Interface for Searching Subject-Related Photographs and Drawings. See Publication Listings at https://mediatum.ub.tum.de/?item=1078037_1.html.

German	German / *English Translation*				*English Translation*
Kollektionen (zu Verzeichnissen)	**Fotografien** *Photographs*	**Übersichten** *Overviews*	**Schemata** *Schemes*	**Diagramme** *Diagrams*	*Collections (to Folders)*
Arbeitswirtschaft	List	List	List	List	*Labour economics*
Bodenbearbeitungstechnik	List	List	List	List	*Soil preparation technology*
Pflanzenschutztechnik	List	List	List	List	*Crop protection technology*
Erntetechnik	List	List	List	List	*Harvesting technology*
Pflanzenanbautechnik	List	List	List	List	*Crop cultivation technology*
Tierhaltungstechnik	List	List	List	List	*Livestock technology*
Ökonomie	List	List	List	List	*Economics*
...	...	...	...	...	...
Forschung und Lehre	List	List	List	List	*Research & Education*

TABLE 33.12

Contents of the Metadata Matrix.

Column Content	Number of Columns	Occupied Cells	Number of Characters	Number of Words
Object IDs	1	61,298	789,215	1
Object Names	1	61,022	8	9
exif columns	89	714,931	715,748	116,995
pdf columns	15	180,419	550,508	34,119
vid columns	25	3,826	293,625	75,182
Columns with other object data	249	401,251	21,838,406	1,192,525
Total	390	1,422,747	24,187,510	1,418,831

convenient choices. In this way, what is being searched for can be found quickly and downloaded for one's own use.

In addition, the column on the left (in German) or on the right (in English) provides a direct link to the respective parent directories in order to view the documents contained therein and in their subdirectories.

AgTecCollection Tomorrow

After more than 22 years since the first digitisation of slides began, a comprehensive agricultural technology archive has now been created with images, documents, initial films/videos, and research data (Auernhammer, 2020). As a result, in addition to the archiving and retrieval software, the digital objects and metadata have become a unique store of knowledge with many new possibilities and with new challenges (Auernhammer, 2022). Important figures may illustrate this as of 31 December 2022:

Digital objects: Agricultural knowledge in visual and written form are documented in digital objects. With 31,765 photographs and 25,669 digitised drawings, this knowledge becomes visible. In words and writing, this is documented in 2,408 documents. There, for 1,609 or 67% of the published objects, 101,626 pages are retrievable in OCR preparation.

Metadata: As indispensable information, metadata describe the digital objects in a matrix of 61,303 rows, with 390 columns in 23,908,170 fields (Table 33.12)

This contains 1,422,747 cells or 6% of information. The "IDs" are the access keys to the digital objects whose names were transferred to the "names" during the upload. The contents "exif columns" describe object parameters obtained during the upload, such as pixel count, length, and width of images. The "pdf columns" represent information about the documents. The large number of the remaining columns contain information about object type, author(s), title, year, and abstract, the characteristics of the technique, and other information.

The total amount of data available can therefore also claim the term "Big Data" for itself, even if this is often understood to mean much more extensive databases and data sets.

In terms of the use of the available digital objects, the metadata are thus the key elements in the system. They are used to enable searches and to describe the selected objects. But objects and associated metadata do not necessarily have to be in the same place (in the same system), because:

1. Independent of the digital objects, the metadata with its available content can be used for direct and differentiated information retrieval.

2. With suitable retrieval software, the metadata can also be used as digital twins via a link to the digital objects at any location in the respective required adaptation to professional and/or regional requirements and needs.

Metadata Analysis

Based on the extensive information in the metadata, two extremely important forms of possible analysis and evaluation are offered. First, it becomes possible to retrospectively identify and understand the evolution of agricultural technology, and second, tools can be derived for the use of digital techniques available today and tomorrow.

Data Mining

Data mining as a sub-concept of knowledge discovery in databases (Knowledge Discovery Data) enables the discovery of existing cross-connections and trends in mass data (Big Data). These are contained in the available metadata when specifically accessing object type, title, author(s), origin, keywords, year, etc.

With the help of the metadata of the AgTecCollection and other digital archives, a variety of approaches are possible, such as:

- Trends in the development of agricultural technology as a whole. Likewise, trends in direct assignment to time periods, to specific technical solutions or, for example, concentrated on animal husbandry or crop production.
- Time intervals between initial development approaches, intensive scientific work, and broad practical application. Additionally, contributions made to this by specific authors or institutions.
- Conclusions with regard to scientific developments and the failed introduction and use of these techniques.
- Work areas and publication activities of the authors over a timeline with conclusions on continuation and reorientation. This also includes publication activities in image and writing, with differentiation according to the publication organs.
- Trends and chronological allocation of the use of media forms of publications with regard to presentation form and layout.

In addition to a much clearer presentation of the achievements of agricultural engineering research and development, conclusions can also be drawn for upcoming and future research approaches. First and foremost, expectations of new research approaches, for example, can be evaluated and classified much more realistically. Conclusions can also be drawn about the sensible orientation, execution, and implementation.

Artificial Intelligence

Compared to the review with a strong focus on documents, the metadata with its image data offers unique opportunities for the use of artificial intelligence. Here, the available keywords come to the fore. They are supplemented by the associated object titles and the available comprehensive image data. With a focus on automatic keywording of digital image objects, today's digital image objects would then be largely automatically inserted and indexed in an archive:

- The basis for this is a selected training set from the collection based on the technique area, the keywords, the title, and the presentation form with colour, type, and exif data.
- An adapted algorithm can then be tested and further developed on a selected training set.
- Finally, new digital image material is processed with it and indexed with the derived keywords. If possible, an image title is also created and inserted.
- Because the corresponding keywords of an object in the AgTecCollection correspond to the classification, an automatic classification in the archive is possible with an adapted script and much less additional manual work.

In this way, existing archives would be automatically updated with the new agricultural technologies of today and tomorrow. Metadata with their digital objects would become a living organism.

Digital Twin

With the integration of the English keywords in 2016, the increased use of the archive became immediately visible. Consequently, this approach can and must be developed further in a targeted manner, for which the tripartition of an archive with IT platform, digital objects, and metadata represent the starting point, because metadata are the basis for digital twins. If they are integrated on a powerful IT platform in a storage and retrieval software available there, then:

- The copy of the original metadata system can be extended with linguistic additions and thus adapted to user requirements or user expectations without changing the original content (in the AgTecCollection, these were the English synonyms to the German keywords). Of course, the copy can also be limited to selected areas of the original system.
- The extension requires the linguistic adaptation for the object group, the type, the title, and the keywords. This can already be done to a large extent automatically with available translation programmes and can produce a very good quality, even for subject-specific terms. The required manual rework is limited to a minimum. Of course, such a metadata extension can also be carried out simultaneously for several languages.
- In addition, it makes sense to make the user interface available to the platform-known user group and other system users with a modified or specifically adapted interface.

- Access to the digitised material (original objects) is always via a link to the original platform. Updates ensure current adaptations and updates, and copyrights allow unrestricted use in accordance with the agreed terms of use.

German Digital Twin

A first digital twin was realised with the image holdings of the AgTecCollection via an integration into the German Digital Library (DDB). This library provides all users with access to Germany's digitised cultural and scientific heritage. Digitised collections are included as digital twins. Predefined forms of presentation provide access to the original objects via links. Searches are performed according to DDB's own algorithms in the adopted metadata (Figure 33.8).

The available search algorithms with integrated filter systems are similar to those of mediaTUM with the platform's own representation. After the user has made a selection, in addition to the object presentation with its characteristic data (a selection of metadata), the related objects are also displayed. This provides the user with much more comprehensive information from related or supplementary representations in the form of photographs, overviews, schematic representations, diagrams, and mappings. Using these, the search can be quickly and reliably expanded and generated with supplementary information.

The example "Digital Twin in the German Digital Library" thus impressively confirms the underlying method and opens up valuable additional information.

Regionally Adapted Digital Twins

However, the digital twin of the AgTecCollection in the DDB remains in the German-speaking area, including the additional use of English keywords. A much more extensive regional adaptation would only result from translations, which can already be performed quickly and with high precision by suitable programmes (e.g. DeepL, Köln, Germany).

For the metadata columns required for this purpose, Table 33.13 shows, as an example, translations into English, French, and Japanese generated with DeepL for five objects, each from the German language.

Further adjustments would then have to be made for problem-free regional use, because:

- It is understandable that these automatically generated texts require review and correction or addition by the specialist on site. However, the effort required for this should be limited.
- Equally understandable is the integration of suitable user interfaces so that the user of this new library information is encouraged to access and use it in their own language and according to their own needs. Only in this way would it be possible to achieve a regional gain in information.
- Related to this, considerations are also needed with regard to the content needed locally. Starting from a general agrarian-technical basis, this concerns above all the cultivation-specific devices and machines and the production procedures available with it.
- Finally, additional considerations are also appropriate to access other archives and to fill technical gaps. This again concerns to a large extent the additions for regional plants and animals in connection with technology adapted for it.

With all these considerations and possibilities, however, regionality must not be considered too restricted. For example, the demonstrated predominant use of AgTecCollection in the northern hemisphere points to uncovered agro-technical conditions and requirements for the central and southern areas of our planet. Why should such adaptations not, for example, integrate all languages spoken on one region or continent (Central and Eastern Europe, Asia, Africa, South America, Australia, and New Zealand)?

Only in this way could a basic version be created that could be comprehensively used by the entire scientific agricultural community. The knowledge gained from previous developments in agricultural engineering would then also be available everywhere regionally where it is most urgently needed.

Conclusions

There is no doubt that the progressive development of agricultural technology has made a beneficial contribution to food production and to relieving farmers of heavy physical labour. But its development has not come to an end; it will build on what already exists and incorporate and integrate the new technologies available. However, nothing fundamentally new will emerge in the process. Rather, it will include techniques, methods, facts, and implementations that have been known for a long time. Therefore, knowledge of these will be essential tomorrow.

However, most of the scientific fundamentals of agricultural engineering are still available today exclusively in analogue form. Moreover, very little attention is paid to the original physical sources. Personnel, structural, and organisational changes, together with rapid decisions, very often lead to the loss of what is already available. Only digitisation can preserve and maintain this knowledge for the future.

With the AgTecCollection in mediaTUM, a successful project was generated for this purpose. In the areas of images, documents, films and videos, and research data, the development of agricultural technology is documented in the collection. The available digital copies are provided with comprehensive metadata. Worldwide access and downloads confirm the need for and use of this repository of agricultural knowledge.

As Big Data, the metadata now available open up a wide range of new possibilities. With data mining, previously unquantifiable correlations and dependencies can be determined. Via the use of artificial intelligence, tools for the integration of today's digital information in picture and script are created.

Moreover, with the established metadata, the available digital knowledge can be used quickly and easily in digital

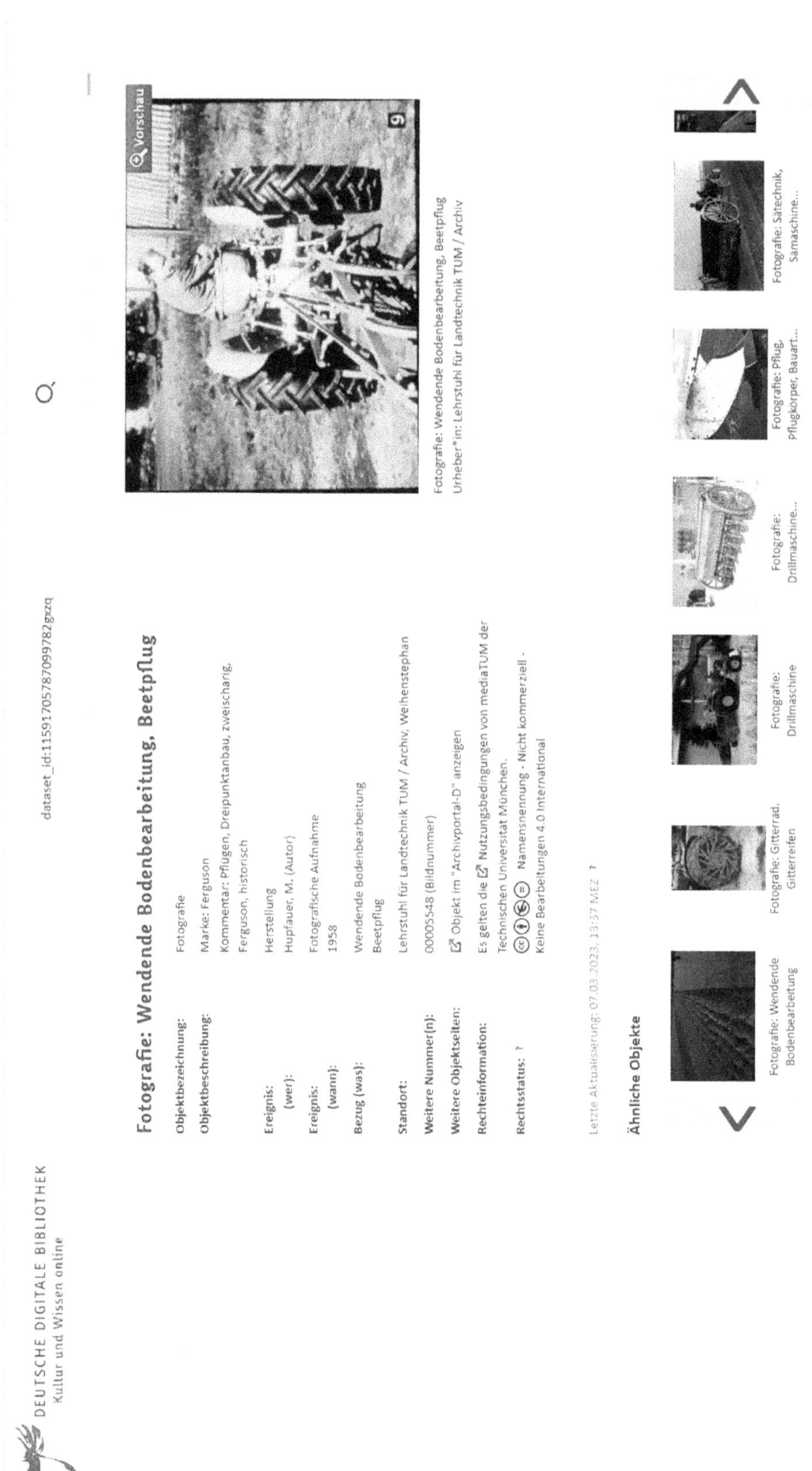

FIGURE 33.8 User interface of the AgTecCollection in the German Digital Library, https://www.deutsche-digitale-bibliothek.de/item/6BMD64NKC5YEFMBWIWNNCFOGWR37EWML?query=dataset_id%3A115917057870997 82gxzq&rows=20&offset=0&viewType=list&firstHit=TNVM37UJESJ5J14JFZHI5AZXKHPGUXWB&lastHit=lasthit&hitNumber=7, 2023-02-03.

TABLE 33.13

Metadata Columns Adapted by Automatic Translation for use in Digital Twins of the AgTecCollection.

		English Adaptation			
ID	**Authors**	**Title**	**Keywords**	**Type**	**Colour**
714785	Schwarzer, J.	AGRIPLAN – Evaluation by tractor	Working-time requirements; Working methods; Labour; Costs	Overview	black and white
722402	Brenner, W.	Piston picker	Maize; Maize picking attachment; Rebler	Photography	colour
724292	N. N.		Braking system; Regulations; Two-axle trailer	Overview	black and white
729048	Schön, H.; Hadlak, M.	Injured body parts due to accidents at work	Workforce; Occupational safety; Occupational accident	Overview	monochrome
729099	Thomé, F.-J.	Dimensions for access openings in cabs	Workplace; Driver's cab; Ergonomics	Overview	monochrome
French Adaptation					
ID	*Auteurs*	*Titre*	*Mots-clés*	*Type*	*couleur*
714785	Schwarzer, J.	AGRIPLAN – Évaluation par tracteur	Besoin en temps de travail; méthode de travail; main d'œuvre; coûts	Aperçu	noir et blanc
722402	Brenner, W.	Cueilleur à piston	Maïs; Cueilleur à maïs; Rebler	Photographie	couleur
724292	N. N.		Système de freinage; Réglementation; Remorque à deux essieux	Aperçu	noir et blanc
729048	Schön, H.; Hadlak, M.	Parties du corps onduit par des accidents du travail	Force de travail; sécurité au travail; accident du travail	Aperçu	monochrome
729099	Thomé, F.-J.	Dimensions des ouvertures d'accès dans les cabines	Poste de travail; poste de onduit; ergonomie	Aperçu	monochrome
Japanese Adaptation					
ID	著者紹介	タイトル	キーワード	タイプ	カラー
714785	Schwarzer, J.	AGRIPLAN – トラクターによる評価	作業時間要件; 作業方法;労働力;コスト	概要	黒白
722402	Brenner, W.	ピストンピッカー	トウモロコシ;トウモロコシ収穫用アタッチメント; レブラー	フォトグラフィー	色
724292	N. N.		ブレーキシステム; 規制;二軸トレーラー	概要	黒白
729048	Schön, H.; Hadlak, M.	仕事中の事故による体の一部の損傷	労働力; 労働安全; 労働災害	概要	モノクローム
729099	Thomé, F.-J.	キャブ内アクセス開口部寸法	職場; 運転台; 人間工学(エルゴノミクス)	概要	モノクローム

twins for a wide variety of requirements. Thus, linguistic and also regional adaptations can be generated, always referring back to the digital copies once created via link. Agricultural knowledge can thus be brought to where it is urgently needed. Because there also applies the maxim: Who does not know the past is condemned to repeat it (Santayana, 1905)!

REFERENCE LIST

Auernhammer, H. (1990). Stallsysteme für die Milchviehhaltung im Vergleich – Methode und Ergebnisse. Habilitation Weihenstephan: Institut für Landtechnik, MEG-No. 182, https://mediatum.ub.tum.de/?id=992683

Auernhammer, H. (2019). Ag-Technology digital in mediaTUM® - State-of-the-Art, ESAlQ-USP Piracicaba (Brazil), 22 pages, https://mediatum.ub.tum.de/?id=1633306

Auernhammer, H. (2020). Preserving Knowledge: A digital repository for biosystems engineering. *CIGR Newsletters* 123, 4–5, https://www.cigr.org/node/739

Auernhammer, H. (2022). Repository AgTecCollection – An Attempt to Rescue Analog Agricultural Engineering Knowledge. XX CIGR2022 World Congress Kyoto (Japan), 44 pages, https://mediatum.ub.tum.de/?id=1691205

Auernhammer, H., Pirkelmann, H., Wendl, G. (Hrsg.) (1985). Prozeßsteuerung in der Tierhaltung – Erfahrungen mit der Milchmengenerfassung, Tiergewichtsermittlung und Bereitstellung von Managementdaten. Schriftenreihe der Landtechnik Weihenstephan, No. 2.

Auernhammer, H., Demmel, M., Rottmeier, J., Muhr, T. (1991). Future Developments for Fertilizing in Germany. ASAE, St. Joseph, Paper-No. 911040.

Auernhammer, H., Demmel, M., Muhr, T., Rottmeier, J., Wild, K. (1994). GPS for yield mapping on combines. *Computers and Electronics in Agriculture*, Vol. 11, No. 1, pp. 53–68.

Auernhammer, H., Seifert, A., Teichert, A., Bernhardt, H. (2013). Digitalisierte Bilder und Schriften Agrartechnik in der "AgTecCollection in mediaTUM®". In Referate der 33. GIL-Jahrestagung 20. -21. Februar 2013, Potsdam "Massendatenmanagement in der Agrar- und Ernährungswirtschaft Erhebung – Verarbeitung – Nutzung" (Hrsg. Clasen, M., Kersebaum, K. C., Meyer-Aurich, A., Theuvsen, B.), Bonn 2013, 23–26.

Braunreiter, C., Auernhammer, H., Steinberger, G. (2008). Potentials of Spatio-Temporal Behaviour Data of Cattle in Alpine Pasture Farming. ASABE, St. Joseph, Paper-No. 08-4897.

Eichhorn, H. (1965). Arbeitswirtschaft, Technik und Gebäude bei der Planung neuer Stallformen für Milchvieh – Dargestellt am Beispiel strohsparender Boxenlaufställe. Frankfurt: Arbeitsgemeinschaft Landwirtschaftliches Bauwesen e.V. ALB, H. 26, https://mediatum.ub.tum.de/?id=992679

Leiss, J., Pretz, E., Seifert, A. (2010). mediaTUM: Der zentrale Medienserver der Technischen Universität München. In Bode, A., Borgeest, R. (eds) *Informationsmanagement in Hochschulen*. Springer, Berlin, Heidelberg, doi: 10.1007/978-3-642-04720-6_29

Oksanen, T., Auernhammer, H. (2021). ISOBUS – The Open Hard-Wired Network Standard for Tractor-Implement Communication, 1987–2020, ASABE: Distinquished Lecture No 42, 2021, St. Josef, MI (USA), 21 pages. doi: 10.13031/913C0121

Santayana, G. (1905). *The Life of Reason*. New York. https://www.gutenberg.org/ebooks/15000

Wenner, H-L., Zeeb, K., Boxberger, J., Kempkens, K., Kirchner, M., Schulz., H., Englert, G., Schwarz, F., Kirchgeßner, M., Pirkelmann, H., Schön, H., Worstorff, H., Sauer, N., Auernhammer, H., Kreuzer, M., Kaufmann, Th.E.G., Paulicks, B.R. (1987). Bericht über das Symposium zum Abschluß des Sonderforschungsbereiches 141 Produktionstechniken der Rinderhaltung. Bayer. Landwirtschaftliches Jahrbuch, 64, 4. https://mediatum.ub.tum.de/?id=1519606

34

A Need to Incorporate Intelligent Food-Drying Systems in ABE Education – A Recipe for Reducing Global Food Losses and Waste

Anthony C. Iheonye, Frank P. Ferrie and Vijaya Raghavan

Introduction

The United Nations Food and Agricultural Organisation (FAO) estimates that the world's population will grow to 9.8 billion people by 2050. This massive number represents an increase of approximately 2.1 billion from the current global population of 7.9 billion (Department of Economic and Social Affairs, 2019). It is projected that most of this growth will occur in Africa and Asia. Interestingly, this huge spike in population comes with unique advantages, such as an increase in labour force and human capital, market expansion, diversity, and innovation. However, just as every coin has two sides, there can also be significant challenges associated with population growth.

An increase of 2.1 billion people will mean more mouths to feed. To prevent food insecurity by 2050, the FAO projects that global food production must increase by at least 60% (FAO, 2018; FAO, 2019c; FAO, 2020). One way to meet this target could be by cultivating more food, however, this option will inevitably add more stress to finite natural resources, such as arable farmland and water resources. This option would also require more financial investment for labour, equipment, rent, insurance, and repairs, to mention a few. However, instead of cultivating more food, there is a better and even cheaper alternative – reducing global food waste and losses.

Globally, over one-third of the food produced for human consumption, or about 1.3 billion tonnes of food, is wasted or lost annually (Environment and Climate Change Canada, 2019; FAO, 2015; FAO, 2019a; FAO, 2019b; FAO, 2019c; FAO, 2020; Julian et al., 2010; Stępień & Dobrowolski, 2017). To put this staggering quantity of waste into context, imagine stacking multiple fully loaded 18-wheeler trucks upon one another, each truck having a carrying capacity of 22 tonnes. To accommodate 1.3 billon tonnes of food waste, the stack of trucks would stretch 945,455 kilometres, which is almost two and a half times the distance between the Earth and the moon. According to the FAO, this quantity of food wasted annually is enough to feed the whole world for one year. What this means is that, going by the quantity of food produced annually, we can ensure food security by 2050 just by eliminating food waste and loss.

One of the popular methods that has been adopted for food preservation is food drying. It is practised both in developing and developed countries. Drying is applied to vegetables, grains, spices, meat, nuts, and seeds. During drying, the moisture content of the food is reduced to a safe level, thereby impeding microbial and enzymatic activities, which in turn inhibits food spoilage and extends the shelf life of the food (Demirhan & Ozbek, 2009; Mayor et al., 2011). However, one may ask why there is still so much food waste and loss with such effective techniques in place. A short response to this question would be that current drying systems lack one major capability that has prevented them from attaining their full potential. Let us discuss this limitation in detail.

As food materials lose moisture during drying, their organoleptic properties (colour, texture, size, shape, aroma) and nutritional properties gradually change. If the drying conditions are not well controlled, these properties are negatively affected, resulting in irregular shrinkage, development of surface cracks, loss of flavour, casehardening, shape distortion, and changes in colour due to enzymatic and non-enzymatic browning. These undesirable outcomes usually occur in food dryers that have their operating parameters optimised to achieve only the desired final moisture content, without considering the visual attributes and nutrient content of the food. Incidentally, current drying systems fall under this group. When such a system is operated by experienced specialists, they always end up with a product that is dried to a safe moisture content. However, when shopping for dried foods, consumers do not look out for what the moisture content of the food is; this information is not even written on the packaging material. Instead, consumers are interested in the visual properties of the food. Before deciding whether to pick up the dried product from the shelf, they first inspect the colour, size, visual texture (how rough the surface of the product is), or shape of the product. If the product is attractive, they pick it up and pay for it, or else it is rejected, thus leading to global food waste.

This problem affects both developing and developed countries. Developed countries such as the US, Canada, and some countries in Europe mainly battle with dry food waste, where the products are dried to the right moisture content but are occasionally rejected due to unattractive appearance. On the other hand, developing countries are ravaged by both food waste and losses. By losses, we mean that the dry product does not even make it to the market. During drying, the product does not attain the right moisture content nor expected visual appearance.

One reason why dry food loss is not a major problem in developed countries is that drying is mainly conducted by

DOI: 10.1201/9780429150111-37

experienced specialists. For instance, in Canada, these specialists account for less than three percent of the population. With the aid of the food's drying curve – that is, the graph of moisture ratio against drying time – these specialists consistently achieve the expected final moisture content. On the other hand, in developing countries, such as Nigeria, subsistence agriculture accounts for over 80% of food production. Food drying is mainly done by farmers or entrepreneurs, who are mostly not specialised in food drying.

For instance, in most West African countries, dried fish is a must-have ingredient in most local soups. A popular variety is dried catfish. Consumers prefer to buy dried catfish that appears golden-brown and is properly dried. The price of well-dried catfish easily quadruples the price of the fresh counterpart. Unfortunately, there are very few specialists that produce such high-quality product. Most times, what customers find in the market are dried catfish that look dry, crispy, and golden-brown on the outside, but are still moist within. Hence, after purchasing the product, a week down the line, the fish is covered by mould, due to the unextracted moisture within it. To avoid spoilage after purchasing the substandard product, consumers continue the drying process either by hanging the product beside a hot stove, by placing the fish on a metallic mesh hung over red-hot charcoal, or by exposing the "dry" fish to direct sunlight.

After purchasing poorly dried fish from a retailer (who, most times, is also the fish farmer and dryer operator), the consumer usually does not buy from them again. To prevent losing a customer, when drying next time, the dryer operator is inclined to extend the drying time to further reduce the moisture content. Unfortunately, while reducing the moisture content to a safe level, most times, they end up with dry fish that is almost black and unattractive to customers. This unpalatable experience is not only encountered when drying fish, but is also experienced with onions, ginger, and many other popular food products. For the non-specialist dryer operator, food drying ends up being a game of trial and error. They are constantly toggling between achieving the expected final moisture content and obtaining an attractive product.

Since the quality of the finished product is dependent on the drying conditions and the initial state of the food, when drying, the non-specialist operator is overwhelmed by numerous unanswered questions, such as:

"What should my initial drying temperature and air velocity be?"

"How long do I need to maintain my starting temperature?"

"As the moisture content changes, do I have to change the temperature? If yes, at what point and to what level?"

"To achieve a specific colour, texture, and size, how should the operating conditions be varied?"

Without answers to such questions, it is not surprising that huge quantities of food harvested annually in developing countries are not always available as dry substitutes during the off-season period. To solve this problem, we need food drying to be as easy as using Google Maps. With this navigation app, we don't need to know the route to our destination before driving out of the garage. All we need to do is type in the destination, and Google Maps shows us the route. It tells you when to turn, what lane to drive on, and how far you are from your next turn. It gets you to your destination, without you having to stop on the road to ask a passer-by for directions.

We can make food drying that way. From the dryer's touch screen, we select the name of the fresh food being dried and the expected values for moisture content, visual attributes, and nutrient content of the end product. Once this is done, the dryer's monitoring system keeps track of these properties. These values will be fed to an intelligent control system that will modulate the drying conditions to ensure we consistently produce dry foods that have long shelf lives, are visually appealing, and retain most of their nutrients. Such a system will significantly reduce, if not eliminate, dry food losses and waste in developing and developed countries, thus improving food security come 2050. However, one may ask why such systems are not readily available if such an advancement in drying technology is possible.

To the best of our knowledge, the absence of such improved drying systems can be attributed to two factors. The first reason can be attributed to current curriculum for teaching drying system design in most departments of agricultural and bioresource engineering. The second reason is that the techniques that now make it possible to monitor and control visual attributes and nutrient content in real-time just came into the limelight in recent years. Let us discuss the second point.

Such an intelligent dryer will be made up of two sub-systems – one for monitoring moisture content, visual attributes, and nutrient content, while the second will be an intelligent control system that knows when and by how much to change the various operating parameters of the dryer in order to achieve the right moisture content, colour, texture, size, shape, and nutrient content.

In the past, there were no mathematical models that linked the operating conditions of the dryer to the moisture content, various visual attributes, and nutrient content of the food. All that was available then was the drying curve and drying equation. With these tools, control systems were developed to obtain the desired final moisture content. Fortunately, with the introduction of faster computing infrastructures and current advances in artificial intelligence (AI), namely deep learning, generative modelling, reinforcement learning, and computer vision (CV) technology, it is now possible to develop effective monitoring and control systems for food dryers. Incidentally, these areas of learning were traditionally taught in departments of computer science and electrical engineering.

However, as we draw closer to 2050, the various departments of agricultural and bioresource engineering (ABE) are presented with the unique opportunity to develop state-of-the-art intelligent drying systems, that could drastically reduce food waste and losses. To achieve this, we need to integrate within our curriculum a pathway for upcoming ABE students to build solid foundation in AI. Usually, ABE students encounter these areas of learning during their graduate studies. Unfortunately, due to the short duration of graduate studies and the resulting steep learning curve, it is challenging to

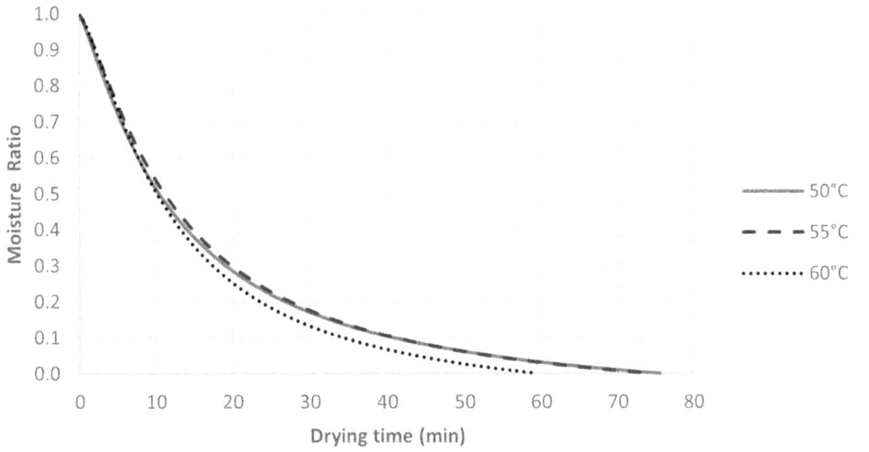

FIGURE 34.1 Moisture ratio vs time for 350 g of peas dried at 50°, 55°, and 60°C.

build a strong foundation in these fields in order to develop such intelligent dryers. However, if students are introduced to these topics during their undergraduate studies, the possibility of revolutionising food drying becomes more realistic.

Within the field of deep learning, students will learn to develop feedforward neural networks (FFNN), convolutional neural networks (CNNs), and recurrent neural networks (RNNs) including long short-term memory (LSTM) and gated recurrent units (GRUs). Learning generative modelling would equip the students with the knowledge and skills needed to develop autoencoders (AEs), variational autoencoders (VAEs) and generative adversarial networks (GANs). The other two topics are reinforcement learning and CV. Each of these techniques plays a vital role in developing this state-of-the-art drying system.

In this paper, we will discuss each of these areas, pointing out why they are needed and how they can be combined to develop an intelligent drying system. We will also discuss the progress that has been made so far within the research community to achieve this goal and conclude by articulating additional works that still need to be done.

Improving the Status Quo

To solve the problem of dry food waste and losses, we need dryers that can control the visual attributes, moisture content, and even nutrient content of the food, in real-time, during the drying process. Unfortunately, it has been almost impossible to achieve this due to the lack of full understanding of the complex underlying mechanisms that take place during food drying. One such mechanism is "heat and mass transfer" – heat moves into the product, and due to this flux, moisture within the product leaves as vapour. Most food dryers are operated solely on this mechanism. However, it is not enough, for it does not account for other processes, such as various chemical and bio-chemical reactions taking place within the product; simultaneous coupled momentum; component migration; several phases of transformations; and sometimes abrupt surface hardening. All these interconnected and non-linear processes affect the visual attributes and nutrient content of the food. To consistently end up with high-quality dry products, we need dryers that account for all these processes.

Current ABE curriculum on drying system optimisation focuses on monitoring and controlling process parameters that are linked to the heat and mass transfer process. These parameters are associated with the drying kinetics of the food; that is, the relationship that describes how the moisture ratio or the moisture content of the product decreases over time. Some examples of these parameters that are monitored during convective drying include the temperature, velocity, relative humidity, and the pressure of hot air. By optimising these parameters, the operator can replicate the drying curve (Figure 34.1) consistently. Once the initial moisture content of the product is known, with the help of the drying curve, the operator can accurately estimate what the moisture content of the product would be at any point during the drying process.

Unfortunately, with these tools, we cannot predict the colour, texture, size, shape, or nutritional quality of the food. To solve this problem, we need models that can learn precise relationships between the inputs and outputs of the drying process. For instance, for convective drying, we will need a model that relates the operating parameters of the dryer to drying kinetics, and real-time changes in visual attributes and nutrient content of the food. Unfortunately, due to the complex and non-linear processes taking place during drying, it has been cumbersome, computationally expensive, and sometimes not feasible to model such relationships using mathematical, numerical, statistical, and analytical methodologies (Aghbashlo et al., 2012; Stępień & Dobrowolski, 2017). Fortunately, with the introduction of high-speed computing technologies and advances in deep learning techniques, it is now possible to model such complex relationships. Using deep learning methodologies such as FFNNs, RNNs, and CNNs, researchers are making substantive breakthroughs.

Feedforward Neural Networks

In the section, we will discuss what FFNNs are, how they work, and how various researchers have applied this technique to advance the state-of-the-art in food-drying technology.

Some of the concepts that will be discussed here are also applicable to other networks, including RNNs, CNNs, and generative models.

The building block for a neural network is called a perceptron (Figure 34.2 A). It is made up of inputs, a neuron, and an output. Each input is associated with a weight. Both the input values ($x_1, x_2, x_3 \ldots x_n$) and their weights ($w_1, w_2, w_3, \ldots w_n$) flow into the neuron, where their weighted sum is computed, and an activation function is applied on the sum to produce the predicted output $h_w(x)$. All the input terms make up the input layer. A bias feature (b) is added to the weighted sum before the activation function is applied. This term is usually represented using a special neuron called a bias neuron. During the training process, the networks continually tweak (optimise) the weights and biased terms so as to minimise the error between the predicted and the actual outputs.

When multiple perceptrons are stacked together, they form a multilayer perceptron (MLP), which is made of an input layer, one or more hidden layers, and an output layer, as shown in Figure 34.2 B. If the signal through this network flows in a single direction, such a network is called a feedforward neural network. Additionally, if the network contains a deep stack of hidden layers, it is referred to as a deep neural network (DNN). The output of a fully connected layer is computed using Equation 1,

$$h_{W,b}(X) = \varnothing(XW + b) \tag{1}$$

In this equation:

- **X** is the matrix of input features (e.g., air temperature, air humidity, air speed, pressure drop, and the different physical parameter of the food).
- **W** is the weight matrix. It contains all the connection weights (w_1 *to* w_n) except the ones for the bias neuron. It contains one row per input neuron and one column per artificial neuron in the layer.
- b is the bias vector. It contains all the connection weights between the bias neurons and the artificial neurons. This vector contains one bias term per neuron in the layer.
- The function $\varnothing$ is the known as the activation function.

Training a Neural Network

The neural network (NN) is trained using the backpropagation algorithm. This algorithm is made up of two paths: the forward pass and the reverse pass. During the forward pass, the input values flow through all the neurons in the input layer, the hidden layer(s), and the output layer. At the output layer, predictions (depending on the number of output neurons) are made.

Once a prediction is returned, the forward pass concludes by computing the error between the predicted value and the actual value. This computation is carried out using a loss function or a cost function, depending on whether the error is computed for a single training example, or for the entire training set. When the error is computed for a single training example, the calculation is done using a loss function, and the result obtained is referred to as the "loss", l, while the cost function is the average of the loss function for the entire training set. Similar to the loss function, the result obtained by the cost function is known as the "cost", J. Equations 5 and 6 are sample functions used for computing the loss/cost for regression problems, while Equation 14 is for classification problems.

For regression problems:

a) Single training example:

$$l(\hat{y}, y) = \left(w^T \boldsymbol{x}^{(i)} - y^{(i)}\right)^2 \tag{2}$$

$$\hat{y} = w^T \boldsymbol{x}^{(i)} \tag{3}$$

b) Entire training set:

$$J(w) = MSE(\boldsymbol{X}, h_w) = \frac{1}{m}\sum_{i=1}^{m}\left(w^T \boldsymbol{x}^{(i)} - y^{(i)}\right)^2 \tag{4}$$

c) The vectorised form:

$$MSE(\boldsymbol{X}, \boldsymbol{h_w}) = \frac{1}{\boldsymbol{m}}(\boldsymbol{XW} - \boldsymbol{y})^T(\boldsymbol{XW} - \boldsymbol{y}) \tag{5}$$

If MSE (mean squared error) is sensitive to outliers, MAE (mean absolute error) from Equation 6 can be used to compute the cost.

$$MAE(\boldsymbol{X}, h_w) = \frac{1}{m}\sum_{i=1}^{m}\left|w^T \boldsymbol{x}^{(i)} - y^{(i)}\right| \tag{6}$$

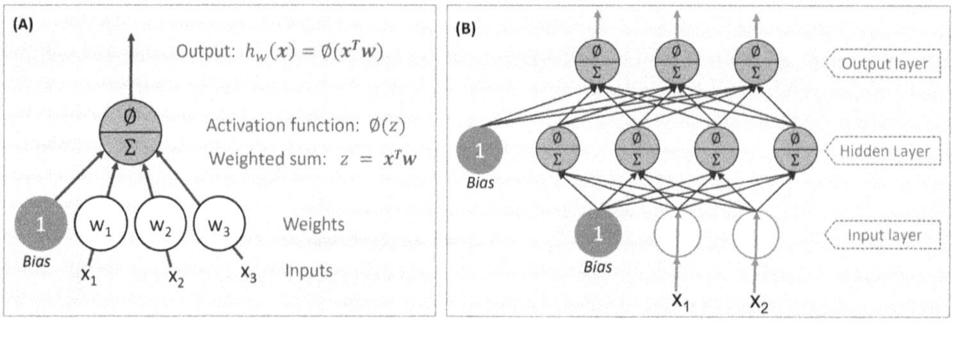

FIGURE 34.2 (A) A perceptron that computes a weighted sum of its inputs then applies an activation function. (B) A multilayer perceptron with two input neurons, one hidden layer containing four neurons and three output neurons.

For classification problems:

a) Single training example:

$$l(\hat{p}, y) = y^{(i)} \log\left(\hat{p}^{(i)}\right) + \left(1 - y^{(i)}\right) \log\left(1 - \hat{p}^{(i)}\right) \quad (7)$$

b) Entire training set:

$$J(w) = -\frac{1}{m} \sum_{i=1}^{m} \left[y^{(i)} \log\left(\hat{p}^{(i)}\right) + \left(1 - y^{(i)}\right) \log\left(1 - \hat{p}^{(i)}\right) \right] \quad (8)$$

Where $\hat{\boldsymbol{p}}$ is computed using Equation 9 (vectorised form)

$$\hat{\boldsymbol{p}} = \boldsymbol{h}_{\boldsymbol{w}}(\boldsymbol{x}) = \sigma\left(\boldsymbol{x}^{\boldsymbol{T}} \boldsymbol{w}\right) \quad (9)$$

In these equations:

- $MSE(\boldsymbol{X}, h_w)$ or $MAE(\boldsymbol{X}, h_w)$ is the cost function (MSE and MAE) measured on the set of training examples using the hypothesis h.
- $\mathbf{x}^{(i)}$ is the instance's feature vector, containing x_1 to x_n.
- σ is the activation function (sigmoid activation function in this case); it is computed using Equation 10, where $t = x^T \boldsymbol{w}$.
- $\hat{p} = h_w(\boldsymbol{x})$ is the estimated probability that an instance belongs to the positive class. It makes its prediction $\hat{y}$ using Equation 11.
- $$\sigma(t) = \frac{1}{1 + \exp(-t)} \quad (10)$$
- $$\hat{y} = \begin{cases} 0 \; if \; \hat{p} < 0.5 \\ 1 \; if \; \hat{p} \geq 0.5 \end{cases} \quad (11)$$

Whether the problem is a regression or classification problem, once the cost is computed, the algorithm begins the reverse pass. The aim of the reverse phase is to tweak the individual weights and bias terms in order to minimise the error between the predicted values and actual outputs. To modify the weights of the model, the algorithm computes the gradient of the cost function (using Equations 12 or 13) with respect to each parameter within the network.

For regression (vectorised form):

$$\nabla_w MSE(\boldsymbol{w}) = \frac{2}{\boldsymbol{m}} \boldsymbol{X}^{\boldsymbol{T}} (\boldsymbol{X}\boldsymbol{W} - \boldsymbol{y}) \quad (12)$$

For classification:

$$\nabla_w J(w) = \frac{1}{m} \sum_{i=1}^{m} \left(\sigma\left(w^T x^{(i)}\right) - y^{(i)} \right) x_j^{(i)} \quad (13)$$

Afterwards, the individual gradients are used to update their respective parameters through an optimisation algorithm, such as gradient descent (Equation 14).

$W \leftarrow w - \eta \nabla_w J(w) \#(14)$ In Equation 14, the term η. Represents the learning rate. Once the updates are made, the forward and reverse phases are continuously repeated till the algorithm converges at a solution. Next, let us discuss some of the activation functions.

Activation Functions

There are various activation functions used in neural networks, ranging from logistic function, hyperbolic tangent function (tanh), rectified linear unit function (ReLU), to various variants of the ReLU. Due to their ability to prevent the unstable gradient problem (to be discussed shortly), the ReLU activation function or its variants have been used by numerous researchers as the activation for hidden layers.

However, one may ask why we need an activation function. Without an activation function, a neural network would just be a chain of linear transformations that ends up producing a linear function. But most times, the relationship between the input data and expected output(s) are complex and non-linear (for example in food drying). Hence, in order for the network to learn the appropriate weights and bias terms that would be used to make accurate predictions, activation functions are inserted between layers to introduce some non-linearity. If not, a deep neural network with numerous hidden layers would still be equivalent to a single layer with limited modelling ability.

Just as using an activation function is important, more so is fine-tuning the hyperparameters of the network. If they are not tuned, the network would experience difficulties converging to a solution.

Fine-Tuning Neural Network Hyperparameters

There are so many hyperparameters that could be modified within a neural network. By tuning these parameters, an already-trained network could be easily adapted to carry out a different task. However, if these hyperparameters are not properly tuned, the model may take a long time to converge to an optimal solution, or in some cases it may never converge. Some of the hyperparameters that need fine-tuning are the number of hidden layers, number of neurons per layer, learning rate, optimisation algorithm, and the batch size, to mention a few. Let us talk about the number of hidden layers.

Number of Layers

How many hidden layers should a network have? For many machine learning problems, a neural network with one hidden layer is a good start. In most cases, the network produces reasonable results. Theoretically, as long as the network has enough neurons, it can model a variety of complex problems (Chollet, 2021; Géron, 2019). But as the problem gets more complex, deeper networks become more attractive. These networks have higher parameter efficiency, meaning that with fewer neurons per layer, they could reach better performance, even with the same amount of training data.

To put this into context, imagine you are told to use a drawing software to paint an apple orchard containing 1,000 trees, but you are also told not to copy and paste. You would have to draw each apple tree leaf-by-leaf and branch by branch; it would be very time-consuming and stressful. On the other hand, if you were allowed to draw one leaf, copy and paste it to form a branch, then copy and paste the branch to form a tree, in a short while, you would have completed the orchard by duplicating numerous trees. This enhanced pattern of painting is similar to the way deep neural networks can learn complex relationships.

The lower hidden layers model low-level features (e.g., lines segments of different shapes), the intermediate hidden layers combine this low-level information to model intermediate-level features (e.g., squares, triangle, circles, leaves), while the much higher layers and output layers combine the intermediate features to model high-level features (such as trees, faces, peas, and tractors). This hierarchical architecture enables DNNs to converge faster to an optimised solution, and it improves their ability to be generalise to new datasets. To determine the optimal number of layers our network should have, during training we can compare various sizes of network using the cross-validation method. This technique would determine the network whose number of layers minimises the cost function.

Now that we have seen how to choose the number of hidden layers, suppose that we have trained a deep neural network (e.g., a CNN) to predict the colour of Red Delicious apple slices during drying. Once a picture containing the slices are fed to the network, it outputs the Lightness (L^*), Greenish index (a^*), and Yellowish index (b^*) colour values. Having developed such a network, if we are later commissioned to build a new network that will be used to predict the colour of different variety of apples known as "Golden Delicious" apples, how would we go about developing the new network? One method would be to build the new network from the scratch, then randomly initialise and train its parameters. This method is good; however, there is a better, cheaper, and faster way to go about it. We could leverage the parameters learned by the previous deep neural network that was trained on Red Delicious apples.

By using this enhanced method, our new network does not need to learn the low-level features (and even some intermediate-level features) that are common to all varieties of apple. All we need to do is go to our previous network, copy the weights and bias terms for its low and intermediate hidden layers, then share them with the new network. This way, our Golden Delicious network will not have to learn these parameters from scratch. It will mainly focus on learning the high-level parameters. At the end of the training process, the low- and intermediate-level parameters may be tweaked a bit, which is better and faster than learning from scratch. This method of training a network using the parameters of a pre-trained network is known as transfer learning. This method can be applied to FFNN, RNN, CNN, and generative models. Now that we have an idea how to choose the number of hidden layers in a neural network, let us discuss how many neurons should be used in each layer.

Number of Neurons Per Layer

The number of neurons in the input and output layers would be dependent on both the type and the number of inputs and output the task requires. If an FFNN is designed to predict moisture content when values of air temperature, air pressure drop, air humidity, and air velocity are supplied to it, then the input layer would have four neurons, while the output layer would only have one neuron for moisture content.

In the past, it was normal to design the layers in the form of a pyramid, with fewer and fewer neurons as you move from the first hidden layer towards the last hidden layer. The rationale was that, as information flowed from the input to the output nodes, low-level features combined to form high-level features (Géron, 2019). However, this practice is no longer popular, because it seems that using the same number of neurons in the hidden layers produces the same results and sometimes even better results than the pyramidic design. Additionally, by using equal-sized hidden layers, we have one less hyperparameter to tune. It sometimes helps to have the number of neurons in the first hidden layer larger than the others.

In practice, it is easier to model with more layers and neurons than needed, then apply early stopping and other regularisation techniques to stop it from overfitting. This method is referred to as the "stretch pant" technique (Géron, 2019): instead of wasting time searching for pants that matches one's size, just pick a larger pant that shrinks to your size. Just like the number of hidden layers, the optimal number of neurons could also be found using cross validation. Next, how do we choose the appropriate learning rate?

Learning Rate

The learning is considered one of the most important hyperparameters. During each epoch, as the learning algorithm goes through each training set, the learning rate determines how fast the algorithm adapts to new information. If this rate is set too high, the algorithm would quickly adapt to new data input; however, it would also loose memory of the previous information that was fed to it. When the learning rate is set to high, during gradient descent, the cost would take huge and random steps across the gradient descent curve. Such huge strides could cause the algorithm to diverge, thus making it difficult for the network to find an optimal solution that accurately models the data.

On the other hand, if the learning rate is set too low, the backpropagation algorithm will go through numerous iterations before it converges at a minimum. During this process, the algorithm would not adapt rapidly to new data. It would keep track of both present and previous information. Although learning would be really slow, it would not be so sensitive to noise in the new data or to outliers.

To converge to a solution faster, it is preferable not to set the learning rate to high, nor too low. An optimal learning rate is about half the maximum learning rate; that is, the learning rate above which the learning algorithm diverges (Smith, 2018). A popular method for obtaining the optimum learning rate is to train the model for a few hundred iterations, starting with a low learning rate (e.g., 10^{-5}), then gradually increase it to a large value (e.g., 10). This increase can be achieved by multiplying the rate by a constant factor (e.g., by $e^{\left(\frac{\log(10^6)}{500}\right)}$. It would keep reducing for a while, then at a point the cost would spike up again (Smith, 2018). The optimal learning rate would be a bit lower than the point where the spike (divergence) occurred. The value obtained would then be used by the optimisation algorithm (such as gradient descent) to learn the optimal values of the weights that minimise the cost function. However, are there other optimisation algorithms other than gradient descent that could speed up the learning process? That's our next point of discussion.

Optimisation Algorithm

Choosing an optimisation algorithm and tuning its hyperparameters would affect how fast our model converges at a solution. There are other much faster optimisers than the mini-batch gradient descent algorithm. Some of those algorithms are Momentum optimisation, RMSProp, AdaGrad, Adam and Nadam optimisation, etc. Just like gradient descent, these optimisers are used to update the model's parameters after the learning algorithm has gone through the batch of training examples and computed the cost. However, before we start sending training examples through our network, it is very important to consider the size of the training batch so that training performance is not negatively affected.

Batch Size

The size of the training batch has a significant effect on a model's performance and training time (Hoffer et al., 2017; Masters & Luschi, 2018). With the introduction of high-speed GPUs (graphics processing units), it has become attractive to train models using large batch sizes which GPUs can process efficiently. However, in practice, models trained with large batch sizes usually experience instabilities at the initial stages of training (Géron, 2019; Hoffer et al., 2017; Masters & Luschi, 2018). Hence, these models would not generalise as well as models trained with a smaller batch size. Thus, Maters and Luschi (2018) advised that an optimal batch size should be between two and 32. On the other hand, it is possible to use a large batch size (up to 8,192) if, during the training process, we integrate techniques such as "warming up the learning rate" (Goyal et al., 2017).

Having considered the major hyperparameters that need to be tuned when designing either a shallow or a deep neural network, let us move on to consider specific problems that could crop up when training deep neural networks (DNNs).

Problems Encountered When Training Deep Neural Networks

When training DNNs, we could experience challenges such as vanishing or exploding gradients, insufficient data to train the network with, the training becoming extremely slow, or the model overfitting the training set. In the upcoming section, we will discuss each of these challenges and outline techniques that could be used to solve them. Let us kick off with vanishing and exploding gradients.

Vanishing or Exploding Gradients

As we discussed previously, during backpropagation, the algorithm moves through all the neurons in a network, starting from the output layer to the first hidden layer. As it journeys backwards, it computes the gradient of the cost function with respect to the parameters in the network. Afterwards, these gradients are used to update the weights and bias term at each connection.

Occasionally, during this backward movement, the gradients keep decreasing significantly. At a point, the gradients are so small that the parameters in the lower layers remain unchanged during the gradient descent update, thus preventing the algorithm from converging at a solution. This significant reduction in gradients during training is known as vanishing gradients.

In some cases, instead of vanishing, the gradients grow astronomically. As a result of these massive spikes, the updated weights become too large, causing the algorithm to diverge. This scenario is referred to as exploding gradients. Just like the vanishing gradient case, exploding gradients also make it difficult for the algorithm to converge at an optimal solution.

In 2010, Glorot and Bengio shed light on why this problem exists. It was attributed to both the logistic activation function and the weight initialisation technique that were mostly in use at that time. Both authors showed that when the sigmoid activation function and the initialisation technique are applied, the variance of the outputs (activation values) coming out of each layer is greater than the variance of the input that flowed into them. As the algorithm moves from the input layer towards the output layer, the variance of the activation values keep increasing after each layer, until it saturates at the top layers (Glorot & Bengio, 2010).

From the sigmoid activation function shown in Figure 34.3, we will observe that when the input is so high (negative or positive), the function saturates at zero or one. At both regions, the gradient is almost zero, hence there may be little or no gradient to propagate backwards to the lower layers. So, when the reverse pass begins and the weights are to be updated using computed gradients, the very minute gradient obtained at the uppermost layer would be diluted further as backpropagation moves towards the lower layers. Hence, the weights in the lower layers would remain unchanged, thereby impeding convergence. The problem of vanishing or exploding gradients can be prevented by applying a couple of techniques. The first technique is the use of an efficient parameter initialisation scheme.

Parameter Initialisation

To solve the problem of unstable gradients, Glorot and Bengio (2010) developed an improved initialisation technique that ensures the variance of the inputs into each layer is the same as the variance of the output coming out of it. This initialisation strategy is known as Glorot initialisation. It speeds up the training process, when combined with either the logistic, tanh, or the SoftMax activation functions (Glorot & Bengio, 2010).

A similar efficient initialisation technique was proposed by He et al. (2015) and it is known as the He initialisation. It is used together with the ReLU activation function or any of its variants. Just like parameter initialisation, another way to alleviate the problem of unstable gradients is by using non-saturating activation functions.

Non-saturating Activation Functions

There are other activation functions that perform better than the sigmoid (which saturates at high input values). One of those non-saturating functions is the ReLU. It does not saturate for positive input values, and it quickens the learning process. However, this activation function suffers from the dying ReLU problem. This happens during training, when a neuron

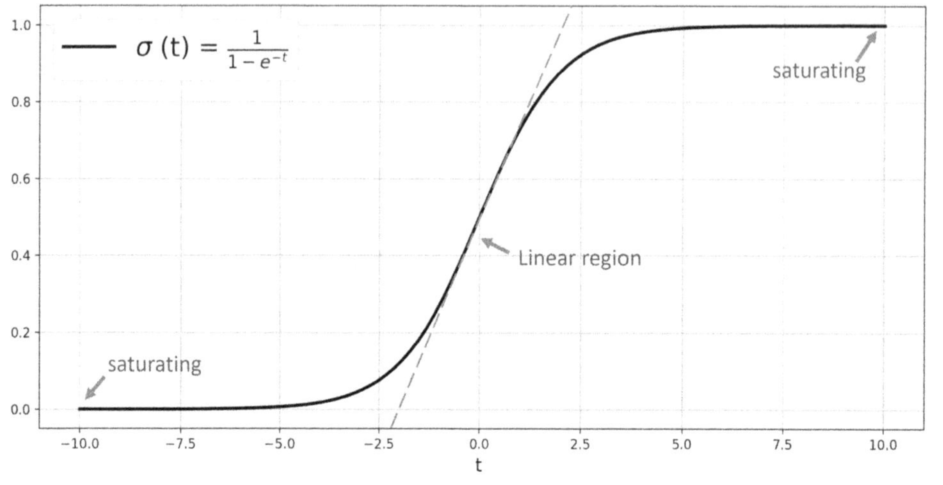

FIGURE 34.3 Sigmoid activation function saturation.

stops outputting results other than zeros. When this happens, you may notice that over half of the neurons in the network are dead. A neuron dies during backpropagation when the weighted sum flowing into it is less than zero (Géron, 2019).

If we observe a plot of the ReLU function, we will notice that once the input is less than zero, the gradient becomes zero, hence, there would be no gradient to update the parameters during gradient descent. To solve the problem of dying ReLU, we can use the Leaky ReLU activation. The Leaky ReLU activation function is very effective for generative models.

The Leaky ReLU (Xu et al., 2015) is a variant of the ReLU activation function, but it has a small slope (leak) to accommodate negative values. Hence, it is non-saturating both for negative and positive inputs. To solve the problem of dying ReLU, it provides a hyperparameter ($\propto$) that denotes by how much the activation function leaks. Hence, when the input is negative, there will still be some gradient that will be used to update the model parameters.

$$LeakyReLU_{\propto}(z) = \max(\alpha z, z) \tag{15}$$

With the hyperparameter α, the Leaky ReLU never dies; occasionally, some neurons may fall asleep, but they eventually wake up after some time (Xu et al., 2015). Aside from the Leaky ReLU, there are other non-saturating variants of the ReLU activation. Those additional ones are:

a. Randomised Leaky ReLU (RreLU) (Xu et al., 2015)
b. Parametric Leaky ReLU (PreLU) (Xu et al., 2015)
c. Exponential Linear Unit (ELU) (Clevert et al., 2015)
d. Scaled ELU (SELU) (Klambauer et al., 2017)

All that said, the ReLU is still one of the most-used activation functions both in feedforward and convolutional neural networks. With the strategies we have discussed so far, we can solve the problem of saturating/exploding gradients by using a combining parameter initialisation and the ReLU activation function (or its variants). However, this does not always guarantee that the problem will not crop up sometime along the training process. There is another popular and very effective technique that can keep unstable gradients at bay. It is called batch normalisation.

Batch Normalisation

Before delving into batch normalisation, let us first discuss how normalisation affects a neural network with one hidden layer. Normalising the input values to have a mean of zero and a variance of one speeds up the learning process. The input values are normalised using Steps 1 and 2 below:

Step 1: Estimate each input's mean and standard deviation (Equations 16 and 17).

$$\mu_B = \frac{1}{m_B}\sum_{i=1}^{m_B} x^{(i)} \tag{16}$$

$$\sigma_B{}^2 = \frac{1}{m_B}\sum_{i=1}^{m_B}\left(x^{(i)} - \mu_B\right)^2 \tag{17}$$

Step 2: Subtract the computed mean from each input, then normalise them using the standard deviation (Equation 18).

$$\hat{x}^{(i)} = \frac{x^{(i)} - \mu_B}{\sqrt{\sigma_B{}^2 - \epsilon}} \tag{18}$$

In the algorithm:

- μ_B is the vector of input means, evaluated over the whole mini-batch B. The vector contains one value for each input.
- The vector $\sigma_B{}^2$ contains input standard deviations, evaluated over the whole mini-batch (it contains one value per input).
- m_B is the number of instances that make up the mini-batch.

- $\hat{x}^{(i)}$ is the vector of zero-centred and normalised inputs for instance i. Epsilon ($\in$) is a smoothing term, usually a very small value (typically 10^{-5}) added to the denominator in Equation 18 to prevent division by zero.

How does normalisation speed up training? Suppose the activation function used in the hidden layer is a sigmoid function. By normalising the inputs, these values would cluster around the linear section of the sigmoid curve, thus providing the initial momentum required for gradient descent, hence, speeding up convergence (Géron, 2019). However, in the case of a deep neural network where data flows through more than one hidden layer, normalising only the training set is not good enough. Yes, the original input is normalised before they get into the first hidden layer, but after the activation function has been applied, the output from the layer most likely would not be normalised. Hence, we may experience a scenario where the variance of the activation values from the first hidden layer is higher than the variance of the input that was fed to it (Géron, 2019; Glorot & Bengio, 2010). This unwanted scenario would set the stage for unstable gradients. To avoid this problem, each batch of activation values should be normalised before they flow into the next hidden layer (Ioffe & Szegedy, 2015). However, they shouldn't all be normalised to have a mean of zero and a variance of one.

If they are normalised the same way as our training set (assuming we are using the sigmoid activation function), all the activation values would cluster around the linear section of the sigmoid function, thus defeating the purpose of having a deep network. Why? Because the network would be hindered from learning much more complex and non-linear relationships.

To take advantage of the non-linearity of the sigmoid function (and similar activation functions), and at the same time guard against vanishing and exploding gradients, two trainable parameters γ and β (in Equation 19) are used to scale and shift (offset) the inputs after batch normalisation has been done.

$$z^{(i)} = \gamma \otimes \hat{x}^{(i)} + \beta \quad (19)$$

In this equation:

- γ is the output scale parameter vector for the layer. The vector contains one parameter per input.
- $\otimes$ represents elementwise multiplication.
- β is the output shift parameter vector for the layer. The vector contains one parameter per input. Each input is shifted by its corresponding offset parameter.
- $z^{(i)}$ is the final output of batch normalisation, after rescaling and shifting has been applied.

In conclusion, to solve the problem of vanishing or exploding gradients, we can apply improved parameter initialisation techniques, use a non-saturating activation function, and apply batch normalisation. The next problem we could run into when developing deep neural networks is insufficient training data. Even when we have sufficient raw data, it may be costly to label them. Towards the end of this chapter, we will discuss how to solve this problem. But first, let us discuss what we can do when training becomes extremely slow.

Extremely Slow Training

Training very deep neural networks sometimes can be a very daunting and slow process. However, there are effective techniques that could make the process go faster. We have deliberated on some of them, such as applying a good initialisation technique, using non-saturating activation functions, applying batch normalisation, and re-using parameters from pretrained networks. As we can see, these strategies also solve other challenges associated with DNNs (in addition to speeding up training); however, there is another vital strategy that could boost the network's training speed. It's the use of much faster optimisation algorithms. These algorithms are much faster and more efficient than the regular gradient descent optimiser. Some of the popular ones are Momentum optimisation, RMSProp, Adagrad, and finally Adam and Nadam optimisation. Table 34.1 compares each of them by convergence speed and quality.

With all these techniques that have been discussed, ABE students will be able to tackle the challenges that may crop up when developing deep neural networks.

Application of FFNNs in Food-Drying Research

When deep neural networks are trained with historical data obtained during drying food products, these systems effectively learn relationships between inputs and outputs of the drying system. For instance, at regular time intervals during convective drying, with the help of a data acquisition system, we can record various inputs and outputs. Some of the inputs can be the temperature, velocity, relative humidity, the pressure of the hot air, and the drying time, while the outputs could be the size and moisture content of the food.

Once the drying runs are complete, the FFNN can be trained to predict moisture content, while product size and other system parameters are set as model inputs. The training could also be done the other way around – the model learns to predict one or more system parameters when presented with moisture content and remaining parameters. Several researchers have used FFNNs to model and predict the drying kinetics of various foods, including apple slices (Khoshhal et al., 2010; Samadi

TABLE 34.1

Optimiser comparison (Chollet, 2018; Géron, 2019).

Class	Convergence Speed	Convergence Quality
Regular gradient descent (SGD)	*	***
Momentum optimisation	**	***
Adagrad	**	* (stops too early)
RMSProp	***	** or ***
Adam	***	** or ***
Nadam	***	** or ***

(bad = *, average = **, good = ***)

et al., 2013), beetroot (Kaminski et al., 2004), carrot slices (Erenturk & Erenturk, 2007; Saraceno et al., 2012), cassava crackers (Lertworasirikul, 2008; Saraceno et al., 2012), cassava slices (Hernandez, 2009; Hernández-Pérez et al., 2004; Kaminski et al., 2004), and ginseng roots (Martynenko et al., 2007). Researchers have also applied this machine learning technique to hazelnuts (Ceylan & Aktas, 2008), mango slices (Hernandez, 2009; Hernández-Pérez et al., 2004; Murthy & Manohar, 2014), potato slices (Islam et al., 2003; Saraceno et al., 2012), thyme (Samadi et al., 2013), and tomato slices (Khazaei et al., 2008; Movagharnejad & Nikzad, 2007). In their in-depth review titled "Application of Artificial Neural Networks (ANNs) in Drying Technology: A Comprehensive Review", Aghbashlo et al. (2015) present various research works where neural networks, especially FFNNs and RNNs, were trained to predict food and process-sensitive parameters.

With all we have discussed so far, we can train a FFNN to predict moisture content, when fed data such as the temperature, relative humidity, and velocity of the drying air. Once this trained network is fed inputs captured at time t, it will return a corresponding value for moisture content. However, due to the non-linear transformations taking place during food drying, it will be very beneficial if, at time t, the deep model has the capability to remember what the moisture content was in the past, and to predict what it is at time t, and what it will be at future time steps. Such a rich pool of data is needed for effective system control. Unfortunately, FFNNs can only predict values for the current time step.

Recurrent Neural Networks

Recurrent neural networks are effective at predicting future outcomes using historical data. Unlike FFNNs that make predictions based on the set of current inputs that were fed to them, RNNs learn to predict future outcomes by remembering inputs from previous time steps. The FFNN will predict moisture content for the 45th minute, using the values of air temperature, velocity, and relative humidity also captured at the 45th minute. The RNN, on the other hand, most times will make more accurate predictions by taking in account what the air temperature, velocity and relative humidity was at the 20th, 25th, 30th, 35th, and 40th minutes. By learning how previous trends affect future outcomes, at the 45th, the RNN can tell what moisture content is at the 45th minute, as well as what it will be at the 50th, 55th, and even 60th minute.

The use of RNNs is very important for monitoring nonlinear processes where the values we are interested in predicting are affected by what they were in the past. Various researchers have applied RNNs to monitor food quality attributes during drying. There are two major challenges RNNs face during training: there are unstable gradients and have a limited short-term memory. Interestingly, there are effective techniques to solve these. Unstable gradients can be solved by applying recurrent dropouts and recurrent layer normalisation, while limited short-term memory is solved using LSTM and GRU cells.

Using FFNNs and RNNs, we can accurately predict system parameters and drying kinetics of the food, in real-time. However, predicting system parameters and drying kinetics is not enough. Consumers are also interested in visual attributes. To train neural networks to predict visual appearance of food products, during training, we need to feed the network historical data on the colour, texture, size, and shape of the food. Ground-truth values of these parameters are generated from images of the food captured and processed during the drying process. Acquiring and processing these images is done using a computer vision system (CVS). The final output from the CVS would be continuous variables representing each of the visual attributes. So, what are computer vision systems and how do they work?

Computer Vision Systems

Computer vision systems are used to obtain the colour, texture, size, and shape of food from images of food captured during the drying process. The system comprises of two major sub-systems – an image acquisition system and an image processing system. The image acquisition system is made up of a digital camera for capturing images of the food, a lighting system to provide uniform illumination, and appropriate software to control the camera. On the other hand, the image processing system is made up of computer hardware and software to process the acquired image and compute or predict visual properties of the food within the image. Apart from food drying, CV systems are also used for ascertaining the quality of agricultural and food products (Du & Sun, 2004; Du & Sun, 2006; Gunasekaran, 1996; Zheng et al., 2006). With CV technology, the quality of the food can be examined nondestructively without touching the product, thereby maintaining product quality and improving the accuracy of the process (Gunasekaran, 1996).

Evolution of Computer Vision Technology

The use of CVS in food drying has been evolving. It started out as an offline system for evaluating food quality during drying. With the offline setup, the CVS was a separate unit from the dryer. At distinct intervals during drying, food samples are extracted from the dryer, then placed in the CVS for image acquisition and processing. The system has been used to study the visual attributes of food products, such as apples (Fernández et al., 2005), carrots (Nahimana & Zhang, 2011), hawthorn fruits (Aral & Beşe, 2016), kiwifruits (Fathi et al., 2011), pumpkins (Zenoozian et al., 2008), and soybeans (Huang et al., 2014).

The challenge with the offline system is that it does not provide real-time feedback. It only provides information corresponding to the time the food was extracted from the dryer. With this system, after retrieving a sample for examination, the dryer operator does not have granular information on how the drying conditions are affecting visual attributes of the food that are still in the dryer. Occasionally, by the time additional samples are retrieved for examination, the damage has already been done.

Another challenge with an offline system is the frequent opening and closing of the dryer during sample retrieval. In the case of convective drying, this practice leads to a lot of heat loss, thus reducing the efficiency of the drying system. The drive to solve this challenge gave birth to the online CV system – the next stage of evolution in CV technology.

The online CV system is an integral part of the dryer. The camera and lighting systems are configured in a manner that makes it possible to acquire images of the whole food in real-time. With this setup, researchers and dryer operators can obtain granular and real-time data on visual attributes of the food. With the online CV system, researchers can monitor visual attributes of food within shorter intervals, even as short as 30 seconds. The colour of the product can be computed as Lightness (L^*), Greenish index (a^*), and Yellowish index (b^*), product size can be computed as Feret diameter, filled area, perimeter, eccentricity, equivalent diameter, and roundness, while surface texture can be computed as contrast, correlation, entropy, uniformity, and homogeneity.

Suppose our data acquisition system collects data every 30 seconds. By the end of a six-hour drying cycle, we would have obtained 720 training samples. Each row of training sample will contain data for visual attributes, system parameters, and moisture content of the food. If we run the same drying experiment two more times, using the same experimental conditions, we would end up with 2,160 training samples. We can go a step further by running two more sets (three replicates each) using new drying conditions. At the end of the drying run, we will have 6,480 training samples. This data can used to train an FFNN or an RNN to predict one or more parameters of interest (e.g., moisture content and visual attributes), while the other parameters will be used as model inputs.

By leveraging the benefits of online CV systems and deep learning methodologies, researchers have developed numerous solutions to monitor visual attribute of food in real-time. With this solution, they can study in granular detail how the drying conditions affect the visual attributes of the food. This CVS makes it possible to detect critical points beyond which the visual appearance of the food becomes less appealing to customer.

Developing off-the-shelf technologies that provide the dryer operator real-time feedback on how the food looks is a vital step towards eliminating dry food waste. To develop such technology, we need to equip ABE students with the requisite knowledge, skills, and experience needed to design and build real-time AI-driven solutions. The amalgamation of deep learning and CV technology have been used by researchers to investigate the visual attributes of various food products, including apples (Nadian et al., 2015; Wang et al., 2017), bell peppers (Romano et al., 2012), blueberries (Saldaña et al., 2013), ginseng (Martynenko et al., 2005; Martynenko, 2006), kiwifruits (Nadian et al., 2017), potato slices (Yadollahinia & Jahangiri, 2009; Yadollahinia et al., 2009), and shrimp (Hosseinpour et al., 2011; Hosseinpour et al., 2013; Hosseinpour et al., 2014).

Online CV systems provide researchers with steady and recurrent stream of data on the product being dried. This consistent and timely flow of data naturally propelled CV technology to its third stage of evolution – the use of real-time imaging to identify vital control points to control and optimise the drying process (Martynenko, 2017). Building on this advancement, Martynenko (2008) developed a control system for drying ginseng roots, using information generated by an online CVS. Having discussed how CV systems improve the quality of dry food, one may ask how visual attributes are computed from images captured by the CV system. Computing the visual attributes occurs during the image processing stage. This stage is divided into two stages – image segmentation followed by visual attributes prediction.

Image Segmentation

The goal of image segmentation is to produce a mask that identifies regions on the original image that are occupied by food materials (the object of interest or foreground) and the background (regions that are not occupied by the food). Suppose we are drying green peas in a convective dryer; the segmentation mask will be a binary image containing ones and zeros, where one represents pixel locations occurred by the food and zero represents pixel locations on the original image that are occupied by the background.

The pixel locations containing intensity values of one can then be used to compute the average size of all the peas, or the size of individual peas. Since the peas have unique values on the mask, the RGB image in conjunction with the mask can be used to compute the colour and texture value for each location containing green peas. Image segmentation method can be based on classification, thresholds, gradients, and regions.

We refer to this two-step process of image segmentation followed by visual attribute prediction as the "two-pass framework". Most research papers where CV systems were used to compute visual attributes of food are based on this framework – after the image is segmented, visual attributes are predicted from the segmented output.

Having mentioned the two-pass framework, the natural question could be whether it is possible to predict visual attributes without segmentation. We will answer this question later in the chapter. Interestingly, bypassing the segmentation step may reduce computational cost and processing time. Such a solution is called the "single-pass technique". We will talk more about it soon.

Aside from the two-pass framework, there is another common link that connects most CV systems that have been applied to food drying. This link has to do with the layout, the state and pattern of motion, and the behaviour of food materials within the image used to predict visual attributes. Let us describe each of them, starting with the layout the food.

By "layout", we are referring to the layout of the food product during the drying process. For most dryers with integrated CV system, the food materials are usually on a surface throughout the drying process. One such dryer is the convective dryer. Within the system, the products are evenly spaced out on the drying tray. With even spacing amongst the products, as well as the contrasting colour of the drying tray, it is easy and straightforward to segment the objects of interest within the captured image. The next common factor is the state and pattern of motion of food products.

For most dryers with CV systems, the food product lays static throughout the drying process. Since the products are static, we can ascertain their distance from the camera, hence it is straightforward to obtain their actual size from the segmented image. Also, since the food materials are static and spaced out, we will not encounter problems with overlapping objects of interest. The last factor is the behaviour of the food material during drying.

As food products lose moisture during drying, the colour and texture on the surface of individual food products may change. For some products like corn and soybeans, the colour and texture on the surface of the individual grains will mostly remain uniform as drying progresses. For other food products like green peas and catfish, the change in colour and texture is non-uniform across the surface of each product. If the products were not evenly spaced out, the non-uniformity in colour and texture would make segmentation a nightmare. Fortunately, this is not the case with most food dryers where CV systems have been used to monitor visual attributes.

Going by the layout and the state and pattern of motion of the food within the image, online CVS can be grouped into static-input online CV systems and chaotic-input CV systems. So far, we have been discussing static-input CV systems; most CV systems used in food-drying research belong to this group. On the other hand, chaotic-input CV systems are used to monitor the visual attributes of products that exhibit the following characteristics during drying (as shown in Figure 34.4):

a. The products are in random (chaotic) motion, jumping up and down within the drying chamber.
b. Due to their chaotic state, the products are at different distances from the camera.
c. The chaotic motion results in products being overlapped and clustered within the image.

These characteristics of chaotic-type images act as a major roadblock to effective image segmentation. Two popular and efficient drying systems where these chaotic behaviours occur are the fluidised bed and spouted bed dryers. They are popular within the pharmaceutical and food industries. Both dryers are renowned for their high rate of heat and mass transfer, high thermal efficiency, uniform temperature distribution, effective material mixing and handling, and their suitability for processing heat sensitive solids (Hosseinpour et al., 2011; Oman et al., 2011; Sampson et al., 2014).

Unfortunately, due to the layout, as well as the state and pattern of motion of pharmaceutical/food products during fluidised drying, segmentation pipelines for static-input CV systems fail miserably when used for segmenting chaotic-type images (Iheonye et al., 2019a). Fortunately, several researchers (Možina et al., 2009; Oman Kadunc et al., 2014; Oman et al., 2011; Watano & Miyanami, 1995) have developed improved classical computer vision solutions to segment and compute visual attributes of pharmaceutical products during granulation and coating. Incidentally, due to non-uniform changes in texture and colour on the surface of individual food products during drying, these classical computer vision solutions encounter challenges when segmenting images of food products captured during fluidised bed drying.

To the best of our knowledge, prior to 2019 there were no published papers that presented a CV solution for monitoring visual attributes of food products during fluidised bed drying. To solve this lingering problem, Iheonye et al. (2019a) developed a novel and improved classical CV approach for segmenting and monitoring shrinkage of peas during fluidised bed drying. Despite the improvements introduced by their methodology, there were still some unsolved challenges. Occasionally, the CV system encountered challenges when the food products were overlapped, clustered, and displayed variation in colour and texture. To solve this problem, deep learning techniques had to be introduced.

Unfortunately, feedforward neural networks are not suited for image classification and segmentation. The network is not invariant to rotation, translation, and scale, meaning that if an FFNN is trained on images of peas, during inference or deployment, the network will succeed at detecting the peas only if they are the same location as they were during training. If the position, size, or rotation of the peas are different in the test image, the model will fail or perform very poorly.

A deep learning solution that is well suited for such images is known as a convolutional neural network. Hence, building on the stride achieved by Iheonye et al. (2019a), the authors of this paper developed two effective CNN-based solutions for

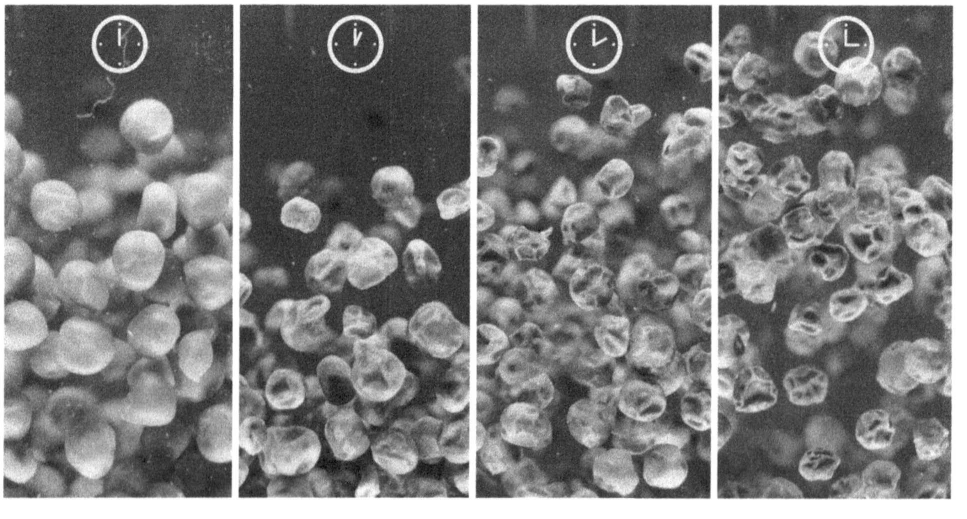

FIGURE 34.4 Image depicting the real-time change in visual attributes of green peas, during fluidised bed drying.

monitoring visual attributes of food during fluidised bed drying. Later in the paper, we will briefly discuss both novel solutions. Before then, let us delvc deeper into CNNs.

Convolutional Neural Networks

Earlier on, we described how FFNNs can be trained to predict drying kinetics when fed inputs, such as air temperature, air humidity, pressure drop within the fluidised bed, and air speed. If this FFNN were to be trained to make the same prediction, but this time with high resolution images of the food, we would likely run into challenges of overfitting and enormous processing cost due to increased data size and the unique architectural pattern of FFNNs.

Speaking of the architecture, the FFNN is a fully connected network; each neuron in a hidden layer is connected to all the neurons in the layers before and after it, hence there would be massive number of parameters to be trained. To describe the challenge further, suppose the size of our input image is 200 × 200 pixels, that is 40,000 pixels in total, and the first hidden layer of the network has 1,000 neurons (which in fact are too few, as they would greatly restrict the amount of information transmitted the next hidden layer). For that first layer alone, the network needs to learn the optimal values for 40,000,000 weights (200 × 200 × 1,000) plus 1,000 bias terms. When we include the extra parameters that need to be trained for other hidden layers, the task becomes even more cumbersome and expensive, as it would be practically impossible to provide enough annotated images to train the network. Interestingly, these problems are solved by CNNs.

One of the unique features of a CNN is that all its neurons are not connected to each other, hence we would need to train fewer parameters. When compared to a FFNN, the risk of overfitting the training set is reduced for CNNs. Before we discuss how we can be use CNNs to develop intelligent solutions for monitoring the quality of food products during drying, let us first examine the building blocks of a CNN that will be helpful to ABE students in the quest to develop intelligent dryers. Some of these concepts include convolutional layers, padding and strides, filters, and pooling layers. What are convolutional layers?

Convolutional Layers

The convolutional layer is the most important building block of a CNN. Unlike the convectional deep FFNN, the neurons in the first convolutional layer are not connected to all the pixels in the input image, but only to the pixels within its receptive field. Also, each neuron in the second convolutional layer is connected only to a small group of neurons in the first convolutional layer. This trend continues as the network progresses.

With this architecture (connection pattern), the first hidden layer (convolutional layer one) concentrates on low-level features within the image, these features are then combined in the next convolutional layer, and so on (Géron, 2019). A neuron located in row i, column j of a given layer is connected to the outputs of the neurons in the previous layer located in rows i to $i+f_h-1$, and columns j to $J+f_w-1$, where f_h is the height of the receptive field and f_w is the width of the receptive field.

Padding and Strides

During convolution, in order for layer l to have the same size as the previous layer $l-1$, before convolving with a filter, the layer $l-1$ is padded with zeros. To reduce the model's computational complexity, it is possible to connect a large convolutional layer to a much smaller layer. We achieve this by spacing out the receptive fields. This is known as striding.

When using a stride of one, the receptive field moves from one neuron to the other, without jumping any neuron. On the other hand, if the stride is set to two, and the size of the new convolutional layer is 3 × 4 compared to the original 7 × 9 (including the padding). A neuron in the new convolutional layer is connected to the output neurons in the previous layer located in row $i \times s_h$ to $i \times s_h + f_h - 1$, columns $j \times s_w$ to $j \times s_w + f_w - 1$, where s_w and s_h are horizontal and vertical strides.

Filters or Convolutional Kernels

A filter or convolutional kernel is a representation of a neuron's weights. It has the same size as the receptive field. To produce the first convolutional layer that comes after the original image, the filter slides through each of the receptive fields, performing convolution operations. As the filter slides over a particular receptive field, the weights at each position in the filter are multiplied by the corresponding pixel intensity in the receptive field; once that is done, the products are summed up to form the value for the resulting position in the convolutional layer.

An image can be convolved by more than one filter. The result of each convolution operation is called a feature map. If we used two filters, the resulting convolutional layer would have two feature maps. Hence, convolutional layers are not always a two-dimensional thin sheet of neurons; instead, they are mostly three-dimensional volumes made up of a pile of feature maps (Figure 34.5).

During training, the network learns the appropriate filters that minimise the cost function. To monitor the visual attributes of food, our input image will most likely be in an RGB image, made up of three channels – red, green, and blue. To obtain the first feature map, the RGB image (input layer) can be convolved with a 3 × 3 × 3 filter. The first two threes are the height and width of the filter, while the last three is the number of layers or channels that make up the filter, which should at least be equal to the number of colour channels (feature maps) in the input image to avoid losing vital spatial and cross-channel information within the image.

When an RGB image is convolved by a 3 × 3 × 3 filter, the result will be one feature map. However, in order to make the CNN capable of detecting multiple features anywhere in the image, the input image is convolved with multiple 3 × 3 × 3 filters. Hence, once a CNN has learned to recognise a green pea in one location in an image, it will also locate it at any other location in the image. However, this is not the case with

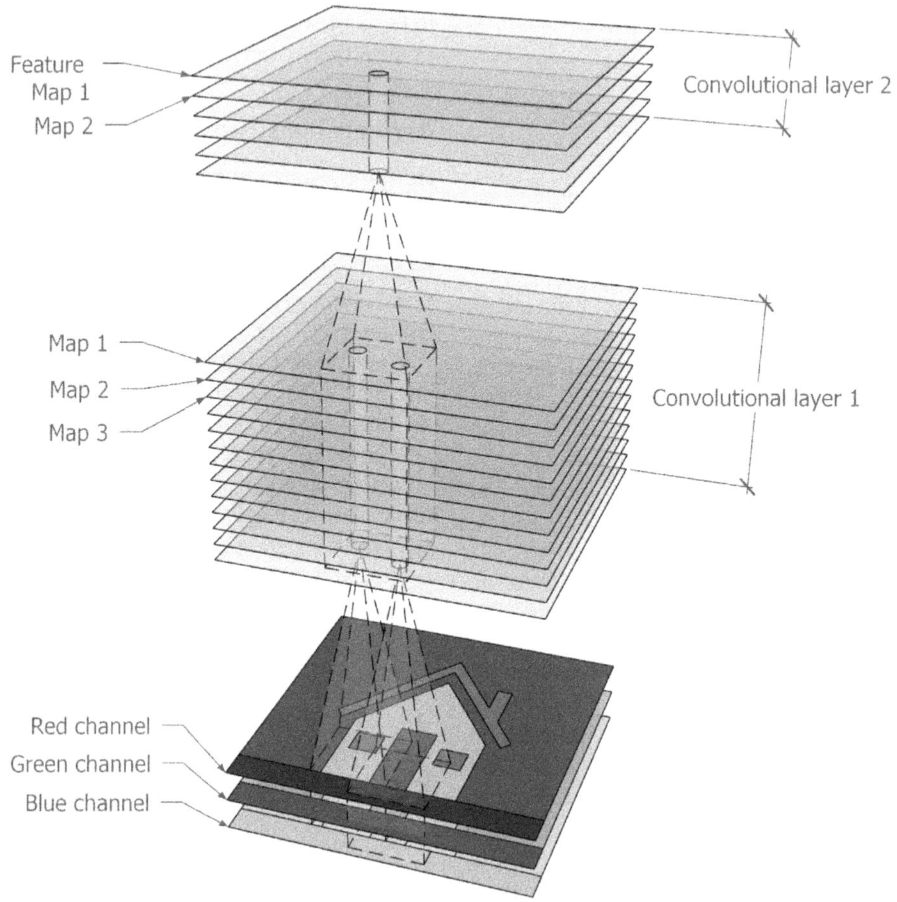

FIGURE 34.5 An image with three colour channels and two convolutional layers with several feature maps.

regular DNNs, like FFNNs. The output of a neuron within a convolutional layer is computed with Equation 26 (Géron, 2019).

$$z_{i,j,k} = b_k + \sum_{u=0}^{f_h-1}\sum_{v=0}^{f_w-1}\sum_{k'=0}^{f_{n'}-1} x_{i',j',k'} \times w_{u,v,k',k} \quad (26)$$

with $\begin{cases} i' = i \times s_h + u \\ j' = j \times s_w + v \end{cases}$

In this equation:

- $z_{i,j,k}$ is the output of the neuron located in row i, column j in feature map k of the convolutional layer (layer l).
- s_h and s_w are the vertical and horizontal strides, f_h and f_w are the height and width of the receptive field, and $f_{n'}$ is the number of feature maps in the previous layer (layer $l-1$).
- $x_{i',j',k'}$ is the output of the neuron located in layer $l-1$, row i', column j', feature map k′ (or channel k′ if the previous layer is the input layer).
- b_k is the bias term for feature map k (in layer l).
- $w_{u,v,k',k}$ is the connection weight between any neuron in feature map k of the layer l and its input located at row u, column v (relative to the neuron's receptive field), and feature map k′.

Pooling Layers

Although CNNs are better and less computationally demanding than conventional DNNs, the amount of RAM they require during training is considerable. During backpropagation, the optimisation algorithm will need all the intermediate values that were computed during the forward pass. For instance, if a 150 × 200 RGB image is convolved by 200 different 5 × 5 filters using a stride of one and "same" padding (adding zeros so that the resulting convolutional layer has the same size as the input image), the resulting 200-layer convolutional layer will have a size of 150 × 200. The number of parameters that will have to be trained for just that convolutional layer is $((5\times5\times3)+1)\times200 = 15,200$. The +1 is for the bias terms (one bias term per feature map).

Each of the 200 convolutional feature maps contain 150 × 100 neurons, and each of the neurons would compute the weighted sum of $5\times5\times3= 75$ inputs; that is 225 million float multiplications $(200\times150\times100\times75)$. Yes, this is not as cumbersome as training 675 million parameters for a FFNN, but it is still considerable. If the feature maps are represented using 32-bit floats, then each convolutional layer would occupy 200 × 150 × 100 × 32 = 96,000,000 bits (12 MB). And this is just for one instance/image. If each training batch contain 100 images, we would need 1.2 GB of RAM. That is substantial!

Pooling layers can be used to solve this problem. They are used to shrink the input image in order to reduce the computational load, the memory usage, and the number of parameters that needs to be trained, thus reducing the risk of overfitting. Pooling layers are similar to convolutional layers; each neuron in the pooling layer is connected to a limited number of neurons within the previous layer, located in a small rectangular receptive field. The neurons of the pooling layer have no weights; they are computed by aggregating the values within each of their receptive fields. The aggregation can be done by either computing the max or mean. Max pooling is more popular. Assuming we are to carry out a max pooling operation using a 2 × 2 pooling kernel, two strides and no padding (valid padding), and the values within the receptive field are one, two, three, and five, the result of the pooling operation will be five.

In addition to reducing computational cost, memory usage, and the number of parameters that need to be trained, pooling layers also introduces some level of invariance to small translations. By placing max pooling every few layers in a CNN, the network also becomes invariant to translation at a larger scale. This type of design also offers small amounts of rotational and slight scale invariance.

This unique feature is helpful in cases where the CNN's prediction should not be dependent on the translation, rotation, and scale of the object(s) in the image. A typical example can be found in images captured during fluidised bed drying, as the food materials are located at different positions in the image, and their respective angles, inclination, and sizes are different.

A downside of max pooling is that it is destructive in nature. For a 2 × 2 kernel with a stride of two, the original image is reduced by half. So, the convolutional layer loses spatial quality. This reduction would not be a problem for classification, but it proves problematic for semantic segmentation. Although max pooling is destructive, the resulting layers preserve the strongest features within the image or convolutional layer, thereby getting rid of insignificant information. Hence, the upper layers in the network would receive cleaner data to work with.

Finally, max pooling can also perform depthwise convolution instead of spatially. This can make the CNN invariant to many features, including thickness, brightness, and skew. This may be of help when segmenting images of food with varying illumination.

If we combine everything that we have discussed so far in this section, we will end up with a CNN that can classify objects of interest within an image. When we feed this trained network an image containing green peas, the model will predict that there are peas within image. However, that is not good enough for food drying, because in our case, we want to monitor size, colour, and shape. To achieve this goal, the CNN should go beyond telling us that there are peas in the image; it should also produce a mask showing us where those peas are located within the original image. This process is known as semantic segmentation.

Semantic Segmentation

There are various state-of-the-art CNN architectures that are used for semantic segmentation. Some of these architectures include U-Net (Ronneberger et al., 2015), DCAN (Chen et al., 2017; Hao Chen et al., 2016), and DeepLabV3 (Liang-Chieh et al., 2018). To produce the segmentation mask, the input image usually passes through two paths – an encoder (or down-sampling path) followed by a decoder (also known as the upscaling path). Both the encoder and decoder are CNNs.

As the image moves through the encoder, the encoder learns important spatial and cross-channel relationships between the image and the object(s) of interest within it. With this information, the encoder can easily identify if our object of interest exists within the image. Due to the use of max-pooling layers and convolutions with strides greater than one, during the encoding process the size of the convolutional layers keeps reducing, while their number of feature maps increases.

At the end of the down-sampling step, the output from the encoder is fed into a decoder (the upscaling path). The overall goal of the decoder during semantic segmentation is to produce the segmentation mask that identifies pixel locations occupied by the object of interest. To produce such a mask, the decoder has to regain most (if not all) of the spatial resolution that was lost during encoding, while retaining the important spatial and cross-channel information it received from the encoder.

To achieve both goals, the decoder is lined up with blocks containing transpose convolutional layers followed by convolutional layers. As the output from the encoder flows through the decoder, the size of the feature maps increases while their numbers decrease. To regain more spatial resolution, at each decoding step, before applying the convolution operation, the upsampled feature maps are concatenated with corresponding equal-size feature maps within the encoder

At the end of the upscaling step, the decoder produces a segmentation mask used to compute visual attributes of the food within the RGB image. With the help of the mask, we can compute the visual attributes of the all the pixel positions on the RGB image that corresponds to the locations on the mask. This computation can be done in two ways. The first method is by using the mathematical equation associated with each visual attribute, while the second method is to train a CNN to do the prediction. We successfully developed the first (Iheonye et al., 2019b) and a second method.

The second method is a "two-pass" deep learning solution we refer to as "Unet-Xception". It is made up of two models – U-Net for semantic segmentation and Xception for predicting visual attributes using RGB image containing only the objects of interest.

Since our research was conducted on peas dried in a fluidised bed dryer, we had to take to cognisance of the layout, the state and pattern of motion, and the behaviour of green peas during fluidised bed drying. During drying, as the peas jump and down within the drying chamber, they keep mixing. Due to the mixing action, there is a consistent supply of peas sliding over the walls of the dryer. We refer to the peas on the wall facing the camera as the "infocus peas", while those behind them are referred to as "defocused peas". Due to the continuous and random up and down motion, each time a new image is captured, different set of infocus peas appears in the image.

Since the peas are mixing while in motion, the infocus peas touch the wall of the dryer (a known distance from the camera), and each captured image contains new set of infocus peas, the visual attributes of the peas at a given time can be computed by calculating the average visual attributes of infocus peas from multiple images captured within that short period. For instance, if we are to monitor visual attributes every five minutes, at the start of each interval we capture 15 images (within ten to 15 seconds) and compute the average visual attributes of all the infocus peas within all 15 images. The value obtained would be representative of all peas for that point in time.

Hence, our goal was to develop a semantic segmentation model that can segment only the infocus peas (our object of interest), despite challenges such as overlapping, uneven distance of peas from the camera, clustering, uniform colour distribution at the onset of drying, and non-uniform distribution as drying progressed. Unet-Xception achieved this goal. Figure 34.7 compares the ground-truth mask to segmentation results obtained by a classical CVS and those obtained using the two-pass deep learning solution. We would notice that Unet-Xception succeeded at identifying even infocus peas that were mistakenly omitted in the ground truth. After segmentation, the results were fed to Xception, which predicted visual attributes. The performance significantly outperformed the results from a classical CVS. The results of this research have been submitted for publication.

The Single-Pass Solution

One may ask if it is possible to predict visual attributes of food without segmenting the captured image. We worked on answering this research question. Our study gave birth to a novel and effective single-pass deep learning solution. With this solution, once a RGB image of the food is fed to the system, it predicts visual attributes without segmenting the image. The single-pass solution was made up of a trained CNN. The model achieved similar results as our two-pass deep learning solution (Figure 34.6), significantly outperforming the classical CV system. The advantage of the single-pass solution is its reduced computational load, since it contains one model, compared to the two-pass solution that has two.

Insufficient Training Data

When developing a model for semantic segmentation, the network is trained using pairs of original images and their corresponding ground-truth mask (Figure 34.7). These masks are produced manually by tracing the outline and/or shading the regions occupied by the objects of interest. The mask can be created using annotation software such LabelMe (Russell et al., 2007). There are a lot of parameters that need to be trained for semantic segmentation models. It would require numerous training examples to converge at the optimal solution. Hence, it is expensive, time-consuming, and almost impossible to produce a sufficient volume of training data manually. On the average, it takes approximately 40 minutes to annotate a single image of green peas captured during fluidised bed drying.

One way to produce an adequate number of annotated images is by manually annotating a few hundred images (for instance, 300), then generating many realistic variants of each training example by applying random transformations to them. This process is known as image augmentation. Some the transformations include up-down flip, left-right flip, random rotation, random crop, and random shear (Figure 34.8). By the time we are done, from just 300 images we can end up with more than 3,000 new images, which is a good start.

In addition to increasing the size of our training set, data augmentation forces the model to be more tolerant to various sizes, positions, and orientations of the object of interest, thus making the network much more robust. By generating various realistic versions of the training set, data augmentation reduces overfitting, making this a regularisation technique. A second method for solving the problem of insufficient training data is through transfer learning. We described the process earlier in the chapter. A third method for generating realistic training data (images and continuous values) is through generative modelling. Some of the generative models include autoencoders, variational autoencoders, and generative adversarial networks.

By exposing ABE students to the deep learning techniques discussed so far, they will be able to develop intelligent systems that can monitor visual attributes, moisture content,

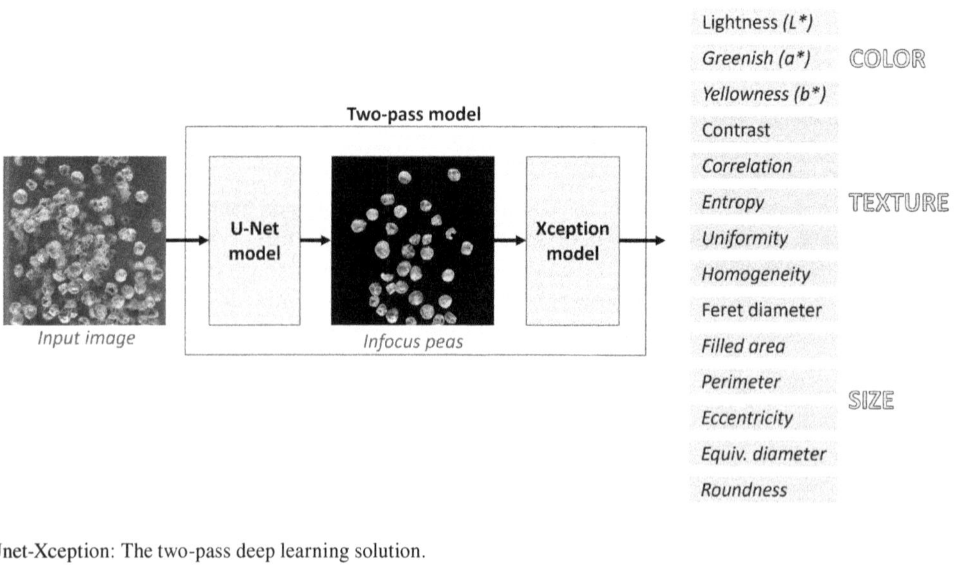

FIGURE 34.6 Unet-Xception: The two-pass deep learning solution.

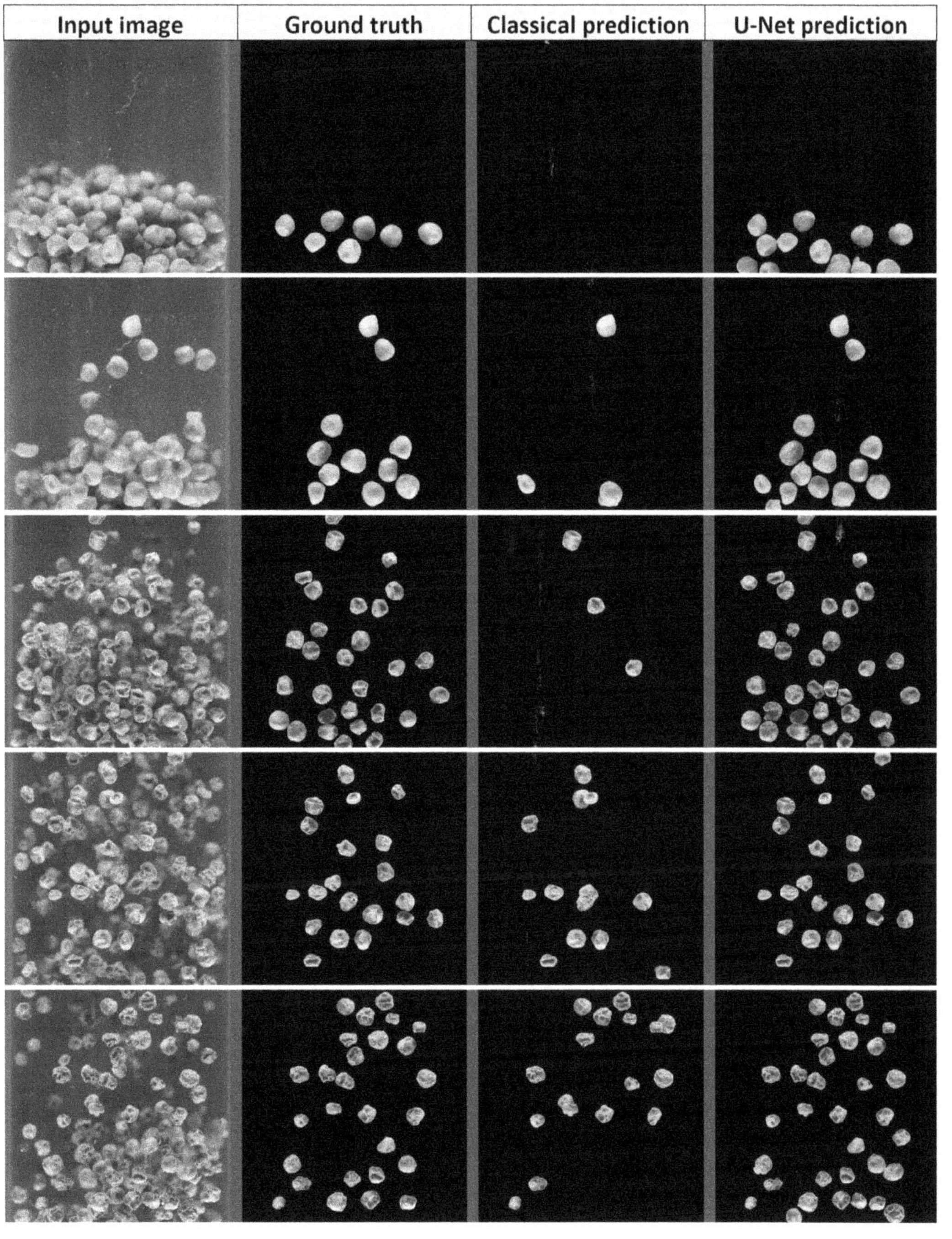

FIGURE 34.7 Segmented results: Input image, ground truth, classical results, and U-Net prediction.

and nutritional quality of foods in real-time, during drying. Interestingly, there is a final piece of the food-drying puzzle we need to discuss, referred to as intelligent system control.

With the monitoring system, we can keep track of food quality parameters as drying progresses. However, to eliminate or significantly reduce dry food waste and losses, we need dryers that will consistently produce dry foods that are visually appealing, that are rich in nutrients, and that have long shelf life. To achieve this goal, there has to be an intelligent system that can control the drying conditions to ensure the visual attributes, moisture content, and nutrient content of the end product is always within the limits we set out to achieve. Due to the non-linear nature of drying, and the lack of full understanding of how the various drying conditions affect food properties, we need a system control technique that can effectively learn this underling relationship. We can achieve this using a deep learning technique known as reinforcement learning (RL).

Intelligent System Control and Future Prospects for ABE Education

Developing a real-time food monitoring system will solve only half of the problem. The food can appear attractive but can still

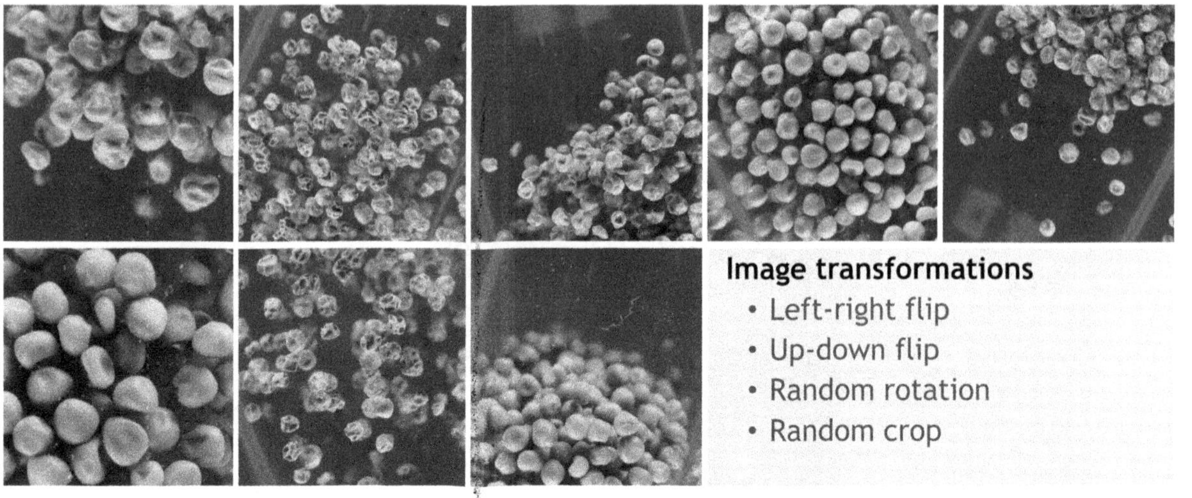

FIGURE 34.8 Realistic variants generated through image augmentation.

be moist inside. At this point, we run the risk of browning or casehardening the product if we keep drying it. However, if we choose not to continue drying, we also run the risk of losing the product to spoilage. Hence, to get out of this dilemma, we need an additional intelligent system that can modulate the drying conditions, based on the real-time feedback it receives from the monitoring system. The modulation is done in such way that we always attain the expected final moisture content, visual appearance, and nutrient content. To attain this optimal balance, this additional system should be able to learn latent relationships between drying conditions and food quality parameters, then building on this knowledge to design an appropriate system control. Such an intelligent control system can be developed using a deep learning technique known as RL technology.

Currently, there is no published literature where RL has been applied to drying. However, the technology has been successfully applied to several industrial processes, including heating, ventilation, and air conditioning systems, robotics, factory automation, machine tuning, smart grids, and autonomous vehicles, to name a few. By modifying the ABE curriculum to include topics in deep learning, computer vision, RL, and generative modelling, we can equip our students with a strong, practical, and theoretical foundation needed to develop intelligent food dryers that have the potential to significantly reduce food waste and losses.

Acknowledgement

The authors acknowledge the doctoral scholarship from the Tertiary Education Trust Fund (TETFund) of Nigeria, and the financial support of the Discovery Grant Programme of the Natural Sciences and Engineering Research Council of Canada (NSERC).

REFERENCE LIST

Aghbashlo, M., Hosseinpour, S., & Mujumdar, A. S. (2015). Application of artificial neural networks (ANNs) in drying technology: A comprehensive review. *Drying Technology, 33*(12), 1397–1462. https://10.1080/07373937.2015.1036288

Aghbashlo, M., Mobli, H., Rafiee, S., & Madadlou, A. (2012). The use of artificial neural network to predict exergetic performance of spray drying process: A preliminary study. *Computers and Electronics in Agriculture, 88*, 32–43. https://10.1016/j.compag.2012.06.007

Aral, S., & Beşe, A. V. (2016). *Convective drying of hawthorn fruit (crataegus spp.): Effect of experimental parameters on drying kinetics, color, shrinkage, and rehydration capacity.* https:////doi.org/10.1016/j.foodchem.2016.04.128

Ceylan, I., & Aktas, M. (2008). Modeling of a hazelnut dryer assisted heat pump by using artificial neural networks. *Applied Energy, 85*(9), 841–854. https://10.1016/j.apenergy.2007.10.013

Chen, H., Qi, X., Yu, L., Dou, Q., Qin, J., & Heng, P. (2017). DCAN: Deep contour-aware networks for object instance segmentation from histology images. *Medical Image Analysis, 36*, 135–146. https://10.1016/j.media.2016.11.004

Chollet, F. (2018). *Deep learning with python.* Manning.

Chollet, F. (2021). *Deep learning with python* (second edition). Manning.

Chen, H., Qi, X., Lequan, Y., & Heng, P-A. (2016). DCAN: Deep contour-aware networks for accurate gland segmentation. Paper presented at the 2487–2496. https://10.1109/CVPR.2016.273 https://ieeexplore.ieee.org/document/7780642 https://ieeexplore.ieee.org/document/7780642

Clevert, D., Unterthiner, T., & Hochreiter, S. (2015). Fast and accurate deep network learning by exponential linear units (ELUs). https://arxiv.org/abs/1511.07289 https://arxiv.org/abs/1511.07289

Demirhan, E., & Ozbek, B. (2009). Color change kinetics of microwave-dried basil. *Drying Technology, 27*(1), 156–166. https://10.1080/07373930802566101

Department of Economic and Social Affairs. (2019). *World population prospects 2019: Highlights.* United Nations. https://10.18356/13bf5476-en

Du, C. J., & Sun, D. W. (2004). Recent developments in the applications of image processing techniques for food quality evaluation. *Trends in Food Science & Technology, 15*(5), 230–249. https://10.1016/j.tifs.2003.10.006

Du, C. J., & Sun, D. W. (2006). Learning techniques used in computer vision for food quality evaluation: A review. *Journal of Food Engineering*, *72*(1), 39–55. https://10.1016/j.jfoodeng.2004.11.017

Environment and Climate Change Canada. (2019). *Taking stock: Reducing food loss and waste in Canada*. Waste reduction and Management Division, Environment and Climate Change Canada.

Erenturk, S., & Erenturk, K. (2007). Comparison of genetic algorithm and neural network approaches for the drying process of carrot. *Journal of Food Engineering*, *78*(3), 905–912. https://10.1016/j.jfoodeng.2005.11.031

FAO. (2015). *Global initiative on food loss and waste reduction*. FAO. Retrieved August 10, 2019, from http://www.fao.org/3/a-i4068e.pdf http://www.fao.org/3/a-i4068e.pdf

FAO. (2018). *The future of food and agriculture – Alternative pathways to 2050*. https://ageconsearch.umn.edu/record/319842/

FAO. (2019a). *Food loss and food waste*. Retrieved August 10, 2019, from http://www.fao.org/food-loss-and-food-waste/en/ http://www.fao.org/food-loss-and-food-waste/en/

FAO. (2019b). *Key facts on food loss and waste you should know*. Food and Agriculture Organization of the United Nations. Retrieved 2019, from http://www.fao.org/save-food/resources/keyfindings/en/ http://www.fao.org/save-food/resources/keyfindings/en/

FAO. (2019c). *State of food and agriculture: Moving forward on food loss and waste reduction*. FAO. https://statistical.proquest.com/statisticalinsight/result/pqpresultpage.previewtitle?docType=PQSI&titleUri=/content/2019/3410-S2.xml https://statistical.proquest.com/statisticalinsight/result/pqpresultpage.previewtitle?docType=PQSI&titleUri=/content/2019/3410-S2.xml

FAO. (2020). *World hunger statistics*. Food Aid Foundation. Retrieved 2019, from http://www.foodaidfoundation.org/world-hunger-statistics.html http://www.foodaidfoundation.org/world-hunger-statistics.html

Fathi, M., Fathi, M., Mohebbi, M., Mohebbi, M., Razavi, S., & Razavi, S. (2011). Application of image analysis and artificial neural network to predict mass transfer kinetics and color changes of osmotically dehydrated kiwifruit. *Food and Bioprocess Technology*, *4*(8), 1357–1366. https://10.1007/s11947-009-0222-y

Fernández, L., Castillero, C., & Aguilera, J. M. (2005). An application of image analysis to dehydration of apple discs. *Journal of Food Engineering*, *67*(1), 185–193. https://10.1016/j.jfoodeng.2004.05.070

Géron, A. (2019). *Hands-on machine learning with scikit-learn, keras, and tensor flow* (Second edition). O'Reilly Media.

Glorot, X., & Bengio, Y. (2010). *Understanding the difficulty of training deep feedforward neural network*. Paper presented at the International Conferenceon Artificial Intelligence and Statistics, 249–256.

Goyal, P., Dollár, P., Girshick, R., Noordhuis, P., Wesolowski, L., Kyrola, A., Tulloch, A., Jia, Y., & He, K. (2017). Accurate, large minibatch SGD: Training ImageNet in 1 hour. https://arxiv.org/abs/1706.02677 https://arxiv.org/abs/1706.02677

Gunasekaran, S. (1996). Computer vision technology for food quality assurance. *Trends in Food Science & Technology*, *7*(8), 245–256. https://10.1016/0924-2244(96)10028-5

He, K., Zhang, X., Ren, S., & Sun, J. (2015). Delving deep into rectifiers: Surpassing human-level performance on ImageNet classification. Paper presented at the 1026–1034. https://10.1109/ICCV.2015.123 https://ieeexplore.ieee.org/document/7410480 https://ieeexplore.ieee.org/document/7410480

Hernandez, J. A. (2009). Optimum operating conditions for heat and mass transfer in foodstuffs drying by means of neural network inverse. *Food Control*, *20*(4), 435–438. https://10.1016/j.foodcont.2008.07.005

Hernández-Pérez, J. A., García-Alvarado, M. A., Trystram, G., & Heyd, B. (2004). Neural networks for the heat and mass transfer prediction during drying of cassava and mango. *Innovative Food Science & Emerging Technologies*, *5*(1), 57–64. https://10.1016/j.ifset.2003.10.004

Hoffer, E., Hubara, I., & Soudry, D. (2017). Train longer, generalize better: Closing the generalization gap in large batch training of neural networks. https://arxiv.org/abs/1705.08741 https://arxiv.org/abs/1705.08741

Hosseinpour, S., Rafiee, S., Aghbashlo, M., & Mohtasebi, S. S. (2014). A novel image processing approach for in-line monitoring of visual texture during shrimp drying. *Journal of Food Engineering*, *143*, 154–166. https://10.1016/j.jfoodeng.2014.07.003

Hosseinpour, S., Rafiee, S., & Mohtasebi, S. S. (2011). Application of image processing to analyze shrinkage and shape changes of shrimp batch during drying. *Drying Technology*, *29*(12), 1416–1438. https://10.1080/07373937.2011.587620

Hosseinpour, S., Rafiee, S., Mohtasebi, S. S., & Aghbashlo, M. (2013). Application of computer vision technique for on-line monitoring of shrimp color changes during drying. *Journal of Food Engineering*, *115*(1), 99–114. https://10.1016/j.jfoodeng.2012.10.003

Huang, M., Wang, Q., Zhang, M., & Zhu, Q. (2014). Prediction of color and moisture content for vegetable soybean during drying using hyperspectral imaging technology. *Journal of Food Engineering*, *128*, 24–30. https:////doi.org/10.1016/j.jfoodeng.2013.12.008

Iheonye, A., Gariepy, Y., & Raghavan, V. (2019a). Computer vision for real-time monitoring of shrinkage during fluidized bed drying. Paper presented at the CSBE/SCGAB 2019 Annual Conference.

Iheonye, A., Gariepy, Y., & Raghavan, V. (2019b). Computer vision for real-time monitoring of shrinkage for peas dried in a fluidized bed dryer. *Drying Technology*, 1–17. https://10.1080/07373937.2019.1649277

Ioffe, S., & Szegedy, C. (2015). Batch normalization: Accelerating deep network training by reducing internal covariate shift. https://arxiv.org/abs/1502.03167 https://arxiv.org/abs/1502.03167

Islam, M. R., Sablani, S. S., & Mujumdar, A. S. (2003). Artificial neural network model for prediction of drying rates. *Drying Technology*, *21*(9), 1867–1884. https://10.1081/DRT-120025512

Julian, P., Mark, B., & Sarah, M. (2010). Food waste within food supply chains: Quantification and potential for change to 2050. *Philosophical Transactions of the Royal Society B: Biological Sciences*, *365*(1554), 3065–3081. https://10.1098/rstb.2010.0126

Kaminski, W., Tomczak, E., & Skorupska, E. (2004). Estimation of the effect of shape and temperature on drying kinetics using MLP. *Drying Technology*, *22*(1–2), 191–200. https://10.1081/DRT-120028227

Khazaei, J., Chegini, G., & Bakhshiani, M. (2008). A novel alternative method for modeling the effects of air temperature and slice thickness on quality and drying kinetics of tomato slices: Superposition technique. *Drying Technology*, *26*(6), 759–775. https://10.1080/07373930802046427

Khoshhal, A., Dakhel, A. A., Etemadi, A., & Zereshki, S. (2010). Artificial neural network modeling of apple drying process. *Journal of Food Process Engineering*, *33*, 298–313. https://10.1111/j.1745-4530.2009.00435.x https://agris.fao.org/agris-search/search.do?recordID=US201301796380

Klambauer, G., Unterthiner, T., Mayr, A., & Hochreiter, S. (2017). *Self-normalizing neural networks*. 31st Conference on Neural Information Processing Systems (NIPS 2017), Long Beach, CA, USA.

Lertworasirikul, S. (2008). *Drying kinetics of semi-finished cassava crackers: A comparative study*. https://https://doi.org/10.1016/j.lwt.2007.09.009

Liang-Chieh, C., Papandreou, G., Kokkinos, I., Murphy, K., & Yuille, A. L. (2018). DeepLab: Semantic image segmentation with deep convolutional nets, atrous convolution, and fully connected CRFs. *IEEE Transactions on Pattern Analysis and Machine Intelligence*, *40*(4), 834–848. https://10.1109/TPAMI.2017.2699184

Martynenko, A. (2017). Computer vision for real-time control in drying. *Food Engineering Reviews*, *9*(2), 91–111. https://10.1007/s12393-017-9159-5

Martynenko, A., Yang, S. X., & Pan, L. (2007). Intelligent computation of moisture content in shrinkable biomaterials. *Drying Technology*, *25*(10), 1667–1676. https://10.1080/07373930701590871

Martynenko, A. I. (2006). Computer-vision system for control of drying processes. *Drying Technology*, *24*(7), 879–888. https://10.1080/07373930600734067

Martynenko, A. I. (2008). Computer vision system for ginseng drying: Remote sensing, control and optimization of quality in food thermal processing. Paper presented at the ADM Verlag, 200.

Martynenko, A. I., Davidson, V. J., & Brown, R. B. (2005). Intelligent computer vision system (SAIF) for automated inspection of ginseng root quality. Paper presented at the CSAE/SCGR 2005 Meeting.

Masters, D., & Luschi, C. (2018). Revisiting small batch training for deep neural networks. https://arxiv.org/abs/1804.07612 https://arxiv.org/abs/1804.07612

Mayor, L., Moreira, R., & Sereno, A. M. (2011). Shrinkage, density, porosity and shape changes during dehydration of pumpkin (cucurbita pepo L.) fruits. *Journal of Food Engineering*, *103*(1), 29–37. https://10.1016/j.jfoodeng.2010.08.031

Movagharnejad, K., & Nikzad, M. (2007). Modeling of tomato drying using artificial neural network. *Computers and Electronics in Agriculture*, *59*(1), 78–85. https://10.1016/j.compag.2007.05.003

Mozina, M., Burmen, M., Tomazevic, D., Pernus, F., & Likar, B. (2009). Automatic visual inspection of pharmaceutical pellets in coating process. Paper presented at the 1–5. https://10.1109/ICIT.2009.4939618 https://ieeexplore.ieee.org/document/4939618

Murthy, T. P. K., & Manohar, B. (2014). Hot air drying characteristics of mango ginger: Prediction of drying kinetics by mathematical modeling and artificial neural network. *Journal of Food Science and Technology*, *51*(12), 3712–3721. https://10.1007/s13197-013-0941-y

Nadian, M. H., Nadian, M. H., Abbaspour-Fard, M. H., Sadrnia, H., Golzarian, M. R., Tabasizadeh, M., Martynenko, A., & Martynenko, A. (2017). Improvement of kiwifruit drying using computer vision system (CVS) and ALM clustering method. *Drying Technology*, *35*(6), 709–723. https://10.1080/07373937.2016.1208665

Nadian, M. H., Rafiee, S., Aghbashlo, M., Hosseinpour, S., & Mohtasebi, S. S. (2015). Continuous real-time monitoring and neural network modeling of apple slices color changes during hot air drying. *Food and Bioproducts Processing*, *94*, 263–274. https://10.1016/j.fbp.2014.03.005

Nahimana, H., & Zhang, M. (2011). Shrinkage and color change during microwave vacuum drying of carrot. *Drying Technology*, *29*(7), 836–847. https://10.1080/07373937.2011.573753

Oman Kadunc, N., Šibanc, R., Dreu, R., Likar, B., & Tomaževič, D. (2014). In-line monitoring of pellet coating thickness growth by means of visual imaging. *International Journal of Pharmaceutics*, *470*(1–2), 8–14. https://10.1016/j.ijpharm.2014.04.066

Oman, N., Možina, M., Tomaževič, D., Pernuš, F., & Likar, B. (2011). A study of a visual inspection technique for in-process monitoring of coating of pharmaceutical pellets. Paper presented at the 2011 IEEE International Conference on Industrial Technology, 239–243. https://10.1109/ICIT.2011.5754380 https://ieeexplore.ieee.org/document/5754380 https://ieeexplore.ieee.org/document/5754380

Romano, G., Argyropoulos, D., Nagle, M., Khan, M. T., & Muller, J. (2012). Combination of digital images and laser light to predict moisture content and color of bell pepper simultaneously during drying. *Journal of Food Engineering*, *109*(3), 438–448. https://10.1016/j.jfoodeng.2011.10.037

Ronneberger, O., Fischer, P., & Brox, T. (2015). U-net: Convolutional networks for biomedical image segmentation. *Medical image computing and computer-assisted intervention – MICCAI 2015* (pp. 234–241). Springer International Publishing. https://10.1007/978-3-319-24574-4_28

Russell, B. C., Torralba, A., Murphy, K. P., & Freeman, W. T. (2007). LabelMe: A database and web-based tool for image annotation. *International Journal of Computer Vision*, *77*(1–3), 157–173. https://10.1007/s11263-007-0090-8

Saldaña, E., Siche, R., Huamán, R., Luján, M., Castro, W., & Quevedo, R. (2013). Computer vision system in real-time for color determination on flat surface food. *Scientia Agropecuaria*, *4*(1), 55–63. https://10.17268/sci.agropecu.2013.01.06

Samadi, S. H., Ghobadian, B., Najafi, G., Motevali, A., & Faal, S. (2013). Drying of apple slices in combined heat and power (CHP) dryer: Comparison of mathematical models and neural networks. *Chemical Product and Process Modeling*, *8*(1), 1–12. https://10.1515/cppm-2013-0009

Sampson, D. J., Chang, Y. K., Rupasinghe, H. P. V., & Zaman, Q. U. (2014). A dual-view computer-vision system for volume and image texture analysis in multiple apple slices drying. *Journal of Food Engineering*, *127*, 49–57. https://10.1016/j.jfoodeng.2013.11.016

Saraceno, A., Aversa, M., & Curcio, S. (2012). Advanced modeling of food convective drying: A comparison between artificial neural networks and hybrid approaches. *Food and Bioprocess Technology*, *5*(5), 1694–1705. https://10.1007/s11947-010-0477-3

Smith, L. N. (2018). *A disciplined approach to neural network hyper-parameters: Part 1 – learning rate, batch size, momentum, and weight decay*. US Naval Research Laboratory Technical Report 5510-026. https://doi.org/10.48550/arXiv.1803.09820

Stępień, S., & Dobrowolski, D. (2017). Loss and waste in the food supply chain: An introduction to the problem. *Progress in Economic Sciences*, *4*. https://10.14595/PES/04/022

Wang, D., Martynenko, A., Corscadden, K., & He, Q. (2017). Computer vision for bulk volume estimation of apple slices during drying. *Drying Technology*, *35*(5), 616–624. https://10.1080/07373937.2016.1196700

Watano, S., & Miyanami, K. (1995). Image processing for on-line monitoring of granule size distribution and shape in fluidized bed granulation. *Powder Technology*, *83*(1), 55–60. https://10.1016/0032-5910(94)02944-J

Xu, B., Wang, N., Chen, T., & Li, M. (2015). Empirical evaluation of rectified activations in convolutional network. https://arxiv.org/abs/1505.00853 https://arxiv.org/abs/1505.00853

Yadollahinia, A., & Jahangiri, M. (2009). Shrinkage of potato slice during drying. *Journal of Food Engineering*, *94*(1), 52–58. https://doi.org/10.1016/j.jfoodeng.2009.02.028

Yadollahinia, A., Latifi, A., & Mahdavi, R. (2009). New method for determination of potato slice shrinkage during drying. *Computers and Electronics in Agriculture*, *65*(2), 268–274. https://10.1016/j.compag.2008.11.003

Zenoozian, M. S., Devahastin, S., Razavi, M. A., Shahidi, F., & Poreza, H. R. (2008). Use of artificial neural network and image analysis to predict physical properties of osmotically dehydrated pumpkin. *Drying Technology*, *26*(1), 132–144. https://10.1080/07373930701781793

Zheng, C. X., Sun, D. W., & Zheng, L. Y. (2006). Recent developments and applications of image features for food quality evaluation and inspection – A review. *Trends in Food Science & Technology*, *17*(12), 642–655. https://10.1016/j.tifs.2006.06.005

35

The Need to Incorporate Ergonomics in Smart Agriculture and Organic Farming in Agricultural and Biological Engineering Education

Masakazu Komatsuzaki, Muhammad Zakaria Hossain and Nugrahaning Sani Dewi

Introduction

The agricultural industry, unlike any other human enterprise, dominates land and water usage, and the agroecosystem plays a crucial role in sustaining human life and activities (Lal, 2004). The need for effective soil management is becoming increasingly apparent, as the agricultural output must be increased to promote equity and maintain global political and social stability. Farmers, consumers, researchers, and policy makers worldwide are beginning to promote and develop regenerative organic agriculture practices (Newton et al., 2020). Organic agriculture, which involves covering the soil surface with vegetation and supplying organic amendments, offers multiple benefits to the agroecosystem (Reganold & Wachter, 2016).

The objective of the Ministry of Agriculture, Forestry, and Fisheries is to attain several goals by 2050; the targets include the elimination of carbon dioxide emissions stemming from the combustion of fossil fuels in agriculture, forestry, and fisheries, a 50% reduction in the risk-associated employment of chemical pesticides via the propagation of integrated pest management techniques and recently devised substitutes, a 30% decrease in the utilisation of chemical fertilisers, and an expansion of organic farming to encompass one million hectares, thereby constituting a quarter of the arable land (MAFF, 2021). Amidst the COVID-19 pandemic, organic food witnessed an upsurge in demand owing to the public's perception of its health-enhancing attributes and potential for fortifying immunity. However, it is noteworthy that demand for organic produce has been on a continuous rise, regardless of the pandemic situation.

Despite its many benefits, organic farming practices can be labour-intensive and offer limited returns to farmers (Durham & Mizik, 2021). In response to labour shortages in agriculture, the concept of smart agriculture has emerged as a means of enhancing productivity while reducing labour requirements (Chetan Dwarkani et al., 2015). This strategy entails employing state-of-the-art technologies such as the Internet of Things, drones, robotics, and artificial intelligence to manage farms, thereby augmenting the quantity and quality of agricultural outputs (Agrimonti et al., 2021). Technologies such as robotic tractors and smart phone-operated water management systems for paddy fields are gaining popularity due to their ability to automate agricultural work and expand production capacity (Noguchi & Barawid, 2011). However, the adoption of smart agricultural technology is still in its early stages for labour-intensive crops such as fruit trees. To enhance agricultural productivity, monitoring tools such as self-propelled transport vehicles, assist suits, and field servers for temperature, ground temperature, and soil moisture are being employed. Among these technologies, grass-cutting robots are considered the most practical smart agricultural technology for fruit tree orchards.

This chapter discusses the ergonomic aspect of organic farming and smart agriculture. Developing smart agriculture for organic farming, which includes farming practices such as no-tillage cover crop and robotics adoption to mitigate climate change, will be an effective way to develop a new organic farming system. We also provide a brief overview of some of our ongoing research interventions aimed at improving productivity and reducing labour in organic agriculture. Finally, the significance of agricultural and biological engineering (ABE) education for smart agriculture and ergonomics is discussed from the perspective of developing a truly sustainable agricultural system.

Robotic Lawnmowers in Orchard Production

Currently, over 100 types of fruit are produced in Japan, with most of them being cultivated over the course of a year for a single annual harvest. Although these fruits are essential to our diets, their production has been declining since peaking in 1979. However, while the wholesale quantity of domestically produced fruit has decreased, the wholesale price has continued to rise. To maximise profits, fruit growers have chosen to focus on high-quality production through labour-intensive techniques, which is resulting in market and consumer recognition, leading to a stable increase in wholesale prices.

DOI: 10.1201/9780429150111-38

On the other hand, the labour-intensive production of high-quality fruit comes at the cost of sacrificing labour productivity through manual labour. Compared to other agricultural products, fruit production requires longer working hours and is more labour-intensive. The labour peak is particularly concentrated during short periods, such as harvest time, which requires securing temporary labour. For example, in Ibaraki Prefecture, Japanese pear production decreased by 27.1% over the ten years from 2009 to 2018 (Figure 35.1). Therefore, addressing labour shortages in fruit production is a significant challenge that must be tackled to maintain Japan's food security.

We evaluated the effectiveness of a robotic lawnmower for grass management in pear orchards compared to other methods, including walk-behind mowers, shoulder-mounted mowers, and ride-on mowers (Figure 35.2) (Hossain et al., 2020; Hossain & Komatsuzaki, 2021). The study was conducted in a pear orchard in Japan, where grass management is critical to maintaining healthy trees and preventing pest and disease infestations. We found that the robotic lawnmower was effective in maintaining grass height at five centimetres throughout the year, resulting in a significant reduction in weed growth compared to other methods (Figure 35.3a). The use of the robotic mower also reduced labour costs (Figure 35.3b). However, the robotic mower had limitations, including the need for boundary wire installation and frequent stops due to obstacles and low battery levels.

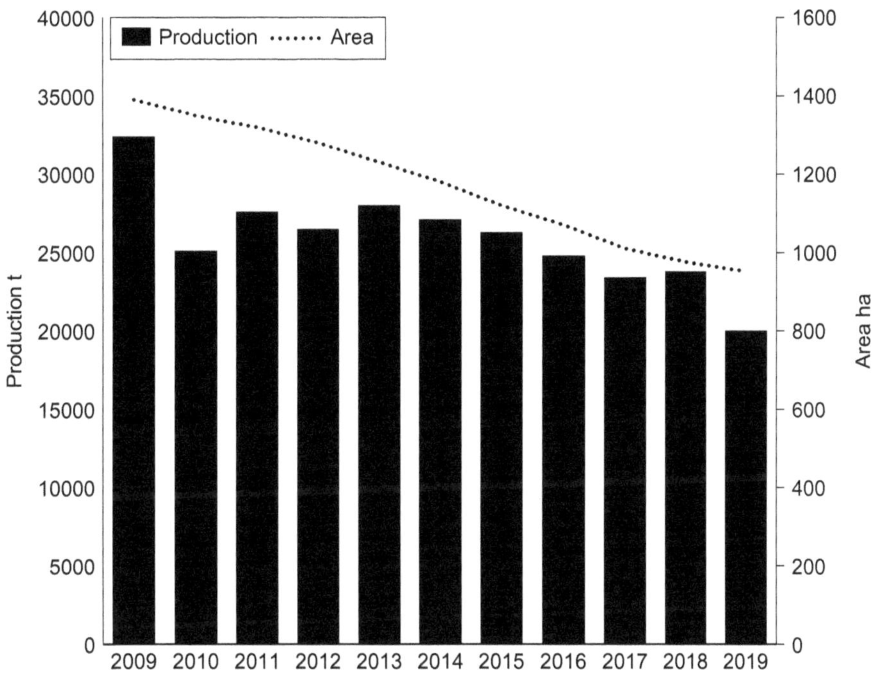

FIGURE 35.1 Trends of Japanese pear production and cultivation in Ibaraki, Japan (Ministry of Agriculture, Forest, and Fisheries, 2023).

FIGURE 35.2 Comparison of weed growth in a pear orchard where a robotic lawnmower (left) and a mechanical weeder (right) was used.

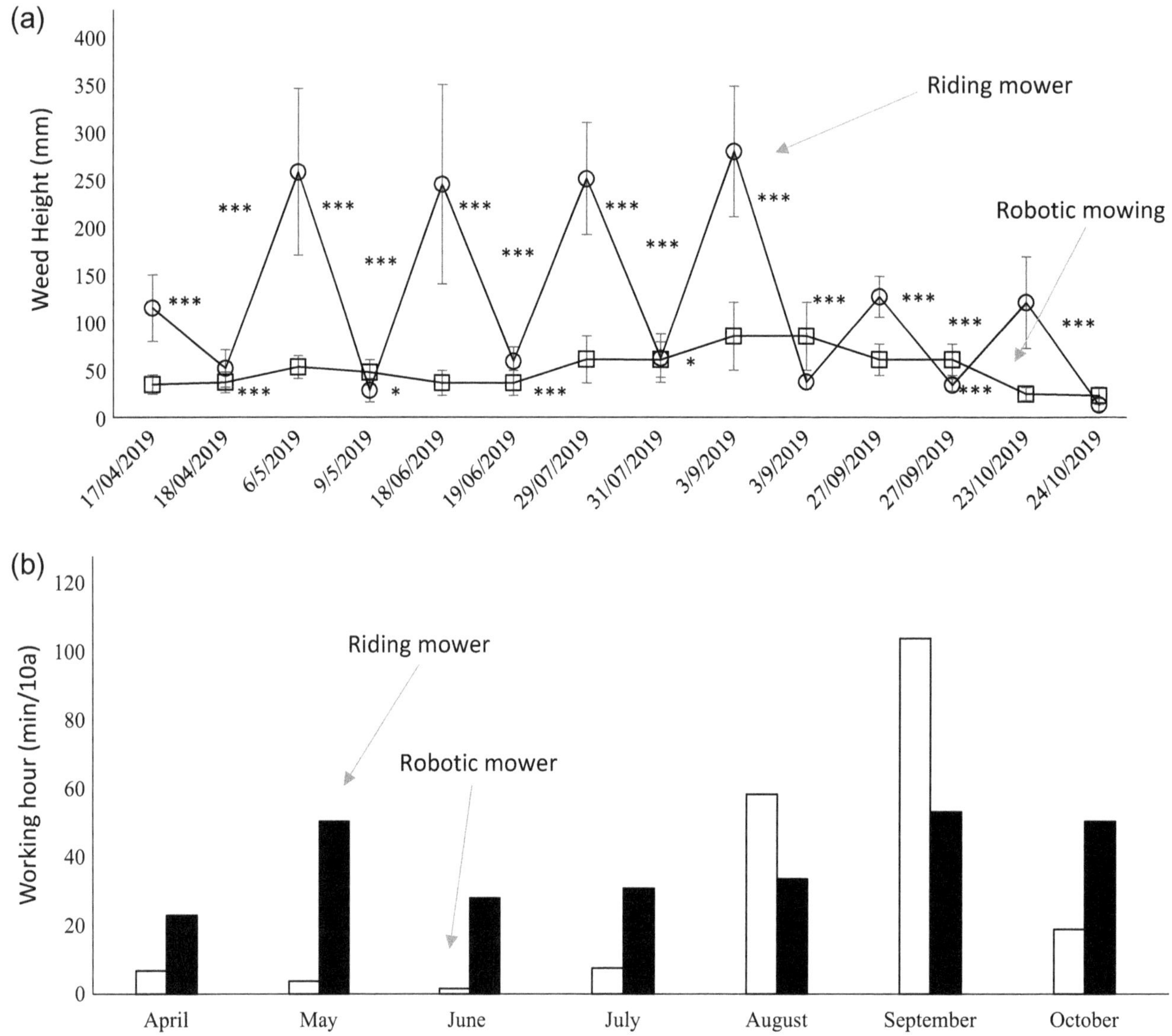

FIGURE 35.3 Comparison of weed management (a) and working hours (b) in Japanese pear production (Hassain et al., 2020, with permission from Japanese Journal of Farm Work Research).

The study also showed that the appropriate size of the area for a single robotic lawnmower was approximately 0.4 hectares, considering the maximum boundary wire length of 800 metres and the need to maintain a distance of 35 metres between the robot and the wire (Hossain et al., 2020). Our findings suggest that the robotic lawnmower is an economically feasible and ecologically sustainable alternative for grass management in pear orchards with a size of up to 0.4 hectares. However, for larger orchards, other methods such as ride-on mowers may be more suitable.

Organic No-till Systems as Climate-Smart Agriculture

Improvements in agricultural productivity through technological innovations are important in positioning organic farming as viable agricultural methodologies that have the potential to fulfil the demands of an expanding populace. Yields in no-till and cover crop cultivation are often less than half of those in conventional tillage. Therefore, there are concerns that the benefits of increased soil carbon sequestration and greenhouse gas reduction in no-till and cover crop cultivation may be significantly reduced if larger areas are needed to meet food demand due to lower yields, and that the greenhouse gas reduction per unit of harvest may be markedly lower. A global analysis of 610 case studies on no-till and conventional tillage found that no-till farming resulted in yield reductions in many cases, indicating that there are limits to the sustainability of food production with no-till farming as a standalone technique (Pittelkow et al., 2015). In this regard, the use of organic material coverings such as cuttings of weeds in no-till farming can increase yields and ensure equivalent harvests to conventional tillage and no-till and cover crop cultivation has an advantage over conventional farming (using chemical fertilisers) in terms of greenhouse gas emissions per agricultural product

(Matsuura et al., 2018). It is very interesting to note that cuttings, a traditional organic farming technique in Japan, may be useful for climate change mitigation. This suggests that it is necessary to combine various techniques to create a new agricultural system, rather than simply introducing no-till farming as climate-smart agriculture.

A comparative field test was conducted at the Ibaraki University International Field Agricultural Centre between the organic farming system that involved tillage and a model field of no-till and cover crop cultivation (Yagioka et al., 2014). The sequestration of soil carbon and the emission of greenhouse gases were measured for both methods, and the results showed that the adoption of no-tillage and cover cropping practices yielded a substantial positive impact on the sequestration of carbon in the soil and the mitigation of greenhouse gas emissions originating from agricultural fields, indicating that it could contribute to climate change mitigation (Yagioka et al., 2015). This result is similar to the results of previous studies on cover crop utilisation and no-till farming.

However, crop productivity with this method is not necessarily high. A comparison of yields of eggplant, bell pepper, watermelon, and tomato between the no-till and cover crop zone and the tilled zone was conducted at the Ibaraki University International Field Agricultural Centre, and the tilled zone tended to have slightly higher yields. No-till farming requires labour for weed management, as the soil is not tilled, and frequent weeding is essential. While no-till and cover crop cultivation have been shown to accumulate soil carbon, increased yield is still difficult due to competition between crops and weeds.

Organic No-till Cherry Tomato Production Using a Robotic Lawnmower

Excessive tillage, which is commonly employed for weed management in organic farming, has been found to lead to both physical and biological degradation of soil. To combat this issue, organic farmers can instead utilise weed residue mulch, a technique that has been shown to enhance soil aggregation, boost microbial activity, and stabilise organic carbon levels in soil (Hashimi et al., 2020). However, the management of weeds in organic no-tillage farming is considerably difficult and often provides poor harvest. The employment of commercial robotic lawnmowers (RLMs) in orchards and gardens has gained significant traction as of late (Hossain & Komatsuzaki, 2021). To determine the optimum weed management, we tested a Husqvarna RLM (Automower 430X) by operating it in an organic no-till cherry tomato production system at Ibaraki University, Ami, Japan (Figure 35.4).

During the tomato cultivation period, a study was conducted to assess the efficacy of two distinct treatments, namely conventional tillage (CT) and no-tillage (NT). Furthermore, a

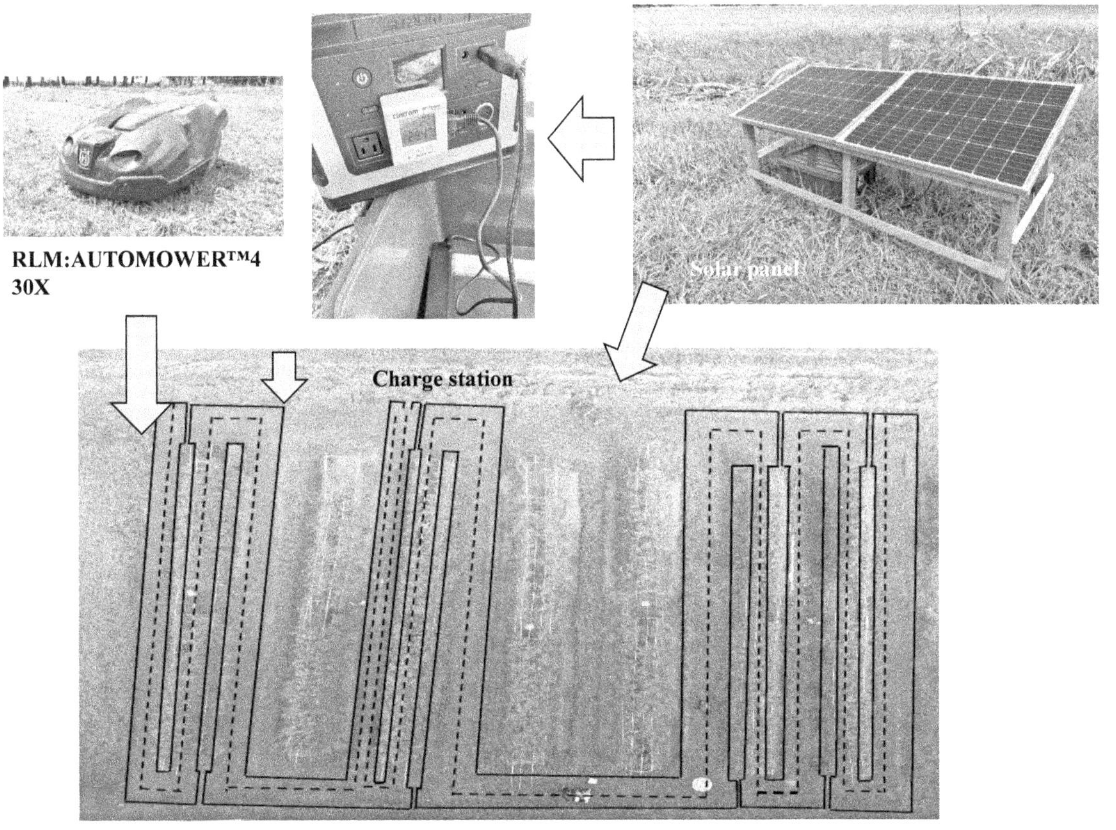

FIGURE 35.4 The experiment design between NT with RLM and CT. The power of RLM was supplied by solar panel, and the mowers run between the boundary wires along guide wire cables in the soil.

robotic lawnmower (RLM) was deployed exclusively in the NT treatment. We investigated the impact of introducing a robot lawnmower for weed management on the yield and quality of cherry tomato cultivation using a no-till, cover crop, and organic approach (Figure 35.4). The robot lawnmower used was bound with a guide wire and a boundary wire in the no-till cover crop area to enable automatic weed management and was equipped with a charging station and solar panels. We used a new cultivation method called no-till with weed mulch for cherry tomato cultivation in the open field during the summer and autumn seasons. The weed suppression degree of the robot lawnmower prevented the proliferation of tall weeds and significantly reduced the labour required for weeding to the extent of weeds growing from the gap in the mulch film. Moreover, the consumption power was sufficient, as the power from the solar panels was utilised. No-till and cover crop cultivation resulted in a 63% increase in yield, with 935 kg per 10 are compared to 572 kg in the tilled area (Fukuda, 2021). In cases where pesticides and chemical fertilisers are used, the average yield of cherry tomatoes is 2,000 kg per 10 are. Our cultivation method has a low planting density and no fertilisers, allowing for relatively spacious cultivation. Considering this, we expect this method to lead to a new environmentally friendly agricultural system that utilises the soil organic matter and solar energy through robots.

On-Body Assistance Suits in Farming

Individuals employed in the agricultural industry, particularly those engaged in activities involving significant manual labour, are at high risk of developing musculoskeletal disorders due to repetitive motions, such as digging. To address this issue, wearable ergonomic assistance devices, such as personal lift-assist devices and personal weight-transfer devices, have been developed to aid farmers in handling objects without compromising their back health. Studies have demonstrated that these devices effectively alleviate the workload of the back muscles during lifting and static forward bending. Despite these advancements, the impact of the recently introduced on-body personal assist suit (PAS) on agricultural tasks remains largely unexplored. Therefore, we are engaged in ongoing research to evaluate the effect of PAS on heart rate, muscle engagement, and trunk movement during digging tasks, while also considering user acceptance. Both male and female Subjects' heart rate, muscle activity, and motion acceleration will be monitored and recorded. PAS may serve as a straightforward safety intervention in field-related applications, and the study posits that PAS use will decrease demand for back muscles during digging (Dewi & Komatsuzaki, 2018). The study will monitor both male and female, and the study hypothesises that wearing a PAS will reduce the demand for back muscles during digging (Figure 35.5).

In this research, we used a device to record heart and muscle activity, as well as back movement, all at the same time. During the last minute of the activity, we made it easier for male participants using a specific method, but the overall difficulty level stayed the same. Male participants showed a more consistent increase in speed when using this method compared to female participants (as shown in Figure 35.6). Because 86% of females experienced discomfort in the crotch area, we recommend that hip straps be modified to increase user acceptance. For a typical agricultural forward-bending task, such as digging, PAS can significantly reduce strain ($p < 0.05$) on lower back muscles (L3) (Dewi & Komatsuzaki, 2018). The measurements of muscle and heart activity showed that using the PAS reduces strain on the lower back while digging. To make PASs more appealing to users, they should be designed differently for males and females.

Organic Smart Rice Production Using Global Navigation Satellite System Tractor and Transplanter

Organic farming is becoming more popular in agriculture, but in Japan it can be very labour-intensive and the amount of crops produced is often less than traditional rice farming (*Oryza sativa* L.). A two-year study was conducted to test organic farming methods in a rice field at the Centre for International Field Agriculture Research and Education, Ibaraki University (Matsuoka et al., 2022). For this farm management, we installed a global navigation satellite system (GNSS) assist tractor for tillage and puddling, and GNSS assist transplanter for rice transplanting in organic rice production (Figure 35.7).

The use of GNSS-assisted tractors for tilling provides high accuracy and straight-line operation even in flooded conditions and can suppress weed growth after tilling. Furthermore, GNSS-assisted transplanters exhibit excellent straight-line operation in soft, flooded soil conditions and facilitate the use of machine weeders afterward. By utilising such guided agricultural machinery, it is possible to enhance the efficiency of organic rice cultivation. Introducing such smart agricultural technologies into organic rice cultivation may have the potential to reduce the labour required for various agricultural tasks that were previously labour-intensive.

Future Prospects for ABE Education

The COVID-19 pandemic has been going on for four years now and has had a huge impact on our daily lives, including on the organic food industry. Many people are looking for healthy and safe food options to feed their families at home, and organic food has become a popular choice. The organic production side, however, still requires intensive labour for farming. Our results suggested that robotics, assisting the operation of the tractor, and personal assist suits showed dramatically reduced labour for organic farming. Organic farming often increases soil carbon sequestration; therefore, this system may provide an appropriate way for both human and environmental conservation.

Agricultural engineering plays a vital role in organic farming by providing specialised knowledge in improving agricultural systems, efficiency, automation, environmental protection, and sustainability. However, the current agricultural engineering

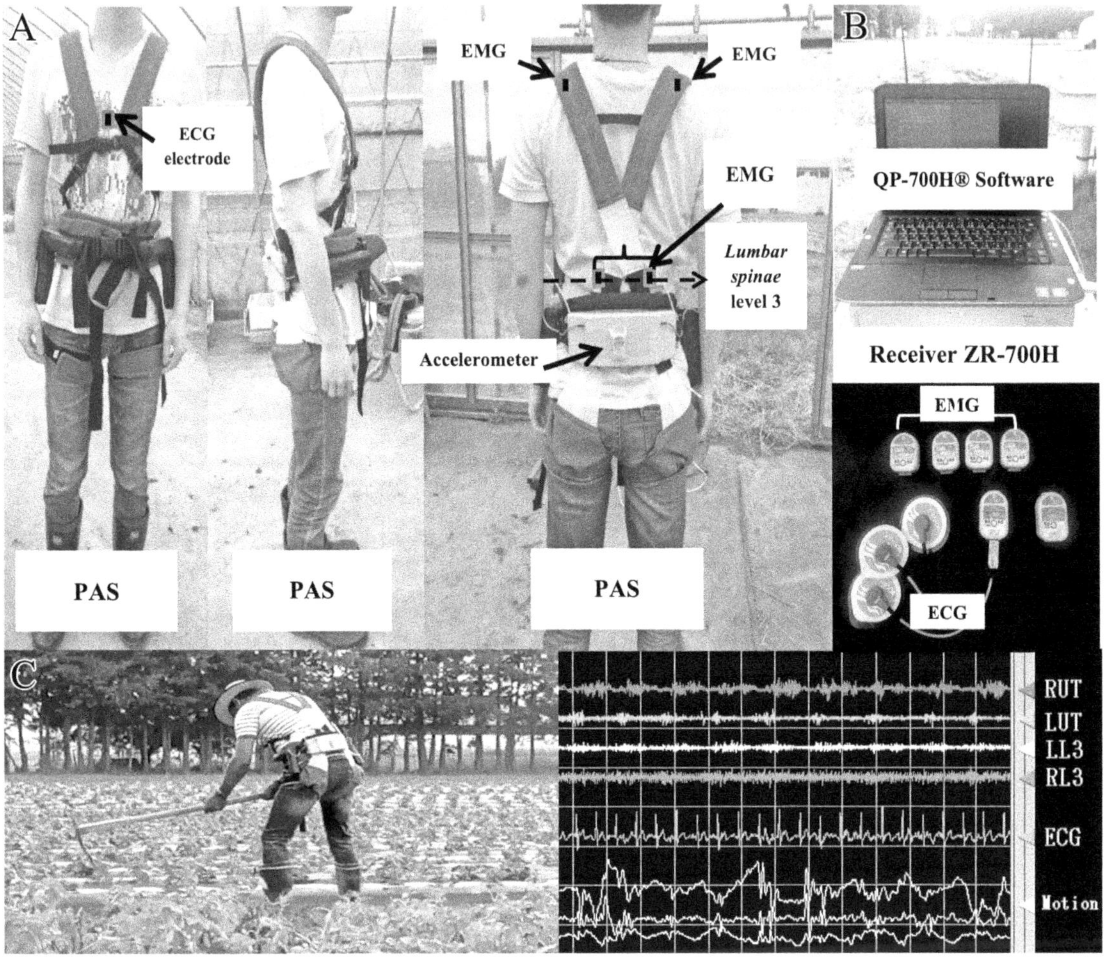

FIGURE 35.5 (A) On-body personal assist suit (PAS, front, side, and back view) and the placement of electrocardiogram (ECG), electromyography sensor (EMG), and accelerometer; (B) multi-channel telemetry system; (C) Subject performs digging activity with PAS and EMG, ECG, and motion acceleration displays (Dewi and Komatsuzaki, 2021, with permission from Elsevier).

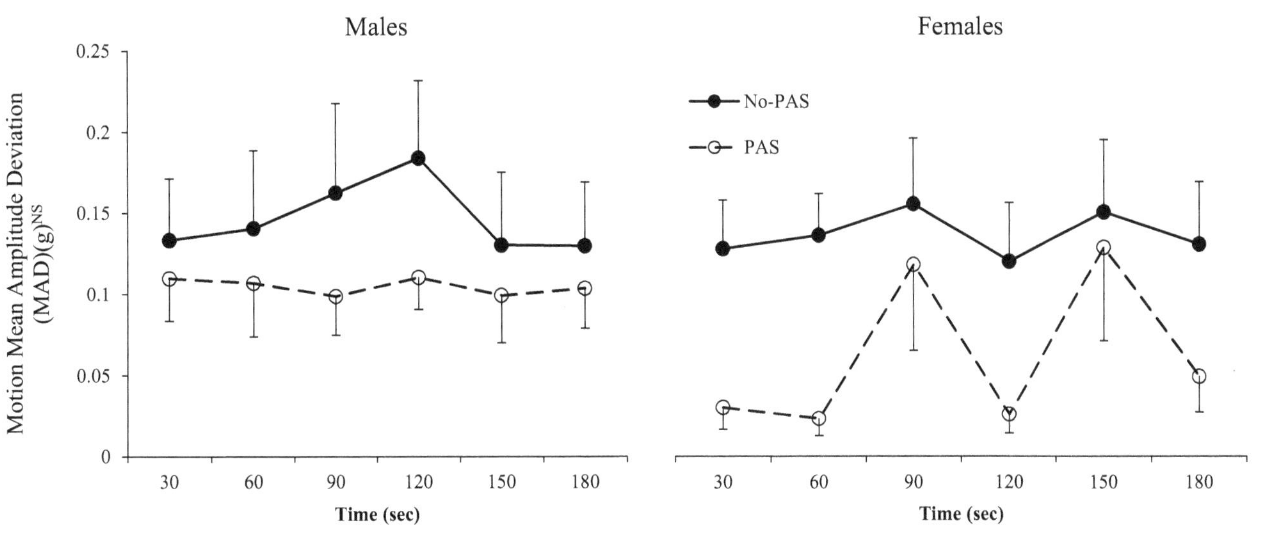

FIGURE 35.6 Mean amplitude deviation of motion acceleration across male and female subjects for no-PAS and PAS conditions during three-minute digging activity. NS: not significantly different at $p < 0.05$ across utilisation of PAS (one-way ANOVA); the error bar indicated standard error (data from Dewi and Komatsuzaki, 2021, with permission from Elsevier).

FIGURE 35.7 Organic farming in paddy rice system using GNSS-assisted tractor and transplanter.

education curricula have mainly focused on reducing labour and labour intensity, which has led to monoculture dominance and a decrease in biological diversity. To enhance sustainability in farming, including organic agriculture, a more site-specific approach that incorporates smart agriculture and ergonomic trials is necessary. Therefore, it is crucial to incorporate this approach into agricultural and biological engineering education.

REFERENCE LIST

Agrimonti, C., Lauro, M., & Visioli, G. (2021). Smart agriculture for food quality: Facing climate change in the 21st century. *Critical Reviews in Food Science and Nutrition*, 61(6), 971–981. https://doi.org/10.1080/10408398.2020.1749555

Chetan Dwarkani, M., Ganesh Ram, R., Jagannathan, S., & Priyatharshini, R. (2015). Smart farming system using sensors for agricultural task automation. *Proceedings – 2015 IEEE International Conference on Technological Innovations in ICT for Agriculture and Rural Development*, TIAR 2015, 49–53. https://doi.org/10.1109/TIAR.2015.7358530

Dewi, N. S., & Komatsuzaki, M. (2018). On-body personal assist suit for commercial farming: Effect on heart rate, EMG, trunk movements, and user acceptance during digging. *International Journal of Industrial Ergonomics*, 68, 290–296. https://doi.org/10.1016/j.ergon.2018.08.013

Durham, T.C., & Mizik, T. (2021). Comparative economics of conventional, organic, and alternative agricultural production systems. https://doi.org/10.3390/economies

Fukuda, M. (2021). Investigation of new no-till and organic cultivation techniques for mini tomatoes utilizing a robot lawn mower and root endophyte. Master Theses, Graduate School of Agriculture, Ibaraki University.

Hashimi, R., Kaneda, S., Kaneko, N., Ohta, H., & Komatsuzaki, M. (2020). Aggregate distribution and substrate-induced respiration under different tillage and mulching management systems in organic farming. *Soil Science and Plant Nutrition*, 66(6), 878–888.

Hossain, M.Z., & Komatsuzaki, M. (2021). Weed management and economic analysis of a robotic lawnmower: A case study in a Japanese pear orchard. *Agriculture (Switzerland)*, 11(2), 1–16. https://doi.org/10.3390/agriculture11020113

Hossain, M.Z., Takahashi, K., & Komatsuzaki, M. (2020). Robotic Lawnmower saves labor and operation costs in a pear (Pyrus pyrifolia) orchard. *Japanese Journal of Farm Work Research*, 55(3), 143–153.

Lal, R. (2004). Soil carbon sequestration impacts on global climate change and food security. *Science*, 304(5677), 1623–1627. https://doi.org/10.1126/science.1097396

Matsuoka, T., Asagi, N., & Komatsuzaki, M. (2022). Response of weeds and rice yield to Italian ryegrass as a cover crop and planting density in organic farming. *Agronomy Journal*, 114(1), 689–699.

Matsuura, E., Komatsuzaki, M., & Hashimi, R. (2018). Assessment of soil organic carbon storage in vegetable farms using different farming practices in the Kanto Region of Japan. *Sustainability*, 10(1), 152.

Ministry of Agriculture, Forestry, and Fisheries. (2021). Strategy for sustainable food systems, MeaDRI. https://www.maff.go.jp/e/policies/env/env_policy/meadri.html

Ministry of Agriculture, Forest and Fisheries. (2023). Crop statistics. Retrieved on April 1, 2023, from https://www.e-stat.go.jp/stat-search?page=1&toukei=00500215&bunya_l=04

Newton, P., Civita, N., Frankel-Goldwater, L., Bartel, K., & Johns, C. (2020). What is regenerative agriculture? A review of scholar and practitioner definitions based on processes and outcomes. In *Frontiers in Sustainable Food Systems* (Vol. 4). Frontiers Media S.A. https://doi.org/10.3389/fsufs.2020.577723

Noguchi, N., & Barawid, O. C. (2011). Robot farming system using multiple robot tractors in Japan agriculture. *IFAC Proceedings Volumes*, 44(1), 633–637.

Pittelkow, C.M., Liang, X., Linquist, B.A., Van Groenigen, K.J., Lee, J., Lundy, M.E., Van Gestel, N. Six, J., Venteea, R.T., & Van Kessel, C. (2015). Productivity limits and potentials of the principles of conservation agriculture. *Nature*, 517(7534), 365–368.

Reganold, J. P., & Wachter, J. M. (2016). Organic agriculture in the twenty-first century. *Nature Plants*, 2(2). https://doi.org/10.1038/NPLANTS.2015.221

Yagioka, A., Komatsuzaki, M., & Kaneko, N. (2014). The effect of minimum tillage with weed cover mulching on organic daikon (Raphanus sativus var. longipinnatus cv. Taibyousoufutori) yield and quality and on soil carbon and nitrogen dynamics. *Biological Agriculture & Horticulture*, 30(4), 228–242.

Yagioka, A., Komatsuzaki, M., Kaneko, N., & Ueno, H. (2015). Effect of no-tillage with weed cover mulching versus conventional tillage on global warming potential and nitrate leaching. *Agriculture, Ecosystems & Environment*, 200, 42–53.

36

CAM-SAM: A Computer-aided Tool for Education and Research on Selective and Sustainable Agricultural Mechanisation Development

Umezuruike Linus Opara

Introduction

Successful agricultural mechanisation in developing regions of the world, especially in sub-Saharan Africa, is still hampered by numerous structural bottlenecks. Agriculture is predominantly small-sized, individualistic, and at the subsistent peasant level. The unit of land cultivated by an average farm family rarely exceeds five hectares, with a degree of fragmentation up to 80%. The other, more general problems facing the growth of agricultural mechanisation in developing countries are embodied in the socio-economic, political, and techno-cultural characteristics of these countries.

In equatorial Africa, for instance, the typical farmer is a hand-tool user. An overview of the situation in Nigeria, which is the most populous and agricultural country in Africa, illustrates the status of agricultural machinery development and mechanisation. For instance, over 90% of farm operation in Nigeria (ranging from clearing to land preparation, planting, weed and insect control, harvesting, and primary processing) are still carried out using hand tools, which depends on human muscle as the major, and sometimes only, source of power. In fact, it is estimated that over 70% of Nigerians whose means of livelihood is farming are rural peasant farmers with two hectares of farmland or less and who mostly rely on hand-tool technology for their agricultural activities. Due to the low power output of the human muscle (which is rated at about 0.1 horsepower [hp]), farm operations depending on hand-powered technology are arduous, inefficient, and characterised by low rates of work which range from 0.017 to 0.059 ha per man-day for mound-making and primary processing operations, respectively (Anazodo et al., 1989a).

In addition to the above characteristic of hand-powered mechanisation technology in Nigeria, the village blacksmith industry has equally been faced with serious problems which adversely affect the availability and cost of hand tools. The blacksmiths who produce these tools have continued to be neglected by both the government and the public, resulting in a lack of good workshop facilities and equipment for the efficient production of tools. Raw materials are in short supply. These inadequacies and neglect have led to a steady and astronomical rise in the price of farm hand tools from about three naira in 1983 to well over ₦45 in 1989, and ₦5,000–7,000 for one hand-hoe in 2023. The lack of technological innovation in local agricultural machinery led, in part, to the emergence of private and commercial hand-tool hiring schemes in some states of Nigeria (Anazodo et al., 1989b).

The introduction and use of animal-draught technology in Nigeria is generally limited to the Sahel and Sudan Savannah ecological zones, where the terrains are relatively flat with sandy soils and where animals can survive. The use of animal-draught power in Nigeria was first tried in Daura in 1922 by the British colonial administrators in a bid to raise the cultivation of the two important cash crops in the Savannah ecologies, namely, groundnut and cotton (Musa, 1988).

Animal-draught implements have not been introduced in many parts of Nigeria, particularly the southern parts, because of a high incidence of tsetse flies in those areas which cause the deadly animal disease called trypanosomiasis. Other factors frequently attributed to causing limited use of draught animals in the central and southern areas of the country include unfavourable soil type and vegetation; unavailability of work animals, animal-based implements, and credit facilities; and poor extension and veterinary support.

Current reliable figures on the total number of farmers who employ animal-draught implements and equipment in Nigeria are not available. Previous estimated figures range from 100,000 to 300,000 work bull owners and less than 100 work camels (Musa, 1979, 1988), with 80% of the work bull owners concentrated in areas north of latitude 12°. The major operation for which animal-draught tools and implements are being used in Nigeria are primary tillage, weeding, and transportation. However, in areas where the appropriate equipment is available, animal power can be further exploited for planting and operation of stationary machines such as water wheels, pumps, threshing machines, winnowing, chaff cutters, and grinders (Hopfen, 1969).

Nearly four decades ago, the predominant animal-drawn implement in use in Nigeria was the Emcot ridger, which was produced locally by John Holt and Company Limited in Zaria. There were a few imported ridging ploughs (Chinese and Italian), mostly of lighter construction, which were also available in some areas (Yadav & Oguejiofo, 1987). Due to the increasing cost of tractors and implements, the use of animals as power sources for farm operations was on the increase, as

DOI: 10.1201/9780429150111-39

was greater attention by various government and private agencies on draught animals in the middle-belt and northern parts of Nigeria.

The use of engine-empowered technology for agricultural operations has continued to receive considerable attention by government, development agencies, and agribusiness. Various efforts have been made to improve agriculture in developing countries, especially in sub-Saharan Africa, using some imported engine-powered agricultural mechanisation technology. These range from power tillers (single-axle tractors) to conventional four-wheel tractors with complementary implements for ploughing, harrowing, weeding, chemical application, and shelling, threshing, drying, and transportation. The introduction of engine-powered agricultural equipment in these countries was initially mostly limited to few state-owned farms and River Basin Development Authorities, as is the case in Nigeria. The use of tractorised machinery on private farms was encouraged through the services provided by the tractor and equipment hiring units of the ministries of agriculture. Even these tractor-working units were poorly managed (Anazodo, 1988). Currently, more private individuals and a few public liability companies have shown interest in large-scale farming, especially in the production of the raw materials needed for their industries.

Several decades ago, studies by Anazodo et al. (1989a) showed that, in general, agriculture machinery was in short supply. However, prices were prohibitive and in most cases beyond the reach of private individuals. During the same period, the results of a national investigative survey research conducted by Anazodo et al. (1986) showed that at the end of 1985, the total number of tractors used for agricultural purposes in Nigeria was 15,906. In view of the difficulties in importing agricultural machinery and other products into the country because of the ongoing structural adjustment programme coupled with the high cost of what little was available locally, as well as the general difficulties in obtaining spare parts, the 1989 number of operational farm tractors, at best, was projected to remain at the 1985 figure as above. Details of the relevant findings on the agricultural machinery situation in Nigeria are presented in the survey report (Anazodo et al., 1986).

Based on a comprehensive inventory of agricultural equipment presented in the national investigation survey research, the status of agricultural mechanisation in Nigeria was summarised and it was shown by the authors that the most predominant size of four-wheeled, two-axle, conventional, or standard farm tractors available in Nigeria was 66 to 75 hp. Disc ploughs, disc harrows, knapsack sprayers, and ridgers were found to be the most numerous modern agricultural equipment available and it was further discovered that other modern agricultural equipment, such as planters, cultivators, harvesters, threshers, shellers, dryers, and irrigation technology, were not commonly in use in many of the establishments visited. The least number of agricultural equipment reported to be available in Nigeria were the root and tuber crop planters and harvesters. It was deduced from the study that about 51.1% of the total number of farm tractors purchased in Nigeria were purchased for agricultural purposes and this figure was used by the authors to determine a tractorisation intensity of 0.024 hp/ha for the country.

In another study reported by Anazodo et al. (1989a), the individual contributions of the various levels of agricultural mechanisation in the overall or national agricultural mechanisation technology were evaluated. The amount of power available to the farmer was considered the most critical factor in his ability to effectively and efficiently adopt improved production equipment to raise his volume of agricultural production and overall farm activities. This amount of power available to the farmer per unit area of land cultivated was then considered as a measure of the level of farm mechanisation. It was determined that the total draft power available on Nigerian farms was 0.1072 hp/ha of cultivated land. Manual labour accounted for over 71% while farm work animals and tractors account for six percent and 23%, respectively.

These data on mechanisation intensity are considered low compared to 0.5 hp/ha, which is the minimum power requirement for necessary cropping function, as recommended several decades ago for developing countries (Giles, 1967). Four decades ago, the overall draft power available in Nigerian agriculture was only about 21% of this minimum requirement and it was thus necessary to increase the number of operational farm tractors in Nigeria as well as the number of available farm power sources. Similarly, the ratio of farm implements to the available operational number of farm tractors was equally adjusted not only to ensure more effective utilisation of the farm tractors but also to achieve timeliness in planting, weeding, and harvesting operations.

The prospects for the development of appropriate agricultural mechanisation technology and its effective utilisation in transforming smallholder agriculture are bright. A high willingness index and favourable level of awareness have continued to increase among farmers, especially on the need to adopt mechanised methods to enhance the productivity and efficiency of the entire crop production enterprise. What is now needed most importantly is the identification and adoption of an effective strategy for agricultural development in which agricultural mechanisation and management are properly integrated.

The search for sustainable agricultural mechanisation to modernise and transform agri-food systems requires the development of optimal models for efficient allocation and effective utilisation of available technology. For agricultural mechanisation to be effectively applied, it must be appropriate to the political, cultural, economic, and technological environment. It must be profitable. It must be sustainable. Indeed, the application of such technologies must be meaningful or selective, as recommended by several investigators over the past half'century (Giles, 1973; Abercombie, 1975; Anazodo, 1975, 1981, 1986, and 1988).

The foregoing review of the status of agricultural mechanisation development in Nigeria, as a case study of the need to conceptualise, design, develop, and adopt an effective strategy in which agricultural mechanisation and management are properly considered, raises some fundamental questions, as previously articulated by Anazodo (1975):

1. What appropriate levels of mechanisation are needed?
2. Are we planning for a selective gradual and progressive mechanisation, or do we aim at a complete mechanisation at a stroke?

3. Do we go for a complete substitution of our more familiar but probably inefficient hand tools with the latest designs in imported tillage, seeding, weeding, harvesting, and processing equipment?
4. Have we sufficiently identified the social, technical, economic, and resource problems that may impede a successful introduction of farm power and machinery?
5. Are these impediments quantifiable?
6. How limiting are they as input factors in a mechanisation programme?

In order to evaluate the optimal levels of mechanisation in any agri-food system, the above fundamental questions were introduced into our system of goals in agricultural mechanisation, and it was considered necessary to provide suitable answers to the following (Anazodo, 1975):

- What is the desired (or optimal) level of production consistent with maximal availability of productive land, capital, and market resources?
- Does the farmer achieve an optimal level of production under his existing level of mechanisation?
- Can the optimal level of production be reached by cultivating more land?
- What level of machinery introduction will achieve the desired level of production?
- What powered system (that is, hand, animal, or engine power) is most effective in achieving the optimal levels of production, etc.?

As part of the continued research endeavours in this respect, Anazodo (1985) reported the results of a study undertaken to evaluate the appropriate systems of maize production in a community farm project. By a graphical approach, traditional and selectively mechanised systems were compared on the basis of timeliness and profitability, and it was possible to determine the optimum farm size for manual labour and the expected increase in the cultivated area if the farmer is encouraged to use a tractor-power system.

The graphical technique was further developed into nomographs to assist in the selection of farm power and machinery management systems for timely and economical crop production (Opara, 2023, in this book). Furthermore, a probabilistic inventory (stock-control) model has been adapted and applied to develop a guideline for systematic replacement and stocking of agricultural machinery spare parts for improved farm machinery use in Nigeria (Anazodo et al., 1989a, b).

These reports effectively summarise the basic considerations, concepts, principles, features, and applications of systems analysis and operations research into the solution of the fundamental question of determining the appropriate level of agricultural mechanisation systems and the strategy for increased agricultural production and productivity. The next stage in these research endeavours is to finally develop a package of computer-based optimal models that will facilitate a more accurate and broad-based solution to the fundamental question. Such a package will provide a systematic basis for the proper selection and effective management of farm machinery at various levels of agricultural mechanisation technology.

The information to be generated from a solution of the models will be useful to the government, agricultural mechanisation development planners, advisers, extension officers, farmers, and other entrepreneurs interested in agribusiness. Indeed, the results of the overall study will provide the rational basis for the selective mechanisation of agriculture and act as a tool in agricultural and biosystems engineering education. The broad aim of this chapter was to develop a computer-aided model for a systematic analysis of the farm power and machinery and labour requirements for different levels of agricultural mechanisation. Specifically, based on timeliness consideration, minimum cost, maximum net economic benefit, and overall optimum financial efficiency, the objectives were to:

1. For a given number of available manual labourers, determine the appropriate farm size.
2. For given farm sizes, determine the farm power requirements for field operations.
3. For the optimally selected farm power levels, determine the appropriate farm machinery for seedbed preparation, planting, weeding, and harvesting.

The entire model was designed to provide a learning tool that will enhance a broad-based solution to the selection of farm power and machinery at all feasible levels of agricultural mechanisation.

Model Development

Background to the Model

As previously pointed out in chapter one of this report, agricultural mechanisation development in many developing countries is still in its first phase of evolution. Despite a growing desire among farmers to adopt improved methods of agricultural production, several bottlenecks have continued to impede the growth of successful agricultural production, including mechanisation. Except in a few government farms and those owned by public liability companies and a few individuals, farm sizes are small with a high degree of fragmentation.

Differences in ecologies and agro-climatology affect the relevance, acceptance, and diffusion of agricultural mechanisation technology into the existing cropping systems. Most farmers are, therefore, only able to apply specific components of the available agricultural machinery technology. Investment in modern engine-powered technology is quite unattractive in most circumstances because of small farm holdings. In some other instances, farmers can only mechanise a fraction of the entire production system due to the lack or unavailability of the technology for other operations; introduction of modern agricultural mechanisation technology into an environment must consider the characteristics of the farmer and the level of technology compatible with the existing cropping system. The growth in farm machinery utilisation must be gradual and progressive. Farmers should be encouraged to adopt only those technologies which justify the extra investment in machinery. Any programme designed to enhance agricultural mechanisation development should consider the various categories of farmers in the country to ensure its relevance and success.

In this study, the call for selective mechanisation of agriculture is because some technologies in farm machinery will be best suited to certain categories of farmers who fall into particular levels of agricultural mechanisation. Agricultural mechanisation should, therefore, only be applied to specific conditions where the ingredients exist or can be easily developed. A basic consideration in the model presented in this report is that there is a need for a systematic basis for adopting any level of agricultural mechanisation. The transition from any level to another should equally be used in a careful analysis of the enterprise system to select the optimal machinery systems needed to minimise costs.

An integrated block diagram showing the components and decision pathways of the computer-aided model for selective agricultural mechanisation (CAM-SAM) is presented in Figure 36.1. CAM-SAM was conceptualised, designed, formulated, and developed for the optimum system of farm machinery and specifies the following:

1. The optimum level of farm power source required in terms of number of farm labourers, number of animals, size of tractors.
2. The optimum size of implements (ploughs, harrows, ridgers) for seedbed preparation and other field operations.
3. The costs (fixed, operational, timeliness, total system cost) and the amount of investment needed for the selected system of farm machinery.
4. The optimum range of farm sizes for each level of mechanisation, given power and farm labour limitations.

The above information is considered particularly useful to farmers, farm managers, and extension agents to help determine whether existing labour and machinery systems are appropriate and efficient in terms of economy and timeliness. It will further be useful to decide whether to "transit" or "switch" over to another level of mechanisation. From a system of the least costs obtained from each level of mechanisation, the final selection will be made based on the appropriateness of each level of mechanisation.

The appropriateness of any level of mechanisation is evaluated in terms of the following:

a. The availability of the production technology associated with that level.
b. The availability of the production technology. This criterion is assessed based on the cost of investment, in which case the farmer's initial capital savings and the amount of loan he can secure are the limiting input factors.
c. The profitability of the entire farming enterprise. This overall criterion is apparently the most sensitive because of the role of financial incentive and/or the role of profit motive in the acceptance and diffusion of new technologies which are expected to cause disequilibrium in the traditionally stable peasant economy (Schultz, 1964).

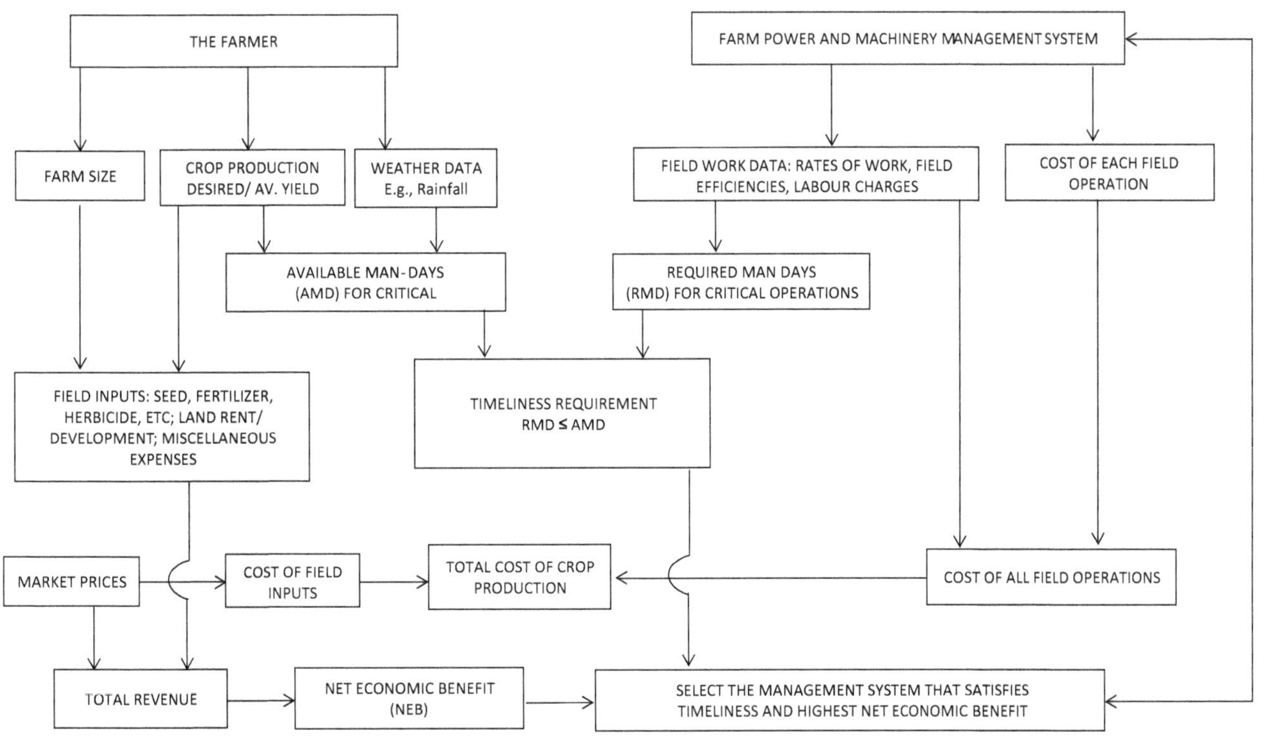

FIGURE 36.1 Block diagram of the relationships in the CAM-SAM model for selective gricultural mechanisation.

The model has also been designed for all crops and agro-climatic zones. Its successful application in any situation will depend on the availability of the requisite input data.

Characterising Levels of Agricultural Mechanisation

The development of agricultural mechanisation over the years has led to a gradual replacement of farm labourers and animal power by engine-powered systems. The three basic levels of technology remain hand-tool technology, animal-draught technology, and engine-powered machinery technology. Due to differences in farm holdings, cropping patterns, and agro-climatology, one major level of mechanisation may not be optimal for all sizes and classes of farmers. Farmers, therefore, often accomplish their farm operations using either one major level of agricultural mechanisation technology or a combination of levels.

For the model presented in this report, five levels of agricultural mechanisation were studied based on current production practices in Nigeria:

Level 1: Where only human muscle and associated hand tools are used for all field operations.

Level 2: Where animal power and associated implements are only used for seedbed preparations, while human muscle and associated hand tools are used for other field operations.

Level 3: Where tractor power and associated implements are only used for seedbed preparations, while other field operations are performed using human labour.

Level 4: Where only animal power and associated implements are used for all field operations where feasible or where the technology is adaptable or could be adequately developed.

Level 5: Where only tractor power and associated implements are used for all field operations.

In the development of CAM-SAM, each of the five levels was analysed for power, implements, and labour requirements. Hence, each level of mechanisation represents a sub-model of the main CAM-SAM model. Thus, CAM-SAM.1, CAM-SAM.2, CAM-SAM.3, CAM-SAM.4, and CAM-SAM.5 are used to represent the constitutive five levels of agricultural mechanisation.

Optimum Calendar for Farm Operations

For a given crop, the optimum time available for farm operations varies significantly due to differences in cropping practices and agro-climatology. Depending on a package of agronomic practices suitable for optimum crop performance, a chart is developed for any cropping system, specifying the optimum calendar dates for various field operations. Oftentimes, farmers are required to perform a combination of farm operations during the same period using the same power source, in which case the actual available optimal time for each of those operations is less than the total time duration specified in the chart. The optimum working hours available for each of such farm operations is estimated from Equation 1:

$$OPWH_i = \left[\sum_{J=1}^{N} \left(\frac{NOD_j}{NOO_j} \right) \right] \times WH_r D_i \qquad (1)$$

Where:

$OPWH_I$ = optimum working hours for i^{th} operation, h.

NODj = number of days available during j^{th} time interval, days.

NOOj = number of operations to be performed by the same power source during j^{th} interval.

WH_rD_i = working hours per day for i^{th} operation, h/day.

Determination of Timeliness Cost

For a given agro-climatology and cropping system, there exists a date to complete specific field operations that results in maximum crop yield. However, there is an optimum period during which crop yield is constant for any given field or primary processing operation. It is assumed that crop yield suffers no loss if farm operations are completed within the optimum period and loss decreases at the same rate whether the farm operation is carried out before or after the optimum period of the operation.

The crop yield function with respect to the completion time of the farm operation is therefore of the form shown in Figure 36.2.

For any given farm operation, the annual cost can be evaluated by relating the crop value for optimum completion of the operation and that for the operation completed before or after the optimum period. For operations in which the total required working hours ($REWH_i$) is less than or equal to the optimum working hours available ($OPWH_i$), the annual timeliness charge ($ANTC_i$) is assumed equal to zero. In general, we can therefore write Equation 2 as:

$$ANTC_i = \frac{\left(REWH_i - OPWH_i\right) \times YILS_i \times FMSZ \times VCRP}{WHPD_i} \qquad (2)$$

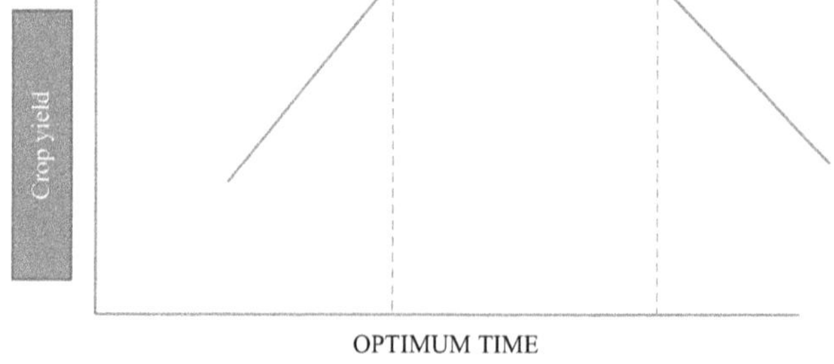

FIGURE 36.2 Crop yield versus completion time operations.

Determination of Farm Machinery Costs

The following costs are associated with the use of farm machinery:

(a) Fixed cost – depreciation, interest, taxes, insurance, and shelter.
(b) Operating cost – labour, fuel, oil, repairs, and maintenance.
(c) Timeliness cost – due to the inability of the machinery to complete an operation within the optimum period of time available.

For machines that do not have a power component, the cost of power source (fractional) is included as part of their operating cost.

The total annual cost (TANC) for a farm machine can, therefore, be expressed in Equation 3 as:

$$TANC = ANFC + ANOC + ANTC \tag{3}$$

Where:
$ANFC$ = annual fixed cost
$ANOC$ = annual operating cost
$ANTC$ = annual timeliness cost

For animal and engine-powered systems, the components of each annual cost equation that depended upon the size or capacity of the machine were expressed in terms of horsepower, width, or capacity rating; farm work animals were assumed to appreciate in value, as they were expected to gain weight at the end of their useful life. The appreciated value of the animals was, therefore, deducted from their TANC as associated implements. Taxes for animal-powered operations were equally assumed to be zero, since the payment of *jangali* (cattle tax) has been abolished in Nigeria. Fuel and repair and maintenance costs per hour were expressed in terms of the fuel cost using five percent of the specific fuel consumption of the machine (Singh & Gupta, 1980).

The prices of farm animals, tractors and other machines vary significantly in Nigeria depending on the make and location of the machinery dealer, and, most remarkably, on the fluctuating value of the naira at the foreign exchange market or *bureau de change*. However, for all power sources and associated equipment, the relationship between price and size was assumed to be linear, as this was found to be true for products from the same company.

The annual cost equation for various farm machines associated with each level of mechanisation was derived following the procedure by Hunt (1963) and Singh and Gupta (1980). Equations for optimum size were developed using simple differential calculus technique to minimise the annual cost, as presented by Hunt (1963) and Chancellor (1968). The minimum cost equation is obtained when the first derivative of the annual cost with respect to the size of the machine is taken as equal to zero, provided the second derivative of the equation is positive.

For manual farm operations, the annual fixed cost is set equal to zero, since the labourer owns the hand tool, the cost of which is reflected in the labour cost for any operation as the annual operating cost. The average rate of work for manual operations was considered critical, as this greatly determines the amount of work that can be accomplished. Manual labour was considered to consist of family and hired labourers. The hirable number of labourers for a given farm size is determined as the difference between an optimum number of labourers required and the number of family labourers available for farm operations.

Despite noticeable variations in the types and sizes of hand tools used in Nigeria, which is expected to influence the size of mounds and overall work output by labourers, the size of manual tools was assumed to be fairly the same for each field operation. Hand-powered implements were, therefore, not selected, since it is assumed that each manual labourer is equipped with a single unit of each piece of equipment.

Similarly, it was also assumed that manual labour is available; however, the use depends on the ability of the farmer to pay. Thus, while human resources are available all year round for farm operations, the minimum number of labourers a farmer can hire depends on the capital available and the overall economics of manual labour operations. On the other hand, for an average farm family, comprising of a man and his wife with full-time farming as their major economic activity, with five dependent relatives and children, there are between three and five family labourers available for farm work at any time of the year.

Model Testing

Model Application to a Selected Area

The optimal selection of any farm machinery system depends on climate, type of crop(s), and the other conditions prevailing in a given area. To illustrate the use of CAM-SAM and the programmes presented in this study, the Nsukka area in the rainforest zone of Nigeria was chosen. Maize crops and a soil type of light texture were also chosen.

Data on manual labour operations are based on studies reported by Anazodo (1976, 1982, 1985), Odigboh (1976), and as assembled by Opara (1987) from relevant publications. Machinery management data on tractors and associated implements are based on research reports (Anazodo, 1982; Anazodo et al., 1986, 1989a, 1989b). Price quotations for equipment were also obtained from reputable agricultural machinery dealers. The price of tractors and other agricultural machinery varies significantly according to their make and quality. For these and farm work animals, the relationship between price and size was assumed to be linear. This is true for other products from the same company.

The calendar of farm operations for maize production is based on that recommended by Ibe (1987). Due to the unavailability of timeliness data on maize crop based on the project area, yield losses from late planting of maize (80 kilogrammes per hectare per day), as reported by Jones (1976) based on planting time studies on maize at Samaru, was used as input data for the model. Data on the performance of farm work animals is based on reports by Musa (1988), Anazodo et al.

(1989a, 1989b), Kline et al. (1969), Hopfen (1969), Munzinger (1982), and Singh and Gupta (1980). Agronomic requirements for maize production are based on AgroSystems (1984) and Anazodo (1985) and a review of the problems and prospects of maize production in Nigeria. Due to time and budget limitations, it was impossible to carry out an extensive survey to assemble all the input data required. The computer programmes developed were used to determine the optimum farm machinery system for each level of agricultural mechanisation.

Results and Discussion

The computer programmes developed were used to determine to farm mechanisation and labour requirements for maize production. For given farm sizes, the sub-model CAM-SAM was used to determine the number of manual labourers required to complete all field operations without incurring timeliness costs. For hypothetical farm sizes 1, 5, 10, 20, 50, 100, 500, and 1,000 ha, the number of manual labourers required are 2, 8, 16, 31, 77, 154, 769, and 1,537, respectively. The computer programme and printouts are contained in the appendix. Also, for a given number of available manual labourers – four and six – it was determined that for zero timeliness charge, the labourers can effectively cultivate two and three hectares, respectively. When the farm size was increased to ten hectares, 16 manual labourers were required to complete all field operations within the optimum time available.

The results obtained for animal-powered field operations are summarised and presented in Table 36.1. The table shows the number of pairs of farm work animals and associated optimum-sized field implements for seedbed preparation only. The additional labour required to accomplish planting, weeding, and harvesting without incurring timeliness costs is also presented. For the one-hectare farm, the optimum-sized animal power required is less than two horsepower, which is the equivalent of a pair of animals. However, this size of power was selected, since it was assumed that the animals work in pairs. The power levels and implements selected for hypothetical farm sizes under the third level of mechanisation are presented in Table 36.2.

TABLE 36.1

Number of Pairs Of Animals, Sizes of Associated Implements, and the Number of Manual Labourers Selected for Given Farm Sizes Under Level 2 of Agricultural Mechanisation.

Farm Size (ha)	Number of Animals (Pairs)	Size of Implements (m)			Number of Labourers Required
		Plough	Harrow	Ridger	
1	1	0.6	0.7	0.6	2
5	1	1.3	1.6	1.3	8
10	2	1.3	2.0	1.3	16
20	2	1.9	2.3	1.9	31
50	4	2.1	2.6	2.1	77
100	5	2.7	3.3	2.7	154
500	12	3.9	4.8	3.9	769
1,000	17	4.6	5.6	4.6	1,537

TABLE 36.2

Power Levels and Implement Sizes Determined for Seedbed Preparation Under Level 3 of Agricultural Mechanisation.

Farm Size (ha)	Tractor Size (hp)	Size of Implements (m)		
		Plough	Harrow	Ridger
50	11	21	2.85	2.15
70	21	3	4	4
100	42	4	6	5

Figure 36.3 shows the relationship between the farm size and system cost for the first three levels of mechanisation. For comparison purposes, the system cost was divided by farm size to give the system cost per hectare (Table 36.3). The graphs were plotted using the machinery system in Table 36.1 and 36.2 and the number of labourers earlier determined for the given farm sizes. From the graph, system cost per hectare is at a minimum using level one of mechanisation for farms up to three hectares where the available number of labourers is four. Further, an increase in the availability of labour up to seven will shift this minimum point to 5.3 ha. For farms varying from three to 56 ha, the cost per hectare is at the minimum using level 2 of mechanisation, and where this level is not available, the farmer is advised to employ the services of a tractor and hiring service for his seedbed operations.

The third level of mechanisation is the minimum for farm sizes above 56 ha. This result suggests that for farm sizes below this minimum, it would not be economical to own tractors and associated implements for seedbed preparations only. The farmer then has a choice of either using farm work animals if available or using a tractor hire service for his field operations.

Results included in the computer printout also indicate the animal operating costs, timeliness costs, and fixed costs for the machinery systems selected. The system cost for a particular farm size used in the graph presented was computed by assuming the annual costs of performing various field operations. The results presented above provided optimal matching of equipment to a power level that was determined for a given farm size. The amount of power determined is a function of the draft requirement for the various field operations and the final selection is based on minimum cost.

Conclusions and Future Prospects

Sustainable agricultural mechanisation is an important technological input for the successful intensification of agri-food systems. The use of human muscle and manually powered tools such as the hand-hoe and cutlass are the major sources of drudgery which characterises subsistence smallholder agriculture and limits the modernisation of agri-food systems. Even where farmers are willing and able to access improved agricultural mechanisation technologies, a common problem is whether to buy/own or rent. To address these problems, farmers seek alternative and improved technologies such as animals and related animal-drawn implements, as well as access

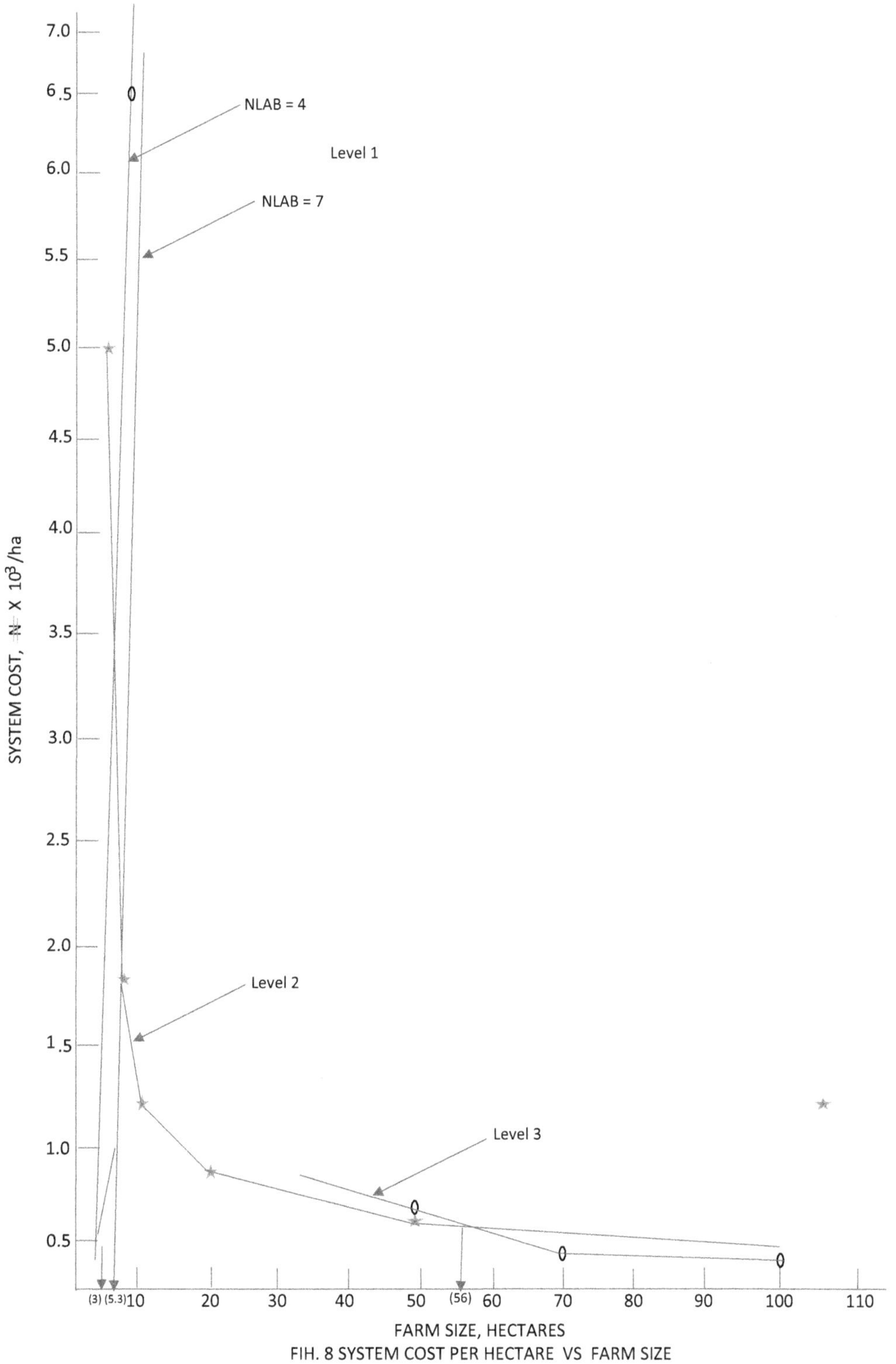

FIGURE 36.3 Relationship between the farm size and system cost for the three first levels of mechanisation (Level 1, Level 2, and Level 3). NLAB = Number of manual labourers.

to mechanical power such as tractors and related tractor-powered and tractor-drawn implements. In this chapter, we present a heuristic computer model as an aid in selecting the level of mechanisation. Each level of mechanisation is characterised by a combination of power source type (and size, hp), and associated types of implements (and size, m). The objective function of the model is the timely operation of agricultural activities and optimum economic benefit. The model represents a simple, yet effective, tool for introductory to mid-level courses in agricultural mechanisation and rural development, especially in regions aiming to transform and modernise subsistence smallholder agriculture. Case study application of the

TABLE 36.3

Total System Costs (₦/Ha) of Selected Farm Machinery for Each Level of Mechanisation. NLAB = Number of Manual Labourers.

Level of Mechanisation		Total System Costs (₦/ha)				
		1 ha	**5 ha**	**10 ha**	**20 ha**	**50 ha**
1	NLAB = 4	607.50	6,790.73	23,567.73	63,166.62	9,098,264.00
	NLAB = 7	347.14	1,000.00	8,373.69	28,963.78	96,848.72
2		4,956.90	1,889.97	1,297.55	915.21	559.41
		50 ha	**70 ha**	**100 ha**		
3		664.28	331.77	322.81		

model provided answers to the appropriate level of mechanisation for selected farm sizes under a maize cropping system in the southeast agro-climatological zone of Nigeria. This model can be adapted to educational programmes at the secondary and introductory tertiary school levels including technical colleges and universities to provide students with the principles and applications of selective agricultural mechanisation and modernisation to enhance productivity, reduce drudgery, and enhance profitable agribusiness.

REFERENCE LIST

Abercombie, K.C. (1975). Agricultural mechanization and employment in developing countries. In *Monthly Bulletin of Agriculture Economics and Statistics*. FAO, Rome.

Agro System. (1984). Proposal for increased food and industrial raw materials production in Nigeria. Prepared by agro systems consultancy unit, University of Nigeria, Nsukka, November 15.

Anazodo, U.G.N. (1975). System approach to farm mechanization in Nigeria. *Nigerian Journal of Technology*, IC1: 7–11.

Anazodo, U.G.N. (1976). A program for selective mechanization of agriculture in the Nigerian rainforest zone. In *Appropriate Approaches for Accelerated Food Production in the Rainforest Zone of Nigeria*, edited by Anazodo & Odigboh, Department of Agricultural Engineering, University of Nigeria, Nsukka, pp. 140–151.

Anazodo, U.G.N. (1981). Selective Agricultural mechanization or massive Food Importation: The choice facing Post-independent Nigeria. Paper presented at a National Conference of Agriculture, Rivers State University of Science and Technology, Port Harcourt, May 3–8.

Anazodo, U.G.N. (1982). Mechanization of family farms in Nigeria: Part I and II. *Agricultural Mechanization in Asia, Africa and Latin America*, 13(3): 59–65.

Anazodo, U.G.N. (1985). A study of traditional and mechanized systems of maize production in Nigeria. *Agricultural Mechanization in Asia, Africa and Latin America AMA*, 3(16): 51–56.

Anazodo, U.G.N., Abimbola, T.O. & Dairo, J.A. (1986). Agricultural machinery inventory, type and condition in Nigeria (1975–85). A national investigative survey report. Federal Department of Agriculture, Lagos.

Anazodo, U.G.N. (1988). The current status of farm mechanization technology in Nigeria. An invited paper presented at AERLS National seminar and exhibition on Intermediate/Appropriate Technology for Nigerian Agriculture. Agricultural Extension and Research Liaison Service (AERLS), ABU, Samaru, Zaria, November 15–16.

Anazodo, U.G.N., Opara, L.U. & Abimbola, T.O. (1989a). Agricultural mechanization study report. Perspective Plan for Agricultural Co-ordinating Unit (FACU), Ibadan.

Anazodo, U.G.N., Ikiriko, C.H. Opara, L.U. & Agunwamba, C.C. (1989b). Projected requirements of farm tractor spare part in Nigeria based on optimum stock control probabilistic model. Paper to be presented at the 13th annual conference of the Nigerian society of Agricultural Engineers, to be held in ABU from September 26–29 (Paper No: NSAE conf. 89/fpm/05).

Chancellor, W.J. (1968). Selecting optimum-sized tractor for developing agricultural mechanization. *Transaction of the ASAE* 11(4): 508–514.

Giles, G.W. (1967). World food problems, basic needs. ASAE paper no. 67-501. ASAE, St. Joseph, Michigan.

Giles, G.W. (1973). Selective mechanization an aid to increased yields. ASAE paper no. 73-512. ASAE, St. Joseph, Michigan.

Hopfen, H.J. (1969). Farm implements for arid and tropical regions. FAO agriculture development paper no. 91. FAO, Rome.

Hunt, D.R. (1963). Efficient field machinery selection of agricultural engineering, Vol. 44, No. 2, pp. 78–79, 88..

Ibe, D.G. (1987). Chief farm manager. University of Nigeria, Nsukka, Farm operations. Personal communication with the author, January 15.

Jones, M.J. (1976). Planting time studies on maize at Samaru, Nigeria, 1970–73, Samaru miscellaneous paper 58. IAR, ABU, Zaria, Nigeria.

Kline, C.K., Green, D.A.G., Donahue, R.L. & Stout, B.A. (1969). Agricultural mechanization in equatorial Africa. Institute of International Agriculture. Research report no. 6. Michigan State University.

Musa, H.L. (1979). Traditional tillage operation and development and use of anima drawn equipment. In *Proceedings of the Appropriate Tillage W/Shop*, January 16–20, IAR, Zaria, pp. 29–36.

Musa, H.L. (1988). Animal power utilization for agricultural production in Nigeria. Paper presented at the pastoralism conference, National Animal Research Institute, ABU, Shika, Zaria, June 26–29.

Munzinger, P. (1982). Animal traction in Africa. GTZ German Agency for Technical Co-Operation.

Odigboh, E.U. (1976). Engineering the mechanization of Nigeria agriculture: Sociological, economic and technological implications. In *Proceedings of the Appropriate Approaches for Accelerated Food Production in the Rainforest Zone of Nigeria*, edited by U.G.N. Anazodo & E. U. Odigboh, pp. 120–129.

Opara, L.U. (1987). *Nomographs to aid in agricultural machinery selection. BEng thesis (Unpublished).* Department of Agricultural Engineering. Nsukka: University of Nigeria.

Opara, U. L. (2023). Nomograph-based models for introductory undergraduate teaching and research in selecting agricultural power and machinery ownership system in developing countries. In this book.

Schultz, T.W. (1964). *Transforming Traditional Agriculture.* New Heaven: Yale University Press.

Singh, G. & Gupta, M.L. (1980). Machinery selection method for farms in North India. *Agriculture Systems*, 6: 93–120.

Yadav, P.P.S. & Oguejiofo, I.O. (1987). Strategies of agricultural mechanization in Nigeria. Federal Agricultural Co-ordinating Unit, Regional Office, Kaduna, Nigeria, March 28.

37

Information and Operational Technologies Curricula in Agricultural Engineering Studies

Francisco Rodríguez, Jorge Antonio Sánchez-Molina, Manuel Muñoz and Manuel Berenguel

IT/OT in Agricultural Engineering

The agricultural sector lacks information technology (IT) and operational technology (OT) development due to a variety of socioeconomic constraints (small facilities, low investment in technology, products with low profit, seasonality, and low-qualification workforce), and technical factors (large variability of agricultural products, different types of farms, climate, and orography).

The FAO predicts that 70% more food will need to be produced by 2050 (FAO, 2018), and, given the rise in prices and transport costs due to increased input prices caused by market uncertainties (without even considering those aspects related to climate change and geopolitics) (FAO, 2022), this means that a technological revolution in agriculture must take place. In such a context, the role of agricultural engineering is fundamental, and training in these types of technologies is essential.

In this scenario, the utilisation of IT and OT is central. A good definition of these can be found in the *Gartner Glossary* (Gartner Group, 2016), which states:

- Information technology (IT) is the common term for the entire spectrum of information processing technologies covering software, hardware, communications technologies and related services (communications networks, Internet of Things, data analytics, decision support systems, etc.).
- Operational technology (OT) is the hardware and software that detects, or causes a change, through the direct monitoring and/or control of industrial equipment, assets, processes, and events (automation, robotics, and others).

Nowadays, we are involved in integrating and converging IT and OT to enable more direct control and complete monitoring, providing easier data analysis of these complex systems from anywhere in the world (Garimella, 2018) and thus facilitating and improving decision-making and process optimisation. Within such a framework, the information will be available in digital form. To make this possible, a digital transformation is required that involves not just the production systems but society as a whole. This requires digital technologies (IT/OT) to enable, improve, and/or transform processes (i-SCOOP, 2020).

The core of the digitalisation process is the datum, which is defined as a piece of information. In an agricultural farm, it is possible to generate many kinds of data, whether acquired automatically or not, from climate sensors, fertigation sensors, crop monitoring, control actions, machinery reports, and camera images (e.g. in drones and satellites), as well as from farm management and the agricultural standards and policies. In the first phase, the data are raw, so they must be processed, organised, structured, and presented in a given context to create useful information. In the second phase, these refined data are analysed to identify patterns, trends, and relationships. From this, knowledge is obtained – a combination of information, experience, and insight that can benefit the individual or the organisation (Zins, 2007), i.e. useful information. When a system has knowledge and expertise regarding a process, it can observe the inputs – the factors, data, and characteristics (information) – to analyse them. Consequently, it can generate decisions and actions based on certain criteria.

There are four steps in the digitalisation process – data acquisition, information gathering, conversion into knowledge, and decision-making. This way, we can divide digitalisation technologies into three categories: data acquisition technologies, data analysis and exploitation technologies, and acting technologies. For example, to control the diurnal temperature inside a greenhouse using natural ventilation, it is necessary to measure different variables (acquisition), check the indoor and outdoor weather conditions and the state of the crop to help make a decision (data analysis) and determine the optimal vent opening to send the corresponding command to the motor that positions it (acting). Hence, we must know the enabling technologies that underpin the agricultural digitalisation process, which, based on our experience, would mainly be computer programming, sensing (including machine vision), communications, the Internet of Things (IoT), automation, robotisation, and data analytics (machine and deep learning).

IT and OT technologies are currently used in each of the main phases to analyse the agricultural production chain (Rodríguez et al., 2013):

- Soil preparation – there are a lot of autonomous tractor prototypes for undertaking tasks such as tillage, weeding, pre-fertilisation, and others (Kondo et al.,

DOI: 10.1201/9780429150111-40

2011). These include commercial developments such as the *Field Robots Catalogue* (Misset, 2023).

- Planting – outdoor seed planting processes are being developed using autonomous tractors with the corresponding implements. In nurseries and seedbeds, machinery for automatic planting in planting trays has systems for monitoring and controlling both the planting process and the seedlings' growth (Ureña et al., 2001). Related to this phase, there are also automatic systems for taking cuttings.
- Production – there are numerous developments in the field for irrigation and fertilisation, weeding, pruning, spraying of plant protection products, and climate control (in the case of closed environments such as greenhouses) (Rodríguez et al., 2015).
- Harvesting – a distinction can be made between continuous and piece harvesting. The first refers to massive crops of cereals, maize, fodder, and others, in which the plants (dry or green) are normally cut at the base by devices with comb-type blades; these crops are collected and packed in a nonstop process. Such operations involve various mechanisms that make up the machine (the harvester), synchronised by the machine's forward movement. The incorporated automation and robotisation are tasked with the self-guidance of these machines. Harvesting by pieces, on the other hand, is of greater interest because it has not yet been suitably resolved. Furthermore, it is how most table vegetables and fruits are harvested, whether grown on plants or on trees. Technical advances in recent years have made it possible to take up the challenge of total harvesting robotisation (Kondo et al., 2011).
- Post-harvesting – in the last phase, once the product has been harvested, a wide range of processes are carried out in which automation and robotics techniques are applied, of which the following stand out: storage, ripening and conservation in refrigerated chambers, inspection and classification, packaging, product transformation, palletisation, and transport. This type of agri-food industrial process works like a factory controlled globally through integrated production management.

These stages are no longer considered as being isolated. All the data produced in each phase must be integrated (e.g., for traceability purposes). Within the framework of the European Union's Internet of Food and Farm 2020 Project (IoF2020) (IoF2020, 2017), an IoT cloud platform has been developed to integrate diverse data from all the actors involved in the greenhouse agriculture supply/value chain – from farm to fork – as can be seen in Figure 37.1:

- Greenhouse data from the local systems (sensors and actuators), field notebooks (labour), lab analyses, and other agents (mainly public agencies), such as weather forecasts, market price fluctuations, energy prices and models, etc., are used to predict the greenhouse climate, diseases, and fertigation conditions, etc.
- Postharvest data from the Enterprise Resource Planning undertaken by the companies include procurement, project management, risk management, compliance, quality standards, product alerts, market models, etc.
- Transport data comprises distance, duration, transport conditions, alarms, etc.

All this information is integrated, processed, and available to different users – the farmers, technicians, and institutions. Moreover, these systems contribute to: a) generating information standards for interoperability, b) adding value through new techniques, new products, new processes, and improving air, soil, and water quality, c) management of the agricultural processes through data, and d) overall, improving social-economic-environmental sustainability.

As one can see, the agri-food sector is currently the focus of special attention in incorporating IT/OT technologies, resulting from the ever-increasing demands on production, product variety, product quality, and product presentation. All of this is required while dealing with the growing problem of labour shortages. In response, numerous international research and development projects have been developed, such as AGRI2 ICT (Information and Communication Technology and Robotics for Sustainable Agriculture) (ICT-AGRI, 2014). Indeed, the main international associations related to agricultural engineering are working on incorporating this type of technology into the agricultural sector, for example:

- The International Commission of Agricultural and Biosystems Engineering (CIGR) is organised into seven technical sections, each focused on different scientific and technical areas of agricultural engineering – for example, section VII: Information Technology, which includes OT technologies such as automatic control and robotics. CIGR also organises congresses dedicated exclusively to these topics, such as the different editions of: The International Congress on Information Technology in Agriculture, Food, and Environment; The International Symposium on Technology Application in Horticulture and Viticulture; or The International Symposium on Machinery and Mechatronics for Agriculture and Biosystems Engineering.
- The International Society for Horticultural Science (ISHS) has numerous working groups, including Automation and Integrated Production Systems in Horticulture, Sensors in Horticulture, Greenhouse Environment and Climate Control, and Modelling Plant Growth. It also organises congresses on these topics, such as the IFAC/ISHS Workshop on Mathematical and Control Applications in Agriculture and Horticulture, Greenhouse Environment Control, and Automation; The Symposium on Glasshouse Construction and Automation; Computers and Automation; Electronic

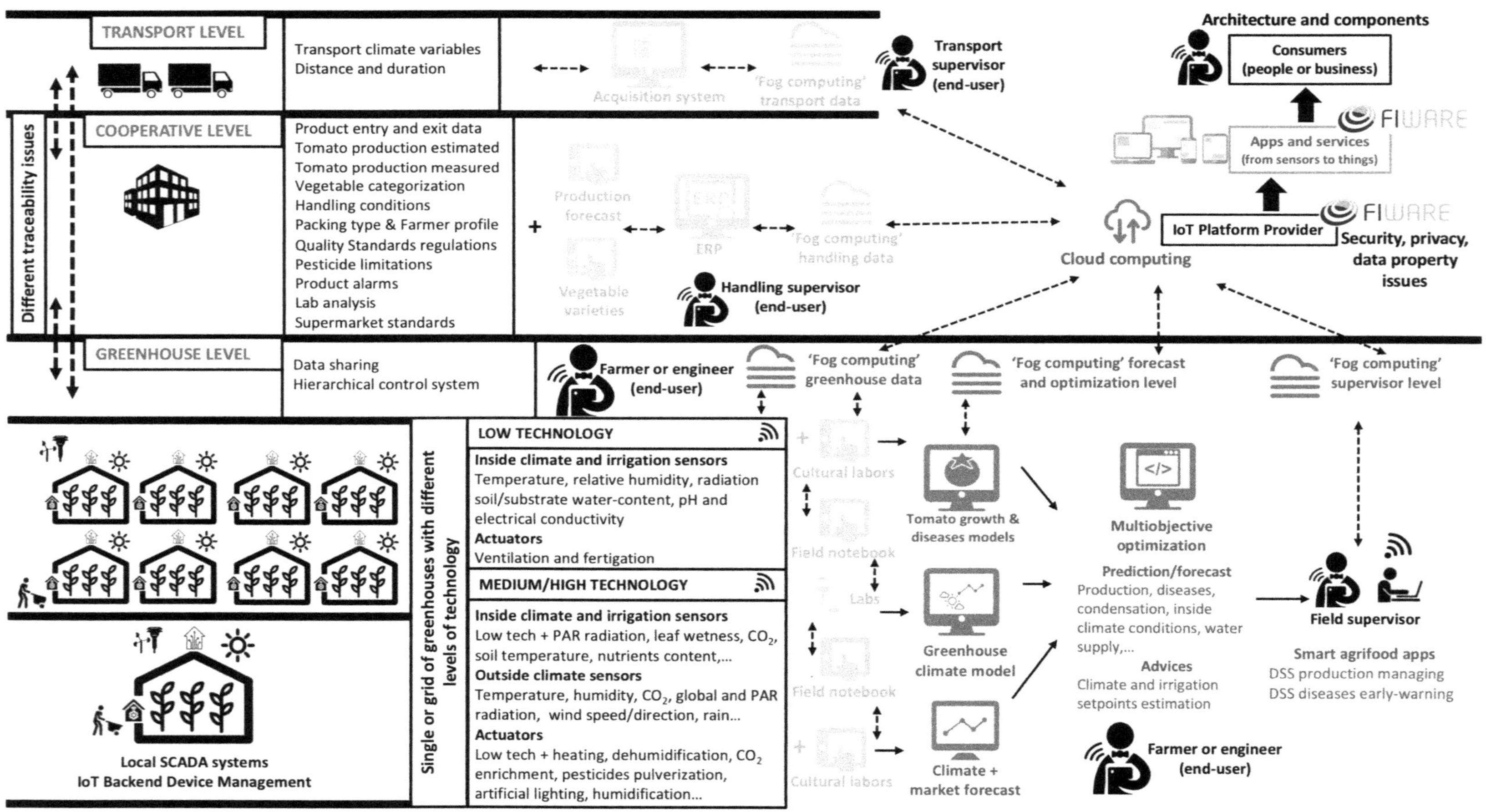

FIGURE 37.1 IoF Architecture: "From farm to fork".

Information in Horticulture; and the International Symposium on New Technologies for Environment Control, Energy-Saving, and Crop Production in the Greenhouse and Plant Factory.

- The European Society of Agricultural Engineers (EurAgEng) is organised into technical working groups related to automation and emerging technologies, IT, and precision agriculture. It organises scientific meetings such as the European Conference on Precision Agriculture; Sensors and Systems for Agriculture and Plant Monitoring; New Methods and Technologies for Agricultural Application; as well as management and technology applications to empower agriculture and agri-food systems.
- The American Society of Agricultural and Biological Engineering (ASABE) is organised into communities representing the various impact areas affecting the membership, for example, IT and sensors and controls. The ASABE Annual International Meeting includes sessions on machine vision for sensing and automation; biosensors and bioinstrumentation for food and agriculture; connectivity, cloud computing; the IoT in agriculture and natural resources; and robotics and AI-enabled robotics for production agriculture.
- The European Federation for Information Technology in Agriculture, Food, and the Environment (EFITA) is an umbrella organisation for the various national associations applying information technologies to agriculture. It organises its own conferences, the topics of which are sensors, data, decisions, and action.

As can be deduced from this introduction, it is necessary to include subjects related to IT/OT technologies, particularly their applications to agriculture, in the agricultural engineering studies curricula, evidently more from a user than from a developer standpoint, given how they will exercise this knowledge in their professional careers.

This chapter will include a summary covering the recommendations for an international agricultural engineering university curriculum related to these topics and a proposal for an IT/OT technologies curriculum for these students. Subsequently, we will share the experiences gained at the University of Almería (Spain), where the authors teach, and summarise with a conclusion section.

International Agricultural Engineering University Curricula Recommendations

Until the mid-19th century, agriculture was conceived from an organic perspective (everything happened naturally and according to a previously established order). However, the emphasis has shifted to precision agriculture (the application of IT/OT technologies to optimise crop productivity, reducing its environmental impact without affecting its quality) through agricultural mechanisation (designing machines to improve the quality of the work and reduce cultivation costs) and agrobiology (conveniently manipulating the characteristics of plants and animals, correcting the soil, and feeding the plants in an optimal way). These changes have led to the farm being conceived as a modern enterprise in which the available resources are combined appropriately to produce what society demands (Maroto, 1998).

Agricultural engineering has unquestionably contributed to this technical progress through the different spheres of its action – agricultural mechanisation, the development of improved plant varieties, the use of fertilisers and pesticides, the introduction of irrigation systems, the grid electrification of rural areas, and improvements to the transport networks are all innovations that have brought about both social and economic change in the countryside. This shows that agricultural engineering professionals must be trained in wide-ranging activities as part of the study subjects to adapt to the new IT/OT technologies being continually applied in this sector while acquiring this new knowledge in short periods. Consequently, the training of agronomists requires specific knowledge of the speciality, and a dynamic mental posture that is open to progress, given that some technologies become obsolete in less time than a university degree lasts.

Since the end of the last century, there have been numerous initiatives – European (USAEE, 2006), American (Sandun et al., 2006), Chinese (YuanHui, 2005), and others – to harmonise agronomy studies, including training in IT and OT.

As a consequence of the meetings held in Gargnano, Italy in May 1991 and in Silsoe, England in May 1992 (attended by representatives from different European Union countries, at which various papers related to the agricultural studies curricula were presented) (Pellizzi & Febo, 1994b; Febo & Ward, 1995), two working groups were created: WG1 within the International Commission of Agricultural and Biosystems Engineering (CIGR) and SIG 12 Education and Communication within EurAgEng. Their work sowed the seeds for harmonising agricultural engineering studies at the European level. Between 1990 and 1994, various surveys were carried out in several countries to analyse the organisation of university study programmes and specialisations related to agricultural engineering (Pellizzi & Febo, 1994a). Between 1996 and 1999, the surveys were updated and extended to countries on all continents (Febo & Sun, 1998). In June 2000, the results of the latest survey were presented in the report entitled "The University Structure and Curricula on Agricultural Engineering". This provided an overview of 36 countries and was published and presented by FAO in July 2000 at the EurAgEng Conference in Warwick (UK) (Febo & Sun, 2000). From this report, the following conclusions related to IT/OT technologies can be drawn:

- Agricultural engineering is a multidisciplinary field that provides knowledge and skills to define and apply engineering systems that can improve, control, and maintain the quality and sustainability of biological and environmental processes related to agriculture, food, natural resources, and rural development. This involves the technical and economic optimisation of the soil, water, energy, labour force, and biological and natural resources, considering the ecological constraints and social considerations.

- The content distribution should be 30–35% basic engineering subjects (mathematics, physics and so on), 20–25% general aspects of agriculture (genetics, botany and so on), and 45% agricultural engineering subjects (irrigation, construction, and so on).
- One of the main basic courses is related to computer science, which should include content on a computer structure, operating systems, programming languages, and skills related to system-modelling tools.
- Agricultural engineering curricula integrate learning outcomes (knowledge, competences, and skills) from different subjects and disciplines so that engineers can develop new technologies to improve the quality and reliability of agricultural products, plan energy-efficient field operations, automatically control agricultural production, and develop efficient technologies for processing agricultural products.

The coming of the third millennium in Europe also brought with it the new study structure (3+2) or (4+1), according to the Bologna Declaration of 1999, whereby most EU countries updated their university study programmes, including those in agricultural engineering, to be in line with the general economic conditions, scientific and technological development, and the need to create competitive study programmes. The EU Socrates Thematic Network for Agriculture, Forestry, Aquaculture, and the Environment (AFANet) was funded within this framework. In 2001, AFANet Working Group 3 published the report, "Towards a European Standard for Agricultural Engineering Curricula" (Briassoulis et al., 2001), which outlined the basis for European curricula and proposed the creation of basic content for all agricultural engineering studies in Europe. The idea was that there should be a degree that would be accepted in all countries, achieved by adopting a general curriculum. This would lead to and improve labour mobility and help develop and specialise the European agricultural sector. In the first phase, an attempt was made to identify new trends and suggestions for realising this common curriculum. The conclusions were laid out in the report (Briassoulis et al., 2001), of which the following are highlighted:

- New terms and concepts related to agronomic agriculture are starting to be used to meet societal demands, even replacing the name for this discipline – bioengineering.
- Different degrees in different European countries lead to a chaotic situation when it comes to homologation. The curricula are very different, even in the first academic year, meaning the contents can vary significantly.
- The contents of other IT- and OT-related disciplines are being introduced into the curricula; these can be divided into two blocks, as shown in Table 37.1.

Based on these initiatives, a project was devised in Europe to develop basic study plans that could be used as reference points to develop agricultural engineering locally and train future agricultural engineers. In 2002, the USAEE (University Studies of Agricultural Engineering in Europe) was created; this was the thematic network which analysed all the previous studies and proposals to draw up the desired curricula. It served as the basis for all the studies in European countries that teach the agricultural engineering degree. It comprised 31 institutions from 27 European countries and was approved and supported by EurAgEng through SIG RD12 (USAEE, 2002). The main objectives of USAEE-TN were:

TABLE 37.1
New Contents in Agricultural Engineering Studies

Classic Engineering Courses	Modern Engineering Courses
Electronics	Computer Engineering
Materials engineering	Sensor technology
Systems engineering	Robotics
Control systems design	Control techniques (predictive, optimal, and others)
	Image processing

- To define and develop core curricula to be benchmarks for European agricultural engineering studies.
- To determine a set of minimum criteria and/or requirements against which any curriculum can be tested to decide whether it complies with them and can thus be recognised as an agricultural engineering programme.
- To define common accreditation procedures, including European Credit Transfer System (ECTS) credits (equivalent to 25 hours of study), and to establish bodies and/or committees to carry out these procedures.

For a core curriculum to be recognised, it must take into account the following requirements (Febo & Sun, 2000):

- The Fédération Européenne d'Associations Nationales d'Ingénieurs (FEANI) recommendations for curricula, which propose a minimum of seven years of study to obtain an engineering degree – this is divided into university studies (a minimum of three years, after which the student is awarded the degree), internships in companies (optional), and a minimum of two years professional experience.
- The EurAgEng recommendations on agronomic curricula, mainly related to a basic knowledge of plants, animals, and the environment, and to ECTS.

In 2005, the USAEE-TN published a draft report – "Core Curricula of Agricultural/Biosystems Engineering for the First-Cycle Pivot-Point Degrees of the Integrated M.Sc. or the Long-Cycle Academic Orientation Study Programs" (USAEE, 2006) – which recognised that the weakness of agricultural engineering studies in Europe is the inadequate engineering base of the relevant curricula. Therefore, the main challenges are:

- To improve the engineering part of the core European curricula to meet the FEANI criteria for engineering curricula.

- To reduce the agricultural and/or biological sciences part of the core curricula.

The main issue is to agree on a set of minimum standards for core curricula and clearly describe the criteria to implement them professionally and academically. Relevant to this is the work carried out by the E4 Thematic Network (Enhancing Engineering Education in Europe), coordinated by the University of Florence in cooperation with the European Society for Engineering Education (SEFI) and other organisations.

USAEE-TN established a strong collaboration with SEFI and E4 TN to pursue common goals (USAEE, 2006). As a first step in developing the agricultural engineering core curricula, the first cycle of study was examined, and two different plans were defined:

- Plan A, which is academically oriented, comprises the core curricula of the five-year integrated degree programmes (M.Sc.) and the core curricula of the three-year first-cycle degree programmes.
- Plan B, which is applied technology-oriented, represents the core curricula of the first-cycle (mostly three-year) professional study programme.

This report contains not only the core curriculum but also the following seven modules (or specialisations) of agricultural engineering, the last being the one of interest for this chapter:

1. Water Engineering
2. Mechanical Systems and Mechanisms
3. Structural Systems and Material
4. Waste Management
5. Bioprocessing
6. Energy Supply and Management
7. Information Technology and Automation

This proposal was sent to the FEANI and EurAgEng Executive Committees for evaluation from September 2004 until the end of 2005. Subsequently, the results were widely disseminated by the EU Directorate General for Education and Culture. As a result, a new Thematic Network ERABEE (Education and Research in Biosystems Engineering in Europe) was proposed in 2007. It was funded by the Lifelong Learning programme to continue and develop the results obtained from the previous project (ERABE, 2009). The main achievements of the ERABEE project were (Febo & Comparetti, 2013):

- To define the new discipline of biosystems engineering in Europe.
- To describe the current outlines and curricula.
- To describe the basic infrastructures necessary to ensure the quality of the studies.
- To describe the tools necessary to improve the image of these studies.

At the same time as the ERABEE-TN project, the POMSEBES (Policy-Oriented Measures in Support of the Evolving Biosystems Engineering Studies in USA-EU) Consortium was formed. This comprised 12 higher-education institutions, of which eight were in the EU and four in the United States. It was jointly funded by the European Commission, the US Department of Education, and the Fund for the Improvement of Post-Secondary Education (FIPSE), as part of the ATLANTIS (Actions for Transatlantic Links and Academic Networks for Training and Integrated Studies) programme, within the framework of the EU-US Agreement 2006–2013 on Higher Education and Vocational Education and Training. Its objectives were (Febo & Comparetti, 2013):

- To provide a platform for the systematic exchange of experiences and ideas between the US and the EU in order to contribute to improving the quality and linkage of education and research and to establish appropriate policy-oriented measures.
- To develop appropriate policies and curricula for biosystems engineering.
- To promote compatible curricula, both within the EU and between the EU and the US.

All these initiatives (AFANet, USUAEE, and ERABEEN) proposed a core curriculum in agricultural engineering studies to be used as a benchmark in Europe (or more widely), specialising in solving problems in systems involving plants (production agriculture, including power and machinery), animals (animal production, including structures and the environment), humans (health and biomedical engineering), and the environment (bio-environmental engineering, including soil and water engineering, waste management, and so on) (USAEE, 2006). It is explicitly stated that all these areas require competences in instrumentation, IT, and many other disciplines. This core is composed of seven agricultural engineering modules (or specialisations), which include courses related to IT/OT technologies, one of them being Information Technology and Automation in Agriculture and Bioprocess Engineering, made up of 28–30 ECTS credits, including seven courses of four to five ECTS each, as can be seen in Figure 37.2.

The contents, objectives, and competences proposed for these courses can be consulted in (USAEE, 2006). These conclusions were reaffirmed in the Europe-US collaboration (Panagakis et al., 2008), which stated that IT/OT should be a compulsory discipline in agricultural engineering curricula, being both a necessity and an opportunity for professionals. The main drawback of this proposal is that there has been an advance in IT recently, especially with the implementation of the IoT, Cloud Computing and Data Sciences, while OT is enabling technologies in the digitalisation process that global agriculture is currently undergoing. Hence, updating some of them and even proposing new ones is necessary.

Although the report is now somewhat dated (Briassoulis et al., 2001), it analyses the topics covered by courses at various agricultural engineering departments worldwide. From this, certain data on courses related to IT/OT can be extracted:

- Regarding basic technical courses, the instrumentation measurements and control topic is taught in about 60% of the consulted universities, the computer

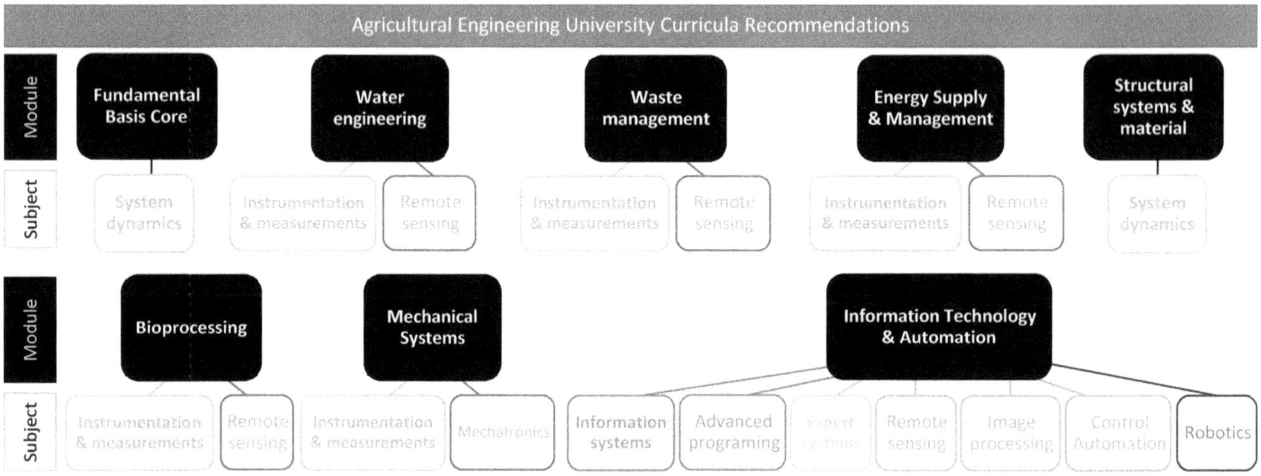

FIGURE 37.2 IT/OT courses proposed in the International Agricultural Engineering University Curricula Recommendations.

applications topic in 55%, and the systems engineering topic in only ten percent.

- Regarding agricultural engineering courses, the topic of automation, sensing, and image processing is taught in 35% of the universities surveyed.

This data shows how important this content is in the agricultural curricula and how the professionals in the sector have become aware of it. It should be noted that by studying the above report, one can observe that, in areas of the world that can be considered more industrialised, such as countries in Europe, North America, and Japan, most agronomy degrees include subjects on IT/OT. The influence of the environment on the university degree curricula is also demonstrated.

With all these considerations and the authors' experience in teaching, research and development and innovation tasks, we propose an itinerary to train agricultural engineers in these technologies in the next section.

A Proposal for an IT/OT Module in the Curriculum Studied by Agricultural Engineering Students

Module Justification

Despite the efforts of the different administrations, agriculture, in general, is still not very technologically developed. Furthermore, growers and technicians are not usually trained in applying technology. To achieve this, the EU Commission has highlighted the Common Agricultural Policy as the core of the EU's policy on agriculture. Its main objective is to consider local conditions and requirements to trigger productivity gains in EU agriculture. These include research and innovation programmes, new technologies, rural development and infrastructure, efficient advisory systems, and continuous training for farm managers (Bachche, 2015; Shamshiri et al., 2018; Yıldırım & Solmuş, 2014; Zhang et al., 2020). For all these reasons, in the training of future agricultural engineers, basic knowledge of computer science, data science, automation, and robotics is necessary since it is the agricultural engineers who will have to confront the reality of the problem and implement new trends in the sector as a whole, not just to advance IT/OT. Efforts must focus on specialising the curriculum, with more emphasis on widening knowledge across specialisations to benefit from the strengths of each, resulting in graduates with more diversified training. This situation will provide them with skills that increase their employment opportunities beyond the agricultural sector. Equally important, it will improve the competitiveness and safety of the agri-food sector.

A paradigmatic example of this transversal curriculum in which the proposed IT/OT knowledge is applied (i.e., programming, IoT, data science, instrumentation, automation and control systems, and robotics) is a farm management system based on cloud computing. In the case of the connected greenhouse (see Figure 37.3), an architecture divided into three interconnected layers is proposed: the sensing and actuation layer, the data storage and processing layer, and the application layer. These layers are closely related to different academic disciplines, such as robotics, programming, IoT, data science, automation, and sensing. Combining these technologies enables farmers to obtain real-time information about their production process and make more informed decisions, improving their efficiency and sustainability. In short, cloud connectivity is a vital tool for the agricultural sector's future.

IT/OT Subjects

Having justified the six subjects proposed for adequate training in IT/OT technologies, this section will describe each in two modules: the first three are grouped into IT technologies and the last into OT technologies. The information is presented in a table indicating the semester, the ECTS, the classroom and lab hours, the objectives, skills, the theory, and the lab programme. It should be noted that the laboratory programme depends on the material available at each university. For this reason, the type of laboratory exercise needed to be carried out is indicated, with examples that could be developed with the software and hardware tools shown.

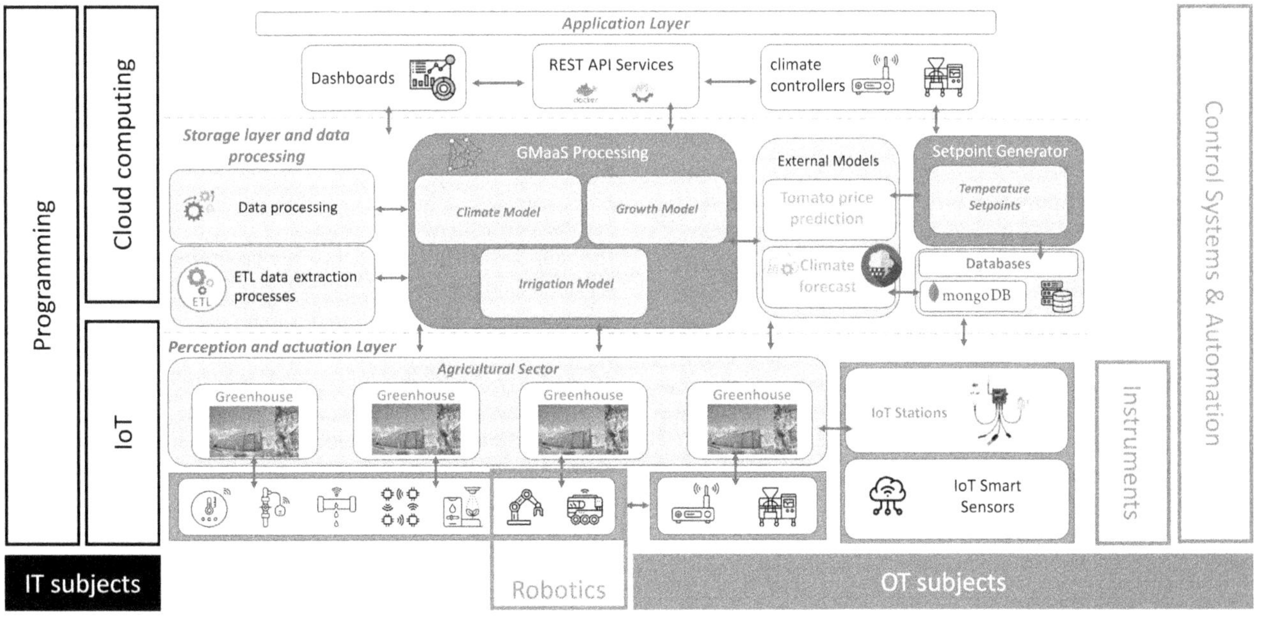

FIGURE 37.3 Conceptual digitalisation architecture in a greenhouse and curricula proposal.

IT Module

The skills that an agricultural engineer will need in the future depend on the trends and technological advances that occur in the industry. However, some skills are likely to become increasingly relevant to an agricultural engineer's success in the future.

In terms of specific skills, here are some that may be important for a future agricultural engineer:

1. Knowledge of technology: New technologies will be increasingly present in the agricultural field, so it is important for agronomists to be familiar with them. These technologies include IoT, Cloud computing and Data Sciences.
2. Data analytics: The ability to analyse large amounts of data is an increasingly important skill for agronomists. Data science can be especially useful for data analysis in the agricultural field, which can help improve productivity and efficiency.
3. Programming skills: Programming can be very useful for an agricultural engineer in terms of data analysis, task automation, and control system design. In addition, the ability to programme can enable the agronomist to develop customised agricultural tools and applications.
4. Communication skills: As technology becomes more complex, it is important for agronomists to be able to communicate effectively with other professionals, both technical and non-technical. Communication skills, including the ability to present information clearly and effectively, are crucial to future success.
5. Flexibility: Finally, programming is a very versatile skill that can be applied in a wide variety of areas and sectors. Having a technical profile that includes knowing how to programme offers great flexibility to adapt to different projects and work environments.

In summary, the agricultural engineer of the future will need to have skills in technology, data analysis, programming, biotechnology, and communication. Programming, IoT, and data science, especially, are skills that are likely to become increasingly important for future agricultural engineers. Specifically, the following three subjects specialised in the IT branch are proposed:

- Computer programming
- Internet of Things
- Data science

The information below shows the main features of each proposed module.

Subject 1: Computer Programming

Semester: 2 ECTS/Hours: 6/150 Classroom and Lab hours: 60

The objectives are:

1. Introduce students to the fundamental programming concepts: A basic programming course should address the fundamental programming concepts, including variables, operators, control flow structures, and data structures.
2. Teach students to solve problems using programming: Programming is a powerful tool for solving problems and automating tasks. A basic programming course should teach students to identify problems that can be solved using programming and to develop solutions using computational thinking.

3. Familiarise students with a specific programming language: There are many programming languages, and a basic programming course should focus on a specific one. Students should learn the basics of the chosen programming language, including its syntax, structure, and most important features.
4. Develop programming skills: Programming is a skill that improves with practice, and a basic programming course should provide students with enough exercises and projects to apply what they have learned and improve their programming skills.

In summary, the objectives of a basic programming course are to introduce students to the fundamental concepts of programming, to teach them to solve problems using programming, to familiarise them with a specific programming language, and to develop programming skills.

Theoretical programme:

Lesson 1. Introduction to programming: What is programming? Why is it important for agricultural engineers? How does a programme work? (two hours).

Lesson 2. Fundamental programming concepts: Variables, data types, operators, conditionals, and loops (eight hours).

Lesson 3. Flow control structures: Conditionals, loops, and functions (eight hours).

Lesson 4. Data structures: Arrays, lists, stacks, queues, trees, and graphs (ten hours).

Lesson 5. Programming languages: An overview of programming languages and their differences, such as Python, R, and MATLAB, and how to choose the right language for different applications (two hours).

Lab Programme:

Possible laboratory material: To develop this didactic unit, a programmable microcontroller such as Arduino will be used, with the possibility of using any other available on the market to perform the exercises for each lesson. This platform is widely used (even in high schools), but one must remember that this is a basic course in structured programming oriented to agricultural engineers, who must know the fundamentals to practise their profession; these types of systems are ideal for the purpose. Basic programming exercises using Arduino can be performed with a simulator or a real microcontroller. If a simulator is used, several free programmes can be found online that allows one to simulate Arduino programming, such as Tinkercad, Fritzing, and Proteus, among others (Figure 37.4). On the other hand, if a real microcontroller is used, an Ardunio started kit is required, that includes an Arduino board, a breadboard, cables, and some electronic components such as LEDs, resistors, buttons, sensors, and so on. In both cases, the exercises are similar, and the goal is the same: to practise basic programming with Arduino.

Lab exercise 1. Basic programming exercises: create simple programmes using variables, operators, and loops.

Example 1.1: Create a programme that flashes an LED connected to the Arduino board. This LED must be connected to an Arduino pin and turn on continuously using a loop.

Example 1.2: Using a temperature sensor, create a programme that displays the temperature in Celsius and Fahrenheit on the serial monitor.

Lab exercise 2. Create functions: create and call functions that perform simple operations, conditionals, or unit conversion.

Example 2.1: Once we have the temperature data from the sensor, we must create a conditional that warns the serial monitor if the temperature is higher than 35 degrees Celsius.

Example 2.2: Use a function that transforms the degrees Celsius to Fahrenheit and displays them on the serial monitor.

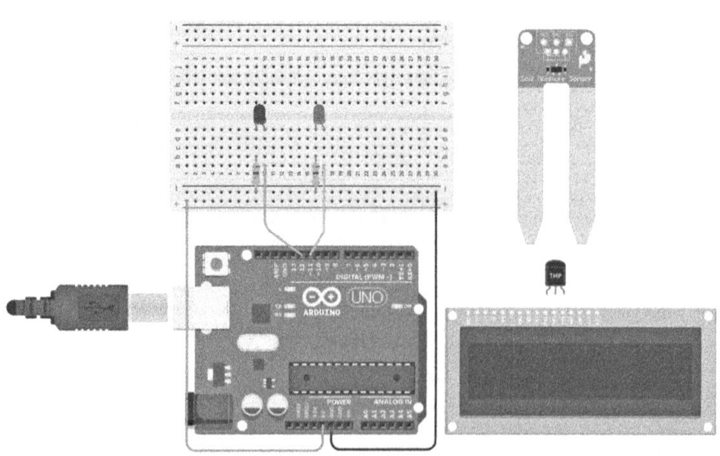


```
// C++ code
//
void setup()
{
  pinMode(12, OUTPUT);
  pinMode(11, OUTPUT);
}

void loop()
{
  digitalWrite(12, HIGH);
  delay(1000); // Wait for 1000 millisecond(s)
  digitalWrite(12, LOW);
  delay(1000); // Wait for 1000 millisecond(s)
  digitalWrite(11, HIGH);
  delay(1000); // Wait for 1000 millisecond(s)
  digitalWrite(11, LOW);
  delay(1000); // Wait for 1000 millisecond(s)
}
```

FIGURE 37.4 Virtual simulator of a programmable microcontroller based on Arduino.

Lab exercise 3. Create a programme to calculate agricultural statistics: develop a programme that reads agricultural data and calculates basic statistics, such as average, standard deviation, and minimum and maximum.

Example 3.1: Suppose we are farmers and want to monitor the temperature on our plantation to know the environmental conditions and improve the crop yield. To do this, we will use a temperature sensor and build a programme in Arduino that reads the data and calculates basic statistics (the average, standard deviation, minimum and maximum, and others). We will then show the results on a display.

Subject 2. Internet of Things

Semester: 4 ECTS/Hours: 6/150 Classroom and Lab hours: 60

1. Introduce students to the fundamental concepts and basic principles of IoT and its applications in agriculture.
2. Students will learn how to collect and transmit data on weather, soil, moisture, plant growth, crop production, and other key factors using IoT sensors and devices, enabling them to collect accurate and useful information about their crops.
3. Students will learn about different connectivity technologies for IoT, such as Wi-Fi, Bluetooth, and MQTT, allowing them to choose the most appropriate connectivity technology for their specific needs.
4. Students will learn how to configure, programme, and control IoT devices using software and hardware tools, enabling them to use this technology effectively in practice.
5. Students will learn how to analyse and visualise IoT data for decision-making in agriculture, enabling them to make informed decisions based on data collected through IoT devices.

In summary, the objectives of a basic IoT course are to introduce students to the fundamental concepts of IoT, including the different types of devices and sensors used, the connectivity technologies, software and hardware tools, data collection and transmission, and IoT application opportunities in different fields to improve efficiency and productivity.

Theoretical programme:

Lesson 1. Introduction to IoT: Basic concepts and definitions (four hours).

Lesson 2. IoT applications in agriculture: Examples of how IoT is used in agriculture to improve efficiency and productivity (four hours).

Lesson 3. IoT sensors and devices for agriculture: Types of sensors and devices and their application in agriculture (four hours).

Lesson 4. Communication and connectivity in IoT: Communication and connectivity technologies for IoT, such as Wi-Fi, bluetooth, Sigfox, and LoRaWAN, among others (eight hours).

Lesson 5. IoT platforms and tools: Tools and platforms used to develop IoT solutions (six hours).

Lab programme:

Possible laboratory material: The visual programming tool based on Node-RED (Figure 37.5) flows will be used for carrying out the practical exercises because it is very useful for programming IoT devices and can be used with microcontrollers and sensors to create customised solutions.

Lab exercise 1. Configure and programme an IoT device, such as an Arduino or a Raspberry Pi, to measure temperature and humidity in an orchard or agricultural field (four hours).

Example 1.1: Create an IoT device with a programmable microcontroller that reads the basic variables in a greenhouse (indoor temperature, indoor humidity, electrical conductivity, and soil moisture).

Lab exercise 2. Use an IoT platform to collect and visualise sensor data in an orchard or agricultural field (six hours).

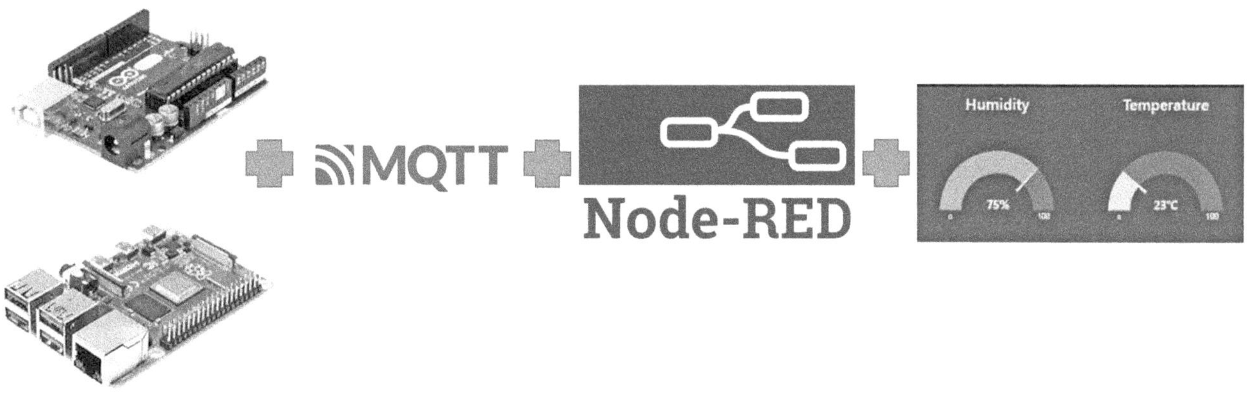

FIGURE 37.5 Workflow for developing an IoT system with Node-RED.

Example 2.1: Once the above IoT system is designed, use the Node-RED tool to perform the data collection flow from the sensors and send it through the MQTT protocol.

Example 2.2: Configure an IoT platform, such as ThingSpeak or Fiware, to receive data from a temperature and humidity sensor in an orchard. Programme the device to send the data to an IoT platform using the MQTT protocol.

Lab exercise 3. Create a programme that reads and writes to files: create a programme that reads a file with agricultural data (e.g., temperature, precipitation, and crops) and writes the results to another file (eight hours).

Example 3.1: Create a Node-RED flow that reads an agricultural data file in csv format and uses processing nodes to extract relevant information, such as temperature and crops. Then, use a file write node to save the results to another csv file for further analysis.

Lab exercise 4. Create an environmental condition alert system using sensors and an IoT platform (four hours).

Example 4.1: Set up an alert system using an IoT device such as an Arduino and a motion sensor to detect intruders in a greenhouse. Programme the device to send alerts to an IoT platform using the MQTT protocol.

Lab exercise 5. Design and develop an IoT solution to automate the irrigation of an orchard or agricultural field (ten hours).

Example 5.1: Set up an automated irrigation system using an IoT device such as a Raspberry Pi and a soil moisture sensor to measure the moisture in the orchard. Programme the device to send irrigation commands to an irrigation valve using the MQTT protocol. This valve can be simulated with a motor.

Subject 3. Data Science

Semester: 5 ECTS/Hours: 6/150 Classroom and Lab hours: 60

1. Understand the basic concepts and principles of data science and its agricultural applicability.
2. Learn how to collect and pre-process data from different sources, such as IoT sensors, historical records, and surveys, among others.
3. Know the different data analysis techniques, such as descriptive statistics, machine learning, and data mining, among others.
4. Learn to use software tools for data exploration, analysis, and visualisation, such as MATLAB and Python, among others.
5. Identify patterns, trends, and correlations in data for decision-making in agriculture.
6. Know the ethical and legal challenges related to data science and learn how to implement measures to mitigate them.
7. Identify opportunities for using data science in agriculture, such as yield prediction, early disease detection, and optimisation of resource use, among others.

In summary, a basic data science course for agricultural engineers in the future should provide students with the skills and tools necessary to use data science to improve agricultural production and decision-making in the field.

Theoretical programme:

Lesson 1. Introduction to data science: Basic concepts and definitions (two hours).

Lesson 2. Data sources: Types of data, ways of collecting them and their relevance in agriculture (four hours).

Lesson 3. Data processing: Data cleaning, transformation, and exploratory data analysis techniques (eight hours).

Lesson 4. Modelling and forecasting: Statistical models and forecasting techniques used in agriculture (ten hours).

Lesson 5. Data visualisation: Tools and techniques to visualise and communicate results (six hours).

Lab programme:

Possible laboratory material: The practices belonging to this didactic unit will be performed with MATLAB software. MATLAB is a powerful and comprehensive data science tool that offers numerous benefits, such as integrated functions and tools, support for large datasets, statistical and machine learning analysis, a user-friendly interface, integration with other tools, and a large community of users and support. Most technical universities licence this tool, although there are also free versions such as Scilab or Octave.

Lab exercise 1. Collect and clean data about a specific crop, such as yield, soil moisture, or temperature (four hours).

Example 1.1: Use MATLAB to import and clean a greenhouse data set to check the crop yield in a specific region. Then, calculate the basic descriptive maximum, minimum, and mean statistics and visualise the data to identify any patterns or trends.

Lab exercise 2. Use exploratory data analysis tools to find patterns and relationships in the collected data (four hours).

Example 2.1: Develop a linear regression model using MATLAB to predict crop yield as a function of variables such as soil moisture, temperature, and amount of irrigation water. Use cross-validation techniques to evaluate the model's accuracy.

Lab exercise 3. Use data visualisation techniques to communicate analysis results and predictions to crop stakeholders (six hours).

Example 3.1: Use MATLAB to perform a cluster analysis on a crop data set to group crops based on their characteristics and yields. Then, use data visualisation techniques to present the analysis results and recommend improving production.

Lab exercise 4. Develop a simple statistical model to predict the crop yield as a function of the collected variables (ten hours).

Example 4.1: Use MATLAB to perform a principal component analysis on a crop data set to identify the most important variables influencing yield. Then, use these variables to create a more accurate predictive model.

Lab exercise 5. Analyse the impact of different factors on crop yield and how it could be improved by adjusting the production process (six hours).

Example 5.1: Use MATLAB to analyse how different factors, such as soil type, time of year, and amount of sunlight, affect crop yield. Use data visualisation techniques to communicate results to crop production stakeholders and make recommendations on improving the production process.

OT Module

The skills that an agricultural engineer will need in the future depend on the trends and technological advances that occur in the industry. Operational technologies are especially important for this type of application because most processes are being instrumented, automated, and robotised. In such processes, there is particular interest in certain aspects, for example, knowledge regarding the state of a heater, the opening/closing of an irrigation valve, or the different instructions carried out by a robot when a sensor detects a piece in a certain position. These systems are characterised by having many variables with a finite number of values, since the possible system states are associated with a finite number of possibilities. Therefore, the system must be equipped with several sensors, to detect events or values, and with actuators that allow one to execute the actions that should be performed at each instant. In addition, a control system is used to manage all these subsystems. Sequential processes are also present in the performed tasks common in these kinds of applications. The most used controllers in these processes are programmable logic controllers (PLCs). As an example, to control the greenhouse air temperature with a water pipe heating system, it is necessary to control this actuator. For this, the fuel inlet must be controlled along with the boiler's water temperature, the water temperature in the tubes that enter the greenhouse, and, finally, when mixing with the return water (which is colder). The process involves an ordered sequence of interventions in the valve apertures, sensors, motors, and others, and continuous temperature or flow controllers. This kind of system is very common in the agricultural sector for transformation products, manufactured products, packing products, or production systems such as greenhouses. Consequently, automation, instrumentation, machine vision, and control systems have become important subjects in many educational fields.

The above aspects have led to the inclusion of automatic control, instrumentation, artificial vision, and robotics (as applied to agriculture) in the agricultural engineering degree curricula. Therefore, a core of six modules (or specialisations) is proposed to combine the IT and automation modules. Specifically, a specialisation module is proposed of between 28 and 30 ECTS (five subjects of three or six ECTS each) called "Automation in Agricultural and Bioprocess Engineering", which includes the following contents:

- Instrumentation
- Control Systems and Automation
- Robotics

The information below shows the main features of each proposed module:

Subject 4. Instrumentation

Semester: 3 ECTS/Hours: 3/75 Classroom and Lab hours: 30

1. Understand the importance and significance of the signals used in automatic control.
2. Understand the fundamentals of measurement and the actuation of industrial instruments.
3. Know how to select the most appropriate measurement and control instrumentation for the process to be controlled.
4. Know how to interpret the Pipes and Instrumentation Diagrams (P&ID).
5. Find a solution to specific problems in the field of instrumentation.
6. Know the accelerated technological evolution in this field.

In summary, the module aims to provide the student with knowledge of the physical elements (sensors and actuators) used in process plants to implement control loops, and to acquire the skills to specify them.

Theoretical programme:

Lesson 1. Instrumentation basics. The measurement process, calibration, and P&IDs. Introduction to electronic instrumentation, metrological characteristics and measurement errors (two hours).

Lesson 2. Sensors: Classification. Electronic sensors for measuring different physical quantities, their characteristics, and conditioning circuits (four hours).

Lesson 3. Actuators: Valves, motors, relays, and others (four hours).

Lesson 4. Signal conditioning: Amplifiers, filters, modulation, and Analog to Digital and Digital to Analog conversion, (four hours).

Lesson 5. Data visualisation. Monitoring and supervisory control and data acquisition (four hours).

Lab programme:

A possible lab exercise consists of using different sensors and actuators (pumps, pneumatic valves, rotameters, water content sensors, load cells, climate sensor, and others) connected to a control system. These elements will be studied, analysing the information provided by the manufacturer in their catalogue, connecting to a data acquisition and control system, and carrying out tests to relate the physical-chemical variable measured by the control system reading. The instrument's linearity will be analysed along with other characteristics, such as range, resolution, or accuracy. The students will be introduced to techniques that allow them to linearise the non-linear static curves of the sensors and actuators (Figure 37.6).

Lab exercise 1. Sensor calibration (four hours).

Example 1.1: Determination of the optimum sensor calibration parameters.

Lab exercise 2. Actuator characterisation (4 hours).

Example 2.1: Determination of the inherent and installed characteristics of a valve. Study of the hysteresis and calculation of the flow coefficient for a pneumatic valve for the range of pressures at which it can operate in the installation.

Lab exercise 3. Measurement of a temperature sensor (four hours).

Example 3.1. Temperature measurement with Resistance Temperature Detector RTD and LM35, including signal conditioning, digitalisation, monitoring and storage processes.

Subject 5. Control Systems and Automation

Semester: 6 ECTS/Hours: 6/150 Classroom and Lab hours: 60

1. Understand the evolution of these types of technologies since they present a great future perspective.
2. Become aware of the need for automation in the agricultural sector.
3. Apply the concepts and techniques learned to real systems that exist in the market.
4. Analyse low- and medium-level control systems, recognising their fundamental modules and the techniques used in their design.
5. Focus on the basic knowledge of agricultural engineering from a process automation perspective.
6. Compare the different control systems in order to study their advantages and disadvantages. Consequently, it will be possible to select the appropriate controller depending on the type of process to be controlled.
7. Introduce a series of basic engineering tools for performing mathematical calculations and simulation, as well as programming tools that are fundamental in process automation.

In summary, this course aims to introduce students to the new technologies related to automation and the specific control

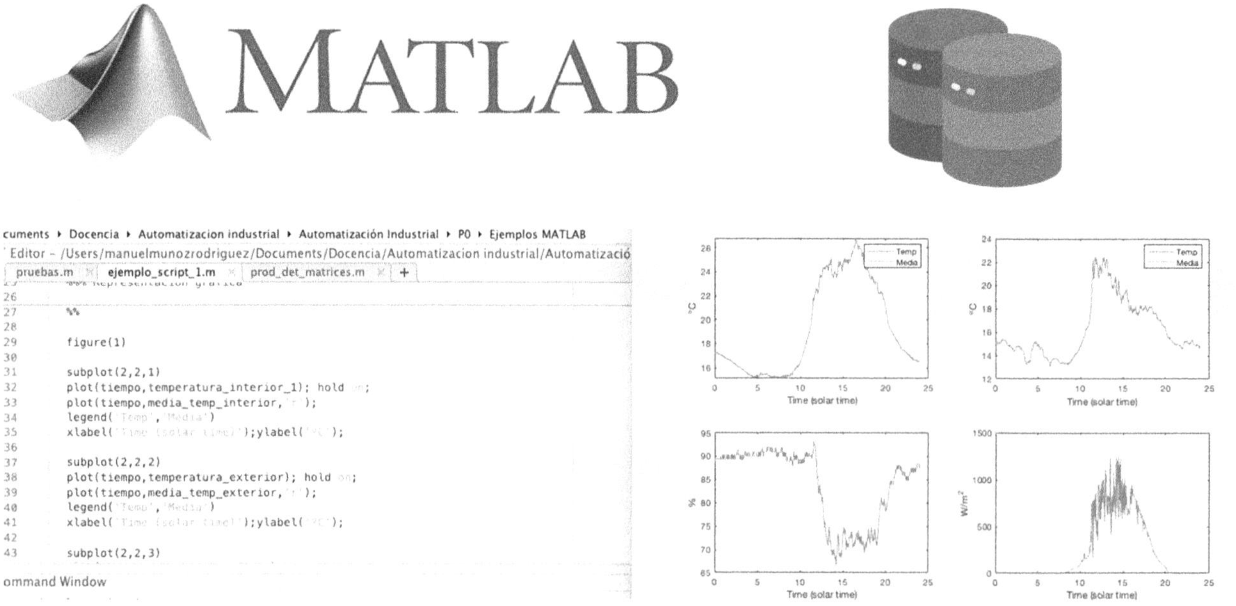

FIGURE 37.6 Arduino with sensors and actuators.

techniques that allow the different systems that make up an automated environment to perform correctly.

Theoretical programme:

Lesson 1. Control systems engineering. Introduction to system engineering. Elements of a control system. Feedback concept (four hours).

Lesson 2. Analysis of continuous agri-food processes. Analysis of agricultural processes, their modelling and simulation, and their analysis (six hours).

Lesson 3. Continuous process controller design. The knowledge of basis control techniques needed by an engineer. Specifications. On/Off and proportional–integral–derivative (PID) type controllers (six hours)

Lesson 4. Analysis of sequential agri-food processes. Automation process modelling through GRAFCET and Petri Nets (six hours).

Lesson 5. Programmable Logic Controllers. Automatic process control using the IEC 1131 standard (eight hours).

Lab programme:

Two possible lab exercises making use of two different platforms:

1. The TCLab platform (the layout and appearance shown in Figure 37.7) provides an excellent testing ground for industrial multivariable control strategies at an industrial level (note that the actuators and sensors are also illustrative of those used in agricultural and general industrial environments).
2. The PLC is any electronic machine designed to control sequential processes in real-time within an industrial environment. Every industrial process consists of sequences of actions that must be controlled. In simple processes, an operator is responsible for this control and for monitoring the correct functioning of the system, but in most cases, this is not possible due to the size of the process.

Lab exercise 1. Simulation of dynamic systems (five hours).

Example 1.1. Using TCLab, step tests are performed to obtain a model based on the reaction curve.

Lab exercise 2. Continuous process control (ten hours).

Example 2.1. Using TCLab, once the model is obtained, the control algorithm is tuned and implemented by means of PID controllers and with different specifications.

Lab exercise 3. Modelling of sequential agri-food systems (six hours).

Example 3.1. Modelling an agri-food processes using Petri Nets. Design of a fertigation controller and distribution.

Lab exercise 4. Control of sequential agri-food systems (nine hours).

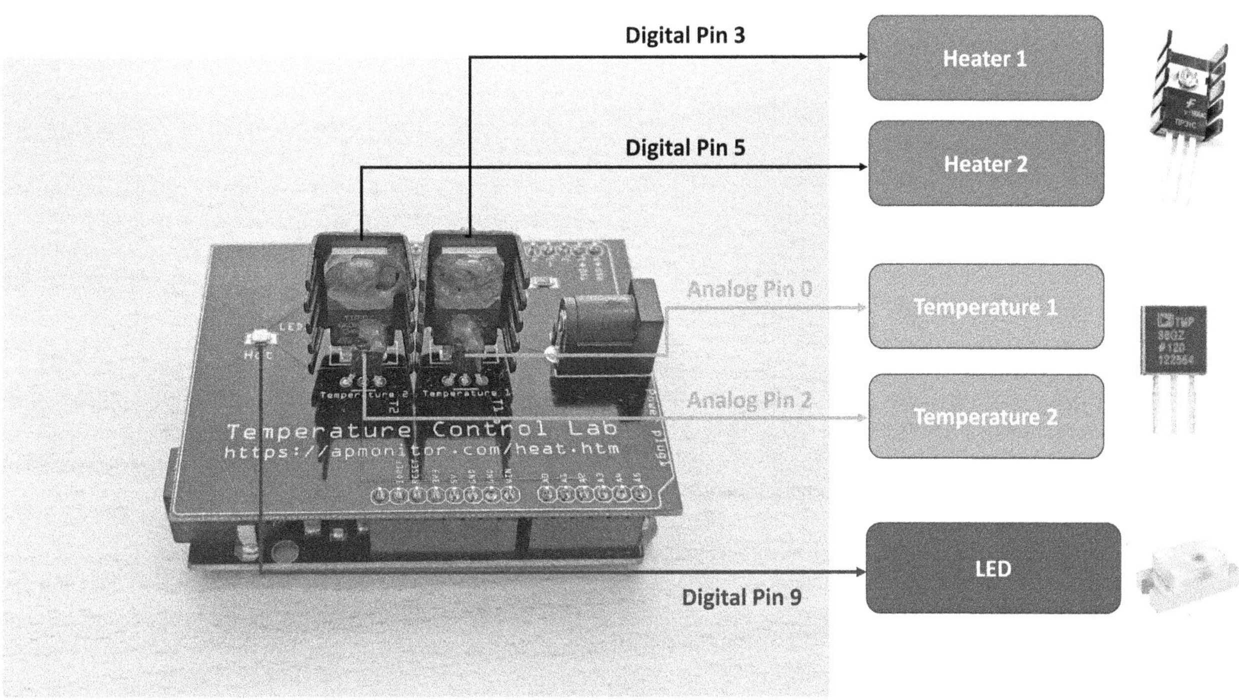

FIGURE 37.7 TCLab used for modelling and control strategies.

Example 3.2. Once the industrial process is modelled, it must be controlled using the PLC programming languages included in the IEC 1131 standard as Sequential Function Charts (SFC) or blocks diagram. Implementation and programming of a fertigation controller.

Subject 6. Robotics

Semester: 8 ECTS/Hours: 3/75 Classroom and Lab hours: 30

1. Introduce the student to the fundamental concepts of manipulation and mobile robotics.
2. Present the criteria, standards, and techniques necessary for the design and implementation of robotic cells for problem-solving.
3. Describe the techniques used in the trajectory and actuator control of a manipulator robot.
4. Describe the teaching methods and robot programming languages.
5. Introduce the student to the field of artificial vision in robotics and algorithms used in mobile robotics for trajectory tracking.
6. Show the student specific fields of application in which automatic control and robotics can be used in the agriculture sector: robotisation of transplanting, transport, harvesting, and post-harvesting tasks.

In summary, the study of robots is outlined in terms of their morphology, kinetics, and basic programming concepts.

Theoretical programme:

Lesson 1. Introduction to robotics. Basic concepts of robots and robotics (two hours).

Lesson 2. Fundamentals of manipulation robotics. Basic concepts, elements, and classification. (five hours)

Lesson 3. Fundamentals of mobile robotics. Basic concepts of mobile robotics (four hours).

Lesson 4. Fundamentals of artificial vision. Concept, processing levels, image capture, segmentation, and recognition (four hours).

Lesson 5. Implementation of a robotic system. Creation of robotic cells, basic concepts, and modelling (three hours).

Lab programme:

A possible lab exercise. Project-based learning plays a key role in teaching engineering and science. Robotic operating systems (ROS) is used as a standard that is gaining momentum in the industry; MATLAB/Simulink with the ROS toolbox makes it easy to implement the prototype using Gazebo as a robot simulation environment and Turtlebot as an example. This scheme allows programming in ROS in a well-known environment such as MATLAB/Simulink with all the possibilities it offers and simulates the robot and its environment in Gazebo, which gives it a dose of realism.

Lab exercise 1. Basic programming of a manipulator robot (two hours).

Exercise 1.1. Basic concepts of robot programming. Develop a pick-and-place process for agriculture objects using manipulation robots using MATLAB.

Lab exercise 2. Fundamentals of mobile robotics (two hours).

Exercise 2.1: Basic concepts of mobile robot programming. Develop of navigation system for greenhouses robotics using ROS and Gazebo.

Lab exercise 3. Fundamentals of artificial vision (four hours).

Exercise 3.1: Objects detection in an agricultural environment. Implementing a machine vision system for tomatoes' colour and size sorting using MATLAB toolboxes.

Lab exercise 4. Programming a robotic cell (four hours).

Exercise 4.1: Programming a robotic cell using a simulation tool. Design a basic robotic work cell for precision fertigation using Gazebo.

Case Study: University of Almería (Spain)

Socioeconomic Context of Almería Province

Almería is a Spanish province located in the southeast of Spain (the Andalusian region) on the shores of the Mediterranean Sea. The province covers an area of 8,774 km^2, with 219 km of coastline and a population of 740,534 people, representing a population density of 84.4 inhabitants/km^2. Foreign residents in Almería account for 19% of the total, so this migratory balance is the main reason for the province's demographic growth (IEA, 2023).

The main economic activity in Almería is the trade and services sector (including tourism), which accounts for 57% of GVA (or gross value added, which at factor cost is the difference between the production value and the costs of the raw materials, intermediate products and services purchased from other sectors or enterprises), compared to 21.7% for agriculture, fishing, and livestock, 6.1% for industry, and 5.4% for construction (INE, 2023).

Greenhouse agriculture is an important driver of the economy. To understand its importance, the surface area covered by plastic greenhouses is 32,827 hectares, producing 3,494,830 tonnes of horticultural produce. In the 2021/2022 season, 2,319,230 tonnes were exported, while the total production value amounted to €2,954.48 million. During this season, pepper was the most produced crop in terms of surface area, volume, and production value, followed by tomato, cucumber, zucchini, and watermelon (Horto-Info, 2022).

The evolution and success of agriculture in this province are due to three basic factors (Galdeano, 1999). Firstly, the natural conditions are favourable – there are numerous aquifers, and the climate is ideal, with an average annual temperature of 19.2°C and approximately 3,000 hours of sunshine per year (the highest in Europe; it also has the lowest rainfall), with an average of 8.3 hours of sunshine per day. Furthermore, thanks to the support of the Spanish Ministry of Agriculture in the 1960s, numerous wells were opened, allowing the use of the sandy soil cultivation technique and the construction of greenhouses. These two techniques made it possible to grow crops during the coldest months when there was no cultivation in the rest of Spain or Europe. This led to the economic launch of the sector. Lastly, the human factor has been a key element in consolidating the sector, with people migrating from inland areas to the coast, acquiring land at low prices, building economic greenhouse structures, and relying on the family ownership structure, which is the ideal type of agriculture, given the amount of labour required. It also highlights the entrepreneurial capacity of these producers (helped by good harvests). In addition, and equally important, are the local professionals, the intermediaries, who have opened and consolidated sales channels for horticultural products that reach the most demanding markets in Europe and North America; markets that are constantly expanding. Therefore, given this socioeconomic context, the demand for agricultural engineering is continuous, with almost non-existent unemployment figures. On the other hand, it is a sector that continually demands new scientific and technical developments to improve production and quality while also bearing in mind the quality of life of agricultural workers.

Therefore, much of the teaching and research at the University of Almería revolves around greenhouse agriculture.

University of Almería Academic and Research Work

The University of Almería (UAL) is a public higher-education institution founded in 1993, although it had previously been part of the prestigious University of Granada since 1972. At present, the University of Almería has more than 14,500 students, 1,152 professors, and 561 administrative and service staff. Its centres distribution comprises eight colleges (experimental sciences, economic and business sciences, health sciences, psychology, humanities, education sciences, law, and international doctorate) and two schools (engineering, and postgraduate studies and continuing education). UAL offers 32 degree courses, 42 official master's degrees and 14 doctoral programmes (UAL, 2022). Engineering studies at UAL began with the creation of the School of Agricultural Engineering in 1987. Subsequently, studies in computer science (1989) and industrial engineering (mechanical, chemical, electrical, electronic, and automatic) were added in 2005, becoming the Higher Engineering School.

The university's main areas of knowledge are agricultural science and engineering, industrial and information technologies, natural resources and the environment, physics and mathematics, health science, social sciences, business and law, and humanities and education. UAL research relates to the socioeconomic environment in which it is located, focusing particularly on agriculture, renewable energy, and water management. Currently, there are more than 150 research groups, of which more than half work in agriculture and bioprocessing.

This chapter's authors belong to the Automatic Control, Robotics, and Mechatronics (ARM) group, code TEP-197, of the Andalusian Government, which comprises more than 20 researchers (ARM-TEP197, 2023). The group is part of the Agrifood Campus of International Excellence (ceiA3) and the Solar Energy Research Centre (CIESOL). Our group's research lines include the application of modelling, automatic control, and robotics techniques to energy efficiency in productive districts, comfort control in buildings, greenhouse agriculture, renewable energy, vehicle dynamics, biotechnology, and bioengineering, as well as innovative techniques in teaching control theory. The group has extensive experience in applying OT technology (automation and robotics), IT technology, and all the digitalisation processes that the greenhouse agriculture and bioprocessing sectors are currently undergoing. In this field, it has participated in and led more than 25 national projects and ten international projects, including the European projects – Internet of Food 2020, Smart AgriHubs, and FutureFarm. At the same time, the group is strongly involved in engineering education, mostly regarding the design of interactive tools, and virtual and remote laboratories. Related to this work line, the group has participated in several projects, amongst which are the Erasmus+ Knowledge Alliances EAC/A02/2019 proposal "Next Generation Training in Intelligent Greenhouses" (NEGTHRA) and "Interactive books for control education", financed by the Institute of Electrical and Electronics Engineers (IEEE). This experience has led to several of its members joining the Control in Agriculture (TC 8.01) and Control Education (TC 9.4) technical committees of the International Federation of Automatic Control. Furthermore, they are members of various IT/OT technologies and agriculture associations. The above academic and research experience has been used to teach these contents in agricultural engineering studies.

Agricultural Engineering Studies in Spain

The title of Agricultural Engineer was created in Spain in 1855, as was the first college in Aranjuez (Madrid). There are now 26 centres where agricultural engineering can be studied. Currently, agronomy studies are made up of two degrees:

- Bachelor's Degree in Agricultural Engineering. The teaching objective is to provide adequate training in the theoretical bases and technologies of engineering, preparing students to apply engineering techniques to agricultural problems. The programme lasts four years, equivalent to 240 ECTS. Students take two courses of basic and common training in agricultural engineering. Then, they specialise in horticulture and gardening, agri-food industries, farming and livestock operations, and agricultural mechanisation for the following two years. Students complete their studies through a company internship and a final dissertation.
- Master's Degree in Agricultural Engineering. The teaching objective is to train students in the advanced

scientific, technological, and socioeconomic aspects of agricultural engineering and to prepare them for professional practice. It comprises between 90 and 120 ECTS, depending on the university, with subjects in technology and planning of the rural environment, technology of plant and animal production, technology of agri-food industries, and management and organisation of agri-food companies. To complete their studies, students complete a master's thesis.

There are also official master's degrees specialising in certain agricultural engineering subjects taught at the University of Almería; examples of these subjects are discussed in the next section.

The international university curricula recommendations (described in section 2) were taken into account in the *White Paper on the Degree in Agricultural Engineering and Forestry Engineering* (ANECA, 2005) to develop the contents of agricultural engineering studies in Spain. In this document, the following considerations were highlighted:

- Instrumental competence in computer knowledge and its application in the agricultural field is considered mandatory.
- A set of professional profiles are established according to the capacities, skills, and abilities of the proposed degrees. For example, Profile 5, "Process control and automation", is defined as competencies to select, assess, and implement the different control mechanisms of process lines, as well as the automatisms that allow their optimisation. On the other hand, Profile 10 includes "Automation in the rural environment".
- Concerning the level of knowledge, it is proposed that a high level of teaching is provided because it will equip future graduates with the necessary competencies to manage the construction, maintenance, automation, and equipment of industries in this sector.
- The following are proposed as basic training contents: auxiliary equipment and machinery for the agri-food industry and automation and control of agri-food processes.
- As far as competencies, skills, and abilities relating to the agro-food industries, the following knowledge areas stand out: electronic circuits and systems used in automatic control; basics of automation in sequential processes; industrial instrumentation; process optimisation, control and simulation; and planning, designing, and executing a complete control project.
- Concerning the operation of agricultural and livestock farms, it is necessary to know the automation systems and how to implement their facilities' agronomic, hydraulic, and automatic control design, such as the irrigation systems.

Based on all these considerations, in 2009, the Spanish Ministry of Science and Innovation published the requirements for verifying official university degrees that enable one to practise as a professional agricultural engineer. Regarding the IT/OT knowledge that should be included in these studies, the following requirements are highlighted:

- Degree in Agricultural Engineering:
  - A basic training module that includes knowledge of using and programming computers, operating systems, databases, and computer programmes applied in engineering.
  - A module common to all specialisations on the transfer of technology, understanding, interpreting, communicating, and adopting advances in the agricultural field.
  - A module common to all specialisations on decision-making, using all available resources for working in multidisciplinary groups.
  - A specific technology module on automating and controlling processes in agri-food industries.
  - A specific technology module on automation in agricultural machinery.
- In the master's degree, a compulsory module on agri-food industry technology, equipment, and systems for process automation and control must be included.

University of Almería Agricultural Engineering Curricula and IT/OT Technology Subjects

UAL offers various degrees related to the agricultural engineering curriculum, which provide the following training in IT/OT technologies:

- Bachelor's Degree in Agricultural Engineering (240 ECTS) – this degree was adapted to the recommendations of the European Higher Education Area in 2010. The 2015 Plan subsequently modified it for this degree course. It is accredited by the European Network for the Accreditation of Engineering Education (ENAEE) and has the International Quality Seal of the EUR-ACE in Engineering. Figure 37.8 shows the IT/OT subjects and their contents.
- Master's Degree in Agricultural Engineering (90 ECTS) – this degree was adapted to the recommendations of the European Higher Education Area in 2014. No subject is exclusively dedicated to these technologies, but their contents have been included in various subjects, such as agri-food process engineering (Module 2. Automation and control equipment), agri-food facility management (Module 3. Information management), biotechnology (Module 2. Bioinformatics), and animal production technologies (Module 2. IT applied to livestock farming).
- Master's Degree in Mediterranean Greenhouse Horticulture (60 ECTS) – this master's is offered only by UAL. It commenced in 2014. Its objective is to provide training for research, innovation, and development of multidisciplinary technical solutions

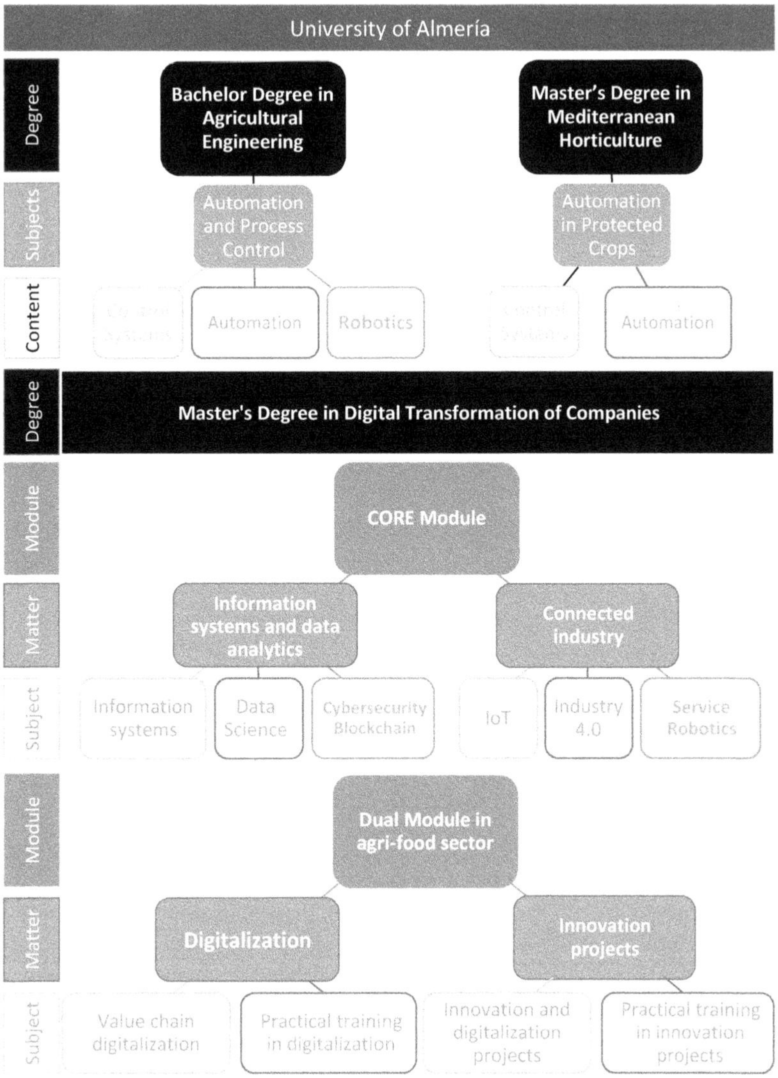

FIGURE 37.8 IT/OT subjects and contents in UAL agricultural engineering curricula.

applied to the engineering, design, management, and equipment in greenhouse horticultural production. Due to the application of technologies in greenhouses, IT/OT training is provided in several subjects. One of the most notable is the automation of protected crops, the contents of which are shown in Figure 37.8.

- Master's Degree in Digital Business Transformation, specialising in Agribusiness (60 ECTS) – an inter-university master's degree for which UAL coordinates the agro-industrial specialisation. Its objective is to provide training in the skills needed to implement digitalisation strategies in companies that promote changes in organisation management. It includes comprehensive training on design strategies in digital environments. This master's degree also includes student internships in companies where they are trained in four of the subjects and where they develop their master's thesis. In the first semester, they are trained in digitalisation-enabling technologies while, in the second, they apply this knowledge to the agribusiness sector. The subjects comprising this master's degree are shown in Figure 37.8.

Conclusions

As discussed several times in this chapter, the agricultural sector has to improve the efficiency of its agricultural processes, reduce costs, and increase productivity. Throughout this improvement process, digitalisation and automation play a predominant role. Specifically, IT refers to using computers, software, and networks to manage, process, and communicate information, while OT refers to using hardware and software to control physical processes and equipment. In the agricultural sector, IT and OT can work together to streamline crop monitoring, irrigation management, and equipment maintenance processes. A drawback to its implementation is the lack of knowledge of the actors involved in the process, from the agricultural workers to the managers, including agricultural engineers.

Different public and private institutions, both academic and professional, advise that training is provided to enable the technologies inherent in this process, given their importance in revolutionising agriculture. The work of agricultural engineers is essential to all the value-chain steps undertaken in this sector; but what role should they play in the process? On the one hand, agricultural engineers are responsible for resolving potential production problems and improving some of the processes. For this, they must know the existing technologies that can help them carry out their work properly, be trained to adapt as these technologies evolve, and even adapt to the appearance of new technologies. Other functions include their making decisions on acquiring and installing these types of systems, so they must grasp the fundamentals to understand the systems available and the budgets.

Nevertheless, these technologies may already be installed on farms or in the agri-food industry, so they must be trained to work on configuring and utilising them. Another area where agricultural engineers might work is in the digitalisation service sector, for companies who seek their advice on developing tools by contributing their agronomic knowledge rather than as programmers or developers of the technologies themselves. In either scenario, agricultural engineers will be part of a multidisciplinary team along with other IT and OT engineers, so they must all be able to understand each other by speaking the same engineering language – and to do this, they must fully understand these technologies.

This chapter proposes that a module should be included in the agricultural engineering curricula that consist of six subjects covering both basic and advanced knowledge of IT/OT technologies – programming, information systems, instrumentation for data acquisition, the IoT, data science, cloud computing, systems analysis, and modelling, automation, and robotics. We know that there are other digitalisation-enabling technologies such as virtual reality, 3D printing, and blockchain but we believe the ones we have described are the most fundamental. Agricultural engineering is based on a knowledge of both engineering and biology, so these subjects should not be eliminated just to include more IT/OT in the curricula. Furthermore, this proposal is based on international recommendations and on the authors' experience, which is mainly in greenhouse agriculture (so it might be slightly biased in this respect). It could be adapted to other agricultural concerns, such as outdoor or livestock farming; hence, modifications could and should be made regarding the allocation of hours or the topics covered.

In summary, this field is in the process of evolution. Consequently, we should be prepared to revise the curricular contents and how they are organised constantly.

Acknowledgements

This work was supported by the Department of Informatics of the University of Almería and the ARM research group. Now the authors are working within the Next Generation Training on Intelligent Greenhouses (NEGHTRA) project, funded by the Erasmus + KA2 programme: Cooperation for innovation and the exchange of good practices – Knowledge Alliances 2020, Proposal: EAC/A02/2019, Application: No 621723-EPP-1-2020-1-EL-EPPKA2-KA

REFERENCE LIST

ANECA. (2005). *Libro blanco del Título de Grado en Ingenierías Agrarias e Ingenierías Forestales.* Agencia Nacional de Evaluación de la Calidad y Acreditación.

ARM-TEP197. (2023). *ARM group – Automatic Control, Robotics and Mechatronics.* https://arm.ual.es/arm-group/

Bachche, S. (2015). Deliberation on design strategies of automatic harvesting systems: A survey. *Robotics*, *4*(2), 194–222.

Briassoulis, D., Papadiamandopoulou, H., & Bennedsen, B. S. (2001). Towards a European standard for agricultural engineering curricula. *AFANet*, 1–54.

ERABE. (2009). Education and research in biosystems engineering in Europe. *A Thematic Network.* Progress Report, 1–19.

FAO. (2018). The future of food and agriculture: Alternative pathways to 2050. Food and Agriculture Organization of the United Nations, Rome, 1–35.

FAO. (2022). Biannual report on global food markets. Food and Agriculture Organization of the United Nations, 1–174.

Febo, P., & Comparetti, A. (2013). Biosystems engineering curricula in Europe. *Journal of Agricultural Science and Technology*, *3*, 1–9.

Febo, P., & Sun, D.-W. (1998). Harmonisation of agricultural engineering university curricula. *Report of CIGR WG1 and EurAgEng SIG*, 1–37.

Febo, P., & Sun, D.-W. (2000). The university structure and curricula on agricultural engineering: An overview of 36 countries. Food and Agriculture Organisation of the United Nations, 1–233.

Febo, P., & Ward, S. (1995). Harmonization of university curricula in agricultural engineering. *International Conference on Agricultural and Biological Engineering: New Horizons, New Challenges*, Newcastle (UK).

Galdeano, E. (1999). El contexto actual del sector hortofrutícola almeriense y la necesidad de estrategias de conjunto. *Anuario de La Agricultura Almeriense 1999*, 12–20.

Garimella, P. K. (2018). IT-OT integration challenges in utilities. *Proceedings on 2018 IEEE 3rd International Conference on Computing, Communication and Security*, ICCCS 2018, 199–204.

Gartner Group. (2016). *Operational Technology (OT); Glossary of Information Technology Achronymns and Terms.* https://www.gartner.com/en/glossary

Horto-Info. (2022). *Las grandes cifras en la campaña 2021/2022 de los invernaderos en Almería; Diario Digital de Actualidad Hortofrutícola.* https://hortoinfo.es/grandes-cifras-resumen-campana-invernaderos-almeria

i-SCOOP. (2020). *The essential Digital Transformation Guide.* https://www.i-scoop.eu/digital-transformation

ICT-AGRI. (2014). *ICT-AGRI Meta Knowledge Base | Information and Communication Technologies and Robotics for Sustainable Agriculture.* ICT-AGRI. https://ict-agri.eu/

IEA. (2023). Boletín Coyuntural Provincial Almería. Instituto de Estadística. Junta de Andalucía, 17.

INE. (2023). Resultados provinciales Almería. Intituto Nacional de Estadística. Gobierno de España.

IoF2020. (2017). *Project Internet of Food and Farm 2020.* https://www.iof2020.eu/

Kondo, N., Monta, M., & Noguchi, N. (2011). *Agricultural Robots: Mechanisms and Practice.* Kyoto University Press and Trans Pacific Press.

Maroto, J. (1998). Historia de la agronomía: Una visión de la evolución histórica de las ciencias y técnicas agrarias. *Madrid, Mundi-Prensa*, 416.

Misset. (2023). *Field Robots Catalogue.* Buyers Guide. Future Farming. https://www.futurefarming.com/dossier/field-robots/

Panagakis, P., Briassoulis, D., Nikopoulos, E., Douzals, J. P., Scarascia-Mugnozza, G., Febo, P., Comparetti, A., Fehrmann, J., Ayuga, F., Mostaghimi, S., Wolfe, M. L., Riskowski, G., Ting, K. C., & Kanwar, R. (2008). Policy oriented measures in support of the evolving biosystems engineering studies in USA-EU. *International Conference on Agricultural Engineering, Hersonissos AgEng08*, 1–9.

Pellizzi, G., & Febo, P. (1994a). The CIGR project for the harmonization of agricultural engineering university curricula. *Journal of Agricultural Education and Extension*, *1*(36), 6.

Pellizzi, G., & Febo, P. (1994b). The University structure and curricula on agricultural engineering: An overview of 25 countries. *CIGR WG Report Series*, *2*, 1–175.

Rodríguez, F., Berenguel, M., Guzmán, J. L., & Ramírez-Arias, A. (2015). *Modeling and Cntrol for Greenhouse Crop Growth.* Springer International Publishing.

Rodríguez, F., Moreno, J. C., Sánchez, J. A., & Berenguel, M. (2013). Grasping in agriculture: State-of-the-art and main characteristics. In: Carbone, G. (ed) *Grasping in Robotics. Mechanisms and Machine Science*, vol 10 (pp. 385–409). Springer, London.

Sandun, F., Murali, N., & Bhushan, S. (2006). The need to reform agricultural engineering curricula in developing countries. *International Journal of Engineering Education*, *22*(1), 79–85.

Shamshiri, R. R., Kalantari, F., Ting, K. C., Thorp, K. R., Hameed, I. A., Weltzien, C., Ahmad, D., & Shad, Z. M. (2018). Advances in greenhouse automation and controlled environment agriculture: A transition to plant factories and urban agriculture. *International Journal of Agricultural and Biological Engineering*, *11*(1), 1–22.

UAL. (2022). *Memoria del Curso Académico de la Universidad de Almería*, 121.

Ureña, R., Rodríguez, F., & Berenguel, M. (2001). A machine vision system for seeds quality evaluation using fuzzy logic. *Computers and Electronics in Agriculture*, *32*(1), 1–20.

USAEE, T. (2002). The newsletter of the USAEE thematic network. University Studies of Agricultural Engineering in Europe, 1, 2.

USAEE, T. (2006). Core curricula of agricultural/biosystems engineering for the first cycle pivotpoint degrees of the integrated M.Sc. or long cycle academic orientation programs of studies. University Studies of Agricultural Engineering in Europe, 77.

Yıldırım, C., & Solmuş, İ. (2014). A parametric study on a humidification–dehumidification (HDH) desalination unit powered by solar air and water heaters. *Energy Conversion and Management*, *86*, 568–575.

YuanHui, Z. (2005). Reshaping curriculum and education for engineering in agriculture - For the centennial anniversary of China agricultural university. *Journal of China Agricultural University*, *10*(4), 7–12.

Zhang, S., Guo, Y., Zhao, H., Wang, Y., Chow, D., & Fang, Y. (2020). Methodologies of control strategies for improving energy efficiency in agricultural greenhouses. *Journal of Cleaner Production*, 274, 122695.

Zins, C. (2007). Conceptual approaches for defining data, information, and knowledge. *Journal of the American Society for Information Science and Technology*, *58*(4), 479–493.

38

Technological-based Resources as Support to Agricultural and Biosystems Engineering Studies

José Luis Guzmán, Juan Diego Gil, Francisco García-Mañas and Ángeles Hoyo

Introduction

Engineering education is mainly characterised by including an appropriate balance between theoretical lessons and practical lab-based experiments. In fact, the main differences compared to other disciplines are the high practical component and the direct connection between theory and practice. In engineering, laboratory exercises and experiments are a fundamental part of the learning process. Moreover, theoretical lectures are usually mixed with practical exercises and simulations to be discussed with students. Thus, for these reasons, engineering studies always require realistic lab equipment and/or software simulators to account for these teaching/learning objectives.

The availability and maintenance of these teaching resources are highly time-consuming and costly for universities. Thus, thanks to the continuous advances in the information and communication technologies (ICT), technological-based resources are becoming an important solution supporting engineering studies, such as virtual/remote labs, interactive tools, or take-home devices. Furthermore, these solutions have also become very relevant during the pandemic for distance teaching/learning activities (Guzmán et al., 2022a).

In developing these educational resources, one should consider that future engineers must be capable of facing any industrial problem and providing solutions for it. So, looking for solutions to teach/train students how to manage all possible industrial situations is really important.

This chapter summarises technological-based resources for agricultural and biosystem engineering studies to support theoretical and practical lessons on modelling, simulation, and control engineering topics. First, virtual and remote labs based on a greenhouse system and an anaerobic digestor process are described as support to practical lectures (Guzmán et al., 2005; Rivas et al., 2012; Guzmán & Joseph, 2021). Then, computer-based simulators are introduced as support to the teaching/learning of complex bioprocesses, such microalgae-based systems (Hoyo et al., 2022). Next, interactive tools are described as support to theoretical control engineering concepts (Guzmán et al., 2008, 2013, 2016, 2023). Finally, the take-home methodology to model and control a temperature-based process is summarised (Guzmán et al., 2022b).

Virtual and Remote Labs (Web-Based Experimentation)

In the last years, new teaching methodologies have appeared in education due to the emergence of new ICT. These technologies have enabled teachers to identify and promote creative and powerful techniques that help to increase students' education and motivation. Digital sliders, multimedia tools, interactive tools, or online platforms for exchanging information between students and teachers represent only a few examples of such techniques. This novel paradigm, commonly referred to as e-learning, has resulted in significant progress for students, who generally exhibit greater involvement in the teaching process, and for teachers in explaining and transmitting complex concepts (Sánchez et al., 2005; Guzmán et al., 2006). However, despite the possibilities provided by these tools, it is often mandatory to complement these teaching methods with content or material featuring a more practical and experimental orientation to incentivise the applicability of the theoretical concepts. In a context of distance education, this fact gives rise to web-based experimentation (Stefanovic et al., 2010), which will be discussed in this section.

In traditional education, the primary means of facilitating practical activities have been through local simulation tools or laboratories, with students constrained by space and time limitations while performing practical exercises on real systems. However, the key services supporting web-based experimentation now offer a more flexible, cost-effective, and integrated approach, enabling remote access to real laboratory resources and simulations, analytical tools, and shared spaces, all capable of real-time data exchange. This more advanced approach, better suited to support collaborative and hands-on learning activities typical of modern educational environments, has led to two new concepts in the distance-education framework: virtual and remote laboratories. The first concept, virtual laboratories, constitutes a novel simulation tool with greater computational power than its traditional counterparts, enabling students to utilise remote simulation modules concurrently. In contrast, remote laboratories facilitate students' execution of primary laboratory activities from remote locations without necessitating their physical presence

DOI: 10.1201/9780429150111-41

at the laboratory site (Potkonjak et al., 2016). This feature allows students to monitor and control physical devices at any time from any place, while interacting with their teachers synchronously. Remote and virtual laboratories, therefore, offer students an opportunity to practically apply their theoretical knowledge while reducing critical aspects such as displacement costs and the limitation of the use of time-limited resources or equipment.

Focusing on engineering degrees, the impact of the tools described above on engineering education has been and continues to be significant. A prime example is the automatic control field, where an engineer's professional success relies heavily on the quality of education received during their academic years. A noticeable feature of automatic control education is the persistent use of experiments. Consequently, university teachers and researchers must continuously devote attention to exploring new teaching methodologies and techniques. Special committees, such as the International Federation of Automatic Control (IFAC) and the Institute of Electrical and Electronics Engineers Control Systems Society (IEEECSS), that research this topic exclusively promote advances and incentives for teaching improvements related, among others, to distance and virtual education and e-learning teaching technologies. These committees' presence in the most important associations on automatic control underscores the significant interest placed on the education area. These efforts in academic advancement have yielded fruitful results, with the availability of numerous virtual and remote labs, such as those presented by Rivas et al. (2012) and Guzmán et al. (2021), which will be briefly described in the following paragraphs to exemplify the possibilities of these tools and their potential.

On the one hand, Rivas et al. (2012) presented an integrated virtual and remote laboratory in their study aimed at teaching greenhouse climate control. This task is particularly challenging due to the nonlinear and multi-perturbed nature of greenhouse systems, making them an ideal benchmark for practising various skills in modelling and control. The virtual lab comprises two components: a greenhouse simulator replicating climate conditions using a nonlinear model implemented in MATLAB/Simulink (Rodríguez, 2002) and a graphical user interface (GUI) programmed in Easy Java Simulations (EJS). The GUI is divided into three main parts: a greenhouse synoptic, which presents a virtual representation of a greenhouse; a panel of parameters related to actuators such as natural ventilation, shade screen, and heating based on a boiler; and a set of graphics that display the evolution of important simulation variables and control signals (see Figure 38.1a). It is important to remark that the lab was designed to be executed by multiple users simultaneously over the internet, with a single copy of the model hosted on a server.

In turn, the real plant accessible remotely, i.e. the remote lab, is a greenhouse scale model equipped with multiple sensors and actuators, which was developed as part of the DAMOCIA project (Guzmán et al., 2005). In this laboratory, the LabVIEW programming environment was employed for input/output tasks, enabling the acquisition of data from sensors and the manipulation of the state of the actuators. A LabVIEW application was created and published on JIL, with input/output parameters automatically available for referencing from EJS. This approach frees developers from low-level programming details, enabling the GUI used for the virtual lab to control the real plant in the same manner, with the same execution options, and by using the same control algorithm parameters (see Figure 38.1b). The only difference is that the remote lab provides visual feedback of the plant's status, thanks to its connection to a video server and an IP camera positioned inside the scaled greenhouse. This setup allows users to observe the actions of the actuators in real time.

Note that virtual and remote labs were integrated into a single software application within a collaborative environment accessible via the internet. This platform allows users to access the tool anytime without spatial or temporal constraints and provides an educational environment where students can work the same way as they would in a traditional laboratory. An advantageous feature of the platform lies in its capacity to facilitate learning GUI through a single experience, as the shared configuration parameters and graphics of both virtual and remote labs foster improved comprehension among users. This, in turn, enables users to effectively apply controllers to both environments, thereby enhancing students' understanding of the basic principles.

On the other hand, Guzmán and Joseph (2021) developed a web-based virtual simulation model of an anaerobic digester (AD) using a flexible Javascript-based platform to support courses in process control and design. The goal of this lab was to provide students with a virtual laboratory experience that could be accessed from any device with internet capabilities, including laptops, tablets, and smartphones. The virtual lab was based on a continuous heated stirred tank design AD, as shown in Figure 38.2a. This type of system involves critical design aspects such as tank size, control of liquid level and digester temperature, and optimisation through set points for level and temperature. These challenges allow students to explore various features related to a complex biochemical system, including design, operation, control, and optimisation.

As shown in Figure 38.2b, the developed virtual laboratory incorporates a user-friendly GUI that enables straightforward specification and testing of operational parameters. Specifically, the GUI provides various tabs, including system synoptic, concentration variables, controlled variables, and all variables, to enhance the user experience. The system synoptic tab offers a graphical representation of the process, as depicted in Figure 38.2b, while the remaining three options provide real-time logs of the system's primary variables. Moreover, the laboratory features a parameter display zone comprising four sections. The first section contains tool actions, including four buttons for play, pause, reset, and start recording. The second section allows users to modify input variables and parameters, while the third section displays the process outputs. The final section of this area concentrates on control loop parameters. The reader is referred to Guzmán and Joseph (2021), who presented a comprehensive collection of exercises that employ virtual labs to teach process design and control. Through these exercises, students can gain practical experience in the field.

The works by Rivas et al. (2012) and Guzmán and Joseph (2021) constitute only two examples of virtual and remote labs used to teach control concepts; however, they illustrate the potential of these tools conclusively. In addition to the

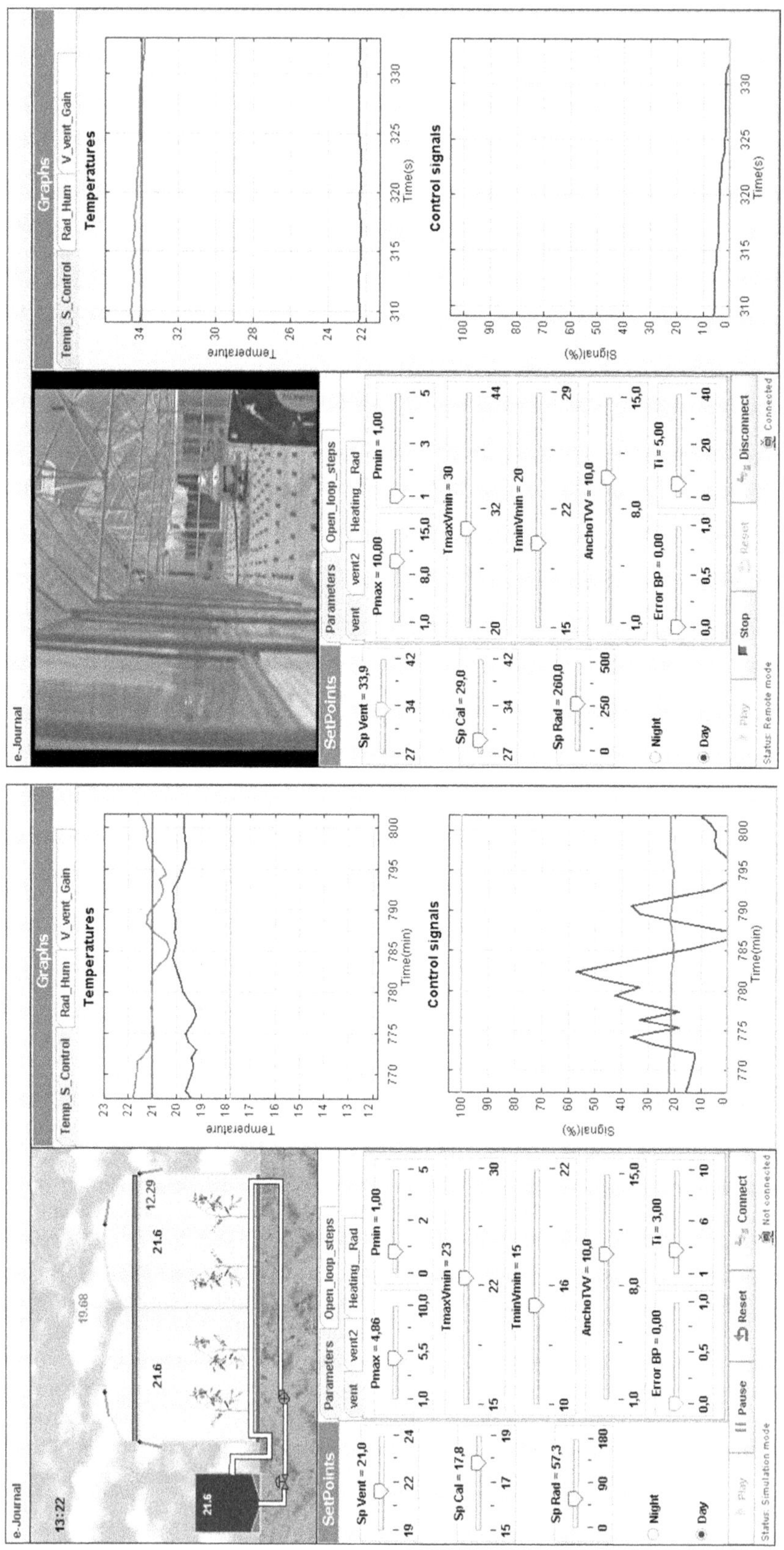

FIGURE 38.1 Illustration of the graphical user interface (GUI) of the virtual and remote lab, adapted from Rivas et al. (2012). (a) GUI of the virtual lab. (b) GUI of the remote lab.

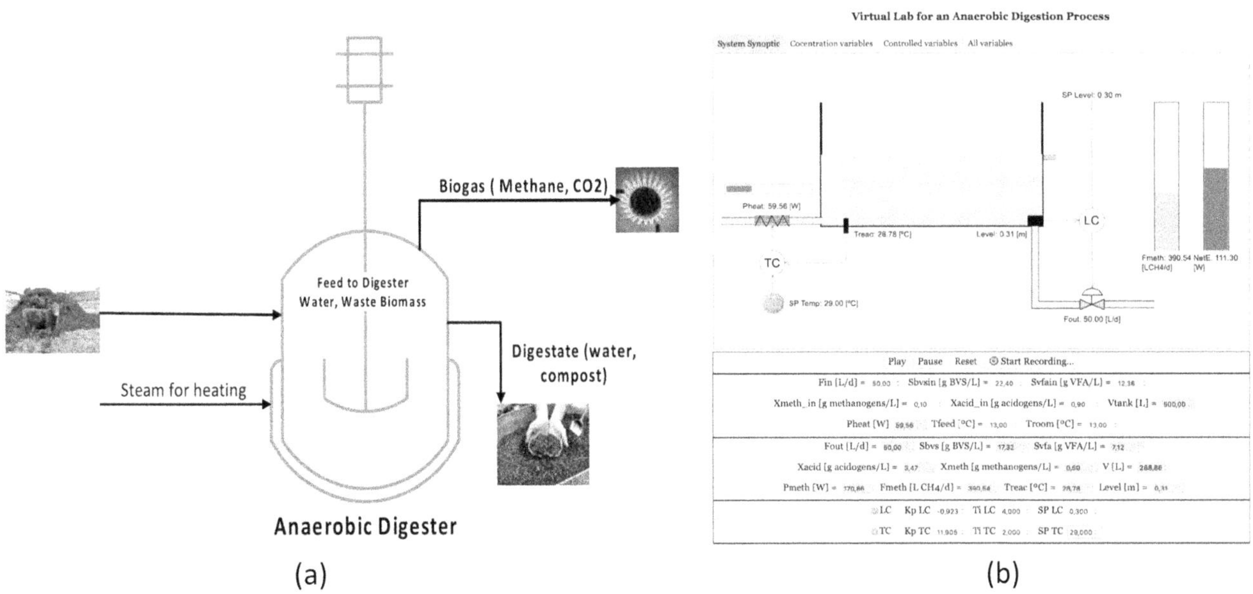

FIGURE 38.2 Illustration of anaerobic digester (AD) and the GUI of the virtual lab, adapted from Guzmán and Joseph (2021). (a) The AD process. (b) The main screen of the virtual lab for AD processes, as viewed from a laptop.

well-known benefits of virtual and remote labs, such as the flexibility to access them without any temporal or spatial constraints, it is noteworthy that the developed labs by Rivas et al. (2012) and Guzmán and Joseph (2021) are designed around socially relevant problems, which can stimulate student motivation and interest. Therefore, it is clear that virtual and remote labs have the potential to revolutionise the way control concepts are taught and put into practice in traditional teaching courses.

Computer-based Simulators

For decades, several strategies using technology have been employed to enhance teaching and learning processes in engineering education. Most of these strategies include the use of computer programs as educational support to ask and answer elemental questions to engage students (Kayssi et al., 1999), as online educational resources for teaching, assessing, and expanding students' skills (Ong & Mannan, 2004, Guzmán et al., 2023), or as virtual reality simulations (Hoyo et al., 2015).

Simulation has been demonstrated to be one of the most potent and appealing computer-based educational tools. For decades, computer simulations have been used as an educational tool and are especially critical for engineering education in laboratory exercises (Balamuralithara & Woods, 2012). Furthermore, computer-based simulators can be used to create experiments and simulations that are dangerous or expensive to perform physically. Thus, they allow students to explore concepts, theories, and phenomena that they might not otherwise be able to experience in a traditional laboratory setting. They can be used to create virtualised simulations of real-world scenarios, offering a complete understanding of a concept than what can be achieved through a traditional lab setting (Tüysüz et al., 2010). Additionally, these simulators can provide personalised instruction and feedback to students and access to data and resources that may not be available in a physical one.

Research has demonstrated that simulation-based educational tools benefit classroom teaching and self-learning. For example, the study by Kabigting et al. (2021) determines that using computer simulation as instructional material in teaching physics is recommended. In (Rodrigues et al., 2010), a computational tool called EMSO is introduced for use in chemical reaction engineering classes. Its structured modelling language allows a more organised implementation, exploring different operating conditions. These facilities enable the users to save time in understanding. Interactive computer simulators have improved student engagement and understanding of complex topics (Ben Ouahi et al., 2022). In line with this, a study by Ratamun et al. (2018) proves the effectiveness of using a virtual laboratory to learn a chemistry experiment over conventional methods.

Computer-based simulators can also be used for research purposes. They can be used to study the behaviour of a system, allowing researchers to observe how various parameters and conditions influence the system's performance. For example, to study the behaviour of epidemics, such as the spread of the coronavirus (Malkov et al., 2020), and to study the behaviour of financial markets (Neri, 2012). Simulations can also be used to study the behaviour of complex systems, such as robot modelling and behaviour (Staranowicz & Mariottini, 2011), or the behaviour of complex systems involving humans, such as social networks (Tsvetovat & Carley, 2005). They are also influential in studying the behaviour of complex biological systems, such as the immune system (Rapin et al., 2011).

This section presents a computer-based simulator for a microalgae-based system as an example of an educational resource. The simulator was developed by Hoyo et al. (2022) and provides a graphical tool to simulate a microalgae

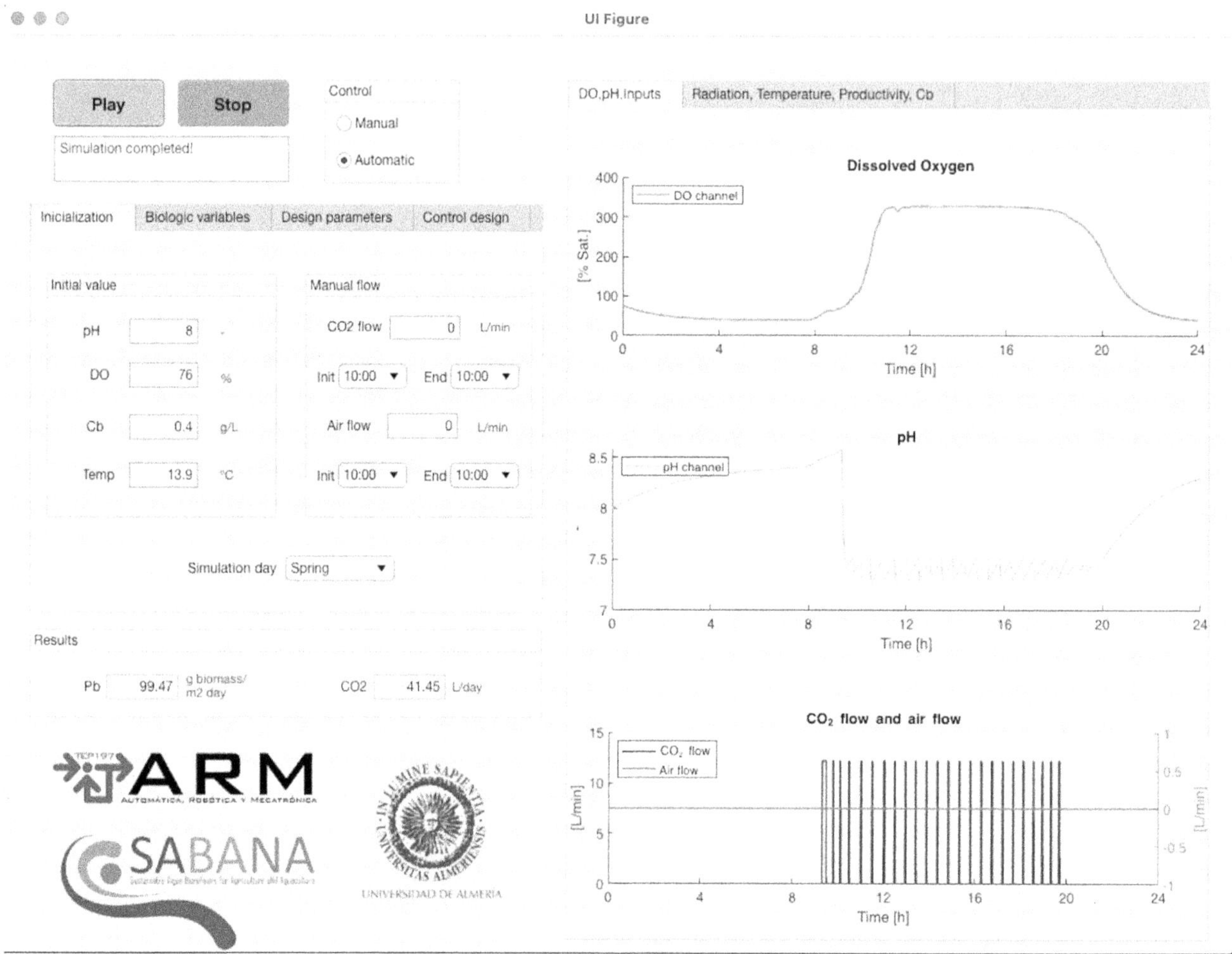

FIGURE 38.3 Raceway reactor simulation tool main screen. Adapted from Hoyo et al. (2022).

production process in raceway reactors. Figure 38.3 shows the main screen of the simulator.

Microalgae cultivation is a biological process that, due to its great potential for energy purposes and other derived products, has become widely recognised today. The reactors mainly used are the raceway reactors, which require less complex equipment and structures and a lower initial investment cost than the tubular ones. This biological system has complex dynamics affected by biological, environmental, and design variables. The pH, temperature, radiation, and dissolved oxygen are the most critical variables influencing microalgae production. In raceway reactors, the system architecture determines the temperature and solar radiation variables. The pH and dissolved oxygen variables strongly depend on photosynthesis, so keeping them close to their optimal operating points is necessary to maximise productivity. To achieve this, it is essential to implement control engineering approaches that actuate automatically to the inputs of the plant. The injection of CO_2 flow modifies the pH value, and airflow modifies the dissolved oxygen value. The automatic feedback controller measures the difference between the desired value of the variable (setpoint) and the actual value in the system, calculates the tracking error, and then gives a control action to minimise the error. Apart from complex control strategies, the most generally control strategies implemented in the raceway reactor are the on/off controller (On/Off) and the proportional-integral-derivative (PID) controller. The complexity of these processes arises from the operation, design, modelling, and control perspective (Guzmán et al., 2021). To simplify the system's behaviour simulation and to modify the principal elements from biological, design, and operational standpoints, a graphical simulator tool was developed by Hoyo et al. 2022 and shown in Figure 38.3. The raceway reactor design parameters and data collection implemented in the tool are based on the real reactor located at the Instituto de Investigación y Formación Agraria y Pesquera (IFAPA) centre in Almería, Spain.

The simulator tool is designed with three different main areas. Play and stop buttons are placed on the top, up to an information text box. On the right part, manual/automatic mode can be selected. It allows the tool to simulate open or closed loops. Four tabs in the configuration area enable users to adjust the essential parameters of the process. The first tab allows user to input initial dissolved oxygen, pH, biomass concentration, and initial temperature values, such as air and CO_2 flow, can be set. It is also possible to select a season to simulate. Real data from different year dates are implemented as input for the simulation.

In the biological variables tab, users can change the most important biological parameters of the microalgae strain implemented. It also allows the user/student to implement any strain in the simulation. The one implemented in the first setup is the *Scenedesmus almeriensis*. The nonlinear dynamic models implemented in the tool are developed by Fernández et al. (2016), Rodríguez-Miranda et al. (2021) and Sánchez-Zurano et al. (2021). These models consider mass transfer, fluid-dynamics, and the biological phenomena occurring in the reactor.

In the third tab, design parameters, the physical size of the reactor can be modified, such as the loop length, width, and height. The default setup consists of two channels, each measuring 50 m in length and one metre in width, connected by a U-shaped bend. A paddlewheel of 1.5-diameter aluminium blades drives the reactor flows along the channels at two metres per second velocity with an electric motor. The reactor is operated with a liquid level of 0.15 m.

In the last tab, different control approaches can be implemented. The most important variables to be controlled are the dissolved oxygen and the pH, being able to set their corresponding setpoints. Control can be done on each variable alone or both simultaneously. It is also possible to control these variables only during the day or day and night. The On/Off control is applied for the dissolved oxygen control, whereas on the pH, it is possible to select between On/Off and PID control (Åström & Hägglund, 2006).

The results of the simulation are displayed in two different areas. On the one hand, microalgae productivity (Pb) and CO_2 flow consumption are displayed numerically on the result box. On the other hand, the results can be visualised with graphics on the right side. On the first tab, the variables represented on the graphs during the simulation day are the progression of the dissolved oxygen, pH, and CO_2, and air injection. The second tab shows solar radiation, temperature, biomass productivity, and microalgae biomass concentration.

Numerous potential scenarios could be used to simulate microalgae productivity. In this way, students from different disciplines can use the tool to understand different concepts of the system. Students from the biological and chemistry areas can use it to understand the different production patterns of the microalgae in several setups. In this way, it can be used to understand the best conditions for the microalgae to grow and produce more biomass, the different growth and production rates of different strains, as well as the effects of different environmental factors on them. The simulation tool can also help learn the system's dynamics: how the microalgae interact with each other and their environment. Students can explore the different effects of a specific parameter on productivity and the interactions between different strains. It would be also useful to understand the effects of different environmental parameters (such as temperature, solar radiation, and nutrients) on the productivity of microalgae. For control engineering students, the simulation tool can be handy in understanding the open- and closed-loop system concepts and implementing different control techniques in a complex nonlinear biological system.

This simulation tool allows students to develop a comprehensive understanding of the dynamics of microalgae production systems and gain experience in applying different control techniques to optimise the system, visualising the effect on the system variables graphically. They can explore different proportional integral (PI) tuning parameters for the controller and analyse how microalgae productivity can be affected by it. Additionally, they can experiment with different parameters and control scenarios to identify the best strategies for optimising and maximising production. It can also help to study the feasibility of its implementation at an industrial scale.

Interactive Tools

Agricultural and biosystems engineering are disciplines of knowledge with significant mathematical content. Consequently, in many cases, the key concepts are difficult for most students to grasp. This is especially the case for those contents where students have only an introductory course in their engineering curriculum about a certain topic, like courses on automatic control. In those situations, teachers look for ways to introduce and make automatic control attractive to the students.

Theory can be acquired through textbooks, stimulating lectures, and dedicated study. A key element in problem-solving is the ability to effectively use theory and to divide large problems into smaller manageable parts. An important objective for engineering teachers is to not only impart the formal and logical structure of the discipline but also, and more importantly, the strategic and intuitive aspects. This latter part can be difficult to express and assimilate for students since it frequently lies in the more subconscious realm of the expert's activity. Interactive tools can be an excellent means of developing a student's intuition. These interactive tools try to make mathematical concepts more understandable by visualising certain examples (Guzmán et al., 2013).

Although visualisation is an important educational tool, interactivity is even more powerful. Interactivity is defined as the interaction between two or more entities which enables the exchange of information. This is different from looking at a figure or reading a book, as interactivity involves a two-way flow of information which helps to enhance the learning capacities of the involved parties. Interactivity in learning is an essential step for gaining knowledge and for developing cognitive and physical skills. To sum up, interactivity increases the users' motivation through more engaging activities (Guzmán et al., 2008; Guzmán et al., 2013).

An interactive tool is a set of graphical windows that can be manipulated simply by using a mouse. Students do not need to write any sentences, as all they have to do is alter the active elements in the windows and the result is immediately calculated and displayed. This helps students to understand the effect of their changes on the displayed results. The use of interactive tools can be a great way to illustrate theoretical concepts during lessons and to enhance students' skills in analysis and design tasks.

In conclusion, the use of interactive elements can be beneficial in enhancing the learning process and making lectures more appealing and engaging for students.

In recent years, the authors have been involved in several engineering education projects to develop interactive tools, particularly in the area of control systems, using Sysquake as a development environment. Two of these projects are briefly described in this section, along with some of the interactive tools that have been created. It should be noted that the interactivity of these tools is not easily conveyed in written form, and the reader is encouraged to visit the website https://w3.ual.es/personal/joguzman/material_docente_itools.shtml to experience the interactive features of the tools.

Interactive Learning Modules. This set of interactive learning modules has been designed to help learners gain an understanding of PID control. It is an extension of a book, providing interactive elements to the key concepts described within. The modules feature menus for selecting process transfer functions and PID controllers, in which parameters can be changed and results can be saved and loaded for use in another session with the interactive tool. A graphical display showing time and frequency responses is also included, and this can be manipulated via dragging lines, points, curves, and sliders. The core module is entitled "PID Basics", while two additional modules, "PID Loop Shaping" and "PID Windup", illustrate loop shaping and windup (see Figure 38.4a, 38.4b, and 38.4c) (Guzmán et al., 2008).

Automatic Control Interactive Tools. The central idea behind this project was to create a suite of applications that would make up an introductory course about classical control theory. It is not necessary for both teachers and students to stick to a complete course based on interactive tools. Each tool represents an independent concept which can be used independently, or the tools can be combined to encompass different concepts depending on their contents. This project is structured in such a way that it can be used by different audiences by means of individual cards. These cards, with interactive tools, are used to explain and understand the essential concepts and to solve exercises based on interactivity. This project consists of learning cards containing interactive software applications, brief theoretical descriptions of the different concepts to learn/teach, a novel kind of exercise for students to learn by discovery, and references to well-known books. More than 35 interactive tools have been developed to support this project. Figure 38.5a and 38.5b shows an example of a learning card and the associated interactive tool (Guzmán et al., 2016, 2023).

These two projects show how interactive tools provide a wide range of possibilities to teachers and students. In the authors' experience, interactivity is a very powerful element that can be considered an excellent resource to support teaching and learning, as it increases students' motivation and participation.

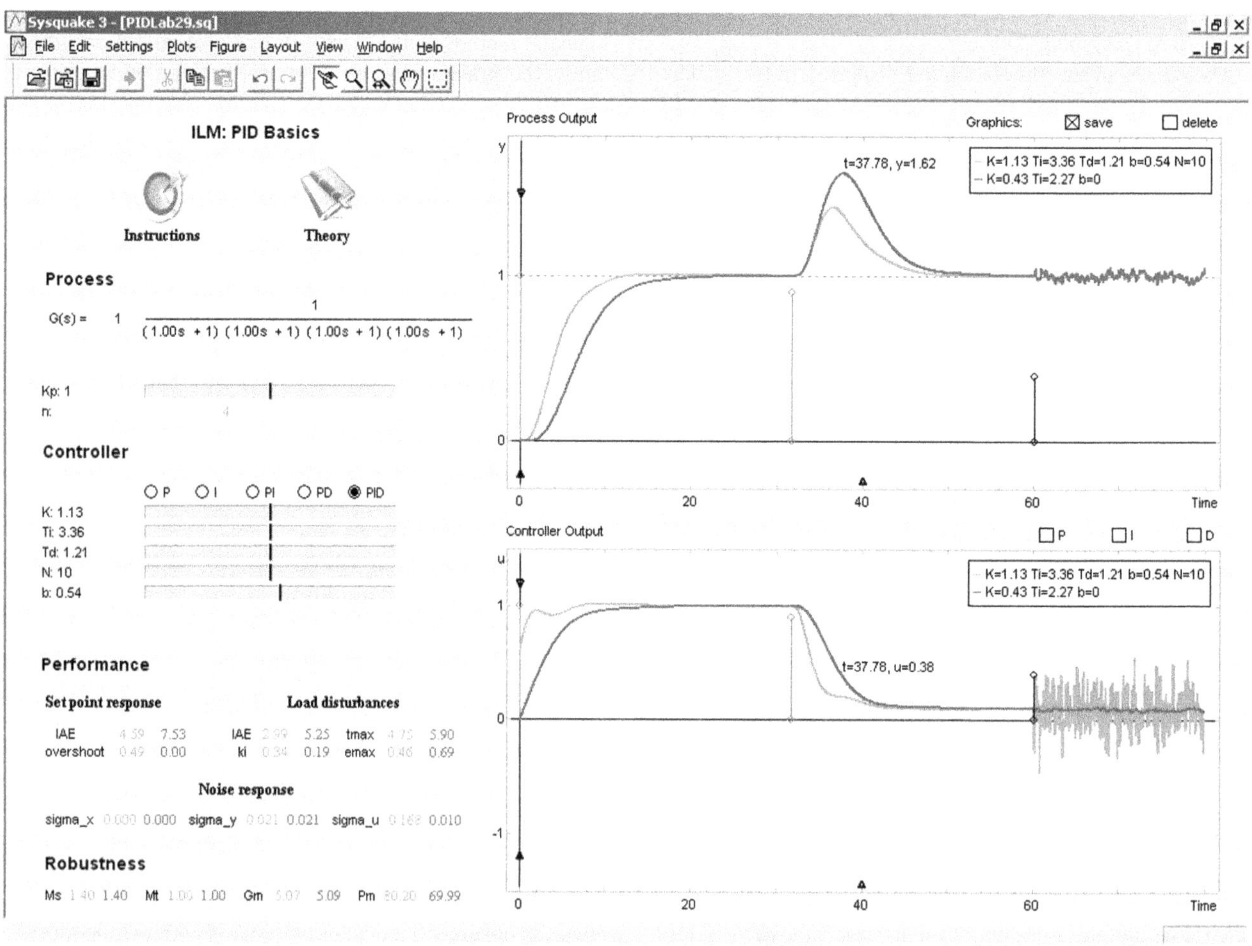

FIGURE 38.4 Screenshots of main interactive tools from the Interactive Learning Modules project. (a) GUI of the PID Basics Interactive Tool. (b) GUI of PID Loop Shaping Interactive Tool. (c) PID Windup Interactive Tool.

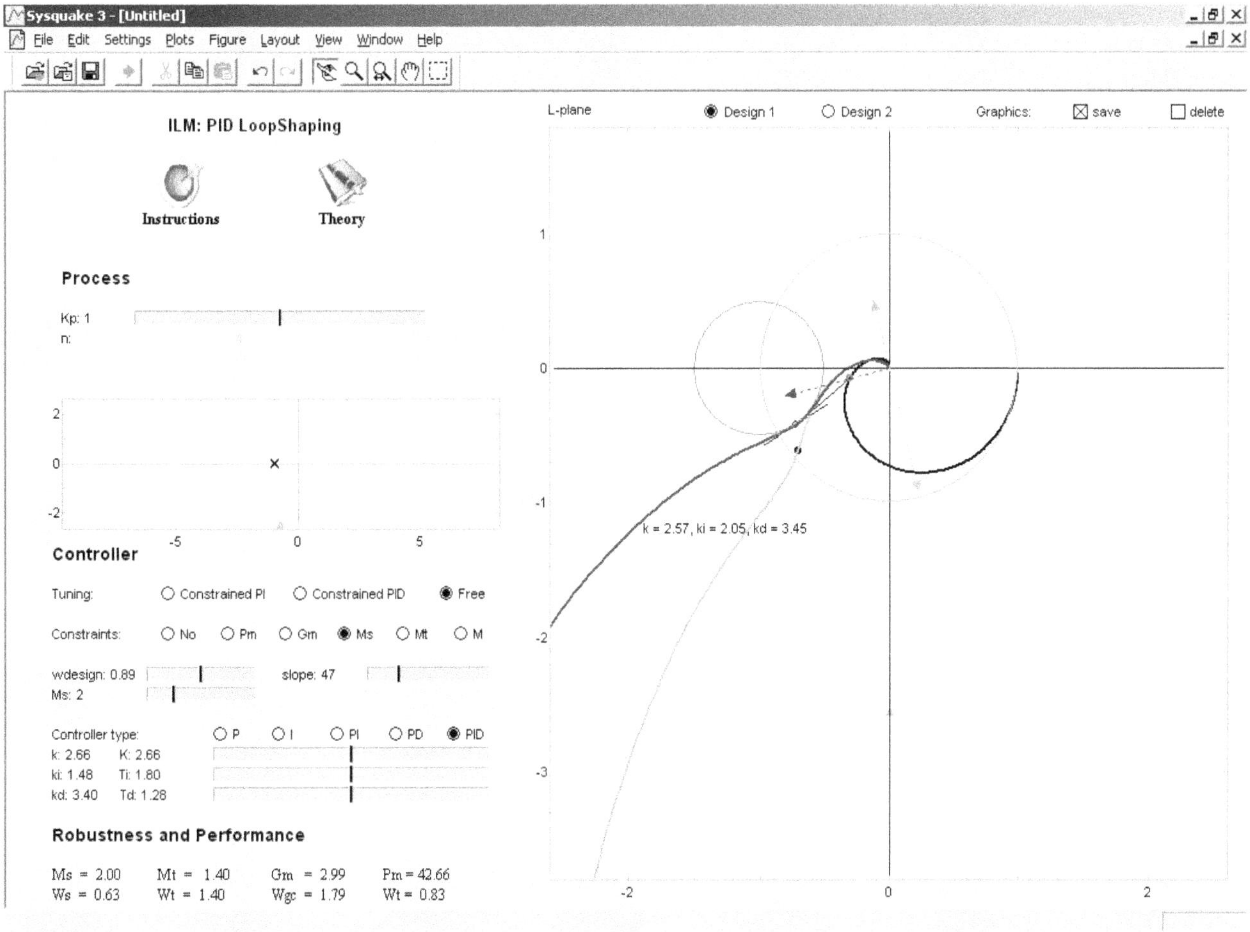

FIGURE 38.4 (b) GUI of PID Loop Shaping Interactive Tool.

Take-Home Labs

Laboratory classes are essential in engineering studies to bridge the gap between theory and practice in the student learning process. Despite the advantages of laboratory teaching, such as consolidating theoretical concepts and increasing student motivation, traditional laboratories have significant drawbacks. For example, in most educational centres, practical classes are normally held in fewer laboratories with sophisticated and expensive equipment. For these reasons, students only have access to traditional laboratories for a limited period, under the supervision of lecturers who also need the assistance of technicians responsible for maintaining the experimental equipment.

As explained in previous sections of this chapter, remote and virtual laboratories can be an alternative to using expensive and scarce laboratory equipment. Nonetheless, there are also some disadvantages to these options. On the one hand, connectivity issues like slow internet speed or loss of connection can occur when using remote labs. In addition, queuing systems must be used to schedule the work time for each student, which can be a problem in large groups. On the other hand, while virtual labs can provide a very realistic simulation of a system, students do not learn how to deal with real hardware problems (e.g., measurement noise or non-measurable disturbances). In this sense, most students prefer to have the opportunity to practise on a real system, as this would be an experience more similar to what they might encounter in real-life situations (de Moura Oliveira et al., 2020a; Guzmán et al., 2022b).

Nowadays, with the reduction in the price of electronic devices and thanks to technological advances like 3D printing, it is possible to provide students with low-cost educational hardware (Castilla et al., 2020; Hoyo et al., 2015). This is a substantial change from an educational point of view because students can have their own equipment for practical experimentation, reducing the need for access to traditional laboratories at fixed hours. This idea is known as the take-home labs paradigm (Durfee et al., 2004), and it consists of providing students with portable devices that they can "take home" and use out of the classroom, at any time of the day and any day of the year. Compared to traditional laboratory experiences, take-home labs are a much more flexible teaching resource, allowing students to work with a real system at their own pace.

In this context, pocket-sized labs, such as those based on Arduino and Raspberry Pi, are interesting devices due to their portability and reduced cost (Afonso et al., 2021; Sobota et al.,

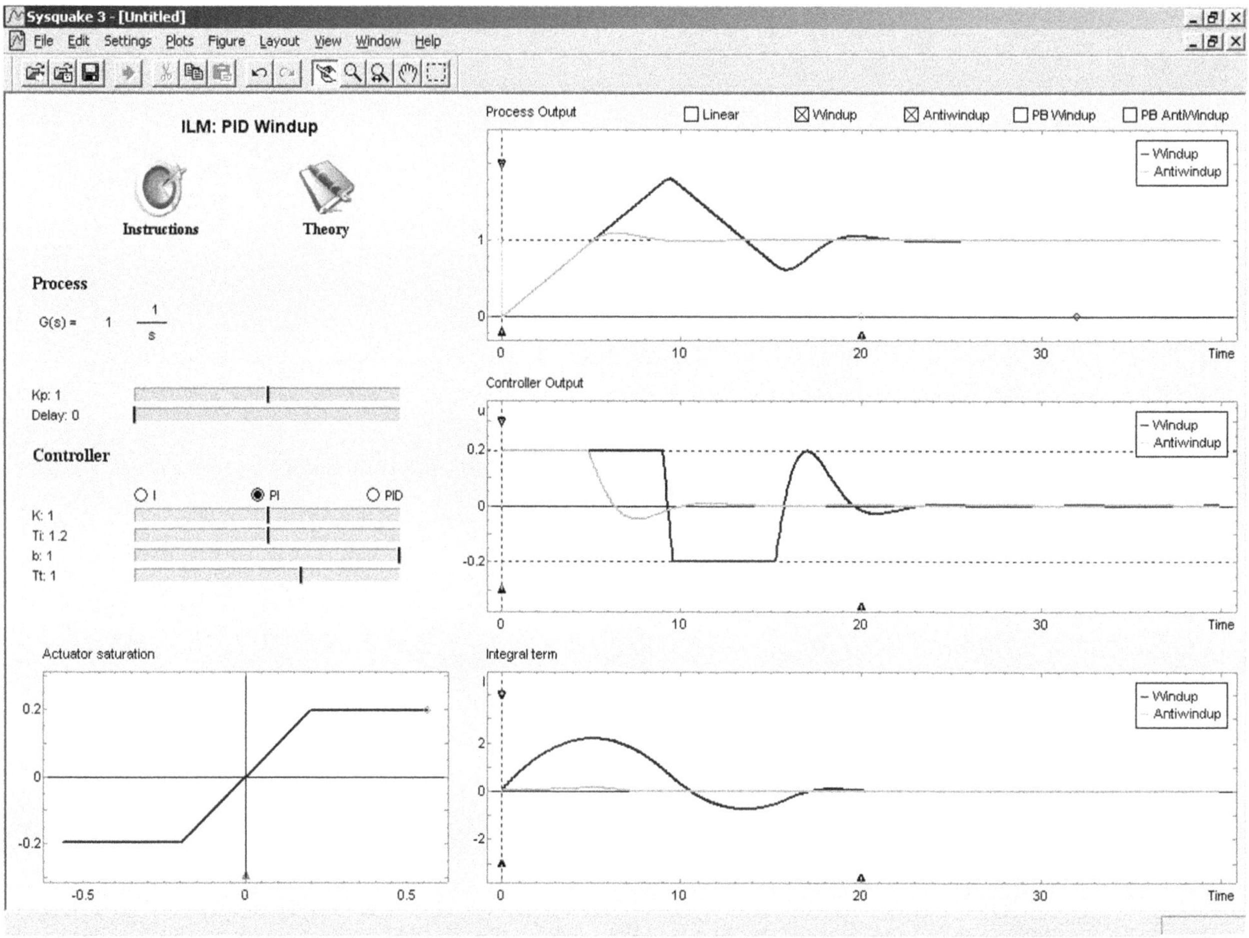

FIGURE 38.4 (c) PID Windup Interactive Tool.

2013). There are several examples of pocket-sized labs for educational purposes, but their use in control engineering studies has been particularly noteworthy in recent years (Docekal & Golembiovsky, 2018; Kalúz et al., 2019). Also, take-home labs have been very useful during the COVID-19 pandemic because lecturers could assign practical activities and students could use portable labs while confined at home (de Moura Oliveira & Soares, 2021).

Among the experiences using take-home labs reported in the literature, the APMonitor Temperature Control Lab (APMonitor, 2016), also known as TCLab, has become very popular for teaching modelling and control concepts in engineering courses (Guzmán et al., 2022b). TCLab is a commercial kit based on Arduino (see Figure 38.6) that consists of two heaters (transistors) and two temperature sensors (thermistors). The control objective with TCLab is to regulate the temperature of each heater by varying the electric current passing through them. As can be noticed, TCLab includes actuators and sensors that represent those used in chemical, agricultural, and industrial environments. This makes TCLab a great take-home lab for students of all engineering disciplines and particularly interesting for agricultural and biological courses, as the temperature is one of the most important variables in bioprocesses (e.g., affecting plant growth).

The suitability of TCLab for teaching and learning automatic control has been validated with real classroom experiences at the university level, as reported, for example, in de Moura Oliveira et al. (2020b) with biomedical and bioengineering students, and in Guzmán et al. (2022b) and Hoyo et al. (2021) with students of different industrial engineering degrees. As reported in Guzmán et al. (2022b), before incorporating the use of TCLab kits in different courses, it is important that the lecturers analyse some criteria to decide which courses can be the most appropriate for the new take-home labs methodology. Some relevant criteria are the number of students, the year, and the previous knowledge of the students. For example, if a high number of students are enrolled in a course and they have no previous experience of laboratory practices or automatic control concepts (i.e., first- or second-year students), the TCLab kits should be used in traditional laboratory sessions before letting the students take them home, for better supervision and support from the lecturers during the learning process. In contrast, TCLab kits can be used directly as take-home labs in courses with a reduced number of students, or if the students are familiar with practical activities and need less supervision from the lecturers.

The following explains an example of how to use TCLab kits as take-home labs for any introductory course related to automatic control, covering concepts of modelling and control

Temporal response of first order linear systems

Theory

This card analizes the time response of linear time invariant first order systems without zeros. These systems constitute the most simple dynamics coming from first order differential equations as $\tau \frac{dy(t)}{dt} + y(t) = ku(t)$, where $y(t)$ and $u(t)$ are the output and input of the system respectively. The corresponding transfer function can be written as:

$$G(s) = \frac{k}{\tau s + 1} \quad (1)$$

where

- k is the static gain,
- τ is the time constant (the system pole is $s = -\frac{1}{\tau}$).

The transfer function represents the system dynamics around an operating point. Usually, the time response is associated to an input excitation, that traditionally is a unitary step ($U(s) = \frac{1}{s}$). Thus, the unitary step time response of a first order system is given by:

$$y(t) = \mathcal{L}^{-1}\left\{G(s)\frac{1}{s}\right\} = k(1 - e^{-\frac{1}{\tau}t}). \quad (2)$$

that can be also obtained solving the differential equation with $u(t) = \begin{cases} 1, t > 0 \\ 0, t \leq 0 \end{cases}$.

The system output performance depends on the sign of τ:

- If $\tau > 0$, the signal is bounded and thus the system is stable.
- If $\tau < 0$, the system is unstable.

Note that if $\tau = 0$ the system is static. The name of the time constant τ, comes from the fact that this parameter gives an idea of the speed of the transient response of the system. The higher τ, the slowre the response. Thus,

- for $t = \tau$, the system output reaches the 63% of the final value,
- for $t = 3\tau$, the system output reaches the 95% of the final value,
- for $t = 4\tau$, the system output reaches 98% of the final value.

The static gain, k, corresponds with the quotient between the system output change and input change in steady state. The slope of the system response at $t = 0$ is $\frac{k}{\tau}$.

References

- Katsuhiko Ogata. **Modern Control Engineering (in Spanish)**. Pearson Prentice Hall. 4th Edition. ISBN: 84-205-3678-4. [Capítol: 5.2, Pàgines: 221-224]

Tool

Link: first_order.exe

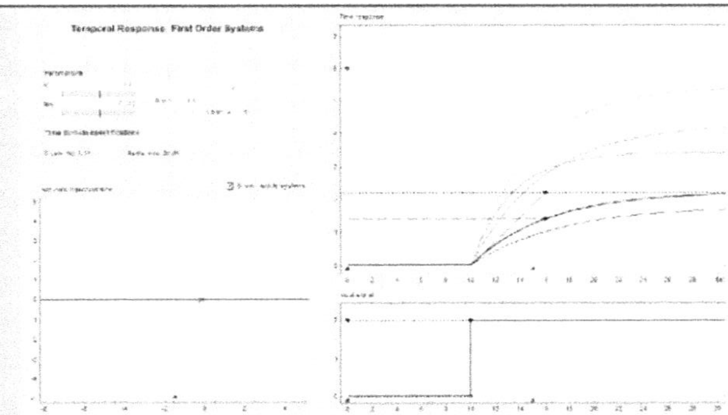

The main objective of this card is to analyze the unitary step time response of a first order linear system. The tool presents in its lower left part the s-plane, *Pole Zero representation*, showing the location of the poles of the represented systems; in the upper right part the time responses are plotted, *Time Response*; in the lower right part the input step is shown, *Input signal*; and finally, in the upper left part the values of the parameters and the transfer function are shown, *Parameters*, as well as the indexes characterizing the time response of the selected system *Time domain specifications*, that in this case are the static gain and settling time (10%-90%).

All these plots are interactive: in the s-plane it is possible to move the pole position; in the time response figure the initial slope and final value of the different responses can be changes; in the input signal graph it is possible to modify the amplitude and activation instant and finally, the system parameters can be changed quantitatively. The different scales of the figures can be easily changed by using the black colour triangles in the x (by pointing on the left and on the right) and y axes (by pointing upper and down the triangles). When the *Show multiple systems* option is selected, the active system shown in the *Parameters* area can be selected both by pointing on its corresponding pole location or on the time response and is changed to bold format. There is also the option of activating the simultaneous visualization of several first order systems, very useful to compare the different responses depending on the defining parameters. In the *settings* menu the different options are:

- *Default* different options for τ and k.
- *Gain effect* the systems have different values of k while τ is constant.
- *Tau effect* the systems have different values of τ while k is constant.

Exercises

1. Move the system pole on the real axis. How is the change in the time response? what happens in left movements?, what happens in right movements?, how long takes the system to reach 63% of the steady state value?.
2. When the static gain is changes, how the pole location changes?.
3. Given a system with the pole in $s = -3$ and which steady state value after a unitary step input is 0.5, which is its transfer function?. Use that transfer function in the tool and verify the results. Which is the value of the time constant?, how long takes the system to reach 63% of the steady state value?. Verify the results using the tool.

FIGURE 38.5(a) Learning card and interactive tool as support to the concept of first-order dynamic systems. (a) Learning card example.

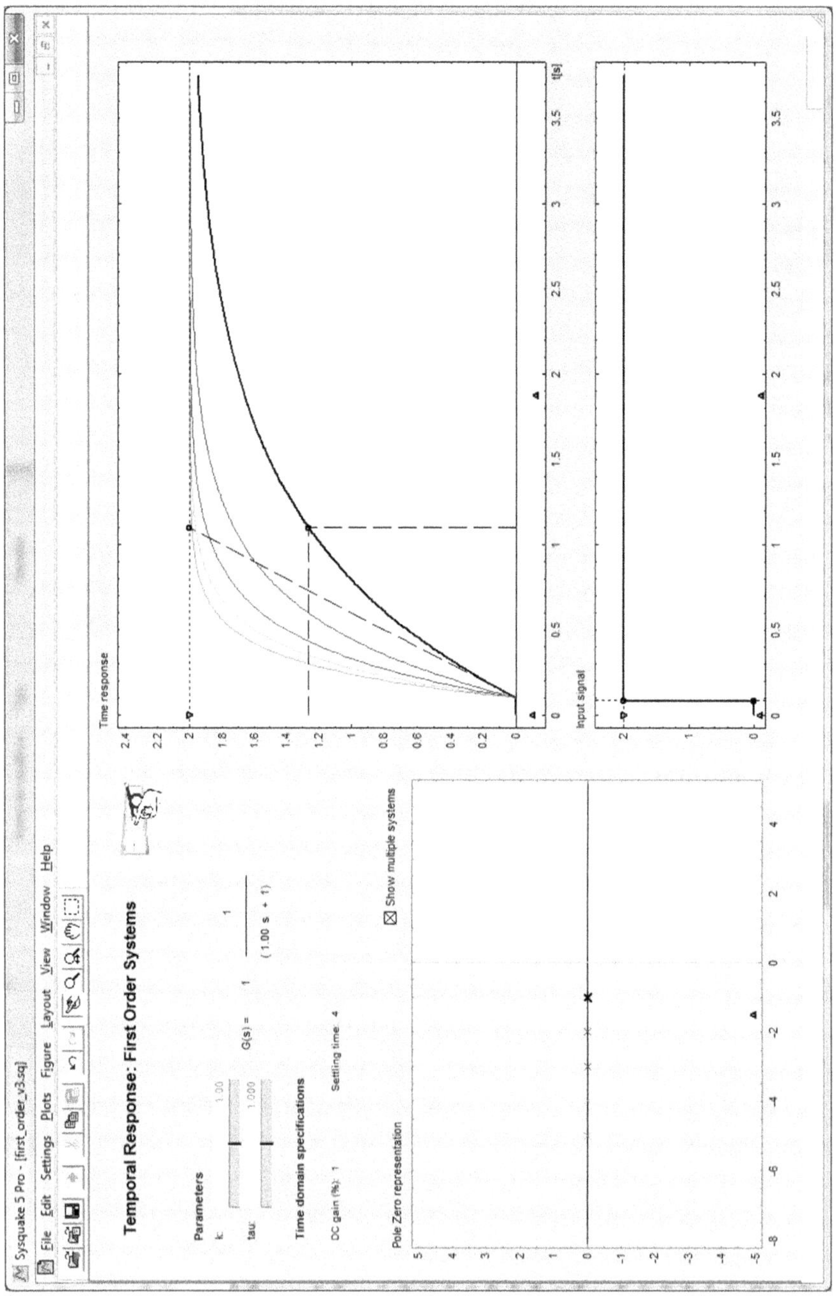

FIGURE 38.5(b) Learning card and interactive tool as support to the concept of first-order dynamic systems. (b) Interactive tool for first-order systems.

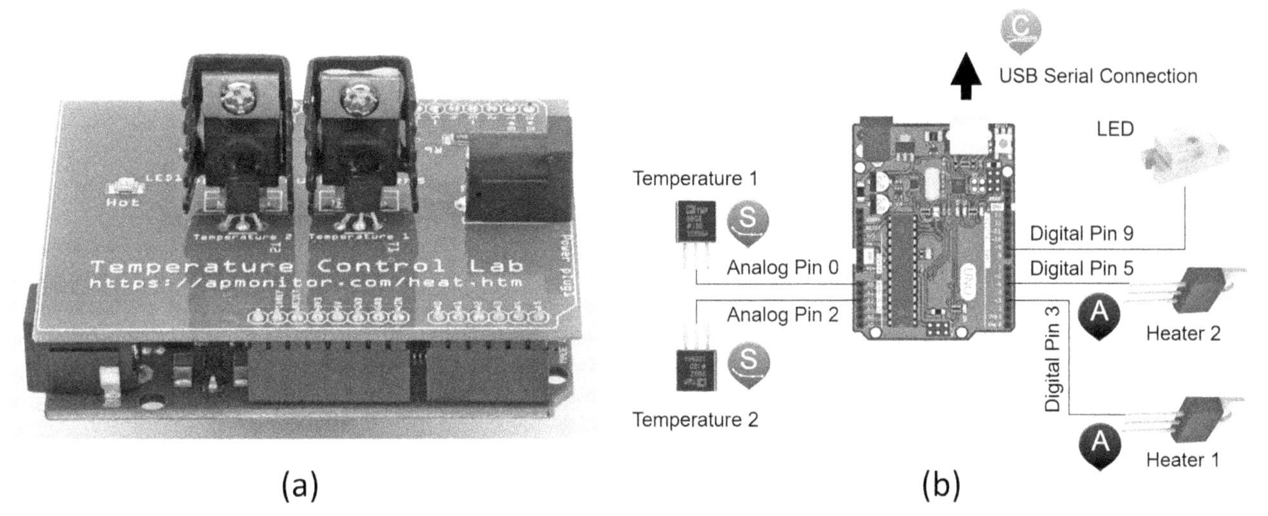

FIGURE 38.6 TCLab platform (APMonitor, 2016). (a) Picture of a TCLab kit. (b) Connections diagram, where "A" indicates actuators, "C" indicates connection and "S" indicates sensors.

of continuous dynamic systems. At the start of the course, students receive TCLab kits with practical assignments to solve at home. Although TCLab is an easy-to-use device, students need a brief tutorial on how to use it to understand the system's behaviour before starting their own trials. It is important to highlight the advantage of using an Arduino-based kit because students can connect the TCLab kits to their computers via USB and use software they are familiar with, such as MATLAB and Simulink (MathWorks, Massachusetts, USA), thanks to the availability of compatible libraries. In addition, the TCLab dynamics are explained according to the convection, conduction, and radiation that affect the temperature of the heaters. As can be noticed in Figure 38.6, TCLab is a multivariable system with two heaters and two temperature sensors. However, for introductory courses, students can focus on controlling heater one, while heater two can be used to produce disturbances in the temperature of heater one. The students receive the simplified differential equation explaining the temperature phenomena of one heater and the parameter values of the model, as presented in Equation 1,

$$mc_p \frac{dT(t)}{dt} = UA\left(T_\infty - T(t)\right) + \epsilon\sigma A\left(T_\infty^4 - T^4(t)\right) + \alpha Q(t) \quad (1)$$

where m is the mass of the transistor-thermistor-sink assembly, c_p is the specific heat capacity, T is the temperature of the heater, U is the heat transfer coefficient, A is the surface in contact with the environment, T_∞ is the temperature of the environment, ϵ is the emissivity, σ is the Stefan-Boltzmann constant, α is a factor that relates the current passing through the heater to the heat produced, and Q is the percentage of heater power.

Once the students understand the TCLab system, they can face some modelling activities to obtain a model that they can use to design a controller. Since TCLab is a nonlinear system, the first activity that can be proposed to the students is to obtain a linear model that relates the temperature of the heater (system output) to the percentage of heater power (system input). They can therefore apply the concept of linearisation around an operating point and use the Laplace transform to obtain a first-order transfer function from Equation 1.

On the other hand, students can perform open-loop tests with TCLab to obtain another model but from experimental data. In this case, they manually regulate the percentage of heater power to reach a desired operating point in temperature. When the heater is at the desired temperature, they increase in a step-form the percentage of heater power and observe the temperature's evolution again. With the data recorded in this test, they can apply the reaction curve method to obtain a transfer function model. They can validate the identified model by performing further tests with TCLab and analysing how well the model fits to the real system (i.e., measured data) in a graphical comparison, as shown in Figure 38.7a.

As a control activity, the students can use the obtained models to design a controller for a temperature setpoint tracking problem. Hence, they have to decide the type of controller, which in introductory courses is usually a PI controller. They also have to decide which tuning method they should use to calculate the controller parameters, depending on the desired behaviour for the closed loop. Normally, they select well-known classical tuning methods, such as pole-zero cancellation, lambda method, or Ziegler-Nichols. Figure 38.7b shows a result of a PI controller tested by students in TCLab. In the proposed activities, they have to study the results to explain the control action performed by the PI controller, which regulates the heater power so that the heater temperature reaches the desired setpoint with the desired behaviour that the students imposed for the closed loop.

This is an example of the methodology for using take-home labs in university-level courses. Comments from students are positive, as mentioned in Guzmán et al. (2022b). It is expected that new pocket-sized devices will be developed in the coming years, such as the Lab-Tec@Home (Sotelo et al., 2022), which proposes a control system for artificial lighting of a greenhouse tomato plant. In this sense, the outlook for take-home labs is optimistic.

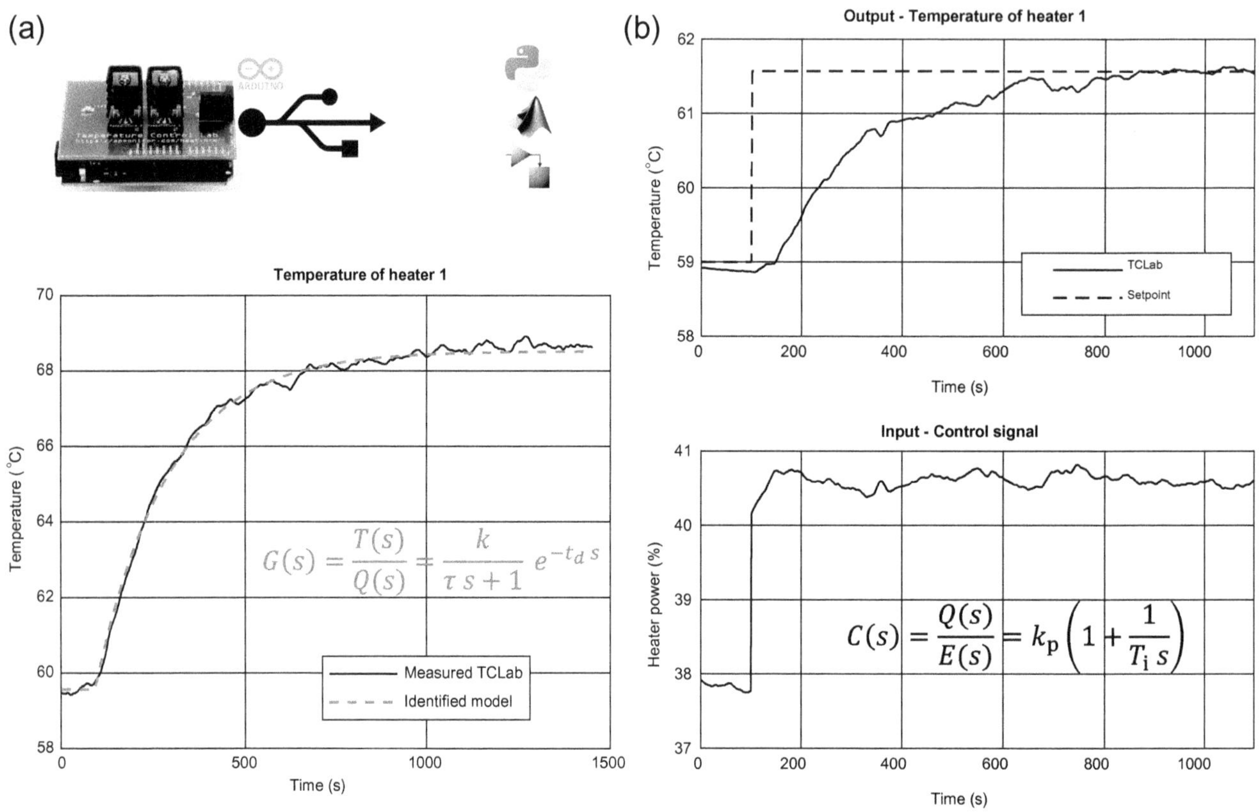

FIGURE 38.7 Example of modelling and control results obtained by the students. Adapted from Hoyo et al. (2021). (a) Open-loop test for model identification. (b) Closed-loop test with a PI controller.

REFERENCES

Afonso, R., Soares, F., & de Moura Oliveira, P.B. (2021). Innovative teaching/learning methodologies in control, automation and robotics: A short review. *Proceedings of the 4th International Conference of the Portuguese Society for Engineering Education*, 1–6. https://doi.org/10.1109/CISPEE47794.2021.9507242

APMonitor. (2016). *Temperature control lab*. http://apmonitor.com/pdc/index.php/Main/ArduinoTemperatureControl

Åström, K.J., & Hägglund, T. (2006). *Advanced PID control*. ISA The Instrumentation. System, and Automation Society.

Balamuralithara, B., & Woods, P.C. (2012). An investigation on adoption of the engineering simulation lab exercise: A case study in Multimedia University, Malaysia. *Computer Applications in Engineering Education*, 20(2), 339–345.

Ben Ouahi, M., Lamri, D., Hassouni, T., Ibrahmi, A., & Mehdi, E. (2022). Science teachers' views on the use and effectiveness of interactive simulations in science teaching and learning. *International Journal of Instruction*, 15(1), 277–292.

Castilla, M., Rodríguez, F., Álvarez, J.D., Donaire, J.G., & Ramos-Teodoro, J. (2020). A hardware-in-the-loop prototype to design benchmarks for automation and control education. *IFAC-PapersOnLine*, 53(2), 17314–17319. https://doi.org/10.1016/j.ifacol.2020.12.1815

de Moura Oliveira, P.B., Hedengren, J.D., & Boaventura-Cunha, J. (2020a). Bridging theory to practice: Feedforward and cascade control with TCLab Arduino Kit. *Proceedings of the 14th APCA International Conference on Automatic Control and Soft Computing*, 23–32. https://doi.org/10.1007/978-3-030-58653-9_3

de Moura Oliveira, P.B., Hedengren, J.D., & Rossiter, J.A. (2020b). Introducing digital controllers to undergraduate students using the TCLab Arduino Kit. *IFAC-PapersOnLine*, 53(2), 17524–17529. https://doi.org/10.1016/j.ifacol.2020.12.2662

de Moura Oliveira, P.B., & Soares, F. (2021). How we turned fully digital due to Covid-19: Two control engineering teaching experiences. *Proceedings of the 4th International Conference of the Portuguese Society for Engineering Education*, 1–5. https://doi.org/10.1109/CISPEE47794.2021.9507219

Docekal, T., & Golembiovsky, M. (2018). Low cost laboratory plant for control system education. *IFAC-PapersOnLine*, 51(6), 289–294. https://doi.org/10.1016/j.ifacol.2018.07.168

Durfee, W., Li, P., & Waletzko, D. (2004). Take-home lab kits for system dynamics and controls courses. *Proceedings of the 2004 American Control Conference*, 2, 1319–1322. https://doi.org/10.23919/ACC.2004.1386757

Fernández, I., Acién, F.G., Guzmán, J.L., Berenguel, M., & Mendoza, J.L. (2016). Dynamic model of an industrial raceway reactor for microalgae production. *Algal Research*, 17, 67–78.

Guzmán, J.L., Zakova, K. Craig, I.K. Hägglund, T., Rivera, D.E., Normey-Rico, J.E., Moura-Oliveira, P., Wang, L., Serbezov, A., Sato, T., & Beschi, M. (2022a). Teaching control during the COVID-19 Pandemic. *IFAC-PapersOnLine*, 55(17), 31–36. https://doi.org/10.1016/j.ifacol.2022.09.221

Guzmán, J.L., García-Mañas, F., Hoyo, Á., Ramos-Teodoro, J., & Donaire, J.G. (2022b). Use of TCLab kits for control engineering curricula at the University of Almería. *IFAC-PapersOnLine*, 55(17), 362–367. https://doi.org/10.1016/j.ifacol.2022.09.306

Guzmán, J.L., Acién, F.G., & Berenguel, M. (2021). Modelling and control of microalgae production in industrial photobioreactors. *Revista Iberoamericana de Automática e Informática Industrial*, 18(1), 1–18.

Guzmán, J.L., & Joseph, B. (2021). Web-based virtual lab for learning design, operation, control, and optimization of an anaerobic digestion process. *Journal of Science Education and Technology*, 30, 319–330.

Guzmán, J.L., Costa, R., Berenguel, M., & Dormido, S. (2016). An interactivity-based methodology to support control education: How to teach and learn using simple interactive tools. *IEEE Control Systems Magazine*, 36(1), 63–76.

Guzmán, J.L., Dormido, S., & Berenguel, M. (2013). Interactivity in education: An experience in the automatic control field. *Computer applications in Engineering Education*, 21(2), 360–371.

Guzmán, J. L., Costa-Castelló, R., Berenguel, M., & Dormido, S. (2023). *Automatic Control with Interactive Tools.* Springer Nature.

Guzmán, J.L., Åström, K.J., Dormido, S., Hägglund, T., Piguet, Y., & Berenguel, M. (2008). Interactive learning modules for control for PID control. *IEEE Control Systems Magazine*, 28(5), 118–134.

Guzmán, J.L., Åström, K.J., Dormido, S., Hägglund, T., & Piguet, Y. (2006). Interactive learning modules for PID control. *IFAC Proceedings Volumes*, 39(6), 7–12.

Guzmán, J.L., Berenguel, M., Rodríguez, F., Dormido, S. (2005). Web-based remote control laboratory using a greenhouse scale model. *Computer Applications in Engineering Education*, 13(2), 111–124.

Hoyo, Á., Rodríguez-Miranda, E., Guzmán, J.L., Acién, F.G., Berenguel, M., & Moreno, J.C. (2022). A computer-based tool to simulate raceway photobioreactors for design, operation and control purposes. *Computers & Chemical Engineering*, 156, 107572.

Hoyo, Á., García-Mañas, F., Ramos-Teodoro, J., Sánchez-Molina, J.A., Guzmán, J.L., & Rodríguez, F. (2021). Uso del paradigma Take-Home Labs para la enseñanza del control automático en estudios de ingeniería. *XLII Jornadas de Automática: Libro de Actas*, 218–225. https://doi.org/10.17979/spudc.9788497498043.218

Hoyo, Á., Guzmán, J.L., Moreno, J.C., & Berenguel, M. (2015). Teaching control engineering concepts using open source tools on a raspberry pi board. *IFAC-PapersOnLine*, 48(29), 99–104. https://doi.org/10.1016/j.ifacol.2015.11.220

Kabigting, L.D.C. (2021). Computer simulation on teaching and learning of selected topics in physics. *European Journal of Interactive Multimedia and Education*, 2(2), e02108.

Kalúz, M., Klaučo, M., Čirka, L., & Fikar, M. (2019). Flexy2: A portable laboratory device for control engineering education. *IFAC-PapersOnLine*, 52(9), 42–47. https://doi.org/10.1016/j.ifacol.2019.08.121

Kayssi, A., El-Hajj, A., El Assir, M., & Sayyid, R. (1999). Web-based tutoring and testing in a computer networks course. *Computer Applications in Engineering Education*, 7(1), 1–7.

Malkov, E. (2020). Simulation of coronavirus disease 2019 (COVID-19) scenarios with possibility of reinfection. *Chaos, Solitons & Fractals*, 139, 110296.

Neri, F. (2012). A comparative study of a financial agent based simulator across learning scenarios. In *Agents and Data Mining Interaction: 7th International Workshop on Agents and Data Mining Interaction, ADMI 2011*, 86–97, Taipei, Taiwan, May 2–6, Revised Selected Papers 7. Springer Berlin Heidelberg.

Ong, S.K., & Mannan, M.A. (2004). Virtual reality simulations and animations in a web-based interactive manufacturing engineering module. *Computers & Education*, 43(4), 361–382.

Potkonjak, V., Gardner, M., Callaghan, V., Mattila, P., Guetl, C., Petrović, V.M., & Jovanović, K. (2016). Virtual laboratories for education in science, technology, and engineering: A review. *Computers & Education*, 95, 309–327.

Rapin, N., Lund, O., & Castiglione, F. (2011). Immune system simulation online. *Bioinformatics*, 27(14), 2013–2014.

Ratamun, M.M., & Osman, K. (2018). The effectiveness of virtual lab compared to physical lab in the mastery of science process skills for chemistry experiment. *Problems of Education in the 21st Century*, 76(4), 544.

Rivas, J.L., Guzmán, J.L., Rodríguez, F., Berenguel, M., & Dormido, S. (2012). Integrated virtual and remote lab for greenhouse climate control. *IFAC Proceedings Volumes*, 45(11), 264–269.

Rodrigues, R., Soares, R.P., & Secchi, A.R. (2010). Teaching chemical reaction engineering using EMSO simulator. *Computer Applications in Engineering Education*, 18(4), 607–618.

Rodríguez, F. (2002). Modelling and hierarchical control of greenhouse crop production. PhD. Thesis. University of Almería, Spain, 366 pp.

Rodríguez-Miranda, E., Guzmán, J.L., Acién, F.G., Berenguel, M., & Visioli, A. (2021). Indirect regulation of temperature in raceway reactors by optimal management of culture depth. *Biotechnology and Bioengineering*, 118(3), 1186–1198.

Sánchez, J., Dormido, S., & Esquembre, F. (2005). The learning of control concepts using interactive tools. *Computer Applications in Engineering Education*, 13(1), 84–98.

Sánchez-Zurano, A., Gómez-Serrano, C., Acién, F.G., Fernández-Sevilla, J.M., & Molina-Grima, E. (2021). Modeling of photosynthesis and respiration rate for microalgae–bacteria consortia. *Biotechnology and Bioengineering*, 118(2), 952–962.

Sobota, J., Pišl, R., Balda, P., & Schlegel, M. (2013). Raspberry Pi and Arduino boards in control education. *IFAC Proceedings Volumes*, 46(17), 7–12. https://doi.org/10.3182/20130828-3-UK-2039.00003

Sotelo, D., Sotelo, C., Ramirez-Mendoza, R.A., López-Guajardo, E.A., Navarro-Duran, D., Niño-Juárez, E., & Vargas-Martinez, A. (2022). Lab-Tec@Home: A cost-effective kit for online control engineering education. *Electronics*, 11(6). https://doi.org/10.3390/electronics11060907

Stefanovic, M., Matijevic, M., Cvijetkovic, V., & Simic, V. (2010). Web-based laboratory for engineering education. *Computer Applications in Engineering Education*, 18(3), 526–536.

Staranowicz, A., & Mariottini, G.L. (2011). A survey and comparison of commercial and open-source robotic simulator software. In *Proceedings of the 4th International Conference on PErvasive Technologies Related to Assistive Environments*, 1–8.

Tsvetovat, M., & Carley, K.M. (2005). Generation of realistic social network datasets for testing of analysis and simulation tools. Available at SSRN 2729296.

Tüysüz, C. (2010). The effect of the virtual laboratory on students' achievement and attitude in chemistry. *International Online Journal of Educational Sciences*, 2(1), 37–53.

39

Present Content and Trial Approaches of Student Experiments for ABEE in Japan

Yukiharu Ogawa, Yasumaru Hirai, Hirofumi Mino, Terumi Sugawara, Kenji Hiyoshi, Shigeki Inaba, Takeshi Shikanai, Takashi Okayasu, Fumina Tanaka and Fumihiko Tanaka

Introduction

Modern agriculture has become more intensive through mechanisation and the use of chemical fertilisers and pesticides, making it possible for a small working population to produce food for people worldwide. The approaches to such intensification and automation of agricultural production by mechanisation vary according to national and regional characteristics, but the mechanisation trend is irreversible. However, sustainable development with environmental protection requires new approaches to the current situation.

Information and telecommunication technologies have been used in social systems in recent decades. Although it has already been applied to a social network system that can communicate between mass audiences, the COVID-19 pandemic has significantly induced digital transformation (DX) in social activities. This trend has also changed universities' education and academic environment drastically. These circumstances would certainly not change even after the COVID-19 disaster. This means a new style of education has been expanding through the employment of DX technology, even though it has been improved by trial and error due to its incompleteness.

The agricultural and biological engineering (ABE) field has also incorporated information technology as a critical factor. Similarly, as automation technologies through artificial intelligence (AI) are advancing by leaps and bounds, so is the need to accumulate relevant digitised knowledge to apply them. Meanwhile, the basic knowledge of power machinery, unit operations, and so on are still essential factors in the education of the students who aim to study ABE from introduction to application. However, it is better that the combination of basic approaches and DX trials for ABE education must fit the circumstances of the nations and areas, because that is the feature of agriculture.

Japan is one of the leading countries in developing basic power machinery and postharvest handling systems, which can be connected to DX technologies. This chapter mainly describes several examples of the content of student experiments taught at Kyushu University and Chiba University. The current status of incorporating DX into education at the universities in Kyushu, one of the main areas for agricultural products in Japan, is also introduced.

Engine Performance Test for Power Machinery

Agricultural machinery is basically powered by an engine to perform various agricultural tasks. The mechanical substance of the power source is piston movement by gas combustion at high temperature and pressure. In other words, this is a practical application of thermodynamic phenomena, and the principles of thermodynamics can be well understood by analysing the phenomena. However, in an actual educational environment at universities in Japan, there were few opportunities to meet actual machines and machine elements. In 2019, a gasoline engine (GX120, Honda), video tutorials for learning the engine mechanism, and a diesel engine (NS75, Yanmar) were assigned from other institutions and could be used for the experiments of undergraduate students at Kyushu University (Hirai, 2022). The faculties in charge first disassembled and re-assembled the engine to acquaint themselves with the engine structure as a preliminary practice. Although this was a preparatory step for the instruction of the student experiment, the assembly work while looking at the component diagrams was a struggle. The students' interest in the actual experiments was great, too. They seemed to enjoy disassembling and re-assembling the engine, even though some had never used tools. This experience in engine-assembly practice deepened their understanding of engine structure. As a result, two more GX120 engines were installed to ensure the opportunity for practical training in smaller groups by 2022. Today, one engine can be used per four to five students for practical assembly work.

In the following year, during assembly training for the student experiments, the understanding of the engine as a power source would not be deepened unless the engine was operated instead of only disassembled and re-assembled. Simultaneously, a concrete image of performance tests could be constructed using a textbook (Iijima, 2018), which made engine performance, thermal efficiency, etc. more easily understood. Based

DOI: 10.1201/9780429150111-42

on the knowledge obtained from the textbook, the engine performance test could be repeated and the characteristics evaluated through trial and error. Simultaneously, the Kistler's sensor-integrated spark plug (in-cylinder pressure sensor) was also attached to the engine. It was found that a pressure–volume (p–V) diagram could be drawn with this sensor. Because the p–V diagram expresses the work produced by the engine, it could be a concrete image of the substance of the work. Therefore, an additional experiment was constructed to evaluate the working of gas combustion by measuring the pressure in the engine cylinder (cylinder pressure) using an in-cylinder pressure sensor and creating a p–V diagram.

This section introduces the method of calculating performance indices for gasoline engines, a system for measuring various types of gasoline engine performance, the data measured in the student experiments, and the analysis results (Hirai, 2022).

Basic Parameters of Gasoline Engine Performance

Work Done by the Combustion Gases on the Piston

The gasoline engine burns a mixture of gasoline and air (gas mixture) in the combustion chamber, and the high-temperature and high-pressure gas work the piston. This phenomenon causes a reciprocating motion of the piston, which is converted into mechanical work by the crankshaft, gears, belts, and other mechanical elements that transmit power to the moving parts of the agricultural machinery or the work equipment. Figure 39.1 A schematically illustrates the work done by combustion gases. It assumes that a small amount of heat dQ (J) is generated by the combustion of the air mixture and that the pressure in the cylinder (in-cylinder) is p (Pa). It also assumes that this pressure p acts on a piston of area A (m^2), causing it to expand by dx (m). Since the force on the top of the piston is $F=pA$, the differential work dW (J) at this time can be expressed by the following equation (1).

$$W = Fdx = pAdx = pdV \quad (1)$$

where $dV = Adx$, representing the volume change. In contrast, the piston work done on the gas in the cylinder is negative work in the compression process.

Creating a p–V Diagram

The p–V diagram is created by plotting the measured values with the cylinder volume on the horizontal axis and the cylinder pressure on the vertical axis. The in-cylinder volume is calculated using equations (2) and (3) from the schematic of the piston-crank mechanism shown in Figure 39.1 B.

$$V = V_c + \frac{\pi}{4} D^2 x \quad (2)$$

$$x = S\left\{\frac{1}{2}(1-\cos\theta) + \frac{r}{8l}(1-\cos 2\theta)\right\} \quad (3)$$

where V: cylinder volume, V_c: clearance volume, D: bore diameter, x: piston displacement from top dead centre, S: stroke (travel), θ: crank angle, r: crank radius ($S=2r$), l: connecting rod length. By measuring the variables through practical engine disassembly and assembly and synchronously measuring the crank angle and cylinder pressure during performance tests, it is possible to create a p–V diagram.

Indices of Engine Performance Evaluation

1) Indicated work, W_i

The indicated work W_i (J) is obtained as shown in equation (4) by subtracting the negative work for exhaust and intake from the positive work in the compression and expansion processes shown in Figure 39.1 C.

$$W_i = W_{i+} - |W_{i-}| \quad (4)$$

The indicated work is the work done by the combustion gas on the piston during one cycle of operation.

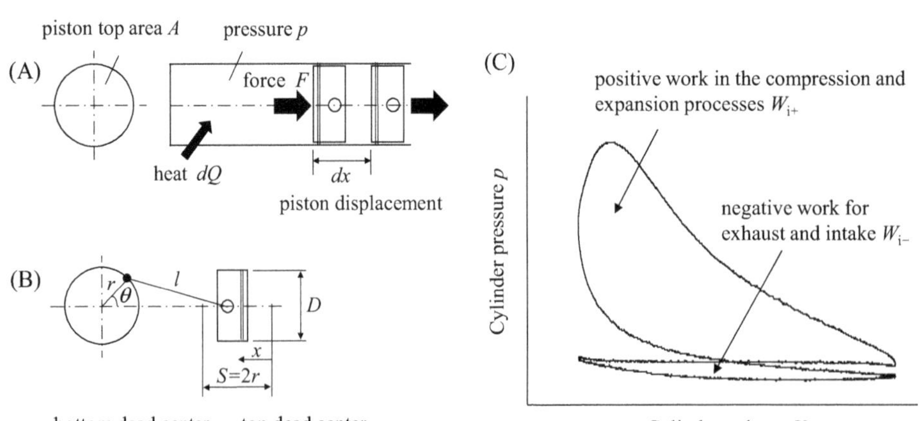

FIGURE 39.1 Work done by combustion gases (A), piston-crank mechanism (B), and negative and positive work shown in the p–V diagram (C) (Hirai, 2022; Iijima, 2018).

2) Indicated mean effective pressure, p_{mi}

By dividing the indicated work W_i by the stroke volume (displacement), the indicated mean effective pressure p_{mi} (MPa) can be obtained, as shown in equation (5).

$$p_{mi} = \frac{W_i}{V_s} \tag{5}$$

Engines with higher p_{mi} can produce more power with lower displacement.

3) Indicated power, P_i

The indicated power P_i (kW) is calculated by equation (6).

$$P_i = \frac{p_{mi} V_s N_p \times 10^{-3}}{60i}. \tag{6}$$

where N_p is the rotational speed (rpm) of the power take-off (PTO) shaft. The engine used in this experiment transmits the crankshaft rotation to the PTO shaft at a reduction ratio of two. Therefore, the value of N_p is half of the engine rotation speed. If the crankshaft and PTO shaft are the same, $i = 2$ for a four-stroke engine.

The engine employed had a reduction ratio of two, and $i = 1$ because the PTO shaft makes one revolution per cycle.

4) Net work, W_e

The net work per cycle, W_e (J), is calculated by equation (7).

$$W_e = 2\pi T i \tag{7}$$

where T: Average torque of the PTO shaft per cycle (Nm).

5) PTO shaft output, P_p

The PTO shaft output P_p (kW) is calculated by equation (8).

$$P_p = \frac{2\pi T N_p \times 10^{-3}}{60i} \tag{8}$$

6) Brake mean effective pressure, p_{me}

The brake mean effective pressure p_{me} (MPa) is calculated by dividing the net work W_e obtained in equation (7) by the process volume V_s, by equation (9).

$$p_{me} = \frac{W_e}{V_s} \tag{9}$$

p_{me} can also be calculated by equation (10) from the PTO shaft output P_p in equation (8).

$$p_{me} = \frac{60 P_p}{V_s N_p} i^2 \times 10^3 \tag{10}$$

7) Efficiency indices

The indicated thermal efficiency η_i is calculated by equation (11) from the P_i in equation (6).

$$\eta_i = \frac{P_i}{\dot{Q}} \tag{11}$$

The brake thermal efficiency η_e is calculated by equation (12) from the P_p in equation (8).

$$\eta_e = \frac{P_p}{\dot{Q}} \tag{12}$$

The mechanical efficiency η_m is calculated by equation (13) from a combination of equations (4) and (7), (5) and (9) or (10), or (11) and (12).

$$\eta_m = \frac{W_e}{W_i} = \frac{p_{me}}{p_{mi}} = \frac{\eta_e}{\eta_i} \tag{13}$$

$\dot{Q}$ (kJ/s) is the heat input per second and is calculated used equation (14).

$$\dot{Q} = 1000 \times m_f H_u / 3600 \tag{14}$$

where m_f (kg/h): mass flow rate of fuel, H_u (MJ/kg): lower heating value of fuel (= 43 MJ/kg).

The brake-specific fuel consumption b_e (kg/kWh) is calculated by equation (15) as an index to evaluate the fuel efficiency of the engine independent of displacement.

$$b_e = \frac{m_f}{P_p} \tag{15}$$

Gasoline Engine Performance Tests

Measurement System

Figure 39.2 A shows the measurement system for gasoline engine performance. This gasoline engine (GX120, Honda) was four-stroke, and its total displacement was 122 cm^3 with a continuous rated output of 2.1 kW (2.9 horsepower) at 3,600 rpm. The in-cylinder pressure was measured by replacing the normal spark plug with a spark plug with an integrated sensor (in-cylinder pressure sensor; 6118CF-6CQ01-4-2, Kistler). The sensor was a charging output type; thus, the signal was detected through a charge amplifier unit (TMR-361, Tokyo Measuring Instruments Laboratory) using a multi-recorder control unit (TMR-311, Tokyo Measuring Instruments Laboratory). Two biaxial strain gauges (KFGS-3-120-D16-11, Kyowa) were attached to the PTO shaft at opposite positions to measure the shaft torque. The strain gauges were connected to a slip ring (Type SR36M, Michigan Scientific Corp.), which was connected to a multi-recorder via a strain 4G unit (TMR-321, Tokyo Measuring Instruments Laboratory). The slip ring was connected to the multi-recorder using a ground wire. TMR-7300 (Tokyo Measuring Instruments Laboratory) was used as a measurement software to set up and operate the multi-recorder and to create files of the recorded data.

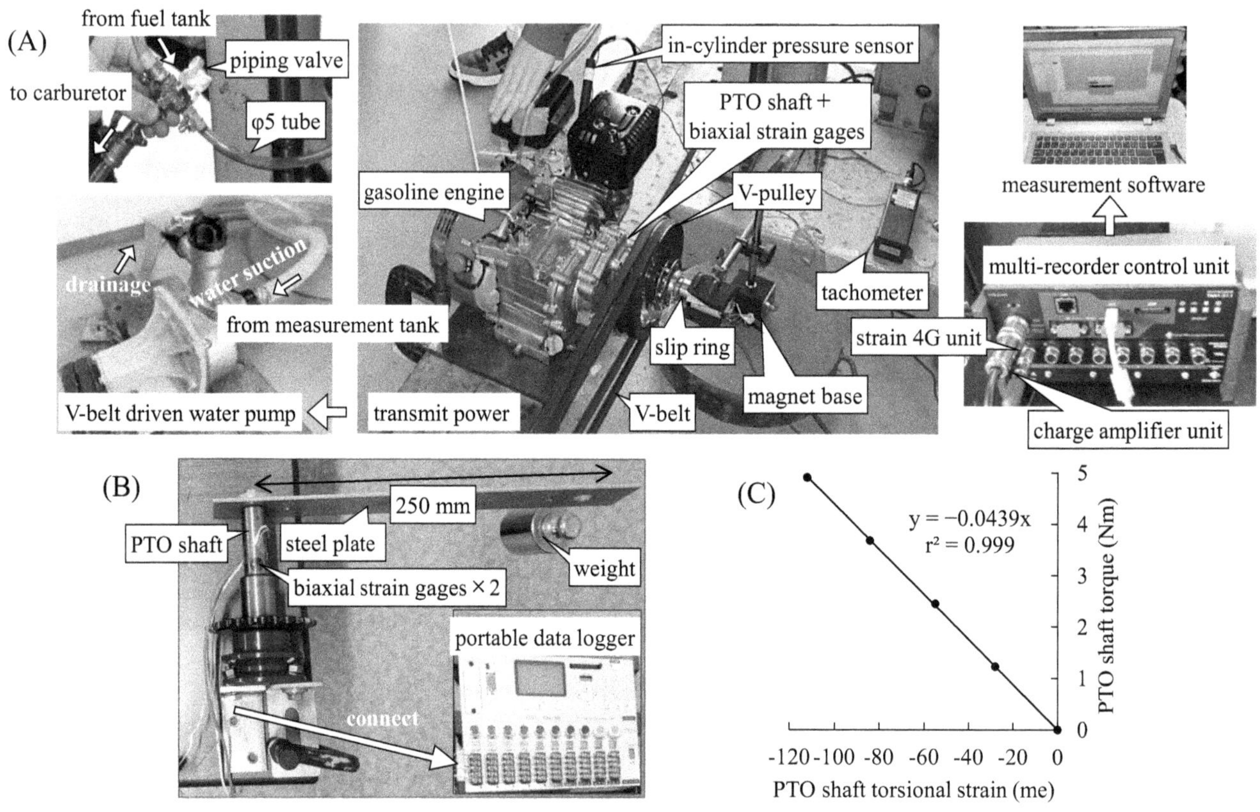

FIGURE 39.2 Measurement system for gasoline engine performance (A), calibration for the PTO shaft torque (B), and calibration line for the PTO shaft torque (C) (Hirai, 2022).

The PTO shaft was loaded by a V-belt-driven water pump (B1, Terada Pump). A JIS V-pulley (170-A-2F-18-BKT-HQ, Nabeya Vitec) was fixed to the PTO shaft, and a V-belt (A88, Mitsuboshi belt) was applied to transmit power to the V-pulley on the pump. The nominal diameter of the V-pulley was 170 mm because the lead wire of the strain gauge had to be connected to the slip ring through the inside of the pulley, for which shape type 30X was selected to allow for this space. The hoses for water suction and drainage were placed in the same water tank to circulate the pump drainage. Gasoline was supplied from the fuel tank and the measurement tank. Tubes from both tanks were connected to a piping valve (BH-5166, Asoh) to supply fuel from the fuel tank during normal operation and from the measurement tank when measuring fuel consumption. Fuel consumption was measured from the length of decrease in the liquid surface of the φ5 tube.

Calibration Factor

The calibration factor of the in-cylinder pressure sensor was –100.4 pc/MPa, a value in the range of zero to 50 bar at 200°C, from the calibration certificate. The calibration factors for the PTO shaft torsional strain (strain) and shaft torque were obtained by screwing a steel plate to the PTO shaft and applying a weight at 250 mm from the centre of the shaft (Figure 39.2 B). A portable data logger (TDS-150, Tokyo Measuring Instruments Laboratory) was used to measure the strain. Figure 39.2 C shows the calibration lines. Table 39.1 summarises the calibration factors.

Measurement Procedure

The experiment was conducted in groups of four to five students. The experimental tasks were preliminarily sorted, as shown in Table 39.2; thus, each student could experience all the tasks. The experiment was conducted according to the flowchart shown in Figure 39.3. First, the fuel consumption measurement tank was filled with gasoline. After setting the measurement conditions, such as the measurement frequency, a zero-point calibration was performed before starting the engine. The measurement frequency in this experiment was set to 50 kHz (0.02 ms). Therefore, the in-cylinder pressure was recorded at intervals of less than one-degree crank angle. The valve on the fuel tank was then opened, the engine was started, and a warm-up operation was performed for approximately one minute. The engine speed was then adjusted to low, medium, or high using the speed control lever. At that time, the rotation speed of the PTO shaft was checked using a pocket tachometer (Model 3632, Yokogawa). After adjusting the rotation speed, the valve on the measurement tank was opened and measured for 30 seconds using a stopwatch after the fuel level dropped to the mark on the tube. Simultaneously, the recording by the multi-recorder was started. After 30 seconds, the valve on the fuel tank was opened to stop the engine. The decreased level of the gasoline in the tube was measured,

TABLE 39.1

Calibration Factors (Hirai, 2022).

Sensor Type	Calibration Factor	Unit
In-cylinder pressure	−100.4	pc/MPa
PTO shaft torque	−0.0439	Nm/με

and the data for cylinder pressure and shaft torque saved in the PC were verified. These measurements were for one set of operating conditions; all students rotated the task to take measures for all rotational speed conditions. The density of gasoline was measured using a specific gravity metre for petroleum (Yokota). Colour, density, and viscosity were evaluated and applied to compare kerosene, light, and heavy fuels.

Measurement of Parameters Used to Create a p–V Diagram

Before the performance test of the engine, the students learned the engine structure, mechanical elements, and power transmission mechanism by disassembling and assembling the same engine (GX120). Since video tutorials are provided for the GX120, students can learn each part's functions and operating principles in detail. In the disassembly and assembly practice, measurement, and observation work, such as the measurement parameters used to create the *p–V* diagram shown in Table 39.3, are incorporated to enhance the students' tasks. For example, the sample data in Table 39.3 and equations (2) and (3) could be used to confirm the relationship between the crank angle and piston displacement or cylinder volume.

Measured Data Processing

Time series data of the output values of the in-cylinder pressure sensor was obtained from the experiment. Since this data has atmospheric pressure as the zero point, the output value was corrected by adding −10.04 pC, corresponding to an atmospheric pressure of 0.1 MPa. The data were then converted to cylinder pressure using the calibration factors shown in Table 39.1. The PTO shaft strain was also converted to shaft torque. To create a *p–V* diagram, it is necessary to specify the top

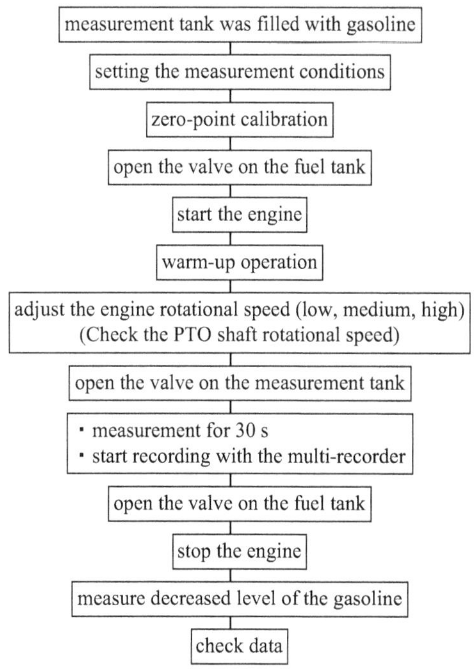

FIGURE 39.3 Flowchart of the experiment (Hirai, 2022).

dead centre (TDC) corresponding to a crank angle of zero degrees. Although an encoder can measure the angle, the TDC was used to determine the pressure peak during the exhaust stroke in this experiment. The rotational speed was calculated from the period of one cycle, measured by the time between the pressure peaks in the expansion process. To create the *p–V* diagram, the data for one cycle from the pressure peak of the exhaust process were used. To calculate the cylinder volume corresponding to the cylinder pressure, the crank angle was firstly determined by multiplying the measurement time by the rotational speed (°/s). The in-cylinder volume was then calculated using equations (2) and (3) and the values in Table 39.3. The indicated work in equation (4) was calculated by dividing the positive and negative work areas in the *p–V* diagram into triangles or trapezoids. The torque T in equation (7) was calculated as the average value of one cycle, including negative values.

TABLE 39.2

Experimental tasks. (Hirai, 2022)

Task	Details
Fuel	• measurement tank was filled with gasoline • measurement for 30 seconds • measure decreased level of the gasoline
Valve	switch the supply of gasoline from fuel/measurement tank
Engine	start, adjust rotational speed
Rotational speed	measure rotational speed of the PTO shaft
Measurement Software	• setting the measurement condition, zero-point calibration, start recording • check data files
Record	record measurement conditions, decreased level of the gasoline, rotational speed etc.
Hose	hold a hose to drain water in the tank (same as the tank for water suction)

TABLE 39.3

Parameters Used to Create a *p–V* Diagram (Hirai, 2022).

Parameters	Value
S : stroke (cm)	4.3
D : bore diameter (cm)	6
l : length of the connecting rod (cm)	7.4
V_S : stroke volume (cm^3)	121.6*
V_c : clearance volume (cm^3)	19.8

*The value is slightly different from one in the specification table due to measurement error.

Results Example

Output Values of In-Cylinder Pressure Sensor and Shaft Strain

Figure 39.4 A shows the output values of the in-cylinder pressure sensor and PTO shaft strain at low rpm for approximately three cycles. Table 39.4 shows the cycle time and engine speed per cycle. The cylinder pressure sensor and PTO shaft strain showed a prominent periodic peak. The peak of the PTO shaft strain appeared later than the peak of the in-cylinder pressure. The output values of the in-cylinder pressure ranged from approximately –80 to 5 pC, and the PTO shaft strain ranged from approximately –1,000 to 200. During the exhaust stroke, small pressure peaks were also observed around 0.055 s, 0.14 s, and 0.22 s. As the rotation speed increased from medium to high, the output value of the in-cylinder pressure sensor and the PTO shaft strain peaks increased.

In-Cylinder Pressure and PTO Shaft Torque

Figure 39.4B shows the relationships between crank angle, in-cylinder pressure, and PTO shaft torque. The peak value of the in-cylinder pressure was 0.67 MPa, then the crank angle at this time was 395°. The negative value of the in-cylinder pressure may indicate that a pressure higher than atmospheric pressure was applied during the zero-point calibration. The peak PTO shaft torque was 34 Nm. The crank angle was 461°, 66° behind the peak in-cylinder pressure. The peak value of cylinder pressure was 1.02 MPa (crank angle 398°) and the peak value of PTO shaft torque was 60 Nm (crank angle 488°) at medium rotation speeds, while the peak value of cylinder pressure was 1.25 MPa (crank angle 389°) and the peak value of PTO shaft torque was 62.5 Nm (crank angle 495°) at high rotation speeds. With increasing rotational speed, the peak values of in-cylinder pressure and shaft torque increased. The delay of the peak phase for the shaft torque increased with increasing rotational speed, from 90° at medium speed to 106° at high speed.

Relationship between the Crank Angle and Piston Displacement/Internal Cylinder Volume

At 0°, 360°, and 720° TDC, the piston displacement and cylinder volumes were 0 cm and 19.8 cm^3, respectively. At 180° and 540° bottom dead centre (BDC), the piston displacement and cylinder volumes were 4.2 cm and 141.4 cm^3, respectively, showing a periodic change (Figures 39.4 C and 39.4 D).

p–V Diagram

Figures 39.4 E–G show the *p–V* diagram at low rotational speeds. The in-cylinder pressure increased gently during the compression process and rapidly at TDC, reaching a peak value of 0.67 MPa. The pressure decreased during the expansion process and dropped to approximately 0.1 MPa at BDC. Negative work was observed in the exhaust and intake processes. At medium and high rotational speeds, the peak cylinder pressures reached 1.02 MPa and 1.25 MPa, respectively, and the positive work area was expanded. The peak in-cylinder pressure approached the TDC as the rotational speed increased.

Fuel Consumption

Fuel consumption was 0.193, 0.386, and 0.634 L/h at low (1,472 rpm), medium (2,140 rpm), and high (2,648 rpm) speeds, respectively (Table 39.5). The values increased linearly with increasing rotational speed.

Engine Performance Index

Table 39.6 shows the calculation results of the engine performance indices. The indicated work, mean effective pressure, and power output obtained from the *p–V* diagram increased with increasing rotational speed. The average torque of the PTO shaft increased with the rotational speed to 1.81, 3.52, and 5.12 Nm at low, medium, and high speeds, respectively. As a result, net work, PTO shaft output, and brake mean effective pressure also increased with increasing rotational speed. The net values calculated from the PTO shaft torque were smaller than those shown in the figure obtained from the *p–V* diagram. The fuel mass flow rate mf increased with the rotational speed with 0.140, 0.279, and 0.458 kg/h at low, medium, and high speeds, respectively. Contrarily, the brake-specific fuel consumption increased with increasing rotational speed to 1.00, 0.71, and 0.65 kg/kWh. The indicated thermal efficiency η_i was the highest at 0.25 for the medium rotation speed and 0.21 for the low and high rotation speeds. Brake thermal efficiency η_e ranged from 0.08 to 0.13, which could be considered low values. The mechanical efficiency η_m was deficient at a low rotational speed, 0.39, and improved significantly to 0.63 at a high rotational speed.

Discussion

In each index shown in Table 39.6, the net values were smaller than those shown in the figures, which is a reasonable result reflecting mechanical loss such as friction in various parts of the engine and drive work of auxiliary equipment. In contrast, the brake thermal efficiency of 0.08 to 0.13 was significantly lower than that of 0.25 to 0.35 (Suzuki, 2020) and 0.25 to 0.3 (Ai, 2007). The engine power was transmitted to the pump using a V-belt in this experiment, but the distance between the pulley shafts was not appropriate, and the loosening of

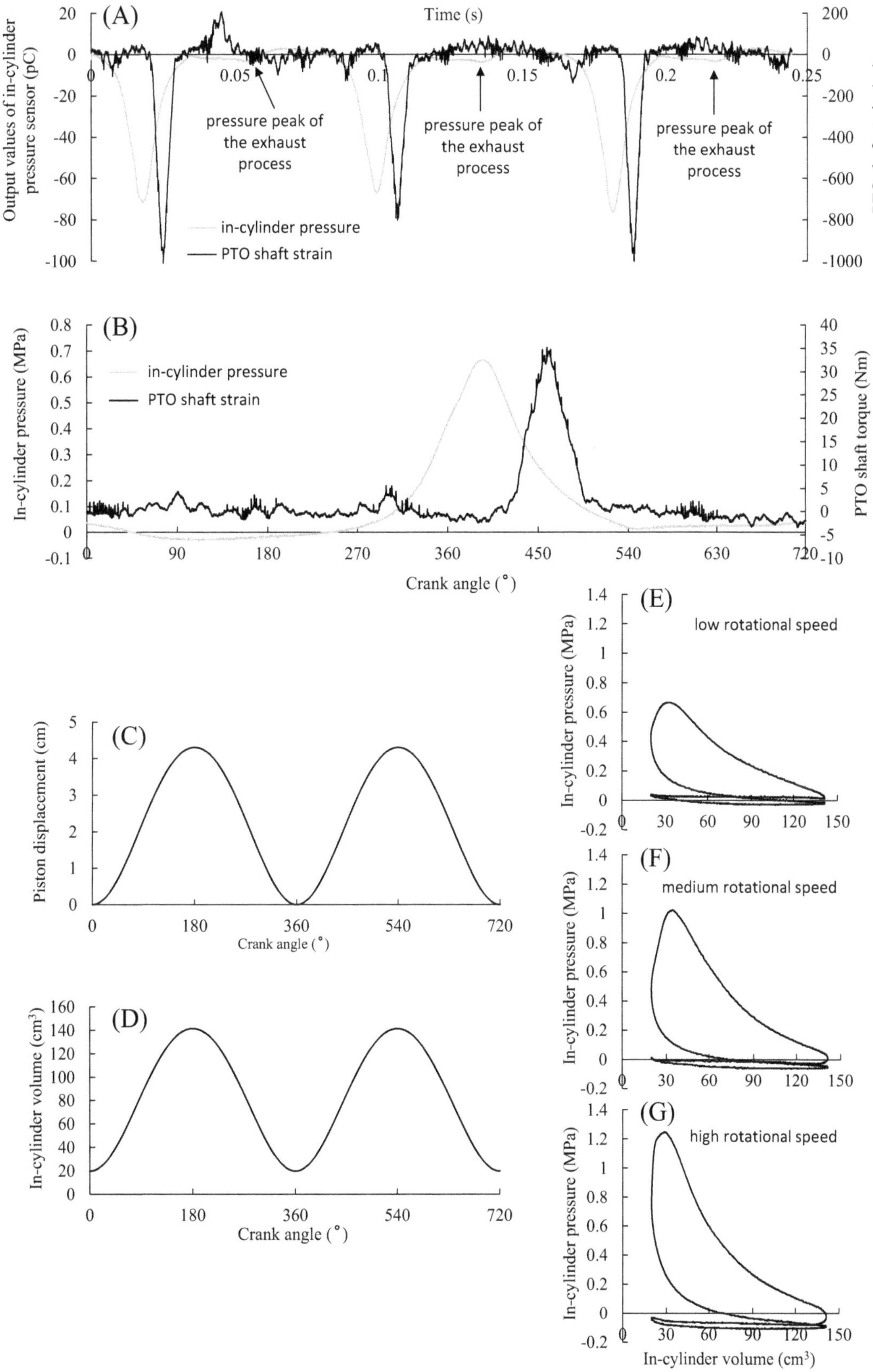

FIGURE 39.4 The output values of the in-cylinder pressure sensor and PTO shaft strain (low rotational speed) (A), in-cylinder pressure and PTO shaft torque (low rotational speed) (B), relationship between crank angle and piston displacement (C), relationship between crank angle and in-cylinder volume (D), and p–V diagram at low (E), at medium (F) and at high (G) rotational speed (Hirai, 2022).

TABLE 39.4

A Cycle Period and Engine Rotational Speed (Hirai, 2022).

	Rotational Speed Level		
Measurements	**Low**	**Medium**	**High**
Period (ms/cycle)	81.54	56.08	45.32
Engine rotational speed (rpm)	1,472	2,140	2,648

TABLE 39.5

Fuel Consumption (Hirai, 2022).

	Rotational Speed		
Measurement	**Low**	**Medium**	**High**
Fuel consumption (L/h)	0.193	0.386	0.634

TABLE 39.6

Engine Performance Index (Hirai, 2022).

	Rotational Speed		
Indicators	**Low**	**Medium**	**High**
Indicated work W_i (J)	29.0	47.2	51.0
Indicated mean effective pressure pmi (MPa)	0.24	0.39	0.42
Indicated power output P_i (kW)	0.36	0.84	1.13
Average torque of the PTO shaft (Nm)	1.81	3.52	5.12
Net work W_e (J)	11.4	22.1	32.2
PTO shaft output P_p (kW)	0.14	0.39	0.71
Brake mean effective pressure p_{me} (MPa)	0.09	0.18	0.26
Fuel mass flow rate m_f (kg/h)*	0.140	0.279	0.458
Heat input per second $\dot{Q}$ (kJ/s)	1.67	3.34	5.47
Brake-specific fuel consumption b_e (kg/kWh)	1.00	0.71	0.65
Indicated thermal efficiency η_i	0.21	0.25	0.21
Brake thermal efficiency η_e	0.08	0.12	0.13
Mechanical efficiency η_m	0.39	0.47	0.63

*The value was calculated from the fuel consumption using measured density of gasoline, 0.723 g/cm^3.

the V-belt may have led to unnecessary fuel consumption and reduced PTO shaft output. As a result, the maximum mechanical efficiency was as low as 0.63. The mechanical efficiency of small gasoline engines ranged from 0.75 to 0.8 (Ai, 2007). It will be necessary to reduce the power transmission loss by the V-belt to bring the various efficiencies closer to practical values in the future. This experiment regarded the pressure peak in the exhaust line as a TDC. This may have caused an error in the position of the peak in-cylinder pressure in the compression/expansion process in the p–V diagram. It would be possible to identify the exact peak position by installing an angle sensor or similar device on the crankshaft or camshaft and synchronously measuring the in-cylinder pressure (Iijima, 2018).

Conclusion

Agricultural machinery transmits the power generated by the engine to the work section through mechanical elements such as gears and belts, which work on the soil and crops to perform practical work. In other words, the engine is the driving force of agricultural mechanisation, and it is essential to learn its power by creating p–V diagrams. With the opportunity to meet machine hardware in education and research, it is fascinating to see how agricultural machines transmit power and work by various mechanical elements. It is essential to convey this fascination to students in an easy-to-understand qY.

Basic Property Measurements for Food Processing

Measurement of Liquid Viscosity

Principles and Fundamentals

In a dynamic fluid, misalignments between the fluid layers in contact with each other cause friction. This is called fluid friction (or liquid friction), and such a property of a fluid is called viscosity. This experiment aims to obtain the viscosity of a liquid using a capillary viscometer following Hagen-Poiseuille's law.

Newton's law of viscosity is expressed by the following equation (16).

$$\frac{F}{A} = \tau = \eta \frac{du}{dy} \tag{16}$$

where τ is the frictional stress per unit area (Pa). Equation (16) indicates that the frictional stress is proportional to the velocity gradient, and the proportionality constant η is called viscosity (Pa·s). Dividing η by the fluid density ρ (kg/m^3), kinematic viscosity ν (m^2/s) is also obtained. Hagen-Poiseuille's law, as shown in equations (17) and (18), holds when the flow is laminar in a capillary tube by the gravity.

$$Q = \frac{\pi r^4 \Delta pt}{8\eta L} \tag{17}$$

$$\eta = \frac{\pi r^4 \Delta pt}{8QL} \tag{18}$$

where r: the radius of capillary (m); L: length of capillary (m); Q: volume of liquid flowing out in t seconds (m^3), i.e., the volume of liquid flowing in t seconds; $\Delta\ p$: pressure difference at the top surface of liquid (Pa), $\Delta\ p = \rho gh$; ρ: liquid density (kg/m^3); h: height of liquid surface (m). Thus, the flow velocity of the liquid, which flows steadily in a capillary tube, is inversely proportional to the viscosity η of the liquid. Since the flow of the liquid is due to gravity, if the volume of the sample liquids is the same, the force by gravity is proportional to the density of each sample, respectively. For example, when water and another liquid different from water are used as samples, equation (18) is valid for both water (equation (19)) and the sample liquid (equation [20]), as follows,

$$\eta_w = \frac{\pi r^4 \Delta p t_w}{8QL} = \frac{\pi r^4 \rho_w g h t_w}{8QL} \quad (19)$$

$$\eta = \frac{\pi r^4 \Delta p t}{8QL} = \frac{\pi r^4 \rho g h t}{8QL} \quad (20)$$

where the viscosity of water is η_w, the density is ρ_w, and the time required for the sample liquid to flow down is t_w, and if the viscosity of water is η, the density is ρ, and the time required for the sample liquid to flow down is t. The ratio of both parameters is expressed as equations (21) and (22).

$$\eta_w : \eta = \frac{\pi r^4 \rho_w g h t_w}{8QL} : \frac{\pi r^4 \rho g h t}{8QL} \quad (21)$$

$$\eta_w / \eta = \frac{\rho_w t_w}{\rho t} \quad (22)$$

By rearranging the equation, equation (23) can be obtained.

$$\eta = \frac{\rho t}{\rho_w t_w} \times \eta_w \quad (23)$$

Therefore, the density of water and the sample liquid is required to obtain viscosity by Hagen-Poiseuilles law. The density of a liquid sample can be determined using a specific gravity bottle (pycnometer) and calculated using equation (24).

$$\rho = \frac{M_s - M_0}{M_w - M_0} \times \rho_w \quad (24)$$

where M_0 is the mass of the pycnometer after washing and drying, M_w is the mass of the pycnometer filled with water, M_s is the mass of the pycnometer filled with the liquid sample, and ρ_w is the density of water.

Methods

In this experiment, an Ostwald viscometer, a type of capillary viscometer shown in Figure 39.5, is employed. Other types of equipment are meth cylinder, 100 cc beaker, pipette, pycnometer, balance, transparent water tank, thermominder, stopwatch, and water bath.

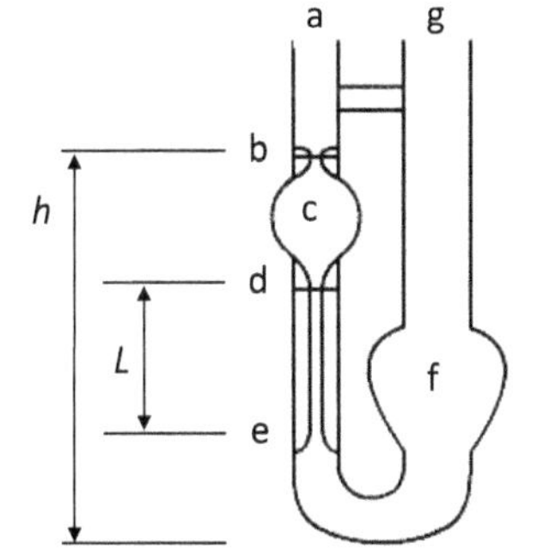

FIGURE 39.5 A schematic of Ostwald viscometer.

For the viscosity measurement, the time required for the liquid to flow down thc volume between the upper and lower marked lines of the Ostwald viscometer is measured. A schematic of the Ostwald viscometer is shown in Figure 39.5. This experiment usually uses distilled water and salt water (ten percent) as sample liquids. First, the measurement is performed on water. Fill the thicker tube (g) with a certain volume of water. A rubber-type suction item is attached to the narrower tube (a), and the water is sucked up until it rises above the line marked b above the sphere (c). When the suction phase is released into the atmosphere, the water begins to fall spontaneously. Use a stopwatch to measure the time from when the liquid level passes the line marked b above the sphere until it passes the line marked d below the sphere. For the measurement preparation, the viscometer is placed in a water tank at a designated temperature and to a depth above the line marked (b) to maintain temperature constant. Then, perform the same operation on the sample liquid. However, the volume of the sample liquid should be measured with a measuring cylinder to be the same volume as that of the water. Simultaneously, measure the mass of water and the sample liquid using a pycnometer. The pycnometer is filled with water and sample liquid under the same temperature conditions, and the density of the sample liquid is obtained by equation (24). The density of water at the same temperature can be referred to in several databases.

The mass M_0 of the cleaned and dried pycnometer is measured using a balance to measure the sample density. Then, fill the pycnometer with water, place it in a water bath at the set temperature for about 20 minutes, thoroughly wipe off any liquid that overflows from the capillary plug, and quickly weigh its mass M_w. Next, dry the pycnometer, fill it with the sample solution, perform the same operation as for the water, and weigh the mass M_s. When the temperature and mass of the water were obtained, the volume can be obtained by dividing the mass by the density. Then, by dividing the mass of a liquid of unknown density at the same temperature by this volume, the density of the liquid can be determined. Our experiment usually sets the water and sample liquid temperature for 30°C and 40°C.

Record the temperature at the time of measurement, the volume of water and liquid samples, the flow time of water and liquid samples, the density of water from a database, the mass of the pycnometer after washing and drying, and the mass when the pycnometer is filled with water and liquid samples. The measurement is typically required to be done ten times.

The requirement for the reporting assignment is as follows.

1) Calculate the mean value and standard error of the measured data.
2) Tabulate the mean value and standard deviation of the density and viscosity of the liquid sample.
3) Divide the viscosity by the density to obtain the kinematic viscosity.
4) Compare the kinematic viscosity obtained by the calculation with values analogous to those in the literature (consult the university library for literature).
5) Discuss the experiment results by referring to relevant literature (search for references on the library's website, etc.).

Measurement of the Specific Heat of the Liquid by Cooling Approach

Principles and Fundamentals

According to Newton's law of cooling, when the temperature difference between the object and the outside is not too significant, the rate at which the object loses heat by heat transfer is proportional to the temperature difference. This is expressed by equations (25) and (26).

$$-Kd\theta = dq = \sigma(\theta - \theta_0)dt \quad (25)$$

$$\frac{d\theta}{dt} = -\frac{\sigma}{K}(\theta - \theta_0) \quad (26)$$

where the temperature of the object is θ, the temperature of the outside is θ_0, the heat capacity of the object is K, the amount of heat loss during a time dt is dq, and the temperature change due to that is $d\theta$. In this regard, however, $\sigma > 0$ is a constant determined by the surface condition of the object. The variable separation of equation (26) gives equation (27).

$$\frac{d\theta}{\theta - \theta_0} = -\frac{\sigma}{K}dt \quad (27)$$

Integrating both sides of equation (27) gives equation (28).

$$\log(\theta - \theta_0) = -\frac{\sigma}{K}t + C \quad (28)$$

Substituting $\theta = \Theta$ into equation (28) under the initial condition $t = 0$ gives equation (29).

$$C = \log(\Theta - \theta_0) \quad (29)$$

This can be transformed to equations (30) and (31).

$$\log(\theta - \theta_0) - \log(\Theta - \theta_0) = -\frac{\sigma}{K}t \quad (30)$$

$$\frac{\theta - \theta_0}{\Theta - \theta_0} = e^{-\frac{\sigma}{K}t} \quad (31)$$

The same thermometer and stirrer are prepared in the same container, filled with a liquid sample of m_1, and the time taken for them to be cooled to θ from Θ by the external temperature at θ_0 is defined as t_1. Perform the same steps for the water of mass m_2 and measure the time t_2. Then, equation (32) is obtained. If $c_2 = 1$ (kcal/kg·°C), equation (33) is obtained.

$$\log\frac{(\theta - \theta_0)}{(\Theta - \theta_0)} = -\frac{\sigma}{cm_1 + w}t_1 = -\frac{\sigma}{c_2m_2 + w}t_2 \quad (32)$$

$$c = \frac{t_1}{t_2}\frac{(m_2 + w)}{m_1} - \frac{w}{m_1} \quad (33)$$

where w is the water equivalent of the vessel, stirrer, and thermometer.

Methods

The experimental items for the specific heat measurement include a cooling unit, heating unit, sample container (are made of copper), two thermometers, stopwatch, burner, balance, beaker, etc., that need to be prepared. This experiment usually uses distilled water and salt water (ten percent solution) as sample liquids.

As an experimental procedure, first wash and dry the sample container and weigh it. Fill the sample container with the liquid sample to a little more than half the total and measure the mass of the liquid sample. The difference between the masses of the empty and filled container is the mass of the liquid sample, which is m_1. The container filled with the liquid sample must be closed by a silicone rubber stopper. Then, the cooling tank is filled with tap water as a cooling medium, recording the temperature as θ_0. At this time, θ_0 should not differ from the atmospheric temperature. In contrast, the heating tank must be heated to approximately 90°C after filling the tap water as a heating medium. The sample container containing the liquid sample is placed in the heating tank and heated until the sample temperature reaches approximately 60°C. The sample container is then transferred to the cooling tank. Then, record the liquid sample temperature every minute for the first 20 minutes and then every five minutes thereafter until it reaches a few degrees above θ_0. Then, perform the same procedure for distilled water, which is the reference liquid. The mass of distilled water is m_2. Because all items including the container are made of copper, w as the water equivalent value can be calculated by equation (34).

$$w = 0.091m_3 \quad (34)$$

where m_3 is the mass of the copper items (kg). 0.091 is the specific heat of copper (kcal/kg·°C).

The requirement for the reporting assignment is as follows.

1) Plot the measured data on a graph.
2) Calculate the water equivalent of the container using equation (34).
3) Calculate the specific heat of the sample from equation (33). In this experiment, the specific heat of water c_2 is 1 (kcal/kg·°C).
4) Discuss the experiment results by referring to relevant literature (search for references on the library's website, etc.).

Drying Characteristic Evaluation of Fruits and Vegetables

To evaluate the drying characteristics of moist solid materials, such as peeled fruits, changes in the mass of the material are measured when the moisture is removed by, for example, contact with hot air. The drying rate can be calculated from a mass change during drying that is recognised as a characteristic drying curve.

Principles and Fundamentals

There are two ways to express the percentage of moisture in a solid, i.e. wet basis and dry basis moisture contents. If the mass of the solid is W kg, if w kg of water is contained in it, and if the mass of the dry matter is W_d kg, the relationship is expressed as equation (35).

$$W = w + W_d \quad (35)$$

Here, the wet basis moisture content m is equation (36), and the dry basis moisture content M is equation (37).

$$m = \frac{w}{W} = \frac{w}{W_d + w} \quad (36)$$

$$M = \frac{w}{W_d} \quad (37)$$

Dividing the denominator and numerator of equation (36) by W_d and substituting equation (37), the relationship between m and M becomes equation (38).

$$m = \frac{\frac{w}{W_d}}{1 + \frac{w}{W_d}} = \frac{M}{1 + M} \quad (38)$$

Also, solving equation (38) for M yields equation (39).

$$M = \frac{m}{1 - m} \quad (39)$$

Generally, moisture content refers to the wet basis moisture content. However, where the moisture is removed and the mass decreases during drying, it is more convenient to use the dry basis moisture content, where the denominator is invariant.

A material has a specific moisture content in the air under equilibrium conditions of constant temperature and humidity, called equilibrium moisture content. This is called the equilibrium moisture content M_e. Therefore, the maximum amount of moisture that can be removed in a drying operation using air is $M - M_e \equiv F$, where F is the free moisture content.

A drying curve is obtained by measuring the change in mass of solid material during drying, converting it to moisture content, and plotting it against the drying time t, as shown in Figure 39.6 A. The period around time $t = 0$ is the period for the preheating of the material, after which the moisture content decreases linearly. This is called the constant-rate drying period.

The curve plotting the drying rate ($-\frac{W_d}{A}\frac{dM}{dt}$) obtained from the slope of the tangent line at each point of the drying curve against moisture content is called the rate of drying curve or the characteristic drying curve (Figure 39.6 B). It is a curve specific to the material and drying conditions. In many materials, the constant-rate drying period is followed by a falling-rate drying period, and finally, drying is completed when the moisture content equilibrium is approached. B in Figure 39.6 B is called the critical moisture content.

The material surface is covered with water film during the constant-rate drying period. Thus, drying is similar to evaporation from the free water surface, and the material temperature is constant T_w. The drying rate in this period, called the constant drying rate, is expressed as equation (40).

$$R_c = -\frac{W_d}{A} \cdot \frac{dM}{dt} = \frac{h}{\lambda}\left(T - T_w\right) \quad (40)$$

where λ is the latent heat of water evaporation at T_w (kJ/kg) and h is the heat transfer coefficient (kJ/ m^2·h·K). For the heat transfer coefficient of air, the following experimental equation (41) can be used, for example, when hot air flows parallel to a plate-like material. (Air velocity: 0.6–8 m, $2{,}500 < G < 15{,}000$)

$$h = 0.013\, G^{0.8} \quad (41)$$

where G is the mass velocity of the air (kg/ m^2·h).

Methods

Experimental apparatus and materials are constant temperature dryer, anemometer, electronic balance, sample fruits or vegetables.

1) Set the temperature of the dryer to 60°C.
2) Measure the air velocity in the dryer using an anemometer.

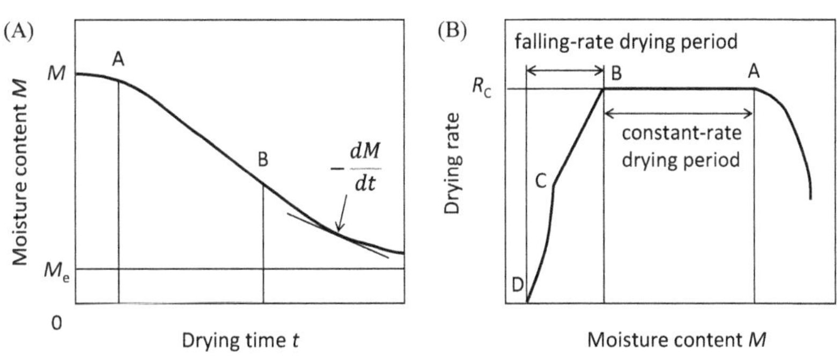

FIGURE 39.6 A drying curve (A) and a rate of the drying curve (characteristic drying curve) (B).

3) Place the electronic balance on the top of the dryer to level it, hang the sample container on the bottom hook of the balance, and set a re-zero.
4) Cut the sample fruit or vegetable into approximately 5 cm × 2 cm × 1 cm pieces. Prepare samples of the same shape for the mass change and temperature measurements.
5) Measure the initial weight of the sample for the mass measurement, measure its size with a caliper, and calculate the sample surface area. For the sample of temperature measurement, thermocouples are attached at the centre and near the surface.
6) The temperature in the dryer is also measured.
7) The temperature and humidity around the dryer are measured before the measurement.
8) The mass change of the sample is measured every five minutes. The measurement should generally be carried out until the equilibrium moisture content is reached, but it could be stopped around three hours after the start of the measure. The surface area of the sample for mass measurement is also measured at the same time.

The initial moisture content of the radish is 95% (wet basis). The dry mass of the sample can be obtained from the initial mass by calculation. The initial moisture content in the wet basis is converted to dry basis moisture content using equation (39), which is 1,900% (dry basis). The change in mass after each time-lapse is expressed as a change in moisture content.

1) Show the drying curve of the sample as a graph.
2) Using equation (40), show the drying curve of the sample in terms of "drying rate" versus "moisture content".
3) Using equation (40), estimate the heat transfer coefficient and compare it with the value calculated from equation (41).
4) Discuss the above results with references.

Online Trials of a Personal Desktop Laboratory at Chiba University

Chiba University, one of the major national universities in Japan, has developed a novel experimental apparatus named the Personal Desk Lab (PDL) to enhance effective physics education (Mino et al., 2017). The PDL is a set of dry experimental items that allow students to conduct physics experiments alone on a desktop, making it possible to conduct experiments while attending lectures in a regular lecture room. The PDL can also simultaneously conduct the student's experiment individually. Meanwhile, remote classes using online tools are also conducted at most universities and schools recently because of COVID-19 circumstances. It has also been revealed that more efficient education can be conducted through remote classes. The most significant advantage of the remote class is being able to offer classes without faculty members having to travel or students changing location, particularly in the universities with multiple campuses. However, the main concerns of remote classes are: 1) whether sufficient measures can be taken to deal with communication equipment problems, 2) whether there is a need for students in distant places, and 3) whether the content is suitable for remote classes. Currently, remote classes are mainly lectures and are primarily delivered one way from the lecturer to the students. Therefore, it is necessary to expand to active learning, which involves primarily two-way interactions. In this regard, it seems that few universities have adequately addressed these issues and successfully conducted remote classes, especially for student experiments.

Chiba University has three distinctive campuses geographically separated from the main campus as extensions. Each campus focuses on applied research and related education. In other words, there are few facilities and personnel for basic science experiments. It is inappropriate for any educational organisation if the university's circumstances cause a disparity in basic education levels among the campuses. For this reason, Chiba University has been considering how to provide classes to geographically different campuses. The PDL system can be used with an online tool, such as a videoconferencing system, and ICT equipment such as electronic blackboards. In that case, there is a possibility that student experiments can be developed at remote locations. If the student experiments in the basic subjects can normally be provided from the main campus to the other location, the disparity between campuses could be corrected. It would also be possible to remotely provide effective student experiments, which can be regarded as one of the practical active learning programmes. The know-how in remote experiments has been continuously accumulated and made to function well. A ripple effect can be expected not only at Chiba University but also in other universities, in high school-university cooperation, and even in an overseas expansion to other universities. This section reports on the pilot practice of remote experiments in basic physics over the last decade and discusses issues for the future of remote experiment education (Mino et al., 2017).

System Configuration for Remote Student Experiments

The following equipment was used for student experiments for the remote experiment from the main campus to the other campus. A videoconferencing system (PCS-XG100, Sony at the main campus, VSX7000s, Polycom at the extension). Each device includes a camera, monitor, projector, screen, and a microphone for capturing class scenes. A portable electronic blackboard (eBeam, Keepad), a clicker (Turning Point, Keepad), a video camera, several laptops (one each at the main and another campus for webcam, Skype or online meeting tools, and portable electronic blackboard), and a PDL experiment kit. Figure 39.7 shows a schematic of the equipment arrangement, configuration, and communication.

Procedure of Remote Student Experiments

PDL experiment kits and textbooks for the number of students were delivered to the target place in advance. The lecturer taught the experiment from the main campus through a videoconferencing system on the day of class. Graduate students

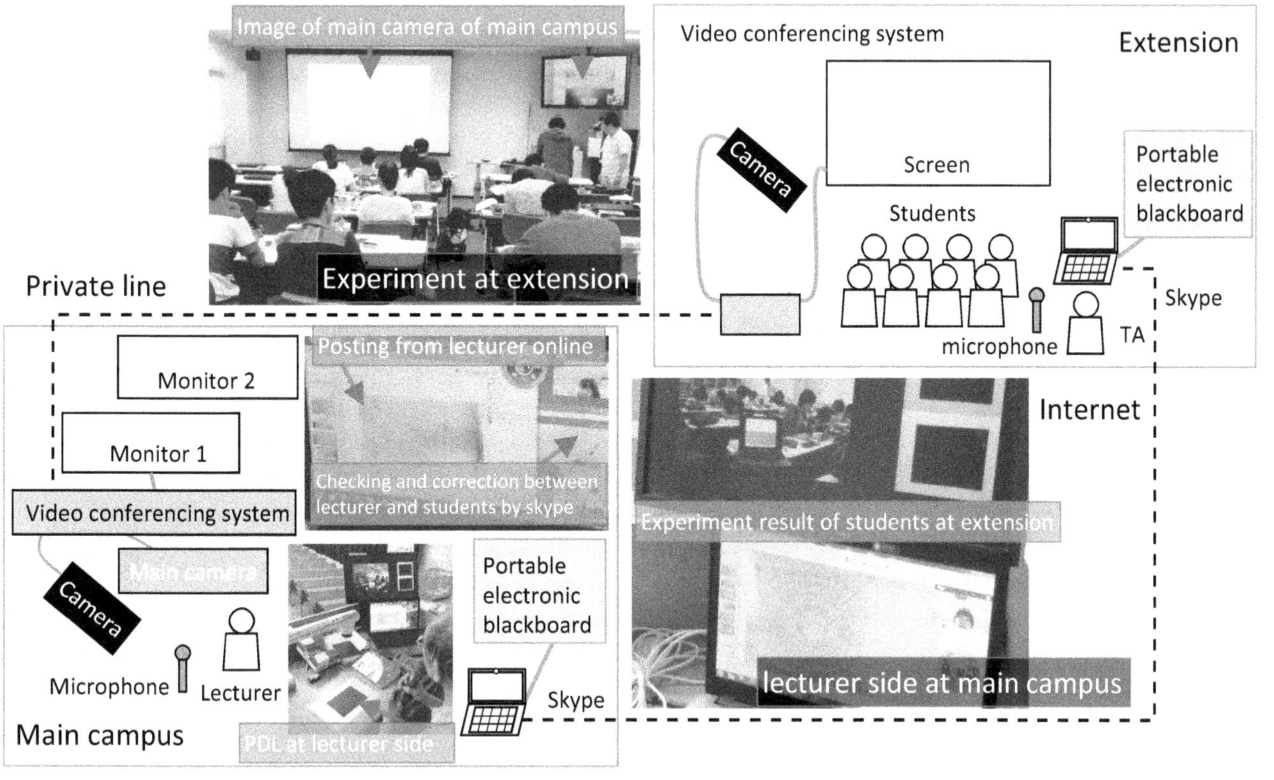

FIGURE 39.7 A schematic of the placement and configuration of remote student experiments at the extension and the lecturer's desktop at the main campus (Mino et al., 2017).

with appropriate skills were employed as teaching assistants (TAs) and provided on-site support for the students in the experiments. As shown in Figure 39.7, the cameras were installed in both rooms. The experimental status in each room was observed on screens captured by the cameras via the videoconferencing system. The students and lecturers could communicate through the audio system. When work required on-site support, the TAs gave practical instructions indicated by the lecturer on the main campus. TAs were given advance instructions on experimental equipment use and the experiments' content. Because of these efforts over the past years, the following issues and areas for improvement were highlighted.

1) It was essential to check the operation of the equipment in advance.
2) In teaching experiments, it was effective to give a broad overview and explain how to use the actual equipment, but this required a system that could stream multiple video images.
3) A separate device from the videoconferencing system was needed to guide individual students on the results of their experiments.
4) Such a separate device was also needed to provide the lecturer with a more authentic visual and audio communication of the students' reactions and activities and to provide real-time instructions to the students.
5) It would be better to have a system to check whether the students understand the instructions and content in real time.

Based on several trials, it was made clear that checking the equipment operation twice was required – once the day before and once 30 minutes before the experiment. In 2016, laptop PCs were installed in both classrooms, and individual responses were made via online meeting tools such as Skype, portable electronic blackboards, and clickers used to check each student's understanding. The following is detailed information on the remote experiment conducted in 2016.

1) Instruction from the lecturer to students via a videoconferencing system and video camera.

The images on lecturer's side at the main campus and the students' side at the extension are also shown in Figure 39.7. The lecturer could check the students' activities through the system. In contrast, the students followed the lecturer's instructions through a monitor in which the lecturer's examples were shown and operated their experimental equipment on the desktop.

2) Individual instructions of the lecturer to the students using a laptop PC, an online meeting tool, and portable electronic blackboard via the internet.

It was necessary to guide each student for each achievement when conducting experiments. Such guidance was performed separately using a laptop PC with an online meeting tool. First, the TA took pictures of the students' experiment results using a webcam. Then, the captured images were uploaded onto a portable electronic blackboard and projected on the computer screens on both sides. The lecturer drew the experimental results on the electronic blackboard in red while giving oral instruction via the online meeting tool (Figure 39.7).

3) Response to questions from each student to the lecturer using a clicker system.

For inquiries from students, a clicker system was used as needed. When a student presses the clicker, the clicker number is displayed on the monitor of the lecturer in red, then green, and so on. Students who had questions were asked to press the clicker, and the TAs of the experiment supported their communication.

4) Reporting assignment and guidance online.

Regarding submitting reports and their instructions, an online system such as Moodle, which could manage the file systems online, was applied.

Results of the Remote Experiment

1) A videoconferencing system could send clear audio and video from a video camera, and all 30 students could receive instruction up to the point where they were producing experimental results.
2) Several students who finished their experiments early could receive individual guidance remotely with corrections through personal advice on a separate line. Meanwhile, the students who completed later could not receive individual tutoring due to a lack of time.
3) When the number of students who had completed the experiment increased, it became hard to provide individualised instruction within the time available. In such cases, several results were picked up as needed and projected on the screen as examples. Explanations were given to the entire group, allowing each student to compare and review the results.
4) Clickers made it possible to grasp questions from the students in real time, which led to actual questioning. However, it also required a certain amount of time and effort.
5) No interruptions by network and device problems occurred during the experiment.
6) A TA handed a microphone and PCs to the students. However, questions from the students were often handled locally by the TA close to them, not by the clicker.

Consideration of Future Issues

In 2016, ICT equipment such as Skype as an online meeting tool and portable electronic blackboards were installed so that one TA familiar with computer equipment and one technical assistant was dispatched from the main campus to the target campus. At the target campus, a TA in charge of remote equipment operation and two caretaker staff were added. Meanwhile, at the main campus, one lecturer for the experiment instruction and one instructor for operating remote equipment were arranged for this remote experiment. Normally, assigning multiple personnel to an experimental class is hard. However, unforeseen problems in remote systems would occur at a specific rate. If it takes much time to solve equipment problems, there is a risk that the remote class will not be successful. For this reason, staff and supporters who can familiarly handle online equipment are indispensable for promoting remote experiments. If communication between mutual classrooms relies solely on the audio function of the videoconferencing system, calls will not be possible during communication errors. The installation of appropriate tools, such as an extension telephone, is essential to keep in touch in such cases. The clicker system used to monitor the status of students in the course needed to be improved, in that it could only be controlled from the main campus. Even though the TAs were monitoring the students and using clickers, there were still students who were not keeping up with the experiments on-site, or the lecturer on the main campus over-explained things without realising that the TA was working with those students. It is necessary to further strengthen the measures taken by the TAs, such as ensuring that audio is exchanged between the teacher and the TA as necessary, discussing the experiment procedures in advance, preparing documents with detailed descriptions of the work, and presenting the overall schedule. Although this initiative did not conduct a class survey limited to remote experiments for students, it would be effective to complete a real-time questionnaire using a clicker system during the class each year to ask students about their level of understanding and their impressions of the experiment.

Concluding Remarks

A remote basic physics experiment class using PDL and ICT was conducted between geologically different campuses, where the internet environment was satisfactory. By dispatching TAs and technical assistants who are familiar with equipment operation to the remote locations and providing staff who can operate remote equipment at both campuses, it was possible to provide experimental classes in a situation similar to that in which the lecturer visited the site. To promote the more widespread use of such remote experiments, the necessary staff mobilisation is a significant issue. The key to this is determining how to minimise equipment problems and simplify operations. These issues have risen to the surface and need to be addressed.

Possibility of DX in ABE Education at the National Universities in Kyushu

The COVID-19 pandemic has influenced human activities, including academia. Such circumstances have meant limited meals with others, restrictions on outings and overseas trips, and so on. Meanwhile, many meetings and discussions are now conducted online instead. In other words, these circumstances have dramatically induced DX in our activities. This trend has also been bringing various changes to universities' education and academic environments. In fact, hybrid lectures that combine the best of face-to-face and online education have been commonplace recently. In particular, online classes have the great advantage of allowing students to attend lectures regardless of their study time and location. These circumstances will probably not change even after the COVID-19 disaster. This means that a new style of education has been expanding through the employment of DX technology. This trend can also increase the diversities of each university's educational content and opportunities for recurrent education if education content and knowledge can be shared online. This section introduces prospective and progressive education using DX technology in the ABE field for the universities in Kyushu, Japan, as well as specific examples of new educational and learning methods developed at various institutions (Technical Series Committee and Human Resource Development Committee, 2022).

Current Status and Issues of ICT Promotion in University Education

Following the first survey conducted in 2021 by the Ministry of Education, Culture, Sports, Science, and Technology (MEXT), the Second National Student Attitude Survey was conducted from 31 January to 28 February 2022, which covered 466,351 second-year university students, 465,813 fourth-year university students, and 25,433 junior college students The participant institutions were 582 universities and 157 junior colleges in the second survey. The specific objectives of this survey were: 1) to use the results to improve education at each university, 2) to deepen public understanding of universities by publicising the results, 3) to provide essential data for formulating national policies, and 4) to make university life more fulfilling for students. The survey results were released in October 2022; thus, everyone can examine the results on the website (Ministry of Education, Culture, Sports, Science, and Technology, 2023) if they are familiar with the Japanese language. The survey was a comprehensive questionnaire constructed with 60 questions. The main questions related to online classes were about the type of online lectures taken by the students, such as the simultaneous interactive or on-demand type, and the pros and cons of each type, which the students evaluated. The survey results revealed that the ratio of the simultaneous interactive online class type taken increased while the on-demand online class type decreased in 2022. This result seemed to indicate the demand for interactive online classes by the students. It was also indicated from the survey result for the pros and cons of the online class type that the simultaneous interactive type could more easily communicate with lecturers than the on-demand type. Meanwhile, more than 50% of the students answered that online classes were more manageable for learning the content than face-to-face classes. Therefore, the survey suggested the students require communication with the lecturers, but the lecture would be preferable to be done online.

Current Status of DX Environment for Researchers at the National Agriculture and Food Research Organisation

NARO, the National Agriculture and Food Research Organisation, is the core institute in Japan for conducting research and development (R&D) on agriculture and food. The mission of NARO is to contribute to the development of society through innovations in agriculture and food by promoting pioneering and fundamental R&D. In other words, NARO is a Japanese version of United States Department of Agriculture (USDA) Agricultural Research Service (ARS). NARO is the largest research institute in Japan's agriculture and food sector, and approximately 1,800 researchers belong to it. Thus, NARO is always regarded as one of the advisable workplaces for graduate students aiming for career enhancement after earning a PhD. The main online tools adopted as internal mission-critical and business systems of NARO are wide-ranging, and researchers with NARO must acquire proficiency in their usage. One online tool is the information-sharing system, namely "Desknet's NEO", used to share schedules and notification documents. Other internal mission-critical and business systems include a personnel and salary system, a financial accounting system, an airline ticket reservation system, a drug management system, an information equipment management system, and an e-learning system. NARO has a comprehensive contract with the Microsoft Corporation for Office365, which includes Office365 Pro Plus (Microsoft Office products such as Word, Excel, PowerPoint, Access, etc.), OneDrive, Yammer, and Teams, so several online meeting tools are available. NARO also provides teleworking services during telecommuting and business trips to cope with the new coronavirus, save electricity, etc. The telework system of NTT East and the Information-technology Promotion Agency (IPA) controls internal terminals from outside the institution through a virtual desktop.. The above online tools can be used in a remote environment. Administrative work has become paperless, and many workflows can now be completed online for greater convenience. Additionally, various training courses, educational training, and seminars are now primarily on-demand and online, making it easier to attend and participate in them. Academic meetings are also held online and can be followed anytime without accommodation or travel expenses. Project meetings with outside parties can now be held and attended flexibly using such online meeting tools. The online system makes it easier to conduct various collaborative activities other than projects because travel expenses are not required. This, however, indicates that various meetings, including urgent appointments, have also increased. In the future, the benefits of both face-to-face and online services must be balanced and considered.

A Case of the Human Resource Development Project on DX at University of Miyazaki

"Development of Highly Specialized Human Resources with Digital Technology to Drive Regional Industrial DX Centered on Agriculture, Forestry, and Livestock Industry Through Agriculture-Industry Collaboration", which was advanced by the Faculty of Agriculture and the Faculty of Engineering at the University of Miyazaki, has been adopted as a project by the MEXT. This project has been implemented since 2022 and supports the renewal of educational facilities and equipment that will serve as the foundation for implementing a curriculum of experiments and practical training that combine digital technology and education in specialised fields through a collaboration between the Faculty of Agriculture and the Faculty of Engineering. This will enable DX to solve the problems of regional industries centred on agriculture, forestry, and livestock, which are the key industries of Miyazaki Prefecture, and foster highly specialised human resources who will lead the DX of the region. This project expects to supply DX human resources to the region in the future and help solve key issues by improving labour productivity in local industries, adding value to products, securing labour, improving safety, and promoting product certification to revitalise the local economy. The Department of Agricultural Engineering has established a new DX-related course linked to existing classes. This course consists of practical exercises applying digital data obtained from intelligent greenhouses and fields that will be installed on a university farm. This will enhance the professional education programme for smart DX. Additionally, by incorporating lectures on the latest DX technology into existing practical training courses, human resources can be developed to respond immediately to problems that arise.

Current Status of DX Use in Classes at Saga University

Online meeting tools, such as Zoom and Microsoft Teams, stream the lectures in classes that are simultaneously face-to-face and remote, called hybrid classes, conducted at Saga University. Both tools are licenced by the university, and the faculty members in charge of the lectures are registered to those licences. In the case of hybrid classes, it is hard to use a practical blackboard in the classroom because camera images are unclear, and a wireless or high-sensitivity microphone is required for the instructor to move. In contrast, using a presentation application such as PowerPoint or the blackboard function of a tablet is clearer and easier to see. The streaming video is recorded simultaneously and distributed on-demand after class using Office365 Stream. This provides a practical situation for students who were absent due to illness or wish to review the class content. However, when the video was played back as a class document, it sometimes resulted in skipped frames and an unsmooth motion on the receiving end. Microsoft Teams has the following functions: 1) group-only bulletin board function, 2) real-time online meetings including video, 3) assignment setting, 4) chatting with individual students or groups, and 5) automatic creation of stream groups. Stream is a video distribution system that uploads MP4s and other videos to the cloud and distributes them. A list of only that subject can be created by creating a channel for the group and associating it with the uploaded video. This makes it possible to provide videos on Teams as well. Office365 Forms is a convenient system for creating surveys and remote exams, which can be linked to Teams assignments and can have deadlines for presentation and revision. If you create multiple-choice questions, you can set the correct answer and score, and the system will automatically grade the questions, including explanations for wrong choices. Scoring results can be presented to the respondent, and the scores and other test results can be output as an Excel file. Saga University has two domains, "@cc.saga-u.ac.jp" and "@edu.cc.saga-u.ac.jp", the latter being the same domain as the student ID. This same domain is essential when sharing information with students. One way to eliminate the complicated login procedure is to use multiple browsers. For example, Edge is used for external communication, and Chrome is used for communication with students. There are many cases where students use their cell phones in parallel to watching videos, e.g., watching videos on their cell phones while doing exercises, etc., on their PCs. Additionally, the automatic subtitle generation function for Stream videos requires skill to generate accurate words, but it may be helpful for students with some disabilities, such as hearing loss.

Corona Disaster Initiatives at University of the Ryukyus and Future Educational DX

In March 2020, the University of the Ryukyus established a guideline for its class policy and syllabus content and then set up a system for conducting remote classes from the new semester, April–August 2020. Additionally, a guidebook with detailed information on online and on-demand classes using the Learning Management System (LMS) was created and distributed to lecturers to conduct remote classes efficiently. Meanwhile, the University of the Ryukyus has been implementing the Inter-University Exchange Project (Re-Inventing Japan Project) since 2018, supported by the MEXT, to strengthen its global development capabilities. Therefore, the lecturers have been using the Zoom platform for online education activities with other countries through COIL (Collaborative Online International Learning). Additionally, as a group member of the Japan Inter-University Consortium for Mathematics, Data Science, and AI Education, the University of the Ryukyus has been working to create content for online classes since 2019. The project has promoted the development and distribution of online teaching materials related to mathematical sciences, and the teaching details of video creation and distribution of class content have been collected. To cope with the environmental changes caused by the coronavirus, an information system for the teaching staff on student management and the web class as an LMS have been introduced. Additionally, Zoom and Microsoft Teams platforms are being used for the online classes. For future educational DX, the development of exchanges with multiple universities using COIL education, implemented under the Inter-University Exchange Project, will serve as a reference. Currently, we are implementing educational collaboration with universities in the Pacific region, including the University of Hawaii and the University of Guam, which have a deep-rooted exchange culture with our university. Moreover, by joining the Japan Forum for Internationalisation of Universities and expanding exchanges with various domestic and international universities, the university is enhancing its educational environment by offering and using COIL online courses to advance human resources for innovation both locally and globally.

Enhancing Data Science and Engineering Education in Agriculture Using DX at Kyushu University

Kyushu University, one of the major national universities in Japan, has formulated "Kyushu University VISION 2030", intending to become a university that drives social transformation through comprehensive knowledge, and has set forth eight specific visions. DX is positioned as the second vision, aiming to develop human resources who can apply mathematics, data science, AI, etc., and contribute to social transformation and implementation through DX. The MEXT has also adopted the proposal by the Faculty of Agriculture for the "Project for the Development of Highly Specialized Human Resources to Drive Industrial DX by Combining Digital and Specialized Fields". The introduction of DX into education and research has been started under the "Digital Native in Agriculture (DNA) Human Resources Development Program to Drive DX in the Near Future Agriculture, Livestock, Fishery, Food, and Biotechnology Industries". In this programme, first, mathematical and data science education for undergraduate students (first year) was reconstructed to create an environment in which students can actively learn the basic knowledge and skills required for data handling. The curricula for specialised higher education (after the second semester of second year) are also redesigned based on the assumption that students have already acquired most of this knowledge and these skills and the curricula are used to develop human resources who can use DX. For example, the "Mechanical Design II (third year)" as a trial class for DX education in the agricultural engineering course acquires basic knowledge and skills related to machine design. Specifically, students draft mechanical parts using 3D CAD Fusion 360 (Autodesk, available free of charge for educational institutions), conduct strength analysis using the structural analysis function (finite element method) provided with the software, and fabricate a simple vehicle robot. By introducing the strength analysis, the students can understand how the "Strength of Materials" (second year) is applied in actual machine design. In the robot fabrication practice, the students can experience a series of processes, from basic design to fabrication of each component, as group work. In this lecture, in addition to drafting the robot components using 3D CAD, the students can understand the structure of the vehicle robot by experiencing the fabrication of parts using a 3D printer and computer numerical control machine while checking the strength characteristics of the actual parts. Driving control programming could not be included in the lecture content due to time constraints. However, some students could immediately check and modify the provided driving control programme because they had already completed one-board microcomputer programming and driving control algorithm using Lego Mindstorms EV3 in "Agricultural Informatics (second year)" and "Bioproduction System Engineering Experiment I (third year)". As mentioned above, it was confirmed that the introduction of DX education and the fusion of related lecture content can improve students' invention and application skills.

Conclusions

Primary content for student experiments, for example, engine tests on power machinery and acquisition of basic characteristic measurement methods on bioprocesses, have still been essential for ABE education; however, ICT and DX education must be incorporated into the course curricula to ensure the fundamental research and practical application progress in ABE. Although it has still been under development in Japan, productive educational curricula incorporating the latest DX into traditional educational content are essential to the steady progress of ABE. Simultaneously, it should promptly assemble content that arouses and develops students' interest.

Acknowledgement

We would like to thank Prof. Masafumi Mitarai, Miyazaki Gakuen Junior College, for providing information on the Second National Student Attitude Survey conducted by the MEXT.

REFERENCE LIST

Ai, F. (2007). Structure and use of agricultural machinery, Rural Culture Association Japan, 60–89. (in Japanese)

Hirai, Y. (2022). Gasoline engine performance test –Introduction of undergraduate student experiments in the educational course of Bioproduction System Engineering, Faculty of Agriculture, Kyushu University–, *Journal of Kyushu Society of Agricultural Machinery and Food Engineers*, 72, 25–32. (in Japanese)

Iijima, A. (2018). Theory, and practice of highly efficient engines from the basics, Grand Prix Book Publishing Co. Ltd., pp. 13–30 (in Japanese)

Ministry of Education, Culture, Sports, Science and Technology. (2023). The second national student attitude survey, https://www.mext.go.jp/a_menu/koutou/chousa/1421136.htm (accessed on February 4, 2023) (in Japanese)

Mino, H., Iida, A., Hayashi, M., Ogawa, Y., & Tozaki, K-I. (2017). A practice of a distance physics experimental class using ICT tools and personal desk lab, *Journal of Liberal Arts and Sciences*, Chiba University, 1, 91–97. (in Japanese)

Suzuki, T. (2020). Thorough investigation of diesel engines, Grand Prix Book Publishing Co. Ltd., pp. 23. (in Japanese)

Technical Series Committee and Human Resource Development Committee. (2022). Online educational materials for transforming university education and strengthening social collaboration, *Journal of Kyushu Society of Agricultural Machinery and Food Engineers*, 72, 16–22. (in Japanese)

40

Practical Engineering Education to Address Sustainability and Food Security

Vijaya Raghavan and Valérie Orsat

Introduction

Food security and sustainability are closely linked to the human resource development, natural resources, developmental status, and geographical location. Sustainability and food security issues have become a global concern, capturing the attention of scientists, academics, socio-economists, lawmakers, and politicians (Paliyath et al., 2018). Food is one of the primary sustainable development goals (SDGs) set by the United Nations for the 21st century. In this context, to "end hunger, achieve food security and improved nutrition and promote sustainable agriculture" is the most important SDG among the 17 SDGs. A significant change in the global agriculture and food system is required to achieve this goal by the target year of 2030. The Food and Agriculture Organisation (FAO) has estimated that we will need to produce 70% more food to feed 9.8 billion people by 2050. It is crucial to enhance sustainability, food safety, food security, and properly implement postharvest management practices. Application of appropriate postharvest technologies would have a decisive role in addressing these issues. It is, therefore, imperative to impart practical agricultural and biological/biosystems/bioresources engineering (ABE) education to the societal system to address sustainability and food security.

In general, countries with high rates of population growth, low income, economic vulnerability, geopolitical crisis, degraded agriculture, poor natural resources, lack of human resource development, low social stability, and lack of governmental stability are susceptible to food security and sustainability crisis. Reducing postharvest losses is essential to realising the immense potential for boosting equitable economic growth, sustainability, and food security. Large-scale practical applications of appropriate postharvest technologies through ABE education would have a far-reaching effect. The ABE education system includes food production, postharvest processing, supply chain, storage, distribution, consumption by end users, waste management, energy utilisation, and environmental aspects. In recent years, agriculture and food systems have given much emphasis on the six E's – namely, education, economics, energy, environment, efficiency, and ecology (Ting, 2010). The integration of practical engineering knowledge with biological, chemical, and physical sciences delivers an effective approach to address food security and sustainability concerns. This chapter focuses on the domains of ABE, the importance of ABE education, and a case study on the project of food security in South India. The case study emphasises institutional strengthening, implementation of appropriate postharvest technologies, business incubators in operation, and pilot projects in adopted villages. This chapter would serve as useful reference material for the implementation of ABE education globally in general and particularly in developing economies.

Domains of Agricultural and Biological/Biosystems/Bioresources Engineering

The ABE education system delivers a holistic approach for teaching, research, and development with the integration of engineering and biosystems for addressing food security and sustainability. ABE is based on a hybrid concept, consisting of engineering elements and biological components. Broadly, the domains of ABE include: (a) engineering (mechanical engineering, chemical engineering, computer engineering, electrical engineering, material engineering, automatics engineering, civil engineering, and energy engineering), and (b) biosystems (vegetal biosystems, animal biosystems, environment biosystems, microorganisms, and human beings) (Roş et al., 2006). Some of the major research and development areas of ABE are: food safety and quality, food processing and preservation, new food products development, bio-processing, food production, land and water resources, biomass and renewable energy, information and precision agriculture, intelligence and automation of agricultural machinery systems, farm machinery and power systems, instrumentation of agricultural systems and controls, agricultural safety and health, and environmental stewardship (Ting, 2010). The key to the effective application of ABE education for global food security and sustainability depends on the teaching, research, developmental, and educational activities in these domains.

Importance of ABE

Over the last few years, the global challenges in food security, food safety and quality, ending poverty, improving nutrition, healthy lives, sustainable agriculture, rural development, and inclusive economic growth have given much emphasis to the

DOI: 10.1201/9780429150111-43

importance of ABE education in the societal system to reinforce education and research activities in order to achieve long-term sustainable development goals. The ABE system requires a multidisciplinary approach to teaching, research, and development activities. The importance of ABE education in addressing major global goals such as food security, employment generation and economic development, and rural agricultural work experience and sustainable rural development are discussed in the following sections.

Global Food Security

During the FAO World Food Summit, 1996, the FAO defined food security as a condition that exists "when all people, at all times, have physical and economic access to sufficient, safe and nutritious food that meets their dietary needs and food preferences for an active and healthy life" (FAO, 1996). This generally acknowledged definition explains the four dimensions of food security – namely, availability, access, utilisation, and stability (FAO, 2006). The American Society of Agricultural and Biological Engineers (ASABE) specified four primary priorities for addressing global food security – namely, consciousness of agriculture, capacity building and higher education and training, international collaboration, and technology adoption (ASABE, 2016). ABE education would play a vital role for achieving food security worldwide by ensuring proper production, handling, processing, storage, value addition, and distribution of foods. The FAO has estimated that approximately 14% and 17% of global food production was lost annually between harvest and the retail market and between the retail market and consumer levels, respectively. Approximately one-third of total food production is lost or wasted, a significant loss of about one trillion US dollars annually. This food wastage signifies a threat to global food security and also causes serious impacts on the environment. The food losses decrease the amount of food available for consumption. Thus, saving more of the produced food is an undeniable priority for all. High postharvest losses due to inadequate postharvest technologies and management (from postharvest processing to storage) contribute significantly to food security crises in developing countries. However, in developed countries, the postharvest losses are mainly due to food expiration or discards (Padi & Chimphango, 2021; Hodges et al., 2011). Reduction in postharvest losses is a key to ensuring food security.

Postharvest losses of food can be due to: (a) metabolic causes (for example, respiration, transpiration, and ageing), (b) developmental causes (seed germination, rooting, and sprouting), (c) mechanical damage (bruising, impacts, breaking, and cutting), (d) chilling and freezing injury (low-temperature injury), (e) physiological deterioration (high humidity, oxygen, and temperature), (f) parasitic diseases (microorganisms and insects), (g) lack of storage facilities, (h) poor packaging concept, (i) inadequate postharvest quality control, (j) lack of infrastructure, (k) poor transportation and supply chain system, (l) wasteful consumption (at home and dining table), (m) lack of market demand (lack of market planning and information and inaccurate production), and (n) lack of awareness among producers, consumers, and policy makers. Therefore, practical postharvest engineering education is imperative in the societal system, including education of technologists, academics, scientists, socio-economists, lawmakers, politicians, and common people. Over the years, ABE education and training activities have been gradually progressing in developing countries. However, the necessary support and follow-up after the education and training programmes need to be improved. Often, due to a lack of locally available supplies, equipment, tools, support, and resources, the training participants are reluctant or find it difficult to implement the technologies, even though they may be convinced to during the training and/or were initially provided with the machineries, supplies, and tools as part of a project and training activities. In most cases, any initial implementations of the postharvest technologies are discontinued or abandoned owing to a lack of locally available services and facilities. Therefore, it is necessary to provide the required resources, facilities, and services at a reasonable cost, even after the training programmes (Kitinoja et al., 2011). Global food security could be achieved by increasing the land under cultivation, improving productivity, reducing pre- and postharvest losses, and controlling the population growth rate. Proper implementation of postharvest technologies and management strategies is one of the most economical and efficient approaches to achieve food security worldwide.

Employment Generation and Economic Development

The International Labour Organisation (ILO) unanimously adopted a "global jobs pact" in the 2009 international labour conference. The government, worker, and employer delegates from the ILO member states agreed to create decent work opportunities. The global jobs pact discusses the social effect of employment crisis and recommends job-centred policies based on national needs. The portfolio of policies includes: (a) employment generation, (b) social protection, (c) fair globalisation, (d) respecting labour standards, and (e) promoting social dialogue (ILO, 2009). It is an urgent call for action regionally, nationally, and globally to promote employment opportunities and protect people by responding to agendas focusing on people and economic growth. Job opportunity is becoming a challenging factor with increasing global population and limited employment openings in government sectors. This issue is particularly dominant in developing and underdeveloped economies and developed economies in general. The migration of unemployed rural youths to cities and towns in search of jobs is becoming a great concern. The world bank (WB) has reported that around 1.6 billion people will reach working age in the next 15 years in low and middle economy countries. Creation of new jobs for this fresh group of working people while supporting the billions who are already in jobs will be a great challenge for all the sectors (WB, 2023). The overall development in the postharvest system would have a far-reaching effect for generating employment opportunities and economic development.

The important areas of postharvest technology where governments, the public sector, companies, groups of people, organisations, and individuals can venture include but are not limited to: food processing; value addition; preservation;

new product development; storage; safety and quality control; packaging; transportation and supply chains; design and development of machineries and tools; automation, instrumentation, and control; teaching, research, and development activities; marketing and logistics; and consultancy. The required postharvest technologies for establishing the above-mentioned food-processing complex are available and need to be adopted and implemented. The scope of employment generation and economic development in the postharvest sector is tremendous. Currently, the postharvest food system provides jobs mostly through individual ventures and wage employment (WB, 2023). The demand for jobs in the postharvest sector will continue to increase with the increase of consumers, per capita increases, increase in health-consciousness, food safety and quality concerns, and shifting of eating patterns. Postharvest operations require large number of human resources to achieve these goals. The job prospects in the food system through postharvest technologies can be strengthened by ensuring positive collaboration among scientists, academics, industrialists, socio-economists, lawmakers, and politicians. Inclusion of more youths and women into the ABE education system and related jobs would increase food system production efficiency, boost economic development, and enhance social life and harmony.

Rural Agricultural Work Experience and Sustainable Rural Development

Agricultural education is imperative for ensuring food security, sustainability, improving productivity, employment generation, economic growth, better social life and equity, and ecological and environmental security. The students of agriculture science and technology are well-equipped with basic concepts and advanced knowledge on agriculture and allied fields. However, students lack practical and field experience and thus it is necessary to gain practical education. Therefore, the rural agricultural work experience (RAWE) programme is introduced as a course curriculum in agricultural studies programmes. The RAWE programme is most commonly followed in Indian agricultural universities and institutes. The RAWE activities help agricultural graduates to become better equipped in the natural setting of rural conditions, interacting and working with farmers, identifying the farming community's problems, and applying the appropriate techniques and methods to solve their problems. Generally, the objectives of the RAWE programme are to: (a) understand rural life and the rural community's difficulties in agricultural activities, (b) gain experience with the socio-economic status of farmers, (c) gain practical knowledge and skill-oriented training, (d) identify the problems of farming communities, (e) solve farmers' problems by applying appropriate tools, techniques, and methods, and (f) provide agricultural students practical hands-on training for better careers in agriculture, agri-business, and allied fields (Lal et al., 2002). Similarly, in-plant training is provided to the students of agricultural engineering and food and dairy technology. The Indian Council of Agricultural Research (ICAR) and the state government encouraged the programme by providing financial support. The programme could deliver four fundamental components of agricultural education – namely, knowledge (what to do), skill (how to do it), ability (how to simplify and translate into a beneficial form), and experience (how to reduce losses and enhance efficacy) (ICAR, 2023). The students may be assigned to the adopted villages adjacent to the institutes, universities, colleges, regional research stations, and Krishi Vigyan Kendras (KVKs) under its jurisdiction. The RAWE programme could uplift both the agricultural graduates and sustainable rural development.

Consolidation of Food Security in South India: A Case Study

Agriculture is a major component of the Indian economy (more than 60%), with over 65% of the country's population living in rural communities. Although India is self-sufficient in food, rural poverty remains high (especially for rural women, who make up the majority of the agricultural workforce), and with a steady population increase, food self-sufficiency is projected to hit a breaking point by 2025. The government of India continues to give high priority to rural development and the needs of the poor, primarily through support for infrastructure and development. The agricultural universities of the country play a major role in this direction by carrying out research and developmental activities catering to local needs. To increase the availability of food in South India, the approach of the last two decades has been that of improving crop yields through cultivation practices (chemical and irrigation) and higher yielding cultivars. The scientific and rural communities have now recognised that this alone is not enough and that the control of losses as well as developing means to harness the value of excess production through value-added processing are vital to the growth and the consolidation of food security and general wealth for rural India.

The programme aimed at addressing India's development priorities in agriculture and food security. The programme involved McGill University, Montreal, Canada with Tamil Nadu Agricultural University (TNAU), Coimbatore, Tamil Nadu, the University of Agricultural Sciences, Dharwad and Raichur Campuses (UAS-D), Karnataka, and the University of Agricultural Sciences, Bangalore (UAS-B), Karnataka. It targeted the strengthening of South India's capacity for development in postharvest technology and food processing to promote rural development, employment opportunities for the rural poor, opportunities for rural women, and the consolidation of food security. The programme consisted of activities at three levels: (1) institutional strengthening through staff upgrading, capacity building, and training in postharvest and food process engineering, (2) implementation of appropriate postharvest and food-processing/preservation methods at the community level through the network of university-affiliated research stations and extension services and participating non-government organisations (NGOs), and (3) pilot projects aimed at improving access to information and technical support in adopted villages.

The Canadian International Development Agency (CIDA) funded project had originally a five-year duration from 12 July 2002 to 30 June 2007. The project was extended until

31 December 2007. The total project value (in Canadian dollars) was $7,734,484 with a CIDA contribution of $4,987,458, a Canadian partner's contribution of $2,148,353, and an Indian partners contribution of $598,673.

Institutional Strengthening

The project facilitated the training of a large number of personnel belonging to the academic, scientific, extension, and professional sectors, as well as farmers, NGO staff, community leaders, entrepreneurs, and social workers. The training provided extensive introduction to issues in postharvest technology and food processing, with a specific focus on the problems, opportunities, and limitations. Also, the project helped curriculum development and infrastructure upgrade in universities, institutes, and training centres.

Human Resource Development

Many of the staff members of the Indian universities who were trained in Canada under the project have been actively participating in strengthening further efforts in India. The activities of these personnel have been reported periodically. These staff members have designed and participated in training programmes for the farming and business communities; their active participation in outreach and agricultural extension programmes has been crucial in the continued success of the project. Overall, more than 10,000 people (40% women) have been trained in the outreach programmes that were initiated by the staff from the three universities, and over 7,000 pieces of postharvest gadgets, technological solutions, implements, and structures have been distributed in the region. The involvement of the university staff in activity related to postharvest and food processing has increased significantly (estimated as the portion of their time in a week spent on these activities as part of their responsibilities), especially in the case of some disciplines such as agricultural extension, economics, physiology, and pathology (where the staff had never worked with a specific focus on such areas). Concepts of postharvest technology, food processing, and value addition to agricultural produce have been incorporated into their regular training/teaching/research programmes. All the returning scientists actively participate in the training and orientation programmes held at the university for rural beneficiaries. Some of the staff members have prepared booklets and leaflets in English and local languages, which are used for distribution during the training programmes. More than 60 scientific papers were published by the scientists in training at McGill University during the project period. The returning trainees have also applied for and received over INR 25 million in funding from the government of India through six research projects and infrastructure development funds. Besides that, some of the staff members are operating five venture capital schemes, where training and production activities in postharvest and food processing are carried out, supported by loans from the universities.

In TNAU, the Postharvest Technology Centre (PHTC) was formed in 2004; the staff members of the PHTC, drawn from various disciplines, have all been active in the field, advising local farmers and interested entrepreneurs. The centre oversees all the outreach activities, acts as a consulting body for postharvest and food-processing issues, and manages the newly created agro-processing complex. The PHTC was established in TNAU for human resource development and helping small-scale farmers (Figure 40.1).

FIGURE 40.1 A glimpse of postharvest technology centre establishment in TNAU to help small-scale farmers (source: author's compilation).

In TNAU, Dr Mohan, an entomologist, has developed a training tool for disseminating information on grain pest removal technologies. He has worked extensively for the transfer of this technology through trainings and workshops. Over 90 technology dissemination centres in India have been trained for providing information (KVKs, colleges, research stations, and NGOs) on this product. Dr Mohan conducted a national level course attended by 25 scientists from different parts of the country. In Tamil Nadu, 124 trainings have been given to 2,793 male and 1,664 female personnel, who in turn serve other beneficiaries and self-help groups (SHGs). More than 35 institutions in India, such as universities, NGOs, research centres, and central institutes, have purchased over 545 pieces of pest traps/kits for further training use.

Dr Angappan, a pathologist, is working on seed and horticultural produce storage technology dissemination. Dr Venkatachalam was trained at McGill University in the use of RETScreen International Clean Energy Project Analysis Software (developed by Natural Resources Canada), which is an innovative and unique energy-awareness, decision-support, and capacity-building tool. Dr Venkatachalam found the experience extremely helpful for his responsibility of teaching courses in the newly started environmental engineering undergraduate programme at the Agricultural Engineering College and Research Institute (AECRI), TNAU. Dr Santhanabosu conducted extensive trials at McGill University to validate the performance of the test and measurement equipment that are used in India. Dr Sekar, an economist, is leading the monitoring and evaluation efforts by the PHTC. He also coordinates most of the survey and decision-making studies concerned with the project. A group of scientists from the department of Bioenergy and the PHTC at TNAU worked with trained field staff from the Development of Humane Action Foundation (DHAN), an NGO, to establish a solar polyhouse dryer for agricultural products in a village in the southern part of Tamil Nadu. The other members of PHTC – Dr Parvathi, Dr Malathi, Dr Bhuvaneshwari, and Dr Geetha – manage the bakery unit, food safety laboratory, and training programmes. Dr Parthiban, Dr Sivakumar, and Dr Bhuvaneshwari liaise with the farmers and entrepreneurs for various queries. Dr Thangavel now heads the PHT Scheme of ICAR, a national-level, networked project.

At UAS, Dharwad, Dr S.M. Mundinamani, an economist, plays a key role in the Karnataka Community Based Tank Management Project. Dr M.B. Chetti, a physiologist, is working on the use of modified packaging material for horticultural storage and Dr M. Palakshappa, a pathologist, is heading a project to develop certification systems in organic agriculture. Dr Ramachandra Naik, a horticulturist, developed simple, cost-effective modifications to potato and onion storage structures that have been found beneficial to the local farmers. Dr Dolli (from the Agricultural Extension) has worked with over 250 women SHG members of the region to develop agro-processing ventures for income generation. Dr Nirmala Yenagi and Dr Rama Naik from the Home Science College have been training and guiding the women SHG members in matters related to safe food production, nutrition education, and family health. At the Raichur campus, Dr Nidoni and Dr Hiregoudar have been leading the efforts to establishing effective processing techniques for local crops. At the Arabhavi campus, Dr Lingegowdaru Jagadeesh and Dr Kumara Swamy are assisting Dr Reddy in establishing training and research programmes in postharvest area.

The staff of UAS, Bangalore, who had already been trained during a previous CIDA-funded project at McGill University during the 1990s, continue to work on strengthening the postharvest-related activities in the institution. The universities have gradually been gaining the attention of industrial collaborators and private sector enterprises. The pest-control products developed by Dr Mohan at TNAU have been licensed to a private company, KSNM Marketing, who has popularised it not only in different parts of the country, but also in Ethiopia. The workshops conducted under the CIDA project strengthened the visibility of the university resources; new long-term collaborations with over 15 companies have been achieved during the past year of activities. Enquiries on postharvest, food processing, and value-addition issues received by the institutions from external sources have increased by over 600%. The universities are now well-equipped to attend to these queries and provide satisfactory services to the interested people.

Curriculum Development and Infrastructure Upgrade

Various graduate and undergraduate programmes at UAS, Bangalore and TNAU on food processing, postharvest technology, and agricultural engineering were evaluated by the visiting scientists from Canada as well as the Indian scientists visiting Canada. The syllabus for the Bachelor of Technology degree in Food Process Engineering at TNAU was revamped in consonance with the syllabus of Bioresource Engineering at McGill University. New areas, such as innovative food-packaging methods and non-thermal preservation technology were introduced. The development of the complete syllabus of a new programme in environmental engineering at TNAU was assisted by staff of the bioresource engineering department of McGill University. Similarly, concepts of postharvest and food processing were incorporated in the teaching material of various disciplines at UAS, Dharwad. The project made major contributions to the development of infrastructure at TNAU and UAS, Dharwad to assist their teaching, training, research, and extension efforts in food-processing sector.

A unique, pilot-scale agro-processing complex consisting of four units has been set up at the Coimbatore campus of TNAU. It consists of: (a) an automated processing plant (100 litres per hour capacity) with heat exchangers (plate, tubular, and scrapped surface), evaporator, spray drier, homogenisers, sterilising retort, fruit pulper, crusher, and juicer with all accessories, utilities, packaging, and storage facilities; (b) a bakery unit with rotary oven, conventional oven, mixers, and sheeters has a production capacity of 1,000 loaves of bread per day; (c) a grain-processing unit with a capacity of three tons per day has grain cleaners, milling equipment, spice grinding facilities, a pasta making machine with dryer, and packaging machines for granular and powder products; (d) a domestic-level training facility has small-scale equipment for training startup ventures.

Apart from the processing equipment, the plant also has an attached quality-assurance laboratory and a walk-in refrigerated storage chamber. This unique facility has been singled out by many international visitors as a blueprint for replication in educational and research institutions around the world. The agro-processing complex is now fully self-sustained; the operating funds and profits are realised through training fees as well as sale of products manufactured in the processing plants and bakery.

Training and production facilities to address small-scale food-processing operations (wet processing) have been set up at the Madurai campus of TNAU, the Arabhavi campus, and the Raichur campus of UAS, Dharwad. These units consist of a domestic boiler, jacketed vessels (kettles) for thermal processing, stainless steel worktables, size-reduction equipment, and pulpers. Additionally, grain-processing facilities (dry processing) have been set up at the Dharwad campus, Arabhavi campus, and Raichur campus of UAS, Dharwad. These consist of grain cleaners, flour mills, dal (pulse) mills, and extruders (pasta making equipment). The most notable feature of these infrastructure establishments is that they can not only be used by the institutions for hands-on training but also serve as production facilities (rental), as part of a business incubator. The trained beneficiaries are encouraged to use the facilities for pilot or initial commercial production activities.

Based on the establishment of the pilot plant and the project activities, the AECRI of TNAU submitted a proposal to the government of India (GoI) to fund the setting up of a food safety and quality-assurance laboratory on the Coimbatore campus of TNAU. This Rs 20 million (C$600,000) proposal has been approved by the GoI. The facility will be first of its kind in South India and is not only a direct outcome of the Tier I Project, but also will complement the project activities and enable is self-sustenance. A training centre in small-scale processing and value addition has been set up at UAS, Dharwad by the returning scientists under the project. This multidisciplinary unit is being managed by the staff of the food science department and is used in short courses, workshops, and orientation programmes conducted as part of the project (Figure 40.2).

Apart from the food-processing facilities, other postharvest-related infrastructure components were provided to the universities that are being used for training, research, and extension. These were created for specific extension and outreach activities under the project; but they continue to be used due to the popularity of the technology (and hence demand for training and adoption) and utility. Some of these are: (a) *Pucca kotis* – permanent, household grain storage structures (masonry) at the Dharwad campus and the Bijapur Research Station of UAS, Dharwad; (b) metal bins for household grain storage on various campuses of UAS, Dharwad; (c) solar tunnel dryers at the Raichur campus, Devihosur, and Matamari Research Stations of UAS, Dharwad; (d) oil extraction and dal (pulse) processing units at the Raichur campus of UAS, Dharwad; (e) heat pump dehumidifier-assisted dryer at the Coimbatore campus of TNA; (f) continuous biodiesel extraction lab-scale unit at the Coimbatore campus of TNAU; (g) pest-control gadgets at all the three universities; (h) various postharvest gadgets (harvesters, size-reduction equipment, groundnut decorticators, etc.); (i) collapsible crates (designed and manufactured in Canada); (j) potato and onion storage structures for demonstration; (k) zero-energy cooling chambers (evaporative cooling chambers) for low-cost, low-temperature storage of horticultural produce; (l) various sensors, data acquisition and control units, audio-visual aids for teaching and demonstration, computers, printers, digital storage devices, glassware, analytical lab equipment, etc., at all three universities. Furthermore, a large number of books, journals and manuals were provided

FIGURE 40.2 Dr G.S.V. Raghavan, McGill University, Canada, Dr B. Ranganna, and Dr P.A. Shankar during the inaugural session of the short course on postharvest processing and APCs for value addition in rural areas (source: author's compilation).

to the PHTC at TNAU and the library at the Raichur campus of UAS, Dharwad.

Implementation of Appropriate Postharvest Technologies

During the initial survey and analysis to determine the major postharvest losses in grains, the following features were revealed:

1. Qualitative and quantitative losses in grains during harvesting and threshing.
2. Quantitative losses during transportation.
3. Qualitative and quantitative losses during storage due to mould infestation (heat- and humidity-driven), pests, and rodents. The combined losses were as high as 30–80%.
4. Filling gunny bags (sacks) and stacking the filled bags in rooms was the common storage method. These are an easy target for rodents which not only cause damage to the stored produce but also destroy the sacks, which need to be replenished each year adding to the storage cost.
5. Use of chemicals such as malathion was common. In fact, the chemical was added to the grains before storage irrespective of the signs of pest infestation (used as an anticipatory measure).
6. Conventional storage methods used sealed, underground storage pits (called *hage*) that had become ineffective. Increased water leakage into the ground due to leaky drinking water pipes caused seepage which leads to mould infestation in grains stored in the subterranean pits. Losses from such infestation were estimated to vary between 40-80% over a normal storage period.
7. The grain stored in underground pits was either inventory for sales or home consumption. The anaerobic conditions rendered the grains unviable for seed use; hence a reliable seed storage system was lacking.
8. The importance of proper drying of grains before storage as well as cooling the grain before filling into storage structures was also not well known among the farmers. Lack of proper drying facilities, overlapping of drying period with seasonal rains and the next growing season caused severe handicap.
9. The underground storage method was not easy to use for the women and elderly. Due to anaerobic conditions, narrow opening, and the deep, dark space, recovery of grains from opened chambers were both time-consuming as well as possible with the help of only a few narrow-framed individuals.
10. With increase in population and subdivision of property, space was unavailable to establish new underground storage structures. Therefore, it is crucial to address postharvest losses with the implementation of appropriate postharvest technologies. To this end, several outreach activities, networking, and awareness programmes were conducted to reduce postharvest losses.

Outreach Activities to Reduce Postharvest Losses

With the help of the GoI's Save Grain campaign, the CIDA project introduced various grain storage technologies and postharvest management measures in selected communities of villages in Dharwad, Bagalkot, Haveri, Bijapur, and Uttara Kannada districts. Following project risk assessment (PRA), selection of receptive communities, intensive training, and demonstrations, the participants were highly receptive to the scientific grain storage methods prescribed by the project. Two types of grain storage structures were provided – a permanent masonry structure made of burnt bricks and concrete with a plastic vapour barrier sheet, called *pucca koti*, was suggested for larger farmers who had the necessary space; smaller households with lower capacity requirements were provided with metal bins shaped out of galvanised sheet metal. Training was provided to the masons as well as the tinsmith; both were assisted by the beneficiaries who provided labour input. Along with the structures, necessary training was also provided to the end users in physiology of grains, the need for proper drying and cooling, use of pest traps and indicator devices developed at TNAU, and the periodical care and maintenance of granaries.

Grain storage structures were established in Jekinakatti, Karadagi, Chikkamallur, Kengapur, Shiggaon, Kamadhenu, Sulla, Turakar Shigihalli, Rakkasagi, Tangadagi, Ittangihal, Siddapur, Kannur, Mavalli, Arbail, Hithlahalli, Magodu, Lokur, and Yadavad villages in the North Karnataka districts of Haveri, Dharwad, Bijapur, Bagalkot, and Belgaum. Over 1,200 structures were set up or fabricated, with a combined capacity of over 600 tons of food grains for household consumption and seed purposes. The beneficiaries have responded as follows: losses in grains stored in sacks due to rodents – 15–25%; losses in new grain storage facilities due to rodents – nil; losses due to mould damage (dampness) in underground structures and sacks – 15–40%; mould damage in new structures – negligible; losses due to pest damage in sacks – 5–15%; damage to grains caused in new structures – 1–3%.

The adoption of scientific grain storage has helped the farmers reduce losses during storage by almost 95%, with economic and social benefits. The new storage structures reduce drudgery for loading/unloading as well as maintenance and usage (saving 60 hours of labour per ton of grain stored). The economic benefits were estimated to contribute to a 15% increase in the income of the farmers (based on the reduction in grain losses, the elimination of the use of chemicals, and gunny bags for storage and labour). The storage facilities freed up 30% more space for the households and with better cleanliness/hygiene. The introduction of these structures had other benefits – the people were more receptive to the importance of proper drying and aeration of grains before storage, the use of inorganic chemicals has been almost completely eliminated, and the new structures are very user-friendly for the aged and women who had to otherwise depend on men to access the grains. Besides, during a storage season, the ease of removing the grains (from an outlet at the bottom of the granary) increases efficiency by over 300% compared to storage in gunny bags and by several thousand orders compared to storage in the underground chambers. The new structures are also useful for seed storage, which is a major advantage perceived by the beneficiaries.

The introduction of storage structures has been accompanied by the provision of pest traps developed at TNAU. These pest-control gadgets are simple to use, relatively inexpensive, reusable, non-invasive, and inert. Based on the behavioural aspects of the pests, an indicator device has also been developed for early detection of latent infestation. By the combined use of the two, the farmers using the grain storage structures can effectively control the pest damage. Reduction in grain loss due to pest infestation is at least 50% – in case of seed storage, the scientific storage coupled with use of pest traps reduces losses by over 75%. Due to the grain storage structures and associated training, better postharvest management is adopted, especially proper drying of grains. Overall, it has reduced the losses by over 90%. In Tamil Nadu, the combined use of drying and storage bins has shown a 60% reduction in losses during the storage of paddy. Farmers have understood the principle and devised their own cheaper designs (using discarded plastic and reusable bottles) with varying success.

In case of horticultural products, the introduction of better harvesting techniques, proper primary grading to remove diseased and damaged produce from the heaps, storing under shade after harvest to remove field heat, proper packing and transportation, and the use of crates and careful handling were helpful in reducing the losses. In the case of potatoes, a simple modification to the traditional storage structures resulted in major benefits for the farmers. For potatoes, the losses are up to 70% during storage over the three months following harvest. Farmers heap the produce on fields in covered temporary structures, leading to the generation of heat and humidity in the core region. The resulting microbiological and fungal infestation can damage up to 80% of the crop if unattended. The farmers, when they can afford to, employ labourers to repeatedly sort the entire stored produce to remove the infested material. However, this solution is neither cost-effective nor efficient. Dr Naik, a horticulturist from UAS, Dharwad, who had been trained at McGill University under the project, introduced a simple solution in the form a network of perforated plastic pipes buried in the heap to provide aeration by natural convection. This simple, zero-energy cooling method was found to be very effective. The prices at the time of storage soon after the harvest are Rs 3,000–5,300 (C$80–132) per ton. This rises to Rs 12,000 (C$300) per ton after two to three months. The losses during storage are between 50 and 70%. The farmers can expect to gain only 12–20% profit by traditional methods. With the utilisation of perforated pipes, the losses are reduced to five to ten percent (the reduction in total losses is over 75%). Combined with the increase in market prices, this practice can potentially increase profits by 80–220%. The field demonstrations in potato storage and onion storage were extremely effective and found widespread adoption. Following the demonstration and the subsequent coverage in mass media, the government of Karnataka came out with a subsidy scheme to encourage farmers to adopt these storage solutions and sought the guidance of Dr Naik, who was nominated as the chief resource person for the scheme.

Chili drying in Ramanathapuram: Ramanathapuram district in the southern part of Tamil Nadu is a drought-prone district with a tropical climate, where agriculture is dependent on rain. Historically, the people of this region have faced complete crop failure three years each decade and managed to harvest only 50% of the potential yield during each four-year period. Hence, the years during which they are fortunate enough to get a good crop are important for their sustenance. The communities are made up of small and marginal farmers growing paddy and chili for their livelihood. Small amounts of millet, pulses, and vegetables are also grown for subsistence. Chili is grown in about 16,500 ha of land, producing about 13,000 metric tons in a good year. The produce is dried on streets and front yards under the sun and sold. There are no infrastructure facilities for processing. The conventional methods result in discoloration of the product, leading to low prices (almost half the potential price). Severe losses in quality as well as losses of produce (10% due to handling; 40–80% due to rains) occur during the drying process due to discoloration, breakage during handling, and mould damage.

The project worked with the farmers of Singerayapuram village under the direction of the DHAN Foundation. 80% of the village is under chili cultivation and it is also the centre for outreach among the five similar villages neighbouring it. The advantages of the dryer are quicker drying (higher temperature reduces the drying time by more than half), full colour retention and markedly improved appearance, the drying process being uninterrupted by rains or cloudy weather, reduced losses during handling, and better quality because of no contamination or dust. As the drying can be accomplished unattended, the time made available is used to earn wages from agricultural labour (Rs 240 to 300 per batch dried, or C$6 to $7.50 per batch). The overall losses were reduced by 95% and the market price for the commodity increased by Rs 2 per kilogramme (5%) due to quality retention. The solar tunnel dryer has been well publicised in mass media. It has attracted interest from different parts of the country, especially from chili traders who export the produce. A second unit was installed with 50% contribution from the farmers (TNAU and the CIDA project each contributed 25% to cover the rest of the cost) after the success of the first demo dryer. The dryer is also used for drying neem seeds.

Networking and Awareness Programmes

A strong network of trained and experienced personnel has been created. The development of the network has been fostered by various collaborative activities, frequent interaction on common forums and close communication encouraged by the project. A large number of activities were carried out in both states due to excellent collaboration with other agencies and organisations that were working in developmental activities. Through such collaboration, the university staff were able to gain access to a larger audience who were potentially receptive to the messages and attempts of the project staff. Simultaneously, they were able to equip the other partners to deal with new challenges, consider novel developmental foci, and explore avenues for development based on food security issues.

In Karnataka, UAS-D and UAS-B were approached by Swa-Shakthi and Stree-Shakthi, two women's welfare organisations supported by the governments of India and Karnataka, respectively. These organisations had worked for the past few years, mobilising women to form SHGs throughout the state of Karnataka and assisting them to acquire different occupational skills, micro-credit, and marketing networks. The SHGs had been networked to form local clusters and state-level federations. The two organisations, in an effort to create a pool of resource personnel in their ranks, joined up with UAS-D and UAS-B to acquire training in food processing and value addition in agriculture. Small-scale processing operations, safety and quality aspects in the food industry, hands-on training in industrial operations, etc., were the areas of interaction. Though the World Bank-funded Stree-Shakti programme was discontinued due to the termination of the donor support, the beginnings made with this collaboration sustained and are continuing in three districts of southern Karnataka (Figure 40.3).

In UAS, Dharwad, the project was able to gain a strong foothold in Uttara Kannada district (Yellapur Taluk), due to the rapport with the directorate of agriculture from the government of Karnataka. The agricultural officer for the district (who is also an alumnus of UAS, Dharwad), took an active interest in strengthening the postharvest activities in the region by facilitating the project personnel to reach a larger audience. Upon training and adoption of the income generation activities, she also assisted the SHGs to obtain financial and material support from the Women and Youth Training and Entrepreneurship Programme of the government of Karnataka.

One of the first activities under the project at UAS, Dharwad involved creating awareness about the project and postharvest issues among the local journalists and reports for mass media. This activity had an excellent ripple effect, drawing a large number of NGOs towards the project personnel. Based on several meetings, orientation programmes, and discussions, selected NGOs were invited to nominate their field staff to attend workshops for basic orientation in postharvest, food processing, and value-addition activities. This led to the development of a close working relationship between four local NGOs – Sustainable Agriculture Development House and Nature Awareness centre (SADHANA) from Yellapur, Adarsha Society for Environment, Agriculture and Rural Development (ASEARD), Bijapur, Karnataka Integrated Development Society (KIDS), Bijapur, and Rehabilitation Assistance for People in Distress (RAPID). Minor collaborations were achieved with Yeshaswini Mahila Mandal, Dharwad for the training of farmers from villages close to the Dharwad suburbs.

SADHANA centre is operating in Yellapur district, where 45 female participants from 16 villages have been actively working to disseminate knowledge. The villages are: Hosalli, Idgundi, Hemmadi, Hitlakargadde, Gerala, Ummachagi, Arbail, Joida, Potoli, Ramapur, Yellapur, Mavalli, Kattige, Shistamudi, Magodu, and Balgimane. KIDS has worked in Rakkasagi, Tangadagi, and Baladinni villages of Muddebihal Taluk, Bijapur district. ASEARD has worked in the following villages of Bijapur district: Ittangihal, Logaon, Siddapur, and Dandaragi. RAPID operates in Dharwad, Hubli and the surrounding peri-urban communities. All the organisations assisted the project personnel in identification of potential target audience, initial contact meetings, PRA for problem identification, feedback, monitoring and evaluation, and facilitation of training programmes, workshops, and demonstrations. Each organisation got trained in specific areas suited to their strengths and focus – SADHANA centre field staff worked on quality improvement issues, packing and labelling, marketing and publicity for the products, grain storage and processing, and small-scale value addition. A federation of SHGs under the name of Sanjeevini Mahila Udyam was formed to assist

FIGURE 40.3 The World Bank team visiting the postharvest technology (PHT) scheme (source: author's compilation).

the marketing of conventional home-made products from the women members and the centre facilitates participation of the women in markets, fairs, and bazaars organised by local banks and other developmental agencies. KIDS and ASEARD organised trainings and awareness programmes focused on surveying indigenous storage practices, postharvest management, control of storage pests, harvesting practices, losses, and training in agricultural produce marketing commission market yards for the staff and vegetable vendors, construction of *pucca kotis*, and the use of metal bins for grain storage, etc. RAPID is providing marketing-oriented training to the SHG members and organises events that provide opportunities for sales of products.

The most significant partnership achieved by UAS, Dharwad was with the officials of the Save Grain Campaign, a unit of the GoI's Ministry of Food and Civil Supplies. The technical officers from the Bangalore branch worked actively with the UAS, Dharwad project staff to promote scientific grain storage in the northern Karnataka region. This extremely useful linkage provided tremendous benefits by boosting the performance of the government body while bringing the benefits of their expertise to thousands of farmers with the help of the CIDA project.

At the Raichur campus of UAS, Dharwad, the participating staff from the College of Agricultural Engineering collaborated with a local NGO, the Society for Women, Agriculture, and Rural Development (SWARD), in a series of contact programmes with the people from the village of Sagamkunta. These farmers were participating in a programme called Eco Net Partnership (ENP) and the CIDA project provided the organised farmer groups training in postharvest management.

TNAU actively worked with SHGs under the Kovai Child Labour Abolition Support Society (Kovai CLASS), a GoI programme. During an initial contact meeting during September 2003, Dr V.V. Sreenarayanan, the liaison person at TNAU for the project, had described the objectives of the CIDA Tier I Project and the possibilities for income generation by the SHGs through various food-processing activities by explaining the scope for developing various food products, like ready-to-serve products, such as jam, sauce, jelly, etc. Members of the SHGs, mostly women, are generating income (these families had children earning for them; alternate income generation activities are required in order to encourage the families to give up child labour and provide them education) by running tiffin stalls and tea stalls, and also by selling idli batter, sambar powder, milk, and milk products, etc. It was agreed that these groups require training on the preparation of jam, squash, pickles, noodles, etc. Accordingly, a training programme was designed and offered to the SHG members during the second week of every month at the Coimbatore campus of TNAU. The SHG members were identified and mobilised by the Kovai CLASS, whose input was be considered for such decisions. Similarly, a monthly training programme with the local SHGs was carried out at the Madurai campus of TNAU during the third week of each month. The project staff received good support from the district collectorates (administrative bodies) for these activities.

A strong working relationship was established between TNAU and the DHAN Foundation, a major NGO operating in South India. The DHAN Foundation works with about 150,000 farmers and 600,000 women in the four South Indian states. Over 100 alumni of TNAU are working for the DHAN Foundation. The collaboration used to be focused on issues related to water and soil management. The project has extended the areas of collaboration and has resulted in a memorandum of understanding (MoU) with a larger number of activities in the postharvest and food-processing areas. One specific technology was identified and disseminated successfully in southern Tamil Nadu during this project.

TNAU has been assisting the farmers trained by another NGO called Mysore Resettlement and Development Agency, operating in the Nilgiris district of Tamil Nadu. The NGO has successfully educated the farmers in high plateaus to conserve biodiversity, reduce soil erosion, and eliminate the indiscriminate use of inorganic chemicals for the production of fruits and vegetables. PHTC of TNAU has guided the farmers to adopt better postharvest management practices that contribute to value addition.

Apart from these long-term collaborations, new partnerships were forged as the opportunities were presented. The agribusiness incubator International Crops Research Institute for the Semi-Arid Tropics was introduced to the project activities during a workshop in Coimbatore in mid-2007. The Agri-Business Incubator (ABI) approached PHTC of TNAU and UAS Dharwad for joint working ventures. Taking up a three-month study of markets and business practices around Dharwad, the team from ABI provided useful input for developing better marketing opportunities by the SHGs working with UAS, Dharwad. It has also proposed to enter into an MoU with the PHTC to complement the technical strengths of TNAU with their skills in business management and marketing. The ABI trained the PHTC staff in incubator management issues in late-2007.

Apart from developing working relationships with other agencies, the linking between the participant universities has also been fostered by the project. The network between UAS, Dharwad and TNAU has been strengthened by the flow of information and technology for adoption in northern Karnataka. The biomass stoves and gasifiers developed at the department of Bioenergy, AECRI, TNAU, had been introduced to the farmers from Hithalahalli, Yellapura Taluk in northern Karnataka. The areca nut processors from the region identified savings in wood fuel for their processing operations by using such stoves. Technicians from TNAU visited the village and have trained the farmers and SHG members of Hithalahalli in the construction of biomass stoves and gasifiers, as well as their installation and operation. Dr Mohan, the entomologist from TNAU, made frequent visits to Hassan (part of UAS, Bangalore), Dharwad and Raichur (UAS, Dharwad) to educate the farmers of the region on the use of gadgets developed by him for the control of pests in food grains during storage. Dr Mohan was also invited by the universities to carry out training in the technology in the neighbouring states of Kerala and Andhra Pradesh in South India. The various workshops, meetings, and field visits organised under the project, as well as living together for short periods during their training at McGill University, provided ample opportunity for the participants to develop professional and personal relationships. These linkages have endured and are being put to good use through constant exchange of information and services.

Concept of KVKs

KVKs are an integral part of the Indian national agriculture research system. KVKs are the centres for research and agriculture extensions formed by ICAR. The centres are fully financed by the GoI and sanctioned to institutes, universities, colleges, research centres, and NGOs working in agriculture. Typically, KVKs are associated with a local agricultural university and oversee the activities. The mandates of KVKs are technology assessment and demonstration, technology applications, and capacity development. The activities to be envisaged in order to effectively implement the mandates are (a) on-farm testing of technologies and methods to evaluate the location specificity under different farming systems; (b) capacity development of human resources, including extension personnel and farmers, with modern technologies and skills; (c) demonstration of technologies and processes to increase the prospects of production in the fields; and (d) serving as a resource and knowledge centre to support and assist enterprises of private, voluntary, and public sectors (KVK, 2023). The KVKs play a vital role in boosting agricultural economy by serving farmers and the community with the extension activities.

The project team of the consolidation of food security in South India has been actively involved in community outreach activities, such as planned field demonstration, village programmes, workshops, and technical sessions for the appropriate transfer of knowhow in postharvest handling and basic food-processing techniques. All Indian partners have been actively working on implementing various training programmes to cater to local needs. Training activities continue to be scheduled periodically in various adopted villages, in KVKs, and in collaboration with NGOs and numerous self-help women groups. The local NGOs have greatly helped in facilitating technology adoption and the popularisation of the training programmes. There has been active participation of SHGs in Krishi Mela activities for the promotion of their processed food production for income generation.

Pilot Projects in Adopted Villages

The major focus of the project was on tackling the challenges on the ground and getting the resources to benefit the people in various parts of the region. The universities identified and selected the communities among which the outreach activities were to be carried out. Various outreach activities jointly conducted by the universities and the NGOs are continuing even after the completion of the project. Some have resulted in noticeable benefits, while some are showing promises of success.

Development of Food-Processing Ventures by Women

To begin with, the dissemination of information about the project and its objectives, as well as the concepts of the postharvest situation, management, technologies, food processing, value addition, income-generating opportunities, etc., was given priority. In order to get wider publicity, some programmes were planned for the university staff and the local media. Following the initial planning, orientation programmes and contact sessions were organised at TNAU and UAS-D to link up with the interested and relevant NGOs. Based on these interactions and the subsequent correspondence, different NGOs were identified for collaboration, and training programmes were carried out for the staff of those NGOs. The trained NGOs identified the stakeholders in their areas of operation and organised training programmes for such potential beneficiaries, either independently or with the assistance of one of the universities. The effects of such trainings were frequently monitored and, based on the feedback and the intensity of response from the participants, the next level of intervention was planned and carried out. In many cases, the trainees having acquired sufficient skills and confidence, organised themselves, and come up with proposals to start their own food-processing ventures. Such proposals were evaluated by the project personnel and the NGOs and, with suitable guidance, when necessary, the applicants were assisted with their efforts.

The implementation of postharvest technologies and outreach activities by women's groups successfully worked to provide avenues of income generation by utilising the local resources such white pepper (Figure 40.4). Members of SHGs from different parts of the two states have been involved in this exercise along with the institutional partners in the project. To date, several small agro-processing centres (APCs) have been

FIGURE 40.4 Prototype white pepper making machine (source: author's compilation).

set up or strengthened in the following places as a direct result of the project activities.

Kuttinagere, Magadi Taluk and Shettihalli and Mutthanallur, Anekal Taluk, Karnataka

UAS, Bangalore has been pursuing the policy of enabling beneficiaries from farming community to set up APCs in their communities following the Bangalore UAS/McGill Post Harvest Engineering BAMPHE project (1990–1996). Mostly funded by ICAR and partially by the beneficiaries themselves, the university scheme is providing technical assistance and training in different aspects of maintenance and management to the participants, with the help of the CIDA Project. Many such APCs have been set up in the vicinity of Bangalore, Karnataka, of which four are completely run and managed by self-help groups of women. Three APCs associated with UAS, Bangalore have received support in the form of materials and training from the CIDA Tier I Project. These are located at Shettihalli and Muttanallur in Anekal Taluk (Figure 40.5) and Kuttinagere in Magadi Taluk.

The SHGs in Anekal Taluk were inspired by Sumangali Sevashrama, an NGO associated with this CIDA Tier I Project. These SHGs have enabled women to obtain training in different commercial activities such as tailoring and animal husbandry; the groups were introduced to the concept of APC and interest is growing.

Ragi (finger millet) is the main crop in the region and is also the staple food of the people involved in agricultural production. The demand for ragi flour is significant, but milling facilities are scarce. People from the villages scattered in the region have to travel up to 15 km to get grain milled. The people of Shettihalli and Muttanallur identified this opportunity and organised to start an APC in October 2004. A three-day training programme on value addition to finger millet, maize-based value-added products, and indigenous milk products was organised for the benefit of SHGs of Shettihalli women during October 2004. Around 54 women participated in this training programme. An APC equipped with ragi-cleaning and milling machines was started subsequently. The centre satisfies the requirements of four SHGs in the village besides catering to the processing needs of a half-dozen neighbouring villages in the vicinity. A small popcorn-making unit has also been set up at Muttanallur, where the SHG members operate a petty shop in the village as part of their activities.

The APC at Kuttinagere evolved out of a successful intervention in the watershed management activities of UAS, Bangalore in the region. The area mainly consisted of marginal lands that were reclaimed by different land and water management programmes conducted by the university. As agricultural production increased, the farmers were encouraged to carry out value-addition activities to further enhance the benefits. More than 200 tons of ragi is produced in the village, but most of the production is sold as unprocessed grain at a low cost. The opportunity to benefit from processing was identified and the 16 APC members were trained in ragi cleaning, milling, and the preparation of value-added products such as beverages, packaged ragi flour, and local delicacies made from ragi. A flour mill, a ragi-cleaning machine, and a mini mill were provided by the university, supplemented by stainless steel equipment required for production provided by the CIDA Tier I Project. The taluk of Magadi produces about 2,500 tons of ragi and there is tremendous scope for ragi-processing ventures to benefit. The demand for ragi-based products has also

FIGURE 40.5 Dr G.S.V. Raghavan, McGill University, Canada, Sri Ramalinga Reddy, Hon'ble Minister, Government of Karnataka, and Mr Narayanaswamy, Member Legislative Assembly, during the inaugural function of APC at Muthanallur in Anekal Taluk (source: author's compilation).

been on the rise in the state and in Bangalore, a major market is easily accessible to the APC. These factors have been repeatedly brought to the attention of the SHG members, who are being encouraged to shift more towards the production of value-added products rather than mere processing services.

Lokur, Near Dharwad, Karnataka

The village of Lokur was selected for intense intervention and monitoring activity under the CIDA Tier I Project. The University of Agricultural Sciences, Dharwad had made initial forays into the village with a programme on integrated pest management and a good rapport had been established with the people of the village. Based on this trust and past success, food security-related programmes were started, and these efforts are bearing results gradually.

Initial orientation meetings and discussions between the village people, scientists from UAS, Dharwad, as well as Canadian visitors inspired the people of the village to form a postharvest committee comprising of 100 members (including five women). This committee selected a panel of 12 members to interact with the CIDA Tier I Project team to identify and resolve postharvest-related problems of the village as well as create food processing related income-generating ventures. The members created a corpus fund that will be used for the development of the committee and its activities. As a direct result of the training programmes, two SHGs in the village showed interest in starting a food-processing venture. Members of one group, Akkamahadevi Stree-Shakthi Sangha, attended a few training programmes conducted by the university and developed plans to establish a vermicelli-making unit. During the second part of 2004, they rented a building in the village and arranged for electrical connection. In February 2005, the group purchased a vermicelli-making machine with help from the CIDA Tier I Project and started operation. The group members prepare vermicelli using raw materials purchased from the Dharwad market and sell the finished product in the same market. They also offer the processing services to local villagers, who bring their own raw material. Members have expressed immense satisfaction with their new venture. They are confident about its continued success and are proficient with the technical and production aspects.

Gullapura, Hithalahalli and Nandolli, Yellapura Taluk, Uttara Kannada, Karnataka

Yellapur Taluk, located about 60 km west of Dharwad, consists of a hilly, forested region with widely dispersed habitations of small to medium size. The undulating geographical profile, extensive forest land, as well as limited road facilities make it a challenging environment to carry out commercial activity. People of the region have to travel to distant markets to obtain household necessities and avail of certain services.

In this district, a collaboration between the directorate of agriculture of the government of Karnataka and the University of Agricultural Sciences, Dharwad has been showing promising results. The agricultural officer (AO) of the district, who was acquainted with the CIDA project and the concerned university staff in this project, launched efforts to mobilise and motivate people from two villages – Gullapura and Hithalahalli. The AO guided the groups to obtain support from the state government funds under a scheme. With the technical guidance of UAS, Dharwad under the CIDA project, the members purchased and installed a vermicelli-making machine, a small flour mill, and a chili-pounding machine. The members were trained in production as well as maintenance of the equipment. The units went into operation in July 2004, with products for sale in nearby markets as well as providing flour milling and vermicelli-making services to the local people who brought their own raw materials. The groups have been successful in overcoming many problems they faced initially and have been confidently running the operations with a frequent display of ingenuity and courage. The local village government has provided a good building to the SHGs for this venture. Overall, the women who have been involved in the APC have found the experience quite stimulating. The increased awareness of the income-generating opportunities as well as the interaction with different government and non-governmental bodies have brought a sense of involvement among the women in the local economic and sociopolitical environment.

At Hithalahalli, after tasting success in food processing and packaging activities, the SHG members have ventured into postharvest processing of paddy, a major local crop (Figure 40.6). This novel venture, where threshing is offered as a service by a self-help group, is expected to serve as an example for the SHGs of South India to take up similar activities for their betterment. Not only was the venture found to be economically profitable, but the availability of a mechanised threshing solution in the region was identified as a major benefit. Due to the lack of such an option (and aided by availability of cheap supply from the government-sponsored public distribution system [PDS]), the farmers of the region had gradually reduced the paddy production activity, even though rice is the staple food in the region. But the introduction of thresher has rejuvenated the paddy growers to take up the activity again. The local cultivation and processing have many advantages – availability of a high-quality, local strain of rice (that is preferred over the PDS supply) and byproducts such as straw.

The intangible benefit that is valued higher than all these other benefits is the return to a preferred agricultural lifestyle involving self-sufficient food production that was vanishing rapidly. This has been widely expressed by the beneficiaries during the feedback surveys and contact sessions. The Hithalahalli SHG also initiated efforts to organise all SHGs of Yellapur Taluk to form a federation that would assist them in creating marketing opportunities. Due to this initiative, a group of women from a nearby village, Nandolli, have started an APC. This APC will be involved in milling flour and grinding pulses and spices. The establishment of such processing units has been found to have a major social benefit in this region, as told to us by the local villagers. The interesting social concept here is that the women of the households control the finances. Normally, the women assign the task of flour milling to the men, by providing them the grains as well as the cash to take the grains to the nearest urban centre where the milling facility is located. In this arrangement, some of the men would sell off part of the grain and use it (as well as the cash given for the milling fee) for the consumption of alcohol. Thus, the household suffered not only the loss of food but also

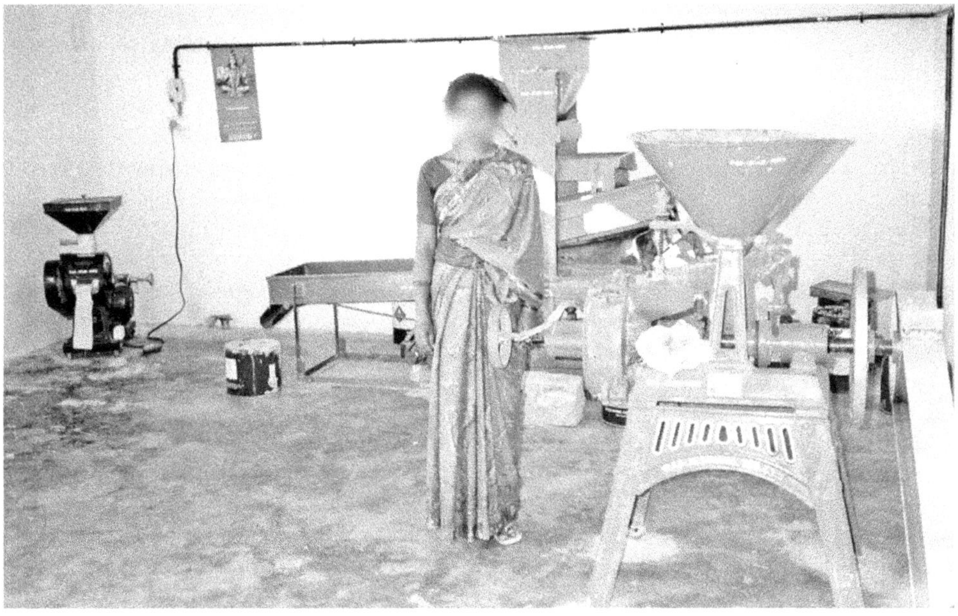

FIGURE 40.6 An APC managed by one of the members of the self-help groups of women (source: author's compilation).

from the contingency that sustained alcoholism. Otherwise, the men would not have access to the cash to indulge in alcohol consumption in their own village. This major social impact was narrated during a monitoring and evaluation event.

Jekinakatti Village, Shiggaon Taluk Near Dharwad, Karnataka

The Home Science College staff of UAS, Dharwad conducted a nutritional education programme in Jekinakatti village for the promotion of traditional processed and value-added foods to improve rural women's economic, social, and health status. The programme involved the use of lectures and demonstrations to encourage the use of locally available nutritious crops that are underutilised, the development of new recipes to enhance marketability, fortification, ease of large-scale production with increased shelf life, as well as adaptability to local resources. The response was tremendous, and the women took up preparing foods incorporating the little millet (*samai*) that is grown locally and is well-suited to the local agro-climatic conditions. The SHGs of the village now prepare foods from the millet for sale within as well as outside the community. This adoption was based purely on the nutritional merits of a local crop that was hitherto ignored or unknown by the people. The local health department conducted tests and found that there was significant increase in the health status of adolescent girls who were receiving these new recipes in their daily diet.

One of the major and simple improvements in the operation of the APCs in northern Karnataka was achieved by training the members in proper packaging of the products. Extensive training had been provided to the members in safe production, but the efforts were not getting proportional returns during marketing. It was observed that the poor packaging negated any value addition and reduced the acceptability of the product. A series of training activities focusing only on proper packaging and labelling resulted in a remarkable improvement in the appearance and appeal of the products, as observed at a university fair where the products were exhibited for sale. The members are now aware of the importance of brand marketing and are working towards having a common identity for their products.

The project staff at the university further assisted the APC operations by providing guidance in marketing and business management. Arrangements were made to provide visibility for the products on the UAS campus as well as in some public places in the city of Dharwad. Permanent display and sales counters were set up on campus for selling the products to university staff, students, and visitors. The sales points were conveniently located in public canteens, cafeteria, and student hostels. The new marketing opportunities created have been yielding a steady income of Rs 600 (C$15) per month for each of the 11 participating groups.

In the eco net partnership programme managed by SWARD at the Raichur campus of UAS, farmers adopted eco-friendly agricultural practices The exercise resulted in the sensitisation of the people to agricultural value-addition opportunities that would strengthen their ecologically responsible approach and make the venture profitable. After many subsequent meetings between the university and NGO staff and the people of the village, plans were drawn up to establish an APC at the university campus that would be accessible to the interested entrepreneurs and users from the surrounding villages. The APC will provide facilities for neem seed oil extraction, drying and dehydration of fruits, vegetables, and spices, as well as a small-scale fruit processing unit.

Pallapalayam, Madhukarai, Velandipalayam and Narasimmanaickenpalayam, Near Coimbatore, Tamil Nadu

As previously mentioned, some members of SHGs collaborating with TNAU under the Kovai CLASS programme required training for alternative income generation activities to avoid relying on child labour to generate income for their families.

In December 2004, 20 SHG members associated with MARIYALAYA (a rehabilitation home for street and working girls) and the Non-conventional Energy and Rural Development (NERD) society approached the PHTC to provide training in income-generating activities. The NERD society group members were already carrying out processing activities at the household level. After acquainting themselves with the latest trends in processing activities at PHTC, TNAU, Coimbatore, the members planned on starting a processing centre on a cooperative basis with a few low-cost machines in the village itself. They asked PHTC for technical guidance and the same was extended to them.

Subsequently, APCs have been set up for three other SHGs operating near Coimbatore and three near Madurai campus of TNAU. These APCs provide an income of Rs 200 per week (C$5.00) to each of the participating members as profits from the venture. In addition, the members contributing to the production labour receive Rs 50 per day (C$1.25) of work and 2% commission for sales. The APCs have been instrumental in providing tremendous economic, social, and emotional support to people with special needs, for example, those with physical disabilities, and single and elderly women without family support. The group activities and social interaction have numerous benefits for these participants, which cannot be measured. By focusing its efforts on selected communities, the project has been successful in creating viable examples that are replicable. The participants in the project have acted as agents of extension. The NGOs and university staff continue to arrange for visits wherein newer participants can adopt similar developmental strategies by interacting with the beneficiaries who are living examples of the development model.

Narrative Summary of Activities

The project addressed some key issues in the developmental efforts of the South Indian regions of Karnataka and Tamil Nadu, principally with a focus on the postharvest and food-processing sectors. The GoI has identified food security as a main challenge to be addressed during the 11th Five-Year Plan. As policy measures, significant amounts of funding have been allocated in the budget for stimulating activities in the food-processing sector. However, the country needs a good human resource base and viable models that can be put into effect to achieve the targets set by the policy objectives. This Tier I project has resulted in the creation of many viable examples that can be replicated successfully on a larger scale by local governments.

The vast amount of experience gained by the training under the project, as well as the establishment of strong ties with the people of the region in these specific areas coupled with the necessary infrastructure facilities, have enabled the universities to establish themselves as leaders in postharvest and food processing. The international experience gained by the Indian staff during training in Canada has already shown remarkable impact through higher acceptance of the proposals and grant applications by the staff members from the three partner universities, the University of Agricultural Sciences, Bangalore; the University of Agricultural Sciences, Dharwad; and TNAU.

Improved grain-storage methods, drying of grains and cereals, pest control with reduced use of chemicals, use of better threshing methods, improved harvesting aids, dehydration methods, seed processing, novel potato and onion storage techniques, the use of crates for the transportation of material, and better grading, sorting, washing, and cleaning of vegetables are some of the practices that were developed and/or popularised under the project in the postharvest management area. In food processing, standardised production methods at small-scale operations with an emphasis on food safety and quality is the major intervention. Training in skills for the use and maintenance of machines, better packaging methods, new recipes from lesser-used grains, and fortification of traditional foods with micronutrient-rich local produce are other practices.

Creation of awareness about postharvest management, its importance, the extent of food losses as well as the means of reducing them, food-processing activities, income-generation ventures in postharvest management and food processing, better grain storage, pest control in stored grains, the introduction of improved gadgets for agricultural processing, nutrition education, and the sensitisation of mass media personnel to postharvest issues are some of the major extension activities that were carried out under the project.

The farmers and rural women involved in SHGs have expressed that they have acquired following new information due to the project outreach activities: (a) The existence of avoidable postharvest losses in grains, fruits, and vegetables, the extent of such losses, as well as the means of controlling and reducing them. (b) Information on grain-storage technologies as well as managing pests in stored grains. (c) Information on postharvest management and technology, processes, and available products in the market. (d) Newer food-processing options for locally available crops, knowledge about the relevant technologies, machinery, and their operation and maintenance. (e) Knowledge of appropriate food-packaging technology, and its utility as well as application. (f) Quality assurance and safety in food processing. (g) Entrepreneurial education related to small-scale food-processing ventures, accounting and bookkeeping skills, marketing strategies for home-made traditional products, as well as knowledge about financial institutions and their products that assist the development of small-scale industrial ventures.

The following are the main research and development activities: (a) Measurement and assessment of postharvest losses in major food grains, fruits, and vegetables in the region. (b) Development of simple gadgets/techniques for pest control and management in food grains. (c) Study on the utility of collapsible crates (developed in Canada) for Indian farming communities. (d) Nutritional status of farm women and their families, benefits of using locally grown minor millets in diets for such families. (e) Effective processing methods and their effect on the quality of Indian herbs, fruits, and vegetables. (f) Processing machinery and equipment for the reduction of drudgery among women involved in agri-food processing. (g) New recipes for the incorporation of underutilised agri-food products in common dietary preparations. (h) Development of food products with health benefits.

The project has brought about many changes in the communities involved. There has been a general improvement in the quality of life, as perceived by the participants themselves. Interaction with the stakeholders during the final days of the project highlighted the following: (a) The knowledge of available opportunities and the necessary skills to utilise them has provided hope and confidence among the people, especially women. This development is crucial in the lives of women who are either single mothers or without children. The project has provided them with sources of financial security, self-confidence, motivation, and, most importantly, self-identity and inclusiveness in the society. (b) Increased disposable income generated from new ventures has the potential to assist the development of families. It is used for educational expenses of children, better health care, and food supplies. (c) Due to the increased income and other resultant benefits, the strong role that women can play in the households has gained more acceptability and respect. Men are increasingly supporting the initiatives that women in the family are proposing. This brings huge human resources potential into the development that was hitherto operating in the background. (d) A strategy that combined income-generation activities with the utilisation of local resources also promises multiple benefits. Food-processing activities based on locally grown crops that were diminishing in importance in local diets has not only provided small-scale business opportunities but also makes available the beneficial qualities of these crops to the consumers. Local crops such as minor millets (rich in minerals and other micronutrients), sorghum, and dicoccum wheat (containing sugars with low glycemic index), which are well adapted to the local agro-climatic region, are receiving higher prices in the market due to this development, assisting the small and marginal farmers who cannot afford intensive agriculture to grow cash crops. Ready-to-eat processed foods made from these crops have been shown to be beneficial to consumers with special physiological needs, such as adolescents, young children, nursing mothers, sufferers of diabetes and osteoporosis, etc. Adoption of such foods in the local diets is assisting these target groups in the villages. (e) Adoption of grain storage technologies is assisting the reduction of grain losses as well as reducing costs of storage. The new storage structures are easy to use for women and the elderly in the villages. Improved harvesting and processing tools have reduced drudgery among the women and elderly involved in agricultural labour. (f) Better methods of pest control have reduced the cost of storage as well as reduced the use of chemicals in grains. (g) The food-processing ventures have provided off-season employment for agricultural labourers. This is especially helpful for landless labourers who have no means of income during the off-season.

New employment opportunities have been created in the sectors of food processing, agricultural processing, grain storage, and handling. In food processing, by incorporating improved production and packaging methods and imparting marketing and business skills, women have been able to gain additional income. By collective action, they have been able to increase the volume of operations with the help of machines. In agricultural processing, opportunities available in better harvesting, threshing, seed preparation, and drying have been shown to have the potential for more employment. Training for building better grain storage structures has resulted in employment for a large number of rural young men who are taking up this work independently.

Establishment of food-processing training facilities has also created the potential to develop entrepreneurs. Unemployed village youth have been provided an opportunity to assist the women in the marketing and sales of the products. By increasing the average income among women SHG members by over 15% as well as creating new opportunities for income for some unemployed youth in the targeted communities, the project has assisted the participants in reducing monetary loans to pay for the education of children as well as unforeseen medical expenses. Overall, almost every participant has continued to earn additional income that has contributed to a reduction in poverty in the targeted communities.

Through enhanced income from off-season employment as well as income security, the economic status has shown improvement. The introduction of better implements and gadgets for operation (e.g., groundnut decorticators, better sickles for harvesting) has reduced the drudgery for women labourers. Education and training in nutrition has stimulated better food habits that have potential to enhance health status. Women have been empowered through economic strengthening – either increasing their earning capacity or creating new opportunities for income generation. In this process, they have also overcome barriers to participation in economic development by newly gained confidence and recognition. It has made small but noticeable socio-cultural changes at the individual, family, and community levels. The women who have benefited from the project have built autonomous and self-reliant organisations to sustain their roles and participation in societal development.

Key Project Beneficiaries and Conclusion

The Indian Ministry of Agriculture (Central Government), which created in recent years a Department of Food Processing, as well as agricultural universities, have identified the importance of postharvest and food-processing practices in establishing and maintaining the stability of food security. This awakening was manifested in some of the developmental schemes launched by the government, which studied the logistics of different postharvest operations. It has now been clearly accepted that progress in this sector of the economy requires intervention at all levels of human resources involved. This can be achieved synergistically through capacity building in postharvest and food process engineering, as well as grassroots training and transfer of appropriate technologies to the lower-income sectors. The educational component at the universities should incorporate necessary improvements in the curriculum towards this purpose. Hence, agricultural universities in Karnataka and Tamil Nadu will coordinate their efforts to develop expertise in and promote the development of agriculture and food to society at large. Strategies will further be developed to minimise losses at the field level, during the harvest and postharvest stages, value-addition and crop processing, food processing and handling of crop residues, wastes, and byproducts. Consolidation of food security will pass by meaningful and deliberate interaction between science and society to create useful technology. Creation of trained human resources that can bring about the interactions is vital to integrating the academic aspects with a multi-dimensional technological approach, leading to self-reliance and long-lasting impacts.

The major programme of this project included activities to strengthen agricultural institutions in the assessment of postharvest losses, adoption of systems to reduce them, development of value-added food-processing practices, creation of an awareness and knowhow within the farming community, and training of human resources at all levels of involvement. The programme involved three major universities of South India – TNAU, Coimbatore, Tamil Nadu, the University of Agricultural Sciences in Dharwad, Karnataka, and the University of Agricultural Sciences in Bangalore, Karnataka. The universities and their network of research stations and farmer training centres are well positioned to carry out need-based research and developmental activities catering to the needs of this project. All these institutions have recognised the need for developmental work in the area of postharvest and food processing and are committed to future successful developments.

McGill University intended to assist the universities in their capacity-building exercise by providing training to find sustainable solutions to the postharvest and food process engineering problems faced in southern India, and to take an active role in transferring the benefits of such solutions to the ultimate beneficiaries – the farmers of the region.

Acknowledgements

This project was funded and supported by the Canadian International Development Agency. The authors acknowledge the support and active participation of McGill University, Montreal, Canada, Tamil Nadu Agricultural University, Coimbatore, Tamil Nadu, University of Agricultural Sciences, Dharwad and Raichur Campuses, Karnataka, University of Agricultural Sciences, Bangalore, Karnataka, the government of Tamil Nadu, and the government of Karnataka.

REFERENCES

ASABE. (2016). Engineering and Technology Innovation for Global Food Security. A Report on the Global Initiative Conference at Stellenbosch, South Africa. https://www.asabe.org/Portals/0/Events/GlobalFoodSecurity/GFSCSummaryPaper.pdf?ver=2018-07-09-174343-937

FAO. (1996). Report of the World Food Summit. World Food Summit, Rome, Italy, 13-–17 November. https://www.fao.org/3/w3548e/w3548e00.htm

FAO. (2006). Food Security. Policy Brief. https://www.fao.org/fileadmin/templates/faoitaly/documents/pdf/pdf_Food_Security_Cocept_Note.pdf

Hodges, R. J., Buzby, J. C., & Bennett, B. (2011). Postharvest Losses and Waste in Developed and Less Developed Countries: Opportunities to Improve Resource Use. *The Journal of Agricultural Science*, *149*, 37–45.

ICAR. (2023). Student Ready. https://education.icar.gov.in/

ILO. (2009). Global Jobs Pact. https://www.ilo.org/jobspact/about/lang--en/index.htm

Kitinoja, L., Saran, S., Roy, S. K., & Kader, A. A. (2011). Postharvest Technology for Developing Countries: Challenges and Opportunities in Research, Outreach and Advocacy. *Journal of the Science of Food and Agriculture*, *91*(4), 597–603.

KVK. (2023). Krishi Vigyan Kendra Knowledge Network, About KVK. https://kvk.icar.gov.in/aboutkvk.aspx

Lal, H., Sharma, S. K., & Khan, I. M. (2002). Manual on Rural Agricultural Work Experience Programme. Department of Extension Education, Rajasthan Agricultural University, Bikaner, Rajasthan.

Padi, R. K., & Chimphango, A. F. (2021). Postharvest Technology for Advancing Sustainable Bioenergy Production for Food Processing and Reduction of Postharvest Losses. In Galanakis, C.M. (ed). *Food Losses, Sustainable Postharvest and Food Technologies* (pp. 281–311). Academic Press.

Paliyath, G., Mattoo, A. K., Handa, A. K., Shetty, K., & Wilson, C. L. (2018). Enhancing Food Security Through Postharvest Technology: Current and Future Perspectives. In: Paliyath, G., Jayasankar Subramanian, J., Loong-Tak Lim, L-T., Subramanian, K.S., Handa, A.K., & Mattoo, A.K. (eds.). *Postharvest Biology and Nanotechnology*, Chapter 1, pp. 1–13.

Roş, V., Mândru, D., Roş, O., & Ghereş, M. (2006). New Trends in Bio-Systems Engineering Education. *Buletin USAMV-CN*, *62*, 138–143.

Ting, K. C. (2010). Development and Perspectives of Agricultural Engineering Towards Biological/Biosystems Engineering. *Journal of Agricultural Engineering*, *41*(1), 1–5.

WB. (2023). The Future of Food: Shaping the Food System to Deliver Jobs. https://www.worldbank.org/en/topic/agriculture/publication/the-future-of-food-shaping-the-food-system-to-deliver-jobs.print

41

Growing Agricultural Engineering in Africa: Students' Attitudes, Perceptions, and Expectations on Agricultural Engineering Education

Umezuruike Linus Opara

Introduction

Agriculture is a critical component of the economy of many African countries, often contributing to over 45% of the gross domestic product (GDP) and providing a source of livelihood for over 70% of the population. Accordingly, the African Union's Agenda 2063 – the Africa We Want, the masterplan for transforming the continent into a global powerhouse of economic prosperity, has identified the modernisation of agriculture for increased productivity and production as a strategic goal. During the launch of the Comprehensive Africa Agriculture Development Programme (CAADP), the former President of Nigeria, His Excellency General Olusegun Obasanjo, noted that "Improving agricultural performance is at the heart of improved economic development and growth, and its role in poverty eradication and in the restoration of human dignity can never be over-emphasised". Agricultural engineering education and research have critical roles to play in achieving the aspirations of Agenda 2063 and the goals of CAADP.

Agricultural engineering (AgEng) is a well-established academic discipline and profession in Africa. It is offered in many institutions of higher education, including universities and polytechnics, and in some countries, such as South Africa, there is a dedicated Institute for Agricultural Engineering as part of the Agricultural Research Council. Earlier academic programmes in agricultural engineering started as part of the faculty of agriculture and later became part of the faculty of engineering, such as at the University of Nigeria, Nsukka. In the 1970s–1980s, student enrolments boomed, but in recent times, students' interest in agricultural engineering has waned and there is concern about the number of undergraduate student intakes in these programmes. To maintain the relevance and viability of both the programmes and academic departments, academic leaders have taken several measures, including renaming both the department and degree programme and reforming the curriculum.

Despite these efforts, problems remain, casting doubt on the future of agricultural engineering education. To adequately address these challenges facing agricultural engineering education, a multi-stakeholder approach is needed, including academia, students, policy makers, the private sector, etc. Understanding the factors which drive student choices and those factors which enhance retention is essential. Studies in New Zealand (Opara, 2003a,b) and the Sultanate of Oman (Opara, 2006; Opara et al., 2006) have provided insights on student perceptions of the AgEng discipline and academic programmes, including curriculum content and attitudes towards professional identity. These studies highlighted the low preference for and poor image of the AgEng academic programme among students and a perceived low profile of the discipline among the public. Students also offered fresh insights on how to enhance the quality of AgEng education, including curriculum reform and implementing dedicated stakeholder awareness programmes to raise the profile of the discipline and profession.

The critical role of agricultural engineering education and research as a key driver in transforming and modernising African agriculture for socio-economic development cannot be over-emphasised. The relevance of agricultural engineering in achieving the United Nations Agenda 2030 Sustainable Development Goals (SDGs) and the African Union's Agenda 2063 – the Africa We Want has been articulated by Fomunyam and Opara (2023). Despite its much-acclaimed success in raising global food production and access during the past century through the impacts of agricultural mechanisation and related technologies, agricultural engineering has continued to face challenges of identity and relevance which have impacted the survival and future of many academic programmes. A case in point is how to grow the pipeline of future agricultural engineering practitioners and thought leaders who are equipped with relevant skills for lifelong learning, technical expertise, and social engineering to drive the agri-food systems transformation agenda. In response to these challenges, AfroAgEng – the Pan Africa Society for Agricultural Engineering (PASAE) initiated a project to grow agricultural engineering in Africa, which was funded by the Royal Academy of Engineering (UK) through the Global Challenges Research Fund – Africa Catalyst programme on "Planning for the growth of Agricultural Engineering in Africa by supporting the establishment of AfroAEng".

DOI: 10.1201/9780429150111-44
This chapter has been made available under a CC-BY-NC-ND license.

In 2018, PASAE organised a workshop on "Strategies to Grow Agricultural Engineering in Africa" as a contribution to the symposium organised by the South African Institution of Agricultural Engineers on "Meeting the Challenges and Growing Agricultural Engineering – Together We Can". In his keynote address, the founding president of PASAE highlighted the need for stakeholder engagements in developing actionable strategies to grow the pipeline of motivated students and well-equipped graduates and to ensure that agricultural education in Africa is tailored to address current and future developmental challenges facing the continent and the world at large (Opara, 2018).

Nearly two decades ago, Opara (2004b) presented a synthesis on the global status of agricultural engineering education and research and prospects for developing countries, which included an outline of the emerging importance of a technological innovation triad – biotechnology, nanotechnology, and information and communication technology. Subsequently, in a three-part series of the special issue of the *International Journal of Engineering Education* on "Trends in Agricultural, Biosystems, and Biological Engineering Education – Transformation and Reinvention", researchers chronicled the evolution of agricultural engineering and emergence of biosystems/biological engineering (Opara & Cuello, 2006a), discussed developments in curriculum reform, assessment, and accreditation (Opara & Cuelllo, 2006b), and highlighted innovative teaching and learning methods in agricultural and biological engineering education (Opara & Cuello, 2007).

Comprehensive studies on student attitudes and perceptions of agricultural engineering education have been reported in New Zealand and the Sultanate of Oman (Opara, 2006). Using a combination of focused group discussion and paper-based questionnaires, students enrolled in the agricultural engineering programme at Massey University, New Zealand, offered their perspectives on their career prospects, critical skills, and discipline name change. The findings showed that the majority of students came from rural farming backgrounds but would prefer jobs in the non-agricultural (farming) sector, including off-farm (downstream) agribusiness (Opara, 2003a). Students also bemoaned the prevailing low attitudes and perceptions among students (especially those enrolled in other disciplines!) and the public and highlighted the critical importance of new efforts to improve this poor "rural" image (Opara, 2003b). In a similar study at Sultan Qaboos University, Oman, students also reported a very low appeal of the agricultural engineering degree programme among students, including a low image and poor public understanding, which altogether contributed to low student enrolments and high attrition rate (Opara et al., 2004a, b). Omani students preferred that the academic home of the degree programme should be the College of Engineering rather than the College of Agriculture and Marine Sciences. At Massey University, both the degree programme in agricultural engineering and the academic department were based in the Faculty of Agricultural and Horticultural Sciences until the department was merged into a new Institute of Technology and Engineering, which became the new home of the academic programme during the period of the study reported here.

To date, there is limited understanding of the perceptions and attitudes of undergraduate students towards AgEng education in Africa and how these can be improved to grow and promote agricultural engineering on the continent. In this chapter, we report the findings of a study aimed at assessing the attitudes of undergraduate students in East, West, and Southern Africa towards agricultural engineering education, their perceptions about the academic discipline and profession, and key priority impact areas where the discipline may contribute towards successful agri-food systems transformation and modernisation for sustainable development in Africa.

Study Design

The study was carried out using SurveyMonkey®, a web-based electronic survey tool that is suited for creating quick, easy, and cost-effective online professional surveys (Collier et al., 2005; Herreid et al., 2014; Herreid et al., 2013; Varela et al., 2016; Symonds, 2011). The questions were placed into categories which explored several issues that were considered relevant to the research objective, including:

- family background of the respondent in relation to agriculture,
- choice of undergraduate degree programme,
- name of the degree programme and department,
- skills areas expected in the agricultural engineering degree programme,
- the ability of the current curriculum to meet the expectations,
- preferred occupation sector after graduation,
- priority impact areas for agricultural engineering education, and
- public understanding and image of agricultural engineering and proposals for improvement.

The questionnaire was administered online to undergraduate agricultural engineering students enrolled in five agricultural engineering departments in Universities in East Africa (Makerere University, Uganda and Sokoine University of Agriculture, Tanzania), West Africa (University of Nigeria, Nsukka, Nigeria and Kwame Nkrumah University of Science & Technology, Ghana), and Southern Africa (University of KwaZulu Natal, South Africa). A total of 160 students responded, with 21.95% from East Africa, 62.17% from West Africa, and 16.89% from Southern Africa. The results are presented as percentages for each question using bar charts and pie charts, where appropriate. Given the relatively small number of institutions and regions covered in the present study (Africa has five geo-political regions – East, West, North, Central, and South, plus the Diaspora), no comparisons were made between regions and institutions. While many institutions offer agricultural engineering education programmes in countries in Southern Africa, only one institution in South Africa could be covered during this preliminary study. Some students did not respond to every question asked; therefore, the results in percentages for some questions may not add up to 100%.

Major Findings and Discussion

Family Background of Students

Most students (51%) reported that they came from a farming family background, with a large majority (42%) identifying their families as smallholder (subsistence) farmers (Figure 41.1). Similarly, a large proportion of the students (48%) reported that they come from non-farming families. Together, these findings highlight some important facts about the current state of agriculture and agri-food systems in Africa, as well as possible implications for agricultural engineering education. First is that smallholder subsistence agriculture still dominates the food systems, and second is that an increasing number of Africans no longer depend on agriculture, particularly farming, for livelihood. For the students who reported that they were from non-farming families, it is presumed that their parents are not involved directly in farming, and this detachment from agriculture is expected to impact the students' perception of agriculture, agricultural engineering, and their future career interests. It is conceivable that during the birth of agricultural engineering as a tertiary educational programme in Africa over half a century ago, nearly all undergraduate students would have come from a smallholder farming family. The growth of the civil service post-independence and the emergence of new industries – from education to health care, transportation, manufacturing, telecommunications, and services, etc., would have contributed to growth in careers and jobs outside agriculture, particularly for the educated middle class.

Together, these findings present challenges and opportunities for agricultural engineering education in Africa. For instance, given the reliance of smallholder subsistence agriculture on the human muscle as a power source and archaic implements such as the hand-hoe and cutlass, questions arise about the relevance of current curricula and practical aspects of teaching and learning. Furthermore, with only nine percent of the students reporting a family background in medium- to large-scale farming, it is clear that a large majority have not been exposed to nor experienced commercial agribusiness characterised by high knowledge intensity and the application of improved inputs and technologies which enhance productivity. As the rate of urbanisation continues to rise in Africa with the increasing demand for a steady supply of food in both quantity and quality, how do we ensure that agricultural engineering education remains attractive to youth – both rural and urban?

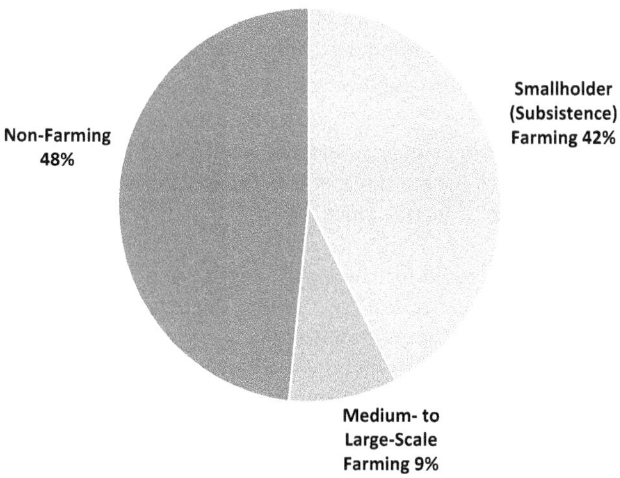

FIGURE 41.1 Family background of the students.

Degree Programme Preference

When students were asked, "Was agricultural engineering discipline your first choice when you applied to study at university?" only 42% agreed, while 58% did not. Among those who did not apply to study agricultural engineering *ab initio*, 76% chose other engineering and related courses, including civil > electrical > mechanical > geological/mining > chemical/petroleum > electronic = telecommunications/computer = quantity surveying/building technology > architecture (Figure 41.2). Additionally, 20% of the students who did not choose agricultural engineering preferred medicine and related health sciences, including medicine and surgery > pharmacy/herbal medicine = nursing > human biology/biological sciences. Some students went further to comment about their lack of knowledge about the agricultural engineering major, saying: "I didn't know about it", "I had no idea that (the) university had such a programme", and "I didn't know much about the programme. I only wanted Mechanical". The preference for other engineering majors and medicine is similar to the findings from previous studies on agricultural engineering education among undergraduate students at Sultan Qaboos University, Oman (Opara et al., 2004a, b). Overall, these findings underline the limited knowledge about agricultural engineering as an academic discipline in engineering among high school matriculants, and possibly including parents, caregivers, and career advisers. The pervasive negative attitude towards agriculture among the youth, parents, and the general public at large could also contribute to the low interest among students to study agricultural engineering (Opara 2003a, b; Opara, 2004a, b; Opara et al., 2004a, b).

Preferred Name for Discipline and Profession

During the past three decades, there has been debate among agricultural engineers and educational administrators about the appropriateness and relevance of the name "agricultural engineering" (Opara, 2002a, b; Opara, 2004b; Opara & Cuello, 2006a, b; Opara & Cuello, 2007) in enhancing the appeal to prospective students and other stakeholders including parents, career advisers, and employers (Opara, 2003a, b; Opara et al., 2004a, b; Opara et al., 2006). Four dominant names in Africa and elsewhere globally were identified and selected, namely agricultural engineering, biological engineering, bioresources engineering, and biosystems engineering, and students were asked to choose one which is most appropriate in the 21st century.

A large majority (40%) of the students preferred the name "agricultural engineering", in the following order: agricultural engineering > bioresources engineering, biosystems

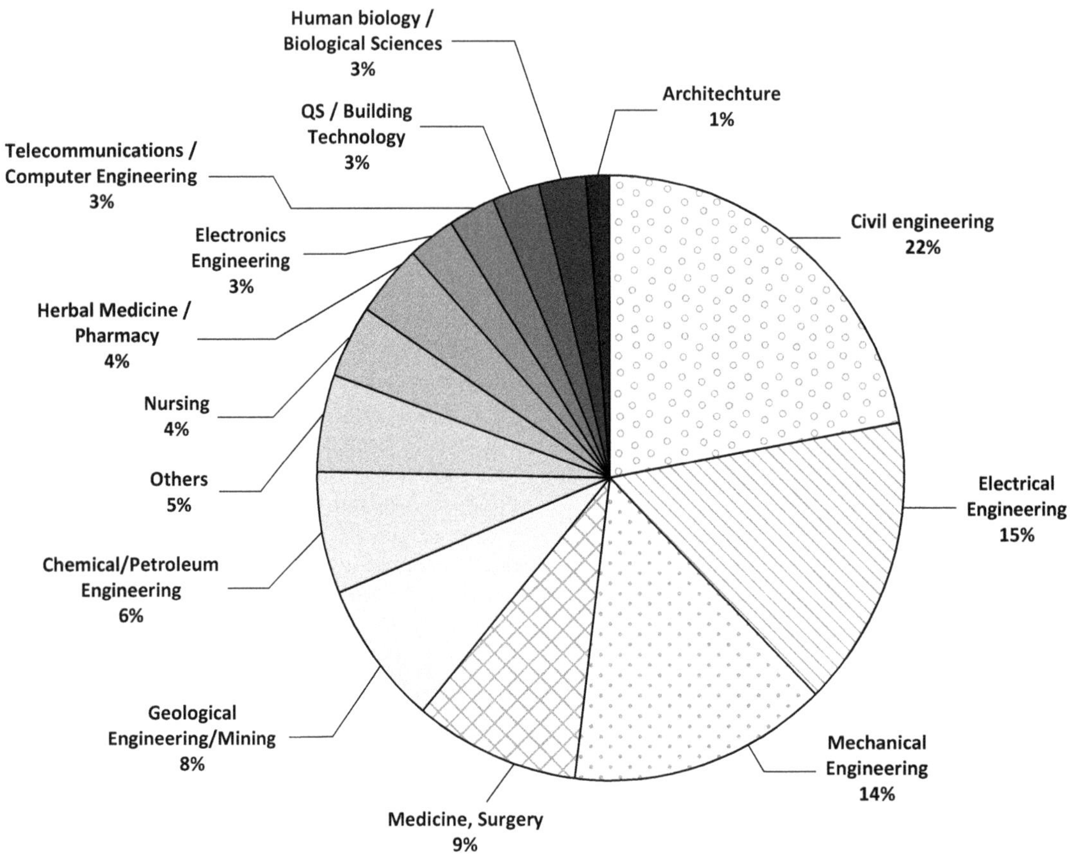

FIGURE 41.2 First choice degree programme among students who reported that they did not apply to study agricultural engineering *ab initio*.

engineering > biological engineering. However, with a higher majority (60%) opting for the bio-inspired names (Figure 41.3), it is clear that the students welcome the proposition and widespread acceptance of biology as the fundamental science underpinning agricultural engineering (Johnson, 2006; Johnson & Phillips, 1995; Johnson & Schreuders, 2003).

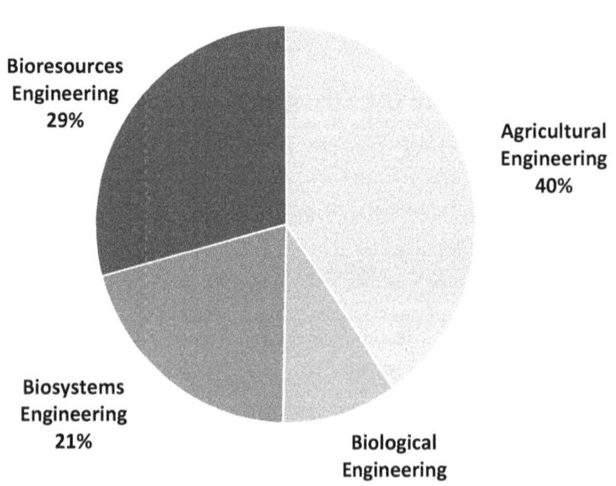

FIGURE 41.3 "There has been a debate about the appropriateness of the name 'agricultural engineering' in enhancing the appeal to prospective students and employers. Which one of the following names do you think is most appropriate in the 21st Century?"

Students also offered alternative names, such as "biomechanical" engineering and compound names, such as "agricultural and bioresources engineering" and "agricultural and biosystems engineering" (Box 41.1). Some of the sentiments expressed against the name "agricultural engineering", an indication of preference for a name change, include the following:

- "People perceive agricultural engineering and agricultural science as one discipline".
- "'Agricultural engineering' comes across as very limited to people who know nothing about it. They immediately assume it's only farming centred".
- "Agricultural engineering makes it sound as if we farmers".
- "A name change would really do the sector good because people consider it to be, just about farming, instead of its true value".

The low preference for "biological engineering" (ten percent) and high preference for "agricultural engineering" are supported by the following comments:

- "Majority of the students in Uganda who apply for this course have not done Biology (or it wasn't one of their favourite subjects), so calling it purely biological would, in my view, bias them. Employers in developing countries can understand directly the

BOX 41.1: KEY COMMENTS PRESENTED BY STUDENTS IN RESPONSE TO THE CHOICE OF A PREFERRED NAME THAT IS MOST APPROPRIATE IN THE 21ST CENTURY.

1 It depends which direction you choose in Ag Eng.
2 People perceive agricultural engineering and agricultural science as one discipline.
3 "Agricultural engineering" comes across as very limited to people who know nothing about it. They immediately assume it's only farming centred.
4 A name change would really do the sector good because people consider it to be, just about farming, instead of its true value.
5 Agricultural engineering makes it sound as if we farmers.
6 Agricultural engineering terms makes people to think you are farmer.
7 In my university it is referred to as Agricultural and Biosystems engineering and I think it should stay that way. Majority of the students in Uganda who apply for this course have not done Biology (or it wasn't one of their favourite subjects) so calling it purely biological would, in my view, bias them. Employers in developing countries can understand directly the need for and agricultural engineer than a Biosystems, Biological, Bioresources engineer.
8 The name agricultural engineering is vast in that in it combines knowledge from almost all other engineering disciplines; civil, mechanical, electrical, etc. This is applicable in agriculture thus allowing work flexibility. The wonder is; potential learners are limited and most people have adopted to specialised fields.
9 Agricultural Engineering is still a suitable name for the 21st century. The agricultural engineers solve problems related to agricultural equipment, water quality and water management, biological products, livestock facilities, food processing, and many other agricultural areas. This disciple combines all the other disciples (Biological, Biosystems and Bioresources Engineering). This is a diverse field of engineering in Agriculture. The future of agriculture depends on the next generation of problem solvers. Creative and skilled individuals, like agricultural engineers, can use their knowledge of agriculture and life sciences, along with the problem-solving skills of engineering, to create new systems and solutions for the 21st Century. Agriculture is changing faster than any time in history. That's why, if you are interested in helping direct the future of agriculture, a degree in agricultural engineering is what you need. When we begin talking of biological, biosystems and bioresources engineering we do contradict with biomedical sciences outlook and thus running away from agriculture context. To my opinion, I wish Africa even in the 21st century outlook should think of Agricultural Engineering in order to enhance appropriate way to students and employers, and as a way to modernising agriculture production, processing and management.
10 The main key issue is food production.

need for an agricultural engineer than a Biosystems, Biological, Bioresources engineer..."

- "Agriculture is changing faster than any time in history. That's why, if you are interested in helping direct the future of agriculture, a degree in agricultural engineering is what you need. When we begin talking of biological, biosystems and bioresources engineering we do contradict with biomedical sciences outlook and thus running away from agriculture context. To my opinion, I wish Africa even in the 21st century outlook should think of Agricultural Engineering in order to enhance appropriate way to students and employers, and as a way to modernising agriculture production, processing and management".

Given these results and comments, it is therefore not surprising that the name "biological engineering" is rarely used in Africa for an academic department and/or degree programme, compared to the United States. Indeed, while the majority of students prefer a name with "bio", the name "agricultural engineering" remains the single most preferred name.

Preferred Career After Graduation

When students were asked, "Where would you like to work after graduation?" based on three broad categories of occupation areas in agriculture and the non-agricultural sector (Figure 41.4), the majority (61%) would prefer to work in off-farm agriculture, comprised of postharvest handling and processing (31.41%) and agricultural support services (29.49%) such as logistics, marketing, research, and teaching. Among the agricultural sectors, the lowest preference was for production/farming (26.28%). Only 12.82% desired to pursue a career in the non-agricultural sector and some of the areas identified by the students include aviation, finance, law, oil and gas, construction, "mechanical engineering companies", automobile, and civic society organisations (Box 41.2). Put together, off-farm and non-agricultural sectors accounted for 74% of student preferences, which strongly underscores the low preference for a career in direct farming/field production. Given that postharvest (post-farmgate) activities can account for up to 80% of both the economic value and profitability of agriculture and given the prevailing drudgery and low productivity of agriculture among smallholder subsistence farmers (who carry

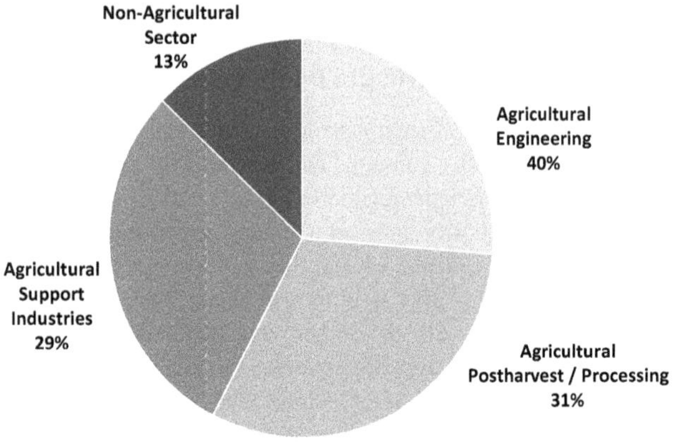

FIGURE 41.4 "Where would you like to work after graduation"?

out the majority of agricultural activities in Africa), it is therefore not surprising that most students prefer careers in off-farm activities and in the non-agricultural sector. These findings are in line with the results reported among undergraduate students in the Sultanate of Oman (Opara, 2004a, 2004b) and New Zealand (Opara, 2003a, 2003b), respectively.

Relevant Skills

A well-trained engineer requires a wide range of skills to succeed in any field, and these are technical and non-technical (soft) skills. Students were asked to select as many options as applicable from a list of key technical skill areas which they would need in order to meet the expectations of future employers and/or the preferred career already identified in the preceding section. The results showed that "machinery management" ranked highest while "field production" was the least (Figure 41.5): machinery management > equipment and/or process design > information and communication > analytical and data management > field production. Students also

BOX 41.2: ADDITIONAL INSIGHTS PROVIDED BY STUDENTS ON TYPE OF WORK/CAREER PREFERENCE AFTER GRADUATION.

1 Machine prototype developing industries
2 Construction companies
3 Aviation or Finance Sector
4 Self-employed engineer
5 Oil company
6 I will be working with computers. I want to be a computer programmer. Didn't know what I loved until I came to school.
7 Academics
8 Irrigation and drainage (hydrology)
9 Material science engineering
10 Design Engineering sector
11 Water and rural community development
12 Water engineering, waste management, or environmental engineering
13 Automobile
14 Mechanical engineering companies
15 Law firm
16 I want to work in both the production and processing sector.
17 Civil society organisation
18 Water sector

Rating of Current Agricultural Engineering Programme Vis-à-Vis the Required Skills

Having examined the preferred career/job after graduation and identified the relevant skills needed to succeed in the workplace, students were asked to assess the extent to which the current agricultural engineering curriculum would enable them to achieve those skills to meet the expectations of future employers/employment. Satisfaction with the current

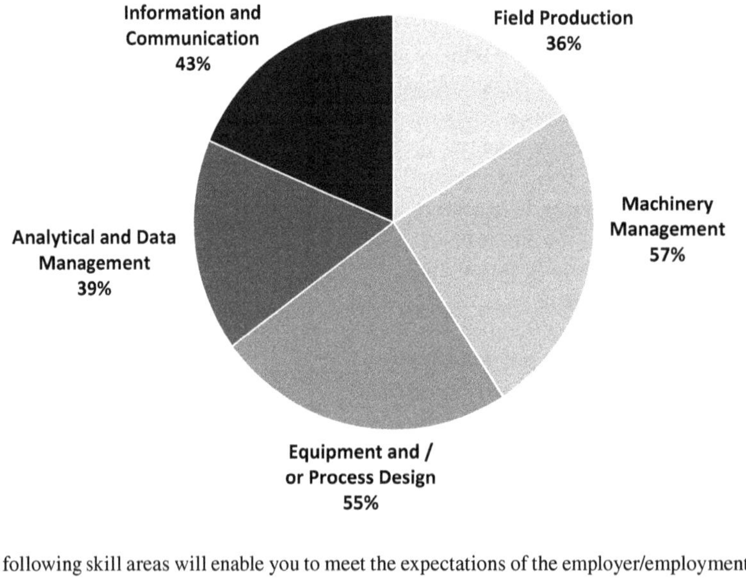

FIGURE 41.5 "Which of the following skill areas will enable you to meet the expectations of the employer/employment you prefer after graduation?"

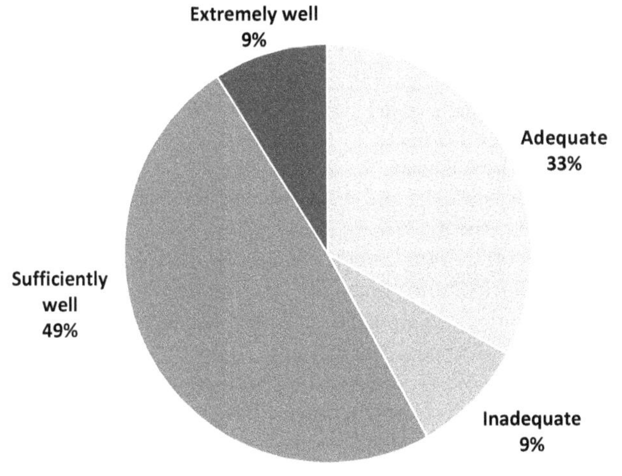

FIGURE 41.6 "To what extent do you think the current agricultural engineering programme would enable you to achieve those skills to meet the expectations of your future employer/employment?"

curriculum/ academic programmes was rated as follows: very good > inadequate to just adequate > excellent (Figure 41.6). Overall, 58% of students considered the programme to be very good or excellent.

Image and Public Understanding

Persistent issues facing the growth of agricultural engineering in Africa and indeed around the world are the low profile, poor image, and limited public awareness and understanding of both the discipline and the profession. These have contributed to both low enrolment and low retention rates of undergraduate students enrolled in agricultural engineering programmes. Even where enrolment has remained fairly high or stagnant during particular periods, students have reported that agricultural engineering was not their first choice of study (Opara et al., 2004a; Opara et al., 2006; Opara & Mullah, 2023, in this book). In this study, and compared to other disciplines, students rated overall public understanding of agricultural engineering and its role in economic development to be very poor to poor > fairly good to good > excellent (Figure 41.7). Students also recommended a wide range of actions to improve public understanding of the discipline (Box 41.3), which can be summarised into the following actionable points:

- Use new and social media platforms.
- Carry out awareness programmes in schools – at all levels.
- Create a special forum to recognise the work of agricultural engineers, including awarding scholarships.
- Promote link with industry in the curriculum, including providing agricultural engineering services.
- Develop agricultural engineering-related industries.
- Improve salary/remuneration.
- Change the name.

A couple of students summarised the poor image and low profile of agricultural engineering and offered solutions as follows:

- "People see Agricultural Engineering as an inferior course compared to Mechanical Engineering, Electrical Engineering, and the rest. There should be awareness created on the importance of Agricultural Engineering to the economy".
- "There is no public education for the other engineering courses, but then they are soaring. This is because they have large markets for employment and also recognised industries. I believe we should rather work hard as agricultural engineers and create an industry of our own. That way, the public has to research into it".
- "Very many people think that agricultural engineering is all about managing tractors, and so most of the engineers in this field are not recognised. This can be improved by increasing[ly] explaining the work done by agricultural engineers through media".

To achieve goals, students specifically single out the role of professional agricultural engineer societies and associations at national, regional, continental, and global levels:

- "Having national and international bodies or associations bringing agricultural engineers together".
- "By setting up a body that unites agricultural engineers in the country. Also, by encouraging the implementation of an agricultural engineering post for each district and by so doing this will improve the publicity of the course and its relevance. Also through frequent seminars both at district and school level".
- "By creating forum for AE recognition, that is by carrying out practicals and creating awareness in the public and scholarships".

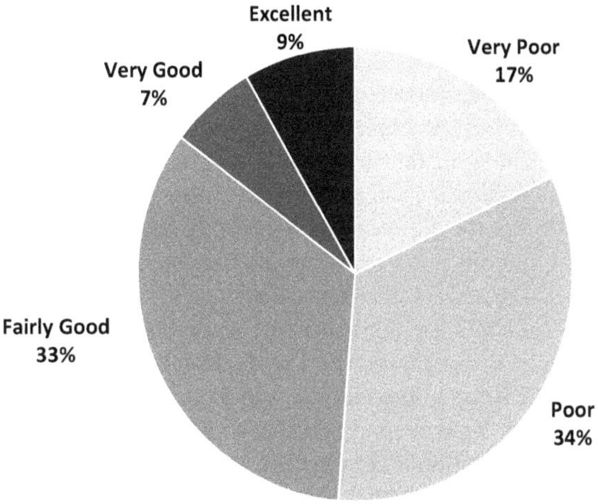

FIGURE 41.7 "Compared with other engineering disciplines, how would you rate the overall public understanding of agricultural engineering and its roles in the economy?"

- "Through setting up seminars to the general public & visiting high schools to let them know about Agricultural Engineering".

These recommendations highlight the critical importance of professional bodies such as AfroAgEng/PASAE in growing agricultural engineering and enhancing its profile in Africa (Opara, 2012; Taylor et al., 2021). This will require working closely with the adhering national and regional associations and in partnership with other stakeholders, using deliberately designed campaigns and other programmes.

BOX 41.3: RECOMMENDATIONS OFFERED BY STUDENTS TO IMPROVE PUBLIC UNDERSTANDING OF AGRICULTURAL ENGINEERING.

1 I think by intensifying education on importance of agricultural engineering aspect.
2 Educating them that agricultural eng are not famers.
3 More "Ag Eng promotions" at high school level nationally and not just at UKZN open days.
4 I think it can be improved by educating people about the importance of agricultural engineering and make it a more popular subject amongst students in high school.
5 Through agricultural extension services.
6 It needs to be promoted on a national level to high school children. Varsity students should also be involved in promoting it because they can relate to scholars as young people.
7 By mutualising/linking the agricultural engineering dept with industries (agricultural). For their design, research, and referral purposes.
8 By creating forum for AE recognition, that is by carrying out practicals and creating awareness in the public and scholarships.
9 One is to improve and promote agriculture in our schools, from primary, secondary to Tertiary institutions, government investing in agricultural resources.
10 Dissemination of information to students who are yet to get admission about the course and letting them see the value and importance of agriculture to humans and the world at large.
11 By the government supporting agriculture and making open agricultural sector to invest and new technological advancement.
12 People see Agricultural Engineering as an inferior course compared to Mechanical Engineering, Electrical Engineering and the rest. There should be awareness created on the importance of Agricultural Engineering to the economy.
13 Through public lectures on the nitty-gritty of agricultural engineering.
14 Through the use of home made goods.
15 Public understanding could be improved by integrating social aspect because engineer should be part and parcel of social change.
16 To have more universities offer agricultural engineering.
17 There should be more articles about this field in newspapers and it should also be offered at more universities.
18 Agricultural policies should be made to improve agricultural practices in Nigeria. Poor policy implementation has been the bane of Agricultural development in Nigeria.
19 It could be improved by high payment of salary to agricultural engineers. This will enable the youths and teens in school look forward to being an Agricultural Engineer hence, making them curious to want to Know more about Agricultural Engineering.
20 Awareness programmes (Ads, campaigns etc).
21 Not being called farmers or the use of the word agricultural.
22 First change the term "agricultural engineering" and make more information available.
23 Having national and international bodies or associations bringing agricultural engineers together.
24 Through education and rebranding.
25 This may be done when agricultural engineers gain relevant and decent paying jobs specifically requiring agricultural engineers. (Job Security).
26 Very poor. Mechanised agriculture should be encouraged and many adverts on prospects should be made available online.
27 Educating the public on the importance/role of agricultural engineering in the 21st century.
28 By changing the name.
29 There is no public education for the other engineering courses but then they are soaring. This is because they have large markets for employment and also recognised industry. I believe we should rather work hard as agricultural engineers and create an industry of our own. That way the public has to research into it.
30 Consultation of various industries dealing with agricultural machinery and equipment and the agro-industrial corporations that are coming up in each country in Africa and a suitable programme develop to suit their needs. Establishment of employment cadres at the government level as well as with the industries. Establishing a centre for a agricultural machinery technology in all countries in Africa even at local levels. Setting up industrial training institutes for agricultural engineers, with proper coordination and cooperation of all facilities effected throughout the country in the complete development of agricultural engineering fields. Proper orientation of agricultural engineering graduates to practical agriculture in all areas of agricultural engineering fields.

31 By teaching the community publicly on what agricultural engineering really means.
32 Through setting up seminars to the general public & visiting high schools to let them know about Agricultural Engineering.
33 By redefining the word Agriculture not to mean only food, tractors but other sectors like water, renewable energy.
34 Communities should be educated on the key roles played by agricultural engineers in developing agriculture and transforming our continent.
35 Very many people think that agricultural engineering is all about managing tractors and so most of the engineers in this field are not recognised. This can be improved by increasing[ly] explaining the work done by agricultural engineers through media.
36 By setting up a body that unites agricultural engineers in the country. Also, by encouraging the implementation of an agricultural engineering post for each district and by so doing this will improve on the publicity of the course and its relevance. Also through frequent seminars both at district and school level.

Priority Impact Areas for Agricultural Engineering in Transforming and Modernising Agri-food Systems in Africa

Using a list of six (6) topical and priority areas contributing to sustainable agricultural intensification and industrialisation of African rural economies, students were asked to select three (3) priority areas where agricultural engineering could contribute towards the successful transformation and modernisation of African agri-food systems. Among these, mechanisation ranked highest (Figure 41.8): mechanisation > biofuel > irrigation and drainage > postharvest handling and processing > conservation agriculture > environmental management. The ranking of mechanisation as the top priority impact area (Figure 41.8) also corresponds to the emergence of agricultural machinery management as the top priority among the desired skills and competencies (Figure 41.5) that would enable them to meet the expectations of future workplaces.

Very low levels of mechanisation in African agri-food systems, symptomised by very limited use of mechanical power and associated implements, improved machinery, in combination with other productivity-enhancing technologies such as irrigation and improved inputs, has contributed to the long-term low productivity of agri-food systems in Africa. By putting mechanisation as the top of the key priority areas for agricultural engineering in Africa, the students have aligned with and corroborated one of the critical challenges facing African agri-food systems. There is a plethora of literature which demonstrates the critical and foundational role of agricultural mechanisation as an effective push-pull factor for widespread rural transformation and agricultural modernisation (Opara, 2023a, 2023b, 2023c, in this book).

A synthesis of agricultural mechanisation approaches in China, including policy, resourcing, and the organisational changes and socio-economic consequences (Hau, 1979), showed positive widespread impacts towards the present-day economic development of the country. These included an improved image of modern farm workers often depicted as "rural folk heroes" and "the impact of rural industries and agricultural mechanisation in creating a 'new type of peasant who is competent in mechanised, scientific farming and in industrial production'". Furthermore, as mechanisation proceeded, the high disparity in income between rural and urban areas diminished, and at the same time, "the interactions between the two areas" increased. To achieve this, and starting

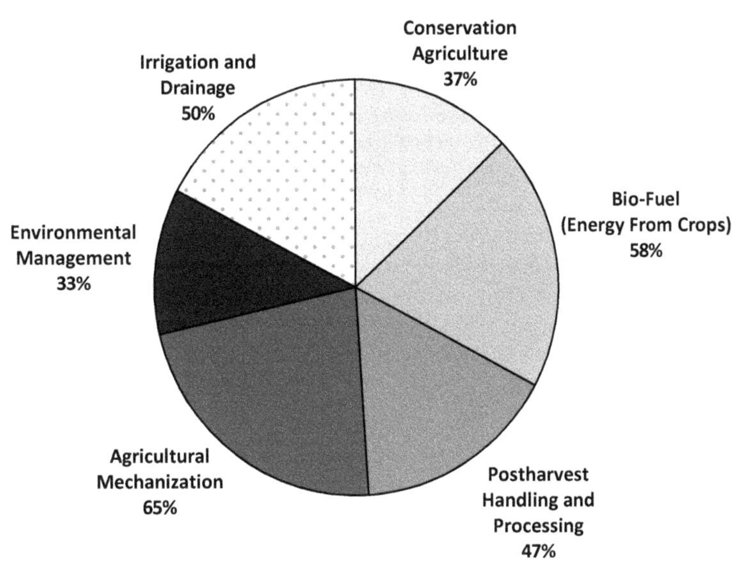

FIGURE 41.8 Priority impact areas where "agricultural engineering" could contribute towards successful agri-food systems transformation and modernisation for sustainable development in Africa.

in the 1950s and 1960s, China pursued a technology strategy that is peculiar to its socio-cultural and political economy, which necessitated "the initiation of indigenous technological advancement – addressing itself to the production and adaption of mechanical techniques to meet the needs of agriculture and local industry" (Matthews, 1987). Similarly, China also pursued what is often referred to as a "gradualistic" approach to economic development push with a "dualistic character", which promoted the advancement of agricultural mechanisation as a key element. In his much-acclaimed presentation on "The Ten Great Relationships" in 1956, the Chinese leader Mao described this dualistic character of development as: "taking agriculture as the foundation and industry as the leading sector" (Singh, 1979). Many would point to a strong link between China's present-day economic stardom to these visionary insights and actions. Again, as Mao aptly put it, "the fundamental way out for agriculture lies in mechanisation" (Hau, 1979).

Studies elsewhere have also shown that combining agricultural mechanisation with the right public policy can bring long-term benefits that increase both individual and social welfare (Martin & Johnson, 1978; Anazodo et al., 1989). While we recognise that mechanisation alone is not the panacea to the socio-economic and development challenges facing Africa, we boldly argue that it is a necessary condition for the radical transformation, modernisation, and industrialisation of Africa's agri-food systems into a global powerhouse for accelerated rural development. Through its benefits in reducing drudgery, achieving timeliness of operations along the value chain, and increasing factor productivity and total production, success in sustainable agricultural mechanisation will also enable Africa to transform and modernise the rural areas, and play its long-awaited role in contributing to a sustainable, competitive, secure, resilient, and equitable global food system.

The African Union has recognised this need and called for the banishing of the hand-hoe to the museum (Opara, 2012; Opara, 2023d, in this book; Fomunyam & Opara, 2023, in this book). The Food and Agriculture Organisation (FAO) of the United States has long been promoting agricultural mechanisation as a key input for sub-Saharan African smallholders (Sims et al., 2016), and subsequently, the FAO and the African Union Commission (AUC) have partnered to provide a framework for sustainable mechanisation of agriculture in Africa (FAO & AUC, 2018). To achieve these goals for agricultural mechanisation in Africa, and indeed elsewhere, the role of agricultural engineering education is important (Kienzle et al., 2023, in this book). Similarly, the AUC, through Agenda 2063 – The Africa We Want, has stated that "Africa's agriculture will be modern and productive, using science, technology, innovation and indigenous knowledge. The hand-hoe will be banished by 2025 and the sector will be modern, profitable and attractive to the continent's youths and women" (AUC, 2014). The African Development Bank (AfDB) has identified "Massively Increase Productivity" with "Agricultural Mechanisation" as a key component as one of the seven (7) enabling actions in fulfilling agricultural transformation in Africa (AfDB, 2017).

In additional comments provided to this section by some of the students, the following issues were listed as areas for potential impact by agricultural engineering: food security, waste management, water, energy, and livestock (breeding, animal tourism, and avoiding poaching).

Additional Comments to Grow and Improve the Future of Agricultural Engineering in Africa

To allow each student to reflect on the objective of this study, the last part of the questionnaire asked for final brief comments with specific actional points to grow agricultural engineering and improve the future in Africa. A total of 116 distinct responses were recorded (Appendix A). Following a detailed review of each comment, those that are similar or seek to achieve the same objective were clustered together, and this yielded the following key results (the percentage in bold represents the proportion to the total comments recorded):

- Enhance the practical (skills) component of teaching and learning – teaching and experimental labs/facilities, fieldwork, vacation work, internships (42.24%); More exposure to the industrial sector ("real world") (6.90%). **49.14%**
- Improve public image and understanding of agricultural engineering and its roles in society – including potential students; Fresh students need enlightenment and an improved understanding of the discipline and what to expect during the course. **23.28%**
- Bear on governments to prioritise agricultural engineering in economic development – including favourable policies. **5.17%**
- Establish and promote academic exchange programmes and international collaboration with the best universities over the world to increase cohesiveness.
- Change the name of the discipline/programme.
- More bursaries/scholarships to support students.

In addition, several students offered some unique insights – often expressing their sentiments and frustrations about their local situations, others focusing on the matters applicable to others (such as mentorship, lecturers that inspire, and standardisation of curricula), while some encapsulate several points into one and thereby provide a more nuanced response:

- "The public is very ignorant about Ag Eng and just assumes it has to do with just farming and thus don't bother to educate themselves about it. My friends often ask me if I'm going to buy a farm after my studies and the youngsters find it to be an 'unglamourous' profession (since they perceive it to be farming), especially those from urban areas. I therefore think that, in addition to vigorous marketing, the name should be changed".
- "Hosting agricultural engineering seminars for African students so they can interact with other successful engineers in the field of study".
- "We rule the world, so we need more labs and workshops, the ones we have are not functional, and we

should be given a large area of land for our field activities. Thank you".

- "Agricultural engineers should establish themselves so that people in the outside world will be attracted to it. Lecturers should be inspiring people".
- "The agricultural engineering discipline should have a standard curriculum. Agricultural Engineering should be understood as being different from Agricultural and biosystems engineering".
- "Agriculture engineering can allow a student to start his/her agric business".
- "It is the most underrated Engineering degree, maybe if the sector could try find ways to promote the Degree, Most of us wanna leave this course because it's unknown to industries mostly and that affect our chances of getting funding from Big companies. Also we struggle to get Vacation work at most, this is due to the little and background of this course. Sector should try at least market our Degree like other Engineering degrees, provide sufficient bursaries for the Agricultural sector. It is very important that we promote this degree as it is future to all the environmental crisis the world is about to face".
- "Don't know about Africa as a whole but for Nigeria, we need only prayer".

Conclusion and Future Prospects

This is the first empirical study to assess student perceptions about growing agricultural engineering and the future prospects in Africa. The study contributes towards achieving the aspirations embodied in the African Union's Agenda 2063 – The Africa We Want, regarding Aspiration 1 – "A prosperous Africa based on inclusive growth and sustainable development", where "modern agriculture for increased production, productivity and value addition contribute to farmer and national prosperity and Africa's collective food security". Agenda 2063's call to action also specifically identifies the need to "Catalyse an education and skills revolution and actively promote science, technology, research and innovation, to build knowledge, human resources, capabilities and skills for the African century" (AUC, 2014).

Similarly, agricultural engineering education, research and related technological innovations have an impact on achieving the United Nations Agenda 2030 for sustainable development with regards to Goal 1 (End poverty in all its forms everywhere), Goal 2 (End hunger, achieve food security and improved nutrition and promote sustainable agriculture), Goal 4 (Ensure inclusive and equitable quality education and promote lifelong learning opportunities for all), Goal 8 (Promote sustained, inclusive, and sustainable economic growth, full and productive employment and decent work for all), and Goal 12 (Ensure sustainable consumption and production patterns), among others. Sub-Saharan African economies remain largely rural, with agriculture being a major contributor to the GDP and source of livelihood. However, the agri-food systems in Africa remain the least productive compared with the rest of the world, and this is to a large extent attributed to the limited applications of modern agribusiness practices and related productivity-increasing technologies across the value chain. Agricultural engineering is central to addressing these challenges, and therefore, Africa needs is a steady supply of endogenous skilled workforce, including knowledge workers, to champion the radical transformation and industrialisation of agri-food systems in Africa. However, agricultural engineering and practice in Africa faces many obstacles – some attributable to the nature of the discipline while others are external (socio-economic) forces impacting it. Stakeholder engagements are necessary to find lasting solutions to these problems.

Using a web-based online questionnaire, this preliminary study assesses the attitudes towards agricultural engineering education and the profession among undergraduate students in selected universities in Africa, including the prospects for impact in realising the Africa We Want (and African agriculture in particular). Low public image among the general public, low profile among other disciplines, especially engineering majors, and limited employment opportunities have contributed to low preference for agricultural engineering education among prospective students. A large proportion of students enrolled in the programme did not wish to study it *ab initio*, but took it when they could not enter their first choice. Nevertheless, students identified key priority actions which can help grow agricultural engineering in Africa, including a name change and related curriculum reform, funding opportunities for scholarships and bursaries, the industrialisation of Africa, including African agriculture to provide new employment and business opportunities for graduates, and aggressive and focused campaigns to raise awareness and enhance the image of agricultural engineering targeting the spectrum of stakeholders including secondary schools, first-year undergraduate students, industry, policy makers, and the general public. Agricultural engineering education must also offer more opportunities for practical experiential learning through functioning laboratories (inside and outside the classroom), and relevant industrial work experiences and internship, which expose students to potential employment and business opportunities.

The mechanisation of agri-food systems was identified by the students as the top-ranking priority impact area for agricultural engineering in Africa. With human muscle and the associated rudimentary tools (hand-hoe and cutlass) still being the major sources of power and equipment, respectively, while the African Union and other stakeholders have called for the banishing of the hand-hoe to the museum in the next couple of years (2025), the stage is set and the stakes are high for novel strategies to urgently realise the mechanisation and automation of Africa's agri-food systems for sustainable and inclusive socio-economic development. While we delay in making the right policy and investment choices to realise these goals, the digitisation of agri-food systems, including the use of mobile networks across the value chain, and the use of drones, robotics, and related site-specific technologies to support precision agriculture, including pest and disease management, are already finding applications around the world and in selected value chains in parts of the continent. Why not? After all, drones are now regular features in many parts of rural Africa for recording events during weddings, funerals, political rallies, and similar social gatherings, even for taking group photographs during local and international conferences.

With increasing rapid advancements in emerging and not-so-emerging technologies that are relevant to agriculture and food, such as biotechnology, nanotechnology, information and communication technologies, the seemingly ubiquitous and perversive emergence of artificial intelligence and the related sciences such as data science, machine learning, and deep learning, etc., the future is here for Africa to lead the next agricultural revolution, having missed out on the Green Revolution. Given its low base in modern technology in the agri-food systems, Africa can leapfrog, just as it has done with the information and communication technology area, such as the use of the internet and mobile telephony. At the same time, the world is confronted with climate change and other grand challenges which have wide-ranging negative impacts on agriculture and are also impacted upon by agriculture.

All signals point to the fact that Africa at large, and Africa's agriculture in particular, will be worst affected at the current rate and future climate change predictions. Africa's agri-food systems remain highly vulnerable to external shocks and lack the resilience needed to absorb the shocks and adapt. It is through quality education, especially agricultural and biosystems/biological engineering education and research, that the workforce with the right skills and attitudes can be developed, mentored, and nurtured. This noble cause makes the case to grow and sustain agricultural and biosystems/biological engineering education in Africa.

Acknowledgements

This study was supported by the Royal Academy of Engineering (UK), through funding the project "Planning for the growth of agricultural engineering in Africa by supporting the establishment of AfroAgEng" through the Global Challenges Research Fund – Africa Catalyst programme.

I am grateful to undergraduate agricultural engineering students who participated in this study by completing the online questionnaire. The support of the following colleagues in publishing the study to students is highly appreciated: late Prof. Dr Noble Banadda (Makerere University, Uganda), Prof. (Sir) Louis Sunday Ezeoha (University of Nigeria, Nsukka), Prof. Tilahun Seyoum and Prof. Jeff Smithers (University of KwaZulu Natal, South Africa), Prof. Emmanuel Bobobee (Kwame Nkrumah University of Science and Technology, Ghana), and Prof. Valerian Silayo (Sokoine University of Agriculture, Tanzania).

I am especially grateful to Mrs Lizanne De Kock (MSc Food Science), who has been my Research Assistant over the years, for her support. Lizanne curated the research data and ensured that this project received my continued attention to the finish line.

REFERENCES

AfDB. (2017). The high 5. Feed Africa: The road to agricultural transformation in Africa. African Development Bank Group, Abidjan, 7 pages.

Anazodo, U.G.N., Opara, L.U. & Abimbola, T.A. (1989). Agricultural mechanization study: Agricultural development plan of Nigeria – 1989–2004. Federal Agricultural Coordinating Unit (FACU), Ibadan.

AUC. (2014). Agenda 2063: The Africa we want, second edition, popular version. African Union Commission, Addis Ababa, 20 pages.

Collier, A.P., Johnson, K. & Dellavalle, R. (2005). Survey monkey: A cheap and easy electronic survey tool for research and program evaluation. *Journal of the American Academy of Dermatology*, 52(3). https://doi.org/10.1016/j.jaad.2005.01.039

FAO & AUC. (2018). Sustainable agricultural mechanisation: A framework for Africa. Addis Ababa. Licence: CC BY-NC-SA 3.0 IGO. Food and Agriculture Organization of the United Nations, Rome, 127 pp.

Fomumyam, K.G. & Opara, U.L. (2023). Top 100 questions of importance to the future of agricultural engineering education, research and practice in Africa (chapter 3 in this book).

Hau, R.C. (1979). Agricultural mechanisation in China: Policies, problems, and prospects. *Asian Survey*, 19(5), 436–449.

Herreid, C.F., Schiller, N.A., Herreid, K.F. & Wright, C.B. (2014). A chat with the survey monkey: Case studies and the flipped classroom. *Journal of College Science Teaching*, 44(1), 75–80. http://www.jstor.org/stable/43631780

Herreid, C.F. & Schiller, N.A. (2013). Case studies and the flipped classroom. *Journal of College Science Teaching*, 42(5), 62–66. http://www.jstor.org/stable/43631584

Johnson, A.T. (2006). The making of a new discipline. *International Journal of Engineering Education*, 22(1), 3–8.

Johnson, A.T. & Phillips, W.M. (1995). Philosophical foundations of biological engineering. *Journal of Engineering Education*, 84, 311–320.

Johnson, A.T. & Schreuders, P.D. (2003). Bringing life to engineering: Biological engineering at the graduate level. *European Journal of Engineering Education*, 28, 37–46.

Kienzle, J., Mpagalile, J. & Houmy, K. (2023). Enhancing the role of agricultural and biological engineering through education: Perspective from FAO (chapter in this book).

Martin, P.L. & Johnson, S.S. (1978). Agricultural mechanisation and public policy. *California Agriculture*, June, 4–6.

Matthews, R. (1987). The role of technology in the process of economic development. *Africa Development*, 12(2), 135–159.

Opara, L.U. (2002a). Agricultural engineering education and research in knowledge-based economy. *Proceedings 30th International Symposium on Actual Tasks in Agricultural Engineering*, Zaghreb, Croatia, pp. 33–47.

Opara, L.U. (2002b). Engineering and technological outlook on traceability of agricultural production and products. *Agricultural Engineering International: The CIGR Journal of Scientific Research and Development*, IV, 1–12.

Opara, L.U. (2003a). Image is everything: Undergraduate students' contributions towards better understanding and improvement of public attitudes and perceptions of agricultural engineering. *World Transactions on Engineering and Technology Education*, 2(3), 457–460.

Opara, L.U. (2003b). From farmers to blue-collar professionals: Agricultural engineering students' perspectives on their career prospects, critical skills, and discipline name change. *World Transactions on Engineering and Technology Education*, 2(3), 399–402.

Opara, L.U. (2004a). Emerging technological innovation triad for smart agriculture in the21st century. Part I. Prospects and impacts of nanotechnology in agriculture. *Agricultural Engineering International: The CIGR Journal of Scientific Research and Development.* Invited Overview Paper, VI, 27 pages.

Opara, L.U. (2004b). Outlook on agricultural engineering education and research and prospects for developing countries. *Outlook on Agriculture*, 33(2), 101–112.

Opara, L.U. (2006). A comparison of students' insights and attitudes towards agricultural engineering education: Evidence from New Zealand and the Sultanate of Oman. *Proceedings of the 34th International Symposium on Actual Tasks in Agricultural Engineering*, Zaghreb, Croatia, pp. 13–19.

Opara, U.L. (2012). The need for a professional network of agricultural and biosystems engineers in Africa. *International Journal Postharvest Technology and Innovation*, 2(4), 426–432.

Opara, U.L. (2018). The future of agricultural engineering and global food security. *Keynote Address, South African Institution of Agricultural Engineers (SAIAE) Symposium and Biennial CPD Event & PASAE Workshop*, 17–20 September. Salt Rock Hotel, Durban, South Africa.

Opara, U.L. (2024a). Chapter 36: CAM-SAM: A Computer-Aided Tool for Teaching and Research on Sustainable and Selective Agricultural Mechanization Development. In: Opara, U.L. (ed.). *Agricultural, Biosystems, and Biological Engineering Education – Global Perspectives and Current Practice.* CRC Press/Taylor & Francis Group.

Opara, U.L. (2024b). Chapter 31: Nomography-based models for introductory undergraduate teaching and research in farm power and machinery selection in developing countries. In: Opara, U.L. (ed.). *Agricultural, Biosystems, and Biological Engineering Education – Global Perspectives and Current Practice.* CRC Press/Taylor & Francis Group.

Opara, U.L. (2024c). Chapter 42. Prospects for agricultural, biosystems and biological engineering education and research for knowledge- intensive, data-driven, climate-smart, and sustainable agriculture. In: Opara, U.L. (ed.). *Agricultural, Biosystems, and Biological Engineering Education – Global Perspectives and Current Practice.* CRC Press/Taylor & Francis Group. (chapter in this book).

Opara, U.L. (2024d). Chapter 4: Top 100 questions of importance to the future of agricultural engineering education, research and practice in Africa. In: Opara, U.L. (ed.). *Agricultural, Biosystems, and Biological Engineering Education – Global Perspectives and Current Practice.* CRC Press/Taylor & Francis Group.

Opara, L.U., Al-Adawi, S.S. & Al-Shukeili, T.Y. (2004a). Students' perceptions of agricultural engineering education at Sultan Qaboos University in Oman, Part 1: Academic program selection. *World Transactions on Engineering and Technology Education*, 3(1), 139–142.

Opara, L.U., Al-Adawi, S.S. & Al-Shukeili, T.Y. (2004b). Students' perceptions of agricultural engineering education at Sultan Qaboos University in Oman, Part 2: Curriculum content and career preferences. *World Transactions on Engineering and Technology Education*, 3(1), 143–146.

Opara, U.L. & Al-Mullah, Y. (2024). Chapter 17: Agricultural engineering education at Sultan Qaboos University, Sultanate of Oman – Historical evolution, curriculum reform, students' perceptions, and future prospects. In: Opara, U.L. (ed.). *Agricultural, Biosystems, and Biological Engineering Education – Global Perspectives and Current Practice.* CRC Press/Taylor & Francis Group.

Opara, U.L. & Cuello, J. (2006a). Guest editors. Special issue series: Trends in agricultural, biosystems, and biological engineering education, Part I: Agricultural engineering and the emergence of biological engineering. *International Journal of Engineering Education*, 22(1), 2. http://www.ijee.ie/contents/c220106.html

Opara, U.L. & Cuello, J. (2006b). Guest editors. Special issue series: Trends in agricultural, biosystems, and biological engineering education, Part II: Curriculum reform, assessment and accreditation. *International Journal of Engineering Education*, 22(6), 1122. http://www.ijee.ie/contents/c220606.html

Opara, U.L. & Cuello, J. (2007). Guest editors. Special issue series: Trends in agricultural, biosystems, and biological engineering education, Part III: Innovative teaching and learning methods in agricultural and biological engineering education. *International Journal of Engineering Education*, 23(4), 630. https://www.ijee.ie/contents/c230407.html

Opara, L.U., Perret, J.S. & Ampratwum, D.B. (2006). The evolution of agricultural engineering education at Sultan Qaboos University. *International Journal of Engineering Education*, 22(6), 1139–1148.

Sims, B., Hilmi, M. & Kienzle, J. (2016). Agricultural mechanisation: A key input for sub-Saharan African smallholders. *Integrated Crop Management*, 23–2016, 44 pages. Food and Agriculture Organization of the United Nations, Rome.

Singh, A. (1979). The basic needs approach to development vs the new international economic order: The significance of third world industrialisation. *World Development*, 7, 589.

Symonds, E. (2011). A practical application of survey monkey as a remote usability-testing tool. *Library Hi Tech*, 29(3), 436–445. https://doi.org/10.1108/07378831111174404.

Taylor, A.J., Opara, U.L. & Smithers, J.C. (2021). Agricultural engineering in Africa: A key driver for transforming agriculture to deliver food security and to support economic prosperity. Pan African Society for Agricultural Engineering (PASAE/AfroAgEng), Pretoria.

Varela, C., Ruiz, J., Andrés, A., Roy, R., Fusté, A. & Saldaña, C. (2016). Advantages and disadvantages of using the website SurveyMonkey in a real study: Psychopathological profile in people with normal-weight, overweight and obesity in a community sample. *E-methodology*, 3, 77–89. doi: 10.15503/emet2016.77.89

Appendix A. Additional final comments to improve the future of agricultural engineering education in Africa.

1	Governments through educational institutions should prioritise agricultural engineering in universities s as to create a bridge between the industry and the farmers. Also Africa is behind on food processing techniques
2	More attention should be given to agricultural engineering because it lies in the faith of our nation
3	We need to start informing people about agricultural engineering and what its all about. Make sure that all our students are well equipped and ready to impact Africa this can be done by educating students on what needs to be done, show students areas that current engineers are struggling with so that we may know what is expected of us in the future.
4	Tell the future generation what is agricultural engineering and its importance to the environment. Because very few people know about this career by so doing we will have more people doing agricultural engineering and more ideas we will get.
5	Agricultural Engineers should be proud of their profession and so be able to reach out to the future generation of potential engineers
6	ESTABLISH EXCHANGE PROGRAMS IN THE STUDY
7	People offering agricultural engineering must feel proud of the discipline and focus on what need to be done. There must be more practical approach to studies in the field than the theory aspect. Also people with brilliant ideas on this discipline must be given support financially to undertake researches.
8	Public enlightment on its advantages and benefits. Also awareness campaign.
9	The public is very ignorant about Ag Eng and just assume it has to do with just farming and thus don't bother to educate themselves about it. My friends often ask me if I'm going to buy a farm after my studies and the youngsters find it to be an "unglamourous" profession (since they perceive it to be farming), especially those from urban areas. I therefore think that, in addition to vigourous marketing, the name should be changed.
10	I think students need career guidance when studying agricultural engineering because most of them have a negative perception about agricultural engineering.
11	Mmmmmmm…I will say that the government should look into agriculture properly…
12	I wanted to see in future, best and hard working agricultural engineering students to enhance improve the above survey.
13	The general public must be educated on what Agricultural Engineering is about.
14	More emphasis should be put on enhancing the public understanding of Agricultural Engineering.
15	I believe there must more practical stuff for the students.
16	Government should place emphasis on agricultural engineering.
17	Practicals should be emphasised more.
18	We rule the world, so we need more lab and workshops, the ones we have are not functional, and we should be given a large area of land for our field activities. Thank you.
19	Hosting agricultural engineering seminars for African students so they can interact with other successful engineers in the field of study
20	Practical aspect of the course should be encouraged
21	Our duties as an Engineer is to provide food everybody.
22	Provision of machines and good tools in Agriculture Engineering farms and schools.
23	We (agricultural engineers) should be united and focus in our quest to archive a healthy agricultural goods and products
24	It should be rather practical than theoretical. And a big part of the program should be spent on industrial training. I feel the 30 weeks are not enough
25	Agricultural engineers should establish themselves so that people in the outside world will be attracted to it. Lecturers should be inspiring people
26	Empower the young innovators
27	Agriculture is the key and basic sector of any African country's economy especially Nigeria.
28	Publicity, in the sense that agricultural engineers should be able to involve the youths in agriculture
29	Africa…should be the world best continent for arable land…and different species of animals…we have lots of youths out there stuck in crossroads of what to study in the university…if top governmental organisation and NGOs can team up…to create Positive awareness about Agricultural Engineering…we will have optimistic youths all over seeking to study…2: Those already studying should be provided with practical experiences, they should have a medium to express their inner thoughts on ways of improving the sector.
30	Agricultural engineering education in Africa should be 60% practical and 40% theoretical.
31	Practical aspect of the course should also be made compulsory.

32	I believe with the help of Agricultural engineering and its branches, erosional problems caused by water can be solved. For instance, Edo state and Anambra state are the most erosion affected states in Nigeria thereby reducing agricultural activities and land in states that are known to be highly agriculturally productive in Nigeria…with the help of soil and water conservationist this can be reduced. If government invest and put interest into agricultural sector, it will help generate huge income and engineering aspect of it can play a big role.
33	taking the student on an excursion so as to appreciate the course more better.
34	I so much believe that the world needs us so much (Agricultural Engineering). We make the green
35	Agricultural Engineering Education in Africa needs to improve more in pratical and Equipments
36	There should be a proper education on what Agric and bio resources can offer so to enlighten our youth
37	It could be improved in so many but I like to point out that the facilities used for teaching are very poor compared to those of other departments.
38	Creating awareness.
39	I think Agricultural Engineering Education should be more practical as well as theoretical
40	Let Agricultural Engineering be like other Engineering profession, by creating job opportunities for student and exposing student to field beyond Agriculture both in design of machinery that can save human labour. THANK YOU
41	Don't know about Africa as a whole but for Nigeria, we need only prayer.
42	Without reliable energy services, farmers and agribusinesses in developing countries are less able to increase food production and engage in value-added processing. New technologies, such as solar food dryers and solar water irrigation, allow farmers living off-grid to replace expensive diesel generators with cheaper and cleaner technologies. This segment of the energy access market is still in the early stages of development but much progress is being made in our work with companies that offer modern energy solutions to boost productivity and economic growth.
43	AGRICULTURAL ENGINEERS UPON GRADUATION SHOULD BE PROVIDED WITH IMIDIATE JOB OPPORTUNITY INTO INTRESTING AGRICULTURAL FIRMS SINCE THE NUMBER OF ASPIRANTS IS USUALLY LOW, THIS WILL MAKE MORE PEOPLE TO ENROLL IN THE PROGRAMME. More should be expected from the graduates of the programme, this will make the student focus more on their studies to improve their abilities. THANKS FOR GIVING ME THIS OPPRTUNITY TO COMMENT. GREATER FUTURE FOR AGRICULTURAL ENGINEERING.
44	There should be change within the education system such that it encourages practical work in the agricultural engineering fields
45	agricultural engineering should be made more of practical, to enable us student know more about the application. THANKS
46	Let us think of innovations that apply in our different countries in Africa to feel the tangible success of agricultural engineering to our societies
47	Special seminars to enlighten secondary school students on the importance of agricultural engineering...
48	Let lecturers be well informed and less borrowing of courses
49	We need innovative minds
50	The agricultural engineering discipline should have a standard curriculum. Agricultural Engineering should be understood as being different from Agricultural and biosystems engineering.
51	Creating awareness and making people know the importance of Agricultural Engineering. This is so because in my country Nigeria many people believe that agriculture is very tedious and stressful. That is why at the mention of Agriculture many people will lose interest in it not knowing what it entails. In my suggestion, if mechanisation of agriculture will be improved more to reduce human labour and also using mechatronics in agriculture.
52	Thank you for this survey. Please I'd suggest you change the name to something more attractive like: Biomechanical engineering. Please read this very important comment which is about agricultural engineering also known as biomechanical engineering. [http://typesofengineeringdegrees.org/agricultural-engineering-degree/]
53	Did the same in India, they call Bachelor of Technology in Agricultural Engineering. Even if it is Bsc.Agricultural Engineering. (The practical oriented course are what important to make the aim of the course successful).
54	The policies, goverment frameworks are not conducive for agricultural engineering education this is the area that should start for future improvement
55	In the attempt to improve the future Agricultural Engineering education in Africa, I would recommend that the course agricultural engineering should be divided according to field of interest; agricultural mechanisation, soil and water resources engineering, Post harvest/Processing engineering, Biosystems engineering, etc.
56	Our curriculum is too theoretical, it would be of great help if heads of education systems make it more practical in Ghana for us… Thank you
57	Please let it be more practical because we still dont know anything. All we can do is solve math questions and memorise stuff
58	Agriculture engineering can allow a student to start his\her agric business
59	Many vocational work for agric students
60	There should be more bursaries for this study program, it seems like most companies they don't have enough knowledge about this whole course.
61	Students of agricultural engineering need to be exposed to bigger companies during their study period.

62	I would like if there could be at least more companies that offer scholarships and bursaries for agricultural engineering student because, there are literally few, like a hand full. e.g SAPPI is an agricultural company but it offers bursaries for other engineering faculties except agricultural engineering, ironic.
63	Improving industrial training of agric engineering students Infrastructural development of Agricultural engineering facilities in tertiary institutions
64	Educating rural farmers on agricultural mechanisation and equipping the educated ones with the machinery.
65	For improvement, more years should be added to learning Agricultural Engineering ie it should be made a professional course. If a year or maximum two is added, students will have ample time in learning both practically and theoretically all there is to Agricultural Engineering.
66	Giving enough information of what is exactly Agricultural Engineering
67	we need to have one campus not howard and pmb please
68	Agricultural engineering is not popular, only a few knows about agricultural engineering in Secondary schools. I think we can try advertise more we can have more engineers in future.
69	Mechanisation of agricultural processes and production would increase the percentage of interest of the youth in the field and would also boost the attractiveness of agriculture generally as a course of study.
70	It is the most underrated Engineering degree, maybe if the sector could try find ways to promote the Degree, Most of us wanna leave this course because it's unknown to industries mostly and that affect our chances of getting funding from Big companies. Also we struggle to get Vacation work at most, this is due to the little and background of this course. Sector should try at least market our Degree like other Engineering degrees, provide sufficient bursaries for the Agricultural sector. It is very important that we promote this degree as it is future to all the environmental crisis the world is about to face.
71	Agricultural engineering is a lucrative discipline of study and impacts a lot in those partaking in the study. It increases their opportunities and possible achievements.
72	This strand of engineering is a little bit of everything and it becomes a concern for the working world
73	We need to spread the word of Agricultural Engineering.
74	there should be more practicals courses involved in the individual courses done from year one to year four especially in the universities.
75	Collaborations between the best universities over the world to increase cohesiveness. The intake of students at Bachelors' level should be more than be doubled.
76	UKZN needs to restructure the program and make its situated on one campus
77	Provide sponsor from previuos disadvantages rural areas student
78	it should be more practical oriented.
79	Pls d program shd be more practical
80	It should be made public to the villages on how to make a quick study into agric engineering by teaching the unknown people who just apply the knowledge of engineering in agric any how while it can be corrected to generate a conservative income from losses and mismanagement or resources and land.
81	Public education should be done to educate the public on the importance of agricultural engineering.
82	Please should work on the issue of public awareness of the proffession. Most People don't understand what the profession is all about.
83	the course should involve more practical aspect and research.
84	More companies are to be set up because agricultural engineers compete with mechanical engineers
85	Mentoring and encouraging more people to have interest in agricultural engineering
86	More industrial training and workshop for agricultural engineers to enlighten us on both current and advanced procedures required by industrial and commercial bodies.
87	I think more attention from our leaders is needed. more work and resources is the most important things. this will help produce more training to we students.
88	I think a name change will do
89	The students from day one should be coached to love the course. More agricultural engineering firms are needed out there to boost interest. Various achievements within the sector should be show cased for the world to see
90	Posting graduates to jobs after completion of the program into various agricultural companies
91	Planning and putting in place measures to conserve, preserve and save specific areas for agricultural work
92	I think it should more of practical work than the abstract we learn all the time. We should understand and practice what is being taught in class
93	Agriculture in itself must be given maximum attention by the state so that new systems that could be implemented to help the various sectors through Agricultural Engineering will be rightly supported and thereby implemented
94	Names of the program should be changed to change people concept about the program
95	specialisation should be introduced at the undergraduate level

96	There should be more investments in the study of agricultural engineering to make it attractive and appealing to students. Students should be made to obtain opportunities after study.
97	Investments!!!
98	More practicals and field tours
99	Increase on the number of practical lessons taught while still in universities
100	More research should be carried out in the field of agriculture.
101	The program has to be funded and supported with research facilities.
102	African should take the disciplines of Agricultural Engineering as a way to modernised and sustainable agriculture. Major focus should be laid on it, through practising extensive training and education activities to producing practical agricultural engineers engaged in agricultural engineering fields. Network and cooperation carried out among the industries, government and institutions in producing agricultural engineering in the 21st Century. More research and technologies should be developed in the agricultural sector for agricultural engineering application. Opening up wide field of jobs for agricultural engineering so that many people can see the good works of agricultural engineers.
103	agricultural engineering should be intergrated with mechanical engineering to make one course
104	The programme should focus more on practical skills atleast 60% hands on
105	Governments in Africa should give first priority to agricultural engineers since most countries rely entirely on Agriculture, they should incorporate them most in processing activities since we are the future as we can do all around engineering.
106	The course should be give students much more hands on experience and practicals during the four years of study other than theory.
107	For the few months i have so far been looking for employment in my profession, i have really come to understand that most people (including potential employers) are not aware of agricultural engineering and do not understand what we can really do. So they end us devaluing us and the whole profession thinking that we can only drive farm tractors. So I think we should first promote our course through advertisements so that people can start recognising our value. Thanks for the survey. I pray that you achieve your goal. Regards.
108	Agricultural Engineering students should be sensitised properly on what the course is all about and what they can expect after successfully finishing the programme. I personally was a victim. I didn't know how great the course was until it second year and this affected my performance.
109	Implore students to like the course as soon as possible because this course could be a core player in this century given the immense challenges we are facing so the earlier students get to realise this, the better.
110	Devote more work on practicals. There should also be exchange programmes between universities offering Agricultural engineering
111	Agricultural engineering should be given the first priority on resource allocation because it can improve lives of very many people within a short period of time. most countries in Africa have got good fertile soils but lack the technology and knowledge of how to fully utilise land.
112	We ought to inform the youth of African about the value that Agricultural engineering plays in the society. I would suggest independent Agricultural engineering workshops that are aimed at exposing all the fields of this profession to the young generation so as to enable them fully understand the necessity of the course. Thank you.
113	Yes. The government should employ an Agricultural Engineer at every district level
114	Student awareness about Agricultural Engineering and its role in society should be done in the earlier stages of education
115	Availability of academic resources and funding since currently the teaching done is so theoretical due 2 limited funds.
116	I think more effort should be put on practical skills rather than theory. Most students graduate and go out in the field and find it so hard to put into practice what was learnt in the classroom.

42

Prospects for Agricultural, Biosystems, and Biological Engineering Education and Research for Knowledge-Intensive, Data-Driven, Climate-Smart, and Sustainable Agriculture

Umezuruike Linus Opara

Introduction

Both agriculture and society are changing more rapidly than ever before. What is assured is that the current rate of change is much faster than ever before. These changes have major impacts on if and how sufficient quantity and quality of food can be produced and delivered in a sustainable, cost-effective, and equitable manner to meet the dietary needs of a growing human population. World population increased from one billion in 1800 to an estimated 2.5 billion in 1950, 7.9 billion in 2020, and at the current growth rate of 1.1% (~83 million) per annum, it is projected to reach 8.6 billion inhabitants by mid-2030, 9.8 billion by mid-2050, and expected to peak at nearly 10.4 billion in the mid-2080s (UN, 2022) with much of the expansion taking place in developing nations. In particular, it is estimated that over 50% of world population growth between now and 2050 is expected to occur in Africa. With Africa having the highest current population growth rate – about 2.37% – compared with major areas of the world, its population of about 1.46 billion is projected to double by 2050. As the economies of these regions of the world develop and consumers have more discretionary cash, the demand for food is anticipated to rise because of nutrition transition and rising life expectancies in these regions as well. In order to satisfy the food (dietary and nutritional) demands of the rapidly increasing population, there is a need for higher agricultural productivity, as the area accessible for cultivation is shrinking concurrently with population growth.

The FAO (2009) has projected that global food production will need to increase by 70% to meet global demand by 2050, while other researchers have presented a much higher increase in global food demand of 100–110% between 2005 and 2050 (Tilman et al., 2011). In a more recent study based on meta-analysis of projected global food demand and population at risk of hunger for the period 2010–2050 across five representative scenarios that span divergent but plausible socio-economic futures, van Dijk (2021) concluded that the total global food demand is expected to increase by 35% to 56% between 2010 and 2050, while population at risk of hunger is expected to change by −91% to +8% over the same period. However, the authors emphasised that by taking climate change into account, the ranges change slightly for total food demand (+30% to +62%) and for population at risk of hunger (−91% to +30%).

On the other hand, major changes are taking place in public views toward agriculture, particularly in terms of its influence on the environment as well as food quality and safety. The eating habits of consumers are increasingly mirroring their way of life – convenience – and personal inclinations consumption patterns are shifting away from high fat, high energy, and high cholesterol meals and towards fresh foods that are more functional, healthier, and nutritious. With increasing awareness of the negative consequences of mechanised agriculture on biological variety, soil degradation, and pollution, a formidable coalition of consumers, scientists, and lobbyists has developed to call into question the sustainability of existing agricultural practices. As seen by the higher pricing of organic products and goods with assurances of quality and safety, these consumers also seek knowledge about the life cycle (systems of food manufacturing and processing) and are prepared to pay extra for it (Opara, 2000). The advent of genetically edited plant and animal products into the food chain for humans has strengthened the demand for better understanding of the synergies between agriculture, the environment, and people, as consumer food preferences continue to evolve. Therefore, it is clear that our roles as researchers, teachers, and engineers of agricultural and biosystems are evolving along with agriculture and the market.

The most effective tools for economic advancement and gaining a competitive edge for industries and nations are innovations in technology (new commercial goods, procedures, and services). The industrial revolution made it possible for factories to produce items in vast quantities. This advancement aided agriculture in a number of ways, including extensive production mechanisation, postharvest managing, and processing operations. The abundance of natural resources in any country or region was crucial to its overall success (land, water, minerals). However, recent technical advancements that support prosperous modern economies depend on the development, wide diffusion, and intensive application of information produced by expenditures in research and education. Therefore, in order to remain competitive, firms must continually push the boundaries of knowledge and information management.

DOI: 10.1201/9780429150111-45

Given this context, a common question that arises is: What are the prospects of agricultural engineering in the emerging knowledge-driven economy, considering that the discipline first emerged and thrived in an era that can be referred to as a machine-based economy, throughout which the mechanisation of agriculture heightened areas of cultivation, factor efficiency, and profits? Information drives the knowledge that advances contemporary, feasible sophisticated agriculture, just as the utilisation of the "mechanical engine" granted the capacity to drive mechanised agriculture and automation eradicated the monotony of muscle-powered duties. In this chapter, we introduce the nexus between knowledge and innovation, highlight some engineering challenges in knowledge-intensive agriculture, and spotlight selected new, emerging, and transformative technological innovations that must be integrated in the education and training of the global agricultural biosystems/ biological engineer.

Connecting Knowledge and Innovation in Agricultural Engineering Education

The impact of technology breakthroughs on business and our lives is changing dramatically. It is now commonly acknowledged that the capacity to produce and deploy fresh information and technology has evolved into a weapon for business advantage and competitiveness, as nations and organisations look for new methods to sustain profitability and competitiveness. Successful organisations are becoming more knowledge-based, and they heavily rely on information technology for the creation, application, and dissemination of knowledge. It is no longer adequate to rely on ongoing production and delivery of outdated products and services that meet market quality standards in order to be relevant in the future market. Businesses and organisations that carry on in this way find themselves abruptly confined to market niches, if not completely shut out of the shifting dynamics of the market (Williamson, 1975). For businesses and organisations looking to adopt new technologies that enhance the value of already-existing goods and services, the capacity to continually develop knowledge has emerged as a significant barrier. Since current knowledge is increasingly across boundaries and most businesses see themselves as knowledge-intensive businesses where intellectual capital is the most important resource, future competitiveness was defined by the availability of knowledge, and each business needs a broad group of knowledgeable individuals with specialised training who can build their own knowledge and work together to produce new knowledge that improves the efficiency of the firm (Ritcher, 2001; Garvin, 1993). This evolution has culminated in a shift in focus from traditional management of innovation, which focused on the generation of specific product and process innovation, to systemic management of technology, which emphasises the creation of knowledge as an underpinning for innovation (Ritcher, 2001; Ritcher & Vettel, 1995).

The growth of new information and technical innovation are key drivers of future economic competitiveness, according to the current worldwide ranking of countries' levels of economic competitiveness, which places the United States, Singapore, and Finland as the top three (IIMD, 1999). The heart of American competitiveness has been connected to the country's exceptional capacity for innovation-driven growth, in which technological advances are transformed into commercial concepts that are then used to generate new economic opportunities (Smith, 1999).

A typical innovation is a new product, technique, or service that is efficient in the marketplace, and it is the activity of converting technological potential developed via research and practice into commercial reality (Lievonen, 2000). Hence, innovation comprises the creation of unique value, which might take the shape of a unique technique, gadget, or service. It is important to new technologies, utilisation, and products that create money and increase the wellbeing of humans (SIAC, 2000). In this view, innovation may be described as the transfer of a concept from the mind to the market, and also from the laboratory to the factory (including the farmland). Sustainable economic growth is characterised, among other things, by ubiquitous innovation and technological advancement supported by an effective innovation system in successful organisations and advanced nations. Thus, dynamic economies, organisations, and disciplines are characterised by innovation, and innovation is increasingly offering the key to these institutions' better competitiveness.

The innovation process has three main components:

(i) Gaining and maintaining skills
(ii) Developing ideas and conducting research
(iii) Hastening the commercialisation of ideas into commodities and services

Education that enables people to gain suitable skills, as well as research that generates ideas and new information, are essential preconditions for the development of innovative goods and services with commercial worth and benefit to society. In truth, research is recognised as the bloodstream of innovation since it promotes long-term creation of ideas.

Some Challenges and Opportunities Facing Agricultural Engineering Education in Knowledge-Intensive Agriculture

The discussion above serves as an example of what engineering may look like in the near future of knowledge-driven agriculture. At the same time, as previously mentioned, the knowledge-innovation nexus has an effect on both agribusiness and business in general. The implementation of engineering and technology advancements in agriculture, however, has extra challenges due to particular agricultural features (such as changing consumer and public attitudes towards agriculture, food, and the environment). The factors must be handled in order to maintain agriculture's competitiveness, stop its declining terms of trade, and increase factor productivity. In a knowledge-based, information-driven economy where advanced biology, information, computer, and related technologies are poised to rule global business activities, what are the practical engineering and technological challenges agriculture faces today?

Several technological obstacles and opportunities exist to change agriculture to transcend the traits it already possesses from the period of the industrial revolution (which are characterised by a heavy reliance on mechanical labour and the use of chemicals per unit of land). The following are some relevant examples which underline these challenges and opportunities for agricultural and biosystems engineering.

How do we manage variability inherent in the biological dimension of agriculture?

In terms of quality and yield, agricultural procedures and the products that arise are highly variable. Hence, it is essential to have precise details about the surroundings, plants, animals, procedures, as well as how they interact to control this variability and produce goods and services that live up to the high standards set by customers and the general public. Future agricultural applications that combine information and communication technology (ICT), mechatronics, and nanotechnology are poised to advance agriculture beyond current levels of mechanisation and automation. The effective integration of expanding technological innovation in agriculture calls for the training of people who are knowledgeable about the requirements for agricultural data and who can take these requirements into account when designing new machinery and procedures. Increased focus is needed when determining what data to collect, the most economical way to collect the data, how to analyse and present the data, and how to incorporate the data into the more expansive system of management for the supply chain for agriculture. Here lies tremendous opportunity for data science and computational thinking in sustainable agriculture and food systems transformation.

What is the future of agricultural mechanisation as applied to small-scale/smallholder agriculture in Africa?

The bulk of the world's population still depends on subsistence smallholder agriculture to support them, especially through producing food, animal feed, and fibre. Given the overwhelming achievement of medium- to large-scale agriculture in reaching high productivity to ensure a plentiful supply of food on a global scale, many parts of the developing world, particularly in sub-Saharan Africa, have been unable to profitably mechanise small-scale agriculture. Genetic engineering and other agricultural biotechnologies, such as those that increase output while exhibiting pest and disease resistance, have the potential to help. Agricultural engineering has the expertise to continue to lead this push for sustainable mechanisation of smallholder agriculture in the face of emerging technologies like information technology, biotechnology, and nanotechnology. Africa cannot continue its reliance on the hand hoe and human muscle (Figure 42.1) to meet the present and future human demands for food, feed, and fibre in sustainable development.

How do we practice agriculture on vulnerable ecosystems and marginal fields?

Since the world's populace, primarily in emerging nations, continues to grow and impoverishment confines their capacity to obtain food and other resources, there exists a pressing need to enhance farm production. Sadly, due to over-cropping and adverse climatic conditions, the majority of these farmlands have marginal fertility for profitable commercial agribusiness. Novel scientific and technological inputs are required to boost farm productivity and increase food production in affected areas.

What is the blueprint for developing new agri-food value chains?

The rising internationalisation of the world agricultural and food supply system necessitates greater vertical integration within and between enterprises in order to manage variability effectively and remain competitive. With ever-changing consumer demand and consumption habits, ensuring stable supply of high-quality, safe, and traceable food goods remains a key priority on the trade agenda. Implementing cost-effective precision agriculture supply chain technologies, site-specific farming, for example, will provide farmers and postharvest personnel with the equipment they need to have greater influence over yields, the quality of the product, and safety. Traceability has become a key indicator of trade in agriculture because of food scares, public concern over genetically modified foods, and the implications of these concerns on lowering public confidence in the safety of the human food chain. The livestock industry, especially in Western Europe, has more developed traceability food chains. The horticultural and agronomical food chains, including fresh produce and bulk foods like cereals and processed foods need to address the gap in assuring safety and traceability in agri-food systems. Agricultural and biosystems engineering technologies will play critical roles in detection, measurement, and control, including data capture, analysis, and transmission among stakeholders.

The desire for traceability from field to plate and the expanding global trade in eco-foods (referred to as "green labelling") are indicators of this (Opara & Mazaud, 2001). Yet, there is worry that new ideas are required in areas where food insecurity and poverty are serious issues in order to increase yields and reduce the prevalence of pests and illnesses using sustainable agricultural techniques. Consumers now rely on product labelling and farm certification by appropriate agencies, as it is difficult to distinguish items produced under various manufacturing methods and sold in the market based on sensory attributes. To distinguish between goods from various production methods, such as organic versus conventional agriculture, new measurement tools are required.

What are the pathways for sustainable agricultural intensification?

The sustainable development goals (SDGs) of the UN's Agenda 2063 and the aspirations of the Africa's Agenda 2063 call for a balance between current production and consumption patterns with the needs of future generations. With increasing evidence on the negative impacts of both intensive agriculture and smallholder farming on the environment, bioresources, and climate change, modern agriculture must now meet consumer demands for waste minimisation, animal care, and sustainable farming practices. As new sustainability indices (such as environmental, water resources, land resources, air, and biodiversity) emerge and agriculture is more closely associated with extensive resource depletion, environmental degradation, and climate change, engineers and the engineering profession will come under increased public scrutiny to incorporate sustainability issues into curricula and adopt sustainability as a professional ethic. While the mechanisation of agriculture has been proven as a necessary engineering input to meet the

FIGURE 42.1 A typical hand hoe used for farming in parts of the middle belt of Nigeria, demonstrated by the author (Prof. U.L. Opara) in Abuja during the Exhibition Session of the 2017 National Conference on Transforming the Livestock Sector in Nigeria.

increasing demand of a growing population for food, feed, and fibre, there is a greater realisation that agricultural mechanisation practices must be sustainable in consonance with the SDGs and aspirations of Agenda 2063. Furthermore, eliminating the drudgery and low productivity of subsistence smallholder agriculture based on the hand hoe and human muscle remains a major stumbling block to the successful transformation and modernisation of agriculture and food systems in Africa. Agricultural and biosystems engineers must embrace these challenges and opportunities through a new education, training, and research, including the reformation of curricula and new research mandates.

What reforms in agricultural engineering curriculum and research agenda are needed?

Developing knowledge-driven technological innovations for agriculture must be incorporated into the syllabus and research schedules of agricultural engineering and related technical programmes serving agriculture and other biosystems in order to take advantage of the potential opportunities presented by these innovations. Prior writings have emphasised the importance of the knowledge-innovation nexus as well as the role that research and education play in the generation of novel ideas. Because the ultimate measure of any profession's worth is how well it serves mankind and how well its graduates may find gainful work, we cannot ignore or give lip service to the technical forces that drive competitive organisations and economies in the twenty-first century. The introduction of curricular changes that provide future agricultural engineers with the knowledge and abilities to expand their employment possibilities into other knowledge and service industries would make the field more alluring to new graduates (Opara, 2001, 2023a, 2023b, in this book; Brumm et al., 2023, in this book).

Where and how will we grow the food we eat?

With increasing demand for protein food, global aquaculture has risen rapidly to supply for fish and seafood products. However, huge deficits still exist, especially in developing countries with high incidences of food insecurity and malnutrition, while the demand for protein foods continues to grow due to a fast-growing middle class. Sustainable harvesting of seas and oceans for food remains an opportunity for feeding and nourishing a hungry world. At the same time, new methods of soilless agriculture have continued to emerge as part of the so-called "urban agriculture". These include hydroponics, aeroponics, and vertical farming. Successful molecular farming to produce novel protein sources as alternatives to industrial livestock farming have been reported and these laboratory-cultured meat are now available in selected market outlets, particularly in Europe and North America. The extent to which these will replace soil-based agriculture remains largely unknown. However, these novel farming techniques present new challenges and opportunities for precision, automation, and resource optimisation. How would engineering technologies shape the future of these new forms of agriculture and food systems? How should we train agricultural and biosystems/biological engineers to take advantage of these rapid transformations of the agri-food system?

Incorporating Emerging and Transformative Cross-Cutting Technological Innovations into Agricultural Engineering Education and Research

Following the establishment of the critical role of knowledge and technological innovation in future competitiveness, allow us to quickly address three knowledge-driven, information-driven technological innovations with the capacity to reshape the nature of engineering and technological contributions to agriculture in the third millennium. The most significant results of the continuing biological and information revolutions to date are these cutting-edge innovations from the modern era.

Information and Communication Technologies

Our daily lives have been transformed by ICT advancements, which have also sparked a new wave of industries. ICTs include databases, application programmes, and hardware-based user interfaces. Computers and new information technologies like the internet, teller machines that are automated, and wireless connections have made it feasible to quickly gather, process, and transmit information. These developments are generally considered the forerunners of the international economy's escalating globalisation. The expansion of the internet (world wide web) and web-based initiatives like online shopping, distance education, and real-time data/information sharing between researchers on various continents have broadened the range and frequency of joint endeavours, which supports an environment that is conducive to inventive innovation. As costs decline along with increasing processing power and the accessibility of new forms of software that enhance current corporate procedures and human activities, this trend is expected to intensify. These advancements are examples of technology-push variables that have agricultural implications across the board. For instance, the trend toward supply chain management and precision farming is only possible with the availability of relevant data regarding inputs and on-farm and postharvest activities.

Automatic diagnosis can be achieved by integrating computing techniques into livestock husbandry, and these improvements may be further reinforced by remote diagnoses as well as services founded on wireless transmission of information (Munack & Speckman, 2001). There are already applications for geospatial technologies like geographic information systems, global positioning systems, and remote sensing, which provide measured position data for agricultural machinery through online data exchange between stationary farm computers and mobile field machinery (Speckman, 2000). The automatic measurement and management of the product environment, field mapping for variations in yield and product quality, online quality and safety assessments of goods, and monitoring and disease epidemiological prediction are some of the beneficial uses of ICT in agriculture. Direct connectivity to the internet or other global networks and databases is necessary for farmers, customers, and other players in the agricultural supply chain to plan and make technical and financial choices.

In an effort to satisfy the diverse consumer demand for products that are produced sustainably, are of high quality, are safe to consume, and can be tracked, it was recently claimed that agriculture constitutes the growing big consumer of ICT and electronic innovations (Sigrimis et al., 1999). To address the different requirements of 21st-century consumers of agricultural goods and services, the employment of artificial neural networks, hazy knowledge-based systems, and ground-breaking algorithms represents a new horizon for agricultural automation and intelligent supervision. "Speaking fruits" during storage and "speaking plants" during cultivation are some of these uses (De Baerdemaeker & Hashimoto, 1999).

Biotechnological Advances

The DNA code preserved and deciphered by living cells serves as the conceptual underpinning of contemporary biotechnology. The Human Genome Project serves as an example of how one of the most ambitious and challenging scientific projects of all time has the capacity to alter the genetic structure and composition of plants and animals. Biochemistry, microbiology, medical sciences, and industrial production technologies have combined to form modern biotechnology (Lievonen, 2000). Historically, biotechnology has been used in agriculture to prepare and preserve food, such as through fermentation. Nevertheless, the success of genetically engineering crops and livestock has created new opportunities and difficulties across the whole agricultural value chain. DNA serves as the theoretical foundation for current biotechnology. As a result of agricultural biotechnology, animals and plants now have unique genetic features for productivity, disease resistance, and insect resistance, which give food products made from genetically modified (GM) plants, animals, and other creatures appealing characteristics.

Let us look at a few of the particular and probable effects of biotechnology (especially genetic engineering) on plants, animals, and the environment, as they relate to agricultural engineers' future roles):

(i) Altering the internal composition, phenology, physiology, and overall design of the plants and animals that serve as the fundamental "factories" for the production of food and fibre. The design, use, and control of future mechanical devices as well as the operational conditions for field and postharvest operations may need to be altered as a result of these consequences.
(ii) Commercial fruits with particular ripening and hardness traits, like the tomato, have already been created by regulating the regulation and expression of genes in plants. This might have an effect on how these plants' products need to be handled and stored.
(iii) With the emergence of new organisms for food processing and conservation, knowledge of novel process kinetics, appropriate parameters for processing, and interactions between the process variables will be required in order to manufacture desired food products.
(iv) For the advancement of biosensors for process classification and innovative devices for site-specific application of biocontrol agents, an integrated knowledge of the fusion of biological and engineering factors driving these processes will be required.
(v) Evaluating the climatic and environmental consequences of GM plants, animals, and organisms necessitates the development of new measures and analytical tools, as well as unique measurements and analytical techniques for assessing the ecosystem and environmental repercussions of GM plants, animals, and organisms.
(vi) To maintain the superiority, safety, pureness, and transparency of GM-based products, new insights in postharvest science and handling of materials technologies will be required.

When the same postharvest handling and/or processing facilities are used, contamination can still happen even when firms can separate their GM- and non-GM product lines. This increases the need for quick detection of products made from GM ingredients. This is an area of considerable competence of the agricultural and biosystems engineering discipline.

Nanotechnological Advances

Nanotechnology is a new area that integrates basic science, materials science, and engineering at the nanoscale (billionths of a metre) to develop and make molecule-sized structures and gadgets. Nanotechnology advancements are the product of the work of scientists and engineers in various domains, including chemistry, biology, photonics, and microelectronics. Despite being least-advanced of the three technical triads to date in terms of practical applications in agriculture, nanotechnology has significant potential in a range of sectors, including health, electronics, and agriculture. Several researchers have proved the potential of nanoprobes, nanorobots, atom-moving machines, nanoresistors, nanowires, and living machinery for industrial use (Stix, 2001), including nanofabrication (Whitesides & Love, 2001). Increasing integration of nanotechnology with other emerging high technologies increases the likelihood of attaining these prospective technological and economic benefits (Lievonen, 2000). The premise behind nanotechnology is that the resultant structures might have better properties, such as chemical, electrical, mechanical, and optical properties. Although investments in nanotechnology have been dominated by biomedical research, defence for fighting cancer, and creating missiles, the development of tools and procedures for characterising and building nanostructures has broad applicability across many industries, including agriculture.

Given the rapid advancements made in various fields over the past decades, what potential benefits does nanotechnology hold for agriculture in the future, and for what reason should agricultural and biosystems engineers pay attention? The following outcomes may result from success in embracing and incorporating nanotechnology in agricultural engineering education and research:

(i) Nanobots, miniature/micro robots that are as tiny as human blood cells or even smaller, might investigate every capillary and even be directed in for close-up examinations of brain complexities in animals during reproduction and unique on-farm diagnostics if they were to be used in their billions. The nanobots would share the knowledge with one another and other computers constructing the scan database utilising high-speed wireless connections (Beijing Review, 2000).

(ii) Nanostructures that can be inserted into plants and animals during maturation to collect and share crucial instantaneous information that includes developmental rates and physiological processes that offer insights into performance, efficiency, and exposure to environmental, chemical, and physical hazards.

(iii) The use of nanowires and nanocircuits will increase the carrying capacity of optical networks, speeding up the transmission and mining of enormous amounts of data. A significant amount of data is naturally produced by agriculture and other bio-industries about environmental factors, agricultural inputs, animal inputs, inputs into processes, and so on (thus the rise of bioinformatics as a subdiscipline in mathematics and the information sciences). Mining these data and incorporating the results into an agricultural system of information facilitates effective business management. Due to the fact that a majority of farms and agricultural sectors are often located in remote areas distant from major communication centres, the development of high-capacity networks of information will make it simpler to capture, analyse, and transfer the critical data that constitutes the foundation of precision agriculture.

(iv) Integrated supply chain traceability may be possible with mechatronic nanostructures that may be implanted into plants and animals and passed on to groups of products created from these components. Such nanostructures include smart nano-cards that gather and store data on product and process history.

Data Science and Computational Thinking

A massive volume of data is generated in agriculture, from the supply of inputs to the production, harvesting, postharvest handling, processing, packaging, storage, distribution, marketing, and utilisation of food, feed, and fibre. Similarly, huge datasets are generated on the impacts of agriculture on the environment, natural resources, climate, biodiversity, and other sectors of the economy. Capturing, storing, analysing, interpreting, and converting these data into information, new knowledge, and novel technological solutions for sustainable agri-food systems' modernisation and transformation presents a tremendous opportunity for education and research in agricultural and biosystems engineering.

Artificial Intelligence and "Unmanned" Agriculture

Rapid advancements in the interconnectivity and interaction between digital electronics, sensors, big data, and computational power has resulted in the creation of artificial intelligence (AI) and its sub-sets such as machine learning (ML) and deep learning (DL). AI is the ability of a machine to carry out cognitive tasks that we connect with human minds, including sensing, thinking, learning, interacting with the environment, solving problems, and even exercising creativity. Applying AI to real-world issues in agri-food systems will have significant effects on agribusiness to become more efficient, profitable, and ultimately, more sustainable. However, in the end, the value of AI lies not in the systems themselves but in how agribusinesses utilise them to help people — and their capacity to communicate this to shareholders and the general public in a way that fosters and gains trust while creating new value propositions along the value chain.

AI in the form of ML is based on algorithms that have been trained on data. Instead of receiving explicit programming instructions, these algorithms can learn by processing data and experience to recognise patterns and make predictions and suggestions. The prospects for applications in agri-food systems, from field to fork, is huge. In order to increase their effectiveness over time, the algorithms also change in reaction to fresh data and experiences, which are the characteristics of agriculture and other land-based industries. ML has both more promise and a greater necessity in agri-food systems because of the volume and complexity of data that is currently being produced – a volume and complexity that is too great for humans to adequately comprehend. In comparison to classical ML, DL can process a larger variety of data sources (such as images in addition to text), requires even less human intervention, and frequently yields findings that are more accurate. Together, the various dimensions of AI present unparalleled opportunities for far-reaching improvements in agriculture by harvesting past, current, and future datasets to improve agricultural productivity while minimising the negative impacts on bioresources, the environments, and the climate. With AI, both "manned" and "unmanned" agricultural machines along the value chain can be trained to perform tasks with greater precision and efficiency, with sustainability as an objective function.

Harnessing the Opportunities for Agricultural Engineering Education and Research

A comprehensive strategy that includes curriculum reform in schools and a shift in the research objective away from its prehistoric concentration on "machines" and toward knowledge, process engineering, and systemic collaboration is required to take advantage of the prospects provided by emergent technologies in agriculture. Young school graduates must find agricultural and biosystems engineering intriguing, and one clear way to do this is to make sure that our curricula allow them to develop abilities that go beyond agriculture. In many nations and organisations, curriculum and educational reforms do not often move as quickly as broader economic changes (Cortez et al., 2001). Yet we must keep in mind that fewer graduates from a discipline area secure employment in the typical sectors the discipline serves. For instance, Prausnitz (2001) discovered that just 20% of contemporary chemical engineering graduates from the University of California, Berkeley, work in the conventional chemical and petroleum businesses. Surprisingly, the author found that the majority of new graduates found employment in fields that either did not exist ten or twenty years ago or, alternatively, did not until recently recognise the value and significance of chemical engineers in their operations.

Agricultural engineering training programmes must expand graduates' employment opportunities beyond agriculture to include other biological and knowledge-driven companies. What worries me about the agricultural engineering profession's future in the knowledge-based economy arises from the undeniable fact that agricultural engineering is intrinsically related to the society it serves through the goods and services it provides. As a result, agricultural engineering must evolve since the world is changing, as a result of rising technology breakthroughs as well as broader economic and socio-political trends. The underlying existence of agricultural engineering, like that of other fields, depends on whether it is in reaction to society's expectations and requirements (Prausnitz, 2001). Professional evolution and transformation must include an acceptance of technical advancements that determine prospective competitiveness in sectors and economies across nations. But, this is insufficient to ensure success in the third millennium. More focus should be placed on sustainable and environmentally friendly agricultural techniques in the profession. The ineffectiveness of numerous farm mechanisation programmes in nations that are developing, particularly in the 1970s and 1980s, with resulting graveyards of heavy-duty machinery and deterioration of soil, is conveniently attributed to "local conditions", but the consequences for public perceptions and attitudes towards agricultural engineering are disastrous. More effort is necessary to minimise pollution from agriculture and to create energy-saving devices. We must evaluate the socio-economic context in which new goods and processes will be deployed in addition to their technological viability.

Expectations of customers and demand for agricultural products and processes have shifted from quantity to quality and safety. Consumers have recently demonstrated a keen interest in the manufacturing technique, livestock welfare, and the genetic composition of food supplies. As a result, the tracking of food commodities and procedures has become crucial across the supply chain. In the increasingly competitive global food commerce, meeting product quality specifications is no longer adequate. Additional work is required to develop traceability systems for rapid and affordable product (or batch) verification and process classification as a component of a unified quality assurance system.

During the past half-century, there has been a strong priority on training agricultural engineering graduates who are eligible to register as professional engineers, with a lack of focus on tutoring graduates with engineering technology skills (often referred to as "gold collar" employees). As a result of this practice, there are fewer technologists in engineering with hands-on project management expertise to occupy the key intermediate technical cadre between expert design engineers and technicians. These graduates are typically more equipped to operate at the interface of new technological advances and industry. A Bachelor's of Engineering Technology (Agriculture/Biosystems) degree may be all that is necessary to guarantee a steady supply of essential technical expertise into agriculture and other bio-industries in numerous economies, among which are New Zealand and South Africa, where there may be a lack of economies of scale for full-fledged agricultural engineering programmes in many universities.

Developing and applying an effective communication strategy is important. The widespread adoption of "biology" as the underlying science of agricultural engineering, which led to the change of many journal names, academic departments, and professional societies to emphasise words like "biosystems" and "biological", is one of the most notable changes that occurred in the agricultural profession at the end of the 20th century. While there has been substantial discussion about these changes within many national and regional professional organisations, I have found that the general public and potential students are aware of these changes. More and focused efforts are required to explain to the public through the means of the media what agricultural as well as biosystems engineers do, why and how we perform it. To tell the public about us, we require the assistance of leaders from far more powerful and larger agribusiness chains as well as the agro-machinery industry. We also need agricultural engineers who have become legislators and policy makers to help in communicating the duties and actions to the wider public. All too frequently, professional and practising agricultural and biosystems/biological engineers are aware of these contributions; however, the great majority of society, including our clients and students, are oblivious to what we perceive them to know about us and the sector we work in.

In research done by undergraduate final-year agricultural engineering students at Massey University in New Zealand (Opara, 2001), the majority of the students reported that people constantly asked them what agricultural engineers intended to achieve (and this included their classmates in other engineering and agricultural science disciplines). In a questionnaire that followed the discussion, one of the students inquired, "Could someone write down what agricultural engineering means and does and give it to us so that we can use it to explain to others when they ask us?" Another asked, "Is there any database or website that has a list of influential agricultural engineers anywhere?" These remarks represent a few of the profession's widely recognised challenges and obstacles. It needs to be emphasised that, notwithstanding modifications to the name and reforming the curriculum, these sentiments are likely to persist among our stakeholders for the foreseeable future unless we properly tackle the concerns of exposure in the community as well as communication with the broader population.

Conclusions and Future Prospects

Agriculture and other biological industries are under a lot of pressure to restore consumer trust in agriculture and the food supply due to factors such as growing globalisation, agriculture's multifunctionality, consumer and public opinions, and the fact that expectations are shifting, as are current food safety concerns and concerns about the need for sustainable agriculture practices. It is essential that agricultural engineering reforms and adapts to respond to these internal and external factors and stimuli. To create novel goods and process technologies that can fulfil site- and process-specific functions, it is necessary to have more knowledge and information. Agriculture will place a greater emphasis on developments in equipment performance prediction and control, process outcomes, and repercussions on commodities, people, and the environment.

Today, it is a common practice in successful businesses to implement quality confidence systems that include measurement, product, and process traceability. Our capacity to use the emerging and transformative cross-cutting technological innovations discussed in this chapter will influence how well we are able to achieve these requirements. Modern technological advancements require highly knowledgeable engineers and researchers who are capable of identifying what data and information to gather, evaluate, interpret, as well as implement to the design and control of apparatus and processes. Agricultural engineering differs from other engineering fields in that it straddles the line between engineering and biosystems (including agriculture), which is where its power rests. Agricultural and biosystems/biological engineers are, therefore, required to be at the forefront of the sustained growth and development and implementation of innovations in agriculture and other bio-industries because innovations are frequently produced at the boundaries of knowledge and disciplines.

To address the significant technical issues encountered by agriculture and other land-based businesses, agricultural engineers face enormous challenges, possibilities, as well as responsibilities regarding research and in the education of an emerging generation of biosystem engineers. For these future engineers to obtain employment outside of the agricultural sector, they will progressively need incredibly advanced skills. Agricultural engineering is a diverse field; therefore, it will always have to balance the depth and breadth of its graduates. To prevent curriculum overcrowding and resource duplication within institutions, it may be necessary to hire new faculty with deeper expertise in the fields highlighted in this chapter and to increase working together with other subject-matter experts in research and teaching. The success of agricultural and biosystems/biological/bioresources engineering education in this new era where information and biology have come closer together depends critically on forging strategic alliances with key business partners and sustaining an efficient communication system with our stakeholders.

REFERENCE LIST

Beijing Review. (2000). Will my PC be smarter than I am? January 25 (Bilingual Club), pp. 1–4.

Cortez,, L., Naas, I. & Braunbeck, O. (2001). Agricultural engineering education programs in Brazil. *Agricultural Engineering International: The CIGR Journal of Scientific Research and Development*, III, 13 pages.

Brumm, T.J. & S.K. Mickelson (2024). Chapter 10: Reshaping agricultural and biosystem engineering degree programs for student success and retention. In: Opara, U.L. (ed.). *Agricultural, Biosystems, and Biological Engineering Education – Global Perspectives and Current Practice.* CRC Press/Taylor & Francis Group.

De Baerdemaeker, J. & Hashimoto, Y. (1999). Speaking fruit approach to the intelligent control of the storage system. *Proceedings of 12th CIGR World Congress*, Milano, Vol. I, 190–197.

FAO. (2009). How to feed the world in 2050. High-level expert forum. Food and Agriculture Organization of the United Nations, Rome.

Garvin, D. (1993). Building a learning organization. *Harvard Business Review*, July–August: 78–91.

IIMD. (1999). *World Competitiveness Handbook.* Lausanne: International Institute for Management Development.

Lievonen, J.T. (2000). Innovation opportunities from fusing high technologies. *Tech Monitor* (September–October), pp. 22–28.

Munack, A. & Speckman, H. (2001). Communication technology is the backbone of precision agriculture. *Agricultural Engineering International: The CIGR Journal of Scientific Research and Development*, III, 12 pages.

Opara, L.U. (2001). What do undergraduate agricultural engineering students think about their discipline? Unpublished Survey, Institute of Technology and Engineering, Massey University, New Zealand.

Opara,, L.U. & Mazaud, F. (2001). Food traceability from field to plate. *Outlook on Agriculture*, 38(4), 239–247.

Opara, L.U. (2000). Market-pull factors influencing perceptions on product quality in agribusiness (or quality assurance for who?). In Highley et al. (eds), *Proceedings of ASEAN/APEC Postharvest Seminar*, September. ACIAR Publication No. 100, Hanoi.

Opara, U.L. (2024a). Chapter 4: Top 100 questions of importance to the future of agricultural engineering education, Research and Practice in Africa. In: Opara, U.L. (ed.). *Agricultural, Biosystems, and Biological Engineering Education – Global Perspectives and Current Practice.* CRC Press/Taylor & Francis Group.

Opara, U.L. (2024b). Chapter 41: Growing agricultural engineering in Africa: Students' attitudes, perceptions, and expectations on agricultural engineering education. In: Opara, U.L. (ed.). *Agricultural, Biosystems, and Biological Engineering Education – Global Perspectives and Current Practice.* CRC Press/Taylor & Francis Group.

Prausnitz, J.M. (2001). Chemical engineering and the postmodern world. *Trans IchemE*, 79(Part A), 777–788.

Ritcher, F.J. (2001). Systemic technology management – An organizational learning view. *Journal of Scientific & Industrial Research*, 60(March), 284–290.

Ritcher, F.J. & Vettel, K. (1995). Successful joint ventures in Japan: Knowledge transfer through organizational learning. *Long Range Planning*, 28(3), 37–45.

SIAC. 2000. New Zealanders: Innovators to the world, turning great ideas into great ventures. A proposed innovation framework for New Zealand. Science and Innovation Advisory Committee, Wellington.

Sigrimis, N., Munack, A., Hashimoto, Y. & De Baerdemaeker, J.. (1999). Prospects in agricultural engineering in the information age – Technological developments for the producer and the consumer. *Agricultural Engineering International: The CIGR Journal of Scientific Research and Development*, I, 20 pages.

Smith, G. (1999). *Australia's Innovative Environment Compared with USA Experience.* Industrial Innovation Working Group, Canberra, 9 pages.

Speckman, H. (2000). Providing measured position data for agricultural machinery. *Computers and Electronics in Agriculture*, 25, 87–106.

Stix, G. (2001). Little Big Science. Nanotech: The science of small gets down to business. *Scientific American*, 285(3), 32–37.

Tilman, D., Balzer, C., Hill, J. & Befort, B. L. (2011). Global food demand and the sustainable intensification of agriculture. *Proceedings of the National Academy of Sciences of the United States of America*, 108, 20260.

UN (2022). Global issues. Population. https://www.un.org/en/global-issues/population. Accessed 25 August.

Van Dijk, M., Morley, T., Rau, M.L. & Saghai, Y. (2021). A meta-analysis of projected global food demand and population at risk of hunger for the period 2010–2050. *Nature Food*, 2, 494–501. https://doi.org/10.1038/s43016-021-00322-9

Whitesides, G.M. & Love, J.C. (2001). The Art of Building Small. *Scientific American*, 285(3), 39–47.

Williamson, O.E. (1975). *Markets and Hierarchies: Analysis and Antitrust Implications.* The Free Press, New York.

Index

N

O

P

For Product Safety Concerns and Information please contact our EU representative GPSR@taylorandfrancis.com
Taylor & Francis Verlag GmbH, Kaufingerstraße 24, 80331 München, Germany

www.ingramcontent.com/pod-product-compliance
Ingram Content Group UK Ltd.
Pitfield, Milton Keynes, MK11 3LW, UK
UKHW050059200926
475702UK00019B/2918
* 9 7 8 1 0 3 2 4 9 6 0 7 8 *